S0-BXL-525

Energy and Man:

Technical and Social Aspects of Energy

OTHER IEEE PRESS BOOKS

Magnetic Bubble Technology: Integrated-Circuit Magnetics for Digital Storage and Processing
Edited by H. Chang

Frequency Synthesis: Techniques and Applications
Edited by J. Gorski-Popiel

Literature in Digital Signal Processing: Author and Permuted Title Index (Revised and Expanded Edition)
Edited by H. D. Helms, J. F. Kaiser, and L. R. Rabiner

Data Communications via Fading Channels
Edited by K. Brayer

Nonlinear Networks: Theory and Analysis
Edited by A. N. Willson, Jr.

Computer Communications
Edited by P. E. Green, Jr. and R. W. Lucky

Stability of Large Electric Power Systems
Edited by R. T. Byerly and E. W. Kimbark

Automatic Test Equipment: Hardware, Software, and Management
Edited by F. Liguori

Key Papers in the Development of Coding Theory
Edited by E. R. Berlekamp

Technology and Social Institutions
Edited by K. Chen

Key Papers in the Development of Information Theory
Edited by D. Slepian

Computer-Aided Filter Design
Edited by G. Szentirmai

Laser Devices and Applications
Edited by I. P. Kaminow and A. E. Siegman

Integrated Optics
Edited by D. Marcuse

Laser Theory
Edited by F. S. Barnes

Digital Signal Processing
Edited by L. R. Rabiner and C. M. Rader

Minicomputers: Hardware, Software, and Applications
Edited by J. D. Schoeffler and R. H. Temple

Semiconductor Memories
Edited by D. A. Hodges

Power Semiconductor Applications, Volume II: Equipment and Systems
Edited by J. D. Harnden, Jr., and F. B. Golden

Power Semiconductor Applications, Volume I: General Considerations
Edited by J. D. Harnden, Jr., and F. B. Golden

A Practical Guide to Minicomputer Applications
Edited by F. F. Coury

Active Inductorless Filters
Edited by S. K. Mitra

Clearing the Air: The Impact of the Clean Air Act on Technology
Edited by J. C. Redmond, J. C. Cook, and A. A. J. Hoffman

Energy and Man:

Technical and Social Aspects of Energy

Edited by

M. Granger Morgan

Departments of Electrical Engineering and of Engineering and Public Affairs, Carnegie-Mellon University

A volume in the IEEE PRESS Selected Reprint Series.

The Institute of Electrical and Electronics Engineers, Inc. New York

International Standard Book Numbers: Clothbound: 0-87942-043-X
Paperbound: 0-87942-044-8

Library of Congress Catalog Card Number 74-27680

PRINTED IN THE UNITED STATES OF AMERICA

For the grandparents . . . who found 110 kwh/capita/day was more than adequate.

For the parents . . . who thought 150 kwh/capita/day was about right.

For the brothers and sisters . . . who have been doing well on 250 kwh/capita/day.

And for the children . . . who will probably feel like they're having to skimp at 350 kwh/capita/day.

Preface

Most people who set out to learn something about the technology of energy and its interactions with society find they must make their way through six to eight shelf feet of material before they begin to feel literate. The objective in editing this book has been to try to draw together many of the highlights of this large body of material. We hope that the resulting volume will help to ease the burden on new students and at the same time provide a convenient reference source for professionals already working in the field.

The book addresses a problem at the technology–society interface largely from the perspective of the technologist. The dominant time scale applied in selecting materials was fairly long, from several years to many decades. For this reason, there is little direct discussion of the short-term disallocations which have contributed to our current "energy crisis." However, most of the material is directly relevant to any long-term solution of the "crisis." Among the various energy technologies, greatest emphasis is placed upon the electric sector. Relatively little space has been allotted to direct fuel use in industrial applications and space heating. There is substantial coverage of some of the fundamental social and environmental impacts and interactions which result from energy use. However, very little material on the subject of "energy policy" has been included. The details of such fields as utility regulation, federal and state energy management structures, energy industry corporate structures, international petroleum, and priorities and management in energy research and development are remarkably complicated and change frequently. There appeared to be no way in which this great body of material could be satisfactorily summarized in 50 or 100 pages of reprints. Rather than provide inadequate coverage, this important area has been largely excluded. Readers who would like some pointers to help them to begin to explore some aspects of the energy policy area are referred to Appendix III, which was prepared by K. Lee of the University of Washington, Seattle.

The book consists of five parts. Part I gives an overview of the basic limits and parameters established by the earth system and explores possible interactions between human energy use and the global ocean-atmosphere system. In Part II, the history and projection of energy use in the United States is explored at a fairly general level. Issues of energy and the environment are introduced and explored at about the same level. The third part discusses the present and future status of most of the major energy conversion, transmission, and storage technologies. These selections are intended to provide sufficient technical detail to allow the reader to develop a broad overview. They are not intended as a primer in the details of the technology. Little or no attention has been given to exploring such underlying scientific and engineering principles as plasma physics, ac rotating machinery and transmission line theory, or engineering thermodynamics. However, some knowledge of such fields has been assumed in making the selections.

The final two parts concentrate on some of the social aspects of energy use. Only a limited amount of serious work has been done on many of the important socially related energy problems, so that the coverage in these sections is not as complete as one would like. This situation is changing and the next few years should witness a considerable growth in this literature. Part IV focuses on the benefits and costs of energy generation and use. In selecting materials, emphasis was placed on nuclear and coal generation technology. Part V contains a few selections on energy prices, demand growth, and energy conservation.

The following five books provide a useful extension and supplement to the material collected in this volume. Student readers and others with tight budgets will find that all five are available in paperback form.

1) A. L. Hammond, W. D. Metz, and T. H. Maugh, II, *Energy and the Future.* Washington, D.C.: Amer. Assoc. Advancement of Science, 1973 (paperback, $4.95).

2) The summary report of the Cornell workshop on energy and the environment (Feb. 22–24, 1972), a committee print of the Committee on Interior and Insular Affairs, United States Senate. Sponsored by the National Science Foundation, RANN Program, May 1972, Serial No. 92–93. Available from the U.S. Government Printing Office, Washington, D.C. ($0.70).

3) *Energy and Power*. San Francisco: Freeman, 1971 (paperback $3.25). Available also as the Sept. 1971 issue of *Sci. Amer.*

4) Federal Power Commission, *The 1970 National Power Survey (Part 1).* Washington, D.C.: U.S. Government Printing Office, 1971 ($4.25).

5) *A Time To Choose: America's Energy Future*, Energy Policy of The Ford Foundation. Cambridge: Ballinger, 1974 (paperback, $3.95).

At the time this book went to press, new materials related to energy were being produced at an unprecidented rate. Most of the papers reprinted in this volume are fairly fundamental in nature and should remain useful for many years. But the reader is urged to remember the fast pace of development in this field and in most cases will find a review of the more recent literature invaluable if he is pursuing a specific problem in any depth.

Acknowledgment

In addition to the authors whose works appear in this volume, I would like to thank the many people who reviewed the contents, including J. W. Baughn, R. Cohen, C. A. Falcone, S. Flajser, R. S. Greeley, K. N. Lee, M. Livine, R. L. Meier, D. Nelkin, P. F. Palmedo, S. S. Penner, D. J. Ragone, M. J. Roberts, P. Schleifer, K. E. F. Watt, and the IEEE Press Editorial Board and staff, for their helpful advice and suggestions.

For encouraging the continued pursuit of my energy interests during my two years at the National Science Foundation, thanks go to J. R. Pasta, P. G. Lykos, and F. W. Weingarten.

I also acknowledge with thanks the assistance and support of Elizabeth N. Morgan.

Contents

Part I
Reference Frame—The Earth System

When studying a system it is important to begin with as clear an understanding as possible of the boundaries and constraints. In a sociotechnical system as complex as the human energy system, this is not an easy task because there is not a simple set of constraints, but rather a whole hierarchy. Indeed some of these constraints change over time as technology, political systems, and human culture evolve. But human activity will be constrained to the near vicinity of planet earth for the foreseeable future, so it is obvious that the most fundamental constraints on the human energy system are in fact those of the earth system.

The earth system may limit human energy use in two ways. First, there are finite amounts of stored fossil and nuclear fuels. Second, there may be limits imposed by the carrying capacity and stability of the oceanic and atmospheric systems which sustain life. We explore the possible resource constraints imposed by the earth system with M. K. Hubbert's article, "Survey of World Energy Resources." Readers who would like to see Hubbert's ideas developed in greater detail are referred to his classic article "Energy Resources" in *Resources and Man* [1]. Beyond this, the many fine publications of Resources for the Future [2] offer one excellent starting place for further readings in this field. Readers interested specifically in U.S. domestic oil reserves are directed to the survey article by Berg *et al.* [3].

The problems of global atmospheric and ocean dynamics, climate change, and the possible impacts which may result from man's activities, are not well understood. In fact, they have recently become the subject of the largest international cooperative science research effort ever mounted. The articles by W. W. Kellogg and W. R. Frisken summarize some of what is known. The probable effects of carbon dioxide are discussed in greater detail by A. D. Watt. Two articles by K. E. F. Watt and R. A. Bryson describe some possible other climatological effects of man's activities. These two articles are controversial. Many geophysicists do not believe that man is pushing quite as hard on the earth system as the Watt and Bryson articles suggest, but at the same time most are uncomfortable about the apparent abandon with which mankind is proceeding and have begun to call for both an accelerated program of research and a serious consideration of earth system problems in the public policy process. Readers interested in further exploring man's impact on the ocean-atmosphere system will find the reports of two major studies available through the M.I.T. Press [4], [5] to be an excellent starting place.

REFERENCES

[1] M. K. Hubbert, "Energy resources" in *Resources and Man.* San Francisco: Freeman, 1969. Study and recommendations by the Committee on Resources and Man, Div. Earth Sciences, Nat. Acad. Sciences, Nat. Research Council.

[2] Resources for the Future is a nonprofit corporation for research and education in the development, conservation, and use of natural resources and the improvement of the environment which was established in 1952 with the cooperation of the Ford Foundation. It is located at 1755 Massachusetts Avenue, N.W., Washington, D.C. 20036.

[3] R. R. Berg, C. J. Calhoun, Jr., and R. L. Whiting, "Prognosis for expanding U.S. production of crude oil," *Science*, vol. 184, pp. 331–336, Apr. 19, 1974.

[4] *Inadvertent Climate Modifications: Report of the Study of Man's Impact on Climate (SMIC).* Cambridge, Mass: M.I.T. Press, 1971.

[5] *Mans Impact on the Global Environment: Report of the Study of Critical Environmental Problems (SCEP).* Cambridge, Mass.: M.I.T. Press, 1970.

■ ENERGY REQUIREMENTS AND RESOURCES

Survey of World Energy Resources

M. KING HUBBERT,
Research Geophysicist,
United States Geological Survey,
Washington, D.C., U.S.A.

ABSTRACT

The present large-scale use of energy and power by the human species represents a unique event in the billions of years of geologic history. Furthermore, in magnitude, most of the development has occurred during the present century. In the United States, the peak in the rate of petroleum production occurred in 1970 and that for natural gas is imminent. The peak in the world production of crude oil is expected to occur at about the year 2000 and that for coal production at about 2150 or 2200. For other sources of energy and power, water power, geothermal power and tidal power are inadequate to replace power from fossil fuels. Nuclear power, based on the breeder reactor and utilizing low-grade deposits of uranium and thorium, has a larger potential than the fossil fuels, but it also constitutes a large perpetual hazard. The largest source of energy available to the earth is solar radiation. This source has a life-expectancy of a geologic time scale, is nonpolluting and is larger in magnitude than any likely requirements by the human species.

In consequence of the large supplies of available energy, the period since 1800 has been one of an unprecedented exponential industrial growth. This also has been accompanied by a world-wide ecological disturbance, including that of the human population. It can easily be seen that such a period of growth must be ephemeral in character and, in fact, is now almost over. One aspect of this transition from a state of exponential growth to a state of nongrowth is the present alarm over an "energy crisis." Actually, the world's present problems are by no means unmanageable in terms of present biological and technological knowledge. The real crisis confronting us is, therefore, not an energy crisis but a cultural crisis. During the last two centuries, we have evolved what amounts to an exponential-growth culture, with institutions based on the premise of an indefinite continuation of exponential growth. One of the principal consequences of the cessation of exponential growth will be an inevitable revision of some of the tenets of that culture.

INTRODUCTION

BY NOW, it has become generally recognized that the world's present civilization differs fundamentally from all earlier civilizations in both the magnitude of its operations and the degree of its dependence on energy and mineral resources — particularly energy from the fossil fuels. The significance of energy lies in the fact that it is involved in everything that occurs on the earth — everything that moves. In fact, in the last analysis, about as succinct a statement as can be made about terrestrial events is the following: The earth's surface is composed of the 92 naturally occurring chemical elements, all but a minute radioactive fraction of which obey the laws of conservation and of nontransmutability of classical chemistry. Into and out of this system is a continuous flux of energy, in consequence of which the material constituents undergo either continuous or intermittent circulation.

Reprinted with permission from *Can. Mining and Metallurgical Bull.*, vol. 66, pp. 37–53, July 1973.

The principal energy inputs into this system are three (Fig. 1): (1) 178,000 x 10^{12} thermal watts from the solar radiation intercepted by the earth's diametrical plane; (2) 32 x 10^{12} thermal watts conducted and convected to the earth's surface from inside the earth; and (3) 3 x 10^{12} thermal watts of tidal power from the combined kinetic and potential energy of the earth-moon-sun system. Of these inputs of thermal power, that from solar energy is overwhelmingly the largest, exceeding the sum of the other two by a factor of more than 5,000.

Of the solar input, about 30%, the earth's albedo, is directly reflected and scattered into outer space, leaving the earth as short-wavelength radiation; about 47% is directly absorbed and converted into heat; and about 23% is dissipated in circulating through the atmosphere and the oceans, and in the evaporation, precipitation and circulation of water in the hydrologic cycle. Finally, a minute fraction, about 40 x 10^{12} watts, is absorbed by the leaves of plants and stored chemically by the process of photosynthesis whereby the inorganic substances, water and carbon dioxide, are synthesized into organic carbohydrates according to the approximate equation

$$\text{Light energy} + CO_2 + H_2O \rightarrow [CH_2O] + O_2.$$

Small though it is, this fraction is the energy source for the biological requirements of the earth's entire populations of plants and animals.

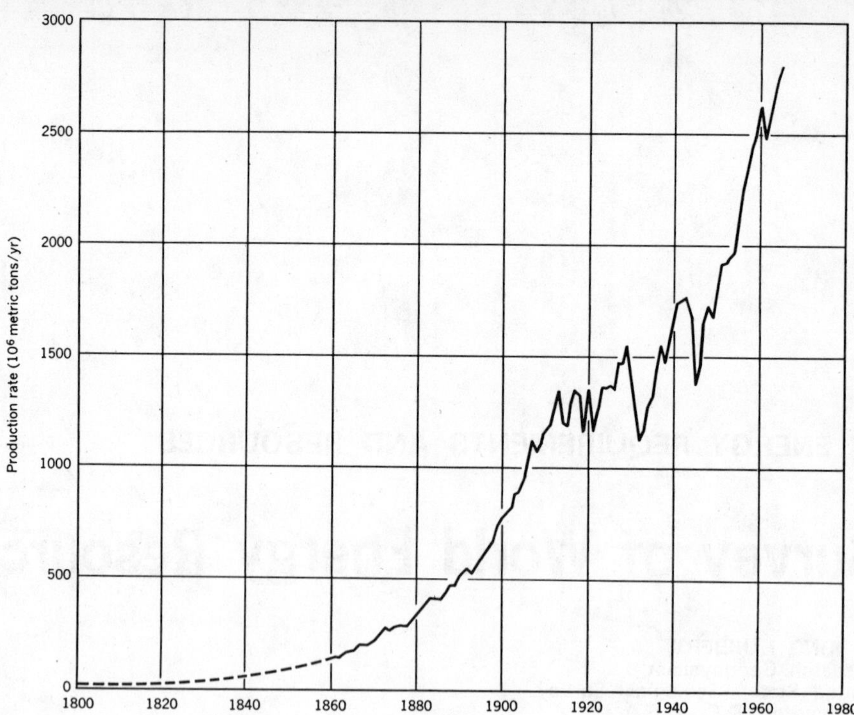

FIGURE 2 — World production of coal and lignite (Hubbert, 1969, Fig. 8.1).

From radioactive dating of meteorites, the astronomical cataclysm that produced the solar system is estimated to have occurred about 4.5 billion years ago and microbial organisms have been found in rocks as old as 3.2 billion years. During the last 600 million years of geologic history, a minute fraction of the earth's organisms have been deposited in swamps and other oxygen-deficient environments under conditions of incomplete decay, and eventually buried under great thicknesses of sedimentary muds and sands. By subsequent transformations, these have become the earth's present supply of fossil fuels: coal, oil and associated products.

About 2 million years ago, according to recent discoveries, the ancestors of modern man had begun to walk upright and to use primitive tools. From that time to the present, this species has distinguished itself by its inventiveness in the progressive control of an ever-larger fraction of the available energy supply. First, by means of tools and weapons, the invention of clothing, the control of fire, the domestication of plants and animals and use of animal power, this control was principally ecological in character. Next followed the manipulation of the inorganic world, including the smelting of metals and the primitive uses of the power of wind and water.

Such a state of development was sufficient for the requirements of all pre-modern civilizations. A higher-level industrialized civilization did not become possible until a larger and more concentrated source of energy and power became available. This occurred when

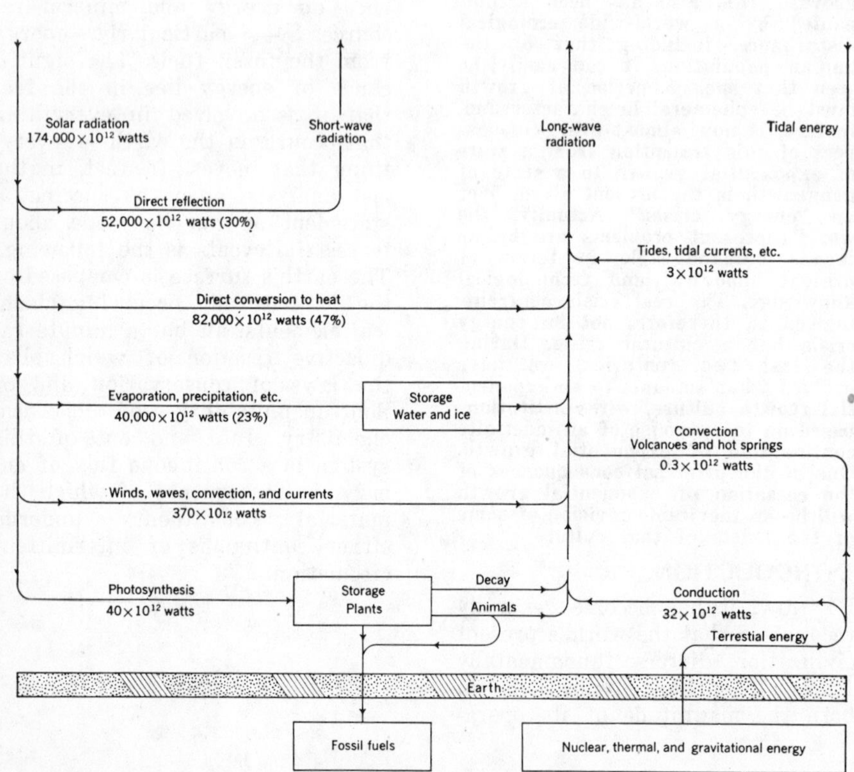

FIGURE 1 — World energy flowsheet.

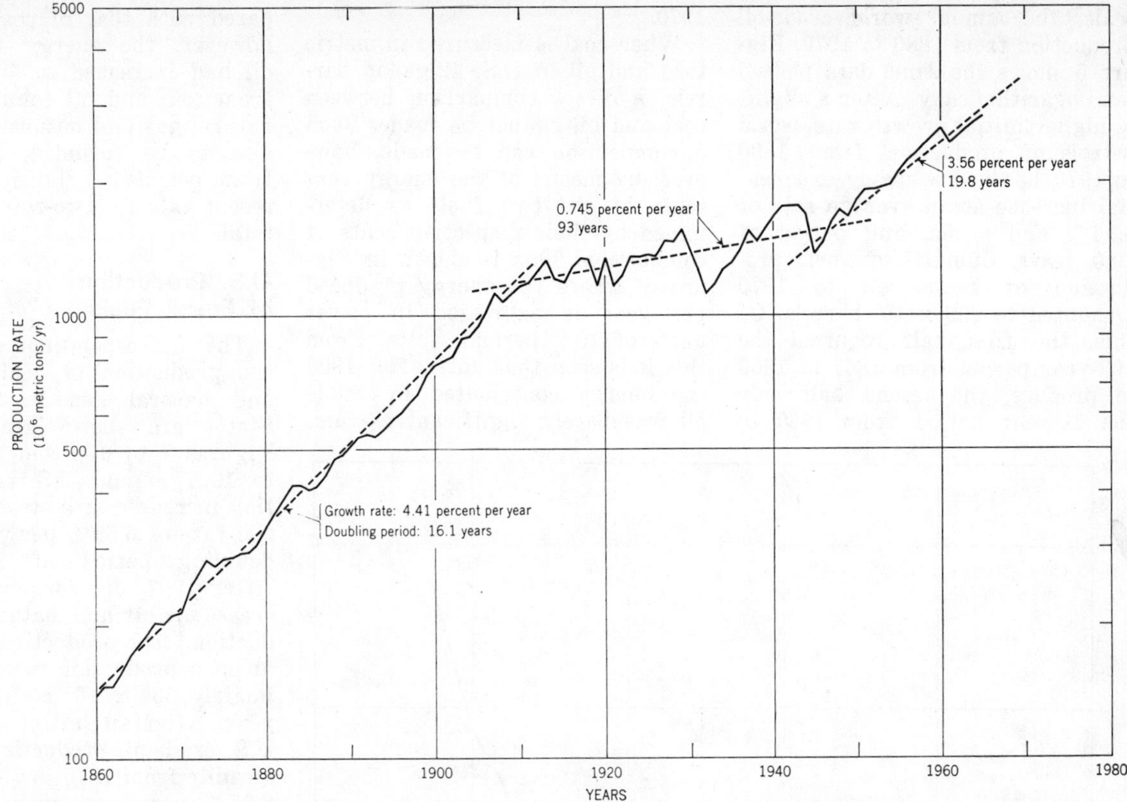

FIGURE 3 — World production of coal and lignite (semilogarithmic scale) (Hubbert, 1971, Fig. 4).

the huge supply of energy stored in the fossil fuels was first tapped by the mining of coal, which began as a continuous enterprise about 9 centuries ago near Newcastle in northeast England. Exploitation of the second major fossil-fuel supply, petroleum, began in 1857 in Romania and 2 years later in the United States. The tapping of an even larger supply of stored energy, that of the atomic nucleus, was first achieved in a controlled manner in 1942, and now the production of nuclear power from plants in the 1,000-megawatt range of capacity is in its early stages of development.

In addition to increased energy sources, energy utilization was markedly enhanced by two technological developments near the end of the last century: the development of the internal-combustion engine, utilizing petroleum products for mobile power, and the development of electrical means for the generation and distribution of power from large-scale central power plants. This also made possible for the first time the large-scale use of water power. This source of power derived from the contemporary flux of solar energy has been in use to some degree since Roman times, but always in small units — units rarely larger than a few hundred kilowatts. With electrical generation and distribution of hydropower, first accomplished at Niagara Falls about 1895, progressively larger hydropower stations have been installed with capacities up to several thousand megawatts.

ENERGY FROM FOSSIL FUELS

To the present the principal sources of energy for industrial uses have been the fossil fuels. Let us therefore review the basic facts concerning the exploitation and utilization of these fuels. This can best be done by means of a graphical presentation of the statistics of annual production.

World Production of Coal and Oil

Figure 2 shows the annual world production of coal and lignite from 1860 to 1970, and the approximate rate back to 1800, on an arithmetic scale. Figure 3 shows the same data on a semilogarithmic scale. The significance of the latter presentation is that straight-line segments of the growth curve indicate periods of steady exponential growth in the rate of production.

Annual statistics of coal production earlier than 1860 are difficult to assemble, but from intermittent earlier records it can be estimated that from the beginning of coal mining about the 12th century A.D. until 1800, the average growth rate of production must have been about 2% per year, with an average doubling period of about 35 years. During the 8 centuries to 1860 it is estimated that cumulative production amounted to about 7×10^9 metric tons. By 1970, cumulative production reached 140×10^9 metric tons. Hence, the coal mined during the 110-year period from 1860 to 1970 was approximately 19 times that of the preceding 8 centuries. The coal produced during the last 30-year period from 1940 to 1970 was approximately equal to that produced during all preceding history.

The rate of growth of coal production can be more clearly seen from the semilogarithmic plotting of Figure 3. The straight-line segment of the production curve from 1860 to World War I indicates a steady exponential increase of the rate of production during this period at about 4.4% per year, with a doubling period of 16 years. Between the beginning of World War I and the end of World War II, the growth rate slowed down to about 0.75% per year and a doubling period of 93 years. Finally, after World War II a more rapid growth rate of 3.56% per year and a doubling period of 19.8 years was resumed.

Figure 4 shows, on an arithmetic

scale, the annual world crude-oil production from 1880 to 1970. Figure 5 shows the same data plotted semilogarithmically. After a slightly higher initial growth rate, world petroleum production from 1890 to 1970 has had a steady exponential increase at an average rate of 6.94% and a doubling period of 10.0 years. Cumulative world production of crude oil to 1970 amounted to 233 x 10^9 barrels. Of this, the first half required the 103-year period from 1857 to 1960 to produce, the second half only the 10-year period from 1960 to 1970.

When coal is measured in metric tons and oil in U.S. 42-gallon barrels, a direct comparison between coal and oil cannot be made. Such a comparison can be made, however, by means of the energy contents of the two fuels as determined by their respective heats of combustion. This is shown in Figure 6, where the energy produced per year is expressed in power units of 10^{12} thermal watts. From this it is seen that until after 1900 the energy contributed by crude oil was barely significant as compared with that of coal. By 1970, however, the energy from crude oil had increased to 56% of that from coal and oil combined. Were natural gas and natural-gas liquids also to be included, the energy from petroleum fluids would represent about two-thirds of the total.

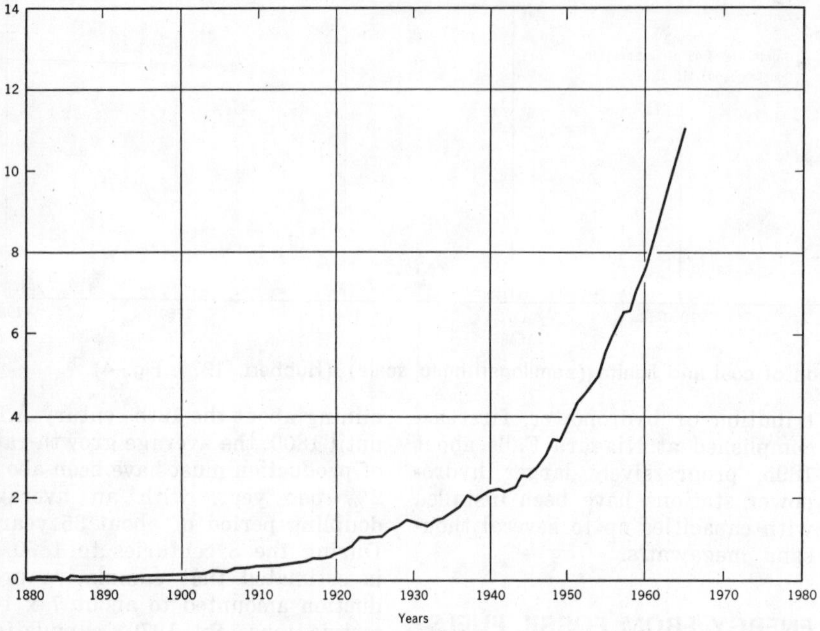

FIGURE 4 — World production of crude oil (Hubbert, 1969, Fig. 8.2).

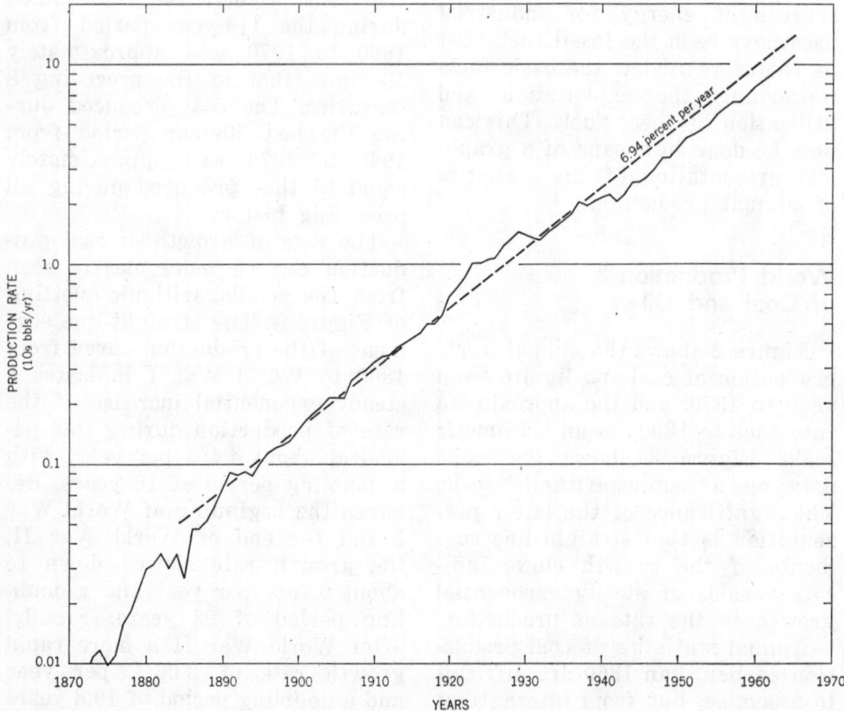

FIGURE 5 — World production of crude oil (semilogarithmic scale) (Hubbert, 1971, Fig. 6).

U.S. Production of Fossil Fuels

The corresponding growths in the production of coal, crude oil and natural gas in the United States are shown graphically in Figures 7 to 9. From before 1860 to 1907 annual U.S. coal production increased at a steady exponential rate of 6.58% per year, with a doubling period of 10.5 years. After 1907, due largely to the increase in oil and natural-gas production, coal production fluctuated about a production rate of approximately 500 x 10^6 metric tons per year. After an initial higher rate, U.S. crude-oil production increased steadily from 1870 to 1929 at about 8.27% per year, with a doubling period of 8.4 years. After 1929, the growth rate steadily declined to a 1970 value of approximately zero. From 1905 to 1970 the U.S. production of natural gas increased at an exponential rate of 6.6% per year, with a doubling period of 10.5 years.

Finally, Figure 10 shows the annual production of energy in the United States from coal, oil, natural gas, and hydro- and nuclear power from 1850 to 1970. From 1850 to 1907, this increased at a steady growth rate of 6.9% per year and doubled every 10.0 years. At about 1907, the growth rate dropped abruptly to an average value from 1907 to 1960 of about 1.77% per year, with a doubling period of 39 years. Since 1960, the growth rate has increased to about 4.25% per year, with the doubling period reduced to 16.3 years.

DEGREE OF ADVANCEMENT OF FOSSIL-FUEL EXPLOITATION

The foregoing are the basic historical facts pertaining to the exploitation of the fossil fuels in the world and in the United States. In the light of these facts we can hardly fail to wonder: How long can this continue? Several different approaches to this problem will now be considered.

Method of Donald Foster Hewett

In 1929, geologist Donald Foster

Hewett delivered before the AIME one of the more important papers ever written by a member of the U.S. Geological Survey, entitled "Cycles in Metal Production." In 1926, Hewett had made a trip to Europe during which he visited 28 mining districts, of which about half were then or had been outstanding sources of several metals. These districts ranged from England to Greece and from Spain to Poland. Regarding the purpose of this study, Hewett stated:

"I have come to believe that many of the problems that harass Europe lie in our path not far ahead. I have therefore hoped that a review of metal production in Europe in the light of its geologic, economic and political background may serve to clear our vision with regard to our own metal production."

In this paper, extensive graphs were presented of the production of separate metals from these various districts showing the rise, and in many cases the decline, in the production rates as the districts approached exhaustion of their ores. After having made this review, Hewett generalized his findings by observing that mining districts evolve during their history through successive stages analogous to those of infancy, adolescence, maturity and old age. He sought criteria for judging how far along in such a sequence a given mining district or region had progressed, and from his study he suggested the successive culminations shown in Figure 11. These culminations were: (1) the quantity of exports of crude ore; (2) the number of mines in operation; (3) the number of smelters or refining units in operation; (4) the production of metal from domestic ore; and (5) the quantity of imports of crude ore.

Although not all of Hewett's criteria are applicable to the production of the fossil fuels, especially when world production is considered, the fundamental principle is applicable; namely, that like the metals, the exploitation of the fos-

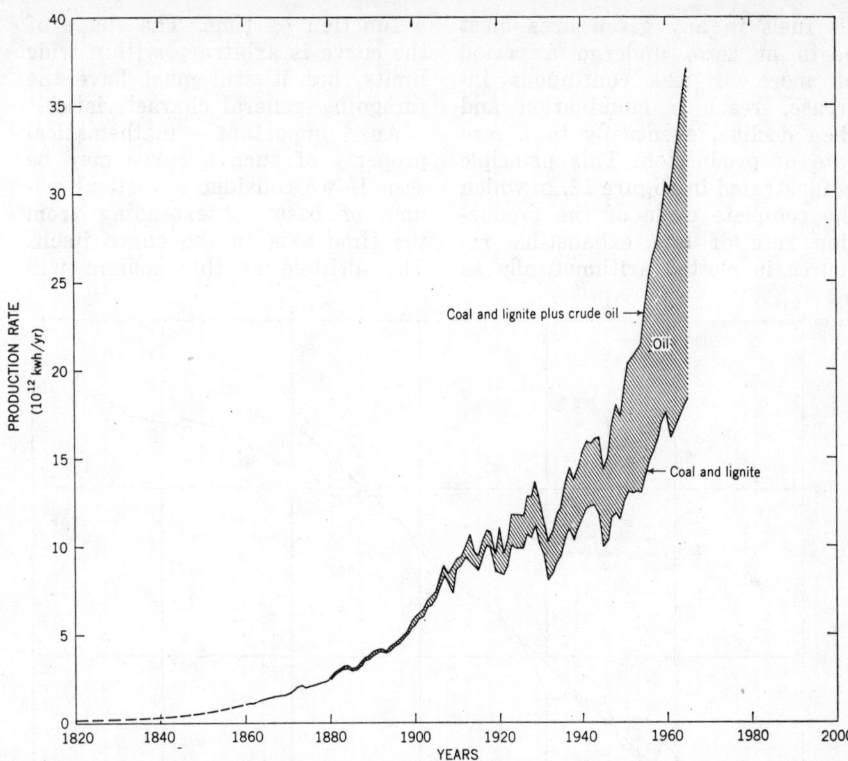

FIGURE 6 — World production of thermal energy from coal and lignite plus crude oil (Hubbert, 1969, Fig. 8.3).

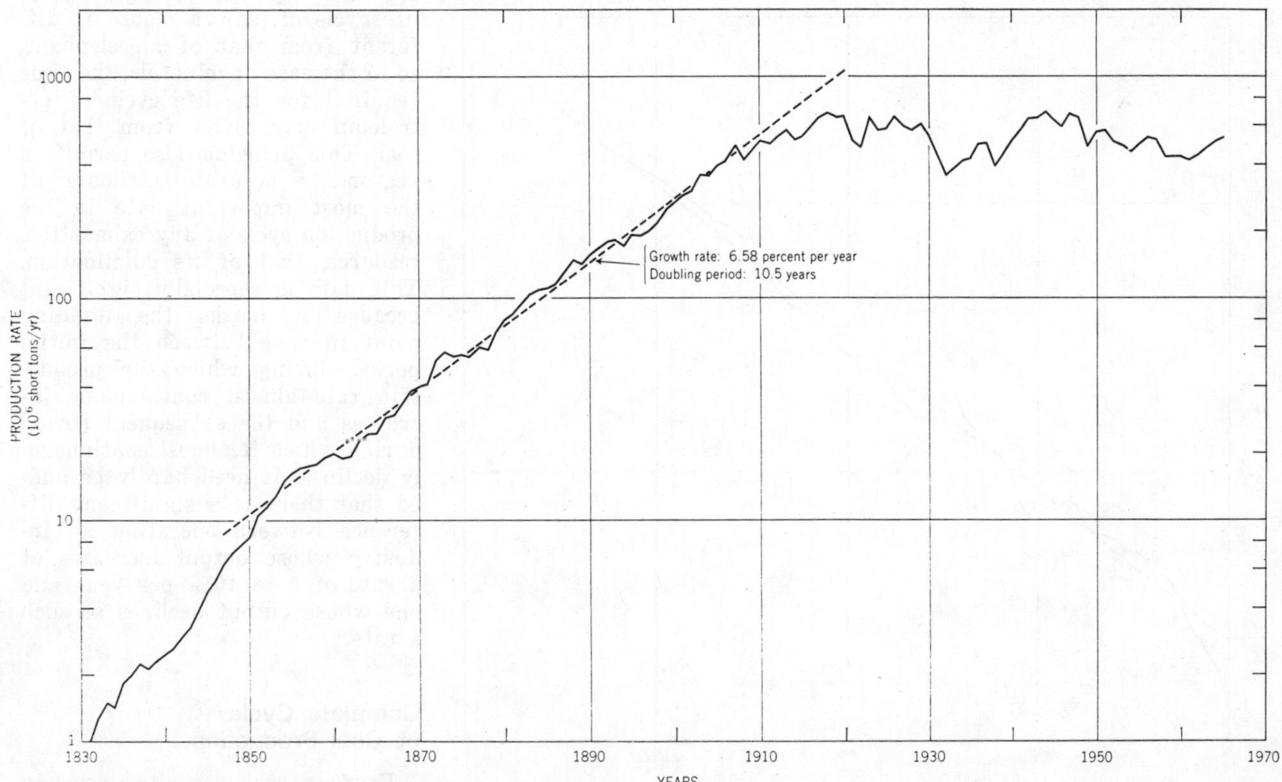

FIGURE 7 — U.S. production of coal (semilogarithmic scale).

sil fuels in any given area must begin at zero, undergo a period of more or less continuous increase, reach a culmination and then decline, eventually to a zero rate of production. This principle is illustrated in Figure 12, in which the complete cycle of the production rate of any exhaustible resource is plotted arithmetically as a function of time. The shape of the curve is arbitrary within wide limits, but it still must have the foregoing general characteristics.

An important mathematical property of such a curve may be seen if we consider a vertical column of base $\triangle t$ extending from the time axis to the curve itself. The altitude of this column will be the production rate

$$P = \triangle Q/\triangle t$$

at the given time, where $\triangle Q$ is the quantity produced in time $\triangle t$. The area of the column will accordingly be given by the product of its base and altitude:

$$P \times \triangle t = (\triangle Q/\triangle t) \times \triangle t = \triangle Q.$$

Hence, the area of the column is a measure of the quantity produced during the time interval $\triangle t$, and the total area from the beginning of production up to any given time t will be a measure of the cumulative production up to that time. Clearly, as the time t increases without limit, the production rate will have gone through its complete cycle and returned to zero. The area under the curve after this has occurred will then represent the ultimate cumulative production, $Q\infty$. In view of this fact, if from geological or other data the producible magnitude of the resource initially present can be estimated, then any curve drawn to represent the complete cycle of production *must be consistent with that estimate.* No such curve can subtend an area greater than the estimated magnitude of the producible resource.

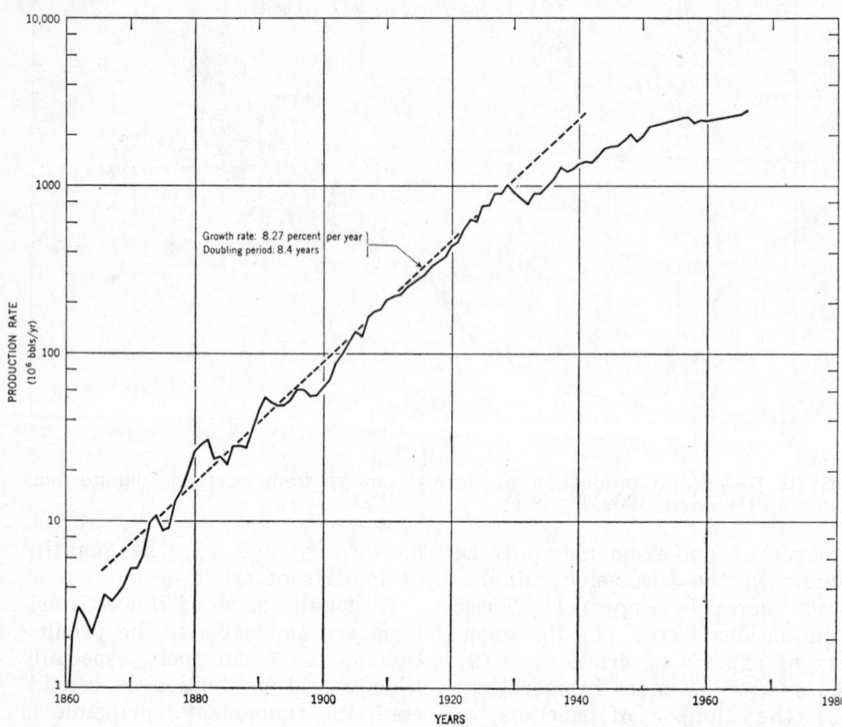

FIGURE 8 — U.S. production of crude oil, exclusive of Alaska (semilogarithmic scale).

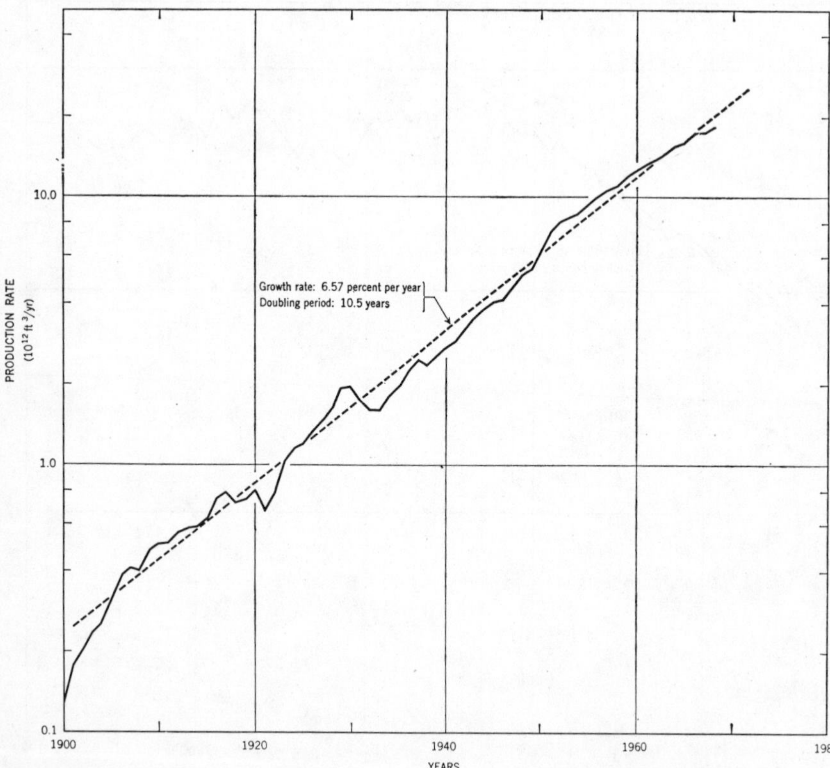

FIGURE 9 — U.S. net production of natural gas (semilogarithmic scale).

Utilization of this principle affords a powerful means of estimating the time scale for the complete production cycle of any exhaustible resource in any given region. As in the case of animals where the time required for the complete life cycle of, say, a mouse is different from that of an elephant, so in the case of minerals, the time required for the life cycle of petroleum may differ from that of coal. This principle also permits a reasonably accurate estimate of the most important date in the production cycle of any exhaustible resource, that of its culmination. This date is especially significant because it marks the dividing point in time between the initial period during which the production rate almost continuously increases and the subsequent period during which it almost continuously declines. It need hardly be added that there is a significant difference between operating an industry whose output increases at a rate of 5 to 10% per year and one whose output declines at such a rate.

Complete Cycle of Coal Production

Because coal deposits occur in stratified seams which are continuous over extensive areas and often crop out on the earth's surface, reasonably good estimates of the coal deposits in various sedimentary basins can be made by surface geological mapping and a limited amount of drilling. A summary of the current estimates of the world's initial coal resources has been published by Paul Averitt (1969) of the U.S. Geological Survey. These estimates comprise the total amount of coal (including lignite) in beds 14 inches (35 cm) or more thick and at depths as great as 3,000 feet (900 m), and in a few cases as great as 6,000 feet. Averitt's estimates as of January 1, 1967, for the initial producible coal, allowing 50% loss in mining, are shown graphically in Figure 13 for the world's major geographical areas. As seen in this figure, the original recoverable world coal resources amounted to an estimated 7.64×10^{12} metric tons. Of this, 4.31×10^{12}, or 56%, were in the USSR, and 1.49×10^{12}, or 19%, in the USA. At the other extreme, the three continental areas, Africa, South and Central America, and Oceania, together contained only 0.182×10^{12} metric tons, or 2.4% of the world's total.

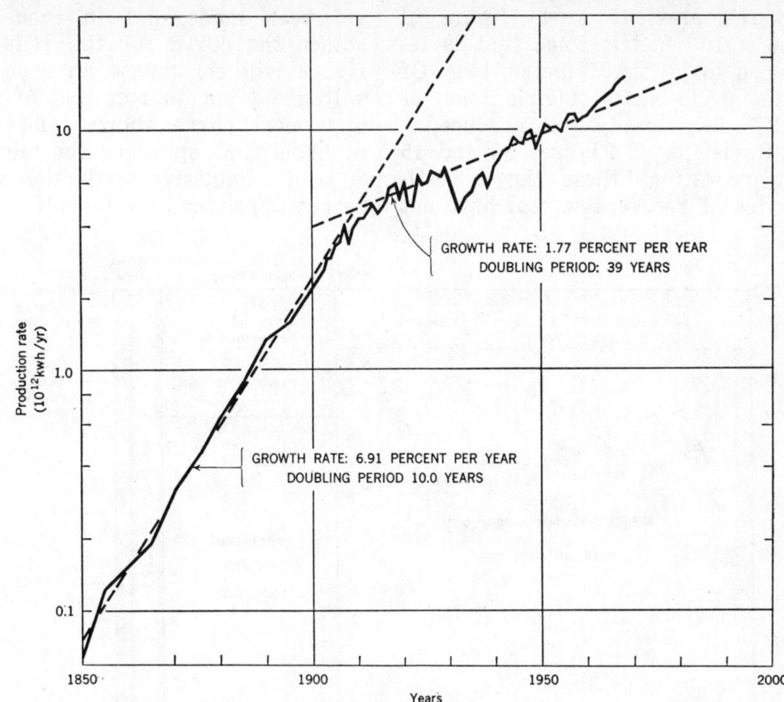

FIGURE 10 — U.S. production of thermal energy from coal, oil, natural gas, water power and nuclear power (semilogarithmic scale).

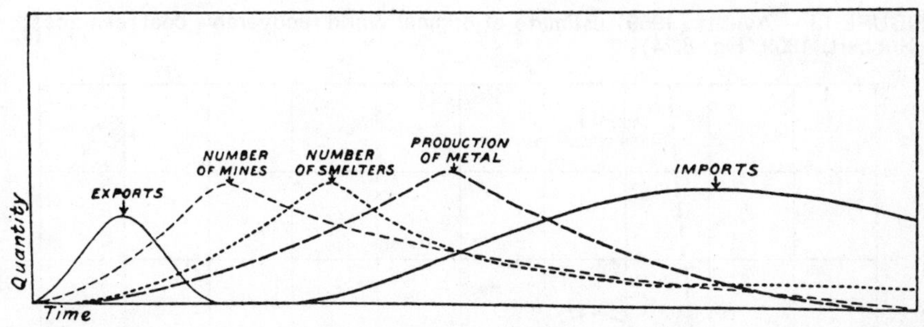

FIGURE 11 — Fig. 7 from D. F. Hewett's paper, "Cycles in Metal Production" (1929).

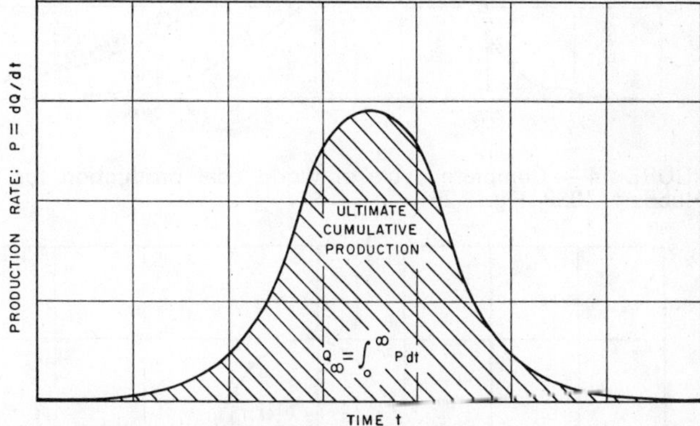

FIGURE 12 — Mathematical relations involved in the complete cycle of production of any exhaustible resource (Hubbert, 1956, Fig. 11).

Figure 14 shows two separate graphs for the complete cycle of world coal production. One is based on the Averitt estimate for the ultimate production of 7.6×10^{12} metric tons. These curves are also based on the assumption that not more than three more doublings, or an 8-fold increase, will occur before the maximum rate of production is reached. The dashed curve extending to the top of the drawing indicates what the production rate would be were it to continue to increase at 3.56% per year, the rate that has prevailed since World War II. For either of the complete-cycle curves, if we disregard the first and last 10-percentiles of the cumulative production, it is evident that the middle 80% of $Q\infty$ will probably be consumed during the three-century period from about the year 2000 to 2300.

Figure 15 shows the complete cycle of U.S. coal production for the two values for $Q\infty$, 1486×10^9 and 740×10^9 metric tons. Here too the time required to consume the middle 80% would be the 3 or 4 centuries following the year 2000.

A serious modification of the above coal-resource figures has been given by Averitt (cited in Theobald, Schweinfurth and Duncan, 1972). Here, Averitt, in February 1972, has given an estimate of the amount of coal remaining in the United States that is recoverable under present economic and technological conditions. This comprises coal in seams with a minimum thickness of 28 inches and a maximum depth of 1000 feet. The amount of coal in this category is estimated to be 390×10^9 short tons or 354×10^9 metric tons. Adding the 37×10^9 metric tons of coal already produced gives 391×10^9 metric tons of original coal in this category. This amounts to only 26% of the 1486×10^9 metric tons assumed previously. Of this, 9.5% has already been produced. If we apply the same ratio of 26%

to the previous world figure of 7.6×10^{12} metric tons, that is reduced to 2.0×10^{12} metric tons. Of this, 0.145×10^{12} metric tons, or 7.2%, has already been produced.

Revisions of Figures 14 and 15 incorporating these lower estimates of recoverable coal have not yet been made, but in each instance the curve for the reduced figure will encompass an area of only about one-quarter that of the uppermost curve shown, and the probable time span for the middle 80% of cumulative production will be cut approximately in half.

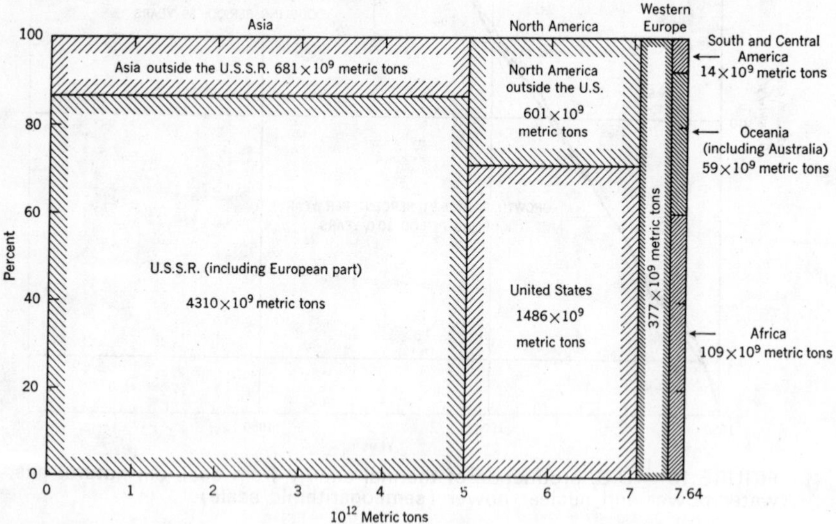

FIGURE 13 — Averitt (1969) estimate of original world recoverable coal resources (Hubbert, 1969, Fig. 8.24).

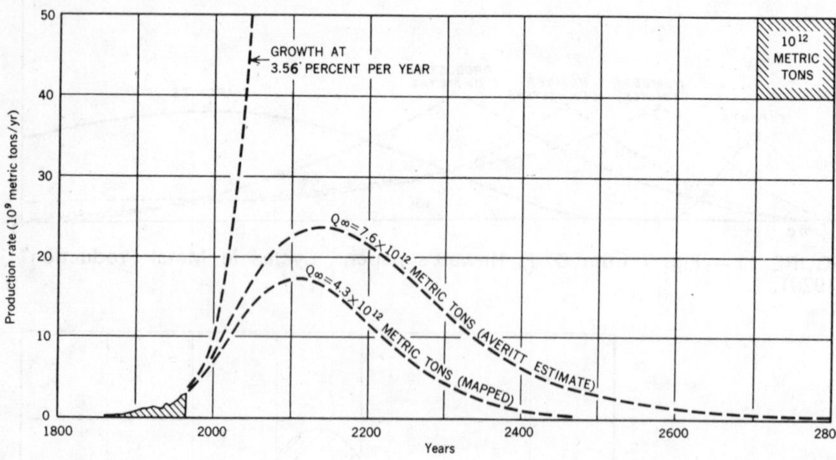

FIGURE 14 — Complete cycle of world coal production for two values of $Q\infty$ (Hubbert, 1969, Fig. 8.25).

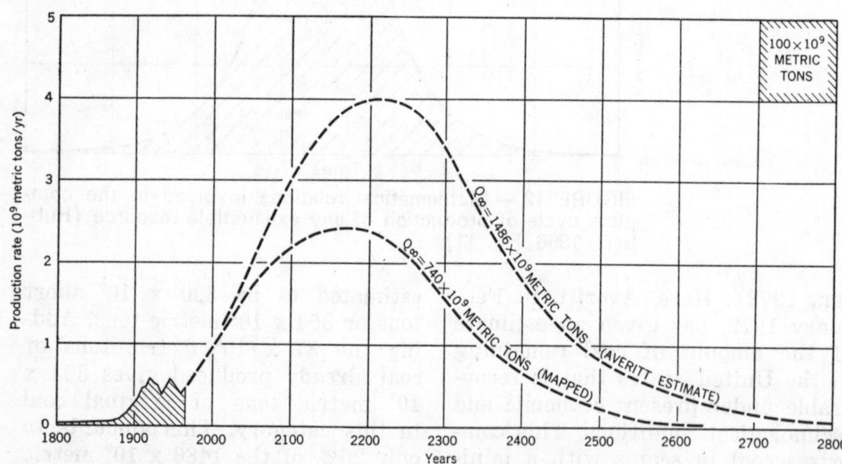

FIGURE 15 — Complete cycle of U.S. coal production for two values of $Q\infty$ (Hubbert, 1969, Fig. 8.26).

Estimates of Petroleum Resources

Because oil and gas occur in limited volumes of space underground in porous sedimentary rocks and at depths ranging from a few hundred feet to 5 or more miles, the estimation of the ultimate quantities of these fluids that will be obtained from any given area is much more difficult and hazardous than for coal. For the estimation of petroleum, essentially two methods are available: (1) estimation by geological analogy; and (2) estimation based on cumulative information and evidence resulting from exploration and productive activities in the region of interest.

The method of estimating by geological analogy is essentially the following. A virgin undrilled territory, Area *B*, is found by surface reconnaissance and mapping to be geologically similar to Area *A*, which is already productive of oil and gas. It is inferred, therefore, that Area *B* will eventually produce comparable quantities of oil and gas per unit of area or unit of volume of sediments to those of Area *A*.

Although this is practically the only method available initially for estimating the oil and gas potential of an undrilled region, it is also intrinsically hazardous, with a very wide range of uncertainty. This is illustrated in Table 1, in which the estimates made in 1953 for the future oil discoveries on the continental shelf off the Texas and Louisiana coasts are compared with the results of subsequent drilling.

TABLE 1 — Petroleum Estimates by Geological Analogy: Louisiana and Texas Continental Shelves

(crude oil, 10^9 bbls)

	U.S. Geological Survey estimates, 1953	Cumulative discoveries to 1971
Louisiana	4	ca. 5
Texas	9	Negligible

In 1953, the U.S. Geological Survey, on the basis of geological analogy between the onshore and offshore areas of the Gulf Coast and the respective areas of the continental shelf bordering Texas and Louisiana, estimated future discoveries of 9 billion barrels of oil on the Texas continental shelf and 4 billion on that of Louisiana. After approximately 20 years of petroleum exploration and drilling, discoveries of crude oil on the Louisiana continental shelf have amounted to approximately 5 billion barrels; those on the continental shelf off Texas have been negligible.

The second technique of petroleum estimation involves the use of various aspects of the Hewett criterion that the complete history of petroleum exploration and production in any given area must go through stages from infancy to maturity to old age. Maturity is plainly the stage of production culmination, and old age is that of an advanced state of discovery and production decline.

In March 1956, this technique was explicitly applied to crude-oil production in the United States by the present author (Hubbert, 1956) in an invited address, "Nuclear Energy and the Fossil Fuels," given before an audience of petroleum engineers at a meeting of the Southwest Section of the American Petroleum Institute at San Antonio, Texas. At that time the petroleum industry in the United States had been in vigorous operation for 97 years, during which 52.4 billion barrels of crude oil had been produced. A review of published literature in conjunction with inquiries among experienced petroleum geologists and engineers indicated a consensus that the ultimate amount of crude to be produced from the conterminous 48 states and adjacent continental shelves would probably be within the range of 150 to 200 billion barrels. Using these two limiting figures, the curves for the complete cycle of U.S. crude-oil production shown in Figure 16 (Hubbert, 1956) were constructed. This showed that if the ultimate cumulative production, $Q\infty$, should be as small as 150 x 10^9 bbls, the peak in the rate of production would probably occur about 1966 — about 10 years hence. Should another 50 x 10^9 bbls be added, making $Q\infty = 200 \times 10^9$ bbls, the date of the peak of production would be postponed by only about 5 years. It was accordingly predicted on the basis of available information that the peak in U.S. crude-oil production would occur within 10-15 years after March 1956.

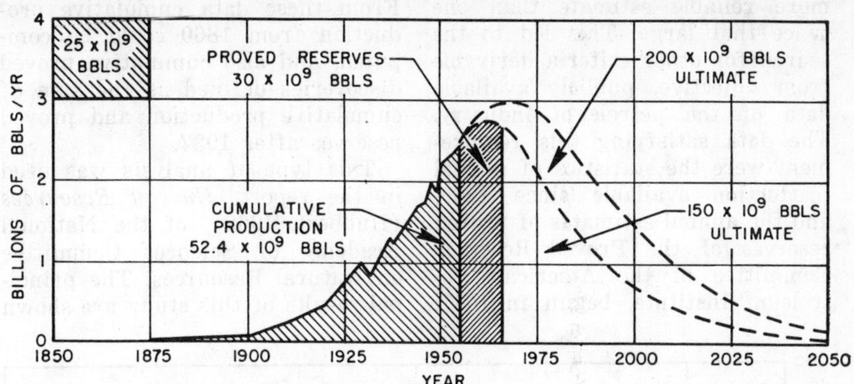

FIGURE 16 — 1956 prediction of the date of peak in the rate of U.S. crude-oil production (Hubbert, 1956, Fig. 21).

This prediction proved to be both surprising and disturbing to the U.S. petroleum industry. The only way it could be avoided, however, was to enlarge the area under the curve of the complete cycle of production by increasing the magnitude of $Q\infty$. As small increases of $Q\infty$ have only small effects in retarding the date of peak production, if this unpleasant conclusion were to be avoided, it would be necessary to increase $Q\infty$ by large magnitudes. This was what happened. Within the next 5 years, with insignificant amounts of new data, the published values for $Q\infty$ were rapidly escalated to successively higher values — 204, 250, 372, 400 and eventually 590 billion barrels.

In view of the fact that values for $Q\infty$ used in Figure 16 involved semisubjective judgments, no adequate rational basis existed for showing conclusively that a figure of 200 x 10^9 bbls was a much

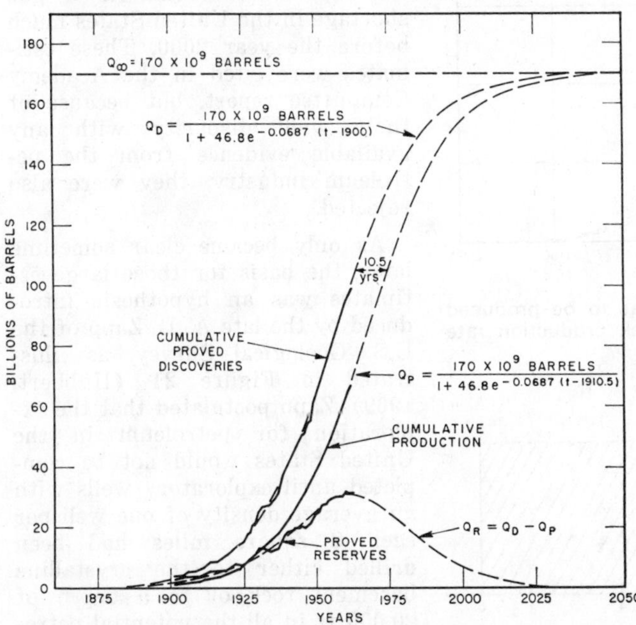

FIGURE 17 — Curves of cumulative proved discoveries, cumulative production and proved reserves of U.S. crude oil as of 1962 (Hubbert, 1962, Fig. 27).

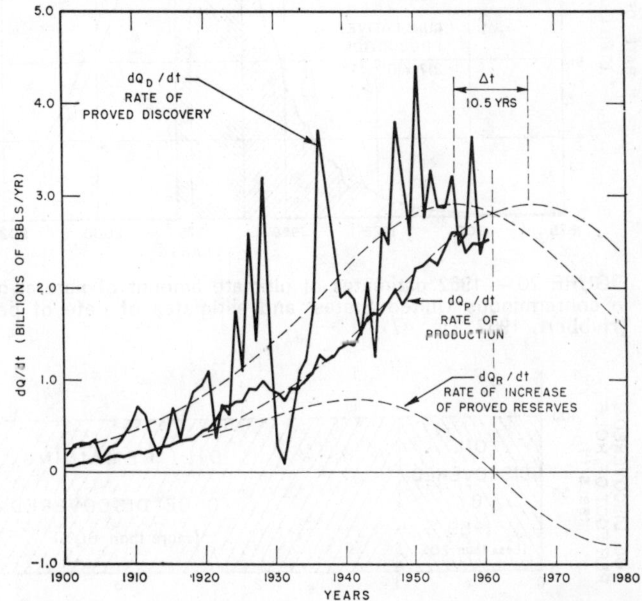

FIGURE 18 — Curves showing the rates of proved discovery and of production, and rate of increase of proved reserves of U.S. crude oil as of 1961. Note prediction of peak of production rate near the end of 1960 decade (Hubbert, 1962, Fig. 28).

more reliable estimate than one twice that large. This led to the search for other criteria derivable from objective, publicly available data of the petroleum industry. The data satisfying this requirement were the statistics of annual production available since 1860, and the annual estimates of proved reserves of the Proved Reserves Committee of the American Petroleum Institute, begun in 1937. From these data cumulative production from 1860 could be computed, and also cumulative proved discoveries defined as the sum of cumulative production and proved reserves after 1937.

This type of analysis was used in the report, *Energy Resources* (Hubbert, 1962), of the National Academy of Sciences Committee on Natural Resources. The principal results of this study are shown in Figures 17 and 18, in which it was found that the rate of proved discoveries of crude oil had already passed its peak about 1957, proved reserves were estimated to be at their peak in 1962 and the peak in the rate of crude-oil production was predicted to occur at about the end of the 1960 decade. The ultimate amount of crude oil to be produced from the lower 48 states and adjacent continental shelves was esimated to be about 170 to 175 billion barrels.

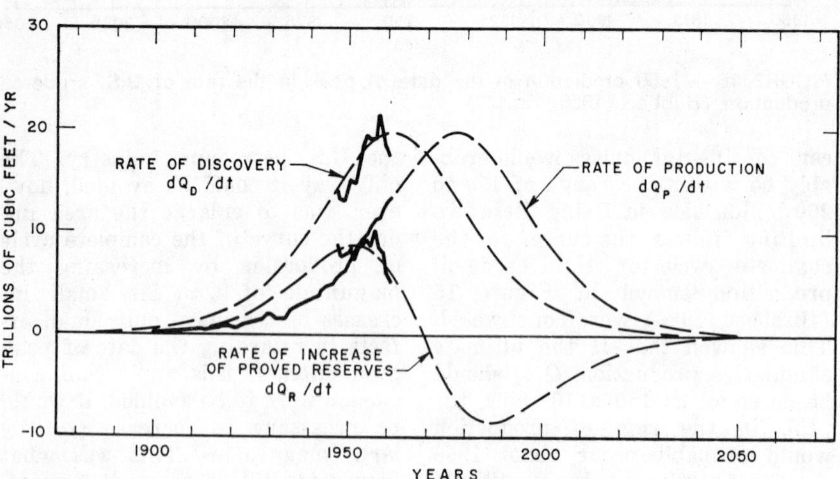

FIGURE 19 — 1962 estimates of the dates of the peaks of rate of proved discovery, rate of production and proved reserves of U.S. natural gas (Hubbert, 1962, Fig. 46).

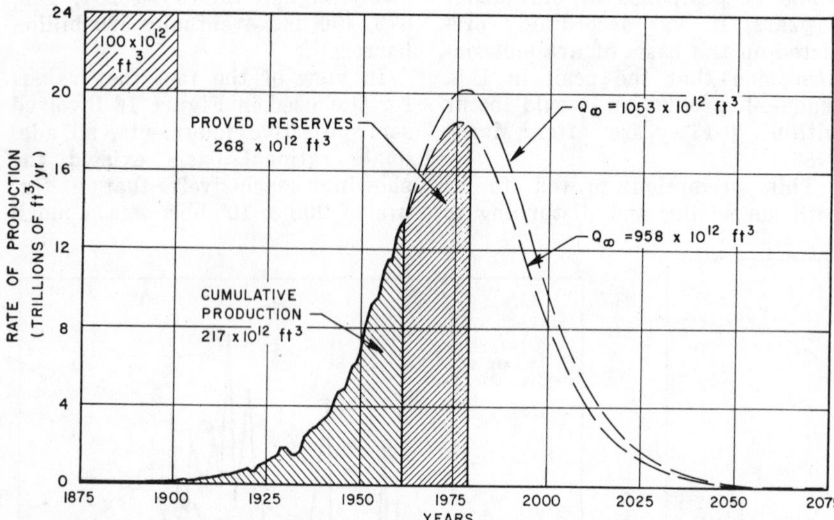

FIGURE 20 — 1962 estimates of ultimate amount of natural gas to be produced in conterminous United States, and estimates of date of peak production rate (Hubbert, 1962, Fig. 47).

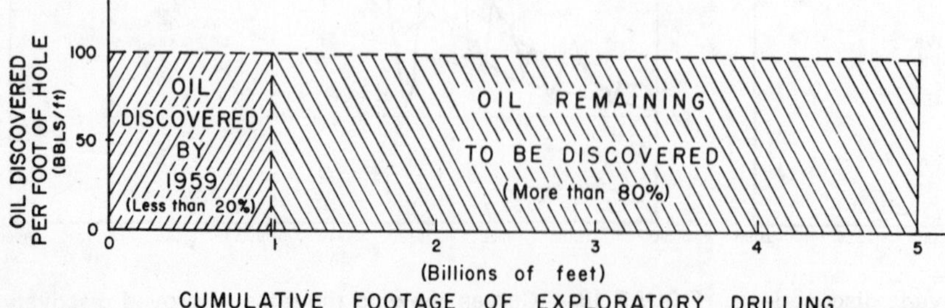

FIGURE 21 — Zapp (1962) hypothesis of oil discoveries per foot versus cumulative footage of exploratory drilling for conterminous United States and adjacent continental shelves (Hubbert, 1969, Fig. 8.18).

The corresponding estimates for natural gas are shown in Figures 19 and 20 (Hubbert, 1962). From these figures it will be seen that the rate of proved discoveries was estimated to be at its peak at about 1961. Proved reserves of natural gas were estimated to reach their peak ($dQ_r/dt = 0$) at about 1969, and the rate of production about 1977.

At the time the study was being made, the U.S. Geological Survey, in response to a Presidential directive of March 4, 1961, presented to the Academy Committee estimates of 590 x 10^9 bbls for crude oil and 2650 ft^3 for natural gas as its official estimates of the ultimate amounts of these fluids that would be produced from the lower 48 states and adjacent continental shelves.

These estimates were, by a wide margin, the highest that had ever been made up until that time. Moreover, had they been true, there would have been no grounds for the expectation of an oil or gas shortage in the United States much before the year 2000. These estimates were cited in the Academy Committee report, but because of their wide disparity with any available evidence from the petroleum industry, they were also rejected.

As only became clear sometime later, the basis for those large estimates was an hypothesis introduced by the late A. D. Zapp of the U.S. Geological Survey, as illustrated in Figure 21 (Hubbert, 1969). Zapp postulated that the exploration for petroleum in the United States would not be completed until exploratory wells with an average density of one well per each 2 square miles had been drilled either to the crystalline basement rock or to a depth of 20,000 ft in all the potential petroleum-bearing sedimentary basins. He estimated that to drill this pattern of wells in the petroliferous areas of the conterminous United

States and adjacent continental shelves would require about 5 x 10^9 feet of exploratory drilling. He then estimated that, as of 1959, only 0.98 x 10^9 feet of exploratory drilling had been done and concluded that at that time the United States was less than 20% along in its ultimate petroleum exploration. He also stated that during recent decades there had been no decline in the oil found per foot of exploratory drilling, yet already more than 100 x 10^9 barrels of oil had been discovered in the United States. It was implied, but not expressly stated, that the ultimate amount of oil to be discovered would be more than 500 x 10^9 bbls.

This was confirmed in 1961 by the Zapp estimate for crude oil given to the Academy Committee. At that time, with cumulative drilling of 1.1 x 10^9 feet, Zapp estimated that 130 x 10^9 bbls of crude oil had already been discovered. This would be at an average rate of 118 bbls/ft. Then, at this same rate, the amount of oil to be discovered by 5 x 10^9 feet of exploratory drilling should be 590 x 10^9 bbls, which is the estimate given to the Academy Committee. This constitutes the "Zapp hypothesis." Not only is it the basis for Zapp's own estimates, but with only minor modifications it has been the principal basis for most of the subsequent higher estimates.

The most obvious test for the validity of this hypothesis is to apply it to past petroleum discoveries in the United States. Has the oil found per foot of exploratory drilling been nearly constant during the past? The answer to this is given in Figure 22 (Hubbert, 1967), which shows the quantity of oil discovered and the average amount of oil found per foot for each 10^8 ft of exploratory drilling in the United States from 1860 to 1965. This shows an initial rate of 194 bbls/ft for the first unit from 1860 to 1920, a maximum rate of 276 bbls/ft for the third unit extending from 1929 to 1935 and then a precipitate decline to about 35 bbls/ft by 1965. This is approximately an exponential decline curve, the integration of which for unlimited future drilling gives an estimate of about 165 x 10^9 for $Q\infty$, the ultimate discoveries.

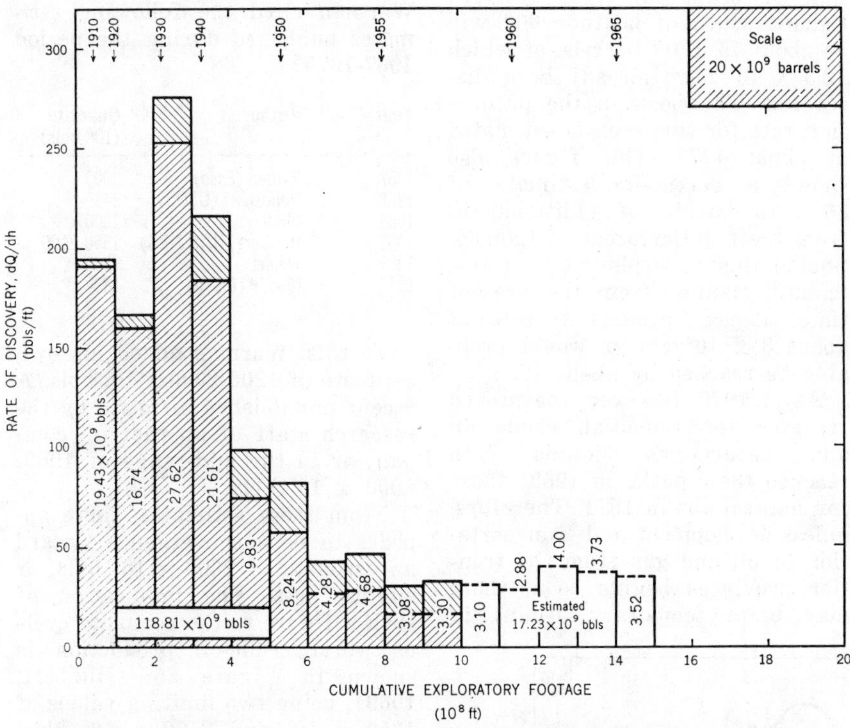

FIGURE 22 — Actual U.S. crude-oil discoveries per foot of exploratory drilling as a function of cumulative exploratory drilling from 1860 to 1965 (Hubbert, 1967, Fig. 15).

The superposition of the actual discoveries per foot shown in Figure 22 on the discoveries per foot according to the Zapp hypothesis of Figure 21 is shown in Figure 23 (Hubbert, 1969). The difference between the areas beneath the two curves represents the diffference between the two estimates — an apparent overestimate of about 425 x 10^9 bbls.

To recapitulate, in the Academy Committee report of 1962, the peak in U.S. proved crude-oil discoveries, excluding Alaska, was estimated to have occurred at about 1957, the peak in proved reserves at about 1962 and the peak in production was predicted for about 1968-1969. The peak in proved reserves did occur in 1962, and the peak in the rate of production occurred in 1970. Evidence that this is not likely to be exceeded is afforded by the fact that for the six months since March 1972, the pro-

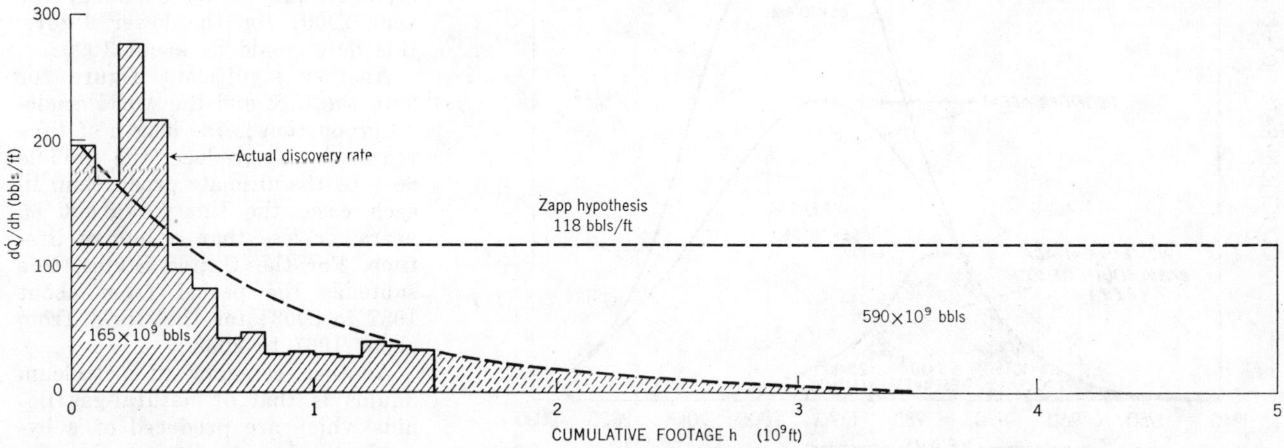

FIGURE 23 — Comparison of U.S. crude-oil discoveries according to Zapp hypothesis with actual discoveries. The difference between the areas beneath the two curves represents an overestimate of about 425 billion barrels (Hubbert, 1969, Fig. 8.19).

duction rates of both Texas and Louisiana, which together account for 60% of the total U.S. crude-oil production, have been at approximately full capacity, and declining.

As for natural gas, the Academy report estimated that the peak in proved reserves would occur at about 1969 and the peak in the rate of production about 1977. As of September 1972, the peak of proved reserves for the conterminous 48 states occurred in 1967, 2 years ahead of the predicted date, and it now appears that the peak in the rate of natural-gas production will occur about 1974-1975, 2 to 3 years earlier than predicted. In the 1969 Academy report, the ultimate production of natural gas was estimated to be about 1000 x 10^{12} ft^3. Present estimates by two different methods give a low figure of 1000 x 10^{12} and a high figure of 1080 x 10^{12}, or a mean of 1040 x 10^{12} ft^3.

Because of its early stage of development, the petroleum potential of Alaska must be based principally on geological analogy with other areas. The recent Prudhoe Bay discovery of a 10-billion-barrel field — the largest in the United States — has been a source of excitement for an oil-hungry U.S. petroleum industry, but it still represents less than a 3-year supply for the United States. From present information, a figure of 30 x 10^9 bbls is about as large an estimate as can be justified for the ultimate crude-oil production from the land area of Alaska, although a figure greater than this is an admitted possibility. Adding this to a present figure of about 170 x 10^9 bbls for the conterminous 48 states gives 200 x 10^9 as the approximate amount of crude oil ultimately to be produced in the whole United States.

Canada's Resources

For the present paper, it has not been possible to make an analysis of the oil and gas resources of Canada. However, Figure 24, from R. E. Folinsbee's Presidential Address before the geological section of the Royal Society of Canada (1970), provides a very good appraisal of the approximate magnitude of Canadian crude-oil resources. According to this estimate, the ultimate production of crude oil from Western Canada south of latitude 60° will be about 15 x 10^9 barrels, of which 12.5 x 10^9 have already been discovered. The peak in the production rate for this area is estimated at about 1977. This figure also shows a maximum estimate of 86 x 10^9 barrels of additional oil from the frontier areas of Canada. Should this be exploited in a systematic manner from the present time, a peak production rate of about 3 x 10^9 bbl/yr would probably be reached by about 1995.

As of 1973, however, the proved reserves for Canadian crude oil and natural-gas liquids both reached their peaks in 1969; those for natural gas in 1971. Therefore, unless development and transportation of oil and gas from the frontier provinces begins soon, there may be a temporary decline in total Canadian production of oil and gas toward the end of the present decade.

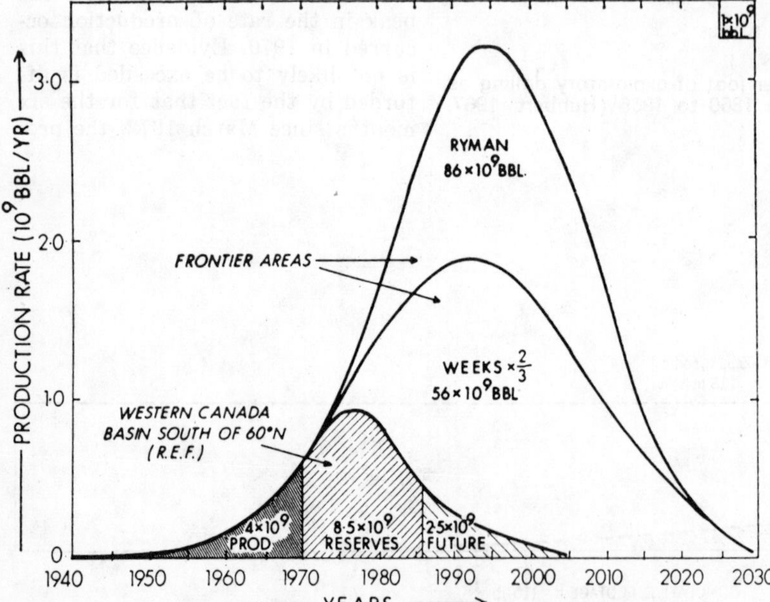

FIGURE 24 — Complete cycles of crude-oil production in Canada (Folinsbee, 1970, with permission).

World Crude-Oil Production

In the present brief review, only a summary statement can be made for the petroleum resources of the world as a whole. Recent estimates by various major oil companies and petroleum geologists have been summarized by H. R. Warman (Warman, 1971) of the British Petroleum Company, who gave 226 x 10^9 bbls as the cumulative world crude-oil production and 527 x 10^9 bbls for the proved reserves at the end of 1969. This totals 753 x 10^9 bbls as the world's proved cumulative discoveries. For the ultimate recoverable crude oil, Warman cited the following estimates published during the period 1967-1970:

Year	Author	Quantity (10^9 bbls)
1967	Ryman (Esso)	2090
1968	Hendricks (USGS)	2480
1968	Shell	1800
1969	Hubbert (NAS-NRC)	1350-2100
1969	Weeks	2200
1970	Moody (Mobil)	1800

To this, Warman added his own estimate of 1200-2000 x 10^9 bbls. A recent unpublished estimate by the research staff of another oil company is in the mid-range of 1900-2000 x 10^9 bbls.

From these estimates, there appears to be a convergence toward an estimate of 2000 x 10^9 bbls, or slightly less. The implication of such a figure to the complete cycle of world crude-oil production is shown in Figure 25 (Hubbert, 1969), using two limiting values of 1350 x 10^9 and 2100 x 10^9 bbls. For the higher figure, the world will reach the peak in its rate of crude-oil production at about the year 2000; for the lower figure, this date would be about 1990.

Another significant figure for both the U.S. and the world crude-oil production is the length of time required to produce the middle 80% of the ultimate production. In each case, the time is about 65 years, or less than a human lifetime. For the United States, this subtends the period from about 1937 to 2003; for the world, from about 1967 to 2032.

Another category of petroleum liquids is that of natural-gas liquids which are produced of a by-product of natural gas. In the United States (excluding Alaska), the ultimate amount of natural-gas liquids, based on an ultimate amount of crude oil of 170 x 10^9 bbls, and 1040 x 10^{12} ft of natural gas, amounts to about 36 x 10^9 bbls. Corresponding world figures, based on an estimate of 2000 x 10^9 bbls for crude oil, would be about 400 x 10^9 bbls for natural-gas liquids, and 12,000 ft^3 for natural gas.

Other Fossil Fuels

In addition to coal, petroleum liquids and natural gas, the other principal classes of fossil fuels are the so-called tar, or heavy-oil, sands and oil shales. The best known and probably the largest deposits of heavy-oil sands are in the "Athabasca Tar Sands" and two smaller deposits in northern Alberta containing an estimated 300 x 10^9 bbls of potentially producible oil. One large-scale mining and extracting operation was begun in 1966 by a group of oil companies, and others doubtless will follow as the need for this oil develops.

Unlike tar sands, the fuel content of which is a heavy, viscous crude oil, oil shales contain hydrocarbons in a solid form known as *kerogen,* which distills off as a vapour on heating and condenses to a liquid on cooling. The extractible oil content of oil shales ranges from as high as 100 U.S. gallons per short ton for the richest grades to near zero as the grades diminish. When all grades are considered, the aggregate oil content of the known oil shales is very large. However, in practice, only the shales having an oil content of about 25 gallons or more per ton and occurring in beds 10 feet or more thick are considered to be economical sources at present. According to a world inventory of known oil shales by Duncan and Swanson (1965), the largest known deposits are those of the Green River Formation in Wyoming, Colorado and Utah. From these shales, in the grade range from 10 to 65 gallons per ton, the authors estimate that only 80 x 10^9 bbls are recoverable under 1965 economic conditions. Their corresponding figure for oil shales outside the United States is 110 x 10^9 bbls.

The absolute magnitude of the world's original supply of fossil fuels recoverable under present technological and economic conditions and their respective energy contents in terms of their heats

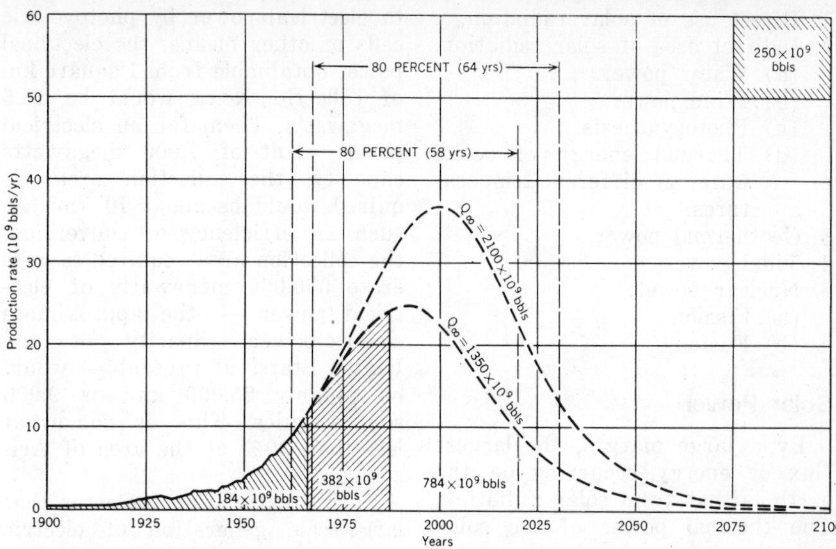

FIGURE 25 — Complete cycle of world crude-oil production for two values of Q_∞ (Hubbert, 1969, Fig. 8.23).

TABLE 2 — Approximate Magnitudes and Energy Contents of the World's Original Supply of Fossil Fuels Recoverable Under Present Conditions

Fuel	Quantity	Energy Content		
		10^{21} thermal joules	10^{15} thermal kwh	Per Cent
Coal and lignite..	2.35 × 10^{12} metric tons	53.2	14.80	63.78
Petroleum liquids	2400 × 10^9 bbls	14.2	3.95	17.03
Natural gas......	12,000 × 10^{12} ft^3	13.1	3.64	15.71
Tar-sand oil.....	300 × 10^9 bbls	1.8	0.50	2.16
Shale oil.......	190 × 10^9 bbls	1.1	0.31	1.32
TOTALS........		83.4	23.20	100.00

of combustion are given in Table 2. The total initial energy represented by all of these fuels amounted to about 83 x 10^{21} thermal joules, or 23 x 10^{15} thermal kilowatt-hours. Of this, 64% was represented by coal and lignite, 17 and 16%, respectively, by petroleum liquids and natural gas, and 3% by tar-sand and shale oil combined. Although the total amount of coal and lignite in beds 14 or more inches thick and occurring at depths less than 3,000 feet, as estimated by Averitt, are very much larger in terms of energy content, than the initial quantities of oil and gas, the coal practically recoverable under present conditions is only about twice the magnitude of the initial quantities of gas and oil in terms of energy content. Therefore, at comparable rates of production, the time required for the complete cycle of coal production will not be much longer than that for petroleum — in order of a century or two for the exhaustion of the middle 80% of the ultimate cumulative production.

To appreciate the brevity of this period in terms of the longer span of human history, the historical epoch of the exploitation of the fossil fuels is shown graphically in Figure 26, plotted on a time scale extending from 5000 years in the past to 5000 years in the future — a period well within the prospective span of human history. On such a time scale, it is seen that the epoch of the fossil fuels can be only a transitory or ephemeral event — an event, nonetheless, which has exercised the most drastic influence on the human species during its entire biological history.

OTHER SOURCES OF INDUSTRIAL ENERGY

The remaining sources of energy suitable for large-scale industrial use are principally the following.

1. Direct use of solar radiation.
2. Indirect uses of solar radiation.
   (a) Water power.
   (b) Wind power.
   (c) Photosynthesis.
   (d) Thermal energy of ocean water at different temperatures.
3. Geothermal power.
4. Tidal power.
5. Nuclear power.
   (a) Fission.
   (b) Fusion.

Solar Power

By a large margin, the largest flux or energy occurring on the earth is that from solar radiation. The thermal power of the solar radiation intercepted by the earth, according to recent measurements of the solar constant, amounts to about 173,000 x 10^{12} thermal watts. This is roughly 5,000 times all other steady fluxes of energy combined. It also has the expectation of continuing at about the same rate for geological periods of time into the future.

The largest concentrations of solar radiation reaching the earth's surface occur in desert areas within about 35° of latitude north and south of the equator. Southern Arizona and neighbouring areas in the southwestern part of the United States are in this belt, as well as northern Mexico, the Atacama Desert in Chile, and a zone across northern Africa, the Arabian Peninsula and Iran. In southern Arizona, the thermal power density of the solar radiation incident upon the earth's surface ranges from about 300 to 650 calories per cm^2 per day, from winter to summer. The winter minimum of 300 calories per cm^2 per day, when averaged over 24 hours, represents a mean power density of 145 watts per square meter. If 10% of this could be converted into electrical power by photovoltaic cells or other means, the electrical power obtainable from 1 square km of collection area would be 14.5 megawatts. Then, for an electrical power plant of 1,000 megawatts capacity, the collection area required would be about 70 km^2. At such an efficiency of conversion, the collection area required to generate 350,000 megawatts of electrical power — the approximate electric-power capacity of the United States at present — would be roughly 25,000 km^2 or 9,000 square miles. This is somewhat less than 10% of the area of Arizona.

Such a calculation indicates that large-scale generation of electric power from direct solar radiation is not to be ruled out on the grounds of technical infeasibility. It is also gratifying that a great deal of interest on the part of technically competent groups in universities and research institutions has arisen during the last 5 years over the possibility of developing large-scale solar power.

Hydroelectric Power

Although there has been continuous use of water power since Roman times, large units were not possible until a means was developed for the generation and transmission of power electrically. The first large hydroelectric power installation was that made at Niagara Falls in 1895. There, ten 5,000-hp turbines were installed for the generation of A.C. power, which was transmitted a distance of 26 miles to the city of Buffalo. The subsequent growth of hydroelectric power in the United States is shown in Figure 27 and that for the world in Figure 28.

In the United States, by 1970, the installed hydroelectric power capacity amounted to 53,000 megawatts, which is 32% of the ultimate potential capacity of 161,000 Mw as estimated by the Federal Power Commission. The world installation, by 1967, amounted to 243,000 Mw, which is 8.5% of the world's estimated potential hydroelectric power of 2,860,000 Mw. Most of this developed capacity is in the highly industrialized areas of North America, Western Europe and the Far East, especially Japan.

The areas with the largest potential water-power capacities are the industrially underdeveloped regions of Africa, South America and Southeast Asia, where combined capacities represent 63% of the world total.

The total world potential water power of approximately 3 x 10^{12} watts, if fully developed, would be of about the same magnitude as the world's present rate of utilization of industrial power. It may also appear that this would be an inexhaustible source of power, or at least one with a time span comparable to that required to remove mountains by stream erosion. This may not be true, however. Most waterpower developments require the creation of reservoirs by the damming of streams. The time required to fill these reservoirs with sediments is only 2 or 3 centuries. Hence, unless a technical solution of this problem can be found, water power may actually be comparatively short-lived.

Tidal Power

Tidal power is essentially hydroelectric power obtainable by damming the entrance to a bay or estuary in a region of tides with large amplitudes, and driving turbines as the tidal basin fills and empties. An inventory of the world's most favorable tidal-power sites gives an estimate of a total

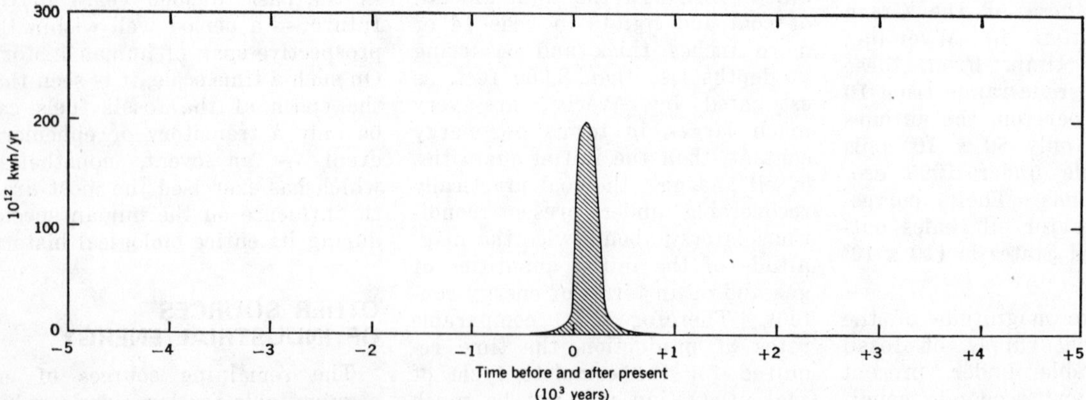

FIGURE 26 — Epoch of fossil-fuel exploitation in perspective of human history from 5000 years in the past to 5000 years in the future (modified from Hubbert, 1962, Fig. 54).

potential power capacity of about 63,000 Mw, which is about 2% of the world's potential water power capacity. At present, one or more small pilot tidal power plants of a few megawatts capacity have been built, but the only full-scale tidal plant so far built is that on the Rance estuary on the English Channel coast of France. This plant began operation in 1966 with an initial capacity of 240 Mw and a planned enlargement to 320 Mw.

One of the world's most favorable tidal-power localities is the Bay of Fundy region of northeastern United States and southeastern Canada. This has the world's maximum tides, with amplitudes up to 15 meters, and a combined power capacity of nine sites of about 29,000 Mw. Extensive plans have been made by both the United States and Canada for the utilization of this power, but as yet no installations have been made.

Geothermal Power

Geothermal power is obtained by means of heat engines which extract thermal energy from heated water within a depth ranging from a few hundred meters to a few km beneath the earth's surface. This is most practical where water has been heated to high temperatures at shallow depths by hot igneous or volcanic rocks that have risen to near the earth's surface. Steam can be used to drive steam turbines. At present, the major geothermal power installations are in two localities in Italy with a total capacity of about 400 Mw, the Geysers in California with a planned capacity by 1973 of 400 Mw and at Wairakei in New Zealand with a capacity of 160 Mw. The total world installed geothermal power capacity at present is approximately 1,500 Mw.

What the ultimate capacity may be can be estimated at present to perhaps only an order of magnitude. Recently, a number of geothermal-power enthusiasts (many with financial interests in the outcome) have made very large estimates for power from this source. However, until better information becomes available, an estimation within the range of 60,000 to 600,000 Mw, or between 2 and 20% of potential water power, is all that can be justified. Also, as geothermal-power production involves "mining" quantities of stored thermal energy, it is likely that most large installations will also be comparatively short-lived — perhaps a century or so.

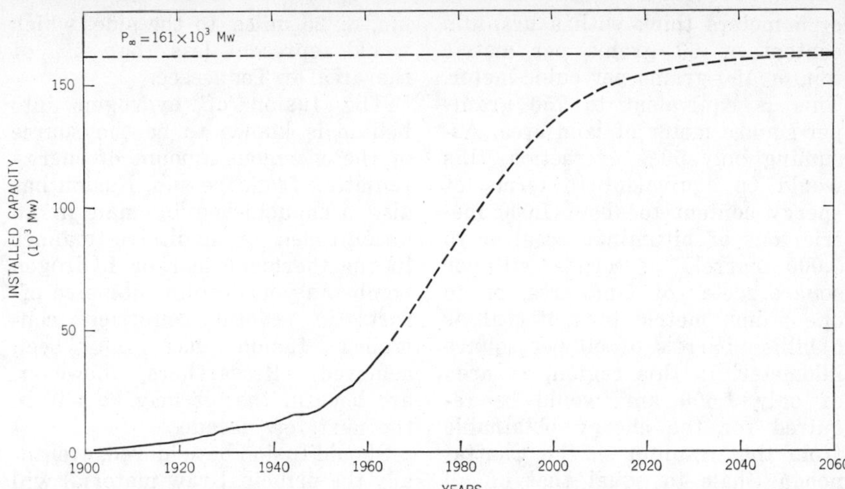

FIGURE 27 — Installed and potential hydroelectric-power capacity of the United States (Hubbert, 1969, Fig. 8.28).

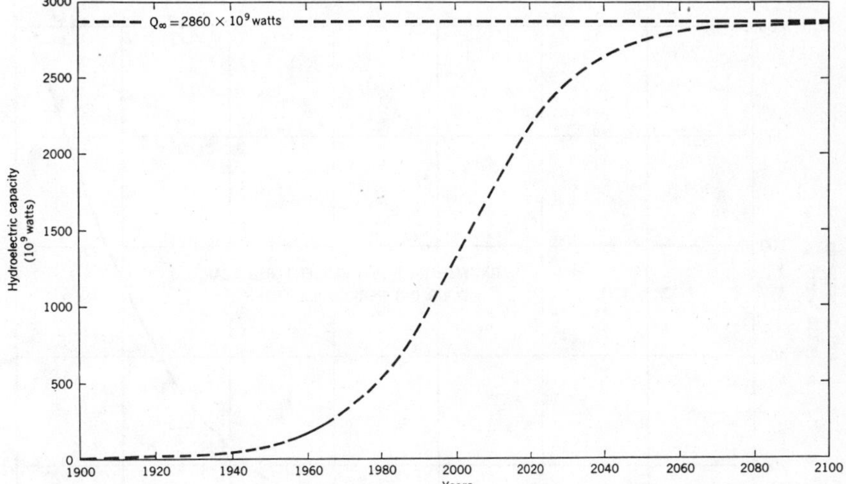

FIGURE 28 — Installed and potential world hydroelectric-power capacity.

Nuclear Power

A last major source of industrial power is that of atomic nuclei. Power may be obtained by two contrasting types of nuclear reactions: (1) the fissioning of heavy atomic isotopes, initially uranium-235; and (2) the fusing of the isotopes of hydrogen into heavier helium. In the fission process, two stages are possible. The first consists of power reactors which are dependent almost solely on the rare isotope, uranium-235, which represents only 0.7% of natural uranium. The second process is that of breeding whereby either the common isotope of uranium, uranium-238, or alternatively thorium, is placed in a reactor initially fueled by uranium-235. In response to neutron bombardment, uranium-238 is converted into plutonium-239, or thorium-232 into uranium-233, both of which are fissionable. Hence by means of a breeder reactor, in principle, all of the natural uranium or thorium can be converted into fissionable reactor fuel.

Uranium-235 is sufficiently scarce that, without the breeder reactor, the time span of large-scale nuclear power production would probably be less than a century. With complete breeding, however, it becomes possible not only to consume all of the natural uranium, or thorium, but to utilize low-grade sources as well.

The energy released by the fissioning of a gram of uranium-235 or plutonium-239 or uranium-233 amounts to 8.2×10^{10} joules of heat. This is approximately equivalent to the heat of combustion of 2.7 metric tons of bituminous coal or 13.4 barrels of crude oil. For the energy obtainable from a source of low-grade uranium, consider the Chattanooga Shale, which crops out along the western edge of the Appalachian Mountains in eastern Tennessee and underlies, at minable depths, most of several midwestern states. This shale has a uranium-rich layer about 16 feet or 5 meters thick with a uranium content of 60 grams per metric ton, or 150 grams per cubic meter. This is equivalent to 750 grams per square meter of land area. Assuming only 50% extraction, this would be equivalent in terms of energy content to about 1,000 metric tons of bituminous coal or to 5,000 barrels of crude oil per square meter of land area, or to one billion metric tons of coal or 5 billion barrels of oil per square kilometer. In this region, an area of only 1,600 km^2 would be required for the energy obtainable from the uranium in the Chattanooga Shale to equal that of all the fossil fuels in the United States. Such an area would be equivalent to that of a square 40 km, or 25 miles, to the side, which would represent less than 2% of the area of Tennessee.

The fusion of hydrogen into helium is known to be the source of the enormous amount of energy radiating from the sun. Fusion has also been achieved by man in an uncontrolled or explosive manner in the thermonuclear or hydrogen bomb. As yet, despite intensive efforts in several countries, controlled fusion has not been achieved. Researchers, however, are hopeful that it may be within the next few decades.

Should fusion be achieved, eventually the principal raw material will probably be the heavy isotope of hydrogen, deuterium. This occurs in sea water at an abundance of 1 deuterium atom to each 6,700 atoms of hydrogen. The deuterium-deuterium, or D-D, reaction involves several stages, the net result of which is:

$$5\,{}^{2}_{1}D \rightarrow {}^{4}_{2}He + {}^{3}_{2}He + H + 2n + 24.8\,Mev;$$

or, in other words, 5 atoms of deuterium, on fusion, produce 1 atom of helium-4, 1 atom of helium-3, 1 atom of hydrogen and 2 neutrons, and in addition release 24.8 million electron volts, or 39.8 x 10^{-13} joules.

It can be computed that 1 liter of water contains 1.0 x 10^{22} deuterium atoms, which upon fusion would release 7.95 x 10^{9} joules of thermal energy. This is equivalent to the heat of combustion of 0.26 metric tons of coal or 1.30 barrels of crude oil. Then, as 1 km^3 of sea water is equivalent to 10^{12} liters, the heat released by the fusion of the deuterium contained in 1 km^3 of sea water would be equivalent to that of the combustion of 1300 billion barrels of oil or 260 billion tons of coal. The deuterium in 33 km^3 of sea water would be equivalent to that of the world's initial supply of fossil fuels.

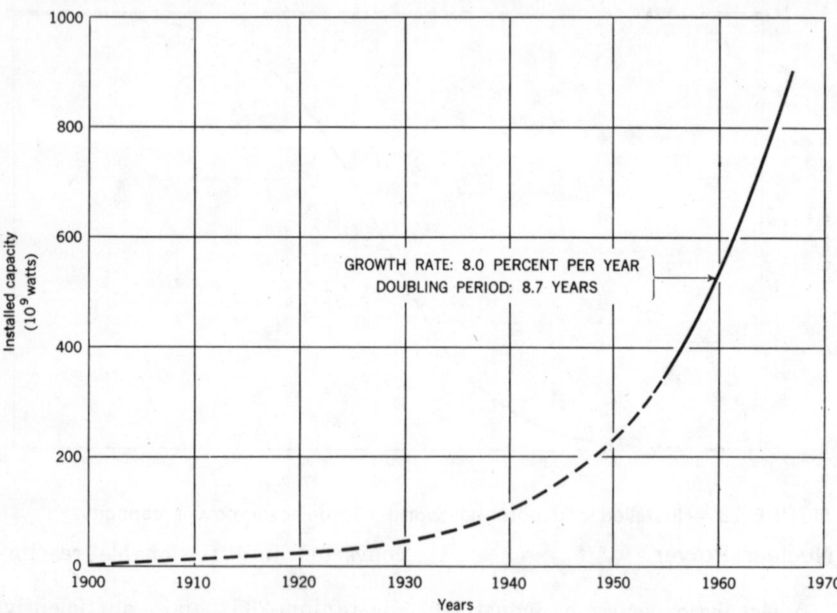

FIGURE 29 — World electric generating capacity as an example of exponential growth (Hubbert, 1971, Fig. 2).

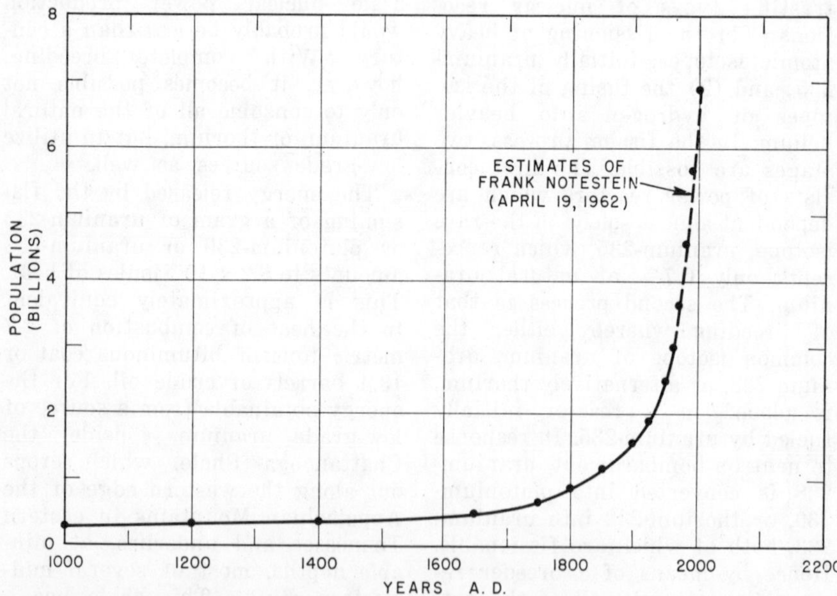

FIGURE 30 — Growth of human population since the year 1000 A.D. as an example of an ecological disturbance (Hubbert, 1962, Fig. 2).

ECOLOGICAL ASPECTS OF EXPONENTIAL GROWTH

From the foregoing review, what stands out most clearly is that our present industrialized civilization has arisen principally during the last 2 centuries. It has been accomplished by the exponential growth of most of its major components at rates commonly in the range of 4 to 8% per year, with periods of doubling from 8 to 16 years. The question now arises: What are the limits to such growth, and what does this imply concerning our future?

What we are dealing with, essentially, are the principles of ecology. It has long been known by ecologists that the population of any biologic species, if given a favorable environment, will increase exponentially with time; that is, that the population will double repeatedly at roughly equal intervals of time. From our previous observations, we have seen that this is also true of industrial components. For example (Fig. 29), the world electric-power capacity is now growing at 8% per year and doubling every 8.7 years. The world automobile population and the miles flown per year by the world's civil-aviation scheduled flights are each doubling every 10 years. Also, the human population is now doubling in 35 years (Fig. 30).

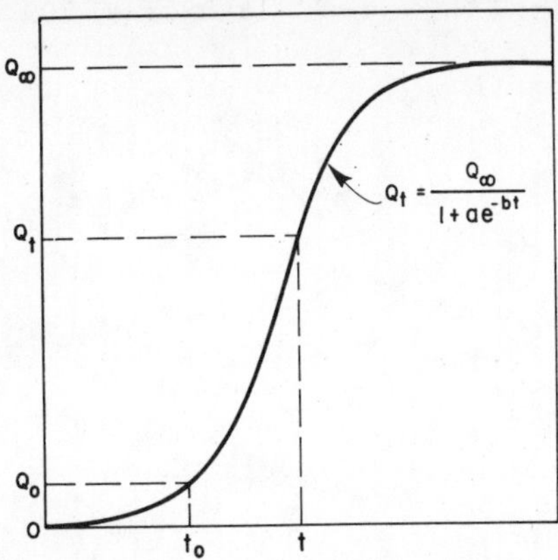

FIGURE 31 — The logistic growth curve showing both the initial exponential phase and the final slowing down during a cycle of growth.

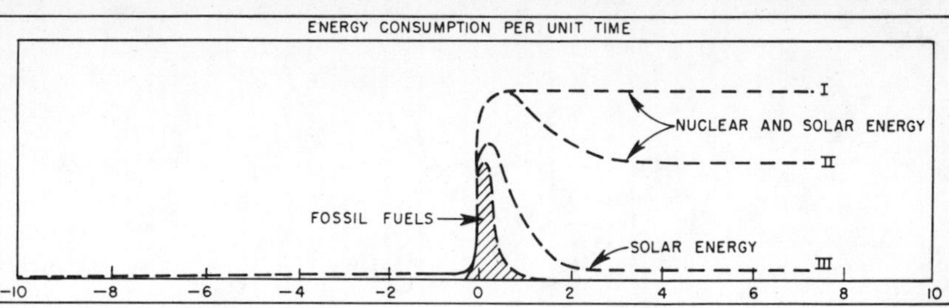

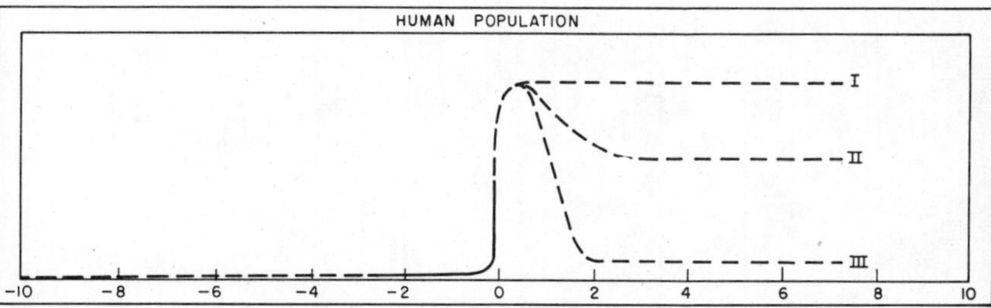

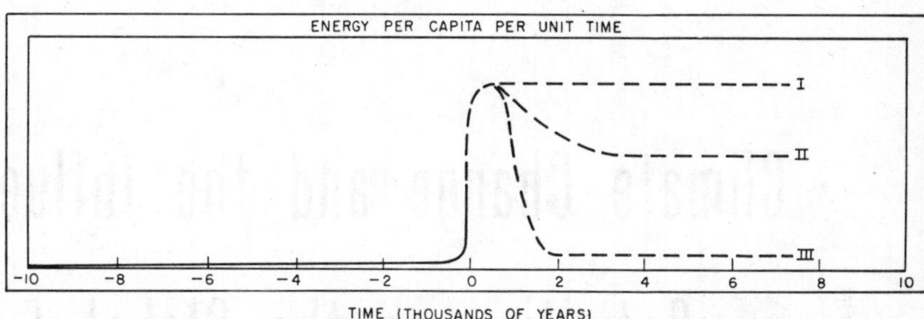

FIGURE 32 — Epoch of current industrial growth in the context of a longer span of human history (Hubbert, 1962, Fig. 61).

The second part of this ecological principle is that such exponential growth of any biologic population can only be maintained for a limited number of doublings before retarding influences set in. In the biological case, these may be represented by restriction of food supply, by crowding or by environmental pollution. The complete biologic growth curve is represented by the logistic curve of Figure 31.

That there must be limits to growth can easily be seen by the most elementary arithmetic analysis. Consider the familiar checkerboard problem of placing 1 grain of wheat on the first square, 2 on the second, 4 on the third and doubling the number for each successive square. The number of grains on the *n*th square will be 2^{n-1}, and on the last or 64th square, 2^{63}. The sum of the grains on the entire board will be twice this amount less one grain, or 2^{64}-1. When translated into volume of wheat, it turns out that the quantity of wheat required for the last square would equal approximately 1,000 times the present world annual wheat crop, and the requirement for the whole board would be twice this amount.

It follows, therefore, that exponential growth, either biological or industrially, can be only a temporary phenomenon because the earth itself cannot tolerate more than a few tens of doublings of any biological or industrial component. Furthermore, most of the possible doublings have occurred already.

After the cessation of exponential growth, any individual component has only three possible futures: (1) it may, as in the case of water power, level off and stabilize at a maximum; (2) it may overshoot and, after passing a maximum, decline and stabilize at some intermediate level capable of being sustained; or (3) it may decline to zero and become extinct.

Applied to human society, these three possibilities are illustrated graphically in Figure 32. What stands out most clearly is that our present phase of exponential growth based on man's ability to control ever larger quantities of energy can only be a temporary period of about 3 centuries' duration in the totality of human history. It represents but a brief transitional epoch between two very much longer periods, each characterized by rates of change so slow as to be regarded essentially as a period of non-growth. Although the forthcoming period poses no insuperable physical or biological difficulties, it can hardly fail to force a major revision in those aspects of our current culture the tenets of which are dependent on the assumption that the growth rates which have characterized this temporary period can somehow be sustained indefinitely.

REFERENCES CITED

Averitt, Paul, 1969, Coal resources of the United States, January 1, 1967: U.S. Geological Survey Bulletin 1275, 116 p.

Duncan, D. C., and Swanson, V. E., 1965, Organic-rich shale of the United States and world land areas: U.S. Geological Survey Circular 523, 30 p.

Folinsbee, R. E., 1970, Nuclear energy and the fossil fuels: Trans., Royal Society of Canada, Fourth Series, Vol. 8, pp. 335-359.

Hewett, D. F., 1929, Cycles in metal production: AIME Tech. Pub. 183, 31 p.; Trans. 1929, pp. 65-93; discussion pp. 93-98.

Hubbert, M. King, 1956, Nuclear energy and the fossil fuels: American Petroleum Institute, Drilling and Production Practice (1956), pp. 7-25.

———, 1962, Energy resources: National Academy of Sciences — National Research Council, Publication 1000-D, 141 p.

———, 1967, Degree of advancement of petroleum exploration in United States: American Assoc. of Petroleum Geologists Bulletin, Vol. 51, pp. 2207-2227.

———, 1969, Energy resources, *in* Resources and Man, a study and recommendations by the Committee on Resources and Man of the Division of Earth Sciences, National Academy of Sciences — National Research Council: San Francisco, W. H. Freeman, pp. 157-242.

———, 1971, Energy resources for power production, *in* Environmental Aspects of Nuclear Power Stations: Vienna, International Atomic Energy Agency, pp. 13-43.

Warman, H. R., 1971, Future problems in petroleum exploration: Petroleum Review, Vol. 25, No. 291, pp. 96-101.

Zapp, A. D., 1962, Future petroleum producing capacity of the United States: U.S. Geological Survey Bulletin 1142-H, 36 p.

Climate Change and the Influence of Man's Activities on the Global Environment

WILLIAM W. KELLOGG
NATIONAL CENTER FOR ATMOSPHERIC RESEARCH
BOULDER, COLORADO 80302

INTRODUCTION

The climate has changed many times in the past. The evidence for this is all around us, written in the shapes of the land, the composition of the ocean sediments, and the structure of the great ice caps that cover Greenland and the Antarctic.

We are now coming to the realization that man can influence the planet's environment on a grand scale. Though he may have influenced the climate already, it has so far probably been in a small way. Our concern is that he may be able to do it in a larger way, and could begin to match Nature's forces with his own in the future.

What can we say about man's influences relative to Nature's? This is the central question that we face when we consider the future of mankind on our little planet, and I will try to explain some of the things that we think we know about it at the end of this talk.

Reprinted with permission from *MITRE Corp. Rep. M72-166*, Sept. 1972.

To a certain extent the conclusions that I will present are those of an international Study of Man's impact on the Climate (SMIC)* that met near Stockholm in the summer of 1971, a study in which I participated, augmented by some additional ideas that have been introduced since then. One overall conclusion that emerged from the SMIC was that we have a long way to go before we can give very many definite answers, but the path to a better understanding of the causes of climate change is now more clearly marked than it was a few years ago.

A BRIEF LOOK AT THE LONG HISTORY OF CLIMATE

Looking at the record of past climatic conditions, it becomes so fuzzy as we turn back the clock beyond 500 million years that very little can be said about the conditions on earth before then. We believe that life came on to the dry land about a billion years ago, and we know that there was a kind of ice age at about 500 million years ago, and then another at about 250 million years ago, and that between these relatively short periods of glaciation of the poles the earth was warmer than it is now. In fact, only about 10% of the time during this 500 million year period, during which mammals evolved on the earth, were the poles glaciated as they are now.

The ice age in which we now live probably started about five million years ago in the Antarctic, and then spread to the Northern Hemisphere two or three million years ago. Figure 1, by H. Flohn gives a rough idea of how the temperature at mid-latitudes may have changed during the past 60 million years. More recently, as shown in this figure, we have had a succession of warming and cooling trends lasting for a hundred to two hundred thousand years each. We have just emerged from the last cold spell, the Pleistocene Ice Age, when a great ice sheet sat over the North American Continent, and another smaller one sat over

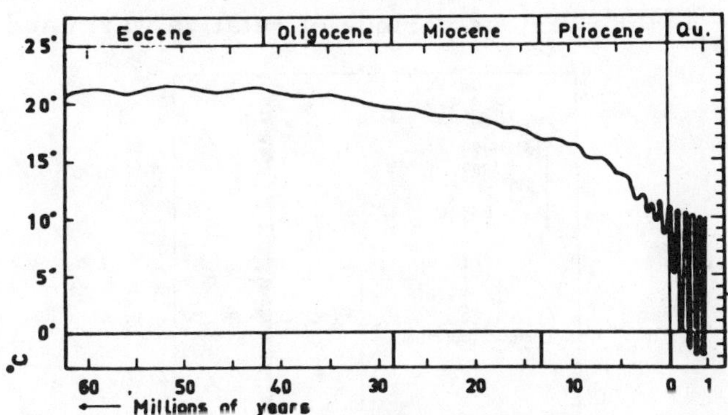

Figure 1. Mean Temperature of the Earth at Mid-Latitudes for the Past 60 Million Years (from H. Flohn)

---

*Inadvertent Climate Modification: Report of the Study of Man's Impact on Climate (SMIC). MIT Press, Cambridge, Mass., 1971.

See also: Man's Impact on the Global Environment: Report of the Study of Critical Environmental Problems (SCEP), MIT Press, Cambridge, Mass., 1970.

And also: Man's Impact on the Climate, ed. by W. H. Mathews, W. W. Kellog, and G. D. Robinson, MIT Press, Cambridge, Mass., 1971.

Scandinavia and Asia. This last period of glaciation only retreated 10,000 years ago, which is but a fraction of a second measured on the scale of time of the earth's history.

Imagine the changes that must have been forced on primitive man by these glaciations of large parts of his continents, and all the changes that must have occurred in the rainfall and temperature patterns of the rest of the land! It is almost certain that many of the deserts that exist now were much smaller or were not even deserts during the last ice age, and the patterns of forests, lakes, and rivers must have been very different from those at present, even though the continents had essentially their present shape.

More recently, only 1,000 years ago, recorded history tells us that the Vikings were able to sail in their small longboats as far as the North American continent, and then at around 1100 A.D. the cooling trend that took place brought the polar ice further south over the Atlantic, and they were no longer able to reach their colonies. Eskimos from Greenland, hunting along the southern edge of the ice pack, were reported in the Faroe Islands during this period of cold.

Even more recently, the cold of the 17th Century caused great hardship in Europe, and has come to be known as "the Little Ice Age." But these climate changes probably corresponded to mean temperature fluctuations of only one or two degrees in northern Europe, whereas the changes that accompanied the major glaciations of the more distant past must have been five or ten degrees at those same latitudes, perhaps more.

With this in mind it is interesting to study the Greenland ice core record shown in Fig. 2, prepared by Dansgaard, Johnson, Clausen, and Langway. Although the quantity plotted is oxygen isotope ratio, it can be considered as roughly proportional to mean temperature at the latitude of Greenland, and the various periods of relative warm and cold that I have just

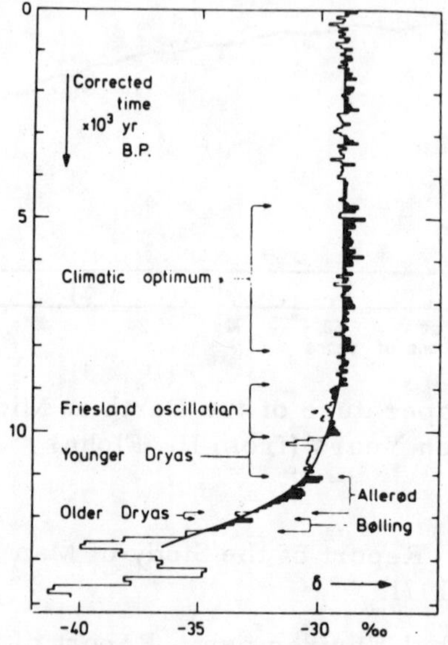

Figure 2. Oxygen Isotope Ratio in Greenland Ice Cores (from Dansgaard, *et al*).

referred to (and others) can be discerned in this 14,000 year record. The total record from this core (not shown here) goes back about 150,000 years.

THEORIES OF CLIMATE CHANGE

Why does the climate of the earth change from time to time? This is of course the question that we would dearly like to answer, and many attempts have been made to do so. The theories that have attempted to account for climate change range from those which invoke a completely external influence, such as a change in the output of the sun, to an almost toss-of-the-coin theory which tries to describe the ocean-atmosphere system as one which has several stable regimes so that it can, as it were, flip from one regime to another. This is the theory of "almost intransitivity" best enunciated by E. Lorenz of MIT. In between are other theories which are undoubtedly all parts of the picture, but we do not yet know the relative importance of such things as volcanic activity (which could darken the sky for decades or centuries at a time), the drift of continents (which on a very long time scale can certainly alter the relative behavior of the oceans and the atmosphere), the fluctuations in the ocean circulations themselves, etc. A complete theory of the climate will have to take all of these various influences into account in order to reconstruct what has happened in the past.

The ocean-atmosphere-land system is a complicated one indeed, and no theoretical model has yet been constructed which pretends to include all the important factors. In fact, even the modeling techniques which have been used successfully to simulate the day to day behavior of the general circulation of the atmosphere will not work for climate study models, a point that is discussed at some length in the SMIC report.

THE CONCEPT OF FEEDBACK MECHANISMS

Even though we really do not have a satisfactory theory or model for the climate yet, we do see several very important sets of relationships that can govern this quasi-equilibrium that I have referred to. In order to understand these, recall the idea of a feedback mechanism in electronics, where one can amplify a small oscillation with an amplifier, or one can damp it out by introducing a resistance in the circuit. Let me show you a couple of examples of feedback mechanisms that are probably operating in our atmosphere right now.

The first of these is the famous example of the carbon dioxide that man has been putting into the atmosphere. The carbon dioxide content of air has indeed increased in the past few decades, and we are reasonably certain that this is due to the large amount of fossil fuels that have been burned during this period, thus releasing the carbon locked beneath the surface of the earth and putting it into the atmosphere and the oceans. Figure 3, by L. Machta, shows the past increase and an estimate of the future trend up to 2000 A.D. One of the effects of this carbon dioxide in the atmosphere is to trap some of the heat which would otherwise escape to space in the form of infrared radiation -- it serves as a kind of blanket.

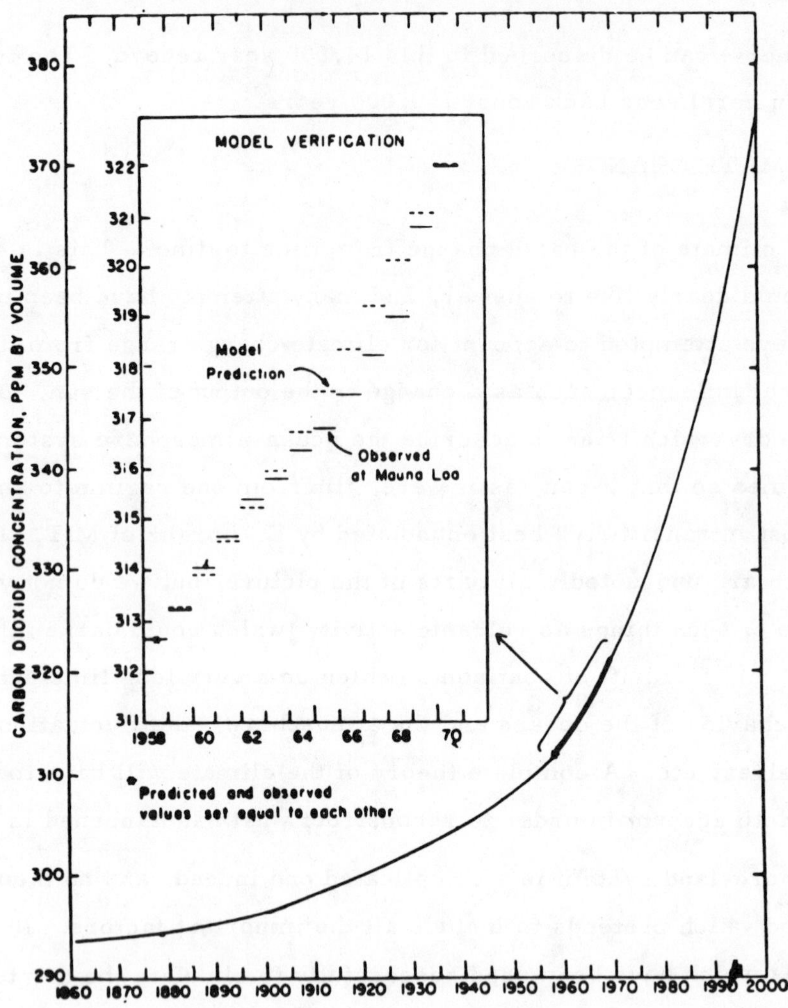

Figure 3. Carbon Dioxide Concentration in the Atmosphere Since 1860 (from Machta).

The sunlight can still come through and heat the surface, but the infrared radiation from the surface cannot escape as easily when there is more carbon dioxide in the atmosphere. This is known as "the greenhouse effect," and adding more carbon dioxide simply increases the temperature at the surface of the earth.

Now, granting this, what other things will happen when we increase the surface temperature? One thing is that more water vapor will be evaporated from the surface of the oceans. Water vapor acts very much as carbon dioxide does with regard to the greenhouse effect, so the extra water vapor added to the atmosphere further enhances the greenhouse effect, and further raises the surface temperature. It seems as though this complementary effect could go on indefinitely, with a small increase in carbon dioxide causing a small increase in water vapor, etc. This positive feedback mechanism, or amplification, has sometimes been referred to as "the runaway greenhouse," and our sister planet Venus is cited as a sad example of the complete end of such a runaway greenhouse effect -- a miserable planet where the temperature at the surface is hot enough to melt lead, and where all the water that might have been in the oceans and all the carbon dioxide that might have been in the rocks has remained in the atmosphere, adding to a super-greenhouse effect.

Another example of a feedback mechanism, one that also happens to be a positive one, is the polar ice. Consider the fact that the polar ice reflects much more sunlight than the land or the open seas, and therefore, if we increase the area of the polar ice, more sunlight will be reflected. A small decrease in temperature of a degree or two (particularly if it is in summer) can cause the polar ice cap to spread towards the equator, this reflects more sunlight, this causes less heat to be available to warm the planet, the temperature falls still further, the polar ice creeps still further towards the equator and so on until (one might expect) the polar ice would reach the equator and the planet would become a solid cake of ice over its entire surface. This mechanism, which has intrigued meteorologists in both the U.S. and the Soviet Union, can be called the "runaway ice cap."

Obviously, the earth has not become frozen, but we can learn from this example how sensitive the polar regions are to a small change in the heat balance of the planet as a whole. We can see that this is so by the fact that, while the equatorial regions seem to have had very little change of temperature during the ice ages, the polar regions undergo very large changes in temperature. During the past century, when the mean temperature of the earth has been changing by something under 1°C, as shown in Fig. 4 by J. M. Mitchell, the polar regions have changed by over one degree Centigrade in just the last decade, as shown in Fig. 5 by H. Flohn.

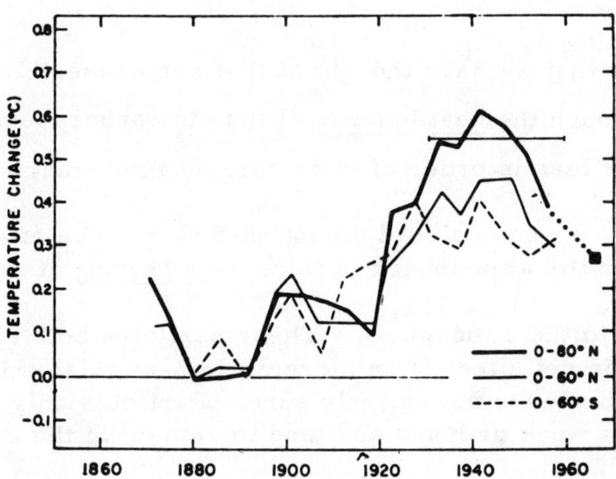

Figure 4. Mean Temperature Changes of the Earth at Different Latitudes since1960 (From J. Mitchell).

The conclusion that one draws from these rather simple considerations is that this ocean-atmosphere system of ours can certainly change its balance, and sometimes quite suddenly. The remarkable thing is that there must be a kind of shakey equilibrium, and that the earth has remained habitable by life forms more or less as we know them for at least a billion years. Somehow, these runaway mechanisms have not been allowed to spoil things irrevocably for the planet earth, as they must have for the planet Venus. There are

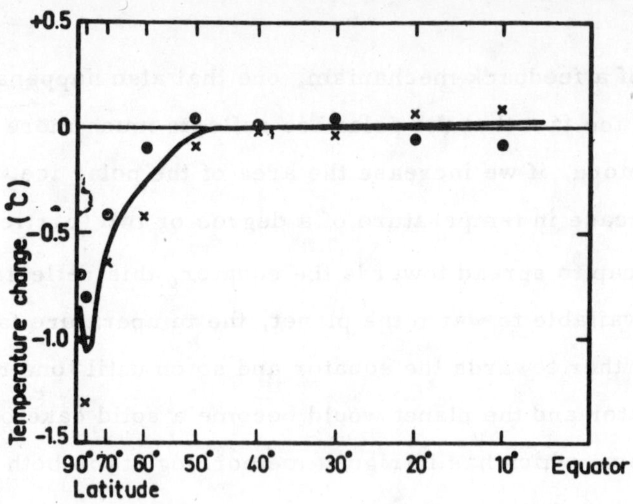

Figure 5. Temperature Change of the Earth at Various Latitudes During the Last Decade (from Flohn).

obviously stabilizing factors that are strong enough to keep our global climate within reasonably narrow bounds, permitting ice ages to come and go but damping out any large fluctuations.

ENTER CIVILIZED MANKIND AS A FACTOR IN CLIMATE CHANGE

Now man enters the scene, and we must ask whether he can reach any of the lever points on this gigantic mechanism and influence it. If there are lever points that he can reach, history has shown that he will probably be tempted to tamper with them, whether for better or for worse.

All the lever points that we have thought of that are accessible to man (except one important one) operate through the heat balance of the atmosphere-ocean system. The main human influences, more or less in order of their current importance as we see them now, are:

- **Carbon dioxide - We have already mentioned this. The effect of increasing carbon dioxide in the atmosphere is to cause a heating at the surface.**

- **Particles added to the atmosphere - These particles both scatter and absorb radiation, and the net effect of an increase in particulates is probably a cooling influence, though we are not entirely sure. Particles only remain in the lower atmosphere for a week or less, and tend to remain in the latitude belt in which they are created.**

- **Changing the reflectivity or absorptivity of the earth -- its "albedo." Most of the things that we can think of, such as plowing grassland, cutting down forests, building cities, etc., cause more heat to be retained by the surface and therefore cause a net heating.**

- **Irrigation of farmland and evaporation of water. Although the immediate and local effect of evaporating water in an irrigated region may be a cooling, the ultimate effect is a warming, since more sunlight is absorbed and the heat is released back into the atmosphere when the water condenses as rain or snow at some other place.**

- **Direct release of heat from the generation of energy - This obviously results in a heating. In the decades to come, this will undoubtedly move to the top of the list if we continue to generate energy at an ever increasing rate.**

Looking back at this list, it seems clear that as we continue to burn carbon fuels the carbon dioxide that we add to the air will continue to increase. So will the other factors, and most of them are in a direction of increasing the surface temperature -- all but one, the dust.

The last one on the list, the direct release of heat to the atmosphere from man's insatiable need for energy, is also continuing to go up, and will do so whether he runs out of fossil fuel and resorts to a nuclear fuel or not. Estimates of the increase in the amount of energy that man will require vary greatly when we go beyond about the year 2000 A. D., as we all recognize, but here are some interesting statistics about the release of heat now and estimates up to the year 2000 A. D., taken from the SMIC Report.

The flux of solar radiation at the earth is 1400 w/m^2, and the average net radiation in the course of a day and night at the top of the atmosphere is about 350 w/m^2, which obviously depends on latitude. Of this, about 30 percent is reflected, and less than half of the remainder is absorbed by the surface, so the solar radiation that can be used to heat the surface and the lower atmosphere is, on the average, about 100 w/m^2.* Now, if we consider that the industrialized area of the world (about 0.5×10^6 km^2) accounts for 75 percent of the world's energy production and release in the form of heat, the average flux of artificially generated heat is about 12 w/m^2. The energy production divided by urban area for some cities is considerably greater than this average, being about 630 w/m^2 for central New York City, 127 for Moscow, 21 w/m^2 for West Berlin and Los Angeles, and 7.5 w/m^2 for all of Los Angeles County, an area of 10^4 km^2. It is thus fair to say that in several areas of 10^4 to 10^5 km^2 the manmade energy production is 5 to 10 percent of the net solar radiation absorbed.

Since a reasonable estimate for the increase of energy production gives about a factor of five increase by the year 2000 A. D. (5.5 percent per year for 30 years), it is likely that over such areas in the developed countries energy production will be 25 to 50 percent of the net solar radiation absorbed, and the average for all the continents if it were to be spread evenly would be nearly 1 percent.

A widely quoted vision of the future by Herman Kahn foresees a world in 2100 A. D. with 20 billion people, each family using an amount of energy equivalent of a U.S. family with a $20,000 per year income (Kahn's "20-20 vision"). One can make an estimate of the total energy production by such a society and compare it with the energy received from the sun, and the human energy flux comes to about 2 percent of the net absorbed solar energy for the planet as a whole -- and 4 to 5 times this for the continents alone. Such an artificial source of heating would cause several degree rises in the mean temperature of the continents, probably more at the poles -- our knowledge of how the non-linear atmosphere-ocean system would respond to this input suggests that it would be highly non-uniform and complex, and still largely unpredictable with our present models. (See below.)

---

*Since the continental land masses are mostly at mid-latitudes, the average for the continents is 67 w/m^2.

There is one part of the system that might be particularly sensitive to such a mean warming of the Northern Hemisphere, the Arctic Ocean. This point is discussed in the SMIC Report, and the essentials are as follows: The Arctic Ocean ice pack is about 3 m thick on the average in winter, and less than 2 m thick by the end of the summer melting season. Although there are leads with open water, especially in summer, they freeze over again quite readily in winter, the freezing encouraged by the low salinity of the top 10 to 30 m of the Arctic Ocean. There is, in other words, a relatively fresh water layer that floats at the surface under the ice cover, not mixed by the wave action that would be present in an open ocean.

We do not know how much mean temperature rise would be required to melt the ice pack to the point that it would not be able to reform the following winter, but it must be just a few degrees. Summer temperatures are more influential here than those in winter. After some appreciable opening of the Arctic Ocean to wave action the fresh water layer would be mixed with the salt water below, and the same wave action would prevent ice from reforming except in bays and estuaries. The lower albedo of the open water would cause more solar radiation to be absorbed, thereby further warming the Arctic Basin. The added evaporation would cause more rainfall in summer and more snowfall in winter.

One study of the regime that would exist with the Arctic Ocean ice removed was made by the RAND Corp., using the Mintz-Arakawa 2-level general circulation model, and Fig. 6 from the report by Warshaw and Rapp shows the temperature difference and the zonal wind difference between the condition with the "ice out" and the "ice in." The circulation patterns were considerably changed by the addition of the oceanic heat source that became available when the sea ice was removed, in particular a weakening of the cold core low vortex over the Arctic Basin. Unfortunately, the model did not, apparently, reveal the altered precipitation patterns.

Thus, the fact is that man by sheer brute force may be able to begin to add enough heat to the atmosphere to be a definite factor in the climate, and in the Northern Hemisphere the Arctic Ocean might boost the effect after it reached a certain limit.

The next thing that should be considered is what would happen then to the ice caps of Greenland and the Antarctic Continent. It is possible that a warming of the polar regions would result in a gradual melting of these ice caps with a corresponding rise in the ocean level. (Melting the Greenland ice cap alone would cause a rise of about 7 m.) However, this outcome is certainly not obvious, since there would probably also be an increase in snowfall over Greenland due to water vapor added by the open Arctic Ocean. In any case, the rate of change of these vast ice caps would be very slow indeed, and the shrinking or growth would be measured in centuries. This is clearly one of the problems that requires more homework.

Some experiments have been done at the National Center for Atmospheric Research by Dr. Warren Washington to see how much the climate does change when additional heat is added to the atmosphere in the populated regions of the earth. As one approaches a point

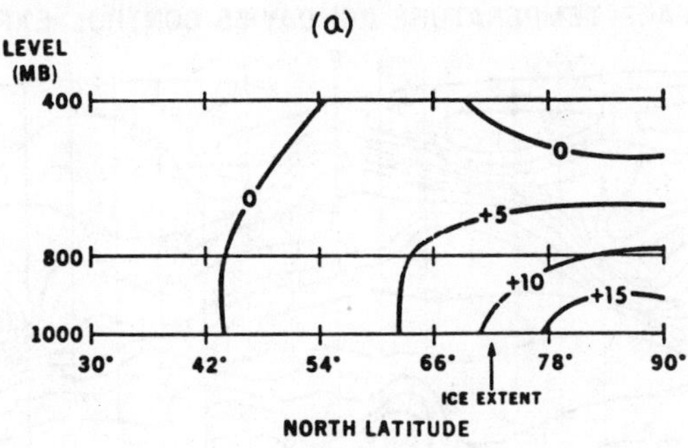

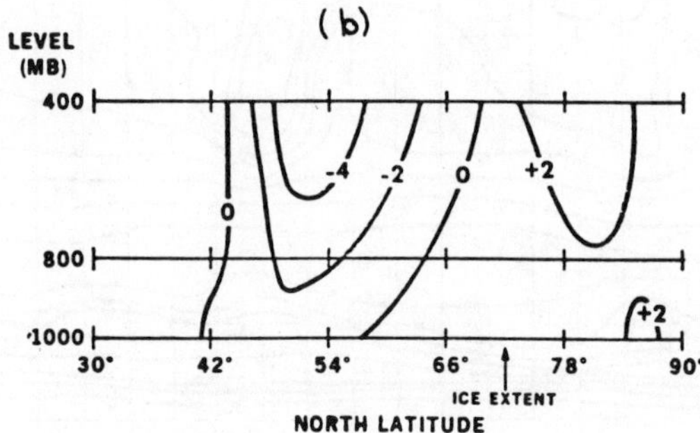

Figure 6. At top (a) are shown the temperature differences between runs of the Mintz-Arakawa general circulation model with the Arctic Ocean Ice out and the ice in, as a function of latitude and height averaged for a latitude zone; and the lower diagram (b) shows the corresponding differences in the west-east (zonal) wind in m/sec. Removing the arctic ice caused a weakening of the mid-latitude westerlies and a slight increase in the higher latitude westerlies.

Source: M. Warshaw and R. R. Rapp, An experiment on the sensitivity of a global circulation model· Studies in climate dynamics for environmental security, RAND Report No. R-908-ARPA, Santa Monica, California, 1972.

where something like 50% of the solar energy is matched by man-made energy over large areas of the earth,(probably an unrealistically large amount, chosen to demonstrate the qualitative changes) one can see definite changes in the circulation pattern of the atmosphere, changes that can be interpreted as changes in the climate. The changes are remarkably nonuniform as shown in Fig. 7a and 7b, but our models are not entirely suitable for this kind of experiment, and we must continue to study the matter further.

The one human environmental factor that I mentioned that was not directly related to the radiation balance is the possible influence that large-scale operations of jet aircraft in the stratosphere might have on the ozone layer. As you all remember, this was quite an important issue in the U. S. during the debate on the SST prototype development appropriation about a year ago. The point was first raised by the Study of Critical Environmental Problems (SCEP), a summer study organized by MIT in 1970, and was further looked at by the Study of Man's Impact on the Climate (SMIC) in 1971, and by a number of other individuals. The idea is that the water vapor and the oxides of nitrogen released by these high-flying jets could react with the ozone in the stratosphere, and in such a way as to reduce the amount of ozone. A reduction in the amount of ozone would cause somewhat more ultraviolet sunlight to reach the surface of the earth.

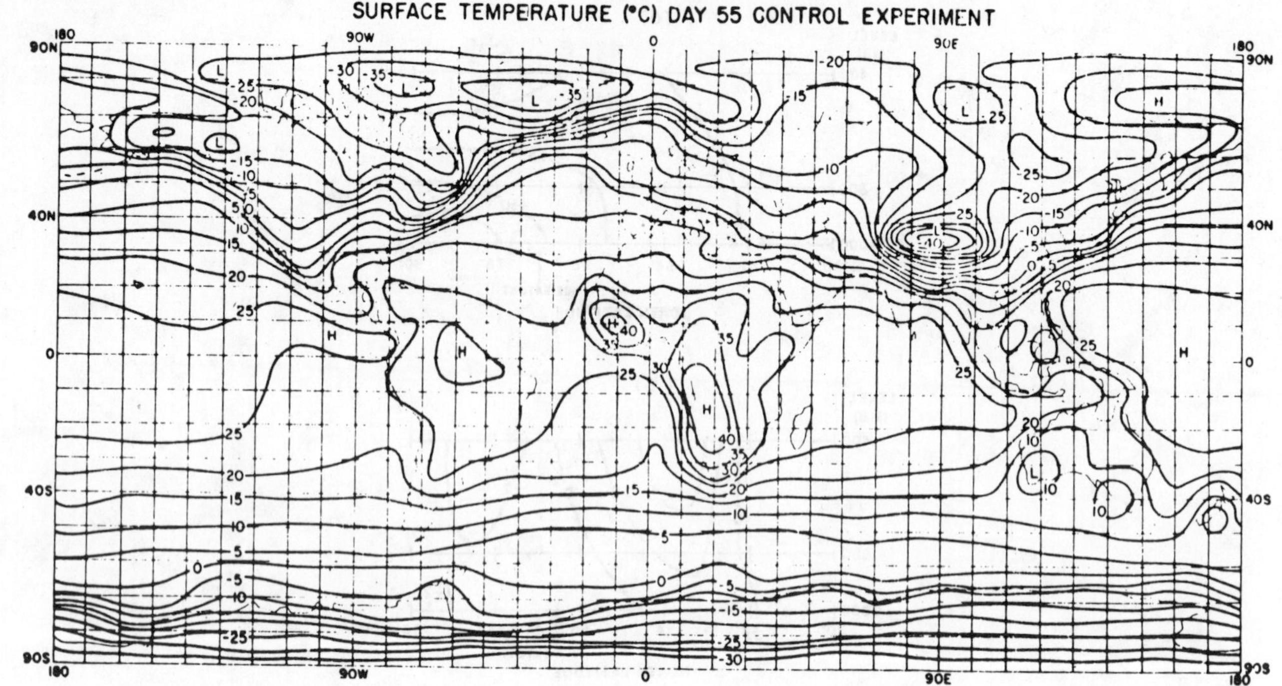

Figure 7a. Global surface temperature distribution calculated by the NCAR six-level general circulation model after the model has run for 55 simulated days. The heat input is just that from the sun alone.

Source: W. M. Washington, On the possible uses of global atmospheric models for the study of air and thermal pollution, Chapter 18 in Man's Impact on the Climate, ed. by W. H. Mathews, W. W. Kellogg, and G. D. Robinson, M. I. T. Press, Cambridge, Massachusetts, 1971.

While this is probably an effect which can eventually be predicted, some of the factors that we need to know in order to make a definite estimate of the effect of SST's are still being sought, and the main hope will be in developing models of the atmosphere and specifically of the ozone region that are adequate to do experiments with. By this I mean that we will use a numerical model, run on a large computer, that will simulate the atmosphere adequately to demonstrate the probable effects of large numbers of SST's. At the moment, we do not know what the answer will be though some early estimates have shown that very large numbers of SST's could indeed cause an appreciable change in the ultraviolet radiation reaching the ground. This would, in turn, be an environmental change on a global scale that would have quite an influence on us. But I must emphasize that we really have not finished our homework on this yet, and the influence of even very large numbers of SST's may turn out to be so small that we should not be concerned about them.

CONCLUSION

From the above one can, and probably should, conclude that man can influence the climate of his planet Earth. The direction that this influence will take in the decades to come, if man continues to demand more energy to satisfy his craving for an ever improving standard

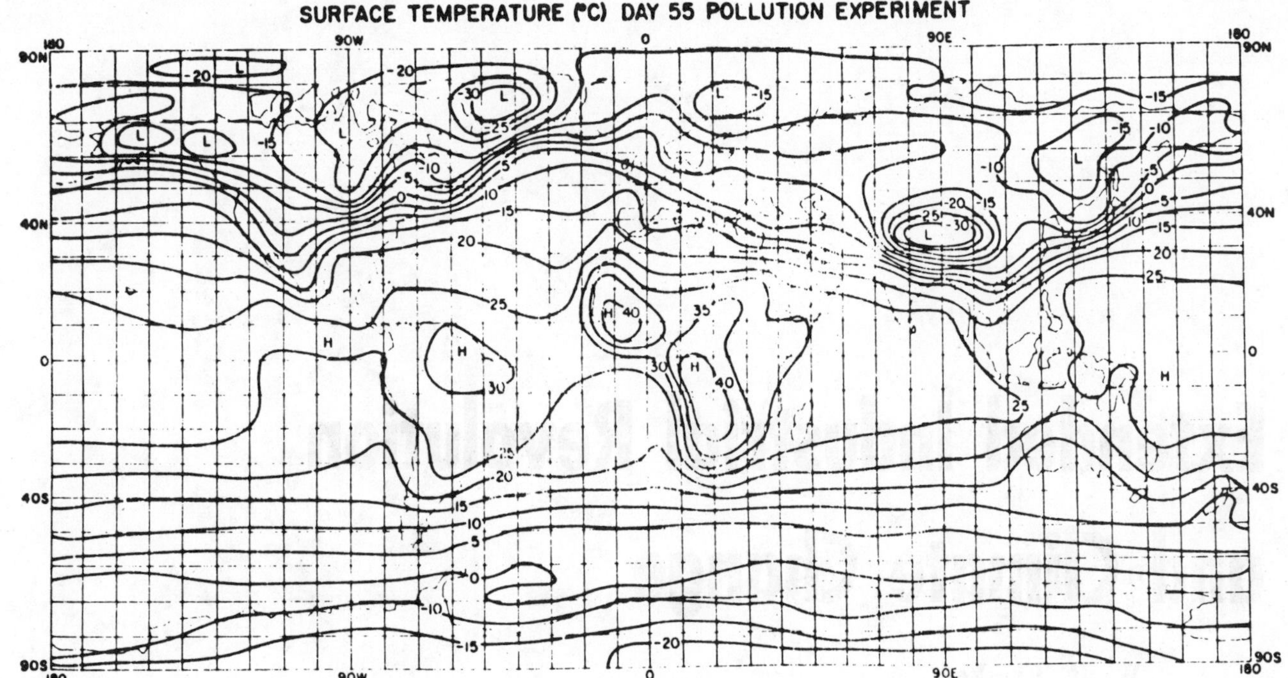

Figure 7b. Thermal pollution is assumed to add another 50% to the heating of the continents. Ocean temperatures are the same in both runs; the conditions are those for January. In this model a complete hydrologic cycle is included as well as mountains.

Source: W. M. Washington, On the possible uses of global atmospheric models for the study of air and thermal pollution, Chapter 18 in Man's Impact on the Climate, ed. by W. H. Mathews, W. W. Kellogg, and G. D. Robinson, M. I. T. Press, Cambridge, Mass., 1971.

of living, coupled with his increasing population, must be that of a warming, especially in the Northern Hemisphere. The Arctic Ocean pack ice represents an unstable part of the ocean-atmosphere system, in that a warming that would remove the pack ice would produce a major one-way transition, and the northern polar regions would then include a large open body of water the year around. The implications of such a transition are very grave for some regions of the earth, but cannot be said (as some have) to spell disaster for mankind.

While we do not know how to predict such a transition, nor indeed whether it will take place at all in the foreseeable future, the distinct possibility that it could occur as a result of human influences warrants every effort to understand the mechanisms that govern climate and climate change. In fact, this is likely to be one of the major future thrusts of atmospheric research, one that has already begun.

Extended Industrial Revolution and Climate Change

W. R. Frisken

The earth's climate has changed noticeably within man's recorded history and much more dramatically during that longer period whose record we must examine in the geology of earth's crustal rock. To plan an intelligent use of our resources, we must frame our plans in a total environment; and the earth's climate is perhaps the determinant factor in this environment. We must understand how the climate is going to change and whether man's activities can influence climate. It is clear that if our activities are of sufficient scale to cause the climate to deteriorate, then they might also be made to improve it, at least for some minority of the earth's population. Understanding is also important on the part of those whose interest might be restraining such experiments.

Fundamentals of Climate

The fundamental physical processes determining climate have been understood for many years. Any object will tend to cool off by radiating electromagnetic energy at a rate proportional to the fourth power of its absolute temperature. This energy is radiated as discrete quanta (or photons) in a spectrum of wavelengths characteristic of the temperature of the source, peaking at a wavelength that is inversely proportional to that temperature. We see then that a hot object emits energy rapidly, peaked at short wavelengths, while a cool one emits energy slowly, peaked at long wavelengths. Figure 1*a* (adapted from *Robinson* [1970]) shows the spectra from the sun and from the earth. Note that the sun's spectrum is populated well into the ultraviolet (shorter wavelength than the visible spectrum which runs from 0.4 to 0.8 microns) and far into the infrared.

If the earth had no atmosphere, the equilibrium temperature of its surface would be just high enough to radiate away as much heat in infrared radiation as it received in the form of 'insolation' or incoming solar radiation. (We neglect here the small amount of heat reaching the earth's surface from its own interior, namely, $10^{-4} \times$ incoming solar radiation [*Stacey*, 1969].) If the earth had a simple atmosphere of molecules which did not interfere with the insolation, but which absorbed infrared radiation, then this atmosphere would absorb some of the outgoing infrared radiation emitted by the surface of the earth and would warm up. This warm atmosphere would now itself become an infrared radiator, and some of its radiation would be directed toward the surface of the earth. This 'greenhouse effect' would clearly cause the earth's surface temperature to rise.

Our real atmosphere is much more complicated than this. The insolation has the high energy (ultraviolet) end of its spectrum absorbed by oxygen in the upper atmosphere (Figures 1*b* and 1*c*), some of its blue light is diffused by Rayleigh scattering from water molecules and dust particles (making the black sky blue, and the white sun red, Figure 1*d*), and all parts of the insolation spectrum suffer some absorption and reflection by clouds and aerosols. So the real atmosphere presents us with a somewhat reduced and considerably diffused solar beam, and in turn it is warmed both by the direct insolation and by the outgoing infrared from the earth's surface. The most efficient absorber of infrared radiation in the lower atmosphere is water vapor, but CO_2 and O_3 are more important in the upper atmosphere. The lower atmosphere or troposphere is mostly heated by contact with the warm surface of the earth

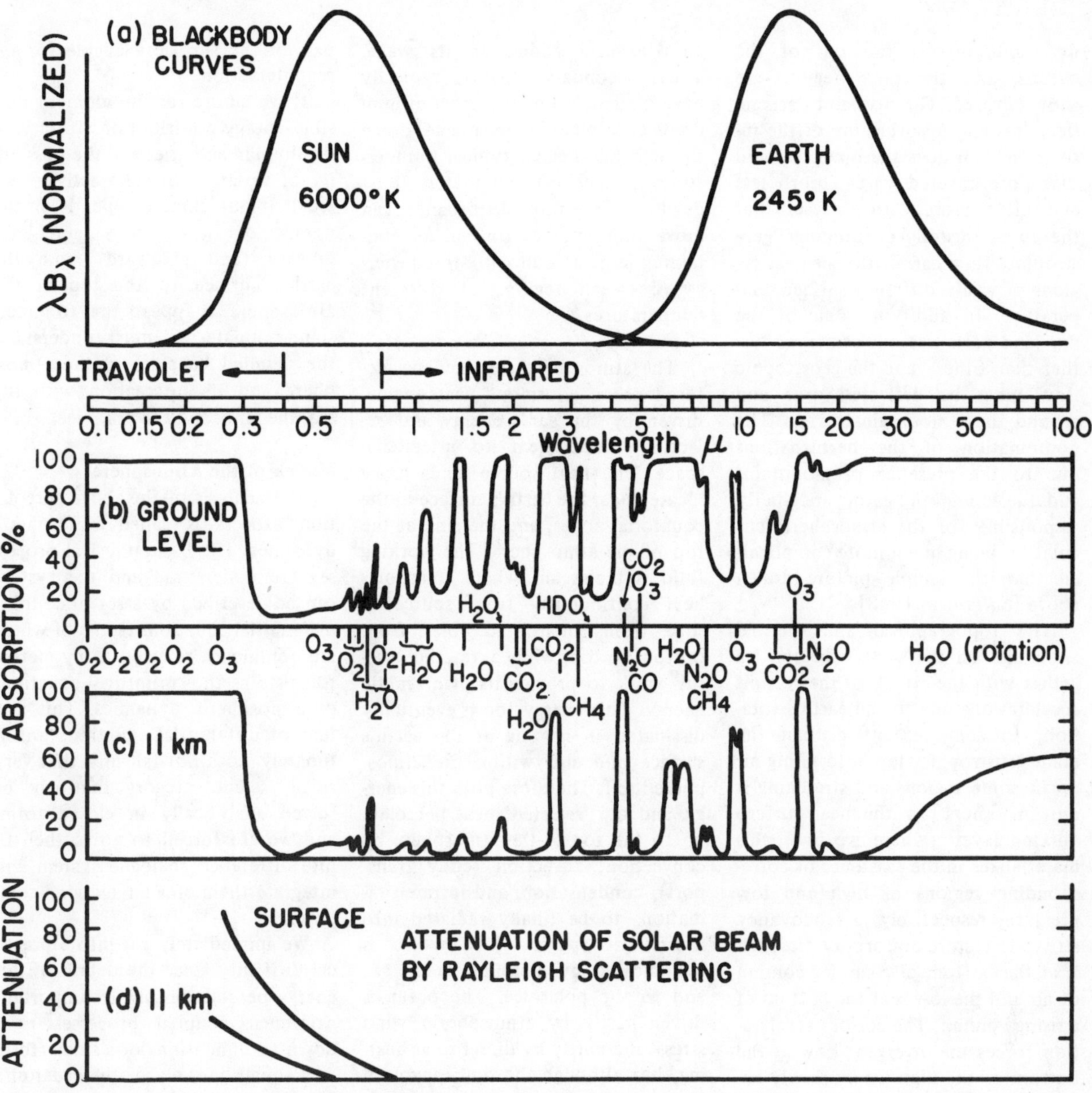

Fig. 1. (a) Blackbody emission for 6000°K and 245°K, being approximate emission spectra of the sun and earth, respectively (since inward and outward radiation must balance, the curves have been drawn with equal areas—though in fact 40% of solar radiation is reflected unchanged); (b) atmospheric absorption spectrum for a solar beam reaching the ground; (c) the same for a beam reaching the tropopause in temperate latitudes; (d) attenuation of the solar beam by Rayleigh scattering at the ground and at the temperate tropopause.

and by water vapor condensing in it, the former having been warmed and the latter having been evaporated by the sun's energy.

In the equatorial zone the solar beam comes from directly overhead, causing the earth's surface and the lower atmosphere (troposphere) there to become preferentially heated. The air in the equatorial troposphere expands and tends to spill over at the top, and this poleward motion in the upper troposphere drives the atmospheric circulation. The earth is a spinning globe, and the poleward flow brings the air closer and closer to the axis of rotation. The associated inertial forces (coriolis forces), from the point of view of an observer fixed on the earth's surface, deflect the poleward flow into an almost *totally circumpolar* upper westerly (eastward) flow. (An observer on one of the fixed stars would see that, as the high level air slides poleward from the equator, its rotation speeds up relative to the surface of the earth.) This gives us a tidy (but simplistic) picture of a troposphere whose slow convective rising at the equator and sinking at the poles is superimposed on a much more vigorous upper westerly flow, giving a toroidal activity in both hemispheres.

However, if we have been describing a fairly real atmosphere, we have not been describing a real earth. The real earth is mostly covered with water of sufficient depth that the ocean circulation patterns can themselves be very complicated and take hundreds of years. Irregular patterns of dry land masses rise out of the oceans, and the polar regions are snow covered. The dry land parts reflect less and absorb more of the insolation than does the open sea, and the snow covered parts absorb less and reflect more, with the result that the upper westerly circumpolar 'geostrophic' flow carries the air over regions of vastly different surface temperature. In addition, some of the dry land parts rise out of the sea further than others, and the geostrophic flow must be deflected over and around these mountain barriers. The combination of the perturbations due to the presence of mountains and due to uneven heating are chiefly responsible for the atmospheric circulation being much more complicated that the simple picture drawn above [*Saltzman*, 1968].

The topographical and thermal anomalies on the surface below, together with the effects of the coriolis accelerations due to the earth's rotation, to some extent distract the main geostrophic flow into piling air up in some regions and stretching it out in others. In the near surface 'mixing layer' (where we live), this materializes in the existence of corresponding regions of high and low pressure, respectively. As buoyancy strives to restore uniformity, the high is at the bottom of a sinking column of air and the low is at the bottom of a rising column. The coriolis acceleration forces the divergent flow at the bottom of the high into a slowly expanding anticyclonic (clockwise from the top) vortex, and the confluence at the bottom of the low into a slowly collapsing cyclonic vortex. (Both directions are reversed in the southern hemisphere.) Cyclonic disturbances of this type usually have dimensions of one or two thousand kilometers.

Although the growth of a large-scale vortex or eddy is often intimately connected with a definite location on the surface of the earth below, once formed it will characteristically move off across the earth's surface under the influence of many factors including the upper westerly flow. As it moves across the surface of the land and sea like a giant vacuum cleaner (for example, in the case of a low), it sets up a series of smaller-scale eddies in its wake. These secondary vortices typically have dimensions of the order of hundreds of kilometers (hurricane) down through kilometers (typical thunderstorm dimension) and meters (dust devil). The entire disturbance can move thousands of tons of air containing large amounts of stored energy between regions of different temperatures.

The atmosphere can then be pictured as an enormous heat engine, driven by the sun's energy and 'rejecting' waste heat to interstellar space. The heat source tends to be located near the earth's surface in the equatorial zones, and the sink at the top of the atmosphere. The working fluid is moist air, which transports heat continuously from source to sink, from equator to poles, from bottom to top, while extracting 'useful' work to provide its own kinetic energy. This energy too is eventually dissipated in heating at the earth's surface and also within the atmosphere itself. Therefore both this energy and the 'rejected' heat percolate up to the top of the atmosphere by convection, advection (eddy transport), condensation, and infrared radiation, to be finally radiated into interstellar space. This picture is complicated by coupling to the sea and to the polar ice. The ocean is driven partly by atmospheric 'wind stress' and partly by direct solar heating, but although the times involved in large-scale atmospheric motions are of the order of a few weeks, the deep ocean circulation is known to be complicated and to take hundreds of years. Melting the polar ice would absorb a very large amount of latent energy and would probably require similar or longer times, although *Fletcher* [1969] thinks fluctuations in the extent of antarctic sea ice may have caused changes in the strength of the general circulation on a much shorter time scale. Despite the long times involved, we are very interested because the temperature of the sea is expected to have a considerable effect on the composition of the atmosphere. The extent of polar ice affects the earth's albedo (reflectivity) and hence the amount of solar energy absorbed; it also covers up part of the surface available for air-sea interaction.

If we ignore for the moment possible energy contributions from various human activities and the possibility of variation of solar activity, we see that our climate depends on the details of how energy percolates upward (and poleward) from the earth's surface to the top of the atmosphere. It appears that to understand climate we must understand the detailed behavior of the atmosphere and its interaction with the sea, the land, and the polar ice.

Models of the Atmosphere

The analysis of the above percolation process is called atmospheric dynamics. The basic physical processes are understood and the system can be described by a set of coupled differential equations (some of which are nonlinear), referred to by meteorologists as the 'primitive' equations of atmospheric dynamics. This system of differential equations unfortunately does not fall into that very small, select category that can be solved analytically in closed form, and we are forced to go to their finite difference analogue system and integrate them on a large digital computer.

We immediately get into a practical difficulty since the details of this energy percolation involve important nonlinear coupled processes right down to dimensions of a small storm or a squall line, say of the order of a kilometer, and beyond. Even the largest computers cannot manage a global calculation with this attention to detail. There are two obvious ways around this difficulty, and both involve specification and/or parameterization of part of the problem. By specification we mean fixing, as in specifying the distribution of ocean surface temperatures which are to be subsequently held constant, and by parameterization we mean expressing as a dependent variable through known physical or empirical relations, as in giving the rate of vertical convective energy transport as a function of the temperature lapse rate.

General Circulation Models. The primitive equations are integrated over a large region, often a hemisphere or the whole globe, using a finite differencing scheme with an integration step as small as is practical. For example, the stepping grid might be 200 kilometers square horizontally and the atmosphere might be divided into as many as 18 layers vertically. Such coarse integration stepping does not allow the simulation of the sub-grid scale eddy transport and diffusion phenomena, and these must be parameterized. Investigators (for references see *Smagorinsky* [1969], *Oliger* [1970], *Mintz* [1968]) variously specify or parameterize conditions of humidity, cloud and snow cover, etc. It is also customary to specify the ocean surface temperature distribution, but *Manabe and Bryan* [1969] have recently done some initial numerical experiments with a joint ocean-atmosphere general circulation model. General circulation models use fantastic amounts of computing time and have always saturated the existing generation of computers. There have always been parts of the problem that have clearly warranted more detailed computation, and this has ensured that the most comprehensive model of any given period has required about one day's computer time to integrate one day's weather. This is especially striking when we see that computing speed has increased three orders of magnitude since 1953, and it will have increased a further two orders of magnitude when the ILLIAC IV comes on [*Smagorinsky*, 1969].

Simple models. Typically, a vertical column of atmosphere at midlatitudes is divided into several layers. Both large *and* small scale lateral transport phenomena are, of course, now parameterized or specified, usually as zonal averages, along with cloud and snow cover, humidity, etc. Such models [*Budyko*, 1969; *Sellers*, 1969; *Manabe and Wetherald*, 1967] do, however, have the virtue of computational speed and lend themselves to initial investigations of the effects of various attempts at climate tinkering, such as that of changing the atmospheric content of aerosols and carbon dioxide. At least one investigator (compare *Manabe and Wetherald* [1967] with the subsequent work, *Manabe and Bryan* [1969]) has used a one-dimensional model to try out schemes of parameterization intended for eventual incorporation into a general circulation model. Since their very nature implies a heavy reliance on specification and parameterization, the simple vertical models tend to suppress synergistic effects, and some of their predictions ('all other things being equal') have been rather extravagant. The climate changes they predict will clearly affect the general circulation of the atmosphere, and tend to invalidate their original parameterization of the lateral transport phenomena. The sharpness of the distinction between the seasons may also change, but they ignore seasonal variations entirely.

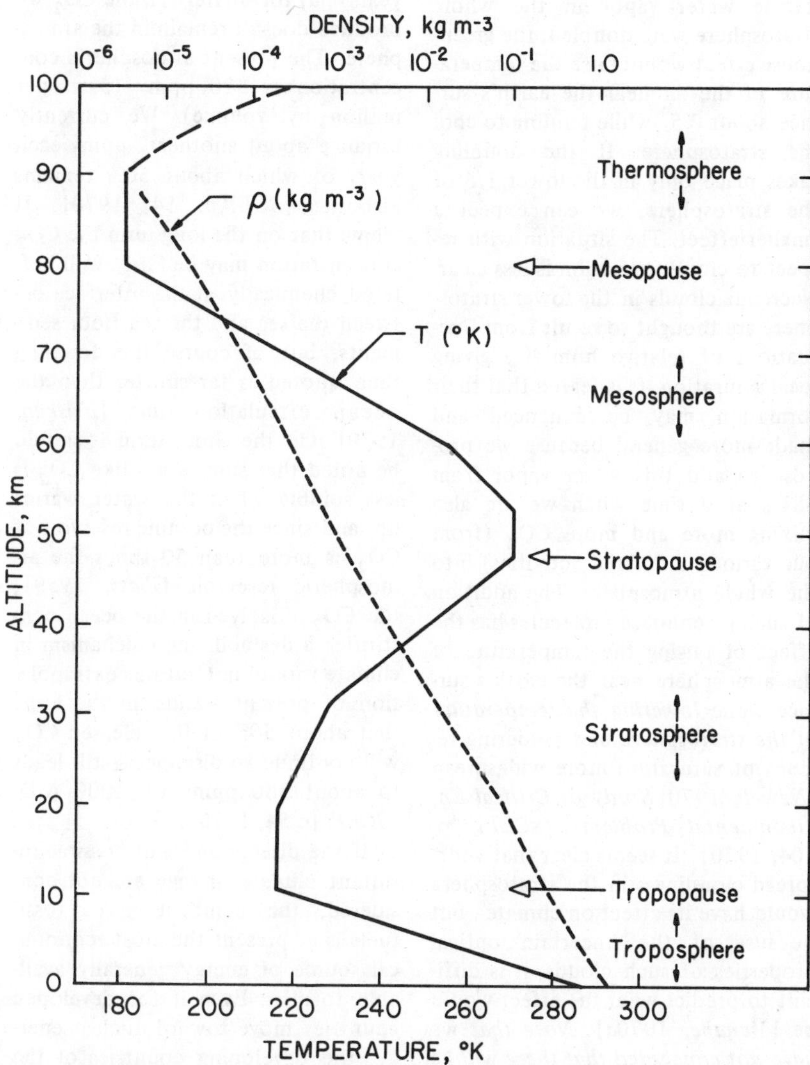

Fig. 2. Temperature and density profile according to the U.S. Standard Atmosphere, 1962, and the IUGG nomenclature, 1960.

Atmospheric Pollutants and Climate Change

Pollution of the troposphere and of the upper atmosphere has been much discussed recently. (Martell, unpublished report, 1970; *Robinson* [1970*b*]; *Kellogg* [1970]). After all the straw men have been introduced and duly knocked down, the following emerge as worth watching: carbon dioxide, aerosols, and stratospheric water vapor; and in a slightly different sense, heat from man's energy conversion. The temperature and density distributions of the atmosphere are given in Figure 2 for reference.

Aerosols. Aerosols are hard to deal with, largely because these small airborne particles have such a variety of sizes, optical properties, and atmospheric residence times. Over most of the size range of interest, they scatter the insolation predominently in the forward direction, and make the atmosphere turbid or hazy. They also tend to scatter some of the light in the backward direction near 180°, and they tend to absorb some of it both on the way down, and again, if reflected from the earth's surface, on the way up. The backscattering tends to reduce the amount of solar energy available to the earth's heat budget, and the absorption tends to warm up the atmospheric layer containing the aerosol. If this layer is high enough, cooling of the near surface environment results, and *Mitchell* [1970*a*] points out that the recent global cooling trends may be caused by the umbrella of fine dust cast into the stratosphere by recent volcanism. *Bryson* [1968] thinks the cooling trend is due to man-generated turbidity, and he cautions against the possibility of triggering another ice age, but a recent calculation by *Mitchell* [1971] shows that typical man-made aerosols in the lower layers of the troposphere lead to net heating of the near-surface environment in most cases. A more detailed discussion with references to the original research is given by *Robinson* [1970*b*]. The residence times of the most noticeable aerosols are short, because of dry fallout and washout, and it seems that natural aerosols from vegetation, dust storms, salt sea spray predominate in regions far from industrial areas.

Water vapor in the stratosphere. The question of increase of water vapor in the stratosphere has arisen during the controversy over the proposed SST, or supersonic transport. Opponents of the program point out that the normal stratospheric water vapor content is very low, essentially because mixing of the troposphere with the stratosphere is weak, and the mixing process requires the water to go through a very cold region (see Figure 2) in which the rising tropospheric air would presumably be dehumidified. It has been estimated that 400 SST's flying 4 flights a day each would introduce 150,000 tons of water vapor to the stratosphere per day or .025% of the total amount naturally present in the altitude range in which the SST's would fly. Since the horizontal large-scale eddy transport processes are well developed in the stratosphere, this water vapor will certainly spread out, and indeed might be fed continously into the tropical tropopause cold trap, and have a relatively short residence time. This is at present a matter of some controversy, and if on the other hand the residence time is very long, it is worth noting that the above rate would lead to doubling in 10 years.

Water vapor in the stratosphere can affect climate mainly in two ways: it can modify the temperature structure of the atmosphere through the greenhouse effect, and it can cause an increase in cloud formation in the lower stratosphere. *Manabe and Wetherald* [1967] estimate that if the water vapor in the whole stratosphere were doubled, the greenhouse effect would raise the temperature of the air near the earth's surface about 0.5° while tending to cool the stratosphere. If the doubling takes place only in the lower 1/6 of the stratosphere, we can expect a smaller effect. The situation with respect to cloud formation is less clear. Nacreous clouds in the lower stratosphere are thought to result from fluctuations of relative humidity giving local saturation. It is feared that their formation may be enhanced and made more general because we propose to add this water vapor from SST's at a time when we are also mixing more and more CO_2 (from our various industrial activities) into the whole atmosphere. The addition of such greenhouse molecules has the effect of raising the temperature of the atmosphere near the earth's surface while *lowering the temperature of the stratosphere* and rendering regions of saturation more widespread [*Newell,* 1970; *Study of Critical Environmental Problems (SCEP),* p. 104, 1970]. It seems clear that widespread cloudiness in the stratosphere would have its effect on climate , but because of the uncertain optical properties of such clouds, it is difficult to predict what the effect would be [*Manabe,* 1970*a*]. *Note that we have not concluded that there will be no effect, but rather that we do not as yet understand the effect.* This suggests we consider the SST program with more caution, not less.

Carbon dioxide. Carbon dioxide is a necessary product of the combustion of fossil fuels, and although it is considerably heavier than air, atmospheric circulation keeps it well mixed, with the result that its concentration is almost constant throughout the troposphere and stratosphere. It is an infrared absorber and contributes to the earth's radiation balance through the greenhouse effect. Its absorption characteristics are known (cf. Figure 1), and the effect of increasing the atmospheric concentration of CO_2 on the earth's surface temperature has been the subject of several numerical modeling experiments. Our combustion of fossil fuels will probably grow at about 4% per year (doubling every 17.5 years) but fortunately all the CO_2 we produce doesn't remain in the atmosphere. The present atmospheric concentration is 320 ppmv (parts per million by volume). We currently produce about another 2 ppmv each year, of which about 50% remains airborne [*SCEP*,p. 54, 1970]. It seems that on the long term the CO_2 concentration may be fairly well buffered chemically at the interface between the sea and the sea floor sediments, but of course the doubling time quoted is far shorter than the ocean circulation time [*Frisken,* 1970]. On the short term it should be noted that since a gas like CO_2 is less soluble when the water warms up, and since the oceanic reservoir of CO_2 is more than 50 times the atmospheric reservoir [*Plass,* 1959], the CO_2 dissolved in the ocean constitutes a destabilizing mechanism in climate variation. Cautious extrapolation of present trends, in the hope that about 50% of the released CO_2 will continue to disappear, still leads to about 380 ppmv by 2000 A.D. [*SCEP,* p. 54, 1970].

If the diseconomies of possible resultant climate change are not considered, the combustion of fossil fuels is at present the most economical source of energy generally available to man. Even if the developed countries move toward nuclear energy, the developing countries of the world can be expected to embrace fossil fuel combustion enthusiastically within the foreseeable future. Even if zero population growth can be achieved fairly soon, and even if at some later point a satisfactory living standard were to be realized for all, the total energy requirements must still increase slowly on the long term. This is because all the good things become used up, dispersed, less concentrated, less retrievable. For example, even the seams of coal will become small and more difficult to work, since we shall have already burnt up the best ones, and we shall get less net energy per ton because we shall have to use more to dig it up. We must make no mistake about it: we cannot afford to fail to understand the effects of increased atmospheric concentrations of carbon dioxide on our climate, because if it doesn't do much harm, we probably want to increase it at least during the foreseeable future. (Coal-Burning is seen to be distinct from SST operation, for which our need seems less than clear.)

Heat from man's energy conversion. We are concerned here with all the energy converted, not just the part 'rejected' as waste heat into the river, because virtually all the 'useful' part is eventually converted into heat also. Except in the very local problem of thermal pollution of small bodies of water, then, we want the total energy converted. In 1967 this was 5.88×10^9 metric tons of coal equivalent (Joel Darmstadter, Resources for the Future, Inc., personal communication), or corrected to 1971 at 4% increase per year, 5.9×10^{12} watts, continuously. On the other hand, the earth intercepts the solar flux of 2.0 langleys/minute for a total of 1.76×10^{17} watts. Since about 50% of this is absorbed at the earth's surface [*Robinson,* 1970*a*], we are now working at about 1/15,000 of the absorbed solar intensity. It has been argued by some of the model builders that 1% may be a noticeable perturbation, and we will achieve this at our present growth rate of 4% per annum in 130 years. The implication is that if this has no noticeable effect on climate, then 17.5 years later we will have the experimental information for 2% of the present surface absorption of insolation, and so on, until 120 years later our industrial activities reach the 100% level.

Sellers [1969] has estimated that by the time we reach the 5% level we should have experienced global warming by more than 10°C, and eventual melting of the polar ice caps. The resultant climatic regimes would presumably be completely different from those we experience today. As was noted earlier and is discussed more fully below, these estimates are crude and likely to indicate only general tendencies. Response of the climate of the real atmosphere is likely to be more complicated than that of the simple model used by Sellers, and less extreme.

Climatic Change and the Numerical Models

As *Mitchell* [1970*b*] points out, climatic change is a fundamental attribute of climate, which means in the first place that climate is not easy to define, and in the second place that the changes induced by man's activities may initially be very difficult to extract from climate's 'autovariation.' If we want to isolate undesirable effects at an early stage, we will have to augment our observation of the real climate with information from numerical experiments with the best available models.

However, despite the known shortcomings of the one-dimensional models, the general circulation models are, at present, time consuming and expensive to run, and most available predictions of climate change have been made using the simpler models. These calculations have yielded what is, at least at first sight, a disagreement among climatologists as to whether the climate is going to warm up or cool down. On closer examination, however, this disagreement appears to be the recorded part of a sometimes heated dialogue between experts in a rapidly developing field as they try out a new way of parameterizing some part of the problem (such as the effects of moist convection), or as they parameterize a previously specified quantity (such as snow cover). In the same paper in which he made his often quoted 'prediction' that doubling the atmospheric concentration of CO_2 would lead to an increase of 10°C in surface mean temperature, *Möller* [1963] makes an almost never quoted disclaimer to the effect that a 1% increase in general cloudiness in the same model would completely mask this effect. Möller was making a first stab at the following synergism: Concentration of CO_2 is increased, causing increased greenhousing and increased temperature. Increased temperature leads to increased evaporation from the sea, and thus to higher absolute humidity (assuming fixed relative humidity), and since H_2O molecules are even more effective infrared absorbers than CO_2 molecules, the warming trend is reinforced. We can easily see why he wanted to make the disclaimer. The very increase in absolute humidity that reinforced the warming trend through infrared absorption might lead to increased cloudiness (or indeed to increased precipitation and winter snow cover) and thus, through reflection of insolation, to a considerable buffering of the warming trend. This paper is also of interest from another point of view, since it is part of an extended dialogue and was not intended to bear alone the bright spotlight of public interest. In this first step, Möller made the approximation of making the radiation balance at the earth's surface. The next step was made by *Manabe and Wetherald* [1967] using a radiative and convective atmosphere temperature adjustment scheme worked out earlier by *Manabe and Strickler* [1964], which enabled them to more nearly treat the atmosphere as the continuum of infrared luminous fuzz that it is. They found that the same doubling of the CO_2 gave them only a 2.4°C temperature rise, and that this could be masked by a 3% increase in low cloud. *Robinson* [1970*b*] gives a historical survey of the various estimates of the effect of increased carbon dioxide concentration in the atmosphere.

The inherent inability of the simple models to deal with more than one parameter at a time has always limited their utility to that of 'educational toys.' With a system as complex and difficult to understand as the earth's climate, this is a considerable utility, but it should be borne in mind that any tendency toward climate change will have a corresponding tendency to change the general ocean-atmosphere circulation. This means that there would be tendencies toward change in the energy transport processes and also, for example, in the amount of cloud cover. The model meanwhile calculates on, with an inappropriate strength for the energy transport and an inaccurate value for the amount of solar energy available.

General circulation models, on the other hand, typically take hours to follow the simulation of one day's weather, and present indications are that at least in some respects they are not sufficiently sophisticated. For example, although general circulation models can presently follow simulation of humidity up to saturation, they then merely dump this immediately as precipitation. They do not generate and transport cloud cover, nor do they consider the associated change in the albedo. Thus we see that even the generation of computers presently under construction will not allow us to evolve climate for hundreds of years by running a general circulation model, even if we could convince ourselves that a unique climate would emerge from integrating weather. Climatologists have worried about the uniqueness of climate for many years, the haunting vision being that an ice-age climatic regime might be just as consistent with the present solar input, atmospheric composition, etc. as is our present climatic regime. *Lorenz* [1968] has worried about this from a mathematical point of view, concerning himself with the 'transitivity' of regions of solution of sets of coupled differential equations, and *Mitchell* [1970*b*] has considered the implications for defining climate and detecting its possible changes.

Some scientists feel it will be possible to get a working understanding of the long-term effects of our activities on climate by varying the input conditions of surface and atmospheric albedo, concentration of infrared absorbing molecules in the atmsophere, level of solar activity, etc. to the general circulation model and merely integrating until (hopefully) the initial transients die away. Smagorinsky (personal communication) and *Mitchell* [1970*b*] suggest that when a comprehensive set of such limited experiments have 'spanned parameter space,' we will finally begin to get some real understanding of what climate changes we can expect to actually take place rather than what the initial tendencies will be caused by a particular facet of man's industrial activities.

The model to be used for this activity should probably be an improved version for a joint ocean-atmosphere general circulation model, but some realistic way of generating and transporting cloud cover must be built into the model before its results will be generally accepted. Computing time will be a problem, as always, but the new ILLIAC IV [*McIntyre,* 1970] should help, particularly when the new intercomputer communication network being designed by the Advanced Research Projects Administration becomes available for routine use by the research community [*Roberts and Wessler,* 1970].

Open Questions and the Future

We are left with some uncertainty on the short term (say, the next 50 years) as to whether carbon dioxide and the greenhouse effect will raise the temperature significantly, or whether increased atmospheric turbidity due to man-made (or caused) aerosols will lower it significantly. There is a possibility that, since the doubling times are presently different, first the former and then the latter will occur [*Mitchell,* 1970*a*]. There is uncertainty about the effect of water vapor from SST's in the stratosphere. There is even the possibility that climate may not be a unique function of the boundary conditions, and that 'the flutter of a butterfly's wings' could trigger a large change from a warm climate to an ice age, for instance, or of course to a much hotter era than the present one. It is worth noting that we have been fluttering fairly hard, and remain uncertain whether we can detect our effect on the climate.

On the longer term (say, more than 100 years) we have the more serious problem of beginning to warm the climate directly with our own energy conversion. This will be with us (in slightly different degree at any one time) whether we derive our energy from coal fires, nuclear reactors, or from imaginary fusion generators. A bare earth (having no atmosphere or oceans) at a uniform temperature of 0°C would have to increase its temperature by approximately 50°C in order to double the rate at which it radiates heat away. We are interested in this number because if we continue to double our energy conversion rate every 17.5 years, in about 250 years it will equal the rate at which we absorb solar radiation at the earth's surface at the present time. A bare earth is ridiculously simple model, and our real earth with ocean and atmosphere would behave in a much more complicated way, but it is hard to see how it would not warm up considerably.

At the present there is apparently a little time to grapple with these problems. We need to conduct some fundamental research into the function of the land and the sea as reservoirs of atmospheric constituents, especially infrared absorbing molecules like carbon dioxide. We need to know what are the precise conditions that lead to cloud formation. We must monitor the effects of various kinds of clouds and aerosol hazes on scattering and absorption of solar radiation and infrared terrestrial radiation. We need to monitor the behavior of the ocean-atmosphere system carefully, watching for secular trends in temperature distribution, in the strength of circulation, in precipitation patterns, and in the concentrations of carbon dioxide, water vapor, and general cloudiness.

We must improve our general understanding of climate behavior through the performance of numerical experiments with climate models. The general circulation models are too time consuming to allow experiments in evolution of climate over extended periods of time, and they are also known to be much too naive in their present forms. The naivite is due in part to the considerable approximations made in the interests of computation speed, and in part to real gaps in the detailed understanding of the basic physical processes involved. New development in computer technology like ILLIAC IV will help with the speed problem, but, although it is clear that an improved general circulation model will yield a superior ten-day forecast, it still does not seem possible that several hundred years of climate can be evolved by integrating weather in 10-minute steps.

The most promising direction would seem to be the development of hybrid models that couple the improved general circulation model with a simpler hydrostatic model. In this sort of scheme the general circulation routine is given realistic input conditions and is made to generate a consistent set of climate statistics complete with appropriate values of parameters such as lateral energy transport, surface albedo, amount and type of cloud cover. Using these values the hydrostatic model takes over and evolves climate for an extended period of say six months, taking into account increased atmospheric concentration of carbon dioxide, increased energy consumption in industrial activities, etc. The ball is then passed back to the general circulation routine, which updates the various parameterizations before the simpler hydrostatic routine is allowed to take the next giant step forward. In this proposal, the spectre of intransitivity rears its ugly head. There is no defense, except to suggest that operation of the hybrid model be tried with a variety of similar initial conditions and rates of introduction of pollutants in the hope that wildly different final states were not generated.

The next generation of predictions of climate change will probably be less spectacular than those we now have from the simpler models, but on the other hand, they will be much more credible and therefore much more compelling. It is clear that if the industrial revolution continues at the present rate, it is only a question of time until man's activities begin to change the earth's climate. We must try to understand the limits this imposes upon us and act accordingly.

Acknowledgments

It is a pleasure to thank Resources for the Future, Inc., for their hospitality and financial support during the summer of 1970.

Many of the scientists whose work is quoted here attended the 1970 Study of Critical Environmental Problems (SCEP) at Williams College. Much of the material they so generously have me in preprint form has now become part of reference *SCEP* [1970].

REFERENCES

Bryson, R.A., and J.T. Peterson, Atmospheric aerosols: Increased concentrations during the last decade, *Science, 162,* 120–121, 1968.

Budyko, M.I., The effect of solar radiation changes on the climate of the earth, *Tellus, 21,* 611–619, 1969.

Fletcher, J.O., The influence of variable sea ice on the thermal forcing of global atmospheric circulation, *Rand Corp. Rep. P-4175,* 21 pp., 1969.

Frisken, W.R., Man's activities and the atmospheric reservoirs of oxygen and carbon dioxide, *Resources for the Future, Internal Rep.,* 10 pp., July 1970.

Kellogg, W.W., *Predicting the Climate,* comments presented to the Summer Study on Critical Environmental Problems at Williams College, Williamstown, Mass., 1970.

Lorenz, E.N., Climatic determinism, in *Meteorol. Monog., 8,* pp. 1–3, American Meteorological Society, Boston, Mass., 1968.

Manabe, S., and R.F. Strickler, Thermal equilibrium of the atmosphere with a convective adjustment, *J. Atmos. Sci., 21,* 361–385, 1964.

Manabe, S., and R.T. Wetherald, Thermal equilibrium of the atmosphere with a given distribution of relative humidity, *J. Atmos. Sci., 24,* 241–259. 1967.

Manabe, S., and K. Bryan, Climate calculations with a combined ocean-atmosphere model, *J. Atmos. Sci., 26,* 786–789, 1969.

Manabe, S., Cloudiness and the radiative convective equilibrium, in *Global Effects of Environmental Pollution,* pp. 156–157, edited by S. Fred Singer, D. Reidel, Dordrecht, Holland, 1970.

McIntyre, D.E., An introduction of the ILLIAC IV computer, *Datamation, 16,* 60–67, 1970.

Mintz, Yale, Very long term integration of the primitive equations of atmospheric motion: An experiment in climate simulation, in *Meteorol. Monog., 8,* pp. 20–36, American Meteorological Society, Boston, Mass., 1968.

Mitchell, J.M., Jr., A preliminary evaluation of atmospheric pollution as a cause of the global temperature fluctuation of the past century, in *Global Effects of Environmental Pollution,* pp. 139–155, edited by S. Fred Singer, D. Reidel, Dordrecht, Holland, 1970*a*.

Mitchell, J.M., Jr., *The Problem of Climate Change and Its Causes,* comments presented to the Summer Study on Critical Environmental Problems at Williams College, Williamstown, Mass. 1970*b*.

Mitchell, J.M., Jr., The effect of atmospheric aerosol on climate, *NOAA Tech. Memo. EDS 18,* 28 pp., 1971.

Möller, F., On the influence of changes in the CO_2 concentration in air on the radiation balance of the earth's surface, and on the climate, *J. Geophs. Res., 68,* 3877–3886, 1963.

Newell, R.E., Water vapor pollution in the upper atmosphere and the supersonic transporter?, *Nature, 226,* 70–71, 1970.

Oliger, J.E., R.E. Wellck, A. Kasahara, and W.M. Washington, Description of the NCAR global circulation model, *NCAR Rep.,* Boulder, Colo., 94 pp., 1970.

Plass, G.N., Carbon dioxide and climate, *Sci. Amer., 201,* 41–47, 1959.

Robinson, G.D., Some meteorological aspects of radiation and radiation measurement, in *Precision Radiometry,* volume 14 of *Advances in Geophysics,* p. 285, edited by A.J. Drummund, Academic Press, New York, 1970*a*.

Robinson, G.D., Long-term effects of air pollution–A survey, *Center for Environment and Man Rep. CEM 4029-400,* Hartford, Conn., 44 pp., 1970*b*.

Roberts, D.G., and B.D. Wessler, Computer network development to achieve resource sharing, in *American Federation of Information Processing Societies Conferences Proceedings, 36,* pp. 543–549, AFIPS Press, Montvale, New Jersey, 1970.

Saltzman, B., Surface boundary effects on the general circulation and macroclimate, in *Meteorol. Monog., 8,* pp. 4–19, American Meteorological Society, Boston, Mass., 1968.

Sellers, W.D., A global climate model based on the energy balance of the earth-atmosphere system, *J. Appl. Meteorol., 8,* 392–400, 1969.

Smagorinsky, J., Problems and promises of deterministic extended range forecasting, *Bull. Amer. Meteorol. Soc., 50,* 286–311, 1969.

Stacey, F.D., *Physics of the Earth,* p. 240, John Wiley, New York, 1969.

Study of Critical Environmental Problems (SCEP), *Man's Impact on the Global Environment,* MIT Press, Cambridge, Mass., 319 pp., 1970.

Placing atmospheric CO_2 in perspective

Contrary to popular belief, increases in atmospheric CO_2 have not drastically altered the world climate; in fact, over the last three decades, rising CO_2 levels have helped to slow down the effects of a steady decrease in world temperature

Arthur D. Watt **Westinghouse Georesearch Laboratory**

Atmospheric carbon dioxide, which amounts to 320 parts per million (ppm) by volume, rather than being a pollutant, is essentially a thread of life woven through the globe on which we live. In the past century alone, the amount of CO_2 in the atmosphere has increased by 40 ppm, with levels increasing at a current rate of about 0.75×10^{10} tonnes per year. Fortunately, man can tolerate CO_2 levels many times present concentrations, and plant life actually grows better at increased CO_2 levels. What does cause concern is the effect that atmospheric CO_2 has on the earth's climate. It appears that the 40-ppm increase over the last century may have contributed to a global temperature increase of the order of 0.2 K. Since 1940, however, the global atmospheric temperature has been decreasing—an indication that other factors (such as atmospheric dust) are of much greater importance in determining the overall heat balance of the world.

Each year, approximately one tenth of the carbon dioxide in the atmosphere is exchanged at ocean and land surfaces, with the biosphere playing a decidedly important role. At present, man is injecting into the atmosphere approximately 2×10^{10} tonnes (metric tons) of CO_2 per year as a result of fossil fuel burning. This is about 10 percent of the annual exchange and 1 percent of the atmospheric total. By far the largest input of CO_2 to the atmosphere is derived from the tropical oceans, amounting to an estimated 15×10^{10} tonnes per year.

Rate of exchange is primarily governed by temperature, *p*H, and wave action, which are the most significant factors in maintaining the level of atmospheric CO_2. Also of great importance is the continual removal of CO_2 by plant growth, believed to consume 7×10^{10} tonnes per year. Despite the delicate balances that contribute to the overall rate of exchange, however, atmospheric CO_2 levels are annually increasing by about 0.75×10^{10} tonnes, or over a third of the total amount introduced to the atmosphere by man in burning fossil fuels.

Approximately 99 percent of the earth's atmospheric materials (5.3×10^{15} tonnes) are contained in the first 30 km of this life-supporting layer surrounding the earth. The major constituents of the atmosphere are gases and water vapor. The gases consist (by volume) of 78 percent nitrogen, 20 percent oxygen, 0.9 percent argon, and 0.03 percent CO_2. In addition, there are numerous other gases present, including ozone with a concentration of the order of 0.000 005 percent. Water vapor amounts to approximately 0.2 percent (by weight), which is equivalent to 2–3 cm of precipital moisture distributed over the whole earth. There is also about 0.008 percent of water droplets and ice crystals, and a small amount of dust with diameters predominant in the region from 0.1 to 1 μm with an effective weight of approximately 0.000 000 1 percent. Atmospheric dust loading varies over a range of several factors, very often resulting from volcanic action.

Gases are rather uniformly distributed up to a height of some 90 km, with the exception of ozone, which is concentrated in a layer 20–60 km above sea level. Water vapor is contained in the first several kilometers above the earth's surface. Fine dust particles (0.1 μm in radius) are found near the earth's surface in the lower troposphere, with increased concentration occurring during windy days in dry areas and near urban centers. Coarser particles (near 1 μm radius) are found in the stratosphere, with a maximum near 18 km above the earth's surface and a concentration that increases after large volcanic eruptions.

There has been considerable conjecture for a number of years as to the effects of the growing amount of atmospheric carbon dioxide on the world's climates. These claims range all the way from rather minor increases in world temperature to several degrees of change with a resultant melting of the ice caps and the inundation of large coastal cities.

Carbon dioxide in the air is one of the essential keys to all life. All primary production of land-based food is dependent upon atmospheric CO_2. In fact, 93 percent of the basic materials used by land plants in the photosynthetic production on carbohydrates come from the carbon dioxide of the atmosphere. The remaining 7 percent is supplied from groundwater. From this we can surmise that, rather than being a contaminant in the air we breathe,

Based on material presented at the First Westinghouse School for Environmental Management, Fort Collins, Colo., June 15–July 10, 1970.

Reprinted from *IEEE Spectrum*, vol. 8, pp. 59–72, Nov. 1971.

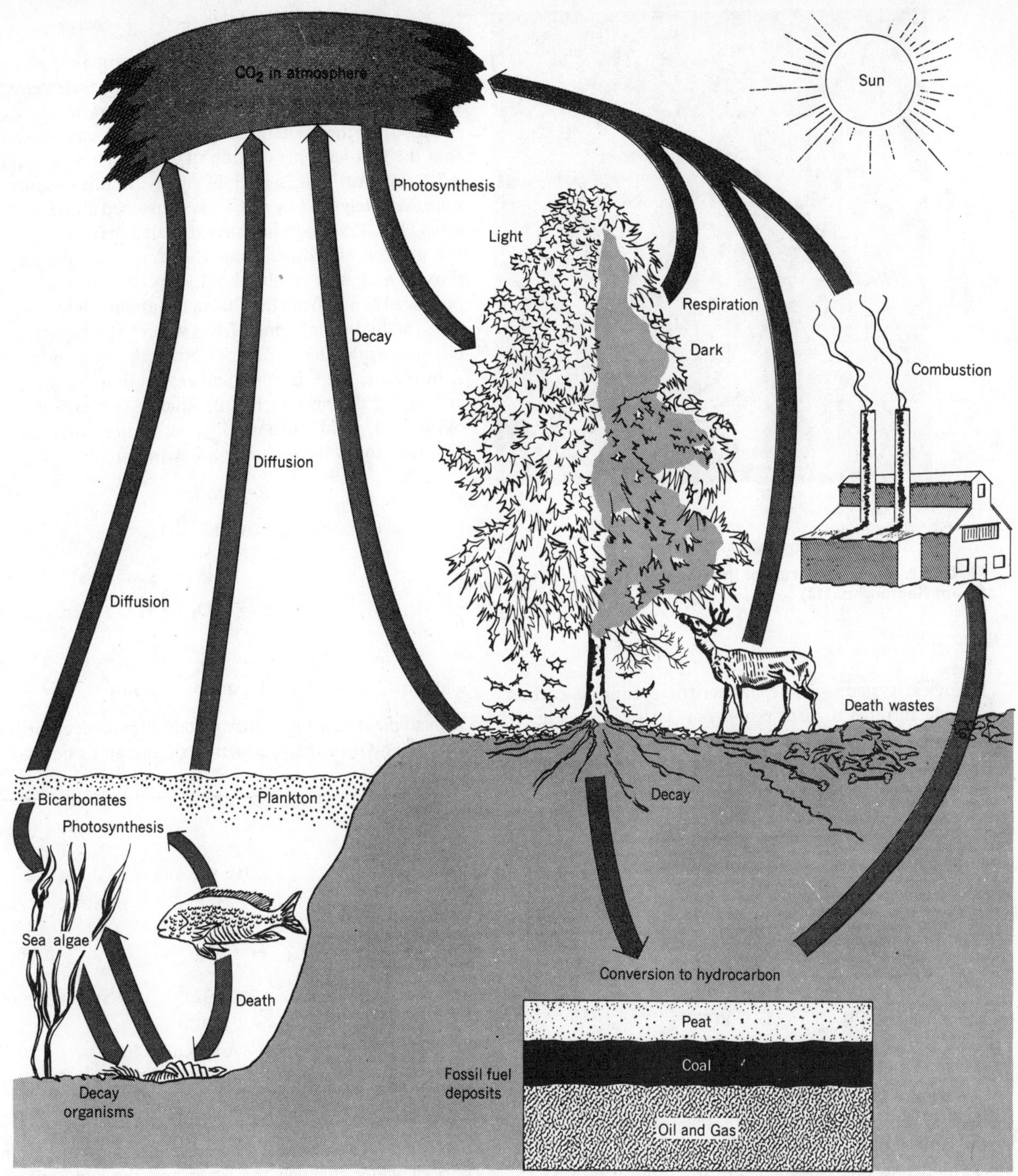

FIGURE 1. The carbon cycle in the ecosystem. (Adapted from Smith,[2] p. 49)

CO_2 is absolutely essential to life on earth.

It is instructive to observe that even if all the world's economic hydrocarbon fuel resources were burned in one year and no greater absorption occurred, the atmospheric CO_2 would only increase to three to six times the present level, i.e., to about 1000 to 2000 parts per million (ppm). By comparison, the human safety limit for prolonged exposure to CO_2 is reportedly[1] about 5000 ppm or 15 times the present level. The upper limit for plant response has not been established, but it is known that plants grow more rapidly with increased levels of CO_2 ranging up to 100 times the present atmospheric level.

In view of these considerations, it appears that the only significant remaining area for possible concern over increasing atmospheric CO_2 lies with its possible effects on global climate. Since existing literature on this subject is controversial, we will first examine the basic CO_2 cycle along with observed spatial and temporal variations of atmospheric CO_2. The probable effects of increasing CO_2 levels will then be examined in the light of experimental and theoretical evidences.

The basic CO_2 cycle

In man's environment, there are a number of highly significant cycles by which both dominant and trace elements are cycled through the combined physical and

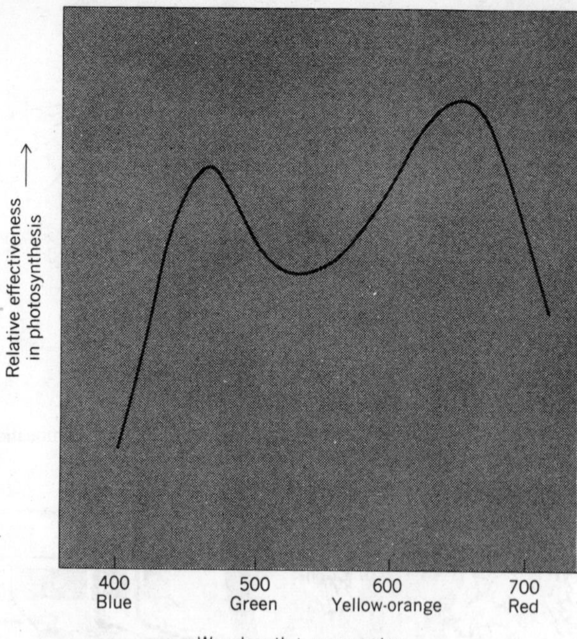

FIGURE 2. Action spectrum of the photosynthesis process. (From Keeton,[3] p. 118)

biological systems. The characteristic cycling of carbon dioxide through a local ecosystem is shown in Fig. 1. Solar energy is a basic requirement for the photosynthesis process, which is expressed in simplified form as

Photosynthesis

		(from the air)		solar radiation		(to the air)	
		↓		↓↓↓		↑	
H_2O	+	CO_2	+	$\xrightarrow{\text{light energy}}$	$\frac{1}{6}\{C_6H_{12}O_6\}$ +	O_2	(1)
18 grams		44 grams		112 kcal	30 grams	32 grams	
water	+	carbon dioxide	+	light energy →	glucose +	oxygen	

The regions of the electromagnetic spectrum that are most effective in photosynthesis are the orange-red portion with a secondary peak in the green region (see Fig. 2).

The actual photosynthesis and respiration process involves various other essential elements and catalysts. Keeton[3] pointed out that "the reduction of carbon dioxide to form glucose proceeds by many steps, each catalyzed by an enzyme." Two key nitrogen- and phosphate-containing organic compounds are involved: ADP (adenosine diphosphate) and ATP (adenosine triphosphate). Energy is added in the reaction

$$\text{ADP} + Ⓟ + \text{energy} \xrightarrow{\text{enzyme}} \text{ATP}$$

where Ⓟ designates the addition of a phosphate group

$$\begin{matrix} & OH \\ & / \\ P & & = O \\ & \backslash \\ & OH \end{matrix}$$

Energy is removed in the reaction

$$\text{ATP} \xrightarrow{\text{enzyme}} \text{ADP} + Ⓟ + \text{energy}$$

From these considerations, it is apparent that nitrogen and phosphorus availability as well as trace elements for enzymes are important factors in plant growth.

Based on these energy relationships, it can be shown that the first-level production of 1 kg of glucose requires 3.7×10^3 kcal = 4.35 kWh of energy. In this production, approximately 1.47 kg of CO_2 is consumed along with 0.6 kg of H_2O. Tracer studies have revealed that the oxygen in the glucose all comes from the oxygen in the carbon dioxide and, as a result, about 93 percent of the glucose produced comes from the CO_2 in the air and only 7 percent (i.e., the hydrogen) comes from water. The oxygen given off, on the other hand, comes from both water and CO_2.

In plants, there is continual respiration going on day and night, during which time glucose is combined with oxygen to yield carbon dioxide, water, and energy. This reaction is shown in the following simplified form:

Respiration

	(from the air)	(to the air)			
	↓	↑			
$\frac{1}{6}\{C_6H_{12}O_6\}$ +	O_2 ⟶	CO_2 +	H_2O +	energy	(2)
30 grams	32 grams	44 grams	18 grams		

Typical plants use up approximately 40 percent of the first-level $C_6H_{12}O_6$ in the production of energy to be used in plant-life housekeeping processes, including the transport of water through the plant and the conversion of first-level glucose to the final carbohydrate forms such as insoluble cellulose ($C_6H_{10}O_5$) within the plant. From this we conclude that the maximum theoretical efficiency expected from plants is 60 percent, and the production of 1 kg of carbohydrate materials in the plant requires 6.2×10^3 kcal = 7.2 kWh* of energy from a spectrally matched source. Since 180 grams of glucose yield 162 grams of cellulose plus 18 grams of water, the production of each kilogram of cellulose removes 1.6 kg of CO_2 from the atmosphere.

The photosynthesis/respiration process has a theoretical maximum efficiency of 60 percent, which requires the assumption that the light spectrum matches the photosynthesis activity curve. Solar energy contains a consider-

* By way of comparison, the heat content of a typical wood is 5 kWh/kg.

able amount of energy outside the action spectrum for photosynthesis and, as a result, the overall carbohydrate production must be modified by a spectral utilization factor that amounts to 10–20 percent for terrestrial solar energy. The result is that we would anticipate an upper bound for photosynthesis production of carbohydrates in the region of 6–12 percent of the total solar energy received. In actual conditions, all the solar energy incident to the earth's surface does not reach plant leaves; for normal growing conditions and spacing between plants, approximate efficiencies of food production from solar energy are 4 percent for algae, 2 percent for sugar beets, 0.2 percent for wheat, and 0.1–1 percent for forest trees.

Returning to the pictorial carbon cycle of Fig. 1, the carbohydrates, once produced by plants, are seen to be eaten by herbivores, or else to decay, in which case most of the carbon dioxide originally consumed is returned to the atmosphere through bacterial action. If appropriate heat and pressure are applied to the carbohydrates, a process of conversion to hydrocarbons takes place and the familiar coal deposits result. In this conversion to hydrocarbons, oxygen is driven off and the heat content per unit mass of the material increases. The net result is carbon withdrawal from the atmosphere and its storage in fuel deposits.

Withdrawal from the short-term cycle can also occur in the marine environment. Phytoplankton take the CO_2 that has diffused from the atmosphere into the upper layers of water and convert it into carbohydrates. This in turn serves as food for marine life, which in some instances produces coral atolls. Such production effectively withdraws carbon from the short-term CO_2 cycle but is available for reintroduction, under proper conditions, to the atmosphere–ocean CO_2 exchange process. Before attempting the formulation of an overall CO_2 cycle, we will examine global carbon deposits and observed levels of atmospheric and oceanic CO_2.

Global carbon deposits and observed CO_2 levels

About 99.9 percent of the earth's carbon is contained in the land. Estimates place the earth's crust deposits at 2.7×10^{16} tonnes, largely as carbonates and with only a small amount in the form of hydrocarbons and a much smaller amount as carbohydrates (Fig. 3).* The world's oceans contain about 0.1 percent of the total, and the atmospheric carbon content in the form of CO_2 is 0.0026 percent or approximately 2.6 percent of that contained in the oceans. The carbonate deposits on land are widely distributed in a nonuniform manner, as are ocean deposits.

At present, the amount of carbon dioxide observed in air has an annual average value of approximately 320 ppm by volume. Since air has a mass of 28 kg per unit volume as compared with 44 kg of CO_2 per unit volume, this is equivalent to 505 ppm of CO_2 by weight. With an atmospheric pressure of 10^4 kg/m² distributed over 5.1×10^{14} m² of global surface, 505 ppm of CO_2 yields 2.6×10^{12} tonnes of CO_2 in the earth's atmosphere. The equivalent amount of carbon is 7×10^{11} tonnes. Observations over the past century have shown that atmospheric CO_2 is increasing according to the trend plotted in Fig. 4. More detailed observations in the last decade have shown

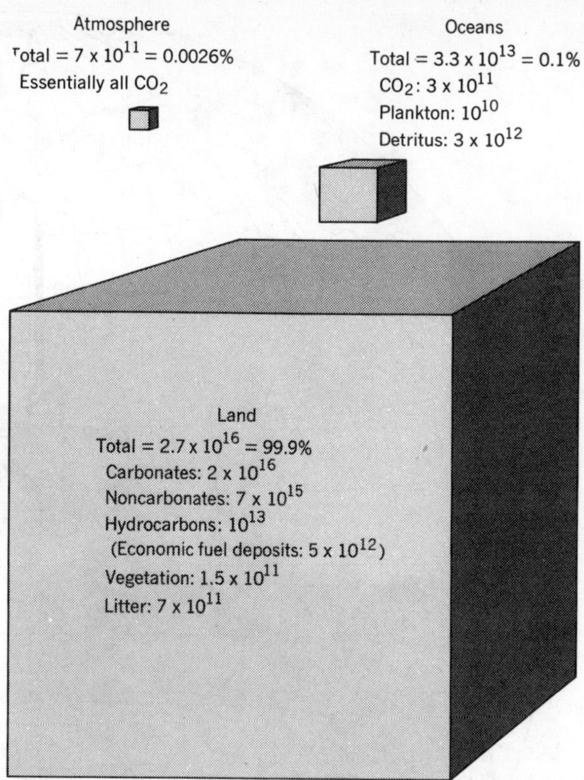

FIGURE 3. Global carbon distribution. Values are given in tonnes of carbon; to obtain equivalent values of CO_2, multiply by 3.7.

FIGURE 4. Trend of atmospheric CO_2 (mean annual data from the northern temperate troposphere).

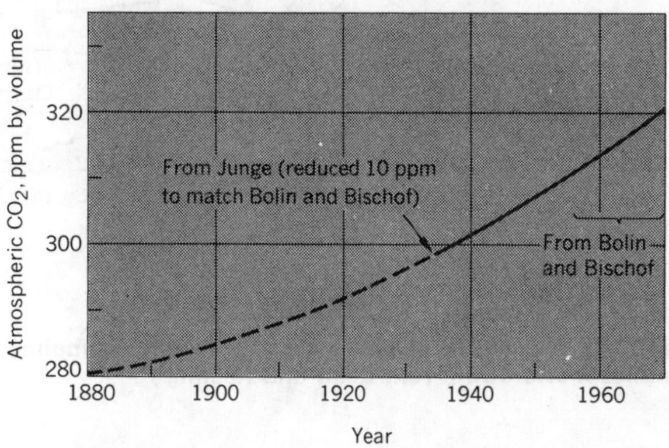

FIGURE 5. Mean monthly atmospheric carbon dioxide concentration at Mauna Loa, Hawaii. (1958–1963 data from Pales and Keeling, J. Geophys. Res., vol. 70, 1965; 1963–1969 data, Bainbridge [Scripps] private communication)

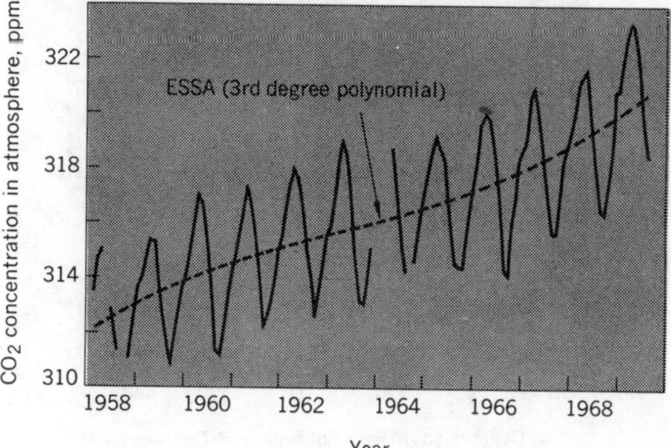

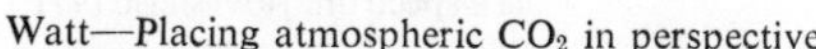

* Much of the data given here for CO_2 deposits and exchange rates were obtained from Lieth,[4] p. 3895.

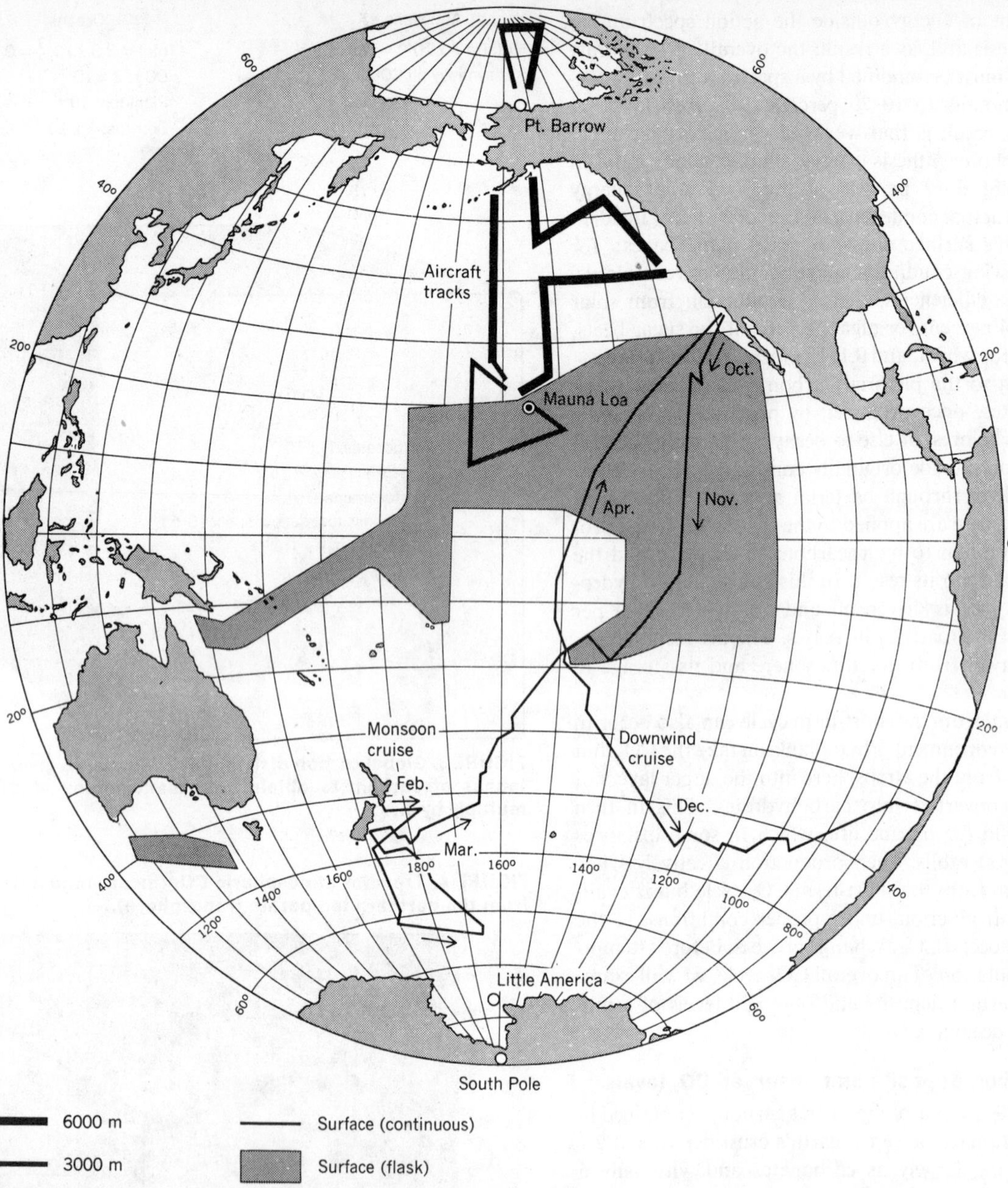

FIGURE 6. Location of stations and tracks for sampling atmospheric CO_2. (From Bolin and Keeling[5])

FIGURE 7. The concentration of atmospheric CO_2 near the earth's surface as a function of latitude in a region near 150°W longitude showing the large changes with season. (From Bolin and Keeling[5])

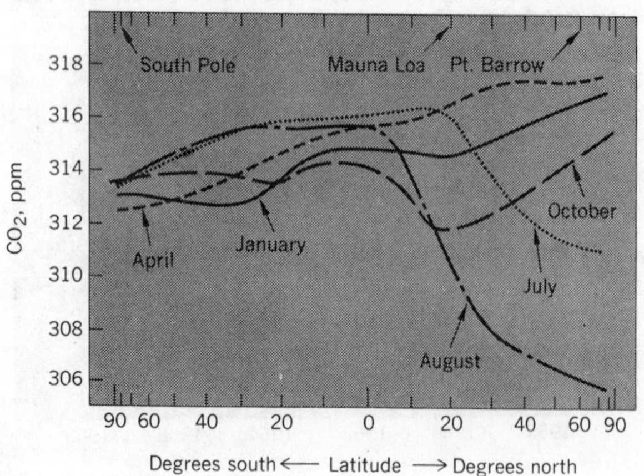

a significant annual variation in the atmospheric CO_2 levels as noted in Fig. 5. Before attempting to explain these temporal variations, it is well to examine spatial variations (with latitude) for the global area (near 150°W longitude) shown in Fig. 6 as described in detail by Bolin and Keeling.[5]

Figure 7 indicates a trend in January from near 313 ppm at the South Pole to nearly 318 ppm at the North Pole. As the season progresses, the trend smooths out and covers a slightly greater range in April. By July, we see a remarkable change taking place—the concentration from 20°N to the North Pole drops drastically to a low of 311 ppm at Point Barrow. This trend, which begins in June, occurs rapidly and indicates the presence of an active scrubbing mechanism operating during the northern-hemispheric summer. An even greater drop is reached in August when large areas of arctic tundra are active and polar ice cover is low. At this time the Point Barrow values reach a low of 306 ppm. By October, the point of lowest concentration has moved down to 20°N latitude with an appreciable recovery of CO_2 levels at Point Barrow. The trend shifts slowly back to the northern-hemisphere winter curve for January.

These rather large spatial variations emphasize the need for careful measurements when attempting to arrive at temporal variations. It can also be seen from these illustrations that temporal variations will differ greatly at various geographic locations.

Atmospheric injection and scrubbing—the global CO_2 cycle

The exchange of CO_2 between the atmosphere and land biosphere has been described, and it was seen that 1.6 kg of CO_2 is removed from the atmosphere for each kilogram of cellulose produced. Correspondingly, the decay or burning of a kilogram of cellulose returns 1.6 kg of CO_2 to the atmosphere. Moreover, the burning of a fossil-fuel hydrocarbon such as coal, which is nearly 80 percent carbon, returns 3 kg of CO_2 to the atmosphere for each kilogram of coal consumed.

The largest amount of atmospheric CO_2 exchange involves the oceans. Data reported by Sverdrup, Johnson, and Fleming[6] (Fig. 8) show the conditions necessary for exchange of CO_2 between the surface waters of the ocean and the atmosphere. For example, with a typical ocean *p*H of 8.1 and temperatures above 20°C, the flow of CO_2 will be to the atmosphere at atmospheric concentrations of CO_2 near our present 320 ppm by volume. For colder water temperatures, the flow is from the atmosphere to the sea. The ease with which this exchange process takes place has been considered in detail by Kanwisher.[7] He shows, for example, that the net CO_2 flux density across the air–sea interface increases with surface roughness. Combining temperature, wind, and area factors, it appears that a large amount of CO_2 removal occurs in the southern oceans.

A conceptual view of the estimated* global annual exchange of CO_2 between the atmosphere, hydrosphere, and biosphere is drawn in Fig. 9. Primary injection is seen to be from the tropical oceans with a large relatively steady scrubbing by the southern oceans. The tundra and Arctic Ocean scrubbing components—each shown as 0.5×10^{10} and 1.6×10^{10} tonnes of carbon per year; i.e., 7.8×10^{10} tonnes of CO_2—are primarily active in the late summer months since in the winter and early spring much of the far north is covered with ice or snow, as shown in Fig. 10.

The large annual fluctuation of CO_2 levels that we have described at high northern latitudes are now seen to be the result of the varying activity of arctic tundra. The low atmospheric CO_2 levels correspond to the time when the snow and ice cover is gone from arctic lands and when the polar ice cover is at a minimum.

The rather persistent bulge in observed CO_2 at mid-latitudes (from 20°S to 20°N) results from warm ocean waters giving off CO_2 to the atmosphere. The persistent low point in the 40°–50°S latitudes corresponds to a rather effective CO_2 scrubbing by rough cold seas and the

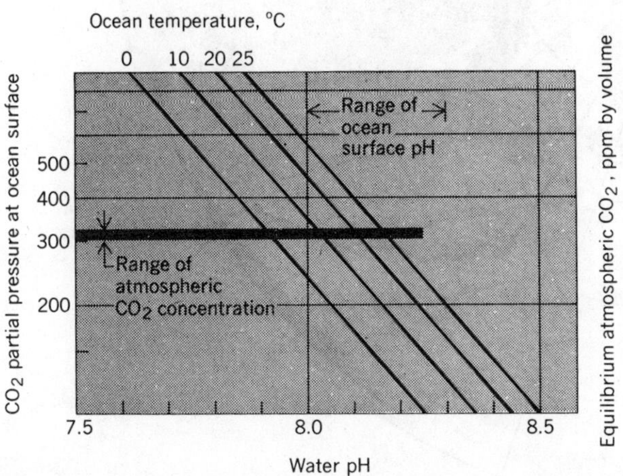

FIGURE 8. CO_2 exchange between ocean and atmosphere. (Based on data from Sverdrup et al.[6])

* Preliminary values based on data from Lieth[4] and others and calculations by the author.

FIGURE 9. Estimated global deposits and annual exchange of CO_2 between the atmosphere, hydrosphere, and biosphere. Flow values are in units of 10^{10} tonnes/year of carbon; to convert to CO_2, multiply values shown by 3.7.

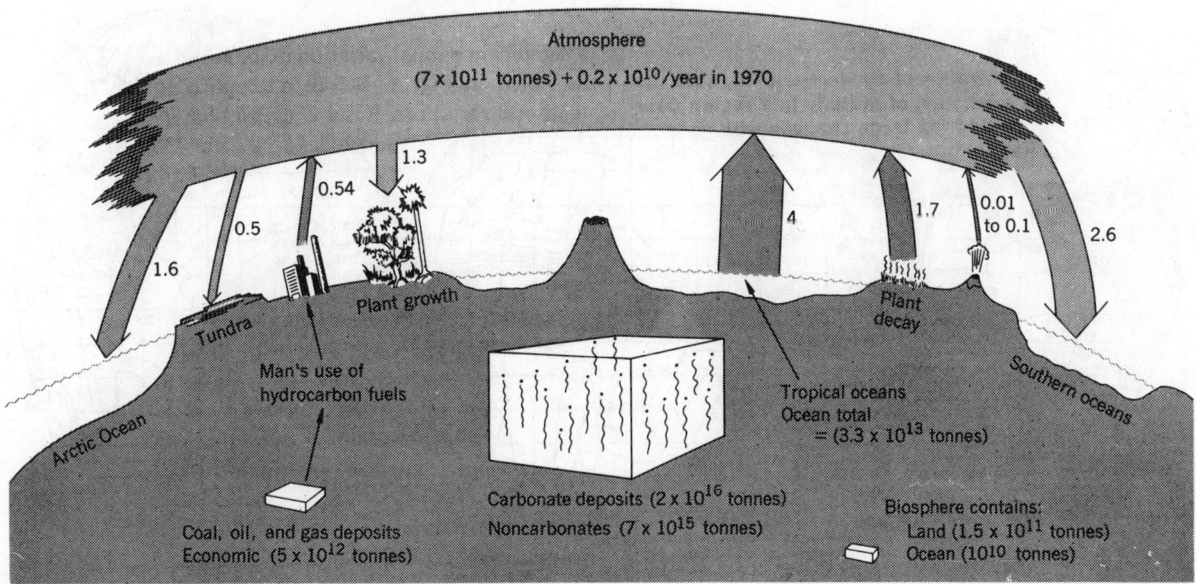

Alaska
Canada
United States
Greenland
U.S.S.R.

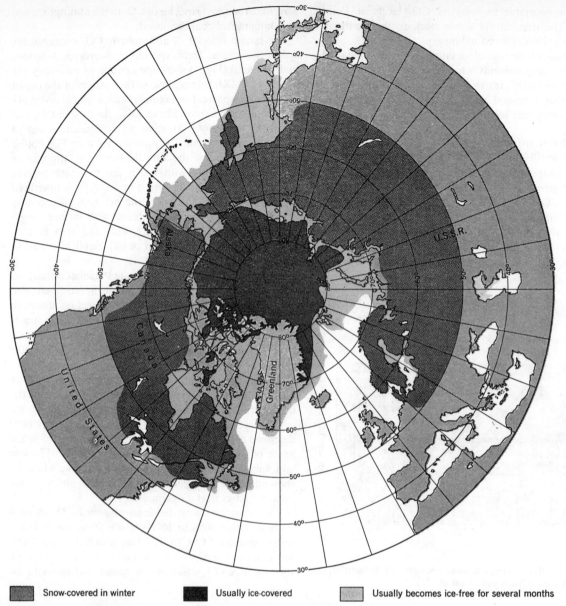

Snow-covered in winter | Usually ice-covered | Usually becomes ice-free for several months

FIGURE 10. Variations in arctic snow and ice cover. (From Sater[8])

FIGURE 11. The distribution of p_{CO_2} of the world's oceans expressed as the departure in ppm from equilibrium with atmospheric CO_2. H indicates high; L, low. Data are largely 1957–1967. (From Keeling,[9] p. 4547)

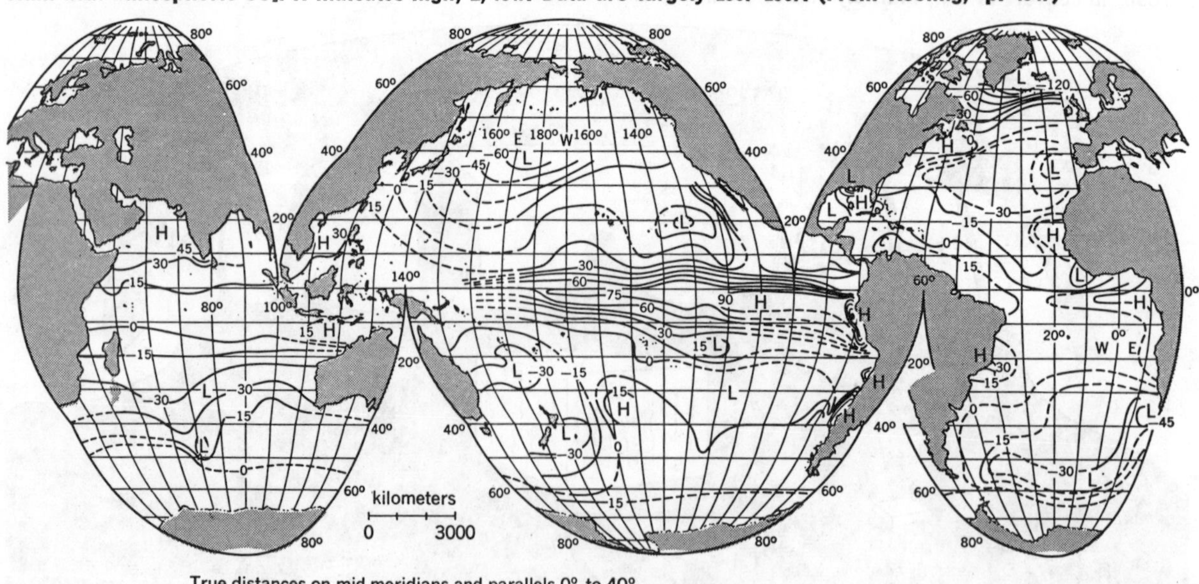

large amount of exposed water in this latitude range.

Some idea of the areas in the ocean that are giving off CO_2 to the atmosphere and those that are taking it from the atmosphere can be seen in Fig. 11. This graph shows the departure in parts per million from equilibrium with atmospheric CO_2. As expected, the high areas are found in the equatorial regions and the lows in the northern and southern oceans. Although there is a considerable amount of data available on CO_2 levels in the atmosphere, as well as partial pressures over surface ocean water, there still are not enough data (pointed out by Keeling[9]) to "clearly establish the oceanic cycle of carbon and the rate of exchange of CO_2 between the atmosphere and the world's oceans." In view of this, we have estimated global gains in the land biosphere and assigned the remainder as the net exchange between the oceans and the atmosphere.

We will now consider the overall balance of CO_2 and man's influence. Additions of CO_2 to the atmosphere calculated on the basis of the amount of hydrocarbon fuels consumed are shown in Fig. 12. In 1970, the level exceeded 2×10^{10} tonnes of CO_2 added to the atmosphere per year. This corresponds to the value of 5.4×10^9 tonnes of carbon indicated in Fig. 9. The observed concentration in parts per million by volume for the past 90 years is shown by Fig. 12A. The dashed curve shows the atmospheric level in 1880 plus man's total addition since then. This is given in tonnes of CO_2 in the atmosphere using the right-hand scale. According to these calculations, the present value would stand at 3×10^{12} tonnes if all man's additions remained in the atmosphere. The actual amount of CO_2 in the atmosphere calculated from observed CO_2 concentrations is shown by the dotted curve as being about 2.6×10^{12}.

Over the past decade, the CO_2 level has increased by approximately 1.4 ppm per year by mass. This amounts to a gain of 7.5×10^9 tonnes per year, which is about one third of man's additions. This means that approximately two thirds of man's input to the atmosphere—i.e., 1.25×10^{10} tonnes/yr—is being removed by some scrubbing mechanism. An estimated annual carbon dioxide budget on a global basis is shown in Table I. Direct data are not yet available to show whether the net oceanic or biosphere CO_2 flows are positive or negative. In this table, the best-known values are the atmospheric gain of 7.5×10^9 tonnes of CO_2 per year and the 2×10^{10} tonnes/yr for man's input of CO_2 to the atmosphere.

I. Estimated atmospheric CO_2 balance (1970)

	Unit Exchange, 10^{10} tonnes/year	
	C	CO_2
Inputs to atmosphere from		
Tropical oceans	4.00	14.80
Plant decay	1.70	6.30
Man (fuel burning)	0.54	2.00
Volcanic action	0.01 (variable)	0.05
Total:	6.25	23.15
Outputs from atmosphere to		
Northern oceans	1.60	6.00
Southern oceans	2.60	9.60
Arctic tundra	0.50	1.80
Other plants	1.35	5.00
Total:	6.05	22.40
Distribution of man inputs to atmosphere		
Atmosphere gain	0.20	0.75
Biosphere (land) gain		
Lumber	0.02	0.07
Forests, soil, and peat deposits	0.10	0.38
Ocean gain	0.22	0.80
Total:	0.54	2.00

The world production of timber and forest products, excluding firewood, is estimated at 7×10^8 m³/yr or near 4.5×10^8 tonnes/yr. Since there are 1.6 tonnes of CO_2 scrubbed per tonne of cellulose produced, this would amount to almost 7×10^8 tonnes of CO_2 removed from the atmosphere. Standing timber is now increasing in the United States and may be increasing on a global basis. Assuming a small amount for forest increase plus an allowance for increasing humus and peat deposits, a total of 4.5×10^9 tonnes of CO_2 per year could easily be accounted for by the land portion of the biosphere.

FIGURE 12. Observed atmospheric CO_2 increase compared with calculations of the potential increase due to man's use of hydrocarbon fuels.

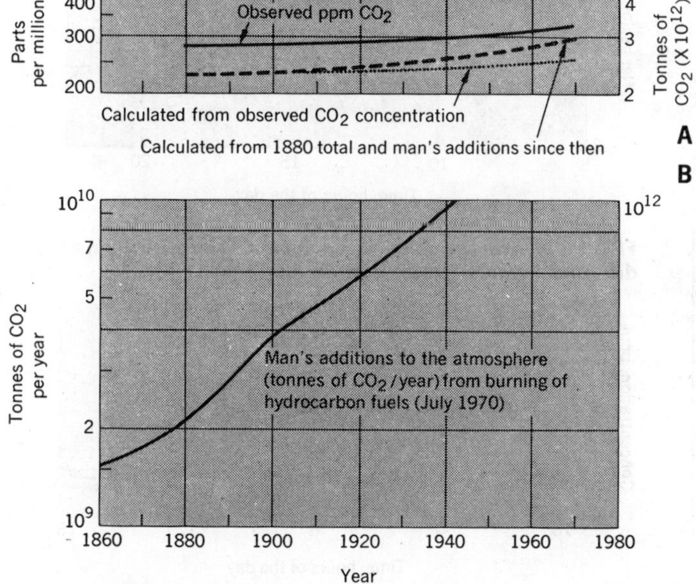

This would leave about 8×10^9 tonnes of CO_2 to be removed annually via the oceans. The details of the oceanic portion of the carbon cycle, spelled out by MacIntyre,[10] are quite complex. The near-surface balancing equilibrium could be produced by a slight decrease in *p*H. In actuality, some near-surface CO_2 is removed via plankton production, which may be transferred to calcium carbonate structures. The cold seas near ice packs also remove CO_2 from the atmosphere, and some is transferred to the tropical oceans via the ocean circulation system for use by phytoplankton and for release to the atmosphere. Some carbon in the form of HCO_3^- is transferred to great depths by the annual sinking of high-salinity waters that result from sea-ice production. This can be trapped in deep-ocean clay deposits, which have the ability to absorb vast amounts of bicarbonate ions.

Effects of atmospheric CO_2 levels and temperature on vegetation

In the preceding section, we saw the importance of plant growth in removing CO_2 from the atmosphere, and noted that the exchange process between the ocean and the atmosphere was dependent upon temperature at the atmosphere–ocean interface.

In a similar manner, we shall see that temperature affects the rate of CO_2 assimilation in plant growth. In Fig. 13, this rate for potato leaves in daylight is given as a function of temperature. Two atmospheric concentrations of CO_2 are included. The lower curve is representative of present-day atmospheric conditions and shows, for example, that for these plants a maximum assimilation rate of approximately 0.2 mg of CO_2 per square centimeter of leaf surface per hour occurs at a temperature of 20°C. It would seem from Fig. 13 that, if the atmospheric CO_2 level were increased to 40 times present levels, the plant productivity for this species would increase by four to one.

Figure 14 gives the relative growth of three different plant species as a function of the partial pressure of CO_2 in millibars. Since there are 505 ppm by weight of CO_2 in the atmosphere, the present partial pressure is 0.5 mbar on this illustration. From these results, it appears: (1) that, at present atmospheric levels, plant uptake increases almost directly with CO_2 concentration, and (2) for temperatures below 20°C, CO_2 uptake increases with temperature.

It can be surmised that if increasing levels of atmospheric CO_2 produce an increase in global temperature, more effective CO_2 assimilation by plants will have a self-regulating effect upon global temperature. Even if temperature effects are small, as will be indicated, increased plant activity as a function of CO_2 level should act as a regulator that will tend to stabilize the level of atmospheric CO_2.

The effect of clouds and rain on plant productivity is demonstrated in Fig. 15. The amount of CO_2 assimilated by plants on clear days is much greater than that assimilated during a rainy day. Another important factor is that, as the sun sets, plant respiration starts returning CO_2 to the atmosphere, as evidenced by the negative assimilation value shown on the lower-right-hand side of Fig. 15.

The effect of plant activity on the near-surface-level concentrations of CO_2 in the atmosphere is shown in Fig. 16. In the early afternoon, atmospheric CO_2 levels measured near the ground are much lower than those at night in vegetated areas. The rapid decrease of CO_2 at low heights during early morning hours is due to the onset of thermal convection of air. The large variations of from 300 ppm during the day to 400 ppm at night show that care must be taken in measuring procedures to allow for this effect when attempting to determine either long-term trends or spatial variations in CO_2 levels.

The actual productivity of land areas varies greatly

FIGURE 13. Rate of CO_2 assimilation in potato leaves as a function of temperature in full daylight at two different CO_2 concentrations. (From Lieth[4])

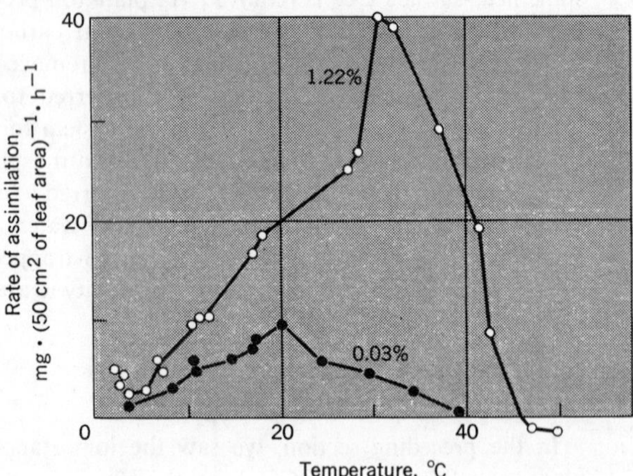

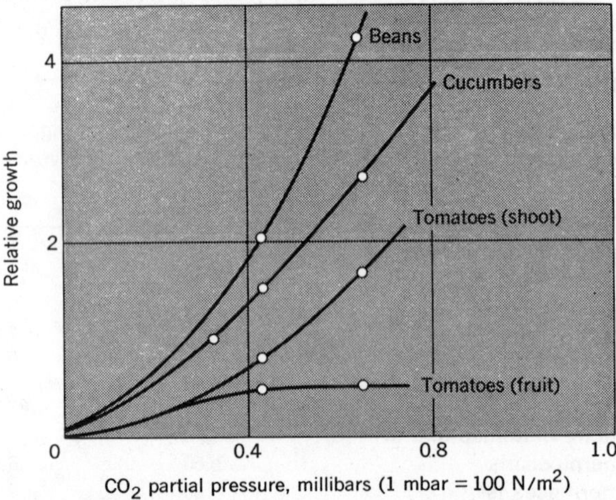

FIGURE 14. Rate of growth of various domestic plants as a function of the CO_2 concentration of the air. (From Lieth[4])

FIGURE 15. The variation of net CO_2 assimilation of corn in a field during both a clear and a rainy day, including soil respiration rate. (From Lieth[4])

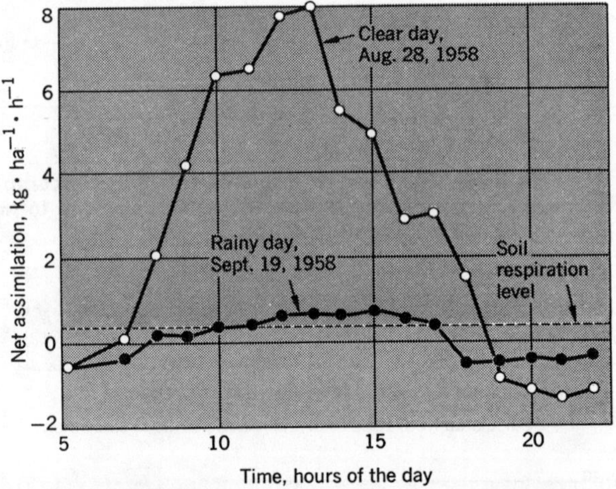

FIGURE 16. Average daily variation of CO_2 concentration at different heights above vegetation. (From Lieth[4])

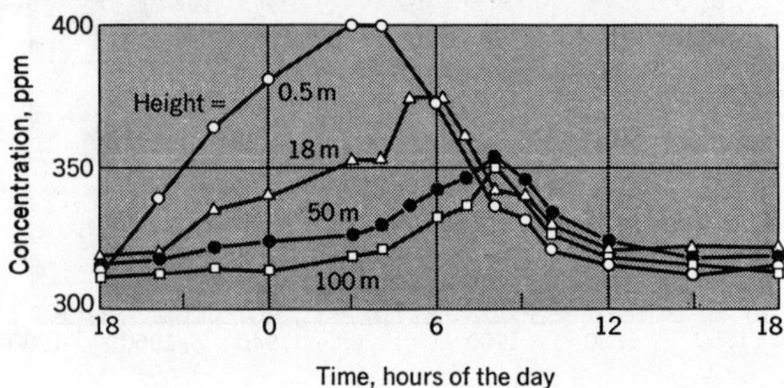

II. Potential productivity of the earth and the population it could support*

North Latitude, degrees	Land Surface, ha × 10^8	Number Months Above 10°C	Carbohydrates, $Mg \cdot ha^{-1} \cdot yr^{-1}$	Square Meters per Man to Support Life: With No Allowance for Urban and Recreational Needs: m^2/man	No. Men × 10^9	With 750 m^2/man for Urban and Recreational Needs: m^2/man	No. Men × 10^9	Percentage Agricultural Land
Column 1	2	3	4	5	6	7	8	9
70	8	1	12	806	10	1556	5	52
60	14	2	21	469	30	1219	11	38
50	16	6	59	169	95	919	17	18
40	15	9	91	110	136	860	18	13
30	17	11	113	89	151	839	20	11
20	13	12	124	81	105	831	16	10
10	10	12	124	81	77	831	11	10
0	14	12	116	86	121	836	17	10
−10	7	12	117	85	87	835	9	10
−20	9	12	123	81	112	831	11	10
−30	7	12	121	83	88	833	9	10
−40	1	8	89	113	9	863	1	14
−50	1	1	12	833	1	1583	1	53
Total:	132				1022		146	

* From deWit,[11] p. 317.

with incident solar energy, soil moisture, and soil nutrients. Lieth[4] has shown, for example, that a typical European forest will yield a yearly production of 10 tonnes of dry material per hectare (1 ha = 10^4 m^2), which results in a scrubbing of 15 tonnes of CO_2 from the atmosphere. In a Malayan forest, yearly production is about 25 tonnes/ha, resulting in a scrubbing of 38 tonnes of CO_2.

DeWit[11] has calculated the potential food productivity of the earth assuming that the only limiting factor is solar energy. Table II shows the land surface available in each 10° latitude band, and the number of months when the temperature is about 10°C. The maximum potential productivity is listed in column 4 in tonnes of carbohydrates per hectare per year. It is interesting to note that maximum productivity is indicated in the 20° latitude regions. This effect in the real world is even more pronounced since in many tropical regions the soils have been heavily leached so that near-surface minerals are not available and as a result productivity, particularly in terms of agriculture, is often very low. The difference between the values quoted in column 4 by deWit[11] and those of Lieth[4] indicate that real-world productivity is perhaps one fourth to one fifth the idealized productivity based on solar energy alone. For many land areas the primary limitation is availability of water.

The world's oceans have 70 percent of the global area and could potentially support a large amount of biological activity and resulting CO_2 scrubbing. In actuality, as shown by Isaacs,[12] large areas of the oceans have low productivity. This results when the near-surface waters (through which the sun's rays can penetrate) are deficient in essential nutrients such as nitrogen and phosphorus. The areas of high productivity are in the upwelling regions shown in Fig. 17, where nutrients are brought to the surface to replace those consumed by the phytoplankton. The heavily shaded areas, which amount to approximately 10 percent of the ocean area, are believed to be the major contributors to the estimated total annual oceanic biological uptake of 1.5 × 10^{10} tonnes of carbon,[13] which is equivalent to 5.5 × 10^{10} tonnes of CO_2 per year. Ocean productivity in some northern-latitude regions has been shown to have major variation throughout the year.[14] Land productivity at high latitudes is, of course, extremely variable from winter to summer.

Probable climatic effects of atmospheric CO_2

There is still considerable disagreement concerning the effect of atmospheric CO_2 on global climates. There is a prevalent misconception that atmospheric CO_2 acts as a blanket that rather uniformly turns back the infrared energy radiated from the earth and thereby increases the earth's temperature. In actuality, CO_2 is effective only in absorbing energy in several very limited portions of the earth's radiated frequency spectrum.

The energy spectrums involved are displayed in Fig. 18. Most of the energy received from the sun is in the region from 0.5 to 5 μm, whereas the major portion of energy radiated from the earth to space is in the 5–30-μm region. Both of these curves are for unattenuated radiation. In the first case, it represents energy from the sun before it encounters the earth's atmosphere, and in the second case it represents electromagnetic energy from the earth at the earth's surface before it passes through the atmosphere on its way to space. Both of these spectral distributions are modified considerably in passing through the atmosphere.

There are numerous atmospheric constituents that influence the radiant energy on its journey through the atmosphere, but those of primary concern are water vapor, carbon dioxide, and ozone. These triatomic molecules, which are very effective in absorbing certain portions of the radiation spectrum, are found in varying degrees at different locations in the atmosphere. Water vapor is concentrated in the several lower kilometers, carbon dioxide is rather uniformly distributed through-

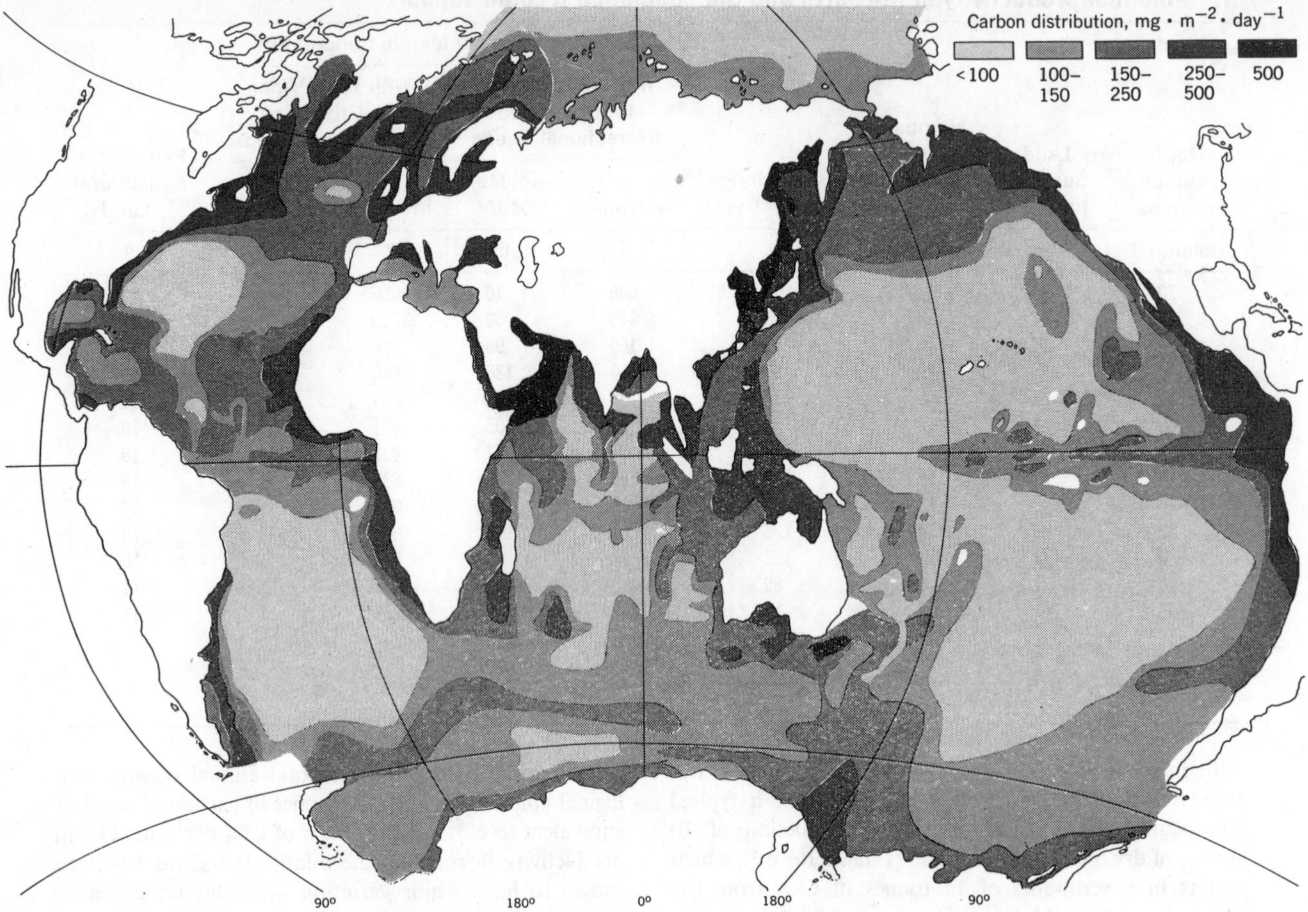

Carbon distribution, mg • m^{-2} • day^{-1}

<100 | 100–150 | 150–250 | 250–500 | 500

90° 180° 0° 180° 90°

FIGURE 17. Distribution of primary biological production in the world's oceans. (From Koblentz-Mishke et al.[13])

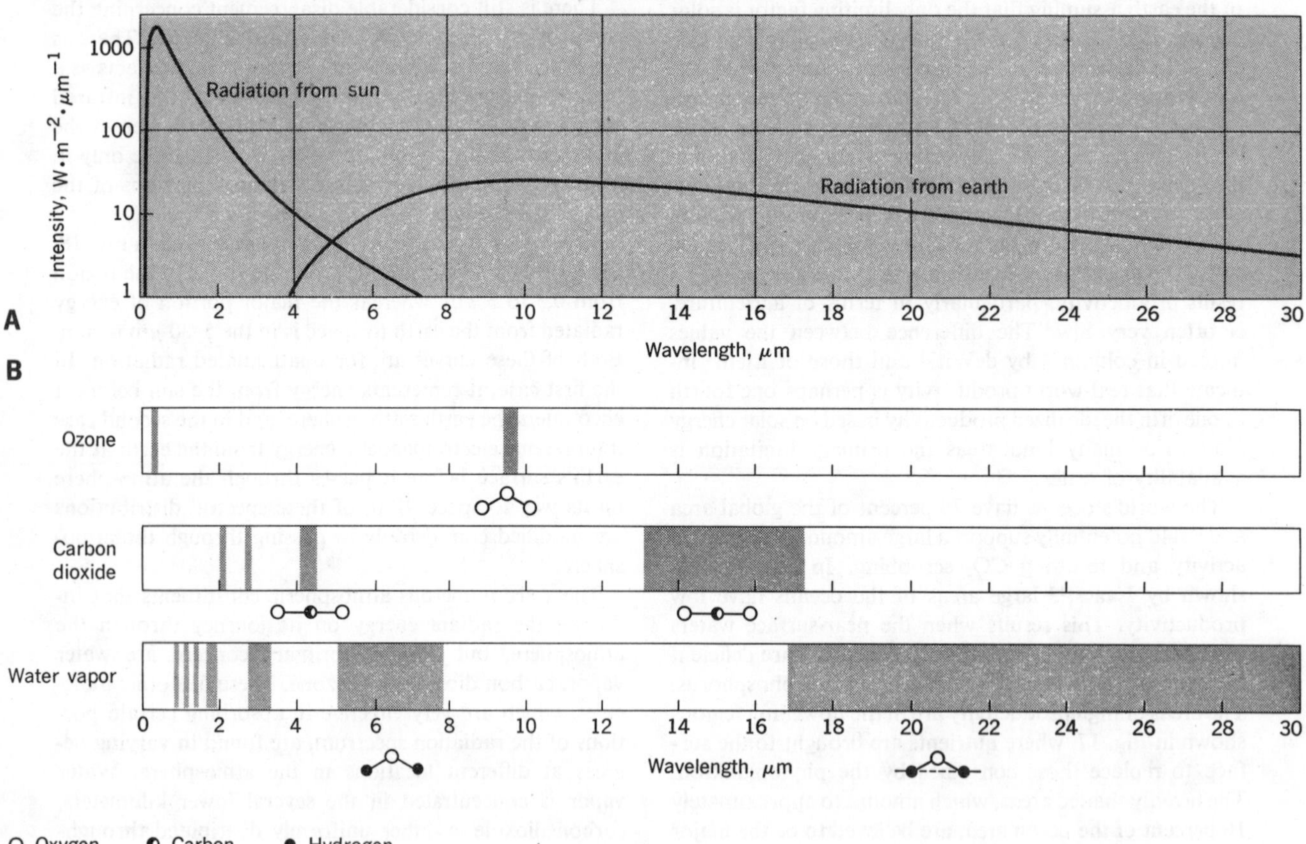

FIGURE 18. Solar and earth radiation spectral distributions. (From Newell[15])

out most of the atmospheric layer, and ozone is largely concentrated in a layer from 20 to 60 km above the earth's surface.

The wavelengths at which each of these gases is effective in absorbing energy are shown in Fig. 18B as shaded bands. The width of these absorption bands is dependent upon the amount of material in the path traversed and also to a great degree upon atmospheric pressure. The so-called "pressure broadening" of spectral absorption bands can be explained as follows: First an absorption spectrum is measured with a particular amount of carbon dioxide. Then an additional amount of gas having no absorption in the region measured is added, which increases the total pressure. The absorption spectrum is again measured and seen to be appreciably broader. Details of such pressure broadening have been described by Plass.[16]

The incoming solar-energy spectrum as observed at the earth's surface and an estimated outgoing energy spectrum for the earth's radiation observed beyond the atmosphere are both shown in Fig. 19. The heavily tinted portions of the outgoing spectrum show where effective CO_2 absorption is occurring. The amount of energy absorbed by CO_2 is seen to be considerably smaller than the energy absorbed by water vapor. The CO_2 absorption bands at 4 and 15 μm are quite deep, and it will be shown later than even doubling the atmospheric CO_2 will only slightly increase the total portion of energy absorbed.

Relative to the heating of the atmosphere, it can be seen in Fig. 19 that a large portion of solar energy is absorbed by ozone and oxygen in the region below 0.3 μm. Theoretical calculations by Pressman[18] have shown that ultraviolet ozone absorption of solar radiation is the main radiative heat source of the atmosphere in the region from 30 to 70 km. Heating rates quoted are in the range of

FIGURE 19. Electromagnetic spectrums of solar and terrestrial radiation. Note that a langley (Ly) is a measure of solar energy density and equals 1 cal/cm² of irradiated surface. (Adapted from Sellers,[17] p. 20)

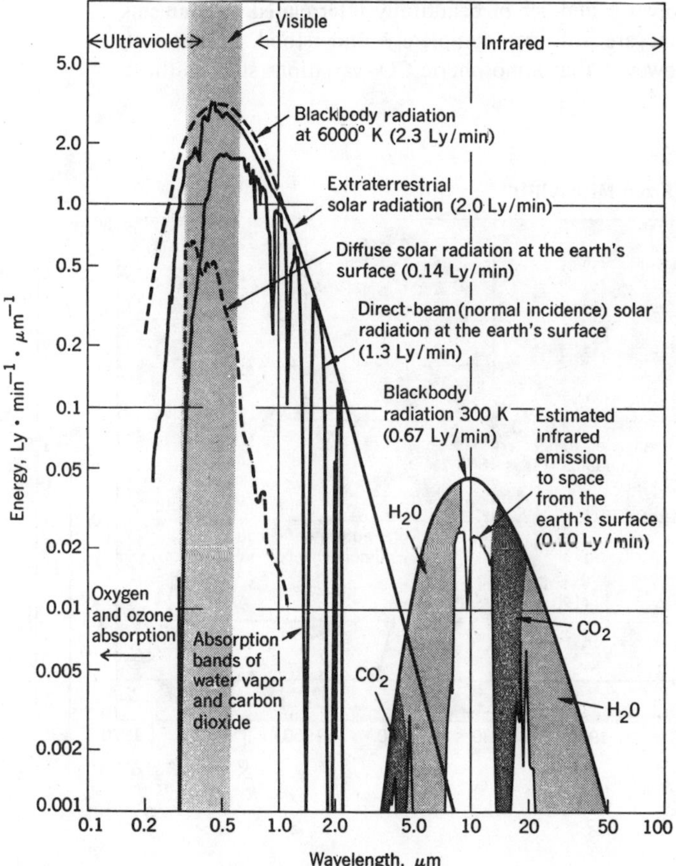

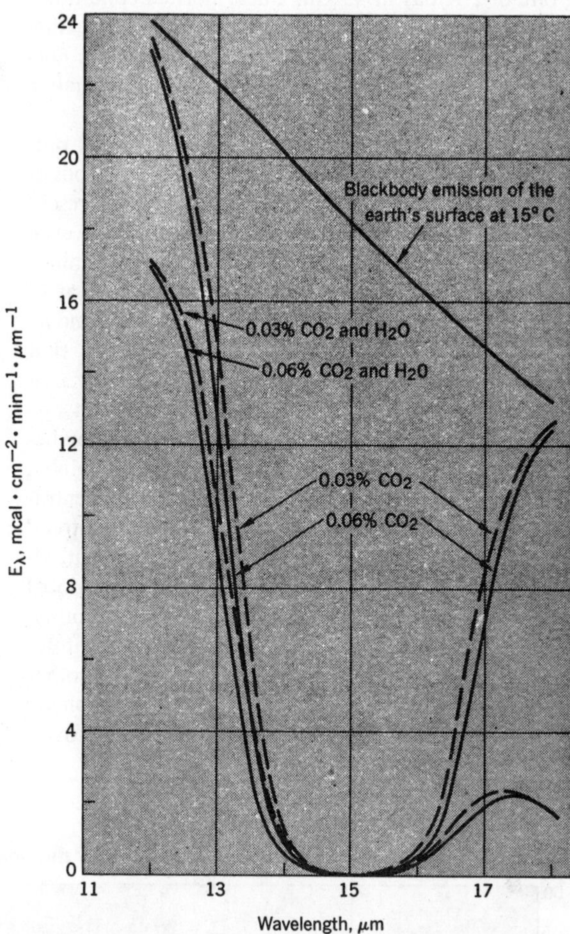

FIGURE 20. Outgoing effective radiation from the earth's surface for two CO_2 concentrations. (From Möller,[21] p. 3879)

FIGURE 21. Trend of atmospheric CO_2 (mean annual data taken from northern temperate troposphere).

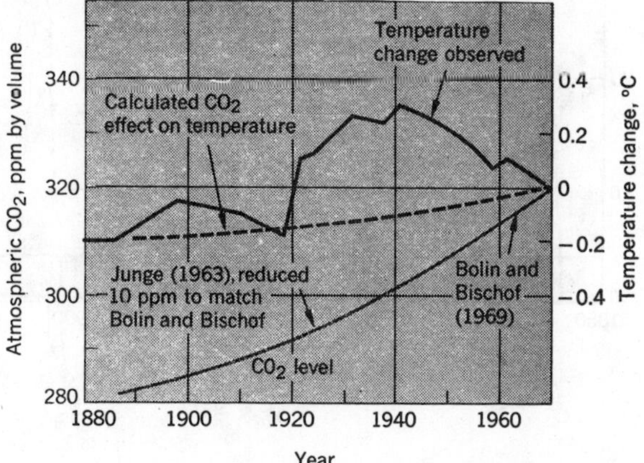

1–10 K per day with a maximum heating usually at a height of 45 km. Below 30 km, the near-infrared bands of water vapor are the primary absorbers of solar radiation. The heating rates at these altitudes of the lower stratosphere may be as large as 4 K/day.

Plass[19,20] has shown that, in addition to the heating effects of CO_2 and O_3 in the atmosphere, the stratosphere is cooled by infrared radiation to space from CO_2 and O_3. The 15-μm CO_2 band is calculated to have a maximum cooling rate of 5 K/day at 45 km for typical conditions and, similarly, the 9.6-μm O_3 band has a maximum cooling rate of 2 K/day at 45 km. Under normal conditions, the sum of these cooling rates nearly balances the heating rates for solar radiation. This would indicate that, normally, the stratosphere is approximately in radiative equilibrium.

Some concept of the magnitude of change in net outgoing radiation from the earth's surface, with variations of CO_2 from 0.03 to 0.06 percent by volume (near present levels), can be seen in Fig 20. It is apparent that even doubling the atmospheric CO_2 levels has only a minor broadening effect on the primary absorption region at 15 μm. The net effect has been discussed by Möller,[21] and he comes to the conclusion that, "thus the theory that climatic variations are affected by variations in the CO_2 content becomes very questionable." At this point one may ask, "But what of the reported good correlation between increasing atmospheric CO_2 and increasing global temperatures?" Figure 21 shows average CO_2 concentrations with a superimposed world mean-temperature trend obtained from Mitchell.[22] It is clearly apparent that, although atmospheric CO_2 levels are still climbing, global temperatures have been dropping since 1941. Obviously, global temperatures do not go up directly with atmospheric CO_2. As a result, we must conclude that other factors in the global heat balance—such as atmospheric dust—are of greater importance and therefore have overridden the smaller, heat-capturing contribution of atmospheric CO_2.

Calculations by Manabe and Wetherald[23] have indicated that a two-to-one increase in atmospheric CO_2 would produce a 1.5°C temperature increase at the earth's surface. With this, one can expect a stratospheric temperature decrease of 10°C at an altitude of 40 km. Using the observed 40-ppm increase in CO_2 levels from 1880 to 1970, an expected increase of 0.2°C is obtained by interpolation of Manabe and Wetherald's results; such a change is shown by the dashed curve of Fig. 21.

Atmospheric dust loading varies appreciably as indicated in Fig. 22. It is important to note that most of the long-residence atmospheric dust particles are in the range of 1 μm in diameter. Such small particles are effective in reflecting back to space the short-wavelength energy from the sun, but they are relatively transparent to the long-wave radiation from the earth. Some of the incoming solar energy is absorbed by the dust particles and some is reflected back to space. For a given tonneage, the smaller particles near 1 μm in diameter are more effective in intercepting solar radiation than are large particles. The height of the dust layer is also *very* important in determining its effect on surface temperature.[24]

Preliminary calculations by the author have indicated possible temperature changes derived from atmospheric dust to be somewhat larger than the reference −1°C shown by Mitchell on Fig. 22. Calculated global temperature changes are of the order of −0.03°C per 10^6 tonnes of high-altitude atmospheric dust in the 1-μm diameter range. Temperature records at numerous locations have shown actual drops in the annual temperature of 2–3°C following large volcanic eruptions.

The earth's temperature-regulating mechanism involves a number of beautifully interrelated mechanisms that are still not completely understood. It appears, however, that atmospheric CO_2 variations such as those

FIGURE 22. Estimated chronology of atmospheric dustload. (From Mitchell[22])

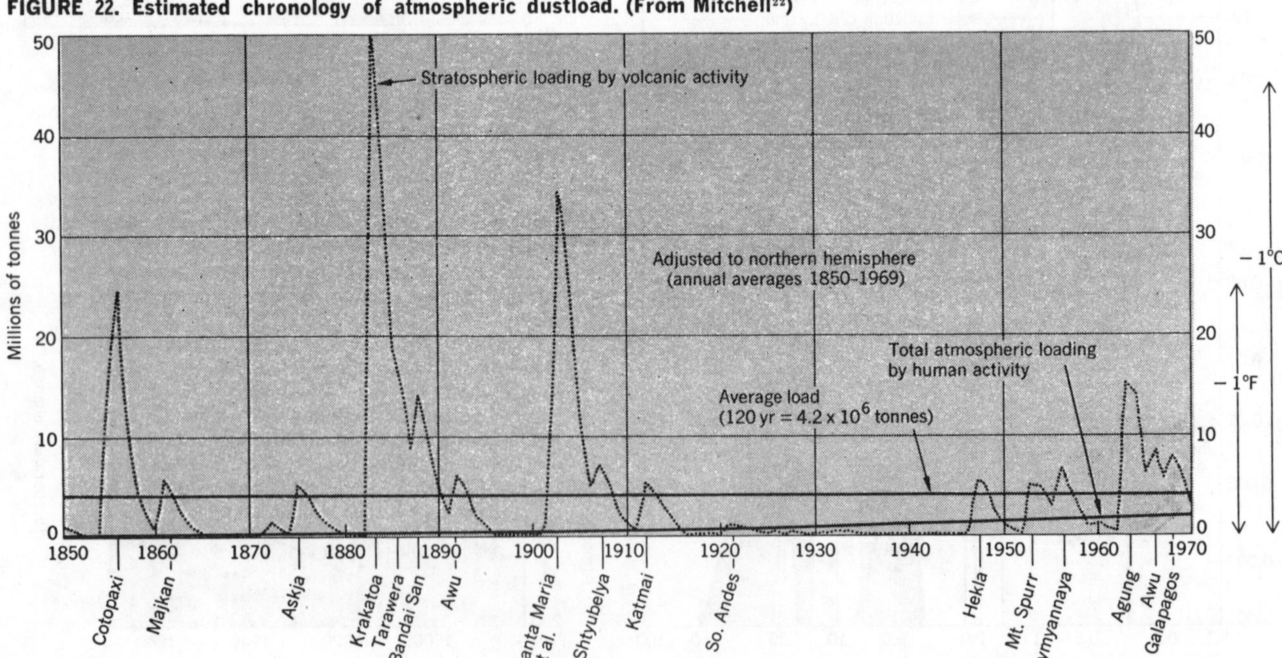

occurring at present are likely to have only a minor effect on the overall global temperature.

Need for a survey article of this type was first suggested to the author by Dr. J. H. Wright. Helpful discussions have been held with H. A. Gunther, F. S. Mathews, D. B. Large, and J. R. Portman. The preparation of the manuscript by Winifred Werth, illustrations by Gary Uridil, and a review of the text by Bob Hill are also gratefully acknowledged.

REFERENCES

1. Peterson, E. K., "Carbon dioxide affects global ecology," *Environ. Sci. Technol.*, vol. 3, pp. 1162–1169, Nov. 1969.

2. Smith, R. L., *Ecology and Field Biology*. New York: Harper & Row, 1966.

3. Keeton, W. T., *Biological Science*. New York: Norton, 1967.

4. Lieth, H., "The role of vegetation in the carbon dioxide content of the atmosphere," *J. Geophys. Res.*, vol. 68, July 1, 1963.

5. Bolin, B., and Keeling, C. D., "Large-scale atmospheric mixing as deduced from the seasonal and meridional variations of carbon dioxide," *J. Geophys. Res.*, vol. 68, July 1, 1963.

6. Sverdrup, H. V., Johnson, M. W., and Fleming, R. H., *The Oceans*. Englewood Cliffs, N.J.: Prentice-Hall, 1942.

7. Kanwisher, J., "The effect of wind on CO_2 exchange across the sea surface," *J. Geophys. Res.*, vol. 68, July 1, 1963.

8. Sater, J. E., (ed.) "The arctic basin," Arctic Institute of North America, Aug. 1969.

9. Keeling, C. D., "Carbon dioxide in surface ocean waters," *J. Geophys. Res.*, vol. 73, July 15, 1968.

10. MacIntyre, F., "Why the sea is salt," *Sci. Am.*, vol. 223, pp. 104–115, Nov. 1970.

11. deWit, C. T., "Photosynthesis: Its relationship to overpopulation," in *Harvesting the Sun*, A. San Pietro, ed. New York: Academic, 1967.

12. Isaacs, J. D., "The nature of oceanic life," in *The Ocean*. San Francisco: Freeman, 1969.

13. Koblentz-Mishke, O. J., Volkovinsky, V. V., and Kabanova, J. G., "Plankton, primary production of the world ocean," in *Scientific Exploration of the South Pacific*. Publication SBN 309-01755-6, Scientific Committee on Oceanic Research, National Academy of Sciences–National Research Council, Washington, D.C., 1970.

14. Parsons, T. R., and Anderson, G. C., "Large scale studies of primary production in the North Pacific Ocean," *Deep-Sea Res.*, vol. 17, pp. 765–776, Aug. 1970.

15. Newell, R. E., "The circulation of the upper atmosphere," *Sci. Am.*, vol. 210, Mar. 1964.

16. Plass, G. N., "Models for spectral band absorption," *J. Opt. Soc. Am.*, vol. 48, p. 690, 1958.

17. Sellers, W. D., *Physical Climatology*. Chicago: University of Chicago Press, 1965.

18. Pressman, J., "Seasonal and latitudinal temperature changes in the ozonosphere," *J. Meteorol.*, vol. 12, p. 87, 1955.

19. Plass, G. N., "The influence of the 9.6 micron ozone band on the atmospheric infra-red cooling rate," *Quart. J. Roy. Meteorol. Soc.*, vol. 82, p. 30–44, 1956.

20. Plass, G. N., "The infrared radiation flux in the atmosphere," *Proc. IRE*, vol. 47, pp. 1448–1451, Sept. 1959.

21. Möller, F., "On the influence of changes in the CO_2 concentration in air on the radiation balance of the earth's surface and on the climate," *J. Geophys. Res.*, vol. 68, July 1, 1963.

22. Mitchell, J. M., "The effect of man's activities on climate," presented at AGU Symp. on The Environmental Challenge, Washington, D.C., April 22, 1970; also "A preliminary evaluation of atmospheric pollution as a cause of the global temperature fluctuation of the past century," in *Global Effects of Environmental Pollution*, Singer, ed. New York: Springer-Verlag, 1970.

23. Manabe, S., and Wetherald, R. T., "Thermal equilibrium of the atmosphere with a given distribution of relative humidity," *J. Atmospheric Sci.*, vol. 24, pp. 241–259, May 1967.

24. Mitchell, J. M., "The effect of atmospheric aerosol on climate with special reference to surface temperature," *J. Appl. Meteorol.* (to be published).

BIBLIOGRAPHY

Bolin, B., and Bischof, W., "Variations of the carbon dioxide content of the atmosphere," Report AC-2, Institute of Meteorology/International Meteorological Institute, Stockholm, Sweden, Dec. 1969, p. 29.

Eriksson, E., "Possible fluctuations in atmospheric carbon dioxide due to changes in the properties of the sea," *J. Geophys. Res.*, vol. 68, July 1, 1963.

Eriksson, E., "The yearly circulation of sulfur in nature," *J. Geophys. Res.*, vol. 68, July 1, 1963.

Georgii, H.-W., "Oxides of nitrogen and ammonia in the atmosphere," *J. Geophys. Res.*, vol. 68, July 1, 1963.

Junge, C. E., *Air Chemistry and Radioactivity*. New York: Academic, 1963.

Kelley, J. J., Jr., "Carbon dioxide in the surface water of the North Atlantic ocean," *Limnol. Oceanog.*, vol. 15, pp. 80–87, Jan. 1970.

Matsushima, S., Discussion of "The vertical distribution of dust to 30 kilometers," by J. M. Rosen, *J. Geophys. Res.* (*Space Phys.*), vol. 73, May 1, 1968.

Rosen, J. M., "The vertical distribution of dust to 30 kilometers," *J. Geophys. Res.*, vol. 69, pp. 4673–4676, Nov. 1, 1964.

Rosen, J. M., Reply to S. Matsushima, *J. Geophys. Res.* (*Space Phys.*), vol. 73, pp. 3088–3089, May 1, 1968.

TAMBORA AND KRAKATAU: VOLCANOES AND THE COOLING OF THE WORLD

BY KENNETH E. F. WATT

You have every reason to be confused as to the possible effects of pollution on the weather. Some scientists have predicted that increasing global air pollution will reduce the amount of solar energy penetrating the atmosphere. This, they say, will cause a decline in the global average air temperature. Other scientists say the earth's weather will warm up. This, it is explained, will come about because of the increase in carbon dioxide concentration in the atmosphere. Since carbon dioxide is a strong absorber of infrared radiation, it will act like the glass roof that prevents loss of heat energy from inside a greenhouse—the infamous "greenhouse effect"—and trap radiation in the atmosphere.

Three types of arguments can be put forth to settle the matter as to which mechanism will be most important. One approach is to argue from mathematical theories of climate determination, using computer-simulation studies. This line of argument is not completely trustworthy, because of the embryonic state of such theories. A second line of argument, pursued by Reid Bryson and Wayne Wendland at the University of Wisconsin, is to discover by statistical analysis the relative strengths of the two processes during recent decades. They found that the impact of the temperature-lowering mechanism will override the impact of the temperature-raising mechanism. Since the temperature has in fact been dropping for more than two decades, their argument is compelling.

Kenneth Watt, who writes regularly for SR, is a systems ecologist at the University of California, Davis.

However, a third line of argument opens up a most fascinating area of research, both in interpretation of history and in climatological prediction. This argument asserts that the most realistic means of assessing the effect of air pollution is to find some gigantic natural event in the past that operated in a way analogous to modern pollution. Fortunately, we are provided with such historical "experiments" by records of a few particularly gigantic and explosive volcanic eruptions.

A close reading of old newspapers gives evidence that people in the early nineteenth century were not aware of any relationship between unusual weather and volcanoes. To my knowledge, most ancient civilizations thought that the earth's weather was almost totally determined by astronomical phenomena, and this belief explains the spectacular development of observational astronomy in those civilizations. The first man known to have recognized that volcanoes could affect the weather was Benjamin Franklin in 1784, following a series of enormous eruptions in Japan and Iceland the previous year. During the twentieth century many climatologists and geophysicists have come to suspect that volcanoes influenced weather and climate and, more recently, that they had effects similar to the global increase in atmospheric pollution.

There are two obvious sources of information on the impact volcanic eruptions could have on weather and hence on crop production. One is the tables of past weather data recently constructed by historical climatologists such as Gordon Manley and H. H. Lamb. The other is old newspapers, which yield detailed information on local weather, crop growing, planting and harvesting conditions, and agricultural commodity prices.

Consider the effect of the eruption of Tambora in Sumbawa, the Dutch East Indies, in April of 1815. This was the largest known volcanic eruption in recorded history: over the period 1811 to 1818 an estimated 220 million metric tons of fine ash were ejected into the stratosphere. Of this, 150 million metric tons were added to the stratospheric load in 1815 alone, mostly in April. Krakatau, which is much better known because it erupted more recently (1883), ejected only 50 million metric tons of ash into the global stratosphere. It is now known that the ash from certain of these very large and explosive volcanic eruptions spreads worldwide in a few weeks and does not sink out of the atmosphere totally until a few years have elapsed. During all of this time the ash is back-scattering incoming solar radiation outward into space and consequently chilling the earth.

The magnitude of the chilling can be ascertained from Manley's tables for central England. The Tambora volcano erupted most explosively in April of 1815; by November the average temperature of central England had dropped 4.5° F. The following twenty-four months was one of the coldest times in English history. Specifically, 1816 was one of the four coldest years in the period 1698 to 1957; the coldest July in the 259-year period was in 1816; October of 1817 was the second coldest October; May of 1817 was the third coldest May. However, these figures convey little sense of the impact

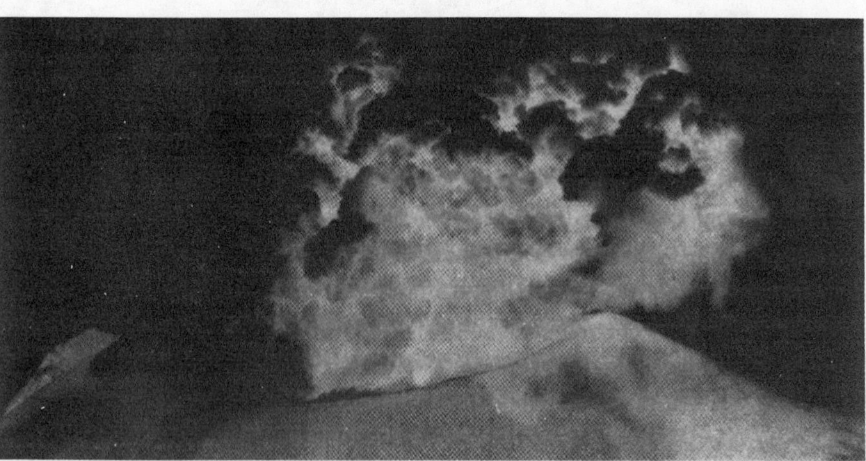

of such chilling on society. For that we must turn to the newspapers of the time. All of the following quotations are from *Evans and Ruffy's Farmers' Journal and Agricultural Advertiser*.

"We had fine mild weather until about the 20th, when it set in cold, with winds at East and North-East, with partial frosts; these together have greatly retarded the operations in Agriculture, and very many cannot purchase seed corn, so that thousands of acres will pass over untilled, and sales of farming stock, and other processes in law, drive many of this useful class in society into a state of despondency. . . . The wheats, late sown . . . have been partially injured by the frosty mornings. . . . Sheep and lambs have suffered from the severity and variableness of the weather. . . . The doing away the Income Tax, and the war duty on Malt, will afford some relief, but are wholly insufficient in themselves to restore this country to its former state of happiness and prosperity." (April 8, 1816)

"From about the 9th or 10th of this month, we have never had a day without rain more or less, sometimes two or three days of successive rain with thunder storms. The hay is very much injured; a considerable part of it must have laid on the ground upwards of a fortnight. . . . Wheat is looking as well as can be expected, considering the deficiency of plants in the ground, and those very weakly . . . but still far short of an average crop." (August 12, 1816)

"Throughout the whole month the air has been extremely cold; there has not been more than two or three warm days, being at other times rather cloudy and dark, and the sun seldom seen. The Oats . . . on high situated ground . . . are the most backward and miserable crop ever seen . . . for the greater part of the Wheat, where the mildew did not strike, has been very much affected by the rust or canker in the head. . . . There have been many seizures for rent this month, and many a farmer brought to nothing, and we hear of very few gentlemen who are inclined to lower their farms as yet; it seems they are determined to see the end." (September 9, 1816)

The preceding quotations all refer to agricultural conditions in England. To indicate that this was a worldwide rather than a local phenomenon, the following quotation, from a letter printed in the same newspaper, suggests the state of affairs in other countries.

"Last year was an uncommon one, both in America and Europe: We had frosts in Pennsylvania every month the year through, a circumstance altogether without example. The crops were generally scant, the Indian Corn particularly bad, and frost bitten; the crops, in the fall and in the spring, greatly injured by a grub, called the cut-worm . . . ." (November 10, 1817).

Thus, in the case of a volcano, which puts an immense load of pollution into the atmosphere suddenly—unlike modern pollution, which builds up gradually—we have a gigantic experiment, the effects of which can be clearly traced throughout all social and economic systems. For example, the volcano of 1815 had a clear-cut influence on world agricultural-commodity markets, as one would expect from the preceding descriptions of the consequences of weather deterioration on crops. Many measures of market conditions could be used to make this point, but the one I have selected is the highest asking price for best-quality flour within a month of trading on the London commodities market. The following table shows how this price changed, from the period before the volcano, to the peak price in June of 1817, to the normal price, which was finally reached again by the end of December 1818.

The table indicates how volcanic

HIGHEST ASKING PRICE FOR A SACK OF FLOUR (IN SHILLINGS)

Year	June	December
1814	65	65
1815	65	58
1816	75	105
1817	120	80
1818	70	70*

(*65 by end of month)

eruptions can serve as the basis for interdisciplinary research, in which a pulse due to a physical event can be tracked through biological, social, economic, and political phenomena.

For example, an interesting feature of this table is the long lag between the time the pollution was introduced into the upper atmosphere and the time that the price elevation was at its peak (April 1815 to June 1817—twenty-six months). These time lags are characteristic of complex systems and indicate why cause and effect are often not connected in peoples' minds: by the time the effect has occurred, everyone has forgotten the cause.

Could the gradual increase in worldwide air pollution concentrations become serious enough to bring crop production to a halt in high latitudes? To answer this question, we must consider the rate of build-up in pollution now and the likelihood that political power to enforce adequate pollution control will materialize.

The worldwide stratospheric particulate loading due to Krakatau and Tambora was 50 and 220 million metric tons respectively. If worldwide man-caused stratospheric particulate loading continues to build up at recently prevailing rates, the following table indicates the likely outcome.

WORLDWIDE STRATOSPHERIC PARTICULATE LOADING DUE TO MAN (MILLIONS OF METRIC TONS)

YEAR	
1970	2
2000	15
2010	30
2020	59
2030	116
2040	228

Thus, we see that by about 2018, if present trends continue, the permanent particulate load in the stratosphere would be equal to that produced temporarily by Krakatau, and by about 2039 the permanent load would be equal to that produced by Tambora.

Will they continue? The reader must make his own judgments. There is ample printed evidence testifying to the difficulty of requiring automobile manufacturers to conform to the limits required by the Clean Air Act of 1975. Also, a casual reading of industry journals does not suggest that industry would like to arrest the rate of increase in sales of oil, gas, coal, or other polluting substances. Computer-simulation studies of trends in the use of fossil fuels do not indicate any significant lessening of the rate of increase in pollution prior to about 2004 unless stringent controls are introduced. Thus, without a really massive political and social change of a type that does not seem likely, judging from present attitudes, the quotations from English farm newspapers in 1816 and 1817 may well be read as a scenario for the future.

There are further complications if these are not enough. What if man continues to build up the pollution load in the atmosphere, and then a volcano of the order of Tambora adds still more pollution to the atmosphere? How likely is this to happen? The answer is: very likely. The period since 1835 has been remarkably free of major volcanic eruptions of the explosive, Vesuvian type. But over the historical record an average of five very large volcanoes has occurred each century. Luckily for us, the modern period of great technological activity has been free from major volcanic eruptions, to an almost historically unique extent. But at some time our luck may run out.

Another complication is what is going on in the minds of people. The world population of 1816 had no idea that there was any relationship between the phenomenally bad weather they were experiencing and air pollution. Will we be any wiser? Can we change in time? Will we be able to take the necessary action to ensure a brighter end to the scenario for civilization? Simple extrapolation of present trends does not lead to an encouraging answer, but history teaches us that sudden, surprising changes in political and social attitudes do occur. Perhaps it will occur once more—in time. □

DROUGHT in Sahelia who or what is to blame?

by Reid A. Bryson

From the 1920s to the 1960s the monsoon rains over the southern Sahara extended well to the north. The rains, together with well-intentioned aid involving mass campaigns to eradicate disease in man and his livestock have brought about an unparalleled surge in numbers of both man and beast in the six countries of the Sahel zone—Mauretania, Mali, Niger, Senegal, Chad and Upper Volta.

That unchecked growth now looks like being checked; for tragedy has struck in a form that can least be countered by man. Over the past decade and in the last five years in particular, the monsoons which bring rain to the Sahel and indeed also to north west India, have not only fallen well below average, but they have also been decreasing year by year. Faced with drought on an enormous scale millions of nomads, farmers and their cattle have been migrating southwards, leaving behind them barren desert and a trail of corpses.

Droughts have obviously occurred before in the Sahel, but never before has the environment been so heavily taxed with people and animals. The present climatic change has therefore been compounded in its effect by over grazing and human mis-management. What can be done? At the recent symposium on drought in Africa held at the School of Oriental and African Studies, London University, Dr. D. Winstanley pointed out that the desert climate will probably continue to shift southwards for a century or more—a devastating conclusion. All solutions would therefore have to be long term. He in fact discussed three solutions: first to modify the climate; secondly, to increase the availability of water in other ways; and thirdly, for the nomads to shift permanently southwards with the rains.

"Even if it were technically feasible to modify climate on this scale", he says "it would be highly irresponsible to do so until we have sufficient understanding of the physical factors and processes at work in climatic change—and this we do not have at present. Research has demonstrated the unity of the general circulation of the atmosphere and that climatic changes in one zone are accompanied by changes in all the other zones.

In contemplating changing climate on this scale, we will probably always be faced with the question 'Can we afford to rob Britain, or Germany, or Russia of rainfall in order to pay Mauritania, or Chad or India?'

"The second solution—that of increasing the availability of water by such methods as building reservoirs, desalinating sea water and constructing huge water pipe-lines—is subject mainly to economic and technical limitations. In trying to increase the availability of water on a continental scale, in a zone where there is liable to be less and less precipitation in the long-run, these limitations could

Reprinted with permission from *Ecologist*, vol. 3, pp. 366–371, Oct. 1973.

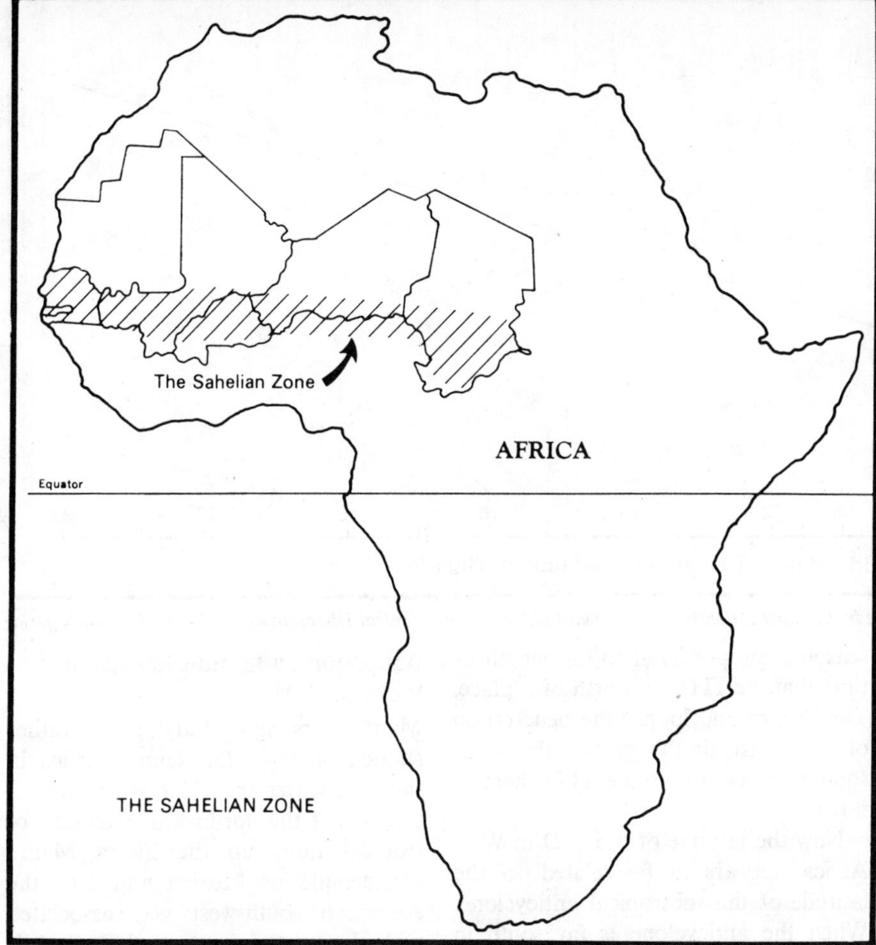

Figure 1: *Most of the population live in the southern region with more than 10 inches of rainfall annually*

prove to be prohibitive. The question that arises in this case is 'Are we prepared to give massive international aid to maintain the independence and economic viability of the Sahelian states, in face of increasingly adverse climatic conditions?'

"The third possibility—that of adapting to the climatic changes and shifting the population southwards with the rains—is the natural solution for the nomads; but is probably unacceptable in the Sahelian states, and to the states such as Nigeria and Ghana in the south, which would have to absorb between about 10 and 20 million people".

Like many of us in the west Dr. Winstanley does not see much hope for the drought-ridden nomads of the Sahel, especially in the face of a natural event beyond man's control. Yet it is unquestionable that western industrialist man, through his programme of aid and development, has been partially and indeed grossly responsible for the extent of the plight in which the nomads now find themselves.

Professor Reid A. Bryson, director of the Institute for Environmental Studies, the University of Wisconsin, believes that industrial man could be responsible for a lot more than having given ill-conceived aid to the Sahelian States. He thinks it possible that the climatic change which has taken the monsoons southwards away from the areas where rain is desperately needed to areas which already have adequate rain, could have been caused through industrialisation. In this article he explains how.

The drought

On May 11, 1973, the Food and Agricultural Organisation in Rome issued the following (abbreviated) communique:

"An appeal for airlifts and for immediate additional aids... for six drought stricken West African countries was made today by the Food and Agriculture Organisation of the United Nations, FAO.

"In making the appeal FAO Director General Addeke H. Boerma stated...:

" 'In some areas there now appears serious risk of imminent human famine and virtual extinction of herds vital to nomad populations.'

"Dr Boerma stated that his special representative for problems of the Sahelian Zone....had reported that ...'the situation is still deteriorating.'

"In order to ensure the survival of populations until the next harvest, it is imperative that the maximum possible supplies be delivered to areas by, at the latest, mid-June when transport is expected to be cut off by the rainy season...

"Sahelian is the term applied to a broad belt of arid land extending some 3,000 kilometres along the southern edge of the Sahara. The zone is shared by all six countries. The countries are Mauretania, Senegal, Mali, Upper Volta, Niger, and Chad (Fig. 1).

"Within the Sahelian Zone, *now in its fifth year of drought*, millions of cattle and other domestic animals have perished in recent weeks. An FAO official for African affairs estimated last week that out of a population of 30 millions in the six countries 'about one-third are now weakened by hunger and malnutrition and some people are dying.' "

According to *Newsweek* (4 June 1973) Mourtada Diallo, a regional director of the United Nations Economic Commission for Africa, said "If the problem is not solved in two months, nearly six million people may die." During June the estimate rose dramatically.

Across Asia the story was similar: the drought-prone regions of India had suffered a harvest estimated at nearly 60 per cent below normal in 1972; Bangladesh suffered a short fall of 2.5 million tons in its rice harvest (enough to just sustain about 10 million people!); in Sri Lanka (formerly Ceylon) drought caused crop losses of

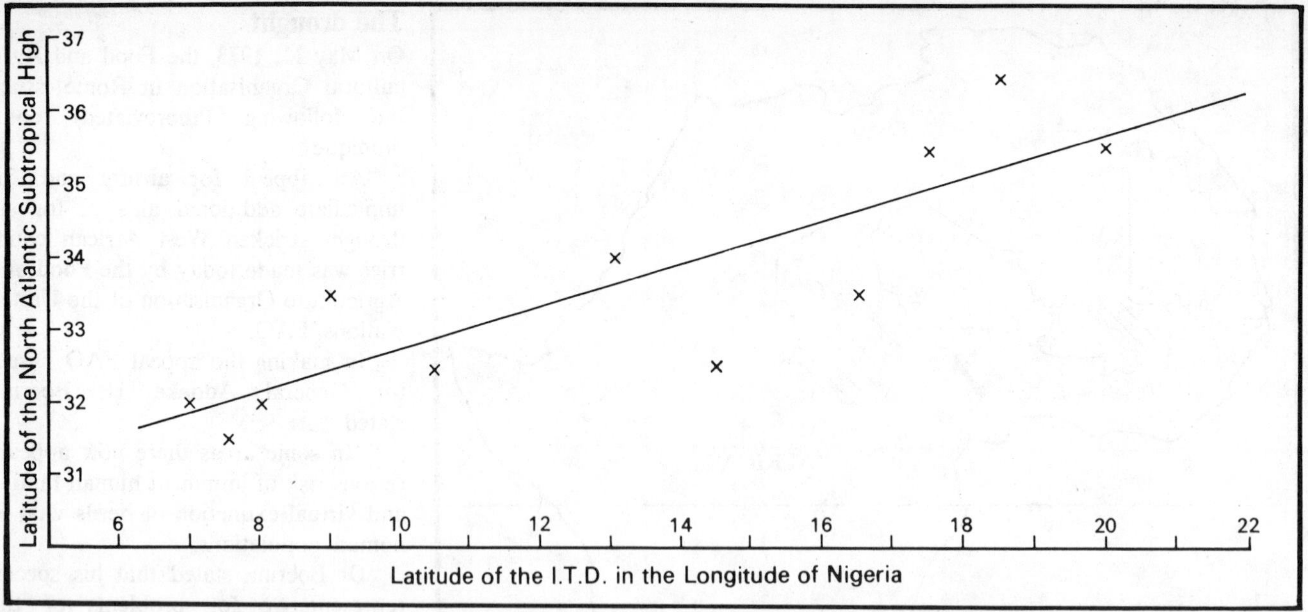

Figure 2: *The latitude of the subtropical anticyclones in the North Atlantic versus the latitude of the Intertropical Discontinuity in the longitude of Nigeria.*

30 per cent of the expected harvest; Chinese newspapers spoke openly of drought and famine.

In the Americas, drought ravaged the Central American maize and Mexico lay in the throes of a drought. In all the monsoon lands the story was similar, as well as in the lands with the monsoon-like climates. The monsoon rains had failed in 1972, but in West Africa the failure was the fifth in a row. What perversity of nature had withheld the life-giving rains in the most hunger-ridden part of the world? Or was it entirely nature?

The monsoon rains

The rains of the Sahelian zone come in summer, when they do come, in common with most regions just south of the subtropical deserts. They are seasonal northward extensions of the tropical rains. During the monsoon, moist air from the equatorial seas enters the continents, replacing the dry, subsident air of the subtropical anticyclones. This occurs when the anticyclones move poleward in summer.

Usually there is a sharp change in the moisture content of the air at the advancing edge of the moist monsoon airstream, which is often called the intertropical discontinuity (ITD) because the origin of the moist air may be traced to the tropics of the opposite hemisphere. In the monsoon regions of the world, the rains occur on the equatorward side of the discontinuity. In West Africa, the monsoon rains increase steadily southward from the position of the ITD for about 800 kilometres, and the length of the rainy season is proportional to the length of time that the ITD lies north of a place. The deeper and longer the penetration of the moist air the greater the monsoon rain. North of the ITD there is no rain.

Now the latitude of the ITD in West Africa appears to be related to the latitude of the subtropical anticyclone. When the anticyclone is far south in winter, so is the ITD, and when one is far north, so is the other (Fig. 2). Year by year, the relationship appears to be true, also, though the data is hard to come by and complicated by other factors. There are some interactions between the pattern of pressure waves in the middle latitude westerlies and the monsoon rains, also, but the dominant relationship is that between the subtropical anticyclones and the ITD. The controls on the latitude of the ITD are a central issue because, as Ilesanmi (1971) has shown, a displacement of one degree in the latitude of the ITD results in a change of seven inches in the annual precipitation in northern Nigeria. When the usual annual total is marginal for food production, a reduction of seven inches is catastrophic! Yet that kind of catastrophic decrease is wnat has happened all across Sahelia since 1957, according to Winstanley's data (1973).

"In some areas there now appears serious risk of imminent human famine and virtual extinction of herds vital to nomad populations."

Monsoon rains and hemispheric wind patterns

Many years ago, studying the rather sudden onset of the summer rains in the south-western United States, I found that the northward extension of tropical rains up the Sierra Madre Occidentale of Mexico and into the American south-west was associated with the concomitant northward shift of the eastern Pacific anticyclone (Bryson and Lowry, 1955). These rains are monsoon-like in character, occurring with a change from dry to humid air. This fact is recognised in the old Zuni proverb: "When the scalp locks on the wall of the Kiva feel moist, the rains will come." The hygroscopic hair on the cool, underground wall metered the advent of moist air. If the anticyclone doesn't move north, the tropical rains do not move north. This relation is complicated a little by the fact that when the anticyclones are unusually far south, the trough pattern between them also appears to change and there is more rain than one would expect on the basis of the latitude of the anticyclones alone.

In India, the monsoon onset has been found to follow the shift of the jet stream to the north of the Himalayas (Yin, 1949). This is equivalent to saying that the anticyclones have shifted, for the jet-stream is near the southern edge of the westerlies and the subtropical anticyclones are at that southern edge. Indeed, there is even a significant correlation between the position of the ITD in India (the penetration of the monsoon) and the latitude of the North Atlantic anticyclone. This is hardly surprising, for the subtropical anticyclone belt is an integral component of the hemisphere wide atmospheric circulation system.

The "Z Criterion"

Some years ago, Fultz (1961) found that he could distinguish two clearly different circulation regimes in his laboratory simulations of the atmosphere. Working with a rotating circular tank of water in which he could produce the equivalent of an equator-to-pole temperature gradient, he found that with small temperature gradients and small rotation rates he produced a direct, vertical cellular circulation. This "convective" type of vertical circulation regime is now widely called a "Hadley regime" because of Hadley's description of it in 1735 (though Halley had hinted at it four decades earlier).

With higher rotation rates and stronger temperature gradients, Fultz found meandering lateral circulations very much like those seen on upper air charts of the polar and mid-latitude regions, and a weaker Hadley circulation. This meandering type of circulation is now called the "Rossby regime" because of Rossby's extensive discussion of it in the late 1940's and early 1950's (though it was described a century earlier by Dove).

Now one can reason that low rotation rates of the earth's surface about the vertical are characteristic of the tropics, and the Hadley regime should be found there, with rising motion in equatorial regions and sinking motion in subtropical latitude. This is observed. High rotation rates are found at higher latitudes, and the Rossby regime is found there. The equator-to-pole temperature gradient is minimal in summer, thus the Hadley regime should extend to higher latitudes in summer than in winter, and this is also observed.

One would expect that the descending branch of the Hadley circulation would thus be found at the equatorward edge of the westerlies of the Rossby regime, moving poleward in summer and equatorward in winter, and such is observed. For dynamical reasons this sinking branch must be associated with high pressure and anticyclonic motion. It produces the major deserts of the world.

In autumn when the equator-to-pole temperature gradient is the same as it is in spring, the subtropical anticyclones, which represent this sinking portion of the Hadley circulation, are found farther north than in spring. A meteorologist would argue that this is due to a greater static stability of the atmosphere in autumn, i.e., the temperature difference between the earth's surface and the air aloft is smaller in autumn than in spring.

All of this summed up says that desert-making climates should be farther south, displacing the monsoons when the equator-to-pole and surface-to-upper-air temperature differences are greater. This was quantified on theoretical grounds by Smagorinsky in 1963. He developed what is called the "Z criterion" which can be viewed as an estimate of the latitude of transition from the Rossby regime to the Hadley regime, i.e., the latitude of the subtropical anticyclones, and in turn the desert climates. This is summarised in Figure 3.

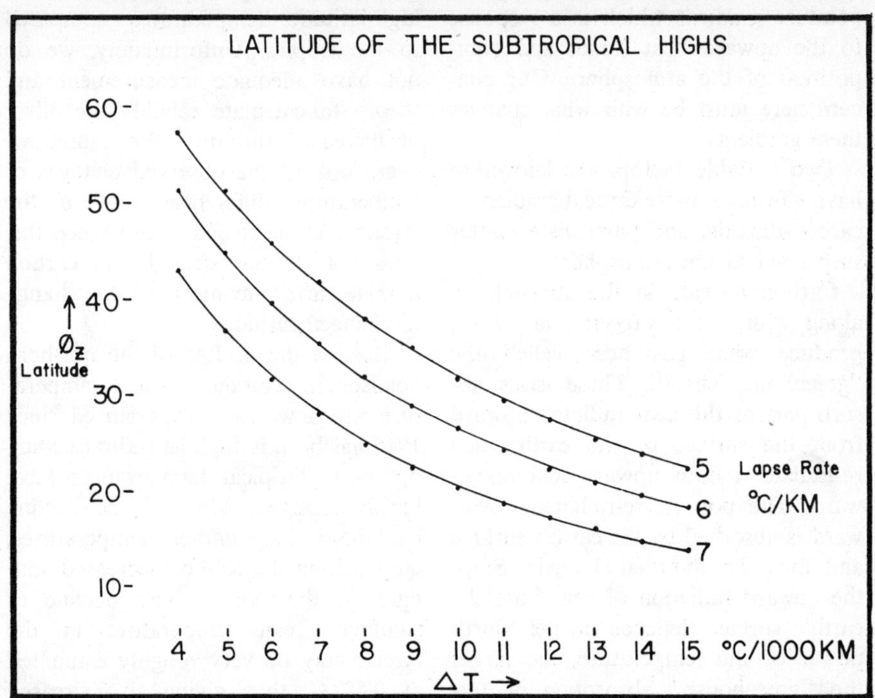

Figure 3: *The approximate latitude of the subtropical anticyclones, estimated from Smagorinsky's Z criterion, as it depends on the north-south temperature gradient, OT and the vertical temperature gradient (lapse rate)*

> According to *Newsweek* (4 June 1973) Mourtada Diallo, a regional director of the United Nations Economic Commission for Africà, said "If the problem is not solved in two months, nearly six million people may die." During June the estimate rose dramatically.

The striking feature of Figure 3 is the sensitivity of the sub-tropical anticyclone latitude to really quite small changes in the world distribution of temperature. We can add one more link to the chain of relations, however. Ilesanmi's data for northern Nigeria (op. cit.) shows that one degree latitude change in the position of the ITD is adequate to explain the Sahelian drought. If we compare the monthly position of the ITD in Nigeria with the monthly position of the North Atlantic subtropical anticyclone, we find that one degree change in the ITD position means only one-third of a degree change in the position of the high—35 km. This is an almost unobservable change! Figure 3 shows that, for summer, this one-third of a degree change can be brought about by a decline of Arctic temperatures of about 0.3° C or an increase of upward temperature gradient of 0.06°C/km. Arctic temperatures *have* declined compared to tropical temperatures, and what data we have so far indicates that the magnitude of the decline is quite adequate to produce the Sahelian effect.

Nature and/or man?

There is a natural equator-to-pole temperature gradient, resulting from the response of the spheroidal earth/atmosphere system to solar radiation. There is also a natural vertical temperature gradient which is a response to the upward heat fluxes and composition of the atmosphere. Our concern here must be with what changes these gradients.

Two variable factors are known to have effects on these critical gradients: carbon-dioxide, and particulate matter suspended in the atmosphere.

Carbon dioxide in the atmosphere, along with water vapour and ozone, produce what has been called the "greenhouse effect". These gases absorb part of the heat radiated upward from the surface of the earth, then re-radiate it both upward and downward. The portion re-radiated downward is absorbed by the earth's surface and must be re-radiated again. Since the upward radiation of heat from the earth's surface depends on the fourth power of the temperature, the larger the "greenhouse" absorption of terrestrial radiation, the higher the surface temperature—but not the temperature aloft. Thus, not only does increased carbon dioxide in the atmosphere increase the surface temperature of the earth, but the vertical temperature gradient as well (Manabe and Wetherald, 1967).

The carbon dioxide content of the atmosphere has increased over the past century, largely from the burning of fossil fuels, and appears to be growing at an increasing rate. The calculated increase of temperature at the earth's surface as a result of this increase in carbon dioxide is on the order of 0.1°C since 1940. If the effect is assumed to be nil at a height of 3 km, then the latitude of the subtropical highs should have moved equatorward by about 15 km and the northern Nigerian rains (perhaps all Sahelian rains) decreased by three inches or so. If so, this decline can be directly attributed to air pollution—man, not nature.

Particulate material suspended in the atmosphere (turbidity) reduces the intensity of sunlight (Bryson, 1972; Reitan, 1971). According to Machta (1972), most of the increase of particulate matter has been at higher latitudes, as shown by the reduced intensity of sunlight at the stations for which data were summarised by Budyko (1969). Even if the increase were uniform, however, the longer path length of the sunlight passing through the higher latitude atmosphere would yield a greater reduction of sun intensity there than in the tropics. This should reduce high-latitude temperatures compared to the tropics. Unfortunately, we do not have adequate measurement and theory to calculate reliably the effect of increased turbidity. We can, however, look at the observed changes of temperature difference between the Arctic and the tropics, and since this does not depend strongly on carbon dioxide, attribute most of the change to changed turbidity.

Most of the decline of the northern hemisphere average surface temperature which we have experienced since 1945 has been in high latitudes (Reitan, op. cit.). Tropical temperatures have hardly changed. Most of the decline has been in summer temperatures, emphasising the role of decreased sunlight in the Arctic. The decline of summer mean temperature in the Arctic may be very roughly estimated at 0.5° C, thus giving an increased equator-to-pole temperature gradient of 0.1° C or so per 100 km. This, according to Figure 3, would however lower the latitude of the subtropical highs and deserts by somewhat over half a degree and decrease the northern Nigerian rainfall by 10–14 inches.

What part of this turbidity contribution to the Sahelian drought and the suppression of the monsoons in general can be attributed to man made pollution depends on the relative contributions of human and non human factors to the production of the particulates. My own analysis suggests that over the past century about 17 per cent of the temperature variance created by turbidity has been due to agricultural, industrial and other human activities. However, it appears that in recent decades the human contribution has been closer to 30 per cent or so, the remainder being largely due to volcanic activity (Bryson 1972).

It seems rather ironic that during the past five years, while drought afflicted the Sahelian zone, the scientific discussion about the climatic effects of carbon dioxide and turbidity has centred on whether one would make the earth warmer or the other make it cooler. There was even the suggestion that neither was important because the two effects would balance. If the analysis in these pages is correct, this either/or argument become irrelevant, for increased carbon dioxide and increased turbidity *both* act to suppress the monsoons of the world!

My own analysis suggests that over the past century about 17 per cent of the temperature variance created by turbidity has been due to agricultural, industrial and other human activities. However, it appears that in recent decades the human contribution has been closer to 30 per cent or so, the remainder being largely due to volcanic activity.

The future of the monsoon lands

If my analysis of the situation is correct, an unpleasant view of the future unfolds.

Prior to 1920 or so there was much more volcanic activity than in the 1920–1950 period. The subtropical anticyclones should have been at lower latitude and drought should have been more frequent in the monsoon lands. The North Pacific anticyclones is the only one for which I have data, but in the 1899–1918 period its summer latitude was 1–1.5 degrees lower than in 1919–1939. Occasions with only half the normal rainfall were half again as frequent in India in the period before 1918. The drought in Sahelia is said to be the worst in 60 years, i.e. since 1913.

Now the carbon dioxide is increasing rapidly to exacerbate the suppression of the monsoon, volcanoes are once more active, and a more-than-doubled human population—rapidly industrialising—provides a "human volcano" source of turbidity.

Will mankind give up burning fossil fuels to aid the people of the monsoon lands? No way!

Will the volcanoes please settle down? Probably not, since they were unusually quiet from 1920–1955.

Will mankind go easy on particulate air pollution by careful pollution controls on factories, slow, careful construction, dust-free agriculture (including no slash-and-burn)? Even the monsoon land nations resist this strongly.

Will the monsoon return? Probably not regularly in this century.

In terms of feeding a rapidly growing population, the present climatic trend is a central fact. In the perspective of history, however, it is not something unusual. What seems to be unusual is what we think of as normal climate. Figure 4 shows the course of

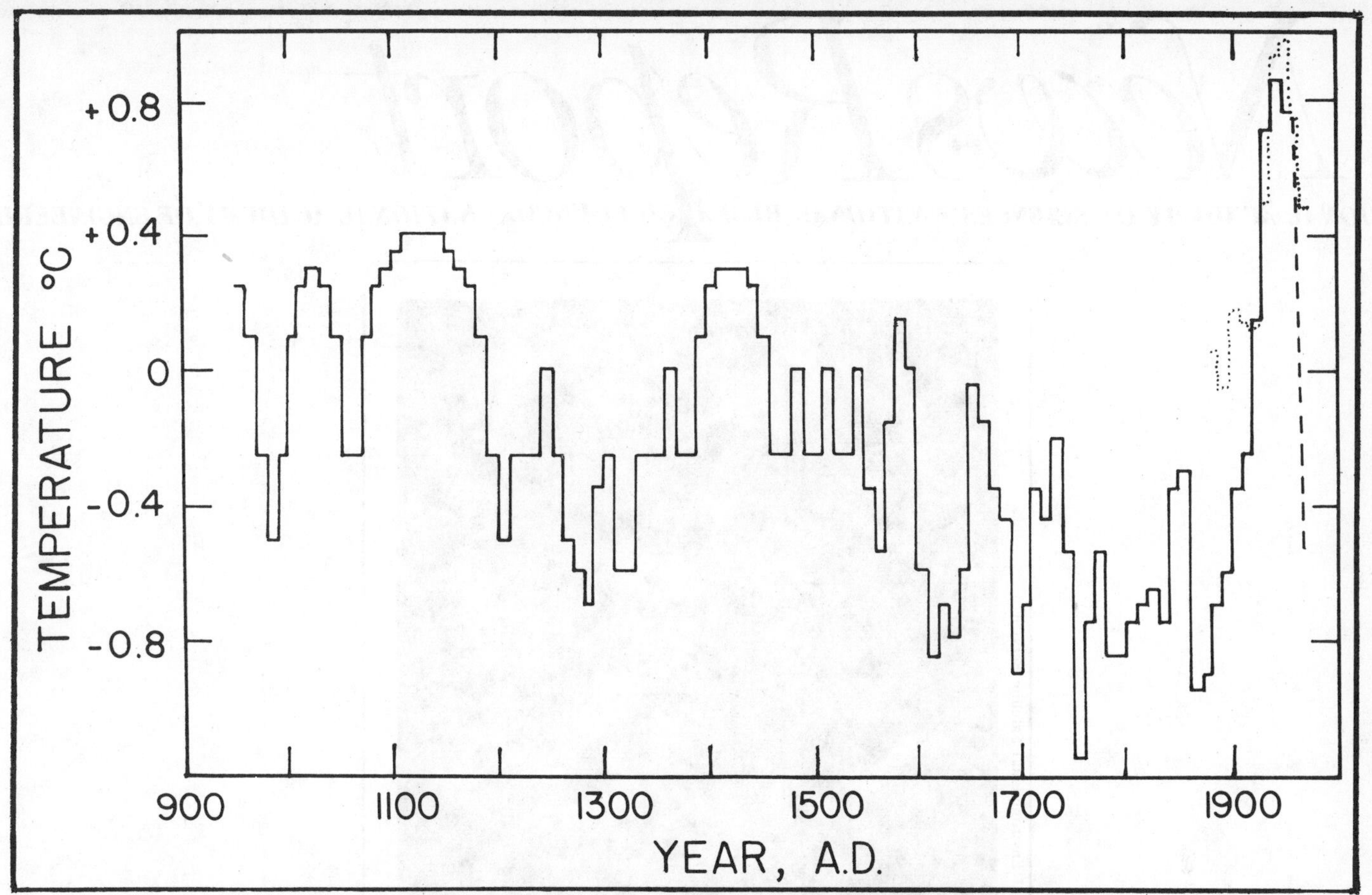

Figure 4: *Mean annual temperature in Iceland over the past millenium (after Bergthorsson). The dashed line indicates the rate of temperature decline in the 1961-1971 period, and the dotted line shows the variation of northern hemisphere mean temperature plotted to the same scale.*

mean annual temperature in Iceland as reconstructed by Bergthorsson (1962). To be sure, Iceland is a rather small country, but it is quite representative of the critical region of main Arctic cooling and warming. The figure shows that the period designated as normal by international agreement, 1930–1960, is the most unusual period in the last millenium. This is a period during which the population of the earth nearly doubled, during which most of the readily arable land was occupied, and during which industrialisation became worldwide. We have become "locked in" to that climate which seems to have been a brief interruption of the "Little Ice Age" of the preceding three centuries. When the earth entered that period between A.D. 1450 and 1600, the Mali Empire collapsed and the magnificent Indian city at Fatehpur Sikri was abandoned as its water supply failed. Do such events lie ahead?

Our climatic data for the Sahelian Zone is rather scanty and our knowledge of the dynamics of the monsoons inadequate. The inter-action of the westerly troughs and the monsoon, mentioned by Winstanley (op. cit.) may perturb the equatorward march of the subtropical deserts. Let us hope so. But aided by the hindsight of climatic history and climatological science let us not assume that the Sahelian drought must soon end as the region returns to an unlikely "normal."

References

Bergthorsson, P., 1962. *Preliminary notes on past climate of Iceland.* Conf. on Climate of the 11th and 16th Centuries June 16–24, Aspen, Colorado (unpublished mimeographed notes) (National Center for Atmospheric Research, Air Force Cambridge Research Labs.). 23 pp.

Bryson, R. A., 1972. "Climatic modification by air pollution", in Polunin, N. (Ed), *The Environmental Future*, London, Macmillan, xiv, 660 pp. (134–174).

Bryson R. A. and W. P. Lowry, 1955. Synoptic climatology of the Arizona summer precipitation singularity *Bull. Amer. Met. Soc.*, 36: 329–399.

Budyko, M. I., 1969. "The effect of solar radiation variation on the climate of the earth", *Tellus* 21: 611–619.

Fultz, D., 1961. "Developments in controlled experiments on larger scale geophysical problems", *Adv. Geophys* 7: 1–103.

Ilesanmi, O. O., 1971. "An empirical formulation of an ITD rainfall model for the tropics: A case study for Nigeria", *Jour. Appl. Met.* 10(5): 882–891.

Machta, Lester, 1972. "Mauna Loa and global trends in air quality", *Bull. Amer. Met. Soc.* 53(5): 402–421.

Manabe, S. and R. T. Wetherald, 1967. "Thermal equilibrium of the atmosphere with a given distribution of relative humidity." *J. Atm. Sci.* 24: 241–259.

Reitan, C. H., 1971. An assessment of the role of volcanic dust in determining modern changes in the temperature of the northern hemisphere, Ph. D. thesis, University of Wisconsin—Madison (unpublished), 147 pp.

Smagorinsky, J., 1963. "General circulation experiments with the primitive equations, I: The basic experiment," *Mon. Wea. Rev.* 91 (3) 99–164.

Winstanley, Derek, 1973. "Recent rainfall trends in Africa, the Middle East and India." *Nature* 243: 464–465.

Yin, M. T., 1949. "A synoptic-aerologic study of the onset of the summer monsoon over India and Burma," *Jour. Meteor.* 6: 393–400.

Will mankind give up burning fossil fuels to aid the people of the monsoon lands? No way!

Will the volcanoes please settle down? Probably not, since they were unusually quiet from 1920-1955.

Will mankind go easy on particulate air pollution by careful pollution controls on factories, slow, careful construction, dust-free agriculture (including no slash-and-burn)? Even the monsoon land nations resist this strongly.

Will the monsoon return? Probably not regularly in this century.

NewsReport

NATIONAL ACADEMY OF SCIENCES · NATIONAL RESEARCH COUNCIL · NATIONAL ACADEMY OF ENGINEERING

National Environmental Satellite Service

This ATS 3 satellite photo in the visible-light spectrum on April 22 reveals a dim haze—from Central America across the Gulf of Mexico—believed to be smoke from burning cane fields.

The Atmospheric Sciences: A Study Of Purpose and Priority

THE ATMOSPHERIC SCIENCES depend upon each other in subject matter and in research timing. A global model of air-pollutant dispersion must rest upon many understandings—atmospheric chemistry and physics, local and regional climate, transport mechanisms, and atmospheric motions. Improved weather forecasting requires improved theoretical grasp, which depends upon improved observations. Weather modification by man raises public questions that in turn influence the course of study and experiment in weather modifications.

This interdependence, the National Research Council's Committee on Atmospheric Sciences believes, has strong implications for atmospheric science and for public policy. Last summer at Friday Harbor, Wash., the committee met to consider objectives and priorities for U.S. atmospheric science in the 1970s. In a report that grew from that study, the committee calls for research-policy emphasis on the "useful" work—especially weather prediction—of atmospheric science. The understanding that must be a part of this, the committee suggests, will bolster not only weather forecasts but also man's ability to avoid influencing the atmosphere by inadvertence.

The Atmospheric Sciences

The report, *The Atmospheric Sciences and Man's Needs: Priorities for the Future* (see "New Publications," p. 8), declares:

"The growth of scientific capability has made possible increased useful application of the atmospheric sciences at the same time that certain environmental problems are approaching critical levels. Enlarged opportunities exist for extending our understanding of the atmosphere, and for applying this understanding more effectively than ever before to human needs."

Human needs with respect to the atmospheric sciences, the committee noted, include more than weather forecasts. They include minimizing weather damage, efficient deployment of resources for pollution control, and possibly the ability to influence the atmosphere to alleviate shortages or excesses of precipitation. And human needs, the committee emphasizes, include the ability to "cope with" the challenges to public wisdom posed by prospects of both purposeful weather modification and accidental influencing of the atmosphere.

The Importance of Prediction

Because "extension of predictive capability is necessary to provide the base which supports much of the fundamental research and the applications of the atmospheric sciences," the committee concluded that the first of the "major objectives" of atmospheric research in the 1970s must be to *"extend the capability for useful prediction of the weather and atmospheric processes."*

National Environmental Satellite Service

This digitized photomosaic of infrared readings from the satellite ITOS 1 is a test display in four shades in an attempt to yield a rough picture of cloud layers. High cloudtops—over 23,000 feet—are seen in white; middle clouds—down to 10,000 feet—are in middle gray; low clouds—down to the surface—are in dark gray. The surface and a wedge for which no data were obtained are in black. The mosaic is from readings made the night of September 11, 1970.

Monitoring Pollutants

Global modeling and other studies essential to improving weather forecasting, the committee said, will make it possible to approach a second major objective: *"To contribute to development of the capability to manage and control the concentrations of air pollutants."* The measurement and monitoring of air pollution contribute not to development of technology for worst-case abatement but rather to assisting the "assessment of alternative pollution control programs." This necessarily involves a detailed examination of urban and regional weather, understanding of which is essential to the committee's third major objective: *"To establish mechanisms for the rational examination of deliberate and inadvertent means for modifying weather and climate."* Depending upon programs to achieve the first three objectives is a fourth: *"To reduce substantially human casualties, economic losses, and social dislocations caused by weather."*

The committee's estimated price tag for U.S. programs to meet these objectives is $453,000,000 over the decade, in addition to current annual funding levels. Of this sum, $180,000,000 would be for capital investment (including $140,000,000 for new computer facilities) and $270,000,000 would be for research operations. Annual funding for research operations in fiscal 1970 was approximately $303,000,000.

Global Atmospheric Research Program

The committee—chaired by Robert G. Fleagle of the Department of Atmospheric Sciences of the University of Washington—recommended that first priority in weather prediction be given to support for the Global Atmospheric Research Program (GARP) "to extend the range of useful prediction into the one- to two-week period." Among GARP requirements stressed by the committee: development of "global observing capability"; "availability of electronic computers with speeds at least 100 times faster than the speeds of the most powerful computers in use today"; and regional field programs to improve the bases for modeling and long-range prediction.

Of special importance, the committee reports, are economical means for measuring wind velocity in the tropics; satellite soundings of temperatures and water vapor, particularly in the tropics; and global measurements of wind velocity near the sea surface, soil moisture, extent of ice cover, extent of Arctic open water, and water content of clouds.

"Improvements in prediction models and the associated increased understanding of atmospheric processes will also contribute to the necessary basis for virtually every other application considered in this report," the committee said.

'Pilot Local Weather Watch'

The committee called, too, for a "pilot Local Weather Watch" for selected urban and rural areas, with "real-time" display, by telecast, in order to achieve better understanding both of local weather and of the economic benefits and costs of weather prediction. The committee said further studies are needed for understanding of climate and for understanding of fronts, jet streams, con-

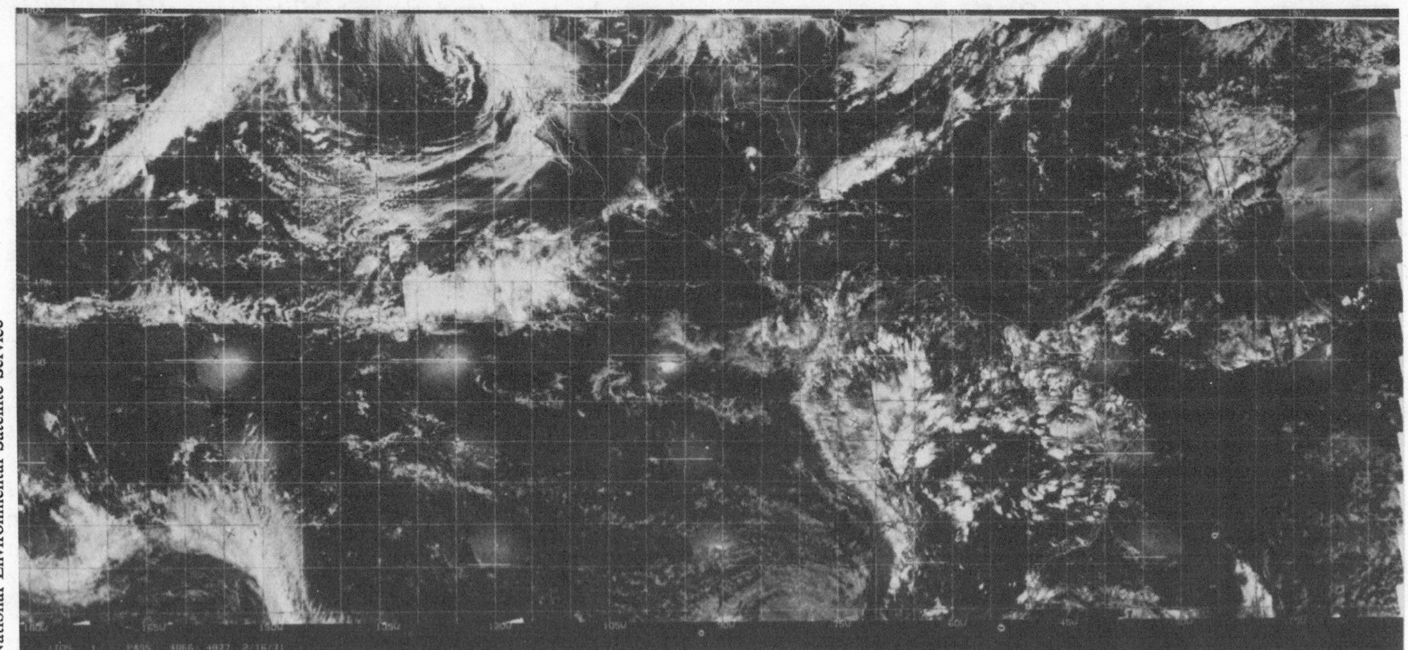

National Environmental Satellite Service

Tropical cloud patterns are evident in this digitized Mercator projection of an April 16 photo in the visible-light spectrum from the satellite ITOS 1. But tropical atmospheric processes, which heavily influence the world's weather, are not well understood. The Global Atmospheric Research Program needs information on tropical winds, temperatures, and water-vapor distribution to improve weather prediction.

vection, and other phenomena of meso-scale—generally phenomena from 100 to 200 kilometers wide and lasting between 2 and 12 hours.

To assist in development of effective air-pollution abatement policies, the committee said, first priority should be accorded development and testing of simulation models of urban atmospheres and pollutant dispersion and a major field program should assess the fate of pollutants from urban sources. A global monitoring system for atmospheric contaminants should be established "promptly," the committee said, and chemical composition of precipitation and dry fallout should be measured in test areas.

"The consequences of deliberate or inadvertent modification of weather phenomena may be so far-reaching that we must know much more specifically what is possible and what is beyond human efforts," the committee warned. "Demonstration of significant weather modification triggers profound ethical, social, economic and political problems which should be given appropriate attention."

Noting increasing, if tentative, evidence that precipitation can be increased or decreased, that supercooled fog can be dissipated, that hail can be suppressed, and that maximum wind speeds of hurricanes can be cut, all by the action of man, and that the atmosphere may be affected as well by the build-up of carbon dioxide and particulate matter, the committee said the "crucial question of what weather modification efforts are in the public interest cannot be settled by atmospheric scientists alone."

Social Implications

This question involves economic, legal, political, and administrative as well as scientific and technical questions, said the committee. "Involved also is the whole complex of environmental issues to which the nation will have to respond . . . : issues of land usage, of the degree of freedom to use and affect the atmosphere, procedures needed to reach equitable decisions in case of conflicting interests, and methods needed to safeguard the future ecology and the climate."

The committee added: "We know so little about possible responses of the atmosphere," the committee said, "that to carry out large scale interventions into cloud processes or energy transfer processes would be highly irresponsible. . . . [N]ational policy should seek to minimize the possibility that any country might undertake to modify weather for military purposes."

Policy Responsibility

The committee said that in order to approach these problems rationally the Executive Office of the President should take responsibility for examining policy issues and developing policy initiatives. Principal Federal responsibility for the national program in weather-modification science should be lodged in the National Oceanic and Atmospheric Administration, the committee said, calling for improved coordination of agency efforts in this field and calling for a major laboratory to implement and coordinate weather-modification research.

"In order to safeguard the life sustaining properties of the atmosphere for the common benefit of mankind," the committee said, "the U.S. Government is urged to present for adoption by the United Nations General Assembly a resolution dedicating all weather modification efforts to peaceful purposes and establishing, preferably within the framework of international non-governmental organizations, an advisory mechanism for consideration of weather modification problems of potential international concern before they reach critical levels."

The committee's over-all hope: that achievement of its objectives for the atmospheric sciences by the end of the decade will provide man with sufficient knowledge of the atmosphere to reduce his vulnerability to it, to protect it, and to influence it more wisely than in the past.

—GERALD S. SCHATZ

Part II
An Overview of Energy Use in the United States

A good but brief overview of energy use by different types of human society ranging from communities of hunters and gatherers to highly industrialized states is offered in the Scientific American book, *Energy and Power* (Freeman, 1971). Another interesting, if perhaps uneven, general discussion of human energy use can be found in Odum's book, *Environment, Power and Society* [1].

In this section we will not attempt to review the full spectrum of human energy use patterns, but will instead focus on the United States, which by almost any reckoning is the most energy-intensive society in the world. The articles in this section outline how the United States is using energy today, how it may use it in the future, and how this use may effect and be effected by environmental quality and the U.S. cultural setting. The articles have been selected to provide a wide overview and to represent a range of opinions which are based on careful analysis.

The section begins with E. Cook's article, "The Flow of Energy in an Industrial Society," which summarizes the ways in which the United States has been using energy. Next, an excerpt from a Brookhaven National Laboratory report provides an overview of the U.S. electric energy system. This is followed by two general reviews of what the future may hold. The first, by C. A. Zracket, summarizes the findings of a series of studies and workshops run by the MITRE Corporation, a nonprofit think tank. The second is from the 1972 annual report of Resources for the Future and summarizes a number of the subjects on which the staff of Resources for the Future has recently done work.

The problem of projecting future energy demand is exceedingly complex. A large number of studies involving various types of modeling have been completed. To date, no study has been sufficiently general and flexible to be really adequate for use in serious policy analysis. A new generation of studies based on considerably more sophisticated models of the energy system is now underway, and research in this area will certainly be intensified over the next few years. A summary of recent energy demand studies is provided in excerpts reprinted from a committee print prepared by the Committee on Interior and Insular Affairs of the U.S. House of Representatives. Because of space constraints it has not been possible to reprint the entire report. Included in the portions of the report not reprinted here are discussions of the difference between "consumption," "demand," "needs," and "require ments" and an extended discussion of forecast methodology. Readers who are interested in exploring this field in greater depth will find the Decision Sciences Corporation's report, "Quantitative Energy Studies and Models: A State of the Art Review," to be a useful starting place [2].

The remaining four papers in this section are by P. L. Auer, P. H. Abelson, R. E. Train, and M. R. Gustavson. These papers address, at a fairly general level, several aspects of the problem of present and future energy use, its impact upon the environment, and some of the implications of alternative values or world views.

A useful framework for thinking about the U.S. energy system is provided by the Brookhaven National Laboratories reference energy system [3]. The following two pages reproduce the reference energy system for 1969 and 1977.

REFERENCES

[1] H. T. Odum, *Environment, Power and Society*. New York: Wiley Interscience, 1971.

[2] D. R. Limaye, R. Ciliàno, and J. R. Sharko, "Quantitative energy studies and models: A state of the art review," Decision Sciences Corp., Jenkintown, Pa., 1973.

[3] "Reference energy systems and resource data for use in the assessment of energy technologies," Associated Universities, Upton, N.Y., AET-8, 1973.

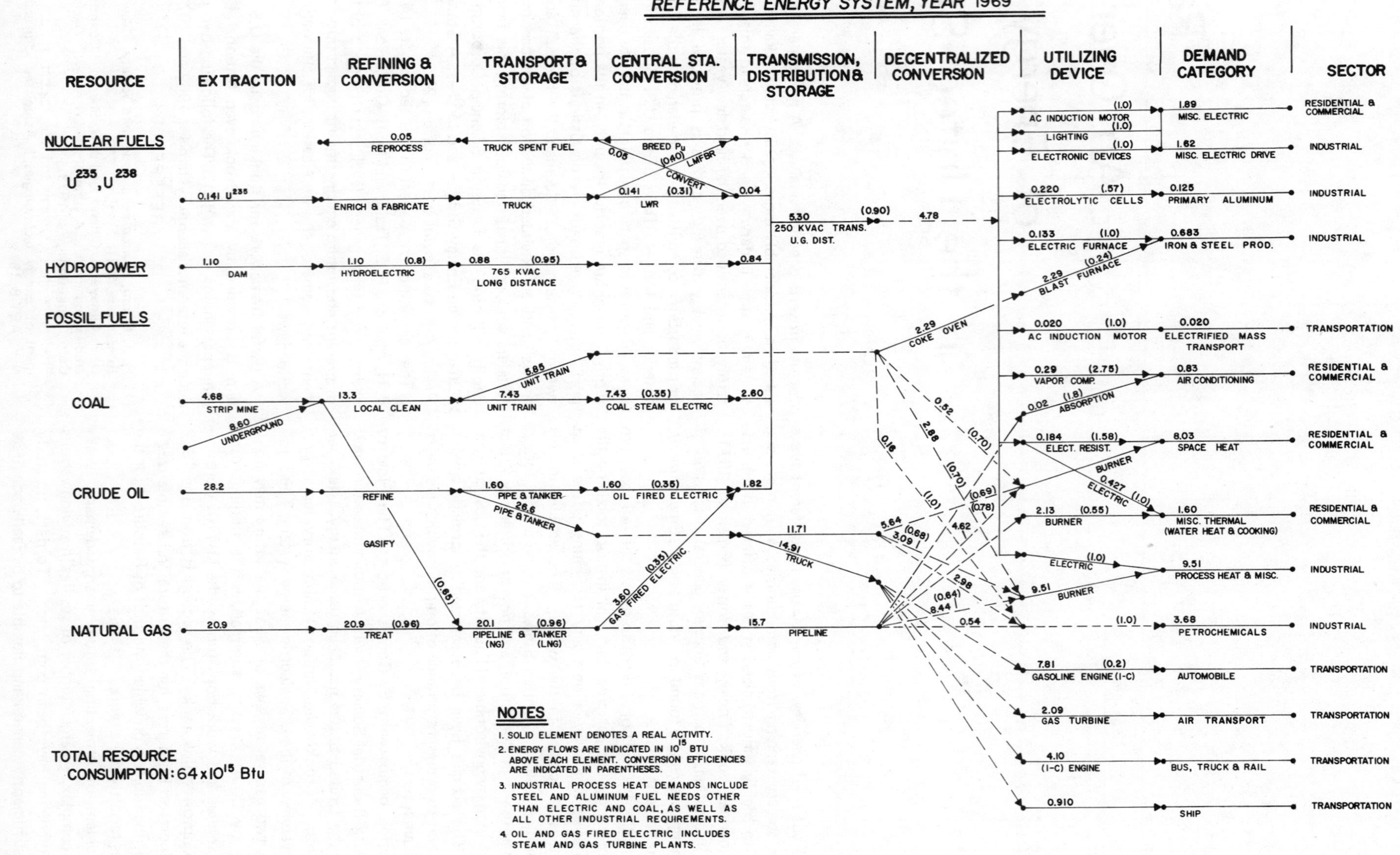

REFERENCE ENERGY SYSTEM, YEAR 1969
RESOURCE
EXTRACTION
REFINING & CONVERSION
TRANSPORT & STORAGE
CENTRAL STA. CONVERSION
TRANSMISSION, DISTRIBUTION & STORAGE
DECENTRALIZED CONVERSION
UTILIZING DEVICE
DEMAND CATEGORY
SECTOR
NUCLEAR FUELS
U235, U238
0.141 U235
ENRICH & FABRICATE
TRUCK
0.141 (0.31)
LWR
0.05
REPROCESS
TRUCK SPENT FUEL
BREED Pu
0.05
(0.40) LMFBR
CONVERT
0.04
HYDROPOWER
1.10
DAM
1.10 (0.8)
HYDROELECTRIC
0.88 (0.95)
765 KVAC
LONG DISTANCE
0.84
5.30 (0.90)
250 KVAC TRANS.
U.G. DIST.
4.78
FOSSIL FUELS
COAL
4.68
STRIP MINE
8.60
UNDERGROUND
13.3
LOCAL CLEAN
5.85
UNIT TRAIN
7.43
UNIT TRAIN
7.43 (0.35)
COAL STEAM ELECTRIC
2.60
CRUDE OIL
28.2
REFINE
1.60
PIPE & TANKER
26.6
PIPE & TANKER
1.60 (0.35)
OIL FIRED ELECTRIC
1.82
GASIFY
(0.65)
NATURAL GAS
20.9
20.9 (0.96)
TREAT
20.1 (0.96)
PIPELINE & TANKER
(NG) (LNG)
3.60 (0.35)
GAS FIRED ELECTRIC
15.7
PIPELINE
11.71
14.91
TRUCK
2.29
COKE OVEN
0.52
(0.70)
2.88
(0.70)
0.16
(1.0)
5.64
(0.68)
3.09
(0.69)
(0.78)
4.62
2.98
(0.64)
8.44
0.54
(1.0)
AC INDUCTION MOTOR
(1.0)
LIGHTING
(1.0)
ELECTRONIC DEVICES
1.89
MISC. ELECTRIC
RESIDENTIAL & COMMERCIAL
1.62
MISC. ELECTRIC DRIVE
INDUSTRIAL
0.220 (.57)
ELECTROLYTIC CELLS
0.125
PRIMARY ALUMINUM
INDUSTRIAL
0.133 (1.0)
ELECTRIC FURNACE
2.29 (0.24)
BLAST FURNACE
0.683
IRON & STEEL PROD.
INDUSTRIAL
0.020 (1.0)
AC INDUCTION MOTOR
0.020
ELECTRIFIED MASS TRANSPORT
TRANSPORTATION
0.29 (2.75)
VAPOR COMP.
0.02 (1.8)
ABSORPTION
0.83
AIR CONDITIONING
RESIDENTIAL & COMMERCIAL
0.184 (1.58)
ELECT. RESIST.
BURNER
8.03
SPACE HEAT
RESIDENTIAL & COMMERCIAL
0.427 (1.0)
ELECTRIC
2.13 (0.55)
BURNER
1.60
MISC. THERMAL
(WATER HEAT & COOKING)
RESIDENTIAL & COMMERCIAL
(1.0)
ELECTRIC
9.51
BURNER
9.51
PROCESS HEAT & MISC.
INDUSTRIAL
(1.0)
3.68
PETROCHEMICALS
INDUSTRIAL
7.81 (0.2)
GASOLINE ENGINE (I-C)
AUTOMOBILE
TRANSPORTATION
2.09
GAS TURBINE
AIR TRANSPORT
TRANSPORTATION
4.10
(I-C) ENGINE
BUS, TRUCK & RAIL
TRANSPORTATION
0.910
SHIP
TRANSPORTATION
NOTES
1. SOLID ELEMENT DENOTES A REAL ACTIVITY.
2. ENERGY FLOWS ARE INDICATED IN 10^15 BTU ABOVE EACH ELEMENT. CONVERSION EFFICIENCIES ARE INDICATED IN PARENTHESES.
3. INDUSTRIAL PROCESS HEAT DEMANDS INCLUDE STEEL AND ALUMINUM FUEL NEEDS OTHER THAN ELECTRIC AND COAL, AS WELL AS ALL OTHER INDUSTRIAL REQUIREMENTS.
4. OIL AND GAS FIRED ELECTRIC INCLUDES STEAM AND GAS TURBINE PLANTS.
TOTAL RESOURCE CONSUMPTION: 64 x 10^15 Btu

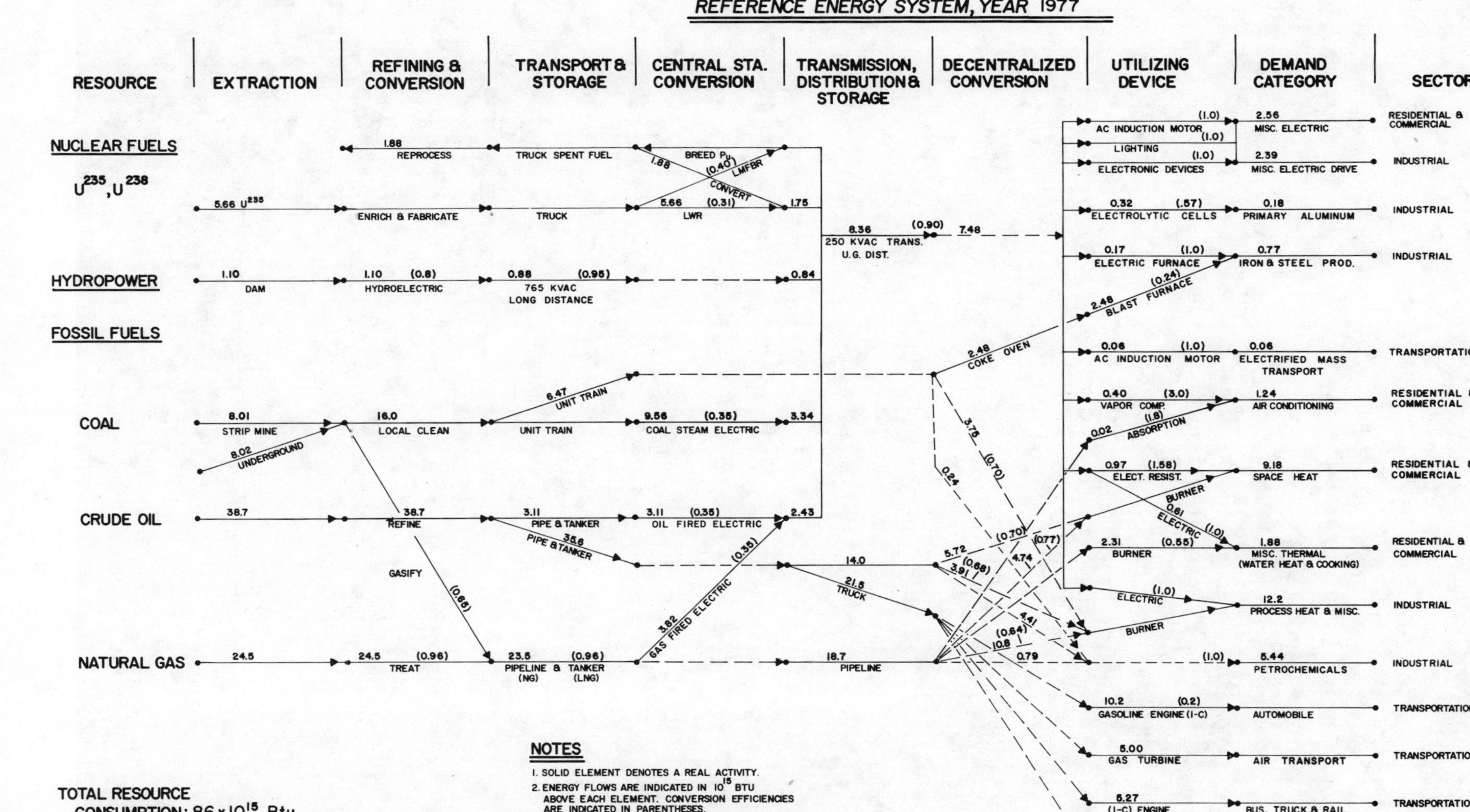

TOTAL RESOURCE CONSUMPTION: 86×10^{15} Btu

NOTES

1. SOLID ELEMENT DENOTES A REAL ACTIVITY.
2. ENERGY FLOWS ARE INDICATED IN 10^{15} BTU ABOVE EACH ELEMENT. CONVERSION EFFICIENCIES ARE INDICATED IN PARENTHESES.
3. INDUSTRIAL PROCESS HEAT DEMANDS INCLUDE STEEL AND ALUMINUM FUEL NEEDS OTHER THAN ELECTRIC AND COAL, AS WELL AS ALL OTHER INDUSTRIAL REQUIREMENTS.
4. OIL AND GAS FIRED ELECTRIC INCLUDES STEAM AND GAS TURBINE PLANTS.

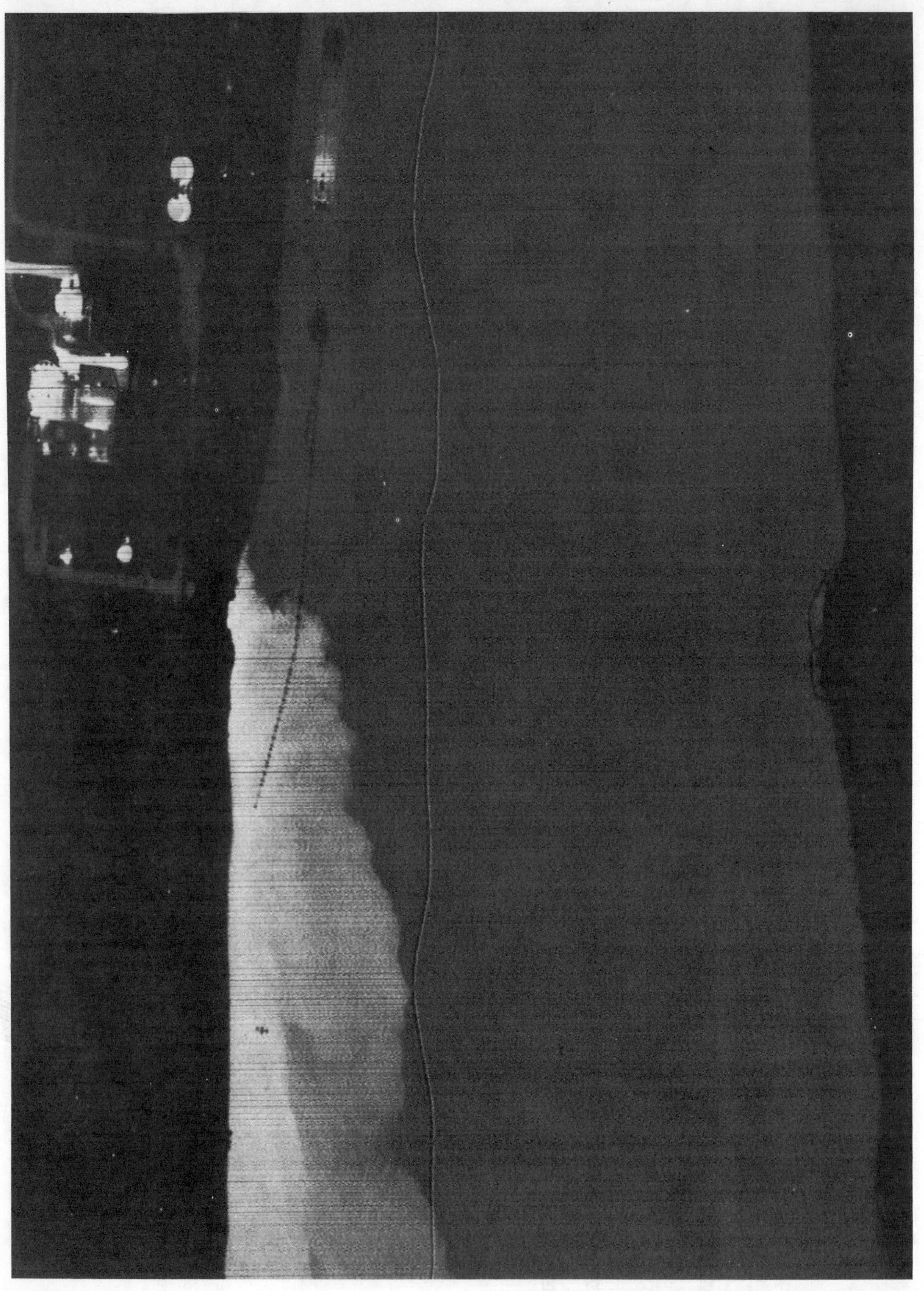

The Flow of Energy in an Industrial Society

The U.S., with 6 percent of the world's population, uses 35 percent of the world's energy. In the long run the limiting factor in high levels of energy consumption will be the disposal of the waste heat

by Earl Cook

This article will describe the flow of energy through an industrial society: the U.S. Industrial societies are based on the use of power: the rate at which useful work is done. Power depends on energy, which is the ability to do work. A power-rich society consumes—more accurately, degrades—energy in large amounts. The success of an industrial society, the growth of its economy, the quality of the life of its people and its impact on other societies and on the total environment are determined in large part by the quantities and the kinds of energy resources it exploits and by the efficiency of its systems for converting potential energy into work and heat.

Whether by hunting, by farming or by burning fuel, man introduces himself into the natural energy cycle, converting energy from less desired forms to more desired ones: from grass to beef, from wood to heat, from coal to electricity. What characterizes the industrial societies is their enormous consumption of energy and the fact that this consumption is primarily at the expense of "capital" rather than of "income," that is, at the expense of solar energy stored in coal, oil and natural gas rather than of solar radiation, water, wind and muscle power. The advanced industrial societies, the U.S. in particular, are further characterized by their increasing dependence on electricity, a trend that has direct effects on gross energy consumption and indirect effects on environmental quality.

The familiar exponential curve of increasing energy consumption can be considered in terms of various stages of human development [*see illustration on next page*]. As long as man's energy consumption depended on the food he could eat, the rate of consumption was some 2,000 kilocalories per day; the domestication of fire may have raised it to 4,000 kilocalories. In a primitive agricultural society with some domestic animals the rate rose to perhaps 12,000 kilocalories; more advanced farming societies may have doubled that consumption. At the height of the low-technology industrial revolution, say between 1850 and 1870, per capita daily consumption reached 70,000 kilocalories in England, Germany and the U.S. The succeeding high-technology revolution was brought about by the central electric-power station and the automobile, which enable the average person to apply power in his home and on the road. Beginning shortly before 1900, per capita energy consumption in the U.S. rose at an increasing rate to the 1970 figure: about 230,000 kilocalories per day, or about 65×10^{15} British thermal units (B.t.u.) per year for the country as a whole. Today the industrial regions, with 30 percent of the world's people, consume 80 percent of the world's energy. The U.S., with 6 percent of the people, consumes 35 percent of the energy.

In the early stages of its development in western Europe industrial society based its power technology on income sources of energy, but the explosive growth of the past century and a half has been fed by the fossil fuels, which are not renewable on any time scale meaningful to man. Modern industrial society is totally dependent on high rates of consumption of natural gas, petroleum and coal. These nonrenewable fossil-fuel resources currently provide 96 percent of the gross energy input into the U.S. economy [*see top illustration on page 137*]. Nuclear power, which in 1970 accounted for only .3 percent of the total energy input, is also (with present reactor technology) based on a capital source of energy: uranium 235. The energy of falling water, converted to hydropower, is the only income source of energy that now makes any significant contribution to the U.S. economy, and its proportional role seems to be declining from a peak reached in 1950.

Since 1945 coal's share of the U.S. energy input has declined sharply, while both natural gas and petroleum have increased their share. The shift is reflected in import figures. Net imports of petroleum and petroleum products doubled between 1960 and 1970 and now constitute almost 30 percent of gross consumption. In 1960 there were no imports of natural gas; last year natural-gas imports (by pipeline from Canada and as liquefied gas carried in cryogenic tankers) accounted for almost 4 percent of gross consumption and were increasing.

The reasons for the shift to oil and gas are not hard to find. The conversion of railroads to diesel engines represented a large substitution of petroleum for coal. The rapid growth, beginning during World War II, of the national

HEAT DISCHARGE from a power plant on the Connecticut River at Middletown, Conn., is shown in this infrared scanning radiograph. The power plant is at upper left, its structures outlined by their heat radiation. The luminous cloud running along the left bank of the river is warm water discharged from the cooling system of the plant. The vertical oblong object at top left center is an oil tanker. The luminous spot astern is the infrared glow of its engine room. The dark streak between the tanker and the warm-water region is a breakwater. The irregular line running down the middle of the picture is an artifact of the infrared scanning system. The picture was made by HRB-Singer, Inc., for U.S. Geological Survey.

network of high-pressure gas-transmission lines greatly extended the availability of natural gas. The explosion of the U.S. automobile population, which grew twice as fast as the human population in the decade 1960–1970, and the expansion of the nation's fleet of jet aircraft account for much of the increase in petroleum consumption. In recent years the demand for cleaner air has led to the substitution of natural gas or low-sulfur residual fuel oil for high-sulfur coal in many central power plants.

An examination of energy inputs by sector of the U.S. economy rather than by source reveals that much of the recent increase has been going into household, commercial and transportation applications rather than industrial ones [*see bottom illustration on opposite page*]. What is most striking is the growth of the electricity sector. In 1970 almost 10 percent of the country's useful work was done by electricity. That is not the whole story. When the flow of energy from resources to end uses is charted for 1970 [*see illustration on pages 138 and 139*], it is seen that producing that much electricity accounted for 26 percent of the gross consumption of energy, because of inefficiencies in generation and transmission. If electricity's portion of end-use consumption rises to about 25 percent by the year 2000, as is expected, then its generation will account for between 43 and 53 percent of the country's gross energy consumption. At that point an amount of energy equal to about half of the useful work done in the U.S. will be in the form of waste heat from power stations!

All energy conversions are more or less inefficient, of course, as the flow diagram makes clear. In the case of electricity there are losses at the power plant, in transmission and at the point of application of power; in the case of fuels consumed in end uses the loss comes at the point of use. The 1970 U.S. gross consumption of 64.6×10^{15} B.t.u. of energy (or 16.3×10^{15} kilocalories, or 19×10^{12} kilowatt-hours) ends up as 32.8×10^{15} B.t.u. of useful work and 31.8×10^{15} B.t.u. of waste heat, amounting to an overall efficiency of about 51 percent.

The flow diagram shows the pathways of the energy that drives machines, provides heat for manufacturing processes and heats, cools and lights the country. It does not represent the total energy budget because it includes neither food nor vegetable fiber, both of which bring solar energy into the economy through photosynthesis. Nor does it include environmental space heating by solar radiation, which makes life on the earth possible and would be by far the largest component of a total energy budget for any area and any society.

The minute fraction of the solar flux that is trapped and stored in plants provides each American with some 10,000 kilocalories per day of gross food production and about the same amount in the form of nonfood vegetable fiber. The fiber currently contributes little to the energy supply. The food, however, fu-

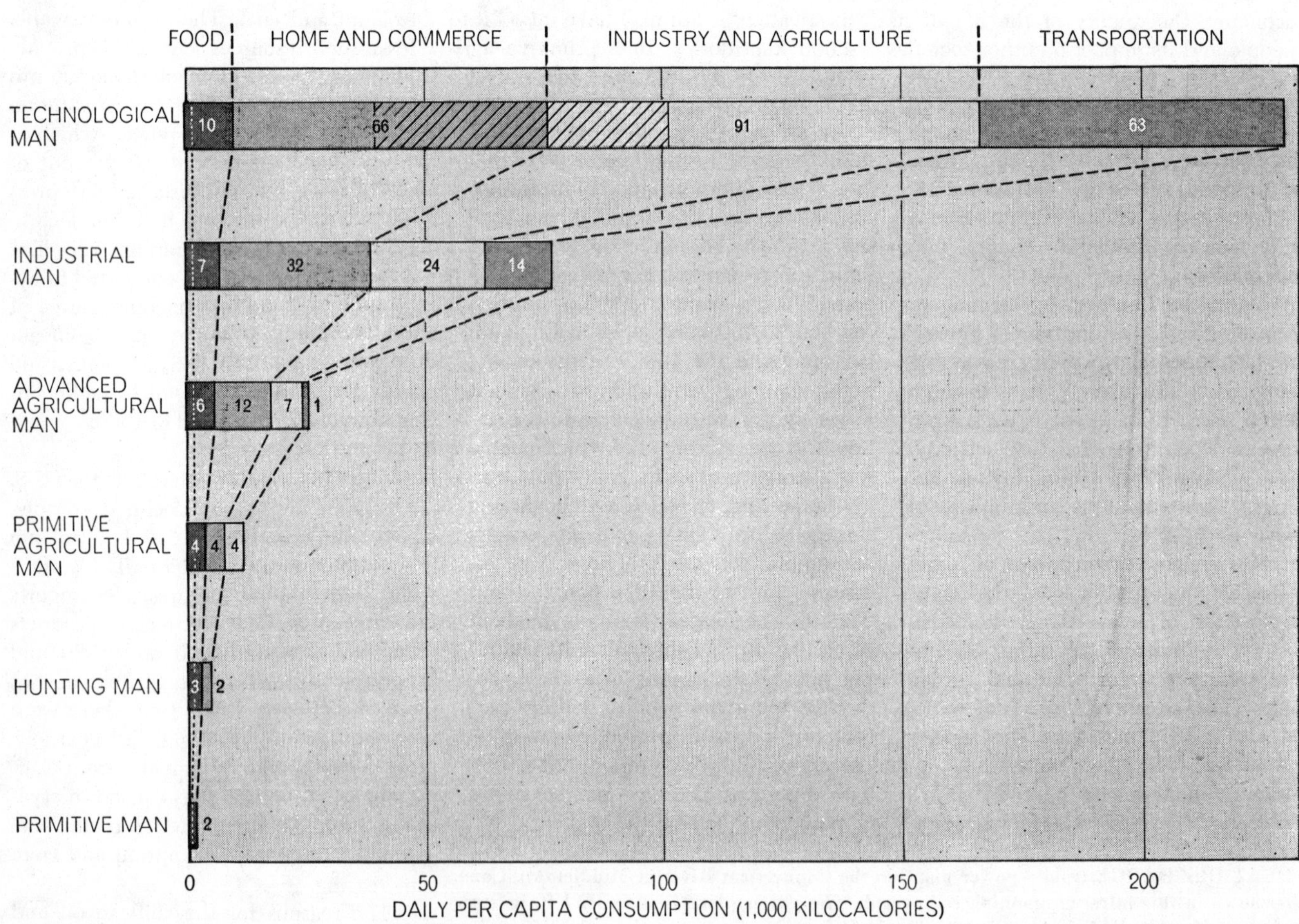

DAILY CONSUMPTION of energy per capita was calculated by the author for six stages in human development (and with an accuracy that decreases with antiquity). Primitive man (East Africa about 1,000,000 years ago) without the use of fire had only the energy of the food he ate. Hunting man (Europe about 100,000 years ago) had more food and also burned wood for heat and cooking. Primitive agricultural man (Fertile Crescent in 5000 B.C.) was growing crops and had gained animal energy. Advanced agricultural man (northwestern Europe in A.D. 1400) had some coal for heating, some water power and wind power and animal transport. Industrial man (in England in 1875) had the steam engine. In 1970 technological man (in the U.S.) consumed 230,000 kilocalories per day, much of it in form of electricity (*hatched area*). Food is divided into plant foods (*far left*) and animal foods (or foods fed to animals).

els man. Gross food-plant consumption might therefore be considered another component of gross energy consumption; it would add about 3×10^{15} B.t.u. to the input side of the energy-flow scheme. Of the 10,000 kilocalories per capita per day of gross production, handling and processing waste 15 percent. Of the remaining 8,500 kilocalories, some 6,300 go to feed animals that produce about 900 kilocalories of meat and 2,200 go into the human diet as plant materials, for a final food supply of about 3,100 kilocalories per person. Thus from field to table the efficiency of the food-energy system is 31 percent, close to the efficiency of a central power station. The similarity is not fortuitous; in both systems there is a large and unavoidable loss in the conversion of energy from a less desired form to a more desired one.

Let us consider recent changes in U.S. energy flow in more detail by seeing how the rates of increase in various sectors compare. Not only has energy consumption for electric-power generation been growing faster than the other sectors but also its growth rate has been increasing: from 7 percent per year in 1961–1965 to 8.6 percent per year in 1965–1969 to 9.25 percent last year [*see top illustration on page 140*]. The energy consumed in industry and commerce and in homes has increased at a fairly steady rate for a decade, but the energy demand of transportation has risen more sharply since 1966. All in all, energy consumption has been increasing lately at a rate of 5 percent per year, or four times faster than the increase in the U.S. population. Meanwhile the growth of the gross national product has tended to fall off, paralleling the rise in energy sectors other than fast-growing transportation and electricity. The result is a change in the ratio of total energy consumption to G.N.P. [*see bottom illustration on page 140*]. The ratio had been in a long general decline since 1920 (with brief reversals) but since 1967 it has risen more steeply each year. In 1970 the U.S. consumed more energy for each dollar of goods and services than at any time since 1951.

Electricity accounts for much of this decrease in economic efficiency, for several reasons. For one thing, we are substituting electricity, with a thermal efficiency of perhaps 32 percent, for many direct fuel uses with efficiencies ranging from 60 to 90 percent. Moreover, the fastest-growing segment of end-use consumption has been electric air conditioning. From 1967 to 1970 consumption for

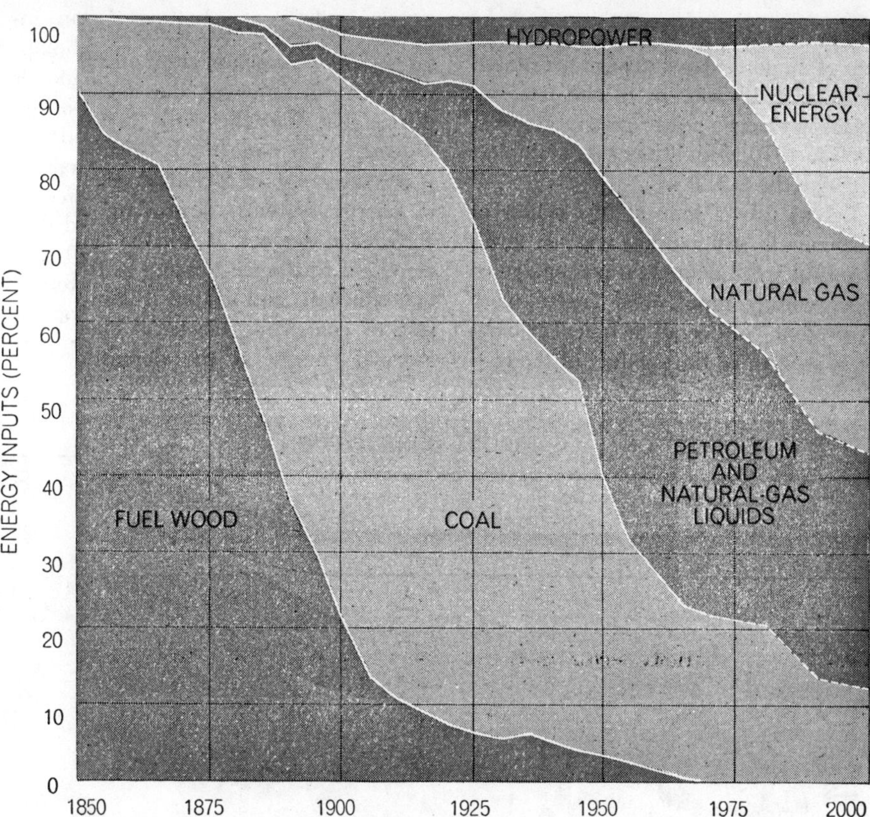

FOSSIL FUELS now account for nearly all the energy input into the U.S. economy. Coal's contribution has decreased since World War II; that of natural gas has increased most in that period. Nuclear energy should contribute a substantial percent within the next 20 years.

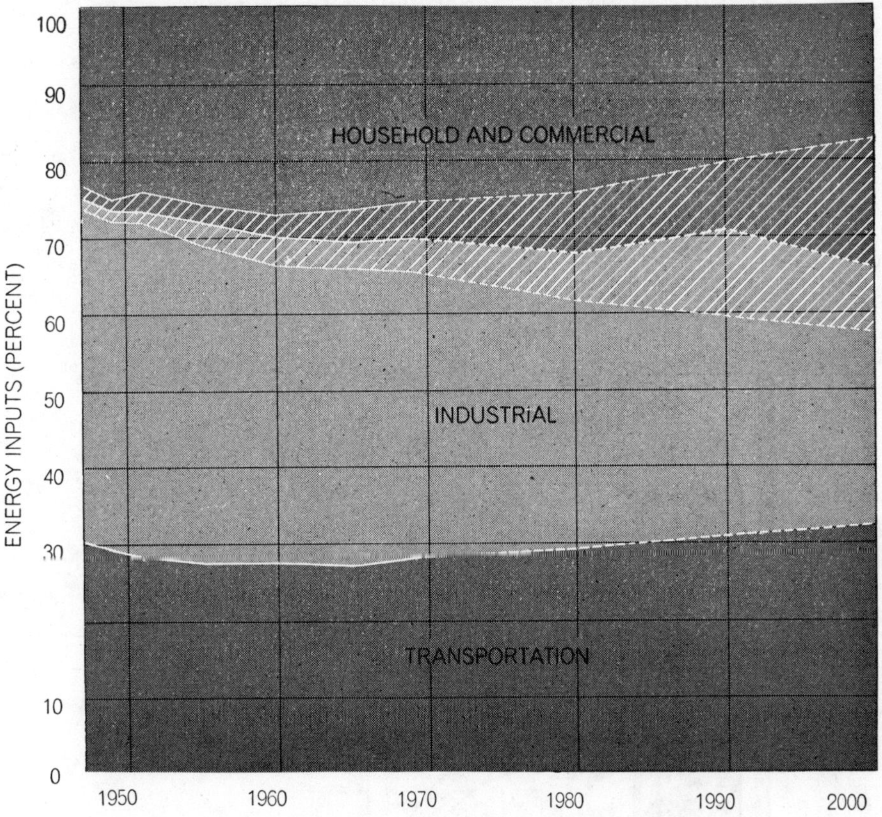

USEFUL WORK is distributed among the various end-use sectors of the U.S. economy as shown. The trend has been for industry's share to decrease, with household and commercial uses (including air conditioning) and transportation growing. Electricity accounts for an ever larger share of the work (*hatched area*). U.S. Bureau of Mines figures in this chart include nonenergy uses of fossil fuels, which constitute about 7 percent of total energy inputs.

air conditioning grew at the remarkable rate of 20 percent per year; it accounted for almost 16 percent of the total increase in electric-power generation from 1969 to 1970, with little or no multiplier effect on the G.N.P.

Let us take a look at this matter of efficiency in still another way: in terms of useful work done as a percentage of gross energy input. The "useful-work equivalent," or overall technical efficiency, is seen to be the product of the conversion efficiency (if there is an intermediate conversion step) and the application efficiency of the machine or device that does the work [*see bottom illustration on page 141*]. Clearly there is a wide range of technical efficiencies in energy systems, depending on the conversion devices. It is often said that electrical resistance heating is 100 percent efficient, and indeed it is in terms, say, of converting electrical energy to thermal energy at the domestic hot-water heater. In terms of the energy content of the natural gas or coal that fired the boiler that made the steam that drove the turbine that turned the generator that produced the electricity that heated the wires that warmed the water, however, it is not so efficient.

The technical efficiency of the total U.S. energy system, from potential energy at points of initial conversion to work at points of application, is about 50 percent. The economic efficiency of

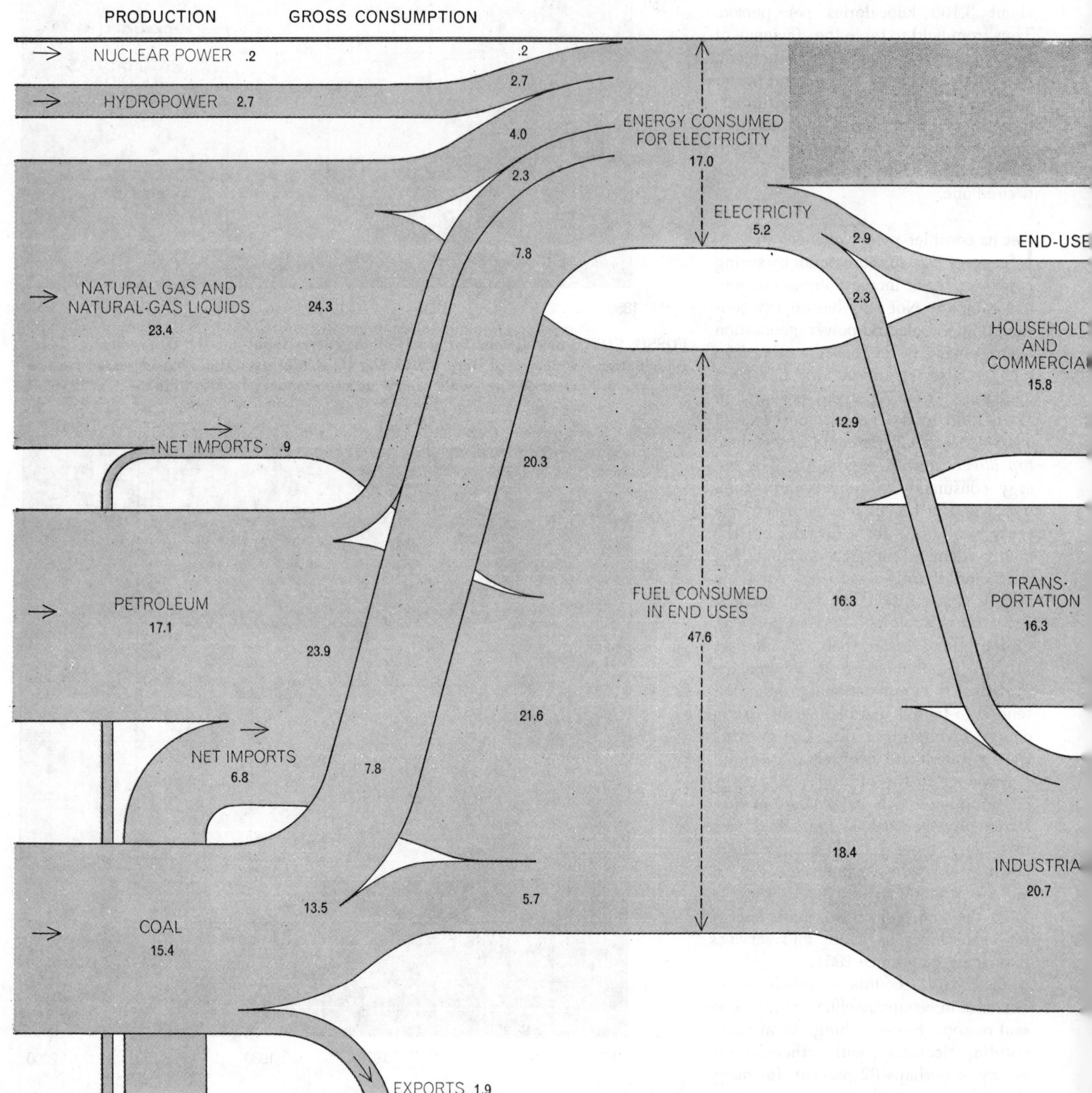

FLOW OF ENERGY through the U.S. system in 1970 is traced from production of energy commodities (*left*) to the ultimate conversion of energy into work for various industrial end products and waste heat (*right*). Total consumption of energy in 1970 was 64.6×10^{15} British thermal units. (Adding nonenergy uses of fossil fuels, primarily for petrochemicals, would raise the total to 68.8×10^{15} B.t.u.) The overall efficiency of the system was about 51 percent. Some of the fossil-fuel energy is consumed directly and

the system is considerably less. That is because work is expended in extracting, refining and transporting fuels, in the construction and operation of conversion facilities, power equipment and electricity-distribution networks, and in handling waste products and protecting the environment.

An industrial society requires not only a large supply of energy but also a high use of energy per capita, and the society's economy and standard of living are shaped by interrelations among resources, population, the efficiency of conversion processes and the particular applications of power. The effect of these interrelations is illustrated by a comparison of per capita energy consumption and per capita output for a number of countries [*see illustration on page 142*]. As one might expect, there is a strong general correlation between the two measures, but it is far from being a one-to-one correlation. Some countries (the U.S.S.R. and the Republic of South Africa, for example) have a high energy consumption with respect to G.N.P.; other countries (such as Sweden and New Zealand) have a high output with relatively less energy consumption. Such differences reflect contrasting combinations of energy-intensive heavy industry and light consumer-oriented and service industries (characteristic of different stages of economic development) as well as differences in the efficiency of energy use. For example, countries that still rely on coal for a large part of their energy requirement have higher energy inputs per unit of production than those that use mainly petroleum and natural gas.

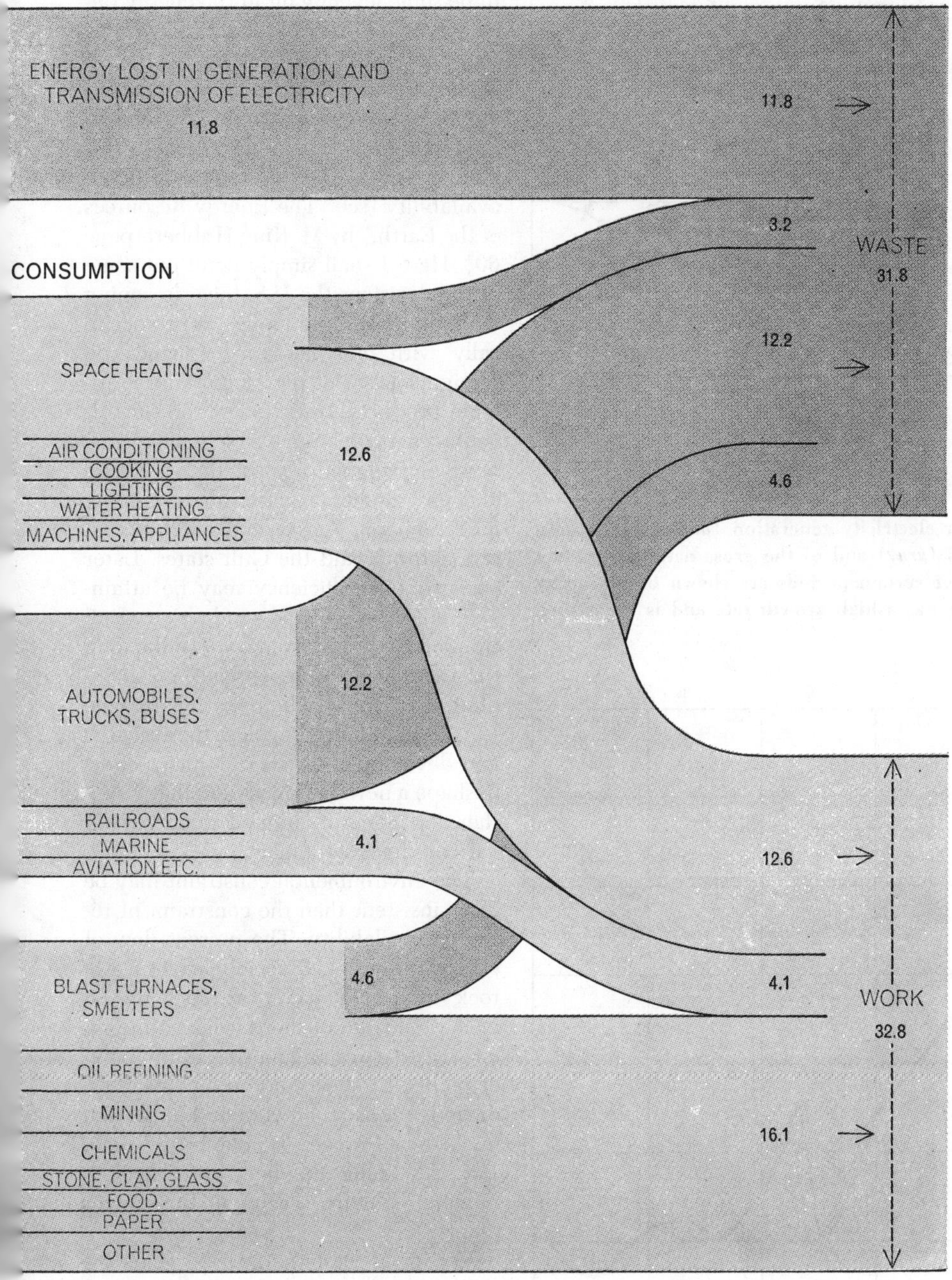

A look at trends from the U.S. past is also instructive. Between 1800 and 1880 total energy consumption in the U.S. lagged behind the population increase, which means that per capita energy consumption actually declined somewhat. On the other hand, the American standard of living increased during this period because the energy supply in 1880 (largely in the form of coal) was being used much more efficiently than the energy supply in 1800 (largely in the form of wood). From 1900 to 1920 there was a tremendous surge in the use of energy by Americans but not a parallel increase in the standard of living. The ratio of energy consumption to G.N.P. increased 50 percent during these two decades because electric power, inherently less efficient, began being substituted for the direct use of fuels; because the automobile, at best 25 percent efficient, proliferated (from 8,000 in 1900 to 8,132,000 in 1920), and because mining and manufacturing, which are energy-intensive, grew at very high rates during this period.

Then there began a long period during which increases in the efficiency of energy conversion and utilization fulfilled about two-thirds of the total increase in demand, so that the ratio of energy consumption to G.N.P. fell to about 60 percent of its 1920 peak although per capita energy consumption continued to increase. During this period (1920–1965) the efficiency of electric-power generation and transmission almost trebled, mining and manufacturing grew at much lower rates and the services sector of the economy, which is not energy-intensive, increased in importance.

"Power corrupts" was written of man's control over other men but it applies also to his control of energy re-

ENERGY LOST IN GENERATION AND TRANSMISSION OF ELECTRICITY
11.8
CONSUMPTION
SPACE HEATING
AIR CONDITIONING
COOKING
LIGHTING
WATER HEATING
MACHINES, APPLIANCES
12.6
AUTOMOBILES, TRUCKS, BUSES
12.2
RAILROADS
MARINE
AVIATION ETC.
4.1
BLAST FURNACES, SMELTERS
4.6
OIL REFINING
MINING
CHEMICALS
STONE, CLAY, GLASS
FOOD
PAPER
OTHER
11.8
3.2
12.2
4.6
WASTE
31.8
12.6
4.1
16.1
WORK
32.8

some is converted to generate electricity. The efficiency of electrical generation and transmission is taken to be about 31 percent, based on the ratio of utility electricity purchased in 1970 to the gross energy input for generation in that year. Efficiency of direct fuel use in transportation is taken as 25 percent, of fuel use in other applications as 75 percent.

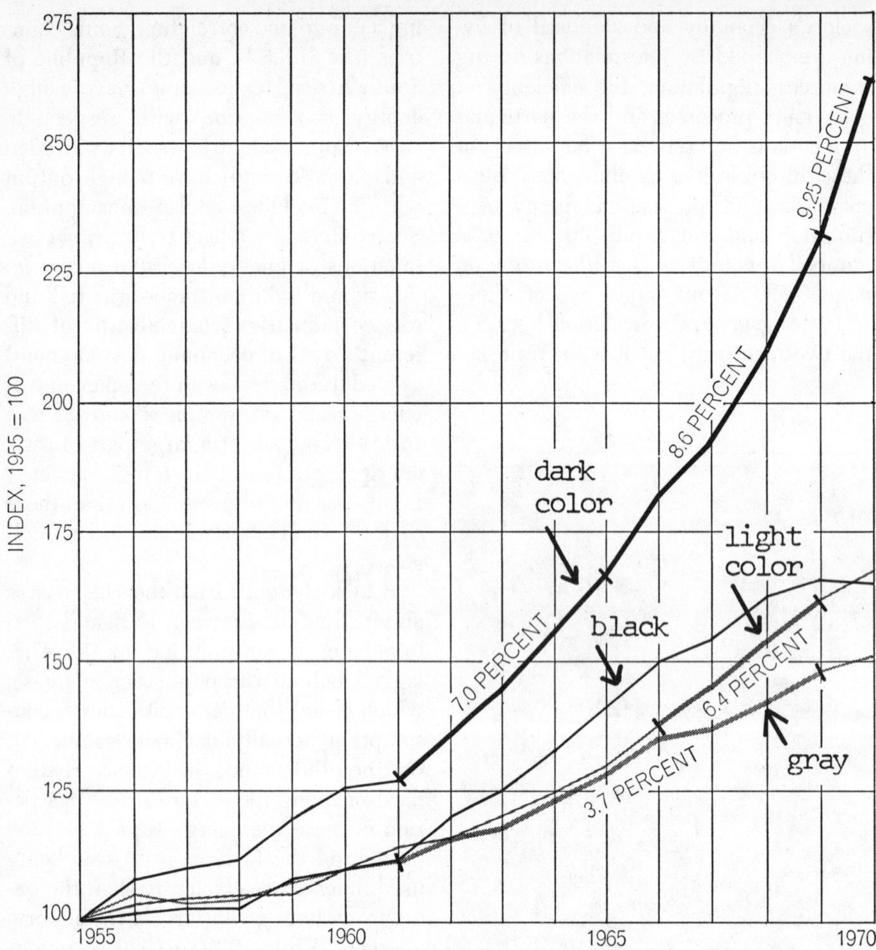

INCREASE IN CONSUMPTION of energy for electricity generation (*dark color*), transportation (*light color*) and other applications (*gray*) and of the gross national product (*black*) are compared. Annual growth rates for certain periods are shown beside heavy segments of curves. Consumption of electricity has a high growth rate and is increasing.

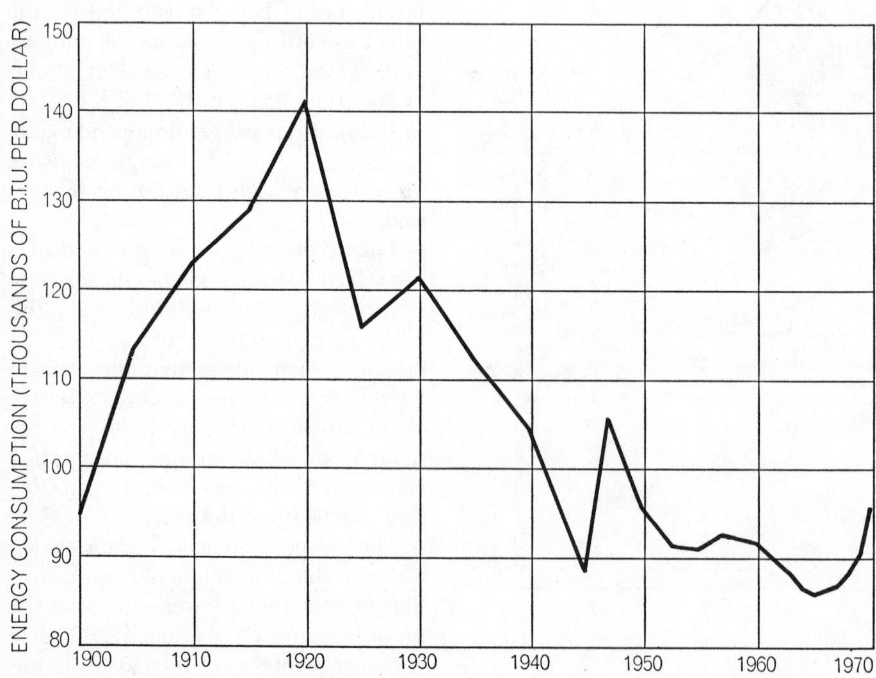

RATIO OF ENERGY CONSUMPTION to gross national product has varied over the years. It tends to be low when the G.N.P. is large and energy is being used efficiently, as was the case during World War II. The ratio has been rising steadily since 1965. Reasons include the increase in the use of air conditioning and the lack of advance in generating efficiency.

sources. The more power an industrial society disposes of, the more it wants. The more power we use, the more we shape our cities and mold our economic and social institutions to be dependent on the application of power and the consumption of energy. We could not now make any major move toward a lower per capita energy consumption without severe economic dislocation, and certainly the struggle of people in less developed regions toward somewhat similar energy-consumption levels cannot be thwarted without prolonging mass human suffering. Yet there is going to have to be some leveling off in the energy demands of industrial societies. Countries such as the U.S. have already come up against constraints dictated by the availability of resources and by damage to the environment. Another article in this issue considers the question of resource availability [see "The Energy Resources of the Earth," by M. King Hubbert, page 60]. Here I shall simply point out some of the decisions the U.S. faces in coping with diminishing supplies, and specifically with our increasing reliance on foreign sources of petroleum and petroleum products. In the short run the advantages of reasonable self-sufficiency must be weighed against the economic and environmental costs of developing oil reserves in Alaska and off the coast of California and the Gulf states. Later on such self-sufficiency may be attainable only through the production of oil from oil shale and from coal. In the long run the danger of dependence on dwindling fossil fuels—whatever they may be —must be balanced against the research and development costs of a major effort to shape a new energy system that is neither dependent on limited resources nor hard on the environment.

The environmental constraint may be more insistent than the constraint of resource availability. The present flow of energy through U.S. society leaves waste rock and acid water at coal mines; spilled oil from offshore wells and tankers; waste gases and particles from power plants, furnaces and automobiles; radioactive wastes of various kinds from nuclear-fuel processing plants and reactors. All along the line waste heat is developed, particularly at the power plants.

Yet for at least the next 50 years we shall be making use of dirty fuels: coal and petroleum. We can improve coal-combustion technology, we can build power plants at the mine mouth (so that the air of Appalachia is polluted instead of the air of New York City), we can make clean oil and gas from coal and oil

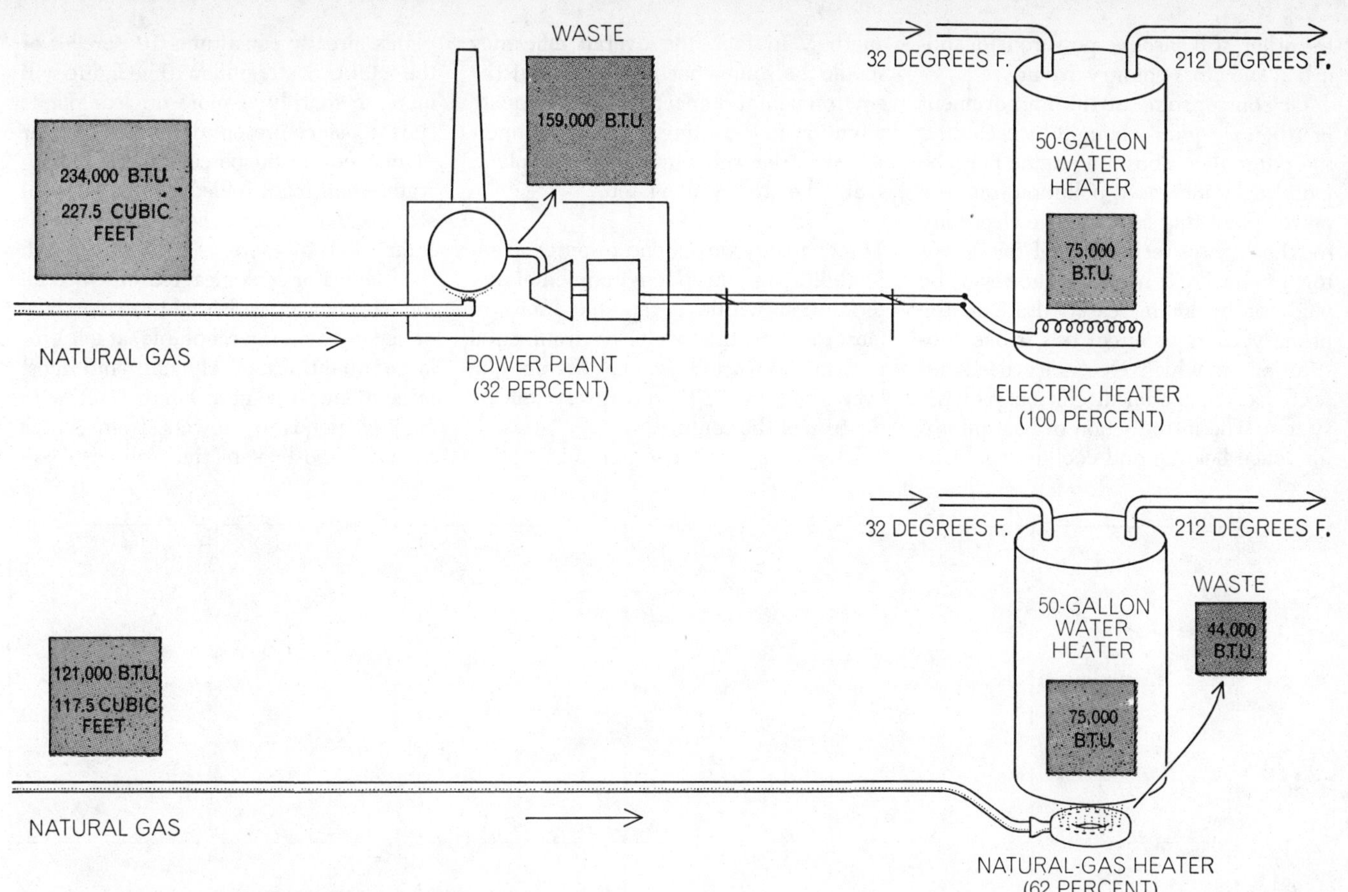

EFFICIENCIES OF HEATING WATER with natural gas indirectly by generating electricity for use in resistance heating (*top*) and directly (*bottom*) are contrasted. In each case the end result is enough heat to warm 50 gallons of water from 32 degrees Fahrenheit to 212 degrees. Electrical method requires substantially more gas even though efficiency at electric heater is nearly 100 percent.

from shale, and sow grass on the mountains of waste. As nuclear power plants proliferate we can put them underground, or far from the cities they serve if we are willing to pay the cost in transmission losses. With adequate foresight, caution and research we may even be able to handle the radioactive-waste problem without "undue" risk.

There are, however, definite limits to such improvements. The automobile engine and its present fuel simply cannot be cleaned up sufficiently to make it an acceptable urban citizen. It seems clear that the internal-combustion engine will be banned from the central city by the year 2000; it should probably be banned right now. Because our cities are shaped for automobiles, not for mass transit, we shall have to develop battery-powered or flywheel-powered cars and taxis for inner-city transport. The 1970 census for the first time showed more metropolitan citizens living in suburbs than in the central city; it also showed a record high in automobiles per capita, with the greatest concentration in the suburbs. It seems reasonable to visualize the suburban two-car garage of the future with one car a recharger for "downtown" and

	PRIMARY ENERGY INPUT (UNITS)	SECONDARY ENERGY OUTPUT (UNITS)	APPLICATION EFFICIENCY (PERCENT)	TECHNICAL EFFICIENCY (PERCENT)
AUTOMOBILE				
INTERNAL-COMBUSTION ENGINE	100		25	25
FLYWHEEL DRIVE CHARGED BY ELECTRICITY	100	32	100	32
SPACE HEATING				
BY DIRECT FUEL USE	100		75	75
BY ELECTRICAL RESISTANCE	100	32	100	32
SMELTING OF STEEL				
WITH COKE	100	94	94	70
WITH ELECTRICITY	100	32	32	32

TECHNICAL EFFICIENCY is the product of conversion efficiency at an intermediate step (if there is one) and application efficiency at the device that does the work. Losses due to friction and heat are ignored in the flywheel-drive automobile data. Coke retains only about 66 percent of the energy of coal, but the energy recovered from the by-products raises the energy conservation to 94 percent.

the other, still gasoline-powered, for suburban and cross-country driving.

Of course, some of the improvement in urban air quality bought by excluding the internal-combustion engine must be paid for by increased pollution from the power plant that supplies the electricity for the nightly recharging of the downtown vehicles. It need not, however, be paid for by an increased draft on the primary energy source; this is one substitution in which electricity need not decrease the technical efficiency of the system. The introduction of heat pumps for space heating and cooling would be another. In fact, the overall efficiency should be somewhat improved and the environmental impact, given adequate attention to the siting, design and operation of the substituting power plant, should be greatly alleviated.

If technology can extend resource availability and keep environmental deterioration within acceptable limits in most respects, the specific environmental problem of waste heat may become the overriding one of the energy system by the turn of the century.

The cooling water required by power plants already constitutes 10 percent of the total U.S. streamflow. The figure will increase sharply as more nuclear plants start up, since present designs of nuclear plants require 50 percent more cooling water than fossil-fueled plants of equal size do. The water is heated 15 degrees Fahrenheit or more as it flows through the plant. For ecological reasons such an increase in water released to a river, lake or ocean bay is unacceptable, at least for large quantities of effluent, and most large plants are now being built with cooling ponds or towers from which much of the heat of the water is dissi-

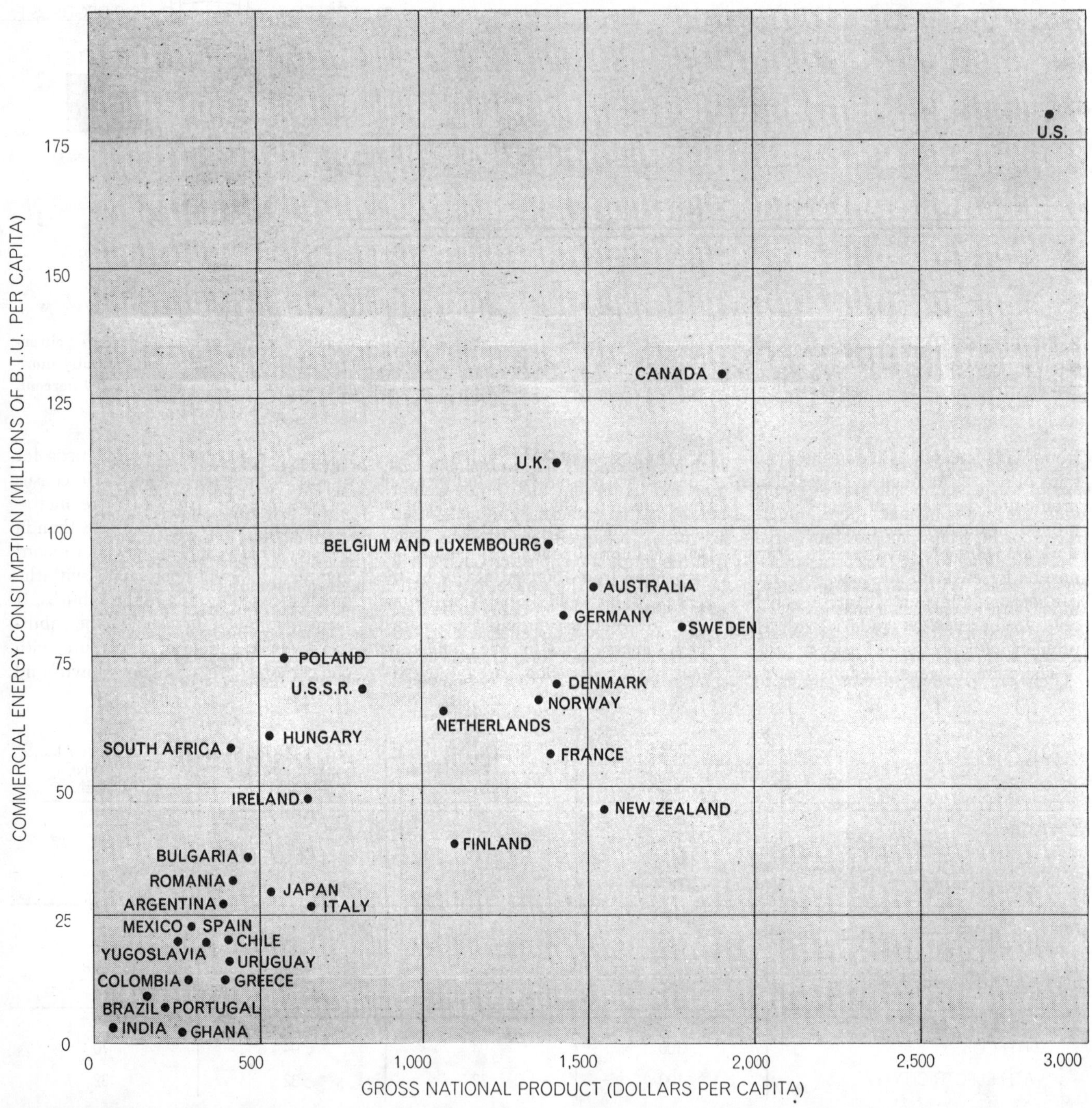

ROUGH CORRELATION between per capita consumption of energy and gross national product is seen when the two are plotted together; in general, high per capita energy consumption is a prerequisite for high output of goods and services. If the position plotted for the U.S. is considered to establish an arbitrary "line," some countries fall above or below that line. This appears to be related to a country's economic level, its emphasis on heavy industry or on services and its efficiency in converting energy into work.

pated to the atmosphere before the water is discharged or recycled through the plant. Although the atmosphere is a more capacious sink for waste heat than any body of water, even this disposal mechanism obviously has its environmental limits.

Many suggestions have been made for putting the waste heat from power plants to work: for irrigation or aquaculture, to provide ice-free shipping lanes or for space heating. (The waste heat from power generation today would be more than enough to heat every home in the U.S.!) Unfortunately the quantities of water involved, the relatively low temperature of the coolant water and the distances between power plants and areas of potential use are serious deterrents to the utilization of waste heat. Plants can be designed, however, for both power production and space heating. Such a plant has been in operation in Berlin for a number of years and has proved to be more efficient than a combination of separate systems for power production and space heating. The Berlin plant is not simply a conserver of waste heat but an exercise in fuel economy; its power capacity was reduced in order to raise the temperature of the heated water above that of normal cooling water.

With present and foreseeable technology there is not much hope of decreasing the amount of heat rejected to streams or the atmosphere (or both) from central steam-generating power plants. Two systems of producing power without steam generation offer some long-range hope of alleviating the waste-heat problem. One is the fuel cell; the other is the fusion reactor combined with a system for converting the energy released directly into electricity [see "The Conversion of Energy," by Claude M. Summers, page 148]. In the fuel cell the energy contained in hydrocarbons or hydrogen is released by a controlled oxidation process that produces electricity directly with an efficiency of about 60 percent. A practical fusion reactor with a direct-conversion system is not likely to appear in this century.

Major changes in power technology will be required to reduce pollution and manage wastes, to improve the efficiency of the system and to remove the resource-availability constraint. Making the changes will call for hard political decisions. Energy needs will have to be weighed against environmental and social costs; a decision to set a pollution standard or to ban the internal-combustion engine or to finance nuclear-power development can have major economic and political effects. Democratic societies are not noted for their ability to take the long view in making decisions. Yet indefinite growth in energy consumption, as in human population, is simply not possible.

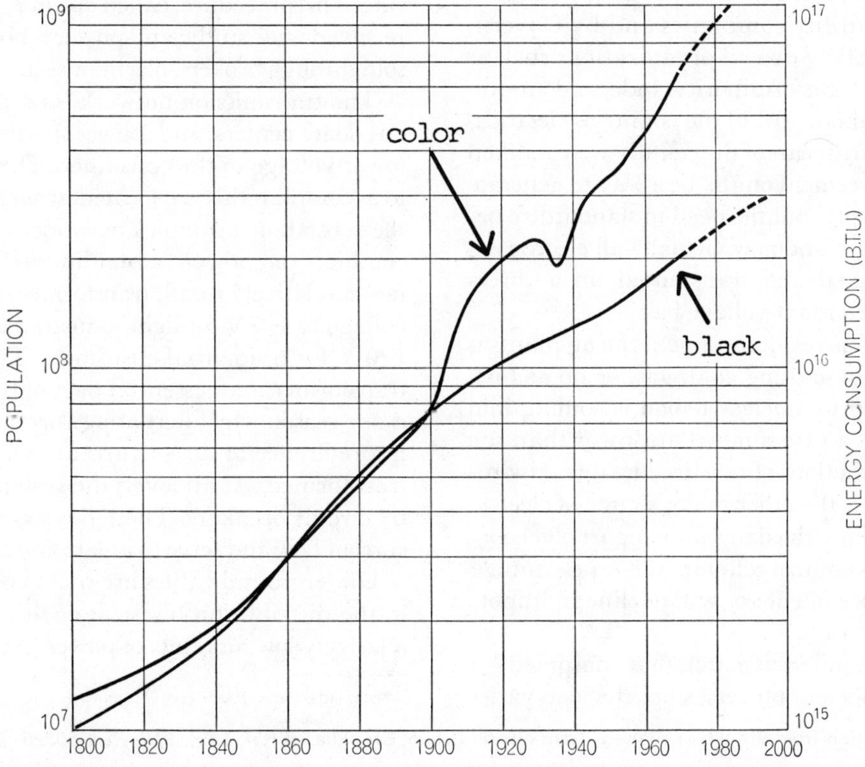

U.S. ENERGY-CONSUMPTION GROWTH (*curve in color*) **has outpaced the growth in population** (*black*) **since 1900, except during the energy cutback of the depression years.**

Excerpts from "Underground Power Transmission by Superconducting Cable"

1.1 CHARACTERISTICS OF THE U.S. ELECTRICAL SYSTEM

The availability of cheap electric power in large quantities is one of the basic requirements of a modern industrial state. It is taken for granted by many dependent consumers as they plan to increase industrial productivity or, in the case of residential customers, their standard of living. The problems involved in the future growth of the electric supply industry are discussed in Chapter II. However, to better understand these problems, some knowledge of the forces that shaped the way the present system evolved will be useful. It is probably true that the average scientist or engineer is not familiar with this background despite the daily impact of electric power on his life and work. In fact, the study of electric energy applied to generation, transmission, and distribution systems has dwindled and few schools of engineering now offer formal courses in these subjects.

Power is generated in the U.S. by hundreds of independent utility companies and by Government and publicly owned organizations such as the Tennessee Valley Authority. Independent utilities produce about 75% of the country's electrical energy. The early years of the century were marked by wide disagreement on the best way to generate and sell electricity, but the need to standardize became paramount and now virtually all electric energy is generated and transmitted on a three-phase, 60-Hz, constant-voltage basis.

The prime mover in most generating plants is the steam turbine using coal, gas, or oil as fuel. Heat produced by nuclear fission is coming into service, but at a rate somewhat slower than the optimistic predictions of two decades ago. Hydroelectric power is the other major source of electric energy; sometimes the dams form a part of a coordinated water control scheme. On a percentage basis this source of energy will decline in importance in the future.

Power is generated to match that consumed – a somewhat inefficient process, since the load varies widely during the day and from season to season. Increasing use is made of gas turbine generators to meet peak load demands, since these generators can be started cheaply and quickly when the load increases. Energy storage schemes such as pumped hydro have been used to a limited extent for the same purpose.

Power from the generating station is stepped up to high voltage by means of transformers. In this way bulk energy can be transmitted efficiently. Transmission line voltages are usually >138 kV,* and the highest blocks of power are now being carried by overhead lines operating at 765 kV. Transmission lines in a region are generally interconnected, a practice that began in the 1920's. The reasons are manifold, the most obvious being an increase in reliability as the network permits energy to reach any point via several paths. A high-capacity transmission system also permits utility companies to exchange power so that they can deliver it to their customers at the most economical rate. This procedure, called optimum dispatch, has received wide study, and power is now bought and sold through brokers, like many other commodities.

The transmission network is tapped at numerous load centers and power is distributed at a lower voltage to the consumer. The transformers to accomplish this are located at substations. From the substation a complex network of cables or lines conducts the power to buildings or streets where more, relatively small, transformers step down the voltage to 440 V for light industrial use and 220 or 120 V for residential customers. The ubiquitous transformer is an essential part of the system and is one reason why alternating current was adopted in preference to direct current, which cannot be transformed. At all levels the system is protected by circuit breakers. These devices can interrupt current flow and serve to isolate faulty components.

Underground cables are quite commonly used in the distribution phase as each branch carries relatively low amounts of power. At the transmis-

*Measured line to line (see Appendix).

Reprinted with permission from Brookhaven Power Transmission Group, Brookhaven Nat. Lab., *BNL Rep. 50325*, E. B. Forsyth, Ed., ch. 1, pp. 1–8, Mar. 1972.

sion level only 1% of present systems is underground, although this small amount probably represents 10% of the capital invested in transmission systems. Besides the financial disincentive there are technical reasons why cables are difficult to incorporate into the transmission system. To keep cables small and flexible and thus relatively easy to install, considerable effort has gone into reducing the thickness of the insulation. About ½ in. of insulation will enable paper-insulated cables impregnated with oil under pressure to withstand 138 kV line to line. Such cables have a high capacitance which is in parallel with the load. The current flowing into the shunt capacitance contributes to heating of the cable, this effect becoming worse the longer the cable. Many cables in service cannot transmit useful amounts of power for more than 20 to 30 miles; this is no drawback in the distribution system but it is a severe restriction for transmission applications. The current flowing into the shunt capacitance, known as charging current, is not a serious problem with overhead lines, and distances of several hundred miles are fairly common. All lines experience another problem, namely voltage drop caused by load current, but the situation is worse for overhead lines because of their greater inductance. Both effects can be compensated by means of external capacitors or inductors, but at added expense.

1.2 REPRESENTATIVE SYSTEMS

Whether or not these restrictions are important often depends on which region of the U.S. is considered. Several typical systems in different parts of the country will be described briefly. The west coast consists of several urban areas situated long distances from convenient energy sources. To the north, and in Canada, are considerable hydroelectric resources. Steam plants are located from San Francisco to the south, mainly on the shoreline. More hydroelectric generators are situated east of Los Angeles, in the Hoover Dam–Mead area. The substations in this area are connected by a 500-kV overhead transmission line to Four Corners, New Mexico. There are many fairly small hydroelectric plants in the mountains of the Sierra Nevada range. Natural gas for steam plants is piped from as far afield as Canada. This region is characterized by very long overhead transmission facilities.

The central region of the U.S. east of the Mississippi River consists of fairly diffused load centers and generating plants. Energy is obtained mainly from hydroelectric resources or steam plants using coal indigenous to the region. Typical transmission facilities consist of complicated networks at different voltages and power levels with extensive interties between pools. Individual lines rarely exceed 150 miles in length.

The east coast is dominated by the power demand of the Boston-to-Washington megalopolis. Regional systems in this area are generally capable of supplying the local load with modest ties to adjacent regions. Power generated in Canada is transmitted on overhead lines. Within the metropolitan areas most power transmission is via underground cables.

1.2.1 Pacific Gas and Electric Company

For a representative utility company, the Pacific Gas and Electric Company (PG&E), the following statistics for 1970* are pertinent.

Number of hydroelectric plants	65
Number of steam plants	12
Hydro plant operating capacity, kW	2.5×10^6
Steam plant operating capacity, kW	7.2×10^6
Available from other sources, kW	2.4×10^6
Gross system output, kWh	56.8×10^9
Under construction or scheduled, kW	5.4×10^6

The service area of this company includes much of California north of Los Angeles.

The main interties for the Pacific coast are shown in Figure 1. Two major transmission installations are the 500-kV ac intertie from John Day to the urban areas of southern California (about 900 miles) and the 850-mile dc line from Celilo to Mead and Sylmar. The dc line has just come into service and is the largest dc transmission system in the world. In certain circumstances, dc transmission is preferable to ac (see Chapter VII).

1.2.2 Tennessee Valley Authority

The Tennessee Valley Authority (TVA) was established by Act of Congress in 1933 primarily to develop water control resources, in particular those of the Tennessee River drainage basin. Whether the Authority was entitled to generate electric power was a question of constitutional law that was not resolved until 1939. Steam generating plants were first built by the Authority in 1940 and it is now the largest single producer of electric power in the U.S. The Authority generates and transmits power but does not sell power directly to

*Reproduced by permission of PG&E.

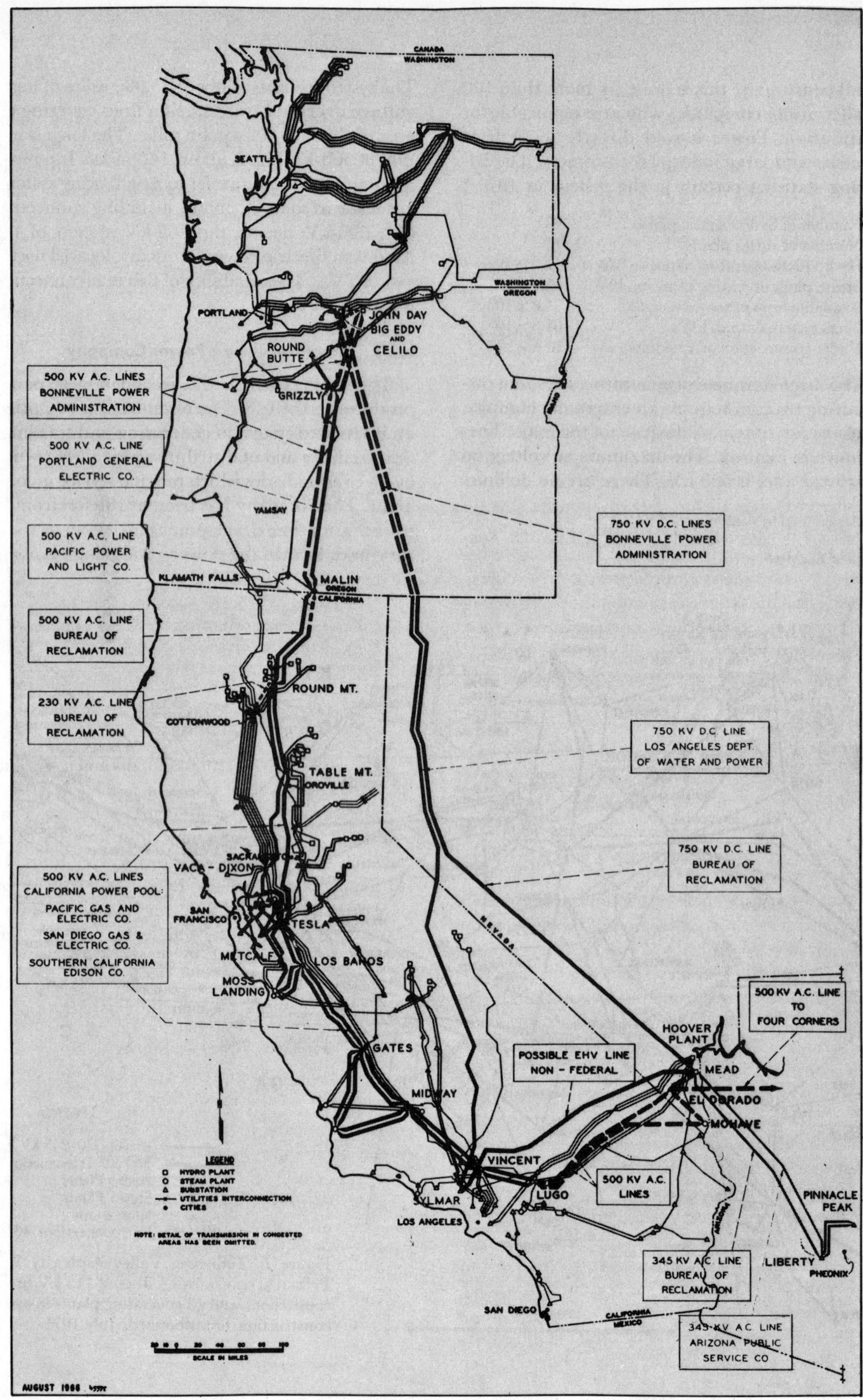

Figure 1. Pacific Northwest-Southwest Intertie and related facilities.

small consumers; this is done by more than 100 smaller utility companies who are responsible for distribution. Power is sold directly to Federal agencies and large industrial customers. The following statistics pertain to the system in 1970.*

Number of hydroelectric plants	48†
Number of steam plants	11
Hydro plant operating capacity, kW	4.3×10^6
Steam plant operating capacity, kW	15.1×10^6
Available from other sources, kW	1.8×10^6
Gross system output, kWh	101.3×10^9
Under construction or scheduled, kW	10.4×10^6

The large numbers of generating sites and distributing utilities require an extremely complex transmission system. A diagram of the major lines is shown in Figure 2. The maximum ac voltage on overhead lines is 500 kV. There are no dc lines.

*Reproduced by permission of TVA.

†Includes 19 dams operated by Alcoa and the U.S. Army Corps of Engineers.

The system consists of about 15,000 miles of high-voltage overhead transmission lines covering an area of about 60,000 square miles. The longest section of 500-kV line is about 140 miles. Interconnections for energy transfer to neighboring systems are made at some 26 points, including connection of a 500-kV line to the 765-kV system of the American Electric Power Company, located northeast of TVA. The capability of this interconnection is 1.5×10^6 kW.

1.2.3 American Electric Power Company

The AEP system is the second largest power producer in the U.S. The organization comprises an integrated system of generating and transmission facilities and of distributing facilities owned by seven subsidiaries which provide service in local areas. The company has been at the forefront of transmission-line development for many years as they have located the power-producing plants as

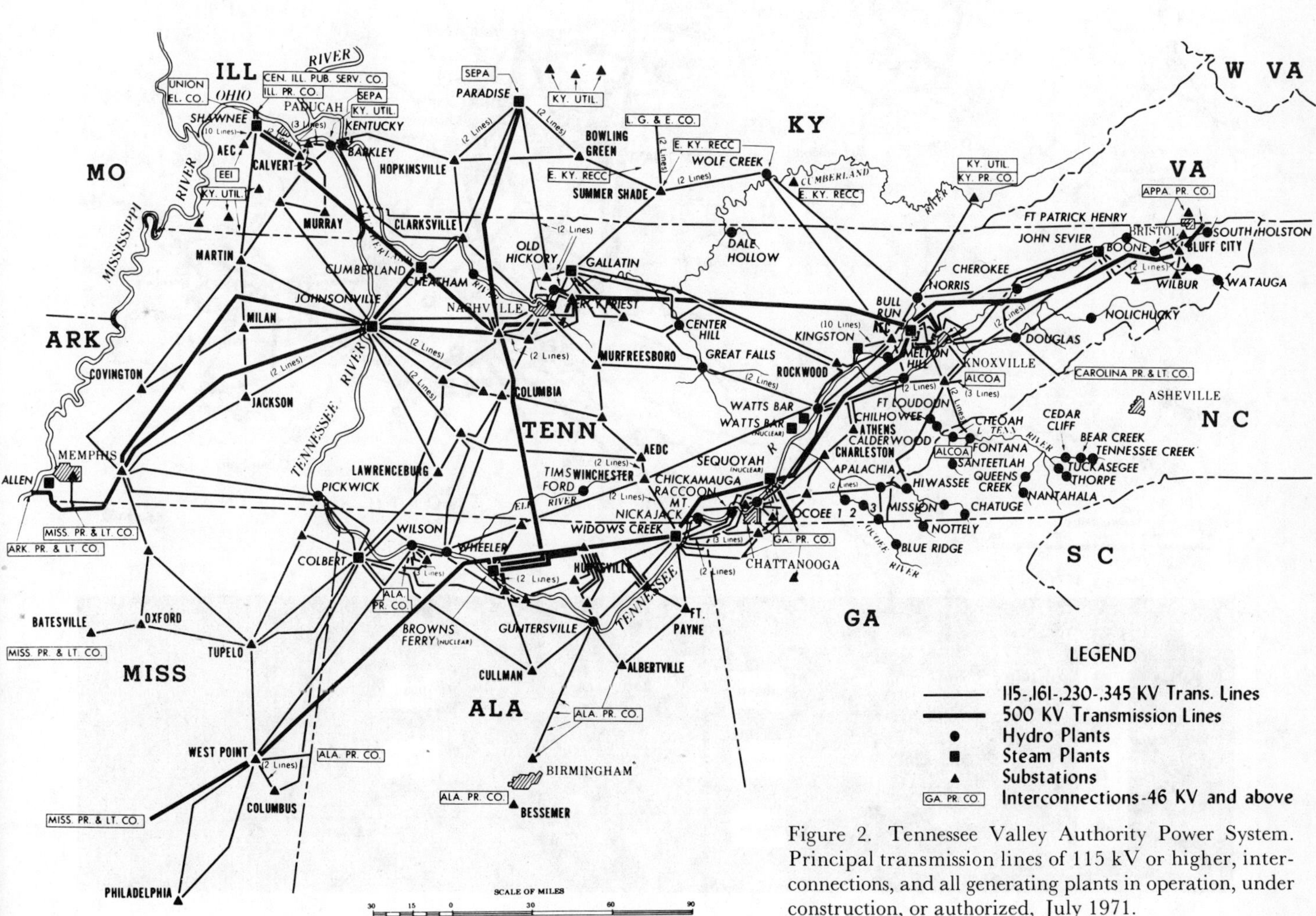

Figure 2. Tennessee Valley Authority Power System. Principal transmission lines of 115 kV or higher, interconnections, and all generating plants in operation, under construction, or authorized, July 1971.

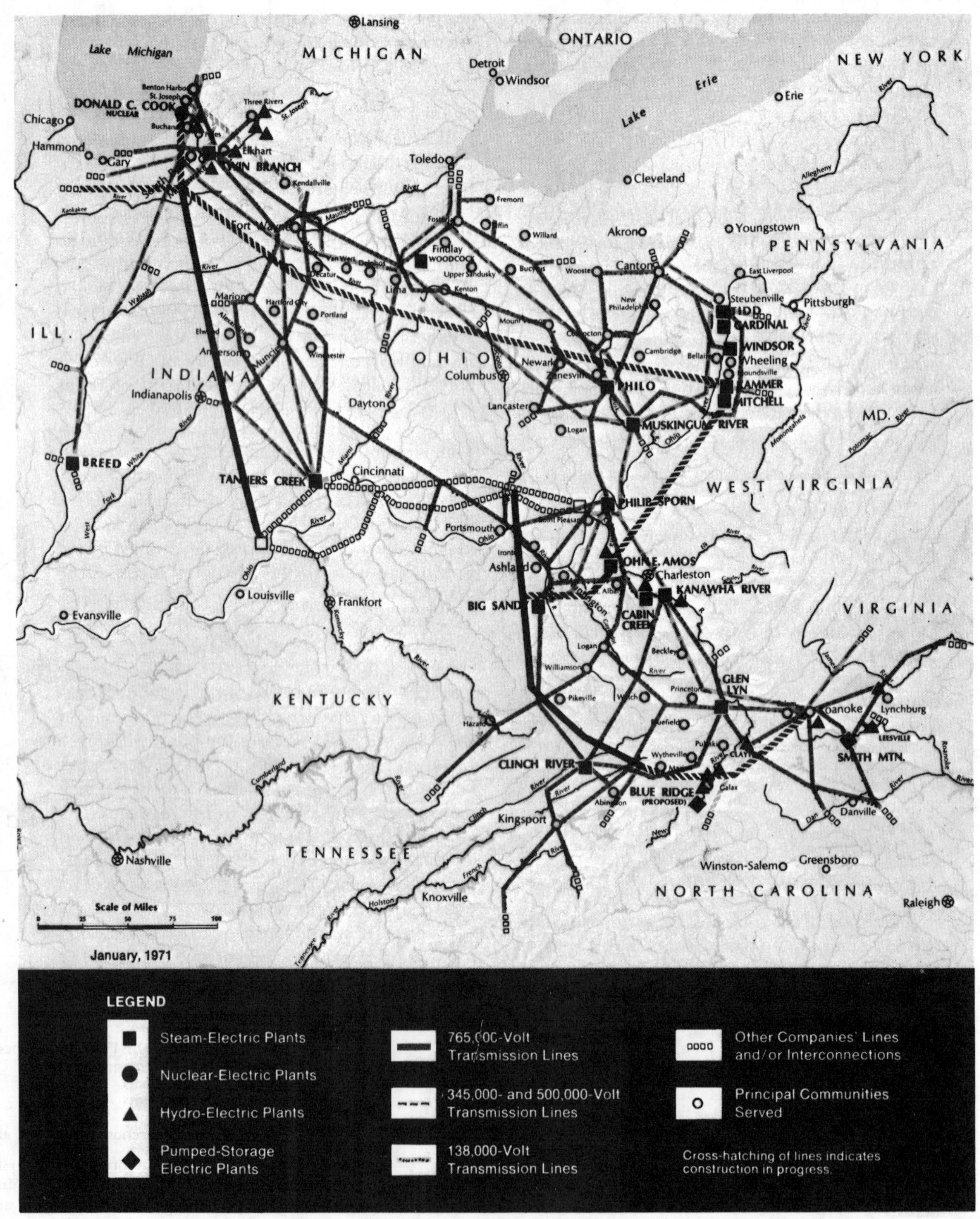

Figure 3. Seven-State American Electric Power System.

close as possible to fuel and hydro sources. In the case of coal-steam plants this is known as mine-mouth operation. Some 1970 operating statistics for this system are*

Number of hydroelectric plants	17
Number of steam plants	19
Total system capacity, kW	12×10^6
Gross system output, kWh	60.1×10^9
Under construction or scheduled, kW	5.9×10^6

The area served by the subsidiaries covers about 41,000 square miles, including portions of Michigan, West Virginia, Kentucky, Virginia, Tennessee, Ohio, and Indiana. A 1200-mile transmission system operating at 765 kV is under construction; about 390 miles are in service. The system is already linked to TVA at 500 kV and Commonwealth Edison at 765 kV. Its major transmission lines are shown in Figure 3.

1.2.4 Consolidated Edison Company of New York

The three systems described so far have high-capacity overhead transmission lines to connect load and generating centers. The use of underground cable for transmission in these systems has been negligible. The fourth system, Consolidated Edison Company of New York, is typical of the densely populated urban areas in which generators are located within the load region and power is transmitted mainly by underground cables. The operating statistics for the company for 1970 are†

Gas turbine, installed, kW	1×10^6
Steam, installed, kW	8×10^6
Other sources, kW	0.5×10^6
Total capacity, kW	9.5×10^6
Gross output, kWh	32.4×10^9
Under construction or scheduled (within 5 yr), kW	4.7×10^6

A diagram of the transmission system showing existing and proposed lines is shown in Figure 4. Within the city boundaries all the feeders are underground except for the Goethals substation connections at Staten Island. This is possible because distances between generating stations and substations are considerably less than in the other systems studied. The service area of the company is 660 square miles. The generating capacity of the system is sufficient to meet the load most of the time. Normally, power transmitted from adjacent utilities does not exceed 5% of the load. However, generator failures have required an increase in this percentage on occasion.

*Reproduced by permission of AEP.

†Reproduced by permission of Consolidated Edison Company.

1.3 SYSTEM REQUIREMENTS

It is clear from these descriptions that hundreds, perhaps thousands, of generators will function satisfactorily when all are properly connected to the same network. To obtain satisfactory performance the system must be supervised and adjusted continuously. The components themselves must also be designed to have static and dynamic stability under all conditions of normal and abnormal loads. Synchronous alternators have the characteristic that, once connected electrically to another generator, they both run at precisely the same speed, or submultiple. Electromagnetic forces are set up within the machines so that this synchronism is maintained automatically.

The situation is analogous to connecting many machines to a common drive shaft through gears and clutches. Once a clutch is engaged an individual machine will run at a speed determined by the gearing ratio; any further change of speed will be in the same percentage as in the system as a whole. The analogy can be carried further. As an individual machine takes up its share of the load, the connecting shaft will twist a little, and a phase difference will appear between the individual machine and a reference point on the main output shaft. As the load increases the phase difference will increase. If the system drives the generator, i.e., puts power into it so that it behaves as a motor, then the phase angle will change sign. The analog of the connecting shaft is the transmission line of the electrical network. A weak line is analogous to a springy shaft that allows a large twist to develop as a torque is applied.

Problems of stability arise in an interconnected network principally because of load changes. The total power consumed by the load must be supplied by the prime movers. As loads change the power difference (plus or minus) between the loads and the prime movers is available to accelerate or decelerate the system. Thus frequency stability is a measure of how well the total input matches the load. The effect of springy connecting shafts, or weak lines, is to allow individual machines to oscillate around the stable phase angle as the continuous small changes in synchronous speed occur. The oscillatory behavior of many coupled bodies is a familiar phenomenon in physics, but the situa-

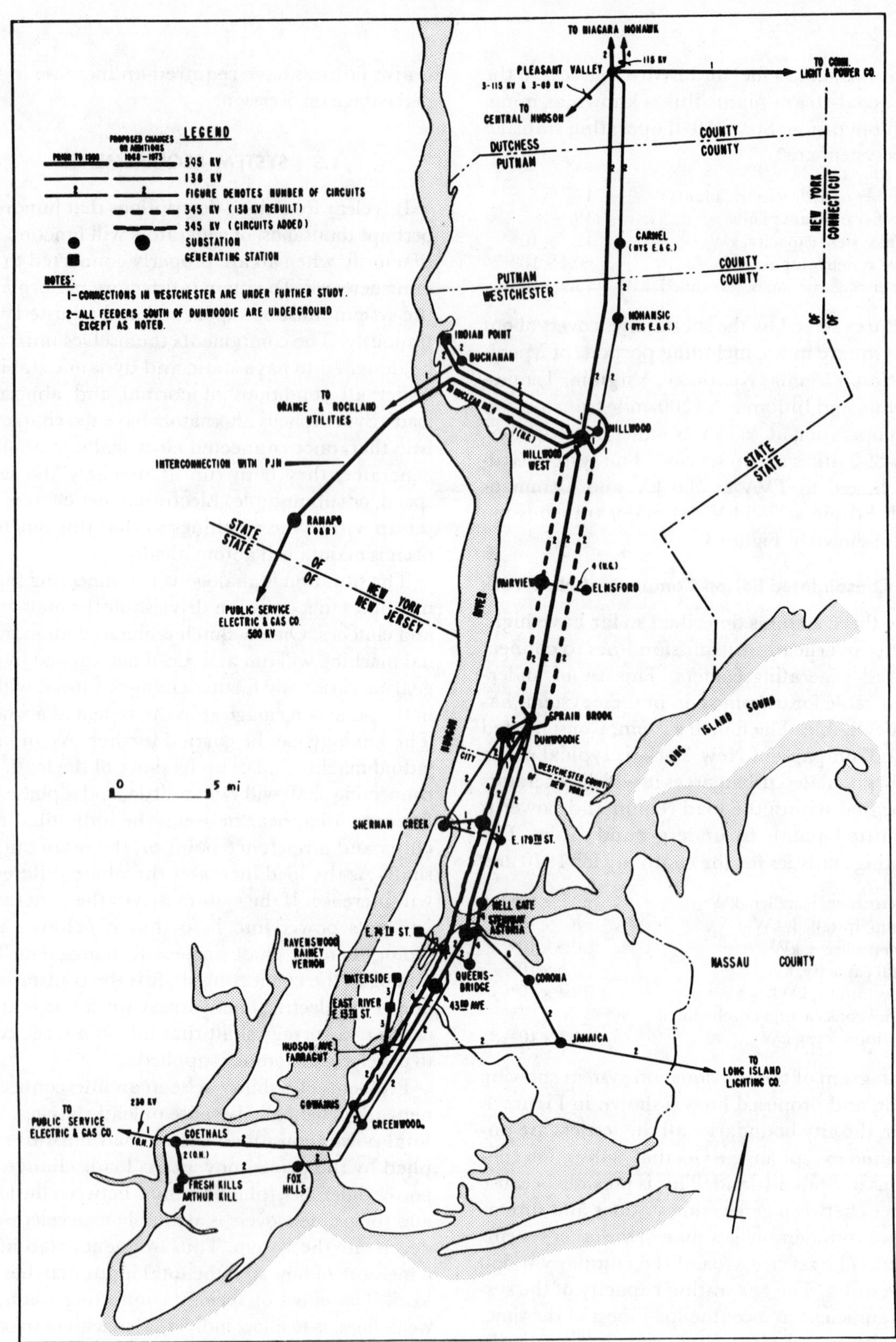

Figure 4. Consolidated Edison Company Transmission System.

tion in electrical power networks is complicated by the grossly nonlinear characteristics of the generators and the regulating mechanisms of the prime movers. An analytical solution can be obtained for a few interconnected machines, but for any practical electrical generating and transmission system the stability must be investigated by using a digital computer and iterative numerical techniques.

1.3.1 Influence of Transmission Lines

If the transmission lines of a system are not designed properly, a large change in load can produce transient oscillations of individual generators which may be large enough to cause them to drop out of synchronism. If this occurs the load on the rest of the generators increases and runaway instability may develop, so that the whole network is closed down.

The average load of an electrical network does not draw current that is exactly in phase with the voltage supplied. The current usually lags a little and thus, if represented vectorially, contains a quadrature component. The magnitude of the quadrature current multiplied by the system voltage is called reactive power. No real work is performed by reactive power. A pure inductance draws lagging reactive current and a pure capacitance draws leading reactive current, referred to the way in which the current vector lies with respect to the voltage vector (see Appendix). An ideal transmission line would not significantly change the reactive power in a network. Sources of reactive current produce power that must be supplied or consumed somewhere else, possibly by expensive components. By controlling the flow of reactive power in a network the voltages at any point can be adjusted and maintained at a constant level as load changes occur. When there is a rapid change of load some generators may be used to absorb reactive power. Maintaining system stability and constant voltage levels during these periods becomes a difficult balancing act.

1.4 SUMMARY

In general, the large power systems developed in the U.S. have been shaped by geographical, technical, and economic forces. The relatively inexpensive overhead power transmission lines have permitted the growth of power pools with much lower reserve capacity than would be tolerable for an individual utility. The interchange of power has encouraged installation of very large generators based on the improvement in economic efficiency with size. Underground transmission cables are not used on a large scale because long-distance transmission by this means requires a large capital investment and, as a technical drawback, large amounts of reactive power compensation. It can be seen that if the technical feasibility of using superconducting underground cables is demonstrated in the future, actual implementation will require very careful integration into the power system as a whole. In Chapter VI the technical problems of integration are discussed and in many areas the prognosis seems favorable. Estimated costs of these cables vary widely and are, in any event, guesses, since no useful length of this type of cable has ever been manufactured. The topic is discussed at greater length in Chapter VI.

RECOMMENDED READING

Underground Power Transmission, Report to Federal Power Commission by Commission's Advisory Committee on Underground Transmission, April 1966.

The Transmission of Electric Power, Report to Federal Power Commission by the Technical Advisory Committee on Transmission for the National Power Survey, October 1970.

Electric Energy Systems Theory: An Introduction, O.E. Elgerd, McGraw-Hill, New York, 1971.

Energy, Resources and the Environment

CHARLES A. ZRAKET

INTRODUCTION

As an outgrowth of Hudson Institute's Corporate Environment Study, Robert Panero suggested to me in June 1971 that MITRE initiate a study on energy, resources and the environment (ERE). We recognized at that time that energy is the strategic commodity with respect to economic growth and development, and that the interrelationships or the interactions of energy, resources and the environment were crucial to the prospects for economic growth and to the prospects for mankind.

MITRE has sponsored eight meetings during the past year on this subject, the first three at MITRE/Washington, a fourth meeting at Lawrence Livermore Laboratories on nuclear and environmental questions and a fifth meeting at Hydro-Quebec in Montreal on Canadian and U. S. issues, and culminating in a Symposium at MITRE/Washington in April 1972 --- the first day of which Herman Kahn of Hudson Institute and Jay Forrester of MIT discussed the subject of the limits to growth --- followed by a Symposium in France in May 1972 where we discussed European issues, two fact-finding trips to Siberia and Australia, and a Symposium in Japan in July 1972 where we discussed Japanese and Pacific Basin issues.

Our purpose, when our current study has been completed, is to have identified and exposed many of the complex issues relating to energy, resources and the environment to a large, worldwide professional community and then to write a summary report for this worldwide audience of government, industry and technical people. We hope that this report will help to provide a substantive, worldwide context and a program outline for detailed planning in E, R and E in the United States and elsewhere.

Reprinted with permission from *MITRE Corp. Rep. M72-180*, Rev. 4, Sept. 1973.

LONG-TERM SITUATION --- THE GROWTH ISSUE

I would like now to present an outline summary of some of the interim conclusions that have come out of our previous meetings on this subject.

The most overriding problem that we see is perceiving, defining, understanding and acting upon the interrelationships, or the linkages, between energy, resources and the environment on an international basis. Here we define resources in the broadest terms --- materials, information, people, ingenuity, and capital. In the environment, we include physical, biological and socio-economic aspects. The overall problem then must be recognized as worldwide in character --- in terms of energy, resources, the environment, and their effect on economic growth and development

In discussing these interrelationships, I would like to do it in the following context: first, with respect to the long term, say post-1985, and then with respect to the intermediate situation, from now until 1985.

The long-term situation is characterized by the "growth" or "Malthusian" issue as illustrated by the current public debate over the so-called limits to growth, and somewhat intensified by some of the work of the Club of Rome and other similar groups around the world. Many of these groups have come to the conclusion that the earth can no longer tolerate the pressures of growth which are driven by the interactions of the increase in population, pollution, the supply of raw materials and resources, industrial processes and food production. A number of questions follow from this thesis. What kind and how much growth should we have? Can we be selective? Should we be selective? How can we implement selective growth on a national and international basis? I think regardless of how we feel about this question of growth, that we cannot deny --- and I certainly felt this force at many of our meetings --- the strong concern that there is about this issue in the United States, in Europe, in Japan and in Australia and many other parts of the world This concern is felt both objectively --- in terms of, for example, potentially catastrophic environmental effects --- and it is felt subjectively --- in terms of social effects such as over-crowding. Responsible political and social planning has to take place on this question --- this is crucial to providing a workable context for this area of energy, resources and environment which we have been studying.

With respect to the energy, resources and environmental area, I feel that the real issue to be addressed is not the possible unavailability of resources and clean energy to meet unconstrained demands, but how much and what kind of economic growth should be achieved, taking the environmental limits of the biosphere into account, with the almost limitless energy resources and the capabilities we have on the planet. In this context then, we believe economic growth is omnidirectional and as Herman Kahn puts it --- "there is an expanding pie for everybody to share in," --- a situation which can contribute to the relief of international tensions and allow for as much internal economic growth in each country as it is capable of and wants.

Also, in the U. S. today, we use about 5% to 6% of GNP on energy generation and distribution, including fuels. If this fraction were to increase appreciably, the cost of all goods would also rise and thereby make it more difficult to raise living standards of the poorer people and to clean up the environment. It is for all these reasons that we believe that more and cheaper energy, with more variety and less pollution, should and can be obtained over the long term through ingenuity and increased efficiency and substitutability in the use of fuels and materials.

THE FUTURE OUTLOOK FOR ENERGY AND RESOURCES

In elaboration on this latter point, although there will be severe dislocations and price changes in fossil-fuel resources over the next thirty years as these resources become relatively scarce in terms of the demand, we believe that adequate clean energy sources for all foreseeable demands can be obtained for the long term through a combination of (1) restraint and conservation (either voluntary in response to or as part of the counter-culture movement or forced through mechanisms such as reversed pricing, taxing, time lags, increased efficiencies in the use of energy, and environmental control), (2) investment money and (3) new choices (e.g., new technology --- breeder reactors, fusion power, coal gasification and liquefaction, solar energy, geothermal energy) and (4) new discoveries (e.g., Arctic oil). The combination of all these items can yield factors of 10^6 to 10^9 amplification in resource availability and energy generation.* The long term problem, therefore, will not be in energy sources but in the energy sinks, i.e., the environment, a subject I will discuss later in this outline summary.

Over the long term we see, then, adequate resources and suitable conversion techniques for energy generation. Also, we see greatly increased conservation (i.e., maximizing performance per unit of materials and energy invested) in the end uses of energy* through, for example, greatly improved insulation of buildings, minaturization of systems which use much less energy per unit task (e.g., computers), systems with high reliability and availability, substitutions of communications systems for some transportation needs, more efficient transportation systems (e.g., mass transit) and vehicles (e.g., smaller, less horsepower cars), fuel switching, recycling of materials, use of new materials, more efficient and cleaner industrial processes through the use of enzymes, for example, and more efficient energy-conversion and distribution through such techniques, for example, as coal liquefaction and gasification, MHD combined-cycle power plants, hydrogen-oxygen fuel cells and the H_2 economy, Li S and Li C batteries, improved distribution networks, cryogenic transmission, breeder reactors, and direct conversion ultimately from fusion power to electricity.

Institutional incentives and structures need to be provided through improved legislative and regulatory action to encourage both vigorous and adequate research and development of these new options and increased efficiency and substitutability of energy and materials. In the U. S. institutional improvements are needed, for example, in government organization, legislative and regulatory policy, environmental standard setting, tax treatment, federal leasing of land and off-shore sites, import/export quotas, and research and development incentives and programs. I will comment on some of these items later.

The number of choices open to us, then, in terms of new discoveries, new technology, new processes, and the restraint and conservation that we could put into the system --- all of these can provide the amounts of clean energy sources that we would need for any foreseeable demands.

INTERNATIONAL CONTEXT

Within a worldwide context, a number of issues will come up with respect to the developed countries who now use almost 90% of the world's energy (North America, Europe, USSR, Japan) and developing countries; resource-rich countries (e.g., Canada, USSR, Australia, Brazil and

*These factors relate to the concept of compound improvements, say 4 to 6 percent per year on the average, which over a period of a few hundred years will yield this kind of amplification and conservation.

to some extent the U. S. and Indonesia) and resource-poor countries (e.g., Japan, Europe); international environment and resource-trade problems; the growing super-industrial society and its urbanization and increased use of energy and resources; and the effect of all of these factors on economic growth and development. I shall only mention them at this point as important issues in perceiving ERE interrelationships, for these interrelationships are affected not only by economic growth per se but also by the so-called "gap" between the developed and the developing countries. We conclude that only in an omni-directional-growth world can all nations and peoples benefit from economic development. Theories which state that the world's resources are fixed and that we have already used up most of them and therefore the developed countries must now give up some of their riches to the developing countries to close the "gap", doom us to a world of tensions and warfare. Such political solutions to equalize wealth inevitably result in totalitarian actions. The more positive action would be to transfer capital and technology know-how obtained from extensive research and development programs to the developing countries in the form of productive systems.

● INTERMEDIATE SITUATION - THE ENERGY "CRISIS"

I would now like to turn to the intermediate situation --- that is the situation between now and 1985 --- where, in the United States, and various other parts of the world, we are experiencing a minor crisis, which may turn into a serious crisis, in being able to furnish the clean energy required by the demand, which is doubling about every ten years (see Chart 1). This exponentially increasing demand, incidentally, is due primarily to the increased per capita use of energy as opposed to an increase due to population growth. Over 90% of current energy use is primarily centered in the North American continent, in Europe, in Japan and in the USSR. Over two-thirds of the increasing demand is due to increased per-capita use in these areas.

At least in the United States, the energy crisis that we are facing today has been brought about primarily by institutional weaknesses and not by a lack of potentially available resources or energy options. These weaknesses include, for example, a long-standing policy that we had energy-plenty, that it was inexpensive and that we really didn't have to do very much to see that it continued that way; a lack of valid data on our resources and on our demand; misdirected incentives with respect to our tax and pricing policies on fuels and electric power, our research and development programs and our environmental-standards; and finally, by government fragmentation in institutions and decision making in, for example, the formulation of a surfeit of environmental standards and controls at the federal, state and local levels, many of which have paralyzed the energy industry on a project-by-project basis with little real effect on the important environmental problems.

These institutional weaknesses have been augmented by a real shortage of natural gas in the United States; by the failure of the nuclear energy sector to meet its projections of the last twenty years, somewhat due to the economics of the situation, to construction costs and delays, and to capitalization needs and somewhat due to the environmental standards that need to be met --- for example, the Calvert Cliff's decision which essentially stopped the building of new nuclear plants until environmental needs are satisfied; and finally by a lack of clean-coal producing capability and our heavy dependence on oil and gas in all of our energy-using sectors. For example, our 1975 shortage of low-sulfur coal for stationary-power sources is now projected to be about three hundred million tons, about half the total coal that is needed in that year. The result of all of these factors and the exponentially increasing demand is that we will have to import more oil. There are currently projections which state that by 1980 we will be importing up to fifty percent of our oil to support all energy sectors of the economy at a minimum cost of twenty billion dollars a year in our balance of payments. I will get back to this point a little later.

All of these factors will have international ramifications. First, there will be increased competition for oil by the United States, Japan, and Europe, in conjunction with opportunity pricing by OPEC. Second, there will be an intensification of the exploration for new sources --- for example, offshore oil, Arctic oil, Southeast Asian oil --- and the attendant ocean oil-pollution problems at these sources and around the world as more and more oil gets shipped to users. Third, there will be an intensification of nuclear-energy development and the need to address as soon as possible questions concerning capitalization needs and also the international nuclear-fuel-cycle problems --- e.g., uranium enrichment plants and their locations, nuclear plant siting and cooling, fuel processing and transportation, and waste disposal. Fourth, there will be a higher visibility of high-energy consumption areas around the world --- the U. S. and Japan and Europe, and the USSR --- versus the low-energy consumption areas. Also, tensions could increase between the resource-rich countries and the resource-poor countries with respect to resource-export issues raised by developing countries. These issues will include possible constraints on economic growth in developing countries as fuel prices increase and possibly the relinquishment of environmental aspirations because of increased costs.

Finally, the U. S. economic situation will be affected significantly. As I mentioned earlier, the possible importation of about twenty to twenty-five billion dollars worth of oil per year by the U. S. will have a significant impact on our national security through lack of a secure supply, on the need for adequate reserve storage, possibly on our balance of trade, on long-term capital inflow from oil-rich countries on U. S. shipping policy, on the environment and on costs through the need for new refineries (about five per year), deep ports (at least two), new distribution pipelines (for resid and crude oil), and oil-spill protection systems. These infrastructure needs will be primarily for the East Coast and will involve a few billion dollars per year in capital investment.

One of the measures that has been suggested to limit imports to about 25% of our needs and to improve our bargaining position is to develop options which can force a reduction of at least one to two dollars a barrel in the projected price of imported oil, a price that some of our more experienced observers say can get as high as six to ten dollars a barrel in the next ten years. With this price reduction, we can project savings of at least $5 billion per year in oil-import costs, so that we could justify easily an investment of a few billions of dollars in a very intensive program of research and development in alternate sources of fuel at prices below $6 a barrel --- e.g., in developing our very extensive oil shale in the western United States, which may take at least a billion dollar investment to make it credible in terms of delivering four-to-five-dollar-a-barrel oil. The environmental problems in developing oil shale need to be researched now --- for example, water needs, local power needs, mine fatalities, and the disposal of the tailings and reclamation of the land.

We can also provide economic incentives to the oil industry to undertake, with appropriate environmental performance-leasing requirements,* an intensive program in secondary recoveries

---

* Federal lands include about 50% of oil, gas and coal left in U. S. and about 80% of the shale and offshore deposits.

and in offshore drilling to get the oil out as fast as possible and then taper off later when the need for this oil has abated as other fuel options are developed; and finally we can both accelerate our nuclear-energy program and we can develop and use new processes for coal cleaning and for coal liquefaction and gasification so as to continue the use of our very extensive coal reserves, if we can develop the water resources necessary to implement this option (e.g., use of mine water) and if we can reclaim land laid waste by strip mining economically.

The viability of coal and oil shale as usable resources from an environmental viewpoint will be extremely difficult to establish and sustain. Much research is needed.

To alleviate current problems, we can relax the compliance schedules and we can differentiate geographically those environmental standards (e.g., SO_X clean-up strategies) which now have little payoff in many areas in terms of improved health of the population as measured by the national primary standards on SO_X control; we can double the price of domestic gas, for example, to stimulate production and to clear the market; we can allocate this clean fuel to the dirtiest emitters until low-sulfur coal from the West and imported oil (resid) are available easily; and we can internalize costs for environmental cleanup of the energy system to restrain energy use if necessary by varying prices, not by rationing. Gas price increases will tend to drive industrial and power-plant users out of the gas market into the low-sulfur coal and oil markets and leave gas supplies for residential use. We can also encourage the use of stack-gas-cleaning processes when these systems are available to operate reliably and economically.

All these options provide a spectrum of choices which will tax our available capital, our skilled labor (e.g., heavy metal industries --- piping, pressure reactors, etc.) and our institutional structures --- federal, state, local and private --- to the limit in providing all the skills and in establishing policies on what and how we get from here to there (1985). This lead-time problem is the most critical one to solve in alleviating potential clean energy shortages.

All these options must be pursued with firm commitment to environmental needs through appropriate investment, design and implementation of resource extraction, land reclamation and energy generation and distribution techniques.

● THE OPTIONS FOR THE LONG TERM

I would like to elaborate now on some of the technical options I mentioned earlier that must be pursued to alleviate the problems discussed above by the 1980's and the 1990's.

Fossil fuels will continue to be an important element in energy, at least until the end of the century when new options will begin to take over. In the United States, our most optimistic projections for nuclear energy in the year 2000 indicate that we will still be at least seventy-five percent dependent on fossil fuels at that time for electric power (see Chart 2), so we feel that great emphasis should be put now on new discoveries of fuel supplies and on synthetic low-sulfur fuels, both liquid and gas, based on the conversion of coal and fuel oil. Those front-end processes which would convert coal or fuel oil into low-sulfur fuels could be used, for example, with advanced gas turbines and steam-topping cycles for high-efficiency, clean power-generation systems. (See Charts 3, 4, 5, 6, 7, 8, and 9 for our projections of price data.)

Over the long term, these front-end conversion processes could be used to make hydrogen fuel which could then be used in a hydrogen economy to feed fuel cells for local power distribution and for transportation vehicles and systems.

Back-end processes, which essentially treat the stack gases after combustion, and fuel-cleaning processes, which essentially desulfurize the coal and oil, can be used in some of the existing power plants along with a more optimum allocation of the available clean fuels and more uniform national environmental standards --- that is over the next five to eight years --- until the front-end processes are perfected.

As I mentioned earlier, there are two other options for the long term that should be pursued much more vigorously than we have been pursuing them. These include solar energy for large-scale power needs and the possible use of hot, dry rock from geothermal sources well below the surface of the earth --- say, 10 to 20 thousand feet --- as a regional supplement for energy where it is available. In the solar energy area, we have studied a solar-energy system, for example, which will make hydrogen fuel, which can then be used in the hydrogen economy. An engineering appraisal is needed for this scheme and for the other various solar-energy proposals being made. Over the long term, the use of solar energy will help to alleviate the earth's heat-balance problems arising from the use of fossil fuels and nuclear energy. In the geothermal area inexpensive exploration techniques are needed now to identify useful sources of energy.

In the nuclear-energy area, which of course is the main strategy that we are pursuing for the long term, in addition to the liquid-metal fast-breeder reactor, we feel that much more emphasis and effort, for both environmental and economic reasons, needs to be put on the high-temperature gas-cooled reactor, the gas-cooled fast-reactor, the heavy-water reactor and the molten-salt breeder reactor as well as, of course, on fusion reactors, especially laser fusion.

Nuclear energy in general will be used for the base load in the overall energy system, including both electric power needs and a source of power for electric-vehicle batteries and electrified mass-transit systems and for the making of hydrogen through the electrolysis of water, for example. An economic advantage probably exists in the future for transmitting energy as electrolytic hydrogen rather than electricity over distances greater than a few hundred miles. The cost of the hydrogen system is much more dependent on the electrolysis process itself rather than the transportation costs.

The international implications of the nuclear fuel cycle need to be addressed now with respect to uranium enrichment and the processing and transportation of fuels and waste. Some of the projections for nuclear-fission energy indicate that by the turn of the century we will be processing about fifteen hundred tons of plutonium per year and at a loss rate of one percent this indicates that there will be something like fifteen tons per year of plutonium that could possibly get lost, and so this kind of problem from a safety viewpoint is obvious and must be looked at.

We also feel strongly that more attention and research and development need to be paid to underground siting of nuclear plants, for both reasons of safety and for reasons of protection against sabotage.

Also more effort is needed on materials and operational procedures for breeder reactors, and on cooling techniques. For example, in the United States, we are planning to install up to about nine hundred new nuclear plants in the next thirty years. This is a rate of about three a month. Some of the questions that must be addressed in such an intensive implementation program include the factors of capital, labor, plant siting and land use, cooling, environmental and safety measures and so forth. In the U. S., for example, $300 to $400 billion of capital will be needed in the next 20 years to implement the planned nuclear energy program.

TRANSPORTATION

In the area of transportation, we believe that the long-term problem in transportation is going to be one of fuel consumption and not air pollution. We think the air pollution problem can be solved readily with an appropriate expenditure of time and money, but that over the long term, we will be facing a problem of oil consumption in transportation as the demand for oil grows exponentially world-wide and the supply levels off relatively. The rate of growth of automobiles in the world today is currently around 6.8% a year. In addition, the use of fuel by aircraft will be increasing exponentially,* especially if we start using large SST fleets. These rates of growth will put great pressure on developing alternate modes of transportation and propulsion to today's automobile, for example.

We see then the trends in transportation to be the replacement of the internal combustion engine, first by modified spark-ignition systems and/or rotary engines, and possibly later with, for example, gas turbines and/or diesel and stirling-cycle engines; the use of electric cars in many urban areas, when and if, for example, lithium-sulfide or lithium-chloride high-power and high-energy-density battery developments are successful and nuclear energy as a source of electric power becomes widespread.

The use of all-electric vehicles initially transfers the source of pollution from the vehicle to the central power station. Vehicle electrification not only shifts the source of pollution but changes the constituents as well, replacing CO emissions with SO_2. Nevertheless, these emissions can take place in non-urban areas and eventually may be eliminated entirely through conversion to nuclear power generation.

Any change which reduces transportation energy consumption is bound to reduce petroleum consumption. However, since transportation energy demand is not likely to decrease in the decades ahead, petroleum consumption can be reduced only through the use of smaller autos and/or alternative fuels.

Electric energy is the most obvious alternate and the soonest to be available. Electric energy can readily supply all guided ground transportation and a large portion of the urban auto sector, presuming successful development of high-energy-density batteries. Electric energy systems will offer no overall efficiency advantage, except for plants which use "free" natural energy (hydraulic or geothermal) or future plants which derive energy from new and highly efficient processes (MHD or thermo-nuclear fusion). In the short term, electric energy will involve an unfavorable emissions trade-off (SO_2 for CO). In the long term, nuclear energy will be a source of electric energy for electrified mass transit systems and for electric vehicle batteries.

Synthetic petroleum produced from shale, coal, or even garbage, will likely be more expensive (shale oil will cost at least 50% more than natural petroleum), may not be available in significant quantities until the mid-80's, and will have major environmental impact related to shale residue disposal and strip mining for coal.

All of the more exotic fuels (alcohol, ammonia, and hydrogen) are considerably more costly than gasoline, and alcohol and ammonia have lower energy density. Except for cost, alcohol would be the most easily implemented petroleum substitute in the short term. For the long term, H_2,

---

* Aircraft fuel will be about 25% of total oil used in transportation in year 2000.

produced, for example, in conjunction with large-scale off-shore nuclear power installations, is a likely alternative fuel. Aside from the different (not more serious, relative to gasoline) safety problems associated with H_2, this gas is in plentiful supply, offers no harmful emissions, and can be used either as a fuel for heat engines (with air or oxygen) or fuel cells.

This discussion of E-R-E effects on transportation seems best summarized by looking into the future to see where the current developments are likely to lead.

With the exception of the use of synthetic petroleum, all changes favor reduced air pollution. Replacement of the modified spark ignition engine with an inherently clean alternative engine and the use of gas turbines in intercity trucks and busses will both serve the cause of pollution reduction only. Several changes do not involve an energy saving but benefit petroleum conservation and cleaner air; in this category are:

- Electrification of rail systems, urban autos, and mass transit (1980's, 1990's).
- Evolution of the electric car into dual-mode systems (year 2000).
- Use of synthetic petroleum and non-petroleum fuels in fleet systems (especially H_2) (1980's, 1990's).
- Use of gaseous fuels (continuing).

Those transportation shifts which favor both cleaner air and reduced consumption of energy and petroleum are the most worth striving for; they are:

- Widespread use of smaller, less powerful automobiles with integral computers to control engine performance (1980's).
- Increased use of urban mass transit and improved high-speed ground systems (1980's).
- Use of H_2 fuel cells for intercity/rural auto use (year 2000).

THE ENVIRONMENT --- CURRENT PROBLEMS

Finally, with respect to the environment, we believe that the currently perceived conventional environmental problems in air, water and land pollution and in waste disposal can be solved with competent design, positive effort and money. We have done some studies that have looked at the cost of cleaning up the environment of these currently perceived problems, and over the next five to seven years, for example, we see a cost in cleaning up the air, in the United States, from both stationary sources and transportation systems, to be about forty billion dollars. This will be the out-of-pocket costs, but there will be a large number of benefits from this clean-up which may far outweigh this direct cost. In the short term, however, we see the consumer's power bill increasing by about twenty-five percent and the cost of his automobile by about three hundred dollars. However, we think that over the long term, with ingenuity, one can design energy-conversion systems and transportation systems which are less expensive than today's and which will still provide a cleaner environment, but this will take a lot of research and development and capital and time.

With respect to environmental standards, the U. S. has established standards in the automotive emissions area and for stationary sources, and in each case, there is an element of a misdirected incentive. In the case of the sulfur dioxide standards for power plants, we have a situation where if the standards adopted by the different states are implemented in the fossil-fuel-burning power plants, this will cost about seventeen billions of dollars for clean fuels and/or cleaning processes and may result in only a small increase in the amount of people in the United

States who would have an ambient air quality within the national primary environmental standards that we have established. This is partly because cleaning up the stationary point sources itself is not the real problem in many of the high-density cities. The real problem comes from the area sources --- industrial combustion and home heating --- and so in many respects, a better strategy, for example, would be to use the available clean fuels like gas in home heating and in industrial combustion resulting in both a better overall effect on the environment and increased efficiency in the use of these fuels in comparison to burning them in power plants. Also, the standards set by the states are different and bear little correspondence to the best strategy for using the available clean fuel supply and to the particular ambient air quality in that state. However, optimum geographic differentiation in standards from a national viewpoint (for example, deferment of state sulfur regulations in those states already meeting federal standards) and efforts to stop the use of gas in power plants are both difficult policies to implement politically and economically at this time. These actions would, though, relieve the immediate problem of the need to import more oil for use in Midwest power plants, for example.

Similarly, it does not look to us like it would be worthwhile to implement completely the clean-car standards in 1975 for the amount of money it is going to cost to do it --- that is, the last ten percent improvement increment which it is going to take to meet these standards over what can be done relatively easily by 1975 does not look like it is worth the expenditure of money that has to be made --- and so we would favor a relaxing, a slight relaxing, of those standards at least in the schedule, say from 1975 to 1978 to allow the automobile manufacturers to develop better long-term solutions. For example, the permissible level of NO_X emission could be increased from 0.4 gram/mile to 1 to 2 gram/mile without materially harming the effort to improve ambient air quality in most regions of the country. This change would require Congressional action, however.

In summary, we need large-scale efforts and an internalization of costs in addressing these areas of air and water and land pollution, radiation effects, nuclear-fuel processing and transportation, nuclear-waste transportation and disposal, power-plant siting (including underground siting of nuclear plants), land use, offshore petroleum exploration and production, refinery siting and deep-water port facilities, fuel distribution, recycling of materials, strip-mining and land reclamation, ocean/estuarine pollution, etc. We need to recognize in our legislation and standard-setting that the environment and energy must be considered as a single entity on a national and an international basis in questions ranging from power-plant siting and environmental standards to worldwide environmental effects and resource trade. This overall planning is needed since we live in a closed system and we can only minimize the effects of pollution on the environments. In the future, the protection of the environment will need to have an ethical and a religious base as well as an economic base for long-term solutions.

THE ENVIRONMENT - LONG TERM PROBLEMS

The environmental problems that we are more seriously concerned about are the barely perceived problems. These include the long-term biological effects of several thousand organic components (chemicals, pesticides, etc.) added to the biosphere every year; the possible long-term radiation effects of krypton, tritium and nuclear wastes; the problem of trace pollutants at high altitudes, for example, nitrous oxides from SSTs; and the interference with ecological cycles, both with the H_2O, O, C, S, N, and pH cycles and with the heat balance of the earth and its effects on local and global climates. Many scientists believe that the heating of the earth is the ultimate limit to growth. All energy outputes (e.g., heat, CO_2) tend to heat the earth except for particulates and this situation will engender both local and global climatic effects for which we don't have yet the kinds of ocean-atmosphere models that can predict accurately what may or will happen. The use of terrestial-based solar-energy systems will alleviate this problem to the extent that solar energy can be used to meet overall needs. Solar energy can be used now, for example, for residential space heating.

Finally, there is the question of catastrophic nuclear-power-plant accidents --- for example, triggered by emergency core-cooling structural and breeching failures or by sabotage --- and what can and should be done to make nuclear power plants even more safe than they currently are. Here, both high-temperature gas-cooled reactor technology and underground siting may help, as well as more adaptive systems and on-line, ensemble testing procedures.

All the above effects may have irrevocable consequences and must be researched heavily if we are to avoid catastrophies and to determine what are the ultimate physical limits to growth.

Overall, we feel that the current energy, resources and environment situation is characterized by confusion, by misunderstanding, by gragmentation and by misdirected incentives. Problems and interrelationships need to be identified and understood --- these are the most critical issues of the future. We need to develop policy alternatives for both decision makers and technicians and to formulate a "lobby for the future" through, for example, intensive research of these barely perceived problem areas.

The questions that need to be addressed include --- what kind and how much of economic growth and development; what kind of institutional and economic incentives do we need; what kind of reallocation and substitution policies for fuels and materi als; what program of lòng-term research and development in resource exploration, extraction and transportation, in new and more efficient energy use and conservation, in conversion and transmission techniques and in the potentially catastrophic environmental effects; and what international policies and cooperation are needed with respect to resources, the environment and the nuclear fuel cycle.

CHART I
U. S. ENERGY CONSUMPTION BY CONSUMING SECTOR

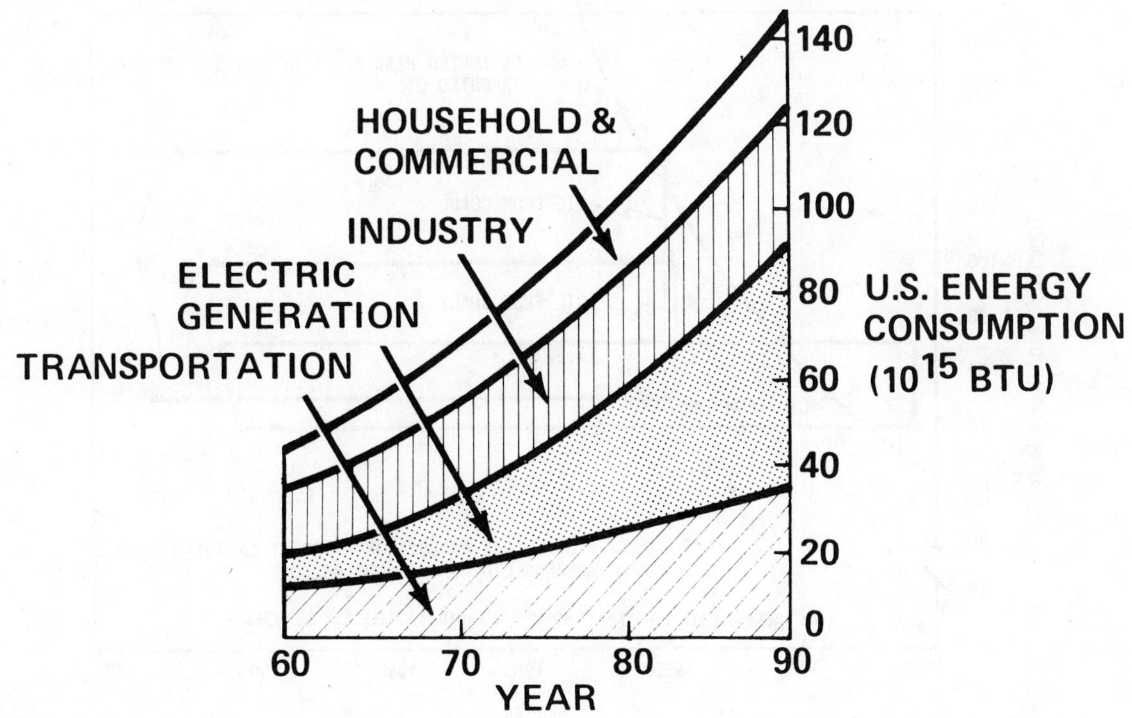

CHART 2
U.S. ENERGY CONSUMPTION SOURCE

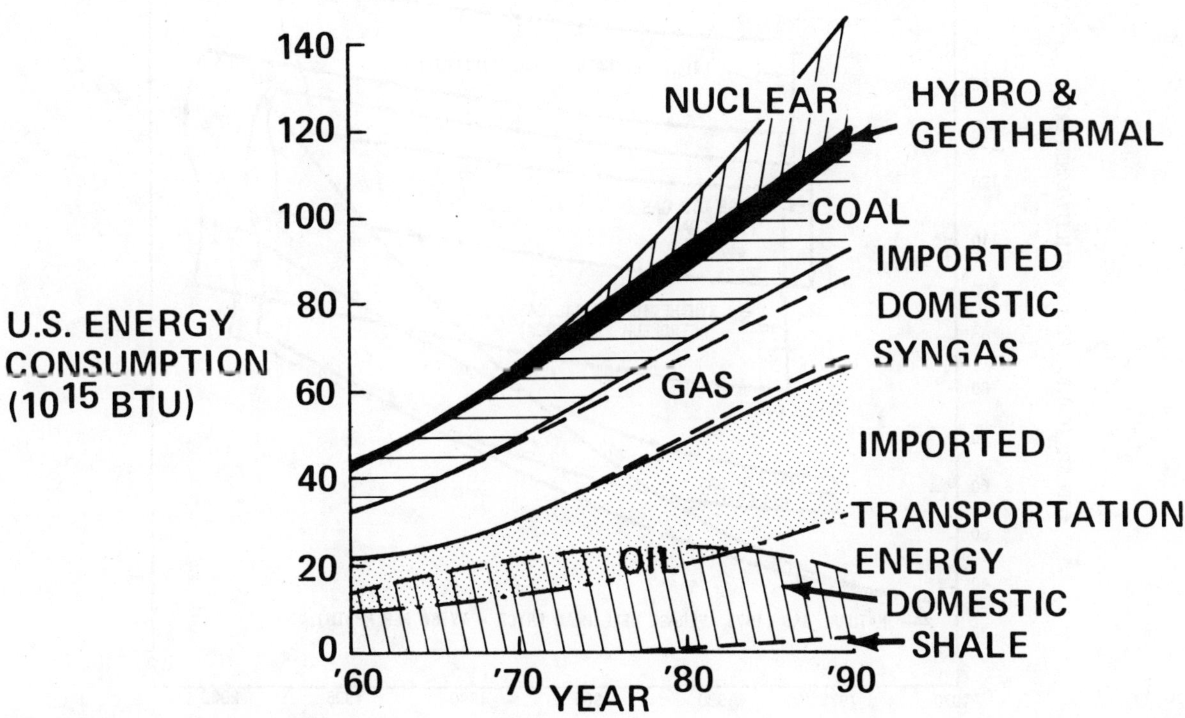

CHART 3
FUEL OIL AVAILABILITY (BY SOURCE) WITH
ESTIMATED PRICE (1970 DOLLARS)

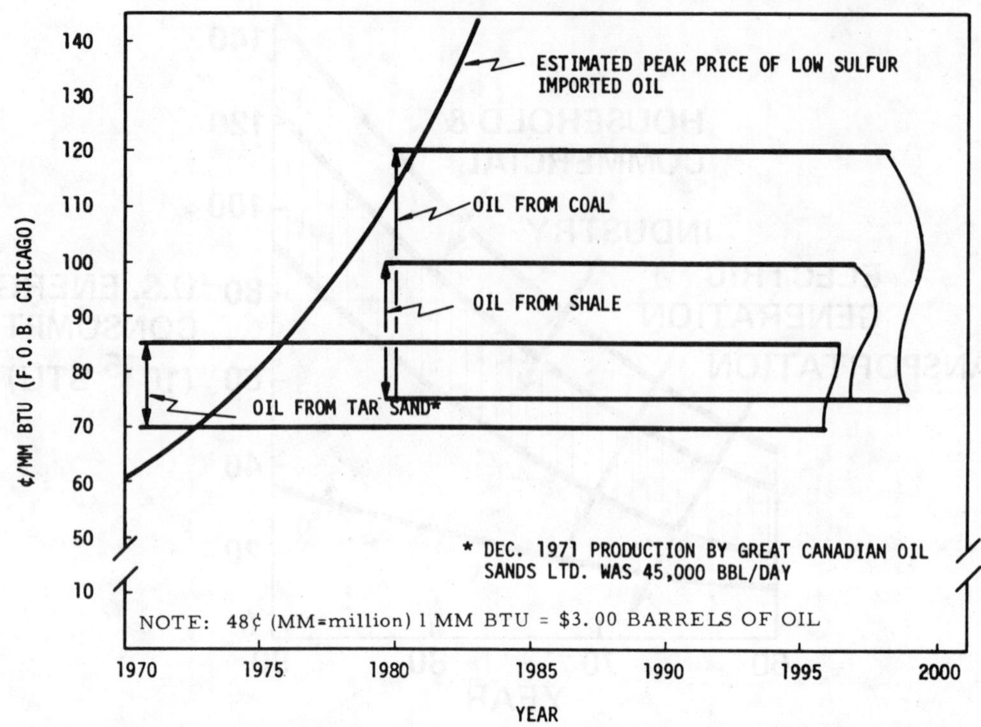

CHART 4
GAS AVAILABILITY BY SOURCE WITH ESTIMATED PRICE (1970 DOLLARS)

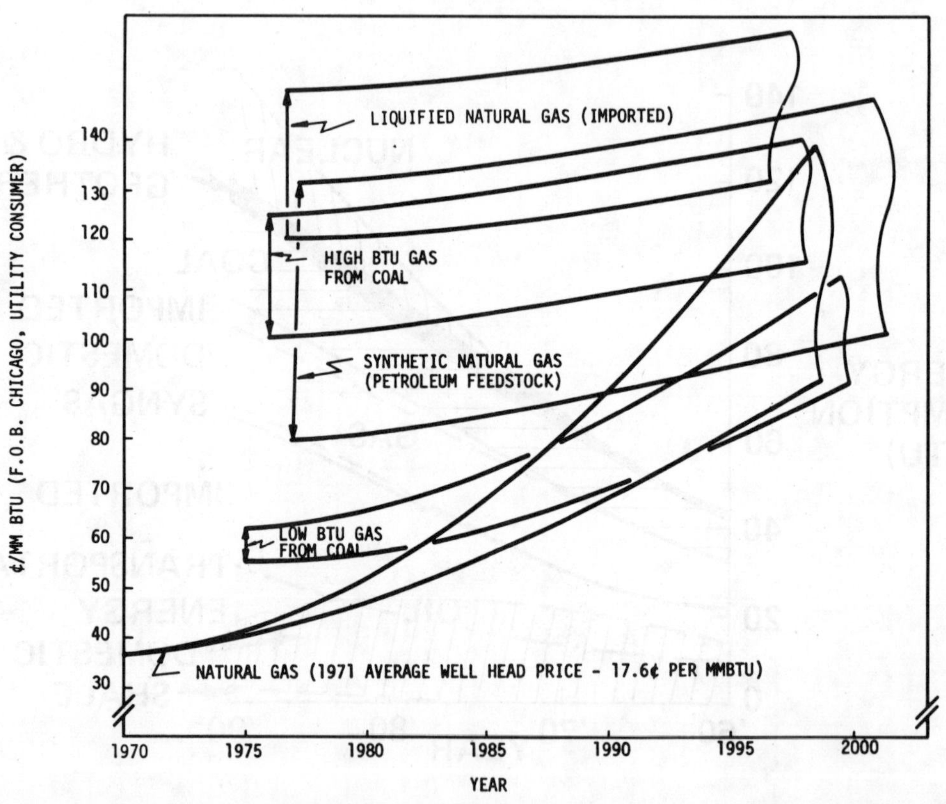

CHART 5
COAL UTILIZATION (LOW SULFUR) POTENTIAL WITH ESTIMATED PRICE (1970 DOLLARS)

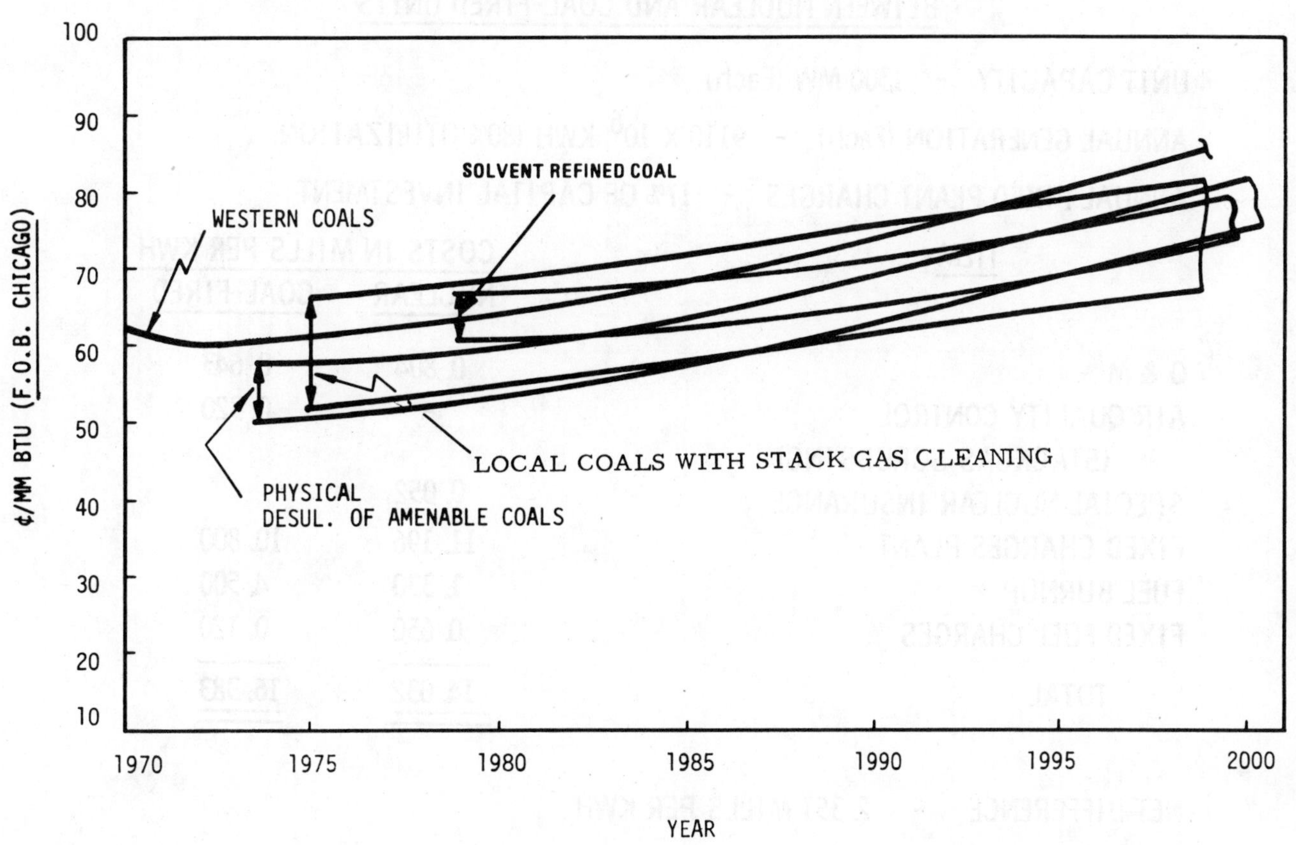

CHART 6
ESTIMATED NUCLEAR FUEL COST (1970 DOLLARS)

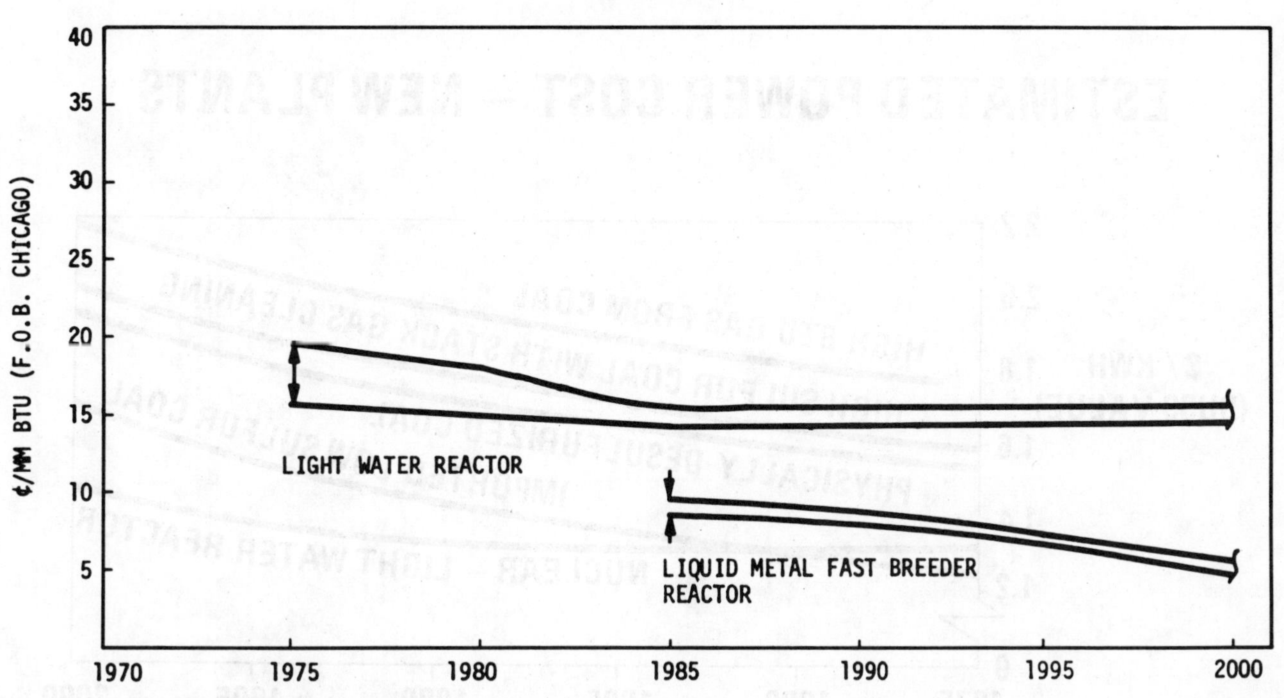

Does not include "subsidies" in uranium enrichment and waste storage.

CHART 7

COMPARISON OF POWER COSTS* (FOR A NEW PLANT TO BE BUILT - 1972 ESTIMATED) BETWEEN NUCLEAR AND COAL-FIRED UNITS

UNIT CAPACITY - 1300 MW (Each)

ANNUAL GENERATION (Each) - 9110 X 10^6 KWH (80% UTILIZATION)

ANNUAL FIXED PLANT CHARGES - 17% OF CAPITAL INVESTMENT

ITEM	COSTS IN MILLS PER KWH	
	NUCLEAR	COAL-FIRED
O & M	0.804	0.643
AIR QUALITY CONTROL (STACKGAS SCRUBBING)	-	0.320
SPECIAL NUCLEAR INSURANCE	0.052	-
FIXED CHARGES PLANT	11.196	10.800
FUEL BURNUP	1.330	4.500
FIXED FUEL CHARGES	0.650	0.120
TOTAL	14.032	16.383

NET DIFFERENCE - 2.351 MILLS PER KWH

ANNUAL COST DIFFERENTIAL - $21.417 MILLION

*THESE ARE COST ESTIMATES OF A UTILITY AND INCLUDE ALL PROVISIONS TO COMPLY WITH POLLUTION LIMITATIONS (INCLUDING COOLING TOWERS, LIQUID WASTE CONTROL, PARTICULATE CONTROL, RAD WASTE, ETC.).

CHART 8

ESTIMATED POWER COST — NEW PLANTS

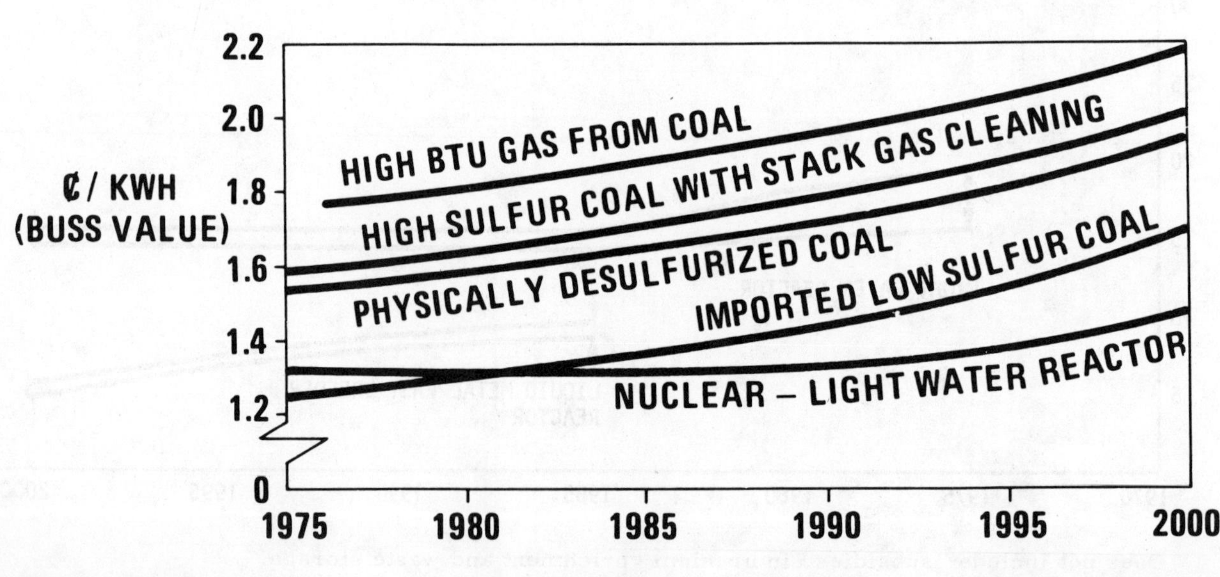

CHART 9

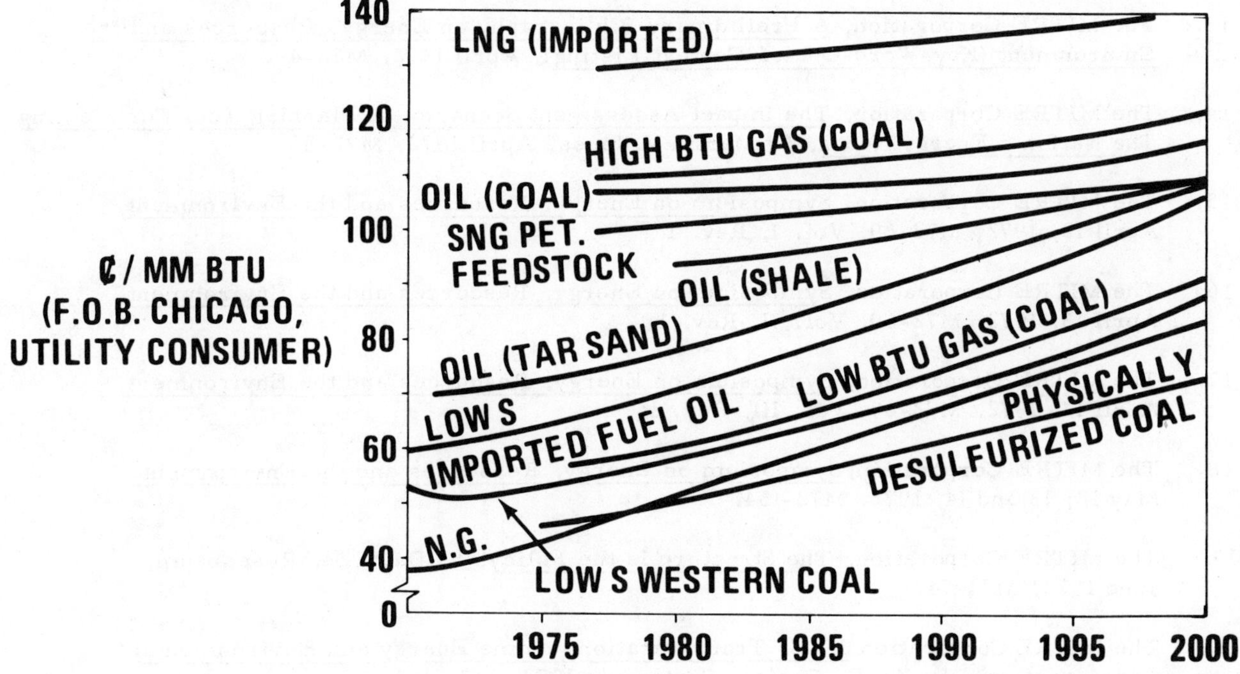

REFERENCE LIST

1. The MITRE Corporation, Symposium on Energy and Resources -- First Organizational Meeting of the Symposium Committee, July 30, 1971, M71-48.

2. The MITRE Corporation, Symposium on Energy and Resources -- Second Organizational Meeting of the Symposium Committee, September 21 and 22, 1971, M71-65.

3. The MITRE Corporation, Symposium on Energy, Resources and the Environment -- Third Meeting of the Symposium Committee, November 8 and 9, 1971, M72-10.

4. The MITRE Corporation, Symposium on Energy, Resources and the Environment -- Fourth Meeting of the Symposium Committee, January 10 and 11, 1972, M72-23.

5. The MITRE Corporation, Symposium on Energy, Resources and the Environment -- Fifth Meeting of the Symposium Committee, February 22, 23, 24 and 25, 1972, M72-50.

6. The MITRE Corporation, Symposium on Energy, Resources and the Environment -- Supplement to the Transcripts of the First, Second and Third Organizational Meetings, December 1971, M72-3.

7. The MITRE Corporation, Symposium on Energy, Resources and the Environment -- Supplement to the Transcript of the Fifth Meeting of the Symposium Committee, February 22, 1972, M72-50, Supplement I.

8. The MITRE Corporation, Dimensions of World Energy, by Dr. M. R. Gustavson, Consultant to The MITRE Corporation, November 1971, M71-71.

9. The MITRE Corporation, Dr. Teller Discusses Energy, Resources and the Environment, by Dr. Edward Teller, September 21 and 22, 1971, M-72-15.

10. The MITRE Corporation, The Availability of Low Pollution Fuels, by Kurt E. Yeager, January 1972, M72-16.

11. The MITRE Corporation, Towards An Energy Ethic, by Dr. M. R. Gustavson, Consultant to The MITRE Corporation, March 1972, M72-43.

12. The MITRE Corporation, Energy, Resources and the Environment, Major U. S. Policy Issues, Discussed by Mr. John F. O'Leary, Consultant to The MITRE Corporation, March 1972, M72-44.

13. The MITRE Corporation, A Preliminary Bibliography on Energy, Resources and the Environment (Key-Word-Out-of-Context Listing), April 1972, M72-49.

14. The MITRE Corporation, The Impact Assessment Scenario, A Planning Tool For Meeting The Nation's Energy Needs, by Martin V. Jones, April 1972, M72-56.

15. The MITRE Corporation, Symposium on Energy, Resources and the Environment, April 12, 1972, M72-69, Vol. I, Rev. 1.

16. The MITRE Corporation, Symposium on Energy, Resources and the Environment, April 13, 1972, M72-69, Vol. II, Rev. 1.

17. The MITRE Corporation, Symposium on Energy, Resources and the Environment, April 14, 1972, M72-69, Vol. III, Rev. 1.

18. The MITRE Corporation, Symposium on Energy, Resources and the Environment, May 12, 13 and 14, 1972, M72-154.

19. The MITRE Corporation, The Structure is the Policy, by David M. Rosenbaum, June 1972, M72-74.

20. The MITRE Corporation, U. S. Transportation - Some Energy and Environmental Considerations, by W. E. Fraize, September 1972, M72-164.

21. The MITRE Corporation, Climate Change and the Influence of Man's Activities on the Global Environment, by William W. Kellogg, as presented at the Symposium on Energy, Resources and the Environment, Kyoto, Japan, July 11, 1972, M72-166.

22. The MITRE Corporation, Large-scale Utilization of Solar Energy, by Gregory M. Haas, as presented at the Symposium on Energy, Resources and the Environment, Kyoto, Japan, July 11, 1972, M72-168.

23. The MITRE Corporation, Magneto-hydrodynamic (MHD) Power Generation Status and Prospects for Electric Utility Applications, by Finn Hals, as presented at the Symposium on Energy, Resources and the Environment, Kyoto, Japan, July 11, 1972, M72-169.

24. The MITRE Corporation, Coal - The Black Magic, by R. P. Ouellette, September, 1972, M72-170.

25. The MITRE Corporation, Environmental Issues and Action Around the World, by Richard S. Greeley, as presented at the Symposium on Energy, Resources and the Environment, Kyoto, Japan, July 11, 1972, M72-171.

26. The MITRE Corporation, Symposium on Energy, Resources and the Environment, Kyoto, Japan, July 9-12, 1972.

Rx for "The Energy Crisis": A Long-Term Policy Base

In an economy geared to the world's most lavish use of energy, the thought of less abundant, higher-priced supplies is disturbing. In recent months there has been much talk of a U.S. "energy crisis." Is this strong term justified? Probably *yes*, if one thinks solely of the short-term supply stringencies that have been piling up in the past few years. Probably *no*, if the long-term possibilities of a well-reasoned, comprehensive set of national energy policies are taken into account. The article that follows is based primarily on the final chapter of a recent RFF study of research needs in the field of energy made for the National Science Foundation. It also draws upon Joel Darmstadter's contribution to an RFF report to the Commission on Population Growth and the American Future and on his background appendix to a new RFF book, *Energy, Economic Growth, and the Environment.*

Many observers see a number of separate recent developments as indications that in the future energy will no longer be in abundant supply at low costs from domestic sources; hence the talk of an "energy crisis."

Reasons for this view include (1) growth in recent years of energy consumption at a rate faster than that of gross national product (contrary to the long-standing downward trend in the energy/GNP relationship); (2) failure of proved reserves of crude oil and natural gas to increase at a rate consistent with the growth in consumption, along with actual shortages of gas at the going price; (3) a recent cessation of improvements in the thermal efficiency of converting fossil fuels to electric energy; (4) rise of concern over the natural environment, implying higher costs at almost every point in the energy supply chain from the extraction of mineral fuels to the facilities in which fuels are utilized to produce heat, electricity, and motive power, and leading to actual prohibitions against the use of high-sulfur fuels, particularly eastern U.S. coal; and (5) doubts that nuclear energy will ever be available at costs as low as had been anticipated earlier.

Do the current difficulties in energy supply foreshadow a fundamental long-term turnabout in the nation's energy position — one

Reprinted with permission from *Resources for the Future Inc., Annual Report*, pp. 11-20, Sept. 30, 1972.

that has in the past been characterized by abundant low-cost supplies, principally from domestic sources? In trying to answer this basic question let us look first at the current situation and recent trends, and then at some projections of future energy consumption in the United States.

■ The expected increase in U.S. energy requirements will start from a level of use that is by far the world's highest. In terms of contained heat, the nation's 1970 consumption of energy resources amounted to about 69 quadrillion British thermal units, about half again as much as was used in all of Western Europe, whose combined population is 75 percent above the U.S. total. The principal components of this Btu consumption were 527 million short tons of coal (20 percent of the total), 22 trillion cubic feet of natural gas (almost 33 percent), and nearly 5.5 billion barrels of oil products (43 percent). Most of the coal and sizable amounts of the gas and oil were used to generate 1.4 trillion kilowatt-hours of electricity. In addition, 245 billion kilowatt-hours were generated by falling water and 19 billion by nuclear energy, together accounting for 4.1 percent of total energy consumption.

In the steady growth of total energy consumption — from 20 quadrillion Btu in 1920 to the 69 quadrillion figure of fifty years later — there were great changes in the relative importance of the different energy sources. The more recent statistics are shown in the figure on page 13.*

Since 1920 the contribution of coal to total energy use has fallen steadily from more than three-quarters to one-fifth. Chief reasons for the drop were replacement of the coal-fired steam locomotive by the diesel electric, loss of most of coal's space-heating market (except in recent years by way of electricity), and the steep rise in automotive transportation in which oil has had no serious competitors. Meanwhile, there was a striking rise in use of oil and natural gas, in both absolute amounts and shares of total energy consumption.

The contribution of hydropower (the statistics include a small nuclear energy component in recent years) has remained fairly steady at around 4 percent. However, nearly every observer expects a marked rise in the nuclear share as it takes on an increasing proportion of electric power generation, which will itself continue to grow. Between 1920 and 1970 the proportion of total energy supplied in the form of electricity increased from 8 percent to 25 percent.

■ America's appetite for energy could triple by the year 2000. There are, of course, many uncertainties about how great the increase will be and how it will be shared among the different energy sources. Will growth in energy use continue to outpace that of gross national product, as in recent years, or drop back to its

*See page 105 of this reprint book.

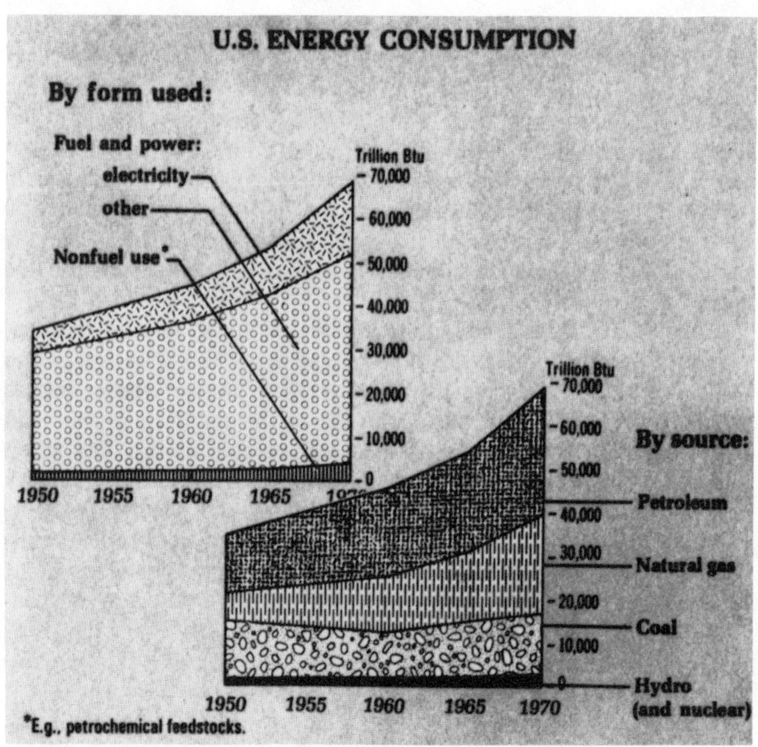

longer-term position of lagging somewhat behind? Would continuation of the trends toward more leisure and toward services rather than manufactured goods necessarily lower the share of energy consumption in national output? At what rate will GNP itself grow? Will population growth hold at recent low levels? Will the share of total energy going into electricity continue to rise as rapidly as in the past? And will nuclear energy take over as large a share of electricity generation as now seems likely?

Despite these and other uncertainties, a formal projection, with assumptions clearly set out and a set of consistent figures to contemplate, is a most helpful device.

The set shown in the figure on page 15* is a composite adaptation of long-term projections by the U.S. Bureau of Mines and the Federal Power Commission. Nationwide energy consumption for the year 2000 is shown as 190 quadrillion Btu, nearly three times the figure for 1969. This reflects a projected annual growth rate of 3.5 percent — slightly above the average rate since 1900 but below the average for the 1960s. Consumption of electric power is shown to be increasing by 6.7 percent a year, a much higher rate than that projected for use of energy in other forms, although it seems quite possible that, as a result of future price increases and environmental constraints, a more dampened electricity growth rate may be in prospect.

*See page 107 of this reprint book.

What are the prospects for our continuing to meet the projected requirements of energy fuels principally from domestic sources?

At present the most widely accepted view — it might almost be called a consensus — is that the chances are not good: that the United States will have to depend much more heavily upon imports. For example, the National Petroleum Council concluded in a 1971 study that by 1985 the nation will be importing almost 60 percent of its oil needs and nearly 30 percent of its gas. (The current import share is around 25 percent for oil and negligible for gas.)

Heavy dependence upon imports would have serious implications for foreign policy and national security. The situation of the United States would become comparable to that of the countries in Western Europe and of Japan. All of the major industrialized nations of the world, except the U.S.S.R., would be drawing much of their vital energy supplies from a handful of countries — mostly Arab — bordering the Persian Gulf and the south rim of the Mediterranean. Moreover, there are new questions about the availability of imports. Factors now at work in international oil point to substantially higher prices than in the past and to grave questions concerning dependable supply — particularly from the Middle East and North Africa, the world's most prolific known oil-bearing regions.

■ *Is heavy dependence on imports inevitable or simply one of several possibilities that will depend upon the policies pursued by the United States? The evidence suggests that choices still are open.*

It is true that recent trends in proved reserves of oil and natural gas, which together account for well over half of the nation's projected energy consumption, are discouraging. Proved oil reserves (crude oil and natural gas liquids) increased only slightly since the mid-1950s from 35.5 billion barrels in 1955 to around 38 billion barrels in 1969. During the same period, U.S. production rose from 2.76 billion barrels per year in 1955 to 3.94 billion in 1969, so that the ratio of proved reserves to production fell markedly. The nation's oil consumption grew still faster. As a consequence, U.S. imports rose from 15 percent to 23 percent of annual use.

For natural gas, the reserves-to-production ratio has dropped in virtually every year since World War II. Since the late 1960s additions to reserves have actually fallen below production and there is mounting evidence that the nation's readily available natural gas supply is declining to critical levels at given levels of demand.

But "proved reserves" is a very restrictive measure, referring only to the amount remaining in the ground that, with reasonable certainty, can be recovered in the future from known reservoirs

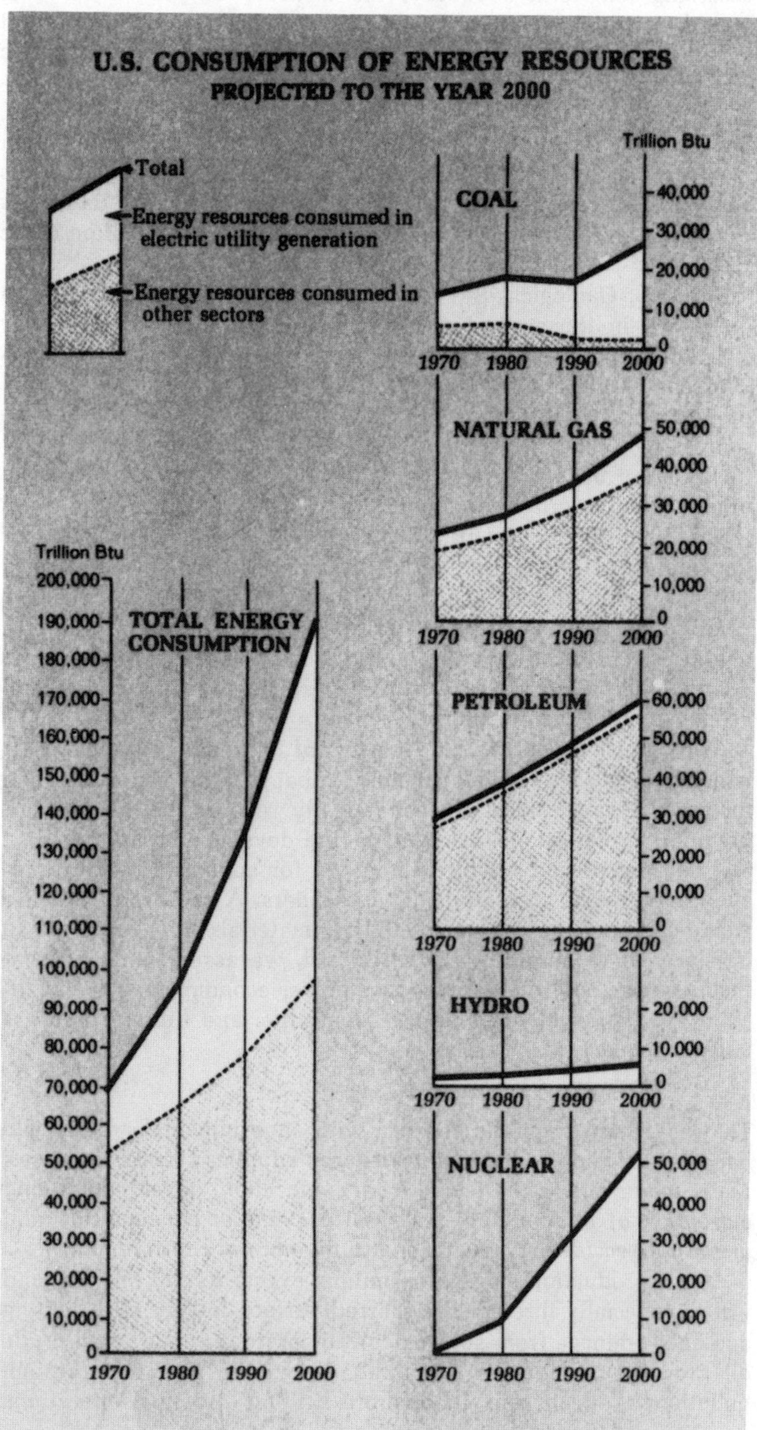
U.S. CONSUMPTION OF ENERGY RESOURCES
PROJECTED TO THE YEAR 2000
Total
Energy resources consumed in electric utility generation
Energy resources consumed in other sectors
Trillion Btu
COAL
NATURAL GAS
PETROLEUM
HYDRO
NUCLEAR
TOTAL ENERGY CONSUMPTION
1970
1980
1990
2000

and under current economic and technological circumstances. In assessing long-term adequacy, the measure of proved reserves must be augmented by estimates of the amount of petroleum that can reasonably be assumed recoverable under future economic and technological circumstances. These amounts can only be approximated on the basis of incomplete geological, geophysical, and other data. Estimates of ultimate U.S. oil resources from which new reserves might be drawn range up to nearly 3 trillion barrels as compared with proved reserves of about 40 billion. Estimates of ultimate resources of natural gas range up to 7,000 trillion cubic feet as compared with proved reserves of 265 trillion.

The U.S. Geological Survey has estimated that coal resources remaining in the ground at the beginning of 1967 totaled over 3,000 billion tons. Half of this amount was based directly on mapping and exploration. The other half was "probable." (Annual production is now around 600 *million* tons.) But these vast resources will mean little if coal, in dwindling demand as a primary fuel, cannot be converted to gasoline or high-Btu gas or some other desirable form of energy.

Then there is the potential of oil shale. The U.S. Department of the Interior estimates that oil shale with a content of at least 15 gallons per ton of rock could yield as much as 1.8 trillion barrels "that could be classed as 'known reserves'." Whatever estimates one uses, it is clear that the quantity is immense and represents a large multiple of proved U.S. reserves. Here again technical and economic problems remain to be solved.

As for nuclear energy, the adequacy of fissionable raw materials is not likely to be a critical factor. Probably more important than the known magnitudes of economically recoverable reserves of uranium (or thorium) ores will be the development in the next few decades of conversion technology for advanced reactors, including the likely development of breeders. A successful breeder would multiply the usefulness of raw materials by a factor of 50 or more. If, looking still further ahead, we assume that nuclear fusion can be controlled to produce power economically, the availability of the required resources (deuterium and lithium) appears still more immense.

■ The recent rise of public concern with environmental quality will handicap development of future energy supplies from all sources. *Every major link in the chain, from exploration, development, and extraction of mineral fuel reserves to eventual consumption and disposal of residuals, is damaging to the natural environment.*

At the production stage, coal mining exacts its toll on land and water, especially through the scarred surface left by strip mining and acid drainage from surface and underground mining. In crude oil production, the most notable damage has occurred through pollution of the oceans in offshore drilling, although operations on land also have created some problems. Exploitation of shale deposits would generate enormous solid wastes and make great demands on water. Uranium mining, like coal mining, results in waste accumulation and defacement of landscape, though on a much smaller scale; its main hazards are from the radioactive effects.

In transportation, spillage from tankers already presents large dangers. Safety factors of shipping liquefied natural gas need to be better understood. Movement of nuclear wastes on a large scale is yet to be experienced but poses obvious hazards. The safety record for pipelines is generally good, but the controversy over the Alaska pipeline may be only a foretaste of the problems of difficult terrain combined with a clash of economic and ecological values.

The processing of crude oil and nuclear raw materials and the use of all fuels in either generation of electricity or combustion for space heating or manufacturing operations, or above all in automobiles, result in pollution of air, water, or both, as well as problems of waste disposal. The siting of large generating plants presents additional problems of amenity, aesthetics, and safety. Most people want more power, so long as the power plants aren't near *them*.

These and other environmental concerns will increasingly affect the outlook for energy. Some resources, such as high-sulfur coal or oil, are becoming much less desirable — in some cases their use is being restricted by government-imposed quality standards. In many instances costs will increase, raising the question of who shall pay them — the industry, the consumer through higher prices, or the taxpayer at large. And ultimately, how shall the choice be made between more energy or cleaner environment?

■ The outlook for domestic energy resources could be one of plenty into a far-distant future. But the emphasis must be placed upon *could*. The presence of vast total resources of fossil fuels and nuclear raw materials does not in itself assure an adequate U.S. energy supply for the future in the forms needed, and consistent with environmental standards. The possibilities will not be realized without conscious effort along many lines.

To give only a few examples:

— There is immediate need to find and prove up additional reserves of crude oil and natural gas. Achieving this end will call for reexamination of incentive, leasing, and regulatory policies, as well as additional research and development on techniques of exploration and extraction. What effects do (and can) price and tax policies have upon exploration for more reserves? How would alternative degrees of import restriction affect domestic prices, production, and the long-range availability of domestic supply?

— To realize the potential of coal and shale resources the technology of transforming them to gasoline or gas must be further advanced and its economic feasibility tested.

— To stretch resources of fissionable materials further, there is an argument for successful development of breeder reactors. Looking farther ahead, more should be done to explore the possibilities of fusion as a source of commercially usable energy.

— There is need to develop better methods of reducing or preventing pollution at the various stages of energy production and use. Some important aspects of pollution require basic scientific research. There is no agreement on the long-range effects of such phenomena as cumulative heat discharge, emission of carbon dioxide, and low-level radiation. Perhaps they are bearable. Perhaps they are threats to the earth's whole life system.

— To guide these and other efforts, there is need for a broad approach to energy problems that not only will take a balanced view of the separate energy sources but also will relate the overall energy picture to such other fields of major concern as economic welfare, national security, international relations in general, and environmental quality.

In sum, there is clear need for a cohesive set of national policies for energy. To shape and carry out such policies will require much study of policy alternatives and their implications. This is not to say that efforts to overhaul energy policies and programs should be shelved until researchers have answered all of the big questions; that time may never come. The only practical course is to proceed as best we can in the light of the knowledge that is available, meanwhile trying to reduce critical areas of ignorance.

■ Scores of important questions call for investigation. Within the next few years it will clearly be impossible to study all of them adequately.* The need for setting priorities under a comprehensive viewpoint is perhaps best exemplified by the current unsatisfactory situation in government technical research and development programs.

The national government's expenditure for R&D has lacked an overall framework. Large sums are being invested in nuclear R&D. Despite some increases in the recent past, the amount of government money spent on other energy technologies has been comparatively small and the selection of projects erratic. From a broad energy standpoint, some of the major research choices appear to have been perverse. The nuclear program was designed to bring a new process for generation of electricity to commercial fruition; yet the nation's resource base of coal, the major fuel for electric energy generation, was known to be great. Resources of crude oil

* However, a sizable effort already is under way. A notable example is the study of national energy policies, sponsored by the Ford Foundation, to which RFF is contributing an analysis of U.S. energy supply options.

and natural gas were known to be far smaller than those of coal. Greater efforts to develop substitute sources for those two fuels rather than for coal would have seemed the rational choice.

Is there some desirable degree of balance in future energy patterns? If so, what role can R&D choices play in achieving such a balance? It is doubtful that an optimum pattern would consist in doling out R&D funds among competing energy technologies evenhandedly. Policy research is needed on the degree to which flexibility and diversity in energy options for the future are desirable.

Consistency with other government policies is equally important. The interrelationship between energy R&D and other energy policies, as well as nonenergy policies, calls for study. For example, if the present pattern of R&D support were to lead the United States to a heavy dependence in the future on very large central-station electricity generation, there could well be certain inherent risks to the nation in terms of systems reliability or vulnerability.

As another example, the government is planning to offer shale oil leases on a test basis. Lessees will be expected to spend large amounts on the commercialization of oil shale technology — quite possibly hundreds of millions of dollars. However, the development of oil shale appears potentially attractive only if oil imports will continue to be restricted for some decades. Once firms have made large investments in oil shale, they can be expected to oppose subsequent relaxation of oil import controls.

In some instances expensive energy R&D is being proposed to bring about results that could perhaps be achieved by other means at less cost. For instance, research on the production of high-Btu gas from coal is being pressed while at the same time field prices of natural gas are controlled at levels much below the anticipated costs of gas from coal. The costs of gas from coal may well be in the neighborhood of 75 cents to $1.00 per thousand cubic feet; until recently, Federal Power Commission regulations have limited natural gas producers to prices of about 25 cents at the wellhead in most areas.

To return to technical R&D: there is no reason for the government to finance projects in fields where privately financed efforts are adequate. But the mere absence of private effort does not necessarily signal the need for government action. Other criteria are needed for deciding when government-financed R&D is justified. The key element in establishing such criteria is the comparison of anticipated costs and benefits of particular lines of research.

This is a complex and often subtle operation. *The relevant costs of a given activity are all those sacrifices resulting directly and indirectly from it, while its relevant benefits are all those gains flowing directly and indirectly from it. Such comprehensive social costs and benefits go beyond those that can be easily identified and expressed in dollar terms.*

In R&D investments the more readily measured private costs and benefits often diverge from social costs and benefits. One reason is that the private initiator of R&D effort often is unable to capture full benefits. The results of R&D are a special type of product — new scientific and technical knowledge. In the absence of special institutional arrangements, it is virtually impossible for the private producer of new knowledge to appropriate his product for exclusive use or sale. The income derived from it therefore may not be commensurate with costs incurred. This peculiarity goes far toward explaining why basic scientific research is confined largely to public and other nonprofit institutions.

The federal government can undertake larger, longer-lived, and riskier R&D projects than even the largest private firm. Governmental resources, although not unlimited, allow it to diversify and maintain flexibility despite the large size and long life of particular projects. In evaluating government R&D outlays, full social costs (including any duplications of effort deemed necessary in R&D strategy) can be compared with expected social benefits. Such a comparison is unlikely to be applied to privately financed R&D.

Still another source of difference between private and social cost-benefit comparisons involves R&D activities designed to maintain or strengthen competition within the private sector of the economy. For government, acting for the entire nation, there could be a benefit in increasing substitutability among different energy sources (thereby strengthening competition) and in general supplementing the R&D efforts of branches of the energy industry disadvantaged by such factors as the small size of typical firms, current economic distress, or lack of a research tradition. Such effects would seldom be considered benefits by a private firm.

Many of the factors in evaluating R&D possibilities on the basis of social versus private cost-benefit considerations are urgently in need of study. With greatly increasing governmental activity in energy R&D on the horizon, it is important that the criteria for intervention be soundly based.

■ To return to the basic question posed in the introduction: Does the United States face an "energy crisis"? The evidence sketchily reviewed here suggests that over the long run the answer is largely up to us. Without decided changes in policies, the set of disturbing developments that have converged in recent years could herald a fundamental change in the nation's hitherto favorable position. But this need not happen. In terms of ultimate resources, an abundance of the raw materials of energy fuels exists in this country. With a coherent set of national policies for energy, the potential of these riches could be realized. The challenge is to devise the means of doing so.

Excerpts from "Energy 'Demand' Studies: An Analysis and Appraisal"

Summary and Conclusions

The forecasts of most so-called energy demand studies are basically just extrapolations of historical trends in energy consumption. The extrapolations are variously labeled as forecasts of energy (1) consumption, (2) demand, (3) needs, or (4) requirements, but usually little recognition is given to the conceptual differences which exist among these terms. Some studies even discuss their forecasts as if they were indications of what future energy consumption "ought to be." There is, however, no established reason to believe that future energy needs are necessarily determined by extrapolations of historical energy consumption trends, and the studies themselves fail to explain why such extrapolations should necessarily be followed in the future.

A comparison of the thirty-five studies reviewed in this document reveals that, in general, they use similar projection techniques and data, they make similar assumptions, and not surprisingly, they obtain similar results. Some studies are even only slight variations of earlier studies while a few give forecasts which are merely averages of existing forecast values. Since the studies closely represent a repetition of the same forecasting effort, the consensus in their results must be cautiously interpreted. One observation, can, however, be made. More recent studies tend to have higher forecast values, reflecting the higher energy consumption rates of the last ten years which earlier studies were unable to incorporate into their calculations of trends.

Most of the studies are based on assumptions of only gradual technological change, constant relative fuel prices, unrestricted fuel availabilities, no major changes in government policy, only moderate swings in the business cycle, and cold-war defense energy consumption. Forecasts of United States petroleum consumption are usually based on an assumption that imports will automatically fill any gap which should arise between projections of oil consumption and future domestic oil production. Growth rates for population are most commonly assumed to be 1.3 or 1.6 percent, and the median assumption for annual growth in Gross National Product is 4 percent. The most recent studies give some mention of the possible impacts on energy consumption of promoting environmental quality, but none of them explicitly accounts for the effects which environmental improvement activities might have on future energy consumption.

The forecasts of most of the studies cannot be interpreted as reliable estimates of what future energy consumption actually will be because they are made without accounting for currently anticipated limitations of supply. Although the forecasts can be used as estimates of energy requirements necessary to sustain projected growth rates of GNP following the trends in the manner with which GNP has been supported in the past, such a use has several limitations. It considers only projected GNP growth rates, it ignores the possibility that a more energy-intensive or less energy-intensive GNP may be socially optimal in the

Reprinted with permission from *Energy Demand Studies*, U.S. House Committee on Interior and Insular Affairs, Sept. 1972.

future, and it fails to recognize the possible merits of future inter-fuel substitutability.

While the studies covered here were undertaken for a variety of reasons, several offer statements which reflect a general attitude about the future energy situation of the United States. These attitudes are particularly interesting because they seem to have changed drastically. Studies made before 1968 seem to anticipate no major energy problems through the end of this century whereas more recent studies express serious concern about the country's future energy situation. The change in attitude most likely results from recent higher energy consumption forecasts, unexpected delays in the development of nuclear sources of energy, and a realization that growth trends in domestic exploration, discovery, and recovery of traditional fossil fuels have failed to keep pace with domestic energy consumption.

In conclusion, an appraisal of the forecasts reviewed here finds that they are of limited usefulness. Most of the studies are fundamentally deficient for their failure to deal with an exact energy concept, their analyses are usually only loosely structured to address some vague notion of energy demand, and little attention is given to a correct interpretation of the forecasts. The crucial issues of price and supply limitations are usually all but ignored. Perhaps most important of all, many studies seem to suggest that their forecasts can be used by policymakers as target levels for future energy consumption. They do so, however, without justifying their implied GNP growth rate objectives and without considering the relative costs and benefits of providing various levels of energy consumption in the future.

Despite the deficiencies of the studies covered here, they represent the best energy studies currently available. Consequently, until a more complete determination is made of socially optimal levels of future energy production and consumption, existing forecasts may have to provide the basic information required for policy considerations. Such forecasts must, however, be adjusted to account for the essential issues generally not addressed by existing studies: the value of various levels of future energy consumption, the costs of supplying them, the price and national security aspects of imports, the advantages and disadvantages of resource utilization, the environmental implications of various energy-related activities, and the interrelationship between energy consumption and the efficient operation of the economy.

Studies Selected for Review

AEC 1960 — Fossil Fuels in the Future
Office of Operations Analysis and Forecasting
United States Atomic Energy Commission
Milton F. Searl, October 1960

SIIC 1962 — Report of the National Fuels and Energy Study Group on an Assessment of Available Information on Energy in the United States
Committee on Interior and Insular Affairs
United States Senate, September 1962

AEC 1962 — Civilian Nuclear Power . . . A Report to the President—1962
United States Atomic Energy Commission
November 1962 (and 1967 Supplement)

BOM 1962 — Patterns of Energy Consumption in the U.S.
Division of Economic Analysis, Bureau of Mines
U.S. Department of the Interior
William A. Vogely, 1962

NAS 1962 — Energy Resources
A Report to the Committee on Natural Resources
National Academy of Sciences, National Research Council
Publication 1000–D
M. King Hubbert, 1962

RFF 1963 — Resources in America's Future
Resources for the Future, Inc.
Johns Hopkins Press, 1963
Landsberg, Fischman, and Fisher

OST 1964 — Energy R&D and National Progress
Interdepartmental Energy Study,
Office of Science and Technology

Energy Study Group, Ali Bulent Cambel
U.S. Government Printing Office, June 1964

FPC 1964 — National Power Survey—1964
Federal Power Commission
U.S. Government Printing Office

AMS 1966 — Technological Change and United States Energy Consumption, 1939–1954 (and Abstract)
Alan M. Strout (unpublished Ph. D. dissertation)
University of Chicago, 1966

AEC 1967 — Forecast of Growth of Nuclear Power
Division of Operations Analysis & Forecasting
United States Atomic Energy Commission, 1967

FRC 1967 — Future Natural Gas Requirements of the United States, Volume 3, 1967
Future Requirements Committee
Denver Research Institute, University of Denver
Under auspices of the Gas Industry Committee

SC 1967 — Energy in the United States, 1960–1985
Sartorius & Co.
Michael C. Cook, September 1967

RRNA 1968 — Projections of the Consumption of Commodities Producible on the Public Lands of the United States 1980–2000
Prepared for the Public Land Law Review Commission
Robert R. Nathan Associates, Inc.
Washington, D.C., May 1968

OOG 1968 — United States Petroleum Through 1980
Office of Oil and Gas
U.S. Department of Interior, July 1968

BOM 1968 — An Energy Model for the United States Featuring Energy Balances for the Years 1947 to 1968 and Projections and Forecasts to the Years 1980 and 2000
Bureau of Mines, IC 8384
U.S. Department of the Interior, July 1968

TETC 1968 — Competition and Growth in American Energy Markets
Texas Eastern Transmission Corporation, 1968

AGA 1968 — Gas Utility and Pipeline Industry Projections 1968–1972, 1975, 1980 and 1985
Department of Statistics
American Gas Association, 1968

CMB 1968 — Outlook for Energy in the United States
Energy Division
The Chase Manhattan Bank, N.A.
New York, October 1968

PWM 1969 — Economic Strategy for Developing Nuclear Reactors
Paul W. MacAvoy
Massachusetts Institute of Technology Press, 1969

AGA 1969 — Gas Utility and Pipeline Industry Projections 1969–1973, 1975, 1980, 1985 and 1990
Department of Statistics
American Gas Association, 1969

SRI 1970 — Requirements for Southern Louisiana Natural Gas Through 1980
Federal Power Commission Area Rate Proceedings
Exhibit in FPC Docket No. AR69-1
Stanford Research Institute, April 1970

EW 1970 — 21st Annual Electrical Industry Forecast
Electrical World, September 15, 1970

WEM 1970 — Energy Resources and National Strength by Warren E. Morrison
Auditorium Presentation, Industrial College of the Armed Forces
(Transcript of lecture and statistical appendix available)
Washington, D.C., October 6, 1970

EBAS 1970 — Energy Consumption and Supply Trends Chart Book, April 1970
EBASCO Services, Inc., 1970

RFF 1971 — Trends and Patterns in U.S. and Worldwide Energy Consumption—A Background Review by Joel Darmstadter, Resources for the Future, Inc.
Forum on Energy, Economic Growth and the Environment
Washington, D.C., April 20–21, 1971

BOM 1971 — Mineral Facts and Problems—1970 Edition
Bureau of Mines Bulletin 650
United States Department of the Interior, 1971

AEC 1971 — Forecast of Growth of Nuclear Power—January 1971
Division of Operations Analysis and Forecasting
United States Atomic Energy Commission, 1971

NPC 1971 — U.S. Energy Outlook, An Initial Appraisal 1971–1985
An Interim Report of the National Petroleum Council, 1971

FPC 1971 — The 1970 National Power Survey
Federal Power Commission
Washington, D.C. 1971

FRC 1971 — Future Natural Gas Requirements for the United States, Volume 4, 1971
Future Requirements Committee
Denver Research Institute, University of Denver
Under the auspices of the Gas Industry Committee

DOI 1971 — Draft Environmental Impact Statement for the Trans-Alaska Pipeline
U.S. Department of the Interior
January 1971

RRFF 1971 — Middle Eastern Oil and the Western World: Prospects and Problems
Studies from a research program of The Rand Corporation and Resources for the Future
Sam H. Schurr and Paul T. Homan
New York, American Elsevier Publishing Company, 1971

PIRF 1971 — Oil Import Dependence and Domestic Oil Prices—A 15-Year Forecast in Oil Imports and the National Interest
Petroleum Industry Research Foundation, Inc.
Henry B. Steele, March 1971

FBD 1971 — The Twentieth Century Fossil Fuel Crisis; Current and Projected Requirements
Ford, Bacon, and Davis, Inc.—Engineers
Gerard C. Gambs, 1971

MOC 1971 — Outlook for the U.S. Economy—Energy—Oil to 1980
Marathon Oil Company
E. R. Heydinger, 1971

Study Forecasts

Fossil fuels, hydropower, nuclear energy, and electricity constitute the major primary and secondary *sources of energy* in the United States. Energy *consuming sectors* purchase either fuels or energy itself in the form of electricity, and finally, the purchased fuels or electricity are applied to various *end-uses*. Thus, there are three aspects of the same energy situation, and each can be used to examine energy consumption. Studies usually choose among the three on the basis of forecast objectives and data availability. Since considerable government and private sector information is available on fuels and electricity production and purchases by major consuming sectors, the sources and consuming sector aspects of energy consumption are usually examined by studies. Relatively little data are available on energy consumption by specific end-uses, and consequently, end-uses of energy are seldom forecast.

Table 2 indicates within standard categories the actual energy items forecast by the thirty-five studies. The table shows that some studies deal with only limited aspects of the entire energy situation while others give a more complete coverage. Within each category comparison of the forecasts is difficult because substantial variation exists among the actual energy items forecast. For example, exports and imports, energy conversion losses, non-energy fuel uses, etc., can be included or excluded before forecasts are made. Forecasts can also differ somewhat according to assumptions made and the specific energy concepts with which the studies attempt to deal.

The three following subsections present the forecasts of the thirty-five studies. Except where noted otherwise, the forecasts are believed to be consumption forecasts.

TABLE 2—ENERGY ITEMS FORECAST

Study	Total energy	Primary sources: Coal	Primary sources: Petroleum liquids	Primary sources: Natural gas	Primary sources: Hydro-power	Secondary sources: Nuclear	Secondary sources: Electricity	Consuming sectors: Industrial	Consuming sectors: Household and commercial	Consuming sectors: Transportation	Consuming sectors: Electric utilities	Regional detail	Other energy items
AEC, 1960	X	X	X	X	X	X							X
SIIC, 1962	X	X	X	X	X	X					X		
AEC, 1962							X						
BOM, 1962	X							X	X	X	X		
NAS, 1962	X												
RFF, 1963	X	X	X	X	X	X	X	X	X	X	X	X	X
OST, 1964	X	X	X	X		X	X						X
FPC, 1964						X	X	X	X	X	X	X	X
AMS, 1966	X												
AEC, 1967						X	X					X	X
FRC, 1967				X				X	X	(1)			
SC, 1967	X	X	X	X	X	X	X	X	X	X	X	X	X
RRNA, 1968	X	X	X	X	X	X	X	X	X	X			X
OOG, 1968	X		X	X									
BOM, 1968	X	X	X	X	X	X		X	X	X	X		X
TETC, 1968	X	X	X	X	X			X	X	X		X	X
AGA, 1968				X				(2)	X			X	X
CMB, 1968	X	X	X	X	X	X		X	X	X	X	X	
PWM, 1969						X	X						X
AGA, 1969				X				(2)	X			X	X
SRI, 1970	X	X	X	X	X	X	X	X	X		X		
EW, 1970					X	X	X	X	X				
WEM, 1970	X	X	X	X	X	X	X	X	X	X	X		X
EBAS, 1970	X	X	X	X	X	X							
RFF, 1971	X	X	X	X	X	X	X	X	X	X	X	X	X
BOM, 1971	X	X	X	X	X	X	X	X	X	X	X		X
AEC, 1971													X
NPC, 1971	X	X	X	X	X	X	X	X	X	X	X	X	X
FPC, 1971	X					X	X	X	X	X	X	X	X
FRC, 1971				X				X	X	(1)	X	X	
DOI, 1971	X		X									X	X
RRFF, 1971	X	X	X	X	X	X							X
PIRF, 1971	X	X	X	X	X	X							
FBD, 1971	X		X										X
MOC, 1971	X		X										

[1] Transportation and other.
[2] Entitled "Industrial and Other."

Sources: Battelle Memorial Institute, Federal Power Commission, individual studies.

TABLE 5.—FORECASTS OF TOTAL ENERGY ANNUAL GROWTH RATES

Study	To base year(s)	1970–1975	1975–1980	1980–1985	1985–1990	1990–1995	1995–2000
AEC 1960		.034				.035	
SIIC 1962		.032					
BOM 1962		.032					
NAS 1962				.0204			
RFF 1963		.027			.025		
SC 1967		.042		.044	.047		
RRNA 1968						.031	
OOG 1968		.031					(Petroleum only)
		.029					(Natural gas only)
BOM 1968						.032	
TETC 1968			.037				
CMB 1968		.038					
SRI 1970		.034	.040				
WEM 1970		.034		.034		.023	
EBAS 1970	.048	.040	.038	.038			
RFF 1971		.034		.035		.035	
BOM 1971				.031 to .043			
FPC 1971			.037				
NPC 1971	.043	.042	.042	.040			
DOI 1971			.042				
RRFF 1971			.035				
PIRF 1971		.046	.041	.040			
FBD 1971			.056				
MOC 1971			.042				

Note: Growth rates generally indicate the compound annual rate of growth from the average value for one period to the average of another period.

Sources: Battelle Memorial Institute; Federal Power Commission; Individual studies.

FIGURE 1.—FORECASTS OF TOTAL ENERGY CONSUMPTION FOR THE YEAR 1980

Forecasts in Trillion BTU's

115,000
110,000
105,000
100,000
95,000
90,000
85,000
80,000
75,000

+ 20% of mean
+ 10% of mean
mean forecast
– 10% of mean
– 20% of mean

1960 1961 1962 1963 1964 1965 1966 1967 1968 1969 1970 1971 1972

Year forecast was made

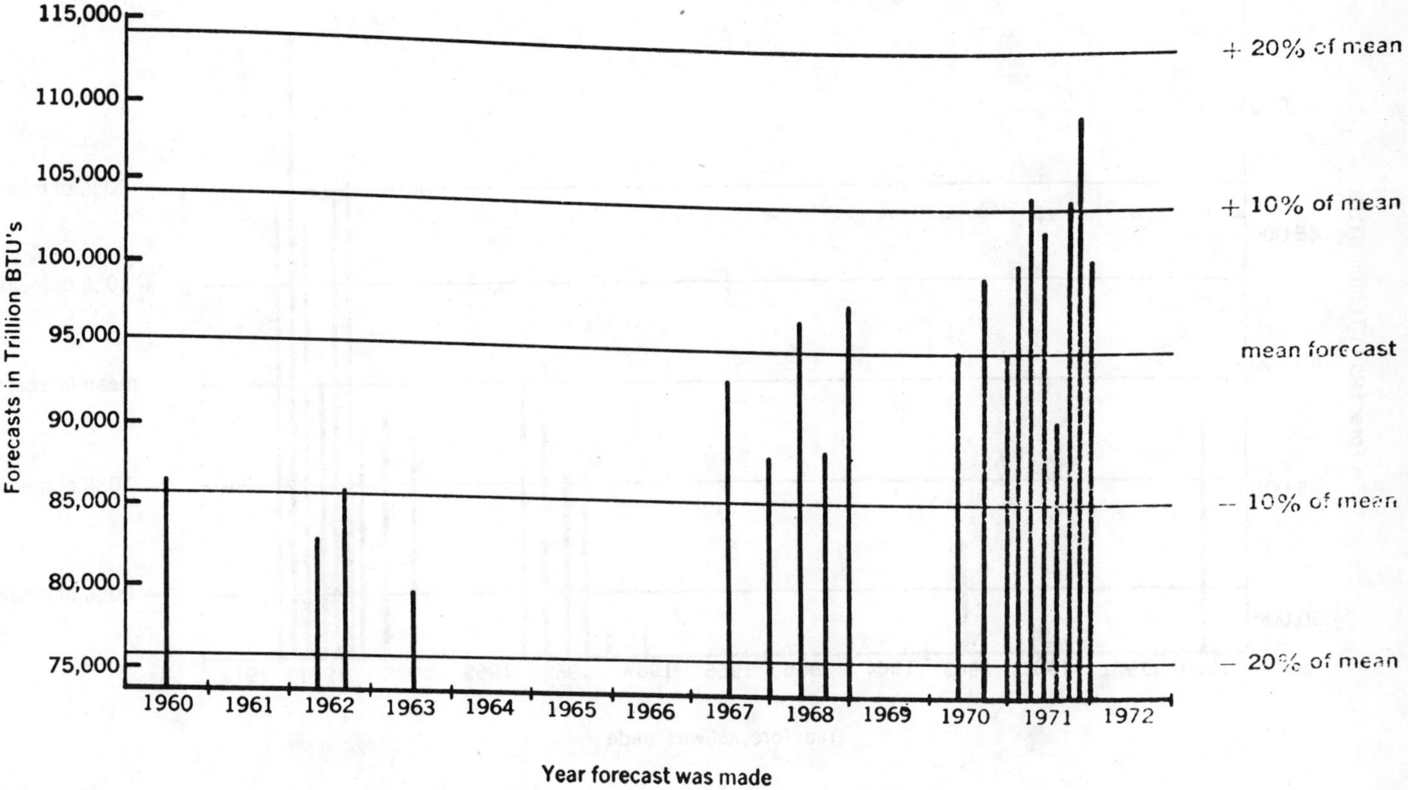

FIGURE 6.—FORECASTS OF UNITED STATES COAL CONSUMPTION FOR THE YEAR 1980

Forecasts in Trillion BTU's

24,000
23,000
22,000
21,000
20,000
19,000
18,000
17,000
16,000
15,000

+ 20% of mean
+ 10% of mean
mean forecast
– 10% of mean
– 20% of mean

1960 1961 1962 1963 1964 1965 1966 1967 1968 1969 1970 1971 1972

Year forecast was made

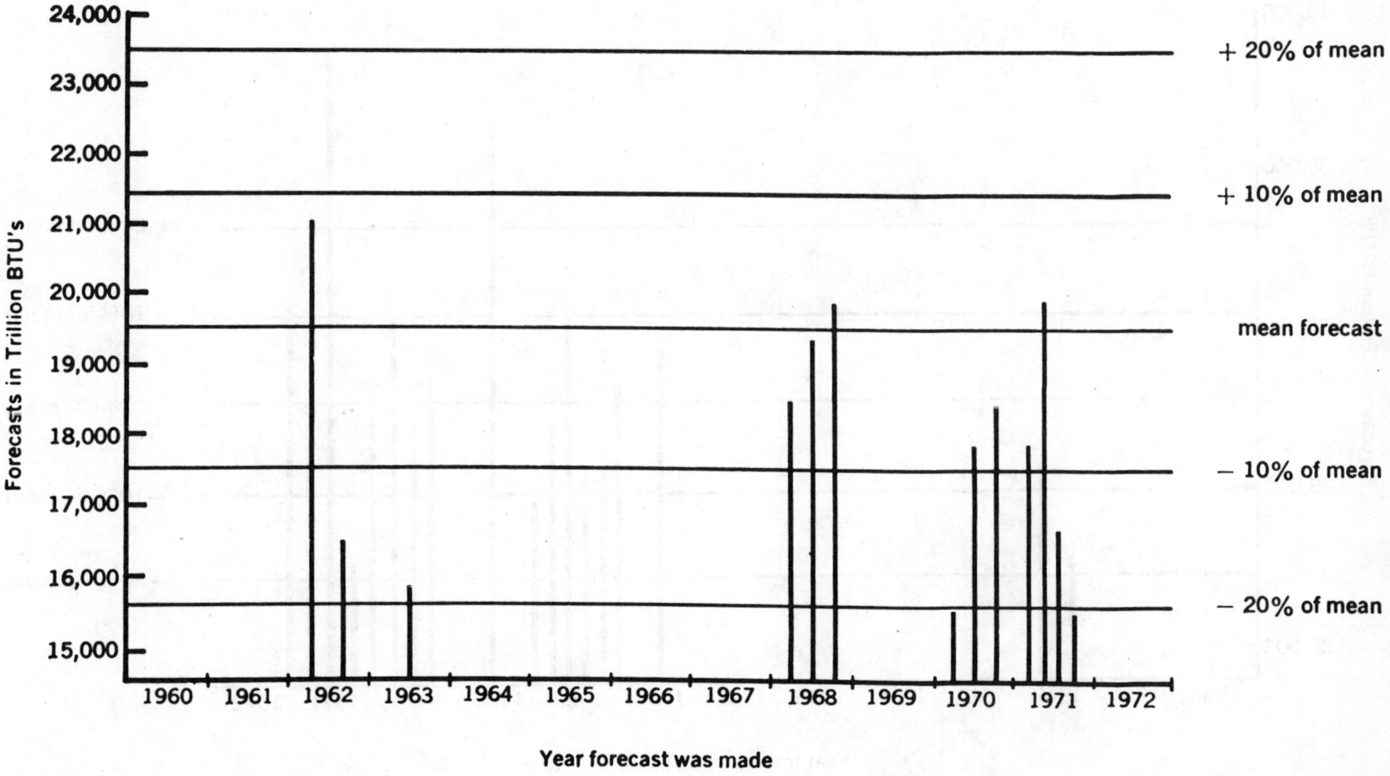

FIGURE 7.—FORECASTS OF UNITED STATES PETROLEUM AND NGL CONSUMPTION FOR THE YEAR 1980

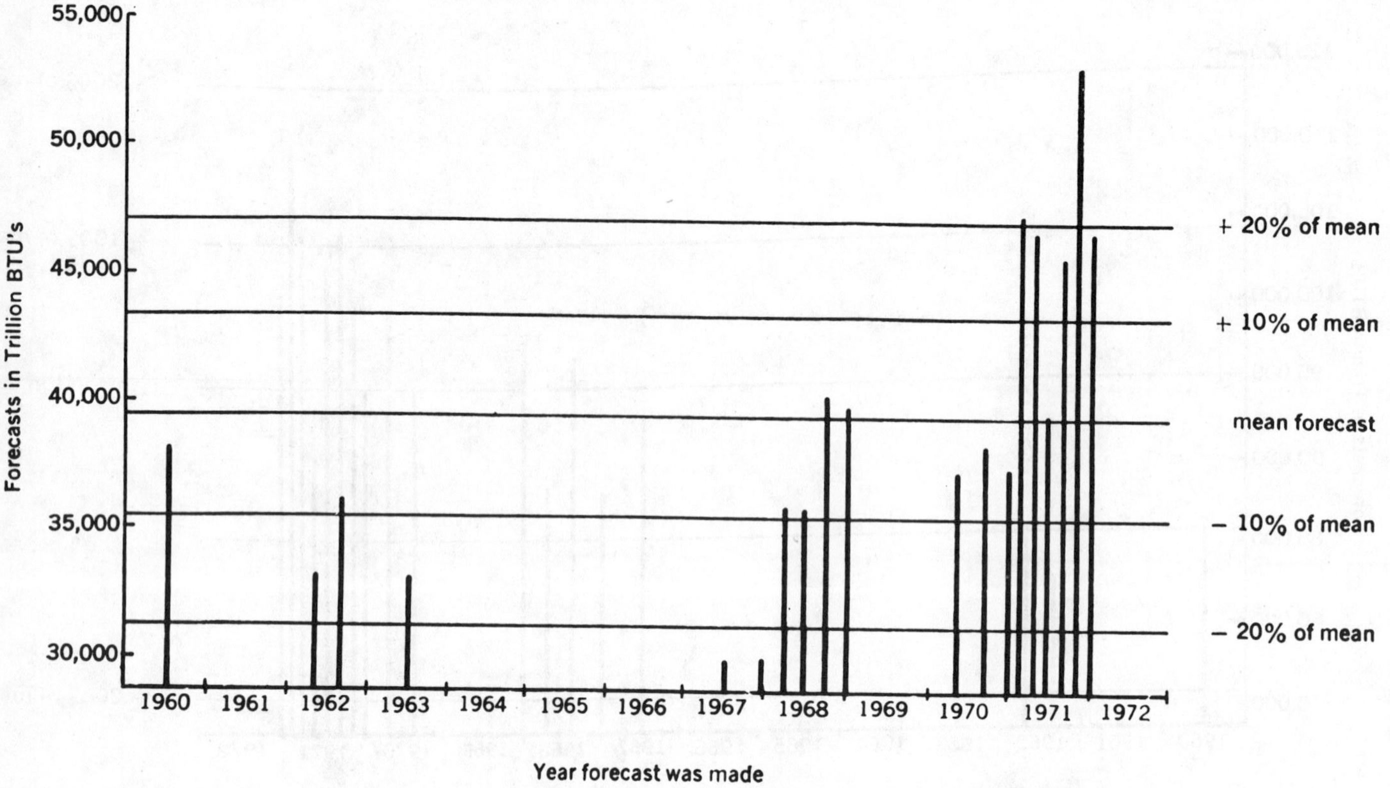

FIGURE 8.—FORECASTS OF UNITED STATES NATURAL GAS CONSUMPTION FOR THE YEAR 1980

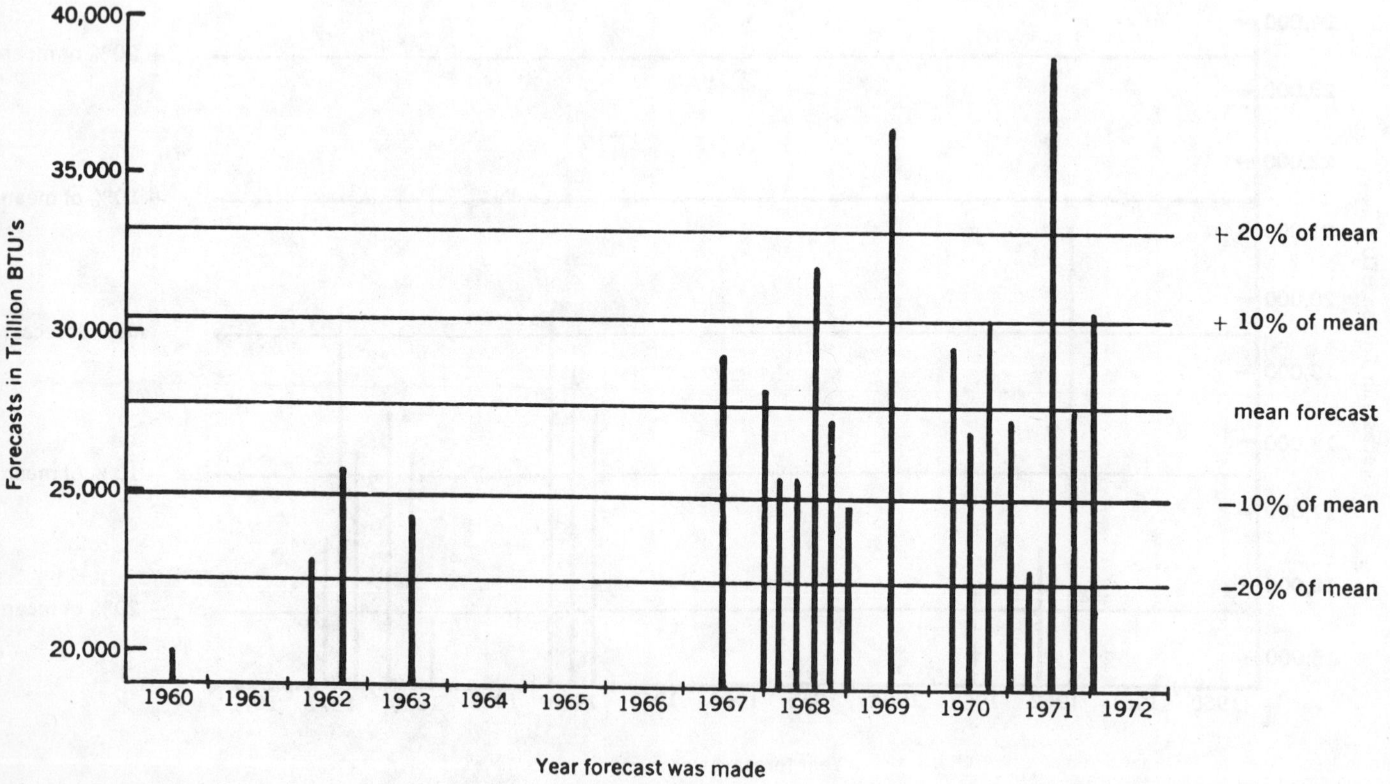

FIGURE 9.—FORECASTS OF UNITED STATES NUCLEAR ENERGY CONSUMPTION FOR THE YEAR 1980

Forecasts in Trillion BTU's

14,000
13,000
12,000
11,000
10,000
9,000
8,000
7,000
6,000
5,000
4,000
3,000
2,000

+ 20% of mean
+ 10% of mean
mean forecast
− 10% of mean
−20% of mean

1960 1961 1962 1963 1964 1965 1966 1967 1968 1969 1970 1971 1972

Year forecast was made

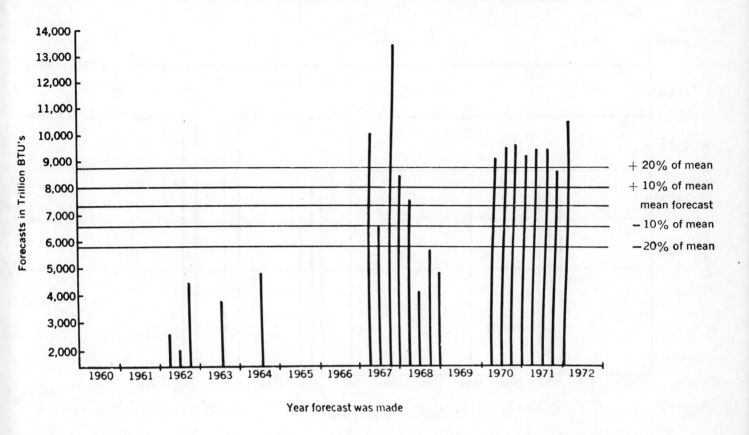

FIGURE 10.—FORECASTS OF UNITED STATES HYDROPOWER CONSUMPTION FOR THE YEAR 1980

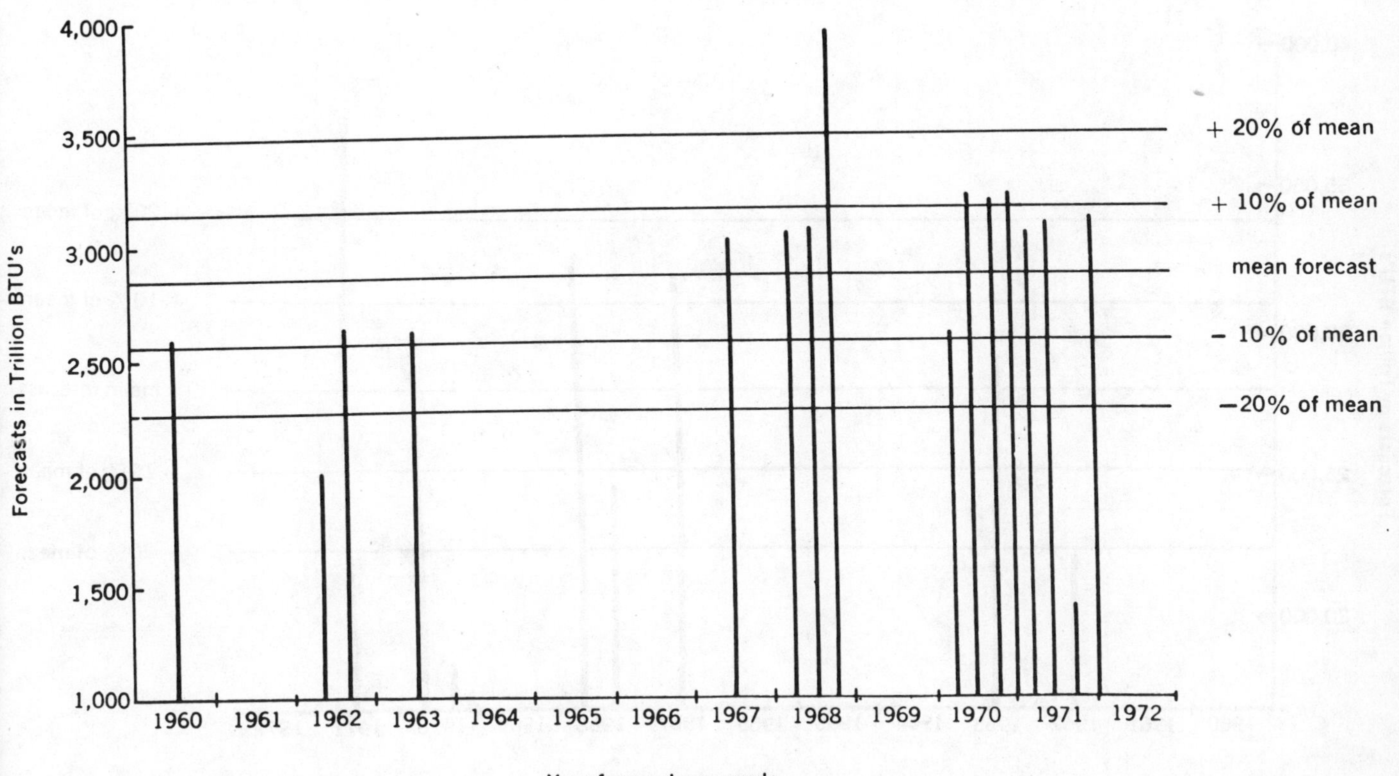

FIGURE 11.—FORECASTS OF UNITED STATES UTILITY ELECTRIC POWER GENERATION FOR THE YEAR 1980

12,000
11,000
10,000
9,000
8,000
7,000
6,000
Forecasts in Trillion BTU's
+ 20% of mean
+ 10% of mean
mean forecast
− 10% of mean
−20% of mean
1960 1961 1962 1963 1964 1965 1966 1967 1968 1969 1970 1971 1972
Year forecast was made

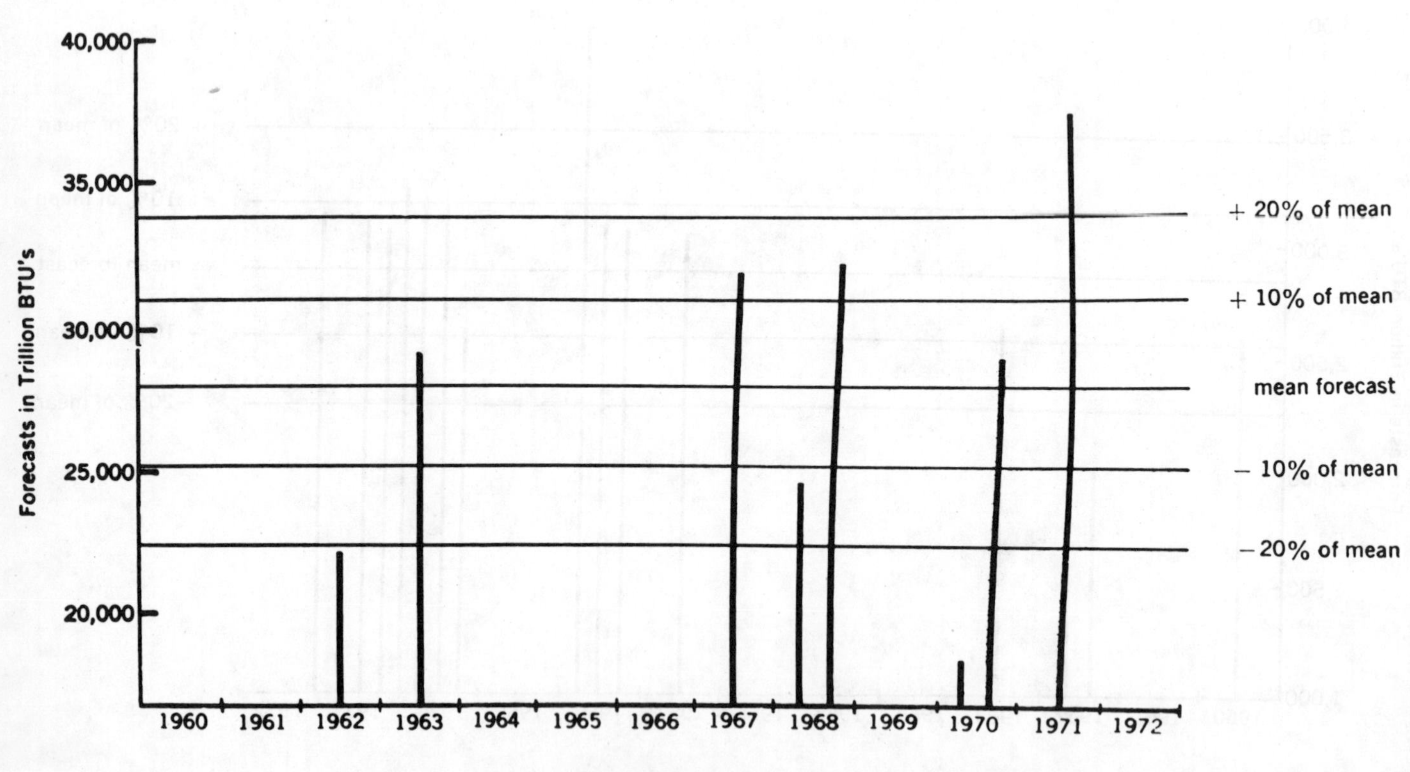

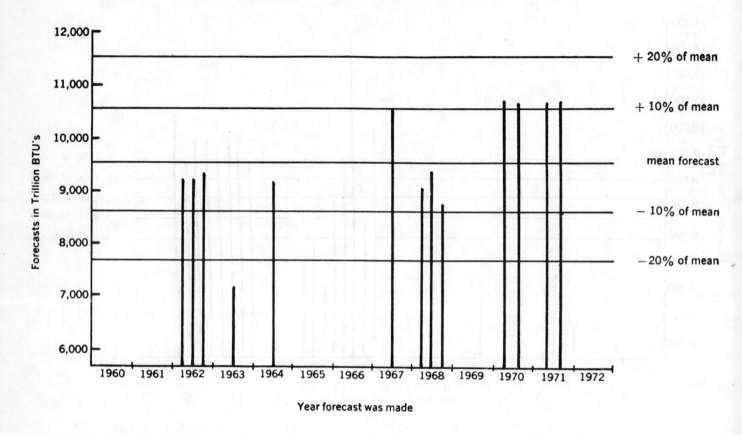

FIGURE 12.—FORECASTS OF UNITED STATES INDUSTRIAL ENERGY CONSUMPTION FOR THE YEAR 1980

FIGURE 13.—FORECASTS OF UNITED STATES HOUSEHOLD AND COMMERCIAL ENERGY CONSUMPTION FOR THE YEAR 1980

Forecasts in Trillion BTU's

40,000
35,000
30,000
25,000
20,000
17,000

+ 20% of mean
+ 10% of mean
mean forecast
− 10% of mean
−20% of mean

1960 1961 1962 1963 1964 1965 1966 1967 1968 1969 1970 1971 1972

Year forecast was made

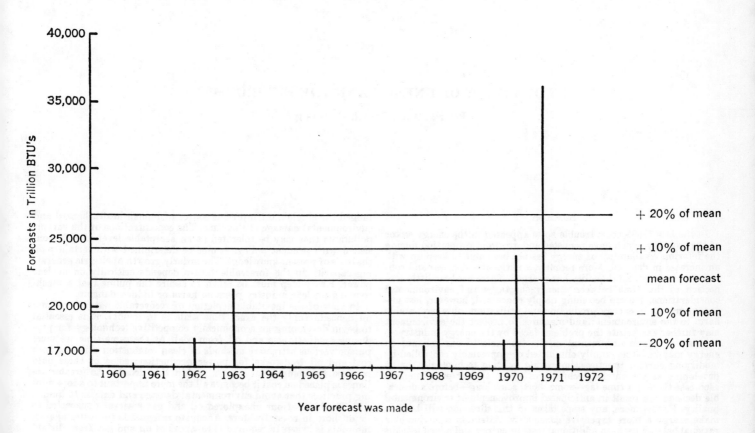

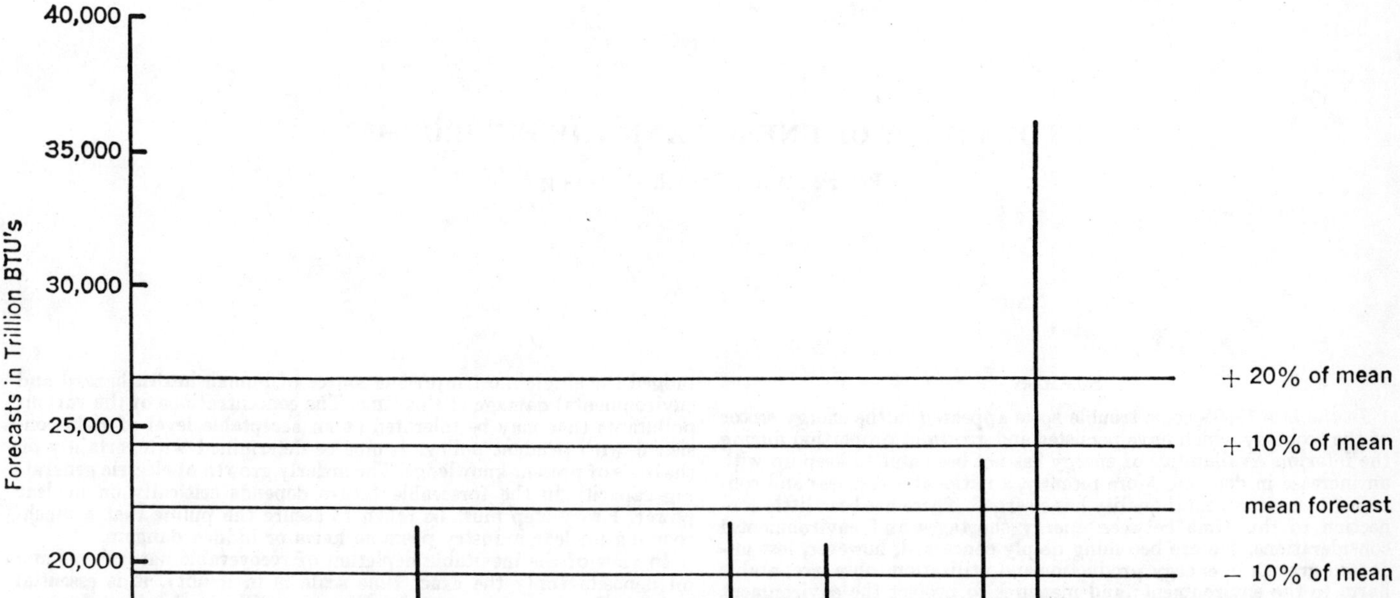

FIGURE 14.—FORECASTS OF UNITED STATES TRANSPORTATION ENERGY CONSUMPTION FOR THE YEAR 1980

Forecasts in Trillion BTU's

26,000
25,000
24,000
23,000
22,000
21,000
20,000
19,000
18,000
17,000

+ 20% of mean
+ 10% of mean
mean forecast
− 10% of mean
−20% of mean

1960 1961 1962 1963 1964 1965 1966 1967 1968 1969 1970 1971 1972

Year forecast was made

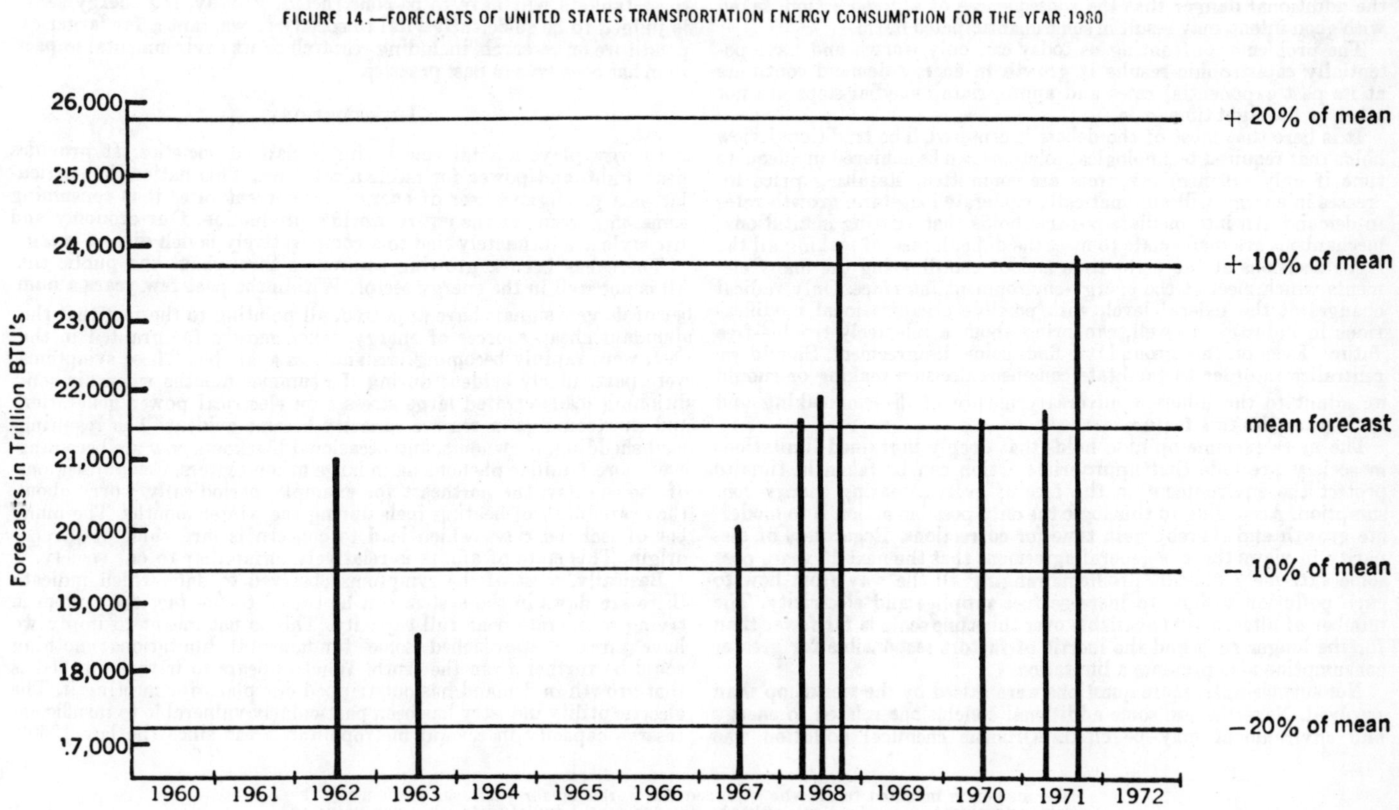

AN OVERVIEW OF ENERGY AND THE ENVIRONMENT

Peter L. Auer, Cornell University [1]

Summary

In the late 1960's some trouble spots appeared in the energy sector of the economy which have persisted and grown in proportion during the interim. Availability of energy has not been able to keep up with an increase in demand. More recently, a national awareness and concern for environmental quality has emerged. There has been little connection to this time between energy shortages and environmental considerations. We are becoming deeply concerned, however, lest unabated growth in energy production and utilization cause irreparable harm to the environment; and measures to protect the environment may further exacerbate the problems faced by the energy industry.

Many of the environmental insults traceable to our reliance on energy may be either totally eliminated or appreciably controlled by modifying current practices, improving existing technologies, or by developing as yet untried technologies which appear promising. In each case there is a time lag—some short, some long—before a desirable decision can result in anticipated improvements of environmental quality. Furthermore, any steps taken in this direction will tend to make energy a more expensive commodity. Attempts to arrive at a rational balance between additional costs in energy and added benefits from an improved environment are hampered by a lack of sufficient quantitative information and scientific data. Because of this there is the additional danger that the consequences of a given action, taken with good intent, may result in some unanticipated harm.

The problems confronting us today can only worsen and have potentially catastrophic results if growth in energy demand continues at its past exponential rates and appropriate remedial steps are not taken on the right time scale.

It is here that most of the debate is centered. The traditional view holds that required technological solutions can be achieved in adequate time if only sufficient resources are committed. Resulting price increases in energy will automatically moderate long-term growth rates in demand. An intermediate position holds that existing institutional mechanisms are inadequate to meet the difficult task of making all the right decisions at the right time and of coordinating the many elements which meet at the energy-environment interface. Only radical changes at the federal level, with possible organizational modifications in industry, as well, can bring about a relatively trouble-free future. Even on this ground one finds some disagreement. Should we centralize in order to facilitate consensus decision making or should we admit to the inherent adversary nature of decisionmaking and institutionalize this feature?

The more extreme opinion holds that deeply ingrained limitations in society preclude that appropriate action can be taken in time to protect the environment in the face of ever-increasing energy consumption. According to this logic the only possible action is to moderate growth and thereby gain time for corrections. Regardless of disparity in views there is general agreement that the next 15 years pose some extremely difficult problems ranging all the way from how to curb pollution to how to increase fuel supplies and electricity. The number of alternatives available over this time scale is far fewer than for the longer run; and the inertia of factors responsible for greater consumption also presents a limitation.

Not surprisingly, more questions were raised by the workshop than resolved. Nevertheless, some additional conclusions related to energy and environment may be cited. Airborne chemical pollution was judged the single most injurious source of human health hazard and environmental damage at this time. The concentrations of the various pollutants that may be tolerated at an acceptable level of risk, consistent with prudent policy, cannot be determined with certainly on the basis of present knowledge. The orderly growth of electric generating capacity in the forseeable future depends critically on nuclear power. Every step must be taken to assure the public that a mushrooming nuclear industry poses no harm or hidden dangers.

In view of the inevitable depletion of recoverable natural gas and oil deposits (only the exact time scale is in doubt), it is essential to begin developing an economically competitive technology for producing synthetic gas and oil from coal. At the same time we must pursue various attractive methods of clean combustion utilizing both coal and oil feedstocks for power production; some of these hold promise for commercial success within this decade. As greater reliance is placed on coal it becomes all the more important to adopt mining practices that avoid environmental damage and danger to human life. The bulk of our unexplored oil and gas reserve is suspected to lie offshore in coastal waters. Adequate safeguards to insure against blowouts is urgently required. Transport of oil and gas from distant sources is best accomplished, from both the economic and environmental viewpoint, in large tankers. Appropriate port facilities for these tankers will have to be constructed. Finally, the energy sector is judged to be sufficiently vital to society to warrant a far larger expenditure on research, including research on its environmental impact, than has been true in past practice.

Introduction

Energy plays a vital role in industrialized societies. It provides heat, light, and power for mechanical work. This nation, in particular, is a prodigious user of energy. At current rates it is consuming some 35 percent of the entire world's production. Our economy and life style are intimately tied to a comparatively lavish diet of energy.

There has been a growing awareness throughout the public that all is not well in the energy sector. Within the past few years a number of danger signals have appeared, all pointing to the fact that the abundant cheap sources of energy, taken mostly for granted in the past, were rapidly becoming less and less available. These symptoms were particularly evident during the summer months when air conditioning loads created large stresses on electrical power generation and shortcomings in reserve capacity became evident. The resulting load shedding, brownouts, and occasional blackouts now are becoming ever more familiar phenomena in large urban centers. Certain sections of the country, the northeast for example, periodically worry about the availability of heating fuels during the winter months. The number of isolated cases which lead to concern is large and of diverse origin. This state of affairs is relatively unfamiliar to our society.

Basically, most of the symptoms observed to date which indicate there are flaws in the system can be traced to the fact the system is trying to operate near full capacity. This is not meant to imply we have already approached some fundamental limitations; nothing could be further from the truth. What appears to have happened is that growth in demand has outstripped our plans for meeting it. The electric utility industry has been particularly vulnerable to insufficient reserve capacity in several metropolitan areas since the late 1960's.

[1] The author is indebted to all who attended the workshop for whatever wisdom may be contained in this section. Particular thanks are due to the individuals who took pains to correct errors in draft. Any misstatements that remain belong to the author exclusively.

Reprinted with permission from *Cornell Workshop on Energy and Environment*, Senate Committee on Interior Affairs, no. 92-23, Feb. 1972.

Unanticipated delays in the construction of nuclear plants due to technical difficulties and labor problems plus isolated cases of system unreliability (for example, "Big Allis" in New York City) were the major culprits. When it appeared that greater reliance had to be placed on coal-fired plants for generating electricity, the apparent shortage of fuel developed because mine operators had not planned on expanding production, were in fact cutting back, and coal cars were in short supply. The threat of heating oil shortages on the East Coast in the recent past was caused more by the lack of shipping than any other factor.

Natural gas is being rationed to some degree at this time: the underlying reason for the shortage, however, appears to be that it is difficult to attract the investment capital necessary for opening up additional production fields and the unregulated intrastate market is far more lucrative than trying to do business under Federal Power Commission (FPC) rules.

Recently a new element entered the energy arena—the rapidly growing concern for environmental quality. While, in a few instances, environmental issues have already affected the energy sector (the delay of "Storm King" is an example, new automobiles that are slightly more expensive and costly to operate because of pollution control devices another), the full consequences of environmental constraints on energy production and use have yet to manifest themselves. There already is ample evidence that conflict of serious proportions can develop in the course of attempting to satisfy both a growing demand for energy and an increasing desire to protect and improve our environment. Indeed, some will question whether these two goals are compatible and many more will contest the wisdom of steps already taken in this direction.

It only takes dramatic events on the scale of the Torrey Canyon oil spill, a coal mine disaster, or the rapture of earth-filled dams from coal waste to serve as stark reminders that energy production and utilization may cause very serious harm to human life and to the environment. Environmental concern took on revolutionary proportions with the enactment of the National Environmental Protection Act (NEPA) of 1969. Many consider this act to be an environmental Bill of Rights. The creation of the Council on Environmental Quality (CEQ) by this act and the creation of the Environmental Protection Agency (EPA) by executive reorganization at about the same time provided a new marshaling ground for environmental matters within the federal executive. Since adoption of NEPA and the establishment of EPA the conduct of business within the energy sector is no longer the same; environmental impact statements and air and water quality standards with the threat of federal intervention have changed many of the old ground rules. NEPA has served to increase the influence of the courts on those responsible for energy policy and planning, a symptom that the principal parties involved are not yet familiar enough with the new way of doing business.

It was on this timely note that the Workshop on Energy and the Environment was organized by Cornell University under National Science Foundation (NSF) sponsorship. While the subjects of energy and environment have led to a proliferation of studies and subjects hardly witnessed before, it was felt desirable now to re-examine carefully the critical interface between energy and environment. A group of leading authorities were gathered together for the purpose of obtaining a balanced, rational view of this matter.

In order to examine the environmental impact of energy production and utilization, the workshop set out to identify the central problems and issues in the realm of energy use and its environmental consequences, ordering them according to priority, and to determine the extent of agreement on these problems and their priorities.

Four panels were formed to address the following set of tasks; (1) weigh the probable social, environmental, and biological costs involved in the various options available for satisfying energy needs, (2) assess the technological options now known or foreseeable for dealing with these problems, (3) gauge the probable course of energy demand and supply over the next several decades, and evaluate the prediction systems now used for this purpose, and (4) examine the existing institutional structures for dealing with energy and environmental problems, to test the fit between newly identified problems and existing mechanisms, and suggest new mechanisms which might require legislative or administrative action.

The four panels which constituted the workshop were provided with opportunities for exchange of views and information during plenary sessions, as well as on an informal basis.

It was the intent of both the organizers and participants of the workshop that the resulting document would help illuminate the pressing energy/environment issue and provide information that would be useful to appropriate committees in both houses of Congress, appropriate agencies of the executive branch, and within the National Science Foundation itself.

The organization of this report is as follows. The first section is an attempt to present an overview of the workshop. It is not meant to represent the consensus nor faithfully reproduce collective opinions. Rather it is the writer's interpretation of what he heard or read. This is followed by a series of reports from each of the panels where information in greater detail may be found. Thus, this report as a whole should not be taken as the viewpoint of the workshop or even that of a majority of those participating; it is a collection of viewpoints, some individual and some representing a group.

General Observations

IS THERE A RATIONAL BASIS FOR PLANNING?

Most of us are familiar with specific arguments advanced by environmental protectionists. Perhaps this view can be best illustrated by the eloquent arguments contained in a recent column appearing in *The New York Times* which is reproduced in its entirety below.

Manuelito Canyon

By TOM WICKER

MANUELITO CANYON, N. M., Feb. 27—Atop one of the rocky promontories that thrust rough fingers into Manuelito Canyon stands the ruin of what may once have been the watchtower of primitive tribesmen. It was built long before the ancestors of the Navajos, who live here now, moved down from the Northwest.

Across the canyon and to the south, clearly visible from the ruins, a cave opens into the rock cliff. The Navajo residents of Manuelito like to use the cave for outdoor feasting around huge fires of piñon wood. Navajo youngsters scratch their initials beside pictographs carved in the rock, depicting hunters on the trail of deer; but these ancient drawings, in the light and shadow of the fires, look much as they must have to the long-forgotten men who carved them there centuries ago.

If the Tucson Gas and Electric Company has its way, the timeless aspect of Manuelito Canyon will disappear. From the watchtower ruins, a number of 100-foot steel towers will be visible, as they carry an electric power line south from the San Juan generating station to run the air-conditioners and can-openers of Tucson. All along the rim of the canyon, crossing it at one point, these towers will rise gaunt and unsightly against the sky, intruding at last the ugly hand of the twentieth century on an area that so far has virtually escaped it.

It is not as if the towers could not be built elsewhere, along an existing power line corridor; but that would cost the electric company more money. The construction can be stopped, apparently, only if the Native American Rights Fund, D.N.A. (Navajo legal services) and other groups prove that, as they contend, T.G.&E. got the Manuelito right-of-way by the same

IN THE NATION

fraud and deception so often practiced on American Indians.

At that, the impending despoliation of Manuelito Canyon is only one small element of what may well be the biggest and most destructive environmental atrocity now afoot on this raped and plundered continent. That is a generous description of the plans by W.E.S.T., a consortium of Southwest power companies, with the collusion and assistance of the Department of the Interior, to construct six giant coal-fired generating stations to meet the so-called "demand" for electric power in the Southwestern states.

No short article can describe all the shocking elements of this scheme, which already has put into operation two of the projected plants. One, Four Corners, near Farmington, N. M., is considered the worst air polluter in the United States, fouling the desert air over 10,000 square miles from Albuquerque far into Arizona and Colorado. It has only one-third of the pollution-control equipment promised by Arizona Public Service, which runs it.

The other operating plant, Mohave, near Bullhead City, Ariz., has electrostatic pollutant cleaners supposed to operate at 97 per cent efficiency. Even so, studies show that each day the plant emits thirty tons of fly ash, 157 tons of sulphur oxides and a hundred or more tons of oxides of nitrogen, much of all this over the Lake Mead National Recreational Area.

A third plant, Navajo, is being built near the Glen Canyon Dam, which also was built to provide power, and which obliterated the Glen Canyon of the Colorado River; now the Navajo plant will ruin the Lake Powell Recreational Area, which the dam made possible. But if Navajo doesn't make the area uninhabitable, a projected fourth plant at Kaiparowits, on the other side of the lake, surely will; it will have a capacity of 5,000 megawatts, twice that of Navajo.

With the other projected plants, these installations will wreak environmental havoc on an almost unimaginable scale. All will be fired by coal strip-mined in the area, a disaster in itself, particularly for the Hopi and Navajo lands on Black Mesa Arizona. The water table at Black Mesa also may be ruined because the Peabody Coal Company extracts up to 4,500 gallons a minute to mix with pulverized coal, which it then pumps through a pipeline 273 miles to the Mohave plant.

Runoff from the mining operations may ruin the Hopi farm lands, on which their ancient way of life depends. Already, the mining trenches and service roads have wrecked much of Black Mesa. Air pollution from the six plants will seriously affect the atmosphere over six national parks, 28 national monuments, two national recreation areas, scores of state parks and historic landmarks, and 39 Indian reservations—but not most of the cities for which the power is to be produced.

But at least five major lawsuits are pending against the grandiose W.E.S.T. scheme to produce 36,000 megawatts —three times the capacity of T.V.A.— by 1985. Environmental organizations are vigorously opposed. Some newspapers and politicans are at last aware, as are the Hopis and Navajos, of the wanton wreckage planned for them in the name of progress and power. The Senate Interior Committee is belatedly stirring. It may be too late to avoid the nightmare altogether, but sufficient public interest can yet save Manuelito Canyon and the essential character of the Southwest.

While the piece is specifically concerned with the production of electricity by fossil fuel-fired plants, most of the dramatic points involving environmental protection on the one hand and the desire for increased energy production on the other are illustrated here. The fundamental dilemma posed by these issues is that many of the objections raised by Wicker and the segment of the public for whom he speaks can be either eliminated or greatly alleviated at a cost. The economist refers to this process as internalization of externalities. The postulated cost of the external damage caused in the process of supplying energy (or any other commodity) can be eliminated at the price of increasing the real cost of energy production; presumably most of the additional cost is passed on to the consumer. As a result of such steps certain benefits accrue to individuals directly affected by the threatened damage and to society at large. The question then is to what extent, if any, do we attempt to balance benefit against cost, assuming reasonably credible values can be assigned to both. In fact let us ask whether this is a promising start toward rational planning, bearing in mind that some groups may wish to have certain environmental features protected at any cost.

The notion that environment versus energy controversies can be settled purely on a quantitative basis by means of benefit-cost analyses is easily challengeable. What price do we attach to aesthetic values? The objection to overhead transmission lines is largely aesthetic. Placing large portions underground would remove the objection but the cost with present technology is prohibitive. Approximately 20 percent [2] of the cost of electricity to residential customers is represented by the cost of transmission. Depending on what voltage and power capacity is involved, underground transmission is some 10 to 30 times more expensive than overhead. Thus, burying half the overhead lines and taking the median value of a 20-fold increase in cost would lead to nearly tripling the price of electricity. Yet, the aesthetic insult of overhead transmission lines can have direct economic impact. At least this is the argument of residents in the Catskill communities of upstate New York who fear that their tourist trade will be adversely affected by the overhead lines carrying electricity from the local pumped storage units of the New York Power Authority to the electricity hungry communities downstate.

With increasing frequency the energy needs of a community can have severe impact on the environment of regions far removed. Pollution and environmental damage are exportable commodities. Therefore, whose costs and whose benefits are we to balance? Obviously, it is difficult to try and balance the ledger sheets unless large enough segments of society are involved. Even if we take our national population and apply rules under the best of democratic principles, the problem is not contained necessarily. Part of the Alaska pipeline controversy involves fears of oil spillage from tankers steaming between Valdez and our West Coast ports. The endangered shorelines are Canadian as well as American. Tall smoke stacks are effective in dispersing SO_2 within the atmosphere and reducing ground-level concentrations. This is a practice which has been tried in England with happy results. But the prevailing wind blows from west to east and the Swedes may not be as happy with this experiment.

Environmental impact may not only transcend geographical boundaries but also time. Certain acts committed by our generation may have delayed consequences for later ones. Depletion of natural resources or inducing genetic damage by man-made radiation are but two examples. These are sources of hidden costs, so to speak; we do not bear them, but those who follow will. Some of today's actions can impose a continuing burden on future generations. As we enter the nuclear age of energy production, for example, we began to accumulate radioactive waste which in some instances must be safely stored for centuries and in some cases for thousands of years. While we are assured that this is technologically feasible and can be done with due regard to safety at an acceptable cost, we nevertheless are requiring future generations to accept the active safekeeping of these wastes until such time when technology finds a more expeditious way for their disposal.

In any attempt to assign costs, it is important to distinguish between environmental damage or consequences that are permanent and those that are not. The unsightly overhead transmission lines may be removed eventually and placed underground once it becomes economically acceptable—just as elevated trains have been replaced by subways in the cities. For the time being perhaps the best solution is to define energy corridors as rights of way which are designed to reach a compromise between the least expensive route and the least offensive. Of far greater impact is the quality of our air. Some long-lasting health effects have already been incurred as a result of air pollution. In extreme cases local pollution levels have reached proportions which represent a danger to human life. But the quality of our air, even if it has worsened in the course of time, has not been irreparably harmed, nor is there now a clear and present danger according to many experts. There is general agreement that air quality must be improved and preserved. If there is disagreement, it is largely on how rapidly we should move and at what sacrifice.

Similar remarks may be made with regard to water quality, although one suspects the time scale of adequate water quality restoration may be long and our understanding of possible irreversible changes in complex aquatic ecological systems is still imperfect. Energy production and utilization, however, is only one of many contributors to water pollution, while it is the principal source of air pollution. Matters are qualitatively different where land is concerned. In modern times each generation has left its indelible mark on the land; whether this is to the benefit or not of succeeding generations is a matter of opinion.

There are already clear indications that as natural gas and liquid hydrocarbon deposits become used up, an increasing dependence will be placed on our far more abundant coal resources. There are vast deposits of coal essentially untapped in the West. Attempts to utilize them have already led to controversy, as described above in Wicker's column. Coal is mined either on the surface or underground. Surface mining leaves scars on the land which make it unuseable unless adequate care is taken to reconstitute it. It has been shown elsewhere, primarily in Germany, that procedures exist whereby successful reclamation of strip-mined land can be achieved providing proper precautions are taken and reclamation steps follow soon enough after stripping. (Good practice requires top soil segregation; open wounds remain as such for no longer than a few weeks at most; restoration follow mining by only a few hundred feet; protection of reclaimed area be provided for up to several years until local flora is fully established). Reclamation of steep grades (usually in excess of 13 percent) stripped by contour cuts in more difficult and complete restoration is subject to some doubt. Cost estimates for adequate reclamation depend on terrain and may vary from $1,000 to $10,000 per acre, with the mean adding $0.25 to $0.50 to the per ton cost of coal in the East and somewhat less in the West.*

Underground mining of coal also entails environmental impact, principally subsidence, the creation of acid mine drainage and solid waste disposal. The classic room and pillar technique employed by United States coal miners does not lend itself to facile control of subsidence. Considerable research is warranted in this area, as described elsewhere in this report, and also in the area of acid mine waste elimination. As vast new tracts of land are mined for coal, it behooves us to employ environmentally acceptable practices in the course of extraction lest irreversible harm be left behind. Policy makers will have to decide what measures of protection should be adopted and at what additional cost to the consumer.

Any attempt to use rational arguments to arrive at energy-environment trade-offs will require a credible data base and wide dissemination of information (public education). There is considerable room for improvement in regard to these matters. While it is generally agreed that at present public health is in greatest danger from air pollution due to SO_2, NO_x, oxidants, hydrocarbon, particulates, and CO contamination, relatively little is known about the epidemiology of these agents in quantitative terms. EPA has projected an annual savings of $14 billion to the public once 1975–1976 ambient air standards are met. By contrast a vast amount of research has gone into the study of radiation effects on humans. This imbalance in knowledge must be redressed by expanding research on chemical pollutants. There are serious doubts in many minds today regarding the scientific basis for deriving air and water quality criteria which serve as a basis for arriving at standards.

HOW IS POLICY MADE?

Steps taken to date to improve environmental quality have sometimes been ruled by expediency and political considerations. The results in some cases may not be in the public's interest. An oft-cited example occurs in the state of California where vehicle emission standards aimed at limiting the discharge of CO and hydrocarbons were recently imposed. As a result engines were forced to operate with leaner fuel mixtures leading to hotter combustion and an increase in NO_x emission. Smog irritation has increased noticeably at the same time. While there seems to be a cause-and-effect relationship here, some groups question this conclusion. Similarly, the decision to tighten and move up federal standards on automobile emission to 1975–1976 has forced manufacturers to meet these requirements with existing engine designs with a variety of pollution control appendages. The resulting additional cost to the purchaser has been estimated at $350 on the average per vehicle and the decrease in performance could

[2] The conventional 10 percent has ben doubled for reasons given by Peterson in the workshop.

*In 1968 the cost of coal at the mine source ranged from $1.42 to $9.02 per ton.

impose a fuel penalty during operation as high as 35 percent. Air quality improvement as a result of these actions may not be significant because of the longevity of old cars, the questionable effectiveness of local enforcement procedures yet to be defined, and the fact that many unabated industrial and residential sources will still be in existence.

On the other hand prospects for a much improved engine in the 1980 time frame is considered high; the gas turbine is currently the leading candidate for the luxury end of automotive vehicles, while the Wankel engine and/or programed combustion engine looks very attractive for the middle and smaller size automobiles. Rankine cycle engines with external combustion are also in contention. It is more than likely these developments would have lagged had the manufacturers not been prodded into action by the threat of government action. The imposition of standards can be a very effective carrot and stick instrument for stimulating environmental improvement. Hasty or arbitrary use of this tactic, however, may lead to undesired results and prove counterproductive.[3]

It is possible to speculate at length along these lines; perhaps the difficult situation facing the electric utilities is worth mentioning briefly. Reserve capacity of several utilities serving metropolitan areas has decreased to alarming levels. Good practice calls for this figure to vary between 15 percent to 20 percent. In some isolated cases the figure is around 5 percent or less. The individual utility's reserve may be augmented if it interconnected with a power pool; this is not always the case. Some of the sources of delay in constructing new plants have already been mentioned; but the role of environmental protection is becoming increasingly apparent.

Part of the difficulty is recognized to be institutional in that a collection of agencies at the federal, state, and local level are involved in certification of nuclear plants. The necessity of independent review by agencies of concern has been underscored by the court's interpretation of NEPA in the *Calvert Cliffs* decision. There are several measures intended to expedite power plant siting procedures now in Congress.

Hopefully, current difficulties regarding nuclear plants will soon pass and measures will be taken which will assure the public that these reactors are properly designed and can be safely operated. In the interim prospects for relief are clouded, since fossil-fired steam plants are endangered by proposed emission and ambient air quality standards. Clean fuel is in short supply and the existing technology of stack gas clean up has yet to be tested on a large scale.

Another cautionary note may be added. If restrictive discharge legislation (for example, Senate Bill 2770) becomes law, it may well prompt utilities to abandon once-through cooling of steam plants and switch to cooling towers or ponds. If this is done precipitously, certain localities may suffer from adverse climatic changes brought on by fogging and icing conditions. Any rush to adopt new technologies in imperfect states of development because of political or any other pressures can be potentially harmful. Finally, the preference of utilities to place generating plants near load centers may have to be forsaken as acceptable sites become less and less available. This brings us back full circle to the earlier discussion on economics and aesthetics of long-range transmission.

The nightmare of the environmentalist is that the horrible mathematical consequences of doubling rates in intervals of anywhere from 8 to 15 years will steadily take place decade after decade until we suddenly become asphyxiated or sink beneath our piles of garbage. By contrast those who are more tradition bound will argue that technological advances and the natural economic forces of the market place will serve to abate and ameliorate. Thus, we are promised that environmentally acceptable energy production will be provided by improved technology; and the increased price in energy due to any added cost and dwindling supplies will serve to damp the fires of growth.

One moment's reflection on the composition of the energy industry—consisting of coal mine operators, oil producers, gas producers, uranium mine operators, electric utilities, and the allied electrical and transportation industries, some regulated and some not, each answerable to one or two or perhaps a dozen different agencies at the federal, state, and local level, each motivated by its own relatively narrowly defined objectives—should convince the casual observer that considerable skill and determination will be required to avert the nightmare and find the right path to the more reassuring view before time gives out.

It should be recognized that there are significant time scales in any major technological development program, that before it may start there must be an adequate research and development base on which to build, that after it is completed there is a time lag between planning and completion of plant constructions, that there is even a lag between finding new deposits of resources and bringing them to market. We may well ask at this juncture whether the institutional mechanisms available at present are adequate to the complex issues facing us as we try to chart a course between Scylla and Charybdis.

To some it appears that the conflict between environmental interests and growing energy demands is a match between unequal opponents. The industries serving the energy sector are well-organized into several interest groups with ample resources and a long tradition of doing business with appropriate offices of government at all levels. The fortunes of the domestic oil industry are strongly affected by oil import restrictions set by the President and coordinated by the Office of Emergency Preparedness, by leasing of public lands under the charge of the Department of Interior, by depletion allowances set by the Congress, by production quotas set at state levels and coordination through Interstate Compacts. The interstate transport of natural gas is regulated in price by the Federal Power Commission (FPC) and everyone must be familiar by now with the growing debate between the gas industry and the FPC over the relationship of price incentive and resource availability.

Similarly, the coal industry is influenced by the health and safety regulatory authority vested in the Department of Interior, which, in addition, must also act on questions affecting any form of energy transport where public lands are in question. Perhaps the electric utility industry best exemplifies the complexity of energy policy and its administration, being responsible to the FPC, the Corps of Engineers, to the AEC when nuclear matters are at issue, state regulatory commissions, and to a variety of additional state and municipal agencies. The EPA with its standards-setting role brings a new element to this array. While not meant to be an exhaustive listing, the above description of dispersed, multiple sources of regulatory authority often find themselves particularly responsive to the very groups they must regulate.

By contrast environmental interests are highly fragmented with very limited resources. Their tactic is to enlist public support through the use of the media, organization of citizen action groups, intervention in administrative proceedings, and selected use of the judicial process. The recent course of legislative action has aided their cause considerably. The result has often been delay and apparent obstruction. Yet, the environmentalists will argue this is necessary and in the long run productive. Thus, for example, delays in the construction of the Trans Alaska Pipeline will assure that it is better designed and less likely to cause ecological havoc; similarly, their outcry on the potential failure of emergency core cooling systems has forced the AEC to make a careful review to assure that major nuclear disasters are truly as unlikely as had been predicted.

Even if the system of confrontations works in the manner expected by its practitioners, one would hope that the unnecessary delays and temporary hardships imposed on society could be eliminated through better organization, objective reasoning, and consensus planning that serves the collective interest of the public at large. Can this be accomplished under existing institutions? Some feel that the basic mechanics of political decision making it characterized by adversary proceedings where special interest forces lobby on their own behalf to influence decisions. If this is accepted as a fact of life, we might improve matters by institutionalizing inherently conflicting forces. For example, advocacy and promotional roles might be separated from regulatory roles.

It is often argued that the AEC cannot perform both these roles simultaneously without bias in the promotion and regulation of the nuclear energy industry. A similar objection might be raised regarding the structure of EPA which must both set standards and promote means for achieving them. Similar internal conflicts exist within the Department of Interior as it tries to balance its licensing and regulatory functions against its promotional role in coal, oil, and gas.

One proposed model would collect all the regulatory functions scattered throughout different offices and agencies of the federal government and place them under a super National Energy Agency, absorbing the Federal Power Commission in the course. This organization would be given broad powers to assure that society's energy needs are adequately met in a manner consistent with environmental safeguards. The super agency in this model is not to be confused with the proposed creation of a Department of Natural Resources, since the advocacy and promotional roles relating to the various segments of the energy sector, the necessary long-range research and development that accompanies this function, would be left in place and no particular need is seen to collect these functions under a single roof.

The argument that you cannot regulate without being informed would be answered by giving the super agency a strong research staff of its own. A logical extension of this model calls for a strong spokesman representing public environmental concern—a kind of environmental ombudsman. By giving the potential intervenors a stronger voice and guaranteed resources, a more orderly process may be achieved. Such a process may go a long way towards consensus decision making ultimately. Were the ombudsman office to be created, it would be best to insulate it as much as possible from interest groups

[3] The Cumulative Regulatory Effects on the Costs of Automotive Transportation (RECAT) Final Report of the Ad Hoc Committee, Feb. 28, 1972, prepared for OST appeared after the workshop.

and executive pressure and to model it somewhat along the lines of the General Accounting Office.

To some workshop participants outside of the Institutional Panel these proposed changes seemed strong medicine and even the wrong kind. They argued that decision making should be highly centralized and based on consensus alone. This presumes some measure of pervasive logic exists—a contention the opposing camp does not accept. In any event, certain tasks growing out of the present concern with energy and the environment will have to be addressed by prompt changes. The amount of money going into energy-related research and development will have to increase substantially immediately. With the exception of nuclear energy, federal support in this area has been negligible up through 1970. If the reader is not already convinced, it is hoped that a further review of this document will show that massive funds are required for research and development and that the magnitudes in question are such that it is incumbent upon the federal government to take a leadership role in stimulating the financing and directing of the effort. Consequently, some degree of coordination and setting of priorities will be required within the executive at a high-enough level to have effective influence on the Congress as well as on the Office of Management and Budget.

A better-informed public and body of decision makers are desirable regardless of what institutional mechanisms are adopted. But there is a scarcity of reliable statistical data and relatively little effort committed to its collection. What is needed is a version of the Bureau of Labor Statistics which would carefully collect and disseminate the agreed upon indices that determine environmental quality. It is not clear that either CEQ or EPA is equipped to perform this function adequately at the present. But, without this information how can we determine whether steps taken to improve environmental quality are having their desired effect? The only problem then would be agreeing on how to determine environmental quality. Most experts believe that knowledge in this area must be greatly expanded through more research, particularly in the area of human health effects, long-term ecological changes, and environmental impact. The call for more research should not be interpreted as justification for delay in reducing the adverse environmental impact inflicted by the energy sector. Objective planning will not only require scientific information regarding cause and effect in environmental matters but also improved analysis of energy demand and resource availability. As for the latter, the Geological Survey provides detailed and comprehensive information on coal; nothing of comparable reliability and detail exists with respect to oil and natural gas, while the uranium and thorium figures furnished by the AEC may also be incomplete. Hopefully, a greater government involvement in research on exploration techniques and mapping can help remove present-day uncertainties.

There are both short- and long-term problems to be faced. Among the short-term problems are finding ways to increase supplies of natural gas and low sulfur fuel, both of which can provide environmentally acceptable energy sources. Efforts to increase domestic finding rates of new deposits is prejudiced by the fact most undiscovered deposits are expected to be on the continental shelf and the oil blowout off Santa Barbara has temporarily curbed enthusiasm for offshore drilling. Import of suitable fuels from outside the Western Hemisphere is encumbered with international political and national security considerations. These and closely related problems, local power shortages for example, fall into a crisis management category and should be under the constant purview of a highly placed office within the federal executive branch.

By contrast there are a set of long-range problems which should be addressed continuously in an orderly manner. Among these is the immediate necessity of starting to convert to coal as the principal fossil fuel base for future energy requirements. (The President's energy message addresses two aspects of this problem, the conversion of coal to synthetic pipeline gas and improving methods of reducing sulfur oxide emissions from high sulfur coals. For purposes of finding ways to burn coal cleanly in order to generate electricity, a number of other approaches have high probability of near-term success and should be pursued vigorously.) As another example, it is one thing to set quality standards on the basis of limited information in the political arena and quite another to pursue a well-structured research effort aimed at a more complete understanding of environmental quality and the underlying cause-effect relationship. An effort should be made to insulate organizations charged with long-term solutions from the day-to-day problems of crises management.

ARE THERE LIMITS TO GROWTH?

Perhaps a significant question to ask at this point is can we support anticipated growth in energy demand without further exacerbation of the environment. The past trend is well-known: electricity consumption has grown on the average between 7 percent to 8 percent annually over the last decade, natural gas at 6 percent, oil at 4 percent, coal at approximately 3 percent, and total energy at better than 4 percent; by comparison our population has increased a little over 1 percent per year. With the exception of coal, whose consumption has been decreasing until the 1960's, all other forms of energy have shown long-term trends of exponential growth in consumption. Within the present limitations of knowledge predictions of future growth patterns in demand cannot be made with any certainty and must be qualified. A fuller discussion of this matter will be found in later sections of the report. Nevertheless, one may anticipate that continued population growth, expansion of the economy as measured in constant dollars per capita, the corollary of increased affluence and redistribution of wealth along with increased productivity all generate pressure for greater demand.

Surely, as limited resources of energy become exhausted, the growth pattern must by necessity attenuate sharply. But in principle our resources are not limited. Both the sun and the promise of nuclear power by means of a breeder technology or controlled fusion represent virtually inexhaustible means for supplying energy. If energy were sufficiently abundant and cheap, we could begin to harvest the oceans and turn to other presently uneconomic deposits to replenish the rapidly dwindling supplies of certain mineral ores. Needless to say, these are visionary statements in the face of hard practical realities.

Roughly three-fourths of today's energy consumption in the United States is supplied by oil and natural gas. Of the two, natural gas has enjoyed the more rapid growth in demand. It has been relatively cheap and is now by far the most favorable fuel from an environmental viewpoint. The often cited shortage of natural gas is apparently real. Consumption and new discovery of natural gas were, respectively, 7 and 12 trillion cubic feet in 1950. Although finding rates of new reserves grew over the next decade-and-a-half, the consumption and discovery figures for 1971 were 23 and 12 trillion cubic feet, respectively. Consumption had tripled, while finding rates decreased suddenly. On the basis of figures supplied by the American Gas Association, the National Petroleum Council predicts an addition of 10 trillion cubic feet annually to reserves in this decade and less in the following one. Proved reserves stood at 289 trillion cubic feet in 1967 and are estimated at 250 trillion cubic feet at this time.

Both natural gas and oil represent resources which are being used up faster than we have replaced from domestic sources; but it would appear we will run out of natural gas well ahead of oil. Nevertheless, we cannot conclude that the energy sector is necessarily resource limited. Historically, fuel substitution has been evident throughout the age of industrialization. As recently as a century ago the bulk of our energy was supplied by burning wood. Coal overtook it during the latter part of the nineteenth century and hydrocarbon liquids overtook coal, in turn, by 1950. The popularity of natural gas is a relatively recent phenomenon. There is every indication that nuclear energy will play an increasingly prominant role in the immediate future.

Nuclear energy in a form compatible with environmental requirements, however, can satisfy growing demand only at some finite rate. Its principal contribution at this time is to the generation of electricity. The state of the art and resources of the industry are such that it would be difficult to supply much more than one-half of the electrical generating capacity anticipated for the year 2000 by nuclear means. Consequently, it is more than likely that still three-fourths of our total energy need will have to be met by fossil fuel sources through the end of the century, unless some unforeseen developments occur in the interim. The production of synthetic gas and oil from coal appears to be the technologically preferred solution for the interim future. Our coal reserves could be adequate for possibly 100 to 200 years if coal assumes a major burden of energy supply along with nuclear fuels and breeder technology.

Unfortunately, the technology of coal conversion to environmentally acceptable fuel is largely in an embryonic state. Optimistic estimates predict that clean fuels prepared from coal will begin to contribute noticeably to energy requirements by 1985. It appears that for most of the remainder of this century we face the difficult task of expanding the availability of natural gas and oil by vigorous domestic exploration, improved recovery techniques, expensive transport from such distant sources as Alaska, and the politically complicated shift to more foreign imports. Unless a number of steps are taken within appropriate time scales, fuel reserve limitations could lead to severe dislocations of the economy by the latter part of this century.

There is every reason to believe that anticipated developments will force the price of energy to increase. Considerable thought is now being given to understanding the detailed quantitative relationships between changes in price and demand. The art is still imperfect and the subject is quite complex; nevertheless, it seems safe to say that direct extrapolations based on past historical trends are fraught with danger if we believe sizeable perturbations will occur in the energy sector. On the basis of a general increase in fuel prices and the cost of conversion, one is tempted to conclude there should be a decreasing trend in the future rate of overall growth. It is equally likely, however, that inequalities will develop within the energy sector in that elec-

tricity will continue for some time to capture a greater share of the market, for example. While it consumes not quite a quarter of all primary energy, electric utilities provide less than 10 percent of what the ultimate energy customer uses (about two-thirds of the initial energy input is lost in the process of electricity conversion, transmission, and distribution). Most predictions project that by the year 2000, the 10 percent figures will rise to 20 to 25 percent.

Assuming technological advances on an appropriate time scale remove any obstacles due to incipient resource limitations or environmental incompatibility (granted that these may be highly controversial assumptions), and assuming the conventional dynamic forces of the economy along with prudent policy are allowed to balance demand against supply, are there still some further limitations to energy consumption? The answer is a qualified yes for the following reasons. Thus far we have passed over the issue of the availability of investment capital, nor do we intend to dwell on it. A significant portion of the total annual capital investment is already being used by the energy sector and it is more than likely that the proportion will have to increase. While not necessarily in the category of a fundamental limit, this matter bears watching.

On a different note, the combustion of fossil fuels unavoidably produces carbon dioxide (CO_2). While not considered harmful to humans in the concentrations produced, CO_2 release to the atmosphere has climatic implications through the so-called greenhouse effect. It partially traps the sun's radiation and consequently leads to an increase in the surface temperature of the earth. There is evidence that a gradual increase of CO_2 content in the atmosphere has occurred in this century; how much of this is due to human activity and how much from such natural events as volcanic erruptions is not known with certainty. Nor do we understand fully the balance between atmospheric CO_2 and what is dissolved in the oceans or tied up in carbonates. The problem is not regarded as being serious at this time but should certainly receive careful attention. Man's contribution to this problem will disappear eventually as he is forced to give up the use of fossil fuels.

Of more immediate concern is the disposal of waste heat. Energy conversion is inherently inefficient and in the process produces waste heat. Ultimately, all waste heat ends up in the atmosphere from where it is radiated to outer space. Should the amount of heat rejected because of energy utilization on earth become a significant fraction of the amount received from the sun, serious climatic changes would result. If waste heat were uniformly distributed over the globe, there would be little need to worry for a long time. But over the northeastern sections of the United States, which account for 40 percent of the nation's energy consumption, heat dissipation already amounts to 1 percent of the estimated solar absorption. It is expected to rise to 5 percent by the year 2000, a level comparable to average conditions in the vicinity of Los Angeles today. It is feared that high local levels of energy dissipation may lead to the formation of heat islands which, in turn, could induce cellular air flow patterns overhead. As a result, free mixing with the rest of the atmosphere could be hampered with significant microclimatic consequences. This could place an ultimate limit on the amount of energy conversion and consumption in a given size locality. With the aid of large-scale computer studies, good field data, and some experimentation, it should be possible to estimate within a reasonable time just how serious a problem heat islands may pose and to what extent it will be necessary to disperse large sources of waste heat.

The essential point, then, is that growth in supply of energy need not be limited because of some fundamental properties inherent to its production and utilization, but only because the complex series of decisions and actions required for orderly progress cannot be made on a timely basis within the framework of society as it exists. This is the essence of the view which holds growth must be checked and moderated in order to find time for corrections to materialize. It is a truism that the severity of energy and environment-related problems can only be aggravated by unbridled growth. The question, however, of whether growth must be moderated in order to safeguard against irreparable harm to the environment and to conserve precious resources belongs in the category on institutional mechanisms. Unless actions designed to correct an imbalance in supply and reduce an adverse environmental impact are taken at the proper time and pursued with sufficient determination, there is little recourse but to moderate growth.

Priorities

Environmental insult due to energy production and consumption is everywhere; some forms are more visible than others. A spectacular oil spill receives wide coverage from the news media and the public's attention becomes focused on miles of fouled beaches and countless oil stained birds. There is great indignation and damage is said to be in the millions. By comparison air pollution is estimated to cause \$10–\$20 billion annual damage to property and agriculture. While the oil spill is a local effect and directly harms a relatively small portion of the population, air pollution affects the entire population in varying degrees. In attacking the seemingly overwhelming environmental impact caused by society's demand for energy, some sense of priorities is desirable.

Air Pollution

There are many forms of environmental impact which may have several local consequences, but air pollution is judged the single most injurious source of damage and potential health hazard presently caused by the energy sector. Its principal source is the combustion of fossil fuels to provide heat or motive power. The chemical composition of air can be traced to still other activities of man and natural processes, but this discussion is limited to major contaminations accompanying energy conversion.

Of the various airborne chemical contaminants, the group sulfur oxide in combination with particulates and nitrogen oxides in combination with oxidants are considered the worst offenders. While there is no medical evidence at this time that sulfur oxide by itself at prevailing ambient air levels is harmful to humans, it is chemically active and harmful to materials and agriculture. In the presence of particulate matter transport of sulfur oxide to the lungs is a health hazard and becomes lethal in large enough concentrations. Sulfur oxide may be regarded as the number one offender in many sections of the country. Due to local meteorological peculiarities, however, some parts of the nation suffer more from nitrogen oxides. In combination with oxidants, the latter is responsible for the all-too-familiar photochemical smog of the Los Angeles basin.

Sulfur oxide emission has increased steadily in the past. The 1965 estimates were 26 million tons discharged annually, while the estimates in 1968 were 33 million tons. The major contributors are stationary power plants followed by industrial processors. The relatively smaller contribution from motor vehicles, residential space heating, and waste disposal can make a somewhat disproportionately higher contribution as observed on the ground because the discharge from these sources is less well dispersed.

Of the three commonly used fuels, natural gas, which consists primarily of methane, is essentially sulfur free. The sulfur content of oil varies according to its source: the high range may exceed 3 percent and the low range is 1 percent or less. Domestic sources are rather high in sulfur, but then virtually all of this supply is refined for more lucrative markets and is not available as a heat source. The bulk of our heating oils, whether for generating electricity or supplying space heat comes from Western Hemisphere imports and these sources also are high in sulfur, averaging about 2 percent concentrations. A principal source of naturally occurring low sulfur oil is Libya, which is not a significant supplier of American markets. Technology exists for reducing the sulfur content of oil up to certain levels at additional cost. It costs approximately \$1 per barrel to reduce 2.1 percent oil to 0.3 percent by hydrodesulfurization. This may represent a 20 percent to 30 percent price penalty. Whether desulfurization of oil can achieve low enough concentrations to meet proposed quality standards has yet to be fully determined, notwithstanding the attendant cost penalty and uncertainty of long-term availability.

The principal contributor to sulfur oxide emission is combustion of coal. As in the case of oil the sulfur content of coal varies from deposit to deposit, from about 0.5 percent to 6 percent in the United States. Much of the sulfur in coal is chemically fixed and not removable by physical processes. Low sulfur grades of coal are to be found mostly in the West, distant from potential customers and in a form which makes it difficult for existing coal burning plants to use. In order to meet proposed air quality standards some form of cleanup will have to be employed by coal burning plants and possibly also oil-fired plants. The major share of the burden will fall on the electric utility industry which, in terms of total tonnage, contributes over half the sulfur oxide emission to air.

There are various avenues of approach to freeing coal combustion of sulfur and other undesirable by-products (fly ash, for example). The one receiving most attention at this time has a number of unattractive features; this is stack gas cleanup. In this method, applicable equally to dirty coal or oil, the fuel is burned by conventional means and the resulting large volumes of rapidly moving gas are scrubbed chemically and passed through electrostatic precipitators before finally escaping to the air via smoke stacks. As an engineering approach stack gas treatment is inelegant; as a technological development its chances of success on a large scale remain to be proved. If it becomes widely adopted, large amounts of solid waste will result, in many cases with no convenient method of disposal in sight. Under the best of conditions, sulfur oxide scrubbing may end up with elemental sulfur or sulfuric acid. While the former may be stockpiled for future use, the value of by-product sulfuric acid is uncertain. Attention should be focused on methods capable of yielding elemental sulfur.

A preferable solution is to burn coal cleanly to begin with and separate sulfur, particulate matter, organic nitrogen, metallic impurities, and other foreign material prior to combustion. Coal gasification is one approach. It is not necessary to produce synthetic pipeline gas for purposes of generating electricity and industrial heat. Modern versions of the old powergas—a method for producing low BTU (British Thermal Unit) gas—appear economically attractive on a reasonably short time scale. Substituting fluidized bed techniques for pulverized fuel combustion is another approach which has been tried with some success in Europe and North Africa but not here. These techniques lend themselves directly to binary cycle power generation in which part of the heat contained in coal is passed into a high temperature gas turbine acting as a topping unit and the remainder is used to generate steam. The combined thermal efficiency of such a plant may approach 50 percent, a marked improvement over the 40 percent efficiency of modern fossil-fired steam plants.

Research and development on clean burning of coal should have the highest priority. Unfortunately, the presently existing programs in the Department of Interior and EPA aimed at solving this problem are woefully inadequate. Nor is there a promising air of activity in the coal industry, while the utility industry is just beginning to stir itself. Money has been but should not be a problem; electric utilities combined represent the single largest industry in our society. Their annual capital outlays now exceed $10 billion and revenues are some three times this figure. Because the utilities are highly fragmented into hundreds of local franchise areas and not organized to conduct high technology research and development and because the coal industry is moribund, the federal government must assume a vigorous leadership role. There is the additional danger that unless we move quickly, our industry will become the captive of foreign technology. We have already started to buy rights to commercial processess from Europe and Japan on syngas production from petroleum feedstocks.

NO_x is the collective symbol for the various nitrogen oxides formed when air is heated to high enough temperatures during a combustion cycle. Clean fuels may produce as much NO_x as dirty fuels. It is a potentially serious hazard to human health and animals. The principal sources are power plants along with other stationary sources and transportation vehicles. Aircraft will become a significant additional source as this mode of transportation increases in use. Effective means for removing NO_x after combustion by means of catalysts or chemical scrubbing have not proved effective. Its formation can be largely eliminated by controlling the temperature and duration of combustion. Promising results have already been obtained in supressing NO_x formation in combustors for small automotive power plants. Additional research is warranted in the study of NO_x formation as a function of combustion conditions and much of this can be done conveniently at the fundamental research level with modest expenditures of money. Much more has to be known about its role in human health as well.

As far as the energy sector alone is concerned, most of the emission of particulates comes from the burning of coal. While technology is well in hand for eliminating the larger size particulates which are responsible for black plumes from smoke stacks, the smaller size fraction is more difficult to remove and may represent a more serious danger to health. The facts are that both industrial precipitators and the human nose are good at filtering large-size objects but allow the small ones to get through. Techniques discussed above for the ultimate removal of sulfur from coal should also help control the problem of particulate emission.

A number of other chemical agents resulting from combustion become airborne and may represent serious health hazards. Among these, carbon monoxide (CO) is a leading candidate. Since CO comes mostly from the incomplete combustion of fuel in motor vehicles, and since the various possible corrective actions with regard to pollution from automobiles have already been mentioned briefly, the reader is referred to more detailed discussions on this and related topics elsewhere in this document.

RADIATION SAFETY

The use of uranium as fuel in nuclear fission reactors to provide energy for generating electricity is increasing rapidly, and within a generation is expected to provide more energy for this purpose than coal. Today, 23 reactors are in operation producing electricity, 54 more are under construction, and many more are planned or on order. By the end of 1971 the total electric generating capacity in the United States was around 350,000 megawatts, of which 10,000 was nuclear. The AEC predicts that the total nuclear capacity will be 150,000 megawatts by 1980 and 800,000 megawatts by the year 2000. This rapid growth is due to the favorable economics of nuclear fuel at many locations, the minimal environmental impact of a properly designed and safely operated nuclear plant, and the assurance of adequate fuel supplies for some time. It is unlikely that near-term demands for increased generating capacity can be met without resource to nuclear fuels.

Although there has been considerable concern about the safety of nuclear reactors, the water-cooled reactors used up to the present have had such an excellent safety record that the companies insuring nuclear plants have had no claims from the public for personal injury or property damage from nuclear causes. (To date there has been only limited experience with civilian fuel reprocessing plants and long-term waste storage.) Despite this excellent record, the large burden of radioactivity contained in each plant, coupled with the anticipated rapid increase in the number of these plants, requires that their safety be subject to the closest scrutiny. The safety question can be divided into two parts, one dealing with routine operation, the other with catastrophic accidents.

Routine operation involves a long chain of steps, which in the case of light water reactors includes uranimum extraction, enrichment, fuel fabrication, reactor operation, fuel reprocessing, and waste disposal. In the course of this cycle the escape of small amounts of radioactive material is unavoidable. Before plants can be licensed and operated they must prove that the amount of radiation escaping satisfies existing radiation standards. The ones given by the Code of Federal Regulations, Title 10, Part 20 were set in 1968 and considered too lax by a number of critics. More recently the AEC has circulated far more stringent standards as a proposed guide for new generating plant constructions and is likely to extend them over the entire industry. Ultimate responsibility for radiation standards now rests with EPA.

Assuming we can reach general concurrence on what levels should be set for radiation standards, nuclear plants operating in compliance, as they must, will be judged safe. That is, whatever risk is represented by their routine operation will be an acceptable one. These remarks also apply to any transport of radioactive material between points along the chain listed previously.

A catastrophic accident is defined as one which breaches the containment vessel and allows large amounts of radioactive material to escape. It is a possible event in large-size nuclear reactors, albeit a highly improbable event. The potential consequences of such an accident in terms of human safety and damage would be very serious. It is estimated that a so-called ultimate nuclear reactor accident could result in as many cancer cases to the exposed population as is occurring now from natural causes. On the probability scale it is more likely that noncatastrophic accidents would take place in which no radioactive substance would escape but routine operation of the plant might be interrupted. Currently, the nuclear industry is under criticism since its provisions for avoiding the one-in-a-million accident have come into sharp question in some quarters. Some of the blame must rest with the AEC which, unfortunately, has not pushed its large-scale testing program on ECCS (emergency core cooling systems) with sufficient emphasis.

Light water reactors are relatively inefficient converters and utilize only a small fraction of the potential energy contained in uranium. High temperature gas cooled reactors using the thorium cycle are somewhat better but have not yet received wide acceptance. The energy supply of uranium and thorium reserves can be extended several orders of magnitude by means of breeders which are capable of producing more fissionable fuel than they consume. Economically attractive deposits of uranium ore are in limited supply and it is predicted that the anticipated growth of the nuclear industry cannot be supported by this base beyond the turn of the century, perhaps earlier, unless breeders come into use by the mid-1980's. The time scales are open to some debate; nevertheless, there is a sense of urgency regarding breeder development as evident in the President's energy message. We eagerly await to learn whether breeders can be made safe and economically attractive.

As envisioned here nuclear energy represents in principle an inexhaustible source of energy which, because of the inherent hostility of its working substance, requires the highest forms of technological skill and human dedication to safeguard its operation. There is little room for human error or miscalculation. The components which constitute the entire system can be designed in principle to operate in a benign fashion, it remains for man to make sure the system performs according to design. There are many implications of societal commitments as a consequence of the nuclear age. Some have been touched upon. Let us look at one more; as breeders enter the energy sector large amounts of fissionable material from which nuclear explosives can be fashioned will become available. Every precaution will have to be taken to safeguard against clandestine diversion of this material into the wrong hands.

A long-term alternative to a nuclear technology based on fission breeders using uranium or thorium is controlled thermonuclear fusion (CTR). According to present concepts this would also be a breeder relying on lithium as one of its fuels. Ultimately, however, controlled fusion could rely entirely on deuterium, a heavy form of hydrogen, which is abundant in the waters of the world and represents an inexhaustible supply of fuel. CTR is still in the laboratory stage of experimentation and even its more ardent proponents do not believe it can become an economically viable force in the energy market before the end of the century. A number of features make CTR highly attractive; there is little reason to doubt that it should enjoy a high priority of research funding.

THERMAL POLLUTION

Rejection of heat to the atmosphere has already been touched upon. Thermal pollution of rivers, lakes, and oceans can result from waste heat disposal by stationary steam power plants. In modern fossil fuel-fired plants for every unit of fuel energy consumed, 40 percent is converted to electricity, 15 percent is lost as heat up the stacks, and 45 percent ends up in steam condensate. In light water nuclear reactors some 67 percent of the energy is retained in steam condensate (a factor 50 percent greater than fossil-fired plants). This latter form of waste heat is usually passed on to a body of water by a once through transit of cooling water. The river, lake, or ocean in question acts as a heat reservoir which subsequently passes the heat on to the atmosphere. There is a growing concern that delicate aquatic ecology may be adversely affected by the temperature rise which results from this practice. Notwithstanding our limited knowledge of ecological damage, the number of available fresh water sites is becoming limited as the number of steam power plants grow.

An alternative to once through cooling is to recirculate the cooling water in a closed-cycle system. In this event the heat reservoir may become a cooling pond or spray pond providing terrain and land availability allow for their construction. A plant generating 1,000 megawatts of electricity may require 1,000 acres of water surface. The atmosphere may act directly as the heat reservoir if cooling towers are used. Cooling towers may be wet or dry and in either case may be mechanical draft or natural draft. With the exception of dry cooling towers, a significant supply of fresh makeup water is required for recirculating systems. This mode of operation consumes more water than once through cooling but avoids direct aquatic damage. Unless local humidity is sufficiently low, adverse climatic changes may also result. Environmentally, dry cooling towers appear to have the least impact; they tend to be large structures, however, particularly if they are natural draft (a structure 600 feet at the base and 500 feet high might be required for 1,000 megawatt nuclear plant), and their aesthetic acceptance has yet to be tested.

In fact, dry cooling towers are four to five times more expensive than alternative choices at the current state of the art and their use is predicted to result in a decrease of plant efficiency. There is virtually no experience in their use with large (namely, 500 MW or greater) power plants.

If the ultimate limitation to waste heat disposal from steam power plants is the availability of fresh water, closed-cycle systems using ponds or wet towers do not promise a lasting solution. A typical 1,000 megawatt nuclear plant using wet tower cooling would require 30 million gallons per day. Once through cooling using ocean water or large lake currents represents an ample heat reservoir as yet hardly tapped. The recent announcement by Public Service of New Jersey that they plan to float nuclear plants three miles off the Jersey shore was immediately met by vocal opposition. The bulk of the adverse reaction seemed to be on aesthetic grounds. Who wants to go to the beach to stare at a string of nuclear plants three miles out?

Any improvement in the efficiency of generating electricity will help moderate the problem of waste heat removal. The breeder, combined cycles using gas turbine topping stages, MHD topping devices, and ultimately direct conversion from fusion reactors all point in this direction. At best, however, we can only expect 50 percent to 60 percent overall plant efficiencies from these developments; and, research must continue on finding effective means for rejecting heat directly to the atmosphere. Equally important is to continue to search for means of employing usefully the low-grade heat contained in steam condensate.

In view of the fact that once air quality and radiation safety considerations are met, thermal pollution may become the principal constraint on power plant siting, the problem deserves very careful scrutiny. Every effort should be made to insure that adequate information on local aquatic ecology has been gathered and to ascertain to what extent environmental harm can be avoided by appropriate design. We may very well have future regrets if stringent uniform standards require that once through cooling be abandoned hastily.

RESEARCH NEEDS

The discussion up to this point has contained numerous references to the need for more research and studies on a variety of subjects. This seems to be an inevitable consequence of reports dealing with energy and the environment. How refreshing it would be to say, "Here is the full extent of the problem, here are alternate solutions and their consequences; let the decision maker pick a choice." At the risk of belaboring the point a number of items requiring particular attention is collected here. The list is by no means complete; but, hopefully, it serves to point us in the right direction.

The first category of items relates to assuring adequacy of energy supply. Repeated mention has been made of the need for coal conversion technology and to find clean methods for burning coal. This matter is worth re-emphasizing. Let us bear in mind that not only current shortages in natural gas are of concern but also anticipated shortages of oil. Stimulation of supply from natural sources by more liberal oil depletion allowances, interstate compacts, import restrictions, or other regulatory instruments can only bring temporary relief. Improved technology in exploration and recovery will help stretch supply and buy additional time. But there is every indication that we are approaching the limit where finding rates cannot keep up with increased demand on a global scale. The rest of the free world is already exceeding our rate of growth in consumption of gas and hydrocarbon liquids. The sooner we establish a domestic industry for producing synthetic gas and oil from coal at a competitive price with natural products, the sooner we arrive at the self-sufficiency required by any wise policy for national security.

This will require a considerable effort in research and development and ultimately in capital outlay. The anticipated total sums are large enough so that no individual oil company may be expected to undertake the burden by itself; consequently, a strong leadership role must be exercised at the federal level. The alternatives of extracting oil from tar sands and shale do not have comparable long-range prospects. Even if technology were to exist for the economic processing of shale containing 10 to 30 gallons of oil per ton, the added cost of land restoration, waste disposal, and water treatment imposed by environmental considerations make this a very unattractive route to take. As for tar sands, the deposits of any significance lie outside our border.

If all of our energy requirements were transferred to coal, we could readily exhaust our known proved reserves inside of 100 years. This implies we must find ways to utilize our nuclear fuel resources to the fullest extent possible. The need for breeders and controlled fusion reactors has already been discussed. Many feel the need for breeder reactors is so vital that the Liquid Metal Fast Breeder development program should be expanded in order to include still other promising candidates.

So far little mention has been made of the prodigious quantities of energy represented by the solar flux of radiation. The problem, of course, is to find ways to concentrate it, store it, and utilize it effectively. There is reason to believe that with modest advances in technology certain sections of the nation could readily utilize solar energy to offset some of their space heating requirements. Large-scale central power generation from terrestial solar collectors is a future possibility which deserves attention. What we need right now is some basic research on collector design and fabrication. If proven practical, such central stations may be rather large, 1,000 megawatts (electrical), and occupy some 25 to 100 square miles of land. We cannot conclude solar power is entirely devoid of environmental impact because of implied land use and extensive requirements for long-range power transmission.

Another possible large source of energy is geothermal power. In select regions of the country it offers immediate relief. If the concept of using hot rocks thousands of feet beneath the earth's surface to supply energy is in fact feasible, geothermal power could have even wider application. There undoubtedly would be some environmental disadvantages in that noxious gases and salt spray may be released to the atmosphere. But learning how to control these is the least of the present problem and the prospects of geothermal power are well worth investigating.

The transportation of energy from source to user is not an insignificant part of the total cost and has environmental consequences of its own. Oil spills from tankers is but one example. One way to minimize this danger is to construct port facilities for giant tankers, thereby reducing the frequency of accidents. Transport of electricity over long distances is another problem. If it could be transmitted economically over arbitrary distances, there would be less of a need to locate generating plants near load centers and one would have greater flexibility in choosing plant sites. An ideal, though distant, scheme would be to use hydrogen to store and transmit energy. Hydrogen can be produced by the electrolysis of water, a process far from being economic with present-day practice.[4] The hydrogen thus formed may be shipped or piped to distant locations where it becomes reconverted to electricity in a fuel cell. Considerable research and development will be required before this can become a practical reality, but it deserves some immediate attention. Fuel cells and energy conversion based on hydrogen are both environmentally benign.

[4] As pointed out by Warren Winsche, even with today's practice one could use off-peak electricity to generate hydrogen and oxygen, receive credit for the latter in use with sewage treatment plants, and utilize the hydrogen as fuel for special urban fleet vehicles economically.

As long as energy was cheap and plentiful there seemed little need to curb wasteful uses of it. Now that our conception has changed a strong research and public education program should be instituted on prudent ways of using energy. Considerable savings should be possible in many areas, particularly in space heating, through improved design.

Research concerned with other than technological issues is urgently required as well. The inadequate state of understanding environmental damage due to chemical contamination of the atmosphere and thermal discharge in water has been repeatedly stressed. Unless knowledge in these areas improve, air and water quality standards will always have a degree of uncertainty associated with them. For the time being general practice has been to set uniform standards; in some cases a latitude for strengthening has been left to the states. It is not at all clear that inflexible standards are desirable in the long run. What we really desire is high quality for the environment and this can not be readily equated to so many parts per million of a particular agent in the atmosphere over the country or so many degrees rise in temperature of water surfaces throughout the nation. We have a great deal to learn about environmental quality.

Every time an environmental impact statement is filed with CEQ or a regulatory pronouncement is issued by EPA difficult decisions and trade-offs must be made on the basis of incomplete information. This is a phenomenon which is not uncommon in the world of politics. Nevertheless, one would like to guarantee that the decision is as sound as possible. To do this we need greater insight into how to balance, for example, the threat of automakers that certain regulations will lead to so many job lay-offs in Detroit against the prediction of public health officials that so many additional respiratory ailments will result unless ambient concentrations of certain chemicals are reduced beyond some level. How real are the threats and predictions and how are they related to each other? To say that more research is needed is an understatement. Much more is required. We—a society composed of many innocent bystanders and potential victims, along with protagonists whose intentions inevitably are subject to question, plus the final arbiter—must all grow wiser together.

Environment: A Delicate Balance of Costs and Benefits

P. H. ABELSON

THE HEALTH AND WELL-BEING OF A PEOPLE is influenced in many ways by material aspects; but it is largely determined by spiritual values and attitudes. A society that is torn by dissension and determined to be unhappy will be unhealthy despite any amount of wealth. The average citizen enjoys much better food, clothing, shelter, and health than did his forebears. Nevertheless, many of our people are unhappy. We have experienced a decade in which so-called intellectuals have labored diligently to discover and proclaim reasons for discontent. In their labors they have been signally assisted by the mass media which historically have found that disaster and disorderly behavior are far more salable than normal activity.

An important area in which most people do not have adequate knowledge or experience to form sound judgment is in matters having a high scientific or technological content. They have little basis for deciding what can reasonably be expected from science and technology, and their attitude often veers from regarding scientists as miracle men to seeing them as creatures of the devil. Correspondingly, public opinion on technological matters is volatile and shifts between admiration and condemnation.

One of the most difficult tasks facing a society or an individual is to form a reasonably accurate estimate of potential role. Ambition and some discontent with the status quo are signs of health. But one of the worst traps that an individual or a society can fall into is that of excessive expectations.

Society's judgment has been inaccurate in its expectations with respect to science and technology. Achievements during World War II and after conditioned our people to assuming that we were guaranteed perpetual world leadership. The successful Apollo landing gave dramatic impetus to this belief and led to the widely adopted view that "if we can go to the moon we can do anything." Soon after the lunar landing, the response to the Apollo triumph took on a sour tinge. "If we can go to the moon, why don't we"—fix our cities, improve transportation, eliminate poverty, or clean up pollution.

The attitude of being able to achieve everything carried over into legislation affecting, for example, our approach to health in the form of the expanded cancer program. The attitude has also affected our policies with respect to foods, drugs, and the environment.

Actually even before the Apollo triumph, it was becoming clear to some that the United States was losing its competitive edge in science and technology. The European countries have long had a tradition of excellence in science. Once they had recovered from the damage of World War II, they began again to display some of their intellectual potential. At the same time, they worked diligently to improve their technology, adapting many of the best American methods. In this they were joined by the Japanese, who both adapted and improved on existing methods. Our superior position in technology has eroded. This is especially demonstrated in our decreased ability to compete in international trade. During the early 1960s we enjoyed a very favorable balance of trade, which was largely based on our exports of high technology items. Recently imports have increased faster

Reprinted with permission from *Carnegie Institution of Washington Year Book 71*, 1971–1972.

than exports, and currently there is a deficit in trade despite the devaluation of a year ago. A continuation of current trends seems likely, for there is no substantial basis for hoping for an improvement in our ability to compete technologically. That is not to say that the matter is hopeless. Rather we have not done much in an organized way about it. There have been some small stirrings in the Office of the President's Science Adviser, but few others have joined him. One major factor that could be improved is in the cooperation of government with business. In that sphere we could take a few lessons from the Japanese and French. A second matter that needs improving is the interface between academia and industry. For decades universities have indoctrinated many of their graduate students in the belief that industrial employment is demeaning. A more cooperative attitude on the part of universities would be helpful.

The environment is another important area in which lack of public understanding of science and technology has led to questionable judgments. The average citizen can easily make a decision as to whether a pulp mill stinks or his eyes water from smog. In trying to gauge the possible toxicity of chemicals that he can not see, taste, or smell, he has little to go on. Many people are predisposed to acceptance of the view that irresponsible devils are trying to poison them. They have been encouraged in this belief by environmental zealots.

There have been scare headlines about running out of oxygen, choking to death on pollution, mercury poisoning from tuna, toxicity of DDT, and many more. The mass media have given great play to almost any shaky claim of an environmental hazard. When there is a problem such as that of mercury release, the story is headlined. When the problem is cured or proven to be nonexistent, little or no attention is given. The average citizen learns of potential hazards. He is rarely informed of constructive measures.

The overwhelming majority of the American people are determined that various forms of pollution shall be abated, and rightly so. The question arises, though, how far down this path do we go. There are those who demand that no emissions or effluents be permitted. This is unrealistic. If no man lived on earth, the air and water would still contain materials we call pollutants. The quantities of sulfur put in the air from natural sources exceed those coming out of stacks. Vast quantities of carbon monoxide are released by microorganisms, and nitrogen is present as ammonia. In addition, nitrogen oxides are formed by lightning discharges. Natural radioactivity is everywhere, and in amounts far greater than that due to man-devised atomic energy. Trace elements such as arsenic and mercury have always been present in river waters, sometimes in heavy concentrations. We could not be entirely free of pollution even if we were willing to pay huge sums for it.

Those dealing with abatement report that cost is sharply dependent on the degree of clean-up demanded. That is, they find that abating the first 50 percent is usually not very costly; after that costs escalate. After the 90 percent level is reached, it often costs as much to achieve a further 5 percent as it did to get the first 50 percent. Perfection is impossible, and the last steps toward it are extremely costly.

It is clear that we are going to spend substantial sums in an effort to improve the environment. The Council on Environmental Quality has estimated that the bill will come to $300 billion for this decade. They point out that this is only a small fraction of the gross national product. This argument, however, neglects many secondary effects, such as a further change in our ability to compete internationally and substantial disruption in our methods of obtaining energy. Already we are experiencing a deepening energy crisis with much more trouble likely and a drastic increase both in the cost of energy and in the imports of foreign oil.

For individual companies the impact is great. Some report that a substantial fraction (as much as half) of their capital expenditures next year will be devoted to pollution control. These expenditures produce no revenue, and indeed the abatement equipment will require continuing maintenance. All these costs will be passed on to the consumer.

The most important impact of our determination to enjoy an improved environment has been on the production and consumption of energy. Principal uses of energy are in transportation (automobiles), electricity, home heating, and industrial processing. Concern about sulfur dioxide pollution has led to increased demand for fuels low in sulfur, notably natural gas and desulfurized fuel oil. It has led to a sharp decline in the use of coal along the Atlantic seaboard. The increased demand for petroleum products has come at a time when this country no longer has sufficient oil production capacity to meet its needs. Our dependence on foreign sources is now rapidly expanding. Prospects are that during the next few years at least, it will continue to expand. This will be due to a persistent growth in use of energy, further conversion of power plants from coal to oil, and an increase in demand for motor fuel. This latter will reflect the fact that the new models with pollution control consume more fuel than their predecessors.

When we seek to increase imports of petroleum and liquified natural gas, we find ourselves in competition for available supplies. The burgeoning economies of Europe and Japan are in even greater need of oil than we. These competing needs have been obvious to the oil producing and exporting countries, and they have not been slow to take advantage of a seller's market. During the last few years the tempo of upward price adjustments has increased. The current doubling time for return per barrel to the oil producing and exporting countries is two to four years. If present trends in our oil demand continue, authorities estimate that by 1985 we will be importing 10 to 15 million barrels of oil a day. Our increased demand and that of Europe and Japan would give the oil producers more leverage even than they have now. Simple extrapolation of current trends suggests that the price of oil could quadruple by 1980. At present prices the cost would be about $15 billion dollars, but with escalation it could be as much as $60 billion. This is a sum greater than the value of our present total worldwide exports.

Production would come largely from a few countries of the Middle East, notably Saudi Arabia and Iran. How would we pay for the oil? Of necessity this would be by exporting goods. But we would be in competition with Europe and Japan, which would likewise be trying to import oil and sell goods. The capacity of the Saudi Arabians to utilize goods is limited. Who will get the oil and how much will they pay?

In one sense the year 1985 seems far, far away. But in terms of the time required for development of new technology and the construction of plants, the time is short.

In addition to pushing us toward a dangerous dependence on foreign sources of petroleum, the environmental standards are leading to sharply increased costs for energy. For example, one effect has been to encourage demand for natural gas because of its nonpolluting characteristics. Until recently, the cost of a thousand cubic feet (Mcf) of natural gas in Texas was less than 20 cents. New contracts are being made at 40 cents per Mcf and plans are being made to import the gas from Algeria at a cost of about $1.00 per Mcf.

Higher costs of energy will be reflected in every aspect of the economy and in the cost of every object that the consumer buys. In a few years his heating bills will double or worse and costs of other forms of energy will rise.

In their enthusiasm for environmental perfection, proponents have apparently assumed that great improvements can be obtained without cost to society. Moreover, some seem to assume that endless lawsuits and delays will affect only the particular companies involved. It is not possible to foresee all the ramifying effects of our environmental program, but it is evident that costs will be far greater than the public dreams. The costs will hit everyone, and especially the poor. Our national security is being placed in jeopardy. The tensions arising from an arbitrary cut-off of fuel oil supplies at the beginning of a heating season are not pleasant to contemplate.

We are going to pay a high price for improvements. What will we get? Two types of benefits have been cited—better health and better quality of life.

In spite of the scare headlines, it would be difficult to prove that any more than a minor fraction of our people had suffered more than minimal effects from pollution. True, some annoyance, as in smog, but not real damage. A perusal of the documents on which our air pollution standards are based indicates that this evidence is flimsy. The documented cases where air pollution has increased morbidity are few. As a rough estimate, the injuries and lethal effects of automobile accidents have been 10 to 100 times as damaging as those due to pollution.

To get additional perspective, consider the air pollution associated with cigarette smoking. Carbon monoxide, a major component of auto exhaust, is also a major component of cigarette smoke. Our present national standard for permissible concentration of carbon monoxide is 9 ppm (parts per million). This level is not to be exceeded for more than 8 hours once a year. The carbon monoxide content of cigarette smoke is more than 20,000 ppm. A chain smoker and his associates are daily subjected to more than the national standards. Similar remarks apply to the nitrogen oxides and particulate matter. The health effects of smoking far, far exceed anything that can be attributed to pollution.

We will continue efforts to abate pollution, and we should continue those efforts. However, it is time to consider more judiciously costs and benefits and to recognize that the costs will be higher than had been estimated and the health benefits not so great as imagined.

As for quality of life, there will be improvements perceptible to some, but most people will not notice benefits. Kenneth Boulding has commented,

"The environment is a luxury of the Sierra Club." This is an exaggeration, of course, and perhaps intended to be. But it has a kernel of truth.

In future, science and technology will play an even greater role in our lives. The centralizing power in government means that many great decisions involving technology will be made by politicians. For the most part, politicians are followers, not leaders. Politicians are sensitive to the winds of public opinion. If the government is to make wise decisions, it will be necessary for the public to make better judgments. And it is especially necessary that the public understand some of the limitations of technology. We can't have everything. We can't do everything, and what we can do usually takes a long time.

Energy Problems and Environmental Concern

RUSSELL E. TRAIN

"There are some who simplistically blame the strong concern over environmental quality as the cause of our energy problems. This assertion is simply not true." Russell E. Train, former chairman of the Council on Environmental Quality, is now administrator of the Environmental Protection Agency. He made these remarks before the American Power Conference on May 8, 1973 in Chicago.

Energy problems are complex and closely related to a wide variety of forces. Prominent among these forces is, of course, the question of environmental quality; but prices, technology, regulatory requirements, international relations, and national security considerations are also integral parts of the problem. There are some who simplistically blame the strong concern over environmental quality as the cause of our energy problems. This assertion is simply not true.

A recent issue of a national news magazine quoted the chief executive of a major international oil company as identifying environmentalists as the major culprits in blocking new generating facilities and new refinery capacity. Such statements obscure the real facts, confuse the issues, and can only serve to delay effective solution of our energy problems.

A recent spate of advertising has sought to convey to the public that auto emission standards will be the cause of major fuel diseconomy—ignoring the fact that comparable fuel losses come from the use of automatic transmissions and air conditioners, among other factors. Moreover, if we assume a 10 per cent increase in gasoline consumption because of emission controls, the public should also know that a 5,000 pound vehicle consumes 100 per cent more fuel per mile than does a 2,500 pound vehicle —ten times the fuel loss from emission controls. Half-truths are not going to help meet our energy needs.

Environmental factors have been given as the reason for nuclear power plant delays. However, Atomic Energy Commission data confirm that the National Environmental Policy Act review process is not the major controlling factor in bringing a nuclear plant into operation. A plant's readiness for fuel loading is requisite to its being licensed. Data submitted to the Council on Environmental Quality (CEQ) by the AEC in March 1973 indicate that final environmental impact statements were available, on the average, 8.2 months prior to fuel loading.

Nuclear Delays

In another study in January 1973, the AEC reviewed 25 operating license applications for 35 nuclear power reactors. With the possible exception of one or two units, the AEC at that time anticipated no delays solely as a result of environmental considerations in the issuance of operating licenses for these reactors. But of those 35 plants then under consideration for operating licenses, 18 had fuel densification problems, and 17 had steam line break problems.

So let's keep the environmental aspect of nuclear power in perspective.

If environmentalists are blamed for power plant siting delays, it should be remembered that nearly 2½ years ago President Nixon first submitted a power plant siting bill to Congress. Should his most recent submission, the Electric Facilities Siting Act of 1973, be enacted, the review and approval process for siting new plants would be simplified. It would also give the public earlier notice and a larger role in the decisions over power needs and how and where to meet them. And although some spokesmen for the power industry publicly lament the difficulties in getting new plants approved, the National Association of Electric Companies' posi-

tion before Congress has been that no new legislation is needed. If this legislation had been enacted, we might be two years closer to the institutional arrangements necessary to deal with some of our crucial energy problems.

Environmentalists have also been charged with hindering the construction of new petroleum refineries. Although some companies have been refused sites for new refineries, by and large the oil industry has been most reluctant to commit large sums of money to new refinery construction because of past uncertainty about government policies, such as oil import policies, and because of a severe shortage of cash from current company earnings. In addition, for the large international oil companies, extreme uncertainty as to their situation in the Middle East vis-a-vis the Organization of Petroleum Exporting Countries has created a wait-and-see attitude. And besides, until the second half of 1972, domestic refinery operations never got above 85 per cent of rated capacity.

Let us not permit our current concerns over energy supply to obscure the fact that the environmental costs of energy production are likewise very real. The high levels of lung cancer and respiratory disease, such as emphysema, in areas with high levels of air pollution is a fact, not emotional imagining.

Those who seek to portray the environmental movement as the cause of our energy problems would seem to prefer confrontation to accommodation. Confrontation can only lead to polarization and irrational responses from all sides. Our energy problems are serious and they are real. Our environmental concerns are likewise serious and they too are real. We need balance and restraint—by both environmentalists and industry—as we pursue both objectives as matters of high priority national interest. Above all, we need full disclosure of all the facts and the broadest possible public understanding of the issues.

The CEQ is committed to find answers to many of the questions raised by our increasing energy demands. Some of our work is contained in a report, "Energy and the Environment," which was released last August. Examination of related questions is just beginning.

Central to these investigations is the deep concern of the CEQ about the impact of the total energy system on the quality of the environment. The exploitation of any energy resource inevitably leads to some environmental damage. At the same time, however, our decision-making with regard to the use of energy or the development of energy resources has not reflected a keen understanding of the specific environmental penalties that each decision forces us to incur. For example, the construction of a power plant which will use coal, oil, natural gas or nuclear energy usually reflects consideration for the environmental impacts at the power plant site, but rarely reflects concern for the environmental impacts of providing the fuel.

Myopic Approach

Generally, we have failed to look systematically at the totality of environmental impacts caused by a single decision. Rather, we have considered only the direct environmental implications of each decision. This myopic approach is not good enough. And it is for this reason that the CEQ has undertaken its extensive effort to determine the environmental consequences of energy systems—from extraction to consumption. As a first step, the report, "Energy and the Environment," quantifies the specific environmental impacts derived from the production and use of electricity—not just at the power plant, but at the mine, the oil well, the pipeline, the tanker, the refinery and the transmission line.

It quantifies the environmental effects of electric energy systems with today's minimal or presently-prevailing environmental controls and looks at the economic costs and environmental gains that result from the use of controls that are now available or soon will be. The report is not only concerned with the costs and effects of sulfur oxide, particulate and thermal discharge controls at the power plant, but also with the results of improved strip mine reclamation, better controlled refineries and cleaner tankers.

Total Look

By looking at the entire electric energy system from the point of energy extraction to the delivery of electricity to the consumer we will be able to more effectively assess the environmental implications of specific electric energy decisions.

The nation's growing demand for more energy makes it clear that for the foreseeable future we will be required to increase oil imports from foreign sources. The most economical way to transport crude oil over long distances is by supertanker. No conventional port in the United States can accommodate supertankers, but if deepwater ports are not developed in the United States, crude oil will be shipped to deepwater terminals in the Bahamas and in the Canadian maritime provinces by supertankers and then transshipped to existing U.S. harbors in smaller tankers.

Concern over environmental implications of supertankers and superports resulted in a CEQ study of superports and their alternatives. Preliminary results suggest that use of smaller tankers would result in greater oil spillage from groundings and collisions than would supertankers using appropriately located and operated offshore, deepwater port facilities. Indeed, the U.S. Coast Guard, which worked with us, concluded that by establishing Federal construction and navigation control over supertankers and offshore superports and by transshipping the oil to shore by pipeline, oil spilled in our coastal waters would be about 90 per cent less than if we continued to use only smaller tankers. Furthermore, adding preventive measures such as double bottoms on supertankers and traffic controls at port locations would further reduce the magnitude and frequency of oil spills. The report also found that vulnerability to environmental damage from supertanker port development is likely to be greatest at inshore locations where the marine ecosystem and recreation areas would be more vulnerable.

Exploitation of one major domestic source of energy—oil and gas on the Outer Continental Shelf (OCS)—has raised great concern over its impact on environmental quality. In his April 18, 1973 energy message to Congress, the President directed the CEQ to undertake a comprehensive study of the environmental impact of possible oil and gas production on the Atlantic Outer Continental Shelf and Gulf of Alaska. The study will be done in consultation with the National Academy of Sciences, the Environmental Protection Agency, and other Federal agencies. Public hearings on the subject are scheduled to be held along the Atlantic Coast in September and October 1973. No drilling will be undertaken in these areas until the environmental impacts are determined.

Although the study is now in its formative stages, I will try to briefly summarize what is planned. In a year's time, we obviously cannot develop much in the way of brand new data. However, we can and will identify and analyze the best information that is available on every aspect of this issue. Our first problem will be to find out with as much confidence as possible where the oil and gas is located and what is the potential environmental vulnerability of these areas to drilling operations and

to oil spillage. We will develop estimates of the frequency and magnitude of oil spills under various developmental scenarios, as well as models of oil spill spread and trajectory. Then, having a grasp of the probabilities of a spill and the likely fate of the oil, we will try to assess the potential effects of spilled oil.

In addition to resource location and the fate and effects of oil spills, our study will rigorously review the state of the art in exploration, production and transportation technology for OCS oil and gas. We will focus on the effectiveness of technological methods to minimize environmental impacts in order to determine how, and at what cost, technology can prevent and control oil spills.

The study will also explore the potential shoreside—or secondary—environmental impacts of OCS development. Where should the pipeline come ashore and what kind of industrial development can be expected as a result? What are the likely environmental effects of new industry, such as refineries, petrochemical plants, and associated commercial and residential development? What are the potential impacts on land use and air and water quality? What demand will induced industrialization and population growth put on water supply, electric power supply, and other public services? These and other questions will be answered in the course of this study, which is expected to be completed by April 1974.

Challenge of Coal

Of course, increased imports and production from offshore oil wells will not satisfactorily quench America's hunger for energy. The nation has turned to its coal reserves to supplement a diet of oil and natural gas. Clearly, coal presents the greatest challenge to our environment. From its extraction to its consumption, coal exacts an environmental price. That price has been paid in many forms of environmental degradation—from strip-mined hills and acid-polluted streams to the sulfur oxides emitted in our air.

At the request of the Senate Interior Committee, we conducted a review of mining and reclamation technology and state regulatory programs. In addition, the CEQ investigated the fuel supply and economic implications of legislated prohibitions against surface mining on steep slopes and submitted its report to Congress for use in considering badly-needed mining legislation.

Our study clearly showed that only the best reclamation techniques can substantially reduce the type of damages that have devastated and are continuing to devastate so much of Appalachia. In a tour of strip mining operations in Kentucky and Tennessee, I saw that even the best reclamation techniques practiced in those areas can still lead to the worst kinds of abuses—landslides, massive erosion and acid mine drainage. I was impressed with the potential of a recently-developed method of contour strip mining called modified block cut or cut-and-fill to significantly reduce the environmental impacts of contour strip mining. The CEQ's cost analyses indicated that the incremental costs of total reclamation using this method is not excessive.

Federal Mining Law

We also discovered, unfortunately, that many states do not have adequate mining regulations and, in addition, are not funding adequate regulatory programs to enforce the laws that they do have. We are convinced that a strong Federal law is required which will establish stringent performance standards for all the states to follow.

Probably the most important contribution of the study was the development of information on strip mine coal reserves and current coal production on steep slopes. The data showed that much of the surface mined coal in central Appalachia comes from steep slopes, in fact from slopes over 20°. Much of that coal is low sulfur coal needed to meet the air quality standards of the Clean Air Act. Utilities are paying premium prices for this coal, averaging over $9.50 per ton at the mine, so there is no economic excuse for not using the most advanced reclamation techniques to minimize environmental damage.

So far this article has centered on more: more coal, more oil, more gas to meet the energy crisis. But unless we as a society take steps to conserve our energy resources, we will exhaust the new supplies in a relatively short time. The President has called upon us as a nation to develop a national energy conservation ethic. Industry and consumers can help by developing and using products which conserve energy.

We in the Federal government have an important leadership role. The General Services Administration, for instance, is constructing a new Federal office building in Manchester, N. H., using advanced energy conservation techniques, with a goal of reducing energy use by 20 per cent over typical buildings of the same size. The National Bureau of Standards is evaluating energy use in a full-size house as a means to develop analytical techniques for predicting energy use for new dwellings. These programs will assist the Federal government and architects and private contractors to design and construct energy-efficient buildings. Current engineering and design of buildings is often outrageously wasteful of energy.

Further, during the past two years, the President has twice directed the Department of Housing and Urban Development (HUD) to upgrade insulation standards in single and multi-family residences financed by the Federal Housing Authority. These revisions can cut heat losses by one-third in new homes, thus conserving energy in the residential sector. Now the President has directed HUD to evaluate the extension of insulation standards to mobile homes. And HUD, with the technical support of the National Bureau of Standards, is pioneering new methods of residential energy conservation. A total energy system, which in effect "recycles" waste heat, is now being evaluated at the HUD Operation Breakthrough residential complex in Jersey City, N. J.

Transportation, too, offers many opportunities for saving energy. Transportation uses about 25 per cent of the nation's energy, and energy efficiencies of various passenger-transporting modes vary greatly. The fastest form of transportation, the airplane, is also the one that uses the most energy per passenger mile. On the ground the automobile, the most flexible and ubiquitous transportation vehicle, uses much more energy per passenger mile than buses or trains. But at least it should be possible to shift to smaller, lighter cars. With the fuel economy characteristics of present small cars—about 22 miles per gallon instead of the current average for all cars of less than 14 miles per gallon—the annual fuel savings could be enormous.

In addition to our use of smaller cars, perhaps by providing alternative forms of transportation we can induce people to leave their cars at home. Congress has passed, and the President signed, legislation that for the first time allows the Highway Trust Fund to be used for mass transit. Hopefully this long awaited action will result in more emphasis on mass transit solutions to urban transportation problems.

Consumers' Role

Automobiles and home appliances account for approximately 20 per cent of current energy demand. Just as with automobiles, similar types of home appliances have widely different energy efficiencies. Some of the new, large, frost-free refrigerators, for example, consume 2½ times more energy than earlier models. Therefore consumers play a significant part in conserving energy. In order to provide consumers with the information needed to make conservation-oriented purchases of energy-consuming appliances, the President has directed the Department of Commerce to work with the CEQ and the Environmental Protection Agency to develop a voluntary system of energy efficiency labels for major home appliances, automobiles and automobile accessories. These labels will provide not only data on energy use but, most important, a rating comparing the product's efficiency to similar products. As a first step toward this goal, the Environmental Protection Agency has just released a comparison of new-car gasoline mileage performance.

In the industrial sector, there are significant opportunities for energy conservation—in plant and process design, and even in the choice of feedstock materials. For example, in many cases significant amounts of energy can be conserved by using secondary materials in place of virgin feedstocks. In the paper industry, the energy consumption to produce pulp from recycled fiber is 70 per cent less than the energy required using virgin wood pulp. Similar figures for the steel industry show a 74 per cent saving in energy when scrap is used to produce steel instead of virgin iron ore. I believe we should explore aggressively the development of incentives, including tax incentives, to encourage greater recycling.

These proposals for government, industry, and consumers represent only a beginning in our efforts to conserve energy. We must explore all alternative ways to reduce energy, weighing the amount of energy conserved with other factors, such as potential regressivity and possible secondary effects. The CEQ will give energy conservation a high priority in its program.

The President has stated that energy conservation is a "national necessity." His energy message has demonstrated that there are alternatives to just responding to increasing energy demands. We can look for ways to reduce our total energy demand. In the future we can look to a mix of energy systems which will supply our energy needs with lower total environmental damage.

The so-called energy crisis stems from the economic forces and complexity of the energy industry, from the difficulty in planning for our voracious energy appetite, and from the need to satisfy social values—other than those that depend on energy. To blame this crisis solely on an increased concern over environmental quality would be a grave failure to face the problem honestly and squarely.

It seems to me that the best way to deal with the difficulties presented by our current energy position is to completely reorient our thinking about energy. In the short run, we are looking for increased energy supplies. But in the long run, we must increasingly shift our efforts from simply finding more energy supplies to concerning ourselves with how to use energy to best meet our many needs. And when we do this, I am sure that the efficient use of energy will profoundly improve the quality of life for all Americans.

Toward An Energy Ethic

M. R. Gustavson

Nationalistic, greedy, myopic, unfeeling, ignorant—these are the words that so frequently enter current dialogs on energy resources. Enter or end them. For often what starts as a dialog ends in an exchange of charges. The issues continue unresolved.

Although important topics frequently lead to heated exchanges, for energy resources there is a deeper set of causes at work. There are disagreements as to the nature or even the existence of the problem. There are differences in the types of consequences that are of primary concern. There is no commonly accepted and suitable framework within which to resolve many of the fundamental issues.

A first step toward resolution must be to make explicit these differences in viewpoints. Implicit or underlying beliefs and concerns must be given explicit statement. This is the primary goal of this paper. A start is also made on constructing an adequate stage for further dialog. Within this framework and on this stage it is hoped that the questions arising from the energy issue may be more clearly perceived and productively investigated.

The balance of this paper first attempts to describe the variety of viewpoints or attitudes toward 'the energy crisis.' The variety of such attitudes is part of the barrier to productive discussion. Even a recognition of the requirements imposed by internal self-consistency of viewpoint is likely to prove salutary.

With this background, some of the numerous issues that interact with energy decisions are next reviewed. This review gives some insight into the variety of primary concerns that motivate those engaged in the energy dialog, another barrier to productive discussion that often goes unrecognized. This review also indicates the enormous potential impact of decisions about energy. It should serve as fair warning to those who have the responsibility for making these decisions.

Finally, a notion is offered as to the nature of the consensual framework that must be constructed as a basis for broadly acceptable decisions on energy policy. It is conjectured that the construction of this framework will require the development of an energy ethic. The need seems clear, but its form can be but dimly perceived.

Attitudes about Energy

As has been widely noted, the annual production and consumption of energy by mankind have been increasing ever more rapidly. This exponential growth is a historical fact, but from this fact observers can draw diametrically opposite views of the future

One observer might say that the best prediction of the future is simply a continuation of the historical trend: increasing rates of production and consumption for the reasonably foreseeable future. Such an 'extrapolationist' sees no problem. A continuation of cheap and abundant supplies of energy seems likely.

To another observer an exponential spells trouble. He feels (knows) that no such exponential growth can long go unchecked, that it characterizes an unstable system, that fundamental factors will impose some limit, that sooner or later a crisis must intervene as production limits consumption, i.e., as energy becomes costly and in relatively short supply.

Quite different reasoning has led others to the same conclusions. For example, there are some who would set no bound on the ability of technology to provide man's material wants. Others total the available but irreplaceable fossil fuels and foresee an early end to the availability of cheap energy. For whatever reason they are held, both views represent extremes and are in this sense close to each other.

It is interesting to use a horseshoe arrangement to depict these extreme views as well as the more popular intermediate viewpoints. In Figure 1, positions A and E represent these extremes. Position A is that energy will be cheap and plentiful. Our concern

Reprinted with permission from *EØS, Trans., Amer. Geophys. Union*, vol. 54, pp. 676–681, July 1973.

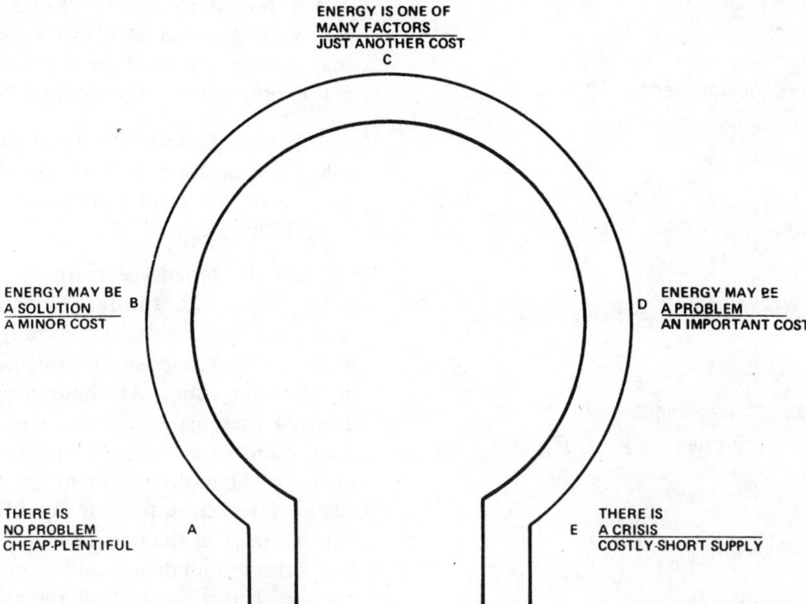

Fig. 1. Attitudes about energy

should be how to exploit this fact. Position E is that energy will be costly and in short supply. A crisis is envisioned. The intermediate position C, at the top of the horseshoe, is a moderate attitude. This viewpoint perceives energy availability as one of many factors, as another cost that has to be taken into account in the relationships that govern men's affairs. The intermediate positions B and D view energy, respectively, as either tending to stay low enough in cost to represent a solution to some problems or becoming sufficiently dear that it will represent an important cost and therefore create problems.

These brief descriptions of attitudes A–E can be given fuller meaning by describing how a person who took each one of these positions might feel about a number of energy-related issues. Such descriptions presume of course that energy is a primary concern and that an internally consistent set of positions is taken. For this purpose the following sample set of key issues has been used:

1. Source development: particularly of the fossil fuel type.
2. Use: as effected by education, advertising, and legislation.
3. Nationalism: in the national security sense as seen by a citizen of a consumer nation.
4. Pollution: as a negative factor in the quality of life.
5. Federal funding: as an element of public support.
6. Fusion reactors: as an example of a possible technological key to abundant energy.
7. Federal lands: concern only with their value for energy production and not with their other possible values.
8. Depletion allowance: assuming that a high depletion allowance encourages resource development.
9. Imports-exports: for fuel, power, and high-energy-content materials as seen by a citizen of a consumer nation.
10. Resource control: assuming that more control can lead to more efficient use.
11. Plowshare: assuming that development of this technology will help to provide a greater abundance of energy.

Figure 2 shows the views of a person with position A attitudes, i.e., that there is no problem and energy will be cheap and plentiful.

A person with type A attitudes feels that major new sources of energy will be easily developed, that its use should be promoted whenever possible, that there is no need for any nationalistic attitude in preserving our sources of supply, that we can give first place to problems like eliminating pollution without worrying about the impact of energy availability, and that we can perhaps even use energy to solve the problem. Since energy abundance is so readily maintained, there is no need to fund research and development in this area. Nor need there be concern, when leasing federal lands, with the development of their power production potential. There is no sense in encouraging people to find fuel resources, and therefore the depletion allowance can be removed. The country should reduce its imports and encourage exports of fuel material, minimize the legislative controls on resources, and, when industry wants Plowshare, let industry pay for it.

Attitudes toward the same set of key issues for positions B–E are summarized below. The trend goes from no concern about new sources to major concern, from promoting use to restricting use, from no nationalistic concern to a concern that national security could be in jeopardy, from a willingness to use energy to remove pollution to a worry that pollution may have to be accepted to meet minimal energy needs, from feeling that federal funding to encourage energy production is not required to giving it a high priority, from lack of concern with new energy production technology to strong federal support, from lack of concern with the power potential of federal lands to close control, from decreasing to increasing depletion allowances to encourage exploration and development, from seeking a high to seeking a low ratio of imports to exports of fuel, power, and high-energy-content materials, from light to tight resource control, and from a hands-off, wait-and-see policy toward Plowshare to government implementation.

Figure 3 shows the views of a person with position B attitudes, i.e., that energy use may be a solution and energy will be a minor cost.

Figure 4 shows the views of a person with position C attitudes, i.e., that energy is one of many factors and is just another cost.

Fig. 2

Position A

1. Major new sources will easily be developed.
2. Promote use wherever possible.
3. There is no need for nationalism.
4. Eliminate pollution.
5. Do not federally fund research and development.
6. Fusion reactors will readily be developed.
7. Freely lease federal land resources.
8. Remove depletion allowance.
9. Reduce imports; encourage exports.
10. Minimize resource control.
11. Develop Plowshare when industry is willing.

Fig. 3

Position B

1. Source development needs no encouragement.
2. Encourage use where reasonably effective.
3. Promote use as national objective.
4. Greatly reduce pollution.
5. Federally fund research and development only where essential.
6. Fusion reactor basic research deserves support.
7. Lease federal land resources at incentive prices.
8. Reduce depletion allowance.
9. Encourage exports.
10. Control only critical resources.
11. Support only Plowshare basic research.

Fig. 4

Position C

1. Source development should proceed as economically desirable.
2. Allow use to be controlled by market forces.
3. Provide for adequate national reserves.
4. Control major sources of pollution.
5. Federally fund research and development for major innovations.
6. Fusion reactors should be encouraged but not rushed.
7. Lease federal land resources with proper return.
8. Adjust depletion allowance.
9. Balance imports and exports.
10. Control resources to achieve efficiency.
11. Federally support Plowshare research and development but not implementation.

Fig. 5

Position D

1. Source development requires encouragement.
2. Prevent waste of resources.
3. Preserve national resources for security.
4. Control worst sources of pollution.
5. Federally fund research and development for all valuable innovations.
6. Strongly support fusion reactor development.
7. Lease federal land resources only under tight control.
8. Increase depletion allowance.
9. Discourage exports.
10. Control resources of all types.
11. Federally support Plowshare demonstration programs.

Figure 5 shows the views of a person with position D attitudes, i.e., that energy use may be a problem and energy will be an important cost.

Figure 6 shows the views of a person with position E attitudes, i.e., that there is a crisis and energy will be terribly expensive.

Thus the broad spectrum of attitudes about the future abundance and cost of energy leads naturally to widely different positions on many other major issues. At times one or more of these issues can become the focal point of a dialog on energy. In many cases a disagreement at this level can be traced to basically different attitudes of the participants as to the future abundance and cost of energy. Unless such fundamentally different attitudes are recognized, the dialog is unlikely to be productive.

Energy Issues

Another source of discord in energy dialogs, in addition to differing views about the relationship of energy supply and demand, is the wide variety of consequences that may be the focus of concern. Energy is an issue that truly impacts on all aspects of our society.

A feeling for the broad range of issues involved can perhaps be gathered from the following lists of some energy-related issues on the international, national, local, and personal levels. Though far from exhaustive, these lists by their diversity

Fig. 6

Position E

1. Major new sources must be developed.
2. Restrict use.
3. National security is at stake.
4. Pollution control may prove unfeasible.
5. Federally fund research and development on large scale.
6. Fusion reactors must be developed.
7. Control federal land resources tightly.
8. Increase depletion allowance in major way.
9. Prevent exports and encourage imports.
10. Control resources tightly.
11. Government implements Plowshare.

clearly suggest that participants in any dialog on energy may be viewing the issue with widely different concerns.

International Issues

- Foreign resource development
- Export-import policy
- National security
- Soviet bloc trade relationships
- Seabed rights
- Nuclear treaty implications
- International health standards
- Monetary agreements

National Issues

- Development of federal lands
- Tax policy
- Interstate commerce
- Research and development support
- Environmental standards
- Antitrust policy
- Safety standards
- Area development plans
- Utility regulations

Local Issues

- Tax policy
- Building codes
- Zoning and plant citing
- Antipollution regulations
- Business development policies
- Mining regulations
- Business competition
- Transportation systems

Personal Issues

- Housing preferences
- Choice of transportation
- Working hours
- Budget priorities
- Environmental quality
- Personal health
- Employment opportunities
- Style of life
- Choice of elected officials

It is not surprising that a person concerned with the national security implications of a rapidly growing volume of oil imports should find it difficult to conduct a constructive dialog with another person who is primarily concerned with the impact of a local power plant on his environment or a third person who is concerned with raising revenues for an urban transit system through increased taxation of utilities. (Relating the interstate flow of gas to the peaking of demand created by common working hours or determining the impact of local building regulations on air conditioning cannot be an easy task.) Nor is a business corporation likely to find much stockholder support for a profit-reducing policy that ignores seabed resources whether or not the man with a home on the beach has his environment affected or another nation feels that its resources are being exploited.

This diversity also has another disruptive effect on energy dialogs, because it offers many techniques for solution. Although this diversity should and does create opportunities, it may also make the discussion more complex by offering many means to the same end. For the same goal of reducing pollution from the burning of sulfur-containing oil for electric power production, one may have a choice of importing more low-sulfur fuel, altering depletion allowances to encourage low-sulfur discoveries, promoting research and development on sulfur control or alternate power sources, relocating power production centers, regulating the use of air conditioning by changes in the building codes, staggering working hours to reduce peak loads, changing property taxes to inhibit additional home construction, altering interstate pipeline policies to make an alternative fuel more available, or increasing consumer cost to encourage reduced consumption. Each technique will have its advocates. Few will be aware of the total implications of the policy that they advocate.

It would be a grave error, however, to conclude that energy policy is merely a complex issue, to acknowledge only that it has many ramifications and that for each problem there will be a diversity of alternate remedial measures proposed. The situation is actually much more difficult than that to resolve. For what is at stake is a whole diversity of value systems that exist, at least in people's minds, as independent and unrelated areas of human concern. When we talk of our disadvantaged and the benefits that more cheap power might confer on them, the question of a potential military confrontation with the Soviet bloc countries is usually far from our mind. It is not commonplace to discuss both the seaward limits of national control over the potential riches of the ocean and urban transport systems at the same time. Our feeling of what is right and proper in one of these areas and what is fitting in another may well lead to conflicting conclusions as they relate to energy. The thread of energy runs through so many areas of human endeavor that it not only creates a conflict between various people but also leads to difficult problems of fairness and equity within each thinking person's spectrum of concerns.

No wonder then that strongly felt and highly emotional terms enter our dialogs on energy. But how is an issue involving such incommensurables to be resolved?

Ethics and Energy

Resolution of the variety of attitudes and the diversity of concerns that make up the energy dialog cannot be achieved through the acquisition of more data or through greater educational efforts. Such measures will help, but they do not go to the root of the problem. They fall short of addressing the question of striking a balance among conflicting and incommensurate 'goods.' Such a balance can be achieved only within a broader framework, within what might properly be called an *energy ethic.*

The fundamental problem is all too easily illustrated. Who among us can perceive a rational calculation that will uniquely define the proper balance between national security and environmental quality, between encouraging private investment and promoting international accords, or between aiding the underprivileged and preserving open lands? Yet these seemingly unrelated issues and many more are all closely interconnected when they are viewed from the perspective of energy policy. To arrive at some solution requires that one make a set of value judgments within a broader moral framework or ethic. When it is shared between people, such an ethic can provide a basis for their common decision making.

Perhaps this problem can be seen more clearly if one starts with a more common phrasing of the issue. Such a frequently heard phrase is whether energy will or should be cheap or dear. Obviously, this phrase is meant in a relative rather than an absolute sense. But even on a relative basis such a decision implies difficult moral valuations. For in reality one is asking what sacrifices it would be worthwhile to endure to achieve a given level of energy cost and what benefits can be expected in return. Stated in this way, the issue can clearly be resolved into two parts: (1) relating costs to each of the other factors and (2) establishing the relative desirability of these other factors.

The first step is clearly amenable to analysis. It will benefit from the collection and evaluation of obtainable data. Education as to these relationships is clearly of benefit.

The second step, judging relative desirability, is not of the same ilk. For this step a different approach is required. This paper contends that the second step will require the development of a consensual energy ethic. This contention seems to be supported by the historical events that have resolved analogous problem areas. But before we pass on to such examples, the breakdown above of the problem into two parts deserves illustration.

Consider, for example, the conjecture that low-cost energy will confer benefits on the underprivileged and that government regulation to achieve this goal is desirable. Here the first part of the problem is to examine whether low-cost energy will really benefit the underprivileged or will merely debase the one commodity that they have to sell, their own energy for doing work. Also one must examine whether government regulation would really lead to low-cost energy, since such a consequence is not obvious and the amount of regulations versus cost must in any case be ascertained. Hence in the first part of the issue both benefits to the underprivileged and degree of regulation are factors that must be related to energy cost.

The second part of the question then arises. If government regulation is good and will reduce costs and if helping the underprivileged is good and can be identified with lowered costs, then the two together optimize this one set of factors. But, although many would agree that benefiting the underprivileged is desirable, an equal number believe that government regulation, even if it can reduce energy costs, is undesirable beyond some level because of other adverse effects. Even more poignantly, the very freedom whose attainment might be sought via cheap energy might be taken away by the regulation required for its achievement. How then are two good goals, such as benefiting the underprivileged and avoiding regulation, to be weighed against each other when they are found to be in conflict.

For some clues we can look at how other problems of similar complexity have been resolved. For example, since nations have existed, debate has continued on the proper procedure for their financing. With the growth of popular democracies, this question has been a matter for public decision. One crucial turning point for this debate within the United States was the institution of a federal income tax. Complicated though this issue was by questions of state's rights and constitutional law, the debate was rich in moral content. Although the fairness of taxing real wealth in land or goods was widely accepted, the desirability of taxing a man's ability to work, his earnings, was hotly contested and declared immoral by many. Ultimately, a national referendum in the form of a constitutional amendment was required after the initial taxation legislation was declared unconstitutional by the Supreme Court. Note also that a progressive or graduated income tax of the type now in vogue was still held in disrepute even after the principle of an income tax was accepted.

In this debate and the many that have followed and continued down to the present day, one can perceive how an issue such as energy policy may evolve. Clearly, there is no demonstrably right way to settle such an issue. The decision changes with time. What determines our policy at any time is a sense of equity or fittingness in the public mind. It is their consensual ethic about taxation that determines the balance of the many benefits and sacrifices at stake in any major decision on tax policy. So, too, must ultimately be the case with respect to energy. Improved analysis and education will make for a more knowledgeable decision. The ultimate decision will be made by the people on the basis of their sense of what is right and fitting.

In the pollution area we have at hand a very young and rapidly developing new ethic in the United States. The developments are in fact so recent that they are more readily perceived because they have not yet become adopted as an almost unquestioned way of life. We see legislation today in the pollution control area that would have been nearly unthinkable just a decade ago. A still growing public consensus has declared that excessive pollution is a fitting subject for legislation and has sanctioned severe punishments for offenders. Here many see the adverse results of inadequate analysis and inadequate education of the electorate. However, few can continue to doubt the power of a widely held consensual ethic.

The disparity between our growing pollution ethic and the almost total lack of an energy ethic leads to some marked contrasts. The neighbor who creates a completely legal smudge is looked down on, criticized, and suspected of being morally corrupt. The society feels that he deserves censure, that 'there ought to be a law against it.' But the neighbor who needlessly leaves his lights burning around the clock is merely considered eccentric and perhaps not too sound in his economic training. There is no consensual energy ethic that suggests that he be condemned or accused of moral turpitude.

There are many other examples where complex relationships between benefits and sacrifices have come to be in the proper bounds of free speech the severity of criminal punishment, the access to ocean shores, the proper constitution of juries, or any other factor that deeply touches men's lives in many ways. Conflicts in these areas can only be guided by impersonal analysis. They will be settled by a personal set of ethical standards held in common by a cross section of our society.

Although the need for such a consensual ethic seems clear, its detailed content cannot of course be determined in advance. This is true in the same sense that the taxation policies of popular democracies could not be predicted with any certainty. One is not seeking *the* answer. One is seeking *a* consensus.

The most that can be done is to discern some of the conflicts in value systems that such an ethic must address. The preceding section on energy issues has attempted to do this. And note that internal resolution at each level is required as well as resolution between levels.

It is not possible to be more explicit about such an energy ethic without risking a charge of prophecy or advocacy. But one can set down a list of energy-related questions against which a suggested ethic might be tested. The following is such a list, based on some issues that currently appear to be of importance and urgency:

Is there some minimal level of energy consumption to which every individual has a right?

Is man's tendency to use even more energy to be regarded as a civilizing trend or as the reflex action of one hooked on a habit-forming drug?

Do nations possess a total freedom to use their energy as they see fit, or are they bound by some world responsibility to mankind?

Does the state have any right to determine how or for what purpose the individual uses the energy available to him?

Should energy be treated as taxable real property, and, if so, what value should be placed on unrealized potential or on fuel types that are perceived to be finite in nature and nearing exhaustion?

To what extent should the direct and indirect costs related to energy production be passed on to the consumer rather than be borne by a broader segment of the public?

Is federal sponsorship of new energy generation techniques a proper use of tax dollars, and, if so, what equity is due those with prior investments?

What is the proper balance between cooperation and self-sufficiency with respect to energy in the international arena?

A comprehensive energy ethic should provide for the resolution of these questions. But I think that most persons will agree that even a personal set of self-consistent answers is difficult to arrive at.

Perhaps these questions can be epitomized in a simple question: Is energy a right, a privilege, or a commodity? To this question no single simple answer would seem sufficient. For on close inspection, energy appears to be more like air, water, and land, i.e., a dimension of human existence inextricably intertwined with all that man chooses to be.

Conclusions

The pejorative opening words of this paper highlighted the nonproductive outcome of many dialogs on energy. This result has been attributed to a special set of causes: (1) a broad set of possible attitudes or views as to the reality of an 'energy crisis,' (2) the tremendous diversity of concerns directly related to energy issues, issues of concern at every level from personal to international and covering most fields of human endeavor, and (3) ethical values spanning many aspects of human existence.

Today there is no consensual energy ethic on which to base a resolution of the differing judgments of the various societal and personal values involved. Analysis and education can but lay a desirable foundation for the development of such a consensual ethic. On this fundamental level any real resolution must await or be accompanied by the development of such an ethic.

Acknowledgment

Research for this article was supported by the Mitre Corporation, Washington, D.C.; however, the views are the author's alone.

Part III
The Technology of Energy

The papers in this section provide a brief look at most of the major technologies used or proposed for use in energy conversion, transportation, and storage. While there is considerable overlap, the papers group roughly into the following six areas: geophysical energy sources, fossil energy sources, nuclear fission and fusion, energy storage, energy transmission, and energy use in transportation.

The term "geophysical energy sources" is commonly used to refer to energy which can be obtained by drawing upon the natural energy *flows* within the earth system. There are two sources for this energy, the solar radiation flux and the earth's own self-heat, which presumably results from the decay of radioactive materials. Falling water was one of the first of these sources which man turned to, and until the last few decades hydropower has been viewed by many as the ideal energy source. Anyone who can visit Hoover, Robert Moses, Grand Coulee, or Hungry Horse without a surge of delight in the face of clean engineering beauty has missed out on one of the most interesting and creative currents in the past 150 years of American social history. However, with the growing realization that additional conventional hydroelectric capacity can have only a modest impact on our energy needs, but a devastating impact on our remaining free running rivers, growth in conventional hydropower has now become widely unpopular.

At the same time, the need for large blocks of flexible power to meet peak electric energy loads has prompted the development of pumped storage hydroelectric facilities and the full or partial diversion of many conventional hydroelectric facilities from base load generation to peak load generation. Because it can be rapidly turned on and off, hydropower is well suited to peak load generation. However, such use frequently draws criticism because of the rapid fluctuations in water level which result. The first paper in this section provides a general look at hydroelectric power in the United States; the second provides a brief case study in river basin development.

Over the years, man has made use of other geophysical energy sources. Hot springs have been used for heating and for baths, the wind has driven mills, ships, and even the occasional land vehicle. But until recently none of the other geophysical energy sources has witnessed the systematic application of advanced technology which has been applied to hydropower. Now, under the pressure of rising demand, and the accompanying stimulus of rising prices, we are turning to a careful evaluation of such other "clean" energy sources as direct solar radiation, ocean thermal gradiants, wind, and hot rock. Much of the support for this new research has come from the federal government, largely through the Research Applications Directorate (RANN) of the National Science Foundation.

Papers 3 through 10 of this section offer a look at several avenues of current work on geophysical energy sources. Because, viewed at least from human time scales, geophysical energy sources draw upon energy *flows* within the earth system rather than upon finite fuel reserves, they are widely viewed as attractive long-term energy sources. It also appears that they can be designed to produce a very low per-unit energy environmental impact—but promises of "no environmental impact" should be greeted with healthy skepticism.

For years poets, and not a few engineers, have talked of harnessing the tides. However, with the exception of a handful of special locations, this is one geophysical energy source that is not being looked at very seriously. There is not much energy available from this source. Doubters should be able to quickly convince themselves with some simple back-of-the-envelope estimates.

Traditional man has depended upon fuels like wood and animal fats which provide short-term biological storage of solar energy. With industrialization, energy demands became so great that the usual small stocks of short-term storage fuels were gradually abandoned in favor of the much larger stocks of accumulated fossil fuel, first in the form of coal, oil, and various types of trapped gas, and now also in the form of oil shale and tar sands.

Paper II in this section reviews the current status of fossil-fueled steam-electric generation, while the following three papers address pollution control problems, largely related to the burning of coal. Of course, a large fraction of the fuels used in the United States are consumed directly and are not involved in electric power generation. The large and important field of combustion technology is only touched on in the reprints of this section.

Coal, our most abundant fuel, has traditionally been the dirtiest. Developing the technology to produce clean power from coal has become a major national research and development priority with most of the federal research support provided by the Office of Coal Research (OCR) in the U.S. Department of the Interior. A useful summary of much of this work is contained in the OCR report "Clean Energy from Coal—A National Priority" [1]. A review of technical approaches to coal gasification is provided in paper 15.

A major effort is also being undertaken to develop the technology to exploit our large reserves of oil shale, that is, rock impregnated with oil-like organic compounds. Oil shale development is a very controversial subject because of the considerable uncertainties at this stage in land and water use requirements and other possible environmental impacts. At the time this book went to press, we were unable to find a brief unbiased discussion of this subject. The interested reader is referred to the four-volume "Final Environmental Impact Statement for the Oil Shale Leasing Program" prepared by the U.S. Department of the Interior [2] and to the report," A Scientific and Policy Review of the Prototype Oil Shale Leasing Program—Final Environmental Impact Statement of the U.S. Department of the Interior," prepared by the Environmental Impact Assessment Project of the Institute of Ecology [3].

Papers 16 and 17 deal with several important aspects of petroleum production, research, and development.

Papers 18 through 22 very briefly summarize nuclear fission and fusion technology and research. There are fundamentally three classes of technologies for converting nuclear fuel to energy through fission reactions: nuclear burners, characterized in this country by the boiling and pressurized light water reactors in commercial service; nuclear converters, characterized by the now commercial high-temperature gas cooled reactor (HTGR); and nuclear breeders. As their name implies, nuclear burners essentially use up their nuclear fuel to produce energy. Converters also use up fuel, but manage to convert a portion of other material into usable fuel. Breeders, by converting other material into usable fuel, actually manage to produce or "breed" more fuel than they consume.

The generation of electric power from fission has been a subject of controversy in this country for over a decade and it is clear that this controversy will continue within various professional communities and the public at large for many years to come. The problems include questions of public safety where one is required to estimate potential costs by multiplying a very small but uncertain probability of accident or damage by a possibly large but uncertain cost of damage. In addition to the risk problem, there are also serious problems of long-term public policy and social philosophy raised by the need to store radioactive wastes for periods long compared with the time scale of human history. Some of the social and technical aspects of this problem are discussed in papers in Section IV of this book. A nice semitechnical discussion of the technology of waste management is provided in the Atomic Industrial Forum's *Nuclear Power Waste Management* [4]. Finally in this area there are questions about the proper rate at which to adopt the various available technologies; for example, should one push rapidly for a large breeder program or proceed more leisurely through a stage of HTGR converters? A balanced reprint book on this subject would be very valuable but could easily fill as many pages as this volume. In our selections in this section and in Section IV we have attempted to hit a few of the highlights, but a comprehensive review has not been possible.

Except in the form of radiation from the sun, man has yet to realize the potential of energy from nuclear fusion. In paper 21, D. Rose gives an overview of this field as of mid-1971. A discussion of some of the more recent research in this area can be found in a *Physics Today* article by Post [5]. The relatively new field of laser fusion is summarized in paper 22.

The ability to store and transport energy is of great importance, and the status of these technologies can have a major impact on the character of the total energy system. Most of the papers selected in this area reflect research on important new technologies. Some brief characterization of the current electric distribution system is given in paper 2 of Section II. No serious treatment has been given to such important technologies as energy transport by tanker, rail, or pipeline, all of which may have enormous carrying capacities when compared with large high voltage electric transmission lines.

The last two papers in this section are on energy use in transportation with emphasis on the problems of efficiency and conservation.

REFERENCES

[1] "Clean energy from coal—A national priority," Annu. Rep. Office of Coal Research, U.S. Dep. Interior, Feb. 1973. Available from the U.S. Government Printing Office, Washington, D.C., Stock No. 2414-00056.

[2] "Final environmental impact statement for the prototype oil shale leasing program," 4 vols., U.S. Dep. Interior, Aug. 1973. Available from the U.S. Government Printing Office, Washington, D.C. Stock No. 2400-00785.

[3] K. Fletcher and M. F. Baldwin, Eds., "A scientific and policy review of the prototype oil shale leasing program: Final environmental impact statement of the U.S. Department of the Interior," prepared by the Environmental Impact Assessment Project of the Institute of Ecology, 1717 Massachusetts Avenue, N.W., Suite 300, Washington, D.C. 20036, Oct. 1973.

[4] J. F. Hogerton, J. G. Cline, R. W. Kupp, and C. B. Yulish, *Nuclear Power Waste Management*, Atomic Industrial Forum, Inc., 1971.

[5] R. F. Post, "Prospects for fusion power," *Phys. Today*, vol. 26, pp. 30–39, Apr. 1973.

CONVENTIONAL AND PUMPED STORAGE HYDROELECTRIC POWER

Introduction

Conventional hydroelectric developments use dams and waterways to harness the energy of falling water in streams to produce electric power. Pumped storage developments utilize the same principle for the generating phase, but all or part of the water is made available for repeated use by pumping it from a lower to an upper pool. At the end of 1970, conventional hydroelectric capacity totaled 52,323[1] megawatts compared to only 3,689 megawatts of pumped storage capacity. During the next 20 years, however, the installation of pumped storage capacity is expected to exceed greatly the installation of new conventional capacity. By 1990, conventional hydroelectric capacity is expected to total approximately 82,000 megawatts and pumped storage capacity about 70,000 megawatts. Existing conventional and pumped storage capacity and projections for 1990 are shown by National Power Survey regions on figure 7.1.

Over the years, hydroelectric plants have provided a substantial but declining proportion of the nation's electric power supply. This trend is expected to continue despite the construction of many large pumped storage plants. Hydroelectric plants, which now account for 16 percent of total generating capacity, are expected to provide about 12 percent of the total capacity in 1990.

Operating Characteristics

Hydroelectric power plants have distinct advantages over thermal plants. Operation and maintenance costs are relatively low, and in many instances, the plants can be designed for automatic or supervisory control from a remote location. The cost of fuel, a major expense in thermal installations, is not an item in the operational costs of hydroelectric plants except for the consumption of pumping energy at pumped-storage plants. Hydroelectric installations have long life and low rates of depreciation. Unscheduled outages are less frequent and downtime for overhaul is of short duration because hydroelectric machinery operates at relatively low speeds and temperatures and is relatively simple. A hydroelectric unit is normally out of service about 2 days per year due to forced outages and about 1 to 2 weeks for scheduled maintenance. The average outage rates of modern steam-electric units are several times greater.

The ability to start quickly and make rapid changes in power output makes hydroelectric plants particularly well adapted for serving peak loads, and for frequency control and spinning reserve duty. If operating at less than full load, they are, in most cases, able to respond very rapidly to sudden demands for increased power. Their ability to supply starting power to steam-electric plants following a major power failure has been demonstrated on several occasions in recent years. There are no emissions that would affect air quality and there are no heat discharges to the receiving waters. Conventional hydroelectric developments do not consume natural fuel resources. Under some circumstances, they can provide a source of replacement power for use when generation at fossil-fired plants might need to be reduced during air pollution alerts. They occupy large areas of land, however, and cause short- or long-term changes in stream regimens, including such items as reservoir drawdowns, so they are often strenuously opposed on esthetic or ecological grounds.

Conventional Developments

Many associated benefits are provided by conventional hydroelectric developments. Reservoirs are used extensively for recreation, and may

[1] Includes 682 MW of industrial capacity.

Reprinted with permission from *Nat. Power Survey Rep. FP1.2:P87/970/PT.1*, ch. 7, Dec. 1971.

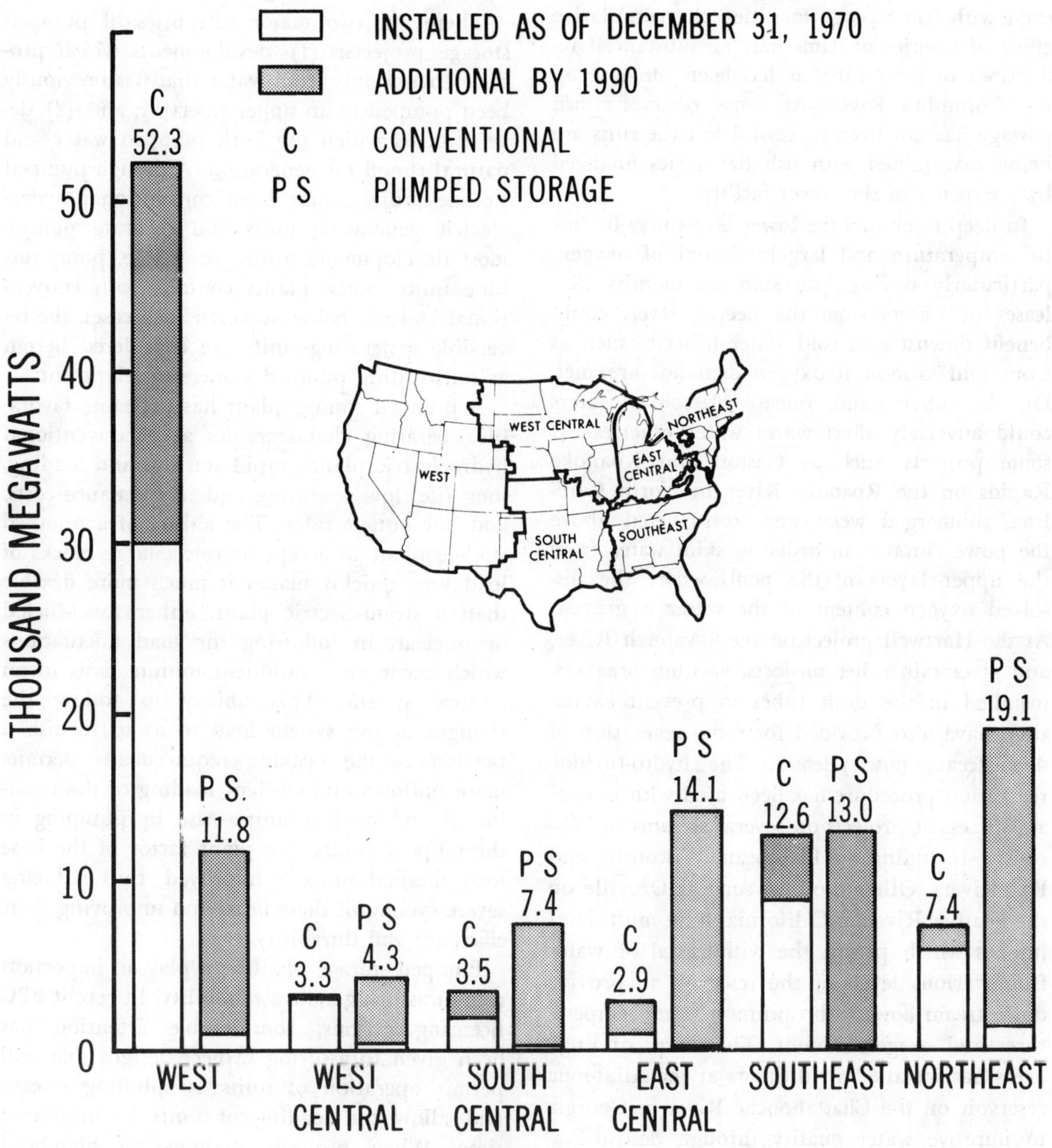

Figure 7.1

provide water for domestic and industrial water supply and condenser cooling water for thermal-electric plants. Some projects provide flood control, and some are operated to supplement natural flows during low-flow periods to provide benefits related to such non-power functions as water quality control, navigation, municipal water, irrigation water, and fish and wildlife.

Streamflow regulation by reservoirs may have adverse as well as beneficial effects, but special measures can usually be taken to minimize adverse effects. For example, a reregulating reservoir may provide uniform flows downstream from a peaking hydroelectric plant. Facilities may be constructed to provide for passage and protection of anadromous fish runs. However, even with fish passage facilities, the cumulative effect of a series of dams may be substantial reductions of such runs, as has been the case on the Columbia River. At some reservoirs fish passage has not been successful but the runs are being maintained with fish hatcheries financed by the owner of the power facility.

In deep reservoirs the lower levels may be low in temperature and largely devoid of oxygen, particularly during late summer months. Releases of waters from the deeper levels could benefit downstream cold water fisheries such as trout and salmon if oxygen demands are met. On the other hand, release of colder waters could adversely affect warm water fisheries. At some projects, such as Gaston and Roanoke Rapids on the Roanoke River in North Carolina, submerged weirs were constructed above the power intakes in order to skim water from the upper layers of the pools where the dissolved oxygen content of the water is greater. At the Hartwell project on the Savannah River, and at certain other projects, vacuum breakers, installed in the draft tubes to prevent cavitation, have also provided for some reaeration of downstream flow releases. The hydro-turbine reaeration procedure has been used with moderate success at projects on several streams in Wisconsin, including the Flambeau, Wisconsin, and Fox Rivers. Other projects, such as Oroville on the Feather River in California, have multi-level intakes which permit the withdrawal of water from various levels of the reservoir to provide downstream flows with optimum water temperatures and oxygen content. The Corps of Engineers has installed air diffusers at the Allatoona reservoir on the Chattahooche River in Georgia to improve water quality through destratification. They have provided significant increases in the oxygen content of water discharged from the reservoir. Some experimental use has been made of special spray type valves for reaeration of downstream releases to improve oxygen content. At some dams, spillway discharges have caused downstream waters to be supersaturated with nitrogen, resulting in adverse effects on fish. Additional research, including comprehensive ecological evaluations, is needed on means of improving the quality of water released from deep reservoirs.

Pumped Storage Developments

There are two major categories of pumped storage projects: (1) developments which produce energy only from water that has previously been pumped to an upper reservoir, and (2) developments which use both pumped water and natural runoff for generation. Although pumped storage projects may have conventional hydroelectric generating units and separate pumps, most developments utilize reversible, pump-turbine units. Some plants contain both conventional and reversible units. In such cases, the reversible generating units are considered herein as constituting pumped storage developments.

A pumped storage plant has the same favorable operating characteristics as a conventional hydroelectric plant—rapid start-up and loading, long life, low operating and maintenance costs, and low outage rates. The ability of a pumped storage plant to accept or reject large blocks of load very quickly makes it much more flexible than a steam-electric plant, either fossil-fueled or nuclear, in following the load fluctuations which occur on a minute-to-minute basis in an electric system. This ability to follow the changes in the system load so as to furnish a portion of the peaking requirements permits more uniform and efficient loading of the fossil-fueled and nuclear units. Also by pumping in the off-peak hours, the plant factor of the base load thermal units is improved, thus reducing severe cycling of these units and improving their efficiency and durability.

Pumped storage plants can play an important role in assuring system reliability. In recent FPC licensing actions, considerable attention has been given to assuring project designs that will permit operation of units as spinning reserve and allow the loading of units in minimum time. When properly designed, a high-head pumped storage unit may operate as a synchronous condenser, with the spherical valve closed, and be fully loaded in about one minute. Units in the lower head ranges may be operated as synchronous condensers without the use of cut-off valves and have an even faster response to load changes. Also, a unit can be operated at 50 percent to 60 percent of full load and have the ability to pick up the remainder of its capacity in about 15 seconds. This provides an ideal source of spinning reserve capacity to protect a system at a time when generating capacity is suddenly lost. In the event of an emergency on the system during the pumping cycle, the system load may be reduced quickly merely by dropping the pumping load. By rapidly picking up full generating load, approximately twice the capacity of the plant can, in effect, be made available to the system to meet a generation deficiency occurring during the pumping cycle.

Operating experience at the Muddy Run project on the Susquehanna River in Pennsylvania shows that, if the units are in a normal ready condition, one of the 100-megawatt units can be started in three minutes and all eight units in the plant can be started in 12 minutes. The changeover time from pumping to generating cycle is 5½ minutes for one unit or 20 minutes for the plant. In a recent test, two units were tripped while in the pumping cycle and two others were immediately started in the generating mode, for a net increase in power of 460 megawatts in approximately six minutes.

In normal operation of the Cabin Creek plant in Colorado, shown in figure 7.2, a period of 10 to 13 minutes is required to synchronize and fully load each of the two 150-megawatt units. In an emergency, it is possible to synchronize and load both machines simultaneously in approximately the same time period. The time required for the plant to change from pumping to generating or vice versa is about 25 minutes.

The design of the eight 250-megawatt units planned for the Cornwall project, New York, provides for changeover from full pumping to full generation in 158 seconds and from full generation to full pumping in 17 minutes. Design for a relatively short turn around time from pumping to generating is usually feasible but, because of pump starting problems, the turn around time in the reverse direction is usually longer, as illustrated by the Cornwall design.

Use in Serving Loads

Hourly loads of a southern utility system for the peak week in August 1968, a week when extremely hot weather was experienced, are shown in figure 7.3.

Figure 7.2—The Cabin Creek project of the Public Service Company of Colorado, a 300-megawatt pure pumped storage development utilizing a gross head of 1,199 feet, went into operation in 1967.

The usual practice is to use new, efficient steam-electric units, either fossil-fueled or nuclear, to serve the base portion of the load. The less efficient steam-electric capacity, usually the older equipment, is used to serve the higher portions of the load and therefore operates at a lower capacity factor. Normally, conventional and pumped storage hydroelectric capacity is used to serve the peak portions of the load, although gas turbines may be used to serve sharp peaks of short duration. Figure 7.3 is an illustration of normal practice.

In some cases, conventional hydroelectric capacity may operate in lower portions of the load. Some plants must operate at high capacity factors because the rate of flow releases must be relatively constant in the interest of navigation or other purposes. Also, run-of-river plants operate at high capacity factors during periods when available streamflows permit.

Because of losses in the pumping-generating cycle, pumped storage plants require approximately three kilowatt-hours of pumping energy to provide two kilowatt-hours of generation. Therefore, the availability of a dependable supply of pumping energy is essential. Equally important is the necessity of constructing reservoirs with sufficient storage capacity to fit the pumping and generation requirements. For the load shown on figure 7.3, steam-electric energy would be available only about 10 hours each weekday

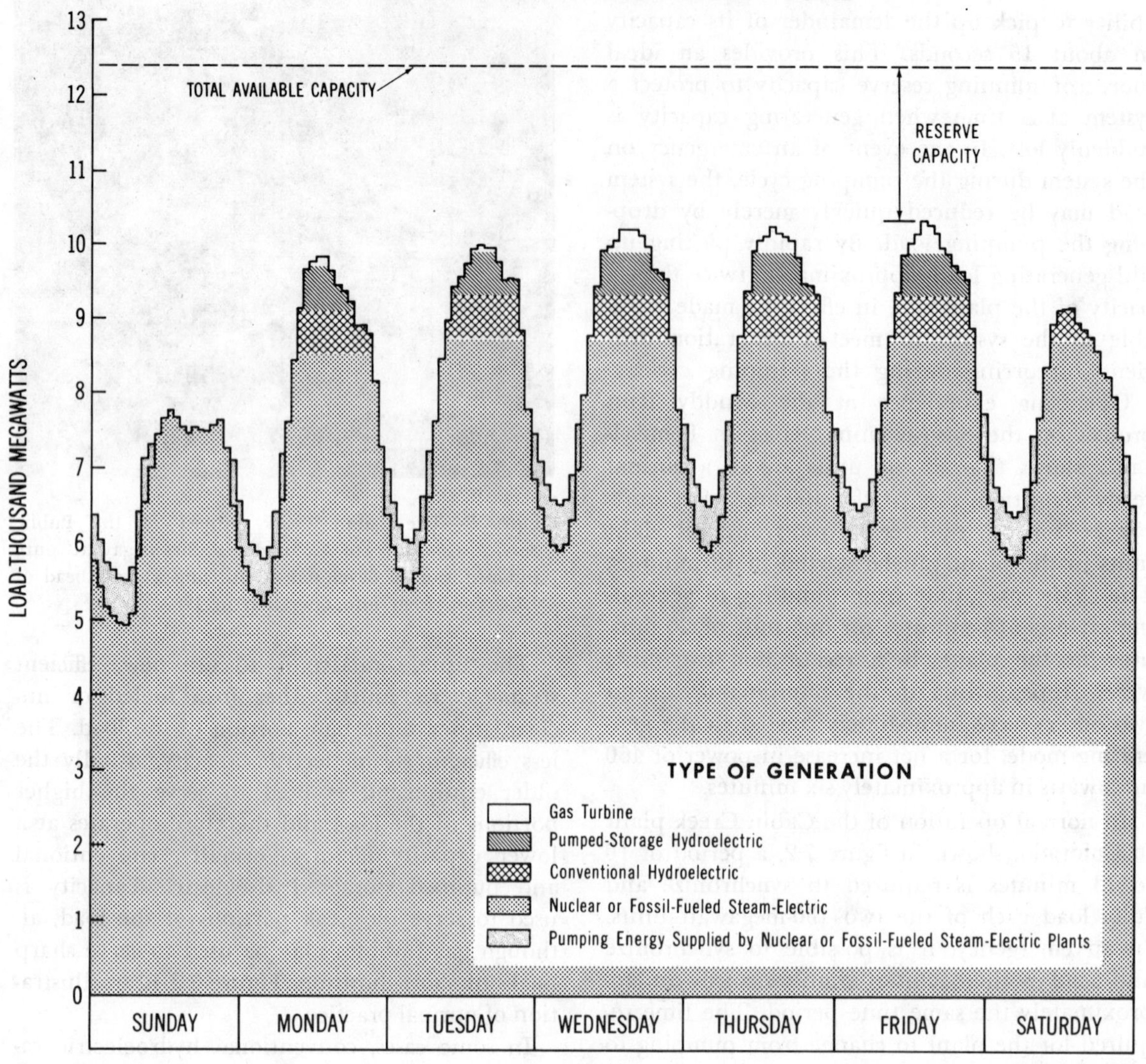

Figure 7.3

night to provide pumping energy. Because the time of the pumping cycle may be as much as 1½ times that of the generation cycle with units operating at full load, a daily pumping cycle of eight or nine hours would assure only about six hours of generation per day. This would be inadequate to serve the load shown on figure 7.3. Normally, therefore, it is necessary to construct reservoirs having adequate storage capacity to permit operation on a weekly cycle with substantial pumping over weekends, as depicted on figure 7.3 Reservoir storage capacities should be sufficiently large to assure dependable operation under most adverse load conditions. In most areas such conditions occur during summer peak periods when long and continuous periods of high temperatures are experienced. Under these criteria, with few exceptions, the reservoirs of pumped storage developments should have sufficient storage capacities to permit from 10 to 20 hours of continuous full load operation.

Trends in Development

The current trend toward construction of very large nuclear and fossil-fueled steam-electric units which operate best at high plant factors has increased the need for plants designed specifically for peak load operation. This has led to the construction of pumped storage and other hydroelectric plants with large generation capacity designed for low plant factor operation. In some cases, existing plant capacities are being greatly expanded to meet system peaking needs. Various studies of the feasibility of increasing the capacity of existing plants are under way, including studies of large plants on the Columbia and Missouri Rivers.

In many cases, conventional hydroelectric developments with reasonably high heads can have their capacities increased substantially by the construction of lower pool afterbays and the installation of reversible units that can be used for both pumping and generating. Such combined developments usually have some advantages in flexibility of operation because of the large upper reservoirs normally available in such projects. The Smith Mountain project in Virginia and the Oroville project in California are examples of projects where the inclusion of reversible units made possible a substantial increase in total project capacity.

Generating units are being constructed in increasingly larger sizes. For many years, the largest conventional hydroelectric units were those of 150 megawatts in the Robert Moses plant at Niagara Falls, shown in figure 7.4. The enlargement of the Grand Coulee power installation now under way will use 600-megawatt units. Units of 270 megawatts are being constructed for the 1,620-megawatt Ludington pumped storage project adjacent to Lake Michigan in Michigan and 382.5-megawatt units are scheduled for the 1,530-megawatt Raccoon Mountain pumped storage project on the Tennessee River in Tennessee. This trend to plants and units of larger sizes is likely to continue, with probable savings in both capital and operating costs.

For developing economically the power potential of sites with heads in the range of 15 to 35 feet, axial-flow turbines of the tubular type have been designed and are now being constructed. Also, bulb units of the type developed in Europe are being considered for installation at certain low-head plants in this country.

Improvements in the design and construction of dams are contributing to the economies of hydroelectric power developments. Also contributing to such economies are advances in the techniques and equipment used in tunneling and other underground excavation. In some instances, underground powerhouses are proving to be economical. In other cases, they may be adopted in order to preserve the natural surface conditions at the plant sites. Many new plants and some existing plants are being equipped for remote control to reduce operating costs.

Figure 7.4—Robert Moses Niagara Power Plant of the Power Authority of the State of New York houses thirteen 150-megawatt generating units. The Lewiston Pumped storage plant shown in the background has 12 units used either as 37,500-horsepower pumps or 20-megawatt generators. The total installed capacity of the two plants is 2,194 megawatts.

Experience has shown that care must be exercised in selecting the site and designing the upper reservoir of a pumped storage development to avoid excessive leakage of water. Initially, excessive leakage occurred at the Taum Sauk (Missouri), Salina (Oklahoma), and Kinzua (Pennsylvania) projects, necessitating various remedial measures. The experience gained at these projects and others under construction should prove helpful in providing better designs for future projects. A unique lining is being provided for the upper reservoir dike of the 1,620-megawatt Ludington project adjacent to Lake Michigan. Beginning from the surface and proceeding toward the foundation, there will be (1) a mastic seal coat, (2) two, three-inch layers of asphaltic concrete, (3) a two-inch asphaltic base course, (4) a drainage zone of 18 inches of crushed rock, (5) a two-inch layer of asphaltic concrete, (6) a two-inch layer of asphaltic sub-base, and (7) a 4½ foot thick layer of calcareous silty sand. Submersible pumps will be embedded in the crushed rock drainage zone and connected with a continuous 12-inch perforated pipe located in the lower portion of the drainage zone and running around the reservoir.

There is considerable flexibility in the design of pumped storage projects. An example is the licensed Kinzua pumped storage project which has two 198-megawatt reversible units and one 26-megawatt conventional unit. The Corps of Engineers' Allegheny Reservoir serves as the lower pool. One of the reversible units, through use of a divided draft tube, has been designed to discharge either into Allegheny Reservoir or into the river below the dam. A sketch of the Kinzua project is shown in figure 7.5.

An FPC license has been issued to Arizona Power Authority to construct the Montezuma project in Arizona with an initial installation of 505 megawatts. Original plans were to use the effluent from a sewage treatment plant serving Phoenix as the source of water. Since, under present plans, the effluent will be used for irrigation and thus will not be available to replace water lost during project operation, groundwater from wells will be used instead. Upper and lower reservoirs of equal capacity will be constructed to provide a gross head of about 1,660 feet. As initially planned, there would have been an underground powerhouse with five conventional 100-megawatt generating units and five, three-stage pumps. Reversible units were not proposed because of the high head. In reviewing the plans for this project, however, it was concluded that reversible pump turbine units could be used and that this would result in a reduction of about $14 million in construction costs. Accordingly, the plans now provide for the installation of four 126-megawatt reversible units.

Because of the relative newness and the physical complexities of large pumped storage projects, the effects they may have on the aquatic environment are not well understood. The power houses are designed with deeply submerged pump-turbine runners to reduce negative pressures which cause cavitation and injure fish. Ob-

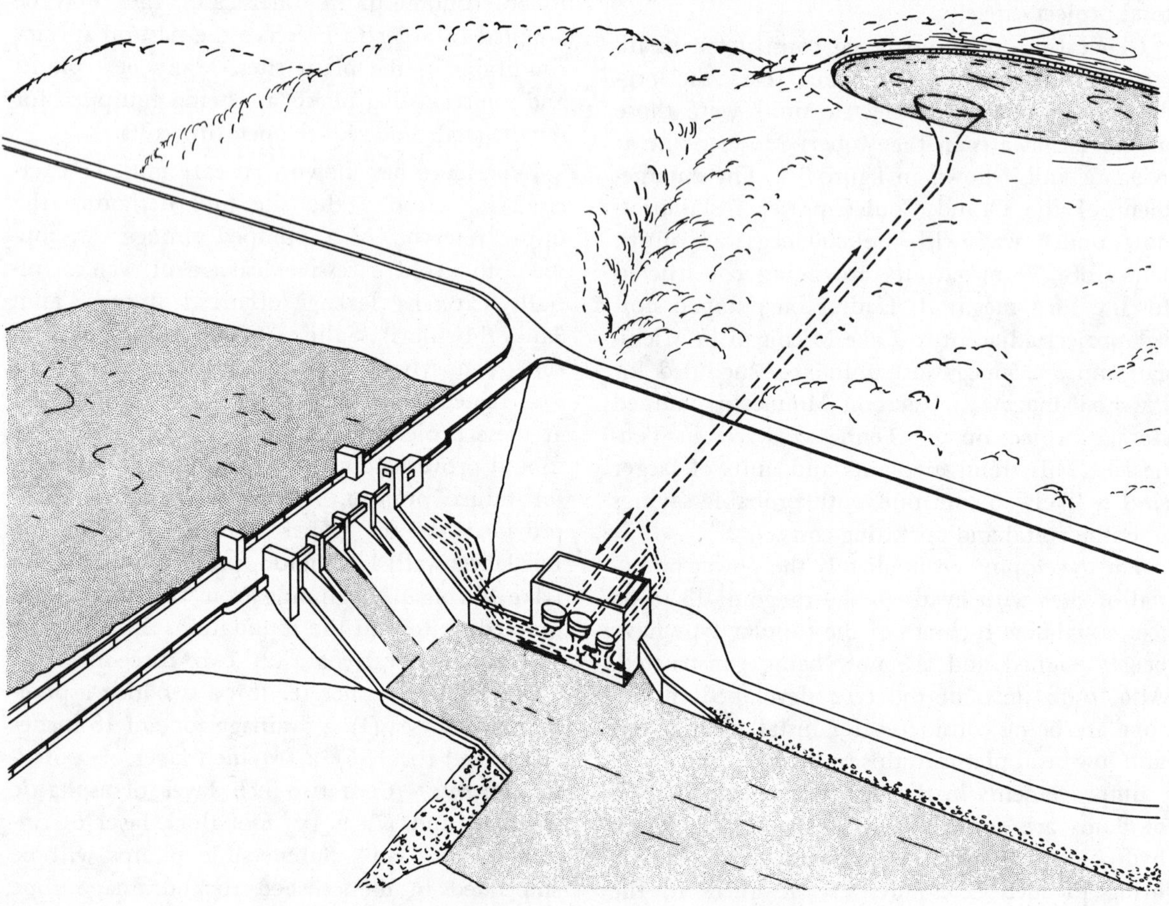

Figure 7.5—Sketch of Kinzua pumped storage project.

Figure 7.6—Philadelphia Electric Company's Muddy Run Recreation Park, developed as part of the pumped storage project, includes a constant level, 100-acre recreational lake in the upper power pool.

servations at some projects, although unconfirmed by quantitative studies, indicate that mortality rates generally are not significantly high for fish in the water passing through such units. The widely fluctuating reservoir surface elevations associated with pumped storage projects may adversely affect the spawning of warm water fish although some studies have shown that these species adapt to the fluctuations.

Trashracks and fish screens at pumped storage projects are regarded as self-cleaning because of the alternating direction of water flows through the units. Therefore, the design of fish protective facilities, when needed, is quite flexible. However, the location of fish screens and the velocities of water approaching them require particular attention. Additional studies are needed on the direct and indirect effects of pumped storage projects on the aquatic environment.

Because of the wide fluctuations in reservoir levels, many pumped storage projects are not suitable for recreational use. However, recreation facilities usually may be provided adjacent to, or in connection with, pumped storage developments. In some cases, sub-impoundments can be provided for recreational use, as shown in figure 7.6.

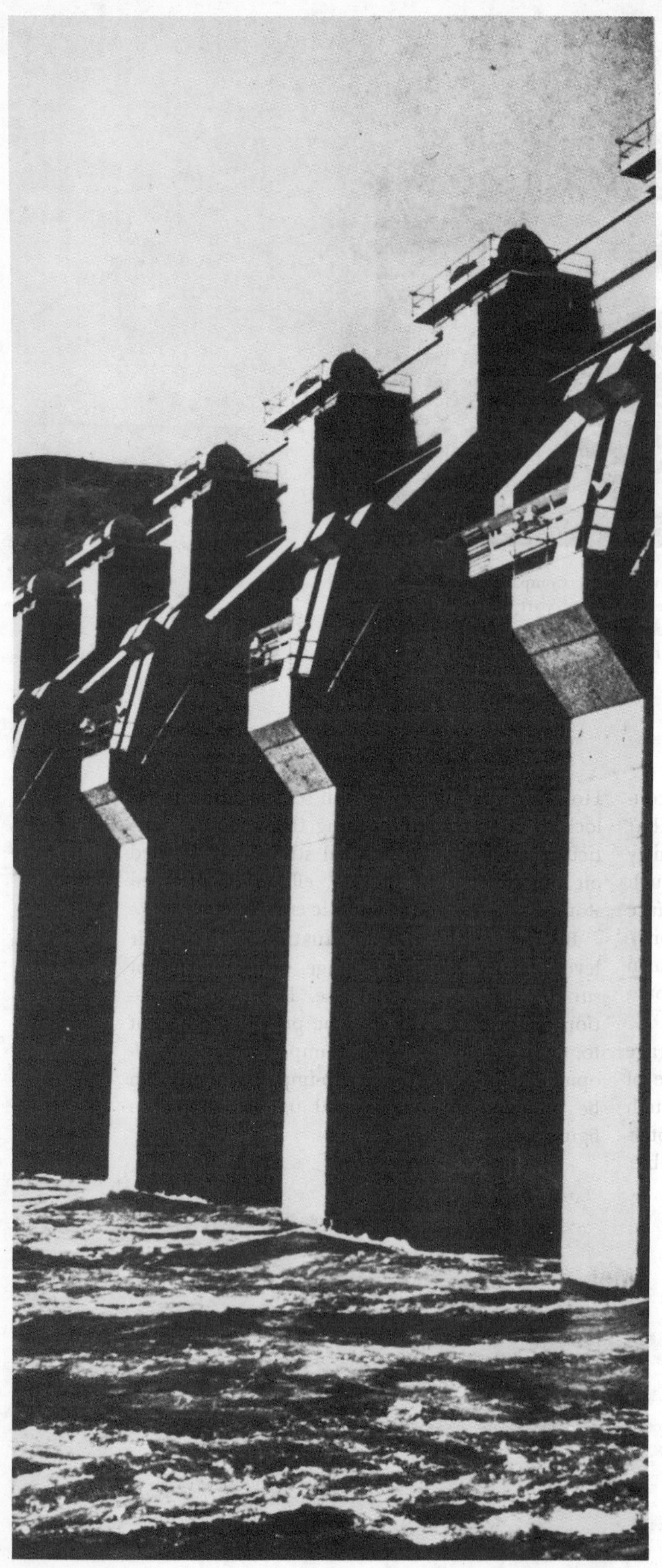

The story of Bonneville Power: 1937–1968–1987...

Dams of the Columbia River Basin

Bonneville Dam, the original "make work" project built in the depression days of the 1930s, was the forerunner of one of the world's most extensive hydro power projects. Today—31 years and 23 dams later—the Columbia River Basin is reaching toward its ultimate development as a vast source of electric energy

Gordon D. Friedlander *Staff Writer*

Realization of the vast hydro power and flood-control potential of the Columbia River, its tributaries, and watershed basin dates back more than half a century. As early as 1923, the U.S. Department of the Interior and the Army Corps of Engineers presented comprehensive testimony to the Congress on the feasibility of such eventual development. And, in 1937, B. E. Torpen, a hydroelectric engineer, presented to the ASCE an outline of the power possibilities of the Columbia River and its tributaries. This paper, entitled "Where Rolls the Oregon," was an accurately prophetic blueprint for the optimum development of the Columbia River Basin over a 50-year period. In it, he anticipated the need for the Canadian Treaty to provide upstream storage and to interconnect the great hydro energy in the future of British Columbia. His dream and his predictions are well on their way toward ultimate fulfillment.

From 1937 to the present time, the population of the Pacific Northwest has increased by 100 percent. Six million people now live in the states of Oregon, Washington, Idaho, and the portion of Montana that is west of the Continental Divide. Thirty-one years ago the aver-

Reprinted from *IEEE Spectrum*, vol. 5, pp. 78–94, Nov. 1968.

age residential consumer in the four-state area required approximately 1200 kWh per year. As of 1967, the average per consumer increased tenfold—to more than 12 000 kWh annually.

The Columbia River—background to power

The mighty Columbia rises in the Selkirk Range of the Canadian Rockies in the Province of British Columbia. Its source is Columbia Lake, a 21-km-long, deep, glacial-fed pool that is situated 800 meters above sea level. From here, flowing southwestward 2000 km to the Pacific Ocean, the initially small stream is joined by more than 150 tributaries large enough to be designated as rivers. And on this long journey to the sea, the Columbia develops into one of the world's most powerful rivers. It is the fourth largest river in North America. Although the Mississippi, St. Lawrence, and the Mackenzie are larger streams, the Columbia easily outpaces them as a source of energy. The river and its tributaries contain 30 percent of the hydroelectric potential of the North American continent.

The Columbia River Basin drains an area larger than France. The watershed includes part of British Columbia and portions of Oregon, Washington, Idaho, Montana, Wyoming, Nevada, and Utah.

The flow of the Columbia and its tributaries, however, seasonally fluctuates over a wide range. For example, at the Canada-United States boundary, the Columbia's largest recorded flow was 15 400 m^3/s; its smallest was 390 m^3/s. The great river enters the United States about 140 km north of Spokane, Wash. At the international boundary, it has descended 410 meters and has traveled 800 km from its source.

The volume of water and its varying flow can create problems in flood control and power production. When heavy springtime precipitation is accompanied by a normal-rate melting of the snowpack on the mountain slopes of the basin, there will be adequate runoff for power production, irrigation, potable water, and fish breeding. But calamitous floods can ensue if warm, torrential rains cause a rapid melting of snows in the watershed area. At the other extreme, a lack of rainfall can reduce the runoff in the basin to that point where power production becomes critically affected; it may also sharply reduce the amount of water that can be diverted for agricultural irrigation.

Historical background

The early dams. The first sizable dams in the Columbia River Basin were built on the Upper Snake River in Idaho by the Bureau of Reclamation of the U.S. Department of the Interior. The Minidoka Dam, a 30-meter-high structure, was the initial project. It began delivering powe to nearby farms in 1909. This dam still serves its area with a generating capability rated at 13 400 kW. In 1912, the Boise Diversion Dam added 1500 kW of hydro power; and, in 1925 and 1927, respectively, Black Canyon and American Falls raised the generating capacity along the Bigwood and Snake Rivers by 38 500 kW. The generators at the earlier dams furnished electric power primarily to operate the irrigation pumps of federal projects.

Not all of the early hydro projects were federally sponsored, however. Private and public utilities in the Pacific Northwest completed the Twin Falls Dam (13 500 kW) on the Snake, in 1912; the Long Lake Dam (70 000 kW) on the Spokane River (Wash.), in 1918; the Bull Run Dam (21 000 kW) on the Bull Run River (Oreg.), in 1928; the Leaburg Dam (13 500 kW) on the McKenzie (Oreg.), in 1928; and the Lewiston Dam (10 000 kW) on the Clearwater, in the same year. The year 1927 saw

FAR LEFT. Spillway section of the John Day Dam, presently under construction. This facility will generate a minimum of 2160 MW in its first-phase service, scheduled to go on the line in 1970.

FIGURE 1. The nonfederal Merwin arch-gravity dam on the Lewis River in Washington was the first structure of this type to be built in the Pacific Northwest. Capable of generating 135 000 kW, electric energy from this facility was put on the line back in 1930.

the completion of the American Falls (27 500 kW) federal dam on the Snake. In 1930, the 100-meter-high nonfederal Merwin arch dam (135 000 kW) on the Lewis River in Washington (Fig. 1) was completed. There were also a number of other notable dams built in the basin area for flood control and irrigation purposes only in the 1910–1930 period. Figure 2 shows the existing and projected federal and nonfederal power dams in the Pacific Northwest, both within and outside of the Columbia River Basin. It may be used as a ready reference throughout this article.

Bonneville: the first of the great federal dams

In that grimmest year of the Great Depression, 1933, the initial large-scale power development of the Columbia River at Bonneville, Oreg., 65 km east of Portland, began

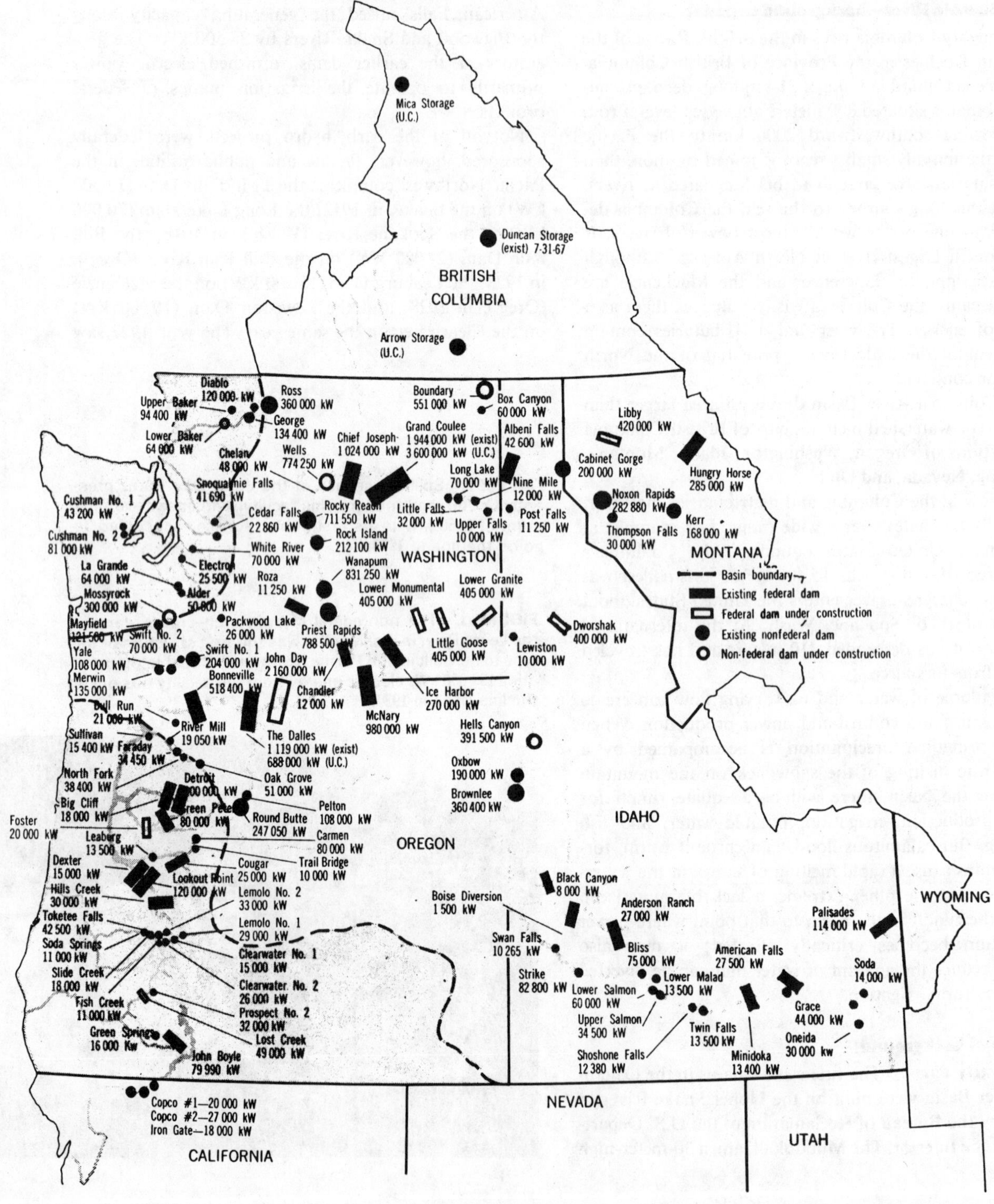

FIGURE 2. Federal and nonfederal hydroelectric dams in the Pacific Northwest, lying both within the outside of the boundaries of the Columbia River Basin. Quantities shown represent both existing generation and that which is being installed (where applicable). Symbols indicate federal and nonfederal projects authorized, licensed, or under active consideration, as well as potential additions.

on September 30, under the provisions of the National Recovery Act. For the short term, it was admittedly an emergency public works project to ease the huge unemployment problem then existing in the Pacific Northwest. The formal authorization for construction was made by Congress under the River and Harbor Act of 1935. For the long term, however, it represented the first step in a 50-year program.

The entire project was constructed—and is being maintained and operated at present—by the U.S. Army Corps of Engineers.

Details and statistics. The Fig. 3 map shows the immediate dam area and appurtenances. The Columbia River reservoir runs east of Bonneville to The Dalles Dam—the next great low dam in the system, situated 75 km upstream.

As may be seen in Fig. 3, the powerhouse portion of the dam is situated between Bradford Island and the Oregon shore. The original construction provided for two hydro generating units (plus a 4000-kW station service unit), each rated at 43 200 kW, for a total generation capacity of 86 400 kW with substructures for four additional units. The subsequent installation, completed in 1943, provided eight additional generators of 54 000 kW each—to bring the present total generating capacity to 518 400 kW.

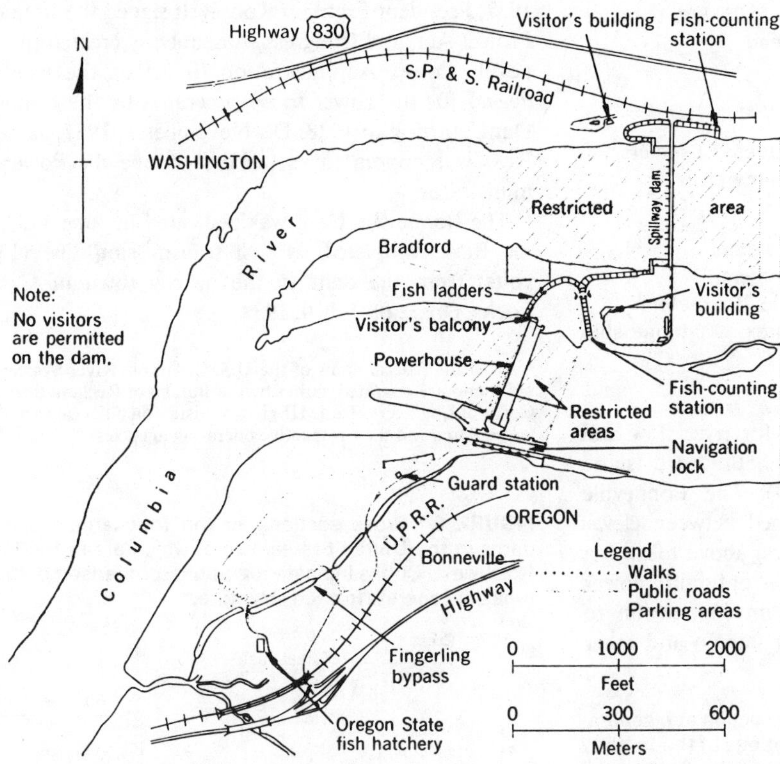

FIGURE 3. Map of the immediate area of Bonneville Dam and its appurtenances.

FIGURE 4. View of the spillway portion of the Bonneville Dam. This section forms a barrage between the Washington shore and Bradford Island.

I. Bonneville Dam statistics

Powerhouse

Length	313 meters
Width	58 meters
Height above lowest bedrock	58 meters
Station service unit	4000 kW
Number of hydro-generating units	2 @ 43 200 kW 8 @ 54 000 kW
Total rated capacity	518 400 kW
Generators	13 800 volts
Generator housing diameter	14.6 meters
Transmission voltage	115 000 and 230 000 volts
Kaplan turbines (adjustable blade)	2 @ 49 000 kW, 15-meter head 8 @ 55 000 kW, 18-meter head
Revolutions per minute	75
Discharge per turbine, in cubic meters per second	386
Propeller	5 blades, 7.1 meters diameter

Navigation Lock

Length	153 meters
Width	23 meters
Lift	9.2 to 21.4 meters
Capacity	approx. 8000-tonne ship

The spillway dam, used primarily for river flow and navigation control, is situated between Bradford Island and the Washington shore (Fig. 4). The Bonneville Reservoir pool is normally maintained between elevations 72 and 74 feet (22.0–22.6 meters) above mean sea level, and it is regulated to provide optimum power and navigation benefits with minimum obstruction to the upstream movement of migrating salmon and other fish. Daily fluctuations in water level are generally small; however, in emergency situations, the drawdown may be as great as 61 cm (2 feet) per day. During the annual spring high water, elevation 74 is usually exceeded at the crest of the flood. In the upper reaches of the reservoir, significant seasonal fluctuations are experienced as the result of varying tributary river flows. The backwater effect of these streams increases with the magnitude of their flow and their distance upstream of the dam. Flood flows and resultant high backwater elevations normally occur during May and June.

Table I gives the quantitative statistics for the powerhouse and navigation locks; and Fig. 5 shows a typical powerhouse cross section at a generator.

Bonneville Project Act: birth of BPA. On August 20, 1937, President Franklin Roosevelt signed the Bonneville Project Act, and Congress subsequently created the Bonneville Power Administration to act as the *marketing agent** for the power to be generated by the Bonneville Dam, starting in 1938. On November 1, 1937, James D. Ross was appointed as the first Bonneville Power Administrator.

The Bonneville Dam was dedicated in June 1938, and the BPA completed its first transmission line (13 800 volts) from the dam to the nearby town of Cascade Locks, Oreg., on July 9, 1938.

* All of the federal dams of the U.S. Columbia River System were built, and are operated, either by the Bureau of Reclamation or the Corps of Engineers. Table III gives statistics on all existing and proposed dams and their respective operating agencies.

FIGURE 5. A typical powerhouse cross section at a generator in Bonneville Dam. With minor variations in the forebay and tailrace construction, this section is applicable to most of the concrete low dam installations in the Northwest.

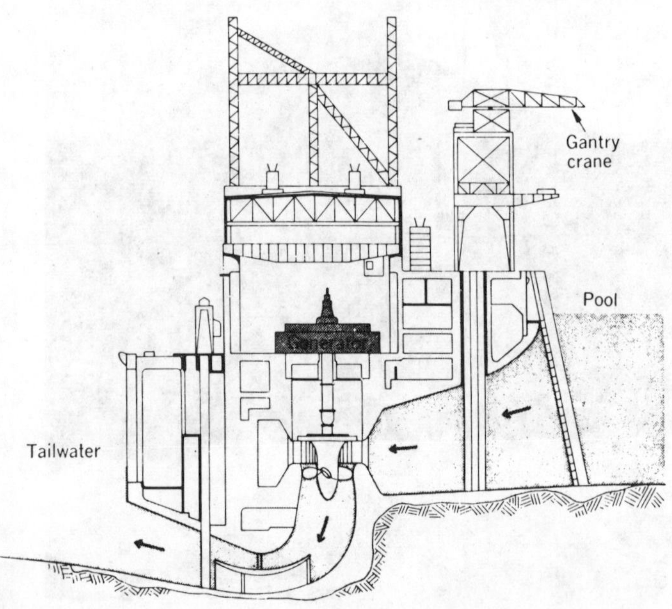

FIGURE 6. Cross sections of the four largest concrete dams in the United States, two of which are in the Pacific Northwest. Cubes indicate a graphic comparison of the volumes of concrete in each structure.

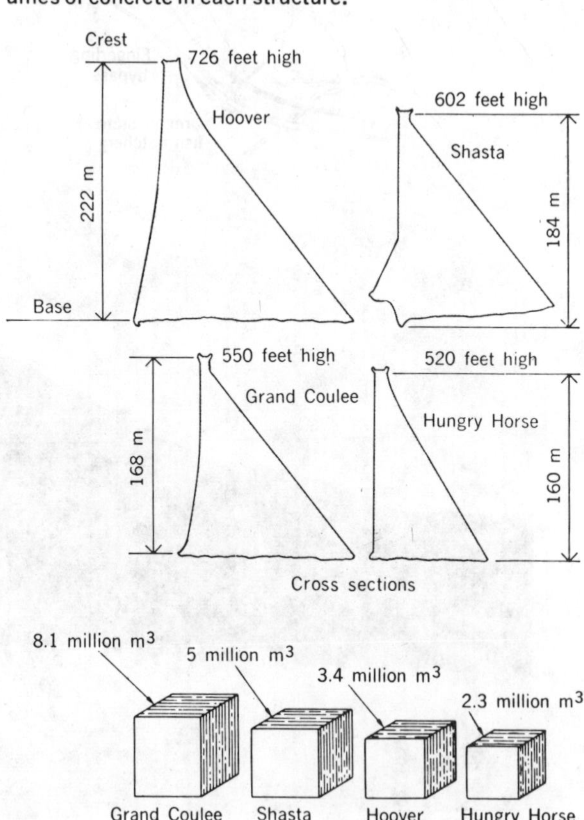

Public vs. private power. Bonneville effectively rekindled the 40-year-old simmering feud between the public and private power interests. In 1937, the industrial, commercial, and domestic consumer loads were adequately served by the existing public and private utilities. Many of the more vocal critics snorted: "What the hell are we going to do with an additional 300 000 kW of electric energy? We won't need that much by the end of the 20th century!" Thirty years later, the power dams of the Columbia River Basin were generating more than 16 000 MW (about 50 percent of this at federal dams)—with no end to the ever-increasing demand in sight.

War, progress, and Grand Coulee

When Administrator Ross died in March 1939, he left to his successor, Dr. Paul J. Raver, a master plan for the future growth of the BPA system. In 1939 and 1940, the first 115 000-volt transmission lines were erected from the Bonneville Dam westward to the J. D. Ross Substation in Vancouver, Wash. (directly across the Columbia from Portland), and from the dam eastward to The Dalles. The Ross Substation subsequently became the "power junction" for southwest Washington, western Oregon, and the Willamette Valley.

II. Grand Coulee Dam statistics

Length at crest	1272 meters
Volume of concrete (including appurtenances)	8.1 million m³
Length of reservoir	236 km
Max. depth of reservoir	153 meters
Total storage capacity	9.5 million acre-feet
Ultimate installed generating capacity	9435 MW

Meanwhile, World War II broke out in Europe and, in late 1940, a state of national emergency was declared in the United States. As part of our accelerated defense effort, work was speeded on meeting the power requirements for the expanding industrial complexes in the Pacific Northwest, and on the completion of the world's most ambitious hydro project up to that time . . .

The Grand Coulee Dam. This immense undertaking, regarded by skeptics as a white elephant and the greatest "boondoggle" of all federal projects, began in 1937 and was completed in 1942. To this day, in sheer volume of mass concrete—10.6 million cubic yards, or 8.1 million cubic meters (including the appurtenance structures)—Grand Coulee stands as the largest straight gravity dam in the world, and it will hold this distinction until the Aswan Dam, on the River Nile, is completed in the 1970s. Figure 6 shows the comparative volumetric quantities of the world's four largest dams and their relative crest heights. Note that Hungry Horse, another dam (to be discussed later in this article) of the Columbia River Basin system, is in this category. Like the other three in the figure, Grand Coulee is classified as a "high dam." Although its height is about 54 meters less than that of Hoover Dam, its crest is almost four times longer. Thus Coulee contains three times more volume of concrete than Hoover. Other construction statistics and reservoir information are contained in Table II.

Grand Coulee stands on a granite base, exposed in early geological times when the cordilleran icecap cut into the lava plateau to form the 500-meter-wide canyon in which the Columbia River flows (Fig. 7). The dam is situated 150 km northwest of Spokane and 390 km east of Seattle. The south end of this reservoir is 960 km from the mouth of the Columbia, and the normal surface elevation of the lake is 1288 feet (393 meters) above mean sea level. The water stored in this huge upper pool is used to irrigate more than 500 km

FIGURE 7. View of the downstream face of the Grand Coulee high dam and powerhouses. This facility, originally completed in 1942, will have an additional powerhouse by 1987, to bring its total generating capacity to more than 9400 MW.

of farmland, to regulate the flow of the Columbia River, and to develop a large block of electric energy that is used for local pumping and irrigation. This energy is also transmitted to distant load centers.

An extensive hydraulic distribution system lifts water needed for irrigation from the Columbia River and transports it through an 8-meter-wide concrete-lined flume for a distance of 96 km to the northerly ridge of the irrigable areas. South of the dam, a 44-km-long equalizing reservoir (which acts as a retention pool to store surplus water from the Columbia during flood season) was formed by earth- and rock-fill dams across an abandoned watercourse called the Grand Coulee.

The major portion of the electric energy produced at the dam is carried over BPA's 230-kV lines to substations near Spokane, Portland, and Seattle for distribution to industrial plants and to private and municipal power utilities of the Northwest Power Pool.

Power from Grand Coulee was first put on the line in 1942, when it was transmitted over 115-kV lines eastward to Spokane and westward, over the Cascade Range, to the Puget Sound area. This historic event actually represented the birth of the Northwest Power Pool, in which all of the power resources of the Pacific Northwest existing at that time were "pooled" for the use of industry in the massive war production effort of the United States. Bonneville and Grand Coulee Dams contributed 26 billion kilowatthours as their share of this task during the war years.

The present power plant was completed in 1951. It contains 18 main hydro generating units, each with a nameplate rating of 108 MW (plus three station service generators, rated at 10 000 kW each), for a total nameplate generation output of 1944 MW. Each of the generators, however, is capable of a short-time maximum output of 120 MW.

A third powerhouse, presently under construction, will boost the existing generating capacity by 3600 MW. This installation will consist of six generating units, each with a nameplate rating of 600 MW. Authorized for future construction is the installation of six more 600-MW generators, plus six pump–turbine units (to be used in a pumped-storage operation) for peaking power that will generate 291 MW, for an eventual grand total generating capacity of 9435 MW at Grand Coulee.

The postwar era begins: McNary and Hungry Horse Dams

McNary. Two years after the end of World War II, ground was broken for the construction of the Corps of Engineers' second Columbia River project, the McNary

FIGURE 8. Hungry Horse Dam, a graceful high dam structure of the arch–gravity type, contributes 285 MW of generated power to the BPA transmission system.

FIGURE 9. Construction plan, upstream elevation, and sections of the Hungry Horse Dam.

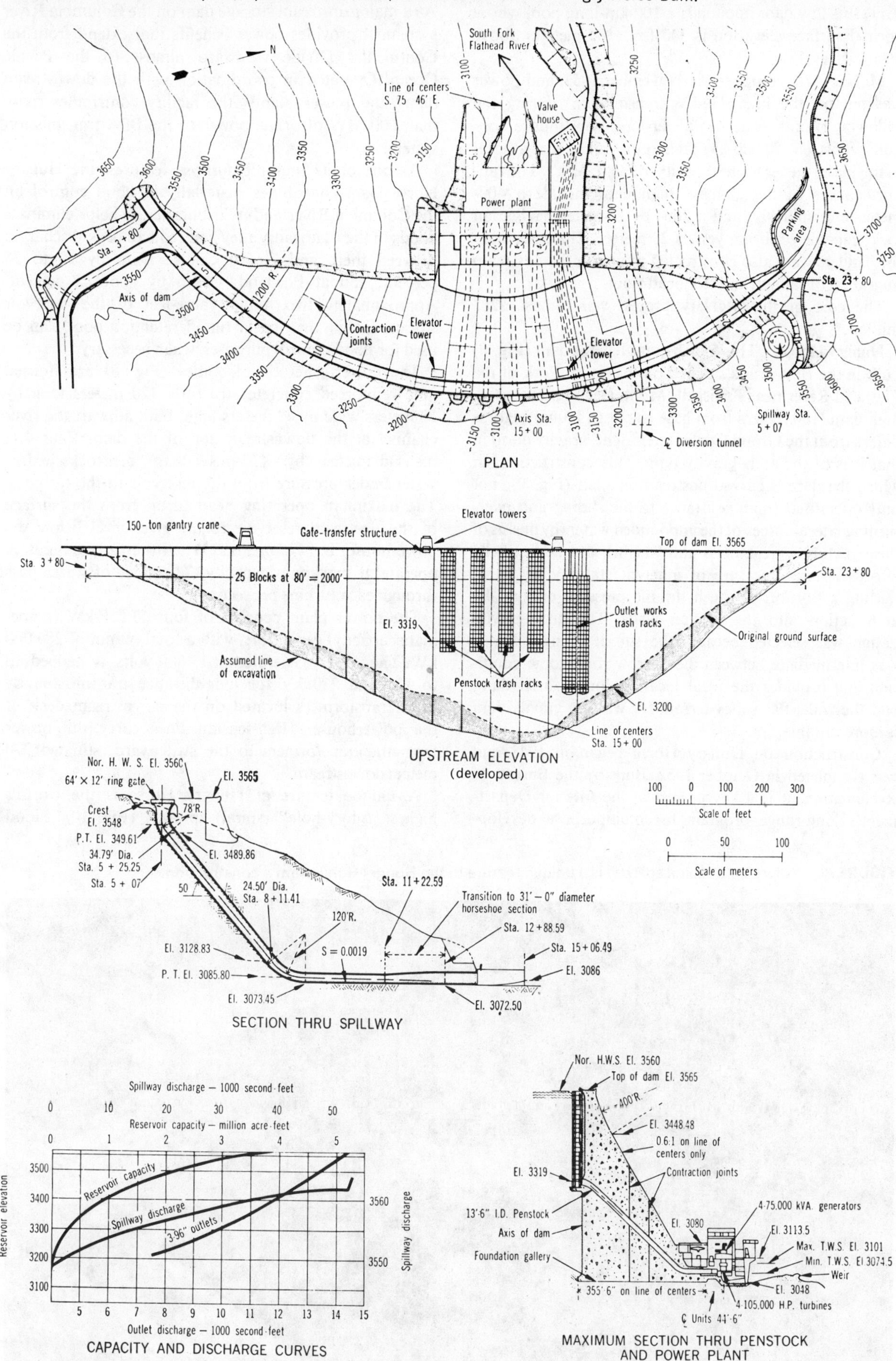

Dam. Situated 470 km upstream from the mouth of the river, this low dam impounds a 100-km-long pool whose normal surface elevation is 340 feet (104 meters) above mean sea level.

McNary was completed in November 1953, and power was put on the line (230-kV transmission) during the following month. The 435-meter-long powerhouse contains fourteen 70 000-kW generator units for a total (nameplate) generation capacity of 980 MW. The appurtenances of the 28-meter-high dam include a 400-meter-long spillway, and a 206- by 26-meter navigation lock, capable of lifting vessels 28 meters. The customary fish ladders were also constructed to permit the upstream migration of salmon and steelhead trout.

The cost of the McNary project was almost $287 million.

Hungry Horse. This fourth great federal dam (Fig. 8) built in the BPA era is situated on the South Fork of the Flathead River near Kalispell, Mont. (see Fig. 2 map). A high dam (160 meters from base to crest), Hungry Horse differs from the Grand Coulee (a straight, gravity dam) in that it is of the arch–gravity type. This construction, in which the dam is curved upstream in plan (Fig. 9), not only offers maximum resistance to the sliding and overturning lateral forces of the impounded water (by means of the structure's great mass and weight), but additionally affords a large measure of inherent stability by transmitting a portion of the hydraulic pressure or load by arch action into the canyon walls. The arch–gravity design was selected because the site at Hungry Horse was intermediate between the narrow rock gorge condition that provides the ideal location for an arch dam, and the wide, flat valley terrain for which a gravity dam is more suitable.

Construction on Hungry Horse began in 1948 and was completed in October 1952. Built by the Bureau of Reclamation, it is a key project in the Interior Department's long-range program for multipurpose development of the water resources of the Columbia River Basin. As a major upstream storage dam on the Columbia River system, it provides power benefits that extend from the Continental Divide westward almost to the Pacific Ocean. Operated in coordination with the downstream dams and power plants, the facility contributes more than 600 MW of prime power to the BPA transmission system.

As one of its multiple-purpose features, the Hungry Horse Dam contributes materially in flood control on the Columbia River and its tributaries. It helps eliminate floods in the Flathead Valley and reduces peak discharges between there and Grand Coulee Dam by 10 to 25 percent, and at Portland, Oreg., by about 5 percent. Approximately two million acre-feet of the reservoir storage (acting as a basin runoff retention pool) can be used for flood-control purposes, when necessary.

The power-generating facilities (Fig. 8) are housed in a reinforced concrete structure, 120 meters long by 23 meters wide by 48 meters high, built athwart the river channel at the downstream toe of the dam. Four 4.1-meter-diameter by 137-meter-long penstocks carry water under pressure from the reservoir to the turbines. The maximum operating head (drop from the surface of the upstream reservoir to the river level below the powerhouse) is 147.5 meters. When the turbogenerators operate at maximum capacity, 73 tonnes of water pass through each turbine per second.

The power plant consists of four 71 250-kW (nameplate capacity) generators, with a total output of 285 000 kW. The power, generated at 13 800 volts, is stepped up to 115 and 230 kV for long-distance transmission by eight transformers located on the downstream deck of the powerhouse. High-tension lines carry the power from the transformers to the switchyard, situated 370 meters downstream.

A unique feature at Hungry Horse is the world's highest "glory-hole" (spiral) spillway (Fig. 10). Flood

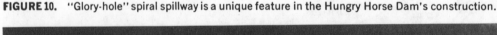

FIGURE 10. "Glory-hole" spiral spillway is a unique feature in the Hungry Horse Dam's construction.

water from the reservoir, cascading over the circular rim of the spillway, drops a maximum of 150 meters through a 10.7-meter-diameter concrete-lined tunnel over a distance of 344 meters to the outlet portal. The maximum discharge capacity of the spillway is 1480 m³/s.

Other interesting dimensional statistics of the world's fourth largest dam are: length at crest, 645 meters; width at base, 100 meters; width at crest, 12 meters. The elevation of the crest of the dam is 3565 feet (1088 meters) above mean sea level. The total cost of the project was $102 million.

The new decade:
Korea, and the demand for more power

The end of the 1940s witnessed the authorization of three new earth- and rock-fill multipurpose dams: Detroit and Lookout Point, in the Willamette Basin of Oregon (100 000 and 120 000 kW capacity, respectively); and Anderson Ranch (27 000 kW), on the South Fork of the Boise River, Idaho. Anderson Ranch was built by the Bureau of Reclamation; the others were constructed under the aegis of the Corps of Engineers. Earth- and rock-fill construction is often specified for smaller dams that are erected across narrow river gorges with a hydraulic head range of 60–90 meters.

In 1950 the Korean War began, and hydroelectric energy requirements for war material production soared to unprecedented levels. In that year, seven new dams were authorized by Congress: the low concrete dams at Albeni Falls, Idaho, and at The Dalles (Oreg.-Wash.). Construction of the Chief Joseph Dam in Washington was also approved. The Congressional action also provided for building the Big Cliff, Dexter, Chandler, and Roza earth- and rock-fill dams.

The 'brownouts' of 1952–1953. With the completion of Hungry Horse in 1952, 220 MW of additional hydro generation was made available for long-distance transmission in 1953. The at-site storage capability of the high dam also increased production at the federal and nonfederal downstream dams by 832 MW. But even this large block of power, plus the generation of all available thermal plants in Utah and hydro power imported from British Columbia, did not fill the power deficiency caused by low winter streamflows and frequent periods of subnormal temperatures.

During the winter of 1952–1953, lights often dimmed during peak-load periods. Thus power to aluminum plants in the Portland area and heavy industry in the Seattle–Tacoma–Everett complex was curtailed to prevent the brownouts from affecting commercial business, farm, and residential customers. And the utilities requested their customers to limit their use of electricity, particularly during the morning and evening peak-load hours.

'Doldrums' of 1954; three
nonfederal dams—and a new administrator

In 1954, new authorizations for federal dam projects came to a halt. Public agencies (public utility districts and municipalities) quickly filled the vacuum created by the lack of projected federal generation by requesting licenses for three large mid-Columbia hydro facilities—Priest Rapids (789 MW), Wanapum (831 MW), and Rocky Reach (712 MW)—to be built along the 450-km stretch between McNary and Chief Joseph.

On January 15, 1954, Dr. William A. Pearl, a former director of the Washington State Institute of Technology, became BPA's third Administrator. In that year a series of opinions from the U.S. Solicitor-General, based upon the power interchange provisions of the Bonneville Act, made it possible for a nonfederal utility to use BPA's interconnected regional power grid to transmit electric energy from an isolated generating facility to distant load centers. This so-called "wheeling" program brought three significant benefits to the Pacific Northwest:

FIGURE 11. Aerial oblique view of the Chief Joseph Dam and powerhouse. Note the 11 openings for the installation of future penstocks and generators.

1. Power from nonfederal projects could be marketed at distant load centers.
2. Costly duplication of transmission facilities was avoided.
3. Power could be transmitted at a lower cost.

In 1955, BPA completed the first EHV transmission system west of the Rocky Mountains, a 345-kV line to carry power from McNary Dam to Portland and Vancouver (Wash.).

Albeni Falls and Chief Joseph

The year 1955 saw the completion of the Corps of Engineers' concrete, gravity-type low dam at Albeni Falls on the Pend Oreille River of northern Idaho. At this relatively small dam, three hydro-generating units, each of 14 200-kW nameplate rating, contribute a total of 42 600 kW of generation—plus downstream prime power storage benefits—to the BPA power system.

The dam impounds the 110-km-long Pend Oreille river and lake reservoir to provide a normal usable storage of 1.15 million acre-feet. The maximum controlled pool elevation is 2062.5 feet (630 meters) above mean sea level.

Chief Joseph. This important Corps of Engineers' dam, named for the famous warrior chief of the Nez Perce Indians, is situated on the Columbia River about 80 km west of Grand Coulee, at a strategic point that takes full advantage of the hydraulic energy release from the high dam upstream. Chief Joseph is an L-shaped structure (see Fig. 11), with a 281-meter-long spillway section built athwart the mainstream flow of the river, and a 622-meter-long intake and powerhouse structure skewed at nearly a right angle (slightly obtuse) to the spillway and almost parallel to the stream flow.

Connecting the two sections is a horizontal arch abutment whose foundations bear on a solid rock outcropping in midstream. Another abutment section, at right angles to the powerhouse (shown in the lower right foreground of the illustration), serves to form an enclosed backwater pool for the sixteen 7.6-meter-diameter penstocks presently installed in the intake section. Provision has also been made for the future installation of 11 more penstocks and a corresponding number of generating units. The total length of the abutments is 410 meters.

The present power-generating equipment consists of 16 Francis-type turbines and 16 generating units, each with a nameplate rating of 64 000 kW, for a total generating capability of 1024 MW. Future plans call for the installation of 11 more units for an additional capacity of 704 MW. The rated hydraulic power head on the turbines is 50 meters.

Chief Joseph impounds an 82-km-long reservoir (Rufus Woods Lake), with a usable storage of 36 000 acre-feet. The normal surface level of the pool is 946 feet (289 meters) above mean sea level.

Transmission lines at 345 kV carry power 360 km westward from the dam (and also from Grand Coulee) to the principal load center in Seattle.

The cost of the Chief Joseph project from its inception to June 30, 1967, totaled $144.4 million.

The Dalles Dam

The Dalles Dam, a major facility authorized in 1950, begun in 1952, and completed in 1957, is an important multipurpose project that provides

1. A 40-km-long slackwater pool for navigation upstream to the site of the John Day Dam (now nearing completion).
2. The necessary additional generating capacity to meet the present and future demands of the Northwest Power Pool, and particularly for the nearby Portland area.
3. Reduction in the pumping lift required for irrigation projects on lands along the reservoir.
4. Additional recreational facilities to area residents.

The design and construction aspects of The Dalles are unique in that the contiguous dam structures are of a heterogeneous type. The 37-meter-high, concrete, 23-gate gravity spillway section (Fig. 12) runs from the navigation lock along the Washington shore athwart the stream flow of the Columbia River to a right-angle intersection return with a 61-meter-high nonoverflow concrete gravity dam section that is integral and monolithic to the powerhouse portion, running parallel to the Oregon shore. A rock-fill closure section runs from the concrete barrage at an open angle, across the main channel, to the Oregon shore.

The Dalles Dam is situated about 145 km east of Portland and approximately 72 km upstream from Bonneville. The normal elevation of the reservoir pool is 160 feet (49 meters) above mean sea level. The pool drainage area (watershed) is 61 000 km². A single-lift navigation lock, measuring 26 by 206 meters clearance in plan dimensions, provides a normal vessel lift of 26.7 meters.

The initial power facilities consisted of 16 generating units with a total nameplate generating capacity of 1119 MW. Presently under way is the installation of eight additional generators, rated at 86 000 MW each, to bring the ultimate generating capacity of The Dalles powerhouse to 1807 MW.

Temporary dislocations, and a consolidation of power

Because new projects—federal and nonfederal—were not coming on the line in an orderly scheduled manner, BPA, in 1958, was confronted by a temporary surplus of firm power that could not be offered for sale; yet, it had to be held to meet the normal load growth of preference customers. And there were no short-term customers for this power. Thus the system began to waste large surpluses of secondary, or seasonal, power for want of a Northwest market. In the course of this crisis, BPA began incurring annual operating deficits for the first time in its history. Its surplus of $80 million, accumulated in previous years, was reduced during this period.

In 1961, Charles F. Luce (presently chairman of the Board of the Consolidated Edison company of New York, Inc.) was appointed as the fourth BPA Administrator. And in that year, BPA, the Corps of Engineers, and nine private and public owners of hydro facilities made power history by the signing of a one-year coordination agreement to ensure maximum power production for all existing plants in the Columbia River Basin. By means of such an agreement, the supply of firm power was substantially increased.

FIGURE 12. Downstream face of the spillway section of The Dalles Dam. Powerhouse portion is visible at extreme right of photo. This Z-shaped contiguous structure features both concrete and earth- and rock-fill construction.

FIGURE 13. Ice Harbor Dam, showing the powerhouse section at the right, spillway in the center, and fish ladders and navigation lock to the left.

In 1962, Congress authorized the construction of two new (Corps of Engineers) major dams—Little Goose (405 000 kW) and Lower Monumental (405 000 kW)—on the Snake River, and approved the BPA proposal for construction of secondary thermal generating facilities at the Hanford (Wash.) nuclear project of the AEC, to be built and operated by the Washington Public Power Supply System. In the following year, BPA became the power marketing agency for all of the Columbia River Basin federal power projects, including southern Idaho's Upper Snake River drainage area.

Four more dams for the sixties. In the six-year period between 1961 and 1967, one major concrete low dam, Ice Harbor (Corps of Engineers), was completed. Situated on the Snake 16 km from the confluence of the Snake and Columbia Rivers, Ice Harbor (Fig. 13) put an additional 270 000 kW from three generating machines on the BPA transmission lines. A projected future expansion will double this power output. The initial project also included a single-lift navigation lock (31-meter rise) to handle the two million tonnes of waterborne freight being shipped up the Columbia and Snake Rivers annually to Lewiston, Idaho.

This period saw the completion of three earth- and rock-fill power dams along Oregon rivers: Hills Creek (30 000 kW), on the Main Fork of the Willamette; Cougar (25 000 kW), on the South Fork of the McKenzie; and Green Peter (80 000 kW), on the Middle Santiam.

The Canadian Treaty. In 1964, The Treaty with Canada for Joint Development of the Columbia River was implemented. Under its terms, three large storage dams, to be built on the upper reaches of the Columbia River in British Columbia, will add 2800 MW of firm power at downstream dams in the United States, and also will provide much-needed flood control. One of these Canadian dams (Duncan Storage) has been completed and the other two —Arrow Storage and Mica Storage—are under construction. The sites of these projects are shown in the Fig. 2 map of the Columbia River Basin area. In accordance with the agreement, half of the power, 1400 MW, goes to each nation. The Canadian share, however, is tem-

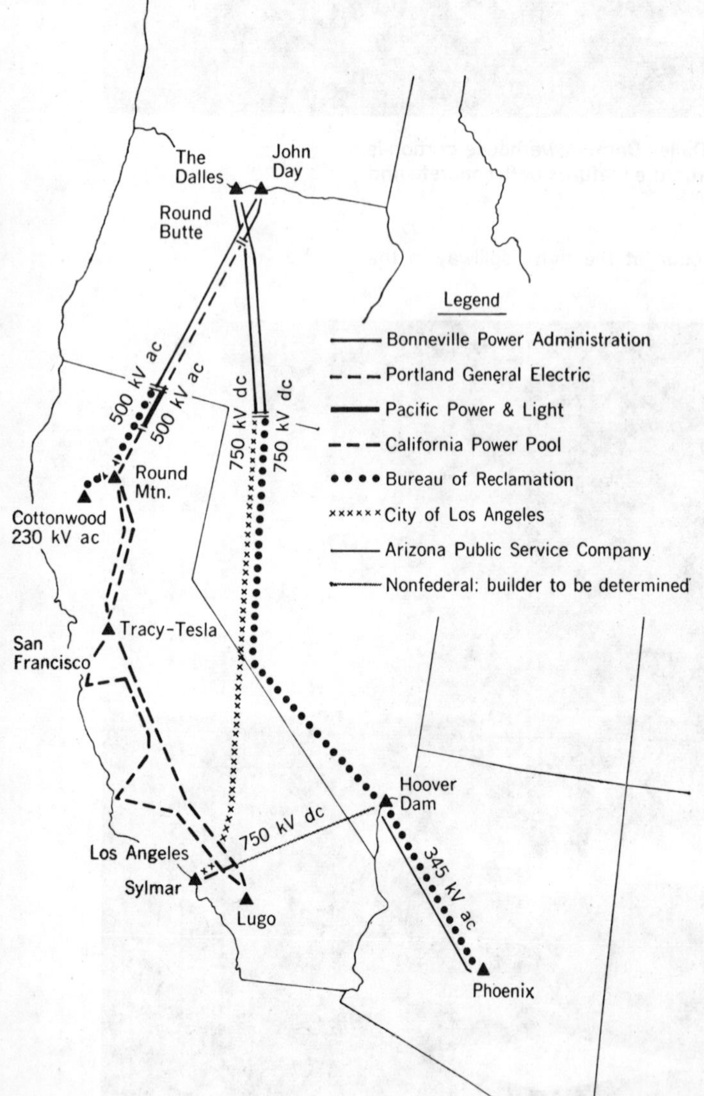

FIGURE 14. Diagram of the eventual Pacific Northwest-Pacific Southwest Intertie, portions of which are presently under construction in a five-year program.

III. Columbia River power system as of June 30, 1967

Project	Operating Agency*	Location	Stream
Bonneville	CE	Oreg.-Wash.	Columbia
Grand Coulee	BR	Washington	Columbia
Hungry Horse	BR	Montana	S. Fk. Flathead
Detroit	CE	Oregon	North Santiam
McNary	CE	Oreg.-Wash.	Columbia
Big Cliff	CE	Oregon	North Santiam
Lookout Point	CE	Oregon	M. Fk. Willamette
Albeni Falls	CE	Idaho	Pend Oreille
Dexter	CE	Oregon	M. Fk. Willamette
Chief Joseph	CE	Washington	Columbia
Chandler	BR	Washington	Yakima
The Dalles	CE	Oreg.-Wash.	Columbia
Roza	BR	Washington	Yakima
Ice Harbor	CE	Washington	Snake
Hills Creek	CE	Oregon	M. Fk. Willamett
Minidoka	BR	Idaho	Snake
Boise Diversion	BR	Idaho	Boise
Black Canyon	BR	Idaho	Payette
Anderson Ranch	BR	Idaho	S. Fk. Boise
Palisades	BR	Idaho	Snake
Cougar	CE	Oregon	S. Fk. McKenzie
Green Peter	CE	Oregon	Middle Santiam
Foster	CE	Oregon	South Santiam
John Day	CE	Oreg.-Wash.	Columbia
Lower Monumental	CE	Washington	Snake
Little Goose	CE	Washington	Snake
Lower Granite	CE	Washington	Snake
Teton	BR	Idaho	Teton
Lost Creek	CE	Oregon	Rogue
Dworshak	CE	Idaho	N. Fk. Clearwate
Strube	CE	Oregon	S. Fk. McKenzie
Libby	CE	Montana	Kootenai
Asotin	CE	Wash.-Idaho	Snake
Total number of projects and installed capa			

*CE—Corps of Engineers; BR—Bureau of Reclamation.
†Nameplate rating.
‡Includes a recommended 3 600 000 kW at the third powerhouse in a

porarily being sold to utilities in the United States.

Finally, the Treaty gives the U.S. the right to build the Libby Dam on the Kootenai River in Montana, the reservoir of which will impound backing water 68 km into Canada. Libby will add 750 MW of firm power at the site and downstream in the United States.

In conjunction with the Canadian Treaty, the original one-year coordination agreement was extended 39 years and expanded to include BPA, Corps of Engineers, and 14 major generating utilities in the Pacific Northwest. This auxiliary agreement was necessary because the Canadian share of the power benefits was calculated on the basis of maximum coordinated international power output. Although voluntary coordination existed in the Northwest for many years, the new joint agreement assured that coordination would be extended and all facilities operated for maximum benefit to the region.

New construction to the present time

In 1965, the construction of power plants and transmission lines reached an all-time high: eight federal and four nonfederal dams were under construction; and, after years of preliminary study and planning, the Pacific Northwest–Pacific Southwest Intertie (500 kV ac and 750 kV dc), approved by Congress in 1964, were under construction. These lines, the largest single transmission program ever undertaken in the U.S., will eventually interconnect the power generation sources at The Dalles, John Day, and Hoover Dams (Fig. 14) with load centers at San Francisco, Los Angeles, and Phoenix, Ariz. This huge program will be discussed in more detail in the second installment of this article.

In 1966, Congress authorized the third power plant for Grand Coulee Dam. Completion of this installation will eventually make the dam one of the world's largest power producers (9435 MW). In the same year, David

	Existing		Under Construction		Authorized		Total	
Initial Date in Service	Number of Units	Total Capacity, kW†	Number of Units	Total Capacity, kW†	Number of Units	Total Capacity, kW†	Number of Units	Total Capacity, kW†
June 1938	10	518 400	—	—	6	324 000	16	842 400
Sept. 1941	18	1 944 000	6	3 600 000	6	291 000	36	9 435 000‡
Oct. 1952	4	285 000	—	—	—	—	4	285 000
July 1953	2	100 000	—	—	—	—	2	100 000
Nov. 1953	14	980 000	—	—	6	420 000	20	1 400 000
June 1954	1	18 000	—	—	—	—	1	18 000
Dec. 1954	3	120 000	—	—	—	—	3	120 000
Mar. 1955	3	42 600	—	—	—	—	3	42 600
May 1955	1	15 000	—	—	—	—	1	15 000
Aug. 1955	16	1 024 000	—	—	11	704 000	27	1 728 000
Feb. 1956	2	12 000	—	—	—	—	2	12 000
May 1957	16	1 119 000	8	688 000	—	—	24	1 807 000
Aug. 1958	1	11 250	—	—	—	—	1	11 250
Dec. 1961	3	270 000	—	—	3	270 000	6	540 000
May 1962	2	30 000	—	—	—	—	2	30 000
May 1909	7	13 400	—	—	—	—	7	13 400
May 1912	3	1 500	—	—	—	—	3	1 500
Dec. 1925	2	8 000	—	—	—	—	2	8 000
Dec. 1950	2	27 000	—	—	1	13 500	3	40 500
Feb. 1957	4	114 000	—	—	—	—	4	114 000
Feb. 1964	2	25 000	—	—	1	35 000	3	60 000
June 1967	2	80 000	—	—	—	—	2	80 000
	—	—	2	20 000	—	—	2	20 000
	—	—	16	2 160 000	4	540 000	20	2 700 000
	—	—	3	405 000	3	405 000	6	810 000
	—	—	3	405 000	3	405 000	6	810 000
	—	—	3	405 000	3	405 000	6	810 000
	—	—	—	—	2	22 000	2	22 000
	—	—	2	49 000	—	—	2	49 000
	—	—	3	400 000	3	660 000	6	1 060 000
	—	—	—	—	1	4 500	1	4 500
			4	420 000	4	420 000	8	840 000
	—	—	—	—	4	540 000	4	540 000
	22	6 758 150	8	8 552 000	16	5 459 000	33	24 369 150

ı to that under construction, and 291 000 kW from six authorized pump–turbine units.

S. Black, a former FPC Commissioner, became BPA's fifth Administrator.

Power system automation. A comprehensive plan for the automation of the U.S. Columbia River Power System was initiated in 1967. The plan proposes to integrate thermal and hydro generation, irrigation, flood control, navigation, and other river functions to optimize the use of the region's total resources for all federal agencies and public and private utilities. System automation is part of a long-range plan, looking ahead to 1975 and 1985. A detailed discussion of the program will appear in Part II.

John Day Dam. To bring this portion of the first installment up to date insofar as dam construction is concerned, the writer must conclude this phase with John Day—the final key link in the important chain of lower Columbia River projects (title illustration).

The $410 million, multipurpose project, now producing more than 300 MW from the first two units in operation, is essentially similar in basic design to McNary, 122 km upstream. Situated 350 km east of the mouth of the river, the dam impounds a 122-km-long reservoir that completes a 525-km slack-water pathway up the Columbia from the Pacific Ocean to a point above Pasco–Kennewick (Wash.), and thence up the Snake River to Lower Monumental Dam. Completion of the four approved dams on the Snake River—Lower Monumental, Little Goose, Lower Granite, and Ice Harbor (second phase)—will add 227 km to this navigable waterway and make slack-water navigation possible above Lewiston, Idaho.

The initial 16 generating units to be installed in the powerhouse each have a nameplate capacity of 135 000 kW, with an overload rating of 155 250 kW. Thus the John Day total capacity in the first phase will be a minimum of 2160 MW. Ultimately, a total nameplate generating capacity of 2700 MW will be available by the installation of four more units.

The 1800-meter-long straight gravity dam consists of a spillway and powerhouse section, running normal to the mainstream flow of the river from the navigation lock along the Washington shore to the fish-ladder appurtenances on the Oregon side. The 206- by 26-meter single-lift navigation lock will raise vessels 34.5 meters.

As previously noted, both John Day and The Dalles Dams have become particularly significant in the BPA system as the northern generating terminals for the 500-kV and 750-kV ac and dc transmission lines of the Northwest–Southwest Intertie.

Industrial growth and power marketing

The history of BPA is essentially interlocked and dependent upon the classical chronicle of supply and demand, coupled with the phenomenal growth of industry in the Pacific Northwest during the past 30 years. Today, it is estimated that 70 percent of the electric power purchased by industries in the Northwest comes from the large federal dams of the Basin. This electric energy is sold and delivered by the BPA, the marketing agent. A portion of this electricity goes to public and private utilities who resell it to their industrial customers. But

FIGURE 15. Diagram of power flows and load center demands in the Pacific Northwest as of January 1967.

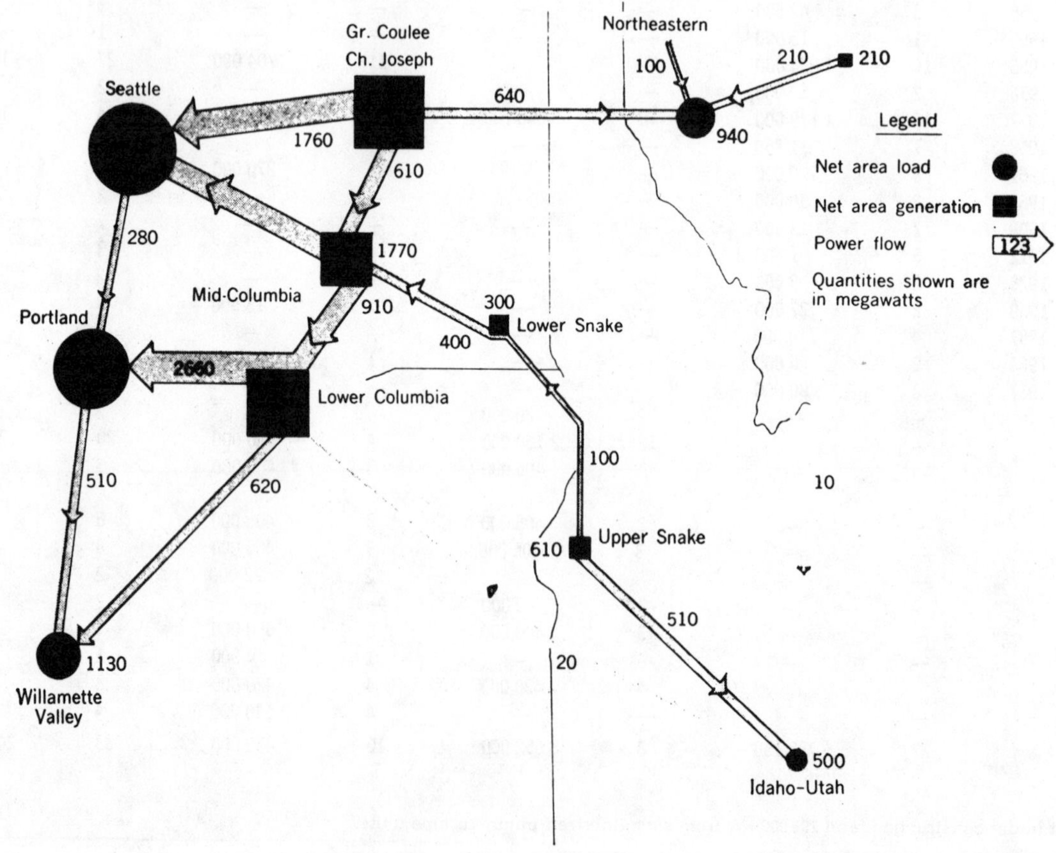

the largest blocks of power go to the heavy electroprocess industries, such as aluminum-reduction plants, who take delivery of the power directly from BPA. Many of these plants were originally located in the Northwest for the primary reason of obtaining low-cost power from the dams as they were completed.

The Aluminum Company of America, for example, became BPA's first industrial customer shortly after power from Bonneville Dam was put on the line. When Alcoa completed its plant at Vancouver, Wash., in 1938, it initially received 26 000 kW—about 5 percent of the dam's generation.

By mid-1941, BPA had signed contracts with six companies to provide a total of 265 500 kW.

Industrial power sales have soared over the years. By May 1967, sales to major industrial customers reached 2140 MW. Of this quantity, 1698 MW are covered by 20-year contracts for firm power. The balance of the power is sold on an interruptible basis, contingent on streamflows. By 1967, 30 percent of the primary aluminum produced in the United States came from Northwest smelters. In addition to aluminum, the products manufactured by the area's electroprocess plants include silicon carbide, ferronickel, calcium carbide, and ferromanganese. Today, 39 percent of BPA's power sales are made directly to heavy industry.

Load growth—present and future

To those who wondered, in 1937, what the Pacific Northwest would do with that 300 000 kW of electric energy transmitted from Bonneville Dam, the example of power flow in the Pacific Northwest (Fig. 15) as of January 1967 may provide some answers! At that date, the load centers in Seattle, Portland, the Willamette Valley, western Montana, and Idaho–Utah were drawing a total of 8350 MW over BPA transmission lines from the hydro and thermal generation capacity of the Columbia River Basin facilities.

Table III shows the nameplate generating capacity of the federal dams of the Basin as of June 30, 1967, the generating capacity presently under construction, and the authorized future installations that will bring these facilities up to their ultimate capacity by the target date of 1987. Note that the deficiency of power generation between Table III and Fig. 15 is met by generation from nonfederal dams and thermal plants.

Figure 16 is a forecast of the power flows in the Northwest by January 1977. It includes 2000 MW in reserve from six new thermal (nuclear) plants that have been recommended for construction west of the Cascades, the transmission lines from which would be built as an extension of the existing BPA system. At this future date, a total of about 17 750 MW—more than twice the present demand—will be required by the same five load centers indicated in Fig. 15.

Finally, Fig. 17 indicates the power flows in the Northwest by 1987, when the projected ultimate generating capacity of the entire Basin (U.S. and Canada) will be realized.

The diagrams, however, should be considered only as

FIGURE 16. Diagrams of power flows and load center requirements in the Northwest anticipated by January 1977.

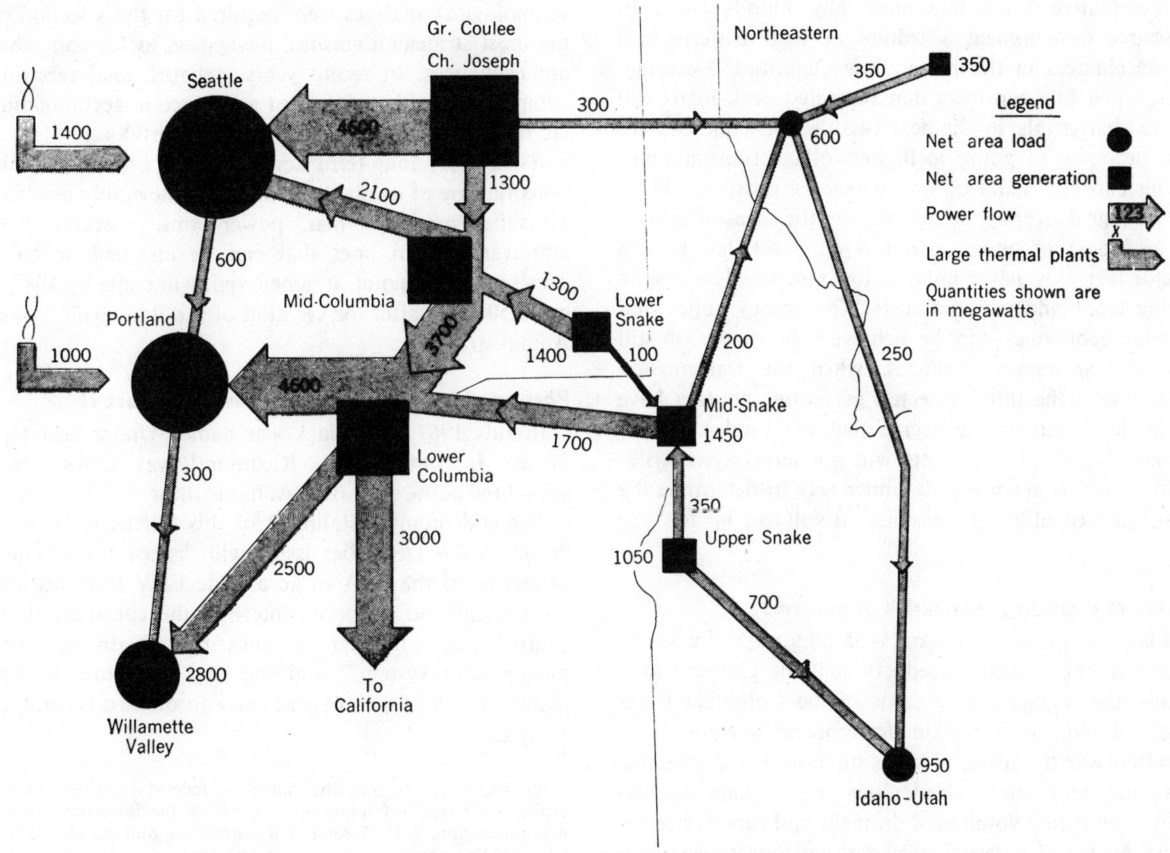

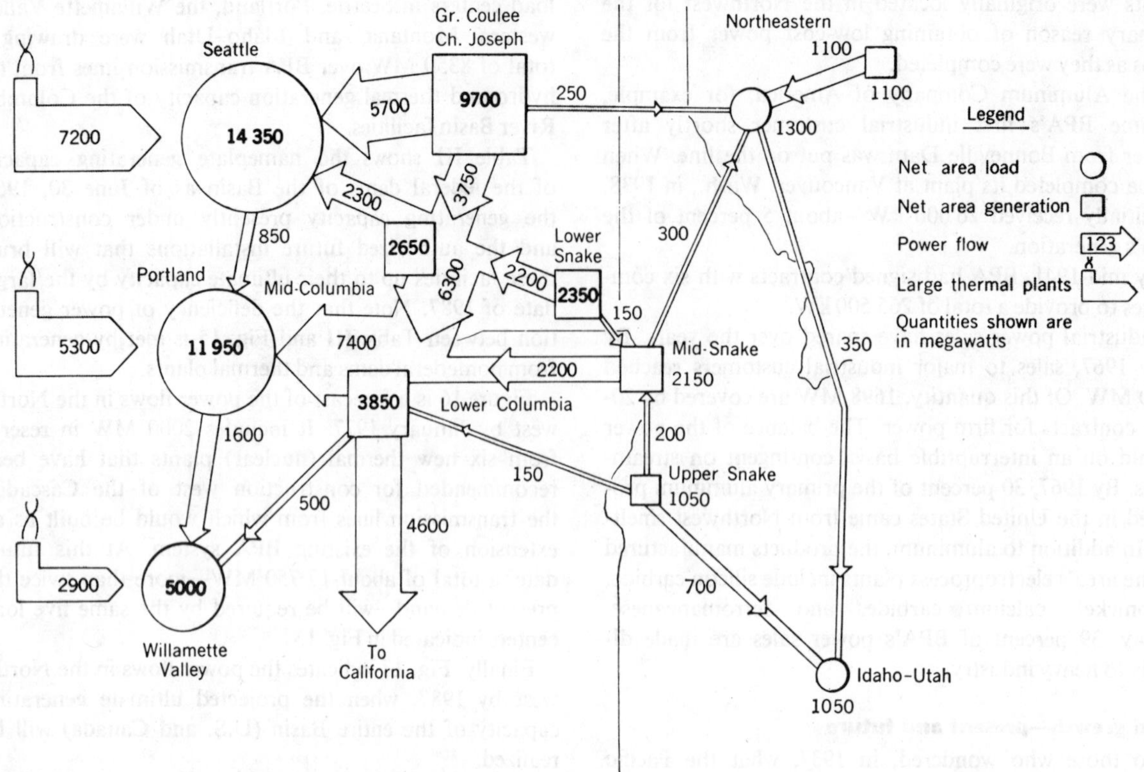

FIGURE 17. Anticipated power flows and load demands in the Northwest as of 1987. Large thermal plant sitings west of the Cascade Mountains are indicated.

representative examples, since any modifications in resource development schedules or load patterns will cause changes in the power flow quantities. Nevertheless, when one considers that projected peak loads will more than triple in the next two decades, one can see the necessity of going to higher-voltage transmissions, using existing rights of way whenever possible. BPA's 500-kV grid overlay, which permits the transmission of large blocks of power at a low unit cost, has been a major factor in maintaining its low rate schedule despite rising labor and material costs. The agency hopes that similar economies can be achieved by means of still higher transmission voltages, when the transmission system and the interconnected generating sources have been developed to the degree that the sudden loss of several thousand megawatts will not affect system stability. Studies are presently under way to determine the feasibility of utilizing transmission voltages up to 1000 kV.

Years of planning—centuries of progress

Like the grand-scale works of antiquity—the Great Pyramid, the Roman aqueducts, and the Gothic cathedrals—the mighty power dams of the Columbia River Basin should benefit mankind for centuries to come. Their creation was the result of years of coordinated scientific planning and study—meticulous engineering calculations concerning volumes of drainage and runoff, streamflows, historical meteorological data and long-range climatic forecasts, future power demands, and the optimum hydro generation development possible in the vast watershed area. In addition, exhaustive geological and seismological analyses were required for the selection of the most strategic damsites, navigation locks, and other appurtenances. In recent years, network analyzers and computers have been of great assistance in accumulating large blocks of data for this gigantic undertaking.

As in every long-term developmental project, the full potential use of available resources is inevitably reached. Thus there are just so many power dams, generator units, and transmission lines that can be installed or built. This saturation point, it is believed, will come by 1987—some 50 years after the creation of the Bonneville Power Administration.

Postscript to the present: a preview of Part II

In July 1967, Mr. Black was named Under Secretary of the Interior; H. R. Richmond was subsequently appointed as the sixth BPA Administrator.

The concluding installment of this article, to be published in the December issue, will discuss the advance programs of the BPA in ac and dc EHV transmission, the present and proposed interties, the construction of control and converter stations, the "hydromet data management system," and the role of future thermal plants as part of a combined and coordinated generation complex.

Photos, diagrams, and illustrations of federal dams used in this article are based on material supplied by the Bonneville Power Administration, U.S. Bureau of Reclamation, and the U.S. Army Corps of Engineers.

Excerpts from "Geothermal Energy: A National Proposal for Geothermal Resources Research"

I. INTRODUCTION AND BACKGROUND

The ever-increasing demand for energy and the recent pressures brought to bear through concern for the environment have led the Nation to consider energy alternatives to the conventional sources.

One of the new alternatives is geothermal energy. There is a great scarcity of reliable data on the geothermal resources of the Nation and this has led planners to ignore geothermal energy in comparison to other, more adequately explored, energy resources.

Nevertheless, other countries, including Italy, Iceland and New Zealand, have utilized geothermal power for some years, and Pacific Gas and Electric has built and is operating geothermal power generation equipment in Northern California.

The natural heat of the Earth is thought to be derived from the natural decay of radioactive core materials and frictional forces resulting from solar and lunar tides as well as the relative motion of crustal plates. Temperature measurements in drill-holes, mines, etc. confirm that, on the average, temperatures increase with increasing depth, reaching values in excess of 1000°C (1832°F) in the molten interior of the Earth.

Theoretically, this energy source can be tapped from any point on earth simply by drilling deep enough holes, providing a passage for heat transfer fluid, and extracting the heat. Practically speaking, this hot mass is much too deep to tap in large areas of the world with existing drilling capability. Yet in other areas the resource is much closer to the surface.

The base of the continental crust, where the temperature is perhaps 1000°C or higher, varies from a depth of 25 to 50 kilometers. When a structural fault reaches deep into the crust and relieves crustal pressures, molten and partially molten mantle materials can move upward toward the surface with great rapidity.

Nevertheless, it is not necessary that the magma break through and flow onto the surface. It may spread out at some depth and mix with rocks of lower freezing points at that level and crystallize. In this way rocks derived from magma of varying composition and varying freezing points may be produced.

The present state of knowledge indicates that such deposits of thermal energy are found at relatively shallow depths in zones of geologically recent volcanism and crustal shifting in areas of above-normal heat flow.

The areas of greatest prospective geothermal resources seem to lie along the tensional environments of the oceanic rises and continental rift systems, and along the compressional environments where mountains are rising and island arcs are forming.

In some areas, surface phenomena, such as geysers and hot springs, suggest that the source of energy is quite near the surface. The United States, particularly in its western region, has an enormous extent of volcanic rocks of recent origin and an abundance of dormant volcanos as well as several active ones. The potential geothermal resources appear to be very large.

It was in this context, in response to a proposal submitted by the University of Alaska with Mr. Walter J. Hickel as principal investigator, that the National Science Foundation, through its RANN Energy Research and Technology Program, granted funds in support of a geothermal resources research conference to be held in September, 1972.

The objective of the conference was to develop an

Reprinted with permission from the V.P. for Research, Univ. of Alaska, *Final Report of the Geothermal Research Conf.*, W. J. Hickel, Conference Chairman, J. C. Denton, Ed., Sept. 18-20, 1972.

assessment of the state-of-the-art of geothermal science and technology and to recommend a research program to provide the requisite knowledge for establishing the proper role of geothermal resources in providing additional energy to alleviate the Nation's impending shortage. A further objective was to investigate the potential of geothermal water to supplement present national fresh water supplies.

The conference was a working conference, as contrasted to one for the presentation of scientific papers, and attendance was by invitation only. The setting for the conference was the Battelle Seattle Research Center in Seattle, Washington, an installation designed for just this type of working conference.

For convenience, the conference was organized into six substantive panels with co-chairmen of each panel responsible for assembling panel members who could speak authoritatively on the substance of the panel.

The panel topics selected were: resource exploration, resource assessment, reservoir development and production, utilization technology and economics, environmental effects and institutional considerations.

Both industrial and university viewpoints were represented among the co-chairmen. Federal scientists were invited to contribute to each of the panels.

In order to complete the planning for the conference, Mr. Hickel held a workshop in Anchorage which was attended by almost all of the co-chairmen. At the Anchorage meeting the co-chairmen discussed the state-of-the-art of geothermal science and technology, specified the topics which were to be explored in each panel, and compiled lists of experts who might be invited to participate.

A preliminary view of a national geothermal research program was also discussed and was subsequently used as one input to the preparation of the report,[1*] "Assessment of Geothermal Energy Resources." The main task of the September conference was to amplify and finalize recommendations for a national geothermal resources research program. The conference drew upon the Anchorage workshop and the report, "Assessments of Geothermal Energy Resources," as resource material.

The objective of this report is to present the results of the Geothermal Resources Research Conference which was held September 18-20, 1972. The sequence of chapters includes one for each of the substantive panels of the Conference.

Inevitably, when concurrent panel meetings are held, there is some degree of overlap between the topical research recommended by each panel. It was not possible to make sharp delineations of the area of responsibility for each panel and this accounted for some of the overlap. In other cases, consideration of a given topic, for example reinjection to maintain reservoir pressure, led naturally into conclusions with regard to another panel, for example environmental effects.

In writing the report, every effort has been made to avoid the overlaps. Thus it has been necessary to rearrange the panel reports to present a consistent program.

For rapid reference purposes, however, each chapter has been written as an individual entity. The repetitive language permits each section to stand on its own.

The national geothermal resources research program was to be presented without regard to the appropriate sector for carrying out the research. In certain cases there is an obvious wisdom in selecting a particular sector to perform, or be responsible for, the proposed research. Clearly the U.S.A.E.C. should be responsible for any nuclear fracturing research. In the same sense it is traditional that industry will be responsible for almost all demonstration plant work.

Nevertheless, it was felt that the Conference, and its report, should address the question of total research and development needed to provide a basis for application of geothermal resources. Thus, it has been necessary to add to some of the panel reports, notably the Utilization Technology and Economics report, in order to assure that each report is presented on the same basis. A substantial part of the additions were covered, at least in part, by duplicate considerations in other panels.

The Geothermal Resources Research Conference attempted to draw together a cross-section of the very best geothermal resources expertise in the country; from industry, government, and universities. Attendance was limited. This report is based on the panel reports which present their findings.

It has been necessary to rearrange, to interpret, and to supplement the panel reports. It is hoped that the rearrangements and interpretations are faithful to the intent of the experts. Any errors are the responsibility of the editor.

II. IMPORTANCE TO THE NATION OF GEOTHERMAL RESOURCES

The potential importance of geothermal resources for contributing to the Nation's supply of energy and fresh water is enormous.

It can be comprehended best when contrasted to some familiar examples of man's outstanding achievements. Grand Coulee will provide 9,771 megawatts of electric power when the third power plant comes on line. The entire electrical power supply system of New England is approximately 15,000 MW. By way of contrast, it has been estimated[1] that geothermal resources have the potential of providing electrical generating capacity ranging from 750 MW in 1975 to 395,000 MW in 2000. These estimates and others that are natural consequences are shown in Table 1.

Using a 90 per cent load factor, which is normal for geothermal plants, the potential electrical energy generated will range from 5.9 million megawatt-hours in 1975 to 3,114 million megawatt-hours in 2000. This amount of energy should be contrasted with the production from Grand Coulee which, utilizing a 60 per cent load factor, in 1975 will be 51.4 million megawatt-hours but in 2000 will still be 51.4 million megawatt-hours. Another view of the significance of the geothermal contribution can be obtained by noting that the present demand for electrical energy in New England is less than 60 million megawatt-hours.

Table I also points out that the amount of oil required to generate these amounts of electrical energy, using a 40 per cent conversion efficiency, ranges from 24 thousand barrels per day in 1975 to 12.6 million barrels per day in 2000. The 1985 figure is 4.2 million barrels per day, which is almost 30 per cent of the anticipated oil imports of 14.8 million barrels per day[2] in 1985. Geothermal energy, therefore, can have a major impact on U.S. national self-sufficiency.

It also follows that geothermal energy can have a major impact on improving our balance of payments posture. Table I shows that the value of oil equivalent to the geothermal capacity is 8.9 billion dollars in 1985 which is over one-third of the anticipated 25 billion dollar deficit[3] in our balance of payments in 1985. In addition it may be possible to enhance our balance of payments posture even further by exporting geothermal energy conversion equipment.

The Bureau of Reclamation has formulated extensive concepts[4] for development of the geothermal resources of the Imperial Valley of California. These concepts include the development of geothermal resources to provide both water and energy, 2.5 million acre-feet per year with 10,500 megawatts (including 2000 MW for use on site.)

Most of the potential geothermal developments which can be brought into operation by 1985 will probably be based on hot-water systems. Thus, in 1985, geothermal resources can provide up to 31.5 million acre-feet of water to augment natural supplies. This is over 15 times the present amount of water for municipal, industrial, and other non-agricultural uses supplied by the Bureau of Reclamation and over 25 per cent larger than the amount delivered for irrigation purposes (25.4 million acre-feet)[5].

Geothermal resources are now and are expected to remain economically competitive with conventional sources of water and energy. The Bureau of Reclamation estimates[4] geothermal water costs in the Imperial Valley to be as low as $85 per acre-foot and to increase to a possible high of $150 per acre-foot, depending upon point of delivery and source of replacement fluids. The cost of electrical energy from geothermal resources was 5.25 mills per kilowatt-hour in 1970[6] for the Pacific Gas and Electric installation at The Geysers in California. The energy costs are expected to be higher for hot water dominated systems and other new systems but are still expected to be competitive because of steadily rising fuel costs as conventional sources grow more inaccessible and suppliers take into account necessary environmental protection.

Comparatively, it is estimated that geothermal resources can be developed to supply power in less time than more conventional power supplies. Measured from the time when the selection of power supply type is final, geothermal power plants should be operative in about two years, whereas fossil fuel power plants may require up to five years and nuclear plants up to ten. These differences result from the smaller scale of the geothermal plants and the different complexities in gaining permits and licenses for the different types of plants.

In summary, geothermal resources, by approximately 1985, can have a potentially enormous impact in supplying the nation's need for energy and augmenting the supply of water in regions with insufficient natural water. These resources promise to be economically competitive and able to be brought on-line rapidly. The development of geothermal resources could substantially increase national energy self-sufficiency and provide a dramatic improvement in the U.S. balance of payments posture.

A vigorous research and development program is urged to assure the timely realization of these advantages to the Nation.

TABLE I. GEOTHERMAL ENERGY RESOURCES POTENTIAL

	1975	1985	2000
Power (thousands of MW)	**0.75**	**132**	**395**
Electrical Energy[1] (millions of MWH)	**5.913**	**1,041**	**3,114**
Oil Equivalent[2] (millions of Bbls/day)	**0.024**	**4.213**	**12.60**
Foreign Trade Impact[3] (billions of dollars)	**0.051**	**8.919**	**26.67**

1. 90 per cent load factor

2. 3,412 BTU/KWH and 5,800,000 BTU/Bbl of oil used at 40 per cent conversion efficiency.

3. $5.80 per barrel ($1.00 per million BTU)

Statement of Robert W. Rex, Director, Geothermal Resources Program, Institute of Geophysics and Planetary Physics, University of California

NATURE OF THE RESOURCE

The heat of the interior of the earth is one of the largest energy resources available to man. It is far larger than the energy available from all of the uranium and thorium combined even using breeder reactors. However, the size of the resoure is not the significant question.

The most important questions are:

1. the size of the resource that can be produced at a particular cost.
2. the technical probability that these cost projections are correct; and
3. the practical constraints such as institutional and environmental problems.

It is evident from present operating geothermal plants in California that certain types of geothermal energy are practical, economic, and technically feasible. Present plants utilize natural steam that exists underground. Hot water geothermal fields are now under test in the Imperial Valley of California and in an identical geological setting in northwestern Mexico. Consequently, geothermal energy can be considered a proven energy source in the U.S.

PRICE-SIZE RELATIONSHIP

The key to understanding the size of the U.S. geothermal potential is to relate size at a given price for thermal energy.

Geothermal plants such as those currently in use at the Geysers in California cost about $120 per kilowatt capacity and it may be expected that this cost will rise to about $140–160 per kilowatt in the next five years. Hot water geothermal generating systems using heat exchangers will probably cost $200–220 per kilowatt over the next five years. In contrast nuclear units cost about $400–600 per kilowatt and coal fired fossil fuel plants with pollution control equipment appear as though they will cost $300–400 per kilowatt over the same time frame. It is evident, therefore, that solving electricity shortages by geothermal energy development would place the lightest capital requirements on the U.S. economy.

The net input of fixed and operating costs on a geothermal plant is to add about 2.3–2.5 mills of the price of geothermally derived electricity. The remainder of the cost is the "fuel" or energy cost which represents the fair market price for a private sector risk-taking energy company that explores for, drills, tests, and produces the steam or hot water and delivers it to the utility.

We have made an assessment of the size of the U.S. potential as a function of price and graded the estimates as to *known reserves*, *probable reserves*, and *undiscovered reserves* (Table 1). The technical basis for the probable and undiscovered reserves are the subjective transfer of our geothermal exploration experience to other areas of the U.S. There has been and undoubtedly will continue to be substantial differences in opinion on how present technical knowledge should be extrapolated into unknown areas.

TABLE 1.—AMOUNT OF GEOTHERMAL ENERGY IN THE UNITED STATES IN UNITS OF MEGAWATT-CENTURIES OF ELECTRICITY PRODUCIBLE FROM THIS GEOTHERMAL ENERGY

Energy price mills/kw.-hr. (1–11)	Known reserves	Probable reserves	Undiscovered
2.90 to 3.00	1×10^3 (12)	5×10^3 (12)	10×10^3.
3.00 to 4.00	30×10^3 (13)	400×10^3 (14)	2×10^6 (15).
4.00 to 5.00	?	600×10^3 (16)	12×10^6 (17).
5.00 to 8.00	?	?	20×10^6 (18, 19, 20, 22).
8.00 to 12.00	?	?	40×10^6 (18, 19, 21, 22).

ASSUMPTIONS

(1) Present costs and prices inflated by 5 percent per year for the next 30 years.
(2) 22 percent depletion allowance for geothermal energy.
(3) Energy price in present dollars.
(4) Development by taxpaying entities.
(5) 10 percent to landowners (Federal or State governments, or private landholders).
(6) No severance taxes.
(7) State taxes at a level no higher than the California corporate rate.
(8) No increase in Federal corporate taxes.
(9) Cost of capital 8.5 percent.
(10) Expensing of intangible drilling expenses.
(11) No change in the depreciation rate for amortizing wells.
(12) Primarily the Geysers area of California.
(13) Primarily the Clear Lake-Geysers KGRA and the various Imperial Valley KGRA's.
(14) Note 13 plus the Jemez area of New Mexico and Long Valley, Calif.
(15) Note 14 plus remainder of Basin and Range area of western United States.
(16) Note 15 plus Hawaii.
(17) Note 16 plus Alaska.
(18) Hot dry rock systems developed based on hydrofracturing or cost equivalent technology.
(19) Present drilling technology. New low-cost deep drilling could substantially improve these economics.
(20) Hot dry rock at less than 20,000-foot depth.
(21) Hot dry rock at less than 35,000-foot depth.
(22) Development of hot dry rock enegy source over 5 percent of the area of the western ½ of the United States.

Reprinted with permission from *Hearings before the Senate Committee on Geothermal Energy Resources and Research*, pp. 48-50, June 15, 1972.

RECOMMENDATIONS

1. Let the private sector finance and develop dry steam and high temperature hot water geothermal systems. (High temperatures defined as those sufficient to yield 20 percent steam on flashing down to 100 pounds pressure.) Federal efforts should be focused on research on basic, environmental, institutional, desalination, and deep basin research.

2. Fund an accelerated program of hot dry rock geothermal research with emphasis on engineering tests.

3. Fund a national deep geothermal drilling program to appraise the total U.S. potential as a base for national energy planning.

The investment of federal money would be returned rapidly and many times over by increases in value of the massive federal geothermal reserves on public lands.

This range of uncertainty, however, is one of the primary reasons that a major federal multiagency effort is needed to appraise the U.S. geothermal potential and to develop the technology for presently marginal geothermal resources.

The units used are megawatt centuries or a megawatt of electrical energy produced for a century. For comparison purposes consider that current U.S. capacity is about 340,000 megawatts. Adding the costs associated with generation to the energy price given in Table 1 we can see that 5.50 to 6.00 mill power at the generating station is available from geothermal plants today. It is this investigator's estimate that 400,00 megawatts capacity for a century could be developed in the western U.S. in 20 years. This assumes that individual geothermal reservoirs are managed to be depleted in 100 years and that reservoir pressure maintenance and water reinjection programs are used whenever needed. The various economic assumptions used in calculating the economics of the steam price are given in the footnotes to Table 1. In summary, these calculations involve continuance of present tax treatment including holding the present depletion allowance of 22 percent, present corporate and state taxes, a 5 percent rate of cost and price inflation.

It is evident from Table 1 that geothermal energy is suggested to be capable of carrying a major portion of future U.S. new electrical energy requirements at costs lower than competing energy systems.

The environmental problems of geothermal energy are primarily *housekeeping* problems for which technology is either now available or can be developed over the next decade if needed. All these problems appear to be technically feasible of solution. They include odor, noise, salt spray emission, residual waste recycling, and subsidence. It is probable that geothermal plants will not be developed in high population density areas, national parks, and in some areas of special environmental concern. However, the size of the resource is so large that, if federal lands are released for geothermal development, then there will be many alternative sites for development.

OCCURRENCE OF GEOTHERMAL ENERGY IN THE UNITED STATES

The western third of the U.S. plus Alaska and Hawaii seem to be the most richly endowed areas where low cost geothermal energy may be expected. There is little or no deep well information on the hot spring areas of Virginia, West Virginia, Georgia, Arkansas and other parts of the eastern and central U.S. The more expensive sources such as the hot dry rock reservoirs of the Appalachians may bevery large but so little is known that it is necessary to project reserves in a very approximate manner. However, present analyses based on fairly conservative projections *assuming hot dry rock systems to be technically feasible* indicate that geothermal energy can be a major source of electrical energy for the entire U.S. (Table 1).

TECHNOLOGY NEEDED

The key to development of the hot dry rock systems is the technology of rock fracturing. A major program of non-nuclear fracturing research is needed to test present hydrofracturing technology and to develop additional more cost-effective methods.

The second key area needed is a national program of deep geothermal drilling in the 20,000 to 30,000 feet depth range with holes fairly evenly distributed across the U.S. to appraise the total U.S. deep potential. This deep geothermal well drilling program would provide data not available from oil wells and would require a substantial amount of new high temperature instrumentation and technology.

The third area of research and development needed is to develop the power conversion technology for low temperature hot water geothermal systems. This would be applicable to hot rock technology as well as the numerous western hot temperature hot water geothermal systems.

A GREAT STATE PLANNING OPPORTUNITY

California's Bright Geothermal Future

STANLEY SCOTT and SAMUEL E. WOOD

The Geysers, Sonoma County

CALIFORNIA is virtually ignoring a major new geothermal discovery—a vast supply of heat and water whose early and carefully controlled development could help alleviate a host of the state's critical environmental and economic problems. Thanks to a rare gift of earth's geology, Southern California has the potential for generating great quantities of cheap electricity, using comparatively pollution-free methods, and for producing a large and virtually inexhaustible supply of distilled water, plus several mineral by-products. Severe water shortages could thus be overcome, substantial power demands met, many struggles over nuclear fission plants avoided, possible radiation hazards reduced, and future air pollution mitigated. But only through comprehensive state and federal planning, now sorely lacking, will such benefits accrue.

The potential for the entire state can be realized by thoughtful and prudent utilization of the vast resources of geothermal energy and water newly proven to lie beneath the Imperial Valley. The technology required to accomplish this near-miracle is yet to be perfected, but scientists and engineers believe that they already see the basic outlines of a large-scale, economically feasible, pollution-free system. To perfect this technology, we must make sure the necessary funds for research and pilot projects are available.

Needed: New concepts and attitudes

Partly because the discovery is so new, it has not yet caused the high excitement and intense public interest that might be expected. Instead, the response in many circles has been quiet, limited and bureaucratic. Leaders in power generation and water supply are presently committed to moving more Northern California water south, and to constructing numbers of nuclear-fueled power plants. Admittedly, the thought and effort invested in such plans make it difficult to shift policy gears as fast as the unfolding of Southern California's geothermal story appears to justify. But the potential of the Imperial Valley field demands a radical rethinking of our whole approach to energy generation and water supply development for Southern California.

The biggest remaining obstacles to realization

Reprinted with permission from *Cry California*, vol. 7, pp. 10–24, Winter 1971/1972.

of the geothermal potential are not primarily matters of science and technology, but of planning, politics, public policy, institutional relations, and resistance of utilities to change. The crucial question is: How do we organize and guide our efforts to insure maximum public benefit through well-planned, expeditious and beneficial use of Southern California's geothermal resource?

Clearly, the first step would be to commit research funds that match the magnitude of the opportunities. The amount of money being spent today is ridiculously small in view of the size of the resource. It is estimated that full development would call for a public-private investment totaling about $5 billion. Prospects are excellent in the Imperial Valley, and the investment is virtually certain to pay handsome dividends. The valley has one of the world's greatest known geothermal resources, and one of the easiest to develop on a large scale. Moreover, what we will learn can be used to develop geothermal energy elsewhere. It is known to be present in huge quantities in the western United States, and in many other parts of the world. In any event, there is urgent need for new funds to support adequate research, field investigation, test drilling and pilot projects in order to tap the Imperial Valley field's tremendous potential.

Global geology

Men have long been fascinated by the look-alike features of the African and South American coasts. Remove the South Atlantic Ocean, and Brazil's great bulge near Recife fits neatly into Africa's Gulf of Guinea. Even in relatively small details, the two coasts mirror one another, like paired pieces in a giant jigsaw puzzle.

We now know that they were joined, until about 140 million years ago, and there was no intervening ocean. Geologists have identified formations in both Africa and South America that are identical in age, origin and composition. Similar rocks formed a few miles apart are now separated by the waters of the South Atlantic. Other continents show analogous former linkages. North America, for example, was once united with Western Europe, and India was attached to Antarctica. In fact, all the world's continents and major islands appear to be derived from two enormous old proto-continents, Gondwana and Laurasia. These have since broken up, and the pieces have drifted thousands of miles, moving over the earth's mantle like great rafts.

What mighty forces are responsible for all of this activity, sundering continents and forming oceans? Only a global mechanism could account for such massive, long-term and worldwide movement. The explanation lies in theories of ocean floor spreading and continental drift that, after many years of doubt and debate, have now taken on firm substance, revolutionizing the staid discipline of geology.

Central to this revolution is a recently discovered global system of rifts or fissures, found principally along the bottoms of the major oceans, as shown in the map on page 14. Over long ages, hot, molten material has welled up from the earth's interior along the length of the fissures, which apparently open outward at rates of an inch or more per year. The slow upwelling gives the rift zones the appearance of ridges or rises in the deep ocean floor.

Named for their locations, the principal elements of the worldwide system are the Mid-Atlantic Ridge, the Mid-Indian Ridge, and the East Pacific Rise. Geologic studies demonstrate that crustal material is youngest near the rises, and oldest close to the continents. In summary, the forces associated with the fissure system and the upwelling hot material have torn the old continents apart and formed the ocean bottoms. These forces continue to cause continental drift.

What does all this have to do with geothermal energy in Southern California? The global background explains the heat source and helps put the Imperial Valley's geothermal phenomenon in perspective. While most ocean-floor spreading occurs along rifts that are usually far from land and under deep water, one remarkable exception is the East Pacific Rise, where it approaches the coast of Mexico. After traversing thousands of miles of deep ocean, it runs under the Gulf of California, where rifting of two or three inches a year began three to four million years ago.

The rift strikes land at the north end of the gulf, and goes under the Mexicali and Imperial valleys. This region in Southern California is one of the few areas in the world where the global fissure system comes ashore and brings hot material up to a position underlying terra firma. Here, too, earthquake fracturing has opened the

---

STANLEY SCOTT is assistant director of the Institute of Governmental Studies, University of California, Berkeley. SAMUEL E. WOOD is a planning consultant and former executive director of California Tomorrow. This article is based, in part, on preliminary research done by Mary Ellen Leary, correspondent for the *London Economist* and the Scripps-Howard Newspapers.

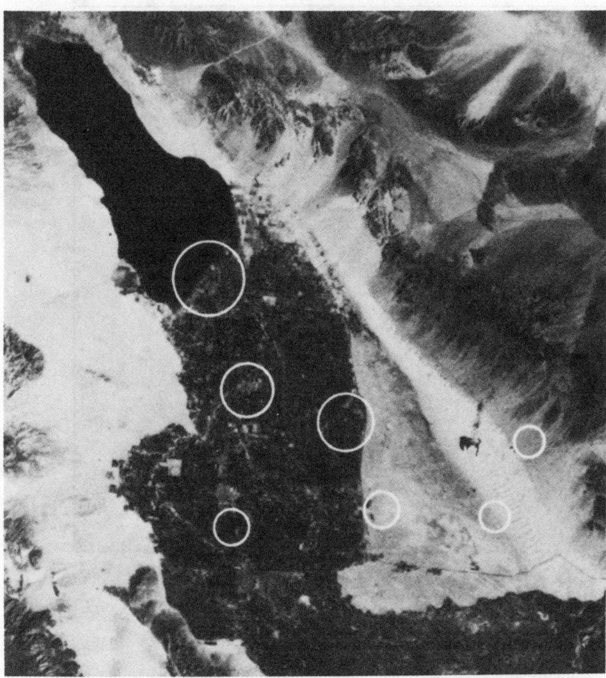

Salton Sea, upper left, and Imperial Valley, center, as photographed from Apollo 9. Circled areas indicate the most promising geothermal fields of the region.

heated rocks to penetration by percolating ground water. The rift has also helped form the Salton Trough, a geologic feature that underlies the Imperial and Mexicali valleys, and is filled to a depth of four miles with porous water-bearing sediments washed down when the Colorado River was eroding out the gorge of the Grand Canyon.

Recent geological investigations led by Robert Rex, professor of geology, University of California, Riverside, and by Alan D. K. Laird, Director, Sea Water Conversion Laboratory, University of California, Berkeley, have confirmed that the Salton Trough contains a huge supply of superhot water—perhaps ten billion acre-feet—varying in temperature from 400 to 600 degrees Fahrenheit, and under considerable pressure. Although the water is moderately saline, much of it is significantly less salty than sea water, thus enhancing its value as potentially usable water, as well as an energy source.

A unique combination

A fortunate combination of circumstances in the Imperial Valley offers Californians a matchless opportunity for tapping this vast supply of heat and water. There are at least seven notable features of the Imperial Valley's power and water potentialities:

- An inexhaustible heat source is found comparatively near the surface and under dry land, rather than far out at sea.
- Geothermally heated and pressurized water is present in great quantities and at comparatively low salinity levels. (The combination of heat, pressure and low salinity makes desalinization of this water more feasible and attractive than sea-water conversion.)
- Southern California's unquenchable thirst for water has given impetus to the first stage of the State Water Plan for diverting Northern California water south, and now threatens to dam the other rivers of Northern California to provide still more water for diversion.
- A local desert environment supports a thriving agriculture that desperately needs new sources of water soon to guarantee its future.
- Great urban industrial concentrations are close by in Southern California, occasioning large future power demands that cannot be met from existing sources.
- A growing concern for the quality of the California environment, and especially dismay with increasing environmental pollution, has prompted a diligent search for pollution-free sources of electrical energy. Geothermal power now emerges as Southern California's best prospect to reduce dependence on fossil and nuclear fuels.
- Official expectations that most of California's future nuclear plants will be concentrated along the Pacific has stimulated concern over radiation safety, thermal pollution and the spreading destruction of California's remaining unspoiled coast.

The fresh-water potential

Knowledgeable observers agree that the geothermal resource underlying the Mexicali and Imperial valleys is gigantic. How much of the power can be economically developed or how long it will last is not yet certain, but the figures presented by Robert Rex suggest that California's sights should be set high. According to Rex, present knowledge supports estimates that the geothermal brine under the Imperial Valley alone should yield at least five to six million acre-feet per year of distilled water. By comparison, the State Water Plan proposes to transfer only 2.5 million acre-feet south of the Tehachapi Mountains annually by 1990.

The Imperial Valley's potential also equals about half the existing water flow in the lower

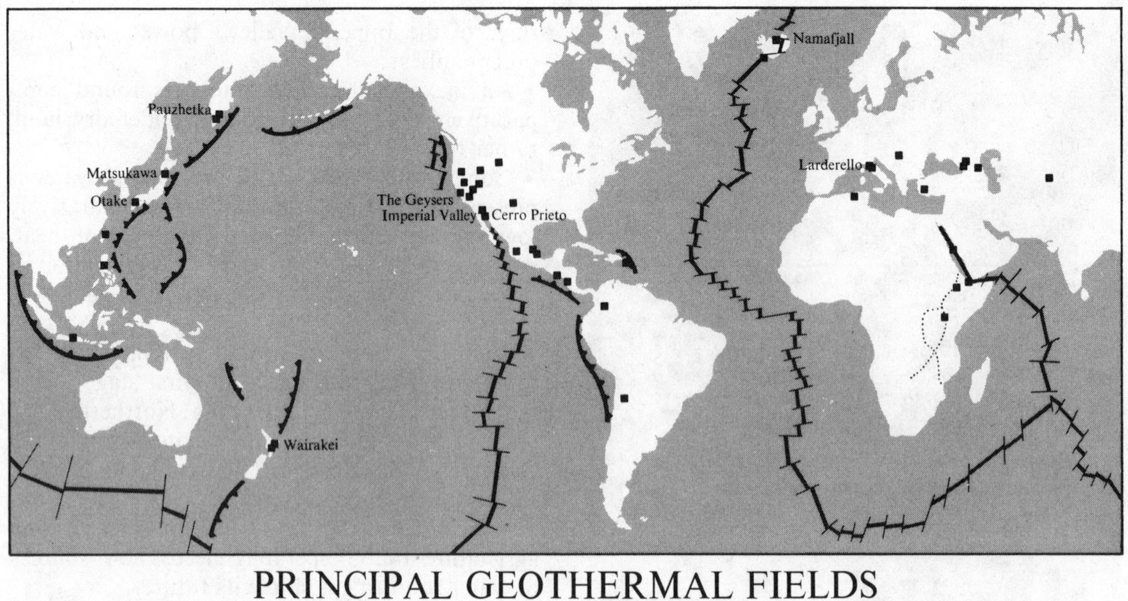

PRINCIPAL GEOTHERMAL FIELDS

Colorado River—a crucial increment in that area of severe water shortage. Because the resource is so large, benefits would not be limited to the Imperial Valley area, but could also have a significant impact on the water-short economies of all the states in the Colorado Basin. Furthermore, the desalted water would enable the United States to meet its treaty commitments to Mexico by reducing salinity levels of Colorado River water flowing across the border.

With proper management, this new source of water can be considered virtually inexhaustible, if present estimates are proved to be accurate. The evidence indicates that hot-water reserves are adequate to last from 100 to 300 years. Moreover, the underground pressure equilibrium could presumably be maintained by injecting sea water to replace the preexisting geothermal brine being removed and desalted. After heating by the underlying rocks, the sea water could in turn be drawn off and desalted, and the process thus continued indefinitely.

The power potential

Paralleling the potential for fresh-water production is the generation of electric energy, which could be part of the same process. University of California research suggests that 20,000 to 30,000 megawatts of power can be produced annually (one megawatt equals 1,000 kilowatts, or one million watts). In 1970, California's total electric-power capacity was 32,000 megawatts (including imports), and is expected to increase to 60,000 megawatts by 1980. The advisability of such an increase is in serious doubt, in terms of both environmental impact and resource depletion, and in the case of nuclear plants, even safety. These considerations require close examination, especially the determination of actual need. In any event, the Imperial Valley's geothermal resource could provide much of whatever increased capacity is required for at least a decade.

Early use of Southern California's geothermal energy could thus "buy time" for perhaps up to ten or 20 years. It could reduce some of the need for increased burning of fossil fuels, and justify deferment or cancellation of plans to construct nuclear reactors at many locations, both along the coast and inland.

Moreover, Rex has estimated the geothermal power potential of the Western states to be between 100,000 and 10 million megawatts. (It should be noted that geologists and students of geothermal phenomena do not accept all of Rex's estimates. In particular, they do not believe that existing information justifies projections showing such huge quantities of *usable* energy in the Western states.) This potential can be compared with the Western area's estimated 1985 peak power demand of 160,000 megawatts. In addition, geothermal possibilities of unknown quantity are thought to exist in Alaska, Hawaii, the Gulf Coast, the Appalachians, the Ozarks and the Ouachita Mountains.

If these estimates are of the right order of magnitude, it seems obvious that the Pacific Gas and Electric Company's highly successful geothermal power development at the Big Geysers north of San Francisco has hardly scratched the surface of a large and potentially cheap power supply that is waiting to be developed, if the necessary technology can be provided.

Unfortunately, the United States seems to be lagging in geothermal research and exploration; many nations have moved ahead of us. Italy has made the most extensive use of geothermal energy, having begun nearly 70 years ago. Other nations with significant geothermal development include Iceland, New Zealand, Japan and the Soviet Union. Other countries which are stepping up their activities include Algeria, Chile, China, Colombia, Czechoslovakia, El Salvador, Ethiopia, France, Indonesia, Kenya, the Philippines, Taiwan, Turkey and Yugoslavia.

Closer to home, Mexico has developed power production in the Cerro Prieto field located in the Mexicali Valley, only about 20 miles south of the international boundary. This ambitious project is well under way, and has already tapped Mexico's part of the same geothermal resource that underlies the Imperial Valley. In moving ahead to develop their portion of this resource, the Mexicans have utilized Italy's and New Zealand's geothermal technology, as well as United States' oilfield practice. Moreover, they are pioneering with original experimentation and are dealing methodically with technical problems. The results of their work are available to help us catch up.

Research and development

Even the more optimistic experts emphasize that many technical problems remain, and at least five to six years of hard work will be required to demonstrate full-scale feasibility. Deep drilling of experimental wells should be accompanied by process testing in pilot plants to find the best ways of producing steam, removing gases, desalting brine, recovering solids, and disposing of the concentrated brine after removal of distilled water.

Reservoir management studies are also essential, if adequate coordination is to be established and the entire geothermal resource developed as a unified or "unitized" field. Such research is imperative if full and long-term public benefits are to be realized, and hazards such as land subsidence and water-table lowering averted. With adequate financial support, these and related studies could be undertaken immediately by the United States Department of the Interior, and by appropriate agencies of the state, including the University of California.

Fitting the institutional pieces together

Innovative planning for new institutional arrangements will be of crucial importance. If one were starting with a wholly clean slate, perhaps the entire job would be handed to a new agency and, in fact, this may come about. But for the time being, proponents of early development must recognize, respect and work with the many institutional cooks who have a responsibility for the geothermal broth.

Both the state and federal governments are moving forward in what has the *appearance* of an orderly approach to developing geothermal resources. But in fact, the effort is disorganized and limited in purpose, being guided principally by the narrow objective of getting geothermal re-

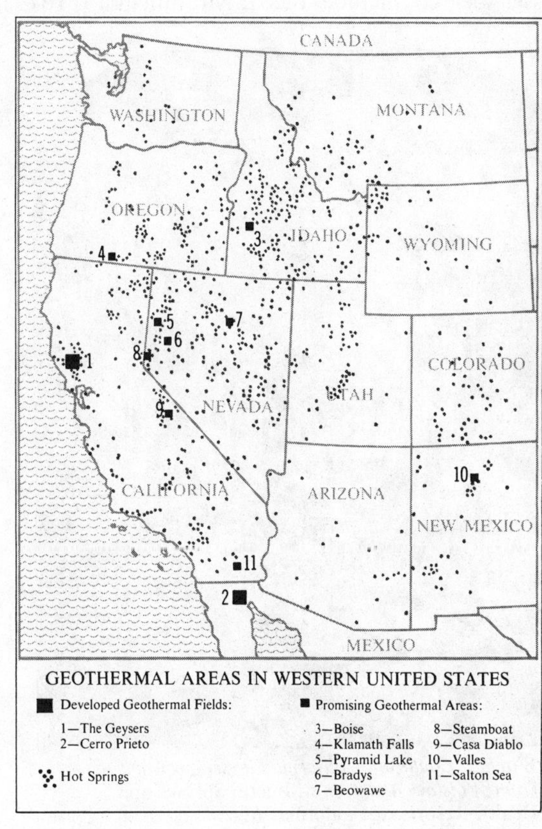

GEOTHERMAL AREAS IN WESTERN UNITED STATES

sources on the market. Thus, the programs are not directed by any overall strategy or plan, such as that called for in the California Tomorrow Plan, which would relate the production of energy and water to state or regional needs. In addition, division of responsibilities between the federal and state governments, and among many units at both levels, is the cause of much confusion and uncertainty. Finally, as we shall see, recent federal and state geothermal legislation is not adequate to the need, and in fact may open the way to injudiciously premature moves, taken without adequate policy guidance for the long-term future of geothermal power and water in California and the West.

Federal law and administration

The Geothermal Steam Act of 1970, and rules proposed for its implementation, give the Secretary of the Interior responsibility for leasing geothermal lands. Within Interior, the Geological Survey is responsible for locating, surveying, exploring and supervising the development of geothermal resources, while the Bureau of Land Management is in charge of preparing and issuing the leases. Other federal agencies involved to a lesser degree include the Environmental Protection Agency, the Department of Agriculture, the Bureau of Fisheries and Wildlife, the Federal Power Commission, the Bureau of Indian Affairs, and the Bureau of Reclamation.

Witches' Cauldron, California Geysers, a woodcut from Winslow Anderson's Mineral Springs and Health Resorts of California, *1892.*

Lands under federal control and within known geothermal resource areas (KGRA) are to be leased on the basis of competitive bidding. Other federally controlled areas are to be leased on a first-come basis. Once a lease has been brought into production, royalties are assessed, based on productive capacity. The geothermal leases will include provisions intended to protect the environment, other resources, and public safety. Requirements are included relating to public access, pollution, erosion control, noise suppression, sanitation and waste disposal, esthetics, wildlife, historic areas, and restoration of disturbed lands. Environmental impact studies for each KGRA are to be prepared by the Bureau of Land Management.

Considered solely as a mechanism for opening geothermal resources for exploration, without destructive environmental and other damage, the federal system seems reasonably well designed, as far as it goes. But many questions remain unresolved, including the following:

- Are adequate measures being taken to forestall premature leasing and other long-term commitments which may later interfere with the orderly development of geothermal water and power in the public interest?
- Will water developed and sold from federally controlled geothermal resources be distributed by the federal government, and if so, for what purposes, to which customers, and under what conditions?
- Will lease agreements permit the Secretary of the Interior to distribute the geothermal power to preference customers over high-voltage intertie systems?
- Will recognized federal preference for power sales to public agencies control the lease arrangements?
- Will federal lands that may be brought into production using the new resources be kept in federal ownership and leased out for development, thus retaining basic federal control of the land use, and recapturing a share of the income for the public benefit?
- If lands in the public domain are released for urban, open-space, and recreational uses, how will these uses be planned for and controlled, on both local and regional bases?
- How will the public and private contributions to the water and power supply be identified and

divided, both as to production and distribution?

• How will geothermal resource development on federal lands be related to state and regional development plans—existing, or impelled by the enormous potential development—in California and other affected states?

California law and administration

Primary state responsibility is lodged in a troika consisting of the State Lands Commission, the Geothermal Resources Board, and the Division of Oil and Gas. The *ex officio* State Lands Commission is responsible for permits and leases of state-owned lands for the development of geothermal resources. Where the land surface is under control of another state agency, the permits are granted with the consent and under conditions set by the other agency. The commission also sets royalty and rental fees.

State law requires that geothermal resources be managed under the principle of multiple use. The commission interprets this policy as meaning that all resources, and both public and private interests, must be considered. Moreover, the commission may not lease any lands until it makes a finding that the lease in question will not have a significantly detrimental effect on the environment, and until it has made an environmental-impact report available to the legislature and the public.

Established by the Geothermal Resources Act of 1967, the Geothermal Resources Board seems largely a paper organization. Real responsibility for administering and supervising the process of exploration, leasing and development rests with the Division of Oil and Gas. Drawing on its 55 years of experience in regulating oil and gas exploration, the division has simply "folded in" the new geothermal resource, and has been applying its petroleum-related expertise to this additional responsibility.

The close ties between the geothermal industry and the operations of both the Division of Oil and Gas and the State Lands Commission make it difficult to distinguish between the public interests of the state and those of the private sector. Perhaps it is worth noting that the Division of Oil and Gas is wholly funded by the industries it regulates, through fees, fines and other revenues generated by the regulatory process.

The State Water Resources Control Board and its regional counterparts are responsible for requirements relating to all waste waters discharged into the surface or subsurface waters of California. The Division of Oil and Gas regulates well

Pacific Gas and Electric Company's geothermal steam wells in Sonoma County are more than a mile and one-half deep. By the end of 1971, 190 megawatts of electrical power will have been generated in this field. By 1975, plans call for a 600-megawatt capacity, the amount equal to the needs of a city the size of San Francisco.

operations when water fluids are reinjected into subsurface formations that do not contain fresh water. Other state agencies with limited geothermal regulatory responsibility are the Department of Water Resources, the Department of Fish and Game, and the Division of Forestry.

The state has disposed of much of the surface of its potential geothermal lands, including water rights, without reservation. Adjudication appears necessary to determine whether these resources belong to the surface owner or to the mineral owner. A further major obstacle to intelligent

The world's largest producing geothermal field is at Lardarello, Italy, where drilling explorations go back more than a century. Today, Lardarello generating plants produce about 360 megawatts of electrical power.

planning and control of the state's geothermal resource development is the absence of basic data on the two million acres of state land. Much of it has not even been properly surveyed and located, and very little of it has had the benefit of land-classification studies. The state simply does not know where its lands are or what their value is. At present, there are at least six separate uncoordinated state land indexes. Without usable information on the location and capability of state lands, adequate management of geothermal resources is impossible.

Another major obstacle to an effective, integrated program is the lack of any viable machinery for coordination. The Geothermal Resources Board does not have the funds, staff or legal responsibility to pull things together effectively. Nevertheless the board attempts to furnish interested parties and agencies with data on research and development. The situation at the state level is chaotic.

Underlying this mess is the lack of a California state development plan. We have a vacuum where guidance is needed, and in the absence of priorities based on such a plan, it will be difficult to determine intelligently what are the best ways to utilize geothermal power and water in the public interest.

But if the state were to take positive action, it could influence geothermal development in several ways. It can, of course, control leasing or disposal policy on lands with geothermal potential which are still in state ownership. Further, the state can facilitate and guide beneficial resource utilization by helping finance the research and pilot studies essential to get started with a planned and coordinated development. Finally, if California had a well-thought-out state plan of development, including water, power and land-use elements, it could employ the plan to influence how geothermal water and power are distributed and used, regardless of the source.

The local and regional roles

As things stand, the cities and counties will have principal responsibility for guiding future land use in the Imperial Valley and other areas opened to new or more intensified development by the availability of new power and water. As long as growth pressures are small, presumably the local governments can do about as good a job as they have in the past. But if great quantities of new water and energy bring large-scale economic expansion, land-use changes and population shifts, local government alone is most unlikely to turn in an acceptable performance.

Before large-scale new resources are made available in the Imperial Valley and adjacent areas, California should be ready with a regional development plan along the lines laid out in the California Tomorrow Plan, to insure that those resources are used in the broad public interest.

Moreover, the plan should contain elements on the resulting land use and economic development of the region, its population growth and distribution, and housing. In addition to a plan, there must be a system for its implementation. Given the size of the growth pressures anticipated, it seems clear that local government will not be able to cope with the situation unaided.

Resource conservation and development

Perhaps it would be helpful to employ the principle of massive yet reimbursable funding that has been used by the Department of Agriculture and the Department of the Interior in aiding rural areas by underwriting development or redevelopment. The framework and processes of the Resource Conservation and Development Project program of the Soil Conservation Service, Department of Agriculture, is one useful model. It offers a proven method of using all available programs for planning, financing and developing urban and rural areas. Under these programs, planning funds are advanced by Congress upon application by the Governor. During the six months or year required for planning, hundreds of citizens and agency representatives explore possibilities, consider alternatives, and arrive at a consensus on what needs to be done and what agencies and programs, federal, state and local, should be involved.

Regional machinery required to administer the program and later to govern the region, along with cities and counties, could be designed as part of the plan. Appropriate authorization procedures and periodic review are, of course, provided in the appropriation process. The Department of Agriculture and the Department of the Interior would assume major roles because of the large power, water and agricultural interests at stake.

In addition, the programs for large new towns administered by the Department of Housing and Urban Development should play an important role in planning, financing and building balanced communities in developing areas. Another prototype for possible emulation is New York State's Urban Development Corporation, which is empowered to plan, acquire land, design, construct and manage, and own projects of all kinds in any part of the state.

Hold the line: New policies needed

Both the state and federal governments would be well-advised to hold the line on geothermal leasing until regional plans have been completed, and the development programs and controls are in being. The necessary local, regional and state-federal policy machinery could be designed as part of the process. In addition, all of the public and private lands needed for the new towns and other regional facilities should be acquired very early in the process, both to insure full acquisition of enough land, and to help guide and channel future growth. At the very least, an emergency leasing policy committee should be formed by the Department of the Interior, to review the implica-

The Great Geyser, Iceland's best-known geothermal display. While Iceland's geothermal resources provide only part of her power needs, much of the city of Reykjavik is heated by steam from geothermal sources.

Mono Lake, one of California's promising geothermal sources. The test rig shown below is part of an ongoing exploratory program at the lake.

Wairakei geothermal power project in Auckland Province, New Zealand. This field currently has an electrical capacity of 160 megawatts.

tions of leasing policy and to devise interim safeguards. The policy committee should have representatives of the Bureau of Land Management, the U.S. Geological Survey, the Bureau of Reclamation, the Environmental Protection Agency, and the Department of Housing and Urban Development.

Until new policies are devised, however, much of the legal theory and regulatory practice that apparently will govern development of geothermal resources will be adapted from another era and based on the administration of quite dissimilar natural resources—groundwater, minerals and petroleum. But it seems clear that the vast, new and different geothermal resource cannot appropriately be treated as a mere retread version of groundwater, or as a mineral. It is a unique phenomenon in its own right, should be so recognized, and demands new policies.

Instead of devising new policies, we apparently

are hoping—through leases to private entrepreneurs for exploitation by bits and pieces—to develop adequately a major resource that requires treatment as a complex total system. To this end, all the following efforts must somehow be meshed: drilling wells and bringing hot water to the surface, making steam and generating power, desalting brine and producing distilled water, extracting useful mineral by-products, removing noxious gases, suppressing pollutants, dissipating heat, reinjecting the concentrated brines into underground formations or otherwise disposing of them, and maintaining subterranean water levels to prevent subsidence. Finally there are the questions of how to distribute the power and water produced, who should distribute it, and for what purpose.

Doing all these things so that they fit together and are mutually supportive will be a monumental task, especially because of the intrinsic complexity of the processes and the multiplicity of the agents and interests concerned. The situation does not give grounds for much assurance that we can "let nature take its course" and muddle through to a satisfactory outcome. Still, the *status quo* is the place where we have to begin, and past experience demonstrates the need for vigilance against the hazards of *ad hoc*kery.

The history of resource development in California is largely a history of *ad hoc*, single-purpose exploitation coupled with a lack of broad guidelines from the state. Instead, the major goals and objectives have been those of the special interests involved in exploiting the resources. The process has been reinforced by state and federal policies that have viewed natural resources primarily as raw material for products to be mass-produced and mass-marketed.

A recommended reorganization

Ad hoc resource development and sale, rather than thoughtful utilization according to a basic resource plan, have been facilitated by the long-term give-away policies of the State Lands Commission. The commission appears to hold as a principal objective the placing of state controlled resources "on the production line."

Accordingly, we urge that California reorganize and enlarge its State Lands Commission to insure that the agency receives direct policy contributions from a wider range of interests than at present. (The commission consists of the Lieutenant Governor, the Director of Finance, and the State Controller who serves as chairman.) A broadened base would help the commission better to represent the public interest in the many land-use issues that come before it.

The commission should be reconstituted by adding to its present membership the director of the Office of Planning and Research, the chairman of the Senate Committee on Natural Resources and Wildlife, the chairman of the Senate Committee on Water Resources, the chairman of the Assembly Committee on Natural Resources and Conservation, the chairman of the Assembly

Committee on Planning and Land Use, the chairman and the vice-chairman of the Joint Committee on the Public Domain, and the secretary of the Resources Agency who would chair the enlarged commission.

The newly reconstituted commission should have two major responsibilities. Its *first* responsibility would comprise the following assignments:

• Within one year it should identify, classify and develop a single index of all state-owned lands.

• Within two years it should review state lands policy and practice, and recommend to the legislature appropriate new policy for state-owned geothermal lands. Policies for other land classes should be recommended, depending on their capability for sustaining development and on the roles envisioned for them in the state plan of development.

• Advise the State Office of Planning and Research with respect to a state-owned land element to be included in the state plan of development.

The commission's *second* responsibility would be a continuing advisory function. It would advise the Resources Agency and the legislature on state-owned land policy—especially with respect to geothermal lands—and on state developmental and planning programs affecting both public and private lands.

The Challenge . . . and the Opportunity

Much of the large-scale planning that has guided past resource development has been both fragmented and controlled by several powerful special interests, such as the water-distribution, land-development and freeway establishments. Although these interests are both influential and entrenched, there is an excellent chance to plan more prudently for the use of our geothermal resource than we have for other resources. Geothermal power and water offer California and the entire West an opportunity to secure maximum long-term benefit to the regions and their inhabitants, unencumbered by past improvidence, oversight and error in resource-development policy. Optimum utilization will profoundly affect the state's future.

"Assuring beneficial development"

The importance of future development based on geothermal resources is stressed by the University of California's Alan D. K. Laird. In a monograph published in August, 1971, by the Department of the Interior's Office of Saline Water, titled "Ranking Research Problems in Geothermal Energy," Laird listed "assuring beneficial development" second in a hierarchy of eight needed research topics. Laird proposes that the Department of the Interior's Imperial Valley program be used as a model by American industries in developing geothermal resources in the southwestern states and elsewhere. He sees this as a first step in large-scale utilization of geothermal resources throughout the world, with great resulting benefits to mankind. But in emphasizing the magnitude of the opportunity offered in the Imperial Valley, he also points to the danger of our failing to rise to the task:

"The availability of abundant power, water, and open land—combined with a desirable climate and the possibility of short-term profits—could precipitate indiscriminate industrialization and urbanization, and cause a degradation of this unique winter agricultural area. Overcrowding, inadequate transportation, pollution, and severe ecological damage could require remedial measures that would cost far more than the total gains from the geothermal resources. Such long-term results in this first big development would have a deterring effect on subsequent developments of the widespread geothermal resources of the western states." To prevent such a catastrophe, Laird recommended that research be initiated early on proposed actions, and on means of influencing the course of development. Careful cause-and-effect studies could help with the formulation of guidance and developmental control policies for use in other KGRA's.

Planned regional growth staged to match the orderly utilization of the geothermal resource appears to Laird to carry the best chance of success. A prototype of such a regional planning and development program is outlined in the California Tomorrow Plan. Laird proposes the testing and evaluating of alternatives at each step, with systems analysis, benefit-cost studies and related analytic processes, some of which are still to be devised. Flexibility needs to be built into the organization and operating procedures, and there must be effective feedback of the results of actions taken to influence the course of the development.

An interim solution: A geothermal task force

Flexibility and feedback are crucial concepts. Keeping options open, while pursuing a watchful and sustained monitoring program, will permit experimentation and progress, while insuring that things do not go irrevocably wrong. Thus at the very outset it is essential to establish the chief criteria for desirable geothermal development, and to take steps for their implementation. The

Cerro Prieto geothermal field, south of Mexicali in Baja California. When fully operational, the field will have a capacity of 75 megawatts.

criteria can form the basis of a plan to develop the Imperial Valley geothermal resource as a single unit. The plan should also include recommended uses and allocations of the energy and water resources made available. Progress toward implementation of the plan could be monitored on a continuing basis, and reserve authority employed when necessary to forestall actions that would contravene the plan.

One way to begin would be the immediate creation of a federal-state intergovernmental Geothermal Task Force to draft a geothermal resource development plan, encourage and review research and pilot projects, and monitor leasing and geothermal field development. The task force should report periodically, indicating desirable state and federal policy changes, and especially suggesting ways of insuring optimum geothermal resource development in the public interest. Their reports would be directed to the attention of the President and Congress, and the governor and legislature of California.

Such a task force would operate as a stop-gap and interim measure, but it would also be a useful first step toward an effective long-term program of planning and development. Given an appropriate authorization, and supplied with adequate resources, a task force could help push along orderly planning and development of the state's geothermal resources.

The pressure of events makes it essential to pursue planned geothermal development forthwith, using diligence, energy, imagination, and foresight. Accordingly, it seems fitting to conclude this article with a highly pertinent and accurate statement, by a key figure in the current state administration, on the need for planning and on the magnitude of the task. Secretary of the Resources Agency Norman B. "Ike" Livermore, Jr., commented in a paper presented at the Imperial Valley-Salton Sea Area geothermal hearing in October of 1970:

"The time is most opportune to fully explore the feasibility of harnessing this vast natural resource in other areas of the State [in addition to the Big Geysers area] such as the Imperial Valley. King-size planning is needed to bring private industry and local, state, and federal agencies together to take a close look at plans and proposals for exploration and development and to consider solutions to the numerous interrelated problems. . . ."

Brighter outlook for solar power

The Sun's relentless flood of energy onto the Earth's surface is a tempting power source in a world that is fast running out of fossil fuels. The United States, which has already been hit by energy shortages, is looking seriously at the prospects for solar power

Graham Chedd
is Washington Editor of New Scientist, and a consultant to the American Association for the Advancement of Science's Communications Programs for the Public Understanding of Science

Although out of phase with the 11-year cycle of solar activity, interest in solar power here on Earth seems to wax and wane with a similar periodicity. But it could be that the Earth-bound cycle is about to be broken. The resurgence of enthusiasm in the United States for harnessing the energy of the Sun has coincided—and has, of course, in large part been brought about by—the age of environmental concern, and more particularly with the year of the "energy crisis".

While in the short-term the energy crisis is increasingly recognised as political and institutional (and, as such, in the words of one well-informed observer, "could be solved by a stroke of the President's pen"), it has alerted the country to the alarming medium- and long-term problems of satisfying the country's voracious energy appetite. With an end to the known domestic oil and natural gas reserves now clearly in sight, and with nuclear power so far signally failing to live up to expectations, even the nation's politicians—President Nixon among them—have begun to cast around in some desperation for energy technologies that will see the United States through to the next millenium. Power from the Sun, while even on the most optimistic of forecasts not expected to provide more than a small fraction of these needs, is suddenly no longer regarded as a quaint, if intellectually appealing, technological backwater.

The solar optimists have recently had two opportunities to display their wares: at a conference on energy held last month at the Massachusetts Institute of Technology; and within the pages of the report of a National Science Foundation/NASA panel devoted to solar energy as a national energy resource. With the drawback of a hefty R & D pricetag —$3500 million spread over the next 15 years —power from the Sun could, in the panel's opinion, economically provide, by the year 2020, 35 per cent of the energy required to heat and cool the nation's buildings, 30 per cent of the gaseous fuel, 10 per cent of the liquid fuel, and 20 per cent of the United States' electricity needs.

The Sun's rays have several notable attractions as an energy source. They are not going to run out (or at least, when they do, mankind will have to face more pressing problems). They come already distributed. They provide as environmentally clean a source of energy as one could wish for. And there is several orders of magnitude more than enough energy from the Sun falling on Earth to cope with all anticipated needs. For instance, assuming a 10 per cent efficiency in the conversion of sunbeams to electricity, the total electricity consumption of the United States in 1969 could have been supplied by the solar energy incident upon 0·14 per cent of the US land area.

The NSF/NASA panel identifies three fields in which solar power could (given enough money) make a significant impact over the next 15 years or so. These are: the heating and cooling of buildings; the production of organic materials and their use directly to provide energy, or alternatively their conversion to solid, liquid, or gaseous fuel; and the generation of electricity. In each case the obviously crucial task is to make solar power economically competitive with conventional energy sources, a task which looks more difficult as one works through the list.

Solar houses

Sun power in the home is unquestionably the prospect most likely to come to something. Already, solar water heaters are commercially manufactured in Australia, Israel, Japan, the USSR and, on a small scale, in the US. Although further behind, space heating in the home by solar power is already technologically feasible, and about 20 experimental buildings have been built. Technologists also have limited experience in the operation of absorption refrigeration systems using heat from solar collectors. The solar energy panel sees no technological and few economic barriers even now to the construction of homes with a combined system providing between about 50 to 80 per cent of all heating and cooling needs. A typical installation would involve a flat, solar collector on the roof, heating water pumped through it. Heat extracted from the stored hot water would be used to warm air in the winter and cool it in the summer. An auxiliary water heating system would have to cope with prolonged cold spells.

Two examples of the use of solar power in houses were presented at the MIT energy symposium, and together illustrate the possibilities and practicalities of exploiting the Sun in the home. The Environmental Quality Laboratory, at California Institute of Technology, is forging a coalition between local energy utility companies, building firms, and solar equipment manufacturers to demonstrate the feasibility of solar water heating in new apartment buildings constructed in Southern California. With a 50 square-foot flat solar collector per unit, an optimised system could reduce gas consumption to 20 per cent of present levels. The major obstacle to the introduction of such systems is not technological or even economic but institutional; the building trade is perhaps the most conservative in the land, and power companies are naturally highly suspicious of any innovation that will reduce the consumption of their output. By bringing these forces into the partnership, the EQL group hopes to overcome these prejudices and hasten the introduction of solar water heating by some five or ten years.

Reprinted with permission from *New Scientist*, vol. 58, pp. 36–37, Apr. 5, 1973.

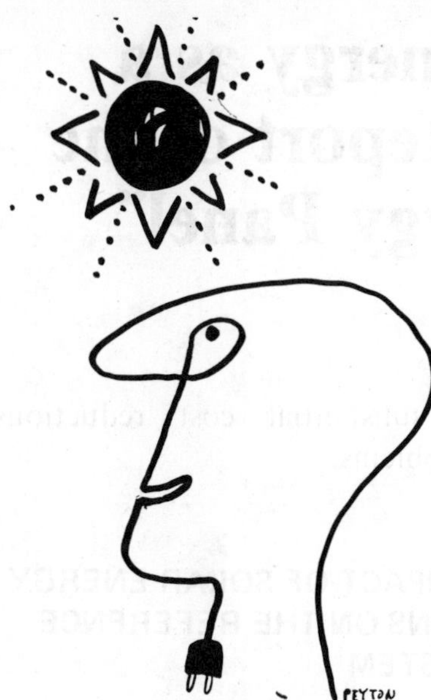

While Southern California has sunshine in abundance, Delaware, in the north-east of the country, has to cope with weather of the English type. Yet it is at the University of Delaware that perhaps the most ambitious feasibility study of solar power for the home is underway. The university is constructing an experimental single-family home which will provide its hypothetical occupants not only with thermal energy but also with a small proportion of their electricity requirements. The home has a total of 40 collectors on its roof and south wall, each employing cadmium sulphide/copper sulphide solar cells to convert sunlight directly into electricity as well as absorbing the Sun's heat to provide hot water.

Growing fuel

Fossil fuels represent a concentrated stored form of solar energy. Because these fuels are now being rapidly depleted, the NSF/NASA panel looked into the possibility of mimicking the processes by which fossil fuels were originally generated as a way of producing renewable, clean, synthetic fuels. The panel points out that if through advanced management techniques, including the exploitation of modern developments in plant genetics, crops of trees, grass or algae could be raised that convert solar power into the stored heat energy of plant materials with an efficiency of 3 per cent, then less than 3 per cent of the land area of the US would produce enough stored solar energy to meet all the nation's expected energy needs in 1985. The plant material could be burned directly to power electricity generators, or converted into convenient fuels. Among the possibilities are anaerobic fermentation of plant matter to produce methane; pyrolysis to produce solid, liquid and gaseous fuels; and chemical reduction to provide oil. Perhaps before the deliberate growing of plants for conversion into fuel (arguably not the best use of land in a world two-thirds hungry) some of the same conversion techniques could be instituted on a large scale to convert organic wastes to useful fuels. Although unlikely to be profitable, the income earned by selling the fuel from municipal pyrolysis plants, disposing of urban solid wastes in this manner, would probably allow the plants to be operated at no cost to the community.

Solar power is most commonly thought of in connection with electricity production, either directly via solar cells or indirectly through the generation of steam for generators. While solar cells offer the simplest, cleanest, and most efficient means of converting solar to electric power, the cost of the silicon solar cells used to power spacecraft must fall by a factor of 100 or so before the method becomes feasible on Earth. But the potential benefits are enormous: solar cells of around 10-15 per cent efficiency could collect from the roof of a house in the north-east of the US more than three times the electricity needs of the home. The energy arrives without transmission lines, and, with a technological breakthrough in electricity storage, could not only run the home but also power the family's electric cars.

Technologically more accessible is indirect generation of electricity from solar power. A number of research and development projects are already underway, and one of the most promising was discussed at the MIT symposium. Under the NSF contract, the Aerospace Corporation is designing—with present day technology—a solar to thermal energy conversion (STEC) system which could be producing about 1000 MW(e) by 1990. Set in the sunny southwest, the system would consist of modular reflectors, focusing the Sun's rays on a pipe carrying either liquid sodium or high-temperature eutectic. To provide a reliable base-load capacity, the plant would have to include some means of storing the hot working fluid so that steam generators could be run through the day and night. One attraction of the system, however, is that electricity demand in the southwestern US, where everything is air-conditioned, also follows the Sun.

The STEC solar farm epitomises the problem of exploiting the Sun, whether it be via the homespun technology of solar water heating or grandiose schemes like giant heat engines exploiting the temperature difference between the ocean surface and its depths, or the literally far-out concept of solar satellites in geostationary orbit. While the fuel for all these proposals is free, their capital costs are high. Yet in the end it is the peculiar economics of solar power that may prove to be its greatest asset. While oil and natural gas have been abundant and cheap, solar power has been unable to edge itself into the market place. If anything is certain, however, it is that oil and gas prices are going to climb steeply over the next decade. Even coal, of which the US still has enormous reserves and to which people are looking with increasing interest as a means of salvation, has very high environmental costs which in future will be reflected in its price. The Sun's rays—clean, abundant, limitless, and free—will become increasingly attractive simply by being there.

Excerpts from "Solar Energy as a National Resource: The Report of the NSF/NASA Solar Energy Panel"

SUMMARY

The following information is tabulated to facilitate an overview of the Solar Energy Panel's findings:

Table 1. Status of Solar Utilization Techniques.
Table 2. Summary of Major Technical Problems.
Table 3. Impact of Solar Energy Applications on the Reference Energy System.
Table 4. Summary of Overall Program Funding.

Comments From the Solar Energy Panel's Consultants

TABLE 1. PRESENT STATUS OF SOLAR UTILIZATION TECHNIQUES

The checkmarks in this table indicate the approximate state of development for the various utilization techniques. The extreme right hand check indicates the progress as of 1972 but some work may still be required in the other areas. For example, water heaters are commercially available but combined water heating, heating and cooling systems are still in a state of development.

TABLE 2. SUMMARY OF MAJOR TECHNICAL PROBLEMS

This table identifies the major technical problems to be overcome. As an example, before any of the photovoltaic systems can become applicable on a large scale, solar arrays costing about \$1.00 per square foot will have to be developed. Before the space power station can be feasible, major advances are required in the deployment of huge lightweight structures in space, and low cost synchronous orbit transportation. In all applications substantial cost reductions are important problems.

TABLE 3. IMPACT OF SOLAR ENERGY APPLICATIONS ON THE REFERENCE ENERGY SYSTEM

This table shows the expected impact that solar energy could have if developed to commercial utilization. The impact is compared with the Reference Energy System's annual consumption figures for the particular application, an estimate in percent of the expected market, the annual savings in fossil fuels based on \$1 per million BTU and an estimate of the significance of the impact based on the specific application and the total U.S. energy needs. A minor impact is considered to range from 0 to 5%, a modest impact from 5 to 10%, and a major impact is equal to or greater than 10%. For example, if by 2020, 30% of the Nation's methane needs are met by the bioconversion of organic wastes, then this would represent a major impact on the gas consumption market and a minor impact on the total U.S. energy consumption (4.2%). At \$1.00 per million BTU, assumed for natural gas costs, the value of the product is \$12 billion per year.

TABLE 4. SUMMARY OF OVERALL PROGRAM FUNDING

The funding recommended by the Panel for the total 15 year program is shown in this Table. Since some systems will terminate at phased decision points in the R&D program, the items have not been totaled. Specific decisions justified by the progress made in a preceding phase are required before progressing to the next phase. In general, the transitions will progress

Reprinted with permission from *The Report of the National Science Foundation/National Aeronautics and Space Administration Solar Energy Panel*, P. Donovan and W. Woodward, Ed., Dec. 1972, and the Dept. of Mech. Eng., Univ. of Maryland, College Park, Md.

Table 1. Present Status of Solar Utilization Techniques

Application	Status					
	Research	Development	Systems test	Pilot plant demonstration	Prototype plant	Commercial readiness
Thermal energy for buildings						
Water heating	X	X	X	X	X	X
Building heating	X	X	X			
Building cooling	X	X				
Combined system	X	X				
Renewable clean fuel sources						
Combustion of organic matter	X	X	X	X		
Bioconversion of organic materials to methane	X	X	X	X		
Pyrolysis of organic materials to gas, liquid, and solid fuels	X	X	X	X	X	
Chemical reduction of organic materials to oil	X	X	X			
Electric power generation						
Thermal conversion	X					
Photovoltaic						
Residential/commercial	X					
Ground central station	X					
Space central station	X					
Wind energy conversion	X	X	X			
Ocean thermal difference	X	X	X			

X indicates effort is underway but not necessarily complete.

from feasibility studies, to component development, to prototype or pilot plant, to demonstration models or plants. It is assumed that some of the funds designated for demonstration plants will be cost shared by the industry and thus reduce the government investment.

COMMENTS FROM THE SOLAR ENERGY PANEL'S CONSULTANTS

To include the concerns from the various other disciplines which would interface with the development of any new energy source, the Solar Energy Panel had representatives from the economic, environmental, psychological, sociological and industrial interests. A summary of their comments follows:

Economic

On close examination, the possibilities for the economic use of solar power, given reasonable R&D support, appear much better than generally realized. In regard to the level of R&D, if the nation is to obtain the maximum benefits

Table 2. Summary of Major Technical Problems

Application	Major technical problems to be solved
Thermal energy for buildings	Development of solar air-conditioning and integration of heating and cooling.
Renewable clean fuel sources	
Combustion of organic materials	Development of efficient growth, harvesting, chipping, drying and transportation systems.
Bioconversion of organic materials to methane	Development of efficient conversion processes and economical sources of organic materials.
Pyrolysis of organic materials to gas, liquid and solid fuels	Optimization of fuel production for different feed materials.
Chemical reduction of organic materials to oil	Optimization of organic feed system and oil separation process.
Electric power generation	
Thermal conversion	Development of collector, heat transfer and storage subsystems.
Photovoltaic	Development of low-cost long-life solar arrays.
Systems on buildings	High temperature operation and energy storage.
Ground station	Energy storage.
Space station	Development of light-weight, long-life, low-cost solar array; transportation, construction, operation and maintenance; development and deployment of extremely large and light-weight structures.
Wind energy conversion	Integration of large wind conversion system with suitable energy storage and delivery systems.
Ocean thermal difference	Large low pressure turbines, large heat exchangers, and long, deep-water intake pipe.

for its energy R&D expenditures, then R&D expenditures on various sources of energy and processes should be carried to the point of equal marginal productivity of the incremental research dollar for each source and process. On the basis of this, as well as other, criteria, it appears that an objective allocation of R&D funds would call for substantially increased R&D support for a number of solar energy opportunities. There are also international benefits in making a viable solar technology available to the world as well as balance of payments and national security benefits in limiting our almost inevitable dependence on foreign energy sources.

Environmental

Solar energy utilization on a large scale could have a minimal impact on the environment if properly planned. It is important, therefore, that a policy of research and review for environmental effects be made an integral part of the R&D process. Continuous feedback into the development program is critically important to prevent the undue expenditure of funds for processes that could ultimately prove unacceptable from a public point of view. One of the major obstacles to public acceptance of new technologies is the fear that there are unknown side effects that have not been adequately investigated or disclosed.

The most environmentally benign solar energy systems might be those of small scale that would fit into space already occupied by buildings. When considering large land based systems, great

Table 3. Impact of Solar Energy Applications on the Reference Energy System[1]

System	Year	Annual consumption[2] (10^{15} BTU)	Percent of total energy consumption in USA	Estimated percent of market captured	$\$10^6$ Annual savings in fossil fuel @ \$1.00/10^6 BTU	Significance[6] of impact on reference energy system by 2020
Thermal energy for buildings	1985	[3] 17	15	<1		Major on building industry Minor on total energy consumption
	2000	[3] 21	12	10	2,100	
	2020	[3] 30	10	35	10,500	
Conversion of organic materials to fuels or energy						
Combustion of organic matter	1985	37	32			Major on electric utility Modest on total energy consumption
	2000	76	43	1	760	
	2020	160	53	10	16,000	
Bioconversion to methane	1985	[4] 27	23	1	270	Major on gas consumption Minor on total energy consumption
	2000	[4] 31	18	10	3,100	
	2020	[4] 41	14	30	12,300	
Pyrolysis to liquid fuels	1985	[5] 50	44			Major on oil consumption Minor on total energy consumption
	2000	[5] 63	36	1	630	
	2020	[5] 80	27	10	8,000	
Chemical reduction to liquid fuels	1985	[5] 50	43			Major on oil consumption Minor on total energy consumption
	2000	[5] 63	36	1	630	
	2020	[5] 80	27	10	8,000	
Electric power generation						
Thermal conversion	1985	37	32			Modest on electric utility industry Modest on total energy consumption
	2000	76	43	1	760	
	2020	160	52	5	8,000	
Photovoltaic						
Systems on buildings	1985	[3] 9	9			Major on building industry Minor on total energy consumption
	2000	[3] 15	9	5	750	
	2020	[3] 21	6	50	10,500	
Ground stations	1985	37	32			Major on electric utility industry Modest on total energy consumption
	2000	76	43	1		
	2020	160	52	10	16,000	
Space stations	1985	37	32			Major on electric utility industry Modest on total energy consumption
	2000	76	43	1	760	
	2020	160	52	10	16,000	
Wind energy conversion	1985	37	32			Major on electric utility industry Modest on total energy consumption
	2000	76	43	1	760	
	2020	160	52	10	16,000	
Ocean thermal difference	1985	37	32			Major on electric utility industry Modest on total energy consumption
	2000	76	43	1	760	
	2020	160	52	10	16,000	

Notes: (1) Each of the above impact estimates assumes the successful development of practical economically competitive systems. However in each case a judgement has been made resulting in estimates that are less than the maximum possible. The estimates are not necessarily additive since not all systems will be carried to commercial readiness.

(2) Nonrenewable fuel consumed to generate the electric power as projected in the energy reference systems and resource data report, AET-8, Associated Universities, Inc., April 1972 [1].

(3) Nonrenewable fuel consumed to generate the projected electric power requirements for buildings, AET-8 [1].

(4) Methane consumed to meet projected energy needs, AET-8 [1].

(5) Oil consumed to meet projected energy needs, AET-8 [1].

(6) Minor, 0-5%; Modest, 5-10%, Major, >10%.

Table 4. Summary of Overall Program Funding*

Applications	Long Range R&D Program (15 years) $ M
Thermal energy for buildings	100
Renewable clean fuel sources	
Photosynthetic production of organic materials and hydrogen	60
Conversion of organic materials to fuels or energy	310
Electric power generation	
Solar thermal conversion	1130
Photovoltaic conversion	780
Wind energy conversion	610
Ocean thermal gradients	530

*See Appendix A for additional funding information.

care must be taken to find suitable areas that would not be of unique ecological or recreational importance or cause serious alterations in local climate or weather.

One can indeed imagine designing "optimum size" environmentally oriented communities which would meet most of its energy needs from direct solar energy and the solar derived fuels from the local waste treatment plant.

Industrial

Before solar energy becomes a major source of clean energy for our nation, it will require the involvement of industrial ingenuity and productive know-how to produce economic hardware and services. Some of the difficulties in achieving industry participation are: (1) most companies are looking for short term projects for their new enterprises with return on their investments in 2 to 3 years; (2) long range projects present great risk, and investment capital is very scarce unless there is a high probability of return in a major line in their business; and (3) companies will undergo major change only under crisis (they will take chances but cannot take failure).

To overcome these problems several things are needed. (1) The item needed must be defined very carefully so that the direction is clear and feasibility shown. (2) The Government must make a long range commitment to assure that the interest will not wane in a couple of years. (3) The incentives must be substantial so that future profit is sufficiently assured over a period of time that makes industrial investment pay off. Further, industries must feel that the system is socially acceptable so that the public will look upon it favorably. Solar energy utilization fits into this pattern extremely well.

Sociological

Research on the social conditions which foster solar energy technology protects against the truncating of a technological policy by the social responses it engenders. Analysis of social problems accompanying solar energy technology development requires a shift of focus from the physical world to the world of social activity. There is a need for more social scientific work to define the social (including economic, political, and cultural) problems presented by solar energy utilization. The establishment of National priorities for the use of solar among other energy forms should recognize the social impacts of the utilization of each energy form.

CONCLUSIONS AND RECOMMENDATIONS

CONCLUSIONS

• Solar energy is received in sufficient quantity to make a major contribution to the future U.S. heat and power requirements.

• There are numerous conversion methods by which solar energy can be utilized for heat and power, e.g., thermal, photosynthesis, bioconversion, photovoltaics, winds, and ocean temperature differences.

• There are no technical barriers to wide application of solar energy to meet U.S. needs.

• The technology of terrestrial solar energy conversion has been developed to its present limited extent through very modest government support and some private funding.

• For most applications, the cost of converting solar energy to useful forms of energy is now higher than conventional sources, but due to increasing prices of conventional fuels and increasing constraints on their use, it will become competitive in the near future.

• A substantial development program can achieve the necessary technical and economic objectives by the year 2020. Then solar energy could economically provide up to (1) 35% of the total building heating and cooling load; (2) 30% of the Nation's gaseous fuel; (3) 10% of the liquid fuel; and (4) 20% of the electric energy requirements.

• If solar development programs are successful, building heating could reach public use within 5 years, building cooling in 6 to 10 years, synthetic fuels from organic materials in 5 to 8 years, and electricity production in 10 to 15 years.

• The large scale use of solar energy as a national resource would have a minimal effect on the environment.

RECOMMENDATIONS

It is recommended that:

• The Federal government take a lead role in developing a research and development program for the practical application of solar energy to the heat and power needs of the U.S.

• The solar energy R&D program provide for simultaneous effort on three main objectives: (1) economical systems for heating and cooling of buildings, (2) economical systems for producing and converting organic materials to liquid, solid, and gaseous fuels or to energy directly, (3) economical systems for generating electricity.

• Research and development proceed on various methods for accomplishing the above objectives and that programs with phased decision points be established for concept appraisal and choice of options at the appropriate times.

• For those developments which show good technical and economic promise, the Federal government and industry continue development, pilot plant, and demonstration programs.

• Environmental, social, and political consequences of solar energy utilization be continually appraised and the results employed in development program planning.

Walter E. Morrow, Jr.
Associate Director
Lincoln Laboratory, M.I.T.

Assuming successful development and exploitation, a $25 billion solar energy industry might provide 25 per cent of all U.S. energy needs by early in the 21st century

Solar Energy: Its Time Is Near

Outside of the earth's atmosphere the sun provides energy at the rate of about 1,400 watts/m.^{2} (4,730 B.t.u./m.^{2}/hr.) normal to the sun. By the time sunlight reaches the earth's surface, atmospheric attenuation, clouds, and earth shadowing have taken their toll; the average of solar energy falling on a horizontal surface in southern New England, taken over a long time period, is about 160 watts/m.^{2}. If the solar energy density is to be measured on a platform which is movable so that it may be maintained constantly normal to the sun, this annual average may be improved by a factor of nearly two—to almost 300 watts/m.^{2} in the case of Southern New England.

Such energy densities are quite low, and this is essentially why solar energy has to date played a negligibly small role in the U.S. However, assuming 30 per cent efficiency, the total present *electric* power demand of the U.S. could be supplied with solar energy plants having a total area of about 2,000 km.^{2}. This is about 0.03 per cent of the U.S. land area devoted to farming and about 2 per cent of the land area devoted to roads; and it is about equal to the roof area of all the buildings in the U.S.

But such statements do not comprehend the very considerable hurdles which stand between this apparently bountiful energy supply and its collection, storage, and use.

There has been extensive development of fixed-orientation, low-temperature collectors for building heating and hot water heating, and some systems are now in use. These collectors are usually faced south at inclinations of 45° to 60° above horizontal, with one or two layers of glass or plastic used to reduce convection and radiation losses. Transmission losses through these windows are commonly of the order of 15 to 20 per cent.

Of the solar energy passing through the windows, 80 to 95 per cent can be absorbed with simple black coatings. Thus, between 70 and 80 per cent of the incident radiation can be collected.

Losses inside the collector are a function of working temperature and occur by reradiation, convection, and conduction. Convection and conduction losses can be made negligible through good design. The reradiation losses are determined by the infrared emissivity of the absorbing surface and the infrared transmission of the windows. Absorbing surfaces with effective infrared emissivities as low as 0.15 are not difficult to achieve. (An emissivity of 0.15 means that the surface will radiate 15 per cent of the infrared energy that would be radiated by a perfect black body of equal area.) The usable heat is usually gathered from the absorbing surface either by water flowing through tubes attached to the surface or by air flowing over the surface itself.

Assuming 75 per cent absorption and an emissivity of 0.15, net collection efficiencies of between 50 and 75 per cent can be proposed, depending on the outlet temperature—the higher the temperature required, the lower the efficiency.

Heat storage for house heating has been accomplished by means of water tanks, bins of rocks, or in hydrated chemicals such as sodium sulfate, whose solid-liquid phase change adds to the heat energy which can be stored.

Assuming a 40° C. temperature change—typical of that required in building and hot water heating applications—water will store 1.6×10^8 joules/m.^{3}; sodium sulfate 3.5×10^8 joules/m^3. Analysis of these figures demonstrates that to achieve the large heat storage required for domestic space heating applications—to provide heat through the night and in bad weather—requires substantial volumes of storage. One to a few days' heat storage is typically used in a solar-heated house, with auxiliary heating sources added to provide heat during extended cloudy periods.

While the temperatures achieved in such low-temperature collector systems as those described above are adequate for heating buildings and water, they are insufficient for high-efficiency production of electricity or artificial fuels by thermal processes; for these applications high-temperature collectors are required. The achievement of high temperatures depends chiefly on concentrating the solar energy from a relatively large area into a small collector, from which it is carried to storage. Conduction and conversion losses can be made small in high-temperature solar collectors by good design, including, if necessary, the use of vacuum insulation around the heat absorber. Indeed, such solar energy concentrators can be very efficient; used in research, they now yield the highest temperatures available in small furnaces for many applications.

The differences between unfocussed and focussed collectors are striking: unfocussed collectors typically have ratios of concentration to emissivity of 10.0 and are limited to output temperatures of 150° C or less; ratios of 300 and temperatures of 600° C. are typical of one-axis-steerable concentrators, and ratios of 10,000 and temperatures of 4,000°C are typical of two-axis-

Reprinted with permission from *Technol. Rev.*, vol. 76, pp. 31–43, Dec. 1973. Edited at M.I.T.

The total present demand for electric power in the U.S. could be supplied—assuming 30 per cent efficiency of collection—by solar energy falling on 0.03 per cent of the nation's land area, or on the roofs of every building in the country

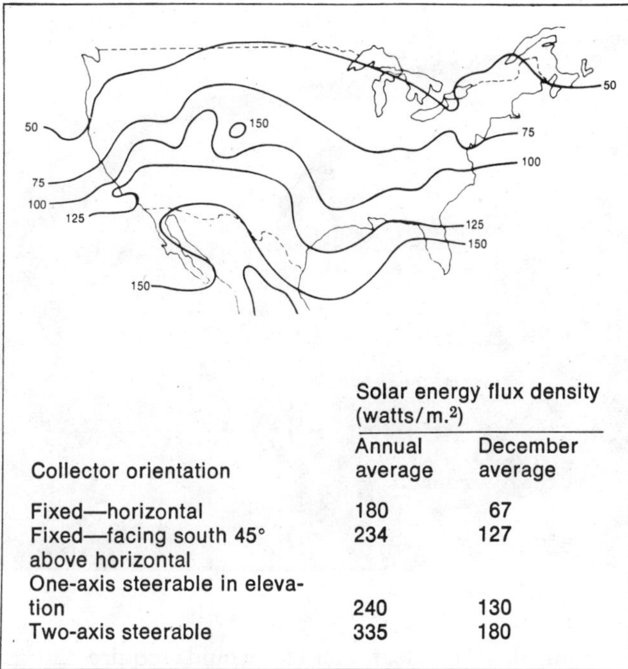

Collector orientation	Solar energy flux density (watts/m.²) Annual average	December average
Fixed—horizontal	180	67
Fixed—facing south 45° above horizontal	234	127
One-axis steerable in elevation	240	130
Two-axis steerable	335	180

In the northern hemisphere the test of solar energy systems comes in the winter, when energy inputs from the sun reach their minimum. The map shows the average solar energy incidence in the U.S. in December in watts/m.^{2}, the table shows solar energy densities, also in watts/m.^{2}, on collectors with different orientations in an average location in the U.S. in December. To obtain B.t.u./hr. from the figures given, multiply by 3.41.

steerable concentrators. These figures suggest that reasonable efficiencies and temperatures can be obtained with a one-axis concentrator, such as shown in the accompanying illustration.

Solar energy can also be utilized through its conversion into combustible fuels by photosynthesis in trees, plants, and algae. Conversion efficiencies have been estimated to be in the range of 0.3 to 3 per cent depending on the vegetation used. It is possible to imagine an energy system built on this conversion: plants grown with sunshine used to fuel furnaces or boilers, for example. Because of the low collection efficiencies, rather large land areas are required to supply significant amounts of energy.

To supply the total current U.S. energy needs at 3 per cent efficiency would require a land area of about 350,000 km.^{2}, about 3 per cent of the total U.S. land area. Soil depletion and the handling of waste products from the combustion of such fuels are likely to be significant problems for large-scale energy systems which may be conceived to utilize solar energy in this way. However, an interesting variation of this plan is that of utilizing waste from forestry operations and municipal trash collections as a fuel to produce power and/or a synthetic fuel such as methanol. For instance, it is estimated that waste from current forestry operations could provide 10 to 20 per cent of U.S. energy needs projected for 1975.

Photovoltaic conversion of solar energy to electricity using silicon solar cells—direct sunlight striking the cells generates current in an n-p junction in the silicon material—has been widely employed on spacecraft. Typical efficiencies are of the order of 10 per cent, and costs for space-qualified systems can be as much as $1 million per kilowatt. Relatively modest efforts have gone into improving efficiency, and these have yielded silicon and gallium arsenide devices with efficiencies of over 16 per cent. Efficiencies of 20 per cent are believed to be achievable, compared with a maximum theoretical efficiency for simple photovoltaic converters of about 35 per cent.

High cost is as much a problem as low efficiency: arrays of currently available silicon cells (with 10 per cent efficiency) cost about $100,000 per kilowatt of peak capacity when engineered for ground installations without solar concentration. Several efforts are under way to reduce the costs of energy from solar cells of this type. Polycrystalline cells would have lower efficiencies than the single-crystal cells now in use in spacecraft, but they should be much less expensive. Use of solar concentrators with special cells designed for high solar intensities would also increase energy output; a system using a one-axis concentrator would require about 1/50 of the area of a non-concentrating system to supply a given amount of energy. If these cells with a concentrator could be provided at the same costs as today's silicon cells, the cost per peak kilowatt might be reduced by 50-fold—to the order of $2,000.

Given these various alternatives for collecting solar energy at various temperatures and efficiencies, what systems can we envision within the realm of engineering feasibility which will utilize solar energy at capital and operating costs which are reasonable relative to those of other energy systems? Three types of systems for three different applications have been proposed:

The University of Delaware's new "Solar ONE" is the first solar house designed to convert sunlight into both electricity and heat. Solar cells in the collector will convert some 5 per cent of the incident solar energy into electricity and 45 per cent into heat (total, 50 per cent efficiency), and the system is expected to provide up to 80 per cent of the total energy demand of the house.

domestic space and water heating systems, total energy systems for commercial and industrial buildings, and large-scale solar electric power generation.

Solar Houses: The First Chance for Reality

A number of experimental solar house-heating systems have been built and there has been substantial production of solar hot water heaters.

A typical house of 1,500 ft.^{2} area requires of the order of 0.7×10^9 joules/day for space heating during December in mid-U.S. locations (40-degree-day). Hot water heating typically might require another 0.1×10^9 joules/day. In central U.S. locations a fixed 45° collector can be expected to receive an average of about 127 watts/m.^{2} in December, of which perhaps 60 per cent can be retained; thus 0.0066×10^9 joules/m.^{2} could be collected in an average 24-hour period. To provide the needed 0.80×10^9 joules/day for space and water heating would require the heat from approximately 120 m.^{2} (1300 ft.^{2}) of collector area. This could be provided by using somewhat more than half the roof area of a single-floor house or slightly more than the roof area of a two-story house.

Four days of heat storage would require 20 m.^{3} (5,000 gal.) of water (or 64 m.^{3} of rock) heated to 65°C. Smaller storage systems could be used if auxiliary heating were provided for cloudy days.

Current costs for fixed low-temperature collectors range between \$20 and \$40/m.^{2}. A 5,000-gal. heat storage system would cost about \$1,000. Thus, the total system costs would be between \$3,500 and \$6,000, not including the usual heat distribution system. If systems of this type were financed as part of house mortgages over a 20-year period at 7 per cent interest, the increase in annual carrying charges would be between \$300 and \$550.

How does this compare with current costs for heating by fossil fuel? Such a house requires fuel or heating electricity equivalent to about 10^{11} joules/year (9.5×10^7 B.t.u.) assuming a total of 5,000-degree-days required. Typical costs in 1972 for 10^{11} joules of heat energy were:

Oil at 75 per cent efficiency (@ 22¢/gal.)	\$280
Gas at 75 per cent efficiency (@ 23¢/100 ft.^{3})	\$275
Electricity at 100 per cent efficiency (@ 1.8¢/K.w.h.)	\$500

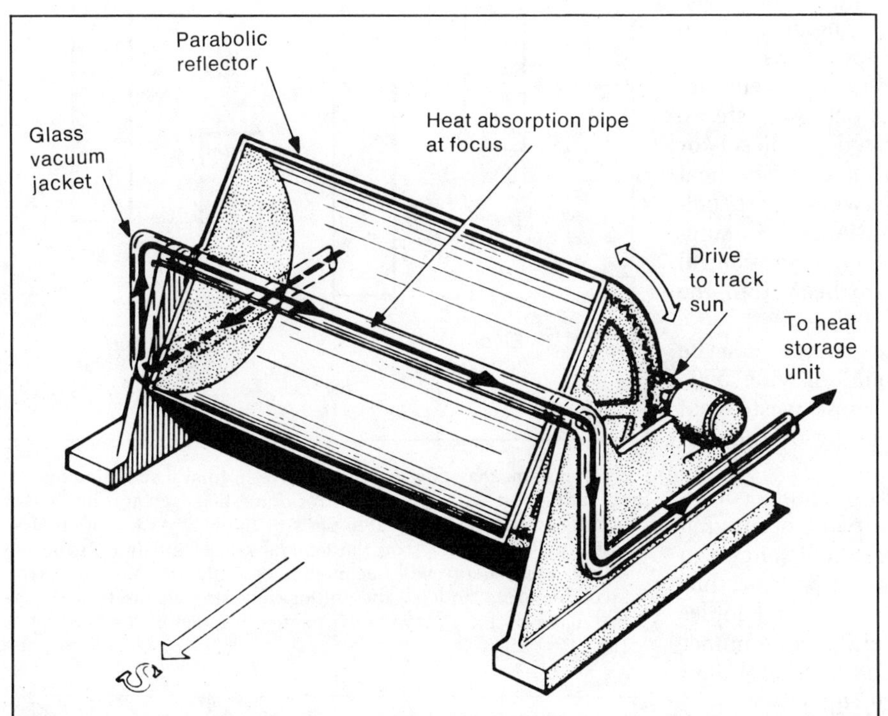

When a simple flat-plate fixed collector provides insufficient efficiency and output temperature, a one-axis concentrator is usually proposed for higher efficiency and temperatures at modest cost. This design incorporates a clockwork drive to track the sun and a vacuum-jacketed pipe at the focus of the reflector through which dry nitrogen is circulated to collect heat at 550° C.

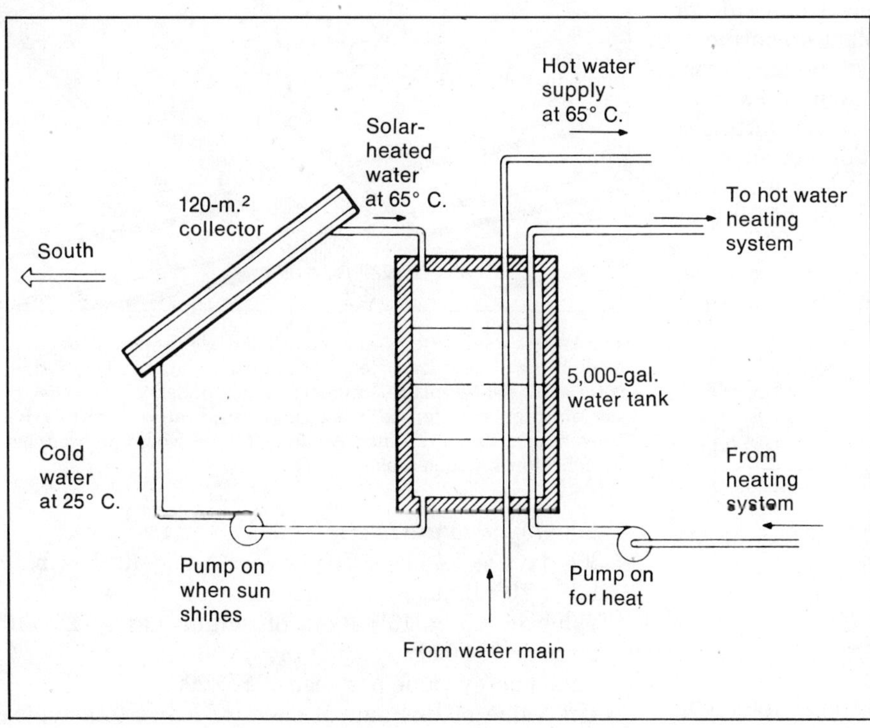

The simplest solar energy systems are those designed for domestic space and hot water heating. In such systems, an inclined, southerly-facing flat collector is typically coupled with a heat storage system from which heat can be drawn when required. This diagram shows a combination system which provides domestic hot water at 65° C. and hot water, as well, for space heating.

This computation suggests that solar heating systems are competitive with electric heating today in mid-U.S. locations. Should fuel prices rise significantly faster than the costs of constructing solar heating systems in the future, solar heating could become competitive with gas and oil at some future time.

Total-Energy Systems for the Industrial Park

Most shopping centers and industrial plants now being built consist of one- or two-story buildings in suburban locations. Such facilities require large amounts of energy for heating, cooling, lighting and operations.

A number of such buildings have recently been constructed with gas- or oil-fueled total-energy systems. In such a system electricity is generated by diesel- or gas-turbine-powered generator units. The waste heat from the engines is used for heating in winter and cooling (by means of absorption air conditioners) in summer. Such systems have the advantage over central power systems of recovering the waste heat from the electric generating process.

A similar arrangement using solar energy can be proposed. Parabolic concentrators could provide 550° steam for a turbine-alternator plant whose waste heat was used for heating or cooling, depending on the season.

Approximately 5,000 m.^{2} of one-axis-steerable collectors can be mounted on the roof of a plant occupying 10,000 m.^{2}. About 70 per cent absorption efficiency can be achieved with an outlet temperature of 550° C; thus the heat collected per day will be about 7×10^{10} joules averaged over the year. In order to average out summer to winter solar energy variations, the order of 100 days of heat storage are required. A storage capacity of 7×10^{12} joules can be obtained with 10,000 m.^{3} of insulated rock heated to 550° C. and located along one side of the building. With a thermal electric plant operating at 40 per cent efficiency (which is typical of modern steam generating plants) for perhaps 12 hrs. out of each 24 that the building is in use, an electric power output of about 670 kw. can be delivered. In addition, an average of 500 kw. of low-temperature waste heat would be available over the full 24-hour period for heating or cooling.

No such system has ever been built on such a scale, so the costs are very difficult to predict. Here is a very rough estimate:

Collector: 5,000 m.^{2} @ \$100	\$500,000
Storage: 6,000 m.^{3} @ \$20	\$120,000
1000-kw. steam turbine-alternator	100,000
	\$720,000

The retail value of the output energy can be roughly estimated as:

Electric power:

250 working days @ 12 hr. $\times$ 670 kw. $\cong$ 2.00×10^6 k.w.h.

115 standby days @ 24 hr. $\times$ 100 kw. $\cong$ 0.27×10^6 k.w.h.

2.27×10^6 k.w.h.

2.27×10^6 @ \$0.018/K.w.h. $\cong$ \$41,000

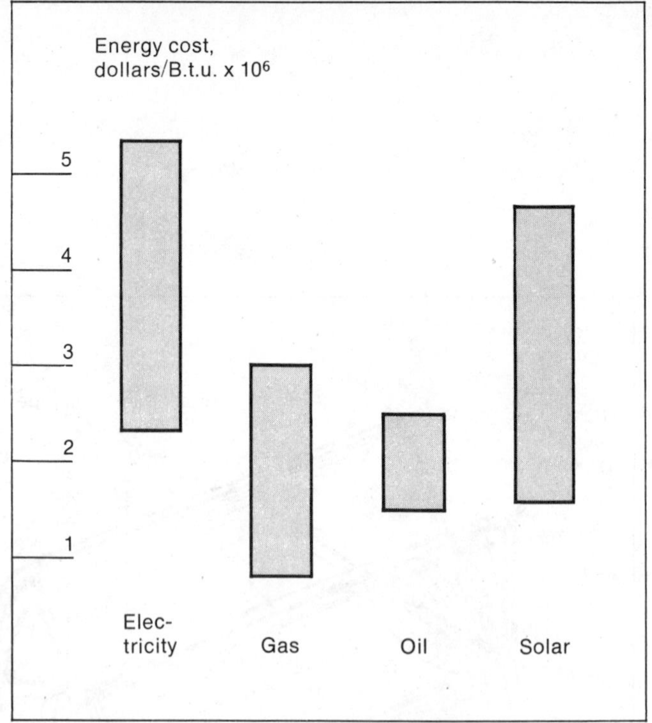

Solar systems may now be competitive in total installation and operating cost with gas and oil for domestic space and hot water heating in the most favorable parts of the U.S. As the price of fossil fuel rises faster than the general rate of inflation in the future, solar energy will become increasingly attractive in less favorable areas. Indeed, the author suggests that the balance may have shifted to favor solar energy in much of the U.S. by 1983.

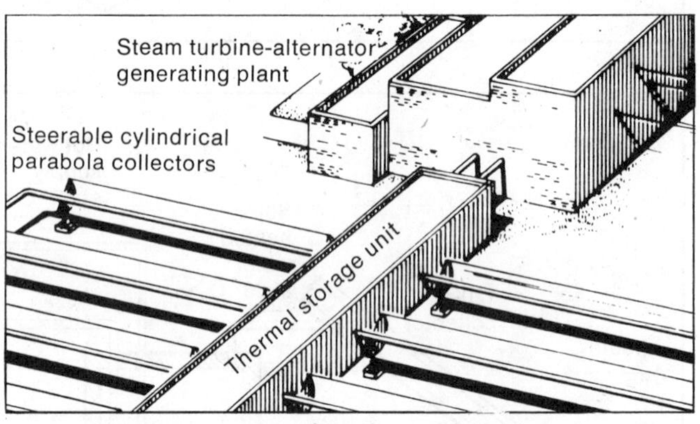

The largest solar systems now envisioned are proposed for medium sized base-load electric power installations. This system is for a 1,000-Mw. plant—utilizing large numbers of one-axis-steerable concentrators to collect high-temperature heat to drive conventional steam turbines. An area 16 km.^{2} would be occupied by collectors for such a plant.

Low-temperature heat:

365 days @ 24 hr. $\times$ 500 kw. $= 4.4 \times 10^6$ K.w.h. $= 1.5 \times 10^{10}$ B.t.u.

Value of 1.5×10^{10} B.t.u. of natural gas @ 23¢/10^5 B.t.u. $\cong$ \$34,000

Total energy value per year $\cong$ \$75,000

The value of this annual energy savings to a typical suburban building can be estimated using typical procedures for calculating industrial plant capitalization. At an 8 per cent discount rate with inflation of 4 per cent/yr., straight-line depreciation, and a 48-per-cent tax rate, this value of the energy savings can be calculated to equal about five times the annual savings or

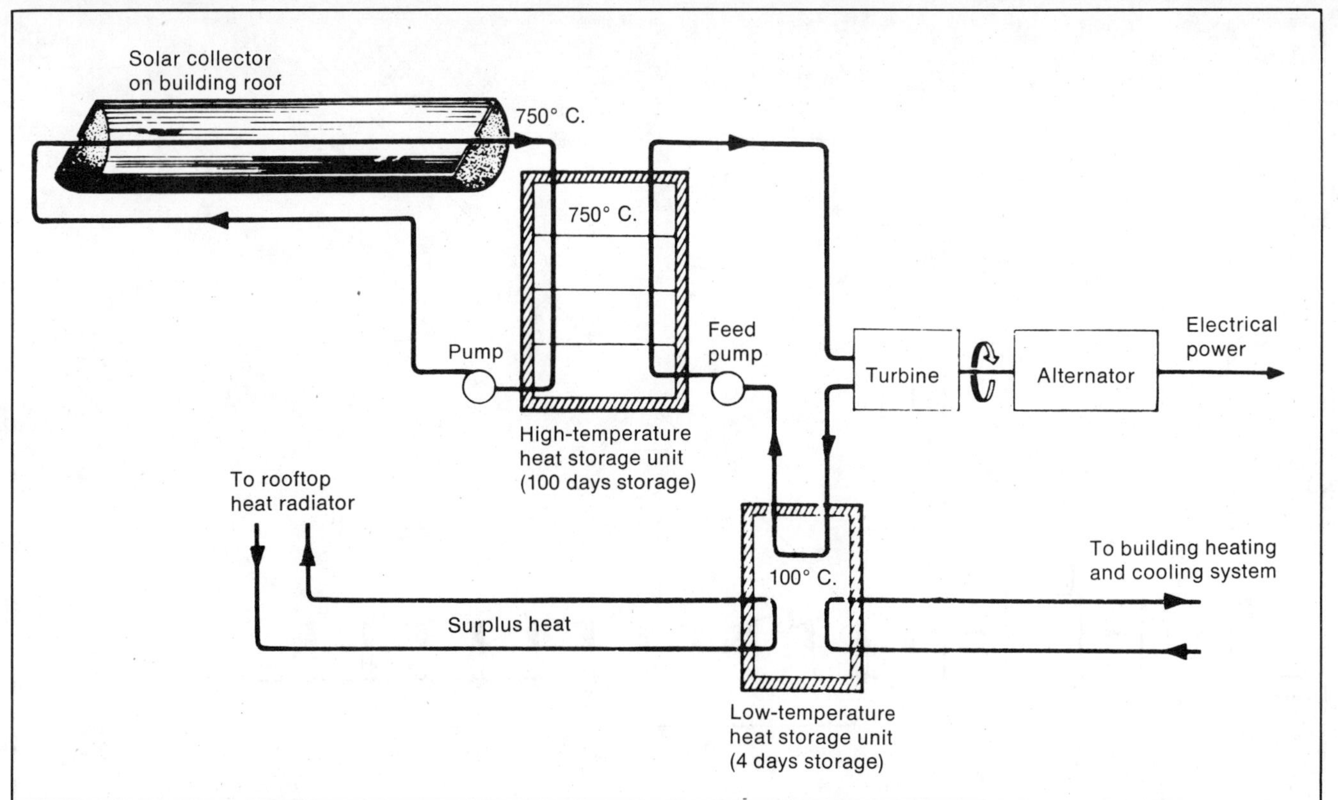

This solar-powered total-energy system is projected for use in suburban shopping centers or industrial plants where space for solar collection is available. One-axis-steerable concentrators would be located on the roof; high-pressure dry nitrogen, steam, or sodium chloride would transfer heat at 750° C. to a high-temperature storage unit of insulated rocks or molten salt. Heat drawn from the storage unit would drive a turbine to produce electrical power and then, held in a low-temperature storage unit, would be available for space heating or air conditioning.

about $375,000 compared with the $720,000 estimated costs.

These calculations suggest that, given the assumptions, such total-energy solar plants are close to being an economic investment for a typical suburban industrial plant in the middle latitudes of the U.S., although the cost uncertainty is considerably greater than for the building heating systems described in the previous section.

Base-Load Electricity from the Sun

A number of proposals have been made for the design of large-scale solar-powered electric generation systems. (Such plants could also generate hydrogen fuel either from electricity through electrolysis or directly from thermal energy by one of several proposed processes.) These suggestions generally fall into one of the following three classes: ground-based thermal conversion systems using conventional collectors, ground-based systems based on photovoltaic arrays, or systems based on photovoltaic arrays mounted on satellites above the earth.

Proposals for large-scale ground-based thermal conversion systems suggest plants similar in design to the total-energy plant described above. One proposal by Drs. A. B. and M. P. Meinel of the University of Arizona includes one-axis-steerable cylindrical parabolas as solar energy concentrators, with vacuum-insulated heat collection pipes at the focal points of the collectors. The energy would be stored as thermal energy in molten salt or in rock, and conventional steam turbines and alternators would be used to produce electricity. Parameters of a 1000-Mw. continuous-output plant would be:

Area of plant		30 km.^{2}
Area of collectors		16 km.^{2}
Outlet temperature of collectors		550° C.
Collection efficiency		~60 per cent
Thermal storage	rock:	2×10^7 m.^{3}
Thermal plant efficiency		~40 per cent
Overall efficiency		~25 per cent

The collector efficiency is reduced from the previous case because of the long distance involved in transferring heat from the collector to the steam plant. Assuming that collector costs can be reduced to $60 per square meter and using the same cost assumptions for the other components as for the total-energy plant, a total system cost of about $1.4 billion is calculated, corresponding to $1,400 per kw. of capacity. Separate studies of this type of solar plant by Aerospace Corp. have suggested capital costs of $1,000 to $2,000 per kw. of capacity for similar designs.

In steady operation over a one-year period, such a 1,000-Mw. plant would produce about 8.8×10^9 k.w.h. At a wholesale value of $0.08/k.w.h., the year's output would have a value of about $70 million. Using the same capital valuation assumptions as in the case of the total-energy plant, a plant investment of about $350 million could be justified. This is about one-fourth of estimated cost. Obviously, substantial reductions in the solar plant cost and/or increases in the value of electricity would be required before such solar plants would be justified.

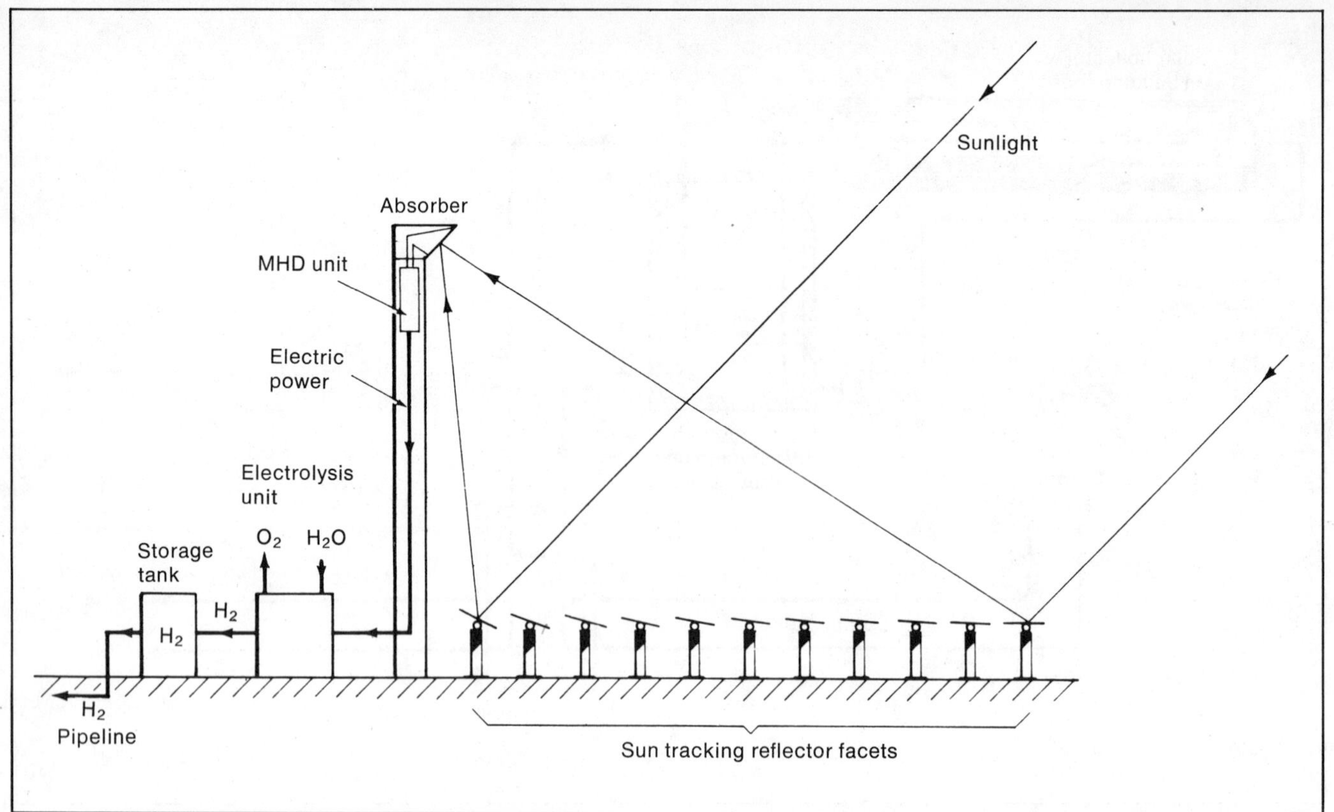

If solar heat can be used to drive a magnetohydrodynamic power generator, the system might look like this. A two-axis concentrator consisting of a large number of movable reflectors focusses energy at extremely high temperature to operate an MHD power unit. Energy from the power unit is used to produce hydrogen by electrolysis, the hydrogen then to be supplied directly to consumers or placed in storage for use when solar energy is not available.

An alternative design for a ground-based thermal plant using a two-axis concentrator which consists of a large number of individually movable facets has been proposed by A. F. Hildebrant, G. M. Haas, W. R. Jenkins, and J. P. Colaco of the University of Houston. Each facet would independently track the sun so as to reflect sunlight on a central collector mounted on a tower at one edge of the array. A magnetohydrodynamic thermal-electric system mounted at the collector would be used to produce hydrogen by electrolysis. Tanks of hydrogen at the plant would provide the energy storage.

Typical parameters for a plant with 1,000 Mw. continuous output might be as follows:

Area of plant	17 km.^{2}
Area of collector	17 km.^{2}
Collector efficiency	~60 per cent
Conversion efficiency to electricity	~60 per cent
Conversion efficiency to hydrogen	~90 per cent
Overall efficiency to hydrogen	~32 per cent

Detailed cost estimates for this class of system are not available, but costs would probably be of the same order as for the previously described ground-based system.

Solar Cells on the Ground and in Space

Large-scale ground-based plants based on photovoltaic collectors have been proposed, using arrays of silicon photovoltaic cells to energize electrolytic cells producing hydrogen in large quantities. The hydrogen would be piped to consumers either for direct use as a fuel or as an input to a fuel cell to produce electricity. A 1000-Mw. plant might have the following parameters:

Land area	~100 km^2
Collector area	~50 km^2
Photovoltaic cell efficiency	~12 per cent
Electrolyzer efficiency	~95 per cent
Net efficiency to produce hydrogen	~11 per cent
Fuel cell efficiency	~80 per cent
Net efficiency to produce electricity at customer location	~9 per cent

The low net efficiency of such units leads to substantially larger land areas for solar energy collection than those required for thermal-cycle solar systems. A much more difficult problem, however, is the high cost of the photovoltaic cells; at present prices, costs would exceed $100,000 per kw. of capacity. Cost reduction by a factor of at least 300 is required to achieve reasonable costs, and many years of intensive development will clearly be required to achieve this goal—if in fact it is attainable. An alternate approach to cost reduction would be the use of a one-axis concentrator, yielding reduction by a factor of 50 in the required solar cell area; but cells for such applications would have to be specially designed to accept energies of up to 50 times the sun's intensity.

Solar power plants mounted in satellites in synchronous earth orbit have been proposed by Peter E.

	Application dates based on comparative cost estimates with present technology only		Application dates with intensive solar energy research and development program	
	First use in favorable areas	Extensive use	First use in favorable areas	Extensive use
Residential space and water heating	1983	1993	1978	1988
Total energy systems in commercial plants	1990	2000	1980	1990
Large-scale power generation	2006	2016	1985	1995

How soon will solar energy become competitive with fossil and nuclear fuels in the U.S.? The chart shows the author's estimates—the left columns based solely on current solar energy technology in competition with increasingly costly fossil and nuclear fuels, the right columns based on improved solar technology which might result from intensive research and development beginning immediately.

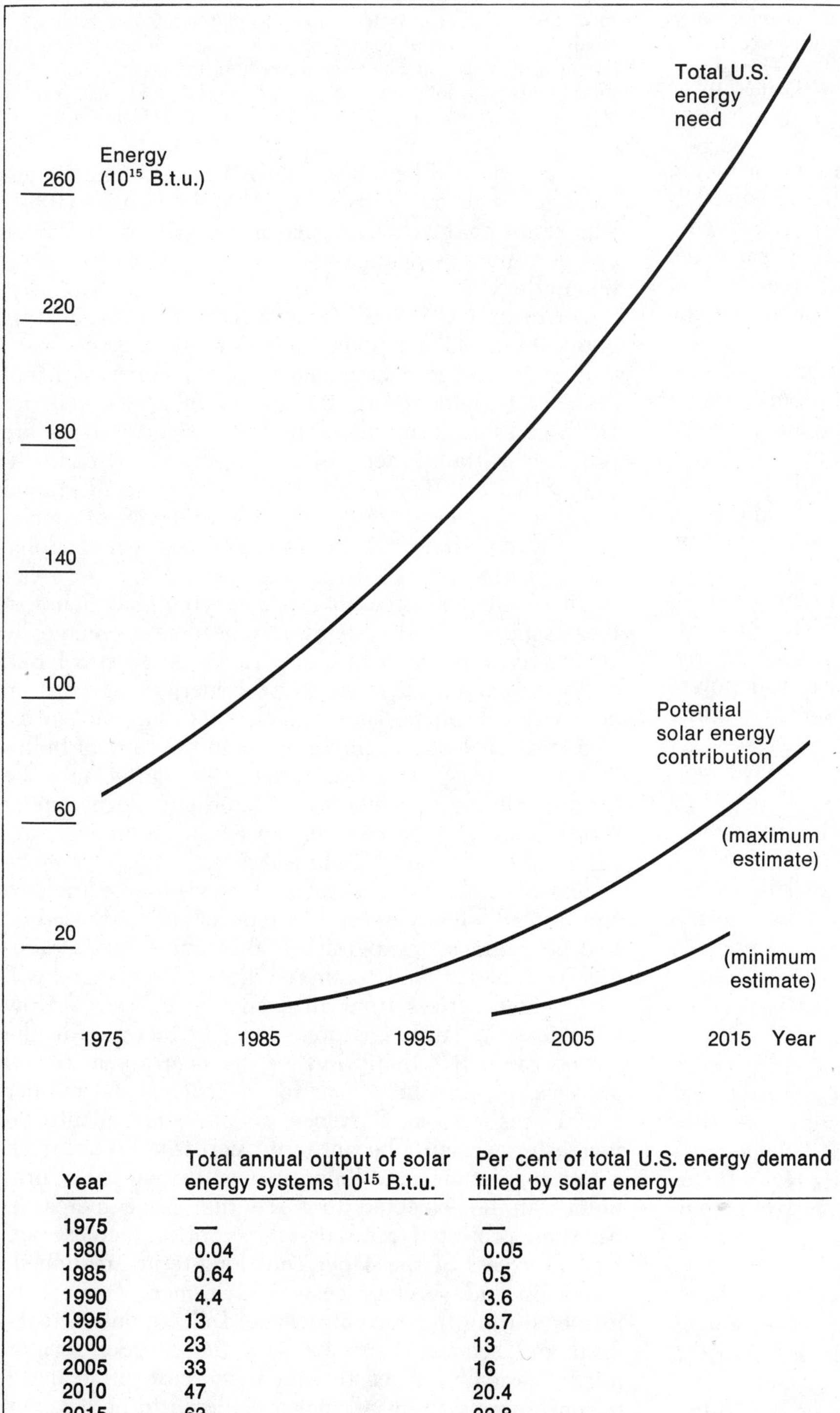

Year	Total annual output of solar energy systems 10^{15} B.t.u.	Per cent of total U.S. energy demand filled by solar energy
1975	—	—
1980	0.04	0.05
1985	0.64	0.5
1990	4.4	3.6
1995	13	8.7
2000	23	13
2005	33	16
2010	47	20.4
2015	63	23.8
2020	78	26

A vigorous successful solar energy research and development program, plus substantial investments in solar energy systems ($25 billion annually by the year 2000) and the facilities for their manufacture (a total of $11 billion by the year 2000) would result in rapid growth of solar energy utilization by the end of the 20th century. Indeed, the author proposes that more than 25 per cent of all the energy which the U.S. requires in 2020 could be drawn directly from the sun.

Year	Solar space and domestic hot water systems		Solar total-energy systems		Solar base-load electric power plants	
	Number of dwellings (millions)	Annual energy savings (10^{15} B.t.u.)	Floor area of buildings (10^6 m.^{2})	Annual energy savings (10^{15} B.t.u.)	Installed capacity (10^6 kw.)	Annual energy savings (10^{15} B.t.u.)
1980	0.3	0.04	—	—	—	—
1985	3	0.4	52	0.24	—	—
1990	8.4	1.2	200	0.92	18.7	1.4
1995	14	1.9	400	1.9	69	5.3
2000	20	2.7	610	2.8	137	10.4
2005	27	3.6	850	3.8	208	15.2
2010	34	4.7	1,090	5	284	22
2015	41	5.9	1,350	6.2	363	30
2020	49	7	1,620	7.6	445	37.6

Assuming the most favorable development, annual energy savings of 52.2 x 10^{15} B.t.u.—over 1,160 million tons of oil—may be realized by the use of solar energy by the year 2020. The estimates for dwellings are based on the use of approximately 10^8 B.t.u. annually and a heating plant efficiency of 70 per cent; those for total-energy systems on consumption of 6.67 million k.w.h./yr. in a plant of 10,000 m.^{2} floor area, with an efficiency of 50 per cent; those for power generation on the construction of solar plants with total capacity of 12.5 x 10^6 k.w.h. each year beginning in 1995, the plants having 40 per cent efficiency.

Glaser of Arthur D. Little, Inc. Such a system would not be affected by clouds, and the orbit proposed is such that the system would receive solar energy continuously except for two six-week periods a year when earth shadowing would occur for about one hour at what would be midnight in the time zone under the satellite.

The proposal is to use photovoltaic cells to produce electricity as direct current, which would then be converted to radio energy at microwave frequency (3,000 MHz.) for transmission to the earth's surface where conversion to a.c. power would occur. All the components of the system would be massive by any scale with which we are familiar: the satellite-mounted transmitting antenna would be 1.4 km.^{2}, and the receiving antenna on earth would cover a square 10 km. on each side. The satellite would carry an array of solar cells about 7 km. square, giving an output power of 10^7 kw. —sufficient, for example, to meet the electrical power needs of New York City. Parameters for such a satellite would be:

Size of satellite microwave transmitting antenna	1.4 × 1.4 km.
Size of ground microwave receiving antenna	10 × 10 km.
Efficiency of solar cells	15 per cent
Efficiency of microwave transmission system	70 per cent
Projected weight in orbit	2.2 × 10^6 kg.

Such a system could be competitive in cost with conventional power plants or even some of the proposed ground-based solar power plants only with very substantial reductions from present satellite launch costs (by a factor of 50 to 1) and solar cell costs (by a factor of 1000 to 1). Very substantial advances in technology are required to achieve such cost reductions.

Economics vs. the Sun: When Solar Energy?

From this range of possibilities for the utilization of solar energy, what applications may we envision for the next few decades?

The first point to be made in answering that question is the lesson of basic economics: no application of solar energy will be made unless it is economically advantageous in comparison with available alternatives. The result of this rule applied in the past is that there are now no significant solar energy systems in operation in the U.S.

For central U.S. residential space and water heating applications, solar systems are currently cost-competitive with electric energy and within a factor of 1.5 of being cost-competitive with gas- and oil-fueled systems. In these same locations total-energy solar systems are probably within a factor of two of being cost-competitive with fossil-fueled systems, and large-scale ground based solar electric plants are within a factor of four of competing with electric plants fueled with conventional energy sources.

There are good reasons for believing that many of these solar-powered systems may become economically attractive in the years to come. This prediction is based on the probability that the costs of energy derived from conventional sources such as fossil fuels and nuclear fission may rise faster in the future than the costs of building solar energy systems. Indeed, the latter may be lowered through a vigorous research and development program which places significant emphasis on economical design and production techniques.

The rationale for the costs of conventional fuels to rise faster than any general rate of inflation is based on two predictions: inexpensive fuel resources will gradually be depleted, and continued increasing demand will force higher prices from all available sources. A Dow Chemical Co. study has projected cost increases in the period from 1974 to 1980 of 8.4 per cent a year for oil and gas, 10 per cent a year for electricity, and 6.7 per cent a year for coal. Current costs of nuclear plants are increasing sharply because of increasingly stringent safety regulations, and limitations on the supply of uranium can be expected to force fuel price increases. All such projected rates of energy price increase are well in excess of the 4 per cent long-term inflation in materials and services costs which most economists foresee during the current decade. Indeed, only a technical and economic success with the breeder reactor might basically affect these energy cost projections; fusion power systems are not considered to be a factor in the current century.

Assuming successful research and development and continued increases in the cost of fossil fuel, a U.S. solar-energy industry may grow rapidly to supply 8.4 million dwellings by 1990 and 69 million kw. of base-load electric power by 1995

Year	House heating	Total-energy systems	Electric plants	Hydrogen plants	Total
	Required annual capital expenditures (billions of 1973 dollars)				
1978	$0.2	—	—	—	$ 0.2
1980	0.8	—	—	—	0.8
1985	3.2	$0.9	—	—	4.1
1990	4.2	1.8	$4.4	$4.4	15
1995	4.4	1.9	8.8	8.8	24
2000	4.5	2.0	9.1	9.1	25

If the author's forecasts of technological and economic developments are fulfilled, solar energy will become increasingly competitive with other energy sources during the last 20 years of the 20th century, and there will be large capital expenditures in solar energy systems. Indeed, by the year 2000 such annual expenditures could total at least $25 billion.

On the basis of an average annual increase in fuel and electricity costs of about 8 per cent and a general price inflation (including solar energy systems) of 4 per cent, one can estimate the dates when various types of solar energy systems might become less expensive than fossil-fueled alternative energy sources. Such calculations suggest that solar energy, now 1.5 times as costly as fossil fuels for residential space and water heating, may be competitive for this purpose in favorable areas of the U.S. such as the southwest—where the climate is moderate and sunshine relatively plentiful—by 1983, and that it may come into extensive use (as a result of a further 50 per cent cost differential) by 1993, when it would be competitive for residential applications in most U.S. climates. Similar calculations suggest that solar-powered total-energy systems for commercial plants might first appear practical in 1990 and come into more extensive use by the year 2000, and that comparable dates for large-scale solar-based power plants are 2006 and 2016.

Two factors may affect these estimates. One is any future development of nuclear technology which may act to reduce the cost of nuclear power. The other is the possible effect of an extensive solar energy research and development program. Cost improvements in solar energy systems by a factor of two seem relatively easy to achieve given even modestly successful technical innovation, and such improvements might be demonstrated in three to five years in installations of modest size. This could lead to the first extensive use of solar heating systems by 1978 and of total energy systems by 1980. Additional intensive development during the next five years could make large-scale solar plants competitive by 1985. All these estimates are summarized in the accompanying tables.

The Growth of the Solar Industry

How rapidly might solar energy systems gain acceptance in the U.S.? The following estimates are made on the basis of two assumptions: that solar energy is applied only to new installations (i.e., that no retrofitting is done); and that rates of investment in new houses, industrial plants, and power plants continue in the future at approximately present levels.

In the case of residential space and water heating systems, a total building rate of 2 million units per year is assumed, of which one-half might be suitable for solar heating. Installation would begin in 1978; the rate would rise linearly to 1 million per year in 1988, and thereafter installations would increase in proportion to population growth. By 1990, with 8.4 million dwellings equipped with solar systems, annual savings could be 1.2×10^{15} B.t.u., equivalent to 30 million tons of oil.

An estimate of the maximum rate of application of solar total-energy systems to commercial and industrial buildings can be made by noting that, on the average, approximately 2×10^7 m.^{2} of industrial floor space are constructed every year. Assuming that half of this construction is single-story and suitable for solar total-energy systems and that such systems are available by 1980, installations associated with 3.5×10^7 m.^{2} of construction per year might be achieved by 1990. By then, annual energy savings of 92×10^{13} B.t.u., or 23 million tons of oil, could be achieved.

In 1970-71, electric generating capacity was being constructed at the rate of about 25×10^6 kw. per year. If it is assumed that large-scale electric power production by solar energy can begin in 1985 and that by 1995 one-half of all new electric power installations are solar-powered, 69 million kw. of base-load electric power generation will then be from solar energy, with an annual saving of 131 million tons of oil or its equivalent.

One may postulate that some of the solar electric power proposed above would be redundant with solar home heating or total-energy commercial units. But one may also propose that many other energy requirements —buildings inappropriate for solar power because of siting or multi-story construction, industrial processes, and transportation—could be supplied with energy in the form of hydrogen or hydrogen-derived fuels produced by large-scale, high-efficiency solar-powered plants.

Though there are no proven methods of directly producing hydrogen at high efficiency (70 per cent or better) from solar energy, a number of multi-stage thermally driven chemical processes seem to have promise. If one assumes that the costs per unit area of collector will be the same for hydrogen plants as for base-load solar-powered electric plants, that such plants are built at the same rate (rated in terms of solar collection area) as electric plants, and that the efficiencies are of the order of 70 per cent, the energy savings will be about 70 per cent of those achieved by construction of solar-powered base-load electric plants. Construction might be feasible by 1990, and projected energy savings by the year 2000 are some 180 million tons of oil.

The estimates of solar energy utilization developed in the preceding paragraphs can be combined to indicate total possible energy savings for the U.S. through solar energy development. As the chart (*p. 39*) indicates, these savings may begin to be significant during the last decade of this century. The growth in solar energy output projected here is set primarily by the capital investment assumptions given earlier. But other factors may also affect the validity of these estimates. If energy demands can be moderated through conservation programs or under the influence of higher prices, solar sources could meet a larger percentage of total U.S. energy need than indicated.

If development funding lags or if opportunities for applications are more limited than assumed, a much lower utilization of solar energy would result—the minimum contribution shown in the chart. But it is clear that in either of these situations solar energy has the possibility of contributing very significantly to the country's energy needs. (A third possibility is that a new energy source—such as nuclear fusion—may turn out to be less expensive and more readily available than solar energy systems. In that case, very few if any solar systems will be built.)

One point needs emphasis: no breakthroughs are needed to make solar energy feasible. All that is required is to engineer solar energy systems which are less costly than current designs by a factor of 1.5 to 4 depending on the type of application—except that considerably greater cost reductions will be necessary to make practical photovoltaic solar plants.

It is fair to say that the engineering problems to be solved in making solar energy practical are considerably simpler than those of the breeder reactor and *far* simpler than those which must be solved in devising a practical fusion reactor. Yet it is also fair to say that a substantial research and development program, together with a vigorous implementation program, will be required to confirm that solar energy may in fact provide a significant portion of the nation's future energy needs.

An Agenda for Research and Development

How much research and development investment can be considered? About 4.5 per cent of sales is typically devoted to research and development in U.S. industry, and such expenditures are usually required about five years in advance of the beginning of production. Hence one may hypothesize that solar energy research and development expenditures should be computed as 4.5 per cent of the sales anticipated five years hence. Data developed in the following paragraphs suggest sales of $200 million in solar energy equipment in 1978, $15 billion (in 1973 dollars) in 1990, and $25 billion in the year 2000. On this basis one postulates research and development expenditures of $10 million in 1973, $670 million in 1985, and $1.1 billion in 1995.

The research and development program should include efforts in the nine areas:

—Development of low-output-temperature (less than 200° F.) solar collectors for space and water heating applications. The emphasis should be on achieving low cost and high collection efficiency in a design compatible with housing construction practices.

—Development of inexpensive low-temperature energy storage techniques capable of holding sufficient energy to heat a house for several days, in designs compatible with housing construction practices.

—Development of low-cost high-efficiency solar collectors for applications requiring high output temperatures (500° C. to 1000° C.) such as electric power

plants using thermal cycles and hydrogen fuel production processes. Areas for study include absorbing surfaces with low emissivity, heat mirrors, insulation techniques, inexpensive solar trackers, heat transfer systems, and techniques for cleaning the collectors.
—Development of economical high-temperature thermal energy storage systems. Areas of possible effort include molten salts, rocks, and metals as heat storage media; insulation techniques; heat transfer systems; and techniques for combining such devices with solar collectors and thermal-cycle generating plants.
—Development of improved photovoltaic solar cells having increased efficiency and lower cost than contemporary designs.
—Development of solar concentrators for use with photovoltaic solar cells.
—Development of inexpensive electrical energy storage techniques for use with photovoltaic solar cell systems.
—Detailed surveys of solar intensity variations in possible areas of application.
—Systems studies on various applications of solar energy with emphasis on system costs, financing methods, and the problems of interfacing solar energy with various processes and systems which now consume energy.

In addition to the above, successful research and development on high-efficiency energy conversion techniques for electric power and hydrogen fuel production would contribute significantly to the early realization of solar energy systems.

$36 Billion for Solar Energy by 2000

Success in these research and development efforts will make possible the rates of application postulated above. But if these rapid advances are to occur, large capital investments will also be required. To simplify the earlier discussion, the cost of the various types of solar energy systems can be projected as follows:

Residential solar heaters	$4,000 each
Total-energy plants	$50/m.^{2} of building supplied, or $100/m.^{2} of collector area
Electric base-load power plants	$700/kw. of electrical output power, or $40/m.^{2} of collector
Hydrogen production plants	$40/m.^{2} of collector

These costs, taken with the application rates developed above, yield the capital expenditure rates shown in the table on page 40—an annual expenditure of $25 billion (1973 dollars) to be invested in solar heating and energy systems by the year 2000.

To these figures must be added the cost of the substantial number of production plants needed to make such large numbers of solar-energy devices and facilities. A crude estimate of this required investment in production facilities can be made by extrapolating from the fact that manufacturing in the U.S. now requires plant investments equal to about 45 per cent of annual output. The plant is depreciated, on the average, over about 14 years. On this basis an investment of $11 billion (1973 dollars) in solar plant production facilities is indicated by the year 2000.

Combining the cost of research and development, production facilities, and the systems themselves gives a total solar-energy investment of about $300 billion in the next 27 years. As the table on page 38 shows, investment at that level would mean that 13 per cent of projected U.S. energy requirements could be filled by solar systems in the year 2000, 26 per cent in 2020.

While substantial collector areas would be required, the total area involved by the year 2020, about 10^4 km.^{2}, would be much less than that used currently for highways. In fact, a substantial fraction of the collector area needed could be accommodated on the roofs of buildings; the rest could be accommodated on land shared with farming or grazing.

The large-scale use of solar energy should have a minimal environmental effect, since such systems operate from an almost inexhaustible energy source external to the earth, produce no pollution products, and can be designed to have minimal effect on the earth's heat balance.

Walter E. Morrow has been a member of Lincoln Laboratory since its founding—the year he completed studies in electrical engineering (S.B. 1949, S.M. 1951) at M.I.T. During this period he has been active in research on a wide variety of radio communications techniques, equipment, and systems; and he has held positions of continually increasing responsibility, becoming the Laboratory's Associate Director in 1972. This article is based on a paper prepared for a technical task force of the Federal Power Commission.

OCEAN THERMAL ENERGY CONVERSION

Robert Cohen
Solar Energy Program
Research Applied to National Needs (RANN)
National Science Foundation
1800 G Street, N. W.
Washington, D. C. 20550

The oceans serve naturally to collect and store (as heat energy) tremendous quantities of solar energy. By operating a heat-engine--utilizing the warm surface water as a heat source in conjunction with cold water pumped from the depths as a heat sink--it was suggested (d'Arsonval, 1881) and shown experimentally (Claude, 1930) that significant amounts of heat can be converted to electricity.

Although renewable "ocean energy" is typically associated by the layman only with ocean tides, there are a number of other ocean energy sources, including ocean waves, currents, salinity gradients, and ocean thermal energy. However, the relative quantity of energy available from ocean thermal energy far exceeds that available from the other sources (Isaacs and Seymour, 1973). For North America, accessible tides, currents, waves, and salinity gradients could each provide several thousand megawatts, whereas ocean thermal might ultimately provide a large percentage of total U. S. energy production.

The ocean thermal energy is low quality heat (i.e., the warm surface water is associated with a high entropy), and the theoretical Carnot efficiency of a thermodynamic cycle utilizing the available temperature difference is about 6 or 7%. The achievable net efficiency of ocean thermal energy conversion can be expected to attain about a third to a half this theoretical value. Thus, the net conversion efficiency is quite low compared to the 30% or greater net efficiency of more familiar power conversion cycles. This means that relatively large amounts of warming and cooling water must be circulated for utilizing the ocean thermal energy. On the other hand, no fuel is required and the amount of available heat energy is very large.

Ocean thermal energy is a form of solar energy collection analogous to hydropower in that nature provides extensive conversion and storage of much of the incoming solar radiation, while smoothing out the intermittence of the source. Thus, since the oceans already act as a natural photothermal collection and storage device for solar radiation, technological solar collectors and storage devices are not required, as in several solar energy utilization options. On the other hand, hardware is required to convert the thermal energy into mechanical energy and thence to electrical energy.

The thermal to electrical energy conversion can be accomplished in several ways. Sea water can be used as a working fluid, as in the experiments of Claude (1930). This method is referred to as the "open cycle", wherein the warm water is flash-evaporated under a partial vacuum, the water vapor then propels a turbine and the vapor is cooled in a condenser using cold water. The conversion cycle tentatively preferred by contemporary workers for producing tens of megawatts and originally suggested by d'Arsonval (1881) would employ a working fluid such as ammonia or propane that is vaporized and condensed by warm and cold water, respectively, in an evaporator and condenser. This "closed cycle" resembles a refrigeration cycle, and is shown schematically in Figure 1, which is adapted from a figure due to Mark Swann.

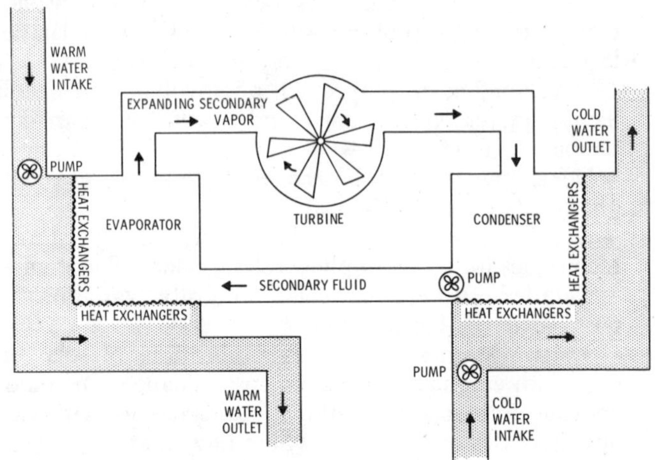

Figure 1--Schematic diagram of a closed cycle ocean thermal power plant (after Mark Swann)

A total ocean thermal system operating at sea will probably be contained in a semi-submersible hull incorporating power-pack modules of optimal size, each containing heat exchangers, one or more turbines, generators, and pumps to circulate the working fluid and the warm and cold water.

Near-shore ocean thermal power plants will likely provide electrical energy directly to land areas. For more remote harnessing of ocean thermal energy, an intermediate chemical product such as hydrogen, methanol or ammonia could be produced, or the power plants could be located where minerals (e.g., bauxite) or other energy-intensive products need to be processed. Although a likely outgrowth of ocean thermal research will be the development of land-based ocean thermal power plants, such plants will probably provide relatively limited amounts of power. This is because the number of suitable locations is determined by the availability of increasingly valuable coastal land combined with a steep off-shore gradient and an adequate, seasonally-steady thermal energy resource. The thermal energy resource of the oceans is thus available mainly at sea, largely in tropical and temperate latitudes. However, relatively large quantities of ocean thermal energy are extractable using large floating power plants located sufficiently close to populated areas so as to be able to transmit electrical energy to shore via submarine cable. An example of this possibility has been described for the southeast coast of the United States by Anderson and Anderson (1973), where the Florida Current could be used to provide power for much of the eastern seaboard.

There has been considerable interest in ocean thermal energy conversion technology both in France and the United States since d'Arsonval proposed the technolog-possibility in 1881. However, although various pioneers such as Claude tried to proceed with experiments, there was little government or private capital available for development in view of the emphasis by governments and industry on fossil and nuclear sources, in a market where fossil fuels became increasingly cheaper and abundant. Now that this competitive pattern is changing, alternative energy technologies such as ocean thermal are being seriously explored.

Research on ocean thermal energy conversion is now being conducted as part of the U. S. solar energy program. That program supports studies in six solar energy technology research areas, where each technology is potentially capable of producing substantial amounts of energy at competitive costs. The objective of the ocean thermal research program is to establish the technical, economic and geopolitical feasibility of large-scale floating power plants capable of converting ocean thermal energy into electrical energy, leading toward the commercial utilization of such plants and the production of energy in various forms.

Ocean thermal studies supported by the U. S. solar energy program began in 1972. The status of the ocean thermal energy conversion program was discussed at a public workshop held at Carnegie-Mellon University in Pittsburgh in June of 1973, and is reported (Lavi, 1973) in the proceedings of that workshop. Other results of the ocean thermal research program are being published in progress reports and technical reports of the University of Massachusetts, Amherst, and of the Carnegie-Mellon University research teams. See, for example, Heronemus et al. (1973, 1974) and Zener et al. (1973 and 1974). Additional research is now being conducted by other university, industry, government, and non-profit organizations.

The technology required for ocean thermal energy conversion systems is relatively low-level technology. That is, no scientific or technical breakthroughs are required (although some are expected), and the basic technology is presently intrinsically available to enable the construction of commercial ocean thermal power plants. However, the state-of-the-art within this body of technology needs to be advanced in order to adapt the technology for this application, especially in order to provide economically viable (i.e., competitive) systems. For the purpose of minimizing costs, study and adaptation of existing relevant technologies are needed. Fortunately, there has been considerable recent experience in such relevant technologies, for example, offshore technology. Present program plans are to work toward the development of an experimental prototype ocean thermal power plant that can be constructed in about 1980. This power plant would be a floating power-pack module, with a power output of about 10 to 25 megawatts. In the meantime, testing and optimization programs are planned for the development of components and subsystems.

Some of the technical problems confronting the development of ocean thermal technology include:

-Optimization of heat exchangers, from the standpoints of cost and performance. Some intertwined problems include biofouling, corrosion, and materials compatibility.

-Design and deployment studies on the pipe required for cold-water intake. (This pipe will be 15 meters or more in diameter, and may extend to depths of 600 meters or greater).

-Perfection of techniques for positioning of ocean thermal power platforms. Possibilities include anchoring, mooring, and dynamic positioning.

-Biofouling of components and subsystems: avoidance and corrective measures.

-Hydrodynamic and structural design of the hull, condenser and evaporator structures.

-Optimization of energy delivery as a function of power plant and market separations and locations, including alternatives for energy storage and transmission.

-Design, testing, and optimization of system components, especially turbines and pumps.

Besides the purely technical problems that confront this technology, there are certain potential environmental and legal problems. The degree of implementation of this technology will probably determine what, if any, environmental and ecological alterations might occur on a regional or global basis. On the other hand, there may be local environmental consequences associated with an individual power plant. However, it is premature to conjecture as to what they might be. Indeed, the ocean thermal energy conversion process could probably be technically modified to avoid or alleviate possible adverse consequences of operation, and there might actually be favorable environmental alterations resulting from the implementation of this technology. Compared to other energy alternatives, the ocean thermal option would appear to be relatively environmentally benign, and any possible adverse effects may be controllable and/or reversible through the character, degree and localization of its implementation. There are siting limitations associated with availability of the thermal resource and of suitable ocean conditions. Also, the amount of available heat will impose a limit on the number of such power plants that can be permitted per unit area of ocean. Regulatory problems, such as freedom-of-navigation and law-of-the-sea considerations, may also impose certain limits to the availability of ocean areas for this application.

Studies are now underway or will shortly be undertaken on all of the technical, environmental and legal problems mentioned above. Other considerations concern resource constraints. The technology for exploiting ocean thermal energy will be capital-intensive, hence resource-intensive. The exact materials requirements remain to be determined, since the optimum choice of materials will depend upon further research. If the heat exchangers were to be made of aluminum, then considerable (but not prohibitive) quantities of that metal would be needed. Even so, the energy required to construct a plant will be repaid in several months.

Ocean thermal power plants will be "fuel-free", yet will need to circulate large quantities of water containing the low-quality heat that is really the fuel on which they operate. They will not require land area, but can utilize the "free" open ocean. Unlike some solar energy applications, resources for providing technological thermal collectors and thermal storage devices are not needed, since the ocean provides them naturally. On the other hand, the heat exchangers represent a significant cost/resource item.

In the mid-1980's, when ocean thermal plant platforms could be constructed, there will probably be a readily available shipbuilding capacity, and the power plants are expected to be of modular design amenable to assembly-line techniques.

The optimal overall size of an ocean thermal power plant or of its power-pack modules has not yet been determined. Preliminary estimates (Heronemus et al., 1973 and 1974) are that the plant size will be in the neighborhood of 400 or 500 megawatts, comprised of modules of about 25 megawatts. A likely size for a first commercial demonstration power plant is about 100 megawatts. A schematic of what such a plant might look like is shown in Figure 2. This figure is largely based upon the conceptual design of Anderson and Anderson (1973).

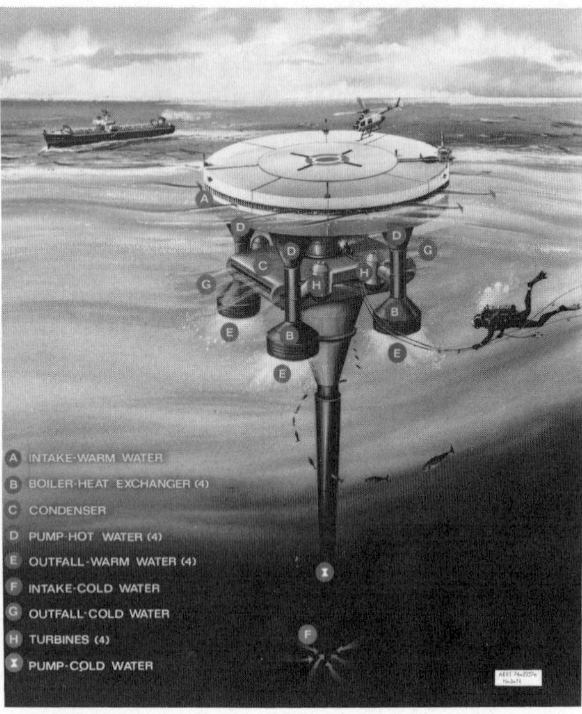

Figure 2--Schematic conceptual diagram of an ocean thermal power plant of 100 megawatt capacity (after Anderson and Anderson, 1973)

Besides hydrogen, a likely product of ocean thermal power plants will be ammonia, synthesized from ocean water and air. Such an energy-intensive product could be transported and utilized for existing fertilizer markets or as a fuel, relieving energy requirements for producing ammonia elsewhere. Thus the offshore production of ammonia or a similar energy-intensive product results in the "virtual" shipment of the energy otherwise required to fabricate it elsewhere.

Other byproducts of ocean thermal power plants include protein, through open-ocean mariculture of fauna such as shellfish, using the nutrients provided in the artificial upwelling of the cold ocean water from depth. Similarly, flora such as kelp could be cultivated at sea, for their food and energy value.

Fresh water is another possible byproduct, especially for land applications. The potential for land-based plants providing energy, fresh water, and protein is discussed by Othmer and Roels (1973). (See also Davitian and McLean, 1974 and Othmer and Roels, 1974).

REFERENCES

Anderson, J. Hilbert and James H. Anderson, Jr., "A summary of the Anderson and Anderson analysis of the sea solar power process, 1964 to 1972", Technical Report NSF/RANN/SE/GI-34979/TR/73/5, University of Massachusetts (Amherst) Energy Program (1973 March).

d'Arsonval, A., "Utilisation des forces naturelles. Avenir de l'électricité.", La Revue Scientifique, 370-372 (1881 Sept. 17).

Claude, Georges, "Power from the tropical seas", Mechanical Engineering 52, 1039-1044 (1930 December).

Davitian, Harry and William McLean, "Power, fresh water, and food from the sea," Science 184, 938 (1974 May 31).

Heronemus, W. E. et al., "Research applied to ocean sited power plants", Progress Report NSF/RANN/SE/GI-34979/73/2, University of Massachusetts (Amherst) Energy Program (1973 July 31). (Available from National Technical Information Service, catalog PB-228 070).

Heronemus, W. E. et al., "Research applied to ocean sited power plants," Progress Report NSF/RANN/SE/GI-34979/73/3, University of Massachusetts (Amherst) Energy Program (1974 Jan. 25). (Available from National Technical Information Service, catalog PB-228 067).

Isaacs, John D. and Richard J. Seymour, "The ocean as a power resource," Intern. J. Environmental Studies 4, 201-205 (1973).

Lavi, Abrahim, ed., Proceedings, Solar Sea Power Plant Conference and Workshop, held at Carnegie-Mellon University, Pittsburgh, Pennsylvania on 1973 June 27 and 28. Report NSF/RANN/SE/GI-39115/73, Carnegie-Mellon University (1973 September). (Available from National Technical Information Service, catalog PB-228 066).

Othmer, Donald F. and Oswald A. Roels, "Power, fresh water, and food from cold, deep sea water", Science 182, 121-125 (1973 October 12).

Othmer, Donald F. and Oswald A. Roels, "Power, fresh water, and food from the sea", Science 184, 938 and 940 (1974 May 31).

Zener, Clarence et al., "Solar sea power," Progress Report NSF/RANN/SE/GI-39114/PR/73/1, Carnegie-Mellon University (1973 October 11). (Available from National Technical Information Service, catalog PB-228 069).

Zener, Clarence et al., "Solar sea power," Progress Report NSF/RANN/SE/GI-39114/PR/74/2, Carnegie-Mellon University (1974 January 25). (Available from National Technical Information Service, catalog PB-228 068).

WINDMILLS

BY JULIAN McCAULL

PICTURE THE GREAT PLAINS OF MID-AMERICA, from Texas into North Dakota, with a forest of giant windmills, each the height of a 70-story building. In contrast to the familiar, multi-vaned windwheels that drove water pumps or grinding stones in years past, these windmills would have clusters of propeller blades similar to those on aircraft, only much larger. The wind-driven blades would turn generators to supply badly needed electricity in the U.S.

This scene is envisioned by William E. Heronemus, Professor of Engineering at the University of Massachusetts, Amherst, whose proposal was detailed early in 1972 at a joint meeting of local sections of The American Society of Mechanical Engineers and The Institute of Electrical and Electronics Engineers.[1] The proposal subsequently was printed in the Congressional Record at the request of Senator Mike Gravel of Alaska.

The Heronemus scheme reflects a recurrent interest in wind power as one of a number of alternatives to meet U.S. energy demands. Whether such an array of towers would be acceptable to those interested in the beautification of America is one question – a particularly serious one considering intensified public pressure to put unsightly electrical transmission lines underground. Another question, that addressed here, is whether wind machines in general are technically and economically feasible. Based on experience in a number of other countries, the answer appears to be a qualified "yes." The scale on which they are feasible, however, is uncertain.

Tapped on a large scale, wind power theoretically could produce prodigious amounts of electricity.

Visions of wind-driven electrical generators have tantalized inventors for some time. The concept is appealing because it is so simple: Wind pressure turns vanes or propellers attached to a shaft. The revolving shaft, through connections to various gears and mechanical or hydraulic couplings, spins the rotor of a generator. The generator creates an electrical current in transmission lines that are tapped for desired uses of electrical energy. The windmill has been used in this way to generate small amounts of electrical energy since 1890, when a Danish machine with patent sails was put into operation. Thousands of streamlined windmill generators, rated at one kilowatt or less each, have been used in rural America to light farms or charge batteries in areas without electrical utilities.

Central Vermont Public Service Corporation

The 1,250-kilowatt Grandpa's Knob windmill being assembled in 1941 in central Vermont. Tower in background carries wind-measuring instruments.

These modest applications of wind power have inspired a number of more ambitious projects to use large wind machines, or wind turbines, to generate substantial amounts of electricity for public use. One problem has been to build machines large enough to utilize sufficient wind power. Another serious problem has been to design a system that can generate electricity reliably despite uncertain winds, since when the wind stops, so does the generator. Though elusive, wind power is nevertheless enticing because it is continuously regenerated in the atomosphere under the influence of radiant energy from the sun, and thus is a self-renewing source of power. It has the further advantage of being a clean source of power, whereas nuclear and fossil-fuel power produce dangerous pollutants.

Tapped on a large scale, wind power theoretically could produce prodigious amounts of electricity. The Heronemus proposal, for example, envisions some 300,000 wind turbines in the Great Plains, spaced as closely as one per each square mile. Each 850-foot tower would carry twenty turbines, consisting of a two-bladed, 50-foot diameter propeller. The Great Plains network could provide the equivalent of 189,000 megawatts of nuclear power plant installed capacity. The wind-powered systems actually would have a considerably larger total installed capacity in terms of potential electricity generating power, but the potential would never be met over a year's time since wind speed variations would make it impossible to keep the generators turning at maximum output. The comparison with nuclear power plant capacity was worked out mathematically by Heronemus to provide a more realistic measure of the electrical production that could be anticipated from wind power. The 189,000-megawatt capacity of the wind network would be significant, nonetheless, since the total installed capacity of electricity generating plants in the U.S. in 1970 was only an estimated 360,000 megawatts.[2]

The Vermont Experiment

Before considering Heronemus' work, however, we must go back three decades to when the groundwork was laid for large, modern, wind turbines. During the period from 1941 to 1945, a 1,250-kilowatt, wind-powered generator with stainless steel blades visible for 25 miles operated intermittently near Rutland, Vermont. This project supplied the only long-term data by which to evaluate current proposals.[3]

The technical and administrative mind behind the experiment was Palmer Putnam, who in 1934 wanted a small wind-driven electrical generator for his

Cape Cod home to reduce the cost of electricity. He discovered that commercial wind-powered generators used for farm lighting were not large enough, so he investigated earlier large-scale wind power experiments. Among those that seemed relevant were German, French, and Russian projects featuring bladed wind turbines. A 1931 Russian project, for example, featured a wind turbine 100 feet in diameter on a bluff near Yalta, overlooking the Black Sea. The windmill drove a 100-kilowatt generator to supplement the electrical output at a steam-power plant in Sevastopol, twenty miles away. Despite limitations in the equipment – the blade skins were made from roofing metal and the main gears from wood – the unit reportedly produced 279,000 kilowatt-hours of electrical energy in one year.

The most economical plan in the U.S. appeared to be the use of large wind machines to generate alternating electrical current that would be fed into existing water- or water-and-steam powered networks. When the wind blew, operators at the main hydroelectric plant could shut down some of the water-driven turbines. With the turbines inactive, water would build up in the reservoir, creating a reserve that would then be used to drive the turbines, thereby supplying electricity, on days when the machines stopped from lack of wind. As explained later, this arrangement would be economical only if the hydroelectric system had enough unused reservoir and water-wheel capacity to take advantage of the supplementary wind power.

A number of engineers, electrical utility officials, and manufacturing experts were drawn into Putnam's scheme. It was agreed that a large test unit would be financed and built by the S. Morgan Smith Company, manufacturer of hydraulic turbines in Pennsylvania. The wind machine would be operated by the Central Vermont Public Service Corporation, a subsidiary of the large New England Public Service Company. The Central Vermont system would provide a good test for Putnam's idea, since it was powered only by water and had insufficient capacity for peak loads. Excess power demands were met under a power-purchase contract with another hydroelectric company, and the hope was that a battery of wind turbines might eventually be built to supplement Central Vermont's existing system and reduce power production and purchase costs. As will be shown, that cheery vision proved overly optimistic in light of what happened during operation of the experimental wind machine.

Some 50 Vermont summits appeared to be promising wind-turbine sites, and one was selected early in 1940. The knoll chosen was 2,000 feet high, well

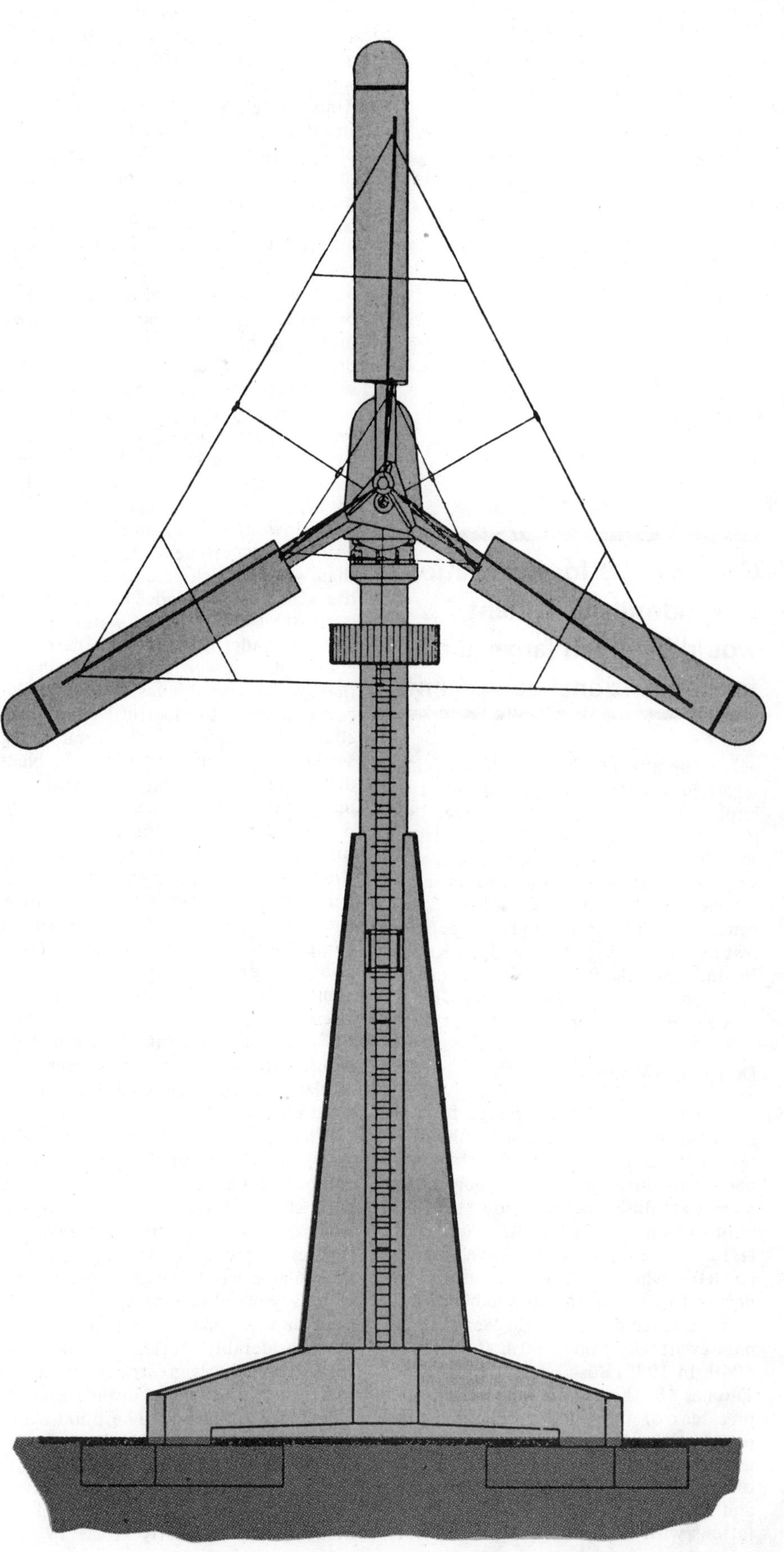

A sketch of the 75-foot wind turbine operating successfully near Gedser, Denmark, since 1957. The 200-kilowatt device produces electricity for the Danish public power system. The equipment, which requires only routine maintenance, generates an estimated 400,000 kilowatt-hours per year — roughly one-third of the output from the much larger installation at Grandpa's Knob in Vermont. The Danish machine can be modified to produce larger quantities of electricity. It also is less expensive than U.S. models.

Capital costs for installation of windmill equipment would be much larger than for conventional equipment.

below the altitude at which heavy icing might have endangered equipment. The knoll had no official map name, but project organizers took their cue from the Vermont farmer from whom it was purchased. He and his family invariably referred to the hill as "Grandpa's." Because of this and the hill's shape, the test site was christened Grandpa's Knob. In due time, the entire project came to be known as the 1939 to 1945 Grandpa's Knob experiment.

Designing a Windmill

In arriving at the final design for the wind turbine, Putnam drew on the professional advice of a number of prominent individuals. An early supporter was Vannevar Bush, better known for his tenure as director of the Office of Scientific Research and Development (OSRD), which mobilized scientific expertise to develop the atomic bomb and tactical radar during World War II. Putnam eventually joined Bush at OSRD in 1940. In 1937, Bush referred Putnam to Thomas S. Knight, commercial vice-president of the General Electric Company in New England. Knight was an avid sailor who was taken with the idea of applying wind power on a large scale, and he helped arrange for design consultations with experts at both the California and the Massachusetts Institutes of Technology. Stress analyses were performed by John B. Wilbur, who became chief engineer of the wind-turbine experiment and later head of the Department of Civil and Sanitary Engineering at the Massachusetts Institute of Technology (MIT). Turbine blade models were tested at the Stanford University wind tunnel. Vibration analyses for the structure were done by J. P. den Hartog of Harvard University. Wind tunnel tests of mountain models were performed at Guggenheim Aeronautical Institute, Akron, Ohio, under the direction of Theodor von Karman. To shed light on uncertainties regarding the behavior of wind over knolls and mountains, a meteorological study program was organized during the operation of the turbine. The program was directed by Sverre Petterssen, director of the Department of Meteorology at MIT, in collaboration with several other prominent meteorologists.

The wind turbine to be constructed at Grandpa's Knob had to be very large to tap wind power economically, because low air density and velocity make wind a less efficient propelling force than water or high-pressure steam. Thus, the wind turbine blades had to be much larger than conventional water or steam turbine blades to drive a generator of comparable power. Calculations showed that with large blades, wind turbines could convert to electrical energy only about 6 percent of the total energy available in winds blowing at the blades during one year. This was because at many times during the year, the wind would be too weak or too strong for turbine operation, and even though the wind was blowing, energy in those winds would be wasted insofar as electrical generation was concerned. By comparison, the most efficient central, coal-powered, electrical generating plants convert up to 36 percent of the latent heat energy in coal into electricity.[4] That is, the central, coal-powered plant extracts up to six times more available energy from coal than does the wind generator from the wind.

This comparison is illustrative rather than technically precise, since we are comparing kinetic energy in the wind with chemical energy in coal. In addition, a very large wind generator can theoretically convert roughly 35 percent of available wind energy into electricity at steady wind speeds of 18 miles per hour or so – small wind generators can do considerably better.[5] Higher wind velocity produces greater electrical output from the very large installations, but at reduced efficiency. The point is that, in addition to the low density and velocity of wind, its uncontrollability makes for poor operating efficiency in wind generators. That is, the wind cannot be made to blow regularly at the 19 to 30 miles-per-hour speed required for satisfactory operation. Another consideration making it difficult to compare the efficiency of a wind versus a steam turbine is that the overall extraction of 6 percent of total wind energy flowing through the windmill blades is only a small fraction of the total energy in wind passing through the general area where the turbine is located. That is, the entire wind stream from ground level up to at least 1,000 feet would have to be tapped by a continuous array of windmill blades to extract a large portion of the energy, an arrangement that would be technologically impractical and which could have serious meteorological side effects. In contrast, coal reserves can be mined systematically, and much more of the total available energy in that fuel is utilized since coal can be fed into boilers at a controlled rate. Neither coal- nor wind-powered generators are as efficient in energy extraction as are hydroelectric generators which capitalize on the greater density of water for driving turbines.

Although some improvement in wind turbine efficiency can be expected with current designs and materials, the Grandpa's Knob turbine proved close to what is theoretically the most economical design for large wind machines. The turbine had a two-bladed propeller nearly 175 feet in diameter. Each blade, weighing eight tons and consisting of a stainless steel skin over stainless steel ribs, was in a streamlined, aerodynamic shape. The pitch of the blades was adjustable to regulate the turning speed of the turbine shaft, and the same system enabled operators to feather (turn edgewise into the wind) the blades to prevent damage during high winds. The blades also were free to move independently of each other forward and backward, through a set angle, to reduce dangerous bending forces that otherwise would have occurred near the base of the blades.

Putting it Together

The blades were designed and fabricated in Philadelphia by the Budd Company, noted for early work in development of stainless steel train cars. The 110-foot tower was constructed at Ambridge, Pennsylvania by the American Bridge Company. Other parts were made in

Wind has the advantages of being a clean source of power and of being continuously regenerated in the atmosphere.

A problem in building wind machines capable of generating substantial amounts of electricity has been to make machines large enough to utilize sufficient wind power.

Cleveland by the Wellman Engineering Company, which also assembled the entire structure at its plant to balance the blades and rotating system before shipping the materials off to Rutland, Vermont in the spring of 1941. A good deal of heavy equipment, including large cranes, was needed to unload, transport, and assemble the wind turbine on Grandpa's Knob. When assembled, the blades were attached by frames to a hub, which was in turn connected to the main shaft. The shaft was linked to the generator by gears and hydraulic couplings. The propeller-generator equipment was mounted on a 40-foot long carriage, or girder, on the top of the tower and inclined downhill to put the blades more nearly at right angles to the wind coming up over the crest. The girder was attached to the tower by means of a vertical shaft and gears. Other machinery automatically rotated the girder about the axis of the tower to keep the turbine aligned with the wind as wind direction changed. The blades were downwind of the control house, which enclosed the generator and its associated equipment. The blades turned the main shaft at a top speed of nearly 29 revolutions per minute. The gears sped up this rotational speed so that the generator rotor turned at 600 revolutions per minute.

After weeks of tests, the turbine unit electrical output was switched into the utility lines of the Central Vermont Public Service Corporation on October 19, 1941. Over the next three and one-half years, the machine operated in winds up to 70 miles per hour, generating as much as 1,500 kilowatts of electrical power. The structure withstood winds of up to 115 miles per hour. The machine was shut down for two years from 1943 to 1945 to replace a main bearing, difficult to obtain during the war, but enough experience had been gained to shift to routine operation in the spring of 1945.

Trouble in the Blade

Although members of the project team thought structural design was generally sound, they were concerned about the strength of the spars, the longitudinal structural shafts attached to the ribs of the blades and leading to the frames, which were in turn attached to the main shaft. The threat of war priorities, which could have precluded the entire project, forced the engineers to have the spars forged early in 1940, even though the overall turbine design was not yet complete. When the design details were finished, they realized that the spars should have been made larger to support the eight-ton blades. By the time this was discovered, it was too late to make alterations, and the project was allowed to continue as a calculated risk. Unfortunately, the risk was increased by field repairs in 1942 that inadvertently weakened the spars still more. These and less serious mechanical problems were taken into account in new wind turbine designs worked out by Putnam and his associates from 1943 to 1945, but it was too late to correct the problems at Grandpa's Knob. Accordingly, the test unit was to be dismantled after some experience had been gained in routine electricity generation.

Early in the morning on March 26, 1945, after three weeks of routine service during which 61,780 kilowatt-hours of electricity energy had been generated in 143 hours and 25 minutes of operation, the test unit project came to a sudden halt. The spar for one of the whirling blades suddenly broke, and the entire 65-foot blade was hurled off into the darkness, landing some 750 feet away from the tower. The remaining blade continued to rotate until halted by automatic controls and efforts on the part of operating personnel. The effect of centrifugal force exerted by the single spinning blade can be appreciated in Palmer's description of conditions in the control house on top of the tower after the one blade broke.

"Harold Perry, the foreman, was aloft, standing on the side of the house away from the control panel and separated from it by the 24-inch rotating main shaft. A shock threw him to his knees against the wall. He started for the controls, but was again thrown to his knees. He tried again, and again was thrown down. Collecting himself, he dove over the rotating shaft, reached the controls, and, overriding the automatic controls which were already functioning, he brought the unit to a full stop in about 10 seconds by bringing the remaining blade to full feather."

Finger in the Wind

Information obtained during the planning and operation of the Grandpa's Knob wind machine shed light on two considerations that are as important today as in 1940. The first is the difficulty in predicting long-term wind velocity and direction in the mountains; the second is the economic feasibility of utilizing this uncertain source of power. As to the first consideration, conventional meteorological data and theory proved faulty in predicting the winds to be expected on Grandpa's Knob. Wind gauge recordings made at the site revealed that the mean annual velocity was only 17 miles per hour instead of 24 miles per hour as had been predicted beforehand on the basis of available meteorological data. Furthermore, prevailing wind direction was from the southwest instead of nearly due west as anticipated on the basis of standard tables showing progressive shifts in wind direction according to altitude. These two miscalculations meant that the predicted level of usable wind energy was considerably lower than expected, with the result that annual electrical generation averaged only 1,200 kilowatt-hours per kilowatt of installed capacity (determined by dividing total kilowatt-hours produced by 1,250 kilowatts, the rated capacity of the generator). This was only 30 percent of the output originally predicted for the Grandpa's Knob turbine.

Subsequent analysis showed that the discrepancy probably arose because the Green Mountain Range acted as a gigantic baffle, or obstruction, affecting air flow far to windward and causing unexpected change in wind direction and velocity in the foothills where Grandpa's Knob was located. Project personnel concluded that it was essential to operate wind gauge equipment on proposed sites ahead of time instead of relying on theoretical calculations. But even a year's data from such equipment would not be sufficient to predict long-term mean annual velocity at the site. This problem could be eased by correcting on-site measurements according to long-term data at some established meteorological station, as long as the station were within 50 miles of the site. Even then, one had to take into account another factor – extreme local variations in wind due to unknown factors of topography (the configuration of the land). For example, a peak near Grandpa's Knob had a similar rounded profile and elevation; yet mean wind velocity over the summit was 8 percent greater. In New Hampshire, the Horn, a 4,100-foot spur, had a substantially greater mean wind velocity than the 6,288-foot summit, Mt. Washington, south of it – even though two ridges

Windmills have been used to generate small amounts of electrical energy since 1890.

upwind of, and higher than, the Horn appeared to have far less wind than did the Horn.

The reason for seeking high elevations for wind turbines is that wind velocity increases with height, since air flow there is less retarded by ground-level friction. On the other hand, mountains act like airfoils to speed up or slow wind flow according to their shape, with high, sharp ridges generally producing the greatest acceleration of wind speed over the summit. Although this was found to be generally true, the local variations were so pronounced that the airfoil-mountain analogy proved unreliable. In fact, one of the most accurate indicators of long-term wind velocity proved to be the progressive deformity of trees, particularly balsam firs, according to wind speed. At an annual mean wind velocity of 17 miles per hour, the firs showed some flagging, a condition in which branches extended downwind, resembling a flag stretched out in the breeze. As the mean annual velocity increased, the deformity passed through four more distinctive stages, the last, at 27 miles per hour, being carpeting, in which growth was limited to a living carpet of branches no more than six inches off the ground. This sensitive ecological indicator defined the path of mountain wind streams that varied sharply in mean annual speed, even within distances of 100 yards.

Costly Power

The economics of wind turbines were analyzed in 1945 on the basis of a design for a 1,500-kilowatt production model incorporating refinements to the Grandpa's Knob machine. The study was based on installation of six of the production models on Lincoln Ridge in the Green Mountains where the altitude was 4,000 feet, and the winds were considerably stronger than at Grandpa's Knob. It was estimated that this battery of turbines, with a total capacity of 9,000 kilowatts, would generate 31.5 million kilowatt-hours per year.

Even today, computing the dollar value of this amount of electrical energy is not a straightforward matter, because the economic value of the energy, except in the large-scale proposals by

Westinghouse Electric Corporation

Wind power and nuclear power coexist in Western Europe.

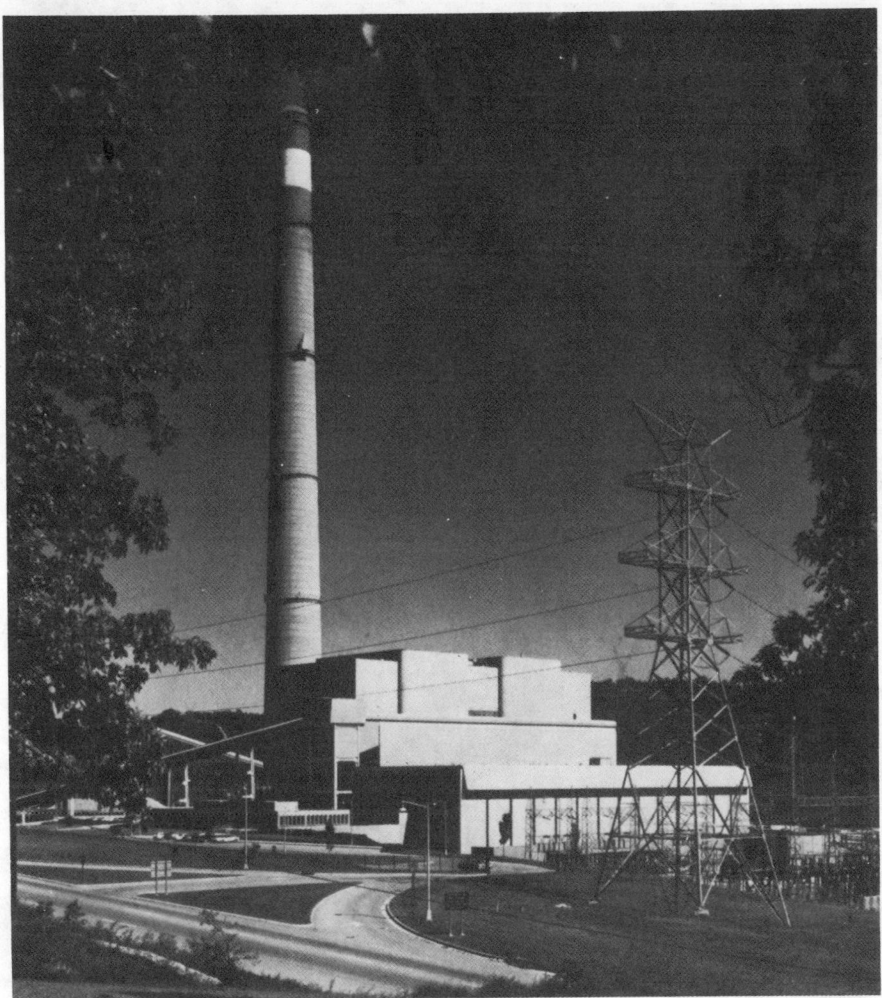

Tennessee Valley Authority

One of the world's largest electrical generating plants near Oak Ridge, Tennessee, consumes a large trainload of coal daily at full operation. The coal is scarcer than wind but can be used with greater efficiency.

Heronemus and some others, is computed in relation to its ability to supplement electrical energy from conventional generating systems. In the case of hydroelectric generation, as mentioned before, this would mean building up a water supply in the reservoir while the wind turbine is running. The degree to which this build-up would bring about an overall increase in generating capacity, with consequent increase in electrical energy to be sold, would depend upon whether the hydroelectric system had enough waterwheels and sufficient storage space in the reservoirs to take advantage of the supplementary electric energy. If the hydroelectric network were already working at full capacity, for example, new waterwheels or reservoir space would have to be added to achieve an increase in overall generating capacity. The costs of the new installations would have to be considered as well as the cost of the wind turbine to determine whether the wind machine would be profitable. If used in conjunction with a nuclear or fossil-fuel generating system, wind power would create no increased capacity, since it would not reduce the size of the conventional plant needed to carry peak loads when there was no wind.

An independent consulting firm took these considerations into account in estimating the value of the 9,000-kilowatt block of power to be supplied by the proposed wind turbines in Vermont. The conclusion was that, in 1945, this supplementary wind power would warrant an investment of $125 per installed kilowatt of power. Unfortunately, the battery of turbines would have cost an estimated $205 per installed kilowatt. This difference was so serious an economic obstacle that the entire project was abandoned, although the S. Morgan Smith Company had already spent more than $1.25 million on the Grandpa's Knob experiment.

A number of methods to reduce the cost of the turbines were suggested by Putnam. These ranged from new production procedures to radical changes in mechanical design, such as elimination or modification of the expensive gear system used to convert wind turbine speeds into higher speeds to operate the generator, a possibility that involved untried technological innovations. If all of the suggestions were realized, a 2,000 to 3,000-kilowatt wind turbine produced in large quantities (on the order of several thousand units) might have been built for about $100 per installed kilowatt in 1945, Putnam estimated.

The World Picture

If mass production techniques and some technical changes had reduced the cost of large wind turbine installation, there would have been a market for such machines in 1945, Putnam believed. But he saw the market as limited to selected areas near centers of heavy electrical demand in windy regions between 30 and 60 degrees latitude in the Northern and Southern Hemispheres. The northern band between these latitudes would take in industrialized nations in Europe, the Soviet Union, and North America but, unfortunately, would not include less industrially developed countries in Africa and in the Asian subcontinent, where the need for cheap power is becoming serious. The band in the Southern Hemisphere would include small portions of Africa, Australia, and South America. Putnam envisioned the largest wind machines as having a 2,000 to 3,000 kilowatt capacity, with electrical output supplementing conventional water and steam power. Medium-sized units rated at 100 to 500 kilowatts could be used in conjunction with small hydroelectric or diesel engine powered installations in windy, isolated regions of the world. Small units of 10 kilowatts or less would be useful in charging batteries for untended airway beacons, as in the Arctic or perhaps in some desert regions, or for farm lighting.

Efforts to harness the wind continued after 1945, but the breakdown at Grandpa's Knob and Putnam's conserva-

Dept. of the Interior

The best ecological indicator of wind conditions at high altitudes is the telltale deformity of trees.

tive estimate of the economic feasibility of large-scale wind power checked further development in the U.S. Feasibility studies were made under the auspices of the War Production Board and later the Federal Power Commission, and in 1951 the Department of the Interior asked for funds to build a prototype high-tower, large capacity machine. Members of the House Committee on Interior and Insular Affairs were unimpressed by the idea, and the project died.

The focus of wind power research then shifted to other countries, where innovators resolutely continued to chase the wind – with some success. Reports on a good deal of this experience were assembled in a 1961 United Nations conference on new sources of energy at which scientists discussed meteorological, technical, and economic aspects of wind power.[6]

As to the meteorological questions, the call by Putnam and others for more information prompted a number of investigations after 1945. At the time of the U.N. conference, there were some 350 sites around the world with instruments to measure wind speed from 150 to 370 feet above the ground, heights corresponding to those at which wind turbines might be built. Observations from these and other locations provided the World Meteorological Organization with enough data to conclude that there were some 20 billion kilowatts of wind power available at favorable turbine sites throughout the world. This figure was 1,000 times the estimate for worldwide wind power potential given by Putnam in 1948.

A number of countries had begun nationwide surveys to determine their potential share of this wind power. Reports were given on recording networks planned or in operation in Spain, parts

The experimental wind machine at Grandpa's Knob illustrated the mechanical problems that can develop in large wind turbines.

of Africa, Argentina, Israel, the British Isles, India, Denmark, Hungary, and the United Arab Republic. In addition to identifying potential sites for wind machines, the surveys produced some new information about wind speed profiles at various heights above the ground. For example, investigators in Israel reported that a key indicator of steady wind power for turbines was the extent to which wind speeds varied at different altitudes up to the level of the proposed turbine. The greater the variation, the greater the wind variability and turbulence, and hence the less suitable the site for wind power. The wind speeds most uniform with altitude, and thus the most satisfactory wind flows, occurred over summits that had smooth, steady slopes near the top. Despite these and other findings, there was general agreement that the theory of wind speed profiles could not yet substitute for long-term surveys at the actual wind turbine sites.

Propellers are Best

A number of U.N. reports dealt with technical advances in wind turbine design. Perhaps the most significant new technique was fabrication of turbine blades from plastic reinforced with fiber glass. These blades were strong, could be moulded in aerodynamic shapes, and were potentially much less expensive to mass produce than metal blades. Experimental plastic blades were installed in a 100-kilowatt wind machine in Germany and a 150-kilowatt unit in France.

Despite possible improvement in the blades, there were no revolutionary ideas to overcome the other major stumbling block in turbine design, namely, the need for expensive gears to translate the slow turbine blade rotational speed into a much faster speed to operate the electrical generator. Furthermore, although a number of novel wind machine designs were tried experimentally, many too complicated to describe here, the most reliable design was still considered to feature propeller blades linked directly to the generator. Units of this type were in operation or under construction in a number of countries. The largest were 200-kilowatt machines in Denmark and Hungary. A 900-kilowatt experimental unit in France was mentioned in a later U.N. report.[7]

The most successful large wind turbine was reported near Gedser, Denmark. The fully automated, 200-kilowatt device had three blades, each nearly 40 feet long, mounted on a 75-foot tower of prestressed concrete. It had been producing electricity for the Danish public power system for more than three years at the time of the 1961 U.N. conference. The equipment generated an estimated 400,000 kilowatt-hours per year, roughly one-third of the output from the much larger installation at Grandpa's Knob. The Gedser turbine cost $56,000 to develop and build in 1957. Costs were expected to drop to about $38,000 for each unit if the machine were mass produced. This would be $190 per installed kilowatt, $15 per kilowatt less than the production machines designed by Putnam in 1945. Although this cost appears to be the best yet achieved, it probably would be higher in the U.S. because of greater labor costs.

Danish engineers contended that in their country, wind-power electrical generation might be economical in its own right rather than as a supplement to conventional generating systems. Coal and oil are scarce in Denmark and expensive to import, they explained. Water power is limited. A wind-power network of 100 Gedser turbines would be economical since it would mean a considerable saving in money otherwise spent for fuel for conventional boiler-operated power plants. In addition, there would be savings in labor. The Gedser turbine required only routine maintenance, and engineers estimated that a team of eight workmen could service a battery of 100 wind-power plants if the machines were in the same general locale. Another advantage was that the strongest winds in Denmark were in the winter when electrical heating demands were also heaviest. Peak winter demand thus would be matched by naturally intensified wind power rather than by use of additional installed conventional generating capacity that would be idle during periods of low demand at other times of the year. A related suggestion made at the U.N. meeting and elsewhere was to develop electronically controlled load distribution systems to shunt wind-generated electricity to different sectors of the public power network as needed. This approach would not make wind power any more reliable, but it would provide for maximum use of the wind when it was blowing at the right speed. An automatic system of this type has successfully matched wind power output to electrical demand in the Scottish Highlands.

Large-scale employment of wind power could possibly generate fuel or fuel-cell electricity at rates potentially competitive in selected power markets.

An additional consideration was that the Gedser windmill design was such that two of the turbines could be stacked on the same tower, one over the other. The tower would have to be about twice as tall as in the single turbine unit, but no taller than the Grandpa's Knob tower. This configuration would house two slightly larger generators that would have a total generating capacity of 600 kilowatts, producing about 960,000 kilowatt-hours of electrical energy each year. Cost per installed kilowatt of power probably would be even less than in the single turbine Gedser unit. In view of these economic and technical possiblities, about 20 percent of Denmark's annual electrical needs could be met by wind power, according to a U.N. report by J. Juul, former sectional engineer, Nordskov, Haslev, Denmark. Few others at the U.N. meeting were optimistic about such wide-scale applications of wind power, however. In particular, there was general agreement that nuclear and fossil-fuel power plants were far more economical in most fully industrialized nations than in developing countries with fewer natural and economic resources.

Windmills on the Plains

Against this worldwide background, the Heronemus proposal for wholesale use of wind power in the U.S. is indeed radical. The individual turbines Heronemus proposes would be similar in size to the Gedser turbine, only with two instead of three blades. The technical design would be a variation of that worked out by a noted wind-power theorist, E. W. Golding. But Heronemus would mount twenty, instead of one or two, turbines, on a tower five times as tall as the one on Grandpa's Knob. The object would be to tap wind power high over relatively flat country – the Great Plains – rather than mountain tops. The terrain would be much easier to build on than mountains, but large allocations of money and metal would be needed to construct the thousands of windmills envisioned by Heronemus. Perhaps the most serious problem is the lack of experience in structural design and equipment maintenance that the giant structures would require. The experience at Grandpa's Knob illustrates the mechanical problems that can develop in large wind turbines. Collapse of part of a

A serious problem has been to design a system that can generate electricity reliably despite uncertain winds.

Great Plains tower or breakage of a spinning blade could bring instant death to persons working in nearby farm fields.

A further difficulty is that commitment to the system also means commitment to a new method of energy storage since the link with hydroelectric power as in Vermont would be impractical. One approach suggested by Heronemus would be to use direct current electricity generated by wind machines to power electrolyzers in which electricity would be used to convert water into its two components, hydrogen and oxygen. The hydrogen would be used on a massive scale for conventional industrial needs, for a fuel in its own right, for hydrogen-oxygen fuel cells that would produce more electricity, or for hydrogenation of coal to produce a synthetic gas to supplement or replace natural gas. This approach would in essence convert wind energy into chemical energy stored in hydrogen (and in oxygen, as discussed later). The hydrogen would be stored or transported by pipeline for use as needed.

This proposal involves many uncertainties. In the first place, use of hydrogen as a fuel is possible, but would involve large-scale changeovers in industrial, domestic, and transportation equipment.[8] Use of hydrogen in fuel cells is attractive, but a good deal of research remains before fuel cell equipment is widely used.[9] Hydrogenating coal to produce a low-sulfur, synthetic gas if feasible, but combustion would still produce nitrogenous pollutants. Finally, linking the unproven wind-power technology to the still formative hydrogen technology poses a hopeless problem for making economic comparisons so crucial to choosing alternative power sources. According to references quoted by Heronemus, fairly recent estimates of the cost of single wind turbines range up to $323 per installed kilowatt of power. By comparison, construction costs for nuclear power plants are on the order of $310 per installed kilowatt, and costs for fossil fuel plants are 20 to 40 percent less than for nuclear plants.[10]

But this is just the start. Mass production of wind turbines is essential to their wide-scale use, and one has to estimate to what degree this would reduce the cost. Based on costs for other production line products, Heronemus suggests that if 20,000 of the 850-foot tall wind machines were produced each year, the cost would drop to $100 per installed kilowatt of power. But since we need 2.74 kilowatts of installed wind power capacity to equal 1 kilowatt of nuclear power plant capacity, as explained earlier, the cost jumps to $275 per kilowatt to allow comparisons with other power generation systems.

Heronemus goes on to analyze the cost according to various uses of hydrogen gas ranging from coal gasification to operation of central fuel-cell electricity generating plants. Depending upon the use to which the hydrogen is to be put, estimated costs per installed kilowatt of wind power would range from $372 to $720, including allowances for associated installations such as water supplies, electrolyzer plants, gas storage facilities, and central fuel plants. The central fuel-cell plant system could provide electricity for public use to be sold at 2 cents per kilowatt-hour; at less, the system would not be economical. For comparison, revenue averaged slightly more than 2 cents per kilowatt-hour produced in the U.S. by privately owned electrical utilities, in 1969, for residential and commercial customers.[11] But 36 percent of total kilowatt-hour sales were not to these customers but rather to industries, for an average revenue of just under one cent per kilowatt-hour. Generally speaking, the more electricity used, as in industry, the more economical the service for the utility, and hence the lower kilowatt-hour price. Wind-power electrical energy at a cost of 2 cents per kilowatt-hour obviously would be competitive with conventionally generated electricity in selected situations only.

Other than this rough outline, there is not space for analysis and critical evaluation of Heronemus' calculations. Furthermore, related considerations need detailed examination in such an evaluation. For example, if only 1,000 wind machines were built each year, instead of 20,000, in order to scale down the scope of the project, the estimated base cost would rise from $100 to $250 per kilowatt of power. Adjustments in costs of associated installations would change, perhaps in different ways. Costs for land acquisition, which Heronemus does not include, would vary according to spacing and location of the turbines. On the other hand, the value of clean fuel such as hydrogen increases with time as stricter air pollution codes put a premium on less polluting fuels. An additional consideration is that the calculations by Heronemus pointedly omit the potential value of oxygen produced by the same electrolytic process that releases hydrogen. Large amounts of oxygen, according to some experts, may have commercial value in various existing industrial processes, including steel production utilizing basic oxygen furnaces.[12]

In summary, large-scale employment of wind power could possibly generate fuel or fuel-cell electricty at rates potentially competitive in selected, but by no means all, power markets. However, capital costs for installation of the equipment – assuming the equipment is technically feasible – would be much larger than for conventional equipment.

Whatever its technological merits, the proposal by Heronemus underscores a potential in wind power that has received only passing attention in our national search for new energy sources, a search that has been taken up primarily by intense efforts to develop nuclear power. The giant wind machine in Grandpa's Knob, though long since dismantled, leaves a legacy of research and potential usefulness that has been neglected. Many other countries less generously endowed with natural resources apparently have been better able to put the problems that developed on Grandpa's Knob into perspective and to proceed with development of improved versions of wind turbines. The Gedser machine in Denmark is perhaps the best example. Faced with dwindling fuel reserves and serious air pollution, the U.S. might do well once again to consider wind power in local situations. □

NOTES

1. Heronemus, William E., "The United States Energy Crisis: Some Proposed Gentle Solutions," presented before a joint meeting of local sections of The American Society of Mechanical Engineers and The Institute of Electrical and Electronics Engineers, West Springfield, Mass., January 12, 1972.

2. **Statistical Abstract of the United States 1971**, U.S. Dept. of Commerce, USGPO, Washington, D.C. 20402, page 497.

3. Putnam, Palmer Cosslett, **Power from the Wind**, D. Van Nostrand Company, Inc., New York, 1948, 224 pages. Discussion of Putnam's work, unless otherwise noted, is based on material in this reference.

4. Kholodovskii, G. Ye., **The Principles of Power Generation**, translated by V. H. Brix, The Macmillan Company, New York City, 1956, p. 94. Grimmer, D. P., and K. Luszczynski, "Lost Power," **Environment**, 14(3): 20, Apr. 1972.

5. **New Sources of Energy, Proceedings of the Conference, Rome, Aug. 21-31, 1961**, United Nations Publications Sales No. 63.I.41 E/CONF 3/8, 7:188, 1964.

6. **New Sources of Energy**, ibid, general.

7. **Science and Technology for Development**, United Nations, New York City, Vol. II; Natural Resources, 1963, p. 169.

8. "Hydrogen Fuel Economy: Wide-ranging Changes," **Chemical and Engineering News**, Jul. 10, 1972, pp. 27-30.

9. Aaronson, Terri, "The Black Box," **Environment**, 13(10):10-18, Dec. 1971.

10. Holdren, John, and Philip Herrera, **Energy**, Sierra Club, San Francisco, 1971, p. 82.

11. **Statistics of Privately Owned Electric Utilities in the United States 1969**, Federal Power Commission, USGPO, Washington, D.C. 20402, Nov. 1970, p. XXXIII.

12. "Hydrogen: Likely Fuel of the Future," **Chemical and Engineering News**, June 26, 1972, pp. 14-16.

FOSSIL-FUELED STEAM-ELECTRIC GENERATION

Introduction

Fossil-fueled steam-electric power plants long have been the mainstay of the electric power industry. They currently account for about 76 percent of total generating capacity and more than 80 percent of total generation. With increased reliance on nuclear power, they are expected to account for only about 44 percent of both capacity and generation by 1990. Nevertheless, the total capacity of these plants is expected to increase from 259,000 megawatts in 1970 to 390,000 megawatts in 1980, and 558,000 megawatts in 1990. The total installed fossil-fueled steam-electric capacity from 1920 to 1970, with projections to 1990, is shown in figure 5.1.

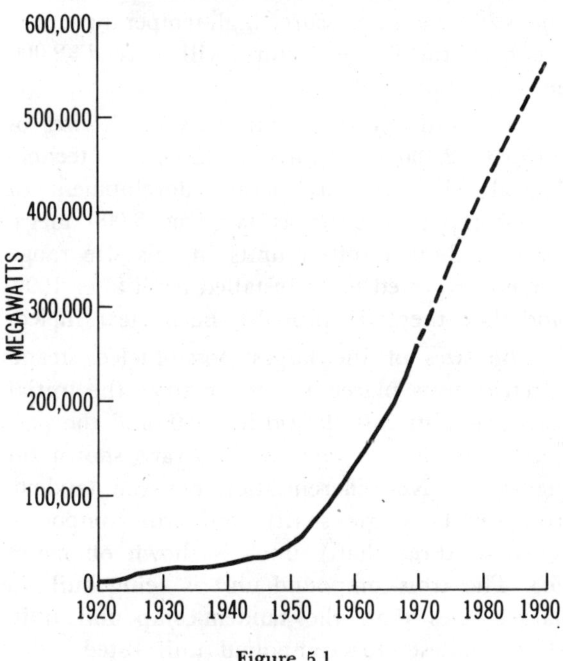

Figure 5.1

The report of the Commission's Technical Advisory Committee on Generation, entitled "The Generation of Electric Power," and published in Part IV of this Survey, includes a comprehensive presentation on the past, present, and probable future position of fossil-fueled steam-electric generation in the total power supply of the contiguous United States. Statistical and other pertinent information on thermal plants are included in the annual publication of the Federal Power Commission entitled "Steam-Electric Plant Construction Cost and Annual Production Expenses."

This chapter summarizes data on fossil-fueled steam-electric plants and discusses those aspects of fossil-fueled steam-electric power supply which relate to its progress and its potential limitations.

Engineering criteria concerned with supplies of cooling water, adequacy of fuel supply, fuel delivery and handling facilities, and proximity of load centers have always been important factors in the selection of power plant sites. More recently, however, environmental factors have gained in influence and now often dominate in the selection of sites.

Environmental problems involved in power plant siting include the discharge of objectionable gases and particulates to the atmosphere, the rejection of waste heat to natural bodies of water, the discharge of chemical and sanitary wastes, and esthetic considerations. These matters are discussed in chapters 10, 11 and 12.

Historically, the improvements in steam electric generating units have been fairly continuous and adequate to meet the needs of the electric utility industry. Although nuclear-fueled and gas-turbine-driven generating units have attracted more attention from the technical press and the public during recent years, there have

Reprinted with permission from *Nat. Power Survey Rep. FP1.2:P87/970/PT.1*, ch. 5, Dec. 1971.

been substantial advances in conventional fossil-fueled steam-electric power supply technology.

The major progress in conventional fossil-fueled steam-electric generation during the past decade has been increases in unit size, reduction in manpower requirements per unit of capacity, automation, better recording of operating data, and improvement in operating reliability of auxiliaries. Reductions in fossil-fueled steam-electric production costs per kilowatt-hour were significant until about the end of 1966, despite continuing inflation. Since that time, however, increases in construction and operating costs have more than offset the gains made through technological improvements.

Figure 5.2—This 1,150-MW unit at TVA's Paradise Steam Plant was the largest generating unit in service in the United States in 1970.

Sizes of Units

In 1930, the largest steam-electric unit in the United States was about 200 megawatts, and the average size of all units was 20 megawatts. Over 95 percent of all units in operation at that time had capacities of 50 megawatts or less. By 1955, when the swing to larger units began to be significant, the largest unit size had increased to about 300 megawatts, and the average size had increased to 35 megawatts. There were then 31 units of 200 megawatts or larger. By 1968, the largest unit in operation was 1,000 megawatts; there were 65 units in the 400 to 1,000 megawatt range; and the average size for all operating units had increased to 66 megawatts. In 1970, the largest unit in service was 1,150 megawatts; three 1,300-megawatt units were under construction; and three additional 1,300-megawatt units were on order. The average size of all units under construction was about 450 megawatts. As the smaller and older units are retired, the average size of units is expected to increase to about 160 megawatts by 1980 and 370 megawatts by 1990.

Capital costs per kilowatt and operation and maintenance costs per unit of energy generated are less for large units than for small ones. This creates incentives to install larger units which will continue until, at some size and some point in time, the incremental savings may be offset by added physical or operational problems. This point is not expected to be reached, particularly for large utilities or those operating in pools, until sometime after 1990. While experience with large units to date has shown some increase in maintenance costs and reduction in unit availability, this is to be expected with prototype units, and the problems should be overcome as later generating units are placed in service.

The maximum size of conventional 3,600 revolutions per minute (r/min), single-shaft generating units expected to be in service by 1980 is approximately 1,500 megawatts. While increases beyond that may occur, it is not expected that the size of high-pressure, high-temperature, single-shaft turbine-generators will exceed 2,000 megawatts by 1990.

Some multiple-shaft units may reach sizes of 2,000 to 2,400 megawatts by 1990, and technological advances may permit development of cross-compound units as large as 3,000 megawatts, although initial units in this size range are not expected to be installed until after 1990, and then they will probably be nuclear fueled.

The sizes of the largest fossil-fueled steam-electric units placed in service since the initial two-megawatt installation in 1900, and the projected maximum sizes to 1990, are shown on figure 5.3. No differentiation between tandem-compound (single shaft) and cross-compound (two or three shaft) units is shown on figure 5.3. The cross-compound unit is being built in larger sizes than the tandem-compound unit. The smallest cross-compound unit listed today in the manufacturer's catalogs is 250 megawatts; the largest is 1,500 megawatts. As a practical matter, very few cross-compound units of less

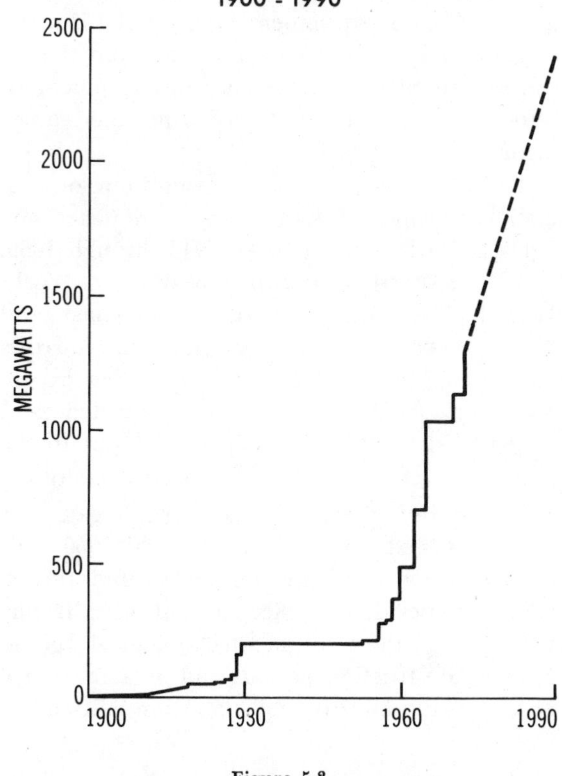

Figure 5.3

Figure 5.4—Dairyland Power Cooperative's Genoa plant south of La Crosse, Wisconsin, has a 350-MW coal-fired unit and a 50-MW boiling water reactor unit.

than 800 megawatts are being ordered because the tandem-compound units are generally more economical in the smaller sizes. The present maximum size of tandem-compound units is approximately 800 megawatts, but larger sizes are expected to accompany improvements in technology.

Sizes of Plants

In 1948 there were only two steam-electric plants in the United States with capacities over 500 megawatts—the 881-megawatt, 12-unit, Hudson plant and the 630-megawatt, 9-unit, Hell Gate plant, both in New York City. Fifteen years later, in 1963, TVA's 9-unit, 1,700-megawatt, Kingston plant was the largest. In 1970, the largest was TVA's 3-unit, 2,558-megawatt, Paradise plant. By 1973, the 4-unit, 3,200-megawatt Monroe plant of Detroit Edison Company, now under construction, will be the largest plant. Some 12 plants will have capacities over 2,000 megawatts by 1975.

The economic advantage of size for individual units also applies to plant sizes. The cost of the components of a plant that are little affected by the number of units, such as office space, shops, docks, and landscaping, can be spread over more capacity. Coal handling equipment, cooling facilities, and other appurtenances can also be operated at less cost per kilowatt-hour at larger installations. The problems of site acquisition and development may be less severe for one large site than for two or more smaller ones. These and other advantages of large installations suggest that plant sizes will continue to increase during the years ahead.

There are factors, however, that tend to limit plant sizes. For example, the amount of land required for a coal-fired plant increases with capacity principally because of requirements for coal storage, ash disposal, and cooling ponds or towers, if required. The amount of land and water required for large plants will preclude the use of many otherwise desirable plant sites. Environmental problems tend to be greater for large plants, and public reactions may limit the amount of capacity that will be permitted at any one location. Plant size is also a factor to be considered from the standpoint of reliability, and there is an important relationship between the maximum size of plants and the capability of system interconnections.

On balance, the advantages of larger plants seem to outweigh the disadvantages. It is expected that plants having 5,000 megawatts of capacity will be in service by 1980, and that by

1990 maximum plant sizes may be as high as 10,000 megawatts.

Average Costs

The average original investment cost of all steam-electric generating capacity in service and operated by Class A and B privately owned electric utilities is shown on figure 5.5 for the years 1950 through 1969. This chart indicates that, despite continuing inflation, the average investment cost per kilowatt of total installed generating capacity decreased during the 1960–1968 period. Price inflation, higher labor costs, increased investment in environmental protection equipment, and a rapidly increasing demand for facilities caused a sharp reversal of this trend, beginning in 1969.

Larger generating units, outdoor type construction where feasible, and unit type construction (one boiler per turbine-generator), have been the principal contributing factors to lower capital costs per kilowatt of capacity. Other factors that have helped to reduce unit costs include: an increase in the number of large single shaft units, reduced weight per megawatt of turbine-generator capacity, standardization of design of major equipment items, prefabrication and assembly of boiler tube panels, and the increased use of heavy construction equipment in building the plant and installing the equipment.

The estimated cost per kilowatt-hour of producing steam-electric energy is depicted by figure 5.6 for each year from 1946 through 1969, and for 1940, the last full year prior to World War II. The "Operation and Maintenance" and "Fuel" expenses are based on recorded costs computed from total reported operating expenses. Unit fixed charges were estimated and added to the production expenses to obtain the estimated total unit cost of electric energy.

The declining trend in costs per kilowatt-hour that prevailed from 1959 through 1966 will probably not recur in the foreseeable future. Cost increases due to inflation and other factors will more than offset technological gains. Longer construction periods and greater control of air quality and thermal pollution will add to future costs.

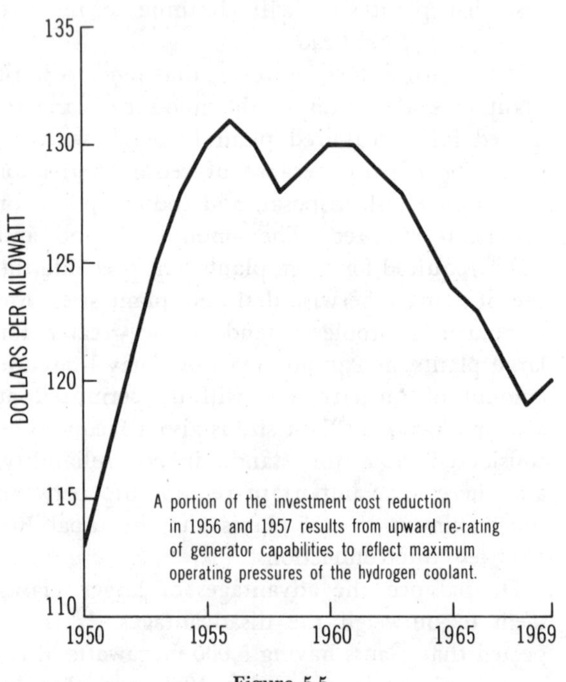

Figure 5.5

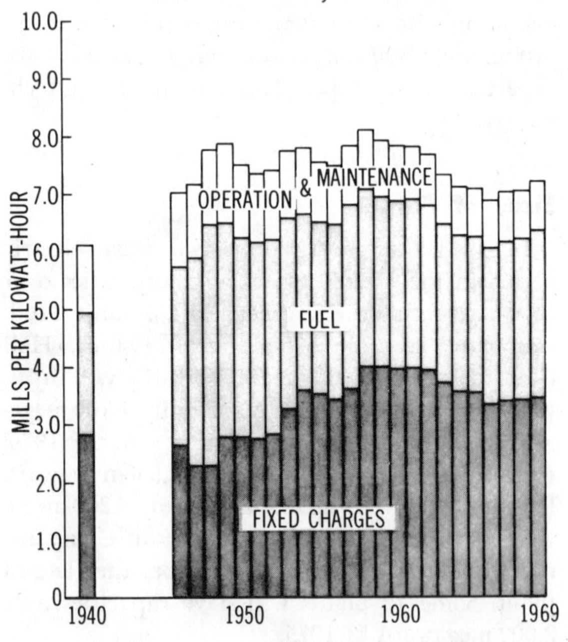

Figure 5.6

Figure 5.7—Unit No 1 in the Asbury plant of The Empire District Electric Company. Coal is transported from mine to plant by conveyor, and cooling water for the 200-megawatt unit is pumped from four deep wells.

MANPOWER REQUIREMENTS FOR OPERATION AND MAINTENANCE OF MODERN COAL-FIRED AND GAS-FIRED STEAM-ELECTRIC PLANTS

(1955 to 1969 Installations)

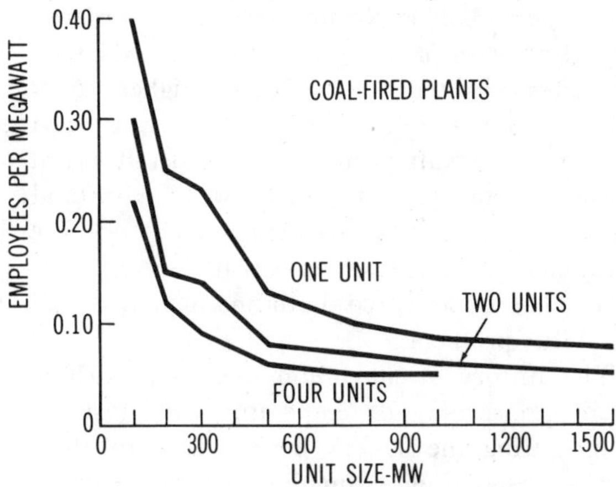

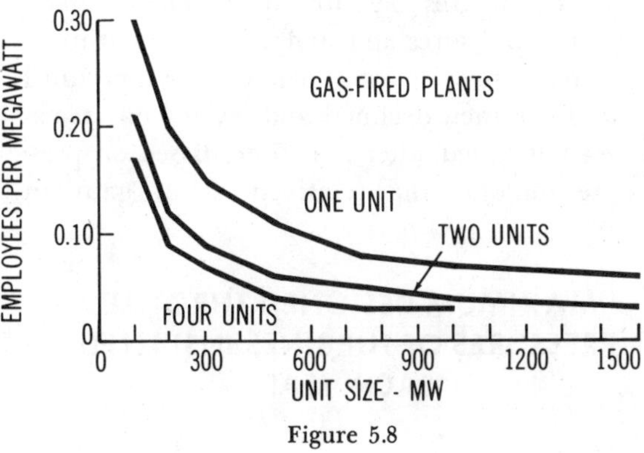

Figure 5.8

Centralized control and increasing automation of plant operations are expected to permit some reduction in the number of plant employees per megawatt of capacity. Other factors tending to reduce manpower requirements are the ever-increasing unit and plant sizes and modernization and mechanization of fuel handling equipment and facilities. Figure 5.8 shows graphically the reduction in manpower requirements per megawatt of capacity as unit size and number of units increase. On the other hand, higher labor costs may negate the manpower savings. By 1990, technical personnel, including those on shift assignment, are likely to require the equivalent of a bachelor's degree to qualify for employment in the highly sophisticated plants that will then be operating, whereas in the past, operating and maintenance personnel have been craftsmen whose skills were developed largely through on-the-job training and experience.

Lower fuel costs during the last decade were due to competition among the suppliers of fossil fuels, adoption of mechanized and strip coal mining, more economical methods of transportation, and improved efficiency of generating plants. The entry of nuclear-fueled generating capacity on an economic basis has introduced an alternative source of power to compete with fossil-fueled electric generation. Despite this, beginning about mid-1969, the cost of fossil fuels for steam-electric power plants has increased substantially. This has been due in part to coal shortages resulting from underestimated demand and increased foreign exports of U.S. coal, and delay in completion of nuclear units. Other important factors have been the enactment of air quality standards, requiring the use of more expensive low-sulfur fuels, higher transportation costs associated with obtaining such fuels, dislocations of transportation equipment sometimes necessitating costly alternative transportation arrangements, and increased mining costs attributable to the Federal Mine Health and Safety Law which became effective April 1, 1970.

Steam Conditions

Since 1964 turbine throttle steam conditions for large new generating units have been relatively constant, with pressures up to 3,500 psi and Fahrenheit temperatures of 1,000° initial and 1,000° reheat. Although some large 2,400 and 2,600 psi units are on order, the 3,500 psi units predominate in sizes of 500 megawatts and larger. A few double reheat generating units have been built with reheat temperatures of 1,025° and 1,050°F. No units have been ordered since 1964 with pressures above 3,500 psi or temperatures (initial or reheat) higher then 1,050°F. These limits on steam pressure and temperature result primarily from metallurgical considerations. Metals that will withstand higher temperatures are expensive and the equipment designed for service at those temperatures has experienced both maintenance and reliability problems.

The history of maximum operating turbine steam pressures and temperatures on installations during the 69-year period, 1900 to 1968, and corresponding estimates through 1990 are shown on figures 5.9 and 5.10. These charts show that pressures and temperatures continued to increase over a long period of years until about 1960, then declined and leveled off, as experience dictated, after that time. Based on present technology, the relatively small gain in

MAXIMUM DESIGNED THROTTLE PRESSURES OF TURBINES INSTALLED EACH YEAR

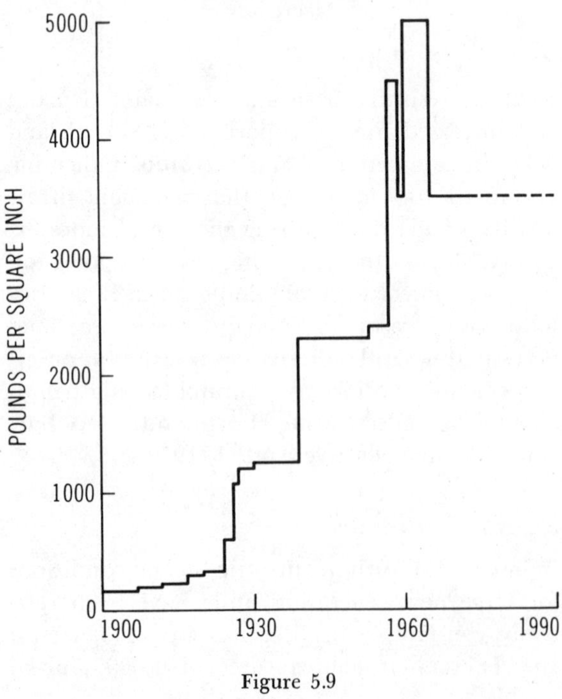

Figure 5.9

MAXIMUM DESIGNED THROTTLE TEMPERATURES OF TURBINES INSTALLED EACH YEAR

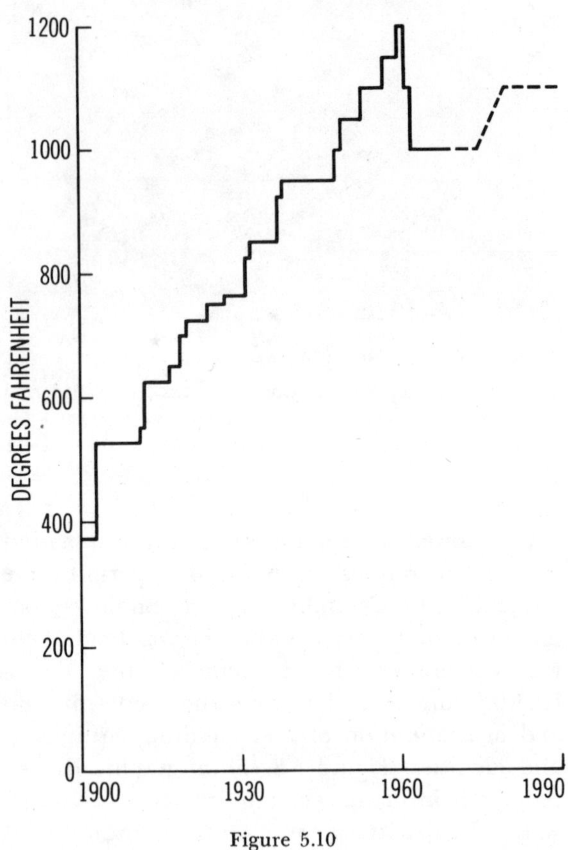

Figure 5.10

efficiency in changing from 3,500 to 5,000 or even 6,000 psi and from 1,050°F. to 1,200°F. is not sufficiently large to justify the substantial additional capital and annual expenditures necessary to install and maintain equipment made of metals required for these steam conditions. Improved metals should be available during the next decade and they may induce some increase in pressures and temperatures. Major increases, however, are not anticipated.

Heat Rates

Heat rate, as the term is used in the electric power industry, is the amount of heat input in British thermal units (Btu) required to produce a net output of one kilowatt-hour of electrical energy. Until improved alloys are developed to permit higher steam pressures and temperatures, the rate of decline in average heat rates will not be as rapid as in the past. Recent improvements in both turbines and boilers, apart from steam conditions, have contributed to somewhat lower heat rates, and increases in unit size have also provided some advances, but better metals and higher throttle temperatures offer the best potential for further major improvements.

The national record annual operating heat rate for a turbine-generator unit is 8,534 Btu per kilowatt-hour. It was established in 1962 by the Philadelphia Electric Company's 350-megawatt Eddystone No. 1 — a 5,000 psi, 1,200°/1,050°/1,050°F., double reheat unit. This is the first and only 5,000 psi, 1,200°F. unit in service in the United States. In its ten years of operation, its availability has been somewhat less than expected and the annual heat rate has varied from the low of 8,534 Btu in 1962 to a high of 8,874 Btu in 1967. The best annual heat rate for a 3,500 psi, double reheat unit was achieved by Eddystone No. 2, the same size as No. 1, also in 1962—8,633 Btu per kilowatt-hour. This unit was placed in service in late 1960. The Eddystone record was bettered in 1969, on the basis of a partial year of its initial operation, by Duke Power Company's Marshall No. 3 unit with a heat rate of 8,617 Btu/kilowatt-hour for about 4,600 operating hours. The unit continued high performance operation in 1970 with a heat rate of 8,638 Btu/kilowatt-hour for the full year. Most of the lowest annual heat rates have been achieved by cross-compound units, but the Marshall No. 3 unit is tandem-compound, with 1000°/1000°/1000°F double reheat.

Figure 5.11 shows that heat rates for the most efficient fossil-fueled steam-electric stations decreased markedly until 1950, and then leveled off. Projected heat rates are shown through 1990. National average annual heat rates will continue to decline as new generating capacity with low heat rates is installed and older, less efficient capacity is retired. The decline may be slowed, however, by the types of fuels which must be burned to meet air quality standards.

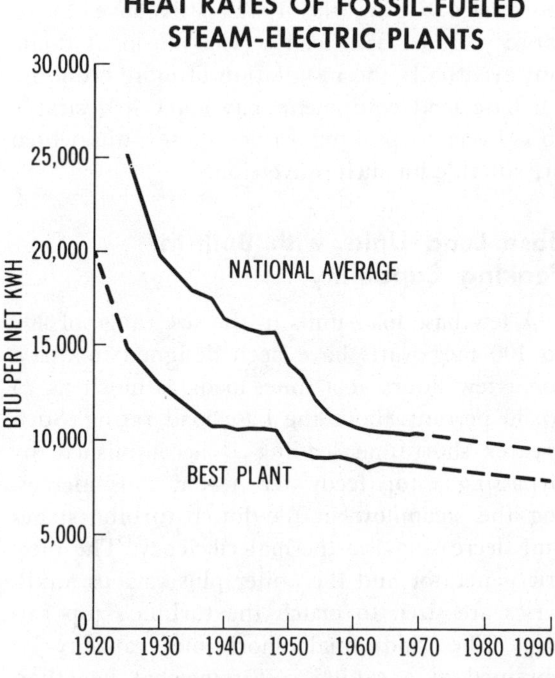

Figure 5.11

Figure 5.12—The 1,606-megawatt Haynes plant of Los Angeles Department of Water and Power, at Seal Beach, California, has six units and uses gas and oil for fuel. Pacific Ocean water is used for condenser cooling.

Base Load Units

All of the high-pressure, high-temperature, fossil-fueled steam-electric generating units, 500 megawatts and larger, have been designed as "base load" units and built for continuous operation at or near full load. Daily or frequent "stops" and "starts" are not consistent with their design and construction and so-called "cycling" or part-time variable generation was not originally comtemplated for these units. However, by the time units having lower incremental production costs become available for base load operation, it is believed that the earlier "base load" units can be adapted and used as "intermediate" peaking units. The units placed in service during the 1960's still have 15 or more years of base load service ahead of them, but eventually the installation of more economical base load equipment may make it desirable to convert to peaking service those units which are suitable for such conversion.

Base Load Units with Built-In Peaking Capability

A few base load units in the size range of 200 to 400 megawatts have been designed to carry, for a few hours at a time, loads as much as 25 to 30 percent above the base load rating. Such greater short-time loading is accomplished by bypassing a top feedwater heater, thus increasing the steamflow in the lower turbine stages and decreasing the thermal efficiency. The electric generator and the boiler plus various auxiliaries are sized to match the turbine's top rating. The additional short-time capacity is obtained at a capital cost somewhat less than that for a conventional steam peaking unit. Operation at loadings in excess of the base load rating results in a substantial increase in the overall heat rate of a unit. A 350-megawatt unit of this type has been operating since 1964 in the North Lake plant of Dallas Power and Light Company. Other units of this type have been constructed and additional units probably will be built in isolated cases, but they are not expected to provide a significant portion of total requirements for peaking capacity.

Peaking Units

Steam-electric peaking units, sometimes referred to as mid-range peaking units, are designed for minimum capital cost and to operate at low capacity factor. They are oil- or gas-fired, with a minimum of duplicate auxiliaries, and operate at relatively low pressures, temperatures, and efficiencies. They are capable of quick start-ups and stops and variable loading, without jeopardizing the integrity of the facilities. Such units are economical because low capital costs and low annual fixed charges offset low efficiency and operation at low capacity factors. The units can, however, be operated for extended periods, if needed, to meet emergency situations.

The first of such fossil-fueled steam-electric peaking units, a 100-megawatt, 1,450 psi, 1000°F., non-reheat, gas-fired unit, was installed in the Arsenal Hill plant of Southwestern Electric Power Company, at Shreveport, Louisiana, in 1960. Two earlier low capital cost fossil-fueled steam-electric plants—the 69-megawatt, single-unit Bird plant of the Montana Power Company (1952), and the 313-megawatt, two-unit Martins Creek plant of Pennsylvania Power and Light Company (1954)—were generally classified as hydro standby; they were not straight peaking installations. The Martins Creek plant was later modified for base load operation.

With increasing loads and the accompanying need for additional peaking capacity, at least 27 peaking units of this general type were on order or under construction at the end of 1970. All are either oil- or gas-fired, because the added costs of coal and ash handling facilities for peaking units are not justified by the small fuel cost saving that might be realized by using coal Eight of the 27 units are in the 250- to 350-megawatt class, fifteen in the 400-megawatt class, and four in the 600-megawatt class. Most of the units are designed for steam conditions of 1,800 psi and 950°/950°F.

The size of peaking units being installed is generally from five to ten percent of system capacity. If this pattern is to prevail for the larger systems of the future, peaking units of 1,000 megawatts capacity, or more, could be utilized. Existing technology is adequate to permit design of such units, but new applications of the technology will be necessary. For example, the problems associated with the need for steam temperature control suggest the possibility of multiple furnace installations with separately fired control surfaces. These and other problems are solvable, and it is expected that steam peaking units in all size ranges will gain in prominence, particularly in areas where hydroelectric potentials are limited.

Some foreign manufacturers have built peaking units using variable pressure boilers which are reported to have rendered satisfactory cyclical service.

Changing Operational Patterns for Older Units

The 25- to 100-megawatt generating units in the 650 to 1,250 psi, 650° to 900°F., pressure-

Figure 5.13—Turbo-generator in the Allegheny Power System's Fort Martin mine-mouth plant.

temperature ranges installed from 1930 to 1950 have been sturdy, dependable, and long-lived. With larger generating units coming into service, many of these older units, especially those of 50 megawatts and larger, are being modified to serve the upper part of the daily load curve. These older units do not have the quick startup characteristics of gas turbine peaking units and they were not designed for automatic operation. Nevertheless, they are available and can operate a few hours per day for peaking and at other times for "hot" reserve, ready to take on load when required.

In several cases where such older units are assigned to peak load service, coal-fired boilers have been converted to oil or oil-gas firing to reduce operation and maintenance labor costs. Maintenance costs are an important factor in determining the extent and type of use to be made of such older units.

During the last half of the 1970's and into the 1980's, the changeover from many hours of continuous generation to a few hours of use with daily or weekly starts and stops will probably be extended to include the 125- to 200-megawatt units installed from 1950 to 1960, or later. This group will include the reheat units with pressures of 1,450 to 2,400 psi and temperatures of 950° to 1,000°F. To make them useful as well as environmentally suitable for such changed operations, modifications and adjustments will need to be made in the units.

Maintenance

Power plant equipment represents a large investment and is expected to have a very high degree of service availability during its 30 to 40 year life. Such performance is not attainable without periodic inspections and continuing maintenance.

Scheduled equipment inspections identify worn, faulty, or otherwise defective equipment that needs overhauling, repairing, or replacing. To make the best and most economic use of sources of power supply, and to reduce exposure to excessive forced outages, scheduled plant maintenance programs are coordinated with maintenance schedules of all important plants on an interconnected system and, frequently, in adjoining systems.

At newer installations, physical inspections of power plant equipment are supplemented by automatic scanning, data logging, and alarm devices which identify actual or potential trouble points. Preventive maintenance resulting from such surveillance eliminates some failures and forced outages. As more experience is gained with scanning and logging equipment, it is expected to provide an improved basis for establishing the optimum period between physical inspections.

Use of contract maintenance for major items of equipment, particularly boilers and turbine-generators, is increasing. Most manufacturers have expert maintenance crews at strategic locations throughout the country. These crews are available to inspect and service equipment on a regularly scheduled basis. There are also independent maintenance firms which specialize in such work. An example of an arrangement for such maintenance is the agreement between the "Keystone Plant Owners Group" and the boiler supplier for the two 820-megawatt Keystone units in Pennsylvania. The agreement provides for an annual boiler inspection and for any necessary repairs and overhauling. The maintenance work is scheduled using the "Critical Path" method—a method which provides a systematic means of scheduling items of work so that the completed task is accomplished in a minimum time.

Present experience with large coal-burning steam-electric power plants indicates that 55 to 65 percent of total maintenance expense concerns the boiler plant. Annual maintenance expense on turbines and generators ranges from 15 to 25 percent of total plant maintenance costs. Some increased maintenance and repair work is associated with very high temperatures and their effects on metals and alloys used in the boiler and the turbine. Because of corrosion and erosion from sulfur compounds and fly ash, coal-fired boilers usually require more maintenance than oil- or gas-fired boilers. In addition, maintenance is required on coal and ash handling facilities, both of which are subjected to rough, heavy duty. There are sometimes added boiler maintenance problems at plants which are converted to burn low sulfur coal, because the boiler systems were designed to work best with coal having other physical and chemical properties.

The problems of scheduling maintenance become more pronounced as larger units are installed, areas of coordinated operation are expanded, secondary system peaks increase, and system load factors improve. Timely and thorough maintenance is essential for system reliability. Some of the difficulties encountered in meeting loads during recent years were attributable, in part, to the fact that planned maintenance was postponed because of capacity deficiencies due to delays in service dates of new equipment, or perhaps in some cases incompletely carried out because of the need to restore units to service.

Performance of New Units

The performance of a generating unit may be rated in terms of its availability for service and its efficiency of operation. The availability for service is usually measured by scheduled and unscheduled outages. The availability of new, larger units with higher operating steam pressures and temperatures has, in general, not been as good as the availability of earlier units. The first of the units in this general category was placed in operation in 1957. Because of the relatively short in-service records of these large, modern fossil-fueled boiler and matching turbine-generator units, with highly complicated control systems and auxiliary equipment, neither their maintenance requirements nor their scheduled and unscheduled outage rates have been firmly established. The so-called "shakedown" period for some of these very large generating units has within recent times been as long as three to five years. It is during this period that initial operating problems are resolved to make the units dependable performers. Current judgment is that the average service life of these very large units will be about 30 years, including the "shakedown" period.

While experience data are being accumulated, manufacturers, utilities, and engineering consulting organizations are cooperating in studies and analyses of forced outages to determine causes, probability of recurrence, and essential steps to improve the availability and reliability of large units. Particular attention is being given to protection against unnecessary tripouts of large units occasioned by transient system instabilities. Some malfunction or untenable condition in a boiler is the most frequent cause of forced outages of generating units. Problems in the turbine-generator are the next most frequent cause. These may include an overheated bearing, water induction into the turbine, low oil pressure, control malfunctions or, in an extreme case, failure of major components.

Trends in Boilers

The size or rating of boilers is in terms of thousands of pounds of steam supplied per hour. The increase in boiler capacity was rather slow until 1955, as indicated by figure 5.14 which shows maximum boiler capacities for 1905 to 1968, with projections to 1990. Prior to 1950, individual boilers were kept small, in large part because boiler outages were rather numerous, so that it was common design practice to provide multiple boilers and steam header systems to supply a turbine-generator. Advances in metal technology since 1950, with associated lower costs of larger units, have made it economical and reliable to have one boiler per turbine-generator.

It is anticipated that the steam output per boiler will continue to increase. Very large boilers, for turbine-generator units of about 1,200 megawatts output and larger, in some cases will be of the double or twin furnace design. Further design and development is indicated on single furnace enclosures, length of soot blowers (gas-side cleaning equipment), forced and induced draft fans, control systems, and pressurized furnaces before a single boiler is used with a turbine-generator larger than 1,300 megawatts.

With the increase in the steamflow from boilers and the raising of steam conditions to the

MAXIMUM CAPACITY OF BOILERS INSTALLED EACH YEAR

1905-1970 Actual
1971-1990 Projected

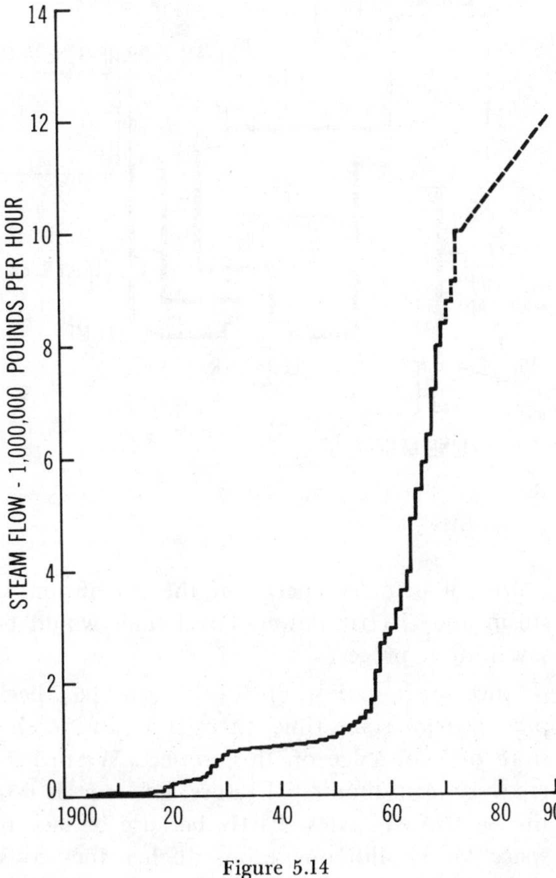

Figure 5.14

3,500 psi and the 1,000°/1,050°F. level, maintenance problems increased and special efforts were needed to maintain service availability and reliability. With high temperatures and high velocity of fuel-air mixtures in the larger boilers, there is an increase in gas-side corrosion and erosion. Sulfur oxides present in the products of combustion produce additional problems of corrosion, and the damaging effects of the corrosive agents are compounded by the abrasive action of fly ash particles. Large single-furnace units experience expansion problems which have resulted in extensive stress cracking of waterwall tubing at buckstays, bustle and windbox attachments, and burner attachments.

Efforts to attain superheat and reheat outlet temperatures significantly in excess of 1,000°F. have met with a difficulty labeled "metal wastage." In coal-fired boilers, this metal wastage form of corrosion is due to a liquid-phase alkali salt which forms at gas temperatures about 1,600°F., and collects on the surface of the boiler tubes under an insulating layer of ash. The metal temperature range in which these salts cause the most corrosion is from 1,100°F. to 1,300°F.

In the newer boilers using once-through flow and supercritical pressures, it is essential that non-corrosive feedwater be used to reduce scale formation and tube failure. In the older boilers, the drum provided a good place to "blowdown" the dissolved solids in the boiler water when the concentrations became too high, but there is no equivalent place in the new once-through systems to perform this function.

Additional design and development work is needed for further increases in size of the boiler, turbine-generator, and principal auxiliaries. Large capacity boilers of today contain over 300 miles of tubing and about 50,000 welds and, because of the pressures and temperatures to which they are subjected, great care must be taken to provide high quality materials and workmanship. Major improvements in metals and metal handling techniques will be required before substantial increases in pressures and temperatures become commonplace.

There are a few installations where two approximately half-size boilers per turbine are being provided in an effort to avoid the very large single boilers which otherwise would be required.

Figure 5.15—Long Island Lighting Company's Northport oil-fired plant has a total generating capacity of 774.2 megawatts in two units.

Prospects for Combined Cycles

In the interest of obtaining electric power and energy at lowest cost, some attention has been given to combined cycles, and they are expected to receive more research and development in the future. An example of a combined cycle is the gas turbine-steam turbine application described below.

Gas Turbine-Steam Turbine

The gas turbine-steam turbine cycle has been proved practical by the 243-megawatt Horseshoe Lake plant installed in 1963 by Oklahoma Gas and Electric Company, by the subsequent 133-megawatt San Angelo installation by West Texas Utilities Company, and by a few other smaller units. The functioning of this cycle is shown on figure 5.17, which is a simplified block diagram of the Horseshoe Lake generating unit. Figure 8.8 shows the San Angelo installation.

Few units of this type are being scheduled for service, probably because of the lack of assurance of a future supply of natural gas or the high cost of other fuels suitable for use in the gas turbine. This type of combined cycle is not likely to be used in large units unless an acceptable coal-burning gas turbine is developed, and prospects for that are not promising.

Combined Steam or Power/Incineration Plants

The electric power industry can play an occasional role in providing a community service through the utilization of municipal waste as a

Figure 5.16—Duke Power Company plans to utilize waste heat from gas turbines, shown at left, to permit retirement of several coal-fired boilers in its Riverbend Plant near Charlotte.

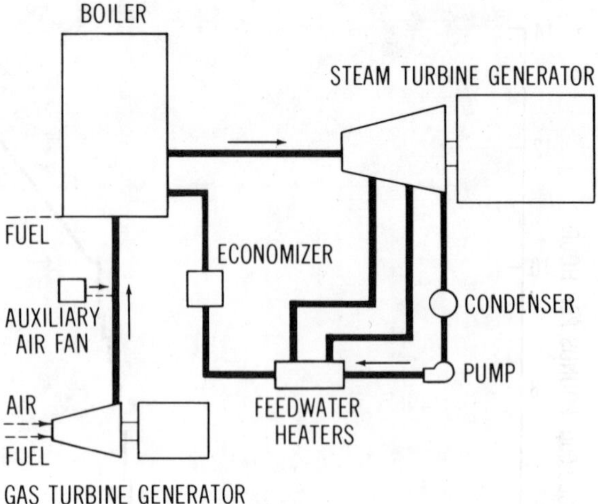

Figure 5.17

source of primary energy in the generation of steam and electric power. Fossil fuels would be saved in the process.

Since incineration of solid waste has been practiced for some time, there is a considerable store of knowledge on this subject. Waste heat recovery from municipal refuse, however, is rare in the United States. Partly because of lack of space for landfill operations, higher fuel costs, and generally lower wages in most European countries, the economics of recovering heat from the burning of wastes are much more favorable and the practice more advanced there than here. Rapid growth in per capita production of refuse in the United States, coupled with such other factors as the growing scarcity and rising costs of suitable landfill areas near urban centers, increasingly stricter environmental control regulations prohibiting open burning and dumping, and the rising costs of fossil fuels, are all making the recovery of waste heat from refuse incineration more economically attractive.

The cities of Chicago, Illinois and Harrisburg, Pennsylvania have started construction of steam-generating incinerators. The Chicago incinerator, consisting of four units, each with a capacity of 400 tons of refuse per day, will have four boilers. Each boiler is expected to generate a steady 110,000 lb/hr of steam at 275 psig and 414°F., using refuse with an average heating value of 5,000 Btu/lb. The furnace sidewalls consist of vertical water tubes with welded fins to assure tight furnace construction. The incinerator will have electrostatic precipitators. The Harrisburg incinerator, which is of the same type, is scheduled to dispose of 720 tons of refuse per day.

A smaller municipal waste-conversion station was built earlier in Braintree, Massachusetts. Here, two boiler-incinerator units complete with water-cooled furnaces, convection banks, and economizers produce a total of 60,000 pounds of steam per hour from 240 tons per day of residential, commercial, and industrial refuse. Plans for the incinerator included electrostatic precipitators for air pollution control.

Potential users of municipal refuse for producing steam and electric power in the United States believe the technology and economics of waste heat utilization could be further improved through research and development.

Recognizing the need to resolve some of the problems associated with steam generation in municipal refuse incinerators, the Solid Waste Management Office (SWMO) of the Environmental Protection Agency is currently sponsoring two projects with somewhat different approaches to waste heat recovery.

In Menlo Park, California the SWMO is testing on a $1/10$ scale the feeding, shredding, combustion, and gas cleaning components of the Combustion Power Unit-400 (or CPU-400). In full scale, the CPU-400 is a gas turbine incinerator designed to burn refuse and generate electric power; it consumes 400 tons of refuse per day and generates 15,000 kilowatts of electric power. In the CPU-400 the refuse is shredded, dried, and burned in a high-pressure fluid bed reactor operating at 1,650°F. The particulate matter is removed from the high-pressure hot gas and the clean hot gas is expanded through a turbine to drive a compressor and electric generator. Heat energy remaining in the gas downstream from the turbine may be used to generate steam or for other uses. The CPU-400 is capable of furnishing ten percent of the electric power requirements of the community supplying the refuse.

In a St. Louis, Missouri experiment, which is to begin on a full scale at Union Electric Company's Meramec Unit No. 1 (125 MW) early in 1972, 10 to 20 percent of the coal will be replaced with prepared municipal refuse. It is assumed that a relatively small percentage of properly prepared refuse mixed with the regular boiler fuel would present few, if any, problems in the operation of the boilers.

If either or both of these experiments prove to be less troublesome and more economical than the European efforts, they may be adopted in many areas of the country resulting in less costly ways to dispose of city refuse, reducing air pollution, and conserving natural resources. In time, 5 to 10 percent of the Nation's electric power requirements could be produced from municipal refuse.

Automation of Steam-Electric Plants

Large, modern high-pressure, high-temperature generating units have many components requiring continuous monitoring and control, and although many problems must be solved before full automation of these plants will be achieved, there has been progress in automation since the first elementary control systems were developed in the early part of the century. Major steps in the evolution have been the advent of the central control room, the introduction of scanning, alarming, and logging instruments and recorders, and the installation of data-gathering systems. Centralization, a major step in automation, consists of bringing the controls of two or more steam-electric generating units into one central control room, in contrast to the remote monitoring and separate control stations used for manual supervision of the boilers, turbines, and auxiliaries during the first three or four decades of this century.

Plants in operation today vary from those under complete manual supervision to a few that are completely automated. Full automation includes scanning, alarming, logging, and recording, plus computer control of plant starts, running periods, and stops, either scheduled or emergency. All modern plants are automated to some degree but full automation is only rarely employed.

The gas-fired Little Gypsy plant of Louisiana Power & Light Company, a 1961 installation, is an early example of the application of fully automated controls to fossil-fueled generating equipment. Significantly more difficult problems are involved in complete automation of coal-fired units and some of them still have not been adequately solved.

Economies gained from computer control vary with plant designs, number of units per plant, and operating practices. The degree of automation that will yield maximum economies is that which optimizes: (1) Reduction in overall maintenance costs; (2) detection of equipment problems; (3) reduction of manpower requirements for operation; (4) reduction in plant outage time; (5) fuel economy; and (6) plant reliability, taking into consideration the cost of the automation system.

Successful operation of highly automated power plants requires competent and trained personnel for instrument and computer maintenance. Reductions in total operating personnel are offset partially by increases in higher paid maintenance personnel.

A potential benefit of complete automation could be a reduction in plant investment, because it would lead to better knowledge of equipment requirements and elimination of "over designed" portions of future generating units. Reduction of design margins might reduce equipment costs by more than the cost of an automation system.

High-Sulfur Coal for Generating Electricity

James T. Dunham, Carl Rampacek, T. A. Henrie

The United States has an abundance of coal. Coal reserves economically recoverable by today's mining technology are estimated at 200 billion tons (*1*), and total domestic coal resources are of the order of 3 trillion tons, or enough to meet a large part of our energy needs for centuries (*2*). We are experiencing an energy shortage in the 1970's, despite such vast amounts of coal, because we have become overdependent on natural gas and oil to supply some of our increasing energy needs, among them that for electrical power.

Electricity provides about 25 percent of our total energy needs. According to a Department of the Interior study (*3*), per capita use of electricity increased from slightly more than 2000 kilowatt-hours in 1950 to 7800 kwh in 1971, and is projected to reach about 32,000 kwh by the year 2000.

Cheap, convenient low-sulfur oil and natural gas are competing with coal as the preferred fuel for the electric utility market (Table 1). While annual consumption of coal for power plants in the northeastern and east northcentral regions of the United States stayed approximately constant in the 6 years from 1966 to 1971, oil consumption has increased by factors of 3 and 25 in these regions, respectively, and gas consumption has increased by up to a factor of 3 (*4–6*). Continued use of petroleum and natural gas at the present rate will aggravate the serious supply problems for these fuels.

Programs under way to augment our oil and gas supplies and to diversify our energy base (*7*), such as coal gasification, extracting oil from western oil shales, harnessing solar energy, wind, and geothermal steam and brines, will have little impact on electricity generating needs for many years. Similarly, although nuclear reactor power plants are expected to provide up to 25 percent of the demand for electricity by 1985 and up to 50 percent by 2000, these optimistic estimates assume the timely development of the fast breeder reactor program and satisfactory solution of environmental problems in siting and operating nuclear reactors. In the meantime, fossil fuel–fired power plants must supply a large part of our electrical power demands, and only coal is available in the United States in sufficient quantity to provide this energy for the next 25 years.

The authors are at the U.S. Bureau of Mines, Washington D.C. 20240. James T. Dunham is staff metallurgist, Division of Solid Wastes; Carl Rampacek is Assistant Director–Metallurgy; and T. A. Henrie is Deputy Director–Mineral Resources and Environmental Development.

Air Pollution from Coal Combustion

An important factor influencing the change in the pattern of energy use in thermal electrical power generation from coal to oil or gas in recent years has been the limitation on the emission of pollutants to the atmosphere. Air pollution regulations affecting power plants are primarily concerned with three pollutants: particulates, nitrogen oxides, and sulfur oxides. Of these three, sulfur oxides are of the most concern from a regulatory standpoint. Ambient air quality standards (Table 2) and emission standards (Table 3) can generally be met only by burning coal containing 1 percent sulfur or less (*8*, *9*).* In coal, sulfur occurs in both the inorganic and organic forms. Substantial amounts of inorganic sulfur, mostly pyrite, can be removed by mechanical cleaning, but the organic sulfur cannot. Because the sulfur in high-sulfur (3 to 6 percent) coal is often about half pyrite and half organic, mechanical cleaning alone does not reduce the sulfur content to the point that the coal can be burned without exceeding the emission standards for sulfur oxides.

Fuel trends in the heavily populated and industrialized regions reflect the impact of air pollution regulations on the use of coal. Burning low-sulfur oil or natural gas has been one method of controlling sulfur oxide emissions, but when switching to oil or gas has not been feasible, utilities have turned to low-sulfur coal, often at great expense because of high transportation costs. Most of the available low-sulfur coal in the United States is in the West, and much of that in the East is captive and used by the metallurgical industry. Accordingly, utilities in Chicago meet air pollution regulations by using low-sulfur coal mined in Montana and Wyoming and pay as much as $8.50 per ton for unit train rail haulage.

In his report to Congress on the energy situation on 23 January 1974, President Nixon urged postponement of the implementation date for air pollution standards to permit conversion of oil- and gas-fired electric generating plants to the use of coal. On 6 February 1974, a report issued by the Federal Energy Office (*10*) cited ten eastern plants that converted from oil to high-sulfur coal and several other plants that were willing to convert if environmental, technical, transportation, and supply obstacles could be overcome. These actions have freed some oil and gas for other uses. In 1973, use of coal for generating electricity increased slightly at the expense of oil and natural gas (*11*). However, widespread conversions by established utilities or construction of new plants designed to burn high-sulfur coal are unlikely unless utilities can be assured that they can ultimately comply with air quality regulations.

Table 1. Sources of energy for generating electricity in 1972 [from (*5*)].

Source	Electricity generated (percent of total)
Coal	42.2
Natural gas	22.1
Petroleum	16.9
Hydroelectric	15.7
Nuclear	3.1
Geothermal	Negligible

Control of Sulfur Oxides at Power Plants

One of the major deterrents to the unlimited and widespread use of coal for generating electricity in the United States, particularly in the Midwest and East, is the quality of the combustion gases released to the atmosphere. If high-sulfur coal is burned, there are three alternatives for producing gaseous emissions meeting air quality standards: (i) Coal can be converted to a sulfur-free fuel; (ii) coal can be burned directly at a rate and under conditions that generate emissions meeting ambient air quality standards; or (iii) coal can be burned and the sulfur oxide gases removed during combustion or before discharge of flue gases to the atmosphere. In the near future, sulfur-free fuels derived from coal by gasification or liquefaction will, at best, have limited application in electrical power generation. The other alternatives are more likely for immediate and short-range use.

Tall Stacks and Curtailment

Before establishment of air quality standards, it was common practice to burn high-sulfur coal and vent the combustion gases to the atmosphere through tall stacks to disperse the sulfur oxides, nitrogen oxides, and particulates. By reducing the rate of coal burning and venting through the tall stacks, ambient air quality standards can be achieved in many cases, but under no conditions can emission standards be met.

The curtailment technique consists of monitoring concentrations of sulfur dioxide at ground level near the power plant and meteorologically forecasting unfavorable atmospheric conditions that

Table 2. National ambient air quality standards. Primary standards are those which protect public health and secondary standards protect public welfare; ppm, parts per million [from (*8*)].

Standards	Concentration $\mu g/m^3$	Concentration ppm	Description
	Sulfur oxides		
Primary	80	0.03	Annual arithmetic mean
	365	0.14	24-hour maximum*
Secondary	1300	0.5	3-hour maximum*
	Particulates		
Primary	75		Annual geometric mean
	260		24-hour maximum*
Secondary	60		Annual geometric mean
	150		24-hour maximum*
	Nitrogen oxides		
Primary and secondary	100	0.05	Annual arithmetic mean

*Not to be exceeded more than once per year.

Table 3. Emission performance standards for fossil fuel–fired steam generation units with heat input of more than 250 million British thermal units per hour (1 Btu = 1.06×10^3 joules; 1 pound = 0.453 kilogram) [from (*9*)].

Pollutant	Fuel	Maximum emission per 10^6 Btu heat input (pounds per 2-hour average)
Sulfur oxides	Liquid	0.8
	Solid	1.2
Particulates	All	0.1
Nitrogen oxides	Gaseous	0.2
	Liquid	0.3
	Solid	0.7

*See correction on page 250.

might force the gas from the stack to ground level. When adverse conditions are indicated, electrical power generation is curtailed to the degree necessary to maintain the ground level sulfur dioxide content of the air below the ambient air quality limits. This control strategy, called the "closed-loop system" or "intermittent control system," has been used for controlling sulfur dioxide emissions at copper smelters in Tacoma, Washington, and El Paso, Texas. The Tennessee Valley Authority has demonstrated the technique at its Paradise steam plant in west central Kentucky (*12*). The Environmental Protection Agency (EPA) has recognized this method of control as supplemental to emission controls and suitable for some power plants (*13*).

Curtailment is simple and cheap, and can be implemented almost immediately for some degree of sulfur dioxide emission control from generating plants. However, extensive use of the procedure would reduce electricity generating capacity. This method of control should be considered a stopgap measure to permit burning of high-sulfur coal until positive methods for controlling sulfur oxides become available.

Removing Sulfur Dioxide from Flue Gases

Research to develop technology for removing sulfur dioxide from gases generated during coal combustion in electric utility boilers covers a span of 40 years in the United States and abroad. These processes include (i) injecting limestone or lime into the combustion chamber to produce a throwaway product; (ii) high-temperature regenerative systems (that is, those that recycle the absorbent) using solid absorbents to concentrate sulfur dioxide gas for conversion to sulfuric acid or sulfur; (iii) direct catalytic oxidation of the dilute flue gases to sulfur trioxide and then to sulfuric acid; (iv) wet scrubbing of the cooled gases with alkaline solutions or slurries to yield throwaway products; and (v) wet scrubbing with regenerative solutions to produce either liquid sulfur dioxide, sulfuric acid, or elemental sulfur. This technology is still controversial, with opinions varying as to the reliability, operating and capital costs, and acceptability of the end products or byproducts of the various processes.

In a 1970 study of available technology for stack gas cleanup made by the National Academy of Engineering (*14*), it was stated that "contrary to widely held belief, commercially proven technology for control of sulfur oxides from combustion processes does not exist." In May 1972, a federal interagency committee responsible for evaluating state air implementation plans formed an interagency task force to evaluate flue gas desulfurization systems. The task force, designated the Sulfur Oxide Control Technology Assessment Panel (SOCTAP), issued its final report on 15 April 1973 (*15*). Having examined the status of stack gas cleaning in the United States and Japan, the task force concluded that the removal of sulfur oxides from stack gases is technologically feasible in installations of commercial size, and that a large number of the nation's coal-fired steam electric plants can ultimately be fitted with commercially available stack gas cleaning systems. Of many processes considered, four wet-scrubbing systems were rated as sufficiently developed for full-scale commercial application within the next 5 years; these processes were as follows: wet limestone or lime scrubbing, magnesium oxide scrubbing with regeneration, catalytic oxidation, and wet sodium base scrubbing with regeneration. Solid absorbent regenerative systems were eliminated as technically deficient or not far enough advanced for application in the near future; several regenerative wet scrubbing systems still being developed were not considered. The method of dry lime or limestone injection in utility boilers fired with powdered coal, in which sulfur oxides are recovered as dry compounds together with the fly ash, was also rejected because flue gases did not meet emission standards for sulfur oxides and the quantities of lime or limestone required were excessive. Serious operating problems also were encountered, including boiler fouling and degraded performance of electrostatic precipitators. Dry limestone injection into a fluidized-bed boiler might capture enough sulfur dioxide to meet emission standards, but the quantity of limestone required is excessive, about 300 pounds per ton of coal burned (*16*) [1 pound = 0.45 kilogram; 1 ton of coal (always short ton) = 0.9 metric ton]. Fluidized-bed boilers are only in the experimental stage and, because of their large size as compared to conventional boilers, are not likely to be readily accepted by utilities.

The EPA supported the SOCTAP conclusions and evaluations in testimony at public hearing (*17*), and added double-alkali sodium scrubbing systems to the list of commercially viable processes. The citrate process, also studied by EPA (*18*), was later added to the list of promising desulfurization systems.

Utility representatives at the public hearing did not agree with EPA's conclusions and testified about operating difficulties with the scrubbers that are installed. They claimed that reliability of units 100 megawatts or larger has not been demonstrated adequately enough to warrant the conclusion that the control systems are commercially available. Another major concern of the utility representatives was the disposal problem attendant with the throwaway control systems. These systems—lime or limestone wet scrubbing and double-alkali scrubbing—produce calcium sulfate and calcium sulfite, which have no market value and must be disposed of in permanent impoundment areas.

The processes considered most advanced have only been tested in a limited number of large-scale demonstration projects, if any, and the controversy continues as to whether the technology is reliable enough for widespread application to coal-fired utility boilers. Large-scale test programs now under way or being planned (*17*) may, in the next 2 or 3 years, solve the engineering design and operating problems to assure fully reliable sulfur oxide control systems. In the meantime, promising new processes now being developed should be available for installation before the end of this decade. Concerned parties are appraising cautiously the more thoroughly researched and advanced processes with regard to their merits and deficiencies.

Wet Limestone or Lime Scrubbing

The wet limestone and lime absorption processes are the most thoroughly studied of all sulfur dioxide control systems. In these systems (*19*), (i) dry lime or limestone is injected into the boiler and the partly reacted material is removed in a wet scrubber; or (ii) slurries of lime or limestone are reacted with sulfur dioxide in scrubbing towers to form calcium sulfates and sulfites which are collected and impounded. In limestone scrubbing systems, efficiency of sulfur dioxide removal depends on intimate contact between solid and gas phases, and it is necessary to install large scrubbers, recirculate large volumes of slurry, and grind the limestone to extremely fine size (– 200 mesh) to achieve an acceptable degree of sulfur dioxide absorption. In addition, limestone utilization is poor, as much as 350 pounds per ton of high-sulfur coal burned. Absorption is more efficient with hydrated lime slurry than with limestone, but construction and operation of a kiln is required for quicklime production.

Wet limestone or lime scrubbing removes particulate matter as well as sulfur dioxide. Although the systems are designed to recycle the scrubbing fluid, the thickened sludges discharged to the impoundment area contain about 50 percent water and require large settling areas for dewatering and stabilization of the solids. Lime or limestone slurry scrubbers are capable of removing up to 90 percent of the sulfur oxides from a typical flue gas containing 0.2 to 0.3 percent sulfur dioxide.

The reliability of limestone or lime scrubbers remains questionable. One lime scrubber in Japan reportedly has operated with near 100 percent availability for 1½ years; in the United States, a scrubber using carbide sludge (calcium hydroxide) has been in reasonably trouble-free operation for 1000 hours.

Several studies have been published on the estimated costs for installing and operating limestone and lime scrubber systems (*17, 20*). Some data are also available on the costs of actual installations. Estimates for capital cost generally range between $27 and $46 per installed kilowatt of capacity. Annualized costs, those which the consumer can translate into the increase in the cost of electricity, range from 1.1 to 1.2 mill/kwh. The as-produced cost of electricity averages about 9 mill/kwh whereas the price to customers averages about 20 mill/kwh. The lower estimates are for new plants of large size, 1000 Mw, and the higher numbers are for retrofitting existing plants of 200-Mw capacity.

With regard to land and water pollution, the purity of the limestone or lime is of considerable concern. Pure limestone is not readily available, because most contains some magnesium. Inasmuch as magnesium sulfate is water-soluble, the throwaway products generated with limestone or lime containing appreciable quantities of magnesium might present a disposal problem, particularly in areas of moderate or excessive rainfall. The soluble magnesium salts might leach and contaminate water at the surface or underground.

The cost associated with disposal of sludge varies appreciably. For some plants, the lack of a nearby sludge disposal site eliminates throwaway systems from the choices of control processes.

Despite the disposal problems, the limestone and wet lime processes are currently the most popular for U.S. power plants. About 28 utilities have selected one of these processes; in seven plants, the process is approaching operational stage (*17*).

Magnesium Oxide Scrubbing

The magnesium oxide scrubbing system is similar to the wet limestone and lime processes (*15, 21*), but it has not been as extensively tested. Magnesium sulfite and sulfate salts are formed by reacting a magnesium oxide slurry with the sulfur dioxide in the flue gas. The scrubber slurry is processed to separate the fly ash, then thickened, followed by crystallization to obtain magnesium salts. The salts are then calcined with carbon at a temperature of 980° to 1090°C to recover 15 percent sulfur dioxide gas. Regenerated magnesium oxide is recycled to the scrubber system. The sulfur dioxide can be liquefied or converted to sulfur or sulfuric acid. Because of the steps involved—thickening, fly ash separation, magnesium salt crystallization, and thermal decomposition—the regeneration is relatively costly.

Long-term reliability of this process has not been demonstrated. Only two units have been installed. One, an oil-fired boiler, was reported to be available 85 to 90 percent of the time during a 2-month period. Capital costs estimates for the process range from $33 per kilowatt of capacity for a new 1000-Mw plant up to $58 per kilowatt for retrofitting a 200-Mw existing plant. The estimated annualized costs are 1.5 and 3.0 mill/kwh for the same plants if no credit for sale of acid is assumed. Marketing the acid would reduce costs only slightly, perhaps 0.3 mill/kwh (*17*). In certain situations it might be necessary to neutralize the acid with limestone or lime and impound the calcium sulfite and sulfate at additional expense.

Advantages of magnesium oxide scrubbing are that the regeneration of the magnesium oxide need not be performed at the power plant site, and a centrally located regeneration facility could service several plants. The process can remove enough sulfur dioxide from flue gases to meet emission standards, but reliability and operating costs must be verified. Current plant tests should provide this information.

Catalytic Oxidation

This process is a variation of the contact sulfuric acid process applied to the extremely dilute gases discharged by utility plants (*22*). The contact process produces 98 percent sulfuric acid from gas containing 3.5 percent sulfur dioxide or higher. There is no technological limitation, other than cost, in treating more dilute gases, but the gases must be thoroughly cleaned and the plants must be designed to treat large volumes.

In the catalytic oxidation process, flue gas, after thorough cleaning in cyclones and electrostatic precipitators, is passed over a catalyst to convert the sulfur dioxide to sulfur trioxide. This combines with the moisture present to form sulfuric acid of about 80 percent strength. This product has limited market value, and large quantities might pose a disposal problem.

In existing power plants, the clean flue gas from the electrostatic precipitators is not hot enough for catalytic conversion and must be reheated to 455°C. The retrofit version of this process has only been tested in pilot plants, but an acceptance test on a 110-Mw coal-fired boiler reportedly achieved 85 percent removal of sulfur dioxide. In a proposed design for new power plants, heat exchangers and hot electrostatic precipitators eliminate the need to reheat the gas. This design, however, has not been tested on a large scale.

Information is not available regarding system reliability, but performance of the particulate cleaning system will influence the percentage of time that the unit is out of operation. Costs are estimated at $41 to $64 per kilowatt, and annualized costs range from 1.5 to 2.6 mill/kwh (*17*). The need to clean particulate matter from the catalyst bed in the acid unit almost continuously, together with reheating and maintaining the large volumes of reaction gases at proper reaction temperatures for effective catalysis, are problems requiring further study.

Wet Sodium-Base Scrubbing

There are several sodium-base scrubbing systems; in the most advanced process, a sodium sulfite-bisulfite solution is used to absorb the sulfur dioxide and convert the sulfite to bisulfite (*23*). In this system, the flue gas must be cleaned thoroughly to remove particulates and must be cooled to about 55°C for effective absorption of the sulfur dioxide. A portion of the liquor is steam-stripped to recover strong sulfur dioxide and is then evaporated to recover sodium sulfite crystals for recycling. The sulfur dioxide can be used to make sulfuric acid or elemental sulfur. Since some oxidation of sulfite to sulfate occurs in the absorber, it is necessary to bleed off part of the solution and make up losses with caustic. Bleeding also controls buildup of particulate matter in the system. The process is capable of removing 90 percent or more of the sulfur dioxide from dilute gases, and has been installed at chemical plants and on oil-fired boilers. Experience in these plants probably has provided more accurate operating cost data than is available for most other advanced processes.

Process reliability greater than 95 percent for more than 2 years has been reported in one instance, for an oil-fired boiler. The process has not been tested at a coal-burning plant, but a demonstration project is planned to begin in 1975. Capital cost estimates range from $38 to $65 per kilowatt with corresponding annualized costs from 1.4 to 3.0 mill/kwh if no credit for byproduct acid or sulfur is assumed (*17*).

No serious technological limitations in the process are apparent. Any reluctance about widespread adoption probably stems from uncertainty about the amount of sulfite to sulfate oxidation and the high annualized costs.

Double Alkali Scrubbing

Although not as well developed as wet limestone or lime scrubbing, this process has potential because it eliminates scaling problems associated with the limestone and lime systems. The scrubbing liquor is an alkaline solution of sodium or ammonium sulfates and sulfites, and efficiency of sulfur dioxide removal is high (*15*). Loaded scrubber effluent is treated with either limestone or lime to recover a throw-away sludge of calcium sulfates and sulfites and to regenerate the solution, which is returned to the scrubber. Development has largely been focused on the sodium system.

Cost estimates for the sodium double-alkali process as applied to utility power plants are encouraging, with capital investment cost ranging as low as $25 per kilowatt for a new 1000-Mw unit. Retrofitting a 200-Mw unit is estimated to cost $45 per kilowatt. Estimated annualized costs for these plants are 1.1 and 2.1 mill/kwh, respectively (*17*). An EPA evaluation has indicated that the double alkali and citrate processes may be up to 20 percent less costly than processes such as wet limestone or sodium-base scrubbing (*18*).

Double alkali scrubbing has the same disadvantages as other throw-away processes, including the need for adding sodium or ammonium salt. However, because of its high efficiency and freedom from scaling in the scrubbing unit, the process is receiving increased attention.

Citrate System

The citrate process is one of the more attractive systems that has emerged in the past several years for flue gas desulfurization (*7, 24*). Developed by the Bureau of Mines to remove sulfur dioxide from nonferrous smelter stack gases, the process has the advantage that elemental sulfur is recovered without the need for intermediate sulfur dioxide regeneration. The system, which is considered among the least costly of the advanced processes (*18*), comprises (i) washing the flue gas to remove particulates and sulfur trioxide, and to cool the gas below 66°C; (ii) absorption of sulfur dioxide in a buffered sodium citrate–citric acid solution in a packed tower; (iii) reaction of the loaded solution with hydrogen sulfide in a closed vessel to form elemental sulfur; and (iv) separation of sulfur from the regenerated solution by oil flotation followed by melting. Hydrogen sulfide for the sulfur precipitation step is generated by reacting part of the recovered sulfur with natural gas and steam.

Recently the bureau began testing the process in a pilot plant with capacity of 1000 standard cubic feet per minute (scfm) at the Bunker Hill lead smelter, Kellogg, Idaho. More than 95 percent removal of sulfur dioxide has been achieved without difficulty from a gas stream containing 0.5 percent sulfur dioxide.

Since June 1973, the process has been tested in a 2000-scfm demonstration unit at a coal-fired steam generating plant in Terre Haute, Indiana (*25*). Tests on gas containing 0.27 percent sulfur dioxide, generated by burning coal containing 3 percent sulfur, have largely confirmed Bureau of Mines findings. Although the citrate process has been proposed for producing elemental sulfur, it also is possible to recover sulfur dioxide for conversion to acid by incorporating a steam-stripping step.

Estimated capital cost of a citrate process desulfurization unit for a 1000-Mw plant burning coal containing 3 percent sulfur is $31 million. Annualized costs would be 1.4 mill/kwh, if no credit for the 214 long tons of sulfur produced daily is assumed.

Summary

We must expand the use of coal for electricity generation as rapidly as possible to help alleviate the immediate oil and natural gas shortage, which threatens to become more acute unless the pattern of energy use is changed.

It is not likely, nor is it proposed, that coal should completely replace oil or gas in power generation; geographic location of plants and ready availability of high- or low-sulfur coal will to some extent dictate the choice of fuel. However, replacing 50 percent of the oil and gas now used in power generation would release more than 200 million barrels of oil and 1.9 trillion cubic feet of natural gas (1 barrel of oil = 0.16 m^3; 1 cubic foot = 2.8×10^{-2} m^3) annually for other uses such as home and commercial heating, transportation, chemical feedstock, and selected industrial and manufacturing uses. Even more important, use of coal instead of oil or gas in new fossil fuel–fired electrical generating plants would go far toward conserving natural gas resources and holding the line on increased petroleum imports.

In recent years, U.S. pollution regulations restricting sulfur oxide emissions from power plants have been one of the major deterrents to the use of the high-rank, high-sulfur coals of the Midwest and East. Reliable flue gas desulfurization processes that permit burning of these coals without adverse environmental effects are approaching full development and should encourage wider use of coal in electricity generation for the next 25 years. Estimates indicate that more than 40 sulfur dioxide scrubbing units will be installed on power plants totaling about 20,000-Mw capacity by late 1976 (*15*). The cost of these units will approach $750 million. Although this is not a significant amount of our coal-fired generating capacity, these installations should give impetus to construction of more and larger ones by 1980 and the next decade; this would refute the tenet that wide use of coal and a clean environment are mutually exclusive. As the choice of proved scrubbing technology broadens, no single process will dominate the market. Individual utilities, in addition to considering the economics, will be faced with making choices on the basis of the type of coal burned; water, land, and air pollution regulations; and the marketability of the end products.

The cost of flue gas desulfurization will be high, ranging from 1.2 to 3.2 mill/kwh. The average increase in electricity cost to consumers is expected to be about 3 to 6 percent, and in some instances as much as 15 percent. However, the added burden may not be as high as that of dependence on foreign oil, both in terms of price and reliability of supply. Combustion of high-sulfur coal followed by stack gas cleanup appears to be the cheapest alternative for meeting our electricity needs in the next few decades.

References and Notes

1. *U.S. Geol. Surv. Bull No. 1136* (1961); *ibid., No. 1275* (1969); *U.S. Bur. Mines Inform. Cir. No. 8531* (1971); Geological Survey, unpublished data; Bureau of Mines, unpublished data.
2. P. Averitt, *U.S. Geol. Surv. Prof. Pap. No. 820* (1973), p. 133.
3. W. G. Dupree, Jr., and J. A. West, *United States Energy Through the Year 2000* (Department of the Interior, Washington, D.C., 1972).
4. *Steam Electric Plant Factors/1967 Edition* (National Coal Association, Washington, D.C., ed. 17, 1967).
5. Data available from the Federal Power Commission and Bureau of Mines, *Bituminous Coal Data* (National Coal Association, Washington, D.C., 1973), pp. 108–109.
6. *Steam-Electric Plant Factors/1972 Edition*, (National Coal Association, Washington, D.C., ed. 22, 1972).
7. E. F. Osborn, *Science* **183**, 477 (1974).
8. *Fed. Reg.* **36**, 8185 (30 April 1971); *ibid.* **38**, 25678 (14 September 1973).
9. *Ibid.* **36**, 24875 (23 December 1971).
10. "East coast shifts to greater use of coal for electric generation, 'borrows' electricity from other utilities (press release and status report, Federal Energy Office, Washington, D.C., 6 February 1974).
11. "U.S. energy use up nearly 5 percent in 1973" (press release, Bureau of Mines, Department of the Interior, 13 March 1974).
12. T. L. Montgomery, J. M. Leavitt, T. L. Crawford, F. E. Gartrell, *J. Met.* **25** (No. 6), 35 (1973).
13. *Fed. Reg.* **38**, 25697 (14 September 1973).
14. National Academy of Engineering–National Research Council, *Abatement of Sulfur Oxide Emissions from Stationary Combustion Sources* (P.B. 192,887, National Technical Information Service, Springfield, Va., 1970).
15. Federal Interagency Committee Evaluation of State Air Implementation Plans, *Final Report of the Sulfur Oxide Control Technology Assessment Panel on Projected Utilization of Stack Gas Cleaning Systems by Steam-Electric Plants* (Report No. APTD-1569, Environmental Protection Agency, Research Triangle Park, N.C., 1973).
16. J. McLaren and D. F. Williams, *Combustion* **41** (No. 11), 21 (1970).
17. F. Princiotta, "EPA presentation on status of flue gas desulfurization technology," testimony at national power plant hearings sponsored by the Environmental Protection Agency, 18 October to 2 November 1973 (hearing records available for consultation at the Environmental Protection Agency, Washington, D.C.).
18. G. T. Rochelle, paper presented at the flue gas desulfurization symposium of the Environmental Protection Agency, New Orleans, 14 to 17 May 1973.
19. A. V. Slack, H. L. Falkenberry, R. E. Harrington, Preprint 39d, 70th national meeting, American Institute of Chemical Engineers, Atlantic City, N.J. 29 August to 1 September 1971.
20. J. K. Burchard, paper presented at the technical conference "Sulfur in utility fuels: the growing dilemma," sponsored by *Electrical World Magazine*, 25 to 26 October 1972.
21. I. S. Shah and C. P. Quigeley, Preprint 39e, 70th national meeting of the American Institute of Chemical Engineers, Atlantic City, N.J. 29 August to 1 September 1971.
22. W. R. Horlacher, R. E. Barnard, R. K. Teague, P. L. Haylen, paper presented at the 71st national meeting of the American Institute of Chemical Engineers, Dallas, Tex., 20 to 23 February 1972.
23. J. L. Martinez, C. B. Earl, T. L. Craig, in *AIME Environmental Quality Conference, Washington, D.C., June 7–9, 1971, Preprint Volume* (American Institute of Mining, Metallurgical, and Petroleum Engineers, Washington, D.C., 1971), pp. 409–420.
24. J. B. Rosenbaum, W. A. McKinney, H. R. Beard, L. Crocker, W. I. Nissen, *U.S. Bur. Mines Rep. Invest. No. 7774* (1973).
25. F. S. Chalmers, L. Korosy, A. Saleem, paper presented at the Industrial Fuel Conference, West Lafayette, Ind., 3 October 1973.

Correction: The Office of Mineral Information, U.S. Bureau of Mines, has provided the following correction for page 246 of this reprinted paper:

"Ambient air quality standards (Table 2) limit the concentration of sulfur oxides in the atmosphere while emission standards (Table 3) limit the amount of sulfur oxides that can be discharged into the atmosphere (*8*, *9*). Emission standards can generally be met only by burning coal containing 1 percent sulfur or less."

National Rural Electric Cooperative Assn.

Nitrogen Oxides from Stationary Sources: Problems of Abatement

Of the major air pollutants by volume, two nitrogen oxides are among the most vexing—both in their effects and in the problems they pose for pollution abatement. The two—nitric oxide (NO) and nitrogen dioxide (NO_2), generally considered together as NO_X—are harmful to health directly and present additional problems because of their role in photochemical smog reactions. Because NO_X is formed in high-temperature combustion in air, greater burning efficiency as a technique to cut carbon-monoxide and hydrocarbon emissions from automobile engines has led to greater NO_X output. Questions of NO_X control in stationary sources are tied tightly to questions of fuel choice, combustion technology, and energy economics.

A National Research Council panel has completed an intensive study of NO_X abatement in stationary sources and has advised the Environmental Protection Agency (EPA) that because electric-utility boilers are the largest single stationary source of NO_X the EPA's abatement-technology research plan "will need to be closely coordinated with national energy policies for optimum use of resources," with "major emphasis . . . on developing combustion-modification techniques that will reduce NO_X formation in coal-, oil-, and gas-fired industrial and utility boilers."

The panel's study and report, *Abatement of Nitrogen Oxides Emissions from Stationary Sources* (see "New Publications," p. 8), follow a request to the National Academy of Engineering for assessment of current control technology, of near-term prospects for control technology, and of research targets for control technology. With the admonition that the pace of control-technology research and development is such that "current judgments may undergo considerable change within a few years," the panel concluded that stack-gas cleaning for NO_X control is not promising and that the most promising approach is in combustion modification. The panel said that new plants burning natural gas might be expected by 1980 to emit about 100 parts NO_X per million parts stack gas compared with current averages of 350-400 parts per million, and that for oil-fired plants a realistic objective is about 150-200 parts per million.

Coinciding Dilemmas

To appreciate the thorniness and the importance of the NO_X-control problem in electric-power generation, consider these coinciding dilemmas:

NO_X production depends on combustion characteristics including temperature and flame geometry, and "the most precise control" of NO_X is achieved when fuel is in a gaseous state, but "natural gas may not be available as a fuel for utility boilers very far into the future." Although coal is abundant as an energy resource, particularly in Midwest and Eastern U.S. regions of high-energy demand, and although it is the most widely used U.S. utility fuel. NO_X-control methods "are not yet established for coal," and for the three fuels (gas, oil, and coal) "least is known about coal relative to minimizing NO_X formation from combustion."

It is no simple matter. It is generally understood, the panel noted, that "NO_X is formed largely by the reactions of nitrogen and oxygen from the atmosphere at the high temperatures existing during combustion," but "[f]undamental differences remain in understanding just how NO is formed in a furnace." NO accounts for all but 5 or 10 percent of the NO_X emitted. Chemistry and physics, particularly the study of combustion kinetics, as well as engineering and economics form a part of the NO_X-abatement problem.

The panel—chaired by Charles N. Satterfield of the Massachusetts Institute of Technology—concluded that the most promising approach to NO_X abatement is to be found in reducing its formation rather than in attempting to remove it from the effluent in utility stacks. The chemical complexity of the problem suggests that stack-gas treatment of NO_X by catalysis will be difficult and expensive, and the relatively unreactive and relatively insoluble nature of NO make absorption, adsorption, and scrubbing processes for cleaning stack gases of NO_X impractical, in the panel's view.

Although NO_X removal from stack gases "may offer potential for control in the future," the panel said, ". . . no proven process is available for substantial removal of NO_X from combustion stack gases." The panel defined "proven process" as a process operated satisfactorily on an industrial scale for one year. Scrubber or adsorption systems proposed primarily for removal of sulfur oxides from stack gases, the panel said, "should also be evaluated for their potential in removing NO_X simultaneously."

The panel preferred what is in effect a twofold alternative strategy taking into account the technological and economic nature of the problem. The report recommends work on combustion modification and it recommends work to make fuels more amenable to combustion control, perhaps with side benefits such as sulfur-oxides reduction resulting from coal gasification. The report recommends concurrently that "[b]oiler manufacturers and utilities incorporate as much flexibility as possible in the design of new boilers to permit taking advantage in the future of increasing knowledge of the factors affecting NO_X emissions in combustion."

Reprinted with permission from the Nat. Acad. of Sciences, Nat. Research Council, Nat. Acad. of Eng. *News Report*, vol. 22, Apr. 1972.

Fossil fuels enter furnaces in these conventional boilers in thermal-electric generating plants. "The most promising prospects for significant early reduction" of nitrogen oxides from such plants, a National Research Council panel says, lie chiefly in combustion modification.

In its discussion of research strategy the panel pointed out:

"The complex nature of control of NO_X, SO_2 [sulfur dioxide], and other pollutants from coal combustion, and to a lesser extent from the combustion of other fuels, fully justifies serious federal attention and support for development of processes for production of 'clean' fuels by such methods as fluidized bed gasification, coal gasification, coal liquefaction, and processing of our vast deposits of oil shale. This requires careful planning and coordination by OAP [the EPA's Office of Air Programs] with other federal agencies, especially the Office of Coal Research and the Bureau of Mines."

Current Technology

Current technology, the panel concluded, is such that "a realistic objective for new plants using natural gas to be placed in operation by 1980" is a reduction in NO_X concentration to about 100 parts per million from current levels, a reduction of about 60 or 70 percent. For oil, the most common NO_X emission range today is about 180 to 280 parts per million for tangentially fired boilers and about 300 to 700 parts per million for horizontally fired boilers.

"A realistic objective for oil-fired plants placed in operation by 1980, achievable by flue-gas recirculation and off-stoichiometric combustion, is about 150 to 200 ppm," the panel said. The panel found insufficient information or experience to warrant a comparable assessment for coal-fired burners.

"Both theory and practice indicate that NO_X emissions from combustion sources can be lowered by: (a) reducing the amount of oxygen present in the flame zone, as by use of staged admission of air (or off-stoichiometric combustion), and (b) reducing the peak flame temperature, as by use of flue-gas recirculation to the flame zone," the panel reported. "The practicality of these abatement techniques has been developed primarily in furnaces burning gas or oil. Little has been done on coal-fired units."

Combustion Modification

The panel urged that combustion-modification studies "be given first priority" in NO_X-abatement research and development. "Studies of coal combustion are especially required. Studies of the effect of fuel nitrogen on NO_X emissions and the potential of flame-temperature-control techniques in oil and coal burning are also needed. A substantial reduction in the amounts of NO_X released to the atmosphere (i.e., of the order of 50 to 80 percent) will come least expensively from modifications of the combustion process rather than from scrubbing or adsorption systems to remove NO_X from stack gases."

Additional work should be funded, the panel said, "on new energy-conversion concepts—such as fluidized-bed combustion, coal gasification for electric-power production, and combined-cycle gas- and steam-turbine generating plants operating in conjunction with such combustors and gasifiers—to develop their potential for reducing NO_X and other pollutants."

The panel recommended that evaluation of all new electricity-generation techniques begin "as soon as practicable," incorporating NO_X control and control-cost estimates. Any economic evaluation of magnetohydrodynamics (MHD) as a future power source "should incorporate as a part of the analysis the economic costs of control of the large amounts of NO_X that will undoubtedly be formed," the panel said.

"Reliable commercial methods for gasification of coal to produce both a synthetic natural gas to supplement existing natural-gas supplies, and a gas of less than pipeline quality to be used as an electric-utility fuel and probably having other commercial, industrial, and residential uses, are of considerable interest," according to the panel. Fuel costs would be higher, but "the advantages of an environmentally acceptable 'clean' fuel and the reduced capital cost and increased efficiency of combined-cycle plants may outweigh the higher fuel costs for power generation."

In 1970, the panel observed, electric utilities placed new orders for generating equipment of about 30,000,000 kilowatts capacity to burn fossil fuels and for equipment of about 15,000,000 kilowatts capacity to use nuclear fuels. The major portion of U.S. electric generating capacity will continue to rely on fossil fuels for several years, "thus ensuring that the generation of NO_X will be a continuing significant problem."

—GERALD S. SCHATZ

Removal of Particulates from Stack Gases

STATEMENT SUBMITTED BY JACK G. HEWITT, JR., ASSOCIATE PROFESSOR AND RESEARCH ENGINEER, ELECTRICAL ENGINEERING DEPARTMENT, UNIVERSITY OF DENVER, DENVER, COLORADO, WILLIAM J. CULBERTSON, RESEARCH ENGINEER, CHEMICAL ENGINEERING DIVISION, DENVER RESEARCH INSTITUTE, DENVER, COLORADO, AND THOMAS D. NEVENS, SENIOR RESEARCH ENGINEER, CHEMICAL ENGINEERING DIVISION, DENVER RESEARCH INSTITUTE, DENVER, COLO.

At present, approximately 83% of electric energy is generated by fossil fuel power plants. In the year 2000, fossil fuel power plants are expected to produce almost 50% of the total electric energy.[1,2,3,4] Although the percentage decreases, actual energy produced will increase from 1.6×10^{12} kwh to 9.5×10^{12} kwh. Most of the fossil fuel power plants are and will continue to be coal fired. If we consider that it takes about 2/3 of a pound of coal to produce one electrical kwh, 3.2×10^{9} tons of coal must be burned in the year 2000, and approximately 3.2×10^{8} tons of ash will be produced. About 15% of this ash would pass directly through the stack into the atmosphere if it were not for collection equipment. It is interesting to note that were it not for collectors, if the fly ash pollution effect in the year 2000 could be equally shared over the 3×10^{6} square mile area of the 48 states, the load would be 20 tons per square mile per year, or 100 pounds per square mile per day. With top quality collection equipment, which is 99% efficient, this load would be reduced to 0.2 tons per square mile per year or 1.0 pound per square mile per day. Of course, the pollution load is not shared equally and depends on meterological conditions, local terrain, characteristics of the boiler, etc. Pollution loads downstream of the stack could be orders of magnitude greater than the numbers cited above.

These data and projections take into account expected advances in nuclear energy technology and advances in fossil fuel technology such as development of gas turbine generators, MHD topping plants, and coal gasification. The severity of the problem of particulate pollution in some places is already apparent. There are some additional aspects of the problem that are not so obvious. First is the biological effect of small (0.1 to 1 micron) particles which are most hazardous to health because they are inhaled and retained in the lungs.[5] Second is the tendency for small particles to remain in suspension and accumulate in the upper atmosphere. This could lead to serious climatic changes.[6] Third is the interaction of particulate pollution control with gaseous pollution control. For example, lowering the sulfur content of fuel or injection of reagents to lower SO_3 stack gas content usually causes a large reduction of electrostatic precipitation efficiency.[7]

There are four basic categories of particulate control equipment: dry inertial or cyclone (mechanical) collectors, electrostatic precipitators, wet scrubbers, and fabric or bag filters.[8,9] Each of these categories can have variations, some of which are proprietary. There are many claims about superior performance of one type over another by manufacturers and the utilities who use them. Some generalizations can be made as follows:

1. Mechanical collectors are the least expensive to install, maintain and operate; however, collection efficiency drops very rapidly for particles below 10 microns.

2. Electrostatic precipitators are moderately expensive to install but inexpensive to operate and maintain. They collect particles of all sizes and have a small pressure drop; however, certain types of coal ash and stack gas sulfur oxide level combinations and possibly other variables cause operational efficiency to drop to an unsatisfactory level. These difficulties are commonly attributed to too high an electrical resistivity of the fly ash.

3. Scrubbers are very expensive to install and expensive to operate. They require a large pressure drop for efficient particulate collection and use fans and stack gas reheat steam that can consume more than 10% of a power plant's output. This, of course, will require consumption of 10% more fuel for the same power output capacity, not a suitable solution with pending fuel shortages. However, collection efficiency is independent of the type or condition of coal ash and scrubbers can be operated to remove gaseous pollutants also. Wet scrubbers can sometimes solves an air pollution problem and create a water pollution problem.

4. Big filters are moderately expensive to install and expensive to operate and maintain. The fabric must be continuously shaken to remove collected ash and allow gas to pass through. This causes fatigue of the fabric material and eventual bag failure requiring frequent bag changes. Pressure drop is higher than an electrostatic precipitator's but lower than a scrubber's.

We believe that electrostatic precipitators and low pressure drops scrubbers offer the most promise for future improvement of particulate and SO_2 and NO_x removal from stack gases in an economical way. One idea is to use electrostatic precipitators to remove about 99% of the fly ash load followed by a scrubber to remove gaseous pollutants. This would enable the scrubber to operate at a lower pressure drop and consume less power.

All too frequently such statements as made by Time Magazine[13] are noted: "In one recent survey of 129 major electric power plants the Council of Economic Priorities found that most utilities have been slow to install *proven* antipollution devices, *even though they are readily available.*" Certain representatives of EPA have also made similar statements at Western states air pollution hearings. Such statements are false, irresponsible and need qualifying.

Scrubbers of commercial size have not as yet been proved in spite of manufacturers' claims, and their economics are questionable, unless consuming 10% more fuel for the same power output is suitable and capital costs need little consideration.

After three years of research and testing at the Denver Research Institute, financed both by members of West Associates (power companies of eleven Western states) and individual Western power companies, it is apparent that no electrostatic precipitator manufacturer can design a unit with any degree of confidence of performance for use on low sulfur Western coals. (Fly ash resistivity measurements have been made at a large number of the major power plants of the Western states by the staff of the Denver Research Institute.) The alternatives to the use of high performance electrostatic precipitators for removal of fly ash from coal fired power plants are indeed bleak.

There are empirical parameters on which the design of precipitators and scrubbers for new power plants are based. Values for the parameters are extrapolated and estimated from performance data of previously built units, usually those using Eastern high sulfur coals. An example is the precipitation rate parameter in the Deutsch-Anderson equation.[10] Although attemps have been made to plot precipitation rate parameter versus sulfur content in coal and gas temperatures, results have limited usefulness because the parameter seems to depend on 1) which coal field the fuel came from, 2) which power plant boiler was used, and 3) the stack gas conditions. This indicates a lack of fundamental understanding of how electrostatic precipitators operate when the fly ash does not behave in the ideal way.

The public utilities and pollution equipment manufacturers take an empirical approach to solve this kind of problem. So far, of course, their approach has not been completely satisfactory. A more basic approach is suggested, that is, to increase the fundamental knowledge pertaining to the precipitation of ash more difficult to precipitate such as high resistivity ash from low sulfur coals. To implement this suggestion, we submitted a proposal to the National Science Foundation entitled, "Factors Affecting Fly Ash Resistivity and Voltage Breakdown in Electrostatic Precipitators," and recently received Grant GI-32610X as part of the energy Research and Technology Program under the Division of Advanced Technology Applications, National Science Foundation.

The Electric Research Council[11] projected environment research and development program gives "fundamental study of particulate behavior in electric fields" a priority of 38 out of 71 recommended research and development tasks. The recommended funding is 1.25 million dollars over 5 years. In our opinion, this priority rating and the dollar amount may be too low. A single precipitator installation costs in the neighborhood of 1 to 5 million dollars, and hundreds of installations will be required for future power plants and to improve existing power plants. Therefore, an expenditure of several million dollars on fundamental research is not disproportionate and would be a good investment to insure air pollution particulate control consistent with future abatement laws. According to Ali Cambel, no one sector of the society or economy (the electric utility companies, electric and pollution equipment manufacturers, the Government, etc.) can do the research to solve this type of problem because of the diversity of interests and economic constraints imposed.[12]

One example of diversity of thinking between Government agencies alone concerns the problem of precipitating fly ash from the low sulfur coals commonly found in the Western U.S. Electrostatic precipitators work quite well with fly ash from Eastern coals which contain more sulfur. An argument presented to us was that this problem was local, was not of national concern and should be solved by local Western industries and utilities. The argument is invalid for several reasons:

1. Some utilities as far away as Illinois are starting to use Western coal and will begin to see a drop in electrostatic precipitator efficiency.

2. Utilities serving populated areas are beginning to build power plants in the less populated Western states near the coal field and send electric energy over transmission lines. Examples are the Four Corners Plant in Arizona, the Jim Bridges Plant in Wyoming and the Kaiparowits Plant in Utah.

3. The requirement to reduce SO_2 stack effluent is causing Eastern utilities to consider removing sulfur from the coal before burning or to obtain alternate low sulfur coal sources.

It would be helpful for the Government to create a research environment useable by all sectors involved in energy production and pollution control to enable more rapid progress.

REFERENCES

1 "Power, Pollution and the Inherited Environment," G. D. Friedlander, *IEEE Spectrum*, Nov. 1970, p. 41.

2 "Energy Utilization," Comments by H. J. Young, Vice President of Edison Electric Institute at Direct Energy Conversion Short Course, ASU, Jan. 1972.

3 "Conventional Power," P. N. Garay, *Science and Technology*, March 1969, p. 20.

4 "Nuclear Power: Suddenly Here," M. Shaw and M. Whitman, *Science and Technology*, March 1968, p. 24.

5 "Pollution From Combustion of Fossil Fuels," P. W. Spaite and R. P. Hangebrauck, paper presented at National Pollution Control Conference and Exposition, San Francisco, April 1970.

6 "Climate Modification by Atmospheric Aerosols," R. A. McCormick and J. H. Ludwig, *Science*, *156*, June 1967, p. 1358.

7 "Review of the Current Status of SO_x and NO_x Control Methods," R. E. Harrington, EPA, Durham, North Carolina, Feb. 1971.

8 *Air Pollution Control*, W. Strauss, ed., Wiley-Interscience, 1971.

9 "Scrubbing Air," A. B. Walker and N. W. Frisch, *Science and Technology*, Nov.-Dec. 1969.

10 "Practical Design and Operating Considerations for Industrial Electrostatic Precipitators," S. Oglesby and H. J. White, *Proc. of Electrostatic Precipitator Symposium*, Feb. 1971, p. 44.

11 "Electric Utilities Industries Research and Development Goals Through the Year 2000," Electric Research Council Pub. #1–71, p. 114.

12 "Impact of Energy Demands," Ali B. Cambel, *Physics Today*, Dec. 1970, p. 39.

13 *Time Magazine*, June 12, 1972, p. 50.

Reprinted with permission from *Energy Research and Dev.*, Hearings before the Subcommittee on Sci., Research, and Dev., House Committee on Sci. and Astronautics, Y4.Sci2: 92/2/24, pp. 497–499, May 1972.

Clean Power from Coal

Recovery of sulfur will be a mere incidental in new systems for burning coal to produce electricity.

Arthur M. Squires

In 1969 the United States power industry discharged to the atmosphere about 7×10^6 tons of sulfur in the form of SO_2. In the absence of controls other than tall stacks, the discharge in 1980 will be about 18×10^6 tons. Not long ago, the industry apparently hoped to rely primarily upon tall stacks to disperse its gaseous wastes (*1*), but the vast ecological experiment implied in the projected emissions is unlikely to occur. In early 1968, New Jersey issued rules that call for sulfur levels in coal and residual oil below 0.2 and 0.3 weight percent, respectively, after October 1971. Fuel suppliers, unable in 1968 to conceive where fuels of such low sulfur levels might be found, were skeptical that the rules would be applied. Other jurisdictions now give signs that they can be expected to follow Jersey's lead, and fuel users are beginning to understand that SO_2 emissions must soon be sharply limited.

What will be the technological response? In a word, I believe the response may work to make power cheaper.

It must be said at once that application of means now generally put forward for controlling SO_2 (*2*) would add significantly to the cost of electricity. Environmentalists would reply that the added cost is a small price for clean air. Yet the attempt to weigh costs of control versus benefits is a premature and probably silly exercise if it is carried out before the technological community accepts the idea that controls are imperative.

There is a suggestive historical parallel. Before 1863, British alkali works poisoned the air with massive discharges of HCl gas. Under pressure from neighbors, managers of the works experimented with water scrubbing to absorb HCl, with indifferent success. The Alkali Act of 1863, passed by Parliament in spite of expert testimony to the difficulties, required a 90 percent reduction in HCl emissions. Soon sophisticated gas-scrubbing towers appeared, exceeding the Act's requirement. For a while much of the hydrochloric acid produced by the towers was taken to sea and dumped, but profitable markets for the acid developed. Most significantly, perhaps, the Alkali Act appears to have stimulated invention. Chlorine gas was a curiosity before Weldon and Deacon filed their patents (in 1866 and 1868, respectively) disclosing chlorine processes which soon turned HCl gas formerly wasted into profits.

In this article I have two purposes:

- to persuade that paths of technological development exist that could lead to suppression of SO_2 from coal and at the same time to a lower cost of power;
- to argue that a massive injection of money into *coal engineering* is the immediate ingredient necessary to open up these paths of development.

The combustion of coal at high pressure in the presence of a desulfurizing agent and the generation of power by a combination of gas- and steam-turbine cycles represent, together, a major opportunity.

The questions arise, What combustion technique? and What desulfurizing agent?

The first question is the more difficult. Efforts that shed light on the question, motivated simply by a desire for cheaper power, have been under way for some time. It will be well first to review these efforts and later to look at agents for the capture of sulfur.

Rethinking Coal Combustion

All of our great power-generating stations based upon coal use pulverized-fuel (PF) firing. Although PF boilers have reached the giant sizes needed for power generation at the 1000-megawatt scale, they are basically quite simple. A number of burners inject pulverized coal and air into a large rectangular box that has walls composed of vertical tubing filled with boiling water. The firing chamber must be huge not so much to allow sufficient combustion volume as to provide sufficient tube surface for transfer of heat to the boiling water.

Much of the inorganic matter in the coal leaves the chamber as fly ash, generally to be collected by an electrostatic precipitator. Some stations in metropolitan settings have installed precipitators to provide collection efficiencies exceeding 99 percent. To handle fly ash from a low-sulfur coal, such a precipitator costs about $10 per kilowatt of capacity (*3*). Although few existing stations are equipped with such precipitators, the power industry will find it hard in the future to escape such costs for fly ash control.

Power engineers adopted PF firing in the mid-1920's, when they began to require steam flows larger than earlier grate-combustion techniques could readily provide. Engineers of the day, accustomed to worrying about grit emissions from the earlier techniques,

The author is professor of chemical engineering at the City College of the City University of New York, New York 10031.

welcomed the finer dust that the PF boiler discharged. Herington (*4*) wrote in 1920:

It is quite true that perhaps 60 per cent of the ash goes up through the stack. This ash is of such light flocculent nature that it is dissipated over a wide area before precipitation occurs and no trouble can be expected from this source, although the amount of tonnage put out through the stack per day seems great.

The engineer of 1920 soon heard from nearby housewives who found "soot" on their wash, but the insults to lung tissue by fine matter were as yet unknown. Would PF firing have seemed so attractive for development if engineers had felt something like today's concern about fly ash?

Schemes to control fly ash and sulfur from PF combustion have a makeshift, tacked-on aspect. The time is at hand to rethink the problem of burning coal, with air pollution as a first rather than a last consideration. If the engineer of 1920 had been as much concerned with fly ash as with grit, he might well have concentrated upon ways to increase the burning capacity of his familiar grate devices.

An idea was at hand. Figure 1 is copied from the specification of Winkler's historic patent (*5*), filed in Germany in 1922 and put into commercial practice there in 1926. Winkler's idea was to increase the rate of gas flow upward through a granular bed to and beyond the point at which each particle in the bed was buoyed by the rising gas. When the pull of gravity upon each particle was canceled by the upward drag of the current of gas, the particles flowed freely, and the bed took on the character of a boiling liquid. A decade later, American engineers coined the term "fluidization" to denote Winkler's procedure.

It does not detract from the simple beauty of Albert Godel's idea to wonder why no one before him thought to fluidize a bed of coal upon a traveling grate. This idea led Godel during the 1950's to his "Ignifluid" boiler, seen in cross section in Fig. 2 (*6*). Godel made the remarkable discovery that the ash of almost all coals is self-adhering at a temperature in the vicinity of 1100°C, no matter how much higher the ash-softening temperature may be. Godel exploits this discovery to burn a wide range of coals in his fluidized bed (see Fig. 2). Coal is supplied in sizes up to 2 centimeters. As a coal particle burns, ash is released. Ash sticks to ash and not to coal, and agglomerates of

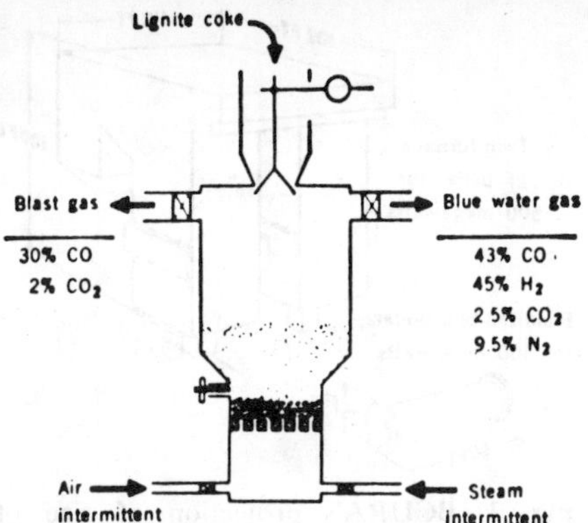

Fig. 1. Winkler's historic idea for partial combustion (that is, gasification) of coal in a "fluidized" bed [from his patent application filed in 1922 (*5*)].

ash are formed. They sink to the grate, which carries them to an ash pit. Godel's bed operates adiabatically, except for radiation from the upper surface. He limits the bed to the desired ash-sintering temperature by maintaining a high inventory of carbon in the bed, so that combustion is incomplete. Carbon appears as CO in the gas leaving the bed, and sulfur appears as H_2S. Godel admits secondary combustion air to the space above the bed, where CO and H_2S burn to CO_2 and SO_2, respectively.

As a result of the high velocity of fluidizing gas (about 3 meters per second) and the low air-to-fuel ratio, the coal-treating capacity of Godel's traveling grate is roughly ten times greater than that of previous grate-combustion devices.

Godel originally thought his Ignifluid system to be useful only in small boilers and for special fuels of low reactivity or high ash content. He believes he lost many years through failure to realize that his system might go into large utility boilers. A mature technique, such as PF firing, tends to become surrounded by an aura of inevitability that inhibits invention and protects it from competitive ideas.

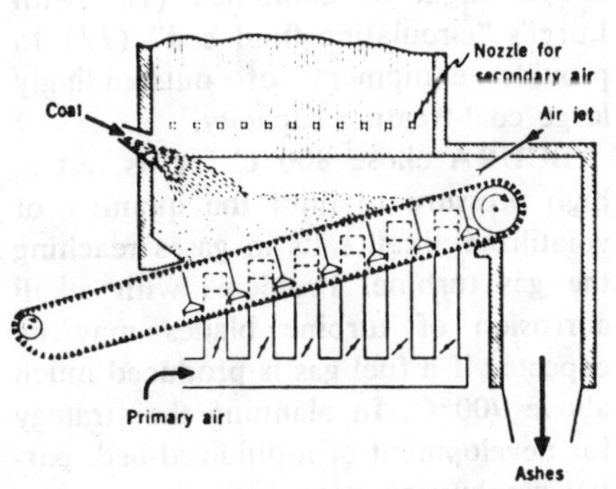

Fig. 2. Cross-sectional view of "Ignifluid" boiler developed by Albert Godel (*6*) and Babcock-Atlantique (*7*).

Recently, Babcock-Atlantique has promoted use of the Ignifluid boiler in large stations (*7*). A 60-megawatt unit is in operation at Casablanca, and negotiations are well advanced for a 275-megawatt unit to burn and remove accumulations of anthracite wastes in northeastern Pennsylvania. The owner of the unit, UGI Corporation, will benefit from a low fuel price, between 12 and 15 cents per million Btu (British thermal units), and the waste supplier, Blue Coal Corporation, will recover valuable urban land. The waste has a high ash content, and Godel's system is uniquely capable of dealing with it.

Another approach to fluidized combustion of coal is receiving worldwide attention. The U.S. Office of Coal Research (OCR) has sponsored large-scale trials conducted by Pope, Evans, and Robbins at Alexandria, Virginia (*8*). The U.S. Bureau of Mines and groups in England and Australia are doing similar work (*9*). The fluidized bed is operated nonadiabatically, the inventory of carbon is very small, and combustion is complete. The bed itself generally comprises the larger particles of ash matter in the coal. The bed is in contact with boiler tubes which hold the temperature to a level where ash does not sinter, generally below about 1000°C.

Although fluidized beds have been built in huge sizes for other purposes, it is nevertheless not yet clear that this work can lead to a boiler that challenges the PF furnace in cost for power generation at the 1000-megawatt scale.

Combustion at High Pressure

A large-scale experiment with fluidized combustion at 6 atmospheres and 800°C is under way at BCURA Industrial Laboratories (formerly the British Coal Utilisation Research Association) at Leatherhead, England (*10, 11*). If the competitive advantage of a fluidized-bed boiler operating at atmospheric pressure is uncertain, there is little doubt that such a boiler at elevated pressure can be much cheaper than a conventional furnace. Figure 3 illustrates the dramatic reduction in boiler size that might be achieved. No electrostatic precipitator would be needed, and the saving might run well beyond $10 per kilowatt.

BCURA's concept is that the hot gases from the bed would be expanded in a gas turbine. Another cost saving would arise from the fact that a gas-turbine power plant costs less than a steam plant by about $30 to $50 per kilowatt. The gas turbine would provide about 20 percent of the power from BCURA's system. Utilities have used gas turbines primarily to supply peak-load power, because the efficiency of an open-cycle gas turbine that operates independently of a steam plant is poor. Such a turbine discharges a hot gas directly to the atmosphere. In a few installations, a gas turbine discharges hot gas to a steam boiler, and such cooperative use of gas-turbine and steam power equipment can provide base-load power at outstandingly low cost (*12*).

The drive for higher performance in aircraft engines will continue, and experience from such engines can be expected to maintain the historic upward trend of temperature of gases at the inlet of industrial gas turbines. As this temperature rises, the proportion of the total power provided by the gas turbine of a combined-cycle system should logically increase. Systems can be envisaged in which the gas turbine would provide more than one-half of the power. With temperatures that might reasonably be achieved within a decade, such systems could provide electricity-generating efficiencies approaching 50 percent (*13, 14*). Adoption of the systems would greatly reduce the quantity of heat rejected to the environment from the steam cycle condenser.

In BCURA's experiment, about 70 percent of the heating value of the coal is transferred to steam in boiler tubes passing through the bed. The remaining heat appears as sensible heat in the hot combustion gases. If the gas turbine is to play a larger role, more energy must be converted to sensible heat in combustion gases. This can be accomplished by substituting a carbon-rich bed for the carbon-lean bed of the BCURA concept, since partial combustion occurring in the carbon-rich bed can provide CO for combustion outside the bed and ahead of a gas turbine.

A problem arises from the fact that dust carry-over from a carbon-rich bed contains a high percentage of carbon, which would represent a serious carbon loss if it is not used. Carbon in the dust carry-over cannot be consumed simply by returning the dust to the carbon-

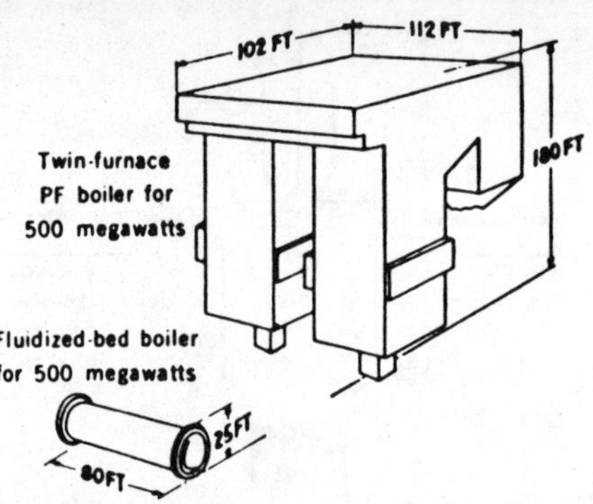

Fig. 3. BCURA's projection of size of fluidized-bed boiler at 15 atmospheres compared with conventional PF boiler (*10*).

rich bed. There is a tendency for the dust to be blown out of the bed again quickly, and, as the dust "ages," its carbon tends to sinter to an inactive coke.

Pope, Evans, and Robbins' success in burning carbon in fines blown from their experimental boiler (*8*) suggests a solution to this problem. They provide an auxiliary bed to which air is supplied in greater than stoichiometric amount for combustion of fines charged to the bed. Figure 4 illustrates how such a carbon burnup bed might cooperate with a partial combustion bed to supply gases at high temperature to a gas turbine. BCURA's temperature, 800°C, is probably too low for a partial combustion bed, and the carbon-rich bed in Fig. 4 should operate at a temperature above 900°C.

Carbon burnup in a partial combustion bed becomes less of a problem if the temperature is raised into the ash-sintering range, above about 1050°C. The Ignifluid traveling grate might be modified for use at high pressure. Jéquier and collaborators at the Centre d'Études et Recherches des Charbonnages de France operated an agglomerating fluidized bed of a design suitable for high pressure (*15*). Jéquier's design might be combined (*16*) with Lurgi's "circulating fluid bed" (*17*) to provide equipment of outstandingly large coal-treating capacity.

BCURA chose 800°C for its test at high pressure to limit the quantity of volatilized alkali salts in gases reaching the gas turbine. Problems with alkali corrosion of turbine blades may be expected if a fuel gas is produced much above 900°C. In planning the strategy for development of a fluidized-bed, partial combustion process for power generation, it will be important to establish upper limits for alkali entering the gas turbines which are expected to become available in the late 1970's. A safe plan would be to provide these turbines with gas that has been cooled and scrubbed free of alkali. Scrubbing with a heavy oil at about 370°C would be preferable to scrubbing with water at a lower temperature, so that heat may be rejected from the scrubber to boiler tubes, raising prime steam. There will be an advantage in keeping the quantity of gas to be scrubbed as small as possible. This will favor schemes in which the coal is devolatilized first, so that the gasification bed must deal only with coke (*18*).

It is hard to escape the feeling that PF combustion might never have been developed, other than for cement-making, if Winkler's work and the chemical engineer's interest in fluidization had appeared sooner. This proposition is, of course, not worth arguing, but the aura of inevitability that has colored much recent thinking about PF combustion and the SO_2 problem should be dispelled by concrete developments now appearing: the Ignifluid boiler to be built in Pennsylvania, BCURA's test of fluidized-bed combustion at high pressure, and Lurgi's application of its historic high-pressure gas producer to power generation (*19*).

A technique as mature as PF firing, practiced on so large a scale, will be difficult to displace. To enjoy the advantages of the combination of gas-turbine and steam power equipment, we must find a way to burn coal at high pressure in equipment of large capacity. A reasonable target would be a technique able to handle the coal for 1000 megawatts in a single unit, or at most in a few units. Development costs are certain to be large, since only expensive large-scale trials could satisfy those responsible for outlays for new power plants. I believe that fluidized-bed art offers the best hope, but I should note that Babcock and Wilcox have faith in a PF partial combustion technique (*20*). Texaco piloted such a technique in the late 1950's (*21*) but published no results. A scale-up of Lurgi's gravitating-bed gasifier would be difficult and uncertain.

Dealing with Sulfur Oxides

Work on alternative techniques for burning coal would be justified simply for the prospect of cheaper power. A special urgency in the effort arises from the opportunity to direct these developments into paths that will lead to a combustion technology in which sulfur is dealt with early in the coal-treating process rather than at its end.

The National Air Pollution Control Administration (NAPCA) has seized this opportunity. It has supported work to explore the possibility of using limestone or dolomite in a fluidized-bed boiler to absorb SO_2, and it has engaged Westinghouse Electric Company to direct a broad effort toward development of nonpolluting fluidized-bed boilers. Under a NAPCA contract, United Aircraft has explored advanced power generation concepts that incorporate gas and steam turbines and coal gasification equipment (*14*).

Atmospheric-pressure tests conducted so far suggest that appreciably more $CaCO_3$ must be injected than just the stoichiometric amount required to react with SO_2 to form $CaSO_4$, if this control technique is to meet New Jersey's requirement for October 1971—that is, a maximum allowable emission equivalent to 0.2 percent sulfur in coal. $CaSO_4$ has a larger molecular volume than either $CaCO_3$ or CaO, and a shallow layer of the first $CaSO_4$ reaction product seals off the interior of a particle (*22*).

If $CaCO_3$ must be used in an amount far greater than stoichiometric, the economics of the operation are improved, particularly in a station of large size, if the resulting sulfur-laden solid is treated for recovery of a valuable sulfur product by a technique that restores the solid to a form suitable for reinjection into the boiler. In atmospheric-pressure tests, Consolidation Coal Company has burned coal in a fluidized bed consisting substantially of particles of calcined dolomite, with good sulfur retention by the bed (*23*). The $CaSO_4$ formed was regenerated to CaO by a roast under slightly reducing conditions. Sulfur was evolved from the roast in a gas containing SO_2 at a concentration adequate for manufacture of sulfuric acid in a contact plant. British Esso (*24*) has conducted similar tests on absorption of SO_2 from the complete combustion of residual oil.

In a system for combustion at high pressure, an agent derived from limestone or dolomite may advantageously capture sulfur during a first coal-processing step. For this step, there are three cases to consider:

- complete combustion, using air in excess of stoichiometric;
- partial combustion, using between about one-third and one-half of the stoichiometric air and yielding a fuel gas containing CO and H_2;
- a carbonization yielding low-sulfur coke as well as fuel gas, heat for the carbonization being supplied by a partial combustion which consumes about 10 to 15 percent of the stoichiometric air (*25*).

Volume of high-pressure equipment would be roughly proportional to the air rate, and carbonization has the advantage that it can provide a low-sulfur fuel product that can be stored and shipped. We will look at this option in more detail shortly.

The CO_2 partial pressure in gas leaving BCURA's fluidized bed for complete combustion at 6 atmospheres and 800°C is sufficiently great that $CaCO_3$ in limestone or dolomite added to the bed would not decompose. Although limestone would not be reactive toward SO_2, half-calcined dolomite would react readily:

$$[CaCO_3 + MgO] + SO_2 + 0.5\,O_2 = [CaSO_4 + MgO] + CO_2$$

Kinetics for this reaction, apparently on account of half-calcined dolomite's porosity, are favorable at temperatures even as low as 600°C (*26, 27*). A way exists to regenerate $[CaCO_3 + MgO]$ from $[CaSO_4 + MgO]$ while liberating H_2S for sulfur manufacture (*27*).

Partial combustion and carbonization have the advantage that sulfur as H_2S can be removed from gas more readily than sulfur as SO_2. It is far easier to prepare sulfur in elemental form from H_2S than from SO_2, and sulfur would be a better by-product than sulfuric acid, since only the former can be stored or economically shipped long distances.

Recent experiments at the City College of the City University of New York (*25*) show that the reaction of H_2S and CaO in calcined dolomite to form CaS occurs homogeneously throughout the particle.

At least three techniques (given below) are available for recovering sulfur values from CaS.

1) The historic Claus-Chance process, introduced about 1880, treated CaS wastes of alkali works with water and a gas containing about 40 percent CO_2.

$$CaS + H_2O\ (\text{liquid}) + CO_2 = CaCO_3 + H_2S$$

In a countercurrent system, the reaction was substantially quantitative, yielding a gas containing 40 percent H_2S, which could be readily oxidized to sulfur in a Claus system. Pintsch Bamag has worked recently on a version of the Claus-Chance method to act in conjunction with partial combustion of residual oil with air at atmospheric pressure (*28*).

2) Consolidation Coal Company (*29*), British Esso (*24*), and FMC Corporation (*30*) have worked on various procedures, each amounting to a controlled oxidation of CaS to release SO_2, which would be converted to sulfuric acid.

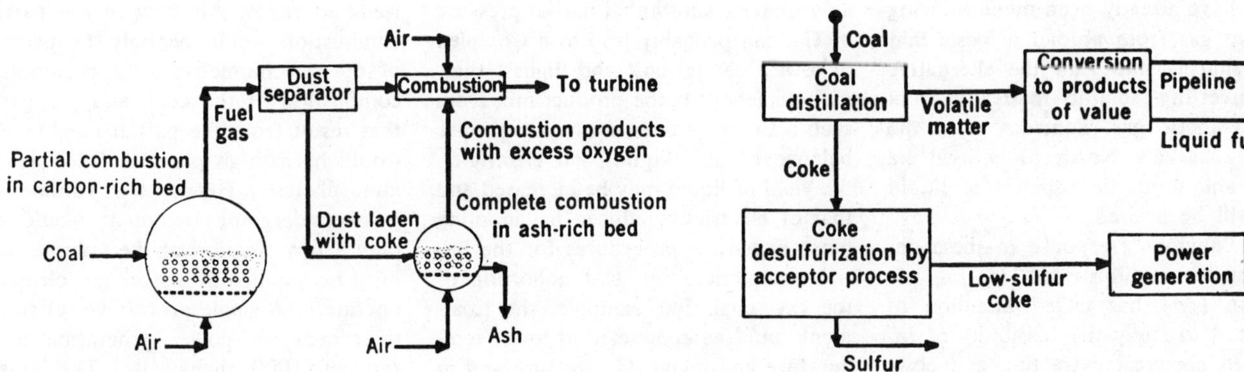

Fig. 4 (left). Scheme to supply CO for combustion ahead of a gas turbine. Fig. 5 (right). A "Coalplex."

$$CaS + 3/2\,O_2 = CaO + SO_2$$

3) The City College is studying a technique for desorbing sulfur from CaS at high pressure by reacting the solid with steam and CO_2 (*31*).

$$CaS + H_2O\,(steam) + CO_2 = CaCO_3 + H_2S$$

The Claus-Chance procedure has the disadvantage that heat from the reaction of CaS with water and CO_2 is wasted; also, $CaCO_3$ is recovered as a wet slime difficult to reuse except at another penalty in thermal efficiency. The oxidative procedures have the appeal of simplicity, if sulfuric acid is an acceptable product or if H_2S is available from another operation in a fuel-treating complex (*30*) to react with SO_2 to yield sulfur. Of the several procedures, the high-pressure desorption is probably best suited to provide a regenerated $CaCO_3$ in a form suitable for repeated cyclic use.

Chemical species other than CaO exist, of course, which are capable of removing H_2S from a fuel gas at high temperature. Consolidation Coal Company has worked on the problem of providing MnO in a form suitable for this purpose (*32*), and the Bureau of Mines has studied sinters containing iron oxide (*33*).

Although operation at high temperature has the advantage of affording better thermal efficiency, the art of scrubbing a fuel gas with an alkaline liquor to absorb H_2S is highly developed, and this alternative may be preferred in the immediate future (*19*).

Perspective for Coal

The country's and the world's reserves of hydrocarbonaceous matter largely reside in coal fields. Our own reserves of natural gas are insufficient for our growing needs, and arrangements have already been made to bring liquified gas from abroad at costs that bring sharply into view the alternative of converting volatile matter in coal into synthetic gas. Someday even the oil of Alaska's North Slope will be gone, and domestic supplies of liquid fuel will be desired.

An "obvious" response to these developments is illustrated in Fig. 5. Behnke (*34*) has called attention to the need to study the feasibility of integrated chemical extraction and power-producing complexes. In recent years OCR has supported work directed toward this goal (*35, 36*). The "Coalplex" depicted schematically in Fig. 5 is a logical choice for study. Volatile matter in coal may be converted by relatively simple procedures into synthetic pipeline gas or liquid fuel. Fixed carbon is converted to products of value only with much more difficulty, but Consolidation Coal Company (*35, 37*) has shown how easily the fixed carbon may be desulfurized through the cooperative action of H_2 gas and a solid acceptor for sulfur in the form of H_2S, such as CaO. FMC Corporation (*30, 38*) has operated a pilot unit demonstrating this procedure for producing a low-sulfur coke. Scientific Research Instrument Corporation (*39*), working under a NAPCA contract, has demonstrated the favorable kinetics for evolution of H_2S from coal distilled in the presence of H_2.

The term "coal distillation" in Fig. 5 should be understood in a broad sense. A pyrolysis at high temperature, preferably under a substantial partial pressure of H_2, can probably lead to a Coalplex of lowest capital cost and highest thermal efficiency. If the product mix from such a Coalplex does not have a proper balance of gas, liquid, and electricity, the yield of liquid may be increased and that of electricity reduced by adopting more expensive procedures for the initial treatment of at least a portion of the raw coal. For example, the treatment could be conducted at lower temperature and higher H_2 pressure and in the presence of catalysts.

Pulverized-fuel combustion is unsuited to handle the low-sulfur coke that will emerge from a Coalplex. Not only would an electrostatic precipitator be expensive, but the coke would need to be supplemented by a volatile fuel to maintain a stable flame. Work to develop fluidized-bed combustion techniques can be amply justified simply for the reason that they will be needed to deal with low-sulfur coke, which is certain to become available in large amounts.

In the near term, a less ambitious complex producing simply power and low-sulfur coke could play a useful role. Figure 6 depicts broadly a scheme under study at the City College (*25*). This pioneering Coalplex would generate base-load power from combustion of volatile matter and would ship low-sulfur coke. Heat to distill volatile matter would be provided from its partial combustion, and these steps, as well as coke desulfurization, would occur within a single vessel housing three fluidized-bed zones. Air flow to the partial combustion would be only 11 percent of the stoichiometric air for complete combustion of the coal, and the gases that result from the partial combustion would be at high pressure (such as 21 atmospheres). Hence, the volume of gases undergoing treatment would be only a tiny fraction of the volume that must be handled in a stack-gas cleaning operation. A single process vessel could treat coal for power generation at a rate of 1000 megawatts. The vessel would be approximately 25 meters in height. Its diameter would be about

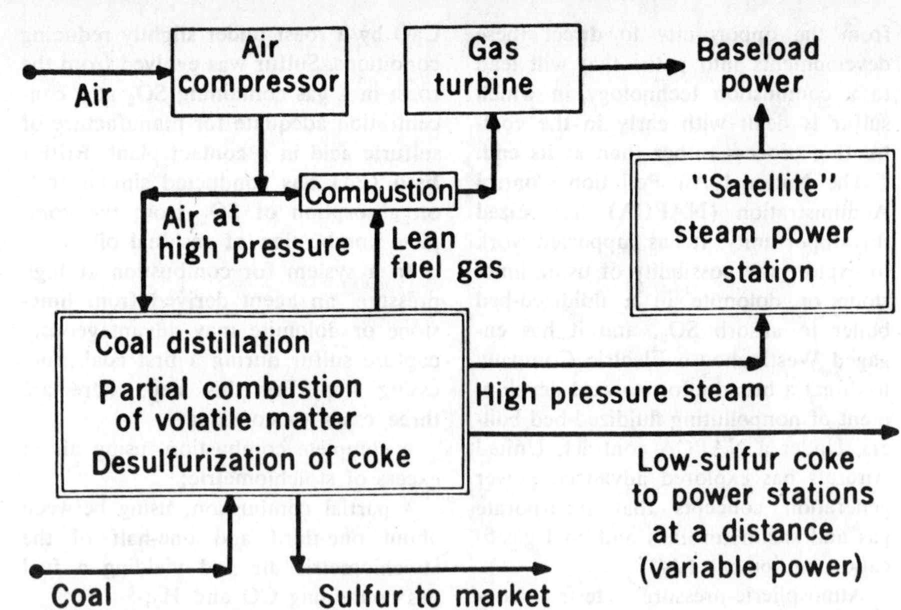

Fig. 6. A pioneering Coalplex (*25*) directed toward recovery of sulfur and generation of clean power.

5.5 meters over roughly 8 meters of the height, and about 3 meters elsewhere. The cost of the coal-treating equipment should be more than offset by revenues from sulfur and savings in cost of power-generating equipment (such as lower boiler cost and lower cost of gas turbines versus steam plant).

The scheme might advantageously be installed at a riverside location to process coal on, for instance, the scale of 13×10^6 tons per year, providing sulfur-free fuel for 5000 megawatts of power and typically shipping 4×10^5 long tons per year of sulfur. Such an installation would enjoy economies of scale in coal processing and sulfur production. Elliott (*40*) has called attention to the role which low-cost fluidized-bed boilers might play in supplying peak-load power, and the availability of low-sulfur coke to such boilers would relieve their owners of the need to install equipment for sulfur recovery.

A natural evolution is foreseen:

- The first Coalplexes would be justified simply for their economy in dealing with sulfur.
- Later, modifications would cream off limited amounts of pipeline gas and liquids from volatile matter.
- Further evolution would increase production of gas and liquids.

Ultimately, in an economy powered principally by breeder reactors, a Coalplex would evolve for which power might be a relatively minor by-product, and fixed carbon would be shipped mainly for metallurgical or electrochemical use.

Reordering Coal Engineering's Priority

There exist paths of technological development that can lead to clean and cheaper power. The missing ingredient is money. Lack of money increases the degree to which an aura of inevitability protects PF combustion from competitive ideas. When money is short, the "practical" man tends to prefer projects aimed at adjusting the mature art, and it is hard to get serious attention for ideas that are not a tack-on to the old.

I envy nuclear engineers in many respects, and not least in the obvious fact that no aura of inevitability will arise to protect the light water reactor. The breeder concept holds out the hope of an efficiency some 50 to 80 times greater than the efficiency of this primitive device, if efficiency is related to our total natural supplies of both fissile (uranium-235) and fertile (uranium-238 and thorium-232) materials. The light water reactor's aboriginality is appreciated if it is recalled that Watt's engine had an efficiency roughly one-tenth that of a modern power station. In a mature power economy based upon breeder reactors, the cost of uranium or thorium will matter very little. The light water reactor is sensitive to the price of uranium, and so also will be the cost of initiating a breeder power economy. I have seen no responsible opinion that the light water reactor, at the commonly projected growth of nuclear industry, can have any competitive standing beyond about 1990, for there simply is not enough low-cost uranium now in sight to fuel this reactor in the 21st cenutry. It is this fact and the uncertainties of the breeder development program that provide powerful arguments for a reordering of priorities for nuclear versus coal engineering. Benedict (*41*) wrote recently:

> Development of the sodium-cooled fast breeder reactor will be difficult and time consuming, and it is not certain that power costs will be low enough to permit them to compete with plants burning fossil fuel at today's price. Nevertheless the potential value of having available a practically unlimited source of energy [is] so great as fully to justify the effort now going into this development.

Equally justified is a vigorous effort to maintain coal's competitive position vis-à-vis the light water reactor, so that some low-cost uranium-235 will remain even if the breeder is delayed beyond present hopes.

Coal engineers need "fun" money, such as nuclear engineers have had, to pursue curiosities. It is a shame that no Godel Ignifluid unit can yet be seen here. An anecdote suggests what we may have missed on account of its lack. In the mid-1930's, M. W. Kellogg Company and Esso were at work on a technique for cracking distillate oils by passing oil vapor together with a fine clay catalyst through a heated coil. Results obtained in a small coil were good, but a larger coil performed badly. Examination of the flow of catalyst and air in a glass coil of the larger size revealed that the greater centrifugal force in the larger coil caused the catalyst to separate from the bulk of the air and to move in a ribbon along the outer diameter of the coil. As it happened, a team of Kellogg and Esso engineers had just returned from a trip to Germany, where the team had visited a Winkler gasifier much like the one shown in Fig. 1. One of the men suggested putting the catalyst in a bed and passing oil vapor upward into the bed to cause it to fluidize. This was done, and within a few weeks a pilot unit was in operation. The tremendous fluid cat cracker of today has a look of inevitability, but, if the men had not seen the Winkler gasifier, how long might it have been before fluid-cracking, so important during World War II, had been invented? How long before the chemical engineering profession awoke to the potential of fluidized-bed art?

It is a shame that no Jéquier unit has been built here, no Szikla-Rozinek unit (*42*), no Secord slagging-grate gasifier (*43*), no Winkler generator of the East German tuyere-blown, pear-shaped design (*44*), no Ruhrgas-Lurgi carbonization unit (*45*), or no plant-scale version of the dilute-phase carbonization unit for agglomerating coals developed by the Grand Forks Station of the Bureau of Mines (*46*). A theme of this article has been the need for a search for equipment of the highest possible coal-treating capacity. Experience with these novelties might have carried us far along a road now only dimly apparent. I have tried to indicate my view of the best paths of work, but it is proper to wonder how many ideas are missing for lack of the chance to see their physical embodiments.

Coal engineers also need the spectaculars, like the nuclear electric power station at Shippingport, which were so important to nuclear engineers before they could offer competitive equipment. The last coal spectacular was the coal-to-oil unit at Louisiana, Missouri, on which something over $100 million was spent after World War II. President Eisenhower canceled the experiment in 1953 for what then seemed proper reasons. In some respects the plant was obsolete even when built, and by 1953 two sad facts were evident: oil from coal would not be competitive for at least two decades, and by that time a far better job could be done. Yet I wonder how many good minds were turned away from coal engineering by Louisiana's closing. Senior men were forced to scramble to keep themselves occupied with responsible tasks, and at least a half-generation of inventors looked elsewhere than at coal's problems.

Spectaculars are important not only for a field's self-esteem and to attract recruits, but also for a reason more subtle. A development engineer is seriously handicapped if he works for years under no great urgency to provide engineering designs for a full-scale plant that he knows will actually be built. There is a coziness in this circumstance hard to resist, which leads to the temptation to resort to dodges convenient for getting on with small-scale work not suited for use in the field. A flaw of some coal development programs in recent years has been too little concern with seeking and testing designs affording coal throughputs per unit volume which are realistic for the commercial scale.

Above all, coal engineers need more coal engineering establishments—and more coal engineers! It has taken a great deal of money to generate the headlines, to produce the fellowships and fine work at colleges, and to support the symposia and demonstrations, reaching even into secondary schools, which have drawn the first-class minds who have carried forward our space and nuclear programs. It will be important to the nation's welfare that coal engineering during the 1970's at last receive its proper share of these inducements (*47*).

Summary

Nuclear engineers have a vision whose fulfillment will make plutonium-239 and uranium-233 the "dirty, cheap" fuels and will make coal the fuel of esteem. It will be valued for derivative chemicals and clean fuels and for metallurgical and electrochemical uses of its fixed carbon. In a coal technology devised to exploit these values, the recovery of sulfur will be a mere incidental. An immediate, properly financed effort to develop means for coping with sulfur can give us clean air with profit, help to conserve our limited supply of vital uranium-235, and take us a large step toward a coal technology for the 21st century.

References and Notes

1. *Hearings on S. 780, U.S. Senate, Subcommittee on Air and Water Pollution of Committee on Public Works* (May 1967), pp. 2043–2047, 2107–2118, 2278–2285, 2287–2316, 2550–2562, 2567–2589.
2. J. R. Garvey, testimony to Joint Committee on Atomic Energy, U.S. Congress (25 February 1970). Stack gases of a 1000-megawatt power station flow at 1300 m^3/sec and typically contain SO_2 at 0.2 to 0.3 percent by volume. Chemical treatment of such a vast throughput for removal of a constituent present in such small amount is bound to be costly. Since the 1930's, development teams have worked upon many ingenious ideas for capturing SO_2 from stack gases. The history of many of these efforts is depressing: initial enthusiasm followed by abandonment when the economic facts became clear. At the moment, some half-dozen or so schemes are alive. Two have reached field trials at a scale greater than 100 megawatts. The Tennessee Valley Authority is conducting trials of a dry limestone-injection technique in the hope of obtaining 40 to 60 percent removal of sulfur oxides through application of twice the stoichiometric amount of stone for absorption of the total sulfur [A. V. Slack and H. L. Falkenberry, *Combustion* **41**(6), 15 (December 1969)]. The hoped-for result would not meet New Jersey's requirement for October 1971. Combustion Engineering, Inc., is testing a wet limestone system which uses substantially the stoichiometric amount of stone. First trials ran into difficulties at the Meramec Station of Union Electric Company. A different design has given better results in preliminary operation at Lawrence Station of Kansas Power & Light Company [*Electrical World* **172**(23), 38 (8 December 1969)], and the technique may represent the best hope for early control of SO_2. The Cat-Ox system of Monsanto Company [J. G. Stites, Jr., W. R. Horlacher, Jr., J. L. Bachofer, Jr., J. S. Bartman, *Chem. Eng. Progr.* **65**(10), 74 (October 1969)] appears ready for a field trial at the +100-megawatt scale. Several other schemes are perhaps also ready for such trials. An overblown heralding of these systems could lead to pressure for their application on an imprudently wide scale. The history of classic disasters of engineering—for instance, the Great Eastern, the post-World War II Fischer-Tropsch synthesis, the Fermi reactor—should teach caution in applying new technology on a giant scale. Too much money and hope committed to multiple installations of inadequately tested systems could make it difficult to fund work on more advanced schemes for control of sulfur oxides. A distinguished panel of engineers recently concluded that efforts to force the broad-scale installation of unproven processes might delay effective SO_2 emission control (Report COPAC-2, prepared by Ad Hoc Panel on Control of Sulfur Dioxide from Stationary Combustion Sources, Committee on Air Quality Management, Committees on Pollution Abatement and Control, Division of Engineering, National Research Council, Washington, D.C., 1970).
3. R. G. Ramsdell, Jr., and C. F. Soutar, paper presented at the Environmental Engineering Conference of the American Society of Mechanical Engineers, Chattanooga, Tennessee, (May 1968).
4. C. F. Herington, *Powdered Coal as a Fuel* (Van Nostrand, New York, ed. 2, 1920), p. 273.
5. F. Winkler, U.S. Patent 1,687,118 (9 October 1928).
6. A. A. Godel, *Rev. Gen. Therm.* **5**, 349 (1966).
7. P. Cosar, *Arts et Manufactures* **196**, 13 (April 1969).
8. J. W. Bishop, E. B. Robison, S. Ehrlich, A. K. Jain, P. M. Chen, paper presented at the annual meeting of the American Society of Mechanical Engineers (December 1968).
9. *Proceedings of First International Conference on Fluidized Bed Combustion*, R. P. Hangebrauck and D. B. Henschel, Eds. (National Air Pollution Control Administration, Cincinnati, November 1968); R. P. Hangebrauck and D. B. Henschel, *BCURA Ind. Lab. Mon. Bull.* **33**(5), 106 (May 1969); *Combustion of Coal in Fluidised Beds: Proceedings of a Symposium Held at Coal Research Establishment*, J. Highley, Ed. (National Coal Board, Coal Research Establishment, Stoke Orchard, Cheltenham, May 1968); P. L. Waters and A. Watts, paper presented at meeting of Institute of Fuel, Canberra, Australia (November 1968).
10. H. R. Hoy and A. G. Roberts, *Gas Oil Power* **65**, 173 (July–August 1969).
11. H. R. Hoy and J. E. Stantan, *Amer. Chem. Soc. Div. Fuel Chem. Prepr.* **14**(2), 59 (May 1970).
12. The gas-fired San Angelo Station of West Texas Utilities Company illustrates the economies afforded by a combination of gas- and steam-turbine equipment [A. R. Cox, L. B. Henson, C. W. Johnson, *Proc. Amer. Power Conf.* **29**, 401 (1967); R. W. Foster-Pegg, *ibid.*, p. 410]. The station cost $78 per kilowatt, a remarkably low figure for its 128-megawatt size. In spite of its small size and modest steam conditions (99 atmospheres and 538°C with a reheat to 538°C), the station's efficiency, about 41 percent on a lower heating value basis, is probably the highest in the world. The high efficiency results from the fact that the gas turbine "tops" the steam cycle—that is, heat rejected by the gas-turbine cycle is employed to raise steam. The proportion of the total power produced by the gas turbine is 20 percent, about the same as in the BCURA concept.
13. I. I. Kirillov, V. A. Zysin, S. Ya. Osherov, L. V. Aren'ev, Yu. E. Petrov, *Teploenergetika* **14**(1), 44 (1967).
14. F. L. Robson and A. J. Giramonti, *Amer. Chem. Soc. Div. Fuel Chem. Prepr.* **14**(2), 79 (May 1970).
15. L. Jéquier, L. Longchambon, G. van de Putte, *J. Inst. Fuel* **33**, 584 (1960).
16. A. M. Squires, in *Power Generation and Environmental Change*, D. A. Berkowitz and A. M. Squires, Eds. (M.I.T. Press, Cambridge, in press).
17. L. Reh, paper presented at meeting of the American Institute of Chemical Engineers, Puerto Rico (May 1970).
18. Through the development of a magnetohydrodynamic (MHD) device, a 50 percent efficiency of power generation might be achieved sooner than may be possible in the expected course of gas-turbine development. A balanced effort will include work on both techniques for topping the steam cycle. Temperatures in a MHD generator are so high as to volatilize most of the ash in coal, and a partial combustion to effect a separation of ash would simplify gas treatment for removal of dust downstream of the MHD device. This device is better suited than a gas turbine to receive gases directly from a partial combustion, since alkali offers no problem. Indeed, the potassium sometimes present in coal ash is an effective "seed" material for rendering the hot gases conductive to electricity. Steam in the gas is undesirable, and magnetohydrodynamics is best suited to generate electricity from coke.
19. P. F. H. Rudolph, *Amer. Chem. Soc. Div. Fuel Chem. Prepr.* **14**(2), 13 (May 1970). A gas turbine will supply 43 percent of the power from Lurgi's 170-megawatt plant for Steinkohlen-Electrizität AG, to operate in 1971. Its remarkably low cost, $90 per kilowatt, includes $19 per kilowatt for coal gasification (that is, partial combustion) but does not include equipment for gas desulfurization, since the plant will use coal low in sulfur.
20. E. A. Pirsh and W. L. Sage, *ibid.*, p. 39.
21. *Chem. Week* **78**(26), 76 (30 June 1956).
22. M. Kruel and H. Jüntgen, *Chem. Ing. Tech.* **39**, 607 (1967).
23. C. W. Zielke, H. E. Lebowitz, R. T. Struck, E. Gorin, *Amer. Chem. Soc. Div. Fuel Chem. Prepr.* **13**(4), 13 (September 1969).
24. G. Moss, in *Proceedings of First International Conference on Fluidized Bed Combustion* (see *9*).
25. A. M. Squires, R. A. Graff, M. Pell, paper presented at meeting of American Institute of Chemical Engineers, Washington, D.C. (November 1969).
26. J. R. Coke, thesis, University of Sheffield, England (1960); R. R. Bertrand, A. C. Frost, A. Skopp, "Fluid Bed Studies of the Limestone Based Flue Gas Desulfurization Process" (report from Esso Research and Engineering Co. to NAPCA, October 1968).
27. A. M. Squires and R. A. Graff, paper presented at meeting of Air Pollution Control Association, St. Louis, Missouri (June 1970).
28. W. Guntermann, F. Fischer, H. Kraus, German Patent 1,184,895 (7 January 1965); A. M. Squires, *Chem. Eng.* **74**(26), 101 (18 December 1967).
29. G. P. Curran, C. E. Fink, E. Gorin, *Advan. Chem. Ser.* **69**, 141 (1967).
30. M. E. Sacks, C. A. Gray, R. T. Eddinger, *Amer. Chem. Soc. Div. Fuel Chem. Prepr.* **13**(4), 287 (September 1969).
31. A. M. Squires, *Advan. Chem. Ser.* **69**, 205 (1967).
32. J. D. Batchelor, G. P. Curran, E. Gorin, U.S. Patents 2,927,063 (1 March 1960), 2,950,229, 2,950,230, 2,950,231 (23 August 1960), and 3,101,303 (20 August 1963).
33. F. G. Shultz and J. S. Berber, *Amer. Chem.*

Soc. Div. Fuel Chem. Prepr. **13**(4), 30 (September 1969)
34. W. B. Behnke, Jr., paper presented at AAAS meeting, Boston (28 December 1969).
35. F. W. Theodore, "Low Sulfur Boiler Fuel Using the Consol CO_2 Acceptor Process: A Feasibility Study" (report from Consolidation Coal Company to OCR, November 1967).
36. J. F. Jones, M. R. Schmid, M. E. Sacks, Y. Chen, C. A. Gray, R. T. Eddinger, "Char Oil Energy Development" [report from FMC Corporation to OCR (January 1967)]; C L. Tsaros, *Amer. Chem. Soc. Div. Fuel Chem. Prepr.* **12**(3), 95 (September 1968); R. T. Eddinger, J. F. Jones, F. E. Blanc, *Chem. Eng. Progr.* **64**(10), 33 (October 1968); C. L. Tsaros, J. L. Arora, W. W. Bodle, *Amer. Chem. Soc. Div. Fuel Chem. Prepr.* **13**(4), 252 (September 1969); N. P. Cochran, *NEREM Rec.* **11**, 134 (1969).
37. E. Gorin, G. P. Curran, J. D. Batchelor, Patent 2,824,047 (18 February 1958); J. D. Batchelor, E. Gorin, C. W. Zielke, *Ind. Eng. Chem.* **52**, 161 (1960).
38. C. A. Gray, M. E. Sacks, R. T. Eddinger, *Amer. Chem. Soc. Div. Fuel Chem. Prepr.* **13**(4), 270 (September 1969).
39. M. L. Vestal and W. H. Johnston, *ibid.* **14**(1), 1 (May 1970).
40. D. E. Elliott, "Can Coal Compete? The Struggle for Power," inaugural lecture, University of Aston in Birmingham, England (20 November 1969).
41. M. Benedict, *NEREM Rec.* **11**, 130 (1969).
42. A. Rozinek, *Feuerungstechnik Leipzig* **30**, 153 (1942); R. H. Essenhigh and J. M. Beér, *J. Inst. Fuel* **33**, 206 (1960).
43. C. H. Secord, U.S. Patent 3,253,906 (31 May 1966); H. R. Hoy, A. G. Roberts, D. M. Wilkins, *Inst. Gas Eng. J.* **5**, 444 (1965).
44. B. von Portatius, *Freiberg. Forschungsh. Reihe A* **69**, 5 (1957); A. M. Squires, *Chem. Eng. Progr.* **58**(4), 66 (April 1962).
45. W. Peters and H. Bertling, *Amer. Chem. Soc. Div. Fuel Chem. Prepr.* **8**(3), 77 (August 1964); R. Rammler, *Erdoel Kohle Erdgas Petrochem.* **19**, 117 (1966).
46. M. Gomez, W. S. Landers, E. O. Wagner, *U.S. Bur. Mines Rep. Invest.* **7141** (1968).
47. There are positive factors in the situation. Many of the present student generation have a bent toward service, and engineering students are growing sensitive to career stability. Word of layoffs from space or military programs reaches campuses quickly, and I predict that enrollments will drop precipitately in non-real-world engineering activities. The space effort attracted thousands of first-class minds whose loss to the real world's business the nation can ill afford. It will be important to decision makers of the future that they possess good data on the degree to which these minds succeed in obtaining retreads and again finding responsible work.
48. Work at the City College on desulfurization of fuels by calcined dolomite is supported by Research Grant No. AP-00945 from the National Air Pollution Control Administration, Consumer Protection and Environmental Health Service. This article is adapted from a paper presented 28 December 1969 at the Boston meeting of the AAAS in the symposium entitled "Power Generation and Environmental Change," arranged by the AAAS Committee on Environmental Alteration.

PETROLEUM IN PERSPECTIVE*

By
HARRY R. JOHNSON†

ABSTRACT

The environment that surrounds the petroleum industry is a complex mixture of economics, technology, and public policy. In the 1950's, the historical relationships between these factors were upset by the widespread application of technology. Because of technology, it became profitable for individual firms to make significant operating changes, which are reflected in a sharply falling balance between domestic supply and demand.

The analysis in this article indicates that a new supply-demand maintenance level will be reached within the next few years. Efforts to maintain this level will be reflected by upswings in exploratory drilling, increased application of secondary recovery technology, and initial attempts to extract liquids from supplemental sources as coal, oil shale, and tar sands. All foreseeable needs for petroleum will be satisfied, but short-run pressure may be created for a higher real price of oil and/or increased imports.

Agitation is high to reform public policies that control a substantial portion of petroleum activity. But policy reform cannot be effective if it is a haphazard, uncoordinated activity. Instead, it must proceed simultaneously along several fronts. Failure to do so may lead to a dangerous over-correction of the future balance between supply and demand.

Early studies of petroleum contain ominous predictions concerning the exhaustion of this resource. Of course, these predictions never came true. The record shows that since the 1930's and up until the very recent past, the supplies of petroleum as measured against its demand have remained nearly constant.

In the 1950's and 1960's, however, significant operating changes were made by petroleum firms. The individual actions of these firms are reflected in a sharply falling balance between domestic supply and demand. The severity of the decline has evoked growing concern among petroleum analysts, and in 1965 a major Government report on petroleum was released that stated "... [W]hat has been done since 1956 to find new supplies of oil ... has not been enough to provide a sound basis for future growth. Additional exploratory effort is needed, ..."[1] But the supply-demand trend continued downward.

Another, more comprehensive, report on petroleum was released 3 years later. The "summary & conclusions" section of the report contains the following:

> It therefore appears that the discovery rate since 1957 will not be sufficient to offset withdrawals from proved reserves between 1965 and 1980 on the basis of anticipated recovery rates. Specifically,
> - either the recovery rate must improve faster than the 0.5 percent annual improvement projected, or
> - Discoveries must be increased above the levels that have prevailed since 1957.[2]

That a fundamental change is or has occurred is not to be disputed. But why were these changes made in the face of huge future demands for petroleum? Have we at last dipped too deeply into our supplies of this non-renewable resource? Can we supply our future needs?

Yes, we will be able to supply all needs into the foreseeable future. The real question relates to the price required to make it profitable to produce these supplies. Why the changes were made is also related to economics. And economics, in turn, is inextricably bound to technology and energy policy. This article explores the interrelationships between economics, technology, and policy to place petroleum in perspective.

I
DEMAND PATTERNS

Private firms, industry associations, and government attempt to estimate future demand-supply relationships for petroleum. Projections of demand provide the basis for rational judgments concerning acequacy of supply and future costs. Common to all projections are assumptions concerning the general economic climate, the political environment, and the pace and direction of technological changes.

Short-run projections of a few years tend to be fairly accurate since petroleum demand has exhibited a long-run growth rate of 3 to 4 percent per year, although this rate in recent years has been higher. In addition, petroleum consumption today is the result of decisions that were made several years ago. The current production runs of

†Petroleum and Natural Gas Engineer, Office of the Chief, Division of Petroleum and Natural Gas, Bureau of Mines, U.S. Department of the Interior, Washington, D.C.

*This article was completed at the University of Washington, Seattle, Wash., under the 1968-69 Career Education Fellowship program granted by the National Institute of Public Affairs. Although the concepts contained herein were shaped by many people, the author feels especially indebted to Professors James A. Crutchfield, Economics, Robert H. Pealy, Public Affairs, and Richard A. Cooley, Geography; all of whom are at the University of Washington.

1. U.S. Department of the Interior, An Appraisal of the Petroleum Industry of the United States, at 17 (1965).
2. U.S. Department of the Interior, U.S. Petroleum Through 1980 at vii (1968).

Reprinted with permission from *Natural Resources J.*, vol. 11, pp. 119–147, Jan. 1971. Published by the Univ. of New Mexico School of Law, Albuquerque, N. Mex. 87131.

automobiles, for example, are the products of a complicated scheduling process that was initiated 3 years earlier. Changes in the price of petroleum or in technology will not, in the short run, drastically alter petroleum consumption.

Long-run projections for any energy commodity are subject to many influences. Among these are price, interfuel competition, consumer preferences, technological changes, government policy, and a growing effort to provide a cleaner environment. The longer range projections have tended to underestimate technological changes and ignore institutional factors.

These limitations cannot be completely overcome, but it is possible to place boundaries on the estimates. One recent study of all of the energy commodities[3] uses both sensitivity analysis and the relatively new concept of contingency analysis to set these boundaries. This study utilizes standard economic techniques to project the "best" or medium estimate of demand to 1980. How sensitive the estimates are to changes in such key variables as gross national product and population growth is then determined. Sensitivity analysis does not alter the percentage of total energy supplied by each commodity, only the absolute amounts.

Contingency analysis attempts to formulate and quantify reasonable events that could change the amounts of energy supplied by each commodity. For example, if low-sulfur oil is not available and technology cannot advance fast enough to permit the economic removal of sulfur from oil, it is reasonable to expect natural gas or other fuel commodities to gradually replace oil in areas suffering from polluted air. The increase in gas usage and decrease in oil was examined for the New York, Chicago, and Philadelphia marketing areas. Other contingencies studied were reduced and increased oil imports. The electric car and other alternatives to the internal combustion engine are not expected to pose a threat to the gasoline market between now and 1980.[4] A contingency not studied was how an increase in the price of petroleum relative to other energy commodities would effect oil consumption. Such an increase is a real short run possibility, as discussed later in this article.

The standard projection of energy use patterns provides the medium projection while sensitivity and contingency analysis serve to bracket this projection with boundary conditions. For petroleum, the domestic oil required to meet future needs is given in Table 1.

The nation in the 15-year period 1965-80 is expected to consume about 80 billion barrels of oil, 56 billion from domestic sources and the balance from imports. Total consumption will be about as much as the nation consumed during the previous 100 years. Concern is evident over the industry's ability to meet these enormous requirements in the face of a falling balance between domestic supply and demand.

TABLE 1

Demand for Domestic Petroleum, 1965 and 1980 Projected

	1965		1980	
	Million bbl/day	Billion bbl/year	Million bbl/day	Billion bbl/year
High			15.1	5.520
Medium	7.8	2.849	12.7	4.645
Low			10.3	3.752

Source: Warren E. Morrison and Charles L. Readling, *An Energy Model for the United States, Featuring Energy Balances for the Years 1947 to 1965 and Projections and Forecasts to the Years 1980 and 2000*, U.S. Department of the Interior, Bureau of Mines Information Circular 8384 (Washington: U.S. Government Printing Office, 1968).

II
SUPPLY PATTERNS

The amount of any commodity that can be made available to any buyer is directly related to the price the buyer is willing to pay for it. In any analysis of supply, therefore, price is a key variable.

A. Price

Price stability of domestic oil has been a characteristic of the petroleum industry over the years. This is readily discernible in Figure 1 where actual and adjusted prices are plotted against time. For the purpose of comparison, the price adjusted for inflation is the more important for it shows how many 1965 dollars[5] would have been required to purchase 1 barrel of oil at any time in the 1951-65 period.

The average price in 1965 dollars over the full 15-year period has been $2.98 per barrel, with price swings around the average confined to within $0.21 per barrel. The extremes, therefore, deviate from the average by only about 7 percent over the entire period.

This remarkable stabilization of prices is the result of many factors. Among the most important of these are the restriction of

Figure 1.
HISTORICAL PRICES OF CRUDE OIL

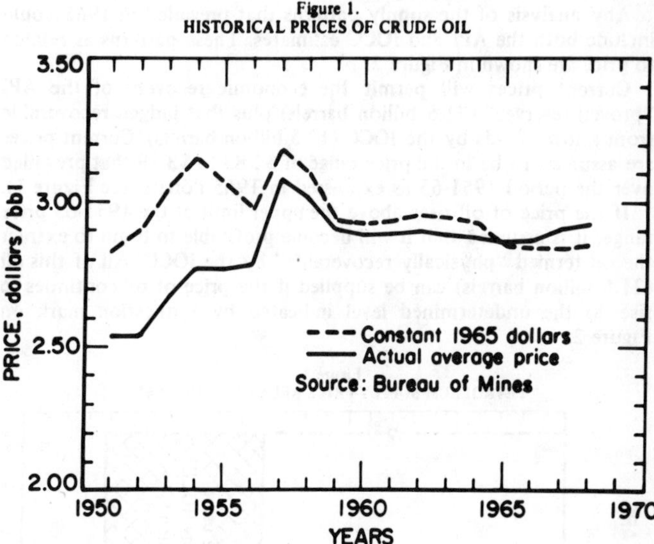

foreign oil imports and compulsory prorationing of oil. Each of these programs limits supply and, therefore, tends to support the price. Another factor is technological advances that have offset large increases in the real costs of labor, oilfield machinery, and casing. The saving due to advances in technology during 1950-65 has been estimated to be about $1 per barrel.[6]

Prices fell slowly between 1962 and 1965 and then began to gradually inch upward. By early 1969, price jumps of up to $0.20 per barrel to the $3.00 to $3.10 level were made, bringing oil close to its 1957 peak price.

B. Supply, 1965

Most discussions of supply seem to center on the "proved reserves" concept as defined by the American Petroleum Institute (API).[7] This is natural since the concept has been used since 1936

3. Morrison and Readling, *An Energy Model for the United States, Featuring Energy Balances for the Years 1947 to 1965 and Projections and Forecasts to the Years 1980 and 2000,* U.S. Department of the Interior, Bureau of Mines Information Circular 8384 (1968).

4. U.S. Department of Commerce, *The Automobile and Air Pollution: A Program for Progress* (1967).

5. Deflated by wholesale price index for petroleum as compiled by the Bureau of Labor Statistics.

6. National Petroleum Council, Impact of New Technology on The U.S. Petroleum Industry 1946-1965, part 1 (1967).

7. American Petroleum Institute, *Reserves of Crude Oil, Natural Gas Liquids and Natural Gas in the United States and Canada, as of December 31, 1966.* Oil Information Bulletin, No. 21 (1967).

when the API first began to publish its annual evaluation of the nation's reserve position.

Underlying the "reserve" concept are a number of assumptions and limitations which have been adequately discussed.[8] The concept applies only to those quantities of oil that have been discovered and can be economically produced using installed equipment or technology that has been pilot tested in that field. In a real sense, "proved reserves" represent the product inventory of the petroleum industry at any given time, just as the amount of goods in a retail store represents the product inventory of that store. This working inventory of petroleum has been nearly constant over the period 1956-65, ranging from 30.3 to 31.7 billion barrels.

The Interstate Oil Compact Commission (IOCC) has extended the concept of recoverable reserves.[9] It periodically estimates the amount of petroleum that could be produced by secondary recovery if such techniques were actually applied to known oilfields.[10] In addition, it estimates the oil that could physically be recovered from those fields if cost were not a factor. This oil could be economically recovered if the price of oil were to increase.[11] The IOCC does not attempt to apply price to its estimates, but reasonable estimates of price can be made since these need not be precise to demonstrate the total United States supply picture.

Any analysis of the supply patterns that prevailed in 1965 could include both the API and IOCC estimates. These patterns as related to price are shown in Figure 2.

Current prices will permit the economic recovery of the API "proved reserves" (31.5 billion barrels) plus that judged recoverable from known fields by the IOCC (17.5 billion barrels). Current prices are assumed to be in the price range of $2.83 to $3.19 that prevailed over the period 1951-65 as expressed in 1965 dollars (see Figure 1).

If the price of oil rises above the upper limit of the 1951-65 price range, it is assumed that it will become profitable to begin to extract the oil termed "physically recoverable" by the IOCC. All of this oil (21.5 billion barrels) can be supplied if the price of oil continues to rise to the undetermined level indicated by a question mark on Figure 2.

Figure 2.
PETROLEUM SUPPLY-PRICE RELATIONSHIP, 1965

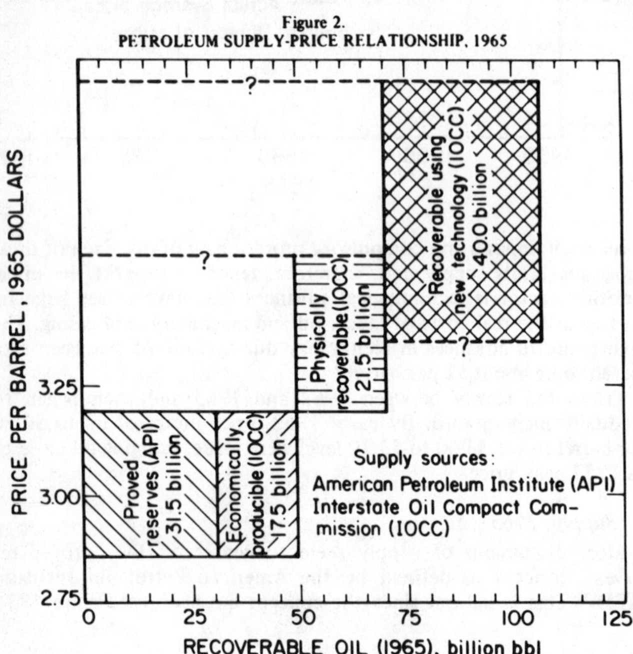

Rising prices will also have another effect. At some point, it will become profitable to begin to apply new technology, such as miscible and in situ recovery, in field operations. Complete application of these methods to existing fields will supply an additional 40 billion barrels of oil.

Figure 2 really shows that supply depends upon price. At current prices, the supply is about 49 billion barrels. If the price rises, an additional 22 billion barrels can be added, making the cumulative supply about 71 billion barrels. New technology will increase cumulative supply to about 111 billion barrels.

The question of physical exhaustion of the nation's petroleum supplies is not relevant to this analysis. The real question is the rate at which new oil can be discovered or old discoveries converted to the "proved reserve" category from the other categories depicted in Figure 2. This can be done from domestic sources in only two ways: (1) new discoveries, and (2) improved technology.

C. *Crude Oil Discovery*

Exploration for oil in the United States reached a peak in the mid-1950's and has fallen since.[12] These trends are quite pronounced, as can be seen in Figure 3. The most important aspects of the situation are stated quite well in Interior's 1968 study:

> The decline in drilling is notable for two reasons. It is not only the longest in the history of the industry, but it has actually worsened over the period 1964-67 which was marked by strong increases in demand, firmer prices for both crude oil and products, and a large gain in wellhead revenue. The result has been one of the more noteworthy *non sequiturs* in the annals of petroleum economics. Within the past three years, production of domestic crude oil has risen by

Figure 3.
CRUDE OIL DISCOVERY TRENDS IN THE UNITED STATES

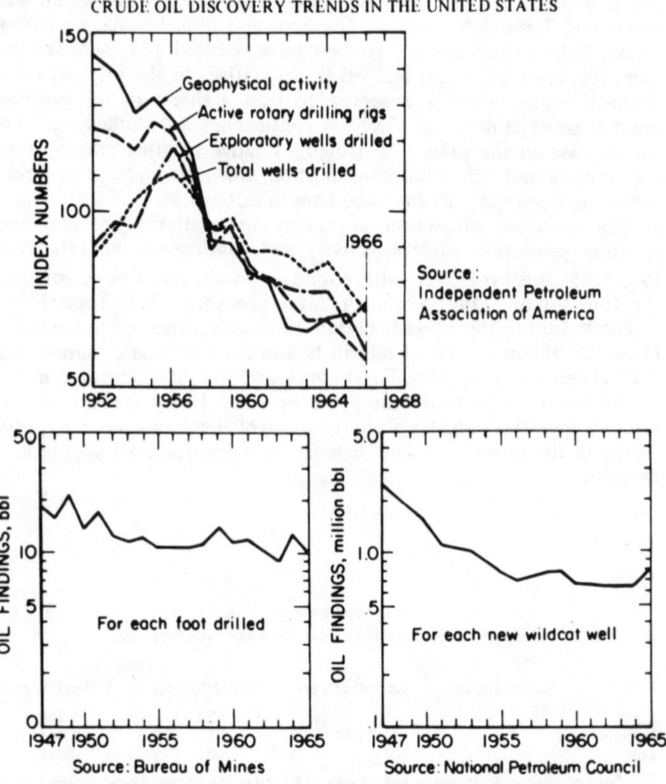

8. Lovejoy and Homan, Methods of Estimating Reserves of Crude Oil, Natural Gas and Natural Gas Liquids (1965).

9. Interstate Oil Compact Commission, The Oil and Gas Compact Bulletin, V. 25, No. 2 (1966).

10. In 1966, the American Petroleum Institute also began to estimate potential recovery from fields where secondary recovery equipment had not been installed; however, the definitions used by these two groups are different.

11. Using the technology that was available in 1965.

12. For a comprehensive analysis of these trends *see*, National Petroleum Council, Factors Affecting U.S. Exploration, Development and Production 1946-65 (1967).

> over one million barrels a day, and this gain would have been almost as impressive even had the stimulus of the 1967 Middle East crisis been absent. Annual revenue to producers from this source alone accordingly increased by over a billion dollars between 1964 and 1967. Yet the response has been the extension of a drilling decline already eight years in length to levels not reached in two decades. Exploratory drilling in 1967 was off by 16 percent from what it had been three years before, while new field wildcats lagged by 20 percent. Total wells drilled declined by 22 percent during the same period.[13]

The rate of oil discovery (Figure 3) is an indicator of the maturity of the petroleum industry. Note that oil findings for each foot drilled or for each wildcat well have gradually declined and appear to have stabilized in recent years. Concurrently, the cumulative oil contained in these deposits reached a peak over 30 years ago and since then has become progressively smaller (Table 2).

TABLE 2
Discovery History, United States

Decade	Giant Fields Discovered (Over 100 MMbbl)*
1870 to 1910 (4 decades)	21
1910 to 1920	32
1920 to 1930	65
1930 to 1940	70
1940 to 1950	43
1950 to 1960	25
1960 to 1966 (6 years)	5
Total	261

Source: Oil and Gas Journal, (August 22, 1966), pp. 150-151.
*Recoverable oil, 1965 conditions

These discovery trends reflect the law of diminishing returns in action. That is, equal investments of capital will produce positive results, but the returns become progressively smaller. In 1948, for example, one wildcat well found over 2.0 million barrels of oil. By 1965, drilling one wildcat well produced a return of only about 1.0 million barrels.

The law of diminishing returns is a characteristic of any extractive industry. For petroleum, it indicates a mature industry that is now well into the decline phase of its discovery cycle. Development of a radically new exploration technique or tool could alter the rate at which oil is discovered. However, exploration technology has been evolutionary over the past 20-year period, and no major changes are to be expected in the near future. Thus, the established discovery trends will probably continue into the foreseeable future. Since there are considerable quantities of oil waiting to be discovered, the cumulative amount of new oil found could be increased sharply by increasing the number of wells drilled. It will just take more drilling to add a given amount of oil than it did during the heyday of the industry, except principally for one new area–Northern Alaska.

1. Northern Alaska: New Oil Province

The big oil news of 1968 was the discovery of oil in the northernmost part of Alaska near the Arctic Circle. This discovery sparked intense interest in this area as a potential source of new oil. Active exploration has established that the potential reserves are huge, some estimate 10 billion barrels or more, which would place these fields high among the world's largest.

Oil in this area has been known to be present since 1944 when the Navy began to develop its Naval Petroleum Reserve No. 4. One gasfield and the Umiat oilfield were found, but the full potential of the area has never been adequately evaluated. Knowing that a large potential exists and actually attempting to define and develop the deposits are two different things. The intensity of the drive to develop this oil will be a function of the costs of extracting and transporting this oil to markets as compared to other supply alternatives.

Man has extended his search for oil into progressively more hostile environments. The price he has paid for this invasion is reflected by drilling costs (Table 3). Drilling on the North Slope presently costs about 3,000 percent more than drilling a well in Texas or Oklahoma. Such investments require that the amount of oil discovered be truly large. It also means that only the biggest fields can be developed, thus the overall recovery will tend to be lower than in less costly areas.[14]

TABLE 3
Drilling Costs, United States

Location	Cost/well, dollars
On shore, Continental U.S.[1]	$ 60,000
Offshore Gulf of Mexico[1]	400,000
Alaska-Cook Inlet[1]	800,000
Alaska-North Slope[2]	2,000,000

[1] Joint Association Survey
[2] *Newsweek*, (January 6, 1969) p. 47

How to move the oil to markets has not yet been fully decided, but submarines and/or tankers able to master ice are still under consideration. Plans have been announced to construct and operate by 1972 an 800-mile pipeline initially capable of transporting 500,000 barrels of oil per day. This will be an engineering feat, since both terrain and extreme temperature swings must be conquered. But the benefits to be gained by this $900 million venture should be carefully weighed against the costs to the natural ecological system of that area.

In the Arctic, the natural balance between the environment and the plant and animal systems is extremely fragile. The environment has permitted the survival of only the hardiest plant and animal species which are highly interdependent on each other for survival. Man's intrusion in this area could upset the natural balance, thus leading to a drastic change in the ecological system. Efforts to prevent this from occurring are being actively pursued.

Many of the popular stories written about Northern Alaska oil assume that significant quantities will become almost instantly available. This is not probable if history is any guide to what can be expected. Consider offshore Louisiana, for example. Oil was discovered here in the late 1940's. The forced development of the technology necessary to operate in this water environment began to become available in the 1950's. In 1957, offshore Louisiana became the most active exploration province in the United States, reaching an alltime high in 1966. By 1968, 8,000 wells had been drilled and $6 billion had been invested, yet the area contributes less than 10 percent of the nation's total production. This is significant, but it has taken 12 years and immense capital expenditures. Furthermore, the return on investment is said to be low by industry standards.

A supply area that will eventually be tied very closely to North Slope oil is the Cook Inlet area of Alaska. This area became an active exploration province in the mid-1950's. Total investment is estimated to be about $1 billion, but production and delivery to west coast markets have been slow because of the lack of transportation. Oil from this area is steadily increasing but accounted for less than 1 percent of the nation's total in 1968. The fact remains, however, that it has taken over 10 years to begin to bring this oil to market in significant quantities.

Cook Inlet oil, large increases in production from California fields due to thermal stimulation, and the expected production from Santa Barbara will all contribute to filling the gap between demand and supply that now exists on the west coast. Inevitably, the transportation system will need to be extended to the mid continent or east coast marketing areas.

Oil from northern Alaska may be needed in the future, but it will not be instantly available or available at extremely low prices. One estimate of wellhead costs[15] (exclusive of taxes) is $1.00 to $1.20 per barrel. Transportation to west coast markets is estimated at $0.60 per barrel and to mid-West and New York markets at $0.80 to $1.00 per barrel. Thus, total production and delivery cost may range from $1.60 to $2.20 per barrel. Others have contended, however, that Alaskan oil will compete in price with Middle East oil. Whatever the price, the rate of development will depend on the actual cost of

13. Interior (1968), *supra* note 2, at 18.
14. In addition, the extreme temperatures of 50°F below zero may make it difficult to economically apply secondary recovery techniques.
15. Jeremy Main, *The Hot Oil Rush in Arctic Alaska*, Fortune Apr. 1969, at 120.

obtaining oil from Alaska as compared with other alternatives. Increased exploration in both offshore and onshore areas is one of these alternatives. Another is to recover more oil from known deposits through improved technology.

D. Crude Oil Recovery

The earliest period of the petroleum industry was characterized by a lack of understanding about how best to produce oil from a field. As a result, many fields were abandoned following the cessation of primary flush production. It has been estimated that the total production plus that which will be produced in the future from the oilfields discovered between 1859 and 1919 will be equivalent to only 25 percent of the total oil that was initially contained in them.[16]

Subtantial improvements in oil recovery were made during the 1920-40 period with the discovery of many of the nation's major natural water-drive fields (East Texas, for example). During this period, principles of underground fluid behavior were established but these principles were not practiced to any great extent in field operations.

Between 1940 and 1960, the overall trend was the discovery of poorer quality oilfields which were not capable of yielding as high a natural oil recovery as those discovered in the 1920-40 period. The decline in reservoir quality, however, was offset by the introduction of formation fracturing and the large-scale application of the earlier established principles of how fluids flow in oil reservoirs.

A constantly improving production technology has permitted the average recovery from a field to increase at an estimated annual rate of about 0.5 percent over the past 20 years.[17] This trend is due mainly to the injection of water into oilfields to supplement or replace the natural energy of the field. Further evolutionary technology is currently, or will soon be, available[18] and is expected to enable a continued improvement in recovery at about the established rate to 1980.

Many wish that this rate could be increased. It can be if the price of oil rises to a point where more efficient methods such as miscible recovery can be economically applied in field operations. However, under the constraint that petroleum must remain competitive with other energy sources, significant increases in the long run real price of petroleum are not likely to occur. This constraint leaves the industry with one main course of action—constant improvement in the recovery efficiency of waterflooding.[19]

Improvements are being made, but the types of oilfields now being found are not as susceptible to this technique as those that were found in the past. Improved technology applied to poorer quality reservoirs will eventually act to slow the 0.5 percent annual improvement in recovery. Like exploration, advancing technology must also respond to the law of diminishing returns.

E. Future Supplies = Discovery + Recovery

Increased discovery and improved technology acting together will determine the rate at which new domestic supplies can be added to oil reserves in the future. Several recent studies of this addition rate have been made by independent groups within the Department of the Interior. They usually start with a projection of future demand similar to that described in the earlier part of this article. Once this demand is established, the investigations generally follow one of three approaches:

1. To develop mathematical expressions that represent established discovery and recovery trends. These are then extrapolated a reasonable distance into the future.
2. To assume that future demands will be met with supply responses as has happened in the past. Using the projection of future demand as a known, historical demand-supply patterns are then used to estimate the unknown, future supplies.
3. To estimate future supply levels by analysis of a number of physical and economic factors. Among these factors are the footage drilled, finding rates, and price of oil.

All three techniques have been used to estimate reserves of petroleum in 1980. These estimates are in surprisingly good agreement (Table 4). The main disagreement occurs when the results of these approaches are interpreted—the subject of the next section of this article.

TABLE 4
Various Estimates of 1980
Domestic Recoverable Reserves

Method[1]	Source of Projected Data	Recoverable Reserves, Billion Barrels Initial	Projected
1	Office of Oil and Gas	31.0 on 1/1/65	29.5 on 1/1/80
2	Bureau of Mines (headquarters estimate)	31.4 on 12/31/65	31.0 on 12/31/80
3	Bureau of Mines (field station estimate)	31.4 on 12/31/65	32.8 on 12/31/80

Source: American Petroleum Institute.
[1] See text

III
DEMAND-SUPPLY BALANCES

Historical demand-supply relationships for petroleum are traced in Figure 4A. A common way of expressing these data is to divide supply by demand (Figure 4B).

Since supply is the API concept of "proved reserves" and domestic demand is production for any given year, the value obtained when supply is divided by production is called the reserve-production (R/P) ratio.[20] It is the R/P ratio that many mistakenly use to predict how many years of petroleum supplies remain to be produced. This ratio really measures the oil that is relatively certain to be produced at current real prices and technology. Increases in the real price or technologic breakthroughs or both will enable the economic recovery

Figure 4A.
HISTORICAL DEMAND-SUPPLY BALANCES

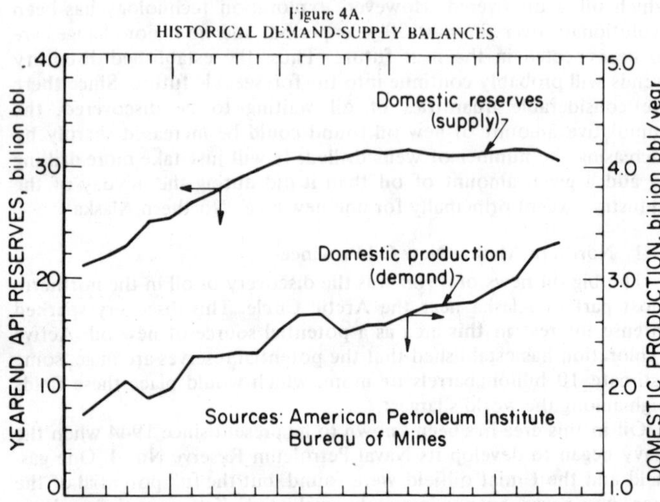

16. National Petroleum Council, *supra* note 6.
17. This increase in recovery due to technology is termed the "recovery factor." It is calculated by adding the cumulative production with proved reserves at a given time and dividing this sum by the total of the original oil in place found by that time.
18. Examples of this evolutionary technology are the injection of water in conjunction with heat, chemical additives, and polymers. Also important are improved exploration, drilling, and production technology applied to deeper producing horizoins or those located offshore.
19. In 1965, oil production from water injection projects accounted for about one-fourth of all of the oil produced in the nation. By 1980, about one-third is expected to be produced by this method.
20. The R/P ratio is calculated by dividing the API reserves reported for the end of any year by the production during that year.

Figure 4B.
RESERVE-PRODUCTION RATIO TRENDS

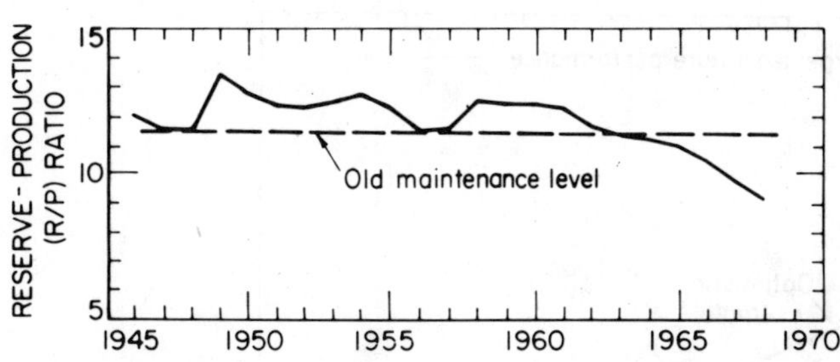

of billions of barrels of additional oil without the need to drill another exploratory well.

The historical trends between demand and supply show that each increased by about the same proportions until the mid-1950's. This balance is reflected by the relatively constant R/P ratio that ranged between 14 and 12 to 1. In the mid-1950's, reserves began to level off at between 30 and 32 billion barrels, but demand continued upward. This divergence caused the R/P ratio to nose downward, breaking the established historical patterns. It is this changing demand-supply balance that has evoked growing concern by those who analyze those trends. Why has it changed? The answer to this question is complicated owing to changes in government regulations and in the character of newly discovered oilfields. However, the single most important change over the past 20 years has been the widespread application of improved recovery technology.

A. Inventory Liquidations

Fracturing the reservoir rocks and supplementing natural reservoir energy has made it possible to produce oil not only more efficiently, but at a much faster rate. This latter trend shows up very clearly in all studies that relate well capacity to other factors. For example, between 1948 and 1965 the number of producing wells, total production, and proved reserves each increased by about 35 percent. The ability to deliver oil to the surface, however, increased more than 120 percent. This increase far outstripped the nation's ability to consume the oil, resulting in enormous amounts of excess capacity (37 percent of the capacity to produce oil was not used in 1965). The growth in capacity is compared to actual production in Figure 5. Note how closely the growth in production capacity is related to the number of operating fluid injection projects. The industry's efforts to reduce the amount of unused capacity are being successful as shown by the rounding and decline in the productive capacity curve between 1964 and 1968. The growth and decline in unused capacity is shown more clearly in Table 5. As indicated, unused capacity grew continuously from essentially zero in 1948 to its peak of 37 percent in 1964. By 1968, it had fallen to 24 percent.

Without fluid injection and fracturing it was probably necessary to maintain an R/P ratio that historically ranged between 14 and 12 to 1. With the application of technology, the need to maintain these historic ratios no longer exists. Like many industries, the petroleum industry was able to make substantial savings by carrying a smaller product inventory. The phenomenon that has reduced the R/P ratio by 20 percent in recent years represents a liquidation of excess inventories which there is no economic reason to maintain.[21]

No one knows for sure when this liquidation will stop. For purpose of this study, however, an answer can be given within reasonable limits.

Figure 5.
PRODUCTION, PRODUCTIVE CAPACITY, AND FLUID INJECTION PROJECTS

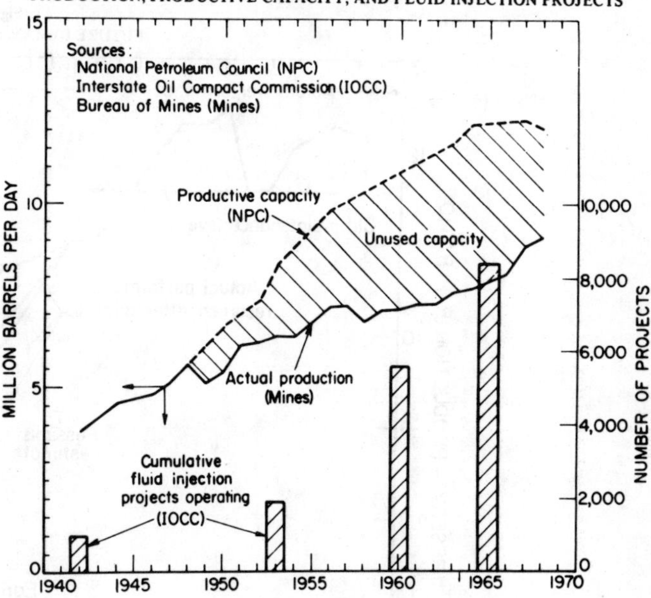

TABLE 5
Production and Production Capacity

Year	Production,[1] million b/d	Production capacity,[2] million b/d	Percent of capacity unused[3]
1948	5.520	5.520	0
1950	5.400	6.727	20
1952	6.256	7.465	16
1953	6.485	8.331	23
1956	7.151	9.867	28
1959	7.054	10.585	33
1963	7.542	11.590	35
1964	7.612	12.107	37
1967	8.810	12.287	28
1968	9.120	12.055	24

[1] Bureau of Mines; production during the year.
[2] National Petroleum Council; end-of-year estimate.
[3] Unused capacity = [(production capacity − production) /production capacity] x 100.

B. The 1971 to 1976 Pivotal Period

For this answer, it is necessary to determine a level where the liquidation should be terminated and the then-existing levels maintained. Some studies of this level have been made, but no one level has been selected as the optimum.

If the R/P ratio were to fall to a level as low as 5 to 1, the nation would have to remove about 20 percent of its reserves in 1 year. This may not be physically possible without causing excessive and irreparable damage to the producing formations. In addition, spare capacity for emergencies would be nonexistent.

Rather than this very low ratio, it is generally agreed that the desirable R/P ratio is about 8[22] or 9[23] to 1, which should leave some spare capacity for emergencies. Accepting 8 or 9 to 1 as a criterion, it is necessary to estimate when the current trends will carry the R/P ratio to these levels. The medium domestic demand estimate from Table 1 (4.645 billion barrels) and the average supply estimate from Table 4 (31.0 billion barrels) have been used to prepare the "best" estimate of future R/P trends (Figure 6). As shown, the desirable R/P ratio will most likely be reached in the period 1971 to 1976. The bounds on the 1971-76 pivotal time estimate, established through sensitivity and contingency analysis, indicate that the desirable level could be reached as early as 1970 or as late as 1985.

21. Other interrelated factors are often cited as the reason for decline. Among the most important of these include increased new oil supplies from imports which reduced the need for domestic production, prorationing regulations which encouraged over development, and a general softening of prices after 1957. All of these contributed to the great growth in excess capacity.

22. Senate Comm. on Interior and Insular Affairs, *An Assessment of Available Information on Energy in the United States,* Doc. No. 159, 87th Cong. 2nd Sess. 302 (1962).

23. Conalez, *How Big a Task is Ahead for the Petroleum Industry,* Oil and Gas Journal 101, May 15, 1967, at 101.

Figure 6.
FUTURE DEMAND-SUPPLY BALANCES

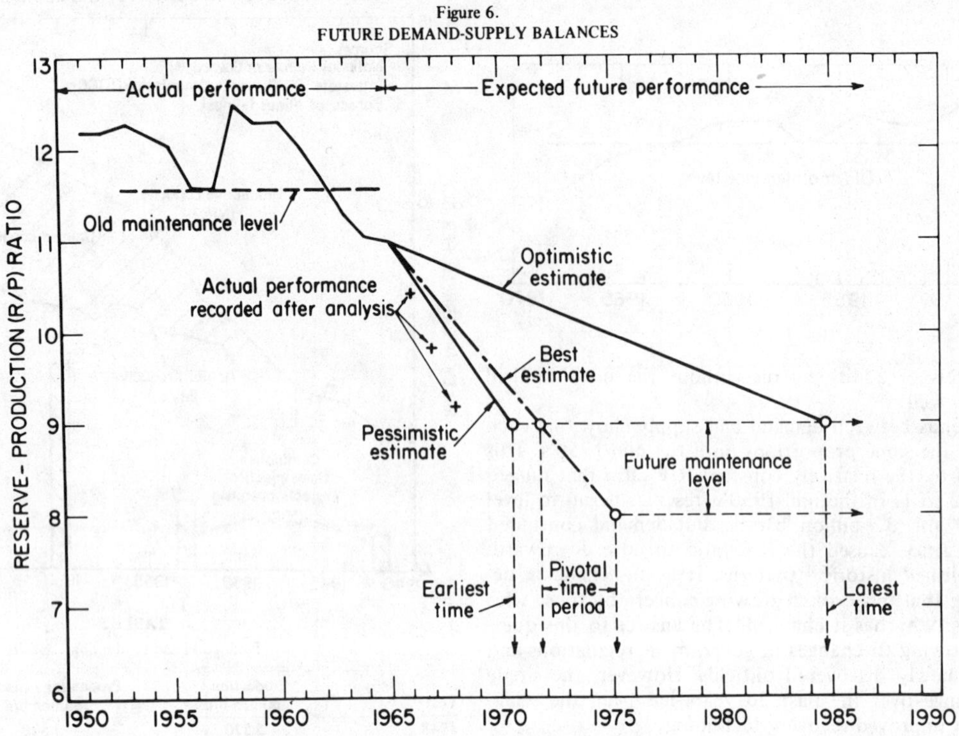

The actual R/P ratios recorded after the cutoff point for historical data (to 1965) are also plotted on Figure 6. The actual data fall below the "pessimistic" estimate. This does not mean that the analysis is incorrect since the R/P ratio does move erratically and seldom follows an ideal straight prediction curve. The eventual reclassification of the Alaskan discoveries from the "probable" to the "proved" category will probably force the R/P ratio to fall within the boundaries of the original prediction. However, these data indicate that projections need constant revision to adjust to changing conditions. This is particularly so during this transition period of rapid changes.

Transition normally implies a change between two other phases, all of which have unique characteristics that can be described. These characteristics cannot be delineated very precisely, but it is possible to begin to grossly classify them using the data previously presented.

C. Three Phases of Change

1. Phase I Pivotal Period (1962-65) This is the period when the petroleum industry's operating changes begin to take effect. Excess productive capacity reached a peak of 37 percent, and the R/P ratio moved below its old maintenance level of about 12 to 1. The allowable production from Texas wells (a bellwether indicator of change), which had fallen from 70 percent during the Korean conflict, begins to firm at about 25 percent. Prices of oil stop eroding and also begin to firm.

2. Phase II Transition Period (1966-70) Increasing production and falling production capacity cause the unused productive capacity to be significantly reduced from 37 percent in 1964 to 24 percent in 1968.[24] The R/P ratio fell sharply, and both prices and Texas allowables advanced. By early 1969, prices were near their 1957 peak and Texas allowables had climbed from 25 to over 50 percent. Exploration and development should soon begin to increase as the R/P ratio nears its new maintenance level.

3. Phase III Pivotal Period (1971-76) The R/P ratio will enter its new maintenance level of 8 or 9 to 1. Efforts to stabilize this new balance will probably include oil from Northern Alaska, but other supply alternatives are available as outlined below. The price of oil will be determined by how well the industry is able to develop these sources.

IV
SUPPLY ALTERNATIVES

Technology, exploration, liquids from solid organic matter, and increased prices and imports are alternative ways of assuring that future oil needs are met. Assuming that petroleum will remain competitive with other energy sources, however, assumes that there will be no significant long run increases in its real price. Variations in the amount of oil imported into this country were treated as a contingency in developing future demand-supply balances, but no major changes in this program can be expected in the immediate future.[25]

Without significant changes in price and the level of imports, oil firms will be faced with a choice of either developing conventional crude oil or extracting liquid products from supplemental sources such as oil shale, coal, and tar sands. Prospects for parallel development routes are excellent.

The application of evolutionary technology in field operations will continue. But without real price increases or technologic breakthroughs, the most efficient of the recovery methods now available (miscible recovery and underground combustion) cannot be economically used in field operations. Therefore, the rate of improvement in recovery efficiency from known reservoirs can be expected to advance in the future no faster than it has in the past.

The cost of increased exploration needed to maintain an adequate reserve position from 1965 to 1980 has been estimated to be $10 billion in drilling and equipping cost alone.[26] The size of this expenditure has prompted the Government to ask: ". . . [j] ust how earnestly the oil industry may be expected to pursue such costly ventures if available alternatives exist . . ."[27]

Each of the supplemental sources has vast quantities of solid organic materials that are convertible to liquid products. The oil con-

24. In addition, the low drilling rates have affected natural gas reserves which in 1968 registered their first decline since records were first kept in 1946.

25. After an exhaustive cabinet level review of the oil import question and extensive public debate, President Nixon, in 1970, decided to retain a modified quota system for controlling foreign oil imports.

26. *Supra* note 2, at 19.

27. *Id.*

tained in these deposits is not measured in billions as is petroleum, but in hundreds of billions and trillions of barrels of oil equivalent. Many factors will govern the use of these sources.

Coal, for example, is found close to major markets and would have a relatively small waste disposal problem. Disposal of waste is a factor in oil shale development, but it probably has the lowest capital investment requirement and highest return on investment of any of the supplemental sources. Tar sands are located primarily in Canada and their development for consumption in the United States must be considered in relation to the oil imports program and balance-of-payments policy.

Most recent studies show that each supplemental source is nearly competitive with crude petroleum using technology currently available.[28] Moreover, these processes involve large-scale material handling which lend themselves to the technological breakthroughs that lead to substantial cost reductions.[29] Scaling up from the current stage of technology to multi-billion dollar industries will require several years to plan both the industry and supporting communities. In addition, enormous commitments of capital will be required.[30]

The alternative that proves to be most economical will depend on the unique positions of individual firms as they evaluate their reserve position and future needs. Many firms will meet their needs through exploration and development, and others will begin to move toward the development of alternative sources.

Potential single-firm entrance into oil shale development was analyzed by Mead.[31] Heading his list was Atlantic-Richfield which has been steadily increasing its holdings of oil shale, coal, and tar sands resources even though it is a leader in Alaskan North Slope development. In March 1969, Atlantic-Richfield became the fourth partner in Colony Development Corp. This move links the financial strength of Atlantic-Richfield and its marketing outlets to that of another oil company, Standard of Ohio, with a firm that specializes in mining (Cleveland Cliffs Iron Co.). The other partner, The Oil Shale Corp., holds patents to one of the three processes thought to be commercially usable for oil shale processing.[32] This combination provides all of the ingredients for the commercial extraction and marketing of shale oil.

By contrast, very large oil companies are absent from Mead's list of potential entrants to oil shale development. This probably reflects the very high reserve-to-production ratios held by these companies.

Thus, parallel routes will be taken to assure future supplies as individual firms analyze their reserve positions relative to future needs. The cumulative actions of these firms will begin to be seen prior to the 1971-1976 period as the R/P ratio enters the 8 or 9 to 1 maintenance level.

The current stage of technology, the size of the capital investments required to develop supplemental sources as compared to alternative investment opportunities (Alaska and offshore drilling), and the planning and the required long lead times suggest that oil production from any of the supplemental sources will not be a significant part of total supplies until after 1975. Thus, the rates at which new oil reserves are found and recovery is increased from known oilfields will determine how well the industry will be able to once again stabilize the demand-supply relationship. These rates will be directly related to the pressure that may be created for higher prices and more foreign imports.

If sufficient quantities of new oil cannot be added to reserves, intense short-run pressure will develop for increased imports and/or for higher real prices to enable the economic recovery of physically recoverable supplies. This is a real possibility because of the evolutionary nature of recovery and exploration technology and the lead time required to develop Alaskan oil. Over the long run, however, oil from supplemental sources will set an upper limit on crude oil prices, or force it lower if its real price has risen relative to that of the other energy commodities.

V
PETROLEUM POLICY

One of the special characteristics of energy that sets it apart from other sectors of the economy is the substitutability of energy sources at the point of use. It matters little to a consumer if his electricity for heat and power is generated from coal, oil, natural gas, hydropower, or nuclear energy. And liquid fuels and gases can be made not only from petroleum, but also from coal, oil shale, and tar sands. Technically, only a few percent of energy use is not susceptible to substitution.

Because of this special characteristic, all energy commodities compete vigorously for a share of the energy market. How well each is able to compete is determined largely by its price relative to that of all other commodities.[33] Since the price of petroleum is indirectly controlled through public policies of the Federal and State Governments, the use of this commodity is also influenced to a great extent. Programs dealing with imports, prorationing, tax treatment, and leasing are all major policies that affect petroleum demand and supply, and the energy sources with which it competes. The wisdom of continuing many of these programs in their present forms is being vigorously challenged on economic grounds.

A. *Current Policies*

1. Foreign Imports

The import program of the Federal Government limits the amount of foreign oil that can enter the country. It has served to assure the nation a secure petroleum supply but only at an added cost since, historically, foreign crude could be delivered to east coast markets for about $1.40 per barrel less than comparable domestic crude. Adelman has calculated that this program costs about $2 billion dollars each year.[34] While there are some economic problems with his analysis,[35] it is true that the American public does pay an added price for secure petroleum supplies from domestic sources.

28. For an overall review, *see* Cameron, *A Comparative Study of Oil Shale, Tar Sands and Coal as Sources of Oil,* 21 J. Petroleum Technology V. 253 (Mar. 1969).

29. The Department of Interior's comprehensive *Prospects for Oil Shale Development,* May 1968, for example, estimates that current technology will enable shale oil production for $2.98 per barrel. This will yield a discounted return on investment of about 12 percent. By 1980, advanced technology will lower this production ost from $2.98 to $1.58 per barrel, enabling a 20-percent discounted return plus substantial returns to the Government through increased land values.

30. Capital investments will range from $2,000 to about $4,500 per daily barrel of production. Thus a 1 million barrel-per-day industry using any of the supplemental sources will require capital investments of from $2 billion to $4.5 billion. This size of industry will be small by future standards. One million barrels per day in 1980, for example, represents only 8 percent of the estimated domestic production.

31. Mead, *The Structure of the Buyer Market for Oil Shale Resources,* 8 Natural Resources J. 609 (1968).

32. The others being those developed by Union Oil Company of California and the Bureau of Mines.

33. Other factors are also important, such as equipment cost and convenience, but price is the overriding determinant.

34. M. Adelman, *Efficiency of Resource Use in Crude Petroleum,* 31 Southern Economic J. 101 (1964).

35. E. Shaffer, Oil Import Program of the United States: An Evaluation, at 219.

2. Prorationing

State prorationing of oil also limits oil supplies and was designed to provide incentives for exploration and to increase ultimate recovery from older productive fields. In practice, this program also supports prices by limiting supplies to meet expected demands. In addition, the oil that would be the least costly to produce is rationed to furnish a market for partially depleted, lower productive wells. Prorationing does prevent physical waste of petroleum, but the annual cost of the program is estimated to be about $2 billion.[36]

3. Taxes

Tax provisions that favor the industry were designed to provide economic incentives to stimulate exploration and development of new oil reserves. However, a study prepared for the Internal Revenue Service (IRS) has concluded that the 27½ percent depletion allowance is a relatively inefficient method to encourage exploration.[37] They have estimated that elimination of this allowance would increase tax revenues by about $1.2 billion each year and result in a decline in crude oil reserves of about 3 percent.

Tax policies that relate to oil shale could serve to greatly stimulate the development of that resource. At present, the depletion allowance is 15 percent applied to the mined ore. This is equivalent to a tax allowance of about 5 cents per barrel of shale oil and would help provide an oil shale industry with a marginally attractive return on investment of about 12 percent. A decision to treat oil from shale the same as crude petroleum (27½-percent depletion allowance on the shale oil) would raise the 5 cent per barrel allowance to about 35-40 cents per barrel, thereby greatly enhancing the overall return on investment.[38]

4. Public Land Leasing

State and federal laws control the access to public lands that contain many of the largest and richest deposits of energy resources that remain in this country. The production history of oil and gas from federal lands makes this abundantly clear. From 1945 through 1955, oil and gas from federal lands accounted for only 2 to 6 percent of domestic production. Production rose rapidly due to the development of the outer continental shelf. By 1968, 16 percent of all the oil and 13 percent of all the gas were produced from federal offshore leases.

In addition to oil and gas, much of the nation's available deposits of oil shale, tar sands, coal, uranium, and thorium are on public lands. Leasing policies can be made to control not only the quantity of energy products that can be extracted, but also which sources will be developed. The rate and types of lands released for development may dominate future energy resource patterns.

B. Reform Movement

It is quite apparent that agitation for policy reform is currently at a fever point. The industry, state, and federal regulatory agencies, and Congress are all calling for different types of policy reforms. Indeed, the percentage depletion allowance appears to have become the major target that must fall before tax reforms in a multitude of other areas are made.

In such an environment, it is quite easy to lose sight of the progress that has been made toward economic efficiency. Lovejoy and Homan, however, recognize this progress and stress that from an economic viewpoint:

> [O]il-conservation regulation has undoubtedly made a significant contribution to oil conservation in the economic sense when compared with what would most likely have happened in the absence of any sort of regulation. We should also like to note that there have been significant advances in the direction of what we call economic efficiency, particularly since 1950.[39]

Many studies of economic efficiency lack validity because they are based on hindsight. They serve to point out what could have been accomplished, but they often are not relevant to the future. Studies such as that prepared for the IRS on depletion allowances are not adequate for use in drafting changes that will affect the petroleum industry in the future. The IRS study was based largely on data covering the early 1960's and the 1960's will not be anything like the 1970's.

This study has shown that petroleum supplies are being liquidated, and that a new demand supply balance will soon be reached. Efforts to maintain this balance will be reflected by marked upswings in exploratory drilling, increased application of recovery technology, and initial attempts to extract liquids from solid organic matter. These will require larger capital expenditures than were needed in the 1960's. In addition, the removal of sulfur from petroleum and application of technology to liquid disposal and oil pollution problems will require increasingly more capital. Yet to be completely evaluated are suggested changes in refining techniques in the manufacturing of gasoline that would be helpful in reducing air pollution caused by internal-combustion engines.

What is needed are not more studies based on past data, but use of these data to realistically reflect future needs of the industry for the capital required to provide both quantity of supplies and quality of the environment.[40] The economic studies of past rewards to this industry do point out the magnitude of the annual savings that may occur in the future, and such savings are worth going after. But it cannot be done in a haphazard, disconnected way. Lovejoy and Homan point out about the petroleum industry that:

> As matters now stand, an extremely complex production and market structure has been built upon the basis of regulatory ground rules which, in many respects, do not conform to the requirements of economic efficiency. There is, however, no easy way to achieve such conformity. An industrial structure is an organic thing, not something that can be taken apart and reassembled to the heart's desire. Policy starts from where you are and is limited by what can be done to modify, without destroying, the organism. When dealing with what Professor John R. Commons called "going concerns," the procedure is to identify the nature of a problem and to devise the practicable means to some definable ends.[41]

The way out of this dilemma is not easy, but neither is it impossible.

C. Policy Reform

Problems of changes are enormous. The vested interests in the status quo in both the private and public sectors and the conflicting goals of regulatory agencies are major hurdles toward any future changes.

The political environment is doing wonders to erode the status quo, but it does little toward the resolution of mutually incompatible goals. For example, security of supply is paid for by limiting entry of low-cost foreign supplies, prorationing supports high prices, and removal of sulfur from oil costs more than leaving it in.

The key to any change is to get all planners to perceive the need for economic efficiency balanced against other desirable goals. If this can be accomplished, the goals to be obtained can be developed in a way that is quantifiable and constrained only by such noneconomic factors as national security and environmental quality. This will require a whole array of data not now available, and new methods of evaluation will need to be developed.

36. Adelman, *supra* note 34.

37. U.S. Congress, House, *Tax Reform Studies and Proposals* U.S. Treasury Department, 91st Congress, 1st Session, 1969.

38. The 1969 tax legislation changed the point of application from the mined shale to 15 percent on the produced shale-oil; increasing the tax allowance to about 12 cents per barrel. The same legislation also reduced the depletion allowance on petroleum to 22 percent.

39. W. Lovejoy and P. Homan, Economic Aspects of Oil Conservation Regulation, at 263 (1967).

40. Such a study has been completed and was released during 1970 as a Bureau of Mines Information Circular 8472, *Changing Investment Patterns of the U.S. Petroleum Industry, 1950-1968,* by Harry R. Johnson.

41. Lovejoy and Homan, *supra* note 39, at 264.

Probably the most difficult problem will be to conceptualize and implement new administrative approaches. Responsibilities for oil and gas matters are now divided between at least a dozen federal departments and agencies, plus about as many congressional committees. In addition, 32 states produce oil, all 50 consume it and exert some influence on demand and supply. Further complications arise when legislative or executive orders are tested in the courts.

All groups who deal with petroleum matters seek to optimize some goal, but no one group has the responsibility for overall coordination of these activities. Such coordination can take several forms, including (1) a new agency concerned only with energy, (2) an effective interdepartmental council to coordinate energy policy, and/or (3) contractual agreements between existing agencies. By realigning role relationships at both the federal and state levels, it would be possible to develop a system where one group emerges able to (1) account for all interrelated energy commodities, and (2) consider all alternatives and directly or indirectly influence all policy decisions related to petroleum. The goal of such a group would be to assure an adequate flow of liquid petroleum into the economy at reasonable costs. The "adequate flow" concept is new and just beginning to appear in some government documents.[42]

This concept can be described as a continuously changing pipeline with one end feeding oil into the national economy. The oil inventory contained in the line would be just enough to assure maximum economic efficiency with adequate surplus capacity to meet emergency conditions. Into the main pipeline are numerous feeder lines that represent alternative means of supplying oil–exploration, technology, imports, and supplemental sources. The valves that control these feeder lines represent public policies–prorationing, import program, tax incentives, and leasing policy. Such a system can only be controlled if one agency is big enough to directly or indirectly adjust all of the valves in sufficient time to enable the main pipeline to remain continuously full. Air bubbles may form if this is not accomplished and lead to temporary shortages of oil, forcing the price of oil upward.

Comprehensive planning does not mean master plans that can be developed and then forgotten. Since no one can adequately project very far into the future, optimum very long term plans may need to be sacrificed for shorter, second best solutions. The objective is not to become permanently locked into one course of action, but to take actions from which one can become disengaged. Flexibility is needed to continuously adjust plans to changing technology and future needs. Of course, a single group that can carry out such a task does not now exist. But steps toward greater economic efficiency can begin to be implemented using existing organizational arrangements.

Lovejoy and Homan list a number of thoughtful suggestions toward this end.[43] These deal primarily with means that could be used to eliminate excess capacity and, range from a soil-bank type program to the future abandonment or restriction of marginal wells. Their work is aimed at the future, not what has happened in the past. It is possible to make the right decision, but at exactly the wrong time. And during this period of inventory liquidation, decisions out of phase with reality could remove capital just when it will be needed to maintain the future balance between demand and supply. This is extremely dangerous since it could lead to excessive inventory liquidations that will draw unused productive capacity below a level adequate to meet emergency conditions. This must not be allowed to happen, and foresight and planning will assure that it does not.

SUMMARY

The petroleum industry operates within the framework of two distinct systems. The first of these is the competitive system where economics and technology largely govern the action of any firm seeking to maximize its profits. Options available to any firm, however, are constrained by a system of public policies superimposed over the competitive system. It is these two systems working together that make analysis of the petroleum industry extremely complex.

Future demand for petroleum forms the basis for rational judgments concerning its adequacy and price. The demand for petroleum is expected to grow at about 3 percent compounded annually. This growth rate indicates that the nation will consume in 1965-1980 about as much oil as it consumed in the 105-period from 1859 to 1964. In the face of such huge future demands, the petroleum industry has made significant operating changes that are reflected by a sharply falling balance between total supply and demand.

These adjustments were made in response to changes in the competitive system. The widespread application of improved recovery technology has made it profitable to hold less of a product inventory, thus eliminating the need to maintain historical supply-demand balances. A new balance will most likely be reached in the period 1971-76. Indications that this level is rapidly being approached at the present time (1969) are (1) a shrinking of excess productive capacity and (2) the upward movement of oil prices and allowable production from wells located in Texas and other states. How well the industry is able to maintain this new supply-demand balance depends largely on advancing technology and exploration.

Petroleum discovery trends are characteristic of a mature extractive industry well into the decline stage of its discovery cycle. Exceptions for this are some offshore areas and Northern Alaska, but oil from the Arctic will require time and large capital expenditures to develop. Recovery technology is evolutionary in nature and can be expected to advance no faster in the future than it has in the past. Liquid products from coal, oil shale, and tar sands are now nearly competitive with crude petroleum. However, the planning and capital required to develop these sources indicates that none of them will contribute significantly to total supplies until after 1975.

Over the short run, the petroleum industry's ability to maintain a new balance between supply and demand will depend largely on its ability to find and develop domestic sources. It is possible that pressure may build for increased imports and/or higher real prices of petroleum to permit more exploration and the economic recovery of the substantial quantities of physically recoverable oil that now exist. Over the long run, oil from the supplemental sources will set an upper limit on crude oil prices, or even force them lower.

Imports, prorationing, tax treatment, and land leasing are major government policies which affect petroleum demand and supply. Many people are calling for reform of these programs, but reform can best proceed by simultaneous adjustment of all programs. This can only be successful if one group emerges that can adequately anticipate future energy needs and is big enough to directly or indirectly influence all major petroleum policy decisions. The goal of this activity is to provide an adequate flow of petroleum into the economy at reasonable costs. Failure to simultaneously reform the present government policies could lead to a dangerous over-correction of the falling balance between supply and demand.

42. C. Mottley, *Strategic Planning*. Management Highlights, U.S. Department of the Interior Release No. 56, at 1 (1967).

43. Lovejoy and Homan, *supra* note 39, at 268-71.

Petroleum-Related Energy Research and Development

STATEMENT PREPARED BY FRANK ICKARD, PRESIDENT, AMERICAN PETROLEUM INSTITUTE, WASHINGTON, D.C.

INTRODUCTION

The U.S. is faced with an energy crisis of deepening proportions. Although rumblings of an impending crisis have been heard for a number of years, only recently has the problem attracted nationwide attention.

The nature of the energy crisis is still not fully understood by many Americans. Some people believe that the crisis results from the fact that reserves of conventional fuels are almost exhausted. Based upon this incorrect premise, they conclude that the only hope of solving the problem would be an immediate and massive crash program to develop new energy sources.. There are also differences of opinion on the timing of the crisis, and when it will reach its most dangerous stage.

In the view of the petroleum industry, the most critical time period will be in the next two decades. This is because sources of low-cost conventional oil and gas are diminishing in the United States. However, if incentives are adequate, oil and gas will become available in greater quantity from beneath the deeper waters of the continental margin and from the recovery of a greater percentage of the original oil in place in onshore and offshore reservoirs. Synthetic fuels from coal, tar sands and oil shale can also begin to become economic and contribute to energy supplies toward the end of this critical period. Also, toward the end of the period there will be a significant use of nuclear fuels to supply energy in the form of electric power. By the end of the two decades, the major adjustment to new sources will have been made.

A brief look at the supply and demand picture indicates the magnitude of the energy supply problem. Energy economists in both industry and government forecast that the nation's energy use will more than double by the year 1990. These same experts indicate that oil and gas, which supply 77 percent of the nation's total energy today, will still have to supply more than 60 percent of these requirements in 1990. If these forecasts turn out to be correct, this would mean that the U.S. would be consuming more than 26 million barrels of oil a day in 1990 compared with just over 15 million barrels daily at the present time.

These demands can and must be met. Our nation has not exhausted its petroleum reserves. The National Petroleum Council estimates that we still have three or four times as much oil and natural gas remaining to be discovered in the U.S. as we have consumed throughout our history. The petroleum industry recognizes the need to find and develop these domestic reserves and make them available at reasonable prices to the nation's consumers. The only alternative is for the U.S. to import the bulk of its petroleum needs. Although it may be necessary to turn to petroleum imports for a larger percentage of our domestic needs, excessive dependence upon insecure sources in the Eastern Hemisphere for vital supplies of petroleum would involve serious risks to our economic health and national security.

These basic facts about the energy crisis are reflected in the industry's steadily-expanding expenditures for research and development. In 1971, it is estimated that approximately $660 million was spent for R&D. This was more than double the $310 million expended 10 years ago. Of this amount, it is estimated that close to $150 million is spent for R&D in exploration and production. This estimate excludes the $1 to $1¼ billion the industry spends for field testing and development to bring new technology to routine commercial use. It also excludes the millions of dollars spent annually on research and development by supporting industries. Petroleum equipment suppliers spend approximately $75 million annually, for example, to develop such innovative equipment as self-propelled vessels for deep water drilling, pre-constructed rigs, and increasingly sophisticated blow-out preventers.

While increasing industry R&D effort is being directed to the longer-run energy needs of the nation, major programs remain focused upon finding and developing the supplies of oil and gas from conventional sources necessary to meet our energy needs during the next few decades.

Much of the oil and gas still to be discovered lies under the continental shelves and slopes off our shores, where water depths may reach 6,000 feet. New technology will be required to extract oil and gas from water depths of more than 600 feet.

The Arctic region of North America is expected to become an increasingly important source of petroleum. The Prudhoe Bay field on Alaska's North Slope is the largest ever discovered in the U.S., and more giant fields may be found in future years. New technology is being developed to find, produce, and transport oil in Arctic regions with minimum impact upon the environment.

Substantial additions to oil reserves can come from improved recovery technology for both existing fields and future discoveries. The average recovery efficiency for domestic fields is presently about 31 percent although an ultimate of 60 percent is predicted.[1]

Gas reserves are especially critical and technology is being developed by both government and industry to recover natural gas from tight reservoir rocks where conventional methods fail.

Better exploration techniques are expected to open up new areas to the search for oil and gas where existing methods have not been successful, because of complex geological conditions, and also to improve discovery efficiency in all areas both on and offshore..

In the exploration and development field, although priorities for R&D funds vary widely among companies, these projects are of major importance.

(1) Exploration techniques to improve finding success in all areas.
(2) Efforts to improve oil recovery rates.
(3) Deep water producing capability.
(4) Reservoir simulation techniques.
(5) Nuclear stimulation of natural gas reservoirs.
(6) Arctic oilfield research.
(7) Research on synthetic fuels.

A brief summary of R&D efforts in each of these areas follows:

Exploration

The most important single exploration tool is probably the seismic reflection technique which has been so successful in mapping subsurface structures that are drilling targets for oil or gas fields. During the past decade large research sums have been expended to improve the resolution of the seismic method for mapping structures. Important developments have been non-explosive energy sources, digital recording of the field data, computer processing of field data and sophisticated interpretation methods aided by computer graphics. Seismic research is continuing throughout industry on a high priority basis to achieve greater accuracy in mapping complex structures, to obtain usable data in so-called "no-record" areas where signals are weak and "noise" levels high and, perhaps most significant of all, to predict lithology and such rock properties as porosity in order to provide seismic capability for finding stratigraphic traps.

A strong research effort also prevails in both geology and geochemistry, the results of which are providing new insights on where to look for new oil and gas resources. New knowledge on the geology of the continental margins has underlined the importance of the shelf and slope as promising petroleum provinces. Significant progress is being made in understanding geological environments and processes most favorable for the generation, migration and entrapment of oil and gas. Techniques are now in use for identification of source rocks of oil and are being developed for source rocks of gas. Computer and computer graphics are being widely used to process geological data, prepare more accurate maps and cross-sections and unravel complex geological situations. Isotopic age dating and zonation of sedimentary rocks has improved the accuracy in mapping traps where petroleum accumulations might be found.

Improved Oil Recovery

The ultimate recovery of domestic known reserves is about 31 percent using existing technology and economics prevailing at year end 1970. The National Petroleum Council reports that recovery is increasing annually as the result of improving recovery technology. Continuation of this trend could add significantly to the nation's oil supplies. For example, on the basis of future domestic discoveries, estimated at 436 billion barrels, the difference between 31 percent and 60 percent ultimate recovery efficiency could amount to an additional 121 billion barrels of recoverable crude oil—equivalent to 24 years of current consumption.

For the foregoing reasons, R&D on improved recovery efficiency can have a major impact on our future inventory of domestic recoverable reserves. Its importance is underlined by the special industry 3-day symposium on recovery methods, held at Tulsa, Oklahoma, in April, and sponsored by the Mid-Continent Section, Society of Petroleum Engineers of the American Institute of Mechanical Engineers. Thirty-three papers reported on the status of secondary-tertiary recovery projects. M. A. Wright, Humble's Board Chairman, stated that the need for tertiary recovery processes was urgent since conventional waterflooding was approaching full maturity and he called for a cooperative research approach by industry.

The target of tertiary processes is the residual oil left after conventional waterflooding but *before* abandonment. Many fields under waterflood are approaching this point which accounts for the urgency.

Most tertiary processes now being developed involve the addition of chemicals with detergent-like properties to the reservoir to reduce the viscosity and surface tension properties of the oil to make it easier for the waterflood to strip more of the oil from the rock surfaces and push it out of the smaller pores where the oil is held more tightly. The major problem is to find chemicals that provide a sufficient incremental volume of new oil to absorb the costs of the chemicals used.

In spite of the work of hundreds of scientists and engineers, the problems of tertiary recovery remain difficult. After years of study and the expenditure of millions of research dollars there has not yet been a major breakthrough.[2]

[1] "Future Petroleum Resources of the United States," p. 104 (1970) A report by the National Petroleum Council.

[2] See *Oil and Gas Journal,* April 24, 1972, p. 50.

Reprinted with permission from *Energy Research and Dev.*, Hearings before the Subcommittee on Sci., Research, and Dev., House Committee on Sci. and Astronautics, Y4.Sci2: 92/2/24, pp. 551-557, May 1972.

There is a growing belief that the best chance for success is by a cooperative effort and the sharing of data.

Another taregt for improved recovery processes is heavy viscous crudes such as are found in California and Saskatchewan. In these areas, fireflooding and steaming are being used successfully in some instances.

Deep Water Production Capability

Rapid progress is being made in developing production capability in water depths of over 600 feet. At least five different systems are under development.

Mobil, with North American Rockwell, has developed and built a prototype seafloor production station, or satellite, for a cluster of up to 18 wells. The system is fully automated and under remote control from a surface facility. The dry chamber may be entered from a diving bell or submarine when necessary for maintenance work. It is designed for an ultimate water depth of 6,000 feet. SEAL Petroleum Company has taken over the system to complete an at sea check-out to be followed by commercialization. R&D costs to date are in excess of $6 million and will reach about $10 million before completion of the first commercial unit.

Shell, along with Lockheed Petroleum Service Ltd., has recently announced an at sea test of a system developed by Lockheed comprising a subsea wellhead cellar, service capsule, manifold center and production station. A dry 1-atm. service capsule will transport personnel to perform work within the wellhead cellar. The system is designed for single well completions.

Humble is developing a Submerged Production System for groups of 5–40 wells completed on the seafloor or on a submerged platform. A robot-type manipulator equipped with TV camera monitors will perform maintenance work under remote control. Humble's investment in the system is reported to be in excess of $5 million.

Other firms, including Standard Oil of California and Fluor Corporation, are also developing sea-bottom completion systems.

These systems add a dimension of safety to offshore production because they are beyond the reach of storms, high seas and shipping.

Reservoir Simulation Techniques

More effective reservoir management can also have a significant impact on improving recovery efficiency. This is being helped in a major way by the rapidly expanding use of digital reservoir simulators, which are now established tools for the design of reservoir exploitation schemes, planning locations for new wells, and predicting producing rates and recovery levels under various operating conditions. Reservoir simulators aid in selecting the best of alternative plans for operating under primary production or the feasibility of various secondary or tertiary processes and their effect on costs and ultimate recovery. Present models are complex and time consuming to design because of the high degree of precision required since long periods of simulated real time are involved, but simplified versions requiring less design time are being developed.

Nuclear Stimulation of Natural Gas Reservoirs

It is estimated in excess of 300 trillion cubic feet of natural gas are locked up in tight reservoirs in Wyoming, Colorado, Utah and New Mexico. The AEC Plowshare program has been established to investigate the feasibility of using nuclear detonation to fracture these tight reservoirs to release the gas in commercial volumes. Various companies are participating in the program and others continue to investigate conventional fracture methods.

The first nuclear test in the Plowshare program was the Gasbuggy project in the San Juan Basin, New Mexico, where a 29 kiloton device was detonated 4240 feet below the earth's surface in December, 1967. The El Paso Natural Gas Company participated in this test. The next was the Rulison project in the Piceance Basin, Colorado, where a 40 kiloton device was detonated 8500 feet below the surface in September, 1969. Industry participants in this test were Austral Oil Company and CER Geonuclear Corporation. Industry funded about 90 percent of the estimated $6.5 million cost. Results of the test were said by those involved to indicate a "significant technological breakthrough in the development and production of natural gas." Each of the test wells had much higher gas production rates and much higher recoveries of gas than wells in the same tight reservoirs completed by conventional methods. However, neither of them produced enough gas to be of economic interest without availability of low cost nuclear devices and a major increase in gas price. Thus, there needs to be additional technical development both in achieving greater fracturing effectiveness of the nuclear explosion and in reducing cost of manufacture, placement, and detonation of nuclear devices. Public acceptance of radioactive content of gas at or below safe limits to be established by state and federal government agencies must be achieved before nuclear stimulation can effectively increase gas supply.

The AEC is now planning two additional tests: Rio Blanco in Colorado, with CER Geonuclear and Equity Oil Company; and Wagon Wheel in Wyoming, with El Paso Natural Gas Company. The Rio Blanco test will use multiple small-diameter, low radioactivity devices detonated simultaneously in an attempt to achieve longer lasting improvement in gas productivity than that obtained by Gasbuggy or Rulison. For Wagon Wheel, sequential detonation is planned.

Arctic Oilfield Research

The severe Arctic cold and the particular surroundings of the Prudhoe Bay oilfield have necessitated a number of special research projects. This work falls into two categories: one concerned with the effects of the severe cold on drilling and engineering operations; and the other with monitoring and protecting the natural environment. In the former, significant work has been done, and is continuing, on the problems associated with drilling and completing producing wells through the 2000 feet or so of permafrost. Core samples in their natural frozen state have been taken from test holes with refrigerated drilling fluids. Laboratory measurements have been made of strains occurring in the sample as they thawed and froze again. The laboratory work has been complemented by measurements of deformations and temperatures taken in test wells drilled through the permafrost in which hot oil was circulated for about a year. At the same time a special "insulated" oil well casing has been developed to minimize the extent of thaw around producing wells. Other engineering tests have been devoted to studying in-situ the load bearing qualities of supporting piles driven into the permafrost.

Environmental research has been carried out to ensure that oilfield operations have no significant effect on the flora and fauna of the area. To this end, the operators have sponsored extensive studies of the Arctic ecosystem. This is apart from, but complementary to, the extremely comprehensive studies made by Alyeska, the pipeline company established by the industry for the purpose of designing, laying and operating a trans-Alaska pipeline and whose submissions have already been made to the Department of the Interior in connection with its Environmental Impact Statement. Alyeska has spent some $35 million on environmental studies covering such items as fish and wildlife, revegetation, waste disposal, permafrost, oil spills, noise abatement, and resistance of materials to earthquake stresses.

Research on Synthetic Fuels

The term "synthetic petroleum" covers five basic extraction or conversion processes: gas from naphtha; gas and petroleum liquids from coal; oil from tar sands; oil from shale; and oil from organic waste. Development of synthetic fuels will require not only large R&D efforts, but massive amounts of capital and the establishment of whole new industries. While the longer-range outlook for the development of synthetic fuels is promising, these industries are not likely to make important contributions to energy needs until the 1990's. Current research efforts are directed primarily to three major areas: oil shale, tar sands, and coal gasification.

Oil Shale—most of the oil shale deposits in North America are in the Green River formation located in Colorado, Utah and Wyoming. Estimates of the potential reserves range from 600 billion barrels for oil shale assaying 25 gallons or more per ton of rock to 1,400 billion barrels for rock yielding 10 to 25 gallons per ton. Some is within economic mining range but much of the resource is deep and will require an in-situ method.

Numerous companies have spent large sums developing technology to exploit the oil shale resource with considerable technological success but costs are still not competitive with conventional crude. Furthermore, present uncertainty over government policies for crude oil imports on the one hand, and leasing of Federal oil shale lands on the other, need to be resolved before industry will risk the large investment required to mine and process oil shale on a commercial scale. The success or failure of efforts to resolve environmental problems, especially the disposal of processed shale, is likely to have a major bearing on government policy decisions.

Union Oil showed the feasibility of the gas retort process, whereby hot combustion gases transfer heat to a moving bed of crushed shale to cause the thermal conversion of kerogen into oil. Oil Shale Corp. (Tosco) developed a modified retort method using ceramic balls to transfer the heat. The United States Bureau of Mines built and operated a pilot plant at Rifle, Colorado using a modified gas-combustion technique. Later a group of oil companies took over the plant and funded a $6 million, four year, R&D project which led to a doubling of the efficiency of the Bureau of Mines retort method and development of practical mining and crushing methods. Participating companies were Mobil, Humble, Phillips, Sinclair, Amoco Production and Continental.

The in-situ combustion-retort method may ultimately be the only feasible way to exploit the deep oil shale deposits, provided that techniques can be developed to obtain economic yields. Practical methods must be developed to overcome the impermeability of oil shale in its natural state to air and oil. Explosive fracturing techniques are among those being investigated to break up the rock mass, before starting the combustion-retorting processes, to establish channels for the high rates of air injection needed to support combustion on the one hand, and subsequent drainage of the shale oil on the other. These techniques show promise but much improvement is still necessary to achieve recovery yields that will bring the price of shale oil to a competitive level.

Tar Sands—the largest tar sand deposits in North America are in Alberta, Canada, where recoverable reserves are estimated to be 430 billion barrels. Most of this, 338 billion barrels, is in the Athabasca deposit. On the basis of the Alberta Conservation Board estimate of resources in-place (626 billion barrels) at Athabasca, recoverable reserves of about 40 billion barrels within economical mining range and nearly 300 billion at depths where in-situ methods must be used are possible.

One mining venture has been operating at Athabasca for several years, and another is in the advance planning and engineering design stage.

Great Canadian Oil Sands Company (Sun), has been operating a mining/extraction plant at Athabasca for about four years with an authorized 45,000 barrels per day (BPD) capacity. Despite an investment of close to half a billion dollars by Sun to commercialize the Athabasca tar sands, this plant has not yet proven financially viable. However, production of crude oil from tar sands is the first synthetic process proved technically feasible and which has been tested on a commercial production scale.

Syncrude Canada Ltd. (Atlantic Richfield, Cities Service, Gulf and Imperial) has authorization for a $500 million project at Athabasca with a plant capacity of 125,000 BPD by 1976 or 1977.

Shell, Texaco, Atlantic Richfield, Mobil, and the Amoco Production Company have conducted experimental thermal in-situ projects. Shell and the Amoco Production Company have applied to the Alberta Board of Conservation for development on a larger scale leading eventually to 100,000 BPD of syncrude production.

Coal Gasification and Liquefaction—coal gasification, which involves the conversion of coal into a synthetic gas interchangeable with natural gas, is seen as an important future energy source. Known minable coal reserves could produce enough fuel to supply the nation's natural gas energy needs for 500 years.

One coal gasification process has existed for 30 years, but it is ten times as expensive as the estimated cost of new processes now being developed. To speed development of these new technologies, the Department of Interior has entered into a joint research agreement with the American Gas Association. This joint venture will cost about $30 million annually over the next 6 to 8 years, and is intended to push current research processes through the pilot plant level to the demonstration plant stage. By 1978, one full-scale demonstration plant is expected to be in operation. Commercial plants are expected to be in operation by the 1990's.

Coal liquefaction research is not as far advanced. It would convert coal into oil, gasoline, or other fuels; or into a low-melting point solid for use in firing power boilers. Pilot plants testing two different liquefaction processes are in operation, and a third pilot plant is planned to test another process.

The commercial development of any of these processes will involve huge capital requirements. Many plants would have to be constructed, and surface mining increased to produce hundreds of millions of additional tons of coal yearly. The estimated investment, during the first decade, would be over $8.5 billion per process. Neither gasification nor liquefaction is expected to contribute significantly to the total energy supply until late in this century.

Conclusion

The petroleum industry, often in close cooperation with the Federal Government, is heavily engaged in research and development efforts designed to make maximum use of our nation's domestic reserves of conventional fuels and to develop new technologies which can meet the energy needs of Americans in the 21st Century.

It must be recognized that technology, by itself, cannot solve our energy crisis. The government must develop policies which will encourage industry to search for and develop new domestic petroleum reserves, and which will permit full advantage to be taken of new technology.

As we have seen, much of the oil and gas still to be discovered lies offshore under the continental shelves and slopes. It will be of little benefit to U.S. consumers to develop new technology to extract petroleum from deep water if the industry is denied access to these offshore areas, or if a highly restrictive leasing policy is followed.

Much of the oil shale in the U.S. is located on public lands, and there is uncertainty concerning future leasing policies affecting this potential energy source.

Nuclear stimulation techniques cannot assist in improving recovery efficiencies if environmental regulations restrict the use of nuclear devices.

The largest oil field in the history of the U.S. has been found on the North Slope of Alaska, and this region may contain other giant fields. However, the long delay in authorizing construction of the trans-Alaska pipeline has almost brought a halt to exploration and development efforts in this promising region.

The search for oil and gas in remote and, hostile environments, such as in the Arctic or offshore, is extremely costly. To bring new synthetic fuels to the commercial stage will involve massive amounts of capital. Government efforts to artificially hold down prices could slow development of new fuel supplies and intensify the energy crisis. In the case of natural gas, it is acknowledged that regulation of prices at the wellhead by the Federal Power Commission has acted to stimulate demand while, at the same time, it has discouraged the development of new supplies, thqs contributing to the energy shortage. As a consequence, natural gas service curtailments have taken place in more than 20 states.

Unless an economic and political climate is developed which will permit industry to take advantage of scientific advances, any level of expenditures for research and development by the industry and its thousands of suppliers all over the nation could fail to prevent a serious energy shortfall.

Social Institutions and Nuclear Energy

Alvin M. Weinberg

Fifty-two years have passed since Ernest Rutherford observed the nuclear disintegration of nitrogen when it was bombarded with alpha particles. This was the beginning of modern nuclear physics. In its wake came speculation as to the possibility of releasing nuclear energy on a large scale: By 1921 Rutherford was saying "The race may date its development from the day of the discovery of a method of utilizing atomic energy" (*1*).

Despite the advances in nuclear physics beginning with the discovery of the neutron by Chadwick in 1932 and Cockcroft and Walton's method for electrically accelerating charged particles, Rutherford later became a pessimist about nuclear energy. Addressing the British Association for the Advancement of Science in 1933, he said: "We cannot control atomic energy to an extent which would be of any value commercially, and I believe we are not likely ever to be able to do so" (*2*). Yet Rutherford did recognize the great significance of the neutron in this connection. In 1936, after Fermi's remarkable experiments with slow neutrons, Rutherford wrote ". . . the recent discovery of the neutron and the proof of its extraordinary effectiveness in producing transmutations at very low velocities opens up new possibilities, if only a method could be found of producing slow neutrons in quantity with little expenditure of energy" (*3*).

Today the United States is committed to over 100 × 10[6] kilowatts of nuclear power, and the rest of the world to an equal amount. Rather plausible estimates suggest that by 2000 the

The author is director of the Oak Ridge National Laboratory, Oak Ridge, Tennessee 37830. This article is the text of the Rutherford Centennial lecture, presented at the annual meeting of the American Association for the Advancement of Science, Philadelphia, 27 December 1971.

Table 1. Estimated total cost of power from 1000-Mwe power plants (mills per electric kilowatt hour). The costs include escalation to 1978. Nuclear fuel costs were taken from (*9*). The coal plant fuel costs are based on average delivered coal price of about $8 per ton in 1971, with escalation to 1978 at 5 percent per year. This leads to about $10.5 to $10.7 per ton in 1978. Estimates for costs of operating SO_2-removal equipment range from zero to about 2×10^6 dollars per year.

	PWR plants		Coal plants			
			No SO_2 system		With SO_2 system	
	Run-of-river	With cooling towers	Run-of-river	Cooling towers	Run-of-river	Cooling towers
Capital cost ($/kwe)	365	382	297	311	344	358
Fixed charges	7.8	8.2	6.4	6.6	7.4	7.7
Fuel cost	1.9	1.9	3.9	3.9	3.9	3.9
Operation and maintenance cost	0.6	0.6	0.5	0.5	0.8	0.8
Total power cost (mills/kwhe)	10.3	10.7	10.8	11.0	12.1	12.4

United States may be generating electricity at a rate of 1000×10^6 kilowatts with nuclear reactors. Much more speculative estimates visualize an ultimate world of 15 billion people, living at something like the current U.S. standard: nuclear fission might then generate power at the rate of some 300×10^9 kilowatts of heat, which represents 1/400 of the flux of solar energy absorbed and reradiated by the earth (*4*).

This large commitment to nuclear energy has forced many of us in the nuclear community to ask with the utmost seriousness questions which, when first raised, had a tone of unreality. When nuclear energy was small and experimental and unimportant, the intricate moral and institutional demands of a full commitment to it could be ignored or not taken seriously. Now that nuclear energy is on the verge of becoming our dominant form of energy, such questions as the adequacy of human institutions to deal with this marvelous new kind of fire must be asked, and answered, soberly and responsibly. In these remarks I review in broadest outline where the nuclear energy enterprise stands and what I think are its most troublesome problems; and I shall then speculate on some of the new and peculiar demands mankind's commitment to nuclear energy may impose on our human institutions.

Nuclear Burners—Catalytic and Noncatalytic

Even before Fermi's experiment at Stagg Field on 2 December 1942, reactor designing had captured the imagination of many physicists, chemists, and engineers at the Chicago Metallurgical Laboratory. Almost without exception, each of the two dozen main reactor types developed during the following 30 years had been discussed and argued over during those frenzied war years. Of these various reactor types, about five, moderated by light water, heavy water, or graphite, have survived. In addition, breeders, most notably the sodium-cooled plutonium breeder, are now under active development.

Today the dominant reactor type uses enriched uranium oxide fuel, and is moderated and cooled by water at pressures of 100 to 200 atmospheres. The water may generate steam directly in the reactor [so-called boiling water reactor (BWR)] or may transfer its heat to an external steam generator [pressurized water reactor (PWR)]. These light water reactors (LWR) require enriched uranium and therefore at first could be built only in countries such as the United States and the U.S.S.R., which had large plants for separating uranium isotopes.

In countries where enriched uranium was unavailable, or was much more expensive than in the United States, reactor development went along directions that utilized natural uranium: for example, reactors developed in the United Kingdom and France were based mostly on the use of graphite as moderator; those developed in Canada used D_2O as moderator. Both D_2O and graphite absorb fewer neutrons than does H_2O, and therefore such reactors can be fueled with natural uranium. However, as enriched uranium has become more generally available (of the uranium above ground, probably more by now has had its normal isotopic ratio altered than not), the importance of the natural ^{235}U isotopic abundance of 0.71 percent has faded. All reactor systems now tend to use at least slightly enriched uranium since its use gives the designer more leeway with respect to materials of construction and configuration of the reactor.

The PWR was developed originally for submarine propulsion where compactness and simplicity were the overriding considerations. As one who was closely involved in the very early thinking about the use of pressurized water for submarine propulsion (I still remember the spirited discussions we used to have in 1946 with Captain Rickover at Oak Ridge over the advantages of the pressurized water system), I am still a bit surprised at the enormous vogue of this reactor type for civilian power. Compact, and in a sense simple, these reactors were; but in the early days we hardly imagined that separated ^{235}U would ever be cheap enough to make such reactors really economical as sources of central station power.

Four developments proved us to be wrong. First, separated ^{235}U which at the time of *Nautilus* cost around $100 per gram fell to $12 per gram. Second, the price of coal rose from around $5 per ton to $8 per ton. Third, oxide fuel elements, which use slightly enriched fuel rather than the highly enriched fuel of the original LWR, were developed. This meant that the cost of fuel in an LWR could be, say, 1.9 mills per kilowatt hour (compared with around 3 mills per electric kilowatt hour for a coal-burning plant with coal at $8 per ton). Fourth, pressure vessels of a size that would have boggled our minds in 1946 were common by 1970: the pressure vessel for a large PWR may be as much as 8½ inches thick and 44 feet tall. Development of these large pressure vessels made possible reactors of 1000 megawatts electric (Mwe) or more, compared with 60 Mwe at the original Shippingport reactor. Since per unit of output a large power plant is cheaper than a small one, this increase in reactor size was largely responsible for the economic breakthrough of nuclear power.

Although the unit cost of water reactors has not fallen as much as optimists such as I had estimated, present costs are still low enough to make nuclear power competitive. I compare the relative position of a 1000-Mwe LWR and of a coal-fired plant of the same size (Table 1).

Water-moderated reactors burn ^{235}U, which is the only naturally occurring fissile isotope. But the full promise of nuclear fission will be achieved only with successful breeders. These are reactors that, essentially, burn the very abundant isotopes ^{238}U or ^{232}Th; in the process, fissile ^{239}Pu or ^{233}U acts as regenerating catalyst—that is, these isotopes are burned and regenerated. I therefore like to call reactors of this type *catalytic nuclear burners.* Since ^{238}U and ^{232}Th are immensely abundant (though in dilute form) in the granitic rocks, the basic fuel for such catalytic nuclear burners is, for all practicaltical purposes, inexhaustible. Mankind will have a permanent source of energy once such catalytic nuclear burners are developed.

Most of the world's development of a breeder is centered around the sodium-cooled, ^{238}U burner in which ^{239}Pu is the catalyst and in which the energy of the neutrons is above 100 $\times 10^3$ electron volts. No fewer than 12 reactors of this liquid metal fast breeder reactor (LMFBR) type are being worked on actively, and the United Kingdom plans to start a commercial 1000-Mwe fast breeder by 1975. Some work continues on alternatives. In the ^{233}U–^{232}Th cycle, on the light water breeder and the molten salt reactor; in the ^{239}Pu–^{238}U cycle, on the gas-cooled fast breeder. But these systems are, at least at the present, viewed as backups for the main line which is the LMFBR.

Nuclear Power and Environment

The great surge to nuclear power is easy to understand. In the short run, nuclear power is cheaper than coal power in most parts of the United States; in the long run, nuclear breeders assure us of an all but inexhaustible source of energy. Moreover, a *properly* operating nuclear power plant and its subsystems (including transport, waste disposal, chemical plants, and even mining) are, except for the heat load, far less damaging to the environment than a coal-fired plant would be.

The most important emissions from a routinely operating reactor are heat and a trace of radioactivity. Heat emissions can be summarized quickly. The thermal efficiency of a PWR is 32 percent; that of a modern coal-fired power plant is around 40 percent. For the same electrical output the nuclear plant emits about 40 percent more waste heat than the coal plant does; in this one respect, present-day nuclear plants are more polluting than coal-fired plants. However, the higher temperature nuclear plants, such as the gas-cooled, the molten salt breeder, and the liquid metal fast breeder, operate at about the same efficiency as does a modern coal-fired plant. Thus, nuclear reactors of the future ought to emit no more heat than do other sources of thermal energy.

As for routine emission of radioactivity, even when the allowable maximum exposure to an individual at the plant boundary was set at 500 millirems (mrem) per year, the hazard, if any, was extremely small. But for practical purposes, technological advances have all but eliminated routine radioactive emission. These improvements are taken into account in the newly proposed regulations of the Atomic Energy Commission (AEC) requiring, in effect, that the dose imposed on any individual living near the plant boundary either by liquid or by gaseous effluents from LWR's should not exceed 5 mrem per year. This is to be compared with the natural background which is around 100 to 200 mrem per year, depending on location, or the medical dose which now averages around 60 mrem per year.

As for emissions from chemical reprocessing plants, data are relatively scant since but one commercial plant, the Nuclear Services Plant at West Valley, New York, has been operating, and this only since 1966. During this time, liquid discharges have imposed an average dose of 75 mrem per year at the boundary. Essentially no ^{131}I has been emitted. As for the other main gaseous effluents, all the ^{85}Kr and ^{3}H contained in the fuel has been released. This has amounted to an average dose from gaseous discharge of about 50 mrem per year.

Technology is now available for reducing liquid discharges, and processes for retaining ^{85}Kr and ^{3}H are being developed at AEC laboratories. There is every reason to expect these processes to be successful. Properly operating radiochemical plants in the future should emit no more radioactivity than do properly operating reactors—that is, less than 10 percent of the natural background at the plant boundary.

There are some who maintain that even 5 mrem per year represents an unreasonable hazard. Obviously there is no way to decide whether there is any hazard at this level. For example, if one assumes a linear dose-response for genetic effects, then to find, with 95 percent confidence, the predicted 0.5 percent increase in genetic effect in mice at a dose of, say, 150 mrem would require 8 billion animals. At this stage the argument passes from science into the realm of what I call trans-science, and one can only leave it at that.

My main point is that nuclear plants are indeed relatively innocuous, large-scale power generators if they and their subsystems work properly. The entire controversy that now surrounds the whole nuclear power enterprise therefore hangs on the answer to the question of whether nuclear systems can be made to work properly; or, if faults develop, whether the various safety systems can be relied upon to guarantee that no harm will befall the public.

The question has only one answer: there is no way to guarantee that a nuclear fire and all of its subsystems will never cause harm. But I shall try to show why I believe the measures that have been taken, and are being taken, have reduced to an acceptably low level the probability of damage.

I have already discussed low-level radiation and the thermal emissions from nuclear systems. Of the remaining possible causes of concern, I shall dwell on the three that I regard as most important: reactor safety, transport of radioactive materials, and permanent disposal of radioactive wastes.

Avoiding Large Reactor Accidents

One cannot say categorically that a catastrophic failure of a large PWR or a BWR and its containment is impossible. The most elaborate measures are taken to make the probability of such occurrence extremely small. One of the prime jobs of the nuclear community is to consider all events that could lead to accident, and by proper design to keep reducing their probability however small it may be. On the other hand, there is some danger that in mentioning the matter one's remarks may be misinterpreted as implying that the event is likely to occur.

Assessment of the safety of reactors depends upon two rather separate considerations: prevention of the initiating incident that would require emergency safety measures; and assurance that the emergency measures, such as the emergency core cooling, if ever called upon, would work as planned. In much of the discussion and controversy that has been generated over the safety of nuclear reactors, emphasis has been placed on what would happen if the emergency

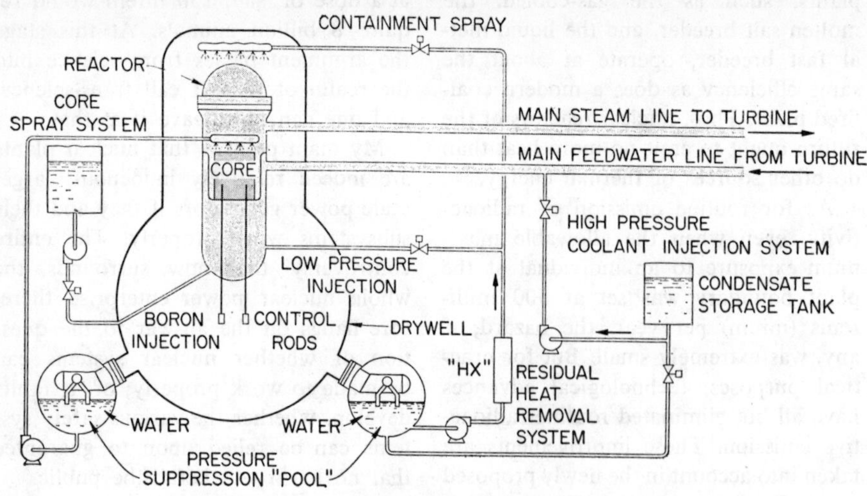

Fig. 1. Boiling water reactor emergency cooling systems.

measures were called upon and failed to work. But to most of us in the reactor community, this is secondary to the question: How certain can we be that a drastic accident that calls into play the emergency systems will never happen? What one primarily is counting upon for the safety of a reactor is the integrity of the primary cooling system: that is, on the integrity of the pressure vessel and the pressure piping. Excruciating pains are taken to assure the integrity of these vessels and pipes. The watchword throughout the nuclear reactor industry is *quality assurance*: every piece of hardware in the primary system is examined, and reexamined, to guarantee insofar as possible that there are no flaws.

Nevertheless, we must deal with the remote contingency that might call the emergency systems into action. How certain can one be that these will work as planned? To better understand the analysis of the emergency system, Figs. 1 and 2 show, schematically, a large BWR and a PWR.

Three barriers prevent radioactivity from being released: fuel element cladding, primary pressure system, and containment shell. In addition to the regular safety system consisting primarily of the control and safety rods, there are elaborate provisions for preventing the residual radioactive heat from melting the fuel in the event of a loss of coolant. In the BWR there are sprays that spring into action within 30 seconds of an accident. In both the PWR and BWR, water is injected under pressure from gas-pressurized accumulators. In both reactors there are additional systems for circulating water after the system has come to low pressure, as well as means for reducing the pressure of steam in the containment vessel. This latter system also washes down or otherwise helps remove any fission products that may become airborne.

In analyzing the ultimate safety of a LWR, one tries to construct scenarios—improbable as they may be—of how a catastrophe might occur; and then one tries to provide reliable countermeasures for each step in the chain of failures that could lead to catastrophe. The chain conceivably could go like this. First, a pipe might break, or the safety system might fail to respond when called upon in an emergency. Second, the emergency core cooling system might fail. Third, the fuel might melt, might react also with the water, and conceivably might melt through the containment. Fourth, the containment might fail catastrophically, if not from the melt itself, then from missiles or overpressurization, and activity might then spread to the public. There may be other modes of catastrophic failure—for example, earthquakes or acts of violence—but the above is the more commonly identified sequence.

To give the flavor of how the analysis of an accident is made, let me say a few words about the first and second steps of this chain. As a first step, one might imagine failure of the safety system to respond in an emergency, say, when the bubbles in a BWR collapse after a fairly routine turbine trip. Here the question is not that some safety rods will work and some will not, but rather that a common mode failure might render the entire safety system inoperable. Thus if all the electrical cables actuating the safety rods were damaged by fire, this would be a common mode failure. Such a common mode failure is generally regarded as impossible, since the actuating cables are carefully segregated, as are groups of safety rods, so as to avoid such an accident. But one cannot *prove* that a common mode failure is impossible. It is noteworthy that on 30 September 1970, the entire safety system of the Hanford-N reactor (a one-of-a-kind water-cooled, graphite-moderated reactor) did fail when called

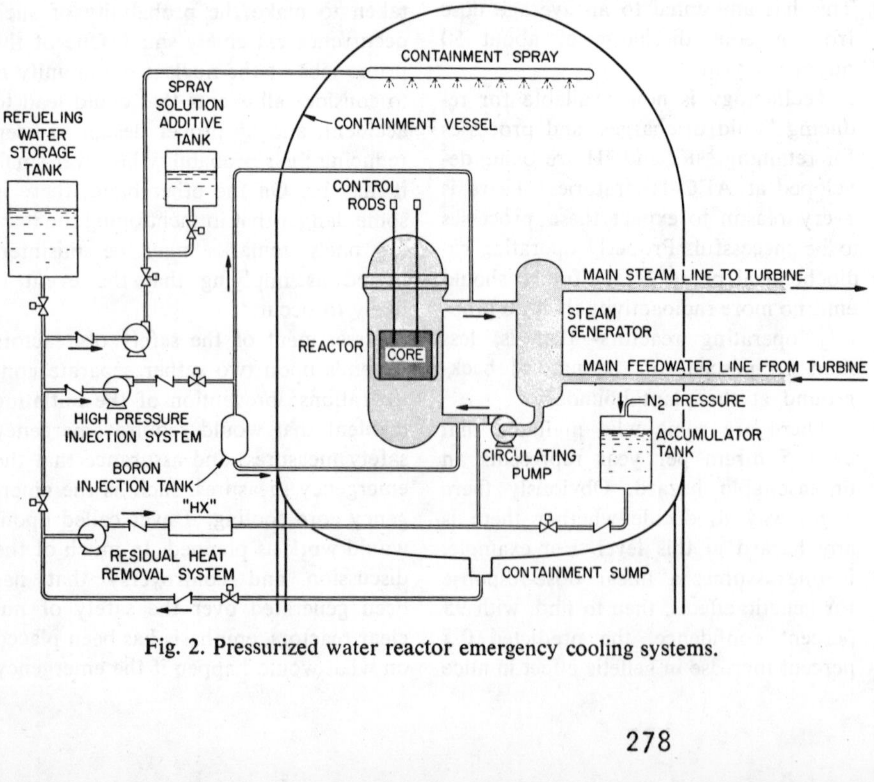

Fig. 2. Pressurized water reactor emergency cooling systems.

upon; however, the backup samarium balls dropped precisely as planned and shut off the reactor. One goes a long way toward making such a failure incredible if each big reactor, as in the case of the Hanford-N reactor, has two entirely independent safety systems that work on totally different principles. In the case of BWR, shutoff of the recirculation pumps in the all but incredible event the rods fail to drop constitutes an independent shutoff mechanism, and automatic pump shutoff is being incorporated in the design of modern BWR's.

The other step in the chain that I shall discuss is the failure of the emergency core cooling system. At the moment, there is some controversy whether the initial surge of emergency core cooling water would bypass the reactor or would in fact cool it. The issue was raised recently by experiments on a very small scale (9-inch-diameter pot) which indeed suggested that the water in that case would bypass the core during the blowdown phase of the accident. However, there is a fair body of experts within the reactor community who hold that these experiments were not sufficiently accurate simulations of an actual PWR to bear on the reliability or lack of reliability of the emergency core cooling in a large reactor.

Obviously the events following a catastrophic loss of coolant and injections of emergency coolant are complex. For example, one must ask whether the fuel rods will balloon and block coolant channels, whether significant chemical reactions will take place, or whether the fuel cladding will crumble and allow radioactive fuel pellets to fall out.

Such complex sequences are hardly susceptible to a complete analysis. We shall never be able to estimate everything that will happen in a loss-of-coolant accident with the same kind of certainty with which we can compute the Balmer series or even the course of the ammonia synthesis reaction in a fertilizer plant. The best that we can do as knowledgeable and concerned technologists is to present the evidence we have, and to expect policy to be based upon informed—not uninformed—opinion.

Faced with questions of this weight, which in a most basic sense are not fully susceptible to a yes or no scientific answer, the AEC has invoked the adjudicatory process. The issue of the reliability of the emergency core cooling system is being taken up in hearings before a special board drawn from the Atomic Safety and Licensing Board Panel. The record of the hearings is expected to contain all that is known about emergency core cooling systems and to provide the basis for setting the criteria for design of such systems.

Transport of Radioactive Materials

If, by the year 2000, we have 10^6 megawatts of nuclear power, of which two-thirds are liquid metal fast breeders, then there will be 7,000 to 12,000 annual shipments of spent fuel from reactors to chemical plants, with an average of 60 to 100 loaded casks in transit at all times. Projected shipments might contain 1.5 tons of core fuel which has decayed for as little as 30 days, in which case each shipment would generate 300 kilowatts of thermal power and 75 megacuries of radioactivity. By comparison, present casks from LWR's might produce 30 kilowatts and contain 7 megacuries.

Design of a completely reliable shipping cask for such a radioactive load is a formidable job. At Oak Ridge our engineers have designed a cask that looks very promising. As now conceived, the heat would be transferred to air by liquid metal or molten salt; and the cask would be provided with rugged shields which would resist deformation that might be caused by a train wreck. To be acceptable the shipping casks must be shown to withstand a 30-minute fire and a drop from 30 feet onto an unyielding surface (Fig. 3).

Can we estimate the hazard associated with transport of these materials? The derailment rate in rail transport (in the United States) is 10^{-6} per car mile. Thus, if there were 12,000 shipments per year, each of a distance of 1000 miles, we would expect 12 derailments annually. However, the number of serious accidents would be perhaps 10^{-4}- to 10^{-6}-fold less frequent; and shipping casks are designed to withstand all but the most serious accident (the train wreck near an oil refinery that goes into flames as a result of the crash). Thus the statistics—between 1.2×10^{-3} and 1.2×10^{-5} serious accidents per year—at least until the year 2000, look quite good. Nevertheless the shipping problem is a difficult one and may force a change in basic strategy. For example, we may decide to cool fuel from LMFBR's in place for 360 days before shipping: this reduces the heat load sixfold, and increases the cost of power by only around 0.2 mill per electric kilowatt hour. Or a solution that I personally prefer is to cluster fast breeders in nuclear power parks which have their own on-site reprocessing facilities (*5*). Clustering reactors in this way would make both cooling and transmission of power difficult; also such parks would be more vulnerable to common mode failure, such as acts of war or earthquakes. These difficulties must be balanced against the advantage of not shipping spent fuel off-site, and of simplifying control of fissile material against diversion. To my mind, the advantages of clustering outweigh its disadvantages; but this again is a trans-scientific question which can only be adjudicated by a legal or political process, rather than by scientific exchange among peers.

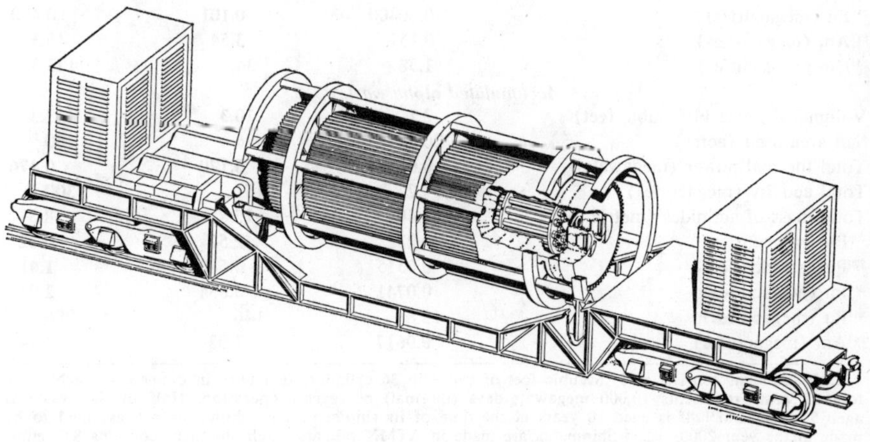

Fig. 3. Liquid metal fast breeder reactor spent fuel shipping cask (18 assemblies).

Waste Disposal

By the year 2000, according to present projections, we shall have to sequester about 27,000 megacuries of radioactive wastes in the United States; these wastes will be generating 100,000 kilowatts of heat at that time. The composition of these wastes is summarized in Table 2.

The wastes will include about 400 megacuries of transuranic alpha emitters. Of these, the ^{239}Pu with a half-life of 24,400 years will be dangerous for perhaps 200,000 years.

Can we see a way of dealing with these unprecedentedly treacherous materials? I believe we can, but not without complication.

There are two basically different approaches to handling the wastes. The first, urged by W. Bennett Lewis of Chalk River (*6*), argues that once man has opted for nuclear power he has committed himself to essentially perpetual surveillance of the apparatus of nuclear power, such as the reactors, the chemical plants, and others. Therefore, so the argument goes, there will be spots on the earth where radioactive operations will be continued in perpetuity. The wastes then would be stored at these spots, say in concrete vaults. Lewis further refines his ideas by suggesting that the wastes be recycled so as to limit their volume. As fission products decay, they are removed and thrown away as innocuous nonradioactive species; the transuranics are sent back to the reactors to be burned. The essence of the scheme is to keep the wastes under perpetual, active surveillance and even processing. This is deemed possible because the original commitment to nuclear energy is considered to be a commitment in perpetuity.

There is merit in these ideas; and indeed permanent storage in vaults is a valid proposal. However, if one wishes to perpetually rework the wastes as Lewis suggests, chemical separations would be required that are much sharper than those we now know how to do; otherwise at every stage in the recycling we would be creating additional low-level wastes. We probably can eventually develop such sharp separation methods; but these, at least with currently visualized techniques, would be very expensive. It is on this account that I like better the other approach which is to find some spot in the universe where the wastes can be placed forever out of contact with the biosphere. Now the only place where we know absolutely the wastes will never interact with man is in far outer space. But the roughly estimated cost of sending wastes into permanent orbit with foreseeable rocket technology is in the range of 0.2 to 2 mills per electric kilowatt hour, not to speak of the hazard of an abortive launch. For both these reasons I do not count on rocketing the wastes into space.

This pretty much leaves us with disposal in geologic strata. Of the many possibilities—deep rock caverns, deep wells, bedded salt—the latter has been chosen, at least on an experimental basis, by the United States and West Germany. The main advantages of bedded salt are primarily that, because salt dissolves in water, the existence of a stratum of bedded salt is evidence that the salt has not been in contact with circulating water during geologic time. Moreover, salt flows plastically; if radioactive wastes are placed in the salt, eventually the salt ought to envelop the wastes and sequester them completely.

These arguments were adduced by the National Academy of Sciences Committee on Radioactive Waste Management (*7*) in recommending that the United States investigate bedded salt (which underlies 500,000 square miles in our country) for permanent disposal of radioactive wastes. And, after 15 years of discussion and research, the AEC about a year ago decided to try large-scale waste disposal in an abandoned salt mine in Lyons, Kansas (Fig. 4). If all goes as planned, the Kansas mine is to be used until A.D. 2000. What one does after A.D. 2000 would of course depend on our experience during the next 30 years (1970 to 2000). In any event, the mine is to be designed so as to allow the wastes to be retrieved during this time.

The salt mine is 1000 feet deep, and the salt beds are around 300 feet thick. The beds were laid down in Permian times and had been undisturbed, until man himself intruded, for 200 million years. Experiments in which radioactive fuel elements were placed in the salt have clarified details of the temperature distribution around the wastes, the effect of radiation on salt, the migration of water of crystallization within the salt, and so on.

The general plan is first to calcine the

Table 2. Projected waste inventories at the permanent repository.

	Calendar year		
	1980	1990	2000
Number of annual shipments			
High-level waste*	23	240	590
Alpha waste†	420	1,200	0
Accumulated high-level waste			
Volume of waste (cubic feet)	3,170	74,200	319,000
Salt area used (acres)	9	200	900
Total thermal power (megawatts)	1.17	24.4	94.9
Total activity (megacuries)	329	7,030	27,700
^{90}Sr (megacuries)	59.0	1,310	5,290
^{137}Cs (megacuries)	83.1	1,850	7,500
^{238}Pu (megacuries)	0.102	2.34	9.88
^{239}Pu (megacuries)	0.00157	0.0368	0.158
^{240}Pu (megacuries)	0.00400	0.101	0.470
^{241}Am (megacuries)	0.151	3.54	15.3
^{244}Cm (megacuries)	1.58	34.1	133.3
Accumulated alpha waste‡§			
Volume of waste (10^6 cubic feet)	2.1	10.3	19.3
Salt area used (acres)	20	96	180
Total thermal power (megawatts)	0.0142	0.170	0.476
Total activity (megacuries)	14.2	151	300
Total mass of actinides (metric tons)	1.40	15.8	38.3
^{238}Pu (megacuries)	0.232	2.57	6.02
^{239}Pu (megacuries)	0.0515	0.580	1.41
^{240}Pu (megacuries)	0.0741	0.834	2.02
^{241}Pu (megacuries)	13.8	146	286
^{243}Am (megacuries)	0.0617	1.03	4.74

* Each shipment consists of 57.6 cubic feet of waste in 36 cylinders (6 inches in diameter). Each cubic foot of waste represents 10,000 megawatt days (thermal) of reactor operation. Half of the waste is aged 5 years, and half is aged 10 years at the time of its shipment. Last shipments are assumed to be made in the year 2000. † Shipments are made in ATMX railcars; each shipment contains 832 cubic feet of waste. Last shipments are assumed to be made in the year 1999. ‡ At end of year. § The isotopic composition of Pu at the time of its receipt is 1 percent ^{238}Pu, 60 percent ^{239}Pu, 24 percent ^{240}Pu, 11 percent ^{241}Pu, and 4 percent ^{242}Pu.

liquid wastes to a dry solid. The solid is then placed in metal cans, and the cans are buried in the floor of a gallery excavated in the salt mine. After the floor of the gallery is filled with wastes, the gallery is backfilled with loose salt. Eventually this loose salt will consolidate under the pressure of the overburden, and the entire mine will be resealed. The wastes will have been sequestered, it is hoped, forever.

Much discussion has centered around the question of just how certain we are that the events will happen exactly as we predict. For example, is it possible that the mine will cave in and that this will crack the very thick layers of shale lying between the mine and an aquifer at 200 feet below the surface? There is evidence to suggest that this will not happen, and I believe most, though not all, geologists who have studied the matter agree that the 500-foot-thick layer of shale above the salt is too strong to crack so completely that water could enter the mine from above.

But man's interventions are not so easily disposed of. In Kansas there are some 100,000 oil wells and dry holes that have been drilled through these salt formations. These holes penetrate aquifers; and in principle they can let water into the mine. For the salt mine to be acceptable, one must plug all such holes. At the originally proposed site there were 30 such holes; in addition, solution mining was practiced nearby. For this reason, the AEC recently authorized the Kansas State Geological Survey to study other sites that were not peppered with man-made holes. The AEC also announced recently its intention to store solidified wastes in concrete vaults, pending resolution of these questions concerning permanent disposal in geologic formations.

Man's intervention complicates the use of salt for waste disposal; yet by no means does this imply that we must give up the idea of using salt. In the first place, such holes can be plugged, though this is costly and requires development. In the second place, let us assume the all but incredible event that the mine is flooded—let us say 10,000 years hence. By that time, since no new waste will be placed in the mine after A.D. 2000, all the highly radioactive beta decaying species, notably ^{90}Sr and ^{137}Cs, would have decayed. The main radioactivity would then come from the alpha emitters. The mine would contain 38 tons of ^{239}Pu mixed with about a million tons of nonradioactive material. The plutonium in the cans is thus diluted to 38 parts per million; since plutonium is, per gram, 700 times more hazardous than natural uranium in equilibrium with its daughters, these diluted waste materials would present a rather smaller hazard than an equal amount of pitchblende. Actually, the 38 tons of ^{239}Pu is spread over 200 acres. If all the salt associated with the ^{239}Pu were dissolved in water, as conceivably could result from total flooding of the mine, the concentration of plutonium in the resulting salt solution would be well below maximum permissible concentrations. In other words, by virtue of having spread the plutonium over an area of 200 acres, we have to a degree ameliorated the residual risk in the most unlikely event that the mines are flooded.

Despite such assurances, the mines must not be allowed to flood, especially before the ^{137}Cs and ^{90}Sr decay. We must prevent man from intruding—and this can be assured only by man himself. Thus we again come back to the great desirability, if not absolute necessity in this case, of keeping the wastes under some kind of surveillance in perpetuity. The great advantage of the salt method over, say, the perpetual reworking method, or even the aboveground concrete vaults without reworking, is that our commitment to surveillance in the case of salt is minimal. All we have to do is prevent man from intruding, rather than keeping a priesthood that forever reworks the wastes or guards the vaults. And if the civilization should falter, which would mean, among other things, that we abandon nuclear power altogether, we can be almost (but not totally) assured that no harm would befall our recidivist descendants of the distant future.

Social Institutions—Nuclear Energy

We nuclear people have made a Faustian bargain with society. On the one hand, we offer—in the catalytic nuclear burner—an inexhaustible source of energy. Even in the short range, when we use ordinary reactors, we offer energy that is cheaper than energy from fossil fuel. Moreover, this source of energy, when properly handled, is almost nonpolluting. Whereas fossil fuel burners must emit oxides of carbon and nitrogen, and probably will always emit some sulfur dioxide, there is no intrinsic reason why nuclear systems must emit any pollutant—except heat and traces of radioactivity.

But the price that we demand of society for this magical energy source is both a vigilance and a longevity of our social institutions that we are quite unaccustomed to. In a way, all of this was anticipated during the old debates

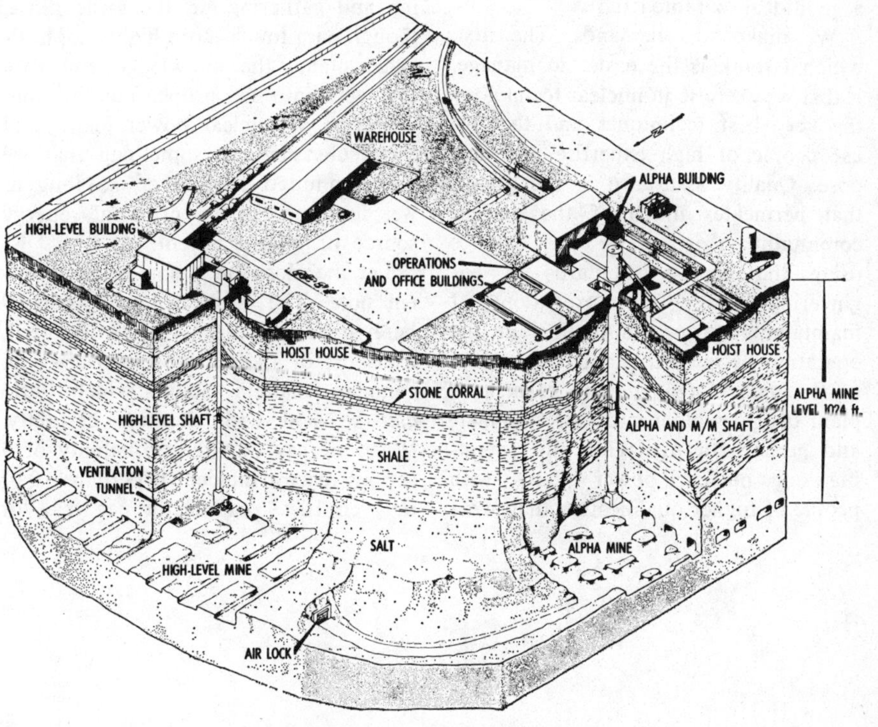

Fig. 4. Federal repository.

over nuclear weapons. As matters have turned out, nuclear weapons have stabilized at least the relations between the superpowers. The prospects of an all-out third world war seem to recede. In exchange for this atomic peace we have had to manage and control nuclear weapons. In a sense, we have established a military priesthood which guards against inadvertent use of nuclear weapons, which maintains what a priori seems to be a precarious balance between readiness to go to war and vigilance against human errors that would precipitate war. Moreover, this is not something that will go away, at least not soon. The discovery of the bomb has imposed an additional demand on our social institutions. It has called forth this military priesthood upon which in a way we all depend for our survival.

It seems to me (and in this I repeat some views expressed very well by Atomic Energy Commissioner Wilfrid Johnson) that peaceful nuclear energy probably will make demands of the same sort on our society, and possibly of even longer duration. To be sure, we shall steadily improve the technology of nuclear energy; but, short of developing a truly successful thermonuclear reactor, we shall never be totally free of concern over reactor safety, transport of radioactive materials, and waste disposal. And even if thermonuclear energy proves to be successful, we shall still have to handle a good deal of radioactivity.

We make two demands. The first, which I think is the easier to manage, is that we exercise in nuclear technology the very best techniques and that we use people of high expertise and purpose. Quality assurance is the phrase that permeates much of the nuclear community these days. It connotes using the highest standards of engineering design and execution; of maintaining proper discipline in the operation of nuclear plants in the face of the natural tendency to relax as a plant becomes older and more familiar; and perhaps of managing and operating our nuclear power plants with people of higher qualification than were necessary for managing and operating nonnuclear power plants: in short, of creating a continuing tradition of meticulous attention to detail.

The second demand is less clear, and I hope it may prove to be unnecessary. This is the demand for longevity in human institutions. We have relatively little problem dealing with wastes if we can assume always that there will be intelligent people around to cope with eventualities we have not thought of. If the nuclear parks that I mention are permanent features of our civilization, then we presumably have the social apparatus, and possibly the sites, for dealing with our wastes indefinitely. But even our salt mine may require some small measure of surveillance if only to prevent men in the future from drilling holes into the burial grounds.

Eugene Wigner has drawn an analogy between this commitment to a permanent social order that may be implied in nuclear energy and our commitment to a stable, year-in and year-out social order when man moved from hunting and gathering to agriculture. Before agriculture, social institutions hardly required the long-lived stability that we now take so much for granted. And the commitment imposed by agriculture in a sense was forever: the land had to be tilled and irrigated every year in perpetuity; the expertise required to accomplish this task could not be allowed to perish or man would perish; his numbers could not be sustained by hunting and gathering. In the same sense, though on a much more highly sophisticated plane, the knowledge and care that goes into the proper building and operation of nuclear power plants and their subsystems is something that we are committed to forever, so long as we find no other practical energy source of infinite extent (*8*).

Let me close on a somewhat different note. The issues I have discussed here—reactor safety, waste disposal, transport of radioactive materials—are complex matters about which little can be said with absolute certainty. When we say that the probability of a serious reactor incident is perhaps 10^{-8} or even 10^{-4} per reactor per year, or that the failure of all safety rods simultaneously is incredible, we are speaking of matters that simply do not admit of the same order of scientific certainty as when we say it is incredible for heat to flow against a temperature gradient or for a perpetuum mobile to be built. As I have said earlier, these matters have trans-scientific elements. We claim to be responsible technologists, and as responsible technologists we give as our judgment that these probabilities are extremely—almost vanishingly—small; but we can never represent these things as certainties. The society must then make the choice, and this is a choice that we nuclear people cannot dictate. We can only participate in making it. Is mankind prepared to exert the eternal vigilance needed to ensure proper and safe operation of its nuclear energy system? This admittedly is a significant commitment that we ask of society. What we offer in return, an all but infinite source of relatively cheap and clean energy, seems to me to be well worth the price.

References and Notes

1. "50 and 100 Years Ago," *Sci. Amer.* **225**, 10 (Nov. 1971).
2. J. Bartlett, *Familiar Quotations* (Little, Brown, Boston, ed. 14, 1968).
3. E. N. da C. Andrade, *Rutherford and the Nature of the Atom* (Doubleday, Garden City, N.Y., 1964), p. 210.
4. A. M. Weinberg and R. P. Hammond, in *Proceedings of the Fourth International Conference on the Peaceful Uses of Atomic Energy* (United Nations, New York, in press); *Bull. Atom. Sci.* **28**, 5, 43 (March 1972).
5. A. M. Weinberg, "Demographic Policy and Power Plant Siting," Senate Interior and Insular Affairs Committee, Symposium on Energy Policy and National Goals, Washington, D.C., 20 October 1971.
6. W. B. Lewis, *Radioactive Waste Management in the Long Term* (DM-123, Atomic Energy of Canada Limited, Chalk River, 13 July 1971).
7. *Disposal of Solid Radioactive Wastes in Bedded Salt Deposits* (National Academy of Sciences–National Research Council, Washington, D.C., 1970); *Disposal of Radioactive Wastes on Land* (Publication 519, National Academy of Sciences–National Research Council, Washington, D.C., 1957); *Report to the U.S. Atomic Energy Commission* (National Academy of Sciences–National Research Council, Committee on Geologic Aspects of Radioactive Waste Disposal, Washington, D.C., May 1966).
8. Professor Friedrich Schmidt-Bleek of the University of Tennessee pointed out to me that the dikes of Holland require a similar institutional commitment in perpetuity.
9. L. G. Hauser and R. F. Potter, "The effect of escalation on future electric utility costs" (report issued by Nuclear Fuel Division, Westinghouse Electric Corporation, Pittsburgh, Pa., 1971).

Statement of John W. Landis, President, Gulf General Atomic Co., Division of Gulf Oil Corp.

Mr. LANDIS. Mr. Chairman, Congressman McCormack, Congressman Seiberling, Dr. Ratchford, Dr. Holmfeld, my name is John W. Landis, as you noted. I am with Gulf Oil Corp., Gulf Atomic Co. We are located in San Diego, Calif.

I have with me Norval E. Carey, who is director of east coast operations for Gulf General Atomic.

We certainly thank you for this opportunity to bring you up-to-date on the national gas-cooled reactor program which is being carried on jointly, as you know, by the Atomic Energy Commission and Gulf General Atomic.

We in Gulf have followed the work of the task force on energy of the Subcommittee on Science, Research and Development with very great interest. We commend the task force for its orderly and discerning approach to the massive problems of energy production, transmission and utilization and for compiling an organized body of extremely valuable information from the vast array of evidence it has received. This information will point the way toward improved utilization of available scientific and technical resources in resolving the prolonged energy crisis into which the Nation has plunged.

As of this date, the Atomic Energy Commission, the electric utility industry and Gulf Oil have invested almost $600 million in the overall U.S. gas-cooled reactor program and almost $450 million specifically in the high-temperature gas-cooled reactor (RTGR), the first major product of that program. The initial powerplant of the HTGR type, the Peach Bottom Station, has been in operation on the system of the Philadelphia Electric Co. since 1967. This plant is now on its second core and its performance has been quite remarkable, both in terms of reliability and in terms of release of radioactive material. The total activity in the primary loop, for example, is holding constant at the unprecedented low level of one-half curie. And that, I should emphasize, is the amount of radioactivity in the entire loop, not just what is being released.

Peach Bottom is the only nuclear station in the world, incidentally, producing steam at modern turbine conditions—that is, a thousand degrees Fahrenheit.

Construction has recently been completed on a 330-MW(e) HTGR for the Public Service Co. of Colorado. This plant, the Fort St. Vrain Station, is undergoing final preoperational testing while awaiting its operating license, which has been delayed because of the *Calvert Cliffs* decision. It is expected to be in full power operation by late fall.

Also, orders have been received from the Philadelphia Electric Co. for twin 1160-MW(e) HTGR's for operation in 1979 and 1981 and from Delmarva Power and Light Co. for twin 770-MW(e) HTGR's for operation in 1980 and 1983. And since we wrote this testimony, Mr. Chairman and members, we have received two additional orders—for twin 770-MW(e) plants of the same type from the Southern California Edison Co. for operation in 1981 and 1982. Southern California Edison has also taken options on two 1160-MW(e) plants for operation at later dates.

During this calendar year we expect to sell, in addition to the two we have already sold to Southern California Edison, approximately four more plants. That would amount to something on the order of 4,000 megawatts. A recent Department of the Interior report forecasts the construction of 181,000 megawatts of HTGR capacity by the year 2000.

The HTGR offers many significant advantages to the electric utility industry, including high efficiency, high conversion ratio, decreased thermal discharge, negligible release of radioactivity, an inert and single-phase coolant, excellent core structural strength at high temperatures, great core heat capacity, large core temperature margins, schedule and siting flexibility, inservice inspection capability, and low fuel and maintenance costs. It uses the thorium-233 fuel cycle and therefore makes the world's thorium resources available for use as nuclear fuels. In addition, it stretches the world's total nuclear fuel reserves significantly by reducing the quantity of fissionable material needed to produce a given amount of electricity.

To assure that the United States reaps the full benefits of these advantages that it has had the foresight to pursue for more than 15 years; the final stages of the long term HTGR development program must be performed on schedule. The bulk of this work is to be done at Oak Ridge National Laboratory. It consists primarily of advanced fuel research, fuel reprocessing development, fuel recycling development, safety research and miscellaneous materials projects. It will cost in the neighborhood of $100 million and will extend over a period of approximately 8 years. We have, of course, strongly urged the Atomic Energy Commission to step up its support of this basic work so that the results are available on time. The benefit-cost ratio is very high, as you have undoubtedly surmised from recent AEC testimony before other Congressional committees.

On top of the many benefits that the present HTGR will bring to the Nation, two promising variations of this reactor will provide important "bonus" benefits. These variations are the "gas turbine HTGR" and the "coal gasification HTGR."

The gas turbine HTGR will help solve two particularly knotty problems: the high capital cost of nuclear power equipment and the tremendous consumption of fresh water by power-generating stations.

The coal gasification HTGR will, if it proves to be economically feasible, open the way to utilization of the vast deposits of high-sulfur coal in the United States.

Because my time is limited here today, I shall not discuss the latter variation except to say that even present-day HTGR's can supply heat at the temperatures required for gasification and that the coal, if liquefied prior to processing, can be converted to pipeline-quality product without polluting the environment in the vicinity of the gasification station—a very important point.

I mentioned two serious problems that the Gas Turbine HTGR will help solve. The first of these is perhaps not as significant from the public's standpoint as the second. Let me therefore address myself chiefly to the second—the cooling-water problem.

In many areas of the country, as you gentlemen know, it is already quite difficult to find sufficient fresh water for powerplant cooling purposes. This problem, I believe, is even more critical than the uranium-availability problem, and it will get far worse in the immediate future.

Some numbers may drive my point home more forcefully than words.

If wet (evaporative) cooling towers are used to dissipate or carry away the waste heat from a 1000-MW(e) nuclear plant of the LWR type, about 25 million gallons of makeup water will be required for each day of operation. This represents about half the total amount of fresh water consumed for all other purposes by the populace and industry served by such a plant.

Two solutions to this problem are on the horizon. The first is to construct nuclear power plants on barges and locate them at protected offshore sites along the coast. Ample cooling water is obviously available at these sites but the costs of the barge, of protecting the site, of transmitting the power to shore and of the required operational precautions may be so high that this approach is not economically viable. Also, it is of course applicable only to our coastal regions.

The second solution is to utilize dry cooling; that is, rejection of waste heat directly to ambient air by means of finned heat exchangers or radiators in either natural-draft or forced-draft towers.

Mr. DAVIS. Excuse me. I wanted to comment on the previous sentence you read.

Reprinted with permission from *Energy Research and Dev.*, Hearings before the Subcommittee on Sci., Research, and Dev., House Committee on Sci. and Astronautics, Y4.Sci2: 92/2/24, pp. 370–390, May 1972.

Would it cause you to be less sure of your previous statement, "also, it is of course applicable only to our coastal regions," if, for example, we should come up with a good cryogenic technology of transmitting electrical power without resistance?

Mr. LANDIS. What I meant there, Mr. Chairman, is that the plant has to be located in the coastal region. If you come up with a good cryogenic technique, or another technique for conducting electricity more cheaply and more conveniently and in a more environmentally-acceptable fashion over long distances, then of course offshore location would be a much more desirable solution to the problem.

Mr. DAVIS. Thank you, sir.

Mr. LANDIS. With respect to the dry cooling method that I mentioned, I want to emphasize that no water is consumed at all, and therefore plants employing this equipment can be built almost anywhere. They do not have to be located near bodies of fresh water, around which people invariably concentrate.

The main disadvantages of the latter approach in current (steam-cycle) plants is the large size and consequent high cost of the required dry cooling facilities. This disadvantage stems from the fact, as Mr. Shaw mentioned, that the temperature difference between the condensing steam and ambient air is very small—only 20 to 40 degrees Fahrenheit.

The high efficiency of the standard HTGR would make it the best choice for dry cooling if we had to use this method of waste-heat disposal study. Still, with the steam cycle, the overall cost would be relatively high compared to other types of thermal discharge. To have our cake and eat it too—that is, to have dry cooling with its siting flexibility and minimal effect on the environment and at the same time hold to acceptable capital expenditures—we must utilize a direct-cycle machine where the reactor coolant also serves as the working fluid for power conversion.

This is what the Gas Turbine HTGR is—a direct-cycle machine. The standard HTGR is convertible to such a machine because it uses helium as its coolant and helium is an excellent working fluid in gas turbines.

The gas turbines will in fact be located inside the reactor vessel, replacing the steam generators and helium circulators. Thus, we shall get usable power from this vessel, not just steam, and the turbine hall and the entire steam and feed-water systems will be eliminated, yielding substantial capital-cost savings.

But that is not the entire story. It so happens that gas turbines reject their heat over a wide temperature range starting at levels 200 to 300 degrees Fahrenheit above normal atmospheric temperatures instead of just 20 to 40 degrees Fahrenheit, as I have noted is the case for steam-cycle systems. This permits a great reduction in cooling-tower size and another important capital-cost saving.

For these reasons—and others that I do not have time to delineate—we in Gulf are confident that the Gas Turbine HTGR will play a vital role in the economy of the United States during the 1980's.

Looking a little farther into the future, we are working on an even more advanced concept—the Gas Cooled Fast Reactor (GCFR). This is the reactor that you, Mr. Chairman, indicated the Committee is especially interested in.

The GCFR is a breeder that uses helium as its coolant—and therefore is based on the proven primary-loop technology of the HTGR. Since its fuel elements are similar to those of the Liquid Metal Fast Breeder Reactor (LMFBR), the incremental cost of developing it should be quite small.

The GCFR has a number of desirable features. Among these are:

(1) Relatively low capital costs, resulting primarily from elimination of the intermediate cooling loop and from the design simplifications made possible by the inertness, transparency, and single-phase nature of helium.

(2) Attractive fuel costs, resulting primarily from the low absorption and moderation of neutrons by helium.

(3) Low operation and maintenance costs, resulting primarily, as in item (1) from the basic properties of helium and from direct access to the secondary containment.

(4) Significant potential for improvement—in both the fuel area and the plant area. (In the latter area, it should be pointed out for the record that the GCFR, like the HTGR, is convertible to a direct-cycle machine.)

(5) Excellent safety characteristics, due chiefly to the impossibility of total loss of coolant and the negligible amount of stored energy in the coolant.

If you would like more information on this subject, I refer you to "The Case for Gas Cooling," by Peter Fortescue, dated April 10, 1972, copies of which I delivered when I was here before.

Mr. DAVIS. Yes, sir.

Now was that included in the record? I forget.

Mr. LANDIS. I don't think it was. I think I merely left it with your staff.

Mr. DAVIS. If not, we will include it at this point.

Editor's Note: The article by Peter Fortescue has not been reprinted here. It can be found in the record of the House hearings or, in somewhat edited form, as "The case for gas cooling," *Power Engineering,* July 1972.

Also, a small but important series of irradiations is being carried out in cooperation with the national laboratories at Oak Ridge and Argonne to test the ability of modified LMFBR fuel to meet certain special requirements of the GCFR.

The results of this and other work have been quite positive and amply justify, in our opinion, the support and interest which the utility industry is according to the GCFR, but funding by the Government far above the present level is essential if this attractive concept is to survive.

Gentlemen, the nonpartisan experts whom I have quoted have said it better than I can, but let me summarize in my own words. It seems reasonable to assume that if the breeder reactor is adjudged to be vital to the future economic well-being of the Nation, we must follow a development path that maximizes our probability of success. One does not need a course in statistics to calculate that pursuing the LMFBR and GCFR approaches to this goal simultaneously more than doubles the probability of success—for an incremental cost in the neighborhood of only 25 percent.

We thank you again for this opportunity to appear before you. I shall be happy to address any questions that you may have—will try to, at least.

Mr. DAVIS. Thank you very much, Mr. Landis, for an excellent statement.

I would like to comment on the general conceptual matters that are touched upon in your statement. It reminds me very much of the debate that went on so long in the space exploration effort of this country as to whether to use liquid or solid fuel. It is interesting to note we went ahead with considerable research in both areas. While we have been using liquid all this time except in some of our military missiles, now the National Aeronautics and Space Administration has reached the conclusion that so far as the space shuttle is concerned, we will use solid fuel. I certainly agree that it is well not to put all your eggs in one basket when you go to develop a new energy source. I want to agree heartily with your position along that line.

Mr. LANDIS. Thank you, sir.

The Fast Breeder Reactor

A Source of Abundant Power for the Future

H. Dieckamp

The fast breeder reactor is essential to meet the future energy needs of our society. This capability derives from three fundamental factors: (1) the fast breeder reactor maximizes the utilization of valuable natural resources and as a direct result can supply energy to an expanding industrial society for many generations; (2) the fast breeder reactor can achieve this goal while protecting the environment better than any other currently available economic method of generating power; and (3) the fast breeder is based upon established technology and is therefore ready for commercial introduction.

Energy Consumption

The discerning observer of the human scene will recognize the importance of energy use as a measure of living standards because energy is basic to most of the requirements of life. The United States with only 7% of the world's population uses more than 35% of the world's energy. This corresponds to about 3 times the per capita energy consumption of Western Europe, about 7 times that of Latin America, and about 20 to 30 times the average energy consumption per person in Africa or Asia. Two countries, Norway and Canada, exceed the U.S. in per capita electrical consumption.

A comparison of the United States and world energy consumption per capita is shown in Figure 1. At present the United States consumption of energy in all forms is at the rate of about 300 million Btu (300 million kilojoules) per person per year or an average rate of about 10 kilowatts thermal per person. The world energy consumption per capita is about 50 million Btu per year or about 1.5 kilowatts thermal per person. The growth rate in energy consumption during the last 15 to 20 years has been about the same for the world as the United States, indicating the increasing industrialization of less developed countries. There is every reason to believe that energy requirements per capita will continue to grow as we increase our standard of living and take added measures to protect the environment.

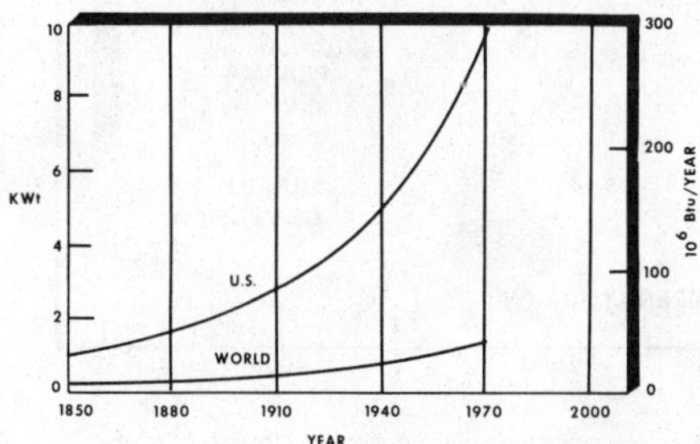

Fig. 1. Energy consumption per capita

This article is taken from a paper presented at the Fifty-Second Annual Meeting of the American Geophysical Union, April 1971.

At the present time the annual consumption of energy resources in the United States is about 0.06×10^{18} Btu (0.06 Q) per year, which is equivalent to an average rate of about 2 billion kilowatts thermal. There appears to be only a small chance that significant changes can be made in the population growth rate during the next 30 years, with the result that by the year 2000 the energy consumption in the United States is projected to grow to 0.14 Q per year or about 5 billion kilowatts. The cumulative total energy requirements for the United States during the next 30 years is estimated to be about 3 Q or about the same as the total energy consumed in the United States during the last 100 years.

Figure 2 shows the growth in en-

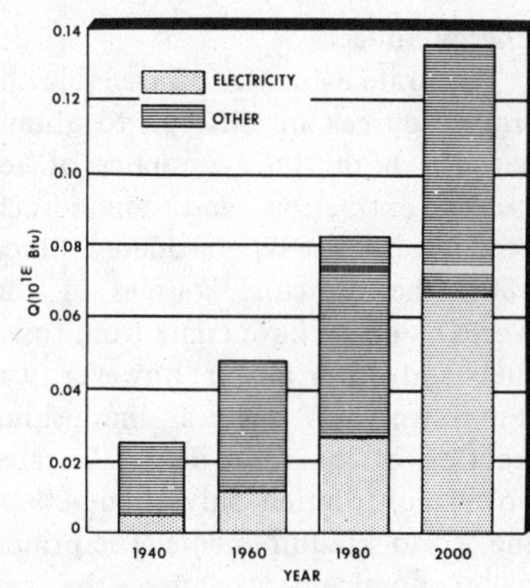

Fig. 2. U.S. energy consumption

ergy consumption from 1940 to the present and projects total energy demand to the year 2000. Energy growth between 1940 and 1970 has been at the rate of 3.5% per year. The figure also shows the percentage of energy used in the form of electricity for the same time periods. At present more than 20% of the United States energy consumption is used to generate electricity. Electric power generation has been growing faster than total energy consumption, about twice as fast, so that in the next 10 years the use of electric power has been projected to be about 6 times greater than in 1970 and will constitute about 50% of the energy consumed in the United States. The current generating capacity in the U.S. is over 300 million kilowatts and the energy consumed in the form of electricity is nearly 1.5 trillion kilowatt-hours per year.

This increase in electricity use is expected to occur because of the convenience of utilizing energy in the form of electricity. Furthermore, electricity is a clean form of energy at the point of consumption where pollution problems can be the most acute. To meet the demands for preserving the quality of the environment, the need for electric power may actually exceed today's predictions. Waste recycle equipment, air purification equipment, and sewage treatment equipment all require electrical energy. The advent of the electric automobile would, for example, create a demand for 10 times today's generating capability.

Energy Sources

Accurate estimates of available energy resources are difficult to obtain because both the economics of resource extraction and commercial utilization must be considered. Since 1800 the principal sources of the world's energy have come from fossil fuels and water power; however, water power has been a diminishing fraction of the total. In 1970 water power contributed only about 2% of the world's requirements. The principal economic energy sources that can supply significant fractions of the energy demands in the United States in the future are fossil fuels and nuclear

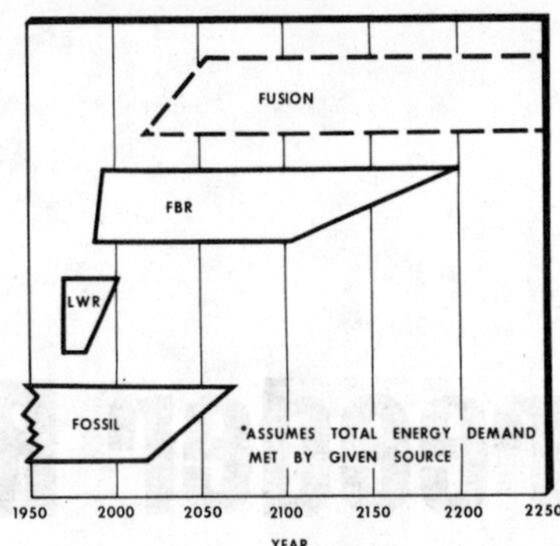

Fig. 3. Economic energy availability

energy. The date of technological availability and possible resource lifetime for fossil fuels, thermal fission reactors, fast fission reactors, and fusion are shown in Figure 3 and are based on the premise that the total U.S. energy requirements in the time period shown for each energy source are supplied by that source. Thus, the figure is a measure of the relative amounts of energy available from each but cannot be construed as an absolute prediction of the lifetime of each fuel source. Furthermore, nuclear energy, as we know it today, can be used chiefly for the generation of electrical energy. Nuclear cannot replace fossil fuels in all their applications. However, since the future use of energy will be primarily in the form of electricity, it is not unreasonable to evaluate nuclear energy sources on the assumption that they supply total energy needs.

Although the short range needs of the United States can be met by existing energy sources (fossil fuels and light water reactors), it is clear that other sources of energy must soon be developed for large scale use. Fossil fuels currently have economic and pollution problems associated with their use which makes light water reactors an attractive alternate for electrical energy generation. However, light water reactors utilize only about 1% to 2% of the latent energy available from uranium, and hence, because of limited economically recoverable uranium, they do not represent a large long-term source of energy.

A fast breeder reactor can economically use up to 75% of the energy content in uranium and thus

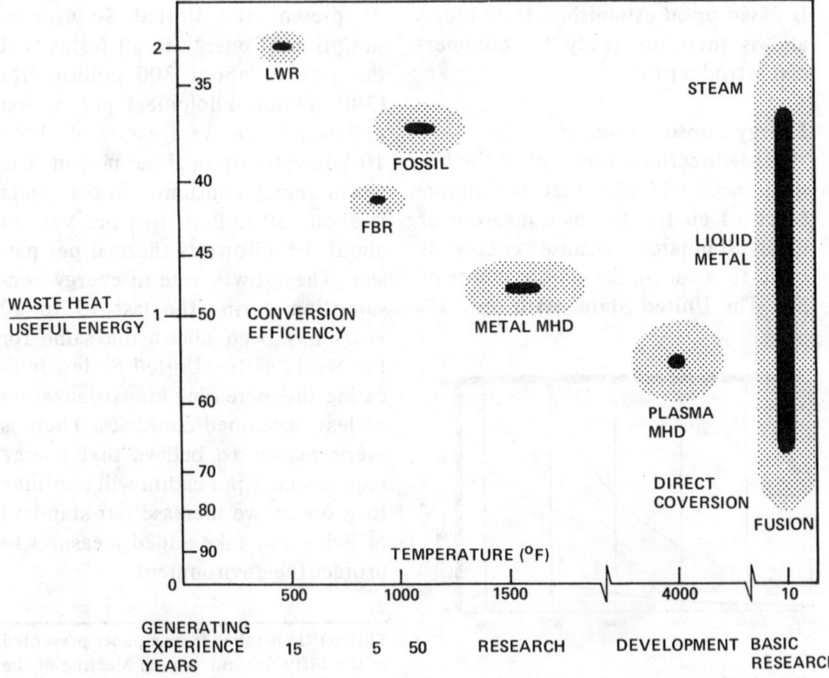

Fig. 4. Electric power sources for the future

achieve a resource efficiency about 30 to 40 times greater than that of current reactors. This multiplication of fuel resources is one of the great attractions of fast breeder reactors. A concomitant advantage is that fast breeder reactors can also afford to use more expensive uranium from lower grade ores than current reactors. As a result, by using fast reactors, the economically available energy from uranium resources in the United States can be expanded 100 to 1000 times as compared to light water reactors. The scientific feasibility of the fission breeder has been demonstrated with the operation of a number of experimental reactors. The building of large demonstration plants is the next step.

Energy can be obtained from a fusion reaction using deuterium, tritium, and lithium. For all practical purposes the energy resources for controlled fusion reactors is essentially infinite–literally millions of times the energy content of all fossil fuels. However, obtaining useful power from nuclear fusion involves solving many formidable problems, not the least of which is the demonstration of scientific feasibility of plasma confinement and stability. No responsi ble scientist has said that fusion is clearly feasible. In spite of the recent successes of the Russian Tokamak system–a temperature of 10 million°F (5 million°C) was held for 1/50 of a second–no one can predict when, or even whether, the development of power from the fusion reaction may be accomplished. The time scale for the first demonstration of the scientific feasibility of controlled fusion has been estimated to be at least 5 to 10 years away, with another 10 to 15 years required for the development of the technology for practical power plants. Add an additional 10 years for the construction and operation of prototype systems and fusion breeder reactors could not be expected to begin commercial operation until after the year 2000.

Other alternate power sources have been proposed as primary power producers. None of these are adequate if we consider the technology involved and the energy requirements. The largest of these is solar energy. The technology has not yet

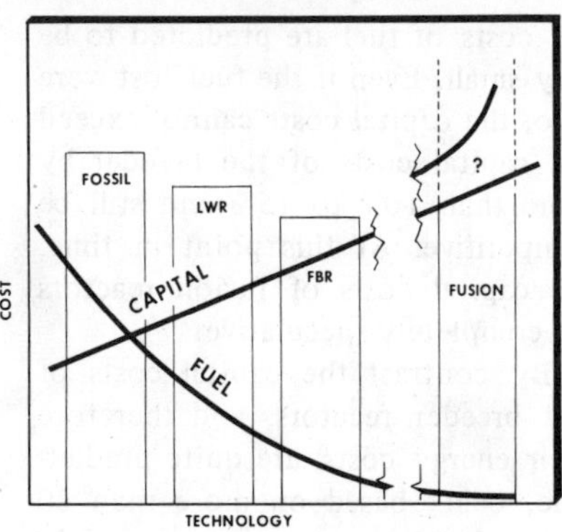

Fig. 5. Energy generation cost

been discovered that would economically and efficiently collect the low radiation energy density necessary for large-scale use of solar energy. Water and tidal power do not have the capacity to serve more than a small fraction of earth needs. Geothermal power from highly localized heat reservoirs can only supply a minor fraction of energy needs. The capacity of wind power is negligible.

Another way to extend energy resources is to increase the efficiency of converting the energy resource into the form suitable for use. The conversion of thermal energy into electric power is a specific example. The efficiency of the process of converting heat into electricity is inherently limited by thermodynamics and is primarily dependent on the maximum temperature of the conversion process. At present, the thermal efficiency of electric power plants is essentially set by the temperature limits of materials, i.e., tube wall in a boiler, cladding of nuclear fuel, blades of a gas turbine, etc. The waste heat is returned to the environment. The ratio of waste heat to usable energy for a variety of power plants is shown in Figure 4. The amount of waste heat decreases as the temperature of the conversion process increases. The materials technology for operation at higher temperatures becomes increasingly difficult and in general there is less generating experience at the higher temperature and efficiencies.

Magnetohydrodynamics (MHD) is considered to offer promise of improving the performance of central station power plants. MHD has never

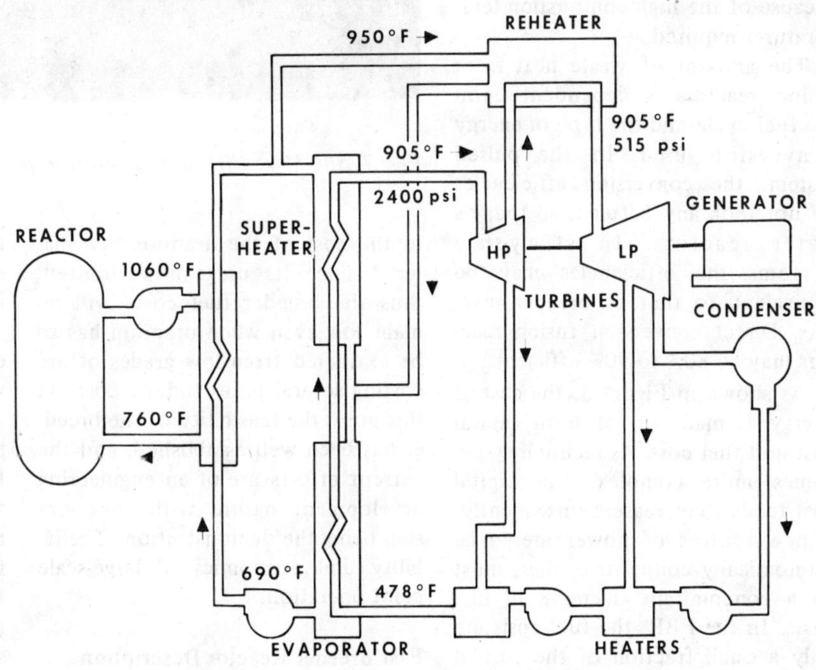

Fig. 6. FBR Flow System Diagram

been seriously proposed to eliminate the Rankine steam cycle in the temperature range in which it is effectively utilized in modern generating stations. It has been considered as a topping cycle in a higher temperature regime. There are two basic types of MHD power conversion methods, depending on whether the electrically conducting fluid is liquid or gaseous. Liquid metal MHD converters operating at temperatures of about 1600°F in conjunction with a steam cycle can increase the energy conversion efficiency to about 45% to 50%. This corresponds to increasing the useful energy available from existing energy resources about 25%. Gaseous or plasma MHD converters operating around 4000°F (2200°C) should achieve efficiencies in the range of 50% to 60%. This corresponds to an increase in the useful electric energy from fossil fuels of about 50%. Thus even with the increases in the efficiency by higher temperature conversion, there is waste heat and the economics of commercial application and the absence of undesirable environmental effect have yet to be demonstrated. One of the more difficult environmental problems of plasma MHD converters could be the excessive production of oxides of nitrogen because of the high combustion temperatures required.

The amount of waste heat from fusion reactors is dependent upon the fuel cycle and the type of energy conversion used. In the pulsed system, the conversion efficiencies do not look any better than today's water reactors. In steady-state systems the efficiencies may be equivalent to that of breeder reactors. Direct conversion fusion reactors may be 80% to 90% efficient.

As shown in Figure 5, the cost of energy is made up of both capital cost and fuel cost. As technology becomes more complex, the capital cost tends to increase. Consequently, if new sources of power are to be economically competitive, there must be a concomitant decrease in fuel costs. In the FBR, the fuel costs are only a small fraction of the cost of electric energy; consequently, even large fluctuations in the costs of fuel cause only minor perturbations in the total cost. In the case of fusion, the costs of fuel are predicted to be very small. Even if the fuel cost were zero, the capital costs cannot exceed the capital costs of the breeder by more than 10% to 15% and still be competitive. At this point in time, the capital costs of fusion reactors are completely speculative.

By contrast the capital costs of fast breeder reactors, and therefore their energy costs, are quite predictable, being based on more than 20 years of sodium cooled reactor technology and component development and an established base of light water reactor fabrication capability. The breeder fuel cycle cost is insensitive to the cost of the uranium raw material since its usage is so limited; thus the breeder fuel costs will remain low even when uranium has to be extracted from low grades of ore costing several times today's cost. At this point the feasibility of the breeder has been well established, and the current efforts are of an engineering development nature with the next step being the demonstration of reliability and economics of large-scale plant operation.

Fast Breeder Reactor Description

A schematic diagram of a liquid-metal-cooled fast breeder reactor system is shown in Figure 6. Heat from nuclear fission is transferred to sodium, the liquid-metal coolant, as it flows through the reactor. A second heat transfer circuit is used to transfer the heat from the coolant to water for the production of steam. The isolation of the reactor coolant from the steam generator is a safety feature of liquid-metal-cooled systems which is possible because of the excellent heat transfer characteristics of the coolant. The turbine-generator is typical of those used in fossil-fired plants. An artist's drawing of the fast breeder reactor demonstration plant proposed for eastern Pennsylvania is shown in Figure 7. The reactor building is shown in the foreground with the steam generator and turbine generator in separate structures at the rear.

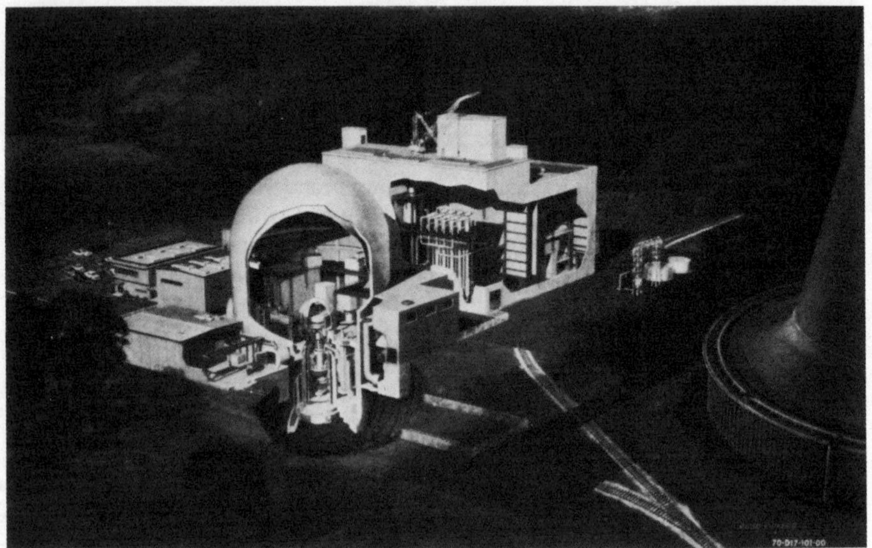

Fig. 7. 500-Mwe FBR demonstration plant

A few words of explanation will clarify how and why the fast breeder works as it does. The bulk of natural uranium, 99.3% of the element, is the isotope uranium-238 which is not fissionable. When uranium-235 fissions (Figure 8) by absorption of a neutron, it produces heat from the fission process and an excess of neutrons. One neutron is required to maintain the fission chain reaction, some of the neutrons escape or are consumed unproductively, and some of the neutrons can be absorbed by uranium-238 to produce plutonium. Uranium-238 is called a fertile iso-

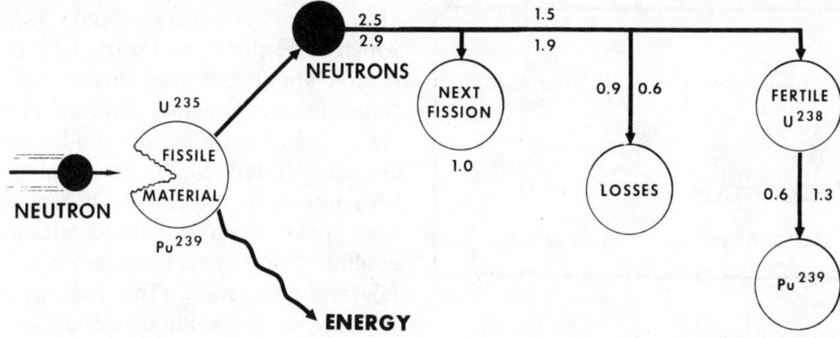

Fig. 8. LWR/FBR neutron balance

tope because this plutonium is fissionable and can be used to produce heat and generate power. The light water reactors of today produce an average 2.5 neutrons per fission and make about 0.6 plutoniums per fission. A fast breeder reactor derives its name from the fact that its neutrons travel faster than they do in a light water thermal reactor and because the efficiency of neutron utilization is such that more new plutonium is produced than is consumed in the production of electricity. In a fast reactor water and other materials that slow neutrons are excluded. As a result 2.9 neutrons are produced per fission and a larger fraction of these are used productively. A typical fast breeder reactor produces 1.3 plutonium atoms per fission.

The impact of this neutronic difference between light water reactors and fast breeders, as it affects uranium requirements, is shown in Figure 9. Over a 30-year period a 1000-megawatt light water reactor requires almost 4000 tons of natural uranium. Only 30 tons is actually used; the remainder is discarded as depleted uranium. During this period more than 5000 kilograms of plutonium will be produced. This plutonium may be used as an additional source of fuel for the system. However, it is used much more efficiently in fast breeder reactors. In addition, the depleted uranium which is discarded from light water reactors is the fertile material for the fast breeder. During a 30-year period, a 1000-megawatt fast breeder reactor requires 2300 kilograms of plutonium to make it self-sufficient and 100 tons of depleted uranium over its life to produce 7700 kilograms of plutonium. The excess plutonium can be used as fuel for other breeders.

Environmental Impact of FBR

The single most important reason for introducing the fast breeder reactor into our industrial economy is the conservation of valuable resources. Today's 1000-Mwe light water reactor uses about 100 tons of uranium per year operating at 80% of capacity, while a 1000-Mwe fast breeder reactor would use only about 1 ton of uranium per year. In Figure 10, this has been translated into the cumulative uranium requirements necessary to meet the projected growth of electricity supplied by nuclear power. It results in an increasing need for uranium, if the electricity is supplied by light water reactors. If electricity from nuclear power is supplied by both light water reactors and fast breeder reactors, with the introduction of commercially competitive fast breeder reactors in 1985, there would be a savings of one million tons of uranium by the year 2000. Furthermore, ore requirements at that time would taper off sharply as fewer and fewer light water reactors are built.

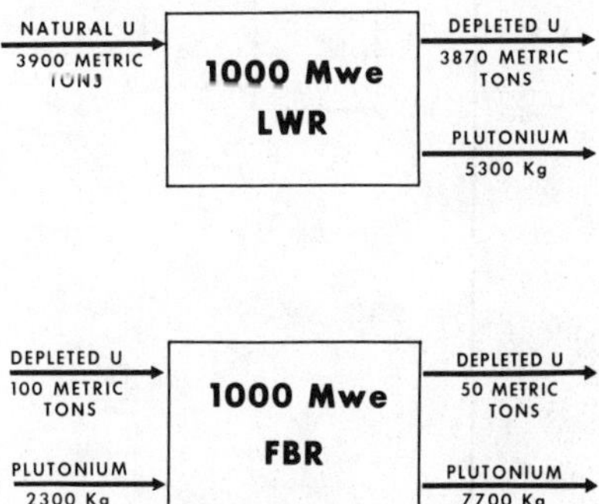

Fig. 9. Fuel balance (30 years)

The FBR minimizes the aesthetics and logistics problems of the mining and transportation of large quantities of raw fuel materials. For example, a 1000-Mwe fossil plant needs about 2.6 million tons of coal or about 10 million barrels of oil per year. In Figure 11 we see that it would require 26,000 railroad cars per year to supply coal to such a generating plant. These figures should be compared with 50,000 tons of uranium ore (0.2% uranium content) per year required for today's 1000-Mwe light water reactor. This corresponds to the shipment of 500 railroad cars per year of uranium ore from the mine to the milling plant. The stockpile of depleted uranium accumulated in

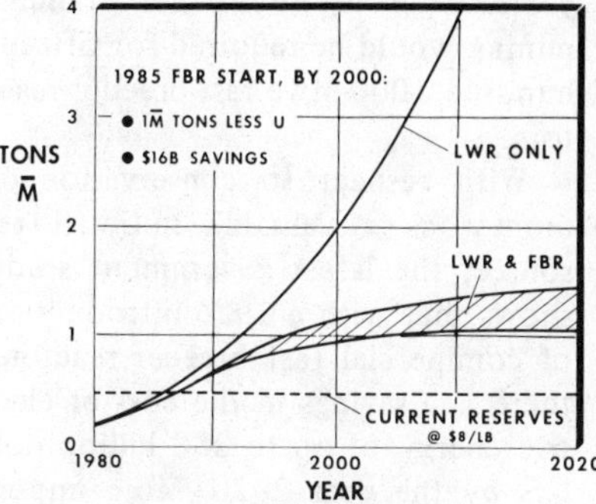

Fig. 10. U_3O_8 requirements

supplying fuel for light water reactors forms the uranium feed for the fast breeder reactor. No further mining of uranium is required for fast breeder reactors for many years.

Over the lifetime of a plant, about 30 years, these figures represent about 75 million tons of coal, 300 million barrels of oil and 1.5 million tons of uranium ore. This corresponds (Figure 12) to the need for mining 50,000 acre-feet of coal and 1000 acre-feet of uranium ore over the plant lifetime. As a hypothetical illustration of the logistics problem, let us assume that the total electric generating capacity of the nation (1.5 billion kilowatts) in the year 2000 was derived solely from coal. Fifteen hundred generating plants, each with a capacity of 1000 megawatts and operating at a load factor of 80% would have to burn 10 million tons of coal per day. This would require the daily movement of the equivalent of 100,000 railroad cars

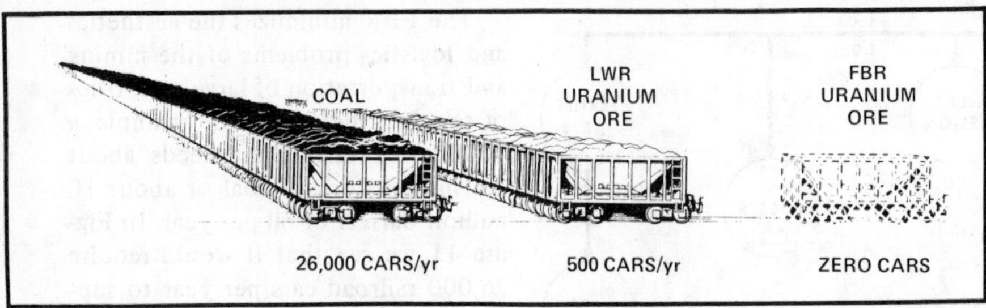

Fig. 11. Fuel mining requirements (1000-Mwe plants)

of coal. By contrast let us assume that the total electric generating capacity was supplied by fast breeder reactors. The existing stockpile of depleted uranium is enough to fuel the projected capacity for the year 2000 and supply all its fuel needs for 100 years. That is, no further uranium mining would be required for fifteen-hundred 1000-Mwe fast breeder reactors.

With respect to conservation of money as a valuable national resource, the latest government study shows that with a 1986 introduction of commercial fast breeder reactors, there is a savings in the cost of electric energy of up to 358 billion dollars by the year 2020. More importantly, the study shows that for each year that the introduction of the breeder is delayed there is a loss in 1971 dollars of more than one billion dollars per year.

The liquid-metal-cooled fast breeder reactor is a practical answer for producing low-cost electricity with minimum pollution of the environment. I should like to discuss three types of pollutants resulting from routine operation of fast breeder nuclear reactors: waste heat, radioactive effluents, and radioactive wastes. The waste heat from various types of power plants is shown in Figure 13. The amount of heat discharged to the environment is dependent on the efficiency of the power plant and the method of heat rejection. In 1969, the average thermal efficiency for all United States fossil plants was about 33%. This results in about 2 units of waste heat rejected for every unit of electricity generated. About 10% of the thermal energy generated is rejected in the stack gases (0.3 units of waste heat) and the remainder (1.7 units) rejected in the condenser cooling water. More modern plants have thermal efficiencies averaging about 38%, which corresponds to a waste heat rejection about 20% less. Present light water reactors operate with an efficiency of about 32%, just under the average United States fossil plant. This results in more than 2 units of waste heat rejected to the condenser cooling water since there is no heat rejected to a stack. The low vapor pressure of the sodium coolant of a liquid metal cooled fast breeder permits operation at temperatures that yield a conversion efficiency better than today's fossil fired plants. The light water reactor has almost 50% greater heat rejection than the fast breeder reactor.

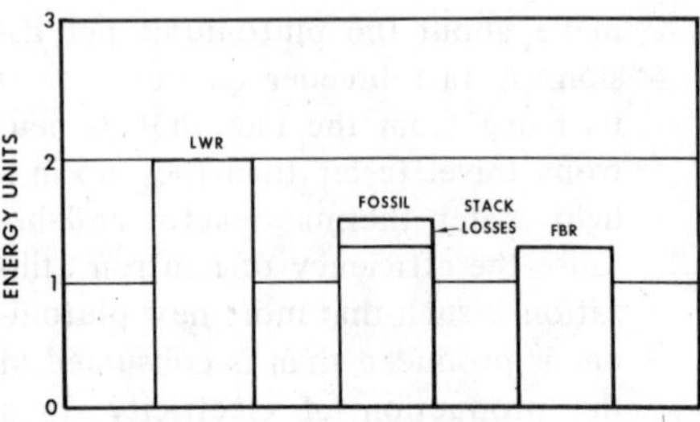

Fig. 13. Rejected thermal heat

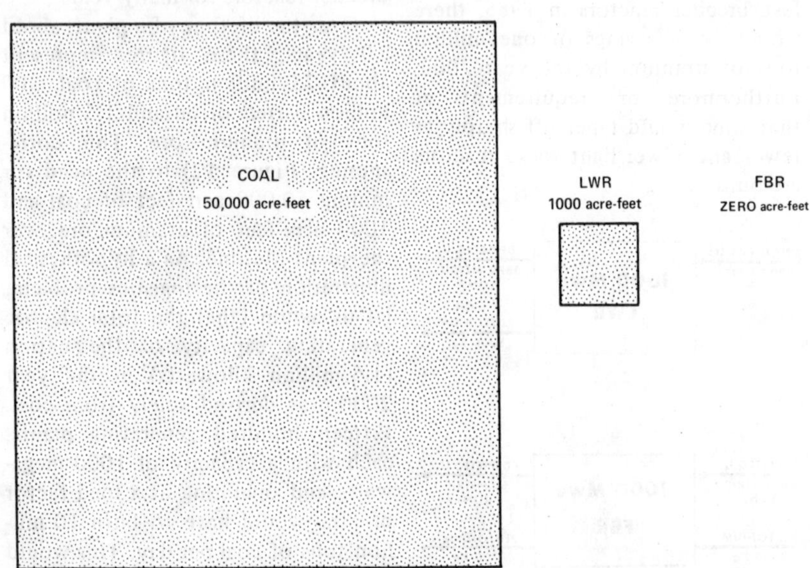

Fig. 12. Life-time fuel mining requirements (1000-Mwe plants)

The fast breeder reactor introduces no unique requirements for control of radioactive discharge to the environment or new problems in waste storage. The quantity of wastes other than fuel generated by liquid-metal-cooled fast breeder reactors is small, and therefore it is easy to design the plant for zero releases of liquid or solid radioactive wastes to the environment. It is also relatively easy to design a sodium cooled reactor so that the exposure at the site boundary from gaseous radioactive effluent leakage is less than 1% of natural background radiation.

The fundamental factors relative to the safety of reactors do not differ significantly between the fast breeder reactors and light water reactors. These are: the use of conservative design margins, inherent safety design features, multiple barrier to fission products, adequate protective systems to prevent accidents, and engineered safety features to mitigate the consequences of accidents. The ultimate safety requirement for all nuclear reactors is the ability to cool the reactor under all conditions. In this regard the sodium-cooled fast breeder reactor has unique advantages due to the excellent cooling properties of sodium and its low vapor pressure. A liquid-metal-cooled fast breeder reactor system operates at low pressures, well below the boiling point of sodium. As a result there is negligible stored energy in the coolant and no loss of primary core cooling or potential for pressurization in the event of a pipe rupture. In addition, the sodium provides good coolant heat transfer even at very low flow and can absorb considerable energy under emergency conditions.

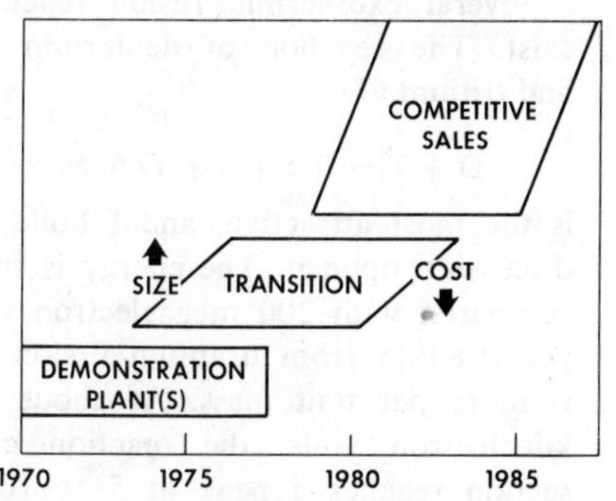

Fig. 14. FBR introduction

Another safety advantage of sodium is its affinity for fission products that tend to inhibit fission product release in the event of an accident.

The reasons for the emphasis that is being placed on the development of the fast breeder reactor in both the United States and Europe as the energy source of the future include: savings on power generating costs, establishment of a premium market for plutonium generated in light water reactors, protection of today's investment in light water reactors, and to the consumer an abundant supply of low cost electricity. What is more remarkable, the fast breeder reactor

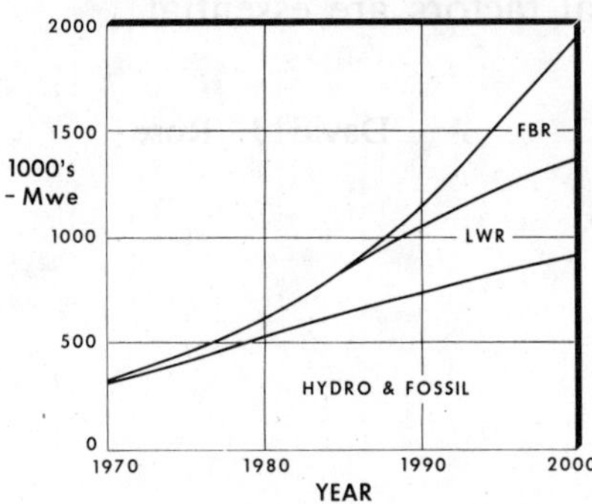

Fig. 15. U.S. generating capacity

can achieve these economic goals while protecting the environment better than any other alternate method of generating electric power currently available to meet rapidly growing power needs. Its potential environmental benefits include: conservation of natural resources, lack of air pollutants, and reduced thermal pollution. The fast breeder reactor can meet the growth requirements of electricity and at the same time provide increased protection of the envi-

the reactor under all conditions. In this regard the sodium-cooled fast breeder reactor has unique advantages due to the excellent cooling properties of sodium and its low vapor pressure. A liquid-metal-cooled fast breeder reactor system operates at low pressures, well below the boiling point of sodium. As a result there is negligible stored energy in the coolant and no loss of primary core cooling or potential for pressurization in the event of a pipe rupture. In addition, the sodium provides good

ronment—at no risk to the general public.

The Future

The immediate step in the introduction of fast breeder reactors is the construction of one or more demonstration plants under a cooperative program between the utility industry, the government, and the reactor manufacturers. Beyond these there will be a transitional period (Figure 14) wherein new plants of increasing size with increasing economic advantage will be built leading to full commercial plants in the mid-1980's. It is expected that by the year 2000 half of all nuclear capacity will be breeders (Figure 15), and they will be the predominant source of all new electrical energy capacity being built.

Controlled Nuclear Fusion: Status and Outlook

Besides plasma confinement, technological and environmental factors are essential.

David J. Rose

The attempt to generate power by controlling nuclear fusion will make an interesting topic for philosophers and historians of science and technology. If such an extravagant statement sounds forced, it is just meant to say at the outset that many factors, not all scientific, and some for the first time, have helped put the state-of-the-art where it is now. I shall try to give some account of these things.

Elements of the Problem

Controlled fusion research has passed through several epochs, the first of which was initiated by four items. First came measurements of reaction energies and rates between hydrogen isotopes and other light elements, which showed that under proper conditions large energy releases would be possible. Second, the well-known laws of single particle physics seemed to show how an assembly of high energy ions and electrons could be confined in magnetic fields long enough to establish the proper conditions. Third, the radioactive ingredients and by-products of fusion appear to be much less hazardous than those associated with nuclear fission: therefore, fusion reactors would be simpler and safer than fission reactors. Fourth, deuterium is a fusion fuel in plentiful supply—one part in 7000 of ordinary hydrogen; and extraction from ordinary water is not difficult. So matters stood in the early days, say up to 1955. Only the first of these items is necessary to make H-bombs. The combination of all four items captured the imagination of a sizable and very competent fraction of the physics community. The ensuing search for controlled fusion—the ultimate power source—has sometimes taken on a moral character, possibly as a reaction to the darker uses to which nuclear energy had been put. Whatever the reason, the efforts exerted by some might be compared to those of an Everest climber who knew that Prometheus was chained to the top. And a good thing, too, for the 1953 worker didn't see the whole field of plasma physics that lay yet to be discovered between his hopes and their realization. Whether it is a field or a gulf is yet to be discovered, and attempts to cross it during later epochs are briefly accounted below.

The present consensus is that, scientifically speaking, controlled fusion is probably attainable. But if fusion reactors are to be truly practical, there are other requisites: producing large volumes of magnetic field at low cost, minimizing the effects of material damage by high energy neutrons, and so forth. All these are equally essential to success; their natural laws being better understood than those of plasma physics, less room exists either for maneuver or speculation.

These phrases introduce the several major topics: how things are now, what is still needed to demonstrate scientific feasibility, what more is needed to make a practical fusion reactor, and how fusion does or does not fit our supposed future requirements.

Several exothermic fusion reactions exist. The reaction of deuterium (D) and tritium (T)

$$D + T \rightarrow {}^4He + n + 17.6 \text{ Mev} \quad (1)$$

is the most attractive, and I build the discussion upon it. The energy is small compared with 200 megaelectron volts per reaction from uranium fission but is more per unit mass. At about 100 kiloelectron volts, the reaction cross section reaches a peak at 5×10^{-28} square meter, which is very large by nuclear standards. Of the 17.6 Mev, 3.5 appears with the ^4He nucleus, and 14.1 with the neutron.

Many difficulties in the way of developing fusion power can be derived from these simple facts. First, consider the nuclear fuel. Deuterium is almost cost-free, but tritium does not occur in nature and hence must be regenerated with the neutrons from the fusion reaction.

The worst problem is presented by the nature of the reaction itself, because the particles must have (about) 10 kev energy or more so that the D and T nuclei can overcome their mutual electrostatic repulsion and fuse. Unfortunately, the cross section for scattering via this repulsion considerably exceeds the fusion cross section at such energies; hence the particles scatter each other several times before reacting. Thus it follows that the fuel will be a randomized collection of ions whose average energy must exceed 10 kev. In conventional terms, this is a gas at a temperature exceeding 10^8 degrees Kelvin. In fact, it will be a fully ionized plasma of D^+ and T^+ ions containing an equal total density of elec-

The author is director of Long-Range Planning of the Oak Ridge National Laboratory (operated by Union Carbide Corp. for the U.S. Atomic Energy Commission), Oak Ridge, Tennessee 37830. He is on leave from the Massachusetts Institute of Technology, Cambridge 02139.

Reprinted with permission from *Science*, vol. 172, pp. 797-808, May 21, 1971.

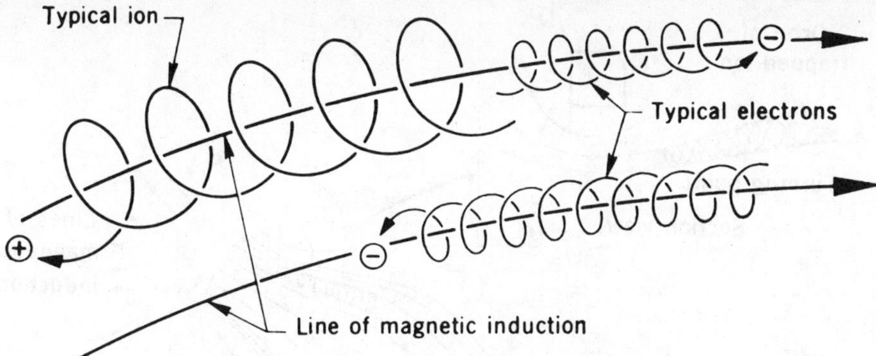

Fig. 1. Orbits of ions and electrons in a magnetic field.

trons to make the medium macroscopically neutral.

As I have implied, the principal difficulty comes in confining this plasma. A D-T nuclear explosive device stays together long enough—less than 10^{-7} second—by inertia alone for the components to react. In the process, the ^{4}He nuclei (and to some extent the neutrons) slow down in the unreacted material and heat it to an "ignition" temperature; transient pressure is millions of atmospheres. For a slower, controlled reaction, the pressure must be something that real structures can withstand; systems that we visualize will have dimensions of the order of 1 to 10 meters, and therefore pressures exceeding (say) a few hundred atmospheres are hardly believable. This restriction, plus specification of temperature already made, determines the density of the ions. Depending on the arrangement, desired D + T ion density turns out to be 10^{20} to 10^{22} m^{-3}, some 7 to 9 orders of magnitude below solid densities, and 4 to 6 orders of magnitude below that in the air around us. Required confinement time for a useful fraction of the nuclear fuel to react is 0.01 to 1 second. The most important parameter is the product of the density by the time, which should be 10^{20} sec m^{-3} or more—the so-called Lawson criterion. Total reacting nuclear mass at any one time would be only about 1 gram, even in a system that operates continuously at several thousand megawatts. All this is remote from any explosive regime.

Present Scientific Program

I will not review in depth the voluminous plasma physics underlying the schemes by which the plasma is hoped to be confined; but some acquaintance is necessary for what follows. The main schemes being developed so far involve use of large volumes of high magnetic fields. Plasma ions and electrons are hindered by magnetic forces from moving across the direction of magnetic fields, but can spiral along the field lines, as in Fig. 1. Thus (naively), confinement in the two directions perpendicular to the field direction is achieved, and one might have to worry only about confinement along the field direction.

From these simple thoughts arose in the first epoch two largely separate categories of device (*1*). In Fig. 2, field lines are curved to form a closed toroidal system; there is no escape except across field lines, and devices of this generic type are called closed systems. In the other generic type of Fig. 3, ions (and electrons) are reflected by increasing magnetic fields at each end. Here, an additional mechanism is required: each ion moving along a magnetic field line has fixed total kinetic energy U—at least until it interacts with the other ions and electrons in the system, or undergoes fusion. The total energy U can be thought of as being composed of two parts, an energy $U_\perp$ of gyrating motion perpendicular to the field line, and a part $U_{\parallel}$ of motion along the field line. That is

$$U = U_\perp + U_{\parallel} \quad (2)$$

Now it can be shown (*2*) that the magnitude of the perpendicular component $U_\perp$ is proportional to the magnitude B of the magnetic field; that is

$$U_\perp = \mu B \quad (3)$$

where μ is a constant (called the magnetic moment) for each particle, depending on details of its orbit. From this we find

$$U_{\parallel} = U - \mu B \quad (4)$$

The consequence of Eq. 4 is straightforward—if the field B becomes high enough in the ends of the device shown in Fig. 3, then μB rises to equal U itself, and no energy $U_{\parallel}$ is left for parallel motion. The particle must be "reflected" from these high field regions, hence contained in the center part. The device is appropriately called a magnetic mirror (*3*).

A difficulty of these "open-ended" systems of Fig. 3 is just that—open ends. An ion or electron whose orbits happen to lie almost along the field direction in the middle of the device has a low value of the magnetic moment. Then the maximum field B at the mirrors is insufficient to reflect the particle, and it escapes out one end. Coulomb interactions continually scatter particles into such directions; hence magnetic mirrors are inherently leaky, even if no worse calamities befall.

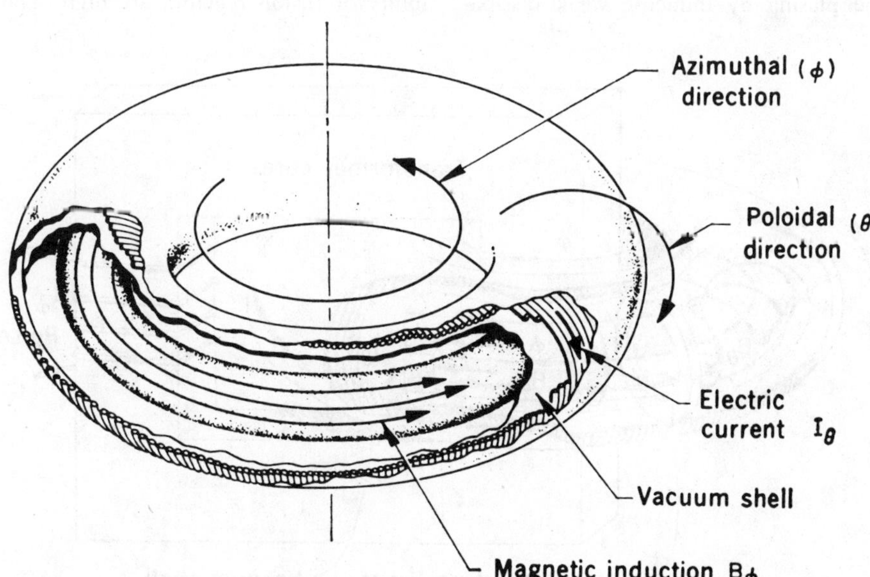

Fig. 2. Toroidal magnetic field B_ϕ made by poloidal electric currents I_θ.

In each case, the confining field might typically have a maximum strength of 8 to 10 tesla (*4*), and an equivalent magnetic pressure $B^2/2\mu_0$ (in meter-kilogram-second units) of 300 atmospheres.

The difficulty with all these truly ethereal schemes is that the plasma turns out to be unstably confined, because a number of electric effects which are negligible for a few isolated particles but important in a large assembly (that is, a plasma) were not included. Thus ended the first epoch of fusion research, a sort of age of innocence. For either the closed or open systems of Figs. 2 or 3, some field lines necessarily bow outward away from the plasma: at such places the plasma tends to develop uncontrolled aneurisms. Modifying the basic configurations (and increasing its cost and complexity substantially) will reduce these unstable growths, but it seems certain that a weak turbulence will remain. As a result, plasma could diffuse toward the surrounding vacuum walls and out the ends at a high rate.

The idea of diffusion is useful for illustrating the situation in the present second epoch of fusion research. If the plasma internal motions can be described by a diffusion theory (there is some doubt about this, which we ignore here), then a diffusion coefficient D can be assigned. The theory then states that the confinement time τ_c in (say) a long cylinder of wall radius r_w should be about

$$\tau_c = r_w^2/6D \tag{5}$$

For long τ_c, we desire small diffusion, but even more importantly large systems. Present custom (*5*) has it that the diffusion coefficient is likely to be some small fraction of the Bohm value D_B for a fully turbulent plasma, where

$$D_B = \frac{kT_e}{e} \cdot \frac{1}{16B} \tag{6}$$

Here, (kT_e/e) is the electron temperature measured in electron volts. Then according to this rubric, we have

$$D = D_B/A \tag{7}$$

where the dimensionless factor A represents confinement quality, measured in "Bohm times." If $A = 1$, the plasma would be lost by diffusion with a coefficient equal to D_B. For adequate fusion system confinement, it turns out that we must have $A \gtrsim 100$ at least, the precise number depending upon the arrangement (*6*).

It is both encouraging and salutory to see where present experimental devices are in relation to these goals. There are many such, but in this summary one example must suffice. The Tokamak, one of the most promising devices today (*7*), is an easy extension of Fig. 2, developed first at the Kurchatov Institute in Moscow, now also appearing in various guises at several plasma laboratories in the United States. Figure 4 shows the arrangement: the strong azimuthal field B_ϕ remains as before; but now the toroidal plasma is itself also the secondary loop of a transformer, which accomplishes two additional purposes. First, a strong current pulse on the primary winding ionizes the gas and generates a secondary plasma current I_ϕ; that current heats the plasma by inducing weak dissipative turbulence—hopefully just enough to heat it but not lose it (Fig. 4). Second, the current I_ϕ produces a new poloidal magnetic field B_θ as shown; the two fields combined, reminiscent of the crossed plies of a tire tread, make up the confining structure. Analysis shows that the plasma should be stable against ordinary hydromagnetic instabilities in the magnetic well so formed. The remaining higher order modes might be too weak to cause excessive diffusion. One penalty for these improvements is abandonment of true steady-state operation, for the device must now be run in long pulses—vide the transformer.

At this time, hopes that a Tokamak device will establish the scientific feasibility of fusion reactors are high. The

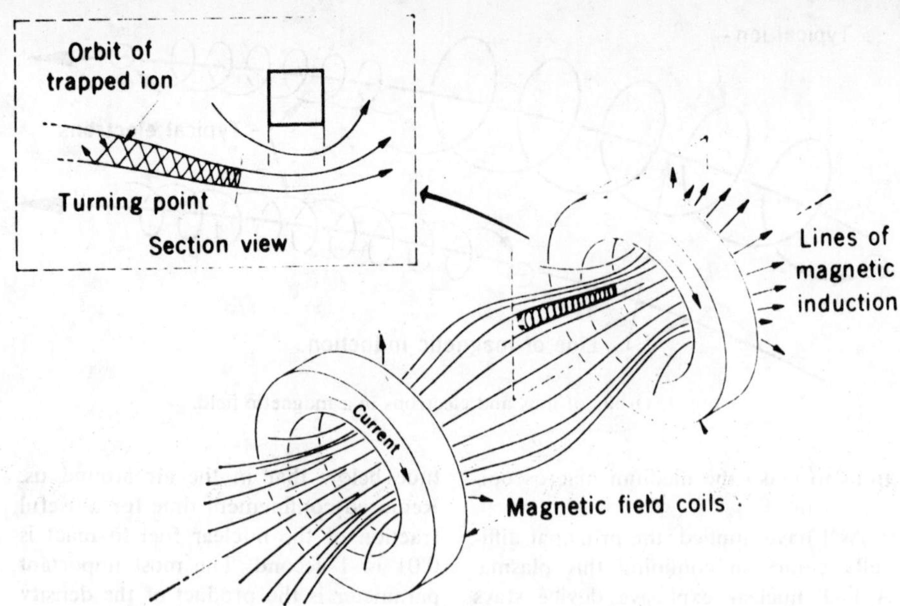

Fig. 3. Magnetic mirror particle (and plasma) confinement configuration.

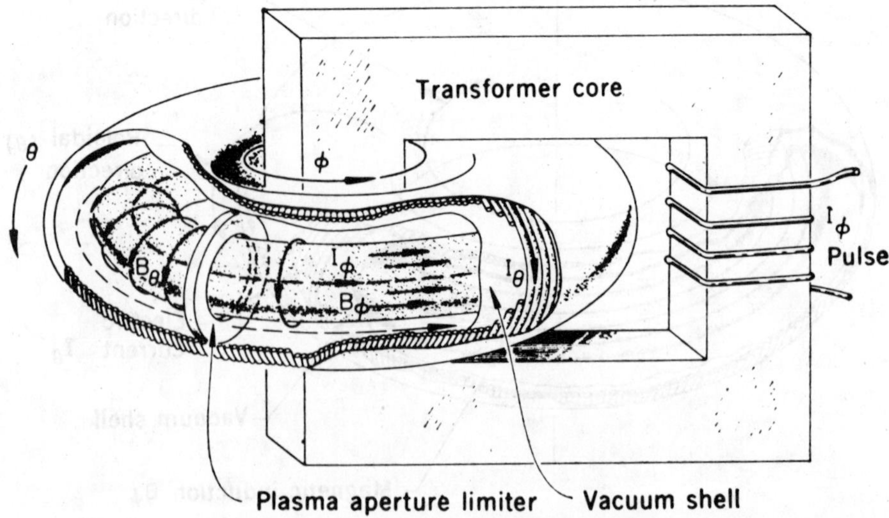

Fig. 4. The Tokamak plasma confinement scheme.

the moderator and vacuum wall, and heat removal.

4) Providing large quantities of high magnetic field and structure to withstand high stress.

5) Radiation damage by the 14.1-Mev neutrons, the consequences of which may be frequent and expensive replacement of much of the structure.

6) Size and cost, which are implicit in many of the above. Other problems are model-dependent; some device concepts seem to require additional developments. The list is long.

Most of the engineering-type problems that are model-independent can be described with the aid of Fig. 6, which shows a stylized fusion reactor as a series of cylinders. The main confining magnetic field is into (or out of) the paper; whether the cylinder is the center section of a stabilized mirror or is wrapped into a torus need not concern us here. The fusion plasma occupies the evacuated center, is surrounded by a neutron-moderating blanket and, at large radius, by a set of magnetic field coils. Here now are summary remarks on the problems listed above, generally slanted to a steady-state (or quasi-steady-state) device (*6, 11*).

1) *The plasma.* How is the plasma heated? What are the equilibrium temperatures and other parameters? The confinement being imperfect, we imagine plasma fuel continually being lost from the ends or sides into some suitable pump, hence also being replaced by some injection process into the center. Thus, the plasma continues in existence, but each ion or electron remains confined only for the period τ_c discussed before. Helium nuclei born in fusion reactions are also trapped for about τ_c, and deliver much (possibly all) of their 3.5-Mev energy to the plasma. Thus, the plasma is at least partly heated by its own reaction. For some fixed τ_c then, a certain throughput of plasma is needed to keep up its density; consequently, a certain calculable fraction f_b of the fuel will be burned per pass through the device; and the helium from the reaction heats electrons and ions (unequally) to temperatures T_e and T_i, respectively. As τ_c is raised, then f_b, T_e, and T_i also go up; the fuel is confined better and is not diluted by so much unreacted throughput. Fractional burnup f_b is a more useful display criterion than is τ_c. Difficulties of replenishing the fusion plasma seem to limit us to $f_b \gtrsim 0.02$; $f_b > 0.1$ would cause too high plasma tem-

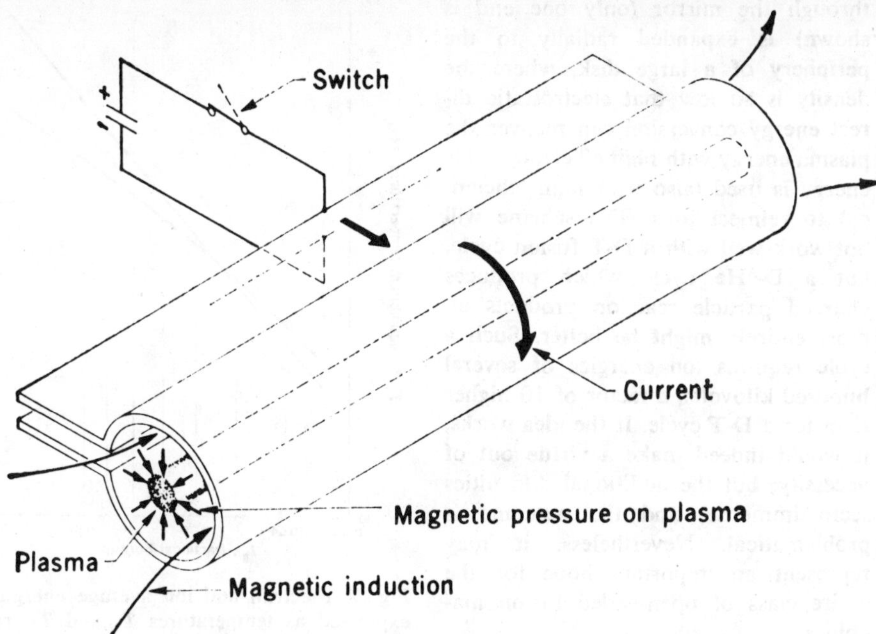

Fig. 5. Pulsed plasma heating and confinement scheme (so-called θ-pinch).

largest device operating ("T-3" at Kurchatov) has a major diameter of 2.0 m, the minor plasma diameter is about 0.3 m, the maximum field B_ϕ is 3.5 tesla, and the current I_ϕ is 10^5 amperes. For these efforts, the results (*8*) are: plasma density is $3 \times 10^{19}/m^3$, confinement time τ_c is 0.03 second, the electron temperature is >1 kev, and the ion temperature is 0.5 kev. Each of these numbers (which has been measured both by the U.S.S.R. and a visiting team from the United Kingdom) is about a factor of 10 too low, but very good by recent standards; and there is more to the story. From Eqs. 5 to 7, we calculate $A \approx 80$; that is, the confinement time of 0.03 second is some 80 times as long as turbulent Bohm diffusion would predict. This bespeaks a fairly quiescent plasma, almost good enough (in these peculiar terms) for a fusion reactor. A respectably optimistic expert could argue that only the small size and relatively low magnetic field prevent the plasma from lasting an adequate number of seconds. Exploring whether larger or higher field devices give a closer approach to fusion reactor parameters is now an exciting activity; the next generation of experiments should tell much.

Analogous descriptions might be made about some magnetic mirror experiments [the so-called 2X experiment at the Lawrence Radiation Laboratory, Livermore, California, for instance (*9*)] or fast shock-heated plasmas [Scylla at Los Alamos, for example (*10*)]. This last device is shown very schematically in Fig. 5. The capacitor discharge through the single-turn coil generates a rapid-rising strong magnetic field ($<10^{-6}$ second, 15 tesla). The field acts as a radial piston, compressing an initially cool plasma into a hot, dense one. In each of these various schemes, the combinations of density, temperature, and confinement time differ. For the Scylla experiment, we find densities up to 5×10^{22} m^{-3}, and temperature ≈ 5 kev, which are nearly satisfactory for fusion; but $\tau_c \approx 10^{-5}$ second is very short: plasma squirts out the open ends of the device. A longer one (Scyllac, 10 m) is being built to reduce these end effects.

General Technological Feasibility

Divinations from plasma physics may permit or deny the possibility of useful power from controlled fusion, but they cannot guarantee it. Some applied problems that are substantially independent of the particular geometric model are:

1) Plasma conditions in imagined practical devices, such as ion and electron temperatures, the fraction of fuel burned up per pass through the reactor, and radiation from the plasma surface. This might be called plasma engineering.

2) Regenerating tritium (for a D-T reactor) in a surrounding moderator-blanket by means of the 14.1-Mev neutrons.

3) Heat deposition, temperature of

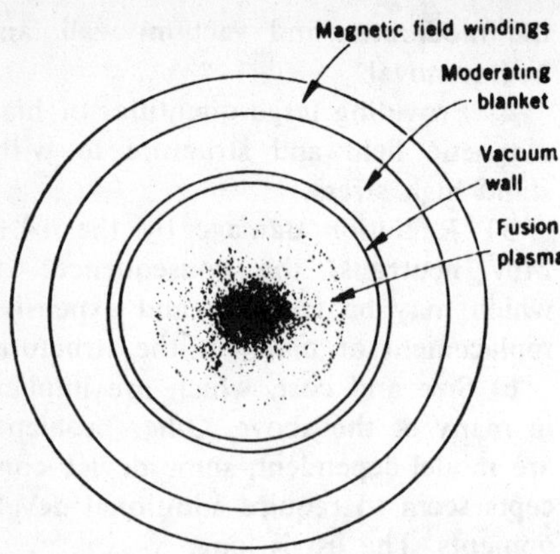

Fig. 6. Schematic controlled fusion reactor.

peratures and also demand unimaginably good confinement.

With some rather restrictive assumptions, these things can be calculated. Figure 7 shows the expected rise of electron and ion temperatures with increasing fractional burnup, for typical conditions expected in a fusion reactor. At high f_b, electron temperature falls below that of the ions. The reasons for this are that energetic electrons radiate energy, and that the ^4He nuclei tend to heat the ions preferentially, if the electron temperature exceeds about 33 kev.

Are these temperatures (once established by some startup scheme) high enough, or must more energy be added? This question lies at the heart of determining energy balance in a fusion reactor. At a given plasma pressure, the highest fusion reaction rate per unit volume occurs at temperatures of 15 to 20 kev. Then Fig. 7 appears to show ample heating if only $f_b \gtrsim 0.03$. For toroidal systems, this may be satisfactory, but an additional problem appears for open-ended systems (mirrors): the ions scatter out of the ends intolerably rapidly unless the ion temperature is very high, perhaps 100 kev or more.

For mirrors, heating the ions (probably by injecting them into the plasma at high energy) appears to be a necessary but expensive step. The expense arises both in additional equipment and in energy. Most of the energy from a D-T fusion reactor will appear as heat, which can be converted to electricity with (at most) about 50 percent efficiency. Then using large amounts of electric power to inject ions could make the system unfeasible.

These objections are serious enough so that a very different energy cycle is being investigated for mirrors (*12*). The field lines of such a device are shown ethereally in Fig. 8. Plasma escaping through the mirror (only one end is shown) is expanded radially to the periphery of a large disk, where the density is so low that electrostatic direct energy conversion can recover the plasma energy with high efficiency. This energy is used (also with high efficiency) to reinject ions. The scheme will not work well with a D-T fusion cycle, but a D-^{3}He cycle which produces charged particle reaction products almost entirely might be better. Such a cycle requires ion energies of several hundred kilovolts, a factor of 10 higher than for a D-T cycle. If the idea works, it would indeed make a virtue out of necessity; but the additional difficulties seem immense, and the outcome is problematical. Nevertheless, it may represent an important hope for the entire class of open-ended fusion machines.

A major difficulty with all these calculations is that they are still nebulous. The hidden assumptions may be unrealistic in serious ways. For example, how are the energy exchange rates inside the plasma affected by the presence of weak turbulence? No one knows. Will the curves of Fig. 7 be affected by inclusion of space charge effects? A subfield of fusion plasma engineering, for lack of a better phrase, needs developing before a fusion reactor can be sensibly designed.

2) *Tritium regeneration*. For a D-T reactor, tritium must be regenerated; the two lithium reactions

$$^7Li + \text{fast neutron} \rightarrow T + {}^4He + \text{slow neutron} - 2.5 \text{ Mev} \quad (8)$$

$$^6Li + \text{slow neutron} \rightarrow T + {}^4He + 4.8 \text{ Mev} \quad (9)$$

are essential and seem adequate.

The general idea in Fig. 6 is, then, to make the vacuum wall and blanket supporting structure of thin section refractory metal. Within it, there would be liquid lithium or a lithium salt coolant, plus an artfully disposed neutron moderator (probably partly of graphite). Leading choice for metals is niobium in that it can be formed and welded, retains its strength at 1000°C, and is transparent to tritium. This transparency helps in two ways: tritium generated in the lithium-bearing coolant is not trapped in the metal; and tritium can be recovered by diffusion through thin section walls into evacuated recovery regions. Some additional neutrons also come from the niobium via (n,2n) reactions, but in this particular respect molybdenum would be a better material.

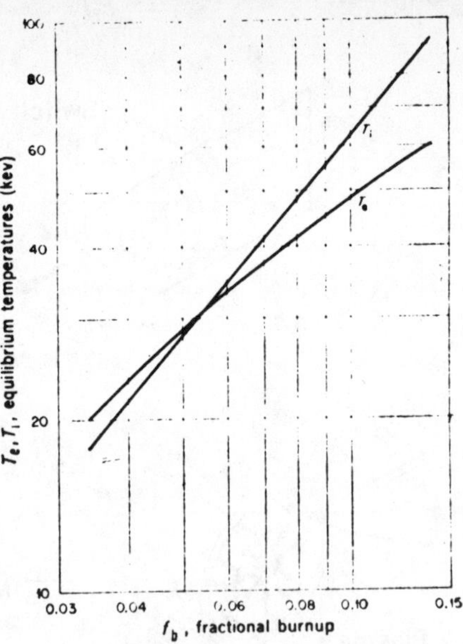

Fig. 7. Electron and ion average energies expressed as temperatures T_e and T_i, respectively, in projected fusion reactors, as a function of the fraction f_b of injected nuclear fuel that is consumed.

Liquid lithium cooling has the advantages of high heat transfer, few or no unfavorable competing neutron reactions; main disadvantage is its high electric conductivity, which makes it hard to pump through high magnetic fields—just how hard is not well enough known. In regions near the vacuum wall where the high lithium flow rate might cause excessive pumping loss, a nonconducting molten salt can be used. The likeliest candidate is Li_2BeF_4; the main penalty for its use is the presence of fluorine, which slows down energetic neutrons unprofitably, hence inhibiting the beneficial ^7Li re-

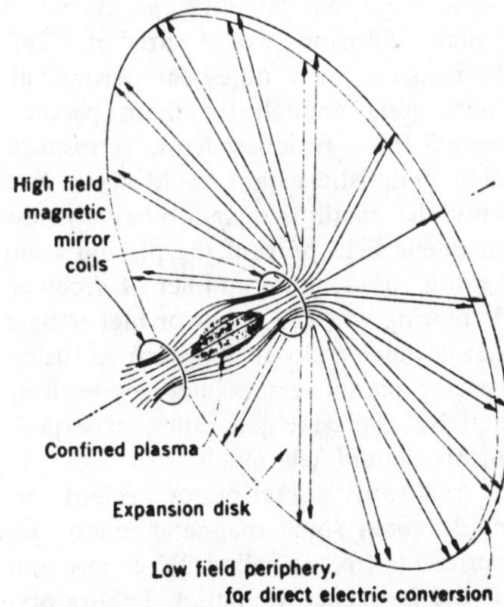

Fig. 8. Magnetic field configuration for a magnetic mirror fusion device with direct electrostatic energy recovery [after Post (*12*)].

action of Eq. 8. That is, using Li_2BeF_4 makes it harder to regenerate enough tritium.

However, with either of these schemes or a combination of them, tritium regeneration seems adequately assured. Calculations with semirealistic combinations of vacuum wall and blanket show that something between 1.1 and 1.5 tritons can be regenerated per neutron incident on the vacuum wall (*13*). Because one triton is used up per neutron generated, we have in fact a tritium breeder reactor, using the raw materials deuterium and lithium. This view of fusion as compared to nuclear fission breeder reactors has not been much emphasized in the past.

In addition to this favorable breeding ratio, present estimates put the tritium inventory in a fusion reactor at only a few weeks' supply—maybe less (*14*). Thus the tritium fuel doubling time in a fusion reactor might be much less than 1 year. Doubling time is an important measure of how quickly new reactors could be built (that is, fueled) either to match expanding power demands or to take over from a prior power-generating scheme. This short doubling time for fusion is in marked and favorable contrast to the situation with fission breeder reactors, where the doubling time tends to be uncomfortably long ($\approx$ 20 years in some designs). Here is one of the predicted large advantages for fusion.

Approximate size of the fusion reactor I have in mind comes directly from these considerations. Fairly simple nuclear calculations establish that the blanket plus a radiation shield (not shown) to protect the outer windings must be 1.2 to 2.0 m thick. This substantial thickness implies not only substantial blanket cost, but also very high magnetic field cost, to energize such a large volume. The only way to make the system pay is to have it generate a great deal of power; but nearly all this power must pass from the plasma into or through the vacuum wall. Engineering limits of power density and heat transfer then dictate large plasma and vacuum wall radii as well —between 1 and 4 m, say. Then overall size will be large, and total power will be high—almost certainly more than 1000 megawatts (electric) and perhaps 5000 megawatts.

3) *Heat deposition and the vacuum wall.* Energy is deposited in the vacuum wall facing the plasma, mainly from three sources: (i) some of the fusion neutrons suffer inelastic collisions as they pass through; (ii) gamma rays from deeper inside the blanket shine onto the back side; (iii) all electromagnetic radiation from the plasma is absorbed there. The plasma itself makes no additional load, being imagined to be pumped out elsewhere. The three sources may constitute 10 to 20 percent of the total reactor power. This is a modest fraction; but the vacuum wall region is thin, and heat deposition (and removal) per unit volume determines the power capability of the whole system. Here is a disadvantage of fusion systems compared to fission reactors; in the latter the energy is more nearly produced throughout the reactor volume and all must not pass through one critical section.

From these considerations, I imagine a total power assignment in the reactor of not more than 15 Mw per square meter of vacuum wall—say 10 Mw/m^2 being 14-Mev fusion neutrons passing through, and the rest consisting of plasma radiation and neutron captures in ^{6}Li. Some (*15*) imagine substantially higher energy fluxes to be possible, with the use of heat-pipe walls—about 30 to 40 Mw/m^2; but the design poses many problems. Even at 15 Mw/m^2, total reactor power is very high, as said before. If the vacuum wall radius is only 2 m, the system of Fig. 6 produces 140 Mw of heat per lineal meter (into the paper) of cylinder. If it is wrapped into a torus, the major diameter can hardly be less than 20 m. Total power of such a device would be 12,000 Mw thermal, or 5000 to 6000 electric, several times that of the largest plant now existing.

One possible way (*16*) out of this and some other difficulties is to run the reactor at substantially lower thermal stress—at ~ 2 Mw/m^2. Total power is conveniently less; and because the plasma density is reduced, so is the magnetic field and the cost of it. Neutron damage (see below) is also ameliorated. Whether this option increases the cost per unit of power excessively has not yet been estimated.

The vacuum wall must support approximately a pressure of 1 atm, which is no small task for a thin-section material in such large sizes. However, preliminary designs indicate that a structure built up in depth of thin sheets (the same principle as in corrugated cardboard boxes) will have the necessary strength, and contain proper passages for coolant flow (*17*).

4) *Magnetic field windings.* Generating even 15 tesla (150,000 gauss) continuously is not the problem; superconducting coils do so routinely at low cost, a dramatic improvement from state-of-the-art 10 years ago. The problem is size: a simple solenoid generating 15 tesla has a magnetic bursting force of 900 atm on its windings. In comparison, contemporary fission reactor pressure vessels are smaller than we imagine here, and are limited to some 40 atm operating pressure. To make matters worse, the magnetic field is not a simple solenoidal one, and stresses arise that cannot be held in simple hoop tension. To be sure, no nuclear excursion impends if the coils fail structurally, but failure would still be an economic calamity. Perhaps also 15 tesla is not required, but no assurance now exists.

Almost all conceptions involve superconducting coils at 4°K, or at least cryogenically cooled ones at 10° to 20°K. This is the reason for placing them outside the blanket, outside a radiation shield; otherwise the refrigeration problem would be intolerable. To make a reinforcing structure for operation at such a temperature, with size and stress loads I have described, is a task yet to be fully contemplated. Titanium is very strong at such low temperatures; but it is also very brittle —as are most other materials under those conditions.

5) *Neutron damage.* This is a very serious problem, for either a fission or fusion reactor. In one way, fusion appears at a substantial disadvantage, as follows. One fission reaction produces 200 Mev and about 2.5 neutrons, each with no more than about 2 Mev. One fusion reaction produces 17.6 Mev, of which 14.1 Mev appears in one high energy neutron. Thus, the "energetic neutrons/watt" is an order of magnitude higher in fusion than in fission, and the structural damage caused by these neutrons is correspondingly high. For the high power levels discussed in the preceding examples, every metal atom in the vacuum wall would be displaced almost once per day (*18*). Many of these displacements anneal out at the high operating temperature; but, even with the delicate choice of materials, design, and temperature, long-term integrity of the vacuum wall against neutron damage will be a major problem facing fusion power development.

In another way of looking at the problem, fusion has an advantage. The damaging neutron flux in this high power fusion reactor is predicted to be about 10^{15}/cm²-sec; but in reference designs for liquid metal fast breeder fission reactors, it will be an order of magnitude higher. We see here a principle of conservation of wretchedness—the fast breeder fuel elements and perhaps the components will require frequent replacement, at substantial expense.

For fusion, this problem translates into the problem of either protecting the vacuum wall (via lower power?) or replacing it. The cost of either of these options may be high; unanswered questions are whether the vacuum wall can be replaced at a cost small compared with the total reactor cost and how often replacement will be required.

Compounding the problem are the facts that probable fusion reactor conditions and materials are not in the fission breeder range of interest. Moreover, no source of 14-Mev neutrons (to test possible arrangements) now existing is intense enough—by a factor ~ 1000.

Within the framework of fusion systems envisaged here, this damage problem cannot be circumvented, cannot be well predicted on the basis of present knowledge, and affects the feasibility of every fusion reactor scheme.

6) *Size and cost.* Size is large for lowest power cost, as I showed earlier. However, over many decades unit size has increased by a factor of 2 to 3 each 10 years. Thus, 10,000 Mw thermal is liable to be quite acceptable before 2000, when fusion might, with good fortune, come into its own.

Cost per thermal kilowatt of capacity makes a reasonable basis for comparison with other generating systems. Components stylized in Fig. 6 are equivalent to the core of a nuclear fission reactor, without some of the nuclear ancillaries (and without any of the turbines and generators of a power station). No definite cost can yet be given for what is shown there; too much is still uncertain. However, outside estimates have been made that the cost might run somewhere between 6 and 20 1970 dollars per thermal kilowatt (*6*). If neutron damage does not require too frequent replacement of the structure, the whole cost range is interesting, and the lower limit is uncontestably attractive.

Such costs warrant continuing development, but they are very perishable commodities, depending on the imperfect and changing state-of-the-art. Designs, costs, trends, and comparisons must be continually reassessed.

Model-Dependent Problems

What of the host of model-dependent problems, more specific than those hitherto listed? I mention just three, to show their kind and importance.

1) *Fuel injection into closed toroidal systems.* Plasma is lost by diffusing toward the vacuum wall and then being absorbed (no mean task, and not well understood) at specific peripheral regions. Implicit in this statement is that something replenishes the plasma at or near the middle (if the device runs on anything like a steady-state basis). Ionized particles will not move across the confining field, so neutral ones must be somehow injected. The trouble now is that the energy flux (of hot electrons) in the plasma is about 10^{14} watt/m², some 10^3 times that of the strongest electron beam made today. Lifetime of a neutral atom or a small cluster of atoms against being ionized in this hostile environment is about 10^{-7} second; upper limit on injected atom velocity is about 10^6 m/sec; otherwise the plasma energy balance is upset. Then the atom penetrates perhaps 0.1 m, a negligible fraction of the way in.

An alternative scheme is to inject pellets so large that they shield themselves by ablation on the way in (as a reentry vehicle into the atmosphere from a space flight). Calculation of what happens here—for example, whether the pellet must be so large that it chokes the fusion reaction—is much more difficult than calculating the fate of atmospheric reentry bodies, and not much has been done (*19*).

2) *Direct energy conversion for open systems.* The necessity for high energy injection and recovery directly as electricity was mentioned in the discussion related to Fig. 8. What cannot be illustrated well is that the diameter of the disklike expansion region may be 100 times the diameter of the mirror confinement region. Can such a structure (albeit with low magnetic field) be built cheaply enough? Can plasma stability and individual particle orbits be controlled well enough throughout this immense region? No one knows.

3) *Fast-pulsed systems.* The scheme of Fig. 5 has advantages of automatic plasma heating, apparently good stability against radial excursions, and some others. But several perplexing complications are as follows. (i) The system requires a substantial amount of stored energy to be delivered in about 10^{-6} second to the coil. At present this is done by capacitors, perhaps at a cost of $100,000 per megajoule. Some cost reduction is clearly possible, but much is necessary. (ii) The fast pulse requires that the magnet coil be next to the plasma in that it forms the vacuum wall. Then the coil must have high strength at high temperature. Electric losses in this coil reduce power output from the system. The coil also slows down and absorbs neutrons, and this process decreases the tritium yield (*20*). (iii) Pulsed operation at (say) 900 atm pressure on a microsecond basis exacerbates problems of mechanical stress failure; yet more reinforcing structure imperils the tritium breeding even more.

Fearless Forecast

To assess the relative merits of many approaches to controlled fusion is a difficult task, and disputatious. But some sort of perspective must be developed from time to time. What follows is partly opinion, partly fact; it is no one's policy but my own.

Figure 9 helps to focus and confine the discussion. In the middle is a level of achievement called Scientific Feasibility: a density-time product of 10^{20} sec/m³ or more, and true thermonuclear temperature—say 15 kev or more, depending on the system envisaged. Whether the device looks like any eventual fusion reactor is immaterial in this context. This level of accomplishment would be crudely the analog of building the Stagg Field fission reactor in 1942: the physics is permissive, but engineering and economics are yet to come. Figure 9 has no absolute scale, but shows where each present scheme is presently situated—all are now below the feasibility waterline. Closest is the Tokamak, but the figure shows two gaps yet to be crossed. These gaps are that it is not yet known whether scaling to larger size really will work (as described earlier) or whether the ions can actually be heated enough in the device, via weak turbulence or some other means. To put some calibrating point on all this, I will bet a modest amount of even money on success of the Tokamak in the next few years.

The stellarator is a related steady-state device, where the toroidal configuration is stabilized not by induced plasma currents (as the Tokamak), but by added helical windings on the periphery of the torus. The big advantage

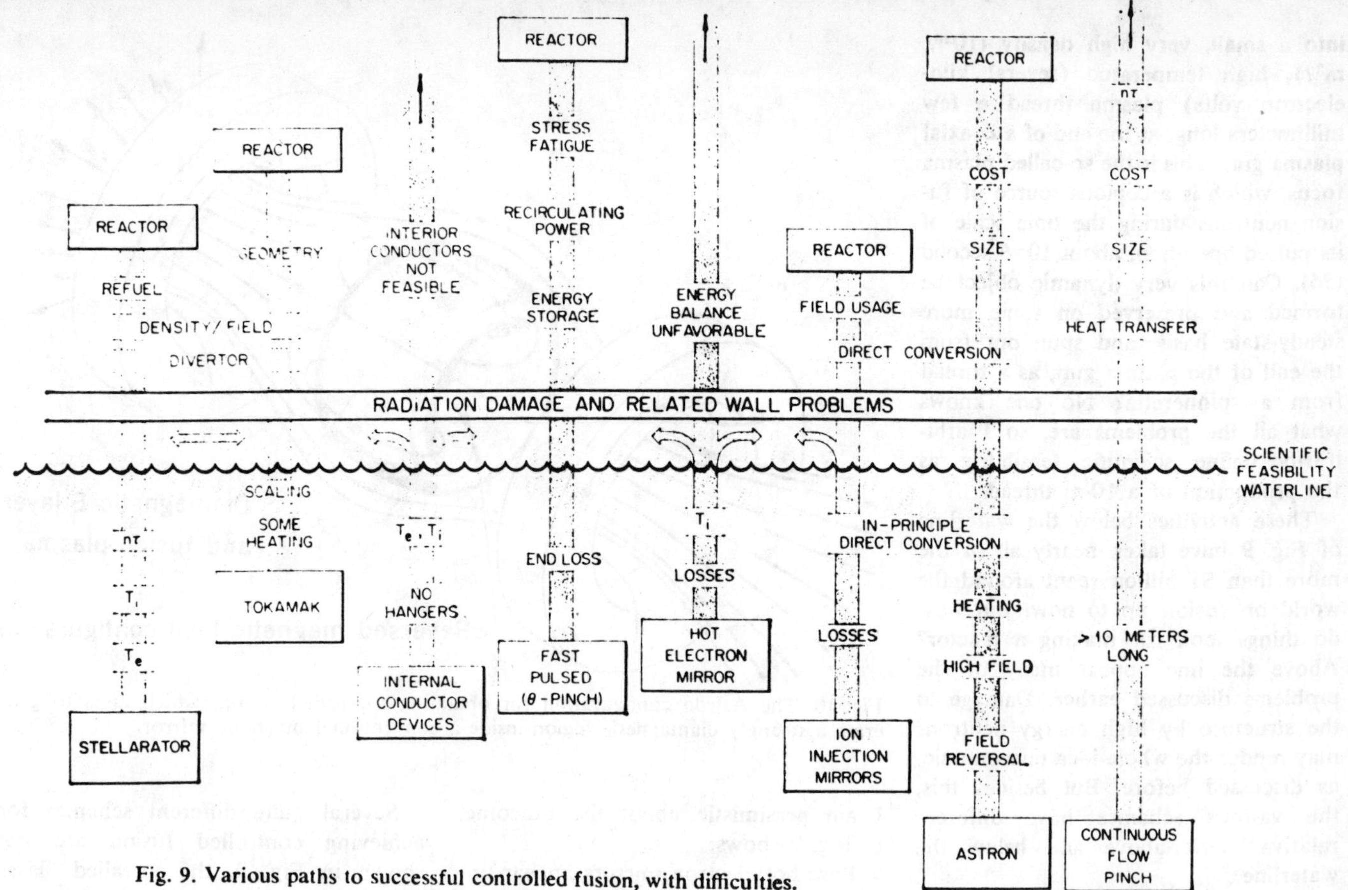

Fig. 9. Various paths to successful controlled fusion, with difficulties.

is steady-state operation. The main disadvantage is that a field configuration made this way seems to give poorer confinement. Thus the density-time product ($n\tau$ in the figure) needs more substantial improvement, and in addition both the ion temperature (T_i) and the electron temperature (T_e) will be harder to raise (*21*). The stellarator lies significantly below the Tokamak at present.

Some toroidal confinement schemes require solid conductors totally surrounded by plasma. The so-called multipoles at the General Atomic Corporation and at the University of Wisconsin, and the spherator at the Princeton Plasma Physics Laboratory are examples (*22*). These internal conductors can be (and are) made superconducting, so true levitation without supports or hangers is possible and has in fact been achieved. On the other hand, no large levitated experiment has yet been performed at high enough field. Thus in the third column of Fig. 9 we see the need to operate without hangers, and to raise both T_e and T_i by some plasma heating schemes yet to be fully developed.

Next in the figure comes the fast-pulsed devices, as shown in Fig. 5. Whether the side losses are now small and whether just reducing end losses will give satisfactory confinement are still questions, but I give the device the benefit of the doubt. One estimate is that the device needs to be 2 km long if linear and the ends are not stopped up (how?); also if wrapped into a torus, new and unresolved questions of plasma stability enter.

All open-ended mirrors suffer from high loss from the ends, and schemes to reduce these losses (by applying high frequency power at the mirrors, for example) seem not to be very effective (*23*). Heating both ions and electrons adequately is an additional problem. The "hot electron mirror" scheme uses large amounts of microwave power to produce an exceedingly dense hot electron plasma, with apparently fair confinement at least (*24*). Ions might be heated (T_i in Fig. 9) by injecting high energy neutral atoms into this "seed plasma." The chances of this scheme making a scientifically feasible fusion plasma are at least fair.

Ion injection mirrors, when the plasma is not substantially aided by hot electrons, face more difficulty. The losses are high; and, as discussed above, it seems that the high losses will require as part of the "in-principle" solution the development of "in-principle" direct energy conversion (see again Fig. 8 and the accompanying discussion).

The Astron at Lawrence Radiation Laboratory is interesting, but hard to describe (see Fig. 10). It starts out generically as a mirror (Fig. 3); but instead of confining a plasma directly there, the aim is to confine a ring of relativistic energy electrons (relativistic protons in a full-scale reactor). This is called an E layer; if dense enough, its diamagnetism actually reverses the magnetic field and sets up a new configuration of closed magnetic field lines: a torus inside the mirror. This configuration holds the fusion plasma. So far, a modest diamagnetic reduction (and no reversal) of a low field experiment has been achieved (*25*). True field reversal in a larger, high field device will be needed to set up the desired magnetic configuration. Beyond that, how the plasma is to be heated is a problem; and high end-losses may also require direct energy conversion.

The continuous-flow pinch is favored in some quarters, particularly in the U.S.S.R. The idea stems from the discovery that plasma can be focused into a small, very high density (10^{25}/m^3?), high temperatue (several kiloelectron volts) plasma thread a few millimeters long, at the end of a coaxial plasma gun. This is the so-called plasma focus, which is a copious source of fusion neutrons during the time scale of its pulsed operation, about 10^{-6} second (26). Can this very dynamic object be formed and preserved on some more steady-state basis, and spun out from the end of the plasma gun, as a thread from a spinnerette? No one knows what all the problems are, so I arbitrarily define scientific feasibility as the production of a 10-m thread.

These activities below the waterline of Fig. 9 have taken nearly all of the more than $1 billion spent around the world on fusion up to now. But how do things look for making a reactor? Above the line appear many of the problems discussed earlier. Damage to the structure by high energy neutrons may render the whole idea uneconomic, as discussed before. But besides this, the various schemes have different relative merit above and below the waterline.

Tokamaks no longer look quite so attractive. Special plasma pumps called divertors have been developed for stellarators, seem necessary for Tokamaks also (where access is more difficult), but must be vastly increased in effectiveness. Plasma stability considerations may demand that the plasma density be uncomfortably low, or the field uncomfortably high [15 to 20 tesla, or more? (27)]. Also, the geometry, inherently pulsed nature, and necessarily large size of the thing are hard to work with.

Some of these problems appear with the stellarator too, but with reduced intensity. Steady-state operation is easier; the additional refueling problem may be no more than moderately serious. Thus, the stellarator tends to look better, *if* we are given scientific feasibility. Stellarator and Tokamak scientific programs support each other extensively, hence the joining arrow on Fig. 9.

The internal conductor devices just will not make fusion reactors, because there is no way of cooling a levitated conductor, especially inside a fusion plasma. This is well understood; no one ever thought otherwise; these experiments are designed specifically for plasma physics and to shed scientific light on other schemes.

The theta pinches have very severe problems, as discussed in the last words of the section on fast-pulsed systems.

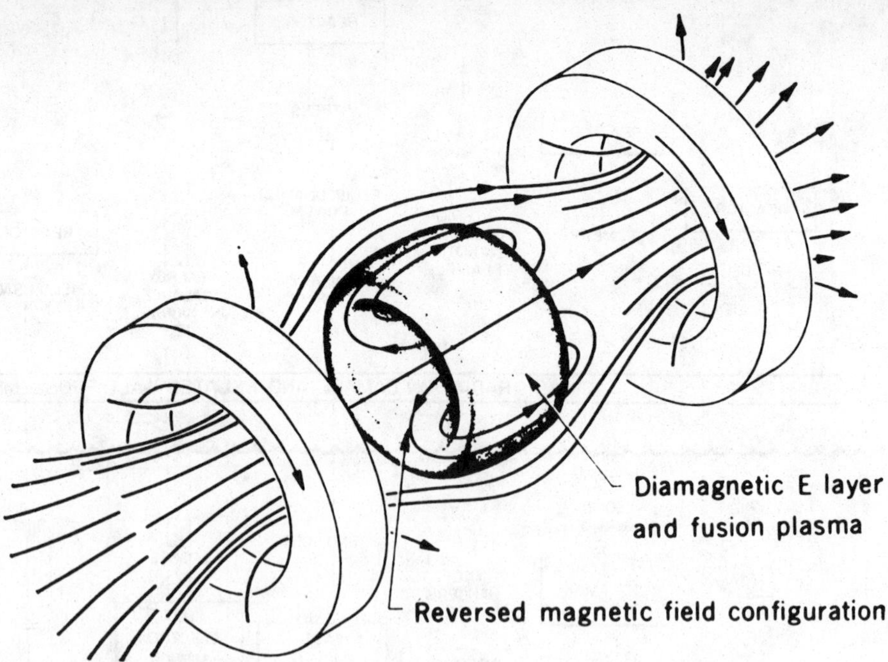

Fig. 10. The Astron configuration for obtaining controlled fusion, which aims to generate a strongly diamagnetic region inside a conventional magnetic mirror.

I am pessimistic about the outcome, as Fig. 9 shows.

Pure hot electron mirrors appear unfeasible for fusion from an energy-balance point of view, but again that is a personal opinion. As with internal conductor devices, the idea is to reach the waterline, not an economic reactor. In addition, *some* electron heating may be valuable for more conventional mirrors.

If conventional mirrors can attain scientific feasibility according to the definition given here, they should be the most likely reactor candidates. The questions are whether direct energy conversion can be developed at a reasonable price; whether the magnetic field is efficiently used (that is, cheap enough); and of course radiation damage.

The Astron seems heir to more difficulties: the size may be very large, and it is not at all clear whether relativistic-energy, high-current guns will be cheap enough. Direct conversion is still a problem.

Even if a continuous flow pinch, 10 m long, can be developed, I doubt that an economic fusion reactor can be made of it. The power density is immense, and presumably an exceedingly high magnetic field is needed to confine the plasma string. Could this ever be done without putting the field coils near the plasma, thus exacerbating heat transfer and tritium regeneration problems? There are more problems besides.

Several quite different schemes for achieving controlled fusion are not shown in Fig. 9; the so-called "laser ignition" scheme deserves mention (28). In that, the pulse from an ultra-high-energy laser is focused on a small pellet of solid D-T and heats it to fusion temperatures before the pellet has time to disassemble. The disassembly speed is about 10^6 m/sec at fusion temperatures, and the pellet size is the order of 1 mm. Thus the main heating pulse must be less than 10^{-9} second long. Even more, the most efficient heating scheme involves using several smaller preheating pulses to set up initial temperature and density gradients in the pellet, and these must be applied with temporal accuracy of perhaps 10^{-11} second. These requirements can be met. About 10^5 joules is the minimum estimated to be necessary for energetic break-even: enough fusion energy out to equal the laser energy deposited. Even these large values are not discouraging; what seems to me very difficult is producing power cheaply enough: for reference, 5×10^7 joules of such "explosive" raw heat deposited in (say) lithium coolant is worth about $0.01; can one do all this repetitively with an expensive and fragile device?

Many of the questions raised above will require systems research, systems development, and systems engineering to answer. These arts have been put secondary to plasma research and experimental device development up to now.

Time Scales

Present pressurized water or boiling water nuclear reactors are satisfactory as interim devices, but their relatively low thermal efficiency and inability to breed much nuclear fuel (from ^{238}U or thorium) condemn them to a brief existence in our society, unless much more uranium is found. The total installed capacity of such devices will be much less than that of fossil fuel plants, so complaints about them are and should be based on relatively local considerations—for example, thermal effects in Biscayne Bay. These words should in no way be taken as denigration of the validity of local complaints.

The view here is broader, and of longer time scales. The real question concerns second-generation fission breeder reactors (for example, a liquid metal fast breeder, or molten salt breeder) vis-à-vis the possibility of controlled fusion. At one time it was thought that fission suffered a relative disadvantage of insufficient nuclear fuel because of lack of uranium in the earth's crust, whereas deuterium is in plentiful supply. This is not true; there are adequate supplies of ^{238}U and ^{232}Th, D, or ^{6}Li for some 10^8 or more years of society based on high energy consumption. Even better, all these are resources for which little alternate use is forecast.

The real questions of fission breeders versus fusion breeders (which have to breed their tritium, as we have seen) involve feasibility, relative cost, time scales, and environmental factors, which all tend to be related. I have discussed the first of these topics and will not return to it in detail. To put the costs in some perspective, I point out that an additional penalty of $20 per thermal kilowatt—that is, doubling the maximum cost mentioned earlier—would add by itself less than $2 per month to the present average residential electric power bill. That is no invitation to adopt expensive options thoughtlessly—as electric power use increases, extra costs hurt more—but it *is* a way of saying that substantial changes could be afforded in reactor cores (fission or fusion) *if* even moderate social benefits were likely to accrue. That view will affect remarks to come later.

With regard to time scale, there is some real misunderstanding. Controlled fusion is *not* an alternative to the first-generation fission breeders, as was at one time thought. The question is whether fusion or some second-generation fission breeder will be preferable. The time scale goes like this: even if scientific feasibility is demonstrated by 1975, basic studies related to topics above the waterline in Fig. 9 will occupy several years beyond. After that, at least one pilot model fusion device would occupy our attention until the mid-1980's; then fission reactor experience shows that the lead time is long for designing and building the economic plants to follow. My own guess is that fusion power will be available in appreciable quantity by 2000, even with a fortunate outcome along one of the paths in Fig. 9. A few optimists propose 1990; pessimists propose never.

This long time before beneficial installation might seem to permit a comfortable period of grace before basic decisions about the overall feasibility and future of fusion need be taken. That is not so: other time scales enter. An important one is the fact that present gas diffusion plants for uranium enrichment may reach the end of their life by about 1990. First-generation fission breeders will have come into service well before then, but large, new, gas diffusion plants will still be needed. The question is in part whether the replacements are for an interim continuation, for a long-term continuation, or something else. Such expensive construction (several billion dollars) and the concomitant commitment bespeak a fairly clear decision by 1980 about what is to be built. For that, relative rank ordering of nuclear power systems will be needed several years earlier. Thus important decisions need to be made about the relative merits and eventual feasibility of nuclear power systems in the next few years. When the decisions start to be made, it becomes increasingly difficult to alter the course of events, because large economic and intellectual investments start to be made in the chosen course, and it usually is easier to stumble forward than to reach back. In truth, controlled fusion must from here on be subject to increasingly detailed technological assessment. To be late or unresponsive in this activity is to risk being irrelevant.

Hazards

Upon the topic of the next two sections, much arrant nonsense has been written, reminiscent of Ben Jonson's *The Alchemist.*

Almost everyone agrees that the most appreciable nuclear hazard of controlled fusion is that of tritium. A 5000-Mw (thermal) fusion plant would cycle about 10^8 curies of tritium through the plasma per day at 0.05 burnup, and actually burn 5×10^6 curies per day. How big will the inventory be? That depends on the rapidity with which unburned tritium can be reclaimed from the plasma pumps and the efficiency with which regenerated tritium can be scavenged from the moderating blanket. What little has been done on the pump problem suggests that something like 1 day's throughput may be held up in transit between exhaust from the fusion plasma and reinjection. For the blanket, more thoughtful analysis (*17*) suggests that 10 or 20 days of bred inventory may be held up in the huge bulk of lithium coolant, graphite, and so forth. At 0.05 fractional burnup, the two inventories would be about equal: a total of 2×10^8 curies.

This is a lot of radioactive material, comparable (in curies) to the amount of the most hazardous fission product ($\approx 10^8$ c of ^{131}I) expected to be found in a fission breeder reactor of the same size. But after that the comparison is not parallel. Per curie, tritium is relatively benign (9 kev average energy β^-) and in the gaseous form is only weakly biologically active. Then to this stage in the discussion, the relative hazards of fusion versus fission are perhaps $1:10^5$; on that basis fusion reactors could be installed anywhere without any containment shells (*17*). Still, extreme care must be exercised.

Complicating this story are the starting-to-be-assessed hazards of tritium being released as T_2O, of tritium leaking through the reactor structure, and the like (*29, 30*). For the first, T_2O enters the life cycle as does water, which increases the relative hazard considerably. For the second, hydrogen (hence tritium) delights in diffusing into and through metals, much more so than does any other element. This is no hazard of critical nuclear accident, but rather the problem of preventing the plant from having radioactive B.O. It can be solved technologically, for example, by placing vacuum barriers at critical places where tritium will migrate. But what will it cost? For example, if the fusion system cost including all such protective arrangements equals the cost of a liquid metal fast breeder plus a carefully prepared hole beneath the city to hold it, any advertised safety advantages of nuclear fusion become hard to see.

These tritium migration and scavenging problems are now starting to receive some attention, and in a few years a lot more can be said. In the meantime, I guess that fusion will retain a substantial advantage, which will be reflected in a price differential of $10 to $20 per thermal kilowatt.

Another nuclear nuisance is that the 14-Mev fusion neutrons will make the basic structure of a fusion reactor highly radioactive (*31*). Fission reactors have the same problem; the components are in no danger of being spread through the environment, so this activation poses more of a maintenance problem than a hazard.

About nonnuclear accident hazards, fusion and fission seem to be a standoff; one uses large amounts of liquid lithium or fused salts; one uses similar amounts of sodium. These hazards seem small, perhaps less than those enjoyed by people who live next to railroads on which many things are transported.

Permanent storage of long-life fission products is an additional problem for fission reactors; the advantage to fusion is modest, because total storage charges are expected not to be severe (on the scale of things discussed here).

Other Environmental and Technology Assessment Questions

Arguments about fossil as compared to nuclear power have often been made in terms of which kind of plant should be installed somewhere remote from population centers. As a corollary, the environment is imagined to be restored by having many nuclear power plants at remote locations producing electricity, which is transmitted to load centers.

That is all very well, but some kind of Sutton's Law (*32*) suggests that we look at the heart of the problem, which is elsewhere. Most people in the United States and other developed countries live in cities. Predictions vary for the energy requirements in (say) 1980, but all agree that even with the trend toward electric power accounted for, the nonelectric energy requirement will exceed the electric energy requirement by nearly an order of magnitude (*33*). Much of this nonelectric demand is for transportation. But even space heating, industrial process heat, and so forth still add up to much more than the predicted electric demand, and all this is now supplied by fossil fuels. Therefore, if fossil fuels are to be substantially traded for nuclear ones, nuclear power plants must be built in or very close to population centers. The question of hazards and the cost of assuring safety discussed in the previous section must be looked at from this point of view.

Analysis of the total social costs and benefits is complicated enough for fission breeders versus fossil plants, and is yet in a primitive stage. Including fusion as an option will make further complications. Either advanced fission breeder reactors or fusion reactors are expected to have good thermal efficiency; some propose 50 percent or more (compared with about 32 percent for present reactors, 41 percent for present fossil fuel plants, perhaps 50 percent for advanced ones). Proponents of fission breeders promote that the total environmental difficulties and social cost of nuclear power are substantially less than those of fossil fuel plants. I agree with this when the various diseconomies—those charges put upon the public sector and not now made a charge on the generating company—are included. That is, the effects of sulfur and nitrogen oxides, and of particulate emissions, place considerable burdens upon us as a whole; the country is taking steps to deal with them, and the curative costs are very large.

Beyond that, many more factors enter; here are some. Strip mining of coal can despoil large tracts of land for long periods. Deep mining of coal or uranium is hazardous; lithium mining also brings problems. Any fission reactor located on the surface in a city probably must have an exclusion area around it. Analyses show that this valuable land can be used for some agricultural purposes, very possibly in combination with some of the reactor's waste heat (*34*). But even if no direct economic use of the land is made, what large city could not do with an internal area having a pleasant vista? It is hard to quantify such social values, but surely they are substantial: recall the view down the Serpentine from Kensington Palace in London. Plant size and tradeoffs between capital cost and fuel cost can and should have substantial leverage on proper urban planning, but so far they do not. For example, large plants with low fuel cost could afford to be run with a policy of very cheap (free to some users?) off-peak power. With such a policy, different activities and living prospects can be stimulated in cities. The well-known positive feedback—via larger plant size, hence lower unit electricity cost, hence increased demand and accelerated technology change toward electricity—involves assessing much of future technology: Can transportation be based on some electric process, for instance?

Even fission and fusion are by no means mutually exclusive choices. They might complement each other, because fusion is predicted to have a large available neutron excess, and some otherwise attractive fission breeder schemes look dubious because the fuel doubling time is too long (*35*). Can fusion reactors then be used to manufacture incremental fissionable material, hence bringing about a useful symbiosis?

Yet all this does not reach the deepest layers of the problem. If we assign importance to the fact that controlled fusion could supply our energy needs for aeons, we should also see what constitutes the energy policy. Just producing more is clearly inadequate; using it sometimes brings difficulties too, such as the summer temperature rise in ghetto streets because of operating air conditioners. Then should we reduce energy dissipation by having better insulated buildings? Perhaps some principle of minimizing the entropy increase needs to be factored in. For fossil fuel utilization, this certainly seems required: jet plane travel is not wholly satisfactory, when almost as much fuel is burned per trip as if each and every passenger drove the distance by himself in his own automobile.

These are not empty phrases; if high speed intercity transport switches from aircraft to tunnel vehicles, substantial switch from fossil to nuclear (electric) power is possible. There is a lot at stake, an adequately broad assessment has not been made, and we are uncertain about what the policy ought to be. Indeed nowhere have problems of this scale—as they really exist in society—been approached in such an integrated fashion hitherto. This comment has broader implication than just to controlled fusion and relates to what appear to be very basic difficulties in how we organize ourselves to solve large societal problems. But that is another story (*36*).

It is in this broad context that controlled nuclear fusion will or will not be brought to fruition. I believe that, for fixed plant requirements, nuclear fission can be made substantially more attractive than can burning coal or oil, for most purposes. As implied in earlier sections, I also believe that the situation could be improved even more with successful fusion power. But these are still beliefs, not yet firm facts.

It would be rash to predict the outcome; not all schemes now being worked on will be adopted, which is the price in technology assessment of keeping options open. Surprises come, not all unpleasant, and a historic parallel occurs to me (*37*). In 1680 Christiaan Hüygens decided to control gunpowder for peaceful purposes, as a perpetual boon to mankind, and set his assistant Denys Papin to invent a controlled gunpowder engine. After 10 years of difficulty, Papin had a different idea, wrote in his diary,

> Since it is a property of water that a small quantity of it turned into vapour by heat has an elastic force like that of air, but upon cold supervening is again resolved into water, so that no trace of the said elastic force remains. I concluded that machines could be constructed wherein water, by the help of no very intense heat, and at little cost, could produce that perfect vacuum which could by no means be obtained by gunpowder.

then invented the expanding and condensing steam cycle, which made possible the industrial revolution.

References and Notes

1. For a very readable summary of the situation up to 1958, see A. S. Bishop, *Project Sherwood: the U.S. Program in Controlled Fusion* (Addison-Wesley, Reading, Mass., 1958); see also T. Alexander, *Fortune* **81** (6) 94–97, 126, 130–132 (1970); T. K. Fowler and R. F. Post, *Sci. Amer.*, **215** (6), 21 (1966), for summaries of the plasma confinement problem.
2. For an account of plasma physics, see for example, D. J. Rose and M. Clark, Jr., *Plasmas and Controlled Fusion* (M.I.T. Press, Cambridge, Mass., 2nd rev. printing, 1965).
3. This scheme is what confines high energy ions and electrons in the Earth's van Allen radiation belts. Particles are reflected above the ionosphere near the north and south magnetic poles, and follow the lines of weak field far from the earth in between.
4. 1 tesla ≡ 1 weber/m², or 10,000 gauss. Earth's field at the surface is about 0.5 gauss.
5. J. B. Taylor, *Nuclear Fusion Supplement*, part 2, 477 (1962).
6. This and a number of feasibility estimates to follow are taken from a review by D. J. Rose, *Nuclear Fusion* **9**, 183 (1969).
7. L. A. Artsimovich, G. A. Bobrovsky, E. P. Gorbunov, D. P. Ivanov, V. D. Kirillov, E. I. Kuznetsov, S. V. Mirov, M. P. Petrov, K. A. Razumova, V. S. Strelkov, D. A. Scheglov, *Nuclear Fusion Special Supplement*, 17 (1969).
8. L. A. Artsimovich, A. M. Anashin, E. P. Gorbunov, D. P. Ivanov, M. P. Petrov, V. S. Strelkov, *J. Exp. Theor. Phys. Lett.* **10**, 82 (1969); M. J. Forrest, N. J. Peacock, D. C. Robinson, V. V. Sannikov, P. D. Wilcock, private communication.
9. A good review is by T. K. Fowler, *Nuclear Fusion* **9**, 3 (1969); see F. H. Coensgen, W. F. Cummins, R. E. Ellis, W. E. Nexson, Jr., in *Plasma Physics and Controlled Nuclear Fusion Research* (Conference Proceedings IAEA, Vienna, 1969), vol. 2, p. 225.
10. E. M. Little, A. A. Newton, W. E. Quinn, F. L. Ribe, in *Plasma Physics and Controlled Nuclear Fusion Research* (Conference Proceedings IAEA, Vienna, 1969), vol. 2, p. 555; A. D. Beach *et al.*, *Nuclear Fusion* **9**, 215 (1969).
11. *Nuclear Fusion Reactors* [Proc. Brit. Nuclear Energy Soc. Conf. on Fusion Reactors, United Kingdom Atomic Energy Authority (UKAEA), Culham Laboratory, 17—19 September 1969], J. L. Hall and J. H. C. Maple, Prod. Eds. (Culham Laboratory, 1970).
12. R. F. Post (chairman) *et al.*, *Preliminary Report of Direct Recovery Study*, Report UCID-15650, Lawrence Radiation Laboratory, Livermore, Calif. (1970).
13. D. Steiner, *Neutronics Calculations and Cost Estimates for Fusion Reactor Blanket Assemblies*, USAEC Report ORNL-TM-2360, Oak Ridge National Laboratory (1968); also S. Blow, V. S. Crocker, B. O. Wade, "Neutronic calculations for blanket assemblies of a fusion reactor," paper 5.5 of reference (*11*).
14. A. P. Fraas, *A Diffusion Process for Removing Tritium from the Blanket of a Thermonuclear Reactor*, USAEC Report ORNL-TM-2358, Oak Ridge National Laboratory (1968); see also reference (*11*).
15. R. W. Werner, "Module approach to blanket design—A vacuum-wall-free blanket using heat pipes," paper 6.2 of reference (*11*).
16. A. P. Fraas and R. S. Pease, private communications.
17. A. P. Fraas, "Conceptual design of a fusion power plant to meet the total energy requirements of an urban complex," part of reference (*11*).
18. D. Steiner, "Neutronic behavior of two fusion reactor blanket designs," paper 5.4 of reference (*11*).
19. D. J. Rose, *On the Fusion Injection Problem*, UKAEA Culham Laboratory Technology Division Report No. 82 (1968).
20. G. I. Bell, W. H. Borkenhagen, F. L. Ribe, "Feasibility studies of pulsed, high-β fusion reactors," paper 3.3 of reference (*11*); also USAEC Report LA-DC-10618, Los Alamos Scientific Laboratory (1969).
21. For example, see M. A. Rothman, R. M. Sinclair, I. G. Brown, J. C. Hosea, *Phys. Fluids* **12**, 2211 (1969).
22. M. Yoshikawa, T. Okhawa, A. A. Schupp, *Phys. Fluids* **12**, 1926 (1969).
23. But for an optimistic opinion about mirror-stopping by electrostatic fields, see A. A. Ware and J. E. Faulkner, *Nuclear Fusion* **9**, 353 (1969).
24. R. A. Dandl, J. L. Dunlap, H. O. Eason, P. H. Edmonds, A. C. England, W. J. Hermann, N. H. Lazar, in *Plasma Physics and Controlled Nuclear Fusion Research* (Conference Proceedings IAEA, Vienna, 1969), vol. 2, p. 435.
25. J. W. Beal, M. Brettschneider, N. C. Christofilos, R. E. Hester, W. A. S. Lamb, W. A. Sherwood, R. L. Spoerlein, P. W. Weiss, R. L. Wright, *ibid.*, vol. 1, p. 967.
26. For an example of state-of-the-art, see J. W. Mather, P. J. Bottoms, J. P. Carpenter, A. H. Williams, K. D. Ware, *Phys. Fluids* **12**, 2343 (1969).
27. I. N. Golovin, "Tokamak as a possible fusion reactor," in reference (*11*); also A. Gibson, "Permissible parameters for stellarator and Tokamak reactors," paper 3.2 of reference (*11*).
28. R. Holcomb, *Science* **167**, 1112 (1970).
29. F. Morley and J. Kennedy, "Fusion reactors and environmental safety," paper 1.3 of reference (*11*).
30. A. P. Fraas and H. Postma, *Preliminary Appraisal of the Hazards Problems of a D-T Fusion Reactor Power Plant*, USAEC Report ORNL-TM-2822, Oak Ridge National Laboratory (1970).
31. D. Steiner, in preparation.
32. After Willie Sutton who, on being asked why he robbed banks, replied, "That's where the money is."
33. *Summary Report: Use of Steam-Electric Power Plants to Provide Low-Cost Thermal Energy to Urban Areas*, USAEC Report ORNL-HUD-14, Oak Ridge National Laboratory (1970), in press.
34. J. A. Mihursky, *Summary of Meeting on Beneficial Uses of Waste Heat*, Oak Ridge, Tennessee, 20 to 21 April 1970, in preparation.
35. L. M. Lidsky, "Fission-fusion symbiosis: general considerations, and a specific example," paper 1.2 of reference (*11*).
36. For a discussion, see D. J. Rose (chairman) *et al.*, "The Case for National Environmental Laboratories," USAEC Report ORNL-TM-2887, Oak Ridge National Laboratory (1970); also *Technol. Rev.* **73** (6), 38 (April 1970).
37. C. J. Singer, E. J. Holmyard, A. R. Hall, T. I. Williams, Eds., *A History of Technology* (Oxford Univ. Press, London, 1954–58), vol. 4, p. 171.

nuclear energy

Laser-induced thermonuclear fusion

Can focused laser pulses in the gigawatt range be used to compress hydrogen droplets by a thousand-fold to create energy-producing reactions?

John Nuckolls, John Emmett and Lowell Wood

Laser-induced fusion has recently joined magnetic-confinement fusion as a prime prospect for generating controlled thermonuclear power. During the past three years, the Atomic Energy Commission has accelerated the national laser-fusion program more than tenfold, to about \$30 million annually,[1] and the Soviet Union has a program of comparable size.

In contemporary nuclear power plants, uranium is the primary fuel and fission reactions provide the nuclear energy. Fusion reactions between the heavy isotopes of hydrogen are another source of nuclear energy, one whose utilization for electricity generation lies in the future. Fusion power is important because, as Richard F. Post pointed out in his recent PHYSICS TODAY article,[2] the deuterium contained in the oceans is a virtually inexhaustible, low-cost, relatively clean fuel.

Fusion reactions are demonstrated in thermonuclear explosions and are thought to be the source of stellar energy. For the past twenty years, the thrust of controlled fusion research has been toward magnetic confinement of plasmas heated sufficiently (to about 10^8 K temperatures) to achieve fuel ignition. Laser-induced fusion, which exploits inertial confinement, has received increased public interest recently with the AEC declassification of important concepts and calculations. The key idea is to use laser light to isentropically implode pellets of deuterium and tritium to approximately 10 000 times liquid density and thereby induce efficient thermonuclear burning. Fusion energies 50–100 times larger than laser input energies of 10^5–10^6 joules have been achieved in sophisticated computer simulation calculations. There is as yet no experimental confirmation, but 10 000-joule lasers are being planned at the Livermore and Los Alamos Laboratories and at the Lebedev Institute in the USSR to explore laser-induced fusion.

The laser fusion implosion system consists of a tiny spherical pellet of deuterium–tritium surrounded by a low-density atmosphere extending to several pellet radii, located in a large vacuum chamber and a laser capable of generating an optimally time-tailored pulse of light energy. Before the main pulse occurs, the atmosphere may be produced by ablating the pellet surface with a laser prepulse. Most of the dense pellet is isentropically compressed to a high-density Fermi-degenerate state and thermonuclear burn is initiated in the central region. A thermonuclear burn front propagates radially outward from the central region igniting the dense fuel.

With 10^5–10^6 joule laser pulses initiating 10^7–10^8 joule fusion pulses, gigawatt electrical power levels may be generated by initiating 100 pulses per second, possibly ten per second in each of ten combustion chambers. The combustion chambers would have a diameter of a few meters and walls wetted with lithium to withstand the nuclear radiation and debris. For economic operation the pellets would have to cost less than a cent each, and they could be fabricated in a drop tower. Electricity would be generated via neutron-heated lithium blankets as in conventional controlled thermonuclear reaction schemes, or by direct conversion in advanced power-plant schemes.

The implosion of bubbles in water was considered by William Henry Besant in 1859[4] and Lord Rayleigh in 1917.[5] A self-similar solution to an imploding shock wave was developed by G. Guderley who made some relevant calculations in 1942.[6] Early work on fission weapons at Los Alamos began by Seth Neddemeyer, John Von Neumann and Edward Teller and others explored spherical implosion systems driven by high explosives.[7] Subsequently, moderately high compressions were experimentally demonstrated and utilized. However, compressions approaching 10 000 fold (relative to liquid or solid densities)—which are required for practical laser fusion power reactors—have not been experimentally achieved.

The invention of the pulsed laser in the early 1960's stimulated further implosion calculations and a proposal of a laser fusion engine for CTR and propulsion applications. Nearly all laser-fusion implosion calculations remained classified until reported recently.[3,8] Experimental work began in 1963, and by the mid-1960's, Ray Kidder and S. W. Mead constructed a twelve-beam implosion-oriented laser at Livermore.[9] In 1972, Nikolai Basov and his colleagues reported implosion of a 100-micron diameter CD_2 microsphere with a few hundred joule, few nanosecond, nine-beam laser pulse.[10]

Implosions and thermonuclear burn

Conditions involving pressure, symmetry and stability must be satisfied

The authors are physicists at the University of California Lawrence Livermore Laboratory. John Nuckolls is associate leader of "A" Nuclear Explosive Design Division, John Emmett is leader of "Y" Laser Fusion Division and Lowell Wood is a member of the Director's Office and Physics Staffs.

Reprinted with permission from *Physics Today*, vol. 26, pp. 46–53, Aug. 1973.

to implode a DT sphere to a state at 10^4 times liquid density, in which both Fermi-degeneracy and thermonuclear propagation can be exploited to achieve maximum gain.

The optimum laser pulse shape generates an initial shock, which is near-sonic ($1/2 \times 10^6$ cm/sec) in the outer part of the pellet. This shock produces an entropy change sufficiently small that subsequent compression to a Fermi-degenerate state is possible. As this shock converges toward the center of the pellet it becomes sufficiently strong to produce significant heating. The pulse shape also generates a maximum implosion velocity of about 3.5×10^7 cm/sec, corresponding to the required average energy density of 6×10^7 joules/gram (see box). The implosion velocity is increased from the initial to the maximum value at such a rate that the hydrodynamic characteristics in the compressing pellet coalesce to form a strong shock near maximum compression, at a distance from the center approximately equal to the range of 10-keV alpha particles in DT. By numerous computer calculations of laser implosions, we know that the optimum pulse shape is approximately

$$E = E_0 \tau^{-s}$$

where E is the laser power, $\tau = 1 - (t/t')$, t is time, t' is the collapse time, and s is approximately 2.[11] No satisfactory analytic derivation of this equation is known. Figure 3 shows how this pulse shape may be approximated with sufficient accuracy by a histogram of 5–10 pulses.

The compression and burn processes that have been described are illustrated in figure 4 by results of a typical computer-simulation calculation of the implosion of a fusion pellet to 10 000 times liquid density, and of the resulting thermonuclear microexplosion. This calculation was carried out at the Livermore Laboratory by Albert Thiessen with a program developed by George Zimmerman.[12] The program includes the following physical processes:

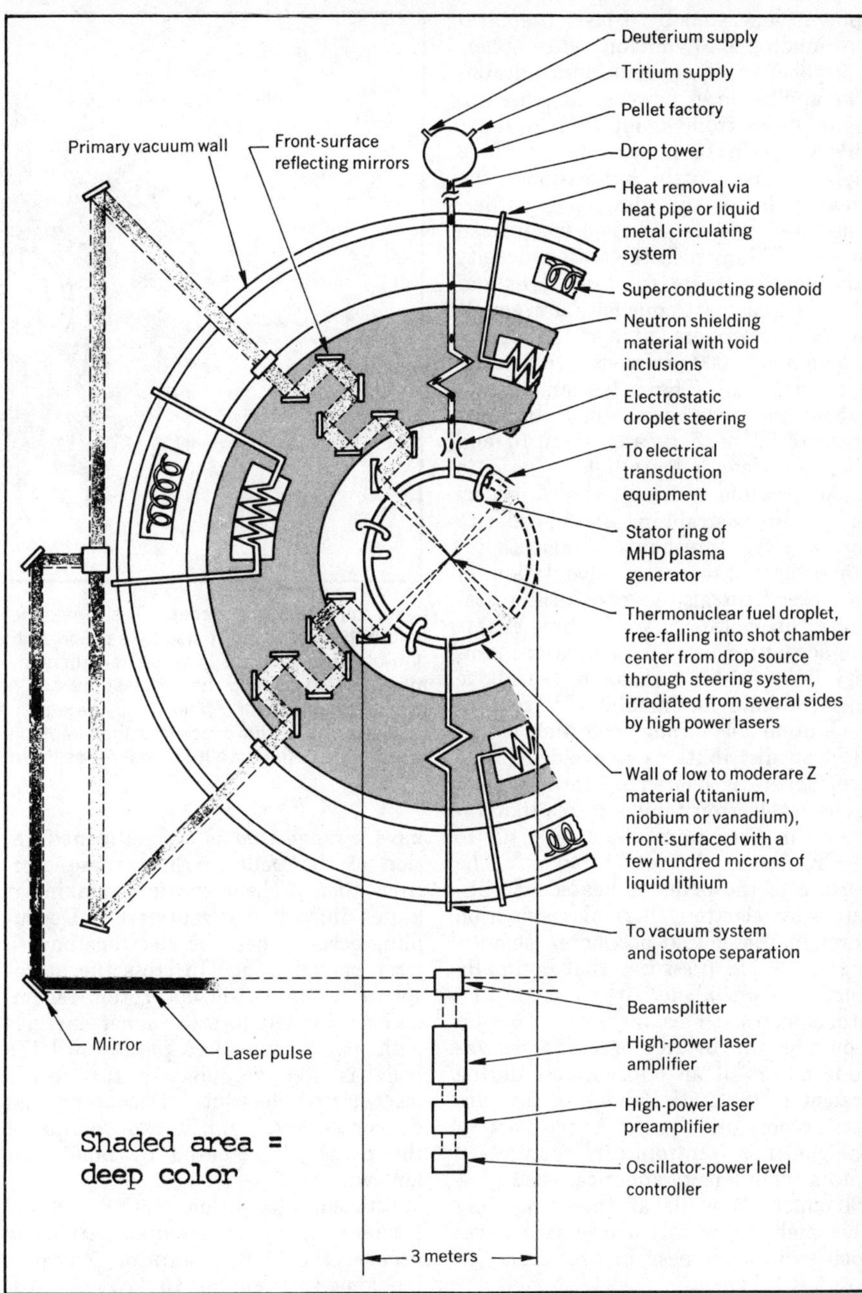

Conceptual design of a laser-fusion power-plant. One hundred laser-induced micro-explosions per second will produce 1000 megawatts of electrical power. The pulse of laser light, which is shown here in deep color, is shaped in time to yield optimum implosion and thermonuclear burning of the pellet. Several processes can be used to convert the explosion energy into electricity, such as thermal and MHD plasma conversion. Figure 1

- Hydrodynamics—Lagrangian; real and generalized Von Neumann artificial viscosities; ponderomotive, electron, ion, photon, magnetic and alpha-particle pressures.
- Laser light—absorption via inverse bremsstrahlung and plasma instabilities; reflection at critical density.
- Coulomb coupling of charged-particle species.
- Suprathermal electrons—multigroup flux-limited diffusion with self-consistent electric fields; non-Maxwellian electron spectra determined by results of plasma-simulation calculations for laser-light absorption by plasma instabilities; inverse bremsstrahlung electron spectrum for classical absorption.
- Thermal electrons and ions—flux-limited diffusion.
- Magnetic field—includes modification of all charged-particle transport coefficients, as well as most of the equilibrium MHD effects described by S. Braginskii.[13]
- Photonics—Multigroup flux-limited diffusion; LTE non-LTE average-atom opacities for free–free, bound–free, and bound–bound processes; Fokker-Planck treatment of Compton scattering.
- Fusion—Maxwell velocity-averaged reaction rates; the DT alpha particle is transported by a one-group flux-limited diffusion model with appropriate energy deposition into the electron and ion fields; one group transport of the 14 MeV neutron.
- Material properties—opacities, pressures, specific heats, and other properties of matter are used, with nuclear, Coulomb, degeneracy, partial ionization, and other significant effects taken into account.

In this implosion-burn calculation, a 10kJ, short-wavelength (1/2 micron), frequency-modulated, pulse of laser light is focussed symmetrically onto a 1200-micron radius low-density atmosphere (generated by a laser prepulse) surrounding a 400-micron radius spherical pellet of liquid deuterium-tritium. The applied laser power is increased in eight pulses from about 10^{11} to about 10^{15} watts in 10 nanoseconds. These eight pulses closely approximate the ideal pulse shape described earlier. The laser light is absorbed via inverse bremsstrahlung near the critical density (the density where the laser light and electron plasma frequencies are equal) in the atmosphere, at a radius of approximately 600 microns, generating hot electrons. The pellet and atmosphere are seeded with small amounts of material of Z greater than 10 and short-wavelength laser light is used to make possible efficient absorption by inverse bremsstrahlung and in order to increase the thresholds of plasma instabilities. Frequency modulation of the laser light also increases the instability thresholds.[14] When these effects are accounted for, the peak laser intensity is less than a factor of ten above the threshold for instabilities—so that generation of strong non-Maxwellian electron distributions is avoided. The atmosphere is heated by the hot electrons to electron temperatures that increase in time from about 3×10^6 to 10^8 K at the absorption radius. The surface of the pellet is heated and ablated by electron thermal conduction through the hot atmosphere, generating implosion pressures that optimally increase from about 10^6 to about 10^{11} atmospheres. This increase in implosion pressure by five orders of magnitude occurs at an optimal rate during transit of the initial shock to the center. Consequently the outer part of the pellet is isentropically compressed into a high-density spherical shell ($p >$ 100 gm/cm^3) while at the same time this shell is inwardly accelerated to velocities that increase in time from 10^6 to 3.5×10^7 cm/s.

As the internal pressure becomes larger than the ablation pressure the rapidly converging shell slows down and is compressed nearly isentropically, at sub-Fermi temperatures, to densities greater than 1000 gm/cm^3. The inner region is compressed by the outer shell to densities approaching 1000 gm/cm^3, and heated to ion temperatures greater than 10^8 K, initiating thermonuclear burn. A thermonuclear burn front then propagates outward. About 1200 kilojoules of fusion energy are produced in less than 10^{-11} seconds. The energy gain is about 20-fold.

There are, however, other effects that may reduce this gain; this is especially so for long-wavelength laser light such as the 10.6-micron CO_2 line. The pellet compression and energy gain might be strongly degraded by electron preheat[15] and decoupling.[16] Preheat occurs when the laser-heated electrons have a range that is a significant fraction of the pellet radius; these electrons then preheat the fuel, making it more difficult to compress. Decoupling occurs when the electrons have a large enough range to cross the atmosphere, re-enter the absorption region, and are heated to still higher energies with longer mean free paths until the pellet is effectively decoupled from the laser-heated electrons. Decoupling can be compensated for if the volume of the pellet is increased by making it hollow.

Efficient absorption of CO_2 light is not possible via inverse bremsstrahlung because the light absorption length is too long ($\gg$1 cm at 10 keV).[17] Absorption is possible via plasma instabilities.[18] However, if the thresholds for these instabilities are greatly exceeded, then plasma-simulation computer codes indicate that non-Maxwellian electron spectra may be generated, with high-energy tails extending beyond 100 times the thermal electron energy.[19] Experiments are needed to determine the electron spectra reliably in such situations. If excessive numbers of superthermal electrons are not generated, then long-wavelength lasers may be suitable for CTR applications, provided that the hollow pellet can be constructed cheaply enough.

In compression of a sphere by 10^4-fold, the radius decreases somewhat more than 20-fold. If, after compression, spherical symmetry is required to within half of the compressed radius—or 1/40 the initial radius—then the implosion velocity (and time) must be spatially uniform and synchronized to about one part in 40, or a few percent. The outer atmosphere may be heated uniformly to 10% to 20% by a many-sided irradiation system, consisting of beam splitters, mirrors, lenses, and other optical elements. This error is then reduced to less than 1% by physical processes occurring inside the atmosphere.[3,20] Asymmetries are reduced during electron energy transport through several scattering mean free paths of atmosphere to heat the surface of the pellet. In addition, since the atmosphere has a large radius compared to the pellet, each point on the pellet surface is heated by electrons coming from almost 2 π steradians of the hot absorbing region in the outer atmosphere. Finally, during most of the implosion, the electron mean free path in the absorbing region is a significant fraction of the absorption radius.

The implosion of the pellet by diffusion-driven pressures generated by ablation is hydrodynamically stable, except for relatively long-wavelength surface perturbations.[3,20] Fortunately, these perturbations grow too slowly to be damaging if the pellet is imploded in one sonic transit time. In part, ablative stabilization occurs because the peaks of surface perturbations are effectively closer to the heat source than are the valleys, so that the ablation-driving temperature gradient is steeper. Consequently, the amplitude of the perturbation is reduced, both because the peak is more rapidly ablated and because the ablation pressure is higher on the peak.

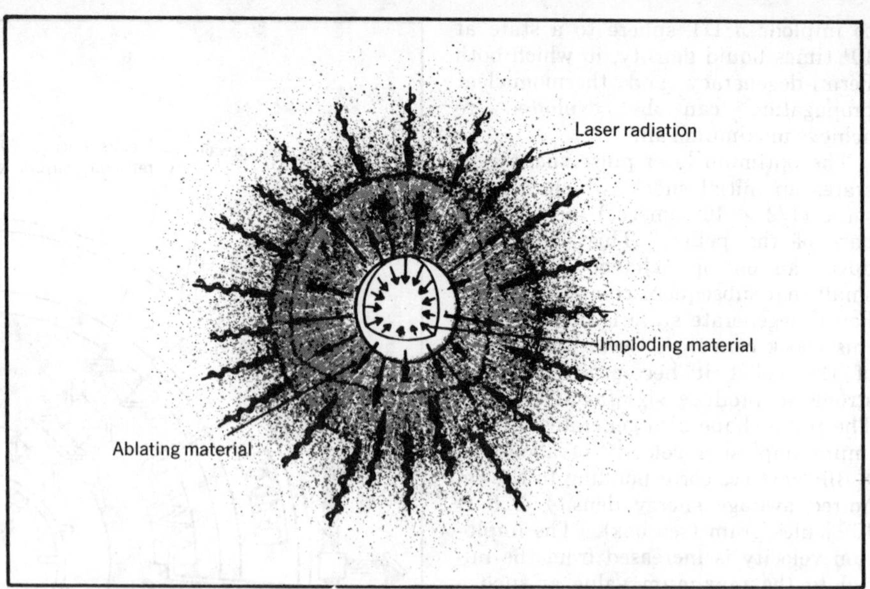

Laser implosion of a pellet. The atmosphere extends to several pellet radii and is formed before the main laser pulse by a prepulse that ablates some of the pellet surface. Absorption of the laser light in the outer atmosphere generates hot electrons. As the electrons move inward heating the atmosphere and pellet surface, scattering and solid-angle affects greatly increase the spherical symmetry. Violent ablation and blowoff of the pellet surface generates the pressures that implode the pellet; the effect is similar to a spherical rocket. The pellet core then undergoes thermonuclear burn. Figure 2

Laser-imploded pellets

Thermonuclear micro-explosions scale as the density–radius product ρR. The rates of burn, energy deposition by charged reaction products, and electron–ion heating are proportional to the density, and the inertial confinement time is proportional to the radius of the pellet. Consequently, the burn efficiency, self-heating, and feasibility of thermonuclear propagation are determined by ρR. If ρR is very much greater than 0.3 gm/cm^2, then only 0.3 gm/cm^2 in the central region of the pellet need be heated to approximately 10 keV to initiate a radially propagating burn front that ignites the entire pellet. In this case, 1.6×10^{10} joules/gm of fusion energy will be released from the central region; one fifth of this energy is in alpha particles, sufficient to heat three times more DT to 10 keV. The alpha particles will deposit their energy in approximately this mass since their range at 10 keV is about 0.3 g/cm^2.

When ρR is approximately 3 gm/cm^2 the fusion energy released is about 10^{11} joules/gm. If we assume propagation of the thermonuclear reaction from the 10-keV central region and compression of the remaining DT to a Fermi degenerate state, all at 1000 gm/cm^3, the minimum average energy of ignition and compression is about 6×10^7 joules/gm. (If the degeneracy condition is not satisfied the compressional energy will exceed the ignition energy.) The gain is then about 1500, but approximately

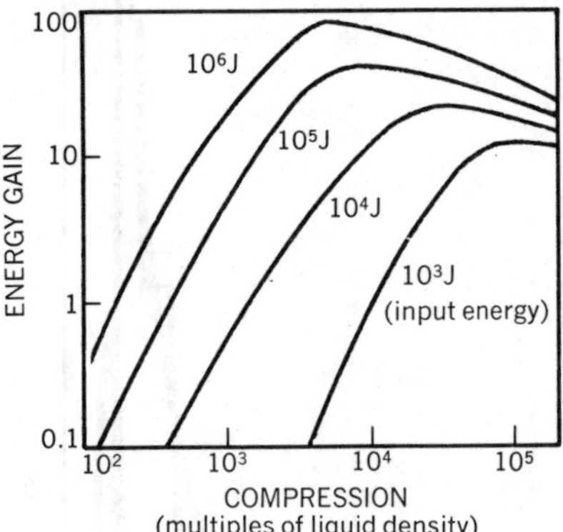

95% of the laser energy absorbed by the pellet during implosion is lost to kinetic and internal energy blowoff. Consequently, the energy gain relative to the laser energy employed is about 75 fold. This is sufficient for CTR applications with a 10% efficient laser, a 40% thermal-to-electric efficiency and about 30% of the electrical energy circulated internally to pump the laser.

The figure shows the variation of gain (relative to laser light energy) with compression and with laser-light energy.[3,11] The curves have been normalized to computer calculations of the implosion and burn. Gains approaching 100 are predicted for laser energies of 10^6 joules. The calculations indicate that less than 1 kilojoule of laser light may be sufficient for breakeven (gain ≈ 1) and 10^5 joules may be sufficient to generate net electrical energy with a 10% efficient laser. These predicted gains are probably upper limits to what can be achieved. Similar gain curves may be generated for D_2 and DHe^3 pellets seeded with a small percentage of tritium to facilitate ignition.

Laser technology

The development of lasers for high-density laser-fusion application poses a special set of problems that have not previously received much attention in the laser R&D community. In the optimum pulse shape, about half the total energy is produced in the final 100 picoseconds. Thus, in terms of a single pulse, a CRT laser has to be capable of producing at least 50 kilojoules in 100 picoseconds. In addition, the optimum plasma heating process may require short-wavelength lasers. Such lasers do not exist at present; however, development of short wavelength devices is receiving increasing attention in the US and several other countries.

The salient characteristics of four laser systems presently under consideration are indicated in Table 1. These systems are representative of the diversity that exists in the laser world. What is interesting to note about these laser systems is the almost total lack of overlap in the technologies required for the development of each. Thus, extensive development effort applied to one of these systems is not usually applicable to another. Also shown in Table 1 are the desired characteristics of a hypothetical laser system that matches the requirements of laser fusion as presently envisioned. The primary characteristics of this hypothetical laser system are high efficiency, high average power, short wavelength and high energy. It is clear that none of the real lasers in Table 1 demonstrates all of these characteristics.

Neodymium-glass lasers develop gain at 1.06-microns wavelength with Nd^{3+} ions in a glass matrix pumped by xenon flash lamps. They have the best developed technology for operation in the sub-nanosecond region. In addition, the high second-harmonic conversion efficiency (60–80%) already demonstrated offers great potential for operation at 0.53 micron. Fourth-harmonic generation (0.265 micron) and stimulated Stokes–Raman scattering (1.9 micron) offer potential for additional wavelengths with efficiencies greater than 20%. Thus, the neodymium–glass laser system provides the best laboratory tool for near-term laser-fusion experiments. However, the extremely low energy of efficiency (0.1%) and the low average power capability (limited by the low thermal conductivity of glass) prevent any consideration of neodymium–glass systems for eventual laser fusion power-generation applications.

The carbon-dioxide laser develops gain at 10.6 microns between vibrational energy levels of the ground state of the CO_2 molecules. This system (actually a CO_2–N_2–He mixture) is pumped by relatively low-power electrical discharges in the gas. This laser has demonstrated efficiencies of approximately 5% for one-nanosecond duration pulses. Operation in the nanosecond regime has yet to be demonstrated. With the addition of high-speed gas flow, CO_2 lasers have the capability to generate high average powers. The major liability of the system is the 10.6-micron emission wavelength. Efforts are currently underway to convert the 10.6-micron energy efficiently to shorter wavelengths, although success has yet to be achieved. Development of high-energy, short-pulse CO_2 lasers continues because of their high efficiency and high average power. The ultimate usefulness of this system remains to be determined.

The iodine laser has recently come under consideration as a possible lower-cost replacement for neodymium–glass. It develops gain at 1.315 microns between the electronic levels of the neutral iodine atom, which is produced in an excited state by photodissociation of molecules such as CF_3I. The emission wavelength and efficiencies are similar; however, the cost of CF_3I is much below that of neodymium–glass. The stimulated-emission cross section of I*, even in the presence of a few atmospheres of a line-broadening buffer gas, is much larger than neodymium-glass. This necessitates an entire different approach to the laser design, in order to control parasitic oscillation within a single laser amplifier section. With the eventual solution to the parasitic oscillation problem and more detailed understanding of the pumping requirements of the system, it is possible that iodine lasers may be built in the 1–10 kilojoule regime.

Recent interest in the xenon laser stems from the short wavelength (1722 Å) and predicted high pumping efficiency (25%).[21] Little detailed information is available on this system, and

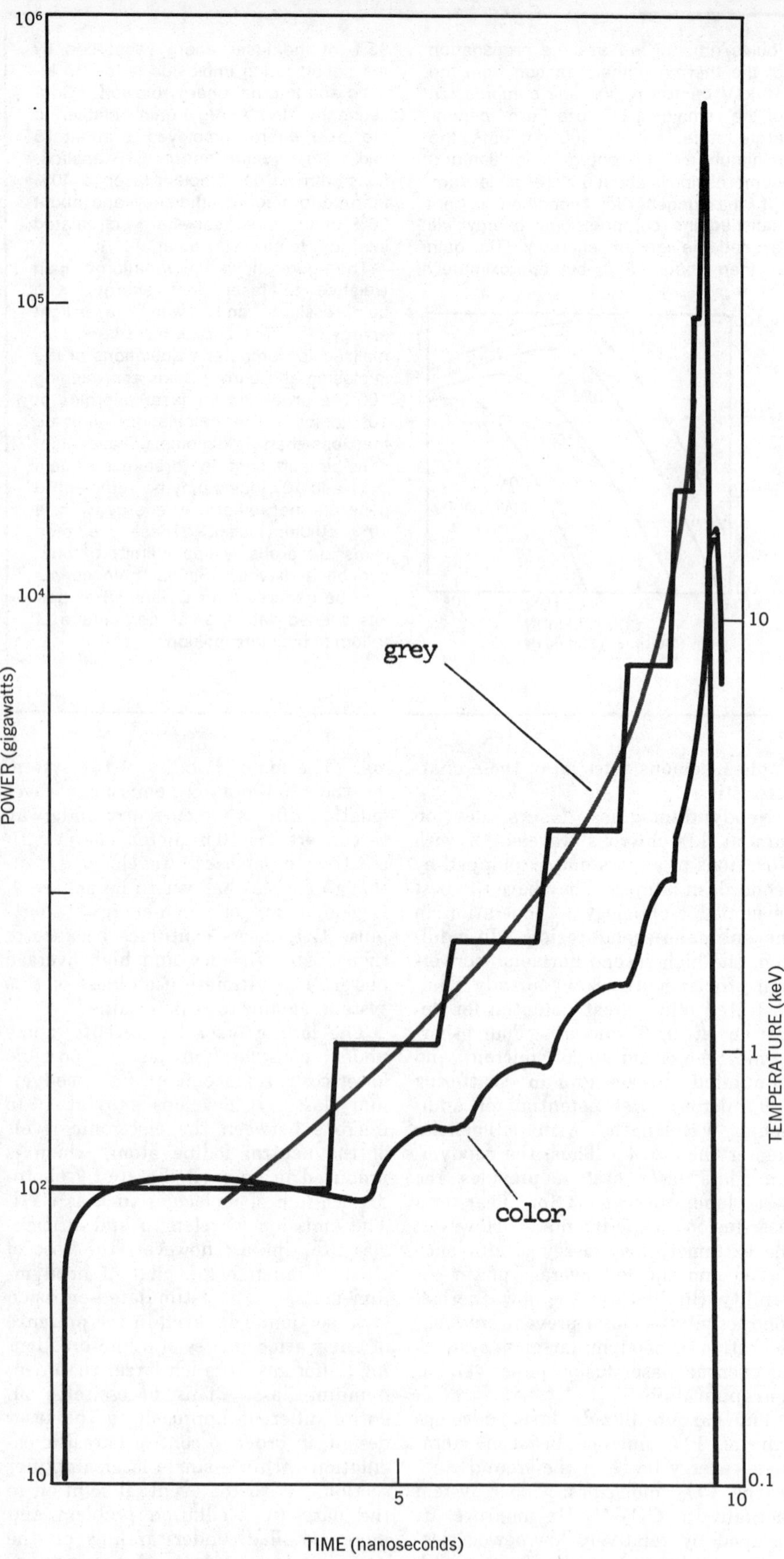

Laser pulse power rise in steps to approximate the ideal shape, which is shown in grey. The corresponding electron temperature in the critical density region is shown in color. These curves were generated by computer calculations and the assumption of a 10 000-fold compression of a fusion pellet. This time-tailored pulse is composed of eight subpulses that form the steps. The first step is a 500-joule pulse that lasts for 4200 picoseconds, and the remaining steps are formed by the following subpulses: 500 J-1600 psec, 1 kJ-1400 psec, 24 kJ-1200 psec, 4kJ-650 psec, 8kJ-330 psec, and finally 30kJ-65 psec. Figure 3

the technology for the generation of subnanosecond pulses has yet to be developed in this region of the spectrum. Two comments are, however, appropriate. First, the stimulated-emission cross section is approximately 3×10^{-18} cm^2. Thus, a large high-energy amplifier will be severely limited in performance by superfluorescence (amplified spontaneous emission) and parasitic oscillation. Second, lasers useful for fusion applications must operate at flux levels of 10^{10} watts/cm^2 (1 J/cm^2, 100 picosec) or greater. At this flux level and 1722 Å wavelength, all transparent materials (window, lenses, coatings) from LiF to Al_2O_3 will exhibit two-photon absorption coefficients in the 1-25 cm^{-1} range. Thus, to use such a short wavelength, a new optical technology of gas lenses and aerodynamic windows will have to be developed.

From the foregoing discussion we may draw several conclusions. Clearly, neodymium-glass laser systems provide the best technology base for the near-term laser-fusion experiments. The wide range of pulse widths obtainable (20 picosec-20 nanosec) and the range of wavelengths (0.265-1.9 microns) render it an almost ideal laboratory tool. For these reasons large multi-aperture neodymium-glass laser systems with energies of 10 kilojoules are in design or construction stages both in the US and the USSR. At the Lawrence Livermore Laboratory, a 10-kilojoule subnanosecond facility is being designed for spherical irradiation. Funds for it have been requested in the President's fiscal 1974 budget submitted to Congress. With this instrument, the important milestone of significant thermonuclear burn and scientific breakeven (fusion energy equals laser energy) will be achieved.

The high efficiency and high average-power capability of CO_2 lasers warrant their further study. The technology of high-energy, short-pulse CO_2 lasers is being aggressively pursued by the Los Alamos Scientific Laboratory. A ten kilojoule, one-nanosecond multiaperture device is being developed to answer the fundamental questions associated with the use of such long wavelengths for high density laser fusion.

The iodine laser system is under intensive development at the Institut fur Plasma Physik, Garching, where the objective is a one-kilojoule, one-nanosecond device for laser-fusion experiments. Additonal low-level investigations of this system are being carried out in many laboratories, both in the US and abroad. However, the only flexibility currently provided by this system is a lower-cost laser medium. This may be more than offset by the increased cost of capacitor banks and

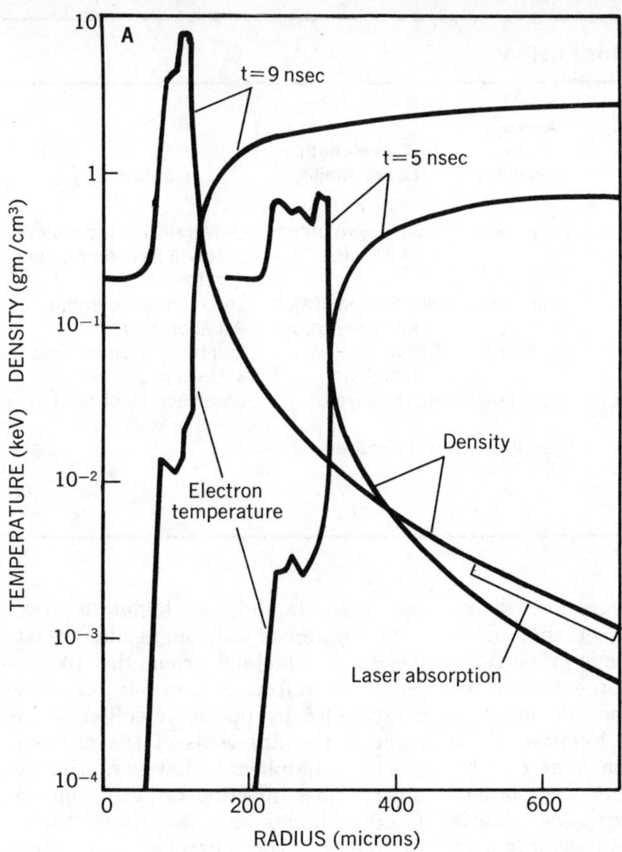

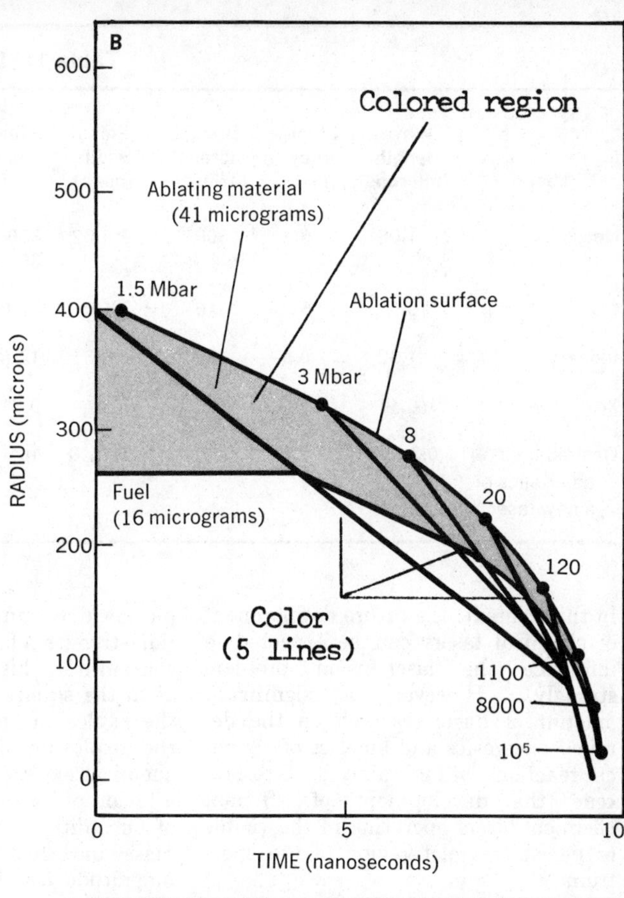

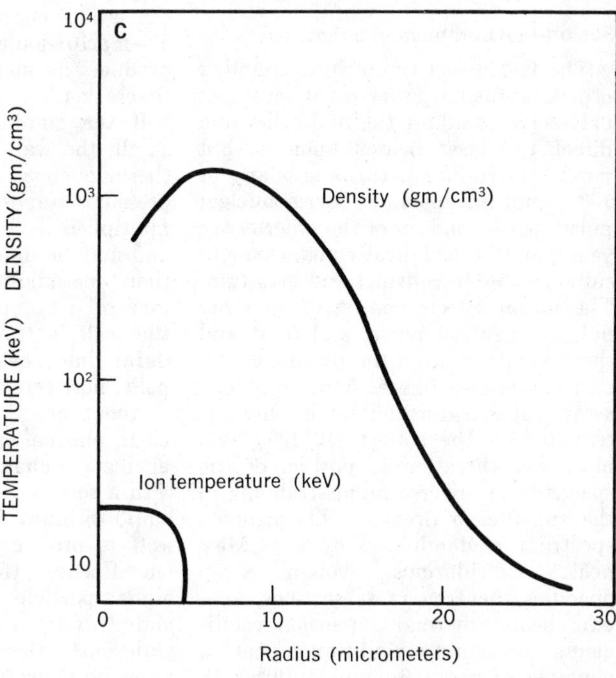

Compression of the fusion pellet from the beginning of the laser pulse and through absorption until thermonuclear ignition. These curves were generated by a computer and the assumption of a 10 000-fold compression. Part A shows the density and electron temperature versus the radius at times, five and nine nanoseconds after the start of the laser pulse but before ignition. In part B, the radius is plotted against time. The colored lines in this figure represent weak shocks from the steps of the pulse power. The inner curve represents that part of the pellet which undergoes thermonuclear burning, and the outer curve represents the outer region of the pellet; all the material in between is ablated away and is indicated by the colored region. The pressures at different stages of the compression are shown in megabars. In part C, the density and electron temperatures are plotted at the time of ignition. Figure 4

flashlamps for the short pulse excitation required.

The xenon laser system is too new to make any meaningful projections as to its ultimate usefulness for laser fusion. It will obviously have significant applications to high-density plasma diagnostics where its short wavelength may be essential.

The development of new lasers is required if laser-fusion power production is to become a reality. The xenon laser represents a class of possible lasers based on the weakly bound or van der Waals molecules. Laser action has already been achieved from Xe_2* and Kr_2*, and it may be expected from some of the similar dimer systems of mercury, cadmium and zinc. Other non-dimer systems such as LiXe or HgXe also look attractive. These systems are pumped by efficient, high-current relativistic electron-beam machines, which have been extensively developed during the last decade. The availability of an efficient pump source and short wavelength of emission makes these systems of great interest for laser fusion. However, it is clear that stimulated emission cross sections smaller than those in the xenon system will be necessary, as will extensive development of a means of reducing parasitic oscillation and superfluorescence.

Probably the most important characteristic of any new laser system developed for laser-fusion applications will be the ability to use energy efficiently.

Table 1. Lasers for Fusion

Laser	Wave-length (microns)	Effici-ency (%)	Energy storage (J/liter)	Pulse width (nsec)	Max. output short pulse (joules)	Average power capability	Wavelength convertability	Laboratory
Nd:glass	1.06	0.2	500	≥0.02	350 (0.1 nsec) 350 (1.0 nsec)	Very low	0.26–1.9 microns (40% eff.)	University of Rochester; Naval Research Laboratory
CO_2	10.6	5	15	≥1.0	17 (1 nsec)	High (flow)	Not demonstrated for high powers	Los Alamos Scientific Laboratory
Iodine	1.32	0.5	30	0.6	12 (10 nsec)	High (flow)	Similar to Nd:glass	Institut für Plasmaphysik, Garching
Xe_2	0.17	<20	300	≈10	0.01 (10 nsec)	High (flow)	Not required	Lawrence Livermore Laboratory
Desired characteristics of a new laser	0.3–0.5	>5	100–1000	0.1–1.0	10^4–10^6	High (flow)	Not required	

In this context, the future development of chemical lasers can be expected to influence the laser-fusion problem strongly. However, a significant amount of basic research on the detailed energetics and kinetics of chemical reactions will certainly have to precede the development of efficient chemical lasers operating in the visible or near-ultraviolet region of the spectrum.

Fusion-fuel combustion chamber

The fusion-fuel combustion chamber of a laser-fusion power plant must not only serve to admit the fuel pellet and direct the laser beams upon it, but must also endure perhaps as many as 100 multimegajoule thermonuclear pulses per second for of the order of ten years, and be technically and economically feasible to contruct and maintain. The fusion effects consist of an x-ray pulse, a neutron pulse, and blast and thermal effects from the plasma explosion debris (see figures 5 and 6). The x-ray pulse is fortunately heavily attenuated in the softest (10–1000 eV), most wall-threatening, portion of the spectrum by inverse bremsstrahlung in the superdense fireball. The neutron spectrum is dominated by a 14-MeV peak. Calculations involving x-ray opacities, neutron cross sections, specific heats, thermal expansion coefficients and compressibilities indicate a chamber of about 3-meter radius with a wall of a few layers of properly chosen low-to-moderate atomic-number materials (for example about 0.01 cm of beryllium backed by titanium, niobium or vanadium) will endure the x-ray and neutron pulses of a 10^7-joule microexplosion. If surfaced with a thin, low-Z liquid layer (for example lithium a few hundred microns thick) by continuous exudation, the plasma pulse of a 10^7-joule explosion may also be repetitively endured by the combustion chamber.[22]

The impulse associated with an explosion determines the size and material strength of a chamber that must contain it. This impulse is proportional to the square root of the product of the explosion energy and the mass of the explosion debris. Relative to a chemical explosion of the same energy, a fusion pulse involves about six orders of magnitude less explosive debris mass and thus about three orders of magnitude less impulse, provided that the surface of the wall is not vaporized. Then a 10^7-joule fusion micro-explosion produces no more impulse than a large firecracker.

If the combustion chamber is too small, the wall will be ablated by the thermonuclear debris. Then the peak pressures imposed on the wall may be multiplied as much as a thousandfold and may be unacceptably high (greater than one kilobar). A crucial advantage of not vaporizing the lithium on the wall is that the chamber pumpdown time does not severely limit the pellet burn repetition rate.

About one joule/cm² of thermonuclear plasma energy may be directed against a chamber wall "moistened" with a several-hundred-micron layer of liquid lithium before significant blowoff is produced. The suprathermal ion fluence (for example 3.5-MeV alpha particles, knock-on deuterons and tritons) associated with a one joule/cm² thermal plasma fluence poses no blowoff hazard, since it penetrates the moist layer relatively deeply and deposits its energy in a large amount of matter. Moistened-wall combustion chambers, rated for ten-megajoule pulses of approximately 3-meter radius would thus be satisfactory from a plasma wall-loading standpoint.

If the combustion chamber wall is shielded from the pellet debris by a minimum-**B** magnetic field, the surface-area requirement for the explosion chamber is determined by x-ray loading considerations. The combustion chamber radius may then be reduced by approximately a factor of two.

The chamber wall might also be satisfactorily shielded from the plasma pulse, as well as from a portion of the x-ray pulse by pulsed injection of gas through the dry walls of the microexplosion chamber. However, the required mass injection rates are uncomfortably large, and the firing rate is limited by the chamber pumpdown time.

A very important problem for laser-fusion reactor design is how to arrange for input of the laser light and the target pellet while at the same time maintaining adequate neutron and x-ray shielding. Laser beams might be admitted through cheap, replaceable windows in the outer vacuum wall, passed through the neutron shield in neutronic-baffling dogleg tunnels on mirror trains, and focused onto the pel-

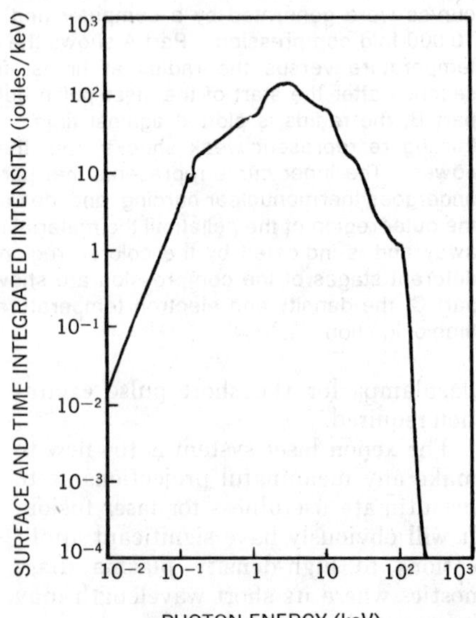

X-ray pulse spectrum of a megajoule DT-fusion microexplosion as calculated by a computer code; note the large self-absorption of the superdense fireball at low photon energies. Figure 5

let atmosphere by aspheric mirrors facing into the explosion chamber through apertures in the inner wall. Continuous, low-Z liquid-metal exudation-surfacing of the mirrors would prevent degradation of the reflectivity by the thermonuclear environment for laser wavelengths greater than 0.2 microns. The fuel pellet, several millimeters in diameter, would free-fall or be electrostatically projected into the combustion chamber.

For tritium breeding and recovery necessary for pure DT burning, a lithium-rich neutron blanket similar to those being considered for magnetic confinement fusion power-plant designs would surround the combustion-chamber wall. A 1% void fraction in the lithium blanket is probably required to permit impulsive neutron heating of the lithium without mechanical damage to the wall.

Fusion energy conversion

Fusion energy pulses, as extremely high-grade energy sources, apparently admit of several very different means of converting their energy into electricity, depending on pellet-fuel composition, the ρR value at which the fuel is burned (product of density and pellet radius, see box) and the combustion-chamber system design and operation. Three types of systems have been identified so far: They are ordinary thermal conversion, MHD hot-gas-generator conversion and MHD plasma conversion.

For first-generation laser-fusion power plants, which would burn DT pellets, ordinary steam-thermal conversion of fusion energy deposited by neutrons in the lithium blanket appears preferable. Such systems would have capital costs of several hundred dollars per kilowatt and energy-conversion efficiencies of up to 40%, which is comparable to conventional and fission-reactor systems.

If the combustion-chamber wall is shielded from the plasma pulse by injected gas, the heated gas might be exhausted from the chamber through a relatively inexpensive, pulsed MHD hot-gas generator. Several atmospheres stagnation pressure at a few thousand degrees temperature could be produced. Such a system might permit electricity generation with higher total efficiency (approximately 60%) for moderate ρR (approximately 10 gm/cm^2) DT or DD pellet burning, or for a 5 gm/cm^2, high-charged-particle-fraction (for example D-He^3) pellet burning. Such ρR's may be obtained with a few-hundred-kilojoule lasers if the pellet is compressed to 10^4 gm/cm^3 densities. However, at these high ρR's, 10–30% of the fusion energy is radiated as x rays. Hence the ultimate efficiency of this approach is limited by the efficiency with which the x-ray energy may be converted to electricity.

The rapidly expanding fusion fireball may be made to do magnetohydrodynamic work on a magnetic field imposed from outside the combustion chamber, transforming its energy into that of a compressed magnetic field. Induction coils suspended from the combustion-chamber walls might be used to transform the compressed-field energy directly into electricity, in a manner basically very similar to the way an ordinary power transformer works. The basic feasibility of such fireball-to-electricity energy conversion has already been demonstrated.[23] Low capital cost, high efficiency (greater than 70%) electrical energy generation may thus be ultimately attainable, in advanced laser-fusion CTR systems. Various estimates also indicate that laser-fusion power plants will be economically feasible.[24]

In the mid 1970's crucial superhigh density laser-implosion experiments will be carried out with lasers now being designed. Edward Teller has recently emphasized the importance of these experiments when he said, "A third of a century ago liquids were considered incompressible for all practical purposes. We are talking now about at least a thousand-fold compression if laser fusion is to be practical. This is a challenge we cannot afford to ignore. I believe that we shall succeed and that the effort will profoundly change our views on how man and matter can interact." Practical power production also depends on the success of programs now underway to develop pulsed lasers with sufficiently high power, energy, frequency and efficiency, and on the engineering of economic reactors. We are excited by the challenge of these difficult and complex tasks, and by the prospect that the mastery of fusion may be more important to Man than the harnessing of fire.

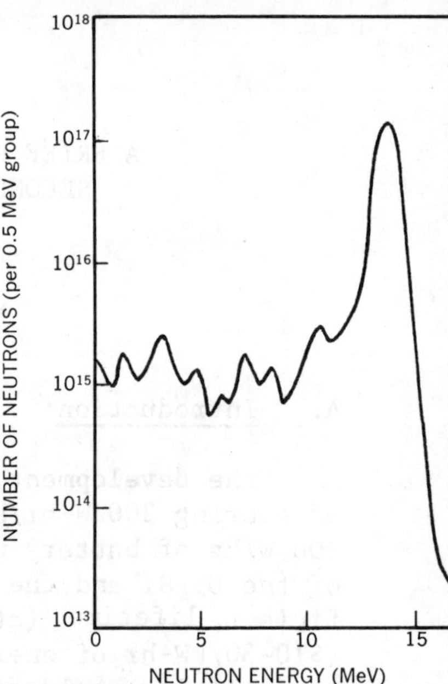

Neutron pulse spectrum of a megajoule DT-fusion microexplosion calculated by a computer code; note the neutron energy peaks produced by single and double 14-MeV neutron scatterings from deuterons and tritions and the 3.5 MeV peak. Figure 6

* * *

This work was supported by the USAEC.

References

1. R. Hirsch, New Scientist **12,** 86 (1973).
2. R. Post, PHYSICS TODAY, April 1973, page 30.
3. J. Nuckolls, L. Wood, A. Thiessen, G. Zimmerman, Nature **239,** 139 (1972).
4. W. Beasant, *Hydrostatics and Hydrodynamics,* Cambridge U.P. (1859).
5. Lord Raleigh, Phil. Mag., **34,** 94 (1917).
6. G. Guderley, Luftfahrtforschung **19,** 302 (1942).
7. F. Hawkins, "Manhattan District History, Los Alamos Project," LAMS 2532 (1961).
8. K. Boyer, Astronaut. Aeronaut. **11,** 28 (1973). W. Daiber, A. Hertzberg, C. E. Wittliff, Phys. Fluids **9,** 617 (1966). J. S. Clarke, H. N. Fischer, R. J. Mason, Phys. Rev. Lett. **30,** 89 (1973). K. A. Brueckner, Trans. IEEE **PS1,** 13 (1973).
9. S. W. Mead, Phys. Fluids **13,** 1510 (1970).
10. N. G. Basov and others, JETP Lett. **15,** 417 (1972).
11. J. Nuckolls and others, Livermore report UCRL-74116 (1972).
12. G. Zimmerman, Livermore Report UCRL 50021-72-1, 107 (1972).
13. S. Braginskii, Rev. Plasma Physics **1,** 205 (1965).
14. S. Bodner, Livermore Report UCRL 74074 (1972).
15. J. Nuckolls, Livermore Report UCRL 74345 (1972).
16. R. Kidder, J. Fink, Nucl. Fusion **12,** 325 (1972).
17. J. W. Shearer, J. J. Duderstadt, Livermore Report UCRL 73617 (1972).
18. P. Kaw, J. Dawson, Phys. Fluids **12,** 2586 (1969). W. Kruer, J. Dawson, Phys. Fluids **15,** 446 (1972).
19. J. Katz, J. Weinstock, W. Kruer, J. Degroot, R. Faehl, Livermore Report UCRL 74334 (1972).
20. L. Wood and others, Livermore Report UCRL 74115 (1972).
21. B. Freeman, L. Wood, J. Nuckolls, Livermore Report UCRL 74486 (1971).
22. L. A. Booth, LASL Report LA 4858MS (1972).
23. A. Haught, D. Polk, W. Fadr, Phys. Fluids **13,** 2482 (1970).
24. R. Hancock, I. J. Spalding, Culham Report CLM-P310 (1972). □

A BRIEF REVIEW OF THE STATUS OF HIGH-TEMPERATURE SECONDARY BATTERY RESEARCH AND DEVELOPMENT

M. L. KYLE, E. J. CAIRUS AND D. S. WEBSTER

A. Introduction

The development of electrically rechargeable batteries capable of storing 200 W-hr/kg of battery weight, and capable of delivering 200 W/kg of battery weight could have a great impact on the economy of the U. S. and the world, provided that the batteries have a sufficient lifetime (at least 3 years and 1000 cycles) and a low cost ($10-30/kW-hr of energy storage capability). Batteries with these capabilities could find many applications, including power sources for high-performance electric vehicles and off-peak energy storage devices for use in the electric utility system. A number of approaches to such a battery have been investigated; however, only those cells operating at elevated temperatures (300-600°C) have shown indications of being able to meet all of the criteria mentioned above. The field of high-temperature batteries is one that has grown rapidly, and has experienced rapid changes. Most of the activity has developed since the mid-1960's when sodium/sulfur and lithium/chalcogen cells were first publicly announced.[1-4]

In general, high-temperature secondary cells and batteries are still in the laboratory stages of development. A number of laboratories have reported on the operation of single sealed cells having active areas of 10 to 70 cm^2, and lifetimes of thousands of hours. Some batteries of limited lifetime (probably hundreds of hours when operated as batteries without special attention to individual cells) having power levels of up to 400 W are known to exist, but have not yet been publicly announced. Plans and activities are under way for batteries of up to about 30 kW for testing in 1973 and subsequently. None of these cells or batteries have yet been optimized for long life, light weight, and low cost. All of these are significant challenges. It will probably require at least 5 years for the development of a reliable (hundreds of cycles) light-weight (200 W-hr/kg, 200 W/kg) prototype battery, if appropriate effort is devoted to the task.

A brief review of the status of all of the known signficant efforts on the development of high-temperature secondary batteries is given in the following sections arranged by country. An overview of the field is given in Table I (following text portion of this appendix).

B. United States

The development of high-temperature batteries was initiated in the United States. The early work was performed by Standard Oil of Ohio,[5]

Reprinted with permission from Argonne Nat. Laboratories, Rep. ANL-7958, Appendix B, pp. 57-70, Mar. 1973.

Ford Motor Co.,[1] General Motors,[6] and Argonne National Laboratory.[2-4] Since the middle 1960's this field has shown rapid, world-wide growth. As shown in Table I, the work has been supported by private as well as government funding.

1. Argonne National Laboratory

The high-temperature lithium/chalcogen secondary cell program at Argonne National Laboratory began in 1966 as a continuation of earlier work (initiated in 1961) on thermally regenerative cells for use with radiosotope or nuclear reactor heat sources and was sponsored by the Division of Research and the Division of Reactor Development of the AEC. This program has continued until recently under AEC Division of Research sponsorship. Funding has been received from various agencies, as indicated in Table II. The references cited provide discussions of the objectives and accomplishments of those programs.

Table II
Sponsorship of Lithium/Chalcogen Cell Development at Argonne National Laboratory

Agency	Period	Reference
USAEC Division of Research	1961-1972	7,16,17
National Heart and Lung Institute	1969-1971	8
NAPCA-EPA	1969-1971	9,10
U. S. Army (Ft. Belvoir)	1969-1972	11-13
National Science Foundation	1972-	14

The current status of the NSF program is as follows. Long-lived (nearly 1 year), small ($\sim$2 cm^2), unsealed cells have been operated successfully. Scaled-up cells (20-50 cm^2) have demonstrated lifetimes of about 800 hr when unsealed, and about 500 hr when sealed. The first battery of sealed cells operated successfully for 120 hr. Lifetime improvement of sealed cells and batteries is a primary objective of the current work.

The present level of effort is 7 staff and 4 assistants. Plans for the future include the development of 1-kW batteries, followed by 5-kW and 20-kW demonstration batteries. The 20-kW battery should be suitable for the propulsion of a full-size automobile, and might be developed in about 7 years, with appropriate effort.

2. Atomics International

Atomics International initiated a program on the development of lithium/sulfur secondary cells in 1969,[15] closely following the work of Argonne.[16,17] The early work at AI was oriented toward off-peak energy storage and vehicle propulsion,[15] and was internally funded.[15,18]

Work was discontinued during 1971, and was recently resumed under joint sponsorship of AI and the Electric Research Council (at about the three-man level). An additional modest effort for the construction of three demonstration cells has been funded by the U. S. Army (Ft. Belvoir).

The status of the AI program has not been presented recently; however, it was reported[18] that cycle lives of 50 cycles and lifetimes near 750 hr had been achieved in sealed cells at capacity densities near 0.1 A-hr/cm^2.[18] Present efforts are believed to be directed toward the development of sealed laboratory cells of improved performance and lifetime.

The application of large Li/S batteries to off-peak energy storage was recently discussed by Herédy and Parkins;[19] projected performance and lifetime figures indicated that Li/S batteries could be attractive for off-peak energy storage, if the necessary performance and lifetime improvements could be achieved.

3. General Motors

The General Motors Research Laboratories and other GM laboratories have been involved in R & D on high-temperature cells since the early 1960's. The major portion of the work which has been reported has been devoted to the primary Li/Cl_2 cell, operating at 650°C with a molten LiCl electrolyte (m.p. = 609°C).[20,21] A number of other systems have been investigated at GM, including $Na/Na_2O \cdot 11\ Al_2O_3/S$, and Li/LiX-MX/S,[22] but no cell performance data have been reported.

The objective of the current programs at GM, which are internally funded, is high-performance secondary batteries suitable for electric vehicle propulsion. At least several people are involved in the program, but detailed effort information is proprietary, as is detailed information on cell performance and lifetime. It is significant to note that GM has a separate electrochemical research department at the research laboratories, which is consistent with the idea of a respectably large effort. Electrochemical power sources for demonstration of compact electric automobiles have been produced in that department (e.g., multikilowatt zinc/air batteries).

4. Ford Motor Company

The Ford Motor Company announced the development of the sodium/sulfur cell in 1967,[1] and has continued to work on improvements in cell performance and lifetime. The size of the effort has not been announced, but is believed to involve several people in at least two departments. Funding has been internal, and little information has been released since the first announcement.

The Ford Na/S cell operates at 300-350°C and makes use of a ceramic electrolyte called beta alumina ($Na_2O \cdot 11Al_2O_3$) and some modifications of this material which conducts sodium ions (5 Ω-cm at 300°C). The ceramic electrolyte acts as a separator (a 1-cm dia tube with an 0.8-mm thick wall), which simplifies cell design. The original design projections[1] indicated an expected specific power of 100 W/lb and a specific energy of 150 W-hr/lb. Some voltage-capacity curves are shown in Fig. 1.

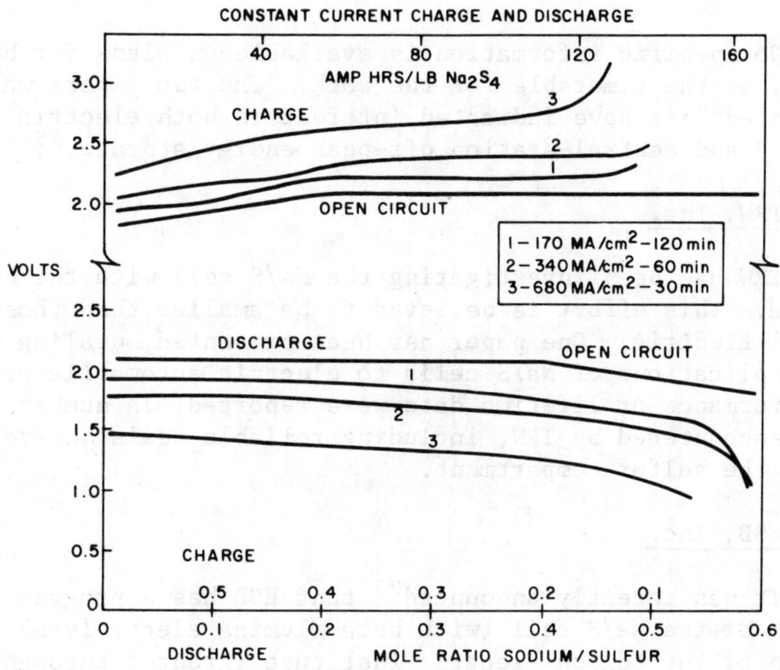

Fig. 1. Terminal Voltage of Small Laboratory Cell as a Function of State of Discharge

Recently, Tischer[23] has reported that a battery of 24 cells with a voltage of 12 V has been constructed, which is capable of delivering 300-400 watts. Exclusive of the weight of heaters and insulation, the battery delivered 210 W/kg and 92 W-hr/kg.[23] Lifetimes of single cells have been improved to 4000-8000 hr.

The current objectives of the program apparently include the development of a multikilowatt battery by 1973, and the continued development of sealed cells having a more rugged design than the present lab cells. Ultimately, this sodium/sulfur battery is intended for application in electric urban automobiles.

5. General Electric Company

The General Electric Research and Development Center has indicated the existence of an internally funded Na/S program of significant size, but has not reported any performance or lifetime data.[24] The problems which have been encountered are the same as those reported by other investigators working on the same cell. These cells exhibit a cycle life of one hundred to a few hundred cycles, limited by cracking of the electrolyte and the formation of sodium deposits within it during the recharge process. Improvements in the durability of the beta alumina electrolyte are being sought.

No specific information is available on plans for battery production, or the timetable for the work. The two papers which have been presented[24,25] have indicated interest in both electric vehicle propulsion[24] and central-station off-peak energy storage.[25]

6. TRW, Inc.

TRW has been investigating the Na/S cell with the beta-alumina electrolyte. This effort is believed to be smaller than those of Ford and General Electric. One paper has been presented, dealing with the possible applications of Na/S cells to electric automobile propulsion,[26] but no performance or lifetime data were reported. A number of problems have been encountered by TRW, including reliable seals and volume changes in the sulfur compartment.

7. ESB, Inc.

It was recently announced[27] that ESB has a program on the development of the Na/S cell (with beta-alumina electrolyte) under the sponsorship of the Edison Electric Institute (granted through the Electric Research Council). No details on the size of the effort or accomplishments are available; however, it is known that the objective is off-peak energy storage.

8. Dow Chemical Company

The Dow Chemical Company, in its Walnut Creek, California laboratory, has for several years been investigating a unique type of Na/S cell. It makes use of hollow sodium-ion-conducting glass fibers as the electrolyte. Several thousand hollow fibers are sealed into a header at one end such that they communicate with a sodium reservoir. Sodium fills the fibers, whereas the outer surfaces of the fibers are immersed in sulfur, and are in close proximity to an aluminum foil current collector which passes among the fibers. This design provides for a very large active area per unit volume of cell, and allows the cell to provide a significant power per unit weight while operating at a low current density (a few mA/cm^2). The cell operates at 300-350°C.

Only small (up to 4 A-hr at the 1-hr rate) cells have been operated to date, and life tests have been of limited duration (a few hundred hours) during cycling. Much longer lives have been demonstrated on open-circuit standby.

One paper has been presented on the Dow sodium/sulfur cell,[28] and two reports have been issued, describing work performed under joint sponsorship of EPA, DOT, and DOD.[29,30] The initial objective of the program is the development of 40 A-hr cells with a cycle life of 150 cycles, to be delivered in 1972.

The problem areas include the development of suitable means for sealing the glass fibers into the header, corrosion of the header, reliable seals and current collectors, and the establishment of long lifetimes.

9. Westinghouse

Westinghouse has been involved in high-temperature electrochemical energy conversion for at least ten years. Recently, a program has been initiated on Li/Cl_2 cells with molten-salt electrolytes. The chlorine is adsorbed on a porous carbon electrode, so that no handling of gaseous chlorine is necessary. It is believed that other types of high-temperature cells are also being evaluated. No information about the size of the program or technical achievements is available.

10. Standard Oil of Ohio (Sohio)

Sohio has had a relatively large effort over a period of at least ten years on the development of cells having a lithium-aluminum alloy anode, a molten LiCl-KCl eutectic electrolyte, and a porous carbon electrode,[31] usually with additives such as tungsten[32] or $TeCl_4$.[33] The porous carbon electrode acts as a chlorine electrode and is capable of storing significant amounts of absorbed chlorine. The cell typically operates at a temperature of 450°C.

The Sohio program is the farthest advanced toward light-weight engineering hardware of any known to the authors. A large number (hundreds) of sealed, metal-encased cells have been life tested for hundreds and thousands of hours,[32] and one 12-cell battery capable of 2-3 kW peak power has been operated.[33] The best specific energy reported for scaled-up cells (~100 cm^2 active area) is about 37 W-hr/lb. It appears that specific energies near 100 W-hr/lb are not likely to be achieved by this type of cell because the cathode reactants (Te, Cl) are too heavy.

Apparently, an effort on the construction of some multi-kilowatt batteries for delivery to the Army will be initiated soon. These batteries are intended to be applicable to hybrid electric Army vehicles.

C. England

England has the largest total effort on high-temperature batteries outside of the United States. This effort is divided among three major laboratories.

1. Admiralty Materials Laboratory

The Admiralty Materials Laboratory is concerned with the development of the Li/S cell (following Argonne's work) for use in classified British Naval programs. The only results reported to date are those on single, sealed primary cells, which exhibit the performance shown in Fig. 2.[34] These cells have demonstrated capacity densities of 0.2 A-hr/cm^2 at 0.5 A/cm^2, and have peak power densities of 1.5 W/cm^2 at 375-425°C.[34] The size of the effort is not known. It is believed that the power goal for these batteries is several kilowatts.

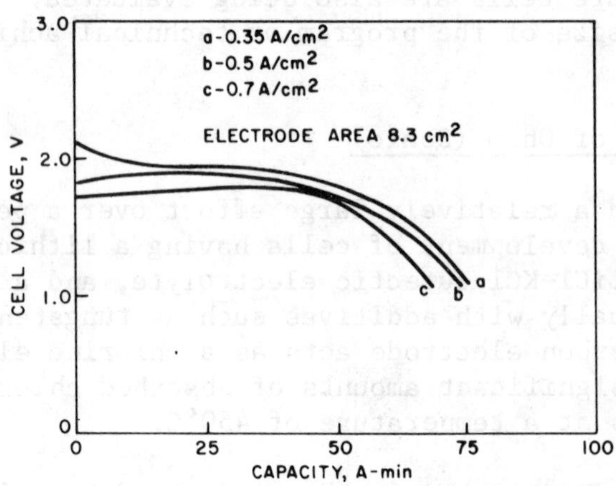

Fig. 2. Discharge Curves for Sealed Li/S Cell Using LiF-LiCl-LiI Electrolyte

2. British Railways Board

The British Railways Board is interested in the possible use of Na/S batteries (with beta-alumina electrolyte) for rail traction.

An effort estimated at 10-12 men has been devoted to the development of sealed cells and small batteries.[35] Many single cells of tubular configuration have been tested, and small multi-tube batteries of up to 1.4 kW-hr energy storage capability have been operated. Typical peak power density values are near 0.25 W/cm^2. Typical cycle lives are near 100 cycles, limited by sodium deposition in, and cracking of, the electrolyte.

Recently, flat-plate cells were constructed and tested in order to compare the results to those for tubular cells.[36] The flat-plate design is preferred for high-specific-power batteries, whereas the tubular design is more appropriate for high-specific-energy batteries. Up-to-date lifetime and cycle life data are not available.

The next step in this program will probably be the construction of a battery of appropriate size (a few kW) for a small demonstration vehicle, perhaps in 1973.

3. <u>Electricity Council Research Center</u>

The Electricity Council Research Center has been developing relatively large tubes of beta alumina, and has been constructing and testing Na/S cells and batteries using these tubes. A unique, rapid, continuous sintering process was developed[37] which permitted the production of high-quality tubes of 70 cm^2 area.

Recent results[38] have indicated that single cells can be operated for at least 4000 hr (at ~350°C), at 0.5 A/cm^2 on discharge and ~0.1 A/cm^2 on charge. Some 20-cell batteries, delivering 350 W, have been tested, but lifetime data are not available.

Work has begun on a 30-kW battery for use in a demonstration electric van. It will contain about 1000 cells of 70 cm^2 each. This effort comprises about 15 people.

This program is apparently the most advanced Na/S program in the world in terms of preparing large cells and batteries. It is not clear who leads in terms of cell lifetime and performance because very little data have been made public by any laboratory.

D. <u>France</u>

France has only one known effort on high-temperature batteries and that is at Marcoussis at the Compagnie Général d'Électricité.

<u>Compagnie Général d'Électricité</u>

This program is centered on the Na/S cell with beta-alumina electrolyte. The initial objective has been the preparation of sealed cells with long lives. A clever method of preparation of beta alumina tubes by electrophoretic deposition of finely divided beta alumina powder onto a metal mandrel has been developed.[39] These tubes are sealed to sodium reservoirs made of glass and are enclosed in glass outer casings which house the sulfur electrode. Typical lifetimes of 10-cm^2-active-area cells have been about 1500 hr of cycling at a temperature of 330°C and capacity density of 0.2 A-hr/cm^2.[40] The discharge has been at 0.2 A/cm^2 for 1 hr (0.3-0.35 W/cm^2); recharge has been at 0.067 A/cm^2 for 3 hr. The capacity density of 0.2 A-hr/cm^2 corresponds to 35% of the maximum value available (based upon Na_2S_3 as the final reaction product). Difficulty has been experienced in achieving complete recharge because of the high resistivity of sulfur. Cell lifetime has been limited by sodium deposition within the beta alumina during recharge, and by cracking of the beta alumina. Some impurities (such as potassium and silicon) have been found to have a deleterious effect on lifetime.[40]

The current objective of the CGE program is to double the cell lifetime, after which a battery of a few hundred watts will be constructed. Anticipated applications include vehicle propulsion and off-peak energy storage. The level of effort is estimated to be about a dozen people.

E. Japan

The Japanese government, under the aegis of the Agency of Industrial Science & Technology and the Ministry of International Trade & Industry is carrying out a National Research and Development program, part of which is the development of an urban electric automobile. The government funding for the electric automobile is $14 million, over a five-year period (1971-1975). This program provides funding to industry for the development of components and the construction of several electric autos. Five types of batteries are being developed, including the sodium/beta alumina/sulfur battery. The Yuasa Battery Company and the Toshiba Electric Company are jointly developing the Na/S battery for this program.

1. Yuasa Battery Company

The Yuasa Battery Company manufactures about 30% of the Japanese lead-acid batteries, and about 30% of their nickel-cadmium batteries, as well as a significant number of dry cells. It has the responsibility for the development of Na/S cells for the government-sponsored electric vehicle program. The beta-alumina electrolyte tubes are supplied by Toshiba.

No recent data are available on cell performance and lifetime; however, a paper was presented in December, 1970[41] which reported that cycle lives (at 300-350°C) of 100 cycles (one cycle per day) at 0.1 A/cm^2 had been achieved with sealed cells having metal casings. The capacity density (for a 30-cm^2 active area) was about 0.2 A-hr/cm^2

at the 2- to 3-hr rate. The usual causes of failure included cracking of the beta alumina.

It is anticipated that a several-kilowatt battery of small cells will be assembled at the earliest opportunity in order to demonstrate a small electric vehicle. This might be attempted within the next year or two. The size of the effort is not known, but probably amounts to several people.

2. Toshiba Electric Company

The Toshiba Electric Company research laboratory has developed expertise in the production of beta alumina tubes, and is being supported by the Japanese government in its activities relating to the electric vehicle program. Beta alumina tubes (∿1 cm diameter and 15 cm long, with 0.1-cm wall thickness) are being produced for use by Yuasa (see above) in constructing Na/S cells. No details are available concerning the method of preparation. It is believed that this activity involves several people.

3. Japan Storage Battery Company

Japan Storage Battery Company is the largest producer of lead-acid batteries in Japan, with about 40% of the market. It also produces about 30% of the nickel-cadmium batteries in Japan. It has been learned that Japan Storage also has a significant company-sponsored effort on the sodium/beta alumina/sulfur cell, and claims to produce superior quality beta alumina. No details are available; no information has been published.

F. USSR

Institute for Power Sources, Moscow

Shortly after Argonne announced its work on lithium/chalcogen cells, N. Lidorenko[42] announced that the Institute for Power Sources was also investigating these cells, and had obtained essentially the same results that Argonne had reported. No other comments were made.

G. Czechoslovakia

Electrical Engineering Laboratory, Prague

Mr. M. J. Lakomy, head of the electrochemical group at the Electrical Engineering Lab., has reported that an effort on lithium/chalcogen cells was to be initiated in 1971. No subsequent information is available to indicate the progress of the work.

H. Germany

As of early 1971, there was no indication of the existence of any high-temperature battery programs in Germany, which is surprising because that country has usually been a leader in electrochemical science and technology.

Battelle Institute, Frankfurt

It is possible that a small effort on high-temperature secondary cells has been initiated recently at Battelle Institute. Representatives of Battelle have recently been following the efforts in this field closely, and have been visiting U. S. laboratories engaged in this work.

I. Summary and Conclusions

This world-wide survey has indicated the existence of at least twenty high-temperature battery efforts in the world, involving about 150 investigators, approximately seventy of whom are in the United States. Only a few systems are being investigated, the main ones being sodium/beta alumina/sulfur, and lithium/molten salt/sulfur. The two main areas of potential application are electric vehicle propulsion and off-peak energy storage. Most of the financial support in the U. S. seems to be private industrial funding.

In nearly all cases, the status of the programs is laboratory cells of 10-to 70-cm^2 active area, with lifetimes of 100 to 1000 cycles, with a few exceptions (such as Ford, Argonne, and Electricity Council Research Center). It is likely that a few multikilowatt demonstration batteries will exist within the next couple of years.

REFERENCES

1. J. T. Kummer and N. Weber, SAE Automotive Engineering Congress, Detroit, Jan. 1967, Preprint No. 670179.
2. H. Shimotake, G. L. Rogers, and E. J. Cairns, presented at the Electrochemical Society Meeting, Chicago, Oct., 1967, Abstract No. 10; see also Extended Abstracts J1 of the Battery Division 12, 42 (1967).
3. H. Shimotake and E. J. Cairns, in Advances in Energy Conversion Engineering, p.951, Proc. 1967 IECEC, ASME, New York (1967).
4. E. J. Cairns and H. Shimotake, presented at the Biennial Fuel Cell Symposium, Amer. Chem. Soc. Meeting, Chicago, Sept., 1967; see also Preprints of Papers, ACS Div. of Fuel Chem. 11 (3), 321 (1967).
5. R. A. Rightmire and A. L. Jones, in Proc. 21st Annual Power Sources Conf., p.42, PSC Publications Committee, Red Bank, N. J. (1967).

6. R. D. Weaver, in Proc. 19th Annual Power Sources Conf., PSC Publications Committee, Red Bank, N. J. (1965).
7. E. J. Cairns, C. E. Crouthamel, A. K. Fischer, M. S. Foster, J. C. Hesson, C. E. Johnson, H. Shimotake, and A. D. Tevebaugh, USAEC report ANL-7316, Argonne National Laboratory (1968).
8. H. Shimotake, A. A. Chilenskas, R. K. Steunenberg, and E. J. Cairns, in Proc. Energy 70 IECEC, Vol. 1, pp.3-61, American Nuclear Society, Hinsdale, Ill. (1972).
9. H. Shimotake, M. L. Kyle, V. A. Maroni, and E. J. Cairns, in Proc. First Internat. Elec. Vehicle Symp., p.392, Elec. Vehicle Council, New York (1969).
10. M. L. Kyle, H. Shimotake, R. K. Steunenberg, F. J. Martino, R. Rubischko, and E. J. Cairns, in 1971 Intersociety Energy Conversion Conf. Proceedings, p.80, Soc. Automotive Engineers, New York (1971).
11. E. C. Gay, L. E. Trevorrow, W. J. Walsh, E. J. Cairns, J. D. Arntzen, and J. G. Riha, presented at the AIChE Meeting, San Francisco, Nov. 28-Dec. 2, 1971, Abstract No. 11C.
12. W. J. Walsh, A. A. Chilenskas, L. E. Trevorrow, E. C. Gay, and E. J. Cairns, presented at the Electrochemical Society Meeting, Houston, May 1972, Abstract No. 146; see also Extended Abstracts 72-1, 375 (1972).
13. E. C. Gay, W. J. Walsh, J. D. Arntzen, and E. J. Cairns, presented at the IECEC, San Diego, Sept. 1972.
14. E. J. Cairns, H. Shimotake, E. C. Gay, and J. R. Selman, presented at the Internat. Society of Electrochemistry Meeting, Stockholm, Aug. 27-Sept. 2, 1972; see also Extended Abstracts, p.432 (1972).
15. L. A. Herédy, N. P. Yao, and R. C. Saunders, in Proc. First Internat. Electric Vehicle Symp., p.375, Electric Research Council, New York (1969).
16. H. Shimotake and E. J. Cairns, in Proc. of Advances in Battery Technology Symp., p.165, Southern Calif.-Nevada Section of the Electrochem. Soc., Los Angeles (Dec. 1968).
17. H. Shimotake and E. J. Cairns, presented at the Electrochem. Soc. Meeting, New York, May 1969, Abstract No. 206; see also Extended Abstracts of the Battery Div. 5, 520 (1969).
18. N. P. Yao, L. A. Heredy, and R. C. Saunders, presented at the Electrochem. Soc. Meeting, Atlantic City, Oct., 1970, Abstract No. 60; see also Extended Abstracts, p.148 (1970).
19. L. A. Herédy and W. E. Parkins, presented at the IEEE Meeting, New York, Jan. 30-Feb. 4, 1972, Paper No. C-72-234-8.
20. D. A. Swinkels, J. Electrochem. Soc. 113, 6 (1966).
21. T. G. Bradley, in IECEC 1968 Record, IEEE, New York (1968).
22. S. E. Beacom, presented at the Argonne Center for Educational Affairs Conference on Materials Limitations in Energy Conversion, Argonne National Laboratory, May 1972.
23. R. P. Tischer, presented at the Electrochemical Society Meeting, Houston, May 1972, Abstract No. 170; see also Extended Abstracts 72-1, 434 (1972).

24. H. A. Christopher, in Proc. 2nd Internat. Elec. Vehicle Symp., Atlantic City, Nov. 1971, Elec. Vehicle Council, New York (1972).
25. F. G. Will, presented at the Electrochemical Society Meeting, Houston, May 1972, Abstract No. 171; see also Extended Abstracts 72-1, 437 (1972).
26. G. H. Gelb, N. A. Richardson, E. T. Seo, H. P. Silverman, and A. Toy, presented at the Columbus Section of The Electrochemical Society, Symp. on Batteries for Traction and Propulsion, March 1972.
27. D. Bird, The New York Times, p. 30M (May 30, 1972).
28. C. Levine, presented at the 25th Power Sources Symp., Atlantic City, May 1972.
29. C. A. Levine, W. A. Taplin, F. Y. Tsang, and W. E. Brown, 1st Quarterly Progress Report on Development of the Hollow Fiber Sodium-Sulfur Battery, Dow Chemical Co., to Naval Underwater Systems Center, U. S. Army ECOM, Dept. of Transportation and Environmental Protection Agency, Contract No. N00298-72-C-0028 (Jan. 1972).
30. C. A. Levine, W. H. Taplin, F. Y. Tsang, and W. E. Brown, 2nd Quarterly Progress Report on Development of the Hollow Fiber Sodium-Sulfur Battery, Dow Chemical Co., to Naval Underwater Systems Center, U. S. Army ECOM, Dept. of Transportation and Environmental Protection Agency, Contract No. 298-72-C-0028 (April 1972).
31. R. A. Rightmire and A. L. Jones, in Proc. 21st Annual Power Sources Conf., p. 42, PSC Publications Committee, Red Bank, N. J. (1967).
32. R. A. Rightmire, J. W. Sprague, W. N. Sorensen, T. H. Haeha, and J. E. Metcalfe, presented at SAE International Automotive Engineering Congress, Detroit, Jan. 1969, Preprint No. 690206.
33. J. E. Metcalfe, E. J. Chaney, and R. A. Rightmire, in Proc. 1971 IECEC, Boston, Aug. 1971, p. 685, SAE, New York (1971).
34. B. A. Askew and R. Holland, presented at 8th International Power Sources Symposium, Brighton, Sept. 1972.
35. J. L. Sudworth and M. D. Hames, in Power Sources, Vol. 3, p. 227, D. H. Collins, Ed., Oriel Press, Newcastle-Upon-Tyne, England (1971).
36. J. L. Sudworth, M. D. Hames, M. A. Storey, and M. F. Azim, presented at 8th International Power Sources Symposium, Brighton, Sept. 1972.
37. L. J. Miles and I. Wynn Jones, in Power Sources, Vol. 3, p. 245, D. H. Collins, Ed., Oriel Press, Newcastle-Upon-Tyne, England (1971).
38. I. Wynn Jones to E. J. Cairns, May 1972, private communication.
39. J. Fally, C. Lasne, Y. Lazennec, Y. LeCars, and P. Margotin, presented at the Electrochemical Society Meeting, Houston, May 1972, Abstract No. 173; see also Extended Abstracts 72-1, 441 (1972).
40. J. Fally, C. Lasne, Y. Lazennec, and P. Margotin, presented at The Electrochemical Society Meeting, Houston, May 1972, Abstract No. 172; see also Extended Abstracts 72-1, 439 (1972).
41. M. Yamaura, S. Kimura, S. Sugaike, and A. Imai, Paper No. 45-17, Battery Technology Committee, Electrochemical Society, Japan, Dec. 14, 1970, translated by H. Shimotake.
42. N. Lidorenko, presented at the IECEC, Washington, D.C., Sept. 1969.

Fuel Cells for the United States Energy System

STATEMENT PREPARED BY CARLE C. MORRILL, MANAGER, FUEL CELL MARKETING, PRATT AND WHITNEY AIRCRAFT DIVISION OF UNITED AIRCRAFT CORPORATION, EAST HARTFORD, CONN.

In order that a meaningful contribution can be made to the process of assessing a national energy research and development policy, a statement of perspective is required. The country is involved in a large scale effort to investigate and establish an overall energy policy; a policy with respect to fuels, a policy with respect to use of the various forms of energy, and a policy with respect to concern for natural resources, ecology, and the environment, both physical and aesthetic. In addressing ourselves to the technological aspects of the problem, there are very few near term answers or solutions and there is confusion and conflict as to the real effectiveness of the contributions of all of the alternatives.

The essential point that should be brought to the attention of the Committee is that of all of the alternatives that are being reviewed and considered, there is a technology requiring research and development effort which can provide timely solutions to today's problems and provide an effective and useful option in all of the years to come that we can foresee. Research and development effort in the fuel cell at this time is an investment that is defensible in consideration of almost any format and configuration the nation's energy policy may take. It is one of the few options that can be of use in the near term and contributes to the critical issues with which we are concerned for the future.

The fuel cell, a clean, quiet electric generator which operates with superior efficiency in both small and large sizes, is an electric power generation concept which could provide significant improvement in each of the critical problem areas associated with our national energy supply.

Operational capability of the fuel cell is being demonstrated currently in a number of small, on-site installations throughout the nation. Its application within the national electric supply system could commence in the 1976-1978 period if the required technology effort is applied in the next three to five years. Support from government as well as increased support from industry is required to assure early application of this concept.

Initial application of fuel cells to the utility systems at those locations having the most severe problems would provide the greatest early relief. Subsequent application to other problem areas would follow. Ultimately the fuel cell, used in modest quantities, compared to the overall national generation capacity, could alter the character of the national system so as to improve the quality of the overall supply.

The fuel cell is a concept that could be brought into use about a decade earlier and require far less technology investment than other well known concepts such as the breeder reactor and MHD (Magnetohydrodynamics). It would provide improvements in many more critical electric utility problem areas than other concepts. The fuel cell has thus far escaped public consideration as a promising new generator, especially in regard to its possible early application. It is understandable that the possibility for early application of the fuel cell should be unrecognized since the entire research and development effort on fuel cells for utility application has been supported by private investment and the results of this effort have not been a matter of public record.

Government support for fuel cells, meanwhile, has been directed either toward development of space hardware (Pratt & Whitney Aircraft developed and produced the Apollo fuel cell for NASA) or towards research at the basic materials and phenomena levels in a large number of small efforts on non-commercial type fuel cells.

Application of fuel cells to the national electric supply system within the 1970's, requires strong support from both government and private sectors. Strongly increased support from the electric and gas utilities is essential to evolve hardware compatible with utility requirements. Public advocacy of private support of fuel cells as well as direct government support for technology advancement is vitally important at the earliest point in time. Government support of a varied nature should be provided also to assure the earliest introduction of fuel cells in substantial numbers. Maximum benefits to society and the economy could thus result from early improvements in the national electric power supply. This support could include government procurement of early units for civilian and military installations and financial assistance to the private sector in some variety of forms. Improved legal, regulatory, tax and environmental inducements are needed to increase utility support for fuel cell technology development as well as for stimulating early fuel cell application in significant numbers. The increased usage of fuel cells could be hastened also by government support for development of clean fuel supplies, including pipeline gas from coal and liquid petroleum.

To achieve the foregoing objectives, government support should be of types which would protect the proprietary rights of private organizations so as to be effective in producing the desired stimulation for the heavy private investment necessary.

CURRENT STATUS OF TECHNOLOGY

Fuel cell research and development programs have been conducted at the Pratt & Whitney Aircraft Division of United Aircraft Corporation since the late 1950's. These programs have been funded by United Aircraft Corporation, NASA, the Department of Defense, a group of natural gas utilities, and a group of electric utilities. The non-government effort has been directed entirely toward the development of hydrocarbon-air fuel cell powerplants for commercial and utility use.

Small scale fuel cell powerplants have been built and have demonstrated in actual field test operation the ability to satisfy completely all of the functional requirements for electric utility service. These 12.5KW powerplants, one of which is shown in Figure 1, operate on air as the oxidant and either gaseous or distillate liquid fuels. Several powerplants are now in operation and shortly a total of sixty powerplants will be operational in some 35 installations including residential, commercial, and industrial buildings.

Data from the initial installations has verified that fuel cell powerplants have minimal air pollution (as shown in Figure 2), can reject their waste heat to air, are quiet, and are highly efficient at both rated power and at part load even in sizes as small as a few kilowatts. These units operate unattended and have a minimum of scheduled maintenance. Multiple fuel cell powerplant operation has demonstrated powerplant paralleling in both AC and DC operation and paralleling of the powerplant with an electric utility distribution network. Both voltage regulation and AC frequency control are equal to or superior to electric utility standards. The response rate to load change from zero to full power has been demonstrated to be less than one cycle. The noise level is no greater than a home refrigerator since the basic energy conversion process is static, and the only moving parts are circulators for process and cooling air.

There are essentially two present obstacles to the widespread introduction of fuel cell powerplants into commercial-utility service: the specific installed cost in dollars per KW, and the length of the economic operating period of the system. The present estimated specific cost of the fuel cell powerplant on a production basis is about $350/KW, and the economic operating period is estimated to be approximately 16,000 hours. A specific cost of about $200/kw and a 40,000 hour operating period would establish the fuel cell powerplant as an economic electric generator for a variety of duty cycles and applications. The limiting factors to achieve lower specific cost are: the power output per square foot of cell area and the amount of cost of materials of construction. The limiting factors in extending the economic operating period are associated with the development of stable and lower cost materials to withstand the chemical and electrochemical environment within the powerplant. Substantial progress has been made in the past five years; specific material cost has been reduced by a factor of nearly 10, and economic operating life increased by a factor of 5 since 1967. Progress in both areas has been limited by the rate of funding available.

Since 1967, more than $50 million has been expended by Pratt & Whitney Aircraft in specific programs to develop fuel cell powerplants for utility service. More than half of the funding has been provided by a group of natural gas and combination utility companies in a program called TARGET. The Institute of Gas Technology has devoted its entire fuel cell activity also to the TARGET program as a subcontractor to Pratt & Whitney Aircraft. In addition, there has been a concurrent program in the last year with ten electric utility companies and the Edison Electric Institute. About one-third of the total investment has gone into the technology associated with cost and operating period. The remainder has been spent in developing experimental fuel cell powerplants, in producing and field testing prototype powerplants, and in building a technology base to make the powerplant concept more versatile (e.g., handle all liquid and gaseous fossil fuels, scaling of cell and reformer component elements to multi-megawatt size, and improved efficiency of ancillary components). These latter efforts are difficult and costly in an engineering sense, but do not require advances in the present state-of-the-art to ensure success.

Initial fuel cell powerplants for electric utility applications will use acid electrolyte systems for the same reason that the first reactor systems used pressurized water; they are the least costly to develop and most easily reduced to practice in a short time period. Once these early powerplants have proven their value as electric generators in practical use, improved cell system concepts will undoubtedly be developed because the assurance of their ultimate acceptance will warrant a greater investment of funds and time—just as the success of the pressurized water reactors has provided much of the confidence for spending the time and funds to develop the breeder reactor. In addition to the progress in the acid cell, sufficient technological progress has been made by Pratt & Whitney Aircraft in the research of two other systems, molten carbonate and high output acid to provide a high degree of confidence in the ultimate success of these concepts.

In addition to the development of commercial powerplant technology, Pratt & Whitney Aircraft carries out the development of special purpose fuel cell powerplants for NASA and the Department of Defense. Because most of these powerplants use pure hydrogen and oxygen as reactants, and because weight rather than production cost is the critical consideration, the technology developed for these units is not applicable to commercial or utility requirements. These efforts, however, have served to demonstrate the inherent ability of fuel cell powerplants to operate unattended and without scheduled maintenance for long periods of time, to operate in hostile environments, and to be highly efficient producers of electric power. Contracts include the development and delivery of fuel cell powerplants for NASA manned spacecraft, the development and demonstration of a 20 KW powerplant for Navy deep submergence application, and technology programs for a variety of Air Force and Army requirements.

POTENTIAL OF THE FUEL CELL CONCEPT

The fuel cell powerplant can contribute to the United States energy system in three formats. One is as a dispersed generator for electric utilities. Figures 3 and 4, where multi-megawatt powerplants are placed within an electric utility distribution system at substation locations. A second is as an on-site transformer of gas to electricity for natural gas utilities customers. The third is as a remote on-site generator for rural electrification and for developing areas such as Alaska. In the first two uses the fuel cell powerplant complements the very large scale fossil or nuclear bulk power generators by permitting a selective attack on critical problem areas. Among its benefits are:

Reprinted with permission from *Energy Research and Dev.*, Hearings before the Subcommittee on Sci., Research, and Dev., House Committee on Sci. and Astronautics, Y4Sci2: 92/2/24, pp. 585–593, May 1972.

FIGURE 1

MINIMUM POLLUTION CONTRIBUTION

POUNDS OF POLLUTANTS PER THOUSAND KW-HR

	← FEDERAL STANDARDS →			
	GAS-FIRED UTILITY CENTRAL STATION	OIL-FIRED UTILITY CENTRAL STATION	COAL-FIRED UTILITY CENTRAL STATION	EXPERIMENTAL FUEL CELLS
SO_2	NO REQUIREMENT	7.36	10.90	0-0.00026
NO_x	1.96	2.76	6.36	0.139-0.236
HYDROCARBONS	NO REQUIREMENT	NO REQUIREMENT	NO REQUIREMENT	0.225-0.031
PARTICULATES	0 98	0.92	0.91	0.00003-0

FEDERAL STANDARD (EFFECTIVE 8-17-71) VALUES CONVERTED TO LB/1000 KW-HR

Pratt & Whitney Aircraft — DIVISION OF UNITED AIRCRAFT CORPORATION

FIGURE 2

J3733-3
R720603

Reduced Transmission Requirements.—By locating powerplants near, or at, the load center the need for additional transmission capacity and right-of-way is reduced. Raw fuel energy is transported to the fuel cell powerplant by gas or liquid pipeline or trucked liquids. Community planning is more flexible because power can be provided at the point of need with minimal lead time and with minimum interference with other communities.

***Improved Site Availability.*—Dispersed fuel cell powerplants are quiet and clean. They do not require a source of cooling or makeup water and can be transported to the installation site using standard trucks. These features increase the number of suitable sites and permit present utility right-of-way, air rights over existing buildings or sites of retired plants to be utilized.**

***Capacity Added in Phase With Growth.*—Fuel cewl powerplants are economic in small scale units; as a consequence they can be installed in blocks as load requirements grow to provide a more effective use of capital and a more responsive generating capability.**

***Factory Assembly.*—Assembly in standard modular units provides mass production economies, reduces lead times and minimizes expensive field construction.**

***System Security.*—Parallel operation of dispersed generation units with the conventional utility system provides enhanced reliability and reduced vulnerability of the generation and transmission system to sabotage, attack or natural disasters. In addition, the multifuel capability of the fuel cell powerplant eliminates reliance on one specific fuel system.**

Initial electric utility fuel cell powerplant installations will provide power in areas with particularly difficult transmission or siting problems. They will be sited in areas experiencing rapid load growth due to new commercial or industrial facilities, in urban areas where renewal projects have increased power demands, and on-site for customers with critical loads such as hospitals, industrial processing plants, and government agencies.

The validity of the fuel cell on-site and dispersed generation concepts is evidence by the fact that it is the gas and electric utilities themselves who have directed Pratt & Whitney to focus their joint program effort toward these ends. Recognition of the importance of immediate development and early application of fuel cells for dispersed generation is contained in the 1971 recommendations of the R&D Goals Task Force to the Electric Research Council.

Fuel cells will make more efficient use of the nation's energy resources. Even

DISPERSED GENERATION
POTENTIAL SITES

J2238-2
711503

Pratt & Whitney Aircraft DIVISION OF UNITED AIRCRAFT CORPORATION UA

FIGURE 3

DESIGN CONCEPT
TYPICAL DESIGN CONFIGURATION
20 MW

J2238-81
R712704

Pratt & Whitney Aircraft DIVISION OF UNITED AIRCRAFT CORPORATION UA

FIGURE 4

with present day technology, characteristic fuel cell powerplant peak efficiency when operating on hydrocarbon fuels is in the range of 40 percent even at power levels as low as 50 to 100 KW, AC. (Figure 5). On hydrogen and air, the efficiency would be about 55 percent, and on hydrogen-oxygen, nearly 60 percent. Efficiency remains high over an extremely broad operating range from approximately 25 percent to 125 percent of the nominal power rating of the powerplant.

When the fuel cell powerplant is located near the point of electrical demand, the combination of the high efficiency of fuel cell powerplants and the elimination of transmission and distribution losses make it possible to use one-quarter to one-third less fuel than conventional fossil fuel systems to produce a given amount of electricity.

The fuel cell powerplant is highly complementary to the concept of distributing synthetic fuel by pipe, whether from fossil fuel sources in the short term, or from non-fossil sources in the more distant future.

Current energy practice involves the delivery of raw fuels to end users for conversion to heat or to electric utilities for conversion to electricity. In both cases the fuel user is responsible for removal and disposal of pollutants. By placing the responsibility for the removal of pollutants contained in the fuel upon the individual user, this practice has created severe economic, waste disposal, and policing problems.

If instead the raw fuel stock is processed to remove pollutants prior to delivery to the user, the economic advantages of large-scale processing are realized and more practical methods for recycling the chemicals removed from the fuel are possible. Central fuel processing is an easier system to police since the responsibility for pollutant removal rests with fewer entities.

Central fuel processing lends itself to the use of underground pipelines for transporting and distributing the clean liquid or gaseous fuel to the user. Pipelines are the most economical, aesthetic, and reliable method of transporting energy over long distances. Using the fuel cell within this type of energy supply system (shown in Figure 6) permits the efficient conversion of fuel to electricity and takes full advantage of the lower costs and improved land use associated with transporting energy by pipeline.

The fuel cell dispersed powerplant concept is greatly enhanced when "synthetic" **rather than unprocessed fuel is distributed to the fuel cell powerplants because the fuel processing part of the powerplant is simplified. Either the**

FUEL ECONOMY/EFFICIENCY

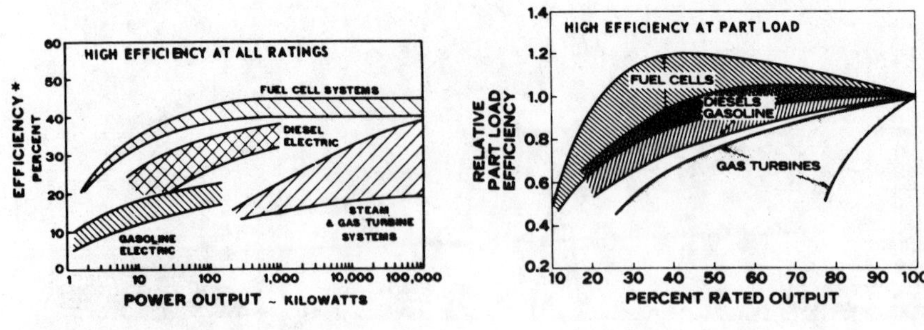

*BASED ON LOWER HEATING VALUE

Pratt & Whitney Aircraft DIVISION OF UNITED AIRCRAFT CORPORATION

FIGURE 5

J3055-14
R722401

FUEL FLEXIBILITY

COAL GASIFICATION

- COAL GASIFICATION UNITS UNDER DEVELOPMENT CAN BE UTILIZED WITH DISPERSED FUEL CELL UNITS

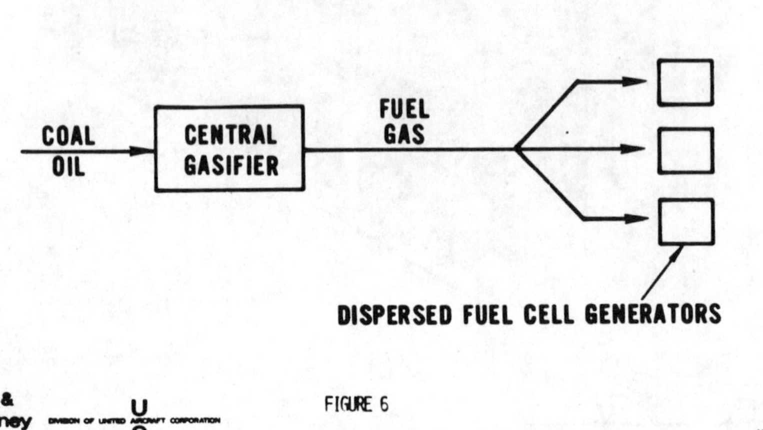

Pratt & Whitney Aircraft DIVISION OF UNITED AIRCRAFT CORPORATION

FIGURE 6

J3750-17
R722505

capital cost of the powerplant is reduced, or the efficiency is increased, or both. In the case where hydrogen itself is the piped fuel, the powerplant efficiency approximates 55 percent and the capital cost is reduced by nearly $25/KW by elimination of an on-site processor. This feature enables the fuel cell to act as a complementary option to possible future nuclear system concepts.

In one approach to future energy needs, the breeder reactor and a more extensive electrical transmission and distribution system would become the source of providing bulk power. The breeder reactor, however, must operate continuously at full power as a baseload generator to achieve maximum efficiency. Therefore, either energy storage systems or a mix of intermediate and peaking non-nuclear plants would be needed to supplement base-load breeder units. Today, pumped hydro-storage systems are used to store energy during periods of low electricity demand and this stored energy is then reused during high demand periods. Pumped storage systems, however, require the use of large land areas to store the water and their siting is inflexible. These problems would be magnified if pumped storage were used in conjunction with large numbers of breeder reactor plants.

As an alternate approach, fuel cell powerplants used as energy storage devices could offer a solution to the problem of storing large quantities of energy. In this concept, off-peak nuclear generation provides electric energy to electrolysis units to convert water into its elemental components of hydrogen and oxygen. Thus the energy from the breeder is stored in the form of hydrogen fuel until periods of high electrical demand. When needed, the hydrogen fuel may be converted back to electric energy by means of a fuel cell powerplant. Studies of this concept show that the energy storage concept employing the fuel cell requires only one percent of the land area needed by conventional pumped hydro-storage systems.

Another future possibility is that the very large breeder and fusion reactors will be located at sites remote from large population centers and that most electrical transmission will be underground. In such cases both economic and environmental considerations would favor the transportation of most of the energy by pipe. As with the above concept, water could be converted to hydrogen and oxygen by high efficiency electrolysis units. Then hydrogen, or a synthetic fuel made with hydrogen as a base, would be transported by pipeline and used as a clean bulk fuel supply for either heating uses or dispersed power generators. The oxygen would be used for industrial processes or for organic waste disposal. In this format the fuel cell is a highly desirable, if not essential, power generator because it can reconvert the fuel energy back to electricity at the highest possible efficiency and in the closest possible proximity to the consumer. In effect, the fuel pipeline system and the fuel cell powerplant form, respectively, the primary and secondary energy distribution system for the breeder or fusion reactor generator.

RESEARCH AND DEVELOPMENT REQUIREMENTS

As a result of sponsorship by United Aircraft Corporation, the member companies of TARGET and the several electric utility companies, the present hydrocarbon-air fuel cell powerplant development effort at Pratt & Whitney Aircraft has a high level of momentum. Subscale demonstrator powerplants are in operation, a 12.5 KW fuel cell powerplant is being developed for specialized applications in the non-utility markets, component building blocks for multi-megawatt units are being fabricated, and there is a sustained effort in developing the technology required to reduce cost, increase performance, and increase the economic operating period. In addition, there is a continuing broadly based research and development effort in hydrogen-oxygen fuel cell systems for NASA and the Department of Devense.

Continued Pratt & Whitney Aircraft investment and participation in the development of fuel cell powerplants is warranted only on the basis of a clear indication that substantial numbers will actually be procured for application in this decade. It is clear that this potential will be realized only if the natural gas utilities, the electric utilities, and the government are willing to:

(1) Support research and development to improve technology.

(2) Share in the costs to design, fabricate, and develop the first production powerplant configuration.

(3) Commit to the procurement and operation of a sufficient number of the first powerplants to establish a production base.

With such support, fuel cell powerplants can enter the commercial utility market for both gas utility on-site and electric utility dispersed generators by 1978. Should the Pratt & Whitney Aircraft effort be insufficiently supported, the introduction of fuel cells into commercial service will most certain.y be delayed at least until 1985 or beyond. Therefore, 1972–1974 is the most opportune and effective time for stong government and utility industry participation with Pratt & Whitney Aircraft.

An optimum program to bring the fuel cell into commercial use by 1978, with the highest probability of continued improvement in cost, operational characteristics, and generating efficiency would consist of the following elements:

(1) Continued funding of fuel cell powerplant technology. This would include evolutionary improvement of present technology, technology evaluation of potentially superior concepts and the extension of operating capability to the complete range of fossil fuels.

(2) Support of a demonstrator powerplant program

(3) Purchase and operation of early production powerplants

The technology program requires a total expenditure of between \$40–\$60 million in the next three years. The demonstrator program requires about the same amount o funds for the design and development of the first on-site and dispersed generator powerplants. A significant number of units must be qurchased to establish a production base to realize viable economics.

Liquid Hydrogen as a Fuel for the Future

Replacement of hydrocarbon fuel for transportation systems by liquid hydrogen is proposed and discussed.

Lawrence W. Jones

I first considered the possibility of using liquid hydrogen as an ultimate replacement for fossil hydrocarbon fuels in vehicular and aircraft transport in casual conversation related to the logistics and use of large quantities of liquid hydrogen in a cosmic-ray experiment. In remarking on the drop in price of liquid hydrogen in recent years, I noted that the cost per liter was about the same as that of gasoline. As other work on this subject came to my attention, I recognized that, although this idea was not original, it had an inherent self-consistency and appeal which warranted broader exposure and discussion. The conclusion I have reached is that the use of liquid hydrogen as a fuel not only is feasible technically and economically, but also is desirable and may even be inevitable.

The amount of fossil fuel (coal, oil, natural gas) is finite, and any extrapolation in our present rate of consumption will lead to the exhaustion of readily available reserves in about 100 years (or somewhere between 30 and 300 years). Singer (*1*) has estimated that we have already exhausted about 16 percent of the earth's readily available hydrocarbon (oil and natural gas) reserves of fossil fuel, and our rate of consumption is approximately doubling every 10 years. In this connection it is academic whether new reserves are found or whether our rate of use increases or remains constant; it is abundantly clear that our rate of consumption so vastly exceeds the rate at which these materials are being laid down that an ultimate crisis is inevitable. As fossil fuels become depleted, their costs will certainly escalate.

Table 1 presents some relevant values for the "energy budget" of the United States. The energy consumed as food is representative of the fraction of the solar energy stored by photosynthesis in farm crops. In the United States our consumption of energy from fossil sources exceeds our consumption of food energy by two orders of magnitude, although it is still very much less than the solar energy input to the earth's surface.

Pollution of the air resulting from the consumption of fossil fuels has been so widely discussed that nothing new can be added here. It is sufficient to note that carbon monoxide, carbon dioxide, and unburned hydrocarbon fragments are major pollutants that are not products of the oxidation of hydrogen.

The Problem for Transportation

It is taken as almost axiomatic that nuclear energy (fission in the immediate future, fusion perhaps in the next century) will eventually supplant fossil fuels as the primary energy source for fixed-station electric power. Plants currently under construction will in several years be able to produce from nuclear energy about 10 percent of this nation's power demands. On the other hand, there seems no serious possibility of using nuclear energy as a direct source of power for vehicles or aircraft. The problems of critical mass, shielding weight, and safety considerations are each formidable obstacles to the use of nuclear reactors as they now exist in any but stationary installations and perhaps ships.

As a consequence, the source of energy for vehicular locomotion in the distant future must be chemical energy synthesized by fixed-station nuclear power. The present options appear to be as follows: (i) the electrochemical storage battery, (ii) the fuel cell, (iii) the internal-combustion engine, and (iv) the external-combustion engine. Chemical and electrochemical reactions are characterized by energies of the order of an electron volt per reaction. Consequently, the most promising energy sources on an energy-per-unit-weight basis are those involving light elements, in particular, hydrogen. At the other extreme lies the lead-acid storage battery (*2*). Electrochemical cells in which lighter metals are used (zinc, sodium, and lithium) are more promising than the lead-acid storage battery, but they are less attractive than hydrocarbon combustion on a strictly weight basis. Exotic storage batteries often involve expensive components and dangerous or corrosive chemicals, and such batteries operate at elevated temperatures. For example, two of the most attractive batteries from the standpoint of energy storage per unit weight are the sodium-sulfur battery operated at 240°C and the lithium-chlorine battery operated at 600°C. Unfortunately, fuel cells do not now appear to have the power-per-unit-weight capabilities, let alone the economic feasibility, to constitute serious possibilites at present. Nevertheless, the fuel cell is a very attractive option, and research breakthroughs in this technol-

The author is professor of physics at the University of Michigan, Ann Arbor 48104.

Table 1. Energy budget of the United States in 1968; 1 barrel = 158 liters.

Energy source	Energy consumption: Conventional units (per year)	Energy consumption: Equivalent number of joules (per year)
Electric power	1.317×10^{12} kilowatt-hours (about 750 watts per person)	4.75×10^{18}
Fossil fuels		
Crude oil*	3.33×10^{9} barrels	1.7×10^{19}
Natural gas liquids*	5.50×10^{8} barrels	0.28×10^{19}
Natural gas*	1.93×10^{13} cubic feet	2.04×10^{19}
Coal*	5.57×10^{8} tons	1.3×10^{19}
Total fossil fuels		5.3×10^{19}
Motor fuel*	1.87×10^{9} barrels (about 7.5 gallons per person per week)	1.0×10^{19}
Food consumption	2000 kilocalories per person per day or about 100 watts per person	6.1×10^{17}
Solar energy	Based on the solar constant of 2 calories per square centimeter per minute over 3.55×10^{6} square miles (area of the continental United States)	1.0×10^{20}

* Figures are for the United States, from (*13*).

ogy will be most important. Hydrogen is currently the most attractive fuel-cell fuel. The situation regarding these various options for automobile propulsion has been discussed in a review paper by Bolt (*3*), from which Fig. 1 is taken.

The Case for Liquid Hydrogen

Figure 1 indicates that the internal- or external-combustion engines appear to be the best choices for vehicular power plants. If we consider a time in the future when fossil fuels are exhausted (or nearly so), it is appropriate to ask what chemical fuel should be synthesized. I believe that liquid hydrogen is the optimum choice.

Our utilization of resources on the surface of the planet is reaching the scale at which we should be prepared to cycle essentially all materials and resources, compatible with the utilization of energy and the second law of thermodynamics. Hence any fuel of the future should be part of a completely closed cycle, wherein its reaction products are identically reconstituted as fuel, while producing no deleterious effects on the environment (for example, pollution) in any portion of the cycle. Thus, while failing on other counts, the rechargeable lead-acid storage battery is ideally cyclic in that its stored energy is used with no effluent and it is later recharged with good efficiency from a source of stationary electric power. Liquid hydrogen likewise is nearly ideal in that its only combustion product is water vapor, and the earth's atmosphere is already in equilibrium with a surface consisting of over two-thirds open water. A fuel economy based on liquid hydrogen would draw water for electrolytic separation by nuclear power, releasing the oxygen and liquefying the hydrogen. The liquid hydrogen would then be transported and distributed as fuel, would be burned in the presence of oxygen from the air, and would then eventually return to the water systems as rain. Virtually any other fuel system would either discharge foreign substances into the environment or be constrained to retain and store its exhaust. Perhaps the only exception would be ammonia, although in this case the nitrogen would not "carry its own weight" in the fuel system, and there is a greater possibility of less desirable substances in the exhaust. A hydrogen-burning system might in some instances carry its own liquid oxygen. This would, of course, eliminate oxides of nitrogen in the reaction products.

Some pertinent physical properties of liquid hydrogen are given in Table 2. A specific comparison between liquid-hydrogen and gasoline on an energy-per-unit-mass and energy-per-unit-volume basis is presented in Table 3. Clearly, liquid hydrogen is an interesting fuel wherever weight is a major factor, as in jet aircraft, for example. As far as I know, there is no other chemical fuel which can equal hydrogen on an energy-per-unit-weight basis. Because of its very low density, hydrogen is about one-third as good a fuel as hydrocarbons on an energy-per-unit-volume basis.

Table 2. Properties of liquid hydrogen.

Boiling point	20.4°K
Liquid density	0.0708 g/cm³
Latent heat of vaporization	108 cal/g
Energy release upon combustion	29,000 cal/g or 2050 cal/cm³ or 1.21×10^{5} joule/g
Flame temperature	2483°K
Autoignition temperature	858°K

It was in connection with a liquid hydrogen fuel system for a hypersonic aircraft that a rather thorough study was made of the large-scale economics of liquid hydrogen production (*4*) by Air Products and Chemicals, Inc., for the National Aeronautics and Space Administration. Prior to about 1958, liquid hydrogen was essentially a laboratory curiosity and was produced only in small quantities. Subsequently, demands of the space program led to the construction of production facilities in the United States totaling over 150 tons (140 metric tons) per day of capacity. The cost of liquid hydrogen (not including marketing and distribution) is currently $0.20 per pound ($0.44 per kilogram) from a plant with a capability of producing 30 tons per day when operating near its full capacity. The Air Products and Chemicals study (*4*) indicates that the cost for liquid hydrogen from a plant with a capacity of 2500 tons per day could be about $0.08 per pound delivered with the production geared to the hypersonic aircraft transportation system. It so happens that, at present, the most economical method of producing liquid hydrogen is not electrolysis, but steam reforming with hydrocarbons. Here the basic reactions may be summarized as follows:

$$a\,CH_n + b\,H_2O \rightarrow c\,CO_2 + d\,H_2$$

with the carbon dioxide removed by solvents. Technological developments could bring the cost of the electrolytic production of hydrogen to 30 percent over the cost of the chemical process, or about $0.11 per pound. With the electrolytic production of hydrogen, about one-fifth as much power would be required for the liquefaction as for the electrolysis. Another estimate of the cost of liquid hydrogen ranges from $0.05 to $0.10 per pound, F.O.B. plant site (*5*). An Allis-Chalmers Manufacturing Company study indicates a projected cost for electrolytic hydrogen produced by large breeder-type nuclear power reactors of $0.20 per 1000 standard cubic feet ($0.07 per 1000 cubic meters), or about $0.04 per pound (*6*). The costs of liquid hydrogen and gasoline are noted on a comparable basis (dollars per calorie) in Table 3 where the figures from the Air Products and Chemicals study are used for liquid hydrogen and the cost of gasoline is taken to be $0.12 per gallon, not including marketing costs, taxes, and other added costs. All figures are normalized to 1968 dollar values.

Table 3. Energy and cost of fuels.

Fuel	Energy/mass (cal/g)	Density (g/cm^3)	Energy/volume (cal/cm^3)	Cost (dollars/cal)
Liquid hydrogen	29,000	0.07078	2,050	6×10^{-9} at $0.08/pound 8×10^{-9} at $0.11/pound
Gasoline	11,500	0.74	8,500	4.2×10^{-9} at $0.12/pound
Fuel oil	10,500	0.96	10,000	

In any discussion of the use of electricity to replace fossil fuels in our economy the figures noted in Table 1 should be borne in mind. The energy consumption of fossil fuels for vehicular transport in the United States was in 1968 more than twice the energy consumption of electric power. Hence the use of electric power to produce fuel, as discussed here, would require that the electric power–generating capacity of the country be tripled. Where electricity is used directly as an energy source for vehicular power, as in electrified rail transport, a rather high efficiency should be realized. On the other hand, a battery-powered vehicle analyzed by Bolt (*3*) was found to have an overall efficiency of only 14 percent (including the efficiency of electric generation in a thermal plant), whereas the overall thermal efficiency of a typical automobile powered by a gasoline engine lies in the range of 13 to 22 percent.

The study of the hypersonic transport system (*4*) was based on a worldwide supply of liquid hydrogen of 8000 tons per day. The liquid hydrogen equivalent of the U.S. gasoline consumption in 1968 corresponds to about 300,000 tons per day.

The State of the Art

Part of the increasing appeal of liquid hydrogen as a fuel arises from the rapid advance of cryogenics technology in recent years. Superinsulated vacuum dewars are able to store liquid hydrogen with loss rates of 2 percent per day for 150-liter containers. A jacket cooled to liquid-nitrogen temperatures can reduce these losses to 1 percent per day. A reduction to zero loss can be achieved with a refrigerator. Larger storage vessels have correspondingly smaller losses, as the ratio of surface area (heat loss) to volume decreases, so that the fractional loss is approximately proportional to (volume)$^{-1/3}$. Modern stationary storage dewars of 5000-liter capacity have a loss rate of 0.85 percent per day (*7*). More dramatic than the storage technology are the recent advances in cryogenic refrigerators. Liquefiers and refrigerators for 20°K service are available in ratings from 1 to 2 watts (at 20°K) or more. For example, a July 1969 survey (*8*) noted 14 commercially produced refrigerators in the temperature range from 12° to 35°K with heat loads of 1 to 10 watts and costs below $11,000. As an example of the evolution of the cryogenics art, the cost of helium delivered in quantity in Denver, for instance, is lower as a liquid at 4.2°K than as a compressed gas. Targets of liquid hydrogen for large accelerators of the Atomic Energy Commission laboratories (Argonne and Brookhaven national laboratories) are now made as closed systems with refrigeration rather than as continuously boiling vessels filled from a reservoir dewar of liquid hydrogen. Liquid hydrogen is presently shipped overland by truck in semitrailer dewar tanks with a capacity of 8300 pounds each, and these tanks are frequently closed during shipment so that the boil-off gas is permitted simply to build the static pressure. Rates of heat loss in such trailers correspond to a boil-off of about 0.5 percent per day. Railroad tank cars with a capacity of 17,000 pounds each are also in current use for the transcontinental shipment of liquid hydrogen. Natural gas (mostly methane) is presently shipped and stored in part as a liquid at 112°K. The rates of boil-off loss are clearly most serious for small units (private automobiles) but even here may be manageable with improving dewars and mass-produced refrigerators.

The use of hydrogen as a fuel in conventional reciprocating engines has been explored. Hydrogen has the desirable feature of burning very efficiently in a lean mixture (more so than gasoline). On the other hand, hydrogen is unfortunately more subject to preignition (knocking) than gasoline.

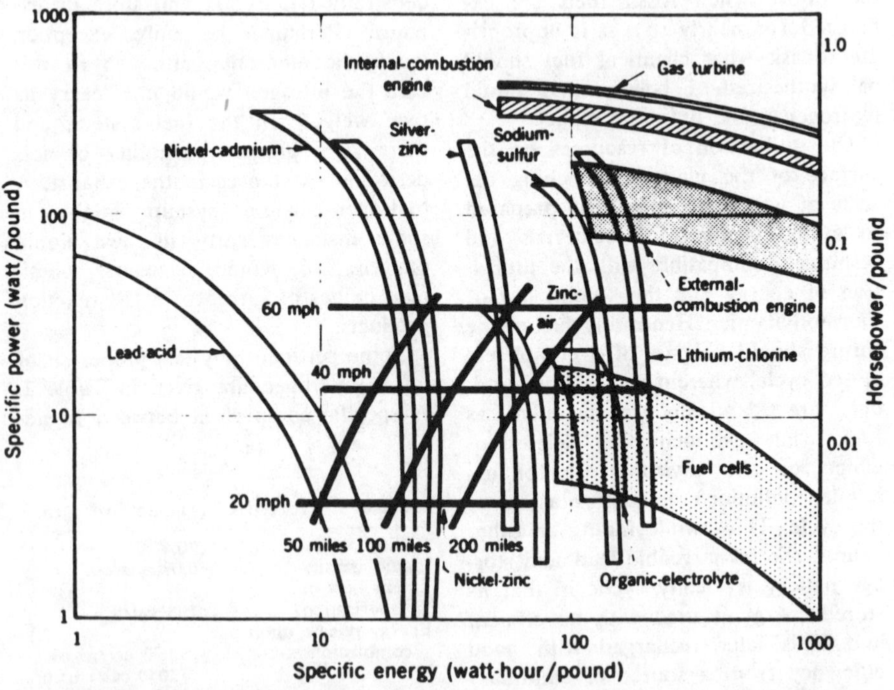

Fig. 1. Vehicle requirements for a 2000-pound vehicle and the capability of power plant systems [from (*3*)]. Solid lines indicate the ranges in miles (1 mile = 1.6 kilometers) corresponding to different constant speeds in miles per hour (mph) transformed on the specific energy–specific power coordinates. [Courtesy of the Society of Automotive Engineers, New York]

King *et al.* (*9*) have summarized work on hydrogen-fueled internal combustion engines. They report that, with a coolant temperature of 60°C and a clean combustion chamber, the correct fuel-air mixture could be used at a compression ratio of 14 : 1 without preignition. A group in Perris, California, has converted conventional automobile internal-combustion engines to run on hydrogen and has operated ordinary automobiles with hydrogen fuel (*10*). They are championing the ambitious objective of adapting the existing worldwide automobile fleet to liquid hydrogen. The use of hydrogen as a turbine fuel should present no problem. A stoichiometric mixture of hydrogen and air contains 2 parts of hydrogen to 1 part of oxygen or 5 parts of air; a typical gasoline-air mixture consists of 1 part of heptane vapor to 11 parts of oxygen (55 parts of air). Since a hydrocarbon such as heptane contains more atoms per unit volume at a given pressure and temperature than hydrogen, the volume of the combustion chamber for hydrogen burning would need to be somewhat larger than that of a gasoline-burning engine for a given power output and compression ratio. The flame temperature of hydrogen-oxygen combustion is 2483°K, comparable to the flame temperatures of gasolines. The peak temperature of the Otto cycle is about 3100°K.

Starkman *et al.* have compared hydrogen to various hydrocarbons, alcohols, and ammonia in terms of the mole fraction of nitric oxide produced in the combustion of these fuels with air in an Otto cycle engine (*11*). On the basis of these studies hydrogen seems to be quite equivalent to isooctane for optimum fuel-air mixtures.

Inevitably, a major question in the use of liquid hydrogen is the fire and explosion hazard. It is well known that hydrogen forms explosive mixtures with air over a broad range of concentrations (4 to 75 percent, by volume), and the use of liquid hydrogen in high-energy physics has been accompanied by one major and several minor accidents. It seems, however, that careful handling of hydrogen could reduce such accidents to a very minimum level in large-scale use. In many ways hydrogen is safer than gasoline in that any escaping hydrogen goes directly into the air rather than remaining as a slowly evaporating liquid. Explosions of hydrogen as opposed to rapid burning are very rare in practice. Apparently in one potentially serious highway accident (*7*) a semitrailer liquid hydrogen tanker went off the road in the mountains and broke apart, spilling its charge. However, no fire ensued and the driver "walked away."

It is logical that the first large-scale use of liquid hydrogen might be in jet aircraft since in such aircraft the boil-off loss and distribution problems would be minimized and the weight advantage over hydrocarbons would be most valuable. Long-haul motor freight and city buses would be the next most effective users; from the standpoint of pollution, the use of liquid hydrogen in city buses would be particularly welcome. The fueling of such vehicles would most logically be through replacement of the entire tank (dewar) with a previously filled tank. Simple, quick disconnects would make it possible to replace these tanks in a minimum amount of time and with almost no loss of liquid hydrogen. Weighing of standardized dewars would then be done with minimum loss at the "service station." Of course, it would be important to adequately vent the ambient boil-off of hydrogen from the fuel tanks of parked vehicles. No discussion is given here of the use of liquid hydrogen by railroads, as it is assumed that trains in the future will be totally electrified.

The private automobile presents the most difficult logistics problem for liquid hydrogen fuel because of its infrequent use, small-capacity fuel system, and the wide spectrum of technical sophistication of the operators. One would not be able to return from an extended holiday and drive off in the family car fueled with liquid hydrogen in view of the boil-off from even the best-insulated tank. Local hydrogen refrigerators could conceivably become economically practical, or, alternatively, hydrogen could be available in "home delivery" by service stations. A potential solution to this problem of the small-scale user, which has been proposed by a group at Brookhaven National Laboratory (*12*), involves the use of metallic hydrides. They point out that Mg_2Cu, Mg_2Ni, and Mg can combine with hydrogen, binding it as Mg_2NiH_4 and MgH_2, and that, in so doing, as much hydrogen is held per unit volume as in liquid hydrogen. These hydrides are stable at the ambient temperature and pressure but dissociate to hydrogen gas and metal at about 260°C. Thus a "fuel tank" of powdered or sintered magnesium or other metal alloy could be charged with hydrogen under the right conditions of temperature and pressure, and the hydrogen could be released through heat from the exhaust of the engine. It may be that an optimum system would include a small hydride reserve tank for long-term, stable storage with the major fuel supply contained in liquid hydrogen dewars.

Conclusion

The use of liquid hydrogen as a long-term replacement for hydrocarbon fuel for land and air transportation seems technically feasible. It is an ideal fuel from the standpoint of a completely cyclic system, serving as a "working substance" in a closed chemical and thermodynamic cycle. The energy-per-unit-weight advantage (a factor of 3) over gasoline or any other hydrocarbon fuel makes liquid hydrogen particularly advantageous for air craft and long-range land transport. As a pollution-free fuel, it must be seriously considered as the logical replacement for hydrocarbons in the 21st century.

References and Notes

1. S. F. Singer, *Sci. Amer.* **223**, 175 (September 1970).
2. H. R. Crane, personal communication.
3. J. A. Bolt, *Soc. Automotive Eng. Pap. 680191* (1967).
4. N. C. Hallett, *NASA Contract Rep. 73* (1968), p. 226.
5. J. E. Johnson, paper presented at the Cryogenics Engineering Conference, Boulder, Colorado, 1966.
6. R. L. Costa and P. G. Grimes, *Chem. Eng. Progr.* **63**, 56 (1967).
7. E. McLaughlin, personal communication.
8. J. G. Daunt and W. S. Goree, "Miniature Cryogenic Refrigerators," Stevens Institute of Technology and Stanford Research Institute preprint.
9. R. O. King, S. V. Hayes, A. B. Allen, R. W. P. Anderson, E. J. Walker, *Trans. Eng. Inst. Can.* **2**, 143 (1958).
10. B. Dieges, personal communication.
11. E. S. Starkman, R. F. Sawyer, R. Carr, G. Johnson, L. Muzio, *J. Air Pollut. Contr. Ass.* **20**, 87 (1970).
12. K. C. Hoffman, W. E. Winsche, R. H. Wiswall, J. J. Reilly, T. V. Sheehan, C. H. Waide, *Soc. Automotive Eng. Trans.* **78**, 981 (1969).
13. L. H. Long, Ed., *1970 World Almanac and Book of Facts* (Newspaper Enterprise Association, New York, 1969).

14. I thank, in particular, D. Sinclair and J. A. Bolt of the University of Michigan, E. McLaughlin of the Lawrence Radiation Laboratory, K. C. Hoffman of the Brookhaven National Laboratory, F. E. Mills and W. E. Winter of the University of Wisconsin Physical Sciences Laboratory, J. E. Johnson of Union Carbide Corp., and W. E. Timmcke of Air Products and Chemicals, Inc., for valuable suggestions and information during the course of the discussion.

Flywheels

Advances in materials and mechanical design make it possible to use giant flywheels for the storage of energy in electric-power systems and smaller ones for the propulsion of automobiles, trucks and buses

by Richard F. Post and Stephen F. Post

In technology, as in women's fashions, old concepts have a way of turning up again, brought up to date and refurbished. The flywheel, which is one of the oldest of human inventions, may be such a concept. Recent developments in materials and mechanical design suggest that the flywheel can help to solve two contemporary problems: the steady increase in the use of energy and the impact of that use on the environment. Specifically, the flywheel offers the prospect of providing (1) an efficient means of storing energy on a large scale to help electric utilities handle peak loads and (2) compact units to power electric vehicles having a range and performance comparable to those of the automobiles, trucks and buses on the roads today.

The principle of the flywheel, which must have been recognized very early in human history, is that a spinning wheel stores mechanical energy—energy that can be put in and taken out, as water is stored in and recovered from a reservoir. Flywheel action plays an essential role in the ancient potter's wheel, keeping it spinning between the occasional kicks of the potter's foot. Considering the age of the potter's wheel (which is mentioned in the Old Testament), it is plain that men understood the basic principle of mechanics exemplified by the flywheel long before Newton formalized it in his laws of motion.

The flywheel also played an essential role in the mechanical developments that underlay the Industrial Revolution. The steam engines that powered the factories and mills of those days could not have functioned without the steadying influence of their flywheels. Even today the internal-combustion engines of automobiles, trucks and diesel locomotives rely on energy stored in flywheels to carry the rotation between pulses of energy delivered by the pistons.

Until recently it was thought that employing flywheels to store energy in a wider range of applications was out of the question because of cost and because not enough energy could be stored for a given flywheel weight to satisfy the foreseeable needs. The picture has been radically changed by advances in materials technology [see "Advanced Composite Materials," by Henry R. Clauser; SCIENTIFIC AMERICAN, July]. Most of these advances have resulted from work in aerospace technology.

To see what developments affecting materials have to do with inertial energy storage, which is another name for what a flywheel does, one must consider the factors that govern the action of a flywheel. Imagine a simple flywheel in the form of a circular rim or hoop connected by thin spokes to a hub.

The amount of energy stored in such a flywheel depends on the mass of the rim and on how fast the wheel is spinning; the storage varies as the square of the rotation speed. In principle one could store as much energy as one wishes simply by spinning the wheel faster, but of course it does not work that way. The limit to the amount of energy stored is

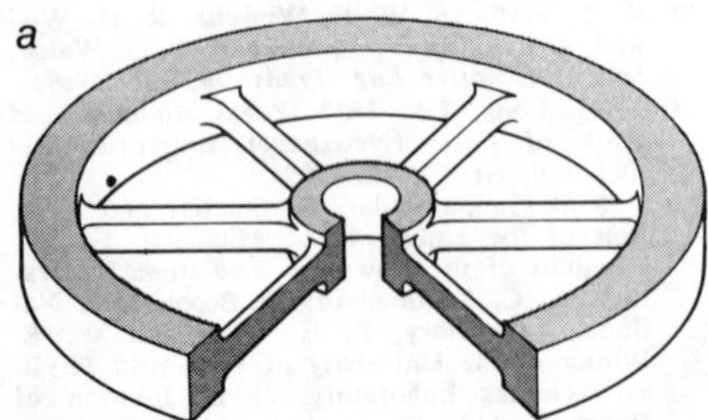

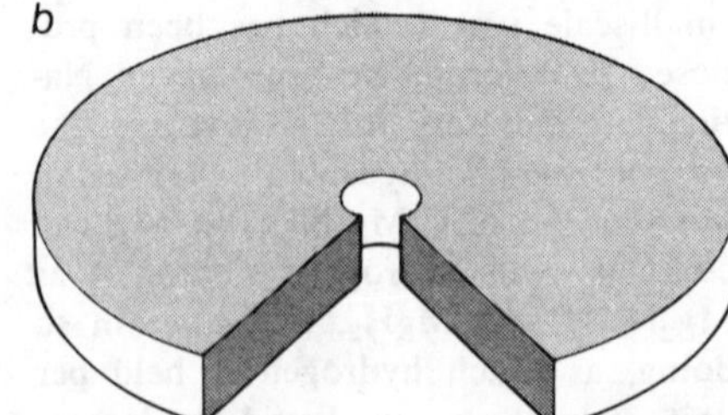

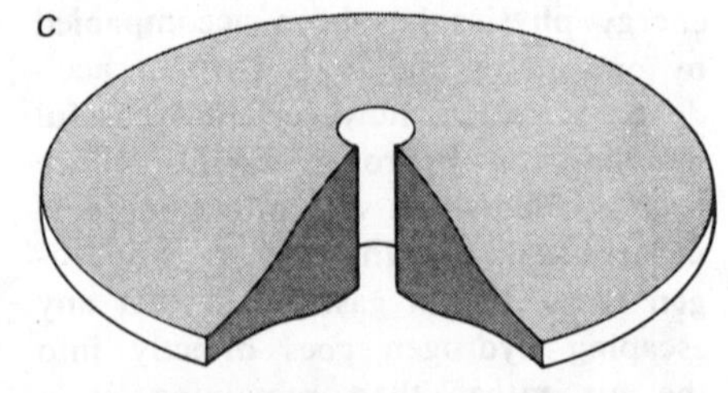

FLYWHEEL DESIGNS for flywheels made of conventional materials include a heavy-rim wheel (*a*), an almost solid disk (*b*), which can store more energy, and a disk that is thickened toward center (*c*) to relieve stress concentration and increase storable energy.

ultimately set by the tensile strength of the material from which the rim is made.

In particular the tensile strength must be sufficient to withstand the "hoop stress" resulting from centrifugal forces, otherwise the wheel would fly apart. As with the energy stored, these forces are proportional to the mass of the rim and increase as the square of its rotation speed. One therefore sees that two properties of the material determine the amount of energy that can be stored in a flywheel: mass density, which provides kinetic energy, and tensile strength, which resists centrifugal forces.

In terms of these properties, what materials are best for storing the most energy for a given weight of material? Contrary to what intuition might suggest, one wants the lightest (lowest density) strong material available.

To understand this surprising fact consider again the simple thin-rim flywheel. Suppose one were to make two such flywheels, identical in dimensions and design but fabricated of two different materials having equal tensile strength but being unequal in density. In other words, one wheel would be made of heavy material and the other of light material.

Now begin to spin up these flywheels, keeping the speeds equal. Under these conditions the heavy flywheel will of course store more energy than the light one, in direct proportion to its greater mass. Note, however, that the tensile stresses from centrifugal forces will also be greater in the heavier rim than in the lighter one, again in direct proportion to the relative mass. As one continues to speed up both wheels, the heavy one will be the first to approach its maximum speed as limited by its tensile strength. In other words, it will be the first to approach the limit of its energy storage.

At the same speed of rotation the lighter flywheel experiences much weaker centrifugal stresses. Therefore it can be speeded up still more before it reaches its limit of tensile strength. When the limit is reached, the light flywheel at its higher speed will be storing the same total amount of mechanical energy as the heavier one at its lower speed. Here, however, is where the advantage of light materials appears. If the light material is, say, 10 times lower in density than the heavy material and both have the same tensile strength, a flywheel made of the light material will require only 10 percent as much mass to store the same amount of energy as the flywheel made of the heavy material. High strength at low density is thus the proper criterion for choosing materials for energy-storing flywheels.

Flywheels have traditionally been made of metal, particularly high-strength steel. Because its density is high, however (eight grams per cubic centimeter), even the strongest alloy steel is not what one wants for making a flywheel capable of storing a large amount of energy for a given weight of flywheel. Per unit of energy stored, steel is too heavy, too expensive and too difficult to fabricate for the demanding applications we envision in this article.

The fiber composites developed initially for aerospace needs have exactly the properties required. They are lower in density than steel by four to six times. Moreover, some of them are far stronger in tension than the strongest steel.

In quantitative terms the limiting amount of energy that can be stored per unit weight of flywheel material is equal to half the tensile stress at the breaking point divided by the density. (The mathematical expression is $U_{max} = 1/2(K_{max}/\rho)$, where U_{max} is the maximum stored energy per unit weight, K_{max} is tensile stress at the breaking point and the Greek letter rho is density.) In metric units this relation is expressed in ergs per gram, but we shall employ a more practically oriented unit: watt-hours per kilogram. The values we discuss are upper limits; practically realizable values will be from 40 to 60 percent of the upper limits. Nonetheless, the maximum value (U_{max}) represents a useful index of the relative ability of materials to store mechanical energy.

Comparing certain fibers with high-strength metals in terms of properties suitable for flywheels, one finds a remarkable showing by the fibers [*see bottom illustration on page 20*]. For example, *E* glass (a commercial fiber produced on a large scale) can store four times as much energy per unit weight as high-strength maraging steel, yet the price of the fiber at the factory is only 15 percent of the price of the alloy steel.

A particularly interesting fiber, which is now coming into production on a large scale for automobile tires and aerospace applications, is PRD-49. This fiber, which was developed by Du Pont, has recently been given the trade name Kevlar. This fiber and similar ones under development by Monsanto are members of a new generation of high-strength polymer fibers: the "aromatic polyamides," descendants of the nylon family. PRD-49

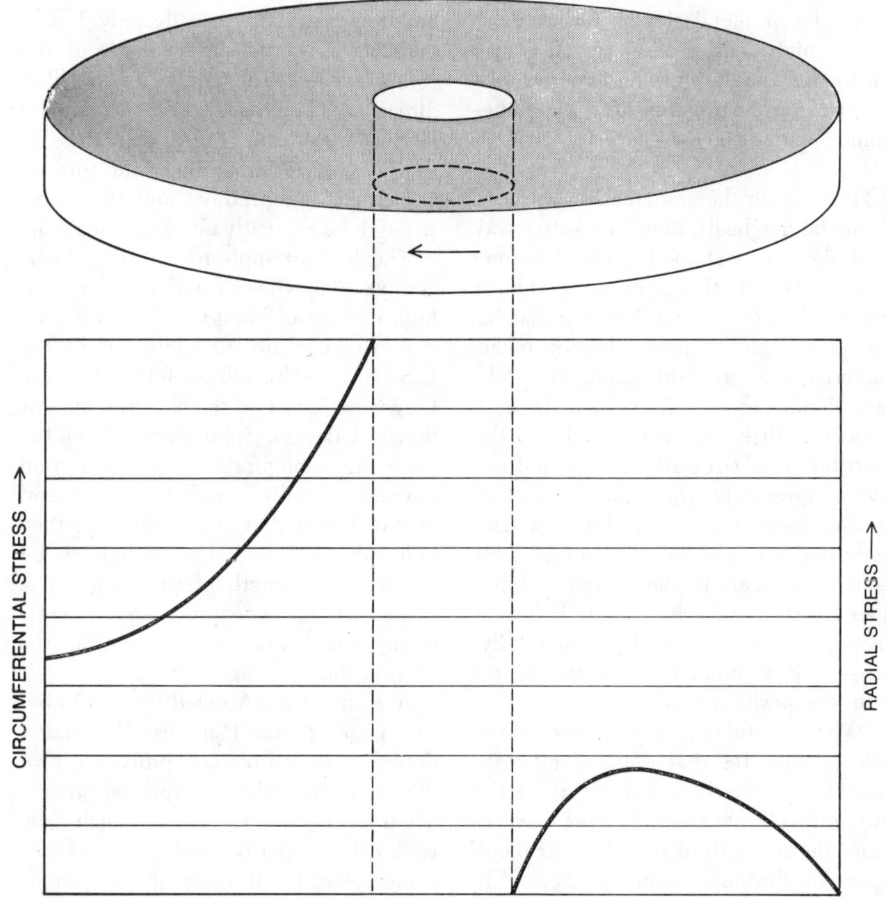

OPERATING STRESSES on a thick-rim flywheel are charted. Circumferential stress is highest at inner hole and radial stress is highest midway between inner and outer edges of disk.

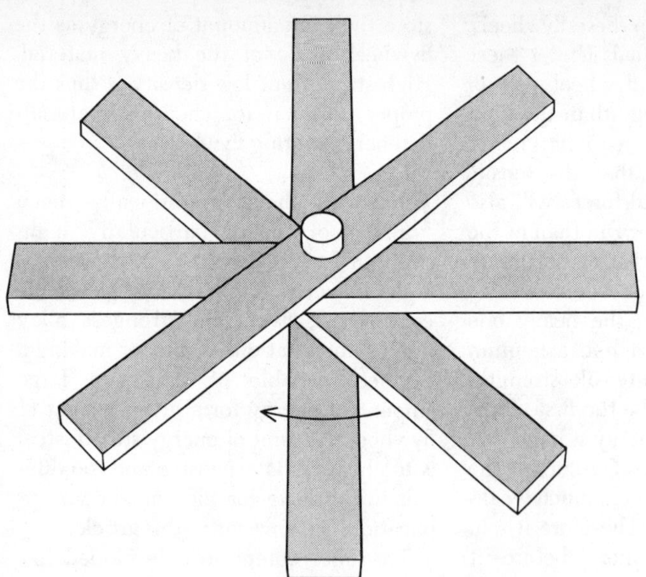

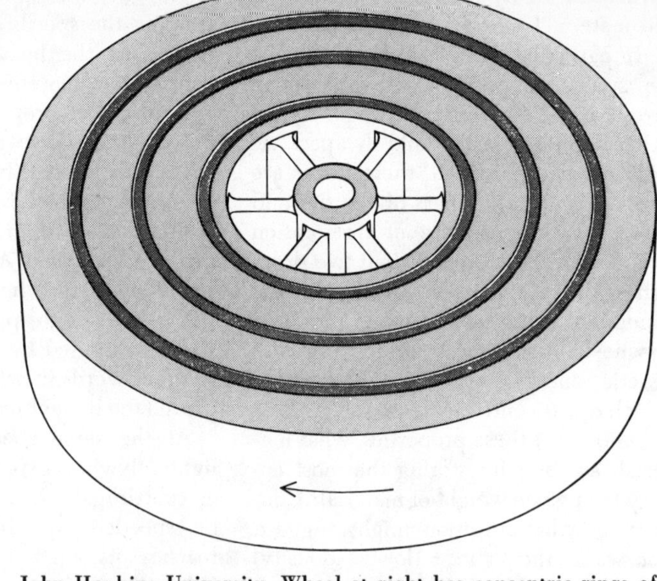

NEW DESIGNS FOR FLYWHEELS have been made with the aim of taking advantage of the light weight and great strength of fiber-composite materials. At left is a rimless, multi-spoke "superflywheel" designed by a group at the Applied Physics Laboratory of John Hopkins University. Wheel at right has concentric rings of fiber composite separated by small gaps filled with bonded resilient material. The design minimizes radial stresses that would cause delamination and makes efficient use of the wheel's volume.

can store seven times as much energy per unit weight as alloy steel.

Fused silica may turn out to be the best of all the fibers for inertial storage. Its estimated maximum energy storage is based on laboratory tests of carefully prepared fibers, so that it cannot be compared directly with fibers already in commercial service. If silica fibers came into use on a commercial scale, however, flywheels incorporating them could store from 10 to 15 times more energy for a given weight of fiber than flywheels made of the best alloy steel.

To take advantage of fiber composites in storing inertial energy, it will be necessary to discard old concepts of how flywheels should be made and to adopt new designs that are tailored to the characteristics of the fiber composites. The problems can be seen by considering again the simple flywheel consisting of a hoop connected by spokes to a hub. Such a flywheel does not have a suitable geometry for the purposes we are discussing; its mass is concentrated in the rim, and the rest of the volume, lying between the rim and the hub, is useless, being mostly empty space.

One's first thought on improving the design might be to fatten the rim, making it into a disk with a hole [*see illustration on page 17*]. Now, however, a new problem appears: since the outer regions of a thick ring necessarily experience greater centrifugal forces than the regions near the center, they will attempt to expand away from the inner portions. This action will set up restraining forces in the radial direction within the body of the ring. These forces will in turn act to intensify the stresses at the inner hole. The resulting stresses, both radial and circumferential, are higher than those in a thin rim rotating at the same speed, and now the highest stress comes at the inner hole. Therefore, because of the uneven distribution of stress, the thick ring is an inefficient shape for a metal flywheel that must store a high density of energy. It is also far too naïve a design for a flywheel fabricated from fiber composites, as will become apparent below.

Pursuing for the moment the matter of metal flywheels, there is a better way to design a disk-shaped flywheel as long as it is relatively thin at its outer edge. It was pointed out a long time ago that for one-piece disks made of a homogeneous material, such as solid metal, the problem of concentrated stress near the center of the disk can be relieved and the distribution of stress made more uniform by progressively thickening the disk toward its center, where the most concentrated stresses appear. Indeed, this concept appears in the design of high-speed turbine wheels. Although the tapered design is optimal for metal flywheels, it is inherently unsuitable for fiber-composite flywheels.

What are the special properties of the fiber composites that demand new approaches to flywheel design? Since the strength of fiber composites derives from the strength of the fibers in them, they will develop maximum strength in tension when all the fibers lie parallel to one another, lined up in the direction of the applied tensile force. The bonding material in which the fibers are embedded (usually an epoxy resin) mainly serves the function of protecting them and transmitting relatively weak forces between adjacent fibers. In such unidirectional fiber composites the tensile strength perpendicular to the direction of the fibers is essentially only that of the bonding material—typically only 1 or 2 percent of the tensile strength of the composite material parallel to the fiber direction. Therefore in order to employ fiber composites at full effectiveness in flywheels it is imperative that this inequality of longitudinal and transverse strength be explicitly taken into account.

The first attempts to construct fiber-composite flywheels capable of storing a high density of energy were made several years ago in work sponsored by the U.S. Navy. The efforts failed because they were based on the old principles of flywheel design. Solid flywheels of the thick-ring configuration were made from circumferentially wound glass fibers and epoxy. When they were spun up, they failed at speeds far below the limit set by the tensile strength of the fibers. The cause of failure was progressively worsening radial delamination within the body of the flywheel.

A group at the Applied Physics Laboratory of Johns Hopkins University, doubtless aware of such problems, proposed a radically different approach. Their energy-storage rotor, which they called the "superflywheel," consists of a rimless wheel with many spokes [*see illustration at left above*]. Each spoke is a bar or rod fabricated of fiber-composite material, with the direction of the fibers

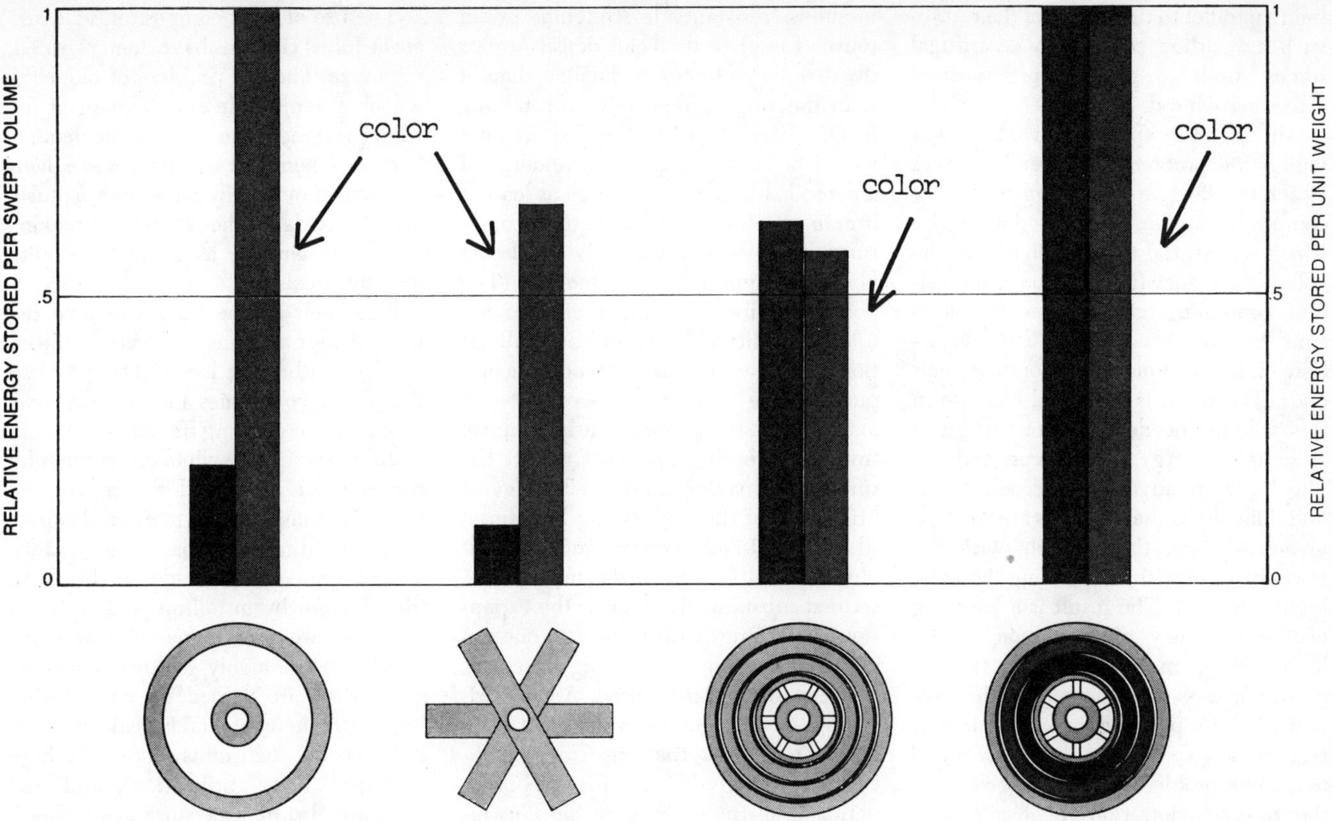

ENERGY STORAGE of fiber-composite flywheels is depicted on the basis of the volume swept out by the outer edge (*gray*) and on the basis of unit weight of fiber (*color*). From the left the wheels are respectively the thin-rim design, the superflywheel, the concentric-ring flywheel and a similar type of flywheel whose inner rings are loaded with weight to distribute the stresses more evenly.

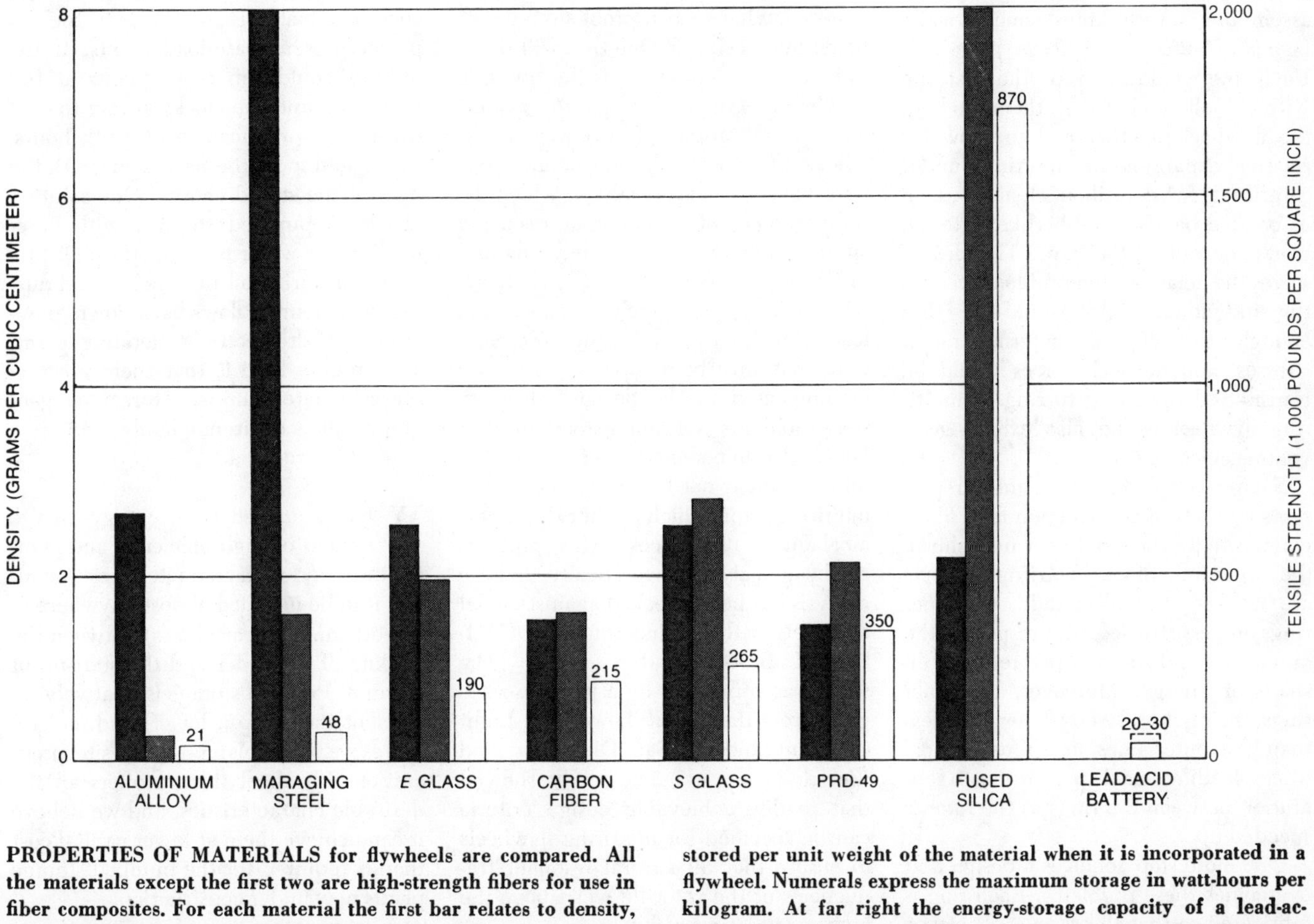

PROPERTIES OF MATERIALS for flywheels are compared. All the materials except the first two are high-strength fibers for use in fiber composites. For each material the first bar relates to density, the second to tensile strength and the third to maximum energy stored per unit weight of the material when it is incorporated in a flywheel. Numerals express the maximum storage in watt-hours per kilogram. At far right the energy-storage capacity of a lead-acid battery per unit of weight is shown for comparison purposes.

being parallel to the length of the rod. In such a whirling pincushion centrifugal forces result in radial (pure tension) forces in each rod.

Although this approach avoids the delamination problem and thereby comes closer to effective utilization of a fiber composite, it has certain drawbacks. First, as with the thin-rim flywheel, the volume efficiency (the fraction of the volume containing useful material) is low, since the rods occupy only a small portion of the volume swept out by their ends. The result is a marked increase in the volume needed to store a given amount of energy, so that cost and applicability are adversely affected. A second difficulty is that the rods are stressed unevenly along their length, with the portions nearest the hub being the most highly stressed. The result is a lowering of the efficiency of utilization of the high-strength material. Finally, the necessity of anchoring the ends of the rods at the hub (or piercing them at their centers to allow space for an axle) gives rise to difficult problems of design because of the stress concentrations involved.

Is it possible to avoid the problem of delamination and still obtain a fiber-composite flywheel that provides for efficient use of the fibers and high volume efficiency? Let us construct a flywheel by assembling several rings concentrically [*see illustration at right on page 19*]. Each ring would fit inside the next one with a small gap between them. To hold the flywheel together and to allow for relative expansion of the rings under centrifugal force, resilient elements, such as bonded bands of rubberlike material, would be put in the gaps. This design solves the delamination problem because the individual rings would be thin enough to minimize internal radial stresses, and no such forces would be transmitted from ring to ring. A multiring flywheel would also exhibit good volume efficiency.

Such a coupled-ring assembly represents a practical construction for a fiber-composite flywheel but does not achieve the optimum. Since centrifugal effects are weaker at smaller radii, the inner rings are less stressed with respect to the outer ones and so do not store their full share of energy. Moreover, the inner rings, being less stressed, expand less than the outer ones under centrifugal forces, leading to a disproportionate separation between the rings as the rotor is speeded up.

One more step achieves a design that overcomes these problems. Each ring is fabricated so that the ratio of its elastic modulus (resistance to stretching under tensile force) to its mean density (mass divided by volume) is smaller than it is in the ring immediately surrounding it. One way to satisfy this requirement would be to add increasing amounts of a dense loading material, such as lead or iron in powder or strip form, to the inner rings. Another way would be to choose fibers with graded elastic moduli. (The polyamide fibers, among others, provide this possibility.) Preferably a combination of the two stratagems would be employed. The main result would be to make fiber stress more nearly uniform from ring to ring and also within the rings, thus maximizing the efficiency of utilization of the high-strength material (the fibers). In addition varying the modulus-to-density ratio in the manner described substantially reduces the expansion of the rings with respect to one another so that simpler coupling structures between rings can be used. A comparison of the configurations we have been discussing shows that the storage efficiency of the graded-ring flywheel is superior to all the others [*see top illustration on opposite page*].

Having arrived at a configuration that is nearly ideal from the standpoint of high density of stored energy and high efficiency in the utilization of flywheel material, what special problems must be taken into account? One that will occur to mechanical engineers is the question of the stability of multiring rotors against internal vibrations. Conventional flywheels and their axle and bearing systems have "critical speeds" at which the rotation speed of the motor, as it is being spun up, momentarily matches a natural vibration frequency of the axle-shaft-flywheel mass system. Such resonances can lead to the buildup of destructive oscillations that must be damped or otherwise circumvented in the design. Although the problem is well understood and can be avoided in conventional flywheel systems, might it not be worrisome with a multiring rotor, which might exhibit new mechanical resonances? This problem has been analyzed theoretically, and the analysis has been checked against model flywheels by the engineering firm of William M. Brobeck and Associates. (The work was supported by a joint research grant from the Pacific Power and Light Company and the San Diego Gas and Electric Company.) The results showed that readily achievable design criteria can be specified for multiring flywheels to ensure that all natural resonant frequencies of the rotor lie well above the highest operating speed.

The use of fiber composites, as workers at Johns Hopkins have demonstrated, minimizes another problem of older flywheel systems: safe containment of flywheel fragments in case of mechanical failure. When fiber composites are overstressed, they fail by shredding or turning to powder rather than by breaking into large chunks, as is the case with steel flywheels.

Flywheels of the kind we have described offer a means for electric utilities to cope with peak loads. The problem the power companies face is that peak demands are growing (for such purposes as air conditioning) while environmental concerns have delayed the moves the companies might make to expand capacity by building new "base load" plants. The companies have tended to deal with the problem by installing peaking units near the areas of demand. The units, which are commonly gas-turbine-driven generators, are turned on only during the hours when demand is heaviest. Capital costs for such units tend to be high (because of low utilization), and fuel costs are also high because gas turbines must burn fuel oil, which is not only expensive but also now increasingly in short supply.

The peaking problem is not basically caused by an energy shortage; except for temporary shortages there is enough coal, residual oil and nuclear fuel to power present base-load plants. If the utilities could run these plants at full output around the clock, storing the extra energy produced in off-peak hours, they could meet the peak loads with the help of the stored energy. One solution has been pumped storage: with base-load power water is pumped uphill to a reservoir during off-peak hours, and during peak hours it flows back down, powering a hydroelectric generating plant. The trouble here is that there are few places where pumped storage is geographically, environmentally and economically practical.

What is needed is an energy-storage system of high efficiency and compactness and with characteristics that allow it to be installed almost anywhere in a short time (a year or less) between the placing of the order and the start-up of the unit. Pumped storage is relatively inefficient, takes up a lot of land and requires years of planning and site preparation. Flywheel storage offers all the desirable characteristics, and we believe it can achieve them at lower capital cost than is required for the pumped-storage plants now under construction.

A flywheel system of energy storage

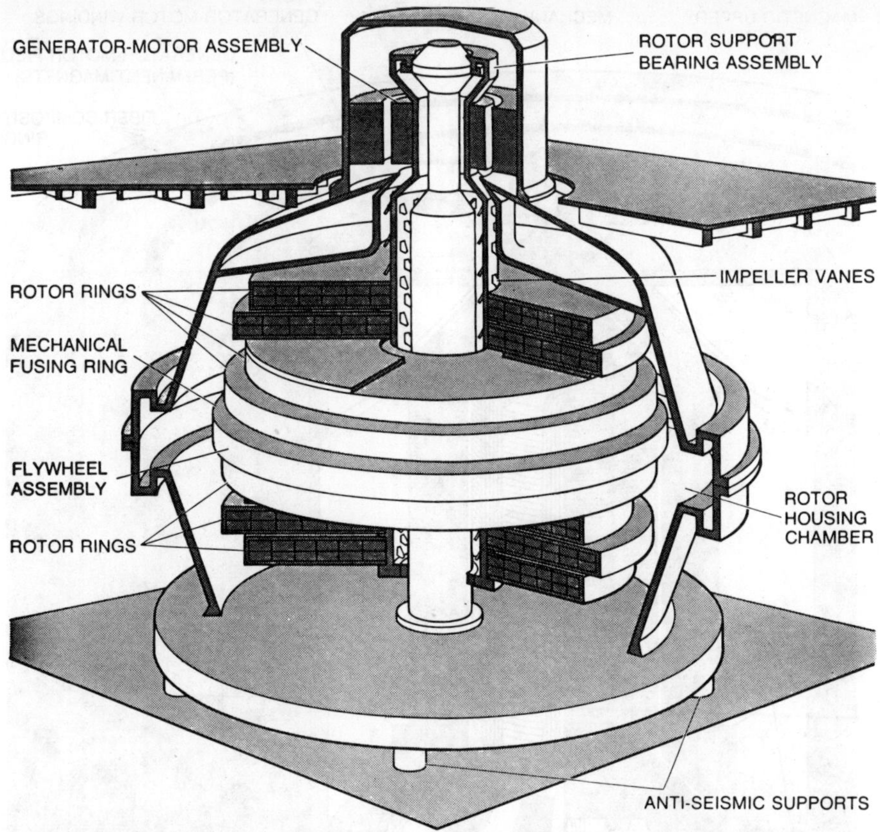

PEAK-POWER UNIT that would store energy from an electric-power plant during off-peak hours and would generate energy at times of peak load is visualized. The fiber-composite flywheel would be some 15 feet in diameter and weigh some 200 tons. It would be coupled to a generator-motor that would function as a motor when the flywheel was being spun up to store energy and as a generator when the system was drawing on the flywheel's stored energy.

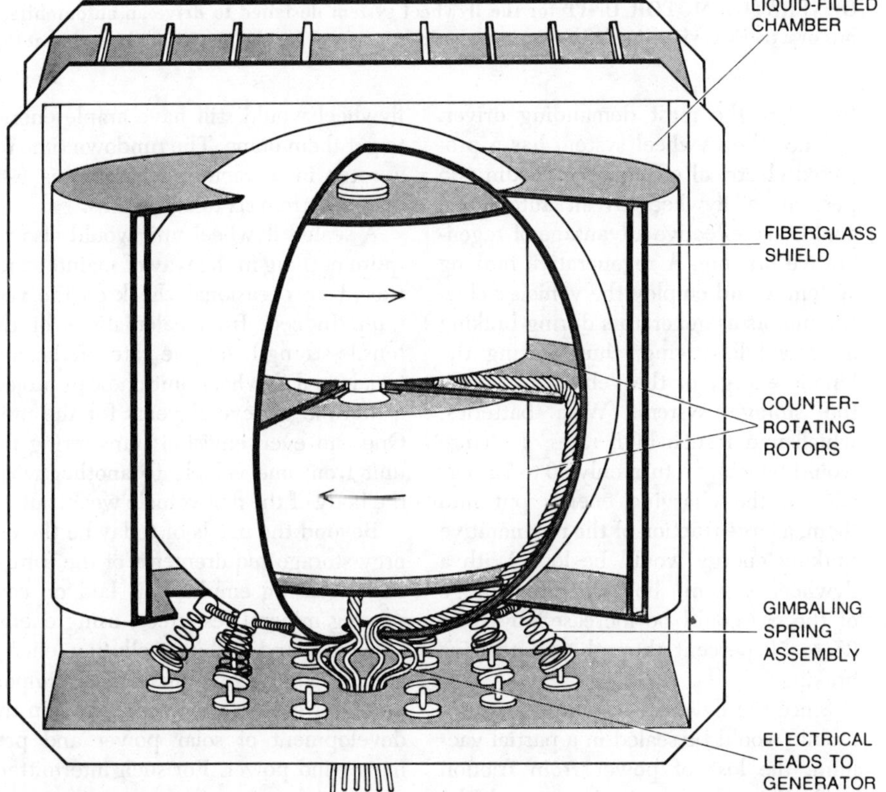

AUTOMOTIVE POWER derived from energy stored in a pair of counterrotating flywheels would employ a system of this kind. The flywheels, sealed in a partial vacuum to reduce air friction, would be coupled to a generator-motor. To store energy in the flywheels the system would be plugged into an electric outlet. Power produced by the system would be delivered to electric drive motors. The gimbal mounting would reduce the gyroscope effect.

for a power company would consist of several individual units. The flywheel in each unit would be housed in a sealed enclosure and coupled directly to a variable-speed generator-motor, which would function as a generator when the system was drawing on the energy stored in the flywheel and as a motor when energy was being stored in the flywheel by spinning it up. The flywheel and the generator-motor would operate in an atmosphere of inert gas (hydrogen or helium) below atmospheric pressure in order to reduce losses from air friction. Each flywheel would be from 12 to 15 feet in diameter and would weigh from 100 to 200 tons. Each unit would store from 10,000 to 20,000 kilowatt-hours of energy at full charge (a rotation speed of about 3,500 revolutions per minute). A utility system might employ 100 or more such units, dispersed throughout the system at substations.

Based on the costs of materials, fabrication of the flywheel and the electrical components and installation, the capital cost of a unit storing 10,000 kilowatt-hours of energy, having a power rating of 3,000 kilowatts and operating at an in-out efficiency of 93 to 95 percent can be estimated at $325,000. The corresponding capital cost per kilowatt would be $110–lower than that of pumped storage. Additional advantages over pumped storage would result from local siting, which would reduce transmission costs, and from the fact that a flywheel unit would take up far less land. (For each 10,000 kilowatt-hours of stored energy a pumped-storage system typically requires from two to four acres of reservoir area; a flywheel unit storing the same amount of energy would occupy a cell approximately 20 by 20 feet in size.) Moreover, the capital costs of pumped storage are rising, whereas one can expect the capital cost of flywheel storage to decline as new and better fibers are brought into large-scale production.

Powering vehicles with flywheels is not a new idea, but so far all the vehicles thus powered have had steel flywheels. Several years ago Swiss engineers developed a bus that ran on flywheel power between stops, recharging the wheel with an electric motor at each stop by bringing an overhead trolley in contact with a power line. Similar buses, which will have alloy-steel flywheels of improved (thickened hub) design, are now being developed for San Francisco.

The real potential of flywheel power for vehicles will appear with the advent of fiber-composite flywheels. Then it will be possible to drive automobiles with flywheels. One would spin up the flywheel with a motor running on power drawn from an electric outlet, and then the energy stored in the flywheel could drive the car with the motor operating as a generator to supply power to smaller motors, one on each wheel.

The economic and social advantages of flywheel-driven automobiles would be considerable because the vehicles would operate at high efficiency and would not give rise to pollutants. The internal-combustion engine in an automobile typically converts only from 10 to 15 percent of the energy in gasoline to motive power, whereas modern electric-power plants burning fossil fuel have efficiencies of 40 percent or better. Moreover, combustion can be better controlled in a central power plant than in an automobile, at least at present. Therefore a high-efficiency system such as a flywheel-driven automobile that stored energy from electric-power plants and used it to drive vehicles would result in greatly improved efficiency in terms of fossil fuel consumed per vehicle mile. In these terms one barrel of crude oil burned in an electric-power plant would be the equivalent of up to five barrels sent to a refinery to produce gasoline. It follows that any new demands for electric power arising from the use of flywheel-electric automobiles could be met readily within the present resources of fossil fuel.

An automobile flywheel system would, like the peak-storage system, include a flywheel and a motor-generator sealed in an evacuated chamber. About 30 kilowatt-hours of stored energy would be needed to provide reasonable performance by a small automobile: a range of 200 miles at 60 miles per hour, which is comparable to the performance of today's internal-combustion automobile on one tank of fuel. Obtaining this amount of energy from conventional lead-acid storage batteries would require a bank of batteries weighing more than 2,000 pounds. A flywheel containing 130 pounds of fused-silica fibers or 280 pounds of PRD-49 could store the same amount of energy. (The weight of the total drive system would be between 500 and 600 pounds.) Another advantage of the flywheel system over a battery system is that energy can be taken out of or put into a flywheel rapidly. A battery cannot deliver the high power needed for rapid acceleration, nor can it take the high rate of charge that would be necessary to make an electric automobile suitable for long-distance driving. A flywheel system, however, could accept a full 30-kilowatt-hour charge in five minutes or less. Furthermore, it could deliver high enough outputs of horsepower to satisfy the most demanding driver.

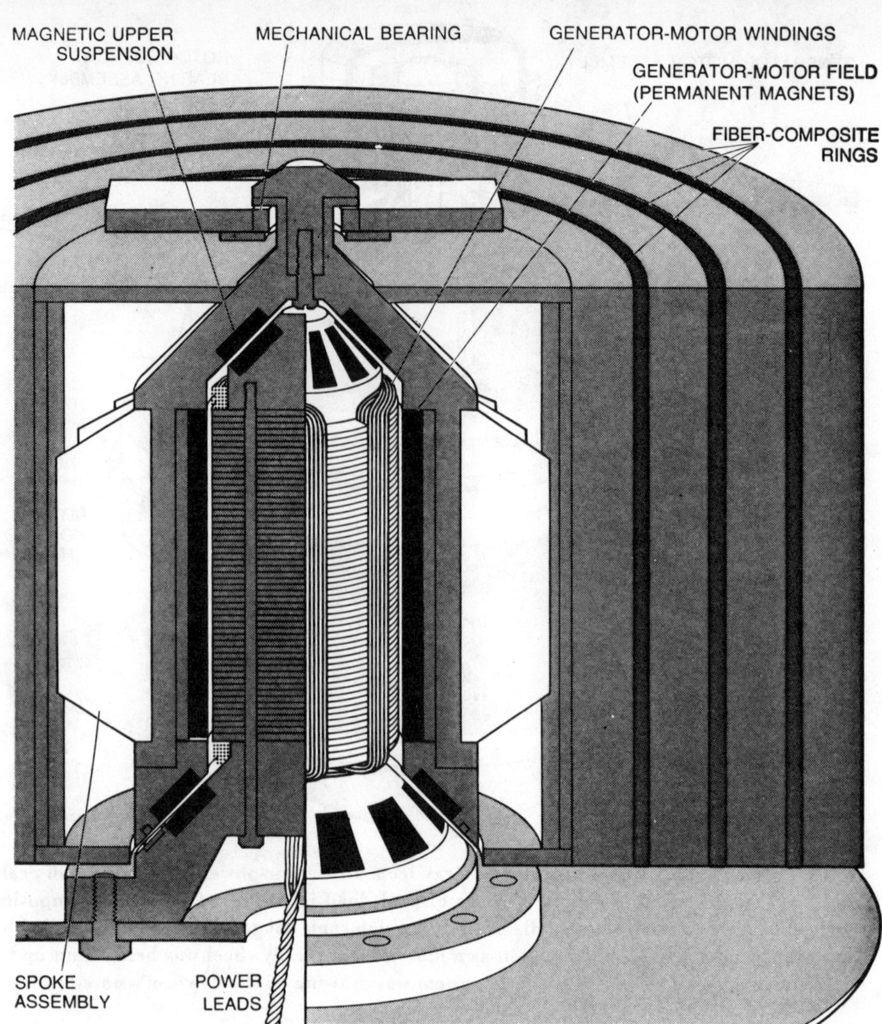

GENERATOR-MOTOR UNIT for the flywheel system designed to drive an automobile, a bus or a truck is visualized. Suspension is magnetic except for the mechanical bearing at top.

Since the flywheel system has a projected electrical efficiency exceeding 95 percent, a flywheel-driven automobile could take effective advantage of regenerative braking. A regenerative braking system would employ the vehicle's electric motors as generators during braking or downhill driving, thus putting the kinetic energy of the vehicle back into the storage system. With batteries, which under such high rates of charge would typically return only 50 to 75 percent of the charging energy put into them, a large fraction of the regenerative braking energy would be lost. With a flywheel system, however, the range of the car could be increased by from 25 to 50 percent through regenerative braking.

Since the flywheel system in an automobile would be sealed in a partial vacuum, the loss of power from friction while the car was not in use could be made very low. For example, a family could leave a flywheel car at an airport for the summer while taking a trip abroad, and when they came back, the flywheel would still have ample energy to get them home. The rundown time for a rotor in a vacuum chamber is estimated at from six to 12 months.

A sealed flywheel unit would also require nothing in the way of maintenance except an occasional check on the vacuum. Indeed, from calculations of the tensile-strength fatigue rate of the materials in a flywheel unit one can project a lifetime of several years for the unit. One can even envision transferring the unit from one vehicle to another when the body of the first vehicle wears out.

Beyond the needs of today lie the energy-storage requirements of the future. As increasing emphasis is laid on conserving mineral resources, using energy efficiently and curbing pollution, energy storage will become increasingly important. For example, one can envision the development of solar power and perhaps wind power. For such intermittent sources of energy to become practical, economical and efficient, means of storing energy must be available. The improved flywheels we have described will meet this crucial need.

Prevention of power failures— The FPC report of 1967

Power demand in the U.S. is increasing at the rate of a geometric progression. Interconnections now cover vast geographic regional areas; hence, reliability of the bulk power supply system is the key criterion for the uninterrupted flow of electric energy

Gordon D. Friedlander *Staff Writer*

Twenty months after the Northeast blackout of November 9–10, 1965, the Federal Power Commission issued its three-volume report calling for the coordinated planning and operation of bulk power supply facilities to ensure maximum possible reliability and the prevention of future cascading tripouts and regional power failures. These and related guidelines are contained in 34 recommendations. As would be expected, the power industry, after evaluating this comprehensive report, may have some disparate reactions. A small sampling of these reactions by three investor-owned utilities—and a federal system—is herein presented.

The FPC's three-volume report, "Prevention of Power Failures," is a comprehensive and painstaking document that has been written and compiled in the best traditions of cooperation between the federal government and private enterprise. This is not the work of some government superagency dictating a set of fiat mandatory recommendations that must be followed, observed, and put into effect by private industry under the threat of punitive regulatory action for failure to comply. Instead, it is a detailed report of the recommendations and conclusions of the FPC for improving present and future electric power service; plus the Advisory Committee's own report on the reliability of electric bulk power supply; and, finally, a presentation of the studies and findings made by the various Task Groups on the Northeast power interruption of 1965 and subsequent outages that affected interconnected systems.

Significantly, the Advisory Committee and the Task Groups were composed of prominent electrical engineers and systems engineers, drawn from public and private utilities, a university, a state commission, and the manufacturers. Thus a cross section of power engineering practitioners and educators participated in the careful studies, investigations, and exhaustive legwork that went into the drafting of this important treatise.

Many improvements made—more needed

During the past two years, federal, state, and city agencies—and private industry—have responded well in eliminating a number of the shortcomings indicated by the massive Northeast power failure of November 9–10, 1965. Actions that have been taken on national, regional, or local levels include

1. A significant upgrading of communications and emergency lighting facilities at airports throughout the United States.

2. The development of emergency plans and procedures to ensure that teletypewriters, PBX phones, and other telecommunications systems be kept in operation and available to the public during service interruptions.

3. The initiation of a nationwide program to ensure that all civil defense and military warning equipment (red

Reprinted from *IEEE Spectrum*, vol. 5, pp. 53-61, Feb. 1968.

phones) have a constant power supply, free from any possible interruption.

4. The implementation of a program to equip federal government buildings with emergency lighting units in stairwells, elevator cars, transformer and switchgear rooms, control centers, and other areas vital to national security.

5. A grant of federal funds, by the Department of Housing and Urban Development, for emergency transportation facility improvements by providing emergency power for the movement of trains; the development of emergency radio, station lighting, and alarm signal equipment, etc.

6. The improvement of standby power sources for communications and control during all critical phases of space flight missions.

7. Formulation of rules by the City of New York to ensure the continuous safe functioning of vital hospital services and facilities during power emergencies.

8. Action by the Port of New York Authority to upgrade communication, lighting, and emergency evacuation procedures for vehicular tunnel and tube train facilities under its jurisdiction.

Although these eight measures represent a significant package of improvements, there are still deficiencies in many critical areas. For example, a recent national survey of statutory provisions for standby emergency power for essential services revealed that 22 states do not have such legislative requirements on their books.

Present status in the Northeast. In the year following the Northeast outage, the major utilities affected invested $20 million for new equipment and improvements to protect existing facilities and to decrease the possibility of future cascading power failure occurrences. Additionally, more than $30 million has been earmarked for further improvements that will be made just as soon as procurement and installations will permit. One problem in this area is that the demands for equipment have temporarily exceeded the manufacturers' and suppliers' production capabilities.

Future development by Northeast utilities

The Fig. 1 map shows the principal transmission network elements in the New York State and New England areas, with interconnections to the PJM pool and to the service area of the Hydro-Electric Commission of Ontario. The map also indicates the lines added since the extensive blackout of 1965, as well as those scheduled for installation through the year 1973. In the eight-year period from November 1965 through 1973, 2250 km of 345-kV circuits and almost 500 km of 500-kV circuits either have been or are scheduled to be placed in service in the Northeast.

The peak load estimated for the Northeast systems—exclusive of Ontario Hydro—for the winter of 1973–1974 will be 31 200 MW, an increase of 35 percent above the area peak load projected for the winter of 1967–1968. Planned generating capacity in the seven-state area by the end of 1973 will be 40 300 MW, or 44 percent more than the capacity in 1967.

Provisions for load shedding

According to Volumes I and II of the report, the best insurance against a major power failure is sound planning plus a well-designed and -operated bulk power supply system. It is possible, however, that an unexpected incident can isolate a utility or power pool from the network; and, if the isolated area created is deficient in generation, loads must be reduced quickly to prevent a total collapse of the power supply system. Therefore, automatic controls are essential for this purpose because manual load shedding is usually too slow to be effective in an acute emergency situation. This fact was dramatically demonstrated during the 1965 blackout when the southeast New York and New England utilities became an "electrical island" five seconds after the disturbance began; but these isolated systems did not collapse until 12 minutes after their separation. In that brief period, system voltage and frequency fell far below normal and the capability of generators that remained in operation diminished rapidly.

At the time of the November 9 interruption, no utility in the CANUSE network used automatic load shedding. This was in contrast to some other areas of the United States in which this procedure is utilized. At the present time, however, load-shedding procedures have been established at all the major utilities in the Northeast, largely through the efforts of the Northeast Power Coordinating Council.

Service to New York City

The Northeast power failure hit New York City particularly hard. At the time of the massive disturbance, the Consolidated Edison Company of New York was meeting a demand of 4770 MW, but the installed capacity of the company's generating equipment was about 7580 MW. The Con Edison system collapsed because of the effects of varied and dispersed deficiencies in the interconnected system of which it was a part. These deficiencies included inadequate coordination between Ontario and

FIGURE 1. Map showing major transmission lines and generating stations, both existing and proposed, in the New York and New England power system (as of January 1967).

FIGURE 2. Map indicating the participating member systems of the Northeast Power Coordinating Council and their respective service areas.
1. Boston Edison Co.
2. Central Hudson Gas & Electric Corp.
3. Central Maine Power Co.
4. Central Vermont Public Service Corp.
5. Connecticut Light and Power Co.
6. Consolidated Edison Co. of N.Y., Inc.
7. Eastern Utilities Associates.
8. Green Mountain Power Corp.
9. Hartford Electric Light Co.
10. Holyoke Water Power Co.
11. Hydro-Electric Power Commission of Ontario
12. Long Island Lighting Co.
13. New England Electric System
14. New England Gas and Electric Association
15. New York State Electric & Gas Corp.
16. Niagara Mohawk Power Corp.
17. Orange and Rockland Utilities, Inc.
18. Power Authority of the State of N.Y.
19. Public Service Co. of New Hampshire
20. Rochester Gas and Electric Corp.
21. United Illuminating Co.
22. Western Massachusetts Electric Co.

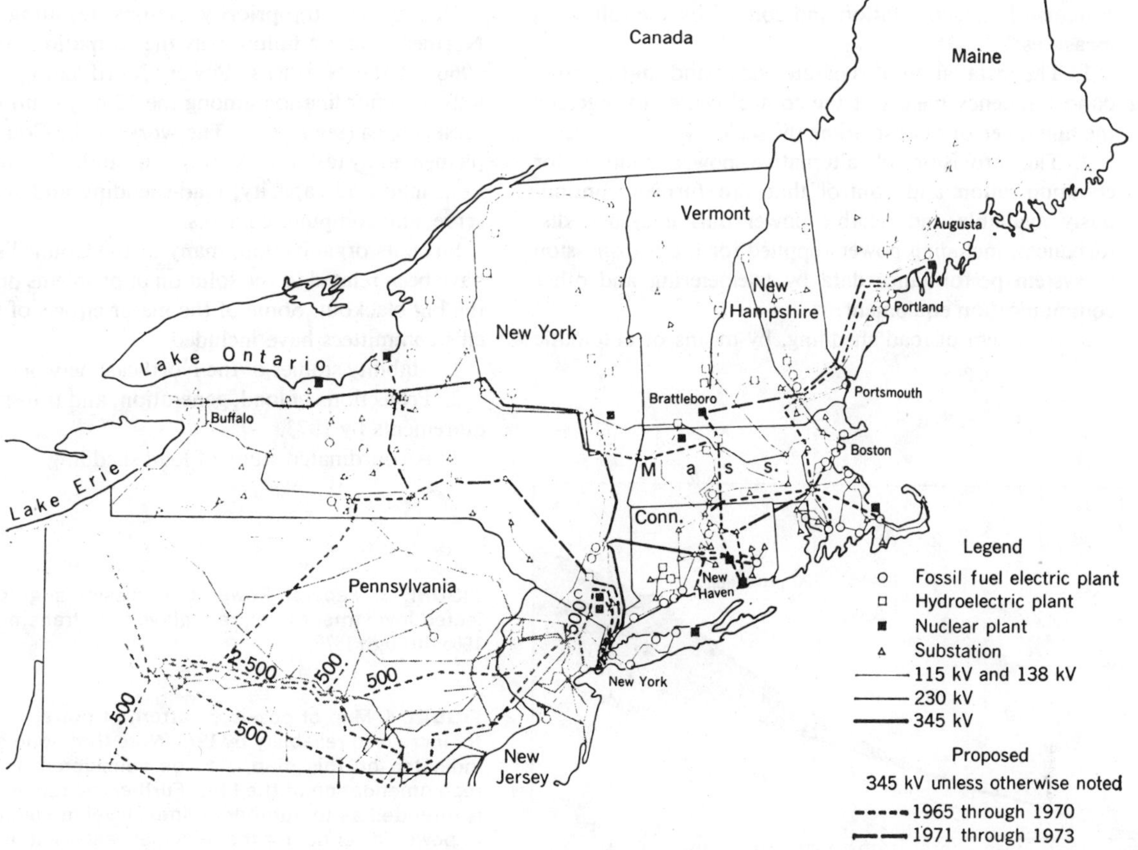

Canada
Maine
Vermont
Augusta
New Hampshire
Portland
New York
Lake Ontario
Portsmouth
Brattleboro
Buffalo
Boston
Mass.
Lake Erie
Conn.
New Haven
Pennsylvania
New York
New Jersey
500
2-500
500
500
500
500

Legend
○ Fossil fuel electric plant
□ Hydroelectric plant
■ Nuclear plant
△ Substation
115 kV and 138 kV
230 kV
345 kV

Proposed
345 kV unless otherwise noted
1965 through 1970
1971 through 1973

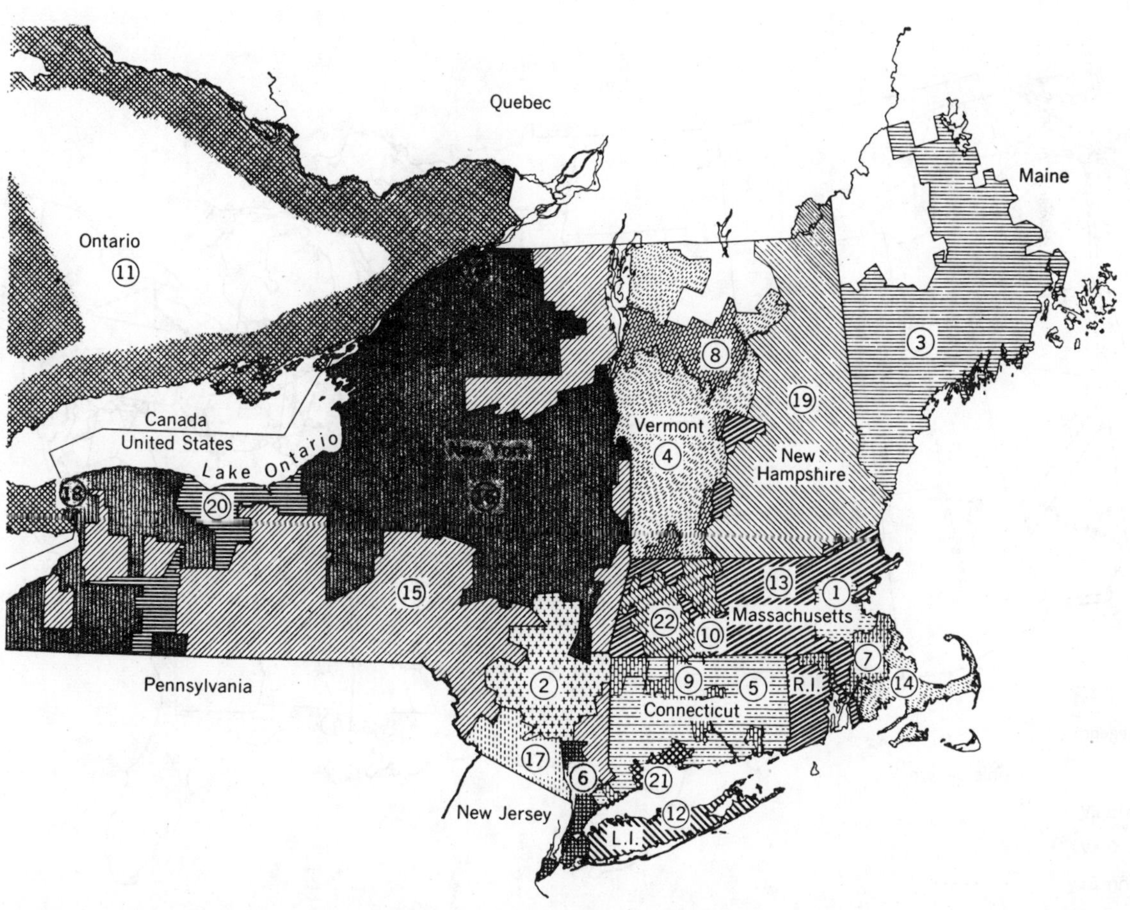

Friedlander—Prevention of power failures

New York, inadequate transmission and interconnection in the overall network, and inadequate coordination in operation among the systems.

During the past two years, Con Edison has improved its systems of instrumentation and control by the following measures:

1. The installation of separate wide-band and narrow-band frequency meters at the control center to preclude the human error of misreading the scale.

2. The provision of alternative power sources for communication and control that can furnish continuously available and reliable power during system disturbances, including power supplied for the transmission of system performance data by telemetering and other communication equipment.

3. A system of load shedding, by means of automatic voltage reduction, that will be applied in two steps by the action of underfrequency relays.

The Northeast Power Coordinating Council

One of the top-priority actions resulting from the Northeast power failure was the formation, in January 1966, of the Northeast Power Coordinating Council to improve coordination among the 22 major utilities in the CANUSE area (see Fig. 2). The work of the Council is supplemented by task forces on system studies, system protection, load and capacity, load-shedding and spinning reserve, and computer controls.

Since its organization, many of the Council's activities have been related to the solution of problems disclosed by the big blackout. Some of the major efforts of the Council's committees have included

1. Stability studies of the Northeast network.

2. Projections of load, generation, and transmission requirements by 1973.

3. A coordinated study of load shedding.

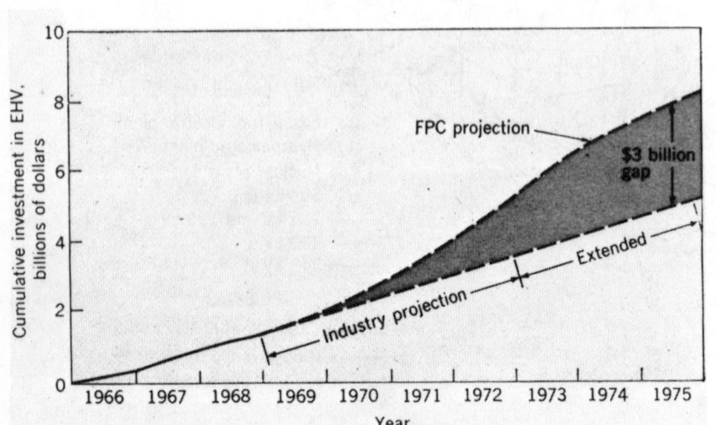

FIGURE 3. Federal Power Commission's graph of projected investment in extra-high-voltage transmission from 1966 through 1975.

FIGURE 4. Map of possible pattern of power transmission for increased reliability by 1975. Note that none of the lines indicated on this map is to be considered as a specific recommendation of the FPC. Further, no recommendation is intended as to number of lines, level of voltage, or type of power (ac or dc) for the principal east–west ties between areas marked with an asterisk (*).

Legend

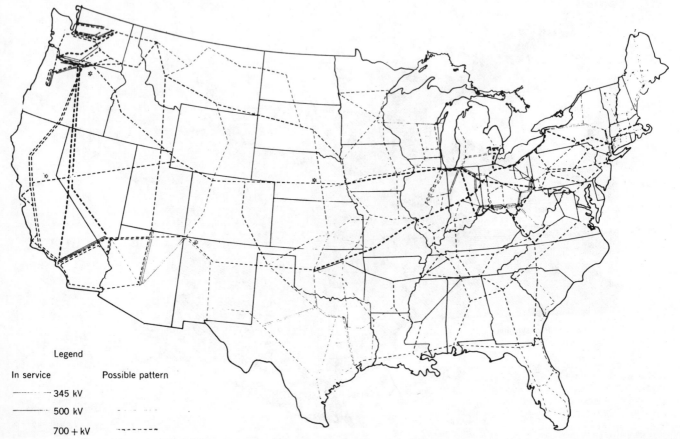

In service — Possible pattern

345 kV

500 kV

700 + kV

The member utilities of the Coordinating Council recently adopted an automatic load-shedding program, which provides that each system will be equipped with underfrequency relays to drop 10 percent of system load if the frequency declines to 59 Hz, and 15 percent more at 58.5 Hz. Each system will also provide for the manual dropping of an additional 25 percent of its load whenever warranted by emergency conditions.

The Northeast Council is one of several similar organizations, either established or now being established throughout the United States, for the purpose of achieving better reliability of bulk power supplies.

Possible pattern of EHV transmission

As outlined by the FPC, the possible pattern of EHV transmission lines required by 1975 would involve an investment of about $8 billion—or 12.5 percent of anticipated expenditures for all power facilities to be built during this seven-year period. This is about $3 billion more than the utilities apparently plan to spend (see Fig. 3), according to the FPC.

In the general pattern projected for 1975 (see Fig. 4 map), the additions in EHV lines beyond those put in service in 1967, include 25 800 km of 345-kV, 34 500 km of 500-kV, 9250 km of 765-kV, and 2680 km of approximately 750-kV dc transmission circuits.

The depicted scheme indicates that the utilities in the Northeast and Southeast will be more strongly integrated with the central body of utility systems in the East. An extension of the 765-kV transmission system, now being constructed in the east–central area, is shown in Fig. 4 to overlay the 345-kV network now under development in New England. This would enable major power flows to occur between the heart of New England and the central eastern section of the United States.

The Southeastern utilities are shown to be interconnected with several 500-kV loops that join an existing system of corresponding voltage that has already begun to span a major section of the Eastern Seaboard from north to south.

Strong north–south interconnections from the heavy industrial load areas of Illinois, Michigan, Ohio, Indiana, and western Pennsylvania to the utilities in the southeastern and Gulf state areas are shown extending through the TVA and CARVA regions. Here, American Electric Power (AEP) has an extensive (2900-km-long) EHV network interconnecting its six subsidiaries throughout a seven-state area from Michigan to West Virginia. A 765-kV overlay network is scheduled for initial operation by 1969 and completion by 1971.

Conclusions and recommendations of the FPC

This major section of the FPC report contains 34 subsections under nine subheads or subjects. A précis of these subsections is presented next in a numbering sequence that conforms to the full text and exact subheads as given in the original document—

Formation of coordinating organizations

1. Strong regional organizations should be established for the coordination of the planning, construction, operation, and maintenance of individual bulk power supply systems. These coordinating organizations should have financial support and representation from all participating utilities.

2. A Council on Power Coordination should be established, composed of members from each regional coordinating organization, to exchange information and to review, discuss, and assist in resolving matters that affect interregional coordination.

3. A Central Study Group, or Committee, should be established to coordinate industry efforts in investigating some of the more challenging problems of interconnected system development. This entails the early coordination of ideas, efforts, and funds required for more effective R&D in planning and operation. Much of this work could be performed under contract with qualified universities and research institutions.

Interconnected system planning

4. Early action should be taken to strengthen the Northeast transmission systems. And the peninsular relationship of New York and New England to adjoining systems demands highly coordinated planning of transmission. For example, the projected 500-kV interconnection to the PJM network is urgently needed. The FPC also recommends an early reinforcement of the Northeast network and more ties to PJM and systems in Ohio and western Pennsylvania.

5. Transmission facilities should be critically reviewed throughout the nation, and planning and construction of needed additions should be accelerated on schedules which provide sufficient capacity to meet the potential requirements of both reliability and economy. Transmission networks and interties between areas are now deficient in numerous locations. Networks should be planned and tested for their ability to remain stable under severe disturbances. The pace of construction should enable transmission capability to lead rather than lag behind emergency requirements.

6. In estimating future loads, full attention should be given to economic trends, potential weather extremes, and growth in specialized uses of electricity (electric heating, air conditioning, etc.) **in each load area.**

7. Lead times for planning and constructing major new facilities should be established that will avoid delays in meeting completion schedules and impairment of system reliability. In comparison with past practice, time extensions of one to two years may be needed for large components. In many cases, it may be necessary to develop relatively firm expansion plans not less than six years in advance of need.

8. Utilities should solicit the participation of interested parties at an early date in the resolution of problems relating to the location and environmental effects of new facilities. Utility planning, in addition to technical factors, involves careful attention to facility location, the satisfactory control of air and water pollution, and the preservation of esthetic values.

9. Special attention should be given to transmission line routing and to switching arrangements in the transmission network to provide maximum reliability in emergencies. Unusual care should be taken to prevent the excessive concentration of critical circuits, which would expose the system unnecessarily to large loss of capability.

10. The size of generating stations, magnitude of area loads, and the capability of the transmission system should be maintained in proper balance. Generating capacity that is too large in relation to the capability of the interconnecting transmission lines and area load concentrations can impair the reliability of supply.

11. Sufficient transmission should be provided to avoid excessive generating reserve margins. Limited transmission capacity tends to result in generating reserve margins larger than those justified by either economic or reliability considerations.

12. A workable number of control centers should be established in each region. The Northeast power failure dramatically emphasized the inability of operators in many centers to have significant communications with each other. Plans are under way to set up two central control points—one in New England, and the other in New York. This simplification should improve the coordination of the Northeast systems for both normal and emergency operations.

13. Relay protection should be continually updated to fit system changes and to incorporate improved relay devices. Relays are key elements in achieving reliability of bulk power supply. Since they are relatively inexpensive components, their adequacy, quality, and periodic readjustment should not be compromised for the sake of negligible economies.

14. Utilities should concentrate on opportunities to expand the effective use of computers in power system planning and operation. Many new applications are being found for digital computers in these areas. Specifically, computers may be used for the collection and printout of operating data, including warning of conditions that are approaching or exceeding safe limitations; the automatic control of generators to meet changing loads economically; increased automation of generating stations; rapid analysis of networks to determine line-load limits, etc. Further progress is needed to use computers reliably for on-line analysis and control of power systems during disturbances.

Interconnected system operating practices

15. System control centers should be equipped with display and recording apparatus to provide the operator at all times with as clear a picture of system conditions as is possible. Desirable displays include narrow- and wide-range frequency indicators, tie-line and principal line flows, lines out of service, switch positions, overload conditions, generators in service, unit and plant outputs, spinning reserve and rate of response, voltages and frequencies at key points, area control error, and appropriate alarms.

16. Communications should be supplied with continuously available power so that information on system conditions can be transmitted to control centers during disturbances. Whenever power supply for communications equipment deviates beyond specified limits, the equipment should be automatically and instantly switched to an emergency power source.

17. Control centers should be provided with a means for rapidly checking on stable and safe system capacity limits. Rapid security checks—now feasible by the use of digital computers—to determine that various elements will be operated within safe limits are essential to prevent hazardous loading.

18. Spinning reserves should respond quickly to a level which can be sustained in meeting emergency power demands. Rapid response normally requires that the reserve be distributed among many units.

19. Coordinated automatic load shedding should be established to prevent the total loss of power in an area that has separated from the main network and is deficient in generation. Load shedding should be regarded as an "insurance program," however, and *should not* be used as a substitute for adequate system design.

20. Plans should be made, and tests conducted, for the quick isolation of generating units to keep them in operation if collapse of system power is imminent. Carefully planned switching procedures should isolate appropriate units quickly during an emergency. Smaller units may be isolated primarily to ensure restarting power for larger units.

21. Emergency power should be available at all thermal plants to prevent damage to turbogenerators during rundown if system power is lost. Pressures must be maintained on bearing lubrication and hydrogen sealing systems. Emergency power should be provided to operate the turning gear and pumps of these systems, to keep the control system operable and to provide lighting in the control room.

22. Auxiliary power should be available to principal thermal plants to enable rapid restarting if system power is lost. Adequate emergency power can save hours of time in service restoration.

23. Thorough programs for operator training should be vigorously administered, with particular attention to procedures for a broad spectrum of potential contingencies. Close coordination among the planning, operating, and maintenance staffs is indispensable.

Interconnected system maintenance practices

24. Programs of system maintenance should be strongly directed toward preventive rather than remedial maintenance.

25. Manufacturers and utilities should promptly disseminate information on troubles or equipment failures. The FPC will also disseminate information on bulk power outages and selected power interruptions.

26. The isolation of any system elements for testing, repair, or replacement should be scheduled by, or receive the clearance of, the operating department.

Criteria and standards

27. Criteria and standards for planning, construction, operation, and maintenance of power systems should be formulated so that each system can be reasonably assured that its own service will not be adversely affected by its neighbors' policies.

Defense and emergency preparedness

28. Although severe damage can be inflicted upon power systems by enemy attack, cascading failures should not ensue. Steps to improve reliability will strengthen utility systems in resisting widespread failures if subjected to wartime attack. Utilities generally have adequate security programs for normal requirements, but many should increase preparedness for the contingencies of enemy attack.

29. All levels of government should establish requirements for emergency power for essential services. More than 50 percent of the states now require auxiliary power for certain critical loads, and this practice should be extended. Thus the FPC urges state, county, and local governments to encourage and direct the planning and installation of auxiliary power facilities to furnish essential and vital services for public safety and welfare.

30. Utilities should cooperate with public officials and customers in planning and maintaining customer standby facilities to ensure service to critical loads in an emergency. Typically, these services include hospitals, police and fire departments, sewage and water plants, transportation systems, communications, and emergency service in buildings that normally contain many people.

Manufacturing and testing responsibilities

31. Manufacturing capacity of electrical equipment suppliers should be expanded to meet future needs. Better liaison on projected requirements between utilities and manufacturers is needed, and this will be aided by improved planning and coordinating procedures.

32. Facilities are needed in the U.S. for more extensive testing of EHV equipment. The Commission urges the early consideration by the electric utility industry of needs, development of plans, early construction of appropriate high-voltage testing facilities, etc., so that the reliability of future power supplies will not be impaired by lack of proper testing.

Increased need for technical proficiency

33. The industry should make young people cognizant of the full challenge of modern power systems engineering. Utilities should work more closely with educational institutions to develop and sponsor appropriate research, to utilize cooperative programs for students and industry assignments of educators, and to exploit opportunities for new and sophisticated R & D.

Power system practices in other countries

34. System design and operating practices in other countries are generally similar to those in the U.S., and foreign power systems are experiencing similar problems in planning and operation. The practice of exchanging technical information in these areas with foreign countries should be continued and expanded.

Synopsis of Volume II . . .

The FPC Industry Advisory Committee report on electric bulk power supply, released as Volume II of the Commission's report, concluded that the sound application of bulk power reliability principles should eliminate widespread or cascading interruptions to service from all "credible" contingencies.

This committee, established by the FPC shortly after the Northeast blackout, advised the Commission that cascading power failures result basically from inadequate transmission facilities within and between power systems, and urged the increased use of properly planned high-capacity transmission. The committee also recommended regional and interregional coordination (as already reported in this article) and proposed a nationwide council on power system coordination.

The Advisory Committee report contains considerable technical information related to various aspects of power system planning and operation—including recommended practices for dependability and reliability of electric bulk power supply systems.

. . . and Volume III

Volume III of the FPC report contains several studies of the Northeast power network, which were made following the 1965 power failure. Selected conditions were studied in detail by task groups under the direction of the Commission's Advisory Panel on the Northeast Power Interruption. The volume is composed of six sections that present the results and conclusions of the several study groups.

Reaction and comments

To provide the reader with a more comprehensive overview of the FPC report than can be presented by a recitation of the Commission's recommendations and conclusions, the writer has attempted to solicit a sampling of reactions and comments from a small segment of the power industry.

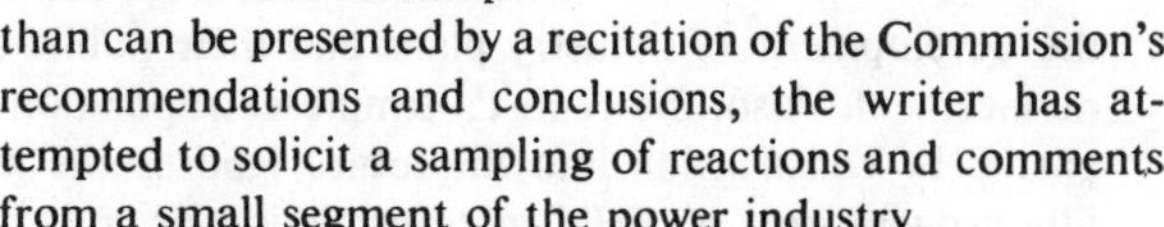

American Electric Power Company (AEP). A spokesman for AEP, the huge, fully integrated utility that provides electric power in parts or all of seven east-central states, believes that the industry can design power supply systems to prevent widespread blackouts, if the various major system components, comprising generating plants, transmission lines, and interconnections with neighboring systems, are planned as an integrated whole, with proper consideration given to their interrelated effects; in other words, if planning is carried out on a truly system basis. If this is done, instantaneous outages can be restricted to small, discrete geographical areas. But this goal cannot be achieved if coordination in system design is ignored because of either an overemphasis on immediate economies, or inadequate attention given to the mutual effects of the actions of other systems.

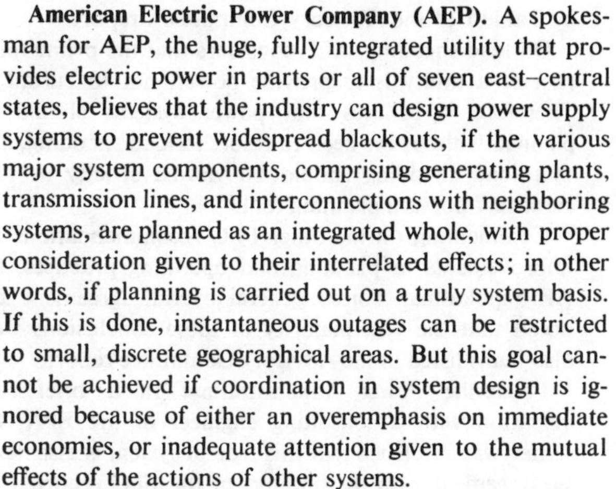

Although agreeing with and subscribing to the concept of regional coordination, AEP felt that the representation and functions of the regional councils, boards, and committees—as proposed in the FPC report—are too broad and diverse. These defects would result in an unmanageable organizational structure, which, in many instances, would attempt to do what the individual systems and pools should be doing and, in turn, would result in an extravagant waste of technical talent.

The regional councils should pursue reliability as their objective and not complicate this primary purpose with economic and other factors. These latter objectives can best be accomplished within the individual systems and pools. Representation on such councils should come from systems that can demonstrate that they have a substantial effect on the reliability of the bulk power network and can contribute to its achievement.

The company feels that to keep planning viable, it must be initiated within the individual utility, or pool of interconnected utilities, and then be submitted to a council for review.

It was also pointed out that some of the industry's problems could best be solved by greater integration of systems under common ownership. Although there are about 3600 individual utilities in the United States, only about 100 of these supply almost 90 percent of all electric power. Thus, to improve overall planning and operation, integration (either by merger or consolidation) to 10 to 20 major systems might be considered as an

eventual goal for greatly improved reliability and service. So far, the FPC has avoided the integration and consolidation issue and, instead, has concentrated on system pooling and coordination.

In regard to future construction of EHV transmission lines, the company believes that the projected FPC estimate is open to question, since it fails to take into account the inevitable time lag between the development of industry plans and their public announcement. Also, the report's simple extrapolation of past EHV construction into the future ignores the rapidly growing role of EHV in transmission as indicated by Table I of the Advisory Committee's report.

On load shedding, AEP considers that the prime function of a utility is to provide power, not to curtail it. Therefore it believes that power systems should be planned without the need to resort to load shedding, and that load shedding should be viewed strictly as "insurance," not as a planning tool.

Commenting on the other FPC recommendations for improvements in planning and operating practices, the AEP spokesman voiced general overall agreement and noted that the vast majority of these conclusions were also items listed in the Advisory Committee report.

Bonneville Power Administration. In a brief conversation with a consulting engineer of the BPA, the writer was informed that this federal system is essentially in agreement with the findings of the FPC report. Like AEP, BPA—although conceding there is a variance of opinion in the industry—regards load shedding as a "last-ditch insurance," but considers load shedding definitely preferable to loss of generation due to sinking frequency. Also, BPA believes that voltage regulators should be kept in service during severe disturbances.

Regarding the EHV "transmission gap," this spokesman agrees with AEP that the projected deficiency (Fig. 3) indicates too wide a margin. Sophisticated design and control is an effective substitute for some aspects of brute capital expenditure.

Finally, maximum credible-incident studies and maximum potential-disaster studies should be simultaneously undertaken by all networks to ensure an absolute minimum probability of a cascading outage.

Pacific Gas and Electric Company. This utility submitted comments on ten of the FPC's recommendations:

On Recommendations 1 and 2, PG&E feels that carefully executed studies by competent and knowledgeable engineers are essential for the proper understanding and evaluation of large interconnected systems. Organizations have been formed, or are evolving, to conduct such studies on a continuing basis in a number of regions.

To function effectively, however, these organizations should be of manageable size, and include engineering representatives from the principal bulk power transmitting and generating agencies. The studies require the best available techniques in system representation, computer programming, and the best computer facilities.

Regarding Recommendation 5, which, in effect, suggests the application of criteria on a national basis for the design of transmission systems, the company believes this suggestion to be of questionable value because it does not adequately recognize the widely divergent conditions (climate, geography, etc.) in different regions of the United States.

In reference to Recommendation 7, PG&E suggests *adequate* lead time for planning and building new facilities. Although it is easy to ask for longer lead time, sound engineering practice also suggests that lead times be kept *as short as possible* to assure that decisions are based upon the latest trends in load growth and state of the art in equipment and facilities. Lead time for major projects should be predicated upon the necessary planning, engineering design, manufacturing, and construction time requirements. Any factors that increase this time may militate against the project in terms of cost and latest available technologies.

The company feels that Recommendation 9, which calls attention to the importance of transmission-line routing and switching arrangements from the viewpoint of reliability in emergencies, is generally sound—but again it does not consider the tremendous range of topographic and climatic conditions encountered in various sections of the country, which are, necessarily, a determining factor in situating facilities.

Commenting upon Recommendation 14, which relates to the use of computers in power system planning and operation, PG&E considers that the stated objectives are sound in principle, and many of them have been, or are being, adopted by progressive utilities. Computer applications presents one of the most challenging technical areas facing the contemporary power system engineer; however, the industry recognizes that some of the more sophisticated proposed applications require considerable R&D to ensure the fulfillment of overall system reliability criteria.

On Recommendation 18, it is felt that emphasis could be added to the thought on spinning reserve. The time-rate of response is vital to effective action of spinning reserve in an emergency situation. More information, based upon field tests, on the pickup rates of actual machines would be desirable, and this can now be factored into system studies.

On Recommendation 19, urging the establishment of automatic load shedding as a backup to other emergency measures, PG&E's views are similar to those of Commonwealth Edison (presented in the next section). The effectiveness of load shedding as a means of providing better service in an emergency has been demonstrated and documented on a number of occasions for systems in California and elsewhere. Nevertheless, system planners generally agree that this practice is not a substitute for adequate system design.

In discussing the last two recommendations, nos. 33 and 34, the utility thinks these are "seemingly self-evident," but many aspects of large regional system analysis, design, and operation involve engineering concepts that are relatively new to the industry. The shifts to higher voltages, direct current, and larger generating unit sizes offer many challenging areas of technological development to the young engineer. Thus the industry must make every effort to present and explain these challenges. Article no. 34 calls attention to the obvious fact that we in the United States do not live in a technologically isolated nation;

other countries also have excellent engineering talent and innovative ingenuity. By active participation in organizations such as IEEE and CIGRE, in contacts with foreign engineers, and in our own travels, we should be alert to these developments in other countries.

Commonwealth Edison Company. This large, Chicago-based utility believes that the Northeast blackout focused attention on the need to improve coordination of planning and operation of the vast bulk power systems that are being built. The advent of EHV transmission (345 kV and higher) and very large generating units (500–1100 MW) has produced a marked change in electric power systems, and a new dimension in power system planning has been introduced. Although the basic principles for designing these systems have been well established, practical experience has been lacking.

Several important lessons were learned from the Northeast blackout and the studies that followed. These were in the areas of—

Regional planning. Working with today's large generators and EHV lines, the planning of an individual utility's needs, without regard to the effects on its neighbors, can lead to dangerous conditions. This situation was recognized before the blackout when regional organizations such as MAIN, MAPP, CAPCO, etc., were organized to coordinate the planning efforts of the bulk power systems in the regions covered by these groups. The Northeast power failure indicated the need for such groups throughout the U.S., and their establishment is now well under way.

Criteria for transmission planning. Although it is axiomatic that a strong transmission system is a basic requirement for protection against cascading outages, the problem is to determine what constitutes an adequate transmission system—after considering all economic factors and system reliability.

Volume II of the FPC report presents guidelines for the tests to be applied for determining the adequacy of a bulk power system, and these are a good starting point in studying the needs for transmission in a particular area. However, conditions affecting bulk power reliability vary greatly throughout the country. Thus it would be unwise to establish uniform planning criteria for transmission design on a national basis. Such variables as load density, climatic conditions, area geography (see PG&E comments), the availability of hydro power, transmission-line concentration on a single right of way, etc., must affect final decisions regarding transmission system design. Further, the need for flexibility in planning criteria also applies to the size of generating units and margin of reserve required.

Although the FPC report indicates that the utilities should invest an additional $3 billion above their present plans for transmission over the next ten years, the Commission has not documented its argument to prove the necessity for this additional transmission increment. Commonwealth Edison believes that utilities constructing the bulk power systems can be relied upon to provide adequate facilities; anything beyond the requirements determined by the utilities will represent an uneconomic investment. Parenthetically, it is entirely possible—based on past history—that the utilities will install even more transmission than that recommended by the FPC. This will depend upon the results of studies to determine the future need for EHV lines.

Load shedding. The utility feels that the use of load shedding to check a condition of frequency reduction is a powerful tool for preventing large-scale blackouts. The basic cause of an area-wide blackout is the shortage of generating capacity, resulting from system separation, which leaves a portion of the system with load in excess of power supply. In such an emergency, the simplest method for restoring stability is to cut off load either by manual switching or the automatic opening of switches by the use of underfrequency relays.

Commonwealth Edison adopted the principle of load shedding many years ago and, in 1963, installed automatic relays for this purpose. On one occasion, a widespread blackout was prevented by manually disconnecting about 10 percent of the system load. In less than one hour's time, the system was stabilized and the load was returned to normal.

The Northeast blackout convinced many power companies that the principle of load shedding—if properly applied—provides good insurance against system failure.

Computers. After the 1965 power failure, a school of thought developed in support of the thesis that bulk power systems had become so complex that it was necessary to automate their operation completely. The utilities generally have been major users of computers, both for system planning and system operation. But experience has indicated that the changeover to automation must evolve gradually over a period of time to "prove in" and coordinate the various automated components. The industry consensus is that available computers do not have the capability and reliability to perform correctly *all* of the necessary functions in the operation of a bulk power system. Also, the communications systems required to provide the input data to the computer are presently inadequate for the successful automated operation of a power system.

The utilities, however, *are interested* in expanding the role of the computer in system operation, and many studies are now under way to develop plans for greater usage of the computer in bulk power systems. The next stage of expanded use of the computer should probably be in assisting system operators, but the final decision on action to be taken in various emergency situations should be left to the operators.

The Commonwealth Edison Company concludes that the FPC report has been helpful to the power industry in forcing discussions and studies of the various factors that affect the reliability of the bulk power systems in the U.S. The company believes that the power industry is well organized to use the tools available for planning and operation, and can be depended upon to provide the adequate electric service demanded in this country, both in quantity and quality.

BIBLIOGRAPHY

"Prevention of power failures, a report to the President by the Federal Power Commission," (in three volumes), U.S. Government Printing Office, Washington, D.C., 1967.

Transmission of Electricity via Superconductors

STATEMENT PREPARED BY JOHN P. BLEWETT, DEPUTY CHAIRMAN, ACCELERATOR DEPARTMENT, BROOKHAVEN NATIONAL LABORATORY, UPTON, N.Y.

INTRODUCTION

Estimates made by industry and its consultants of the demand for electric power by the end of this century yield factors varying from four to ten of increase over present power production. Extrapolation at the rate of 7.5% per year which has been the average rate between 1920 and 1970 yields a factor of about 7. The present capacity of about 340 million kilowatts will thus increase to about 3400 million kilowatts.

The transmission system for this increased supply of power may well increase in size and cost by a factor larger than 7 as power stations are located at greater distances from the centers where the power is consumed. This trend is already visable, for example, in the Pacific Northwest-Southwest interties.

CONVENTIONAL TRANSMISSION LINES

The electrical system in the United States is based on generation and use of three-phase current alternating at 60 cycles per second (now referred to as 60 Hz).

The cheapest method for transmission of power is to transform to high voltage—69 or 138 kV on short lines and up to 765 kV on long lines—and to transmit on overhead lines hung from towers. Typical costs of such lines are about $100 per million volt-amperes (MVA) per mile exclusive of the cost of the 100 ft. or more right-of-way required.

In urban areas where real estate costs are prohibitive, or in scenic regions where strong public pressure forbids overhead lines, power must be transmitted underground in buried cables whose cost per MVA-mile may be ten to twenty times higher than that of conventional overhead lines. The standard cable consists of three paper-insulated conductors in a welded steel pipe filled with oil at a pressure of about 200 pounds per square inch. This is known as high-pressure, oil-filled (HPOF) cable.

The power-carrying capacity of underground cable is much less than that of overhead cable for two reasons. First, heat which is generated by resistive losses in the cable is presently removed by conduction through the oil into the surrounding earth. Although buried cables are usually surrounded by special high-conductivity sand, heat is not removed at a rate that can compare with the rate for an air-colled overhead line. Second, out-of-phase currents are generated due to the close proximity of the ground; thus the input power is diverted into a form that is useless to the customer. Unless expensive compensating equipment is added, this effect limits the useful length of a high capacity underground cable to about 20 miles.

Some improvement in capacity has been achieved by circulation of the oil through cooling stations located every half-mile or so along the line. The power limit for a conventional HPOF cable is about 900 MVA at 550 kV; by forced cooling this can be increased by a factor of about two.

Several advantages are gained by conversion ("rectification") of the generated alternating current to direct current, transmission as direct current, and "inversion" at the point of use back to alternating current. DC lines are free from the out-of-phase current effect that limits the length of underground ac lines. The dc transmission line can be built for about half the cost of a comparably rated ac line. However, the rectification and inversion equipment is expensive and dc transmission is economically justified only in an overhead line several hundred miles long or in special cases like interties between ac systems where the alternations of the current cannot be synchronized. An example of this is the dc tie between the British and French electrical networks.

At present the electric power industry is experiencing two conflicting public pressures—one to increase production and distribution of power and the other to replace overhead transmission by underground transmission. The conclusion is that underground lines must be evolved with much greater capacity for power transmission. This requirement has led to very serious studies of operation at low temperatures and of the possible use of superconductors. As a result of developments during the past decade it now appears that superconducting power lines may provide a satisfactory solution.

SUPERCONDUCTIVITY

At a temperature about 460 Fahrenheit degrees below Fahrenheit zero, or 273 Centigrade degrees below Centigrade zero, Nature seems to have erected an impassable barrier. No lower temperatures can be achieved. Physicists use this point as the zero for a new scale called the absolute scale having degrees of the same size as Centigrade degrees. Temperatures on this scale are in degrees absolute or in °K (K in honor of Lord Kelvin). On this scale, room temperature is about 300°K.

As temperature is lowered toward the absolute zero all substances that are liquid at room temperature, freeze, and all that are gases, liquefy, and all but one freezes Nitrogen, for example, liquefies at 77°K and freezes at 63°K. Hydrogen becomes liquid at 21°K and freezes at 13°K. The last to liquefy is helium, at 4.2°K. Apparently it does not freeze except under high pressures.

The resistance of metals to the passage of electric current, the property that causes losses and results in generation of heat in power lines, diminishes with reduction in temperature and appears to be headed for zero at absolute zero. In 1911 a Dutch physicist, Heike Kamerlingh Onnes at Leiden was investigating this behavior, using mercury, which he had been able to make very pure. The resistance which was decreasing as the temperature was lowered, began to level off at about 14°K then, at 4.2°K it dropped suddenly and discontinuously to zero. It was really zero. A current induced in a loop of frozen mercury will still be circulating a year hence provided the temperature is not allowed to rise above 4.2°K, its so-called "critical temperature."

Other metals were soon shown to become superconducting each at its own critical temperature, mostly in the range between 1 and 10°K. There was a flurry of hope that superconducting electromagnets might be developed requiring no power for their operation. But it soon developed that superconductivity is destroyed by too high a magnetic field—and the critical fields of the then-known superconductors were too low to be of practical interest.

For half a century, superconductivity remained a laboratory curiosity. It was not even understood until 1957, when a group at the University of Illinois produced a quantitative theory. Activity increased in the field and shortly thereafter came a major breakthrough. At the Bell Telephone Laboratories it was shown that a compound of niobium and tin (Nb_3Sn) could retain its superconductivity to a magnetic field of over 200 kilogauss, ten times as high as the peak field in most conventional electromagnets. Moreover, Nb_3Sn proved to have a relatively high critical temperature, 18°K.

During the past decade activity in the field of applied superconductivity has been increasing rapidly. Many new high-field superconductors have been discovered with critical temperatures up to 21°K. High-field superconducting electromagnets have been built for high-energy physics research. These magnets have been used in very large particle detectors and promise to substantially reduce the cost of multi-billion volt particle accelerators. This work has been supported mainly by the Research Division of the Atomic Energy Commission. Design studies and models have been made of superconducting rotating machines. Indeed, a 3000 horsepower pump motor with superconducting windings has been tested at a power station in England. Superconducting motors and generators can be half as large as conventional machines for the same power. They are being studied intensely by the electrical industry and by the U.S. and other armed services.

SUPERCONDUCTING POWER TRANSMISSION

During the past five years work has begun on superconducting power transmission, at first somewhat desultorily with the utilities or with the electrical manufacturing industry. At present the two most advanced groups in the United States are at the Linde Division of Union Carbide, supported by the Edison Electric Institute and the Department of the Interior, and at the Brookhaven National Laboratory, supported by the National Science Foundation with, it is hoped, supplemental support from the Atomic Energy Commission. Proposals for new studies have been, or soon will be submitted by the Oak Ridge National Laboratory, the Los Alamos Scientific Laboratory, and Stanford University. Studies, some relatively advanced, are in progress in the U.K., France, Germany, Austria, Japan, the U.S.S.R., and possibly elsewhere. It was recently reported (New Scientist, May 4, 1972) that an experimental 390 megawatt superconducting cable is in operation in Russia.

Several interesting features of superconducting power lines already are clear:

1. They are peculiarly suited to the transmission of large blocks of power. The large fixed costs of refrigeration prevent them from being competitive at low power levels. But at power levels which are about the ceiling for conventional lines, the costs become comparable. At this point superconducting lines are still capable of power levels a factor of ten higher. In the range above 2000 MVA superconducting cable cost per MVA-mile should be about half, or less, of the cost of conventional underground cables operating in the range around 500 MVA.

2. Superconducting lines will operate, for the same power, at higher currents and lower voltages than those suitable for conventional underground lines. This mode of operation will give an important reduction in the generation of

Reprinted with permission from *Energy Research and Dev.*, Hearings before the Subcommittee on Sci., Research, and Dev., House Committee on Sci. and Astronautics, Y4Sci2: 92/2/24, pp. 585-593, May 1972.

out-of-phase current and will increase the length of underground line possible without compensation. Instead of a maximum length of about 20 miles the superconducting line will have a maximum length of about 200 miles.

3. When superconducting cables are operated with alternating current, the losses in the line are no longer zero. At 60 Hz for relatively high currents they are finite but still very small. Although presently available superconductors give promise of adequate electrical performance, there has been very little superconductor development aimed at this application. Further work on the electrical and mechanical properties of superconducting materials and insulation for cable applications is necessary.

4. To guard against overheating and loss of superconductivity under fault conditions the line should be operated well below its critical temperature. This means almost certainly that the coolant must be helium. This fact should be kept in mind when the Government's program for conservation of helium is considered.

The materials receiving most attention for use as superconductors are pure niobium and the compound niobium-tin. AC losses in pure niobium are lower than those in niobium-tin but the latter can be operated at a higher temperature and is more suitable for carrying the heavy overload currents which occur occasionally in actual power transmission systems.

A typical 3000 MVA cable, designed at Brookhaven, would be housed in a pipe about 17 inches in diameter. This pipe would contain 15 flexible superconducting coaxial cables (five per phase). These cables would be about two inches in diameter. A typical HPOF cable of this size would weigh about eight times more per foot and yet carry only one third of the power. The refrigeration stations would be spaced about every six miles: their size and energy consumption are modest compared with the amount of power transmitted. These stations would fit on a quarter-acre lot or could be located underground.

The necessary technology of refrigeration for use with superconducting power lines is essentially all at hand as a result of developments in the space program. Large quantities of liquefied gases, including helium, were used and a cryogenic industry was built up. Development of low temperature, insulated containers reached the point where liquid helium is now shipped across the country in tank trucks and sold at prices lower than can be obtained by buying gas and liquefying it oneself.

Many engineering problems remain to be solved. No object as long and thin as a power line has yet been refrigerated. Here the experience and facilities of the National Bureau of Standards at Boulder will be used extensively. At each stage of the development the electrical utility industry must be continually consulted. During its 1971 study, Brookhaven had three consultants who know little of superconductivity but who were experts in power station design and siting, cable design and manufacture, and network system performance. Advice of such consultants is essential if the developed cable is to be tailored to the needs and demands of the utilities.

Extensive modeling and testing will be necessary during the coming years. Power lines must have extraordinarily reliable performance if they are to be accepted by the utilities and the public. They must recover rapidly from faults and transient phenomena. And they must be designed for installation at the lowest possible cost.

We feel considerable optimism about the possibility of achieving these goals. Given support we believe that we can have a model superconducting power line in operation within five years. Perhaps a major utility may be using a superconducting link within a decade.

REFERENCES

1. Brookhaven's work to date is summarized in their December, 1971, report to the National Science Foundation, entitled "Report on Superconducting Electrical Power Transmission Studies." (BNL 16339). This report is now out of print: it has been replaced by "Underground Power Transmission by Superconducting Cable," BNL Formal Report 50325, March 1972.
2. A forward-looking viewpoint is presented in "Electric Utilities Industry Research and Development Goals through the Year 2000." This is a report to the Electric Research Council by its "R&D Goals Task Force" and has the designation ERC Pub. No. 1–71. It appeared in June 1971.
3. A convenient reference is the article "Superconductors for Power Transmission" by Donald P. Snowdon, published in the April 1972 issue of "Scientific American."
4. A compact presentation of the field of applied superconductivity will be found in "Superconductivity, A Technology That is Coming of Age" which was presented to the President's Science Advisory Committee on February 21, 1972. Copies can be obtained from Dr. B. W. Birmingham, Head of the Boulder, Colorado laboratory of the National Bureau of Standards.

Needs for Research and Development in Direct-Current Transmission

STATEMENT PREPARED BY BONNEVILLE POWER ADMINISTRATION, PORTLAND, OREG.

I. MOST PROBABLE FUTURE APPLICATIONS OF DIRECT CURRENT TRANSMISSION IN NORTH AMERICA

The principal applications of d–c transmission throughout the world have been for:

1. Crossing bodies of water wider than 20 miles.
2. Interconnecting a–c systems having different nominal frequencies or between systems operating at the same frequency to control power flow.
3. Transmitting large amounts of power over long distances by overhead lines.
4. Underground transmission in congested urban areas.
5. Combinations of the foregoing factors.

In North America, the need for transmission lines crossing wide bodies of water is exceptional, the frequency is universally 60 Hz, and the trend is toward complete interconnection and integration of the transmission networks. For these reasons, applications 1 and 2 listed above will be infrequent. This leaves applications 3 and 4 as those which are likely to be most widespread.

Long Overhead Transmission Lines

Long distance transmission from large remote hydro sites and from remote fossil fuel sites, such as the Montana-Wyoming coal fields, will be required to serve load centers either to the west or east. The feasibility of utilizing remote energy sources would depend, among other things, on the cost of transmission. In most cases, the transmission distance would be 1000 km or more. For such distances, d–c transmission would be less expensive than a–c transmission. In addition, the problem of stability of long radial lines would be much less acute.

Another application of long overhead d–c lines may be for interregional ties with potential diversity between regions, of which the existing 846 mile (1370 km) Celilo-Sylmar line is an example.

Underground Transmission in Congested Areas

As the power demands of large cities continue to increase, sites for new generating plants cannot be found in or near these cities, and it becomes increasingly difficult to acquire the routes for bringing power into the city. Right-of-way problems for overhead high-power lines are becoming almost insurmountable and even routes for underground transmission cables are becoming increasingly difficult to obtain.

Direct-current underground high-voltage cables have several advantages:

1. D–C cables are smaller and cheaper than a–c cables for the same power transmission capability by a considerable factor. 2. D–C cables have no steady state charging current; hence the feasible distance of transmission is not limited by charging current, nor are the generators in danger of self-excitation at times of light load.

3. It becomes feasible to transmit directly from generating stations to distribution substations without going through several intermediate voltage levels, as is the practice in a–c systems. Thus there is a savings of investment in transformers.

4. It is possible to reinforce the supply to existing substations by d–c lines without increasing the short-circuit current capability of these substations and, consequently, having to replace circuit breakers whose interrupting rating would be exceeded.

A pilot installation of this type is now under construction in England to bring 640 mw of power from the Kingsnorth therman generating plant on the south shore of the Thames River Estuary into two substations in London at distances of 59 and 82 km.

It is likely that in the future, d–c links for the same purpose may be introduced into large cities in the U.S.A., such as New York, Chicago, and Philadelphia.

II. AREAS IN WHICH RESEARCH AND DEVELOPMENT ARE NEEDED

Extensive effort must be expended in the miniaturization and development of "mini d–c terminals." Research is required to solve the arc-back problems with mercury arc converters and to reduce the number of components in solid-state valves. In addition, efforts should be made to discover and develop new types of valves and new concepts of conversion techniques.

Substantial development is needed to reduce space requirements for harmonic filtering equipment required for interference reduction.

At present, all d–c transmission lines are two-ended systems in which the terminals control the power flow and interrupt faults .but if d–c transmission lines are to be widely applied and if they are going to be tapped successfully, d–c circuit breakers must be developed.

Conventional circuit breaker technology cannot be applied to these systems.

Extensive research is needed for overhead lines up to voltages of ± 1500 kv and similar research effort needs be expended on high-capacity cable transmission. On overhead d–c transmission, further basic work is required to evaluate insulation requirements, physiological and psychological effects of UHV d–c fields as well as such factors as audible noise, radio and telephone interference and associated corona discharges.

A new family of cables is needed for d–c application because of the fundamental differences in electrical stress distribution between d–c and a–c cables. Included should be an evaluation on the degree of stress inversion with temperatures for different insulation materials, and effect of harmonics on the cable insulation structure.

More knowledge is needed about the role of d–c transmission in integrated a–e networks, that is, how the whole system will work together. More sophisticated controls will make possible the use of d–c lines as controllable dampers during system disturbances.

One effective approach would be to install a scaled-down version of d–c transmission system as part of an existing a–c network. It would be a good test site for various converter elements, d–c breakers, control schemes, filter designs, multi-terminal d–c line operation and other associated functions.

The Department of Interior is currently engaged in underground cable research through the Electric Research Council (ERC). ERC has a number of projects underway both of an a.c. and d.c. nature. The National Science Foundation (NSF) through their RANN program has sponsored several projects pertaining to many areas of national interest including d–c transmission. Their research is intended to promote development of new ideas and concepts which could be of substantial benefit to ERC and others, in future development of d–c transmission.

The Bonneville Power Administration (BPA) recently entered into a three-year research program jointly with ERC which will utilize BPA's HVDC test facility located at The Dalles, Oregon. This program will include studies of insulation requirements, evaluation of radio interference and audible noise levels, and study of clearances and safety aspects for voltages up to ± 600 kv, the limit of the test source.

In another area of on-going research, BPA and the Bureau of Reclamation are providing support for an ERC sponsored research effort by the Hughes Aircraft Company directed toward the development of technology for a practical HVDC circuit breaker.

While there is some research on direct current transmission in progress today, much more needs to be done. In summary, research and development is needed in the following general areas:

1. Conversion technology.
2. Mini d–c terminals.
3. D–C breakers.
4. Overhead lines up to ± 1500 kv.
5. D–C cable development.
6. Converter system performance tests.

Miscellaneous

The report issued in June, 1971, by the Electric Research Council (of which the U.S. Department of the Interior is a member) on "Electric Utilities Industry Research and Development Goals Through the Year 2000" recommended research and development on a number of items pertaining to d–c transmission. We are in agreement with those recommendations.

Reprinted with permission from *Energy Research and Dev.*, Hearings before the Subcommittee on Sci., Research, and Dev., House Committee on Sci. and Astronautics, Y4Sci2: 92/2/24, pp. 421–423, May 1972.

by Richard A. Rice
Professor of Transportation
Carnegie-Mellon University

System Energy and Future Transportation

The United States in a typical year of the late 1960's used some 55,000 trillion B.t.u. of energy for all major activities. Transportation accounted for about 24 per cent of this total energy, or about 100 billion petroleum gallons if 130,000 B.t.u./gal. are allowed. This represents more than half of the 174 billion gal. of petroleum consumed in the U.S. It is a sufficiently large component of our total energy use to justify careful analysis, in a time when our overall energy supply-and-demand system is coming under ever-sharper scrutiny.

Transportation is responsible for about 20 per cent of the gross national product (G.N.P.)—in 1965, about $140 billion. It could be said that to "generate" $140 billion of transport user payments and charges, some 88 billion gal. of fuel (or the electrical equivalent) were used, and that each gallon burned thus "generated" some $1.60 of transport activity. With fuel and power generally available to the typical transport operator at 11 to 23 cents/gal., not including taxes, fuel—or energy cost—constituted between 10 and 15 per cent of all transport user costs for most transport modes.

A second way of looking at this is that some 1,300 billion local and intercity passenger-miles and 2,000 billion local and intercity cargo-ton-miles were moved in 1965. The total movement of 3,300 billion capacity-miles by 88 billion gal. of fuel shows a total national propulsion efficiency of 37.5 net payload unit-miles/gal. used. Some 2,400 billion unit-miles of intercity transportation were accomplished with perhaps 41 billion gal. of fuel—a rate of 58 unit-miles/gal. The urban sector—with autos, transit, and regular trucks—may have generated 800 billion transportation miles with 40 billion gal. (20-miles/gal.). The balance, some 200 billion unit-miles not easily identified, must represent specialized services consuming the remaining 15 billion gal. of fuel at a yield of only 13 unit-miles/gal.

In the United States, in terms of gross-ton-miles moved (4,280 billion), the railways account for half the operation of nonurban transport. In the mid-1960's they moved over 2,200 billion gross-ton-miles annually with 4.0 billion gal. of diesel fuel; their gross efficiency, including passenger trains, was 550 ton-miles/gal. In contrast, the airlines used 2.0 billion gal. of fuel to fly only 60 billion gross-ton-miles. Highway vehicles moved 1,320 billion gross-ton-miles using 35.2 billion gal. for a general efficiency of 37.6 ton-miles/gal. Intercity transportation as a whole accounted for 41 billion gal. of petroleum—two-thirds used in private automobiles—and had a gross propulsion efficiency during the 1960's of 104 ton-miles/gal. The total *net* intercity propulsion efficiency was 58 unit-miles/gal. For freight, the figures (gross and net) were 290/140; for passengers, 33/26.

The figures for urban transportation are, of course, very different. Using 1965 estimates, all suburban railway transportation—6 billion passenger miles—required some 70 million gal. of petroleum or the equivalent in electric energy; this represents some 85 passenger-miles/gal. Other urban public passenger carriers—buses, limousines, taxicabs, etc.,—used 400 million gal. of fuel. Private automobiles in urban service used 35 billion gallons of fuel, 60 per cent in trips of 2½ miles or less. Add to these the totals for urban trucking and parcel delivery (5.4 billion gallons) and institutional vehicles (1.8 billion gal.), and the re-

Reprinted with permission from *Technology Rev.*, vol. 74, pp. 31-37, Jan. 1972. Edited at M.I.T.

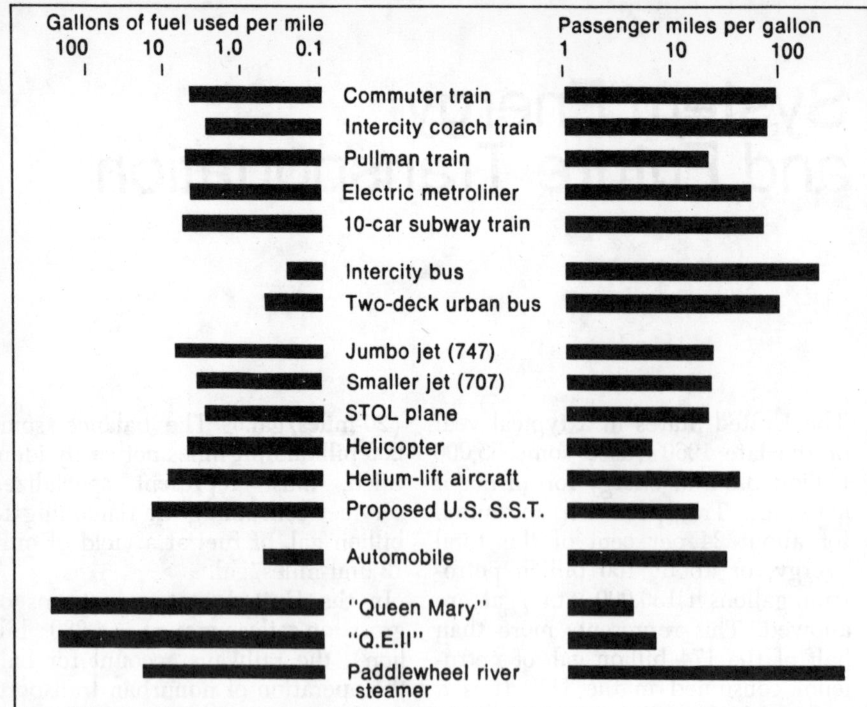

In passenger transit, the high performers in terms of net passenger miles moved per gallon of fuel are buses and commuter trains; the more exotic, faster means of transport are lower in efficiency, and so are such hard-to-die luxuries as superliners and Pullman (overnight) trains. A trend to the most efficient forms of passenger transport, writes the author, would considerably increase the U.S. national propulsion efficiency. Note that the horizontal scales are logarithmic.

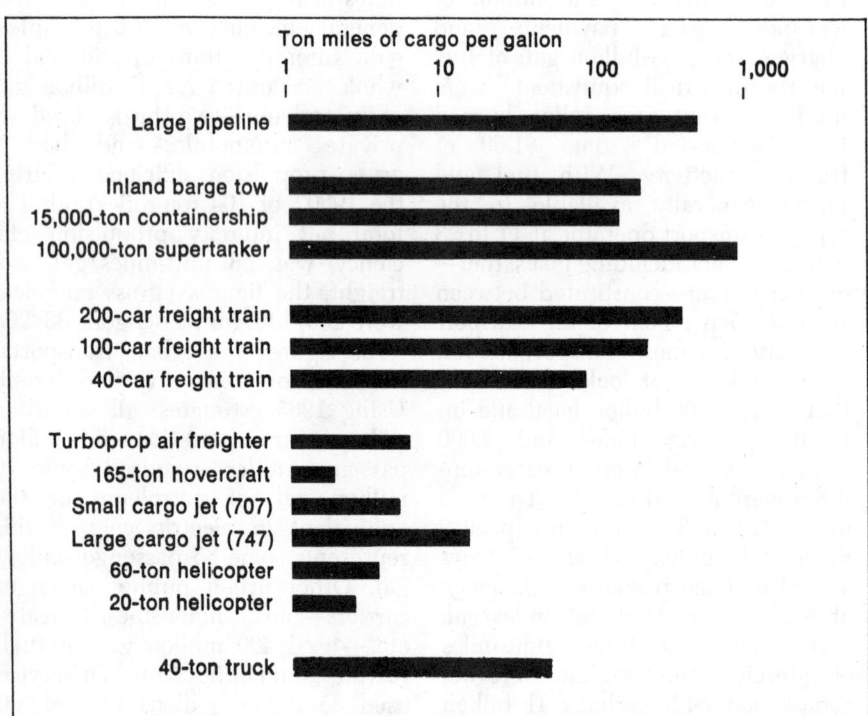

In freight service pipelines, inland waterways, and railroads do not use significant amounts of energy in relation to goods moved. In fact, writes the author, in 1965 these three forms of transport used only about 5.5 billion gal. of petroleum (7 per cent of U.S. transport energy) to provide 1,250 billion cargo-ton-miles of service and 60 per cent of all gross ton-miles moved in the U.S. Note that the horizontal scale is logarithmic.

sult is 42.7 billion gal. of petroleum or equivalent annually devoted to U.S. urban transportation in the mid 1960's.

In addition to the above, a small part of the fuel recorded as used in intercity traffic is in fact used within urban areas at the start and end of each intercity trip. Thus one may hypothesize that perhaps as much as 47 billion gal. of petroleum or equivalent energy was consumed for transportation within the 2 per cent of the U.S. continental area classified as urban; 41 billion gal. were burned in the other 98 per cent of the country.

Future Petroleum Commitments

Since the typical single automobile runs 10,000 miles per year and in so doing uses 670 gal. of fuel, it is easy to see that it is consuming about two tons of petroleum annually. In other words, the auto just about consumes its own weight in fuel every 12 months. In 1960 there were in the world a total of about 150 million autos, using 300 million tons of petroleum.

It is revealing to compare this with the annual energy consumption of a few other transport craft units, as shown in the table on page 33. A bus consumes six times its weight. A DC-6 uses 70 times, and a 707, 175 times its weight. The proposed 375-ton S.S.T. would have swallowed 240,000 tons of fuel, or 640 times its takeoff weight, per year. The old "Queen Mary" superliner of 160,000 h.p. used almost 400 gal./mi. of "Bunker 'C'" fuel oil to drive her across the Atlantic: in a year, she used up to 200,000 tons of petroleum —compared to her gross weight of 80,000 tons. But a single Concorde S.S.T. aircraft, weighing 180 tons and flying 11 hrs./day, might consume 25 million gal. of fuel annually; and the present jumbo-jet—at 320 tons—21.5 million gal. A fleet of 900 of the latter—which has been forecast—makes our future petroleum commitment look alarmingly sizable.

Indeed, if all present air carrier transport plans should materialize, some 88 billion additional gallons of fuel might be required per year for transportation by 1985. Of this increase, automobiles are estimated to need 28 billion gal. and larger air transport craft, 43.5 billion gal. Together, they account for 80 per cent of the projected increase.

Another large increase, 4 billion

gal., would be required by intercity trucking, while smaller private planes would need 4.7 billion gal. more. Railroads, buses, pipelines, and waterways combined would require 3.7 billion gal. more. The latter four carriers are thus not critical contributors to the projected soaring petroleum demand, and some extension of railroad piggyback service could, in addition, alleviate the 4-billion-gal. increase that trucks otherwise might need. This leaves automobiles and aircraft as the targets for further study if we are compelled to revise future fuel commitments.

Urban Transport Energy Needs

In anticipating fuel requirements and in planning the directions for future technological development, it is instructive to compare the true energy efficiencies of various transport modes. This is done in the accompanying tables on the basis of their net propulsion efficiencies (N.P.E.'s), the number of cargo-ton-miles or passenger-miles moved per gallon of fuel.

In passenger transport, the high performers seem to be buses and commuter trains with N.P.E.'s of 100 or over. These provide the highest yield, and a trend to these modes would increase our national propulsion efficiency.

The automobile with an N.P.E. around 30 to 40 is about average for passenger movement. Air transportation is slightly less efficient. Such now-largely-discontinued methods as the ocean liner and the Pullman sleeper may owe their decline in part to their low N.P.E. deriving from their high weights per passenger. However, three proposed modes for future overland service—the S.S.T., the helicopter, and the hovertrain—all seem to have lower N.P.E.'s than even conventional autos or air travel.

In 1960, domestic trunk air-route service required approximately 2 billion gal. of fuel (in 1955, it was only about one billion). By 1965, domestic routes required about 4 billion gal. and in 1970, 8 billion. Between 1955 and 1970, the regular domestic-route airlines thus increased their fuel consumption about eight times, while the traffic carried in terms of passenger-miles rose only about six times.

Pipelines, inland waterways, and railroad freight do not use significant

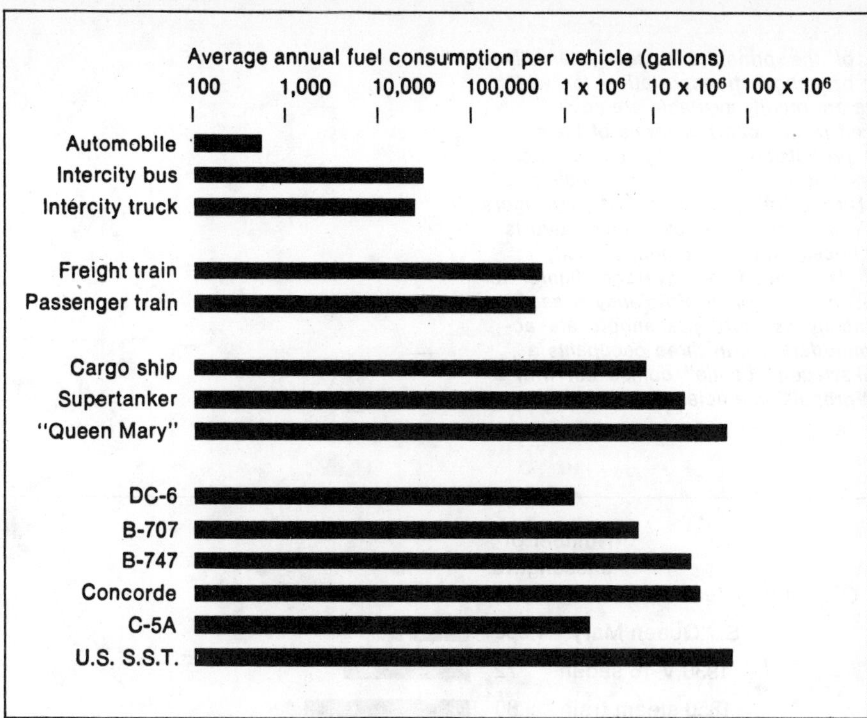

If and when absolute fuel consumption becomes a primary issue in transportation planning, engineers will discover how extravagant are today's more exotic (and faster) forms of transit. The figures given above are based on the current average service required of each vehicle, ranging from less than one hour per day for the automobile to 20 hours per day for a supertanker. Though freight cars move an average of only about one hour a day, locomotives (the figure above) are estimated at between 5 and 6 hours' travel per day. Note that the horizontal scale is logarithmic.

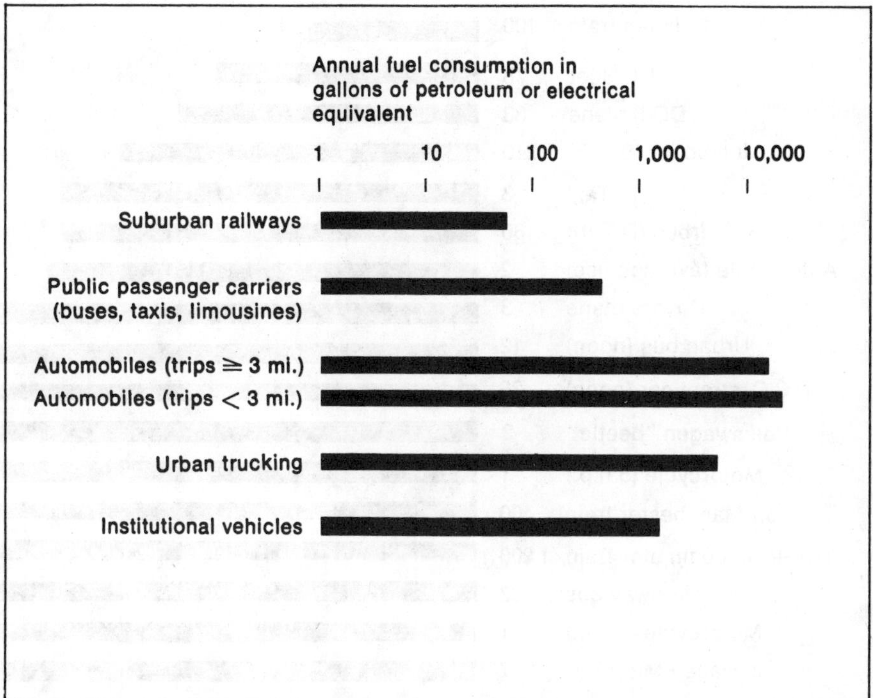

Automobiles have by far the greatest claim on energy resources used for urban transportation. A major reduction in energy expended—and presumably in urban pollution problems—could be achieved if automobiles could be eliminated for trips of three miles and less.

While an automobile accomplishes some 30 to 40 passenger-miles of transport per gallon of fuel used, a bicyclist obtains about 1,000 passenger-miles for the same energy input, about 130,000 B.t.u. or 34,000 calories (kcal.)

All of the principal system alternatives for passenger transportation for which data are readily available are compared in this chart in terms of their net propulsion efficiency, the number of passenger-miles moved per gallon of fuel. Note that the number of passengers on which the efficiency is calculated is not necessarily a maximum capacity but is instead an average figure for present experience. Efficiency rises dramatically as more passengers are accommodated; with three occupants a Volkswagen "beetle" comes out with a net propulsion efficiency of 100.

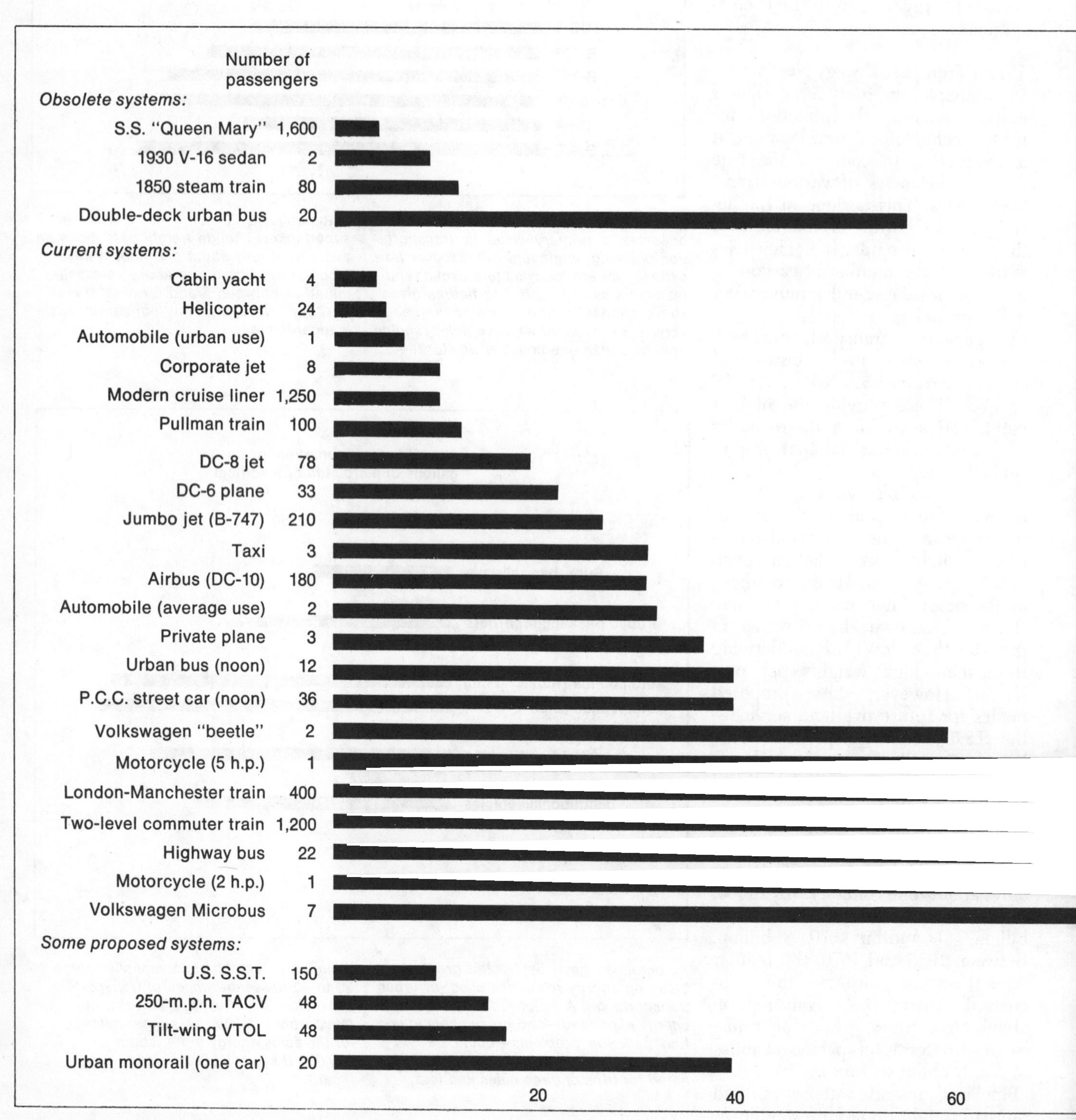

Net propulsion efficiency

80 100 120 140 160

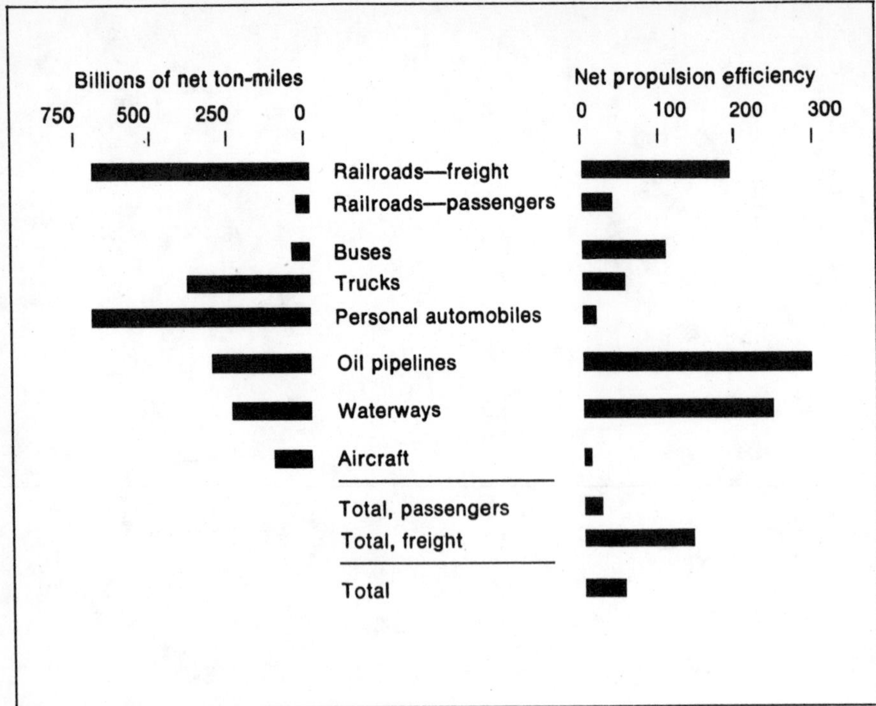

Intercity transport of passengers and freight in the U.S. varies markedly in efficiency, as measured in terms of the net payload (pounds or individuals) carried per gallon of fuel consumed. If our goal is to increase a total national propulsion efficiency, we will concentrate on those methods of moving goods and people which show the highest net propulsion efficiency.

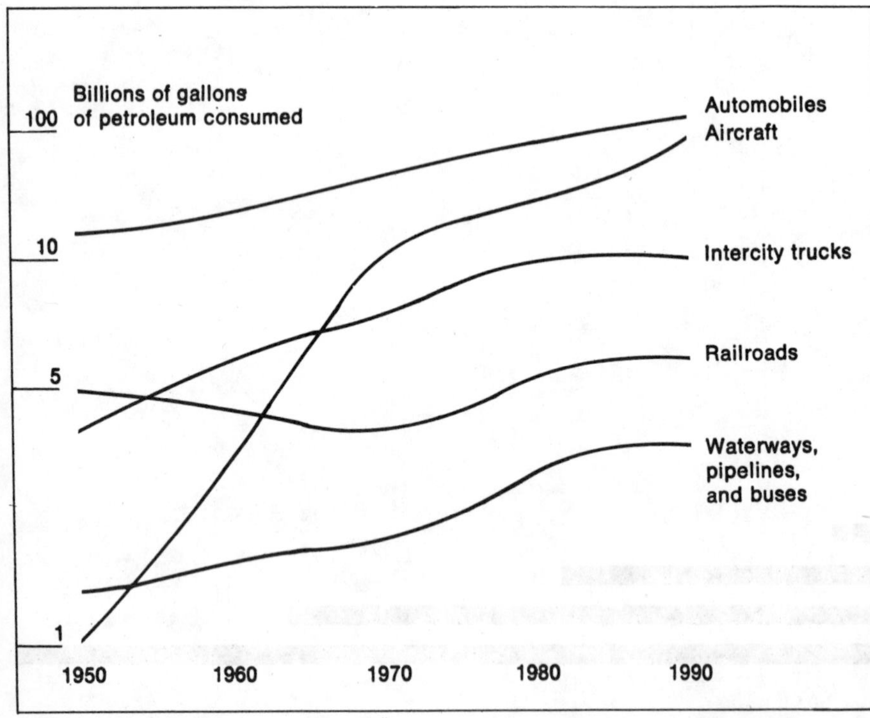

Current government policy favors the development of air and highway transport. If planning subsidies continue to encourage use of these methods, automobiles, aircraft, and trucks will maintain their growth as transportation methods. But the propulsion efficiency data presented by the author suggest that changes in national policies and priorities may in fact be overdue, and he suggests that the balance between transportation methods may be altered before 1990.

amounts of energy in relation to goods moved. In fact, at about 1965 levels, these three carriers used only 5.5 billion gal. of petroleum (or about 7 per cent of U.S. transport energy) but provided some 1,250 billion cargo-ton-miles or roughly 40 per cent of all useful overland passenger and ton-miles generated. In terms of gross ton-miles moved, these three carriers accounted for 60 per cent of all transport.

Trucks achieve approximately 50 cargo-ton-miles/gal., less than half the efficiency of intercity freight as a whole. Airfreight transport, which apparently yields only about five to ten cargo-ton-miles/gal., thus appears suitable only for goods requiring more expedited service.

Projected Total Energy and Transport Outlook

Long-range forecasts indicate that total U.S. energy use may continue at about 100,000 B.t.u. per dollar of gross national product (true G.N.P. in terms of 1958 dollars). Taking median government forecasts of G.N.P. of $970 billion in 1980 and $1,300 billion in 1990, the corresponding energy forecasts for trillions of B.t.u. are 108,000 and 145,000 respectively.

Inserting the most optimistic air transport fleet data and high auto-use projections, we find that in 1980, aircraft and autos might consume 77 per cent of transport energy and in 1990 an estimated 82 per cent. In 1960 and 1970, these figures were 60 and 68 per cent, respectively.

Of the projected growth of 163 billion gal. of petroleum consumption from 1960 to 1990, air and autos would thus consume 150 billion gal. or 92 per cent. This increase by 1990 in air and auto usage alone is far more than the entire 1960 transport energy use.

The impact of such an increase attributed to the two relatively high-energy transport modes is hard to anticipate. But it is probably fair to suggest that unless air travel can deliver more than the current 20 to 25 passenger-miles/gal., the increase projected will in fact not be fulfilled. The prospect for burning from 50 to 90 times as much petroleum in the sky as in 1950 does not look realistic in view of current concern over air pollution; and the same concerns may also serve to stimulate alternatives to the motor vehicle for intercity and urban transport. Fur-

The human being and the bicycle are uniquely efficient as transport, and man's increasingly innovative devices are uniquely inefficient. As energy becomes more costly, propulsion efficiency may become a criterion of wide concern.

thermore, if fuel supplies grow short, efficiency will become an issue of greater concern; and public attention is likely to focus on the vastly more efficient rail, water, pipeline, and even large truck systems.

Human Energy as a Transport Standard

While the chemistry and metabolic rates associated with human activities—especially walking, cycling, canoeing, and hiking—have long fascinated scientists, it is only recently that consideration has been given to the environmental desirability of encouraging human propulsion as a substitute for the urban auto. Human propulsion energy needs are quite small.

The Sierra Club advises that hikers planning their daily trek should allow one hour for each 2½ miles along level trails, plus one hour for each 1,000 ft. of vertical ascent. Assuming a 160-lb. man, the continuous "hiking thrust" at a 2.5-m.p.h. walk works out to 12.3 lb. or 0.082 h.p. for a smooth trail. Ascending at the rate of 1,000 ft./hr. (160,000 ft.-lb./hr.), we get 2,670 ft.-lb./min. or also 0.082 h.p. These rough estimates would indicate that a slow walk requires around 1/12 h.p.

The author's studies of bicycling indicate that, assuming 40 lb. for the vehicle, a thrust of 6.7 lb. is required to average 15 m.p.h. on the level, or about 0.27 h.p. At 20-m.p.h., a 10-lb. thrust indicates power up to 0.5 h.p. Leisurely cycling at 10 to 12 m.p.h. with five-lb. thrust thus is estimated to use about 1/8 to 1/6 h.p.

Many sources confirm that a person hiking or cycling five to six hours daily will actually add some 1,500 to 1,800 calories to his average daily need. Thus a very rough estimate of human propulsion efficiency can be made by noting that 1,800 "extra" calories of input in a moderately active male can yield at least ⅛ h.p. for five to six hours of measurable output—a total of 0.75 horsepower-hours.

For the 1,800 calories (7,000 B.t.u.) that a cyclist uses, he is travelling some 72 miles at 12 m.p.h.: figuring a 200-pound craft (cyclist and cycle), he is performing 7.2 gross-ton-miles of transport work. This implies a propulsive efficiency of over 45 per cent for a male cyclist in good condition, while most engine-powered units have a net propulsive yield of about 25 per cent. In terms of petroleum units, a cyclist can reach the equivalent of over 1,000 passenger-miles per gallon, per 130,000 B.t.u., or per 34,000 calories.

If bicycle and pedestrian journeys were substituted for two-thirds of the 2,050 trips of two miles or less that most urban autos make per year, the savings per household (@1.3 autos) would be 1,800 trips involving 270 gal. of fuel (about 35 million B.t.u. per household). For this there would presumably be substituted the requisite 1,800 personal one-way trips, each of which by cycle or walking would consume only about 500 B.t.u. of human energy from 130 calories. The annual household total for these 1,800 trips is thus about 1 million human B.t.u., which replaces the 35 million automobile B.t.u. Extrapolation of this 35-fold saving to a city of 1 million population owning 300,000 autos and using 80 million gal. of fuel (around 10 x 10^{12} B.t.u.) yields a significantly lower fuel consumption.

Long-Run Energy Implications

Though projections for increased U.S. energy use, and especially petroleum use, can be made based upon G.N.P., the limitations on both petroleum supplies and energy-caused atmospheric pollution make the forecasts for both total and transport use appear questionable. The consumption of fuel per delivered passenger or per delivered ton will assume increasing importance, and here the high-yield systems continue to be the waterways, the pipelines, the railroads, and passenger buses. All of these have propulsion efficiencies in the 100 to 300 range.

Heavy reliance on highly innovative systems in the near future seems doubtful because of the low yields proposed for most of them—hovercraft and helicopters, for example. Even the automobile (30 to 40 passenger-miles/gal.) and the intercity truck (50 to 60 ton-miles/gal.) may have to be curtailed or drastically changed to improve the overall national propulsion efficiency. Continued quantum enlargement of air travel systems yielding only 20 to 30 passenger-miles/gal. (let alone the S.S.T. at 10 to 15) will so increase total petroleum use and so reduce the N.P.E. as to certainly cause early review. The very optimistic jet, jumbo-jet, and S.S.T. fleet projections of just a few years ago carry with them implications for total fuel consumption which begin to appear unrealistic.

These calculations of energy-efficiency suggest modest redirections for intercity transport development, but urban transportation methods appear to be candidates for more drastic changes. Perhaps as much as a 50 to 70 per cent reduction in urban motoring—and a substitution in even amounts of walking, cycling, and mass transit—will be needed to produce a noticeable effect on urban transport energy consumption.

Transportation Energy Use and Conservation Potential

ERIC HIRST

"Recent history shows a steady growth in transportation energy use at a rate more than double the population growth rate. . . . However, oil scarcities and increasing dependence on petroleum imports, coupled with rising environmental concern, could reverse these historical trends. It is technologically feasible to slow transportation energy growth by increasing transportation energy efficiency. Policies to achieve such goals would involve some life-style changes and important institutional decisions, but they do not imply a return to 'caves and candles.'" Eric Hirst is a research engineer studying energy-use patterns in the environmental program at the Oak Ridge National Laboratory.

There are a number of reasons why an examination of transportation energy use is timely and important. First, transportation of people and goods consumes fully one-fourth of the total U.S. energy budget [1]. For example, transportation in 1970 required 16,500 trillion Btus, equivalent to 3 billion barrels of oil. Between 1950 and 1970, energy consumption for transportation had an average annual growth rate of 3.2 per cent, more than double the U.S. population growth rate.

Second, world oil reserves appear to be quite limited. According to Hubbert [2], 90 per cent of the world oil supply will have been consumed by about 2025 and 90 per cent of U.S. oil reserves within 30 years.

Third, U.S. transportation is almost entirely dependent on oil as a fuel [1]. In 1970, imports accounted for 23 per cent of the domestic oil budget, and this fraction is rising. Imports will probably account for more than half our oil budget within 10 years [3]. This dependence on imports poses a balance-of-payments problem (roughly $20 billion a year in 10 years) and national security problems (keeping oil supply routes open during wartime and maintaining good relations with oil-producing nations).

Fourth, exploration, production, transportation, refining and use of petroleum present serious environmental problems. These include oil tanker accidents and intentional spillage, oil well fires and blowouts, disposal of brine at oil wells, refinery air pollution emissions and transportation equipment air pollution and thermal emissions. To some extent, the severity of these problems is directly proportional to the magnitude of petroleum consumption.

Finally, transportation contributes to a number of other environmental problems including urban congestion, inefficient land use and noise. Methods

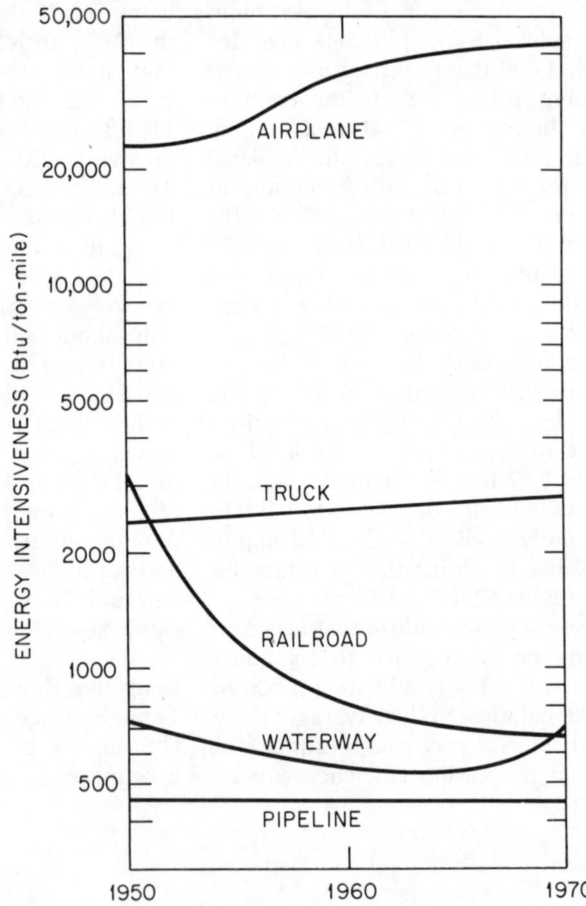

Fig. 1—Historical variation in energy intensiveness of inter-city freight modes.

Table 1

Historical Energy Consumption Patterns for Transportation

Year	Total Traffic	Per Cent of Total Traffic						Total Energy (10^{12} Btu)	Average Energy Intensiveness
		Air	Truck	Rail	Waterway & Pipeline	Auto	Bus[a]		
		Inter-City Freight Traffic							
1950	1,350[b]	0.02	13	47	41	—	—	2,700	2,000[c]
1960	1,600	0.05	18	38	44	—	—	1,800	1,100
1970	2,210	0.15	19	35	46	—	—	2,400	1,100
		Inter-City Passenger Traffic							
1950	500[d]	2	—	7	—	86	5	1,700	3,400[e]
1960	800	4	—	3	—	91	2	2,700	3,400
1970	1,120	10	—	1	—	87	2	4,300	3,800
		Urban Passenger Traffic							
1950	310[d]	—	—	—	—	85	15	2,100	7,000[e]
1960	430	—	—	—	—	94	6	3,300	7,700
1970	710	—	—	—	—	97	3	5,700	8,000

[a] Inter-city bus or urban mass transit.
[b] Billion ton-miles.
[c] Btu/ton-mile.
[d] Billion passenger-miles.
[e] Btu/passenger-mile.

for increasing transportation energy efficiency would generally tend to reduce the severity of these problems.

This article examines traffic, energy consumption and energy intensiveness for inter-city freight and passenger traffic and urban passenger traffic. (For a detailed examination of transportation energy use patterns for 1950 to 1970 see Hirst [4].) Urban freight traffic is neglected in this article and in [4] because it is carried almost exclusively by truck and because accurate data concerning traffic levels is not available.

Energy intensiveness is defined as Btu per ton-mile for freight and Btu per passenger-mile for passenger traffic. It is the inverse of energy efficiency; the latter being the product of two factors: (1) technical efficiency, e.g., seat-miles per Btu, and (2) load factor (per cent of capacity utilized), e.g., passenger-miles per seat-mile.

Freight Traffic

Inter-city freight is moved by railroad, truck, waterway (boat and barge), pipeline and airplane with considerable variation in energy intensiveness; airplanes are almost 100 times as energy intensive as pipelines (Fig. 1).

Pipelines and waterways are the most energy efficient modes; however, they are limited in the kinds of materials they can transport, in the flexibility of their pickup and delivery points, and by their slow speed.

Both pipeline and waterway freight traffic grew steadily between 1950 and 1970. Boats and barges carry primarily raw materials, basic agricultural products, chemicals, semi-manufactured goods and heavy machinery. Pipelines are generally limited to crude oil, refined petroleum products and natural gas.

Railroads are about as efficient as waterways. After World War II, railroad energy intensiveness decreased by a factor of 5 because of the shift from coal-burning steam locomotives to diesel engines. During this period, the fraction of freight carried by rail declined sharply (Table 1). Apparently, the railroads were unable to meet the increased competition from other freight modes, especially trucks. Nevertheless, trains still carry one-third the freight traffic, primarily grain, food products, raw materials, paper, chemicals, metal products and motor vehicles.

Trucks—faster and more flexible than the preceding modes—are only one-fourth as efficient as trains. The amount of freight carried by trucks increased sharply between 1950 and 1970, probably because of increased highway construction and geographic expansion and distribution of industry and commerce. Trucks also offer speedy delivery, flex-

ible routing and scheduling and door-to-door service.

Because of these factors, trucks are particularly useful for short hauls of small, high-value shipments: meat and dairy products, rubber and plastic goods, furniture, fabricated metal products, machinery, and instruments (Table 2).

Airplanes are the fastest mode, with the highest energy intensiveness—60 times that for trains. Between 1950 and 1970, it doubled as the airlines traded energy for speed. Today, airplanes are more than 10 times as fast as other freight modes with an average speed of 400 mph (Table 2).

Because air freight is expensive, only high-value products such as clothing, communication equipment, instruments and small machinery are shipped by air.

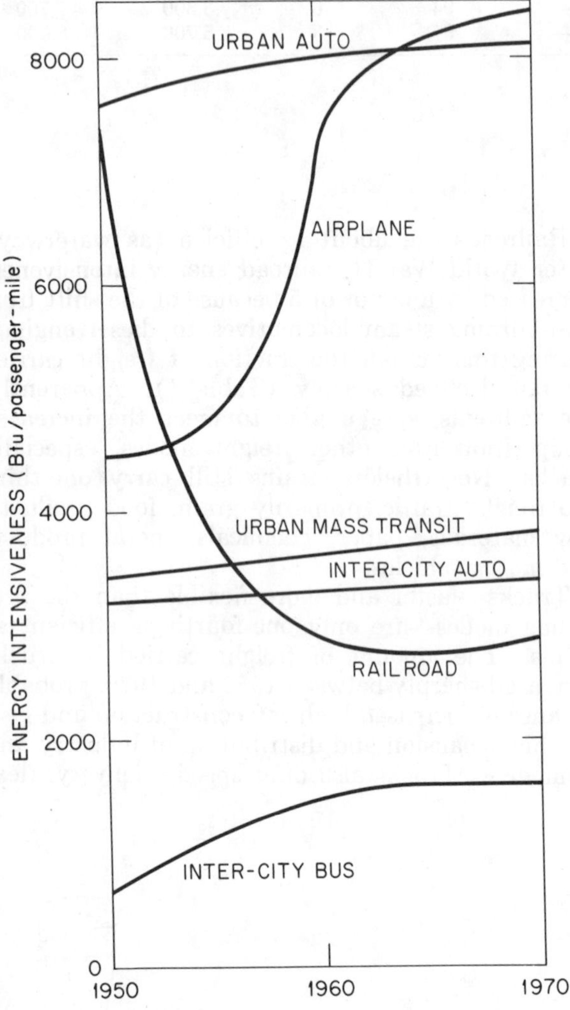

Fig. 2—Historical variation in energy intensiveness of passenger modes.

Between 1950 and 1970, energy consumption for inter-city freight fell by 12 per cent, in spite of a 64 per cent increase in total traffic. Overall energy intensiveness declined by 46 per cent because for trains it decreased by about 80 per cent. Were it not for this sharp drop, overall freight energy intensiveness would have increased as freight traffic shifted to higher modes with (and growing) energy intensiveness.

Table 2

Inter-City Freight Transport Data for 1970

Freight Mode	Energy Intensiveness[a]	Revenue[b]	Haul Length (miles)	Average Speed (mph)
Pipeline	450	0.27	300	5
Railroad	670	1.40	500	20
Waterway	680	0.30	1,000	—
Truck	2,800	7.50	300	~40
Airplane	42,000	21.90	1,000	400

[a] Btu per ton-mile.
[b] Cents per ton-mile.

Inter-City Passenger Traffic

Inter-city passengers are carried primarily by automobile and, to a lesser extent, by airplane, bus and train (Table 1). The variation in energy intensiveness of passenger modes is large, but is not as great as for freight modes (Fig. 2).

Buses are the lowest energy intensive mode and also the least expensive (Table 3). Post-World War II highway construction helped buses as well as cars and trucks. The ability of bus routing to respond to changing demographic patterns, plus low fares, enabled bus traffic to remain nearly constant between 1950 and 1970. The load factor for inter-city buses declined from 51 per cent in 1950 to 46 per cent in 1970 while energy intensiveness rose, probably because of declining load factors and increasing highway speeds.

Railroads are the second most efficient passenger mode. As noted earlier, railroad energy intensiveness dropped sharply during the 20-year period considered. The rather high energy intensiveness for passenger trains, relative to buses, is due to low passenger load factors (Table 3) and the weight of passenger trains. The average passenger train (exclusive of mail, express and baggage cars) weighs over 4 tons per passenger, compared to less than 1 ton per passenger for buses, airplanes and autos.

Rail passenger traffic fell dramatically between 1950 and 1970, in both absolute and relative terms. This decline was probably due to restrictive government regulation, lack of sufficient capital, increased competition from other modes and conservative railroad management.

Automobile energy intensiveness is more than double that for buses. During the period considered it increased slightly, probably because of greater vehicle weight, larger engines, higher speeds and the use of additional equipment such as air conditioners. Autos had an average inter-city occupancy of 2.4 passengers, which implies a 48 per cent load factor, assuming a 5 passenger capacity.

Automobiles account for more inter-city passen-

Table 3

Passenger Transport Data for 1970

Passenger Mode	Energy Intensiveness[a] Actual	Load Factor 100%	Load Factor (%)	Revenue[b]	Fatality Rate[c]	Haul Length (miles)	Average Speed (mph)
Inter-City							
Bus	1,600	740	46	3.6	0.10	100	45
Railroad	2,900	1,100	37	4.0	0.09	80	40
Automobile	3,400	1,600	48	4.0	3.25	50	~ 50
Airplane	8,400	4,100	49	6.0	0.13	700	400
Urban							
Mass transit							
Bus	3,700	650	18	9.5	—	3	—
Electric	4,100	1,100	26	6.2	—	4	—
Average	3,800	760	20	8.3	0.26	3	~15
Automobile	8,100	2,300	28	9.6	2.11	6	~20

[a] Btu per passenger mile.
[b] Cents per passenger mile.
[c] Deaths per hundred million passenger miles.

ger traffic than all other modes combined. The growing use of autos is related to rising personal incomes, shifting demographic patterns, highway construction and the comfort, privacy and freedom of routes and schedules available with autos.

Airplanes are the most energy intensive passenger mode, with energy intensiveness five times the value for buses. Between 1950 and 1970 it nearly doubled while air speed more than doubled. In spite of their high energy intensiveness and high price, airplanes are the most important intercity common carrier, having replaced buses and trains during the past two decades. Air traffic grew rapidly because of technological advances, government subsidies, comfort, high speeds and imaginative promotion.

Between 1950 and 1970, the fraction of inter-city passenger traffic carried by airplane climbed rapidly at the expense of trains and buses (Table 1). During this period, energy consumption increased by 155 per cent as a result of a 124 per cent increase in traffic and a 14 per cent increase in overall energy intensiveness. The latter increase was due to increases in energy intensiveness for individual modes and the shift from buses and trains to airplanes.

Urban Passenger Traffic

Urban passengers are carried almost exclusively by car, with only a small and declining fraction carried by mass transit. Figure 2 shows that mass transit is more than twice as energy efficient as autos [4]. Walking and bicycling, not shown in Figure 2, are an order of magnitude more efficient than autos, based on energy consumption to produce food. Urban values for energy intensiveness are more than double comparable inter-city values because of poorer vehicle performance (fewer miles per gallon) and poorer utilization (lower load factors) in cities.

Mass transit includes gasoline-powered buses and electric vehicles (subways, elevated trains, trolleys, surface trains). The decline in mass transit traffic was due, in large part, to the factors accounting for increased auto use cited earlier. As traffic shifted to cars, transit revenues declined, service and equipment worsened and the spiral continued. Because of higher costs and lower revenues, funds were not available for modernization, experimentation and research. Mass transit load factors declined from 30 per cent in 1950 to 20 per cent in 1970. Partly for this reason, energy intensiveness increased during this time for mass transit.

Buses have a slightly lower energy intensive value than do electric vehicles (Table 3), but the difference is minor. However, because buses have a lower load factor, they are *potentially* more energy efficient than electric trains.

Between 1950 and 1970, energy consumption for urban passenger traffic grew by 165 per cent, as traffic increased by 132 per cent and energy intensiveness by 14 per cent. This increase in energy intensiveness was due to increases in energy intensiveness for individual modes and the shift from mass transit to automobiles (Table 1).

Tables 2 and 3 show 1970 data and minimum values for energy intensiveness, load factors, revenue, fatality rates, haul length and average speed for these modes. The minimum energy intensive values in Table 3 assume a 100 per cent load factor, with no energy penalty for increased passenger weight per vehicle. This column demonstrates the importance of "social efficiency" in determining actual transportation energy efficiencies.

Variations in unit price, fatality rate and speed are positively correlated with variations in energy intensiveness. Thus a shift to low energy intensive modes would imply lower costs to users, fewer deaths and slower speeds as well as energy savings.

Between 1950 and 1970, the fractions of energy used for autos, trucks and airplanes grew steadily —with airplanes showing the largest percentage growth. These increases were offset primarily by the sharp drop in the fraction of energy used by railroads. The fractions of energy devoted to intercity buses and mass transit also fell.

During this 20-year period, the fraction of energy devoted to passengers increased steadily while the freight fraction fell. In 1970, about 60 per cent of the energy budget was for passengers and 15 per cent for freight; in 1950, the figures were 40 per cent and 32 per cent, respectively. The decline in relative freight energy use may reflect the shift from a goods-producing economy to a service economy, as well as income growth that allowed people to buy and use more automobiles. The traffic discussed here accounts for about 75 per cent of the transportation energy budget. The remaining 25 per cent is used for urban freight, military transportation, general aviation, pleasure boating, school buses, non-freight truck uses, greases and lubricants.

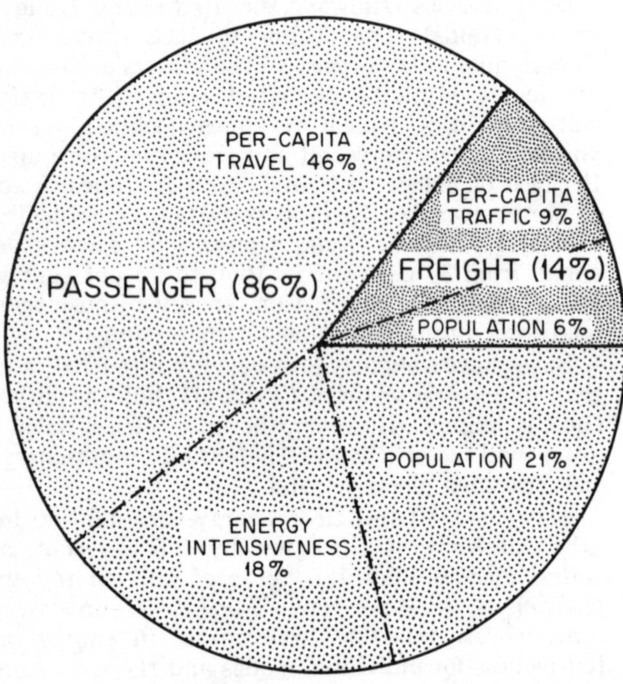

Fig. 3—Factors accounting for the increase in transportation energy use between 1960 and 1970. (Numbers do not add up to totals because of rounding.)

The factors accounting for transportation energy growth during the 1960s are shown in Figure 3. Growth in per capita transportation (especially passenger travel) accounted for more than half the decade's energy growth. Population growth accounted for one-fourth of the rise, and increasing energy intensiveness for one-fifth. Thus, transportation energy growth is due primarily to rising traffic levels and secondarily to shifts toward high-energy intensive modes and increases in energy intensiveness for individual modes.

Life-Style Changes

One obvious way to improve energy efficiency is to shift traffic from high to low energy intensive modes (Figs. 1 and 2). In general, such shifts require no new technologies; however, they may involve "life-style" changes (walking or riding the bus in cities rather than driving) and major institutional actions (massive funding to revitalize mass transit systems). Table 4 illustrates potential energy savings for a shift of one billion passenger-miles (or ton-miles) from one mode to another, although historical trends have been in the opposite direction.

Transport energy intensiveness can be lowered by using existing equipment more fully. Urban autos, mass transit and trains have particularly low-load factors. Table 4 shows potential energy savings per billion passenger-miles for a 10-percentage point increase in load factor for these modes.

Achieving such load factor improvements requires no new technologies; however, life-style changes associated with increased car pools, consolidation of auto trips and greater use of mass transit and trains with existing routes and schedules would be needed.

Technological Changes

Application of existing and emerging technologies can increase vehicle energy efficiency and also improve transport system comfort, speed and service. These latter improvements (e.g., the new $1.5 billion San Francisco Bay Area Rapid Transit System) might help to improve system load factors and induce a shift to low energy intensive modes.

Since auto energy intensiveness is almost directly proportional to vehicle weight [5], the use of smaller, lighter cars would lower it. Several engines are currently under consideration which may meet new emissions standards and provide improved fuel economy. The diesel engine, for example, appears to offer a 40 per cent reduction in auto energy intensiveness, relative to 1973 internal combustion engines [5]. Improved drive-train design, use of low-loss tires (radials reduce energy intensiveness by about 10 per cent), and aerodynamic drag reduction offer additional energy conservation opportunities.

Electric cars are not considered here because they appear to offer no significant energy efficiency advantage over gasoline-powered automobiles. Also, the 1975-76 auto emission-control requirements are not considered, because of the controversy over the energy impacts of such controls. Environmental Protection Agency tests indicate an 8 per cent fuel economy loss due to emissions controls between 1968 and 1973 [5]. Available data are insufficient to show any fuel economy trends for the 1975-76 control systems, but some sources project increases in auto energy intensiveness of 15 to 30 per cent between 1968 and 1976.

Jumbo jets are more efficient than conventional jets. For example, the Boeing-747 consumes only 53 per cent as much fuel per seat-mile as does the Boeing-707 [6]. National Aeronautics and Space Administration programs to reduce aerodynamic drag could provide additional energy savings. Reduced aircraft speed would yield major energy savings because fuel consumption is roughly proportional to the square of speed. For example, a 10 per cent reduction in speed (from 400 to 360 mph) would reduce fuel consumption by almost 20 per cent.

As noted earlier, trains are much heavier per passenger than other modes. Use of lightweight metals and plastics to reduce train weight would lower railroad energy intensiveness. Improved roadway design might also yield fuel savings.

Table 4 shows the energy savings possible (per billion passenger-miles) for 33 per cent reductions in energy intensiveness for autos, airplanes and trains. As indicated above, such reductions are technologically possible. However, other technologies, such as the supersonic transport, vertical short take-off and landing aircraft, high-speed trains and air-conditioned autos, would increase transport energy intensiveness.

Table 4

Transportation Energy Conservation Strategies

From 1970 Situation	To Energy-Efficient Alternative	Energy Savings[a] 10^{12} Btu
Passenger traffic: modal shifts		
Inter-city auto	Inter-city bus	1.8
Airplane	Inter-city bus	6.8
Urban auto	Mass transit	4.3
Passenger traffic: load factor increases[b]		
Urban auto (28%)	Urban auto (38%)	2.1
Mass transit (20%)	Mass transit (30%)	1.3
Trains (37%)	Trains (47%)	0.6
Passenger traffic: technological changes[c]		
Inter-city (auto (3400)	Inter-city auto (2300)	1.1
Urban auto (8100)	Urban auto (5400)	2.7
Airplane (8400)	Airplane (5600)	2.8
Train (2900)	Train (1900)	1.0
Freight traffic: modal shifts		
Truck	Train	2.1
Airplane	Train	41.3

[a] Energy savings are computed on the basis of a one billion passenger-mile (or ton-mile) effect, about 0.05 per cent of 1970 passenger traffic (or inter-city freight traffic).

[b] Energy savings given are for a 10 percentage point increase in load factor; numbers in parentheses are load factors.

[c] Energy savings given are for a 33 per cent reduction in vehicle energy intensiveness; numbers in parentheses are energy intensiveness values in Btus per passenger-mile.

A Scenario

Using the information in Table 4, one can devise various energy consumption scenarios for transportation. The following fanciful scenario for 1970 reduces transportation energy use by 50 per cent with no reduction in total travel:

- Half the inter-city freight carried by truck and by air is shifted to rail with no load factor or technological changes.
- Half the inter-city passenger traffic carried by air and one-third the traffic carried by car is shifted to bus and train. Railroad load factor is increased 10-percentage points, and technological improvements in autos, trains and airplanes (Table 4) are incorporated.
- Half the urban auto traffic is shifted to mass transit. Both mass transit and urban auto load factors are increased 10 points. Urban auto design is changed (Table 4).

The purpose of this totally unrealistic scenario is to show that significant reductions in transportation energy intensiveness are not impossible. The time-scale for such changes, however, is probably a decade or two. Whether or not we improve system efficiency depends more on our collective will and judgment than on scientific and technological breakthroughs.

There are at least three complementary routes through which transportation energy conservation measures could be implemented: economics, public education and government action.

Increasing affluence is undoubtedly an important determinant of transportation decisions. Over the past two decades both gasoline and automobile prices fell by about one-half relative to disposable personal income [7]. Thus a gallon of gas or a new car accounts for a smaller and smaller fraction of the American family budget each year. No wonder so little attention is given to automobile fuel economy.

However, the long-term downward trend in gasoline prices has changed because of the growing shortage of domestic petroleum supplies, the increasing need to import oil and the establishment of environmental standards which internalize the social costs of oil production, refining and use. Oil prices will quite possibly increase by 100 per cent during the next decade.

Perfectly Informed Consumer

As the social costs of oil production and oil scarcities are internalized, market prices will rise, and the demand for oil will decrease. Thus the traditional market system, working through the price mechanism, can encourage greater energy-efficiency of transportation. For example, fuel price increases are likely to inhibit growth in air freight because of the high energy intensiveness for airplanes. The effect on other freight modes is likely to be minor [8].

Public education offers another way to increase energy efficiency. According to one expert, discussing ways to reduce energy demand, "The first thing I would do would be to strive to develop what I call the perfectly informed consumer. That is the consumer who understands what is best for him. This person does not exist now" [9].

People need to understand the environmental costs of the existing transportation system, particularly of the automobile, and the relative benefits and costs of alternative modes such as buses, trains and mass transit. Public transportation must do a better job of advertising its routes, schedules and advantages over higher energy intensive modes. The only two modes that are currently well advertised are autos and airplanes; they are also the two highest energy intensive modes.

The federal government influences transportation decisions through regulations, taxes, subsidies, research and development and educational programs. The Interstate Commerce Commission and the Civil Aeronautics Board regulate inter-city freight —but in an uneven manner. All railroad and airplane freight traffic is regulated, but only 40 per cent of truck and 10 per cent of water freight traffic is federally controlled [10].

For passenger traffic, automobiles are essentially unregulated, although common carriers are regulated. However, increasingly strict air pollution and safety standards are beginning to regulate autos.

The federal budget provides other clues to federal transportation priorities. In 1968, the $6 billion transportation budget [11] was allocated as follows:

Highways	70 per cent
Air travel	15
Waterways	14
Mass transit	1
Railroads	0.2

Federal research and development funds were allocated in a similar manner.

These spending patterns are shifting in response to emerging transportation needs. Estimates of the 1972 federal budget show large increases in funding for mass transit and railroads and a relative decline in highway funding [12]. Estimates of the 1973 federal transportation research and development budget show 18 per cent for urban mass transit and 9 per cent for railroads [10]—compared with 5 per cent and 3 per cent, respectively, for 1969. Increased funding for air travel is a strong exception to the trend toward greater emphasis on low energy intensiveness for all modes.

Federal policies toward oil exploration and production and management of federal energy holdings also influence transportation decisions (through the price of oil). The depletion allowance and increased federal oil leasing tend to lower oil prices, thereby increasing oil consumption. Limiting imports has the opposite effect. Federal and state pollution controls, by internalizing the social costs of the petroleum cycle, serve to raise oil prices, reducing the demand for oil. Reduced demand is likely to be reflected in a shift toward low energy intensive transportation systems.

Recent history shows a steady growth in transportation energy use at a rate more than double the population growth rate. Current trends show continued growth in traffic levels, shifts toward high energy intensive modes and general increases in energy intensive modes.

However, oil scarcities and increasing dependence on petroleum imports, coupled with rising environmental concern, could reverse these historical trends. It is technologically feasible to slow transportation energy growth by increasing transportation energy efficiency. Policies to achieve such goals would involve some life-style changes and important institutional decisions, but they do not imply a return to "caves and candles."

To some extent, incentives for greater energy efficiency may come from the traditional market system as oil prices increase during the coming years. Additional incentives may come from governments if they seek to reduce the environmental and social costs of our transportation system. Finally, individual action—more car pools, greater use of small cars and mass transit, more walking and bicycling, less travel—could reduce transportation energy growth.

Theoretically, energy growth rates can be easily slowed and adverse impacts on the physical environment and energy resources reduced, but so far we, as a nation, have shown little will to do so.

REFERENCES

1. U. S. Dept. of Interior, Bureau of Mines, "U. S. Energy Use at New High in 1971," news release (Washington, D.C.: The Bureau, March 31, 1972).

2. M. K. Hubbert, "The Energy Resources of the Earth," *Scientific American,* 225 (1971), 60.

3. National Petroleum Council, *U. S. Energy Outlook: An Initial Appraisal 1971-1985* (Washington, D.C.: The Council, 1971).

4. E. Hirst, "Energy Intensiveness of Passenger and Freight Transport Modes: 1950-1970" (Oak Ridge, Tenn.: ORNL-NSF-EP-44, April 1973).

5. U. S. Environmental Protection Agency, Office of Air and Water Programs, *Fuel Economy and Emission Control* (Washington, D.C.: EPA, Nov. 1972).

6. R. A. Rice, "Historical Perspective in Transport System Development," in *Advanced Urban Transportation Systems* (Pittsburgh: Carnegie Mellon University, 1970).

7. American Petroleum Institute, *Petroleum Facts and Figures,* 1971 (Washington, D.C.: The Institute, 1971).

8. W. E. Mooz, *The Effect of Fuel Price Increases on Energy Intensiveness of Freight Transport,* Report R-804-NSF (Rand Corp., Dec. 1971).

9. S. F. Singer, Statement to U. S. Senate Committee on Interior and Insular Affairs Hearings, National Goals Symposium, Oct. 1971 (serial 92-11).

10. Transportation Association of America, *Transportation Facts & Trends* (9th ed.; Washington, D.C.: The Association, 1972).

11. U. S. Congress, House, *The Budget of the United States Government,* FY 1970, 89th Cong., 1969 (serial 91-15).

12. R. S. Benson and H. Wolman, eds., *Counterbudget,* The National Urban Coalition (New York: Praeger Publishers, 1971).

Part IV
Social Issues—
The Benefits and Costs of Energy Use

he benefits of a
ɔably more diffi-
ould be obvious
basic light, heat,
r value than the
le economy, our
re, is based upon
pects that a com-
y is probably im-
sales will grossly
repared to believe
plus" would yield
arent that such a

ılt to estimate the
rgy, and this after
incremental social
n can be made, at
ther the new unit

neral energy litera-
positive correlation
of energy use. Of
ndicator of social
bserved correlation
tant component of
is section explores
tion and economic

nmary report of the
ronment, provides a
the problems en-
benefits of energy
that such estimates
ative level, it should
this paper that the
in developing both
put data for use in

excerpt of several
Council on Environ-
mental Quality, entitled "...gy and the Environment: Electric Power." Not included in these excerpts are an introductory chapter that covers material on energy demand and use patterns, and an extended technical appendix that contains much useful data on the environmental impacts of the U.S. energy system. Readers who find this material interesting may also wish to try to locate a copy of the report, "Environmental Considerations in Future Energy Growth," prepared for the Environmental Protection Agency (EPA) by the Battelle Columbus Laboratories [2]. A very valuable set of basic data in this area is provided by the "Hittman Report" [3], a revised and expanded form of which exists as a computer data-base at Brookhaven National Laboratory.

The next three entries are on the incredibly complicated subject of nuclear energy. L. A. Sagan's paper, "Human Costs of Nuclear Power," discusses the human costs associated with operating conventional nuclear power plants. A. Kneese, in his paper "The Faustian Bargain," discusses some of the long-term questions associated with nuclear energy use. And finally, the paper by D. Nelkin provides a case study of a nuclear plant siting controversy. The Nelkin paper is a condensation of her book, *Nuclear Power and Its Critics: The Cayuga Lake Controversy* [4]. One interesting extended discussion of the nuclear power issue is contained in "Nuclear Power and the Public," which consists of papers and brief discussion transcripts from a meeting at the University of Minnesota [5].

The final five papers in this section treat several aspects of the generation of electric power from coal. The reports by Barrett and Waddell are analytical pieces which draw on a wide range of original studies in order to estimate the cost of air pollution. Because of space limitations, only brief excerpts from these reports are reproduced here. In its earlier forms the Barrett and Waddell analysis served as the (usually uncited) basis of many EPA and other federal estimates of the cost of air pollution, including the numbers which appeared in "The President's 1971 Environmental Program," prepared by the Council on Environmental Quality. These studies have thus played an important role in shaping the national discussion of air pollution.

Space limitations preclude the inclusion of any of the primary literature on air pollution environmental and health effects. The interested reader is encouraged to begin his further readings in this area by sampling the 1971 paper by Lave and Seskin [6] and Lave's paper, "Air Pollution Damage: Some Difficulties in Estimating the Value of Abatement" [7]. Paper 8 in this section uses some of the available data to make a first-order estimate of the social cost of producing electric power from coal. The final three papers, 9, 10, and 11, deal with several of the important social and environmental problems of strip mining. One serious problem in this field has been the scarcity of hard data both on the effects of mining, especially strip mining, and on the costs of land recovery. For this reason the work summarized by Schmidt-Bleek and Moore, in a paper prepared especially for this book, is a welcome contribution.

REFERENCES

[1] A. M. Okum, "Should GNP measure social welfare?", *Brookings Bull.*, vol. 8, pp. 4–7, summer 1971. Published by the *Brookings* Institution, 1775 Massachusetts Avenue, N.W., Washington, D.C. 20036.

[2] *Environmental Considerations in Future Energy Growth—vol. 1: Fuel/Energy Systems: Technical Summaries and Associated Environmental Burdens*, report prepared by Battelle Columbus Labs. for the Office of Research and Development of the Environmental Protection Agency under Contract 68-01-0470, Apr. 1973. As of early 1974 this report was not widely available. The editor has been informed, however, that this report should become available through the National Technical Information Service (NTIS).

[3] "Environmental impacts: Efficiency and cost of energy supply and end use, phase I—Draft final report," prepared by Hittman Associates, Inc., Columbia Md., for the Council on Environmental Quality, HIT-561, 1973.

[4] D. Nelkin, *"Nuclear Power and Its Critics: The Cayuga Lake Controversy* (Sci. Technol. Soc. Ser.). Ithaca, N.Y.: Cornell Univ. Press, 1971.

[5] H. Foreman, Ed., *Nuclear Power and the Public*. Minneapolis, Minn.: Univ. Minnesota Press, 1971.

[6] L. B. Lave and E. P. Seskin, "Air pollution and human health," *Science*, vol. 169, pp. 723–733, Aug. 21, 1970.

[7] L. B. Lave, "Air pollution damage: Some difficulties in estimating the value of abatement," in A. V. Kneese and B. T. Bower, Eds., *Environmental Quality Analysis: Theory and Method in the Social Sciences.* The John Hopkins Press, 1972, pp. 213–242, Baltimore, Md.: published for Resources for the Future.

Patterns of Energy Consumption and Economic Growth and Structure

*By Peter E. de Janosi and Leslie E. Grayson**

SUMMARY

The role of energy utilization in economic growth has received much attention in recent years. The distinctive characteristics of our analysis are that this role is dealt with in a quantitative way and that both cross-section and time-series approaches are utilized. We confirm earlier findings that energy consumption tends to be more responsive to economic growth in less developed than in advanced countries. We conclude, however, that the relationship between energy and economic growth activity is affected by a variety of other factors. Multivariate tests suggest that the industrial structure of the economy and the composition of energy consumption are especially significant additional variables.

In recent years economists and policy makers have become increasingly concerned with the nature of the relationship between economic growth and the utilization of inanimate energy sources—coal, oil, natural gas, hydroelectricity and nuclear power. There are sound reasons for this interest as the expansion in energy consumption concomitant with economic development has in some countries created serious problems. In selected instances, the growth in industrial production has placed a severe strain on available inexpensive domestic energy supplies which in turn necessitated an increase in imports and in this way contributed to balance of payments difficulties. At times also, energy shortages have been blamed as a cause for failures to meet planned national growth targets.

The analysis presented is designed to provide some information on a number of broad issues relevant to how the growth and structure of an economy affect energy consumption. The primary emphasis of the paper will be limited to the following two specific questions:

(1) What are the quantitative responses of energy consumption to changes in national income?
(2) What quantifiable economic explanations for these responses can be found?

While the analysis will build on earlier work of others, it is worth noting that much of the past research has been qualitative, or was based on limited statistical analysis and techniques. These limitations have been primarily of two sorts. In some cases only one country's experience was analysed from a historical point of view without any reference to the observed record of other nations. Research, however, that followed an international comparative approach tended to be restricted to one single year. Both approaches—the time series and the cross-section—have provided useful insights into why energy consumption differs in different stages of economic

* Program Officer, Higher Education and Research, Ford Foundation, and Professor of International Business, University of Virginia.

Reprinted with permission from *J. of Development Studies*, vol. 8, pp. 241-249, Jan. 1972.

development. Yet, being limited to either historical experience or geographic coverage reduces the generality of the findings and conclusions. Our research is designed to partially overcome these limitations by bridging the cross-section and the time series approach. It is worth noting, however, that this bridge is only one of many possible ones and much further work will be needed to test its particular validity.

The paper will be organized as follows: Part I will review the pertinent literature. In Part II the response of energy consumption to Gross National Product will be estimated for a series of countries for the period 1953–65. Part III will introduce simple and multiple regression analyses of the inter-country differences in the relationship estimated in the previous section, and Part IV will summarize the major conclusions.

I

In an important study Schurr and Eliasberg [*1962*] investigated the long-term relationship between energy consumption and real G.N.P. in the United States.[1] They conclude that from about 1880 to 1915 there was a steady rise in energy requirements per unit of G.N.P. From 1915 through 1955, however, the use of energy per unit of G.N.P. has been declining steadily. As the decline in the latter period was less than the increase in the former, the total 75-year period witnessed a 55 per cent rise in the amount of energy used per unit of G.N.P.

The decline in the output of energy per unit of G.N.P. is not obvious from an *a priori* point of view. Indeed, increasing industrialization of the United States since 1915 might have resulted in a relative rise. The primary reasons for the decline according to Schurr and Eliasberg are the following:

(1) Increased efficiency of energy production resulting in dramatic increases in the useful energy output obtained from the raw energy input.

(2) A more rapid increase in electricity than in total energy consumption. Generation of electricity has experienced especially significant technological improvements that have resulted in greater efficiency.

(3) The decade of the 1910s which marked the turning-point in the input of energy per unit of real G.N.P. also marks the turning-point in capital and labour productivity.[2]

(4) Finally, and probably most important, changes in the structural composition of G.N.P. partially explain the decline in relative energy output. Prior to World War I the industrial sector of the economy increased faster relative to total G.N.P. than it did in the following decades. At the same time the services sector, as a share of total G.N.P., increased faster in recent years. The impact of these structural changes is self-evident; industry is an energy-intensive sector of the economy and the services sector is, of course, less so.

Attempts to view the relationship between energy and economic growth in a broader geographic context have involved cross-section analyses that measure the level of energy consumption of countries at different stages of their economic development. For example, Mason related *per capita* income and *per capita* energy consumption in 52 countries for the year 1952. He concluded that 'no country can enjoy a high *per capita* income without becoming an extensive consumer of energy' [*Mason, 1955*]. Also, Robinson and Daniel not only successfully related *per capita* income with *per capita* fuel consumption in 1949 but also related the changes in primary fuel consumption and manufacturing production, 1939–50, for the principal countries of Western Europe and North America [*Robinson and Daniel, 1956*].

In addition, a number of other scholars attempted to make broad estimates of the relationship between *per capita* G.N.P. and *per capita* energy consumption, though the conclusions they have reached are not always consistent. For instance, Kindleberger [*1958, pp. 22–25*] suggests that the rate of energy consumption accelerates as development gets beyond $200 *per capita* income. At the other end of the scale, Graham [*1962*] states that 'as consumption rises above about four tons of coal equivalent *per capita* the rate of (energy) growth is invariably slow'.

Particularly interesting work relevant to this paper has been done on the possible use of energy consumption as a proxy variable for capital. One of the first to give this possibility some attention was Frankel [*1955*] whose primary objective was to explain productivity differentials in the United States and Great Britain. He put forward the use of horsepower/worker figures to provide a substitute for the lack of capital data for these comparisons. Similar use of horsepower/worker data was made by J. B. Heath [*1957*] for Anglo-Canadian comparisons of productivity.

Frank [*1959*], using measures of capital data available for the United States and the United Kingdom, correlated energy consumption and the industrial capital stocks for these two countries. In both cases he found very high correlation for the years 1880–1948. Oliver [*1959*], while objecting to some of Frank's conclusions as too ambitious also supports Frank's general findings that 'an index of energy can be used as an index of capital'. Warren [*1964*] suggests that for a less developed country 'statistics on energy consumption represent, with certain qualifications, a broad gauge of aggregate economic activity in the country'. Finally, Olson [*1948, p. 515*] experimented with a formulation of a Cobb–Douglas production function that substitutes energy for capital. He concludes that 'energy is an index of a certain kind of capital; namely, energy consuming capital of all kinds—both measuring different ways the productive apparatus of society'.

II

The initial step taken in estimating the relationship between energy consumption and economic development was the selection of the countries on which to base the first part of the study—the time series analysis. Countries that met the following criteria were selected:

(1) Representative of a wide spectrum of stages in economic development;
(2) Consistent energy and G.N.P. data available for the period 1953–65;
(3) Characterized primarily by a prevailing market economy. Excluded, therefore, were the Soviet Union, Eastern Europe, and China (Mainland).

Altogether thirty countries met all the criteria adequately. In terms of 1964 *per capita* Gross Domestic Product they range from $78 in India to $3,002 in the United States. They also differ greatly in terms of economic characteristics such as degree of industrialization, rate of economic growth, and importance of foreign trade sector.

The basic gross national product data and energy data used in the research are respectively published in the United Nations *Yearbooks of*

National Account Statistics, New York, and United Nations, *World Energy Supplies*, Series J, New York. In all cases G.N.P. data are measured in local currencies, but have been adjusted for price changes. The energy data are measured in millions of metric tons of coal.

We have decided to follow the practice of previous work and define energy consumption as a derived concept. It is obtained by taking a country's production and imports of energy and subtracting from this total exports, additions to inventories and bunkers. As the various energy sources are measured in different units of differing calorific values, it has been found necessary to convert these into a common unit. The standard method of conversion is to express all energy in terms of metric tons of coal equivalent [*United Nations, n.d.*].

In order to make a quantitative estimate of the response of energy consumption to G.N.P., an equation in logarithmic form was fitted for each of the thirty countries as follows:

$$\text{Log E} = a + b \log \text{G.N.P.}$$

where

E = Energy consumption in millions of metric tons of coal
G.N.P. = Gross National Product in price adjusted local currencies.

The results of this statistical exercise are shown in Table I. The first conclusion that can be drawn is that the G.N.P. seems to be, without exception, highly correlated with energy consumption. R^2 is in all but two cases above 0·9; in the two exceptions it is above 0·8. The t-ratio (b/σ) in all cases is well above the 99 per cent confidence level ranging from a low of 7·5 to a high of 43·4.

Even though the statistical reliability of the regression is uniformly high, the striking substantive result is the wide range in the income elasticities. As can be seen from Table I in which countries have been arranged according to their income elasticity, the Philippines has the highest and the United Kingdom the lowest elasticity. At the top of the range the elasticity is four times larger than that at the bottom of the range. The average elasticity is 1·34.

Cursory examination of Table I confirms one finding of some previous studies, namely, that the income elasticity of energy consumptions is in part influenced by a country's stage of economic development. However, such conclusion is confounded by the high ranking of the Scandinavian countries and Switzerland, on the one hand, and the low position of Brazil, Taiwan, and two European less developed countries—Portugal and Greece. Thus a simple interpretation will not do and in the next section alternative and more complex explanations of the wide disparity of income elasticities will be explored.

III

As the initial step we explored simple, one variable, 'explanations' of the calculated elasticities. The first and most obvious one is based on the hypothesis that the income elasticities of energy consumption are related negatively to the level of economic activity—a relationship already noted by Graham [*1962*]. *Per capita* energy consumption [*United Nations, n.d.*] and *per capita* Gross Domestic Product [*United Nations, 1966*] are used as proxy variables for the level of activity.

TABLE I

RELATIONSHIP BETWEEN ENERGY CONSUMPTION AND GROSS NATIONAL PRODUCT

Country	b	σ/b	R^2
Philippines	2·0696	0·0764	0·986
Thailand (a)	1·9341	0·1223	0·958
India (a)	1·8925	0·0974	0·972
Finland (a)	1·8739	0·0939	0·972
Iraq (a)	1·8462	0·1777	0·908
Italy	1·7666	0·0687	0·984
Puerto Rico	1·7442	0·1479	0·933
Switzerland	1·7246	0·0851	0·974
Chile	1·6671	0·1671	0·901
Colombia	1·5758	0·1291	0·931
Sweden	1·4985	0·0862	0·964
Norway	1·4142	0·0691	0·974
Argentina (a)	1·3939	0·1340	0·908
Canada	1·3723	0·0484	0·986
Denmark (b)	1·3526	0·1087	0·933
Mexico (a)	1·3328	0·0798	0·962
Greece	1·2585	0·0832	0·954
Brazil (a)	1·1930	0·0554	0·976
Australia (b)	1·1926	0·0338	0·992
China (Taiwan)	1·1639	0·0484	0·982
Portugal	1·0823	0·0532	0·974
Netherlands	1·0486	0·0798	0·941
Austria (b)	1·0025	0·0698	0·949
Israel	0·9649	0·0776	0·933
Japan	0·9605	0·0300	0·990
United States	0·9588	0·0221	0·994
France	0·8778	0·0637	0·945
Belgium	0·8155	0·0962	0·867
Germany	0·6612	0·0478	0·945
United Kingdom	0·4780	0·0637	0·837

(a) Gross Domestic Product at factor cost.
(b) Expenditure on Gross Domestic Product.

An alternative hypothesis tested is that income elasticities are related to the industrial structure of the Gross Domestic Product. Schurr and Eliasberg's work related to this hypothesis; namely, that the structural composition of G.D.P. partially explains the nature of energy consumption [*Schurr and Eliasberg, 1962*]. For a measure of this structure we use the share of mining and manufacturing (expected relationship negative) and the contribution of agriculture (expected relationship positive) in the Gross Domestic Product [*United Nations, 1966*].

The final hypothesis formulated is that the structure of energy consumption exerts an important influence on the elasticities depending on the efficiency and price of the predominant energy sources. Two measures are employed: the share or coal, and the share of hydroelectricity in total energy consumption [*United Nations, n.d.*].

The rationale for selecting the share of coal was that it is a domestic fuel source, where price in most cases is maintained at an artificially high level. This was, however, not always so and coal, historically, was available in practically all industrialized countries and was probably used inefficiently. Thus, the greater the share of coal in total energy consumption, the less energy consumption is added per unit of increased G.N.P. as there is room for increased efficiency of energy utilization. We, therefore, expect it to be inversely related to the elasticities. Hydroelectricity on the other hand, while also a domestic fuel source, is usually inexpensive wherever it is available, and, therefore, the greater its share in energy consumption the greater we expect the income elasticity to be.[3]

For all six simple regressions representing these hypotheses, the dependent variable was the calculated income elasticities of energy consumption for 28 or 30[4] countries as calculated in section II and the independent variables were the appropriate measures based on 1964. (Not all data were available for 1965, the last year of the time period under analysis.)

The first six equations in Table II summarize the statistical results of the regressions. In all cases the parameters had the 'correct' sign and all but equation 4 were significant at the 0·05 level.[5] Thus, the hypotheses appear to be sensible, but, of course, the important and significant question still remains how much a multivariate analysis can improve on these results.

In view of the fact that the share of hydroelectricity in total energy consumption did not turn out to be significant in the simple regression this variable is omitted from the further analysis. For experimental purposes we combined the remaining five variables (Table II, Equation 7) although the results bore out our expectations that the coefficients would be unstable because of the high intercorrelations among some of the independent variables. There was, however, a sizeable improvement in the correlation coefficient compared to the simple regressions.

In the remaining computations, alternative combinations of variables were used. One of the most interesting findings is that the level of economic activity variables were not statistically significant, except marginally in Equation 9 (X_1) and 10 (X_2). It is also true in these equations that the measure of industrial structure was the share of mining and manufacturing in the G.D.P. (X_5) rather than the share of agriculture.

The two variables that appear to maintain the most persistently significant relationship were primarily the share of coal in total energy consumption and the share of agriculture in G.D.P. In all cases where coefficients were found to be statistically significant, the signs were also consistent with the stated hypothesis, which was that income elasticities of energy are related both to the industrial structures of G.D.P. and to the structures of energy consumption.

IV

The results obtained in our analysis confirm the earlier substantive finding reached in various studies of energy consumption based on the United States experience. We have also found overwhelming evidence in other countries that in fact a similarly strong relationship exists between economic growth and energy consumption, but that this relationship differs widely among countries.

While relative income levels do have some power in explaining the differential response of energy consumption to economic growth, we have also discovered two other significant associations. The first of these consists of measures of the structure of the economy as indicated respectively by the share of agriculture and the share of mining and manufacturing

TABLE II

RELATIONSHIP BETWEEN INCOME ELASTICITY OF ENERGY CONSUMPTION AND SELECTED MEASURES OF ECONOMIC DEVELOPMENT AND STRUCTURE

$\mu = \alpha_0$	$+10^{-3}\alpha_1 X_1$	$+10^{-3}\alpha_2 X_2$	$+10^{-3}\alpha_3 X_3$	$+10^{-3}\alpha_4 X_4$	$+10^{-2}\alpha_5 X_5$	$+10^{-2}\alpha_6 X_6$	R^2
(1) 1·568	−0·218 (0·090)						0·173
(2) 1·559		−0·091 (0·032)					0·226
(3) 1·585			−8·694 (2·738)				0·265
(4) 1·292				+7·913 (9·302)			0·025
(5) 2·084					−2·639 (0·871)		0·261
(6) 0·945						+2·254 (0·599)	0·353
(7) 1·228	−0·028 (0·285)	+0·022 (0·089)	−8·555 (2·514)		−0·220 (1·015)	+2·303 (1·094)	0·602
(8) 1·126		+0·017 (0·037)	−8·710 (2·308)			+2·478 (0·724)	0·602
(9) 2·151	−0·173 (0·087)		−6·537 (2·444)		−1·612 (0·814)		0·518
(10) 2·136		−0·052 (0·031)	−6·397 (2·513)		−1·734 (0·820)		0·500
(11) 1·121	0·044 (0·118)		−8·619 (2·281)			+2·474 (0·794)	0·601

( ) Standard error of regression coefficient.
X_1 *Per capita* Gross Domestic Product (U.S. dollars).
X_2 *Per capita* energy consumption (Kgs of coal equivalent).
X_3 Share of coal in total energy consumption (%).
X_4 Share of hydro-electricity in total energy consumption (%).
X_5 Share of mining and manufacturing in Gross Domestic Product (%).
X_6 Share of agriculture in Gross Domestic Product (%).

in G.D.P. The second set of variables that seems to affect the differential response relates to the composition of the sources of energy consumption. That is, the smaller the proportion of one fuel source (coal) in total energy consumption, the more elastic the response of energy consumption to economic growth. Thus, the response of energy consumption to economic growth is not a simple one, and is also influenced by factors other than the level of income.

In conclusion we must emphasize that the findings presented here are to be interpreted with caution. We are especially concerned about the inadequacies of the currently used index of energy consumption and the possible biases introduced by its use in both developed and underdeveloped countries. In the latter countries an important, but decreasingly so, part of energy consumption is satisfied from non-commercial sources. Because non-commercial energy is not measured in the official statistics, the on-going substitution of commercial for non-commercial energy has probably introduced an upward bias into the calculated elasticities for the less developed countries [*Guyol, 1961*]. A bias in the other direction centres on the developed countries and is the consequence of the substitution of oil and gas for coal [*Adams and Miovic, 1968*; *Turvey and Nobay, 1965*]. (Most developed countries were originally coal-based energy economies; the less developed countries usually satisfied their fuel requirements from oil.) The conversion of all fuels into coal equivalents assumes an average relative efficiency of energy sources even though in reality some fuels are very much more efficient, in some applications, than are others. In inter-country comparisons it may make a difference if the substitution of oil and gas (efficient fuels) for coal (relatively inefficient fuel) is proceeding at a differential pace. This may account, in part, for the low observed elasticities for such originally coal based countries as the United Kingdom, Germany, Belgium and France.

Regrettably, as yet, no satisfactory answers have been provided to the problems of measurement. In spite of this—and we do not believe that this is crucial—we hope that our analysis has taken steps toward answering the basic questions posed at the beginning of the paper and has been successful in offering some new insights to energy consumption and economic growth and structure.

NOTES

1. See also Wardwell [*1961*].
2. Some of the measurements of this point were disputed by E. F. Renshaw [*1963*]; see also du Boff [*1964*].
3. We used the share of coal and hydroelectricity, respectively, as a proxy for relative prices because, contrary to suggestions we have received, such prices are not available on a comparable basis.
4. Sweden and Switzerland do not publish G.D.P. by industrial origin, and therefore had to be omitted in some of the calculations.
5. A possible explanation as to why the hydro variable performed so poorly is that virtually all of hydro is publicly owned and, more likely than not, uneconomically priced. For an interesting institutional account of such pricing see Tendler [*1968*].

REFERENCES

Adams, F. Gerard, and Miovic, Peter, 1968, 'On Relative Fuel Efficiency and the Output Elasticity of Energy Consumption in Western Europe', *The Journal of Industrial Economics*, November.

Boff, R. B. du, 1964, 'Comment' on E. F. Renshaw [*1963*], *Journal of Political Economy*, April.

Frank, A. G., 1959, 'Industrial Capital Stocks and Energy Consumption', *Economic Journal*, March.

Frankel, Marvin, 1955, 'Anglo-American Productivity Differences: Their Magnitude and Some Causes', *American Economic Review*, May.

Graham, M. G., 1962, 'Factors Affecting the Future Pattern of the World Energy Market', *Sixth World Power Conference Proceedings*, Melbourne, October.

Guyol, N. B., 1961, *India's Energy Balance, 1959*, New Delhi: National Council of Applied Economic Research.

Heath, J. B., 1957, 'British–Canadian Industrial Productivity', *Economic Journal*. December.

Kindleberger, Charles P., 1958, *Economic Development*, New York: McGraw-Hill.

Mason, E. S., 1955, *Energy Requirements and Economic Growth*, Washington: National Planning Association.

Oliver, F. R., 1959, 'Comment' on A. G. Frank [*1959*], *Economic Journal*, December.

Olson, Ernest C., 1948, 'Factors Affecting International Differences in Production', *American Economic Review*, May.

Renshaw, E. F., 1963, 'The Substitution of Inanimate Energy for Animal Power' *Journal of Political Economy*, June.

Robinson, E. A. G., and Daniel, G. H., 1956, 'Need for a New Source of Energy', *The World's Requirements for Energy*, New York: U.N. Publications.

Schurr, Sam H., and Eliasberg, Vera F., 1962, *Energy and Economic Growth in the United States*, Washington D.C.: Resources for the Future Inc.

Tendler, Judith, 1968, *Electric Power in Brazil: Entrepreneurship in the Public Sector*, Cambridge: Harvard University Press.

Turvey, Ralph, and Nobay, A. R., 1965, 'On Measuring Energy Consumption'. *Economic Journal*, December.

United Nations, n.d., *World Energy Supplies*, Series J, New York: U.N. Publications.

United Nations, 1966, *Yearbook of National Account Statistics*, New York, U.N. Publications.

Wardwell, C. A. R., 1961, 'Energy Output and Use Related to the G.N.P.', *Survey of Current Business*, Washington, D.C., February.

Warren, J. C., 1964, 'Energy and Economic Advances', *The Philippine Economic Journal*, First Semester.

Risk Estimates and Analyses of Cost-Benefit and Cost-Effectiveness Relationships

GENERAL

The general goal of balancing social utility versus total social costs should be a part of all planning in our society. Traditionally, technical performance as a function of monetary cost is always involved in engineering design decisions. There is the additional need, however, for inclusion of all societal costs and all measures of utility, indirect as well as direct. Existing socio-technical systems, over a period of many years, have developed an empirically acceptable balance between utility and social costs. In addition we have examples of national decision-making involving future socio-technical systems which contain implicit predictive trade-offs of societal benefits versus societal costs.

As the number of our socio-technical systems increases and the impact on the individual becomes more apparent, concern with achieving a planned balance of utility of these systems versus their societal costs has also increased. It has become apparent, therefore, that greater insight, analysis, and predictive planning are essential for the future development of the important socio-technical systems.

These problems originate from certain general assumptions inherent in the operations of our society. First, it is traditionally accepted that everyone should have the oportunity for a natural death, that is, a death from old age and eventual wearing out of the human body. Second, it is now commonly accepted that every individual should have the opportunity to use and enjoy the fruits of our centuries of technological development. Third, more recently we have emphasized our responsibility to assure the best environmental and genetic inheritance for succeeding generations that we can provide. And fourth, it is the philosophy of an egalitarian society that where the activities of an individual infringe on others in an undesirable way, society may intervene to control individual activities in order to achieve a balance between group well-being and the privileges of the individual.

In more specific terms, regarding energy as well as most of man's activities, decisions have to be made at all levels from highest policy (for example, how research resources should be allocated) to day-by-day operations (for instance, the necessity for a given plant to have cooling towers to reduce thermal pollution). Most of these decisions are made by combinations of intuitive judgment and logic. The more data available, the better but these intuitive judgments will remain a large factor because decisions will have to be made before there are complete data available. To aid in decisionmaking we propose that to the extent possible consideration be given to risk estimates and analyses of cost-benefit and cost-effectiveness.

There is great need for extensive discussion within our society of the risks and benefits associated with particular policies and decisions regarding the utilization of our natural resources and our environment. For all major decisions associated with the production and use of energy, estimates should be made of the risk or harm that may be done to man and his environment. There then arises the need to assess the acceptability of this risk to society. The first criterion is comparison of the risk with the benefit associated with it. It is not only necessary that the benefit outweigh the risk, account must also be taken of the feasibility or cost of reducing the risk. A risk is not acceptable to society if it can be easily reduced. In this respect cost-effectiveness can be important from two points of view. First, the risk must be reduced if it is possible to do so at low comparative cost. Second, large amounts of resources should not be wasted to reduce small risks even further while larger risks go unattended. These types of analyses need to be made for all alternative options of achieving the same objective in order to permit a choice.

Some of the advantages and disadvantages of cost-benefit and cost-effectiveness analyses can be indicated as follows. (1) The focus on the biological and environmental cost from technological developments make the need for specific information apparent. (2) Public understanding can be increased with indications of just how and why decisions have been made. (3) Attention can be directed toward effective means of altering the risks and benefits. (4) There can be facilitation of the comparison of the optional modes of accomplishing the same ends.

There are many serious difficulties that arise over and above an inadequate data base in the use of cost analyses, most of which relate to the basic unit, the dollar. Some of these problems are as follows.

(1) Although effects on health can be assessed in terms of the cost of ill health and such factors as the days lost from work or productive life lost, there tends to be an emotional objection to the expression of human suffering and misery in financial terms.

(2) It is difficult, if not impossible, to express in financial terms many of the features that make up the so-called quality of life. This is also true in regard to the description of environmental impacts.

(3) The costs and the benefits very often do not accrue to the same individuals. For example, individuals who are exposed to the environmental risks of power plant emissions may not be the same persons who benefit from the availability of energy. In the case of genetic effects it may be that the individual who has to pay the biological cost is not even of the same generation as the individual who reaps the benefits.

(4) It is extremely difficult to establish a single cost system that would take into account differences in the willingness of individuals to accept various types of risks.

(5) There may be a tendency for heavier weighting of any numbers derived, since greater reliability may be assigned to them than is warranted by the data base that produced the numbers.

Despite these serious difficulties and uncertainties, there seems to be some merit in attempting to use cost-benefit analysis in a reasonable way. It must always be emphasized that the qualifications should be recognized and the results should be used to complement the overall intuitive judgments that are finally made.

APPROACH TO COST-BENEFIT AND COST-EFFECTIVENESS ANALYSIS

The general approaches to balancing societal utility of benefit against societal cost seem to fall into three categories: (1) actuarial or monetary relationships, (2) traumatic or maximum single catastrophes, and (3) philosophic or cultural biases. The first approach, actuarial, requires a systematized quantification of the social good arising from a technical system, such as energy supply, and then an equally complete quantification of the environmental costs, such as public health impacts, life style constraints, ecological damage, and aesthetic penalties.

The second approach relates to the public response to large, infrequent catastrophes such as earthquakes, explosions, oil fires, and possibly extreme nuclear power plant accidents. In the second category, the actuarial importance may be low, that is, the average effects of the infrequent events taken over many years may be much less than the average of the effects of the frequent small events. Nevertheless, the public abhorrence of specific catastrophes may result in relatively large investments to prevent or ameliorate them regardless of the quantitative importance.

The third category, philosophical biases, are not subject to rational argument or analysis. In this category, most pertinent to environmental matters, are the relationship of the individual to the group or the state, and the issues of aesthetics or environmental ethics. It is in this area of societal ethics and cultural values that our political system functions as a most important arbiter.

Returning to the first category of actuarial benefit-cost analysis, the public has shown great concern with the societal health costs of risks arising from proposed or existing energy systems. There is strong indication the the public has emphasized the energy problem because this is a readily identifiable societal problem of the environment which may well play a catalytic role in a process of national reappraisal of our

Reprinted with permission from *Cornell Workshop on Energy and Environment*, Senate Committee on Interior Affairs, no. 92-23, pp. 35–42, 1972.

attitude toward our environment. Recognizing that it appears to be impossible to create a zero risk energy system the most important question then becomes the level of risk which is socially acceptable. The importance of this issue arises from the general consideration that at any given time decreasing the public risk of any system requires an increased cost to society. In a resource limited society the allocations should presumably be made where they will do the most social good. Thus, adding costs for the purpose of decreasing the risk of any one system much below an acceptable level is a misuse of resources when other societal activities with unacceptable risks are not being attended to with sufficient emphasis.

The determination of a level of risk which may be socially acceptable may be done in a variety of ways. If one assumes that the historical behavior of society can be taken as a guide, however, it appears that a quantitative statement of an acceptable risk level can be arrived at for any technical system such as energy supply. Our society appears to gauge acceptable risk levels for involuntary exposure of the public by two natural yardsticks. The upper boundary appears to be the statistical risk of death from internal life span forces (old age and disease) of about 10^{-2} per year per person. The lower boundary appears to be the risk of death arising from external catastrophes, such as floods, hurricanes, lightning strikes, insect and snake bites, and other such events uncontrolled by man, which amounts to about 10^{-6} per year per person (a risk 10^{-6} per year per person means that there is a chance of one in a million of the effect occurring to an individual within one year).

Thus, one can establish a rough scale as follows: greater than 10^{-2} to 10^{-3} is high; 10^{-3} to 10^{-5} is medium; to $^{-5}$ to 10^{-7} is low; and less than 10^{-7} is negligible. Traditionally, society has treated low and negligible risks as acts of God and, conversely, has only focused attention on the excessive and high risks. Presumably, a rational society would accept high risks only for high benefits and demand low risks for low benefit activities. It is interesting that the private automobile is in the high risk range averaged over the nation. It is also pertinent that the public health risk arising from the generation of electricity by both fossil fuel and nuclear power plants has been placed in the low or negligible range which includes accidents, again averaged over the nation.[3] It is thus important to emphasize that public concerns with these power sources arise not from the actuarial risk analysis but from either the issues of episodic trauma or philosophic biases. In the case of fossil fuel combustion the contribution of effluents to air pollution in meteorologically adverse areas has been the primary public issue. In the case of nuclear power plants it is the public conception of the accidental catastrophe which is an important issue. There is also public concern relating to the reprocessing, transportation, and long-term storage of high-level radioactive wastes.

In the case of power plants the public risk issue could, in principle, be removed by geographically separating the physical generating plant from the centers of population. This would provide the benefits of electricity to one area and export the environmental disbenefits (penalties) to another, presumably less populated area. Such practices raise many questions of social ethics which must be extensively discussed on the national level. It should be pointed out that inevitably this type of separation is inherent in any fossil fuel energy system. Coal and oil are rarely found at load centers, so some of the disbenefits of fuel supply (strip mining of coal, oil tanker leaks) are likely to be imposed on a population not receiving the full benefits of electricity generated.

The public controversy over nuclear power plants raises many important policy issues. Because of the low level of public risks so far demonstrated by existing plants, the usual collection of accident data with which to prove public safety is not available now and is likely to be absent for several decades. Thus, safety must be inferred from technical analysis and public acceptance will depend on the image of credibility of our national technical expertise. This is a delicate area of public relations, but it also requires a strong base of in-depth technical analysis. With regard to the lesser public concerns for handling nuclear radioactive wastes, this is a matter where technology is not the issue. Rather the issue is social philosophy. The amount of wastes likely to be generated in the next 30 years by nuclear power is small in volume. Techniques for compactly packaging these wastes are already available and continuously being improved. What is required here is safe management and storage for an indefinite time in a complete recoverable manner. This will permit the next generation to manage these wastes as they choose. Options include continued storage, as with gold in Fort Knox, perhaps separation and productive use, or even (esoterically) transport by rocket to the sun for external disposal.

ATTEMPTS AT ANALYSIS

Cohen and Higgins[4] have summarized various estimates of the biological cost of radiation (the dollar value of the risk of one man-rem) and these range from $10 to $250 per man-rem. As an illustration, these estimates were used to compare the biological costs of tritium and krypton–85 from nuclear power production and the benefit-risk ratios of energy produced from nuclear stimulated natural gas and from power reactors. It seemed clear that the biological cost from ^{85}Kr per unit of electricity produced was about 1,000 times that from tritium. It was also demonstrated that gas stimulation would be a better environmental choice than power reactors based on the biological cost from contaminating radioactivity.

Sagan[5] has attempted to assess the human costs of nuclear power through the use of estimates of the value of human life, lost productivity, and potential radiation effects. Human costs of two kinds were considered: (1) accidental injuries and deaths, usually occupationally incurred, and (2) potential health hazards incurred by those persons exposed to radiation produced throughout the fuel cycle (in-plant and elsewhere). It was recognized that charges of insensitivity could be attached to the use of dollars for measuring such costs but no other workable method is currently in use. The biological cost of nuclear industry radiation was estimated to be $30 per man-rem. Based on a present nuclear power industry of 10,030 MWe, the biological cost to the average member of the population was estimated to be ten cents per individual per year; the cost per year to occupationally involved individuals was estimated as follows: uranium mining and milling $673; manufacturing $45; reactor operation $63; reprocessing $115. Parallel data for the fossil fuel generation of electricity are not available but we can get an inkling of how such comparisons may be made in the future as the data base becomes more reliable and adequate.

Terrill[6] has made estimates of the comparative cost of reducing population exposures by collimation of medical X-ray machines and by modifications of nuclear power plants to provide releases as low as practical. It was concluded that money now spent on improved X-ray equipment would accomplish from thousands to tens of thousands times more exposure reduction per dollar than money spent on reactor waste treatment systems in the current exposure ranges.

Brodsky[7] has proposed a systematic procedure for apportioning the radiation exposure from consumer items according to some prior estimation of benefit and risk.

Another interesting example[8] has been the attempt to compare the cost-benefits of various health programs. This analysis goes right to the operational point of how best to spend money. For several types of programs estimates were made of the cost of the program per death averted. For example, a national program to persuade people to use seat belts is estimated to cost less than $100 for each death averted. In connection with cancer, typical programs showed costs of about $4,000 to $40,000 for each death averted; some of these programs consisted of early detection and treatment, whereas a lung cancer program was preventive, an attempt to persuade people not to smoke.

Garvey, in a forthcoming study,[9] has attempted to identify and quantify the externalized effects of energy in the market place. He has considered such factors as depletive wastes, pollution, and seral disturbances (or interference with the cycles of the ecosystem affronted by the fuel use in question). An estimate of these costs is shown in the following table.

[3] Chauncey Starr, "Benefit-Cost Studies in Socio-Technical Systems," paper presented at the Colloquium on Benefit-Risk Relationships for Decision-Making, Washington, D.C., April 26, 1971.

[4] Jerry J. Cohen and Gary H. Higgins, "The Socioeconomic Impact of Low-level Tritium Releases to the Environment," paper presented to the Tritium Symposium, Las Vegas, Nevada, August 30–September 2, 1971.

[5] Leonard A. Sagan, "Human Costs of Nuclear Power," *Science* (forthcoming).

[6] James G. Terrill, "Cost-Benefit Estimates for the Major Sources of Radiation Exposure," paper presented to the Radiological Health Section, American Public Health Association, October 11, 1971.

[7] Allen Brodsky, "Balancing Benefit versus Risk in the Control of Consumer Items Containing Radioactive Material," *American Journal of Public Health,* Vol. 55, No. 12, pp. 1971–1992.

[8] Elizabeth B. Drew, "HEW grapples with PPBS," *The Public Interest,* No. 8, Summer 1967, pp. 9–29.

[9] Gerald Garvey, "Energy—Economy—Ecology," Norton, forthcoming.

TABLE 1.—ESTIMATED COSTS OF PECUNIARY EXTERNALITIES FOR 1970

[In billions of dollars]

	Coal	Oil	Gas	Nuclear	Total
Extractive phase	0.8	0.028		0.024	0.852
Transportation		.124	0.006		.130
Air pollution	10.2	10.9	.4	.280	21.780
Thermal pollution	.147	.010	.037	.026	.220
Total	11.147	11.062	.443	.330	22.982

The importance of air pollution (95 percent) should not exclude the possibility of much higher contributions from other areas. Since so many of the externalities are not recognized, their costs are hidden and unavailable for full accounting. Without better access to the missing energy-associated externalities, it is very difficult to appraise the real cost-benefits of energy systems.

LEGAL REQUIREMENTS FOR COST-BENEFIT ANALYSIS

In 1969 the courts ruled that the Atomic Energy Acts of 1954 and 1946 intended to confine the responsibility of the Atomic Energy Commission to scrutiny and protection against hazards of radiation. It was generally recognized that there needed to be a broader jurisdiction for AEC, especially in connection with development of the nuclear power industry. The Ninety-first Congress responded both to this matter and to the entire question of environmental management with broad and sweeping legislation known as the National Environmental Policy Act (NEPA). The major feature of the act appears in Section 102.

Sec. 102. The Congress authorizes and directs that, to the fullest extent possible: (1) the policies, regulations, and public laws of the United States shall be interpreted and administered in accordance with the policies set forth in this act, and (2) all agencies of the Federal Government shall—

(A) utilize a systematic, interdisciplinary approach which will insure the integrated use of the natural and social sciences and the environmental design arts in planning and in decisionmaking which may have a impact on man's environment;

(B) identify and develop methods and procedures, in consultation with the Council on Environmental Quality established by title II of this Act, which will insure that presently unquantified environmental amenities and values may be given appropriate consideration in decisionmaking along with economic and technical considerations;

(C) include in every recommendation or report on proposals for legislation and other major Federal actions significantly affecting the quality of the human environment, a detailed statement by the responsible official on—

(i) the environmental impact of the proposed action,

(ii) any adverse environmental effects which cannot be avoided should the proposal be implemented,

(iii) alternatives to the proposed action,

(iv) the relationship between local short-term uses of man's environment and the maintenance and enhancement of long-term productivity, and

(v) any irreversible and irretrievable commitments of resources which would be involved in the proposed action should it be implemented.

Prior to making any detailed statement, the responsible Federal official shall consult with and obtain the comments of any Federal agency which has jurisdiction by law or special expertise with respect to any environmental impact involved. Copies of such statement and the comments and views of the appropriate Federal, State, and local agencies, which are authorized to develop and enforce environmental standards, shall be made available to the President, the Council on Environmental Quality and to the public as provided by section 552 of title 5, United States Code, and shall accompany the proposal through the existing agency review processes;

(D) study, develop, and describe appropriate alternatives to recommended courses of action in any proposal which involves unresolved conflicts concerning alternative uses of available resources;

(E) recognize the worldwide and long-range character of environmental problems and, where consistent with the foreign policy of the United States, lend appropriate support to initiatives, resolutions, and programs designed to maximize international cooperation in anticipating and preventing a decline in the quality of mankind's world environment;

(F) make available to States, counties, municipalities, institutions, and individuals, advice and information useful in restoring, maintaining, and enhancing the quality of the environment;

(G) initiate and utilize ecological information in the planning and development of resource-oriented projects; and

(H) assist the Council on Environmental Quality established by title II of this Act.[10]

The responsibilities of the AEC in connection with the environmental effects of nuclear power were established by the *Calvert Cliffs* decision. In summary, the AEC is required (1) to consider environmental issues at every important stage of the decision-making process, on their own initiative and with the same rigorous procedures they employ to consider other matters, (2) to make independent determinations on environmental matters, (3) to systematically balance environmental amenities along with economic, social, and technical considerations on a case-by-case basis, and (4) to consider environmental impacts on projects initially approved before January 1, 1971, but not yet completed.

In compliance with the regulations, the AEC now requires applicants for nuclear power plant permits and licenses to submit comprehensive environmental reports which include a detailed guide for cost-benefit determinations. The outline of the cost-benefit guide is included with the working papers appended to this report. It is hoped that these requirements will stimulate efforts at cost-benefit analysis. It is doubtful, however, that data and experience are adequate to permit meaningful analyses as detailed as those called for.

Another approach to this subject is contained in the information matrix developed by the U.S. Geological Survey for evaluating environmental impact. This system calls for a designation of magnitude and importance (each on a scale of one to ten) for 8,900 items. The advantage of this system is that it provides a fairly comprehensive check list for many items that might otherwise be overlooked.

[10] National Environmental Policy Act, sec. 102.

Editor's Note: Not included in these excerpts are an introductory chapter which covers material on energy demand and use patterns, and an extended technical Appendix which contains much useful data on the environmental impacts of the U.S. energy system.

Excerpts from "Energy and the Environment: Electric Power"

Chapter II Electric Power and Environmental Quality

Electric Power Alternatives: Their Environmental Impacts

Because energy consumption is a major contributor to environmental degradation, decisions regarding energy policy alternatives require comprehensive environmental analysis. To the extent that it is practical, environmental impact data must be developed for all aspects of energy systems and must not be limited to separate components.

Analysis of the environmental impacts of alternative energy systems delivering an equivalent amount of energy is, unfortunately, difficult and time consuming. There are many systems to be evaluated—for such diverse uses as transportation, heat, and light—and they include many energy sources, many processing and delivery technologies, and many end uses.

For this reason, the present study does not look at all energy systems and end uses; rather, it considers only alternative systems for providing electricity. Because the impacts of extracting and processing these fuels are the same for direct use as for electric power, the environmental impacts of electric power systems are also relevant to energy systems that use coal, oil, and gas directly. Further, although this limited focus on electric energy omits many energy systems, it is important because electricity is the most rapidly growing form of energy use. Indeed, supply and demand trends indicate that within 50 years electricity will be the dominant form of energy use.[56] This rapid growth will require the construction of many new facilities, and we face major policy decisions concerning which electric power systems to build in the coming decades and to what extent energy conservation measures can help lessen demand. One projection of future electric power demand and the fuels necessary for meeting it is seen in Figure 7.

This study identifies the environmental impacts of electric energy systems by comparing the systemwide impacts of electricity production from different fuels. The comparison is based on 1,000-megawatt powerplants operating at 75 percent capacity for 1 year. Such a plant size is typical of many electric powerplants now being built. Each plant generates 6.57 billion kilowatt-hours annually, enough to meet the demands of 900,000 people—a metropolitan area roughly the size of Rochester, New York.[57]

The environmental impacts are those pertaining to systemwide air emissions, water discharges, solid wastes, land use, and occupational health and safety. This study does not quantify environmental effects because the relationship between impacts (i.e., emissions, discharges, etc.) and their effects is complex and not always well understood. Information is often lacking regarding the pathways of emissions from the source to the receptors and regarding the effects of the emissions that reach man and his environment. Regional differences based on geographic, demographic, and meteorological factors are not considered. Even limiting our analysis to total impacts of electric power systems was difficult because accurate data are not available for all steps in a system. Further, it

Reprinted with permission from *Energy and the Environment*, Council on Environmental Quality, U.S. Government Printing Office, 4111-00019, ch. 2, 3, and 4, pp. 12–39, Aug. 1973.

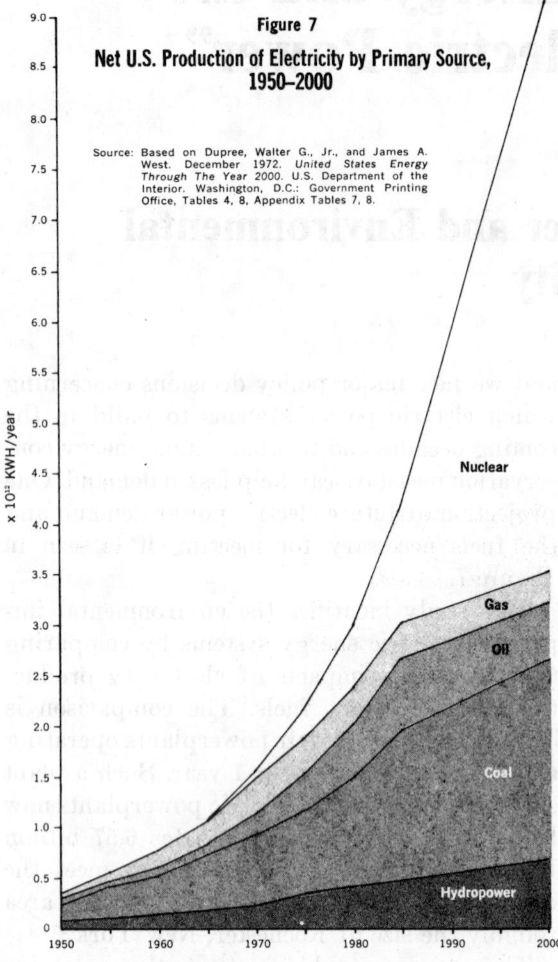

Figure 7
Net U.S. Production of Electricity by Primary Source, 1950–2000

Source: Based on Dupree, Walter G., Jr., and James A. West. December 1972. *United States Energy Through The Year 2000.* U.S. Department of the Interior. Washington, D.C.: Government Printing Office, Tables 4, 8, Appendix Tables 7, 8.

was often necessary to use aggregate national data as the basis for the emissions and other impacts presented in this report. For example, all coal is assumed to have similar properties based on national production averages, even though the sulfur content, ash content, and heating value, and consequently the emissions, vary widely depending on the source of the coal.

Because the present level of emissions control for energy systems varies—from none to relatively high—this report emphasizes the differences between systems with minimum environmental controls, or controls that are now prevalent, and systems with the most effective controls available today, or those required by law within the next several years. The analysis is further restricted to the major existing electric power systems, those based on coal from underground and surface mines; onshore, offshore, and imported oil; natural gas; and nuclear power from light water fission reactors. Hydroelectric power, now meeting almost 16 percent of electricity needs,[58] is not included because of the limited number of sites for future hydroelectric plants. It should be noted, however, that hydroelectric systems also cause environmental damage, such as destruction of natural scenic values, fish and wildlife habitat, and deterioration of water quality.

Although seven energy systems based on different fuel sources for producing electricity are discussed, strictly speaking, they are not alternative systems. Regional factors such as fuel price variations, availability of low sulfur coal, and limited gas supplies in many parts of the Nation restrict the economic alternatives.

Each electric power system produces a unique set of environmental impacts. In an attempt to improve our understanding of the relative impacts of each of these systems, this study aggregates emissions, land impacts, and occupational injuries associated with various system components. This aggregation reduces the number of variables being compared and highlights some significant differences among systems.

There are drawbacks to this treatment, however. First, when similar environmental impacts from several system components are summed, differences in geographical location or proximity to population centers are obscured. For example, the acreage impacted by a coal-fired energy system may include both the land mined, perhaps in unpopulated areas of the West, and a powerplant site in the heart of Chicago. Obviously, the effects on the environment are quite different. There are also drawbacks in aggregating the pollutants. For example, sulfur oxides have different and perhaps more severe effects on health and vegetation than similar quantities of nitrogen oxides or carbon monoxide. Further, the data do not distinguish between inadequately controlled solid wastes, causing environmental damage and solid wastes that are disposed of adequately.

Recognizing these limitations, the Council concluded that a comparative discussion of the aggregated environmental impacts—air emissions, water emissions, solid wastes, land impacts, and occupational health factors—of the alternative electric power systems is useful for illustrative purposes and provides a basis for further analysis. Table 3 presents the aggregated data for each electric power system as it would operate with a low level of environmental controls or with controls that are prevalent today. The disaggregated data for each energy system may be found in the Appendix.

Because of the difficulty in interpreting the data in Table 3, a qualitative estimate of the severity of the impacts is also provided. The index varies from zero to five; five indicates a serious impact and zero indicates no impact. The index is an admittedly subjective assessment by the Council. Although we feel that it can assist the reader in evaluating the quanti-

Table 3

Comparative Environmental Impacts of 1,000-Megawatt Electric Energy Systems Operating at a 0.75 Load Factor With Low Levels of Environmental Controls or With Generally Prevailing Controls [1]

System		Air emissions			Water discharges				Solid waste			Land use		Occupational health		Potential for large-scale disaster
		Tons ($\times 10^3$)	Curies ($\times 10^3$)	Severity	Tons ($\times 10^3$)	Curies ($\times 10^3$)	Btu's ($\times 10^{13}$)	Severity	Tons ($\times 10^3$)	Curies ($\times 10^3$)	Severity	Acres ($\times 10^3$)	Severity	Deaths	Workdays lost ($\times 10^3$)	
Coal	Deep-mined	383	......	5	7.33	......	3.05	5	602	......	3	29.4	3	4.00	8.77	Sudden subsidence in urban areas, mine accidents
	Surface-mined	383	......	5	40.5	......	3.05	5	3,267	......	5	34.3	5	2.64	3.09	Landslides
Oil	Onshore	158.4	......	3	5.99	......	3.05	3	NA	......	1	20.7	2	.35	3.61	Massive spill on land from blowout or pipeline rupture
	Offshore	158.4	......	3	6.07	......	3.05	4	NA	......	1	17.8	1	.35	3.61	Massive spill on water from blowout or pipeline rupture
	Imports	70.6	......	2	2.52	......	3.05	4	NA	......	1	17.4	1	.06	.69	Massive oil spill from tanker accident
Natural gas		24.1	......	1	.81	......	3.05	2	......	......	0	20.8	2	.20	1.99	Pipeline explosion
Nuclear		......	489	1	21.3	2.68	5.29	3	2,620	1.4	4	19.1	2	.15	.27	Core meltdown, radiological health accidents

NA=Not available.

Severity rating key: 5=serious, 4=significant, 3=moderate, 2=small, 1=negligible, 0=none.

[1] See the Appendix for details.

tative information presented, we also recognize that the perceptions and values of informed individuals vary.

Table 3 points up important environmental tradeoffs and highlights the difficulty of identifying the most desirable system when the impacts of each system are very different. The following sections detail the components of the various electric energy systems summarized in the table and briefly compare the environmental impacts of the components.

Coal

Coal systems are based on either surface or underground mining. Surface mining disturbs large amounts of land and often leads to acid mine drainage and silt runoff, both of which degrade water quality. A common belief is that adverse environmental impacts of mining coal result only from surface mining, but underground coal mining also results in acid drainage and causes land subsidence over mined-out areas. Underground mining is a dangerous occupation, resulting in a high rate of fatalities, injuries, and disease.

In comparing surface and underground mining, one can see that although land disruption decreases by a factor of two and solid waste is also reduced, occupational health hazards increase significantly. The ecological and aesthetic damages from surface mining, however, are very severe unless high levels of reclamation are employed. Further, data in the Appendix tables indicate that some forms of water pollution also increase.

Although the chemical properties of coal from different areas do vary, the differences are not a function of its being surface or underground mined. Once the coal is mined, the systems are the same.

Approximately 30 percent of all coal is not mechanically cleaned but is transported directly from the mine to the user.[59] The remaining coal is first cleaned to reduce the inorganic sulfur and ash content. Solid wastes are piled onto waste banks, and processing waste water often pollutes streams with suspended solids.

Most coal moves to powerplants by rail. Considerable amounts of land are devoted to railroad rights-of-way, and some fatalities occur from railroad accidents.

At the powerplant the coal is burned to produce electricity, causing several pollution problems. If uncontrolled, and depending on the characteristics of the coal, a 1,000-megawatt powerplant emits large quantities of air pollutants—primarily sulfur oxides, nitrogen oxides, and particulates—and produces solid wastes and thermal discharge to water. Finally, the electricity is carried to consumers by transmission lines whose rights-of-way occupy large amounts of land.[60]

In the absence of stringent environmental controls, our most abundant fossil fuel resource, coal, causes the most serious environmental damages. It produces the most air pollution—in the form of particulates, sulfur oxides, nitrogen oxides, and carbon monoxide. And it produces the most water pollution—in the form of siltation and acid mine drainage. Powerplant discharges of heated water are comparable to those from other fossil fuel systems and are about 40 percent less than those from nuclear powerplants. Land disruption from surface and underground mining is also considerable. Coal systems cause more occupational deaths and injuries than other systems, chiefly from underground mining and rail transport.

Oil

Almost 80 percent of residual fuel oil burned in powerplants is imported;[61] the rest is refined domestically from crude oil. Petroleum extraction involves drilling through overburden to the oil-bearing strata and removing the oil, with environmental impacts dependent on the location of the oil. Onshore production affects land needed for oil rigs and related equipment. There are also problems of oil spillage and the disposal of large quantities of brine brought up with the oil. Offshore production may result in oil pollution from spills and blowouts.

The crude oil is generally brought to the refinery by pipeline, which in some instances have ruptured and spilled oil. In addition, the pipeline rights-of-way occupy large amounts of land. Refining crude oil is a series of complex processes leading to many marketable products, one of which is residual fuel oil. The refinery and storage facilities contribute to air and water pollution.

After refining, the residual oil is usually transported directly to a powerplant by barge or tanker. Imported residual fuel oil is brought directly to the powerplant. Transfer operations can result in oil spills, and water contamination results if tankers discharge oil during bilge and tank cleaning operations.

When the residual oil, regardless of its source, is burned at a powerplant, it causes air pollution—primarily sulfur oxides and nitrogen oxides—and results in thermal discharges to water. Finally, large amounts of land are used for transmission line rights-of-way.[62]

The residual fuel oil system is generally less damaging to the environment than the coal system. It causes much less air pollution than coal—by a factor of three. Of the three residual fuel alternatives, imported oil is the cleanest for the United States, not because of different sulfur or ash content but because the oil is extracted and refined abroad. (From a global perspective, of course, this factor is not relevant, and the extra transport, with attendant oil spills, would make it less desirable.) Both the importation of oil and drilling offshore can lead to water pollution; at the same time, however, they use less land. Occupational health impacts for all oil systems are much lower than for coal. Of the oil sources, imports have the lowest injury and death rates for U.S. workers; onshore and offshore activities have rates comparable to each other. Solid wastes from oil-fired systems are also negligible regardless of the fuel source.

Gas

Natural gas extraction is in many ways similar to oil production. Indeed, both fuels are often taken from the same well. Gas extraction on land affects some acreage through the use of drilling rigs and associated equipment, and it produces considerable amounts of brine, posing a disposal problem. Pipelines, having extensive rights-of-way, then transport the gas to processing facilities. Here impurities are removed, and some air pollution results.

After processing, the gas is again moved by pipeline to the powerplant. The combustion of gas causes minor amounts of air pollution from carbon monoxide and oxides of nitrogen but leads to thermal discharges to water. And like the other electric energy systems, the transmission line rights-of-way are a major, if not especially damaging, land use.[63]

Natural gas is by far the least environmentally damaging of the fossil fuel alternatives. There is essentially no water pollution other than thermal discharge, and total air pollution is less than 5 percent of the emissions from a coal system and is significantly less than from an oil system. There are almost no solid wastes. Although land use is considerable, the environmental effects are relatively small compared to surface mining of coal because gas requires only small amounts of land for extraction but larger amounts for collection and distribution by pipeline.

Nuclear Fission

The fuel cycle for light water nuclear power systems is more complex than that for other electric energy generation systems, and for this reason it is discussed in somewhat more detail. Present light water reactors use a uranium-based fuel. Uranium ore is extracted from both surface and underground mines. Thorium, which is used in conjunction with uranium as a fuel for gas-cooled reactors, is mined similarly. A series of physical and chemical processes called milling separates the uranium from the excess rock and other material in the ore and usually concentrates the uranium in the form of the compound U_3O_8. Because of the high potential energy content of nuclear fuels, relatively little land is affected in supplying the annual needs of a 1,000-megawatt powerplant. For the same reason, the mining process is not extensive, and few occupational injuries result. Milling releases some radiation to air and leads to a very small amount of radioactive liquid and solid wastes.

After milling, the U_3O_8 is chemically converted to another uranium compound, UF_6, which can be easily gasified. In nature, uranium is primarily composed of the fertile isotope, U-238 (99.3 percent), and the fissile isotope, U-235 (0.7 percent). The UF_6 is transported to a gaseous diffusion plant where the uranium is enriched to a higher concentration (between 2 and 3 percent for light water reactors) of its fissionable, or fissile, isotope, U-235. Very little radioactivity is released in the enrichment process.

After enrichment, the UF_6 is shipped to a fuel fabrication plant where it is converted to uranium dioxide, UO_2, and formed into metal-clad fuel elements which are then shipped to the nuclear powerplant as original or replacement fuel.

At the light water reactor (LWR) powerplant, fission energy is released in the form of heat and is transferred to a conventional steam cycle which generates electricity. Because of coolant temperature limitations in LWRs, their thermal efficiency is lower than modern fossil-fueled plants. This lower efficiency, as well as the absence of hot gaseous combustion products released through a stack, mean that a LWR powerplant discharges over 60 percent more heat to receiving waters than its fossil fuel counterpart. Very small quantities of short-lived radionuclides are also routinely released to water bodies and to the atmosphere. Because most components of the nuclear system require relatively little land, the transmission line rights-of-way are the primary land use for the system.

The spent fuel, containing highly radioactive fission products, is stored at the reactor for several months while the radioactivity dies down. It is then transported to a reprocessing plant where the fuel is chemically treated to recover the remaining uranium and some plutonium that is produced during the fission process. Other fission products are also removed and concentrated. In the entire fuel cycle, radioactive emissions to air, measured in units of curies, are greatest from the reprocessing plant.

The concentrated fission products cannot be discharged to the environment but must be monitored and stored indefinitely because of their biological hazards and long half-lives. The liquid wastes must be converted into solids within 5 years, and U.S. Atomic Energy Commission regulations require that shipment to a Federal repository take place within 10 years. The plutonium separated from the spent fuel may eventually be used to power either LWRs or advanced reactors now under development. Because of its extreme toxicity and long half-life, it is necessary to ensure that plutonium is carefully handled, stored, and monitored constantly. Similarly, plutonium-containing scrap requires careful handling and permanent storage.

In addition to operational release of low levels of radioactivity at several steps within the fuel cycle, radioactive releases, although highly unlikely, could theoretically occur from accidents in powerplant operations or in the transport of radioactive materials.[64]

Relative to the fossil fuel systems, the nuclear system causes little land disruption, air pollution, and water pollution and leads to few occupationally related injuries and deaths. The nuclear system discharges low levels of radioactivity, however, and it is known that high levels of radioactivity are harmful to all forms of life. Further, the nuclear system is unique in producing radioactive and extremely long-lived solid wastes. In particular, the wastes from fuel reprocessing must be stored and monitored indefinitely. And thermal discharges to surrounding waters are considerably higher for a nuclear plant than for a fossil fuel equivalent. Finally, the potential accident from nuclear powerplants and processing facilities ranges from the very small, intermittent release of radioactive materials to major reactor malfunction. Should the most serious reactor failure occur, which is exceedingly improbable and which nuclear plants are designed both to avoid and to contain, the potential for large-scale damage to human health and the environment is vastly greater than for fossil fuel systems.

It should be noted that the standards and regulations applicable to nuclear systems reflect significant research efforts to prevent nuclear accidents and to understand the environmental effects of radiation. Further, there is far greater Federal control over nuclear systems than over other fuel systems. Organizations involved in all important aspects of the nuclear energy system require Federal licenses, and radioactive emissions, as well as other aspects of operation, are closely monitored.

Environmental Controls for Existing Systems

Reducing the environmental damage from electric energy production is possible, and in many cases controls have been mandated by Federal, state, and local legislation and by administrative action. The 1970 Amendments to the Clean Air Act, Environmental Protection Agency and state regulations to implement that Act, the 1972 Amendments to the Federal Water Pollution Control Act, and the pending powerplant siting and surface mining legislation are but a few examples of environmental regulations now affecting or that could affect electric energy systems.

Environmental damage can be lessened by modifying existing energy systems or by developing alternative systems with fewer inherent environmental penalties. The analysis in this section considers only the effects of introducing additional controls to the systems discussed. Subsequent sections discuss new electric energy systems, increased efficiencies for existing systems, and energy conservation as alternative approaches to curbing environmental damages.

Because some pollutants, such as oxides of nitrogen, are not yet considered economically controllable and because the desired level of control for other pollutants has not been clearly established, it is not possible to assess definitively the costs and effectiveness of all possible environmental controls. A preliminary analysis is possible, however, based on control technology that is now available or that can be expected in the very near future. Table 4 summarizes the environmental changes and increased costs that would result from these controls. Detailed information is presented in Tables A–3, A–6, A–9, and A–12.

The coal-fired system, which is the worst polluter when uncontrolled, can be improved significantly, but the cost of producing electricity may rise by 30 percent. As noted in Table A–3,

Table 4

Cost of Controls and Changes in Environmental Impacts of 1,000-Megawatt Electric Energy Systems Operating at a 0.75 Load Factor With a High Level of Environmental Controls [1]

[In percent]

System		Air		Water [2]		Land		Solid waste [3]	Total cost increase [5]
		Tonnage change	Cost increase	Tonnage change	Cost increase	Acreage change [4]	Cost increase	Tonnage change	
Coal	Deep-mined	−81.3	23	−96.2	4	+1	0	+159	28
	Surface-mined	−81.3	23	−92.4	4	−37	4	+29	31
Oil [6]	Onshore	−73.0	31	−38.8	5	0	0	0	36
	Offshore	−73.0	31	−38.8	5	0	0	0	36
	Imports	−39.4	28	−77.2	5	0	0	0	34
Natural gas		0	0	0	5	0	0	0	5
Nuclear		[7] −29.3	1	0	4	0	0	0	5

[1] See the Appendix for details.
[2] Costs of cooling tower construction and operation are considered with water controls.
[3] Solid waste costs are included in air and water pollution controls.
[4] Land impacts are reduced by reclamation but are increased for solid waste disposal.
[5] The total percentage cost increase may differ from the sum of the media percentage cost increases because of rounding.
[6] In the case of oil-fired systems, price increases for desulfurized oil are the primary cost effect.
[7] Radioactive emissions from the nuclear system are reduced by controls at the powerplant.

most of this cost is incurred by reducing the powerplant sulfur oxide emissions by 85 percent. In addition, 99 percent of total particulate emissions can be eliminated, but the less well-controlled, smaller particulates cause much of the health and visibility problem. With controls, most major water pollutants—the silt runoff, acid mine drainage, and suspended solids from mining and processing operations—can be eliminated. Thermal discharge to water at the powerplant can also be controlled through the use of wet natural draft cooling towers. This type of cooling tower discharges waste heat to the atmosphere by evaporative cooling of water, which can modify local weather by causing fogging or icing. It is also a consumptive user of water. Land damage is reduced through adequate mining reclamation, although some additional land must now be used for the disposal of solid wastes caused by sulfur oxide and other air and water controls. Even with the best controls, however, loss of wildlife habitat, diminished recreational opportunities, and aesthetic blight will continue while the energy system uses land, for example, during coal mining. These losses can be reduced but not eliminated.

The environmental damage from oil-fired systems is also controllable to a great extent, with a concurrent 36 percent increase in the cost of producing electricity. As noted in Table A–6, the largest single control cost for this system is the use of low sulfur residual oil, increasing fuel prices by over 50 percent. The use of low sulfur oil reduces sulfur oxide emissions by two-thirds. Cooling towers at the powerplant eliminate thermal discharges to water. Other major environmental gains result from improved controls at the refinery. Environmental damage from imported residual fuel oil can be reduced significantly by use of tankers with load-on-top capability to cut discharges from cleaning operations. Improved navigation systems, tankers with double bottoms, and offshore ports servicing supertankers rather than many smaller tankers would reduce the chance of spills, yielding further benefits.

The natural gas and nuclear systems require relatively few additional control measures, as indicated by projected cost increases of 5 percent each. Control costs and effectiveness for the natural gas and nuclear systems are presented in Tables A–9 and A–12 respectively. The major control cost is for cooling towers, a cost that is somewhat higher for nuclear plants because they discharge more heat than fossil fuel powerplants. Additional control devices at the nuclear powerplant can reduce radioactive emissions to air by 99 percent.

The Effect of Implementing Environmental Controls

The preceding sections evaluated some of the differences in environmental impacts of electric power systems—with minimal controls, or those prevalent today, and with application of control technology that is now economically feasible or is likely to be in the near future. For clarity, a limited number of alternatives have been considered, and their environmental impacts have been discussed only on a highly aggregated basis. The Council recognizes the limitations of this analysis but believes that it provides valuable insights into the differences among electric power systems. Although more comprehensive analyses are needed, the approach used here can be expanded to create a valuable tool for use in making energy policy decisions.

Table 4 shows that costs tend to rise most for those systems causing the most pollution today. Of course, these production cost increases are unlikely to result in a comparable increase in the price of electricity. Production costs constitute half the total costs of a utility company.[65] And for residential users, production costs are less than 40 percent of the price of electricity.[66]

The cost estimates in Table 4 do not take into account current regional differences in fuel prices. Further, the data do not consider the dynamic nature of energy supply and demand forces or the time phasing of control requirements. Hence, they may overstate the effects of environmental control costs or underplay the competitive advantages of the various energy systems in certain parts of the Nation. For example, in areas such as the Midwest where coal is abundant and other fuels are not, the competitive position of coal is much better than this aggregate analysis indicates. In these regions, coal-fired powerplants, and especially new units, with stack gas cleaning devices to remove sulfur oxides would likely remain cheaper than systems burning other fuels.

Despite these moderating factors, the future mix of energy systems will be influenced by environmental control costs. And because environmental quality is itself strongly affected by the mix, it is important to determine the nature and extent of this influence. Further, as more stringent environmental controls are required, emerging energy systems that are now expensive but less damaging environmentally compared to existing alternatives may become more economically attractive.

Changes in the mix of energy systems are desirable to the extent that they reflect internalization of the environmental costs of energy production and use. Internalizing costs of environmental control should stimulate use of energy systems that are the least damaging to the environment, taking into account all system costs—those being paid by consumers and those now borne by society in terms of environmental degradation. Not only will the system mix change, but depending on the elasticity of demand, the rate of increase in total energy use may be reduced.

Emerging Electric Power Generation Systems

This Nation's already large and rapidly rising electricity use creates much pressure for new energy systems and system components. Systems will be developed to use our limited energy resources more efficiently, to exploit energy resources that we cannot now use economically, and to lessen the environmental damage of present systems.

A host of electric power generation components, systems, and combinations of systems are now being developed. Some may be used commercially within 10 years, and others will take much longer. Because many of these systems have yet to be tested even on a pilot project scale, their economic viability is unknown. Some are likely to be considerably more costly than existing electric power systems, but they may compare more favorably as the costs of existing systems reflect increasing fuel costs and more stringent environmental controls. New systems and components range from new ways to use coal for power generation to harnessing the radiant energy of the sun. Each will have environmental impacts that must be evaluated before full-scale development begins.

Assessing the environmental impacts of emerging systems is, of course, difficult. There is no significant operating experience, even with those that may be applied commercially in the early 1980's. Bench-scale testing and pilot plants, although useful for developing information, are not adequate to assess accurately the impacts of large-scale use. For many systems, the processes to be employed have not yet been fully designed.

Despite these severe limitations, systematic analysis of the known, predictable environmental impacts of emerging systems is essential if near-term decisions are to be made. Because it is impossible to be comprehensive, this section identifies areas needing further analysis in order to determine the environmental desirability of the systems. For the most part, it is based on extrapolation from bench-scale systems and experience with other systems having comparable components. It is not a comprehensive environmental assessment; rather, it tries to identify the issues to be considered while these systems are still in their formative stage.

This section does not attempt to review all emerging energy systems and new components of existing energy systems. Rather, it discusses those technologies for which major policy decisions must soon be made if the systems are to be operational in this decade or the next.

There are many techniques under development for exploiting new energy resources, for improving the efficiency of use, and for reducing the environmental impacts of energy resources now being exploited. Those which seem to offer the greatest promise for the short term include tapping geothermal energy and oil shale, gasifying coal, and developing several advanced nuclear power systems. Also of considerable importance but not discussed in this section are such technologies as coal liquefaction, fuel cells, magnetohydrodynamics, cryogenic transmission lines, and stimulating natural gas production from tight gas sands. Nor does it consider such longer-term technologies as nuclear fusion and solar energy systems, which hold promise of significantly freeing us from reliance on the earth's limited fossil and nuclear fuel resources and whose environmental impacts may be substantially less than alternative technologies.

Coal Gasification (Low Btu)

Coal has been and is likely to remain the dominant fossil fuel for generating electricity.

But because coal usage causes environmental problems, means to use coal with less environmental impact are highly desirable.

One such system would.convert coal to a gas (with lower energy value than natural gas) for combustion in powerplants. After extraction, the coal would enter a coal gasification electric power system. Sulfur would be removed during the gasification process. Because of the inefficiencies of converting coal to a gas, a conventional gas-fired power generator would make this system less efficient than the direct combustion of coal. But if a combined-cycle gas turbine is used, the system may have a greater thermal efficiency than present coal-fired systems, which would help reduce thermal discharges as well as emissions of sulfur and other materials, although nitrogen oxide emissions might rise.[67]

The combined-cycle coal gasification electric power system may significantly curb much of the environmental damage from coal use—if it reaches projected efficiency levels and if emissions from the gasification process can be kept low. If high efficiencies are not obtained, air pollution may still be reduced but coal consumption would rise, with attendant magnification of mining impacts as well as solid waste disposal problems.

Oil Shale

Oil shale deposits are found in abundance in parts of Colorado, Utah, and Wyoming, where an estimated 600 billion barrels of oil could be extracted from thick oil shale seams to yield 25 or more gallons per ton of shale.[68] Considerably more oil exists in less economic concentrations. In comparison, this Nation's current proved reserves of liquid petroleum total only 39 billion barrels, although total resources are estimated at several hundred billion barrels.[69]

Tapping oil shale will require its extraction by underground or surface mining, more like mining coal than drilling for petroleum. This is followed by retorting, or heating, to produce crude oil. Alternatively, the shale could be heated in situ—in place underground—and then withdrawn by drilling, as with crude oil. After retorting, the oil would be refined like other domestically produced oil to yield various oil products.

At midpoint in the lifetime of a 1,000-megawatt powerplant, surface mining of oil shale would have disrupted an estimated 1,000 acres. If underground mining without backfilling were employed, about 2,000 acres would be affected, but if 60 percent of the spent shale were returned to the mine, the possibility of subsidence would be greatly reduced and less than 300 acres would be affected by the excess spent shale.[70] If the in situ process were used, only a little land would be disrupted, just that occupied by the oil wells. In any case, mining oil shale would affect less acreage than mining coal with an equivalent energy content.

The retort process would produce significant amounts of hydrogen sulfide.[71] It would also produce extensive solid waste byproducts, whose volume before compaction is 50 percent more than the in-place shale; after compaction, the volume remains about 15 percent more than the in-place shale.[72] Because the solid waste is far less dense than the shale and cannot be readily compacted, only 60 to 80 percent of the wastes could be returned to the mine.[73]

Retorting, spent shale disposal, and shale oil upgrading would use large amounts of water in a region where water is scarce. To produce enough oil for a 1,000-megawatt powerplant would require perhaps 190 million cubic feet of water per year—primarily for spent shale disposal.[74] Surface disposal of the spent shale could lead to the leaching of salts which would enter surface waters. In addition, the retorting releases water containing organic and inorganic pollutants, presenting a water pollution problem.[75]

One major benefit of the residual oil that could be produced from shale crude would be its low sulfur content, which is well below present sulfur levels in residual fuel.[76] As a result, air pollution at a powerplant would be reduced, although the retort could be a significant new pollution source. Development of shale oil holds promise for dramatically increasing domestic oil resources, but the overall environmental impacts depend heavily on the technologies used to extract and process shale, to dispose of spent shale, and to reclaim mined lands. If control measures are inadequate, significant long-term environmental degradation would result.

Geothermal Energy

Geothermal steam or superheated water is produced when the earth's heat energy is transferred to subsurface water from rocks in the earth's crust. Where the pressure and temperature are adequate, the steam output may be used in turbines for conversion to electricity. These heat reservoirs are very large, but were geothermal steam produced in commercial quantities, a reservoir would eventually be depleted.

Exploitable geothermal sites tend to be located along the edge of continental plates. Some are being tapped in Italy, New Zealand, Iceland, Japan, Mexico, and California for electric power generation, space heating, and industrial purposes.[77] In most parts of the world, however, this energy resource is not economic.

Although data on U.S. geothermal resources are sparse, making an accurate assessment of the full potential for exploiting geothermal energy impossible, it may be less than 50,000 megawatts with presently envisioned technology, and full use is unlikely.[78] It is possible that much greater geothermal steam and hot water resources could be exploited by enlarging the reservoirs through the use of hydrofracturing or explosives, but this could cause undesired seismic effects. Further, the use of nuclear explosives could contaminate the steam and possibly release radioactivity to the environment.

Most of the environmental impacts of a geothermal resource are highly localized because it must be used near its source. At the site, up to several thousand acres would be used for extraction wells, although farming, recreation, and wildlife habitat may be compatible land uses.

Geothermal wells drilled in the United States have released gaseous pollutants. For example, at the Geysers geothermal field in Northern California, a 1,000-megawatt powerplant annually would emit 18,000 tons of hydrogen sulfide—an odorous and toxic compound.[79] At other sites in California, the emissions could be much higher, more sulfur emissions, in fact, than from a comparable coal-fired powerplant.[80] These potentially high sulfur emission levels would necessitate the use of abatement devices to conform with state air quality standards.

Wastewater discharge may be a problem because the geothermal steam powerplant condensate contains trace quantities of harmful chemicals.[81] They are of even greater concern for geothermal wells primarily producing hot water because the discharge is often highly saline—sometimes containing more than 20 percent salts. The wastewater discharge and the unwanted chemicals are likely to be disposed of by reinjection into the heat reservoirs. Although the reinjection of water into the faulted areas would relieve the stresses and reduce the possibility of land subsidence caused by the removal of large quantities of fluid, induced seismic activity may result. Because the heat source has a relatively low temperature, geothermal energy systems may be inefficient. The waste heat may be twice that of a conventional fossil fuel unit.

In addition to exploiting geothermal energy in the form of steam or hot water, it may eventually be possible to exploit the thermal energy in magma—the dry heated rocks in the earth's crust. Exploitation would necessitate drilling to the rock formation, fracturing the rock, and then introducing water to be heated. The resulting steam would then be used for power generation. Exploitation of dry geothermal resources is not as advanced as the use of naturally occurring geothermal steam or hot water, but it has significantly greater potential and may avoid many of the adverse environmental impacts of the steam and hot water systems.

Geothermal energy sources are now being exploited, and they could generate significant amounts of electricity in some regions, especially if additional heat reservoirs are found and if stimulation techniques are implemented. Although environmental impacts from current operations at the Geysers indicate that geothermal energy in the form of dry steam can be compatible with other land uses, further study of the environmental impacts of geothermal energy and the effectiveness of environmental controls must still be made.

High Temperature Gas-Cooled Nuclear Reactor

High temperature gas-cooled nuclear reactor (HTGR) technology is well advanced. A 40-megawatt HTGR pilot plant has been in operation since 1967, a 330-megawatt demonstration plant may begin operating in 1973, and orders have been placed for six powerplants of about 1,000 megawatts each.

The HTGR uses helium as the coolant, rather than water as in the current light water reactor (LWR). Because of the thermodynamic properties of helium and the physical characteristics of the graphite core, the coolant can be heated to a higher temperature, yielding an overall thermal efficiency up to about 40 percent—comparable to modern fossil fuel powerplants—as opposed to about 31 percent for the LWR.[82] As a result, the waste heat discharged to the environment by the HTGR is significantly less than for the LWR. The high heat capacity of the HTGR core and the coated particle type of fuel reduce the likelihood of an accidental release of radioactivity from the fuel should there be a sudden loss of coolant or other unexpected temperature rise.

The fuel mixture of the HTGR consists of highly enriched uranium and thorium. The HTGR uses nuclear fuel more efficiently than the LWR, resulting in use of less fuel and in disturbance of less land in mining the ore.[83] HTGR operations produce more fissionable uranium (U-233) than plutonium (Pu-239), as opposed to LWRs that produce significant quantities of plutonium. The uranium (U–233) produced can be recycled in the HTGR, and the HTGR's fuel cycle leads to reduced handling and potential recycling of plutonium, which is extremely toxic and has a long radioactive half-life. Reprocessing technology for HTGR fuel is still being developed, but it is expected that environmental impacts will be comparable to existing facilities for LWRs. Although many of the environmental impacts—radioactive emissions, transportation accidents, and indefinite storage and monitoring of high-level radioactive wastes—are similar for both the HTGR and the LWR, the former may offer some significant environmental benefits; hence it deserves careful consideration.

Fast Breeder Reactor

Several other advanced nuclear reactor systems are being developed for commercial application. Among them are fast breeder reactors with various heat transfer fluids, including gaseous helium and liquid sodium metal. Of these very different systems, the liquid metal fast breeder reactor (LMFBR) is now the lead effort for development,[84] and it may become available for commercial application in the late 1980's.

In an LMFBR, molten sodium is heated in the reactor core. Because the sodium becomes highly radioactive, it is put through a heat exchanger with a second and isolated sodium circulation system in which there is little radioactivity. This second sodium system is used to produce steam for the turbine.

The significant feature of breeder reactors is that they produce more nuclear fuel than they consume.[85] The LMFBR reactor core, for example, converts the most common form of uranium (U-238), which cannot be used directly as a fuel, into plutonium (Pu-239), which is a useful fuel. Compared with the light water reactor that utilizes less than 2 percent of the available energy from the uranium fuel that it burns, the LMFBR could utilize 60 percent or more of the total energy from uranium and could use our extensive reserves of higher-cost uranium without significant economic penalty.

Because the LMFBR uses a much more abundant uranium isotope than a light water reactor, much less mining and milling are needed for the system. In fact, the tailings already accumulated at enrichment plants in the process of preparing fuel for LWRs can be used as a source of uranium for the LMFBR.

The LMFBR, like other nuclear systems, requires the transportation of relatively small quantities of material. Nevertheless, the possibility of an accident remains. Shipping containers are designed to withstand potential accidents. It should be noted, however, that in LMFBR operations there is an economic incentive to shortening the fission product cooling times prior to transport to the reprocessing facility. As presently contemplated, this shortened cooling time will mean the shipment of fission products with greater radioactivity than those from a LWR.[86]

Radioactive emissions from the LMFBR fuel processing plant are expected to be similar to those from a LWR fuel processing plant. At a powerplant, however, environmental impacts from the LMFBR, particularly waste heat discharges, are reduced because the thermal efficiency approaches 40 percent rather than the 31 percent for light water nuclear reactors. The LMFBR would also be expected to discharge less radioactivity to the air and water than a comparable light water reactor.[87]

The projected performance of the LMFBR electric power generating system shows significant reductions in the major impacts of nuclear powerplants—radioactive emissions and thermal discharges. By reducing or eliminating uranium mining, as well as the fuel conversion and enrichment steps, we avoid other adverse impacts of the nuclear system.

Several other factors warrant careful attention. Theoretically, the power level can rise more quickly in an LMFBR than in a light water fission reactor,[88] possibly leading to melting of fuel and release of radioactivity. Offsetting this possibility are other characteristics of the nuclear fuel which inherently stabilize the LMFBR's fission process. In addition, the sodium coolant in an LMFBR is operated at a lower pressure, just above atmospheric pressure, as opposed to the LWR water coolant, which is operated at very high pressures. Although the LMFBR will be designed to avoid reactor failure, significant environmental contamination and damage to human health could result. Further, because other nuclear powerplants would be refueled with the plutonium that the breeder produces from the fertile U-238, the quantities of plutonium will increase significantly with the growth of a breeder industry. Safe handling and avoidance of release of plutonium to the environment will remain a continuing challenge. Despite these potential problems, the LMFBR holds promise as a relatively clean source of energy with a fuel supply that would last for centuries.

Other New Energy Systems

Nuclear fusion, solar energy, and other advanced energy systems also promise abundant electricity generation with minimum environmental damage. In particular, controlled thermonuclear fusion is receiving increasing research and development funds in the United States and several other nations. It would make use of light-element fuels which are sufficiently abundant to supply electrical energy needs almost indefinitely.

Solar energy, also receiving increasing attention as a virtually pollution-free source of energy, may be useful in meeting more of our space heating needs in the near term. In the long term, solar energy has considerable potential for generating electricity, especially in the Southwest.

Chapter III Energy Conservation, System Efficiency, and Environmental Quality

Chapter II examined two options for reducing environmental damages caused by electric power generation—additional environmental controls for present systems and use of new systems. Another approach is simply to conserve energy. This may be achieved by improving the energy-producing efficiency of existing systems or by reducing individual energy consumption. Much of the environmental damage from our use of energy comes from the systems that provide the energy to the consumer. If the systems for providing energy were to function more efficiently, then the adverse environmental effects of energy production would be reduced. Similarly, if the consumer were to use energy more efficiently, that is, if he could expend less energy while achieving his desired ends, then both energy production and environmental damage could be reduced. The consumer could use less energy if he changed some habits that are wasteful of energy or if products used energy more efficiently.

In addition to reducing environmental damage, conservation will enhance the reliability of future energy supplies. Our present energy resources are finite, and increasing amounts must be imported. By slowing the rate of growth of energy demand, we can improve the longevity of our supplies, allowing more flexibility in developing systems for meeting our long-term needs. At the same time, we can reduce projected dollar outflows for fuel imports.

System Efficiencies

Energy is provided to the consumer by several energy systems—each with a different overall efficiency. It may then be used in several ways, with varying efficiencies, to achieve a given purpose. If these efficiencies are very different, a given commodity or service using less energy in its production may conserve energy resources and reduce pollution. For example, in terms of emissions, increasing the conversion efficiency of a powerplant from 38 percent to 50 percent without installing any environmental controls is equivalent to not improving overall efficiency but instead installing a control device that removes 24 percent of the pollutants. Improved powerplant efficiency would reduce pollution from extraction, transportation, and processing because less fuel would be used. Installation of controls at a powerplant would have no such systemwide effects.

Efficiency may be viewed as the ability to produce the desired output with minimum input. The efficiency of an energy system is equal to the output energy of its last stage in consumable form divided by the total input energy. The input energy is the sum of the potential energy in the incoming fuel to the first stage and any external energy needed to run the various stages.

Systems for producing and consuming energy are, for the most part, inefficient. Table 5 estimates that efficiencies of producing and delivering electricity range from 10 to 25 percent. In addition, the consumption of electricity is sometimes inefficient. Systems for providing fuels directly to the consumer, as for automobile gasoline and home heating oil, although usually more efficient than those for producing electricity, are open to major improvement. Further, the consumption of these fuels is often grossly inefficient.

Both energy production and consumption offer tremendous potential for conservation. For example, only about 30 percent of the oil in a reservoir is being extracted from onshore wells; offshore extraction is somewhat more efficient. Should the price of crude oil rise, there would be more extensive use of secondary recovery techniques which conserve oil resources and reduce some of the damages of oil extraction. Similarly, in the deep mining of coal, less than 60 percent of the resource in place is recovered, and over 10 percent of the energy in coal can be lost in cleaning. The use of continuous longwall mining equipment is one way to reduce coal extraction losses. For electric power systems, a major source of inefficiency is the powerplant itself, and there are a number of promising techniques, including magnetohydrodynamics and combined cycles, to increase powerplant efficiency from 38 percent to 50 percent and beyond. Further, by locating powerplants near industrial complexes, powerplant heat that is now wasted through discharge into the environment could be used in industrial processes. Although many of these techniques for conserving energy are economically feasible today, possible increases in the cost of energy would spur improvements in energy recovery rates and in utilization efficiencies.

System Selection

The greatest potential for energy conservation is often in the selection of the right energy system for a particular need. For example, natural gas and electricity may be used interchangeably for most residential needs—heating and air

Table 5

Efficiencies of Electric Energy Systems [1]

[In percent]

System		Component					Total
		Extraction	Processing	Transport	Conversion	Transmission	
Coal	Deep-mined	56	92	98	38	91	18
	Surface-mined	79	92	98	38	91	25
Oil	Onshore	30	88	98	38	91	10
	Offshore	40	88	98	38	91	13
Natural gas		73	97	95	38	91	24
Nuclear		95	57	100	31	91	16

[1] See the Appendix for details.

conditioning homes, heating water, cooking, and drying clothes. Column 1 of Table 6 shows that for all these functions the gas-powered appliance consumes more energy directly than its electric counterpart. For example, an electric stove needs only 49 percent of the energy input of a gas stove.

But this is only part of the story. The conversion of gas and electricity into useful work is the final step in two extensive energy systems that begin when the fuel is extracted and end at the point of use. When the total system is considered, as may be seen in column 2 of Table 6, all the gas-fired appliances are more efficient than their electric-powered counterparts; 1.4 times as much energy is needed to run an electric stove than a gas stove. For home heating, gas is more than twice as efficient as electricity. The low overall efficiency of the electric-powered heating system results from the poor conversion efficiency at the powerplant and the significant transmission line losses.

At the same time that we inefficiently heat our homes and operate many applicanes with electricity produced from natural gas, we are also inefficiently using gas to produce light. Almost 4 million gas lamps are lit in the United States, each using about twenty times more energy than its electric equivalent, a 25-watt bulb.[89] The natural gas savings that could be realized by replacing gas lamps with electric bulbs would heat over 600,000 homes annually.[90]

Home appliances consume large amounts of energy, particularly electricity, and there are big differences in the many products designed to do the same job. Yet information on these differences is not available to the consumer when he purchases these products. If it were, it could stimulate savings in energy and in energy bills.

Perhaps greater potential for energy savings can be found in the devices we use for space heating, water heating, and air conditioning. Raising home insulation standards would also be effective.[91]

One way to improve energy use in the home, in commerce, and in industry is to identify total energy needs for each application and then develop a unified approach or total energy system. For residential and commercial application, the modular integrated utility system (MIUS) generates electricity on site and makes use of "waste" heat for space heating, hot water heating, and air cooling. This system can save between 22 and 35 percent of energy demand relative to using on-site fuel burning equipment for providing heat and buying electricity from an off-site utility.[92] If 65 percent of the solid wastes generated by MIUS users were burned for energy, then it may be possible to realize an additional 5 to 10 percent energy savings.[93]

In addition to new energy-saving technologies, another point to be considered is the high price, in terms of efficiency of energy use, that we often pay for convenience. For example, a frost-free refrigerator uses over 60 percent more electricity than its conventional counterpart;[94] a gas stove pilot light uses about half the total gas that the stove consumes.[95] Indeed, pilot lights are particularly wasteful in view of the fact that inexpensive automatic devices can ignite a burner conveniently and safely.

Table 6

Relative Efficiencies of Residential Use of Gas and Electricity

	Relative electric energy requirements [1]	
	At point of use	For total system [2]
Space heating	0.79	2.2
Water heating	.70	2.0
Cooking	.49	1.4
Clothes drying	.82	2.4
Central air conditioning	.60	1.8

[1] Gas=1.

[2] Based on Table A-7. The efficiency of delivery to the consumer of electricity vs. gas is 35 percent.

Source: Stanford Research Institute. 1972. *Patterns of Energy Consumption in the United States.* Prepared for the President's Office of Science and Technology, Energy Policy Staff. Menlo Park, Calif.: S.R.I., pp. 59, 153.

Transportation too provides examples of energy traded for convenience, comfort, and time savings. Table 2 shows that the energy efficiencies of vehicles for transporting people vary greatly. The fastest form of transportation, the airplane, is also the one that consumes the most energy per passenger mile. On the ground, the automobile uses much more energy per passenger mile than its competitors, the urban transit bus, the intercity bus, the passenger train, rapid transit, and commuter trains. This high rate of energy consumption per passenger mile, coupled with our extensive use of cars, has led to their using 13 percent of all energy.[96]

It is tempting to consider the large energy savings to be realized if larger shares of transportation demand could be satisfied by the more energy-efficient common carriers. For example, if one-half of the intercity air traffic and one-quarter of the intercity automobile traffic could be shifted to passenger trains (requiring almost a hundredfold increase in railroad passenger miles) and if railroads could operate at 70 percent capacity instead of the present 25 percent, we could save about 11 billion gallons of fuel annually.[97] This figure is over 8 percent of all the energy used for transportation in 1971.[98]

Energy savings from transportation shifts would probably take decades to achieve because the present transportation system evolved along with our life styles. A similar evolution to a substantially different transportation system would take time. For example, the automobile, which is one of the most dynamic and rapidly changing components of the transportation system, turns over in about 10 years. Upgrading of railroads and construction of other new transit systems would take many years and large amounts of money. And unless employment and residential siting decisions reverse the trend toward greater dispersion, even car pooling to raise the average occupancy of commuters' cars above the present 1.4 persons per vehicle mile may be difficult to implement.[99]

Because the automobile is the largest single user of energy within the transportation sector, it is a center of efforts to conserve energy. In addition to shifting to public transportation, it may be possible to persuade the consumer to shift to smaller, lighter cars. If just one-half of the 113 million cars expected on the road by 1980 [100] were as economical of fuel as the small car, which uses about 22 miles per gallon contrasted to the current average of less than 14 miles per gallon,[101] the annual fuel savings would be about 17 billion gallons.[102] At about $0.25 per gallon before taxes, this would be an annual saving of about $4 billion annually.

Chapter IV Evaluating Energy Policy Alternatives

Environmental Impacts of Alternative Energy Mixes

In presenting a preliminary analysis of electric power generation systems, this report provides a framework for developing a refined environmental assessment of existing and emerging energy systems. It is a framework that can be an extremely useful tool in helping to evaluate the energy policy alternatives that face the Nation. The following section, using the preliminary data developed, illustrates the environmental analyses possible with more refined data.

A baseline forecast of total environmental impacts for electric power generation was developed from the summary environmental impact data shown in Table 3 and the projected fuel mix to meet total electric power demand shown in Figure 7. Based upon present supply patterns, it is assumed that one-half the coal is surface mined and 80 percent of residual fuel is imported.

If this mix of systems is extrapolated to the year 2000, environmental impacts of electric energy systems with minimal or present prevailing environmental controls would increase, as shown in Figure 8. The figure also shows the impacts of the same mix of energy systems if additional environmental controls that are technically feasible today were installed during a 5-year period and if all new facilities used comparable equipment.

As the controls are applied, air and water emissions are initially reduced considerably but solid wastes increase; consequently, more land is needed for their disposal. This result almost offsets the reduction in the acreage impacted by mined area reclamation. Although the damages from air and water pollution are much less severe with controls, we must recognize the need to avoid unintentionally shifting environmental problems from one medium or location to another.

Many other strategies for improving environmental quality can also have secondary environmental impacts. Reducing air pollution from automobiles, although necessary, may well increase fuel consumption, with attendant increases in environmental damage from the extraction, transport, and refining of crude petroleum.

Figure 8 also indicates that with the rapidly growing demand for electric energy, air and water emissions will eventually become more extensive than today—even with implementation of current control technology. At some point, then, more stringent controls will be required just to keep environmental quality from declining further.

The technique used in Figure 8 can also illustrate the environmental impacts of various policy alternatives. For example, one of the difficult decisions that both Government and industry face is the allocation of limited research and development funds among the large number of potentially useful projects. The development of technologies to provide clean energy is a prime example. Although it is always difficult to estimate total R&D costs, it is perhaps even more difficult to anticipate the benefits of success. Even so, the environmental impacts of new

Figure 8

Annual Environmental Impacts of Electric Energy Systems With Minimal or Presently Prevailing Systemwide Environmental Controls and With Technically Feasible Systemwide Environmental Controls Installed Between 1975 and 1980

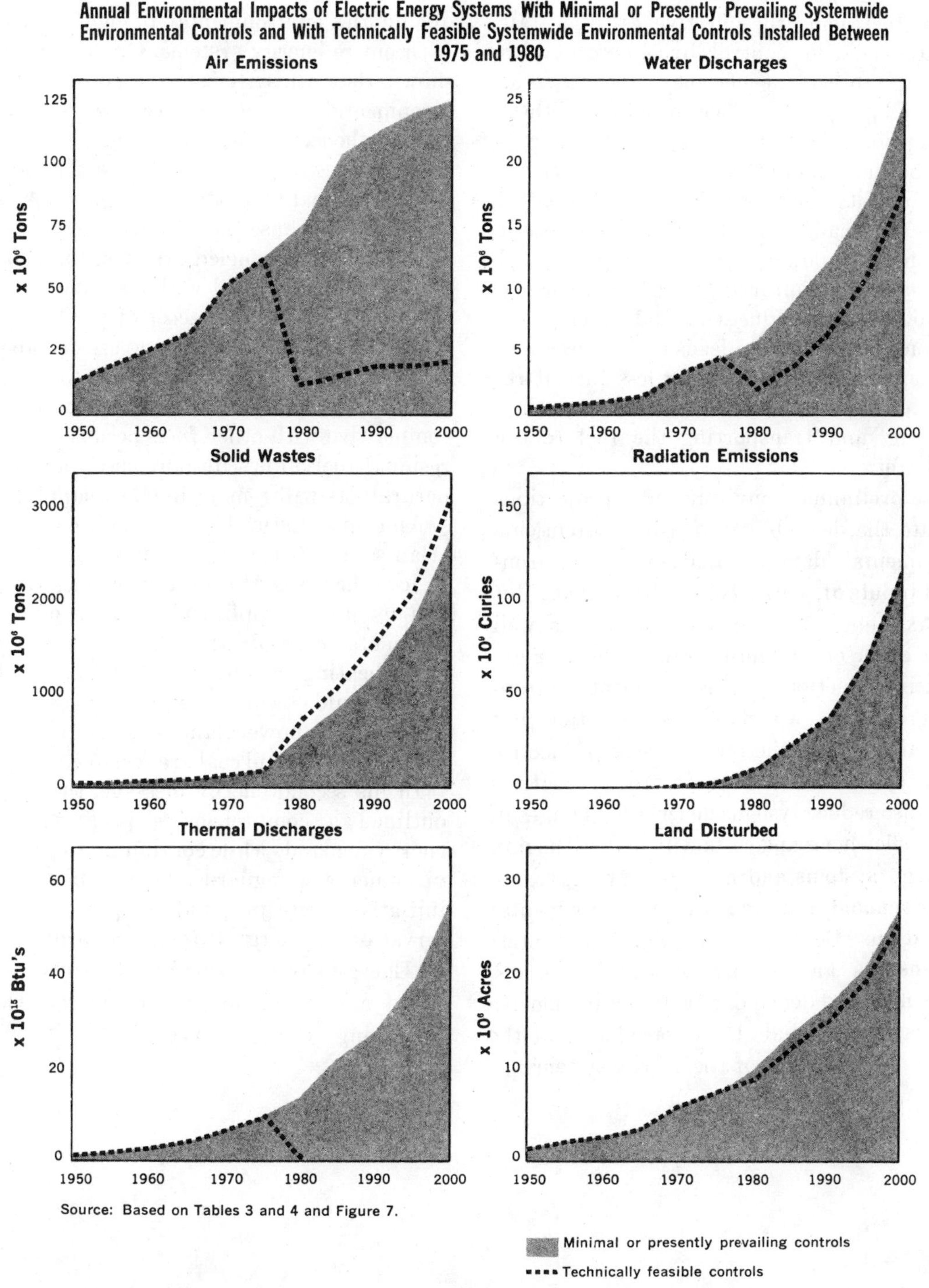

Source: Based on Tables 3 and 4 and Figure 7.

energy technologies can be estimated and compared using the approach developed in this report.

Two policy options worthy of consideration are the development of more effective pollution controls and the development of more efficient electric power generation systems. Figure 9 contrasts the aggregate national environmental impacts of increasing the effectiveness of all air and water pollution controls by 24 percent with increasing powerplant efficiency from 38 to 50 percent. Clearly, achievement of either of these policy options would have significant environmental benefits, but each affects overall environmental quality quite differently. If control equipment is made more effective, air and water pollution at the powerplant is reduced but solid wastes increase. Improved generating efficiency, on the other hand, reduces air and water pollution from all sources and leads to no increase in solid wastes. And by requiring less fuel, it reduces the impacts associated with extracting, processing, and transporting the fuel to the powerplant.

These preliminary environmental projections illustrate the desirability of using systemwide environmental data to analyze the environmental inputs of energy policy alternatives. But for maximum effectiveness, secondary as well as first order effects must be included. For example, introduction of more stringent environmental controls would reduce pollution but would also change the relative costs of electric energy systems. In some instances, controls would also reduce system efficiencies; changes in system efficiencies and costs will change the mix of energy systems and may modify aggregate energy demand, with consequent environmental implications. Because of the significant interrelationships among environmental controls, energy mix, and demand, a truly comprehensive analysis must include these variables and the environmental effects of the energy systems.

In assessing the environmental impacts of electric energy systems, this report has significant implications for several aspects of a national energy policy. First, it is apparent that energy conservation measures, in that they reduce demand for the primary fuels—coal, oil, and gas—as well as for electricity, will directly benefit the environment by quantitatively reducing the environmental damages associated with entire energy systems. Our analyses have shown that all systems have considerable environmental impacts. Hence energy conservation, to the extent feasible, is the most environmentally desirable way to meet our needs.

Second, coal-fired electricity systems, which are shown to cause significantly more pollution than other fossil-fueled systems, could be controlled to a high level without impairing their economic viability. The cost of producing electricity from coal would increase by about 30 percent (with an increased cost to the consumer of about 12 percent), but coal would remain competitive with other fuels because low sulfur residual fuel oil is still more costly and because natural gas will remain in short supply.

And third, natural gas was shown to be a very clean source of energy for powerplants as well as for other uses. However, it is in short supply, and its use in applications where other environmental controls are not practical—such as space heating for homes and commercial buildings—would seem a better alternative than burning it in powerplants where environmental controls for oil and coal are feasible.

In his second energy message, the President outlined a comprehensive program to meet energy demands while continuing to provide environmental safeguards. Among the President's initiatives were proposals to promote the conservation of energy through creating an Office of Energy Conservation, labeling major appliances and automobiles for energy use, and increasing energy conservation in Federal programs.

Figure 9

Annual Environmental Impacts of Electric Energy Systems With a 24 Percent Improvement in Systemwide Air and Water Pollution Controls Occurring Between 1975 and 1980 and With Powerplant Efficiency Rising from 38 to 50 Percent Between 1975 and 1980

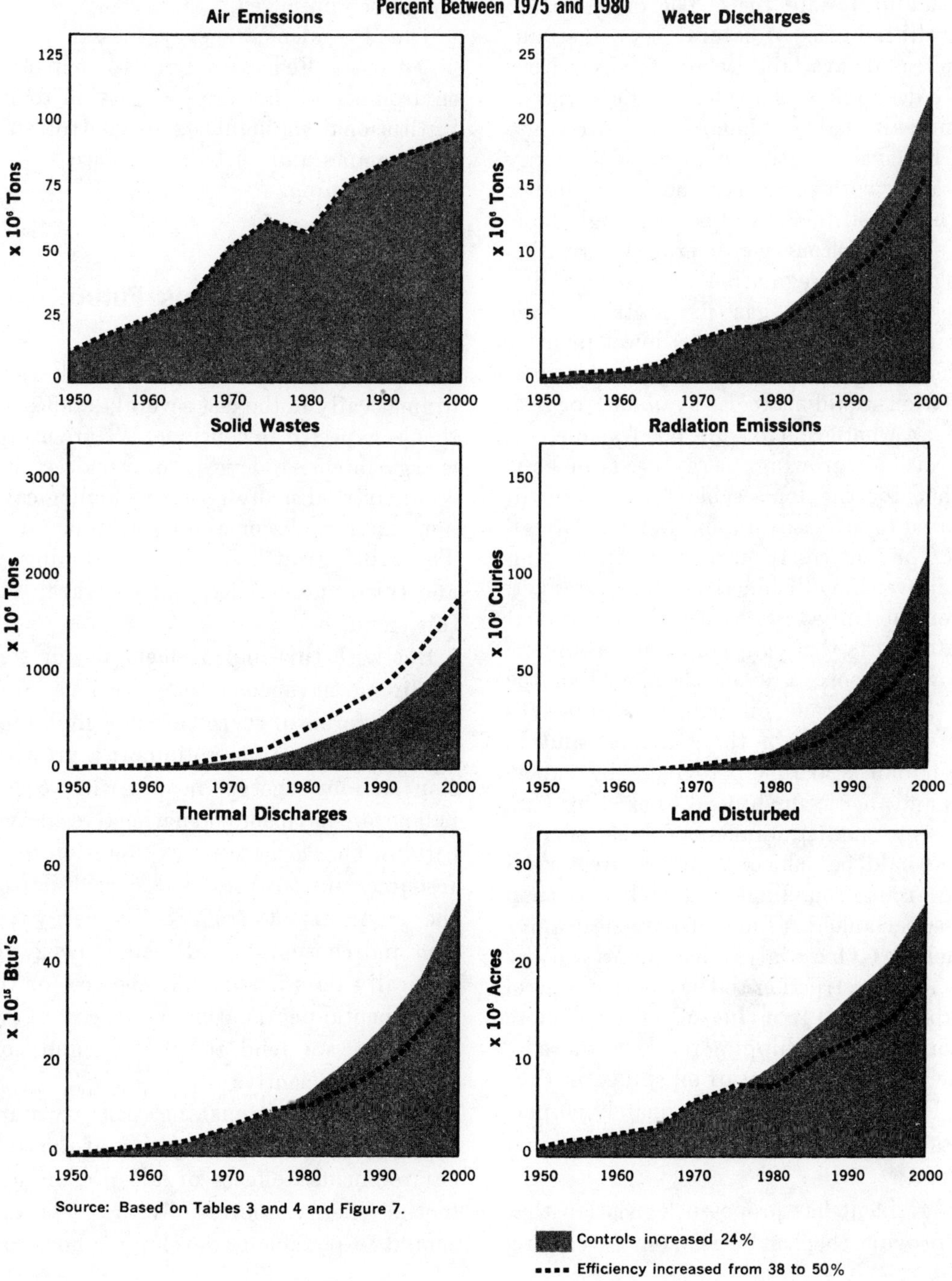

Source: Based on Tables 3 and 4 and Figure 7.

The President called for a number of programs to stimulate the availability of domestic sources of energy and to match energy uses to the most appropriate fuels. The deregulation of the wellhead price of natural gas will stimulate increased availability of this environmentally desirable fuel for high priority needs in meeting air quality standards. The direct use of gas for space heating requires less energy than heating with electricity and is far cleaner than other fossil fuels. And because higher gas prices should encourage careful use of this desirable fuel, deregulation would also discourage the use of gas in electric power generating stations and other lower-priority uses where environmental controls are feasible; deregulation would make gas available for residential use where controls are not feasible.

The Nation's growing energy needs make it clear that for the foreseeable future we will be required to increase oil imports from foreign sources. The least costly means for transporting crude oil over long distances is by supertanker. However, the United States has no conventional deepwater ports that can handle supertankers. If deepwater ports are not developed in the United States, crude oil will be shipped to deepwater terminals in the Bahamas and in the Canadian Maritime Provinces by supertankers and then transshipped to existing U.S. harbors in smaller tankers. These smaller tankers would be subject to much greater oil spillage from groundings and collisions than would supertankers using offshore deepwater port facilities. One analysis has shown that by establishing construction and navigation control over supertankers, receiving oil at an offshore superport, and transshipping the oil to shore by pipeline, the total amount of oil spilled in U.S. coastal waters would be approximately 90 percent less than that spilled under conventional conditions.[103]

The President has proposed legislation that would provide the legal basis for developing U.S. offshore deepwater ports. This legislation would require Federal licensing of these ports and assurance that environmental hazards would be minimized.

The President's energy message also proposed other Federal actions to help minimize environmental damage—legislation to provide institutional mechanisms to control siting of powerplants and curbs on surface and underground mining.

Energy and America's Future

The Nation's energy demands have grown dramatically in the last several decades, growth that is expected to continue. We are using more energy-intensive devices in every sector of the economy, and many recent technological developments have become integral to our life style. The rapid growth in air conditioning, second and third automobiles, and jet travel indicates this trend.

But with this improvement in our standard of living have come some serious problems. Certain forms of energy are now in limited supply, and we continue to import a growing portion of our energy needs—with consequent balance-of-payment and national security implications. Our large-scale exploitation of energy resources now causes widespread impacts on the environment from both energy production and consumption. Because energy pricing generally does not include the cost of this environmental degradation or the cost of possible scarcities, we tend to produce and consume energy inefficiently.

Energy decisionmaking can be improved through the careful evaluation of the costs and environmental effects of all present and potential energy sources. With these data, it should be possible to develop methods to assess the effects of energy supply and demand on environmental quality and, vice versa, the effects of environmental controls on energy supply and demand. Addressing these complexities is indeed a difficult task, but one that demands attention. This report presents a methodology for evaluating the environmental impacts of energy alternatives—a necessary input to such evaluation. Unfortunately, the data and the energy mix forecasts are not sufficiently refined for immediate use in policy development. But without a framework for more comprehensive analysis, the implications of alternative energy strategies will remain uncertain and error is possible.

The energy problem is multifaceted—to provide energy for a rising standard of living while limiting environmental damage and meeting our national security and balance-of-payment concerns. Long-term resolution of this problem will require both an institutional framework and the development of analytical techniques such as those discussed in this report to determine the optimal mix of systems and the effect of energy pricing policies. It will also require effort to develop new ways to conserve energy. Fortunately, the National Environmental Policy Act provides a framework to assess the environmental impacts of major Federal actions with regard to energy. By consolidating Federal energy programs under the proposed Department of Energy and Natural Resources, we will have a comprehensive institutional base from which to apply more sophisticated analytical tools so that the Nation can reconcile its demands for dependable energy supplies with demands for a higher-quality environment.

Footnotes

[1] In 1850, the United States used 2.357 million billion Btu's of energy. Schurr, S.H., B.C. Netchert, et al. 1960. *Energy in the American Economy, 1850–1975*. Baltimore: Johns Hopkins Press, p. 145. In 1971, the United States used almost 69 million billion Btu's. Dupree, Walter G., Jr., and James A. West. 1972. *United States Energy Through the Year 2000*. U.S. Department of the Interior. Washington, D.C.: Government Printing Office, Table 1.

[2] Unfortunately, most energy resources are not directly usable by man, and much energy is lost as it is transformed from a natural resource into useful forms. In particular, electricity is a derivative of other energy sources. Thus, it is most meaningful to measure the total energy potential of the resources that we exploit in the process of providing energy. For coal, oil, natural gas, and nuclear fuel, this is the energy value of the resource as it is extracted. For hydroelectric power, it is the potential energy of water prior to entering the turbogenerator. These are the energy resources that are irretrievably used.

[3] Dupree and West, *op. cit.*, Table 1.

[4] *Ibid.*, Appendix Table 1.

[5] U.S. Bureau of the Census. 1971. *Statistical Abstract of the United States: 1971*. Washington, D.C.: Government Printing Office, p. 497.

[6] *Ibid.*, p. 5.

[7] *Ibid.*, pp. 496–99.

[8] Federal Power Commission. 1971. *Typical Electric Bills, 1971*. Washington, D.C.: Government Printing Office, pp. 157–58.

[9] Stanford Research Institute. January 1972. *Patterns of Energy Consumption in the United States*. Prepared for the President's Office of Science and Technology. Menlo Park, Calif.: S.R.I., p. 50.

[10] InterTechnology Corporation. November 1971. *The U.S. Energy Problem*. Vol. II. Warrenton, Va.: I.T.C., pp. F–39, F–41.

[11] Total energy consumption of the Houston Astrodome in 1971 was 52,100,843 kilowatt-hours. Information provided to the Council by James Garner, Chief Engineer, the Houston Astrodome.

[12] Calculated from *ibid.* and *Typical Electric Bills, 1971, op. cit.*, p. 156.

[13] Hirst, Eric. 1972. *Energy Consumption for Transportation in the United States.* Oak Ridge, Tenn.: Oak Ridge National Laboratory, p. 3.

[14] *Ibid.*, p. 27.

[15] Department of Transportation, Federal Highway Administration. *News.* June 24, 1972. The 1971 national automobile registration was 92,752,515.

[16] Transportation Association of America. 1972. *Transportation Facts and Trends.* Washington, D.C.: T.A.A., p. 8.

[17] Dupree and West, *op. cit.*, Tables 4, 13, 14, 15.

[18] *Statistical Abstract of the United States: 1971, op. cit.*, p. 224.

[19] Dupree and West, *op. cit.*, Table 4.

[20] *Ibid.*

[21] Averitt, Paul. 1969. *Coal Resources of the United States, Jan. 1, 1967.* U.S. Geological Survey Bulletin 1275. Washington, D.C.: Government Printing Office, p. 2.

[22] Dupree and West, *op. cit.*, Table 4.

[23] *Ibid.*

[24] *Ibid.*

[25] *Ibid.*

[26] *Ibid.*, Table 16.

[27] U.S. Department of the Interior, Bureau of Mines. December 1972. *Mineral Industry Surveys. Petroleum Statement, Monthly.* Washington, D.C.: D.O.I., Table 15.

[28] Committee on U.S. Energy Outlook. 1972. *U.S. Energy Outlook—A Summary Report to the National Petroleum Council.* Washington, D.C.: National Petroleum Council, p. 63.

[29] *Ibid.*, p. 64.

[30] U.S. Department of the Interior. 1972. *United States Energy—A Summary Review.* Washington, D.C.: D.O.I., p. 46.

[31] Dupree and West, *op. cit.*, Table 4.

[32] *Ibid.*, Tables 4, 17.

[33] *Ibid.*, Table 17.

[34] *Ibid.*, Table 5.

[35] *Ibid.*, Table 4.

[36] *Ibid.*

[37] *Ibid.*

[38] *Ibid.*

[39] *Ibid.*

[40] U.S. Council on Environmental Quality. 1972. *Environmental Quality—1972.* Washington, D.C.: Government Printing Office, p. 6. Based on U.S. Environmental Protection Agency estimates.

[41] U.S. Council on Environmental Quality. 1971. *Toxic Substances.* Washington, D.C.: Government Printing Office, p. 11.

[42] U.S. Department of Health, Education, and Welfare, Public Health Service. 1969. *Preliminary Air Pollution Survey of Vanadium and Its Compounds.* Raleigh, N.C.: P.H.S., p. 29.

[43] Delson, Jerome K., and Richard J. Frankel. April 1972. "Residuals Management in the Coal-Energy Industry." Manuscript to be published by Resources for the Future, Inc. Washington, D.C.: R.F.F., p. II–12.

[44] U.S. Department of the Interior. 1967. *Surface Mining and Our Environment.* Washington, D.C.: Government Printing Office, p. 117.

[45] U.S. Department of the Interior, Federal Water Pollution Control Administration. 1968. *Industrial Waste Guide on Thermal Pollution.* Corvallis, Oreg.: D.O.I., F.W.P.C.A., Northwest Region, Pacific Northwest Water Laboratory, p. 5.

[46] *Surface Mining and Our Environment, op. cit.*, pp. 39, 42.

[47] *Ibid.*, p. 42.

[48] U.S. Department of the Interior, Bureau of Mines. "Environmental Effects of Underground Mining and Mineral Processing." Washington, D.C.: Unpublished manuscript of the Bureau of Mines, pp. 37, 53. Also see Zwartendyk, Jan. 1971. *Economic Aspects of Surface Subsidence Resulting from Underground Mineral Exploitation.* Ph. D. dissertation for Pennsylvania State University, University Park, Pa., p. 4. Available from U.S. Department of Commerce, National Technical Information Service, Springfield, Va.

[49] "Environmental Effects of Underground Mining and Mineral Processing," *op. cit.*, p. 42.

[50] *Statistical Abstract of the United States: 1971, op. cit.*, p. 164.

[51] U.S. Department of the Interior and U.S. Department of Agriculture. 1970. *Environmental Criteria for Electric Transmission Systems.* Washington, D.C.: Government Printing Office, p. iii.

[52] *Ibid.*

[53] U.S. Council on Environmental Quality. 1970. *Environmental Quality—1970.* Washington, D.C.: Government Printing Office, p. 107.

[54] U.S. Atomic Energy Commission, Directorate of Licensing, Fuels and Materials. 1972. *Environmental Survey of the Nuclear Fuel Cycle.* Washington, D.C.: A.E.C., p. G–6.

[55] *Ibid.*, p. G–10.

[56] InterTechnology Corporation. November 1971. *The U.S. Energy Problem.* Vol. I. Warrenton, Va.: I.T.C., p. 18.

[57] Federal Power Commission. 1972. *Annual Report—1971.* Washington, D.C.: Government Printing Office, p. 11; and *Statistical Abstract of the United States: 1971, op. cit.*, pp. 6, 20.

[58] Federal Power Commission, *Annual Report—1971, op. cit.*, p. 11.

[59] U.S. Department of the Interior. 1971. *Minerals Yearbook–1969.* Vols. I–II. Washington, D.C.: Government Printing Office, pp. 309, 347, 348.

[60] See the Appendix for details on the environmental impacts of coal-fired electricity systems.

[61] U.S. Office of Emergency Preparedness. 1972. "Weekly Fuel and Energy Data Reports." A weekly memorandum for the Joint Board on Fuel Supply and Fuel Transport. Washington, D.C.: O.E.P.

[62] See the Appendix for details on the environmental impacts of residual fuel oil-fired electricity systems.

[63] See the Appendix for details on the environmental impacts of a gas-fired electricity system.

[64] See the Appendix for details on the environmental impacts of a nuclear energy system.

[65] Federal Power Commission. 1971. *The 1970 National Power Survey.* Part I. Washington, D.C.: Government Printing Office, p. I–19–2.

[66] *Ibid.*; and *Statistical Abstract of the United States: 1971, op. cit.*, p. 501.

[67] Squires, Arthur M. "Capturing Sulfur During Combustion." *Technology Review.* December 1971, pp. 52–59.

[68] U.S. Department of the Interior. 1972. *Draft Environmental Statement for the Proposed Prototype Oil Shale Leasing Program.* Washington, D.C.: D.O.I., Vol. I, p. I–2.

[69] *United States Energy—A Summary Review, op cit.*, p. 45.

[70] A 30-year powerplant lifetime is assumed. Acreages are calculated from data in *Draft Environmental*

Statement for the Proposed Prototype Oil Shale Leasing Program, op. cit., pp. I–7, III–11.

[71] *Ibid.*, p. III–48.

[72] *Ibid.*, p. I–24.

[73] *Ibid.*, Vol. I, p. III–15, and Vol. III, p. III–9.

[74] *Ibid.*, Vol. I, p. III–26.

[75] *Ibid.*, p. I–21.

[76] *Ibid.*, pp. I–19, I–28.

[77] Goldsmith, M. 1971. *Geothermal Resources in California—Potentials and Problems.* Supported in part by the National Science Foundation (RANN) under Grant No. GI–29726, EQL Rep. No. 5. Pasadena: California Institute of Technology, Environmental Quality Laboratory, p. 3.

[78] Estimates of U.S. geothermal capacity range between 30 and 100,000 MWe, according to Bowen, R. G., and E. A. Groh. "Geothermal—Earth's Primordial Energy." *Technology Review.* October–November 1971, p. 48.

[79] *Geothermal Resources in California—Potentials and Problems, op. cit.*, p. 31.

[80] For an example, see *ibid.*

[81] *Ibid.*, p. 29.

[82] *The 1970 National Power Survey*, Part I, *op. cit.*, p. I–6–21.

[83] U.S. Atomic Energy Commission, Office of Planning and Analysis. 1972. *Nuclear Power 1973–2000.* No. WASH–7139. Washington, D.C.: A.E.C., pp. 9, 10.

[84] The President's Energy Message to the Congress. June 4, 1971.

[85] U.S. Atomic Energy Commission. 1972. Environmental Impact Statement. *Liquid Metal Fast Breeder Reactor Demonstration Plant.* No. WASH–1509. Washington, D.C.: A.E.C., p. 2.

[86] *Ibid.*, p. 214.

[87] *Ibid.*, p. 225.

[88] Resources for the Future, Inc., in cooperation with MIT Environmental Laboratory. 1971. *Energy Research Needs.* Prepared under contract No. NSF–C644 to the National Science Foundation. Washington, D.C.: R.F.F., p. VI–43.

[89] "PUC Bars Gas Post Light Use; One Utility Calls Them 'Wasteful.'" *Electrical World.* May 1, 1972, p. 50.

[90] Based on data in *ibid.;* American Gas Association. 1972. *Gas Facts 1971 Data.* Arlington, Va.: A.G.A., p. 73; and Stanford Research Institute, *op. cit.*, p. 42.

[91] US. Department of Commerce, National Bureau of Standards, Institute for Applied Technology, Building Research Division, in cooperation with the Office of Consumer Affairs. 1971. *7 Ways to Reduce Fuel Consumption in Household Heating . . . through Energy Conservation.* Washington, D.C.: Government Printing Office; U.S. Department of Commerce, National Bureau of Standards, Institute for Applied Technology, Building Research Division, in cooperation with the Office of Consumer Affairs. 1971. *11 Ways to Reduce Energy Consumption and Increase Comfort in Household Cooling.* Washington, D.C.: Government Printing Office.

[92] Leighton, Gerald S. August 26, 1972. "Estimated Performance of Total Energy Demonstration (Jersey City Site)." Prepared for the Federal Council for Science and Technology, Energy Research and Development Committee. Washington, D.C.: U.S. Department of Housing and Urban Development, p. 1; and U.S. Department of Housing and Urban Development, Office of Research and Technology. December 1972. "Modular Integrated Utility System (MIUS) Program Description." Washington, D.C.: H.U.D., p. 12.

[93] *Ibid.*

[94] Stanford Research Institute, *op. cit.*, p. 50.

[95] Derived from *ibid.*, p. 47; Hittman Associates, Inc. 1971. *Program Review Meeting.* Conducted under U.S. Department of Housing and Urban Development contract No. H–1654. Columbia, Md.: Hittman Associates, Inc.

[96] U.S. Office of Emergency Preparedness. 1972. *The Potential for Energy Conservation.* Washington, D.C.: Government Printing Office, p. 13.

[97] Developed from data in Table II of this report, as well as U.S. Department of Transportation. 1972. *1972 National Transportation Report.* Washington, D.C.: Government Printing Office, p. 75; and American Petroleum Institute. 1971. *Petroleum Facts and Figures.* Washington, D.C.: A.P.I., p. 589.

[98] Dupree and West, *op. cit.*, Table 4.

[99] U.S. Department of Transportation. Federal Highway Administration. April 1972. *Nationwide Personal Transportation Study. Automobile Occupancy.* Report No. 1, Washington, D.C.: D.O.T., p. 8.

[100] U.S. Department of Transportation, Federal Highway Administration, Highway Statistics Division. 1972. "Projections of U.S. Totals of Population, Motor Vehicle Registrations, Highway Use of Motor Fuel, and Relationships Among These Factors, 1971–2020." Washington, D.C.: D.O.T. Based on Series D Census Projections, there will be 112.8 million automobiles in 1980.

[101] U.S. Department of Transportation, Federal Highway Administration. May 3, 1972. "Analysis of Energy Consumption of Highway Vehicles Related to Transportation Service, 1969." Washington, D.C.: D.O.T. Small car gasoline consumption was estimated at 22 miles per gallon. *Ibid.* and U.S. Department of Transportation, Federal Highway Administration. 1972. *Highway Statistics, 1971.* Washington, D.C.: D.O.T. Average automotive gasoline consumption was estimated at 13.6 miles per gallon.

[102] Based on 113 million cars in 1980 and 1.544 trillion vehicle miles in 1980, of which 80 percent is automotive vehicle miles. See note 100.

[103] Statement of Russell E. Train, Chairman, Council on Environmental Quality, before the U.S. Senate Commerce Committee. March 6, 1973. Based on data supplied to the Council by the U.S. Coast Guard.

Human Costs of Nuclear Power

L. A. Sagan

As public attention has focused on a deteriorating environment, economists have noted that a major contributing factor has been the failure to charge industry for emitting wastes into water and air. They point out that the consequences of this failure are not only excessive pollution, but artificially low prices and higher levels of consumption. These economists (*1*) have proposed a remedy with which to optimize pollution levels, namely, an emissions tax. If such a tax were levied, it would properly be based on some objective estimate of the effects of emissions on human health, as well as the economic, social, and environmental costs of emissions.

This article is concerned with the human costs of producing and utilizing nuclear fuel to generate electricity and with the question of whether these costs are equitably compensated for and represented in the price of such electricity. The analysis is based on estimates of the value of human life, lost productivity, and potential effects of radiation. My conclusion, based on certain assumptions, is that major inequities do indeed exist.

Traditionally, cost-benefit ratios have been the province of economists whose major interest is in engineering the costs of a particular project, with which they compare the savings (benefits) anticipated from that project. An example might be the cost-benefit estimates for a flood control project, in which the costs of a dam are compared to the commercial benefits to navigation and the savings in property damage from the prevention of floods.

Recently the concept has been altered somewhat to include a consideration of human lives, as well as property. An example of this is the calculation of costs to the nation of specific diseases, for example cardiovascular disease and cancer (*2*). This has sometimes taken the form of comparing the benefits to patients with the costs of treatment or the costs of technological advances in medical care (*3*).

There is nothing to suggest that there are greater distortions of human costs in the nuclear industry than in other industries. On the other hand, a cost-benefit evaluation is especially suitable to the nuclear industry, for the following reasons.

1) The International Commission on Radiological Protection (ICRP), in proposing radiation dose limits, recommends that all exposure to radiation be considered potentially damaging to man (*4*) and that industrially produced exposure of the public to radiation be permitted only if it can be justified in terms of risk-benefit ratios. The maximum permissible exposure for an individual in the general population under any circumstances is 500 millirems per year (a rem, the usual measure of radiation dose, is defined as the deposition in tissue of 100 ergs of energy, multiplied by an appropriate modifying factor, which will be specific to the particular type of ionizing radiation and the biological effect produced.) Unfortunately, the ICRP has provided no guidelines for calculating risk-benefit ratios. The methodology I use here implies that one can assign a dollar value to each human exposure incurred and then compare these costs to the costs of reducing such exposures. Whether the exposures are justified (benefits) requires separate consideration and will not be attempted in this article; nor will I attempt to compare these costs with the costs incurred in generating electricity from other fuel sources, although such comparisons are clearly relevant.

2) The nuclear industry is new, still relatively small, and in its formative stages; therefore, practices and capital investments have not yet reached such proportions that changes would cause substantial political, economic, or social dislocation. Thus, alterations suggested by cost considerations are more likely to be implemented in this industry than in a mature industry, where large capital investments have already been made.

3) Nuclear power has entered a phase of rapid growth at a time when public interest in and concern about the environment have become intense, resulting in a demand for extensive analysis and justification of the use of nuclear power.

4) Our knowledge of the biological effects of radiation very probably exceeds our knowledge of the effects of any other chemical or physical agent. We therefore have at our disposal a fairly sophisticated estimate of the human risks involved in radiation exposure. Although far from complete, this knowledge may well be greater than our knowledge about most of the other environmental hazards that may require similar attention.

5) The National Environmental Protection Act of 1969 requires government agencies to consider alternatives to any proposed action that would affect the environment. Under the provisions of the Atomic Energy Act, the U.S. Atomic Energy Commission (AEC) is charged with the responsibility of regulating the radioactive discharges of nuclear reactors. The AEC first attempted to meet its obligation by requiring that a utility applying for a license to construct a reactor facility prepare and submit an environmental report describing alternatives. After judicial review (*5*), the AEC issued a revised requirement, dated 9 September 1971, which read in part: "The environmental report shall include a cost-benefit analysis which considers and balances the environmental effects of the facility and the alternatives available for reducing or avoiding adverse environmental effects, as well as the environmental, economic, technical and other benefits of the facility" (*6*). Although not specific, these instructions would clearly seem to necessitate consideration of the human costs of building, fueling, and operating an electric generating plant, whether it be nuclear or conventional in design.

6) Examining the effects on health of each segment of an industry would seem to have some merit in rationalizing public health efforts. Surely, the magni-

---

The author is associate director, department of environmental medicine, Palo Alto Medical Clinic, 300 Homer Avenue, Palo Alto, California 94301.

Reprinted with permission from *Science*, vol. 177, pp. 487–493, Aug. 11, 1972.

tude of preventive and research expenditures should bear some resemblance to the risks involved in the activities to which they are directed. Unfortunately, public health efforts are often stimulated more by emotionally derived public attitudes toward a problem than by any objective consideration of the problem's real costs to society.

The nuclear industry shows glaring examples of such imbalance in its safety and preventive expenditures.

Human costs of two kinds can be considered: accidental injuries and deaths, usually (but not always) occurring among individuals whose occupations are involved with the nuclear fuel cycle; and potential health hazards incurred by those who are exposed to radiation generated throughout the fuel cycle, both in-plant and elsewhere. In order to make comparisons, both of these costs are assessed in dollars. By doing so, I risk the charge of insensitivity, but no other workable method is now in use.

This article represents a "first cut" at assessing the human costs of generating nuclear power. No consideration will be given here to the other portions of the equation: environmental costs and human benefits. Since a great many assumptions and evaluations were, of necessity, arbitrary, the results are presented more as a suggested working model than as a precise estimate of these costs.

Accidents

The very word "accidents" implies the unexpected and even unacceptable. There is, in this society, a myth that we consider life priceless and that no price is too great to pay if it will avoid an accident. Yet, an examination of our practices reveals that we do indeed accept what has been a relatively constant rate of accidents. The record of the past 50 years demonstrates that the death rate in industrial accidents has gradually fallen as automobile death rates have risen, the total showing a gradual, but only slight, decline (*7*).

Accidental deaths, among both industrial and nonindustrial populations, resulting from nuclear power–generating activity have been small, compared to deaths resulting from other industrial activities (*8*). The Department of Labor reports that both the frequency and severity of accidents in the nuclear industry are lower than the national average for manufacturing (*9*). This springs partly from the fact that the hazardous nature of radiation was recognized early and has led to strict regulatory control.

Occupational Injuries: Morbidity and Mortality

The assumptions underlying the following assessments are, to some extent, arbitrary and will undoubtedly be contested. Individuals with more precise data are invited to refine these numbers. In any case, I make the following assumptions.

1) The loss of one day's productivity as a result of injury is assumed to approximate $50. In August 1971, gross weekly earnings for nonsupervisory employees were $173.43 for mining, $220.23 for contract construction, and $141.69 for manufacturing (*10*); beyond this, the employer bears expenses such as vacation, pension, sick leave, and other administrative costs.

2) In addition to loss of productivity (direct costs), medical expenses (indirect costs) of the injury must also be assessed and allocated. Accepting the ratio of indirect to direct costs that obtains nationally for accidents (*11*), I consider these costs equal and estimate $50 per day as the indirect cost of injury.

3) The Department of Labor, in its scale of time charges (*9*), assesses a fatality as 6000 working days (20 years) lost. That assessment is accepted here. At $50 per day for lost time, a fatality would be charged at $300,000. Since death is inevitable, no indirect costs are assessed: that is, society ultimately pays the medical costs of all deaths, whether natural or otherwise. Furthermore, accidental deaths are, by their nature, far less costly in medical terms than are deaths resulting from chronic disease.

Estimates of the economic value of human life vary widely, depending on a number of variables. One example may serve to illustrate the methodology: Fromm estimated the value of a life lost through an airplane accident (*12*). In addition to the loss of the victim's future earnings and personal consumption, he considered the loss in contributed community service time, employer's recruiting and training costs, and accident investigation costs. On the basis of these factors and the income and age characteristics of the average individual killed in an aviation accident in 1960, a total value of $373,000 was assigned. The $373,000 is the sum of the following economic losses resulting from the individual's death: to himself, $210,000; to his family, $123,000; to the community, $28,000; to his employer, $4000; to the government, $4000; and to the airlines, $4000. The value in 1960 of the individual's future earnings and assets was computed from an average salary of $13,000, a yearly increase of 2.5 percent, assets of $25,000, an interest rate of 6 percent, and 40 as the average age at death. The assumption is also made that the individual is paid the full value of his labor and is not exploited.

Another estimate is that of Dublin, Lotka, and Spiegelman, who calculated in 1946 the worth of gross future earnings of a person, age 40, earning $3500 per year. They deducted income tax, estimated savings at 2.5 percent interest, and calculated a future worth to the victim and his dependents of $54,005 (*13*).

Carlson (*14*) used an indirect method of estimating value of life based on Air Force expenditures for development and maintenance of an ejection system for the B-58 bomber. Since these yearly costs were estimated to be $9 million, and since it was anticipated that one to three lives per year might be saved by this system, the implied value of life would lie between $3 million and $9 million.

Uranium Mining

The United States is the largest producer and consumer of uranium in the world today (*15*). Before World War II, demand was small and uranium was used principally by the ceramic industry; however, the uranium-containing ores, of which pitchblende is one, were mined extensively in the earlier half of the century for the radium found in association with uranium. Following the demonstration of the fission process in 1942, U.S. demand for uranium rose rapidly, primarily to meet AEC needs for weapons development and production. As the use of nuclear power for the generation of electricity increases, uranium will be in demand more for reactor fuels than for weapons.

A 1000-megawatt (electrical) nuclear reactor of current design requires an average reload equivalent to 0.140 metric ton of uranium oxide (U_3O_8) in concentrate per megawatt per year (*16*); this is aside from the initial core, which remains as a constant plant inventory.

Averaging employment and production data for the 3 years 1967 to 1969 (*17*), one finds that the mining and milling of 140 metric tons of U_3O_8 would require, at the rate of 2.3 metric tons per man, 62 man-years. Assuming that there are 1760 hours of employment per man per year, the total number of hours at risk per reactor per year would be 109,120. Since the rate of fatal accidents was 0.892 per million man-hours (1969 and 1970 averaged) (*18*), 0.1 fatality per year can be allocated to each 1000-megawatt reactor, or one fatality per year for the 10,030-megawatt capacity in the United States as of 1 December 1971 (*19*).

In addition, nonfatal injuries accounted for the loss of 1065 days per million man-hours worked. Charged at $100 per day, total injury costs, both direct and indirect, would be $11,700 per 1000-megawatt reactor. Together fatal and nonfatal injuries resulting from mining and milling activities would cost the nuclear industry $417,000 per year.

Fuel Manufacture and Reactor Construction

After the mining and milling of the raw uranium ore, it is necessary that the proportion of the ore which is fissionable and therefore useful as reactor fuel, the isotope uranium-235, be increased relative to the nonfissionable isotope uranium-235. This is accomplished by converting the uranium to a gas (UF_6), in which state the increase in uranium-235 can be most easily accomplished. The fuel is then converted to a metal, uranium dioxide, and is formed into small pellets, which are, in turn, encased in long metal tubes, or cladding. Large numbers of these tubes are assembled as bundles and constitute the basic fuel element within the reactor core, which consists of many of these bundles. Before fuel rods are irradiated in the reactor core, they do not produce penetrating radiation; therefore, no significant exposures are encountered in this stage of the fuel cycle.

The Department of Labor maintains injury statistics for each of these stages of manufacture, as well as for design, engineering, and construction of the reactor itself (*9*). These are shown in detail in Table 1 and as assumed total costs in Table 2. The data are for 1969, the most recent year available. Since those rates had been stable, as compared with previous years, it can be assumed that they are still valid today.

Rates of the frequency and severity of injuries are not reported separately for fuel reprocessing, but for fuel fabrication and reprocessing together. Because there are no specific data, the 800 employees involved in reprocessing were removed from the larger group shown in Table 1 and the rates for the larger category were applied to them, thereby producing an estimate of 517 days of injury for the 800 employees.

Table 1. Accidents in fuel and reactor manufacturing—1969.

Activity	Employees (No.)	Deaths (No.)	Injury* (No. of days lost)
Production of feed materials	1,482	0	193
Production of special materials used in reactors	1,439	0	2,281
Fuel element fabrication	2,905	0	1,876
Reactor design and manufacturing	15,572	1	3,122
Design and engineering nuclear facilities	4,793	1	89
Nuclear instrument manufacturing	2,771	0	1,463
Private research labs, including reactor test facilities	1,257	0	114
Miscellaneous (nuclear activities not classified elsewhere)	2,705	0	56
Total	32,924	2	9,194

* Excludes deaths.

Radiological Effects

Although effects on human beings of single, high doses of radiation have been identified and quantified, no effects from the levels of radiation encountered in the nuclear power industry are known or detectable. (An exception, uranium-mining activities, will be discussed.) Nevertheless, it cannot logically be, and has not been, assumed that effects do not occur. The assumption of ICRP, as well as of other groups, is that a maximum estimate of risk at low levels of exposure can be made by presuming linearity—that is, by presuming that risk is in a consistent proportion to dose, whether the dose is high or low (*20*). That assumption will be accepted here in order to assess radiation damage to the individual.

Still another assumption that will, of necessity, be accepted here is that dose rate has no influence on effect. Because dose rate affects human beings in almost all exposures to chemical or physical factors (including radiation), this assumption introduces into estimates of risk a safety factor that lies somewhere between zero and infinity. Studies of risk of human carcinogenesis from radiation exposure are based on effects of radiotherapy and atomic bombs, cases in which the dose rate is on the order of 100 rems or more per minute; therefore, estimates extrapolated to dose rates associated with reactor operation, rates that are on the order of millirems to a few rems per year, are clearly likely to be inflated.

Another concept, in which both linearity and absence of a dose rate effect are implied, is that of the man-rem, a measure of both radiation exposure and numbers of people exposed. Specifically,

Table 2. Total yearly costs to society from 10,030-megawatt (electric) nuclear industry.

Industry	Occupational costs		Public radiation ($)	Total ($)
	Injuries ($)	Radiation ($)		
Uranium mining	417,000	46,200		463,200
Manufacturing	1,519,300			1,519,300
Reactor operation	9,890	72,000	14,790	96,680
Reprocessing	51,700	40,020	1,500	93,220
Long-lived nuclides			3,120	3,120
Total	1,997,890	158,220	19,410	2,175,520

Table 3. Risk to individuals involved.

Activity	Persons involved (No.)	Cost ($)	Annual cost per person ($)
Uranium mining and milling	620	463,200	747.09
Manufacturing	33,724	1,519,300	45.00
Reactor operation	1,290	81,890	63.00
Reprocessing	800	91,720	115.00
Public near reactor	33,841,000	19,410	0.0004
Total U.S.	200,000,000	2,175,520	0.10

a man-rem represents the effect of 1 rem of exposure, whether delivered to one individual or fractionally to a larger number of people.

A number of estimates of the dollar value of the risk of 1 man-rem have been made. Cohen's estimate is the highest, $250 (*21*). Other estimates are $100 (*22*), and "a few pounds sterling" (*23*).

On the basis of all available scientific evidence, and relying on the conservatism of both the theory of linearity and the disregard of any dose rate effect, ICRP has established an upper level of radiation risk. In 1965, they estimated (*20*) that 1000 millirems of radiation received by each of 1 million people at any time will result in approximately 15 cases of leukemia and a total of 15 cases of all other types of cancer during the lifetimes of the exposed population, in addition to the approximately 250,000 cases that would normally occur in a nonirradiated population of 1 million persons. In other words, 1000 millirems would increase the risk of cancer by about 0.01 percent. Accepting that estimate of radiation risk, and accepting the assumptions both of linearity and of the absence of a dose rate effect, one can calculate the risk of cancer to persons exposed to the maximum levels near a nuclear plant and to those persons living within the vicinity of the plant who receive typical exposures.

The risk estimates used in these calculations were reviewed by ICRP in 1969 (*24*). Reference was made to the growing incidence of cancer among patients receiving radiation therapy for rheumatoid spondylitis. Making the assumption that, among these persons, all cancers found in numbers greater than are found in the general population resulted from radiation, the ICRP estimated that cases of other types of cancer may occur six times more frequently than leukemia as a result of radiation. However, in the paper that originally reported this study of spondylitis (*25*), as well as in the 1969 ICRP report, it was carefully pointed out that many of the cancers occurring among spondylitic patients could be the result of factors other than radiation—for example, drug treatment, excess smoking, or a spontaneous effect (not caused by treatment) associated directly with the disease.

Based on the assumption that 1 rem produces 100 cases of cancer per million persons exposed (that is, per million man-rems) and that the cost per life is $300,000 (derived purely from economic considerations), then risk-cost per rem per person would be $30, the estimate to be used here. Implicit in that estimate is the assumption that death caused by radiation-induced cancer would shorten life as much as death caused by accidental injury would—that is, by 6000 working days. Precise data on the latent period between radiation exposure and malignancy do not exist. The data that do exist are conflicting and, furthermore, are based on single exposures and high doses. Latent periods for leukemia, which are the best documented, are relatively short. Among the Japanese who survived the atomic bombs, leukemia began to appear early, reached a peak in about 1953, and declined after 1958 to the rates found among nonexposed groups in the mid-1960's (*26*). In the British study of spondylitics treated with radiation (*25*), leukemia rates rose to a peak 2 years after exposure and then slowly declined.

The latent period for types of cancer other than leukemia may well be much longer. Cancer among persons exposed as children to the atomic bomb is only now beginning to appear (*27*), and types of cancer other than leukemia have a distinctly longer latent period among the spondylitic population. Evans has noted that the latent period is inversely related to body burden of radium among radium dial painters (*28*).

An estimate of $30 can also be reached by extrapolating from the known lethal dose for a single exposure, 1000 rems. Assuming a risk reduced by a factor of 10 for long-term exposures (based on a single lethal dose of 1000 rems and an estimate of 100 cases of cancer per million man-rems), then

$$\frac{\$300,000}{(1000) \times (10)} = \$30$$

Still another approach would be to consider radiation-induced life shortening. Storer has recently reviewed studies relating to life shortening and concludes that the best estimate for man is 1 day life shortening per rem of exposure (*29*). This life-shortening approach to an estimate of radiation risk therefore produces a value of $50 per man-rem, which is, considering all of the variables and unknowns involved, fairly close to the $30 estimate arrived at through a consideration of cancer induction.

Missing from consideration in the above estimate of radiation risk are the genetic effects that have been amply demonstrated in studies of animals. To date, however, and in contrast with the somatic effects described above, no genetic effects have been demonstrated in irradiated human populations. Because it is not known whether such effects might occur at low doses, or what form they might take, no attempt is made here to quantify them in economic terms.

Uranium Mining

It has been demonstrated beyond any reasonable doubt that exposure to radiation at high dose rates in uranium mining leads to lung cancer (*30*). The source of the radiation is not uranium, but the radium and products of radium decay that are found in uranium ores. Radium decays to radon, a gas that is radioactive and that, in turn, decays to a number of other radioactive materials, generally called radon daughter products. Daughter products quickly become absorbed on dust particles that can be inhaled and deposited on lung surfaces, where they can further decay, generating highly energetic alpha particles.

Estimates of radiation dose from this form of exposure cannot easily be expressed in rems because of the great number of physiological and physical variables involved, variables that are only poorly understood. For this reason, estimates of dose, and epidemiological studies based thereupon, are based on concentrations of radon daughter products in the air. The unit is known as a working level (WL), which is any combination of short-lived radon daughters in 1 liter of air that will result in the ultimate emission of 1.3×10^5 mega electron volts of potential alpha energy. Occupational exposure to 1 WL for a period of 1 month is known as a working level month, or WLM. Since 1968, standards for underground uranium mines have limited exposures to 12 WLM per year; as of 1 July 1971 exposures were limited to 4 WLM per year (*31*).

Of the men involved in mining and milling from 1967 to 1969, 31 percent were in milling; of those who were engaged in mining, only 67.9 percent were underground rather than on the surface or in open-pit mining. Therefore, of the 62 men required to do the mining and milling for nuclear fuel to supply one reactor per year, only 29 would be underground and exposed to radiation. In their recently published monograph, Lundin, Wagoner, and Archer have made refined estimates of the anticipated number of lung cancers to be found among 10,000 miners working at 1, 4, and 12 WLM per year, beginning at age 20 and continuing to age 50 (*30*). They predict 353 lung cancers to age 80 for men exposed to 0 WLM per year, 512 for 4 WLM per year (average exposure of 0.3 WL) and 684 for 12 WLM per year (average exposure of 1 WL). From these numbers, it can be easily calculated that one man working for 1 year at 4 WL incurs an additional risk of lung cancer of 0.00053 case. For the 29 men required to work underground at 4 WLM per year to supply one reactor, total projected cases would be 0.0154, at a cost of $4620 for one reactor, or $46,200 for the currently operating 10,030-megawatt U.S. industry.

Reactor Operation

Radioactive materials are released to the environment both as gases that leave the stack and form a plume and as soluble and insoluble materials that are diluted and released into the water of the cooling stream. These materials could find their way into the human food chain through many pathways (*32*), but in practice this has been found to be a negligible source of exposure, compared to the clouds of radioactive gases. These gases are, for the most part, noble gases; they produce external exposures only, and do not enter into any biological processes. Gamertsfelder (*33*) has estimated that the total exposure of the general population to radioactive gases from U.S. reactor operations is 483 man-rems per year.

Since exposures to radiation among employees of reactor plants occur primarily among contract employees, particularly during shutdown for refueling, reported exposures are undoubtedly underestimated.

Goldman (*34*) estimated the total U.S. in-plant exposures in 1971 to be 2400 man-rems. Actual film badge readings of exposures for 1970 totaled 2039 man-rems (*35*). These estimates include only utility employees, not contract personnel. The former estimate would produce a cost of $72,000.

Fuel Reprocessing

Following their removal from the reactor, the "spent" fuel rods, which are intensely radioactive, are first stored at the reactor site to allow for preliminary decay; the rods are then transported to a reprocessing center, where they are both physically and chemically treated to reclaim uranium and other materials; the fission products are separated and prepared for final storage elsewhere.

The sole reprocessing plant for fuel rods from operating nuclear power plants is located in West Valley, New York. Although the radionuclide mixture released from this facility differs considerably from that of a power reactor, the dose limits are the same as those established for a reactor facility. A U.S. Public Health Service survey shows this facility to be operating well within these limits (*36*); therefore, it will be assumed that total exposure to radiation of the nonemployee population living adjacent to the plant is 50 man-rems from this source, at a cost of $1500.

Occupational exposures for fuel reprocessing were 1334 man-rads in 1970 (*35*). This calculation, from film badge readings, excludes film badge readings between 0 and 125 millirems, the great majority of which are 0.

Long-Lived Nuclides

In addition to exposing populations near nuclear facilities to radiation, long-lived fission products generated from both reactor operation and from reprocessing (particularly the latter) accumulate in the environment and are distributed, through natural processes, over very large areas. The importance of these products derives not from the magnitude of dose to individuals, which is small, but from the large number of people exposed. The two nuclides of greatest significance, in terms of curies released and half-life, are krypton-85, a noble gas, and tritium in the form of tritiated water. Neither of these can be easily contained by presently available technology, and they are liberated to the environment. Estimates of individual exposures from current nuclear operations are 0.005 millirem from krypton-85 (*37*) and 10^{-6} millirem from tritium (*38*). This would produce total U.S. man-rem exposures of 100 and 4, respectively, or a cost of $3000 and $120, respectively.

Injuries to the Public

No accidental injuries or radiation exposures beyond permissible limits to members of the public are known as a result of the U.S. nuclear power industry. Clearly, some risk of accidental release of radioactivity exists. Otway estimates that, if the total U.S. power demand were met by 200 reactors located near urban areas, total deaths caused by reactor accidents would be expected to be 0.02 per year (*39*). Since this risk is small, I have ignored it, and I assume that this neglect will not affect my conclusions.

Summary and Interpretation

Approximately 10,030 megawatts of nuclear power are produced by plants now operating in the United States (*19*). According to the estimates developed here, the human costs of generating that electric power are approximately 0.026 mill per killowatt hour. In terms of risk per person exposed (Table 3), the mining of uranium is the costliest portion of the entire fuel cycle. This is particularly significant because these effects are real, in contrast to the hypothetical effects of low-level

Table 4. Cost of dose reduction from Brown's Ferry nuclear plant.

Equipment	Reduction in external dose (man-rem)	Cost ($)	Cost per man-rem of radiation reduction ($)	Incremental cost per man-rem of radiation reduction ($)
Recombiners only	3,600	6,000,000	1,700	1,700
Recombiners and 6 charcoal beds	4,305	9,000,000	2,070	4,250
Recombiners and 12 charcoal beds	4,350	10,500,000	2,400	30,000

radiation exposure, and would clearly deserve far more attention than the off-site exposures from reactors, which have been receiving the greatest amount of public attention. Nor is it clear that the miners themselves or their political representatives have perceived accurately the magnitude of the risk. Compensation for this risk might be reflected either in adequate death benefits (that is, \$300,000) or in wages. Death benefits in the state of Colorado average \$17,000. Starr (*40*) has argued that wages for various soft- and hard-rock mining activities reflect the degree of risk involved. This view has been challenged by Connelly and Mazur (*41*). Whether wages in industrial activities generally in our society do contain some element of compensation for risk has not yet been carefully studied.

In a study carried out for the Federal Radiation Council, the economic effects of reducing radiation levels within uranium mines from the former requirement of 1.0 WL to 0.3 WL (as now required) were considered (*17*). The investigators concluded that additional production costs, primarily for ventilation equipment required to achieve this level, would be 24 cents per pound of U_3O_8. The total production costs for the industry would rise to about \$7 million per year, or an increase in cost to the consumer of 0.18 percent.

Again using the model of Lundin, Wagoner, and Archer (*30*, table 44, p. 109), I find that a reduction of radon levels from 1.0 to 0.3 will reduce the future lung cancer risk for the 2700 U.S. underground miners by 1.54 cases per year of exposure and therefore reduce the cost of uranium mining by \$462,000 per year in return for the \$7 million annual increase in operating cost. To implement such a program of risk reduction would imply an assumed value for human life of \$4.5 million.

Based solely on the controversial theory that some excess risk may exist at 1.0 WL, the Environmental Protection Agency required that the 0.3 WL be mandatory as of 1 January 1971 (*31*). Unfortunately, the \$7 million that the industry must spend to reach 0.3 WL does not benefit the miners, but the ventilation equipment manufacturers. Deaths that do occur among uranium miners as a result of lung cancer or trauma are compensated at a maximum of \$24,492.25 for a widow with three or more children (*42*)—a small fraction of what is assumed here to be the just value of a man's life, and an amount too small to provide an economic incentive to improve safety conditions. Death benefits vary among states, but average a legal maximum of about \$20,000 (*43*).

It cannot be assumed that the much larger costs of nonfatal injuries are entirely compensated. Analysis (*43*) of workmen's compensation provisions for injury, whether permanently or temporarily disabling, shows that income during disability is typically at two-thirds of full employment levels.

With respect to radiation exposures from reactor operation alone, a far greater total exposure is incurred by plant employees than by other persons. Off-site exposures are very small in comparison with radiation from natural sources or medical exposures. However total exposures are distributed between the public and industrial employees, these costs (risks) should be compensated; and, from both an economic and a biological point of view, it is the total exposure that is of concern. For this reason, occupational and public exposures should be considered together, particularly since technology, or standards influenced by public pressures, may reduce public exposures while producing for reactor employees inordinately greater exposures to radiation. Indeed, Goldman has recently argued that the use of additional equipment within the plant to reduce off-site exposures may actually produce a greater occupational exposure because of the additional maintenance requirements (*34*).

The AEC, which regulates reactor emissions, requires that the level of those emissions be maintained as low as technologically practicable (*44*). Under intense public pressure, it has pursued this goal beyond economic justification. The following example, based on data developed by the Tennessee Valley Authority (TVA) for the three reactors at the Brown's Ferry plant (*45*), illustrates the point. Radioactive gases generated by the fission process in the reactor core will decay to very low levels of radioactivity if there is substantial delay before they are released. Delay is influenced by the volume of gas generated, and since large quantities of hydrogen and oxygen are produced by radiation effects on cooling water, hydrogen recombiners can reduce radiation dose by a factor of 6. In addition, charcoal beds, by absorbing radioactive xenon and krypton, can further delay release and thus reduce exposure. Cost of equipment, dose reduction, and incremental cost per man-rem of dose reduction are shown in Table 4 for recombiners and for recombiners with one and two sets of charcoal beds.

Because a reasonable estimate of the cost of biological damage from 1 man-rem is \$30, the addition of any of this equipment could not be justified in economic terms. Nevertheless, TVA has decided to add both the recombiners and six charcoal beds to the plant in order to meet AEC requirements. Ideally, in a society with unlimited resources, no expenditures would be spared to reduce risk, regardless of cost; but practically, radiation protection must compete with other pressing societal needs.

Studies of radiation protection alone indicate that there are far greater economies in reducing public exposure from other sources of radiation than in reducing public exposure from nuclear plants. For instance, Terrill (*46*) has presented a comparative cost-benefit analysis for radiation dose reduction from medical and from reactor-produced exposures. He found that, from the use of automatic collimators on diagnostic x-ray equipment, costs per man-rem reduction are about \$7, compared to his estimated cost of \$10,000 to \$100,000 per man-rem for reducing reactor-produced radiation. He points out that current exposures of the U.S. population to radiation are 430 man-rems from nuclear plants versus 18.7 million man-rems from diagnostic x-rays.

What is left to be explained is why occupational injuries, particularly deaths, are not adequately compensated, whereas the risk to the public of radiation exposure is uneconomically over-regulated, researched, and financed. To an extent, this reflects the general attitude within our society that certain occupational groups are expected to accept higher risks than the public. This attitude can be demonstrated in a number of ways, both within the nuclear industry and without. Standards for permissible levels of exposure to radiation are tenfold higher for employees than they are for the general public. It is also significant that federal support for research in occupational health in all areas (through the National Institute for Occupational Safety and Health) is less than $5 million per year, while research funds for radiation biology are $90 million annually from the Atomic Energy Commission alone. Undoubtedly an argument can be made that those individuals whose jobs are involved with radiation should take greater risks than others, but it is difficult to conceive of an argument that would allow the risks to remain less than fully compensated, other than the undemonstrated thesis that undercompensation serves as a deterrent to accidents. In these days of egalitarianism, such a discrepancy is an anachronism. Hopefully, the Department of Health, Education, and Welfare's Commission on Workmen's Compensation will recommend reforms in compensatory practices to correct this abuse, which is not peculiar to the nuclear industry.

In contrast, public and political interest in protection from radiation exposure has been intense and has led to technological restrictions far out of proportion to what can be justified on an economic basis. This undoubtedly reflects a number of widespread biases arising from the knowledge that nuclear energy is used in weaponry, and that radiation in high doses is associated with cancer.

Although radiation-induced lung cancer is compensated among uranium miners, there is extreme difficulty in compensating others occupationally exposed to radiation because of the problem of establishing a causal relation between their past exposure and disease. In the absence of techniques for distinguishing radiation-induced disease from other causes, a "no fault" insurance against leukemia and cancer, contributed to by both employer and employee, would seem an equitable solution.

Conclusion

This analysis provides some useful insights into the magnitude and distribution of the human costs of generating electricity from nuclear fuels. Our society is able to maintain low prices by evading environmental costs and, as shown here, by failing to pay the costs of occupational injuries. At the same time, the price of nuclear energy is maintained at an artificially high level by an overprotective governmental policy that restricts the public's exposures to radiation to a far greater extent than can be justified in terms of risk reduction or the costs of reducing other (that is, medical) exposures to radiation.

References and Notes

1. J. H. Dales, *Pollution, Property and Prices* (Univ. of Toronto Press, Toronto, 1968); L. E. Ruff, *Public Interest* **19**, 69 (1970); R. M. Solow, *Science* **173**, 498 (1971).
2. D. Rice, *Economic Costs of Cardiovascular Disease and Cancer, 1962*, Health Economics Series No. 5, PHS Publ. No. 947-5 (Government Printing Office, Washington, D.C., 1965).
3. B. Weisbrod, *Economics of Public Health* (Univ. of Pennsylvania Press, Philadelphia, 1961).
4. *Recommendations of the International Commission on Radiological Protection: Report of Committee II on Permissible Dose for Internal Radiation* (ICRP Publ. No. 2) (Pergamon, London, 1959).
5. Calvert Cliffs Coordinating Committee Inc. *v.* U.S. Atomic Energy Commission and U.S.A., 449 Fed. Rep., 2nd ser. 1109 (D.C. Cir. Ct., 1971).
6. Subsection A.3, appendix D, part 50, AEC 10 CFR.
7. R. A. McFarland, *Arch. Environ. Health* **19**, 244 (1969).
8. F. D. Sowby, in *Environmental Aspects of Nuclear Power Stations* (International Atomic Energy Agency, Vienna, 1971), pp. 919–925.
9. U.S. Department of Labor, *Work Injuries in Atomic Energy*, Bureau of Labor Statistics report No. 385 (Government Printing Office, Washington, D.C., 1969).
10. U.S. Department of Labor, Bureau of Labor Statistics, *Empl. Earn.* **18** (No. 5), 93 (1971).
11. D. Rice, *Estimating the Cost of Illness*, Health Economics Series No. 6, PHS Publ. No. 947-6 (Government Printing Office, Washington, D.C., 1966).
12. G. Fromm, *Measuring Benefits of Government Investments* (Brookings Institution, Washington, D.C., 1965).
13. L. I. Dublin, A. J. Lotka, M. Spiegelman, *The Money Value of a Man* (Ronald, New York, 1946).
14. J. W. Carlson, thesis, Harvard University (1963).
15. J. A. DeCarlo and C. E. Shortt, in *Mineral Facts and Problems, 1970*, Bureau of Mines Bull. No. 650 (Government Printing Office, Washington, D.C., 1970), p. 219.
16. *Forecast of Growth of Nuclear Power*, Wash-1139 (U.S. Atomic Energy Commission, Washington, D.C., 1971).
17. *An Assessment of the Economic Effects of Radiation Exposure Standards for Uranium Miners*, report to the Federal Radiation Council (Arthur D. Little, 1970).
18. *Mineral Industry Surveys* (U.S. Department of the Interior, Bureau of Mines, Washington, D.C., 1970).
19. *Status of Central Station Nuclear Power Reactors: Significant Milestones* (U.S. Atomic Energy Commission, Washington, D.C., 1971).
20. International Commission on Radiological Protection, *Health Phys.* **12**, 239 (1966).
21. J. J. Cohen, *ibid.* **19**, 633 (1970).
22. A. Hedgran and B. Lindell, *ibid.*, p. 121.
23. H. J. Dunster and A. S. MacLean, *ibid.*
24. International Commission on Radiological Protection, *Radiosensitivity and Spatial Distribution of Dose* (ICRP Publ. No. 14) (Pergamon, Edinburgh, 1969).
25. W. M. Court-Brown and R. Doll, *Brit. Med. J.* **2**, 1327 (1965).
26. O. J. Bizozzero, K. G. Johnson, A. Ciocco, *New Engl. J. Med.* **274**, 1095 (1966).
27. S. Jablon, J. L. Belsky, K. Tachikawa, A. Steer, *Lancet* **1971-I**, 927 (1971).
28. R. D. Evans, *Health Phys.* **13**, 267 (1967).
29. J. B. Storer, *ibid.* **17**, 3 (1969).
30. F. E. Lundin, Jr., J. K. Wagoner, V. E. Archer, *Radon Daughter Exposure and Respiratory Cancer, Quantitative and Temporal Aspects*, National Institute for Occupational Safety and Health and National Institute of Environmental Health Sciences joint monograph No. 1 (National Technical Information Service, U.S. Department of Commerce, Springfield, Va. 22151).
31. *Fed. Regist.* **36** (No. 132), p. 12921 (1971).
32. L. A. Sagan, *Arch. Environ. Health* **22**, 487 (1971).
33. C. C. Gamertsfelder, "Regulatory experience and projections for future design criteria," paper given at the Southern Conference on Environmental Radiation Protection at Nuclear Power Plants, St. Petersburg, Fla. (1971).
34. M. I. Goldman, "New developments in nuclear power plant waste treatment," paper presented at the American Nuclear Society winter meeting, Miami Beach, Fla. (1971).
35. C. Eason, personal communication (1971).
36. B. Shleien, *Northeastern Radiological Health Laboratory Technical Report BRH/NERHL 70-1* (Bureau of Radiological Health, Public Health Service, Rockville, Md., 1970).
37. C. E. Larson, statement for the record, Joint Committee on Atomic Energy, *Hearings on Environmental Effects of Producing Electric Power* (91st Congr., 1st sess., 1971), p. 214–284.
38. D. G. Jacobs, *Sources of Tritium and Its Behavior Upon Release to the Environment* (Division of Technical Information, U.S. Atomic Energy Commission, Oak Ridge, Tenn., 1968).
39. H. H. Otway and R. C. Erdman, *Nucl. Eng. Des.* **13**, 365 (1970).
40. C. Starr, *Science* **165**, 1232 (1969).
41. T. Connolly and I. Mazur, *Proc. 6th Annu. Health Phys. Soc. Trop. Symp.* **1**, 401 (1971).
42. H. E. Kneeland, personal communication (1972).
43. *Analysis of Workmen's Compensation Laws* (Chamber of Commerce of the United States, Washington, D.C., 1970).
44. U.S. Atomic Energy Regulations 10 CFR, part 50, appendix D.
45. O. M. Derryberry, personal communication (1971).
46. J. G. Terrill, Jr., paper presented at American Public Health Association annual meeting, Chicago, Ill. (11 October 1971).

The Faustian Bargain

In its original form, this statement bore the somewhat abstract title, "Benefit-Cost Analysis and Unscheduled Events in the Nuclear Fuel Cycle." The Atomic Energy Commission had asked for comments on one of its documents, noting that environmental statements for a power reactor should contain a cost-benefit analysis which, among other things, "considers and balances the adverse environmental effects and the environmental, economic, technical and other benefits of the facility." In response to the invitation, Allen V. Kneese, director of RFF's program of studies in the quality of the environment, submitted the following remarks.

I AM SUBMITTING this statement as a long-time student and practitioner of benefit-cost analysis, not as a specialist in nuclear energy. It is my belief that benefit-cost analysis cannot answer the most important policy questions associated with the desirability of developing a large-scale, fission-based economy. To expect it to do so is to ask it to bear a burden it cannot sustain. This is so because these questions are of a deep *ethical* character. Benefit-cost analyses certainly cannot solve such questions and may well obscure them.

These questions have to do with whether society should strike the Faustian bargain with atomic scientists and engineers, described by Alvin M. Weinberg in *Science*. If so unforgiving a technology as large-scale nuclear fission energy production is adopted, it will impose a burden of continuous monitoring and sophisticated management of a dangerous material, essentially forever. The penalty of not bearing this burden may be unparalleled disaster. This irreversible burden would be imposed even if nuclear fission were to be used only for a few decades, a mere instant in the pertinent time scales.

Clearly, there are some major advantages in using nuclear fission technology, else it would not have so many well-intentioned and intelligent advocates. Residual heat is produced to a greater extent by current nuclear generating plants than by fossil fuel-fired ones. But, otherwise, the environmental impact of routine operation of the nuclear fuel cycle, including burning the fuel in the reactor, can very likely be brought to a lower level than will be possible with fossil fuel-fired plants. This superiority may not, however, extend to some forms of other alternatives, such as solar and geothermal energy, which have received

Reprinted with permission from *Resources*, no. 44, pp. 1–8, Sept. 1973.

comparatively little research and development effort. Insofar as the usual market costs are concerned, there are few published estimates of the costs of various alternatives, and those which are available are afflicted with much uncertainty. In general, however, the costs of nuclear and fossil fuel energy (when residuals generation in the latter is controlled to a high degree) do not seem to be so greatly different. Early evidence suggests that other as yet undeveloped alternatives (such as hot rock geothermal energy) might be economically attractive.

Unfortunately, the advantages of fission are much more readily quantified in the format of a benefit-cost analysis than are the associated hazards. Therefore, there exists the danger that the benefits may seem more real. Furthermore, the conceptual basis of benefit-cost analysis requires that the redistributional effects of the action be, for one or another reason, inconsequential. Here we are speaking of hazards that may affect humanity many generations hence and equity questions that can neither be neglected as inconsequential nor evaluated on any known theoretical or empirical basis. This means that technical people, be they physicists or economists, cannot legitimately make the decision to generate such hazards. Our society confronts a moral problem of a great profundity; in my opinion, it is one of the most consequential that has ever faced mankind. In a democratic society the only legitimate means for making such a choice is through the mechanisms of representative government.

For this reason, during the short interval ahead while dependence on fission energy could still be kept within some bounds, I believe the Congress should make an open and explicit decision about this Faustian bargain. This would best be done after full national discussion at a level of seriousness and detail that the nature of the issue demands. An appropriate starting point could be hearings before a committee of Congress with a broad national policy responsibility. Technically oriented or specialized committees would not be suitable to this task. The Joint Economic Committee might be appropriate. Another possibility would be for the Congress to appoint a select committee to consider this and other large ethical questions associated with developing technology. The newly established Office of Technology Assessment could be very useful to such a committee.

MUCH HAS been written about hazards associated with the production of fission energy. Until recently, most statements emanating from the scientific community were very reassuring on this matter. But several events in the past year or two have reopened the issue of hazards and revealed it as a real one. I think the pertinent hazards can usefully be divided into two categories—those associated with the actual operation of the fuel cycle for power production and those associated with the long-term storage of radioactive waste. I will discuss both briefly.

The recent failure of a small physical test of emergency core cooling equipment for the present generation of light-water reactors was an alarming event. This is in part because the failure casts doubt upon whether the system would function in the unlikely, but not impossible, event it would be called upon in an actual energy reactor. But it also illustrates the great difficulty of forecasting behavior of components in this complex technology where pertinent experimentation is always difficult and may sometimes be impossible. Other recent unscheduled events were the partial collapse of fuel rods in some reactors.

There have long been deep but suppressed doubts within the scientific community about the adequacy of reactor safety research vis-à-vis the strong emphasis on developing the technology and getting plants on the line. In recent months the Union of Concerned Scientists has called public attention to the hazards of nuclear fission and asked for a moratorium on the construction of new plants and stringent operating controls on existing ones. The division of opinion in the scientific community about a matter of such moment is deeply disturbing to an outsider.

No doubt there are some additional surprises ahead when other parts of the fuel cycle become more active, particularly in transportation of spent fuel elements and in fuel reprocessing facilities. As yet, there has been essentially no commercial experience in recycling the plutonium produced in nuclear reactors. Furthermore, it is my understanding that the inventory of plutonium in the breeder reactor fuel cycle will be several times greater than the inventory in the light-water reactor fuel cycle with plutonium recycle. Plutonium is one of the deadliest substances known to man. The inhalation of a millionth of a gram—the size of a grain of pollen—appears to be sufficient to cause lung cancer.

Although it is well known in the nuclear community, perhaps the general public is unaware of the magnitude of the disaster which would occur in the event of a severe accident at a nuclear facility. I am told that if an accident occurred at one of today's nuclear plants, resulting in the release of only five percent of only the more volatile fission products, the number of casualties could total between 1,000 and 10,000. The estimated range apparently could shift up or down by a factor of ten or so, depending on assumptions of population density and meteorological conditions.

With breeder reactors, the accidental release of plutonium may be of greater consequence than the release of the more volatile fission products. Plutonium is one of the most potent respiratory carcinogens in existence. In addition to a great variety of other radioactive substances, breeders will contain one, or more, tons of plutonium. While the fraction that could be released following a credible accident is extremely uncertain, it is clear that the release of only a small percentage of this inventory would be equivalent to the release of *all* the volatile fission products in one of today's nuclear plants. Once lost to the environment, the plutonium not ingested by people in the first few hours following an accident would be around to take its toll for generations to come—for tens of thousands of years. When one factors in the possibility of sabotage and warfare, where power plants are prime targets not just in the United States but also in less developed countries now striving to establish a nuclear industry, then there is almost no limit to the size of the catastrophe one can envisage.

It is argued that the probabilities of such disastrous events are so low that these events fall into the negligible risk category. Perhaps so, but do we really know this? Recent unexpected events raise doubts. How, for example, does one calculate the actions of a fanatical terrorist?

The use of plutonium as an article of commerce and the presence of large quantities of plutonium in the nuclear fuel cycles also worries a number of informed persons in another connection. Plutonium is readily used in the production of nuclear weapons, and governments, possibly even private parties, not now having access to such weapons might value it highly for this purpose. Although an illicit market has not yet been established, its value

has been estimated to be comparable to that of heroin (around $5,000 per pound). A certain number of people may be tempted to take great risks to obtain it. AEC Commissioner Larsen, among others, has called attention to this possibility. Thus, a large-scale fission energy economy could inadvertently contribute to the proliferation of nuclear weapons. These might fall into the hands of countries with little to lose, or of madmen, of whom we have seen several in high places within recent memory.

In his excellent article referred to above, Weinberg emphasized that part of the Faustian bargain is that to use fission technology safely, society must exercise great vigilance and the highest levels of quality control, continuously and *indefinitely*. As the fission energy economy grows, many plants will be built and operated in countries with comparatively low levels of technological competence and a greater propensity to take risks. A much larger amount of transportation of hazardous materials will probably occur, and safety will become the province of the sea captain as well as the scientist. Moreover, even in countries with higher levels of technological competence, continued success can lead to reduced vigilance. We should recall that we managed to incinerate three astronauts in a very straightforward accident in an extremely high technology operation where the utmost precautions were allegedly being taken.

DEEPER MORAL questions also surround the storage of high-level radioactive wastes. Estimates of how long these waste materials must be isolated from the biosphere apparently contain major elements of uncertainty, but current ones seem to agree on "at least two hundred thousand years."

Favorable consideration has been given to the storage of these wastes in salt formations, and a site for experimental storage was selected at Lyons, Kansas. This particular site proved to be defective. Oil companies had drilled the area full of holes, and there had also been solution mining in the area which left behind an unknown residue of water. But comments of the Kansas Geological Survey raised far deeper and more general questions about the behavior of the pertinent formations under stress and the operations of geological forces on them. The ability of solid earth geophysics to predict for the time scales required proves very limited. Only now are geologists beginning to unravel the plate tectonic theory. Furthermore, there is the political factor. An increasingly informed and environmentally aware public is likely to resist the location of a permanent storage facility anywhere.

Because the site selected proved defective, and possibly in anticipation of political problems, primary emphasis is now being placed upon the design of surface storage facilities intended to last a hundred years or so, while the search for a permanent site continues. These surface storage sites would require continuous monitoring and management of a most sophisticated kind. A complete cooling system breakdown would soon prove disastrous and even greater tragedies can be imagined.

Just to get an idea of the scale of disaster that could take place, consider the following scenario. Political factors force the federal government to rely on a single above-ground storage site for all high-level radioactive waste accumulated through the year 2000. Some of the more obvious possibilities would be existing storage sites like Hanford or Savannah, which would seem to be likely military targets. A tactical nuclear weapon hits the site and vaporizes a large fraction of the contents of this storage area. The weapon could come from one of the principal nuclear powers, a lesser developed country with one or more nuclear power plants, or it might be crudely fabricated by a terrorist organization from black-market plutonium. I am told that the radiation fallout from such an event could exceed that from all past nuclear testing by a factor of 500 or so, with radiation doses exceeding the annual dose from natural background radiation by an order of magnitude. This would bring about a drastically unfavorable, and long-lasting change in the environment of the majority of mankind. The exact magnitude of the disaster is uncertain. That massive numbers of deaths might result seems clear. Furthermore, by the year 2000, high-level wastes would have just begun to accumulate. Estimates for 2020 put them at about three times the 2000 figure.

SOMETIMES, analogies are used to suggest that the burden placed upon future generations by the "immortal" wastes is really nothing so very unusual. The Pyramids are cited as an instance where a very long-term commitment was made to the future and the dikes of Holland as one where continuous monitoring and maintenance are required indefinitely. These examples do not seem at all apt. They do not have the same quality of irreversibility as the problem at hand and no major portions of humanity are dependent on them for their very

existence. With sufficient effort the Pyramids could have been dismantled and the Pharaohs cremated if a changed doctrine so demanded. It is also worth recalling that most of the tombs were looted already in ancient times. In the 1950s the Dutch dikes were in fact breached by the North Sea. Tragic property losses, but no destruction of human life, ensued. Perhaps a more apt example of the scale of the Faustian bargain would be the irrigation system of ancient Persia. When Tamerlane destroyed it in the 14th century, a civilization ended.

None of these historical examples tell us much about the time scales pertinent here. One speaks of two hundred thousand years. Only a little more than one-hundredth of that time span has passed since the Parthenon was built. We know of no government whose life was more than an instant by comparison with the half-life of plutonium.

It seems clear that there are many factors here which a benefit-cost analysis can never capture in quantitative, commensurable terms. It also seems unrealistic to claim that the nuclear fuel cycle will not sometime, somewhere experience major unscheduled events. These could range in magnitude from local events, like the fire at the Rocky Mountain Arsenal, to an extreme disaster affecting most of mankind. Whether these hazards are worth incurring in view of the benefits achieved is what Alvin Weinberg has referred to as a trans-scientific question. As professional specialists we can try to provide pertinent information, but we cannot legitimately make the decision, and it should not be left in our hands.

One question I have not yet addressed is whether it is in fact not already too late. Have we already accumulated such a store of high-level waste that further additions would only increase the risks marginally? While the present waste (primarily from the military program plus the plutonium and highly enriched uranium contained in bombs and military stockpiles) is by no means insignificant, the answer to the question appears to be no. I am informed that the projected high-level waste to be accumulated from the civilian nuclear power program will contain more radioactivity than the military waste by 1980 or shortly thereafter. By 2020 the radioactivity in the military waste would represent only a small percentage of the total. Nevertheless, we are already faced with a substantial long-term waste storage problem. Development of a full-scale fission energy economy would add overwhelmingly to it. In any case, it is never too late to make a decision, only later.

WHAT ARE THE benefits? The main benefit from near-term development of fission power is the avoidance of certain environmental impacts that would result from alternative energy sources. In addition, fission energy may have a slight cost edge, although this is somewhat controversial, especially in view of the low plant factors of the reactors actually in use. Far-reaching clean-up of the fuel cycle in the coal energy industry, including land reclamation, would require about a 20 percent cost increase over uncontrolled conditions for the large, new coal-fired plants. If this is done, fission plants would appear to have a clear cost edge, although by no means a spectacular one. The cost characteristics of the breeder that would follow the light-water reactors are very uncertain at this point. They appear, among other things, to still be quite contingent on design decisions having to do with safety. The dream of "power too cheap to meter" was exactly that.

Another near-term benefit is that fission plants will contribute to our supply during the energy "crisis" that lies ahead for the next decade or so. One should take note that this crisis was in part caused by delays in getting fission plants on the line. Also, there seems to be a severe limitation in using nuclear plants to deal with short-term phenomena. Their lead time is half again as long as fossil fuel plants—on the order of a decade.

The long-term advantage of fission is that once the breeder is developed we will have a nearly limitless, although not necessarily cheap, supply of energy. This is very important but it does not necessarily argue for a near-term introduction of a full-scale fission economy. Coal supplies are vast, at least adequate for a few hundred years, and we are beginning to learn more about how to cope with the "known devils" of coal. Oil shales and tar sands also are potentially very large sources of energy, although their exploitation will present problems. Geothermal and solar sources have hardly been considered but look promising. Scientists at the AEC's Los Alamos laboratory are optimistic that large geothermal sources can be developed at low cost from deep hot rocks—which are almost limitless in supply. This of course is very uncertain since the necessary technology has been only visualized. One of the potential benefits of solar energy is that its use does not heat the planet. In the long term this may be very important.

Fusion, of course, is the greatest long-term hope. Recently, leaders of the U.S. fusion research effort announced that a fusion demonstration reactor by the mid-1990s is now considered possible. Although there is a risk that the fusion option may never be achieved, its promise is so great that it merits a truly national research and development commitment.

A strategy that I feel merits sober, if not prayerful, consideration is to phase out the present set of fission reactors, put large amounts of resources into dealing with the environmental problems of fossil fuels, and price energy at its full social cost, which will help to limit demand growth. Possibly it would also turn out to be desirable to use a limited number of fission reactors to burn the present stocks of plutonium and thereby transform them into less hazardous substances. At the same time, the vast scientific resources that have developed around our fission program could be turned to work on fusion, deep geothermal, solar, and other large energy supply sources while continuing research on various types of breeders. It seems quite possible that this program would result in the displacement of fission as the preferred technology for electricity production within a few decades. Despite the extra costs we might have incurred, we would then have reduced the possibility of large-scale energy-associated nuclear disaster in our time and would be leaving a much smaller legacy of "permanent" hazard. On the other hand, we would probably have to suffer the presence of more short-lived undesirable substances in the environment in the near term.

This strategy might fail to turn up an abundant clean source of energy in the long term. In that event, we would still have fission at hand as a developed technological standby, and the ethical validity of using it would then perhaps appear in quite a different light.

We are concerned with issues of great moment. Benefit-cost analysis can supply useful inputs to the political process for making policy decisions, but it cannot begin to provide a complete answer, especially to questions with such far-reaching implications for society. The issues should be aired fully and completely before a committee of Congress having broad policy responsibilities. An explicit decision should then be made by the entire Congress as to whether the risks are worth the benefits.

Scientists in an Environmental Controversy

DOROTHY NELKIN
Program on Science, Technology and Society, Cornell University

We have restricted our inquiries so as not to preempt the responsibilities of public decision makers. . . . The group does *not* attempt to comment upon the broader questions of public policy involved. Administrative, economic, political, or psychological questions [are] put aside in order to objectively evaluate the physical, chemical and biological . . . impact of the proposed Bell Station.[1]

This paper reflects the collective knowledge and judgements of persons . . . whose professions are in aquatic sciences, geology, resource management, and engineering. It results from their conviction that such professionals should contribute to public decisions on the management of natural resources.[2]

The present context of public concern for the implications of technological advance brings special pressures to bear on the scientist. He may be called on in an advisory capacity on the basis of his technical expertise; or he may himself feel that the nature of his work obliges him to take a position on public issues. Involvement of scientists in political activity has long been a sensitive and divisive issue within the scientific community, and scientists have traditionally approached political issues with reluctance. 'Politics has been considered an alien element, essentially destructive of scientific endeavour.'[3]

Much of the discussion concerning the political role of scientists has focused on the relationship between science and the federal government following the Second World War, as scientists were drawn into public service, and in some cases took issue with national weapons policy.[4] But even before the war, J. D. Bernal noted the dilemma of scientists who try to extend their skills to the political arena.

Any attempt on the part of the scientist to think for himself outside his own field exposes him to severe 'sanctions'. . . . It is therefore argued that in the interests of science it would be far better for him not to do so. Nevertheless, things can arrive at such a state that neutrality can compromise the very existence of science itself as a living force.[5]

The discussion presented here reveals the perpetuation of this dilemma in a contemporary environmental controversy.

Several questions are raised when scientists, using their technical expertise, engage in political activity. Is science a politically neutral activity with the scientist responsible only for the quality of his work?[6] Or does his vocation, 'circumscribed in a framework of political decisions', throw him, 'whether he wishes it or not, into the political arena'? These questions, which are controversial enough when scientists participate in decisions concerning foreign policy and weapons development, have recently assumed new significance when scientists turned their attention to environmental issues. Such issues are charged with conflicting public values and uncertain technical dimensions, and these are reflected in ambivalent policy. Decisions must often be made despite conflicting technical advice.[8]

When called on for their technical expertise, some scientists recoil from environmental controversies, taking refuge in the 'neutrality of research' position. Others, concerned about the unintended and undesirable consequences of technological application, become 'involved', but often find themselves without channels through which to express their concern. They are then forced into the role of advocates, or political contestants. Defining the role of the scientists in public decision-making was an important aspect of the controversy over the siting of a nuclear power plant on Cayuga Lake in upstate New York. Since the controversy involved a number of scientists at Cornell University, strains developed within the scientific community. As one analyst has suggested:

The vision of responsibility of the scientist . . . is fortified by the confusion between technologist and strategist and by the related notion of the scientist as specially endowed, a seer or a prophet. The notion . . . encourages schismatics and the feuds among scientists have been intolerant, and implicitly rather bloody.[9]

This paper arises from a case study of the Cayuga Lake nuclear power plant controversy, *Nuclear Power and Its Critics*, which has been published by the Cornell University Press, 1971. The author gratefully acknowledges the support of the Cornell Program on Science, Technology and Society (in which she is a Senior Research Associate), and the National Science Foundation. The editorial assistance of Sharon Bryan is appreciated.

[1] Ray T. Oglesby and David J. Allee (eds.), *Ecology of Cayuga Lake and the Proposed Bell Station*, Publication 27 (Water Resources and Marine Science Center, Ithaca, New York, September 1969), Introduction.

[2] Alfred W. Eipper *et al.*, *Thermal Pollution of Cayuga Lake by a Proposed Power Plant* (Ithaca, New York, 27 May 1968).

[3] Joseph Haberer, *Politics and the Community of Science* (New York, Van Nostrand Reinhold, 1969), 1.

[4] For analyses of this relationship, see Robert Gilpin, *American Scientists and Nuclear Weapons Policy* (Princeton University Press, 1962); and Warner D. Schilling, 'Scientists, Foreign Policy, and Politics, in Robert Gilpin and Christopher Wright (eds.), *Scientists and National Policy Making* (New York, Columbia University Press, 1964), 144–73.

[5] J. D. Bernal, *The Social Function of Science* (London, Routledge and Kegan Paul, 1939).

[6] See, for example, the views of Percy Bridgeman: 'Scientists and Society', *Bulletin of the Atomic Scientists*, **4** (March 1948), 66 (discussed in Gilpin, *op. cit.*, 26–7).

[7] Jean-Jacques Salomon, 'The *Internationale* of Science', *Science Studies*, **1** (January 1971), 23, 40. Warner D. Schilling notes: 'The American political system is not one that insulates its experts from the politics of choice', and the important problem for the scientist has been 'how to engage in politics without debasing the coinage of his own expertise' (*op. cit.*, 109, 149).

[8] Schilling, *op. cit.*, 148.

[9] Albert Wohlstetter, 'Strategy and the Natural Scientists', in Gilpin and Wright, *op. cit.*, 185.

Reprinted with permission from *Science Studies*, vol. 1, pp. 245–261, 1971.

Scientific activity

In April 1968 the New York State Electric and Gas Company (NYSE&G) began site clearance for the construction of an 830 megawatt nuclear power plant on Cayuga Lake in the heart of New York State. NYSE&G, with a franchise extending to about 35% of New York State's land area, provides a primarily rural residential service to about 10% of the state's population. The company operates as a part of the New York State Power Pool, through which there is a continuing exchange of power with other companies. It was planned, for example, to sell about 70% of the power generated by the Cayuga Lake plant during its first year of operation to other public service utilities.

The Cayuga Lake site is near Cornell University, which includes a large community of scientists and engineers; many of these became involved in various ways with the problems raised by the proposed power plant. The principal technical issues in the case turned on the potential effect on Cayuga Lake of the heat generated by the power plant's nuclear reactor, and of the pumping activity required for its cooling. Cayuga Lake is a deep, thermally stratified lake, with a natural mixing pattern which limits the distribution of oxygen and nutrients, and minimizes biological activity. The lower layer of the lake remains at about 45°F all year round, and this availability of a continuous source of cooling water is a considerable economic advantage. This, and the location of Cayuga Lake near the centre of the New York State transmission grid, were the main reasons for selection of this particular site.

The first announcement of the proposed construction, in June 1967, met with favourable response from local political leaders, who viewed the development as a potential economic opportunity for the region. There was little publicity concerning the company's plan, and no community reaction at that time. NYSE&G proceeded to purchase land[10] and to apply for the required construction permit from the United States Atomic Energy Commission (AEC). Site clearance began in April 1968.[11]

Just a few weeks earlier, in March 1968, a fishery biologist, Alfred Eipper,[12] first expressed his concern about possible damage to the lake from the proposed power plant. He formed a committee of seventeen scientists from the Departments of Conservation, Limnology, Botany, Biology, Geology and Engineering, and, with the help of this group, wrote a position paper describing in detail the potential damage.[13] It was feared that the heated water discharged from the plant (at a rate of about 1,225 cubic feet of cooling water per second) would upset the thermal balance of the lake and accelerate eutrophication—a process in which a lake ages with increased biological activity. The position paper emphasized the dangers of eutrophication, and warned that it is an irreversible process. To minimize the danger, alternative methods of cooling the discharge water were recommended.

A second position paper written by an associated group of scientists, several of whom were also on the Eipper Committee, urged design modifications to minimize the radiological wastes produced in the normal operation of the plant.[14] But the initial position paper, with its focus on thermal pollution, voiced what was to become the central issue in the subsequent controversy.

Eipper went on to express his continuing concern by helping to form the Citizens' Committee to Save Cayuga Lake (CSCL) in September 1968. David D. Comey,[15] a research associate at Cornell, became interested in the issue and assumed leadership of the organization. The purpose of the citizens' group was 'to inform citizens of the Cayuga Lake region about the potential sources of bacterial, chemical, thermal and radioactive pollution of the lake, and to coordinate the efforts of all concerned organizations and individuals to prevent and eradicate any pollution endangering the foremost natural resource of the region.'[16]

Two problems were emphasized by the CSCL. The first was the danger of thermal pollution. The position of the CSCL was not one of opposition to nuclear power as such; rather its main concern was over the design of the power plant's cooling system, which it was felt afforded inadequate environmental protection. Cooling towers were specifically recommended as a means of minimizing damage to the lake. The second problem was the adequacy of the existing regulatory system for protecting water resources. The ambiguity of thermal standards, the structure of the regulatory agencies, and the extent to which conflicting pressures were brought to bear on power plant licensing procedures became major points of contention. Position papers, 20,000 copies of which were distributed, publicized the CSCL position.

The CSCL became the focus of local activity aimed at protecting natural resources. By the fall of 1969 it had attracted a membership of 854 residents in the Lake Basin area, and raised over $16,000 in dues and contributions. Other local clubs in the lake region became affiliates.[17] The general membership provided a political base by their numbers; the activist core of the organization consisted of a small group of about

10 As a public utility, NYSE&G may exercise the right of eminent domain: 725 acres, including 8,000 feet of lake frontage, were purchased from twenty-five property owners.

11 Three major permits are required to construct a nuclear plant: a construction permit and an operating licence from the United States Atomic Energy Commission, whose sphere of authority is limited to radiological safety; and a discharge permit from the New York State Department of Health, concerned with thermal discharge into state waters. Before receiving permits, a company may proceed with construction to the point of pouring concrete.

12 Alfred Eipper is an Associate Professor of Fishery Biology in the Department of Conservation at Cornell. But he is employed in this capacity by the federal government as a unit leader of the Bureau of Sport Fisheries and Wildlife, US Department of the Interior. His teaching in the area of water resource management first stimulated his interest in the Cayuga Lake issue.

13 Eipper, *op. cit.*

14 Clarence A. Carlson *et al.*, *Radioactivity and a Proposed Power Plant on Cayuga Lake* (Ithaca, New York, 22 November 1968).

15 David D. Comey's professional field was Russian Studies. His previous interests had not included problems of nuclear power generation, nor had he been particularly interested in environmental problems. But the issue fascinated him to the point of total involvement, and he subsequently became a professional environmentalist in a Michigan-based environmental defence group.

16 The Citizens' Committee to Save Cayuga Lake, *Cayuga Lake Handbook* (Ithaca, New York, 1969).

17 These included local sportsmen's groups, the Cayuga Lake Preservation Association, the Seneca Lake Waterways Association, and other citizens organizations which had developed over many years primarily as local home-owners' interest groups. The CSCL Board of Directors, a group of well-established civic leaders, included attorneys, physicians, university professors, and insurance and real estate brokers. Several groups with interests in the labour and tax advantages brought by new development supported the NYSE&G plan, though they were not very active. These included the Building Trades Association and the Taxpayers Association. In February 1968, the Tompkins County Chamber of Commerce honoured NYSE&G as 'Company of the Year'.

twenty scientists who were members of the CSCL Scientific Advisory Committee. These scientists gave speeches, wrote testimony, and took part in the public hearings which were then being held throughout the state on issues relating to thermal standards and power plant siting.[18] It was their active participation, skilfully coordinated by David Comey, which gave strength to the organization.

The press was highly sympathetic to the CSCL, and the statements of scientists were widely distributed in the local newspapers. During the year of controversy, the *Ithaca Journal* printed seventy-five unsigned and forty signed articles relating to the case. Only about a quarter of this coverage focused on the NYSE&G position, and these were mostly press releases. The press covered all CSCL meetings but few of the Company information programmes, and editorials supported the citizens' group. The CSCL encouraged publicity; their strategy was to use the pressure of public protest to delay construction.

But the same activities which prompted public concern also generated tension within the scientific community. This tension reflected not so much substantive disagreement between groups of scientists, as disagreement over the proper presentation of scientific data, the appropriate behaviour of scientists with respect to public issues, and the effect of publicity on an objective approach to technical issues.

The CSCL environmentalists saw a natural resource threatened by a power company acting within inadequate limitations. They considered this a moral issue, requiring scientists to take a philosophical and political position. Another group, however, operated with a different definition of scientific responsibility:

> Members of the public, members of legislative bodies, will have to make the judgements . . . we're not going to try to prove any points, we are testing hypotheses, and trying to be objective as we can, which is the reason why the group engaged in this study has avoided most of the public controversy.[19]

This second group was brought together in June 1968 by the Cornell University Water Resources Center to carry out a research contract provided by the NYSE&G.[20] From July 1968 to April 1969, they collected data from seven sampling stations on the lake. These observations were intended to provide information on oxygen distribution, temperature, the currents and hydrology of the lake, and on plant nutrients. A report,[21] edited and organized by Ray T. Oglesby, a professor of Aquatic Science, was issued in September 1969, with caveats as to the difficulty of drawing definitive conclusions. 'Limnologists know so little about the ecological significance of some of these environmental parameters that prediction of biological effects would be highly conjectural, even if exact description of physical changes were available.'[22] But, despite the reservations, the report included an estimate that, at the time of greatest thermal effect, algae would increase by only 5%, and it was predicted that the combined effects of a higher temperature and longer stratification period caused by the heated effluent would be insignificant. This prediction was based largely on the argument that, in view of the total residential and agricultural situation in the lake drainage area, it is difficult to isolate single effects on the nutrient content of the lake. Agricultural and dairy industries dump tons of fertilizer and animal waste into the soil; the effect of a power plant would be relatively negligible.

Once the issues had been defined, other groups of scientists and engineers became involved. The Engineering College at Cornell undertook a study in the summer of 1968. Six engineers, directed by Franklin Moore, wrote a theoretical analysis dealing with the physical consequences of the process of heat transfer which underlies the biological and ecological phenomena.[23] The Moore report emphasized consideration of the total parameter of the lake, and thus took issue with the NYSE&G-sponsored research, which considered only the area near the site.

Similarly, a nuclear engineer, K. Bingham Cady, proposed a broader perspective on the question of the thermal standards by which the New York State Department of Health regulates the permitted heated water discharge from power plants.[24] Cady felt that these standards were inadequate, since they were based on temperature changes near the point of discharge. Like Moore, Cady was concerned with the ecology of the lake as a whole, and he and five others, including Eipper, wrote a letter to this effect to the State Department of Health.

At the same time, a more inclusive research strategy was proposed by James Krumhansl, a solid state physicist, and two engineers, Donald Belcher and Simpson Linke. This group looked at the input of the power station as a part of the total lake system. Since the lake had already been damaged by a variety of pollution sources, they asked, 'Is it possible to use a major energy source to *beneficiate* the lake, i.e., in some sense to combat the forces (biological and chemical, etc.) which are now driving it?'[25] They proposed several alternative schemes; one would use the plant as an energy source to cool the upper layer of the lake, and in this way decrease biological activity; another would use a pumped water storage facility, which would both ease the thermal problem and remove nutrient-laden water from the lake.

These alternative suggestions were largely ignored by the CSCL and its associated scientists in their public statements. They also ignored the raw data gathered by the Oglesby group, with its implication that the effect of the plant would be insignificant relative to other influences on the lake. These data were of little concern to the CSCL, for it was not the extent of damage, but rather the possibility of *irreversible* damage on which they based their arguments. Thus, the opponents of the power plant remained committed to a position that the only acceptable solution would be the construction of cooling towers.

This posed a dilemma. Though the cost of cooling towers was less than 10% of production costs, the NYSE&G president claimed that information about their possible environmental effects (such as fogging) was inadequate,

18 Statements concerning the Cayuga Lake case were submitted at the following hearings: The US Senate Subcommittee on Air and Water Pollution, April 1968; the NY State Joint Legislative Committee on Consumer Protection, September 1968; the NY State Joint Legislative Committee on Conservation, December 1968; the Senate Committee on Public Utilities, January 1969; and the NY State Water Resources Commission, February 1969.

19 New York State Joint Legislative Committee on Conservation, Natural Resources, and Scenic Beauty, *Hearings*, 22 November 1968. Afternoon question and answer period.

20 NYSE&G awarded contracts totalling $500,000. $320,000 was awarded to the Cornell Aeronautical Laboratory, a Cornell-owned but independent research contract organization in Buffalo. The Cornell Water Resources Center received $135,000.

21 Oglesby and Allee, *op. cit.*

22 *Ibid.*, 452.

23 Franklin K. Moore *et al.*, *Engineering Aspects of Thermal Discharges to a Stratified Lake* (College of Engineering, Cornell University, Ithaca, New York, 1969).

24 Letter from K. Bingham Cady *et al.* to Dwight Metzler, New York State Department of Health, 2 October 1968.

25 James A. Krumhansl, *Ideas for Discussion of 'Thermal Pollution' and Cayuga Lake Protection or Beneficiation*, 1 October 1968 (mimeographed).

and he was not prepared to consider alternatives. Conflict hardened the position of company management.[26] They anticipated that even if they agreed to modifications, the CSCL would launch an attack on some other issue. Moreover, at this time there was uncertainty concerning new criteria being proposed in New York State on water quality standards. NYSE&G would have to meet new standards, thereby raising the possibility of considerable additional construction cost. Estimates had already increased from $135 million in June 1967 to $170 million in March 1968. By April 1969, projections including possible design changes were $245 million. By this time about ten million dollars had been expended for engineering, land, site preparation and research, but the unpredictability of future costs and potential delays from public pressure were compelling. In April 1969, one year after the initial site clearance, the President of NYSE&G announced the indefinite postponement of construction of the nuclear power plant on Cayuga Lake.

Context of scientific activity

To clarify the role of scientific activity in the decisions relating to this case, at least two factors must be distinguished: the technical uncertainty concerning the characteristics of ecological systems, and the ambivalence in public policy relating to the siting of power units.

Crucial aspects of the scientific studies relating to this case were the absence of conclusive data and the lack of an accepted theoretical framework from which to draw definitive quantitative conclusions. There were little existing data from which to evaluate the significant normal variation over time in the thermal characteristics of the lake. The characteristics of heat transfer and the system of nutrient exchange in a stratified lake are poorly understood. Also, it is difficult to separate the effects of a power plant on the nutrient content of the lake from the effects of agricultural development and of other pollution sources in the lake drainage area. Finally, there are aspects of the process of eutrophication—especially with respect to its reversibility—about which little is known.

The question of the extent to which damage to the lake is reversible includes a consideration of the cost of preserving a resource. How does one evaluate a natural resource? Estimates of the total value of a resource must consider its uniqueness, as well as changing tastes which may determine future uses. What is it worth today to have the option to use the lake in unknown ways in the future? How does one trade off various uses of a resource equitably? These are non-measurable variables which contribute to the difficulty of establishing a workable policy for handling such issues as the siting of nuclear power plants.[27]

The present system for regulating the activities of the power industry reflects these difficulties. The United States Atomic Energy Commission (AEC), for example, has two functions which are often in conflict. It is responsible both for promoting nuclear development[28] and for regulating it through the granting of licences. This dual responsibility, originally conceived as the basis of a vigorous federal civilian power programme, is now being reexamined.[29]

Similarly, the New York State Health Department, which regulates the discharge of power plants into state waters through its licensing authority, is subject to conflicting pressures, as revealed in the following statement on the Cayuga Lake case: 'The State Health Department looks on Cayuga Lake as a valuable asset to the State and therefore is interested in protecting it. On the other side of the coin, the State Health Department is involved in the industrial development of the State.'[30] This ambivalence in the regulatory system allows the power industry a great deal of autonomy. Thus, in response to an enquiry from a concerned citizen, a Health Department officer replied: 'Since the company has begun construction it apparently seems that it can meet our requirements concerning its discharge.'[31]

This response demonstrates the conflict between different values in public policy. On the one hand, policy is based on the assumption that maximum economic growth is optimal and should be encouraged. On the other, a major thrust of the environmental movement in the United States is that unrestrained growth threatens scarce environmental resources. This creates a profound dilemma which one writer has called a 'confrontation' between 'the insatiable appetite for energy to run factories, commercial establishments, transportation systems, air conditioners, electric toothbrushes, and the whole gamut of labor saving gadgetry and modern conveniences' and the 'environmental movement which seeks to save mankind from smothering in the waste products that result from the generation of energy.'[32] This context of technical uncertainty and ambivalent public policy contributed to the controversial nature of the scientists' involvement.

Limits and liabilities of scientific activity

The lack of a well-defined political role for scientists[33] is seen in acute form in the case of contemporary ecologists. They have been thrust into the thick of the environmental crisis at a stage in the development of their discipline when there is no accepted theoretical structure to provide criteria for the interpretation of scientific data. In the Cayuga Lake controversy, as soon as questionable data were interpreted in terms of their significance for practical problems, strains developed between those who openly opposed the construction of the power station and those who were doing

26 NYSE&G is a hierarchical organization in which decisions are made centrally by a few people. The technical staff has little to do with the decision-making process. The limited adaptive capacity of hierarchical organizations in dealing with uncertain and rapidly changing problems has been noted by Victor Thomas, *Modern Organizations* (New York, Alfred Knopf, 1963); and Warren Bennis and P. E. Slater, *The Temporary Society* (New York, Harper and Row, 1968).

27 John V. Krutilla, 'Conservation Reconsidered', *American Economic Review*, **57** (September 1967), 785.

28 The total sum of $2.3 billion has been invested to date by the AEC in research to develop commercially feasible atomic power.

29 See, for example, the objections of John W. Gofman and Arthur R. Tamplin, two scientists from the Lawrence Radiation Laboratory, who have suggested that radiation standards are too lax. 'Environmental Effects of Producing Electric Power', *Hearings Before the Joint Committee on Atomic Energy*, Part I, 91st Congress, 1969, 604–746. AEC standards have also become an issue in the state of Minnesota, which is demanding the right to establish its own standards.

30 Letter from Dwight F. Metzler, Deputy Commissioner of the New York State Department of Health, to William Ward (28 August 1968).

31 Letter from Ronald S. Bratspis, New York State Department of Health (on request of Governor Nelson Rockefeller), to Mrs John H. Lehman (17 October 1968).

32 Philip M. Boffey, 'Energy Crisis: Environmental Issue Exacerbates Power Supply Problem', *Science* (26 June 1970).

33 Joseph Haberer (*op. cit.*) notes the absence of an established institutional ethic governing the relationship between science and the political environment. (See also Richard A. Rettig, 'Science, Technology and Public Policy', *World Politics*, **23** (January 1971), 273ff.)

research under New York State Electric and Gas Company contract. Their disagreements were subtle and complex; all participating scientists were clearly concerned with protecting the lake.

Both the opponents of the power station and the company-financed group, however, were criticized for bias and for premature interpretation of their data. The first group was criticized for taking an overt public position, and the other for assuming that it was possible to take a detached, objective position. These criticisms and the responses to them are summarized in the table overleaf.

This conflict had considerable bearing on the scope of inquiry concerning potential damage to Cayuga Lake. To pursue this, it is useful to explore the limits and liabilities of the roles assumed by scientists in the case. Their activities were of three types: substantive research, policy analysis and political action.

With the exception of the Moore report, the substantive research on potential effects of the power plant was performed under NYSE&G contract. This source of funding exposed the researchers to a great deal of criticism, but they claimed that their objectivity would not be affected. For applied scientists and engineers, industrial support is often the basis of their research. Much of the scientific work which underlies decision-making within regulatory agencies is supported by parties with vested interests in the ultimate decision. Scientists and engineers consider themselves able to cope with the pressures of client demands, and to maintain the validity of their work through the critical judgment of professional peers. One of the assumptions of scientific investigation is that if science is conducted according to rules established outside the profession, it can no longer be considered objective. 'The fact is that the scientist gives his allegiance to the scientific community, in particular to the "invisible college" of his specialist field of study.'[34] Outside influence is strongly rejected. 'What I am concerned about, and very much concerned about, is any pressure on us at Cornell—the group that is doing this work—that we have a responsibility to provide *you* with the kinds of specific answers that you want.'[35]

But, in the context of controversy, the ideals of scientific objectivity are difficult to implement. In the organization of the NYSE&G-sponsored research, precautions were taken against bias. For example, to clarify responsibility, the various studies included in the final report were not combined. Each author signed his name to his own work; and the group insisted that the results be published without company editing. Yet certain constraints were unavoidable. The selection of the research group was the first problem. Several individuals who were logical choices because of their relevant scientific expertise were unacceptable to NYSE&G on the basis of their public activities related to environmental problems. The company demanded that researchers be selected only if their activities outside work revealed no bias on the issue. A stipulation of this sort was unacceptable to the university authorities who would not formally sign the contract; the research was therefore performed on the basis of a letter of intent.

In order to meet the company's restrictions, a biologist from outside Cornell, Thomas D. Wright, was brought in as a Research Associate; his main task was to collect the bulk of the data through sampling in the lake at fortnightly intervals. He was ultimately the author of fourteen of the twenty chapters in the final report. Wright's selection posed particular problems of coordination and control in such an interdisciplinary study, for he was unfamiliar with the Cornell research set-up. And when Wright, himself, joined the Citizens' Committee to Save Cayuga Lake, this became a source of aggravation within the research group.

The sampling data was intended to cover a complete annual cycle of the Lake's biological activity. It was, however, terminated three months early due to 'contract, personnel and equipment limitations'. Wright, who had been principally responsible for data collection, left Cornell before the major editing and summarizing of the report; this task fell to Ray T. Oglesby, who had joined the Cornell faculty in October 1968, well after the NYSE&G study was under way.

*Criticism and Response**

GROUP CRITICIZED	CRITICISM	RESPONSE
Opponents of the power plant	Taking a position prematurely destroys scientific credibility and is inappropriate.	Taking a position is necessary to provoke discussion. If scientists don't interpret their own data and thereby take a position, others will, with their own interests in mind.
	Reports were emotional and overstated. Subjective speculations based on selected data were presented as fact. They reflected personal interests in the lake.	This is a philosophical issue where emotionalism is appropriate.
	Taking a position on a public issue is all right, if it is clearly distinct from one's research.	Report, a position paper, was intended to raise issues, not to provide definitive answers. It was based on informed opinion. Accurate data on amount of eutrophication is not available, but is in any case, not relevant if the effects are irreversible.
Water Resources Center Study Group under NYSE&G Contract	Report was bought by the company and was, therefore, necessarily biased and partisan.	All precautions were taken to ensure objectivity; e.g., report not shown to company until after publication, and each individual signed his own articles. In any case, review by scientific peers controls the honesty of scientists, who therefore cannot be bought.
	Outcome of report can determine future funding; for if the evidence discourages the siting of the project on the lake, there will be no further funding.† This makes a bias inevitable, especially considering present scarcity of research funds.	Research was a continuation of on-going work: source of payment is irrelevant.
	The sampling was an inadequate basis for ultimate decisions made by the company, since its timing excluded part of the growing season.	Analysis was cut short for reasons of time and personnel, but solid technical data were carefully collected, and are therefore valid for future analysis.

* This list of criticism and response was compiled from comments by many different persons; therefore, they are not always consistent, nor are they all necessarily derived from what one group of scientists actually and publicly said about the other.

† Employed on the project, and paid out of NYSE&G funds were: eight laboratory assistants and technicians, five research assistants, one labourer, three typists, one programmer, six student assistants, two full-time research associates, and summer salaries for two professors.

Another constraint, imposed by the NYSE&G as a means of keeping company engineers informed, was a directive that the Cornell research group should not be allowed to communicate directly with another research group doing a similar study at the Cornell Aeronautical Laboratory, except

[34] John Ziman, *Public Knowledge* (Cambridge University Press, 1968), 130. See also Derek J. de Solla Price, *Little Science, Big Science* (New York, Columbia University Press, 1963).

[35] This statement was made by Professor Leonard Dworsky, Director of the Cornell Water Resources Center, during the New York State Joint Legislative Committee *Hearings*, 22 November 1968. 'You' is used to refer generally to government decision-making bodies.

through the company's own design engineers. This posed difficulties in the scheduling and organization of research which hampered the collection of data, and some participants saw it as a means of control. NYSE&G engineers were present at the regular research meetings at which the findings and their interpretation were discussed. Their presence was described as 'irritating at times', though it was said not to have influenced the work. These details illustrated how, paradoxically, NYSE&G's attempt to avoid bias helped to establish a context of research in which just such accusations were made.

The second type of activity involving scientists was well within the tradition of scientific dissent following the Second World War, when scientists questioned the American nuclear weapons policy.[36] There is, however, little precedent in the scientific community for questioning the use of technology in domestic areas.[37] Eipper's group, stating its conviction that scientists should contribute to public decisions, proceeded to raise issues concerning a potential threat to the lake which they felt would otherwise be ignored. They were most successful in calling attention to these issues in the position papers; in their eagerness to raise public concern, however, the authors tended to overreach the existing data, and they failed to differentiate between certain and potential dangers. They were criticized by other scientists for their 'non-scientific approach': for allowing their political focus to lead them to selective perception of data to the point of ignoring results of ongoing research which might contradict their political position and allay public concern.[38] Researchers sponsored by the NYSE&G claimed that the available raw data from lake samples were neither examined nor discussed. Public statements neglected to include material from current research on the possible reversibility of eutrophication. Technical alternatives were not seriously considered in the information disseminated by the CSCL, whose concern was political expediency.

This brings us to the third type of activity involving scientists, namely the overtly political. Scientists kept the issue before the public through newspaper releases, public speeches and participation in citizens' committee activities. The effectiveness of these activities is evident in the success of the CSCL, which was started and sustained by scientists. At the same time, however, the political approach of the CSCL scientists was criticized as being single-minded, leading to public misinformation, and contributing to unnecessary polarization over the issue.

The tactic of seeking wide publicity in order to mobilize a base of support on a technical issue is quite different from merely stating a dissenting position. The more active approach was criticized on the grounds that it limited consideration of constructive measures to 'save the lake'. It was argued that the CSCL focused on too narrow a target, failing to take into account the total lake environment. The unexplored dimensions of the problem, and the complexity of the many sources of possible damage to the lake, indicated the need for a very broad research approach. Moreover, the predictable long-term demand for energy requires consideration of alternative technologies to resolve problems of thermal pollution. One of the liabilities of political activity is that the selective focus necessary to make a political point may limit the scope of discussion and inquiry. This is especially the case with environmental issues, where popular sentiment tends to foster quick generalization, and to discourage the time-consuming process of weighing many alternatives.

In this case, serious attention directed to long-range matters was restricted by the immediacy of the political issue, the tentative character of the funding, and the intermittency of the research. Those doing substantive research were dependent on short-term contracts awarded as a response to the particular crisis, and to the need for a specific kind of information. Those who questioned the plans of the power company were forced to move quickly and dramatically if their immediate goals were to be met. In the context of crisis, they had no time to consider alternatives. Had local political activity been less dramatic, few questions would have been raised concerning the adequacy of the plant's cooling system.[39]

In sum, the Cayuga Lake controversy reveals the many problems created by the conflict between the ideals of scientific objectivity and those of public responsibility. The dynamics of the controversy suggest a tendency noted by Don K. Price, that when confronted by the uncertainties that harass politicians, the human mind is tempted to reduce 'the complexities of politics to the simplicities of moral feeling' on the one hand, or to take shelter 'in the purity of research' on the other.[40] One group of scientists engaged in political activity, seeing this as the only way to bring about countervailing consideration of environmental hazards in the absence of an adequate policy concerning nuclear power plant siting. But for others, such political involvement was unjustified, given the context of technical uncertainty. Both positions have their limitation, for a scientist 'operates as much on the basis of his non-technical assumptions as on his command of a body of scientific knowledge'.[41] This places a heavy burden of responsibility both on the scientists who avoid public policy in order 'to objectively evaluate' the technical dimensions of a problem, and on those who feel their actions belong in the public domain.

36 Lawrence S. Wittner, *Rebels Against War* (New York, Columbia University Press, 1969).

37 Haberer (*op. cit.*), in fact, characterizes the relationship of scientists to government as 'prudential acquiescence'. Some of the dangers of criticizing public policy are illustrated in a letter received by Eipper from his federal government employer questioning him for taking issue with public policy on the grounds that it was 'somewhat removed from the competence and expertise of a fishery scientist'.

38 Gilpin (*op. cit.*, 19) has noted how scientists select, emphasize and interpret 'those specific scientific facts which they believed to support their own political position on the wisest course for American foreign policy'.

39 While there were public hearings on the Cayuga Lake case (held on 22 November 1968 by the New York State Joint Legislative Committee on Conservation, Natural Resources and Scenic Beauty), these hearings were not a part of the regulatory process and took place only in response to the local public concern.

40 Don K. Price, 'Purists and Politicians', *Science* (3 January 1969).

41 Gilpin, *op. cit.*, 23. See also James B. Conant, in *Modern Science and Modern Man* (New York, Doubleday, 1952), 114, who talks of the ease of 'clothing personal convictions in technical language'.

The Economic Costs of Air Pollution Damages

Excerpts from the Work of Larry B. Barrett and Thomas E. Waddell

Editor's Note: For a number of years prior to 1973, Larry B. Barrett and Thomas E. Waddell of the U.S. Environmental Protection Agency produced a series of manuscripts in which they performed an analysis of all available air pollution damage cost data in order to estimate total U.S. damages normalized to the year 1968. More recently, Waddell has extended this earlier work and produced a new series of cost estimates normalized for the year 1970.

Because these papers are lengthy, it has not been possible to reprint them. However, they contain valuable discussions of the available data and careful descriptions of the methodologies used to obtain the cost estimates. Serious students of air pollution costs, and *anyone* planning to make use of the cost estimates that are presented, will find it important to read the full papers.

The following pages contain excerpts of summaries and cost estimates, first from Barrett and Waddell's 1971 paper, "The Costs of Air Pollution Damages: A Status Report", and second, from Waddell's 1974 paper, "The Economic Damage of Air Pollution." References for these reports are as follows:

1) L. B. Barrett and T. E. Waddell, "The costs of air pollution damages: A status report," Appendix I-J of "Cumulative Regulatory Effects on the Cost of Automotive Transportation (RECAT)," a report prepared for the Office of Science and Technology, Executive Office of the President, Feb. 28, 1972.

A similar manuscript with the same title is available through Environmental Protection Agency, Research Triangle Park, N.C., Publication number AP-85, Feb. 1973.

2) T. E. Waddell, "The economic damages of air pollution," National Environmental Research Center, Office of Research and Development, U.S. Environmental Protection Agency, Research Triangle Park, N.C. 27711 (Program Element 1AA004), May 1974.

Brief Excerpts from "The Economic Damages of Air Pollution"

PREFACE

This paper was conceived and undertaken in order to: (a) survey efforts-to-date in estimating the cost of air pollution damages; and (b) based on this survey, develop a national annual cost of air pollution damages.

Fulfillment of these purposes entailed a survey of the literature on reported studies, as well as a review of unpublished and ongoing studies. For the first time there will be an opportunity to critically review and compare the many fragmented studies on air pollution costs.

Studies on the cost of air pollution damages that survive critical review will be synchronized to prepare a national estimate of damages. A by-product of this process will be the identification of informational shortages. Although it is not the purpose of this paper to suggest further research topics, such needs may be more readily apparent.

The authors acknowledge that the scarcity of work in several effects areas necessitates assumptions which weaken the conclusions. These conditions are evaluated in each section of the report. While recognizing these limitations, the authors believe that the damage estimates are reasonable. The reader is cautioned against accepting the values as proven; they are reasoned, national estimates. . .

Summary and Conclusions

There are relatively few studies on the costs of air pollution. Of the approximately 36 studies, about a third have been published, another third completed and unpublished, and a third in progress.

Studies have estimated the costs of air pollution for effects on health, vegetation, materials, residential property values, soiling, and aesthetics. Costs have not been studied for effects on animals.

National total annual costs were developed for 1968 as $16.1 billion. This is the sum of $5.2 billion for residential property, $4.7 billion for materials, $6.1 billion for health, and $0.1 billion for vegetation.

The cost of each effect was distributed among the several pollutants considered responsible for that effect according to their relative emissions. The same cost was distributed among the sources by their relative emissions.

The national total annual cost for 1968 may be considered as a function of the pollution levels in that year. No cost function relations could be developed for various levels of pollution. As a consequence the pivotal functions of marginal benefit are precluded.

Considering all factors, the $16.1 billion national total annual costs of air pollution is believed to be a reasonable, conservative estimate.

The subject of the cost of pollution would benefit from further efforts in all aspects. The scope of estimates should be broadened to include more pollutants and more effects. There is a need to relate costs to more specific pollutants or pollutant synergisms and to more specific sources or source categories. Functions on total cost of pollution require the application of more sophisticated analyses.

TABLE J14
Estimates of Nationwide Emissions, 1968*

Emission Source	Type of Emission (10^6 tons/yr) CO	Particulates	SO_x	HC	NO_x
Transportation	63.8	1.2	0.8	16.6	8.1
Fuel combustion in stationary sources	1.9	8.9	24.4	0.7	10.0
Industrial processes	9.7	7.5	7.3	4.6	0.2
Solid waste disposal	7.8	1.1	0.1	1.6	0.6
Miscellaneous	16.9	9.6	0.6	8.5	1.7
Total (1968)	100.1	28.3	33.2	32.0	20.6

**Source:* Division of Air Quality and Emissions Data, BCS, NAPCA.

TABLE J15
National Total Annual Costs of Pollution for Types of Pollutants and Effects in 1968

Effects	Cost of Emissions (millions of dollars) SO_x	Particulates	Oxidant	NO_x	Total
Residential property	2,808	2,392	–	–	5,200
Materials	2,202	691	1,127	732	4,752
Health	3,272	2,788	–	–	6,060
Vegetation	13	7	60	40	120
Total	8,295	5,878	1,187	772	16,132

A dash indicates no known value.

TABLE J16
National Total Annual Costs of Pollution for Types of Sources and Effects in 1968

Effects	Cost of Emissions (millions of dollars) Stationary Source Fuel Combustion	Transportation	Industrial Processes	Solid Waste	Miscellaneous	Total
Residential property	2,802	156	1,248	104	884	5,200
Materials	1,853	1,093	808	143	855	4,752
Health	3,281	197	1,458	119	1,005	6,060
Vegetation	47	28	20	4	21	120
Total	7,983	1,474	3,534	370	2,765	$16,132

Brief Excerpts from "The Cost of Air Pollution Damages: A Status Report"

Summary and Conclusions

The cost of air pollution damage in the United States in 1970 is estimated to fall within the range of \$6.1 billion and \$18.5 billion. The "best" estimate for measured effects for that year is determined to be \$12.3 billion. These estimates are based on: (1) a survey of the literature on environmental economics; (2) an extrapolation of studies that have attempted to estimate air pollution damages and that passed a critical review; and (3) prevailing air quality levels in 1970.

An evaluation is also made of the methods that can be employed to estimate the damages of air pollution. These methods are: (1) technical coefficients of production and consumption; (2) market studies; (3) opinion surveys of air pollution sufferers; (4) litigation surveys; (5) political expressions of social choice; and (6) the delphi method. It is concluded from such a review that some combination of the methods surveyed will ensure the most accurate assessment of the economic damages resulting from air pollution insults. Such damages, in turn, when properly translated, become the benefits of abating air pollutant emissions.

It is shown in this report that only the technical coefficients and market study approaches have been used with measurable success in assessing the benefits of controlling air pollution. The technical coefficients method was utilized in estimating air pollution damages to human health, man-made materials, and vegetation. The "best" (unadjusted) estimates for these effect categories for 1970 are \$4.6 billion for health, \$2.2 billion for materials, and \$.2 billion for vegetation, and total to \$7.0 billion. A market study method, the site value differential or property value approach, yielded a "best" (unadjusted) estimate of \$5.9 billion. This figure represents the value in 1970 of the negative insults of air pollution that are capitalized in the residential, urban property market. It is argued in this report that capitalized in this estimate are primarily those costs associated with aesthetics and household soiling.

Since it is likely that there is some overlap in the \$7.0 billion and \$5.9 billion estimates, they can be considered additive only with minor adjustments. By making such adjustments, any double-counting will be minimized. With such adjustments, the \$7.0 billion determined via the technical coefficients method becomes \$6.5 billion and the \$5.9 billion determined via the property value method becomes \$5.8 billion.

The estimate of \$12.3 billion for 1970 developed here, differs from the 1968 estimate of \$16.1 billion developed by Barrett and Waddell because of the following reasons: (1) the 1970 estimate is based on information that wasn't available in the 1968 study; (2) the levels of air pollutants being worked with in the 1970 study are generally lower than the levels for those same pollutants in 1968; (3) a re-evaluation of the available data has forced the modification of certain assumptions in this report.

The information surveyed in this report establishes that \$12.3 billion is the "best" estimate for 1970. Given the lack of conclusive information to indicate that what is estimated in the \$5.9 billion does not significantly overlap with what is estimated in the \$5.9 billion does not significantly overlap with what is estimated by the \$7.0 figure, the option is left for the reader to use the \$7.0 billion as a measure of air pollution damages in 1970. While the evidence is far from clear, it is reasoned that as interpreted in this study, the estimates determined via the site differential and technical coefficients methods should be considered additive, with only minor adjustment for obvious areas of overlap.

While it is known that air pollution causes losses of domestic animals and wildlife, such losses were not quantified in this report because of data limitations. Air pollution is also believed to cause pervasive effects in the biosphere and on geophysical and social processes. These effects are not without some economic consequences, but until the relationships can be more clearly identified, large-system economic analysis is somewhat premature.

The cost estimates for aesthetics and soiling, health, materials, and vegetation are distributed among the several pollutants considered responsible for the effect. The pollutants considered are sulfur oxides (SO_x), particulates, and oxidants (O_x). Damages in 1970 attributable to SO_x are estimated to fall within the range \$2.8–\$8.0 billion, with a "best" estimate of \$5.4 billion. Particulate damages are estimated to fall between \$2.7 and \$8.9 billion, with a "best" estimate of \$5.8 billion. Oxidant-related damages are estimated to fall in the range \$0.6–\$1.6 billion, with a "best" estimate of \$1.1 billion. Every attempt is made in this attribution process to identify where data deficiencies precluded the generation of estimates. For example, health costs associated with oxidant-related air pollutants are not estimated because of the lack of data.

The same costs are distributed among sources on the basis of the relative level of pollutant emissions. Damages of \$6.1 billion in 1970 are attributed to the general source category, fuel combustion in stationary sources. Damages of \$4.0 billion are attributed to industrial process losses, \$1.1 billion to transportation, \$0.4 billion to both the agricultural burning and the miscellaneous categories and \$0.3 billion to solid waste disposal.

Although estimates are obtained and presented, the reader is cautioned concerning their use. The estimate of air pollution damages of \$12.3 billion is not to be taken as absolute, but is to be considered as indicative of the seriousness of the air pollution problem. The range of $6.1 to $18.5 represents the significant uncertainty in which the "best" estimate of $12.3 billion should be couched. Limitations of gross damage estimates are spelled out in greater detail in the paper. There is certainly at least one significant limitation: many benefits to be gained from air pollution control are not yet amenable to quantification in dollars and cents. Thus, the decision framework set up in the paper is designed to take this limitation into consideration.

While these estimates provide some basic justification for environmental policies and programs, aggregate point estimate offer little policy information for setting environmental standards. The research identified in this report needs to be extended to determine more accurate dose-response relationships, i.e. damage functions, and the economic value of the receptor response over a range of pollution levels. Such information would be very useful for decision-making in matters relating to environmental management.

TABLE 19
NATIONAL ESTIMATES OF AIR POLLUTION DAMAGES (UNADJUSTED) 1970
($ billion)

Effect	Range of Damages Low	High	"Best" Estimate
Aesthetics and soiling[a]	3.4	8.4	5.9
Human Health[b]	1.6	7.6	4.6
Materials	1.3	3.1	2.2
Vegetation	0.1	0.3	0.2

[a]Property value estimator.
[b]Does not include estimates of losses attributable to oxidant-related air pollutants because of data limitations.

TABLE 20
ESTIMATES OF NATIONWIDE EMISSIONS, 1970*
(millions tons/year)†

Source Category	CO	Part.	SO_x	HC	NO_x	Total
Transportation	11.0	0.7	1.0	19.5	11.7	143.9
Fuel combustion in stationary sources	0.8	6.8	26.5	0.6	10.0	44.7
Industrial process losses	11.4	13.3	6.0	5.5	0.2	36.4
Solid waste disposal	7.2	1.4	0.1	2.0	0.4	11.1
Agricultural burning	13.8	2.4	Neg.	2.8	0.3	19.3
Miscellaneous	4.5	1.5	0.3	4.5	0.2	11.0
Total	148.7	26.1	33.9	34.9	22.8	266.4

**Source:* J. H. Cavender, D. S. Kircher, and A. J. Hoffman, "Nationwide air pollutant emission trends 1940–1970," Publ. No. AP-115, Environmental Protection Agency, Research Triangle Park, N.C., Jan. 1973.

†*Editor's Note:* The original entry at this location appears to have been incorrect because of a misprint. I have replaced it with what I believe to be the correct entry.

TABLE 21

NATIONAL COSTS OF AIR POLLUTION DAMAGE, BY POLLUTANT AND EFFECT, 1970

($ billion)

Effect	SO_x			Particulate			O_x[a]			CO	Total		
	Low	High	Best	Low	High	Best	Low	High	Best	Best	Low	High	Best
Aesthetics & soiling[b,c]	1.7	4.1	2.9	1.7	4.1	2.9	?	?	?	*	3.4	8.2	5.8
Human health	0.7	3.1	1.9	0.9	4.5	2.7	?	?	?	?	1.6	7.6	4.6
Materials[c]	0.4	0.8	0.6	0.1	0.3	0.2	0.5	1.3	0.9	*	1.0	2.4	1.7
Vegetation	*	*	*	*	*	*	0.1	0.3	0.2	*	0.1	0.3	0.2
Animals	?	?	?	?	?	?	?	?	?	*	?	?	?
Natural environment	?	?	?	?	?	?	?	?	?	?	?	?	?
Total	2.8	8.0	5.4	2.7	8.9	5.8	0.6	1.6	1.1		6.1	18.5	12.3

Notes:

[a]Also measures losses attributable to NO_x.

[b]Property value estimator.

[c]Adjusted to minimize double-counting.

?Unknown.

*Negligible.

TABLE 22

NATIONAL COSTS OF POLLUTION DAMAGE, BY SOURCE AND EFFECT, 1970

($ billion)

Effects	Transportation	Stationary source fuel combustion	Industrial processes	Solid Waste	Agricultural burning	Misc.	Total
Aesthetics & soiling	0.2	3.1	2.0	0.1	0.2	0.2	5.8
Human health	0.1	2.2	1.7	0.2	0.2	0.2	4.6
Materials	0.6	0.8	0.3	*	*	*	1.7
Vegetation	0.2	*	*	*	*	*	0.2
Total	1.1	6.1	4.0	0.3	0.4	0.4	12.3

*Negligible.

The Social Costs of Producing Electric Power from Coal: A First-Order Calculation

M. GRANGER MORGAN, BARBARA ROSE BARKOVICH, AND ALAN K. MEIER

***Abstract*—A methodology is discussed for quantitatively computing the social costs, or external diseconomies, which result from the production of electric power in conventional coal-fired steam electric plants. With the available data, and our present level of understanding, it is possible to obtain preliminary numbers which place the social cost for the technology of the mid and late 1960's at $\geq 11.5 \pm 2$ mills/kWh, somewhat more than the price of bulk power at the plant bus bar. In applying controls to limit the social costs, control costs are incurred. If the optimum level of control is taken as that level at which the sum of the social costs and the control costs is minimum, then we estimate the total social and control costs with optimum control as $\geq 4.5 \pm 1.5$ mills/kWh and the costs of controlling to that level as $\geq 3 \pm 1$ mills/kWh. These numbers will probably be reduced, and the optimum levels for control increased, as new technologies are developed. The paper is limited to a straightforward development of social costs. No attempt is made to develop policy implications or to draw broad conclusions on the basis of the costs which are derived.**

Significantly most accurate estimates of the social costs of coal-fired electric generation will require large improvements in data, especially in the areas of air pollution and land use costs, as well as some fundamental conceptual progress in techniques for treating value orientation.

Manuscript received November 27, 1972; revised May 8, 1973. This paper had its origin in an undergraduate research seminar in the social costs of electric power held at the San Diego campus of the University of California.

M. G. Morgan was with the Department of Applied Physics and Information Science, University of California at San Diego, La Jolla, Calif. He is now with the National Science Foundation, Office of Computing Activities, Washington, D. C. 20550.

B. R. Barkovich was with the Department of Applied Physics and Information Science, University of California at San Diego, La Jolla, Calif. She is now with the State University of New York, Program for Urban Policy Sciences, Stony Brook, N. Y.

A. K. Meier was with the Department of Applied Physics and Information Science, University of California at San Diego, La Jolla, Calif. He is now with the University of California at Berkeley, Berkeley, Calif.

Introduction

RECENT YEARS have witnessed a growing public awareness that the market price of many of the goods and services we consume does not adequately reflect the societal costs or external diseconomies of their production. Most environmentalists, and indeed many others, would place electric-energy production high on the list of such socially costly activities. However, efforts to quantify the social costs of producing electric power are almost non-existent.

As part of a recent research seminar at the University of California at San Diego we attempted to quantify as many of the social costs as we could which result from the production of electric power from coal. The results are shown in Table I. While these results are tentative, we were surprised to discover how much progress is possible even with today's incomplete data.

We limit ourselves to coal-fired plants for several reasons. While the societal costs of nuclear-power generation may not yet be fully understood or recognized, and while the future value of scarce petroleum fuels as a source of organic materials may be underestimated, our present belief is that

Reprinted from *Proc. IEEE*, vol. 61, pp. 1431–1442, Oct. 1973.

TABLE I

SUMMARY OF THE SOCIETAL COSTS OF PRODUCING ELECTRIC POWER WITH CONVENTIONAL COAL-FIRED STEAM ELECTRIC PLANTS

	Cost to Control to Optimum Level	Total Societal Costs with Optimum Controls	"Uncontrolled" or Baseline Costs
Strip mining and other extraction related land use costs	0.1–0.4	≥(0.4–0.8)	>1
Mine health and safety (a)	≤0.26	~0.26	≥0.26
Other social costs such as "Appalachia"	unknown	unknown	unknown
Sulphur oxides (a)	~1.8	~2.3	~6
Particulates (a)			~(2–3)
Nitrogen oxides (a)	~(0.07–0.1)	~(0.16–0.26)	~(0.3–1)
Cooling (a)	~(0.4–1.0)	~(0.4–1.5)	≥1.0
Heavy metals: Hg, Ra, Th, etc.	unknown	unknown	unknown
Land use, esthetics, global effects, etc.	unknown	unknown	unknown
	≥3±1 mills/kWh	≥4.5±1.5 mills/kWh	≥11.5±2 mills/kWh

Note: The right-hand column shows the estimated societal costs in mills per kilowatthour without major control. With the exception of recently controlled plants, these numbers are typical of performance in the late 1960's. The left-hand column shows the estimated costs $C_c(l_{cont})$ to achieve optimum control. The center column shows the estimated total societal costs $C = C_s(l_{cont}) + C_c(l_{cont})$ which are incurred with this level of control. A discussion of the limits of confidence which we place on most of these numbers is available in the text. All of these numbers could be significantly affected by an improved understanding of damage effects and by future developments in control technology. As noted in the text, several entries (a) involve our lower bound estimate of the "costs" of human death, sickness, and injury. Many readers may wish to factor in their own additional costs above these estimates.

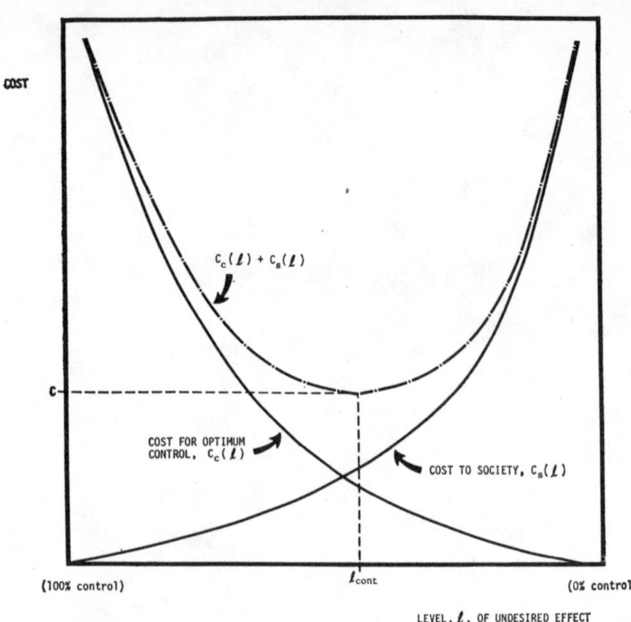

Fig. 1. The optimum level of control l_{cont} is that level for which $C_c(l) + C_s(l)$ is minimum [4]. The total cost of achieving that control is C. $C_c(l)$ is drawn as continuous but may not be in some cases, because of alternative discrete technologies. Ordinarily, $C_s(l)$ will go to zero as l goes to zero and $C_c(l)$ will go to zero as l becomes very large. However, there are processes whose low-level effects can be viewed as a societal benefit so that in some cases $C_s(l)$ may be negative for small l. Likewise $C_c(l)$ can in some cases go negative for large l.

coal-fired electric plants represent the most costly of the current major generation technologies and as such describe an upper limit for the societal costs of electric-power generation. We will exclude from our discussion such global problems as the impact of CO_2, particulates, NO_x, and heat dissipation on atmospheric circulation, although these effects may potentially prove far more costly to man than the relatively localized effects which we consider. We will also exclude a consideration of the "cost" imposed by the impetus to further growth and development which is provided by the widespread availability of low-priced energy. Finally, no consideration will be given to questions involving resource depletion and nonenergy uses of coal.

Approximately 60 percent of the installed electric generating capacity in the United States today is in the form of coal-fired fossil plants and approximately 46 percent of the electricity generated comes from coal. Estimates of installed capacity for the year 2000 are still as high as 40 percent [1], [2]. These facts, coupled with the severe political difficulties which have characterized the nuclear-power industry in recent years [3], lead us to believe that conventional coal-fired plants will continue to supply a significant portion of our total demand at least through the end of the century.

Ideally, in studying the societal cost of some technology, one would like to develop a description of the societal cost function $C_s(l)$ and the control cost function $C_c(l)$ as a function of the level l of the effect being considered [4]. In many cases a description of the control costs may also involve a consideration of the initial capital investment. Processes involving low initial investment may frequently involve relatively high unit control costs ("cheap and dirty") while at least some more capital intensive technologies may involve somewhat lower unit control costs. Hence one must optimize expenditure between initial plant costs and later control costs in deriving $C_c(l)$. If a description of $C_s(l)$ is available, the optimum level of control l_{cont} can be found by solving $d/dl\{C_c(l) + C_s(l)\} = 0$ for l, see Fig. 1. Of course, this operation assumes that $C_c(l)$ is continuous. With real control technologies this may in fact not be true, although the data available in this paper suggest that it is a reasonable assumption for the computations we will perform.

The total societal cost of a technology at the optimum level of control is $C = C_s(l_{cont}) + C_c(l_{cont})$. In order to justify implementation we must require $C \leq B$, where B represents total societal benefits. This analysis is, of course, complicated considerably in cases where C and B represent the sum of a number of effects which do not apply to all persons with equal magnitude. Society must essentially develop some weighting function which can normalize individual costs or benefits to societal costs or benefits. In its simplest form such a function might weight the cost or benefit enjoyed by each individual by one over the total population. We have a legal and moral tradition in this country to protect the rights of the individual, and have developed a body of law which places limits on the range of discrepancy in costs or benefits which will be tolerated. Thus while certain costs in our analysis, such as coal mine health and safety costs, will appear to be quite small on an overall basis, they cannot be ignored because they represent sharp deviations from the desired uniform distribution of individual costs.

We will be unable to answer the ultimate question . . . "are the total societal costs less than or equal to the total societal benefits, for a new increment of installed coal-fired capacity?" For consumers with highly elastic demand curves the answer is clearly no. However, because the computation of elasticities for electric-power consumption is still very much in its infancy we are unable to reach any general conclusion [5].

Computations have been done only for the extraction and conversion phases of the generation process. Transportation and differential transmission costs are not treated, nor do we address the relatively minor problem of ash disposal. Computations are normalized to a 1000-MW plant at a 70-percent plant factor. Fixed costs are taken as 14 percent [6]. We have attempted not to include the backlog of unsupported social costs left from previous production. Our baseline or "unregulated" social cost calculations are based on the technology of the mid and late 1960's. Because of recent environmental controls newer plants are, of course, cleaner.

As a point of reference for the cost numbers derived, the average bus bar price for power from coal-fired electric plants in 1970 was between 7 and 8 mills/kWh [7]. Since 1970, the price has increased significantly.

Extraction Costs

A number of social costs are incurred in the process of extracting coal from the ground through surface or underground mining. We will discuss these costs under the three categories of land use costs, health and safety costs, and human environmental costs.

Land Use and Related Costs of Mining

It is difficult to determine even in a qualitative way the value of the pleasure, peace of mind, inspiration, and similar benefits derived by an individual from a piece of land. This return is a strong function of individual and cultural background, or in the vocabulary of Kluckhohn and Strodtbeck [8], the individual's "value orientation" function. We know intuitively that things that we do to land that are quickly and cheaply undone will not have too great a societal cost for at least most value orientation functions. Mow down a beautiful field of buttercups and, while some people may be sad, most will not view it as a great loss because they know that next year they will grow back again. The things that we do to land that are not so quickly or cheaply undone have potentially far greater societal costs.

As with most environmental changes of this type we can represent the recovery or reclamation process as a family of recovery curves on a two-dimensional cost–time surface. For example, Fig. 2 shows our *very crude* estimate of the potential cost curve for lumbering off a mature western softwood forest. Large initial reclamation expenditure can speed recovery but not past some minimum recovery period. Lesser expenditures result in progressively longer recovery periods. Full recovery is reached with the establishment of a new, essentially mature, ecosystem. Similar plots can be drawn for most such recovery processes, but since some processes are exceedingly difficult to reverse, there is no assurance that the family of curves will converge to a potential cost function for "complete" recovery within societally finite limits of time or investment.

The actual societal cost of any environmental change can be represented by weighting the potential cost function by an appropriate value orientation function. For any given situation, the potential cost function is essentially fixed, given fixed technical capabilities, but the weighting applied by the value orientation function may be very different for different social groups or individuals. Thus for example, the value of Black Mesa as coal far outweighs its value as a mesa if we weight with the value orientation function of some Anglo power officials, while the reverse is true if a "traditional" Indian value orientation function is used in the weighting.

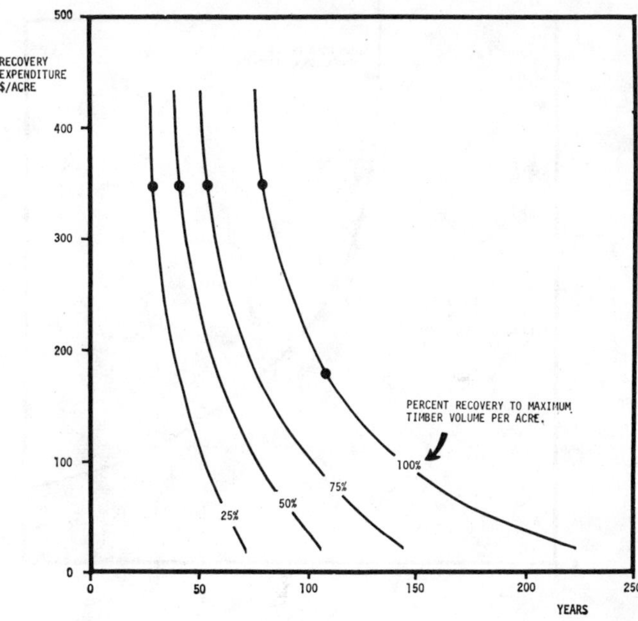

Fig. 2. An example of the recovery or potential cost function description of an environmental change. This family of curves represents our estimate of the recovery of a typical western pine forest to the state of maximum timber volume. They are based on published data [9] and conversations with forest service officials and forest management experts. But because there are great variations with location, climate, soil, and other factors, they should be viewed as at best approximate.

Since value orientations differ, and since they may change in time, we must be careful about proceeding with operations whose potential cost curves lie far out from the origin, since what is acceptable today may not be acceptable in the future.

A classic example is provided by the case of hydraulic mining for gold in the west. When this technique came into use more than a century ago virgin hillsides and clear running streams were perceived as essentially infinite resources. The wasted hillsides and streams which hydraulic mining produced were considered a small and very acceptable price to pay for the resulting gold. Today, after a considerable shift in value orientation functions, this is no longer generally true. Indeed, in most areas the technique has been totally forbidden. However, because the potential cost function for this process lies far out from the origin on the cost–time surface, the unrecovered heaps of gravel will remain for many years to come.

Unfortunately, the development of conceptual techniques for treating value systems, in even semiquantitative ways, is a seriously underdeveloped field of research. One way in which we can approximate the calculation we propose is by assigning value to land in the form of a rental price which is charged for holding it in an unreclaimed status. But, as we will see in the discussion that follows, this approach leaves much to be desired.

Data on the recovery costs for strip-mined land are sketchy, but appear to fall into two categories, recovery to agricultural land and recovery to forested land. German experience in the Rhineland brown-coal lignite fields suggests

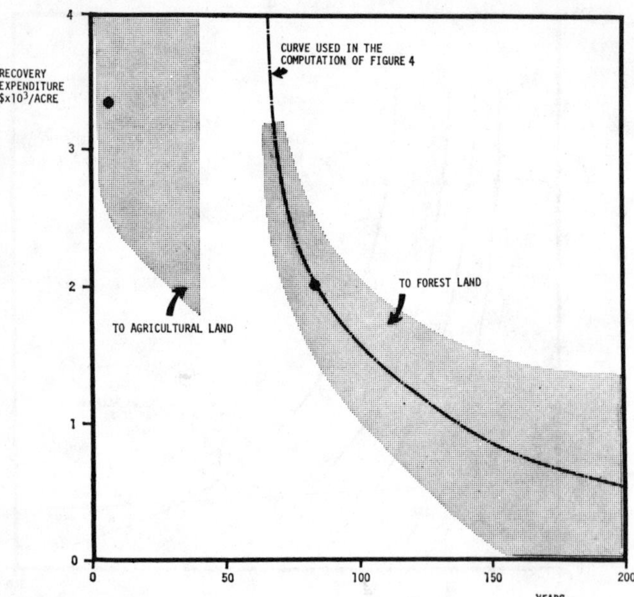

Fig. 3. The shaded curves show our estimate of the recovery or potential cost functions to return strip-mined land to agricultural or forest uses. The curves are based on the data which are summarized and referenced in the text. Note that for very low levels of recovery expenditure some types of land may fail to recover at all in societally finite periods of time.

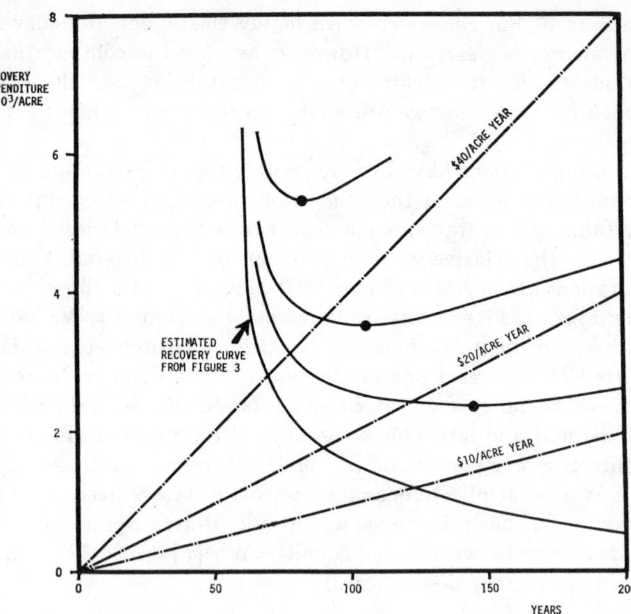

Fig. 4. Computations of the optimum recovery expenditures and times for strip-mined land obtained by combining cumulative rent curves with the recovery curve to forested land from Fig. 3. Note that this optimization procedure is not the same as that described in Fig. 1 and carried out in Figs. 6 and 7.

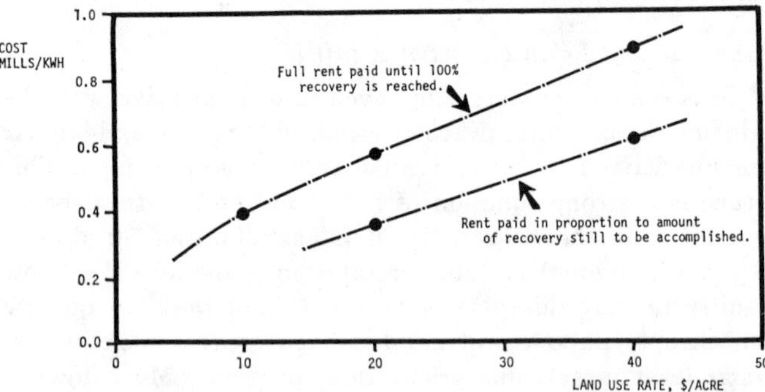

Fig. 5. Recovery costs for strip-mined land as a function of annual land rental rate. The upper curve is derived from Fig. 4. A slightly more complicated derivation, based on the assumption that rent is paid in proportion to the amount of recovery still to be accomplished, yields the lower curve.

a recovery figure of $3000 to $4500 per acre to obtain agricultural land in about five years [10]. Grading costs in the United States appear to run $500 to $3000 per acre [11]–[13], reforestation planting costs average around $300 per acre [11], [14], mulching runs between $800 and $1000 per acre and hydro-seeding between $60 and $100 per acre [15]. These numbers depend very critically upon the local environment. The recovery processes for Appalachian hardwood forests and southwestern sage country are clearly very different. Good data for recovery of arid western lands are not yet available.

Using these very approximate recovery numbers and a bit of basic forestry [9], [16], we have produced two rough estimates for recovery curves as shown in Fig. 3.

If we could agree on a rental price to be charged while strip-mined land is held in its unreclaimed state we could combine the cumulative rent curve and recovery curve to obtain the cost of optimum control. Until recently, Federal grazing right lease rates have been a few dollars per acre and rates for mineral rights have been of the order of $10 per acre [17]. These prices are paid only for the period during which the land is in use. Suppose that we arbitrarily say that $10 per acre is to be charged indefinitely until strip-mined land is recovered. If this rent is future discounted at 4 percent the minimum costs are incurred at times greater than 100 years, which says the companies would rather pay the rent and do no recovery. This is of course the result one would expect and indeed the land has historically not been recovered.

Today a marked shift in value orientations is under way. Land of all types is being increasingly perceived as a scarce and very valuable resource. There is now a sizable block of people, characterized by leaders such as Congressman Ken Heckler, who are undertaking serious efforts to totally forbid all strip mining. At the same time, Federal mineral lease prices have undergone a spectacular rise since 1965. As of late 1970 successful bids in Wyoming were running in excess of $500 per acre [18].

Finally, the situation is further complicated by factors such as the high degree of correlation between western coal fields and American Indian lands.

In order to obtain at least an approximate estimate of the cost of optimum control for strip-mined land we will perform the calculation over a range of present valued rental rates from $10 to $40 per acre. Adding the curves for cumulative rent payments and performing the appropriate computation as shown graphically in Fig. 4 we arrive at an optimum control cost for the $10 per acre rate of 2.5×10^3 per acre which translates to just under 0.4 mill/kWh as shown:

$$\frac{(2.4 \times 10^3 \text{ \$/acre})(640 \text{ acres/mi}^2)}{(1.8 \times 10^6 \text{ tons/mi}^2 \text{ [19]})(2.2 \times 10^3 \text{ kWh/ton})}.$$

Performing a similar computation for several higher land rental rates yields the upper curve in Fig. 5. If instead of charging the full rent for the land until it is completely recovered we charge rent in proportion to the amount of recovery still to be accomplished, a slightly more complicated series of computations yields the lower curve in Fig. 5.

These computations are based on recovery to forested land. For present value rental rates significantly in excess of \$50/A the recovery process becomes exceedingly expensive and is dominated by the finite recovery time of forest land. A similar set of numbers could be generated for agricultural land such as is encountered in Illinois and Ohio. Because of both the higher land values and higher recovery costs the resulting numbers would be larger.

As a first approximation we estimate control costs for strip-mined land as 0.1–0.4 mill/kWh, total costs with optimum controls as $\geq$(0.4–0.8) mill/kWh, and cost without control as $>$1 mill/kWh. We have good confidence in the lower limits on these numbers. The upper limits are uncertain.

Several other types of environmental impact from mining include land subsidence, acid water drainage, and refuse banks.

Since the start of coal mining in this country approximately seven million acres of land have been undermined, of which at least one third have subsided [20]. Surface subsidence may occur at any time from within a few days to many years after the coal is extracted. Land has subsided even when mines are as much as 2400 ft below the surface [21]. While careful engineering, including the practice of leaving substantial amounts of coal in the ground as supporting columns, can help to reduce subsidence, there is at the moment no certain technical solution for this problem short of refilling the mine with material of adequate bearing strength.

The social cost of subsidence is highly dependent on the present and potential uses of the undermined land. To date, most has been grazing and forest land with no current major alternative use, but at least 200 000 acres of urban and suburban land have been undermined [21].

A minimum cost estimate is available from a Bureau of Mines study of a twelve-county area of Western Pennsylvania [22]. This study attempted to compute the total costs due to subsidence, both internal (i.e., reflected in the price of coal because of Pennsylvania's subsidence control laws) and external (i.e., societal). Based on the year 1968 the study obtained internal costs of 0.04 \$/ton and external or societal costs of 0.05 \$/ton. If we generalize from this result and apply the 0.05 \$/ton number to the rest of the country, recognizing that 53 percent of the coal mined underground is used for steam electric generation, we compute a total cost of 0.01 mill/kWh. Since not all states have Pennsylvania's control laws, this may represent an underestimate.

Another estimate of the costs of subsidence can be obtained by looking at the cost of insurance formerly sold in the State of Pennsylvania. Here the property owner paid to have half of the coal left in the ground under his property at rates which ranged from 0.35 \$/ton for homes, churches, and schools, up to 1.50 \$/ton for some commercial properties. Since 1959 the State of Pennsylvania has offered similar coverage to homeowners (\$25 000 coverage at 60 \$/year) [21]. Using the 0.35 \$/ton number we obtain a cost of 0.16 mill/kWh.

Mine drainage, from both active and abandoned mines, also contributes to the social costs of coal mining. Mining activity exposes pyritic materials which become oxidized. Water, often introduced through the fractures created by mine subsidence, leaches out acid and metallic salts which are then carried into the local ground water system and on into the streams. In Appalachia the water quality of 10 500 mi of streams is affected; 5700 mi are continuously acidic [23]. Inactive mines produce 78 percent of the total acid drainage with underground mines contributing the major share (52 percent of the total). Active and abandoned surface strip mines contribute a rather modest 12 percent.

It is virtually impossible with our present capabilities in the fields of aquatic ecosystems and human value systems to estimate either the potential cost function or the actual societal costs of mine drainage. However, a few pieces of incomplete data are available.

The Appalachian Regional Commission has estimated the direct economic damages of acid drainage to municipal water supplies, navigation, and other public and private facilities in Appalachia. Their value is 3.5×10^6 \$/year [23]. No effort was made to place a cost on the ecological and esthetic damages or the restricted options for future growth. These are costs borne very heavily by a few rather small regions.

In sharp contrast to this number is the estimated cost of control. In 1969 the U. S. Department of the Interior estimated that to reach 95-percent control of all acid drainage a capital investment of \$$6.6\times10^9$ over a period of 20 years would be required [23]. Presumably some of this would go into one-time costs such as mine sealing and runoff diversion, but much would have to go into treatment plants which have significant operating costs. The Appalachian region suffers about three quarters of the acid drainage in the United States [21]. Using the Appalachian Regional Commission's number of 5700 mi of stream we can estimate the required capital expenditure for control as 8.8×10^3 \$/mi. While we cannot draw the appropriate potential cost curve, clearly it lies far out from the origin. As with the results of hydraulic mining, acid drainage will be with us for many years. Because it is so difficult to control once it exists, it is clear that in the future considerable efforts should be made to plan mining operations so that acid drainage will not later develop.

As of 1970, 73 billion tons of coal had been mined from United States underground mines. Approximately 2.2×10^3 kWh of electric power can be generated from a ton of coal. With the Department of the Interior estimate we can thus conclude that the control cost for acid drainage would run something of the order of 0.04 mill/kWh. On a kilowatt-hour basis damage costs are probably smaller . . . but they are borne very unequally by a few relatively small regions.

Refuse banks are another land use problem which results from underground coal mining. One analyst [20] has estimated that all of the coal-mine refuse accumulated to date would cover 23 000 acres if piled 75 ft deep with a density of one ton per cubic yard. Of course, in terms of total land use a few tens of thousands of acres is a trivial amount of land. The problem is made more severe by the fact that many of these unsightly refuse piles are located close to population centers. One study of the Pennsylvania anthracite region located 863 banks within a few miles of traffic centers which have a total population of 625 000 people [24]. In addition to being unsightly the banks are subject to fires. In the study just cited, 27 of the 863 banks surveyed were burning. A 1963 national survey located a total of 495 burning banks [25]. A 1968 study found 292 [26]. To date, 55 deaths have been attributed to burning refuse banks and considerable air pollution and nuisance problems are created. Techniques for controlling and extinguishing fires have been developed and cost of the order of 0.40 \$/cubic yard [25].

Finally, refuse banks can pose very severe dangers, as in the case of the recent West Virginia flood disaster, when they block natural drainage and inadvertently become earthen dams of uncertain integrity.

We will make no effort to estimate a dollar value for the social costs of refuse piles. Because they occupy a relatively small land area the per kilowatthour costs, based on any reasonable land use rate, are small. But these costs are borne in a very unequal way, principally by the residents of the Appalachian coal district, and for this reason can probably not be ignored.

Because refuse piles and subsidence generally occur in close proximity one obvious control approach for both problems is to replace the refuse in the underground voids. In a study conducted last year in northern Pennsylvania the MITRE Corporation [27] made a detailed investigation of this approach in the context of the Lackawanna Valley. A total of 80 banks with a volume of 76.6×10^6 cubic yards were investigated. Some banks had been previously burned and thus were marketable as a building aggregate. Some also were appropriate for use as a low-grade fuel in a newly planned steam electric plant. The balance, 24.5×10^6 cubic yards in 48 piles, was identified for removal. A study of alternative transportation modes led to the conclusion that slurry pipelines were most economical with a cost of about 0.06 \$/ton·mi. Average moving costs worked out to about 0.20 \$/cubic yard. MITRE did not estimate the costs of the required underground exploration, drilling, crushing, loading and unloading, water, and right of way. Data in a recent Bureau of Mines report [28] suggest prices of 0.03 to 0.20 \$/cubic yard for just spreading and compacting. For this estimate we will take the total costs of the process as 0.50 \$/cubic yard. Using the MITRE estimate that only 32 percent of the total volume must be placed underground, we obtain:

$$\frac{(2.78\times10^9 \text{ yd}^3 \text{ of accumulated refuse to date [20]})(0.32 \text{ percent to go underground})(0.50\ \$/\text{yd}^3)}{(7.3\times10^{10} \text{ tons of coal mined to date})(2.2\times10^3 \text{ kWh/ton})}$$

or a cost of 2.8×10^{-3} mills/kWh.

The cost entries in Table I for all extraction related land use are essentially the strip-mine costs since these appear to dominate all other land costs.

Coal Mine Health and Safety

Coal miner's pneumoconiosis or "black lung" disease is the most serious occupational health hazard associated with underground coal mining. Estimates of the number of working and retired coal miners suffering from this disease range from 38 000 [29] to 125 000 [30]. Complicated pneumoconiosis causes disability and seriously shortens life expectancy. There is no cure. Simple pneumoconiosis may not be disabling but continued dust exposure often leads to the complicated form. Epidemiological studies indicate that symptoms are related to coal variety and thus show geographic dependence.

Incidence of pneumoconiosis is highly correlated with levels of dust exposure. It is essentially absent in mines with dust concentrations of less than 2 mg/m³ [31]. The postwar period witnessed a marked increase in the use of automated continuous mining equipment which has resulted in significant increases in dust output. A Bureau of Mines study of 29 mines concluded in April of 1969 found that average dust levels for persons working close to such automated equipment are about 8 mg/m³ [31]. The exposure may be significantly higher in some mines.

The Federal Coal Mine Health and Safety Act of 1969 should have the long-term effect of essentially eliminating pneumoconiosis as a major occupational health problem. Under this act a maximum dust level of 3 mg/m³ was established, and this value has dropped to 2 mg/m³ since December 31, 1972. Improved ventilation and wetting should allow most mines to reach these levels without special exotic technologies. In one very unusual mine, U. S. Steel is now successfully using cryogenic respirators for men in high exposure zones [32], [33].

Lucille Langlois of the Appalachian Commission has performed a careful and clearly conservative computation of the societal costs of pneumoconiosis [34]. She obtains a social cost of 5×10^7 \$/year which translates to about 0.029 mill/kWh as shown:

$$\frac{(5\times10^7\ \$/\text{year})(0.53 \text{ fraction of coal production used for electric power})}{(2\times10^{12} \text{ kWh/year})(0.46 \text{ portion of electric power production from coal})}.$$

In a similar way her computed control cost of 3×10^7 \$/year yields a value of 0.017 mill/kWh.

That these numbers are conservative can be seen from the following computation. Data on the number of active miners are available [35]. There are no hard numbers for retired miners although estimates have been made for 1968 which range from 120 000 to 400 000 [29].

A 1963–1964 Public Health Service sample study [36] shows a 3-percent incidence of complicated pneumoconiosis in working miners (ages ranging from 35 to 64) and a 9.4-percent incidence in inactive miners (aged 45 to 64). A 1969 Public Health Service sample study [18] indicates a similar percent occurrence of complicated pneumoconiosis, 3 percent in active and 9 percent in retired workers. From these data we compute an approximate estimate of the number of retired miners by displacing the active miner curve horizontally until it intersects an estimated 1968 value of 250 000. The required displacement turns out to be about 15 years, certainly shorter than most working lifetimes, but probably not an unreasonable number if one includes early deaths, transfer to other employment, and similar factors. From these two curves, using the incidence rates previously cited, we obtain $N(t)$, the total number of black lung cases as a function of time.

Each year, $N_n(t)$ new cases are added, and some old cases die. We have rather arbitrarily chosen a life expectancy of fifteen years for persons entering the complicated black lung population. The number of persons in this population at time t_0 is thus given by

$$N(t_0) = \int_{t_0-15}^{t_0} N_n(t)dt.$$

Expanding $N_n(t)$ in a Taylor series expansion and evaluating the definite integral, we obtain

$$N(t_0) = 15\left[N_n(t_0) - \frac{15}{2}\frac{dN_n(t_0)}{dt}\right]$$

where higher order derivatives have been taken as zero, that is, $N(t)$ is approximated by a straight line.

This simple computation estimates 800 new cases in 1972. There is obviously no available way to convert death or disability numbers into a dollar cost which will satisfy most people. Econometricians typically use the sum of direct costs plus foregone earnings. Perhaps a better measure is what people would be willing to pay to reduce morbidity and mortality, yet even this approach would be objectionable to many. We will not attempt to establish the total "cost" of deaths and injuries in this paper. Instead we will compute a lower bound with which most people can agree and then state costs as greater than or equal to this lower bound. In the case of black lung victims we estimate the lower bound by reimbursing each man or his dependents with a lump sum of 15 years' pay at an average rate of \$11 000 per year. The lower bound thus comes to 0.08 mill/kWh. Less than 25 percent of this cost is now being supported in the form of disability payments.

Pneumoconiosis is by no means the only disease to which coal miners show increased susceptibility. There is evidence of increased incidence of chronic bronchitis and emphysema, pneumonia, heart disease, pleurisy, and other diseases caused by the breathing of coal dust though quantitative data are limited. We feel justified in stating the health costs of underground mining as ≥ 0.08 mill/kWh.

Another hazard of the coal mining industry is mine accidents. In 1968 there were 136 300 working miners in all types of coal mining. In that year 311 were fatally injured, and 9495 nonfatally injured [37]. They produced 561 million tons of coal of which 53 percent was used for the generation of 6.6×10^{11} kWh of electric power. Thus there were 1.4×10^{-9} nonfatal injuries per kilowatthour produced in 1968. If the families of those fatally injured were compensated with a lump sum of \$220 000 (20 years at \$11 000/year) the lower bound on the cost would be 0.05 mill/kWh. Similarly, a two-year compensation for nonfatal injuries (which can range from minor to totally disabling) produces a cost of 0.16 mill/kWh. As with black lung, actual compensation rates lie far below these numbers. This compensation varies considerably from state to state and is only partly reflected in the cost of coal.

The cost entry in Table I for mine health and safety is computed as follows:

$$\geq [(0.08)(0.62 \text{ fraction of coal mined underground}) + 0.05 + 0.16] \quad \text{or} \quad \geq 0.26 \text{ mill/kWh.}$$

Human Environmental Considerations

Coal mining in this country is intimately connected in the public mind with the economic stagnation and human misery of Appalachia. The development of these conditions has had a long and complicated history. It is not clear to us how using coal today affects these conditions. It is clear that if all coal mining in the area were replaced by other equivalent income sources, a major problem would still exist. Solving the problems of Appalachia represents one of our major societal responsibilities, we believe, but while these problems are connected with coal, and their solution will be costly, we are unable to say what if any portion of these costs should be assessed against electric power.

Conversion Costs

Air pollution effects are the largest of the several costs associated with the conversion from coal to electric power. The most comprehensive study of the costs of air pollution currently available is an internal EPA study done by Barrett and Waddell and available as Appendix I–J in the 1972 OST *RECAT Report* [38]. This work is essentially an integration of a wide range of previous studies including the important work of Lave and Seskin [39] on health effects. We will draw upon it extensively in the analysis which follows.

Barrett and Waddell have attempted to normalize all of their data to 1968. Their two most striking conclusions, that total air pollution damages in 1968 were $\$16 \times 10^9$ and SO_x damages were $\$8.3 \times 10^9$, have been widely used (often without citation) in government publications, including the President's 1971 Environmental Program compiled by the Council on Environmental Quality [40].

The Barrett and Waddell work is a careful and honest effort to do the best job possible with the data available. It suffers from the incomplete nature of the data. For example, contributions from health effects account for about 40 percent of the total costs computed but as Lave and Seskin explain, their dollar estimates "are surely underestimates of the relevant costs. The relevant measure is what people would be willing to pay to reduce morbidity and mortality . . . it seems evident that the value used for foregone earnings is a gross understatement of the actual amount. An additional argument is that many health effects have not been considered in arriving at these costs" [39].

Sulfur Oxides

Sulfur oxides are the single most damaging air pollutant produced by coal-fired electric plants. By combining the estimated 1968 total damage number of $\$8.3 \times 10^9$ with a knowledge of the distribution of coal use and power generation technologies in the country, we estimate the damage from sulfur oxides which result from coal-fired electric generation at 6 mills/kWh as follows:

$$\frac{(8.3 \times 10^9 \text{ \$/year national } SO_x \text{ damage})\left(\begin{array}{c}0.467 \text{ percent of all } SO_x \text{ which came from}\\ \text{coal-fired steam electric plants in 1968 [5]}\end{array}\right)}{(0.46 \text{ portion of electric power generated from coal})(1.4 \times 10^{12} \text{ kWh generated in 1968})}.$$

The cost of optimum control is more difficult to estimate. Clearly it is ≥ 0.26 mill/kWh; however, several simple computations based on the few hard numbers we can find leave us reluctant to cite an upper bound that is smaller than this.

There are several options available to reduce the level of sulfur oxides emitted from coal-fired electric plants. These include: 1) the use of low-sulfur coals, 2) the employment of special combustion methods and stack gas cleaning methods, 3) the preprocessing of coal to remove sulfur, and 4) the conversion of coal to some cleaner burning fuel such as gas or oil or the conversion to a fuel suitable for fuel cell use. Only the first two of these options appear to be reasonable for most of the industry for at least the next decade, although some washing and cleaning will be used. Today the use of low-sulfur coals adds roughly $1.50 to $3.00 per ton to the fuel price or something like 1 mill/kWh [1], [41]. However, reserves of low-sulfur coal are severely limited and in great demand not just in the United States but also in Japan and Europe. It is clear that the simple use of natural low-sulfur coal does not offer a major solution to the problem of sulfur oxides emissions.

Several techniques for removing sulfur oxides during and after the combustion process have been widely discussed, usually in qualitative terms. These include fluidized bed combustion and other advanced combustion technologies, limestone injection, limestone wet scrubbing, catalytic oxidation, magnesium oxide scrubbing, alkalized alumina absorption, copper oxide absorption, charcoal absorption, steam and ammonia injection, and several others [1], [7], [42]–[44]. Data on all but a few of these processes are extremely sketchy. To our knowledge, none has yet been proven entirely satisfactory in a major long-term demonstration plant. In Fig. 6 we have plotted what we believe to be the most reliable of the economic data in order to produce a rough estimate of the control cost curve. If we assume that the societal costs are a linear function of level we can then compute the desired optimization level. This computation indicates that at 1968 levels of sulfur oxides and using control cost data based on experience during the last couple of years, optimum control occurs somewhere between 90 and 95 percent. Total costs with optimum control are about 2.3 mills/kWh. And the control costs themselves are about 1.8 mills/kWh. All of these numbers are critically dependent on damage costs of SO_x, the assumption of a linear social cost curve, and possible technological changes in the status of SO_x control technology.

The present damage curve may not be the appropriate curve to use in considering the level to which one should control a plant which will still be operating in the year 2000. At least as appropriate an approach might be to use the estimated unregulated SO_x numbers for the year 2000 [1] and develop an optimization on this basis. The optimum level of control in this case appears to lie somewhere above 95 percent but neither our control cost curve nor the assumption of a linear damage curve is good enough to allow a meaningful estimate.

Nitrogen Oxides

Remarkably little is known about the nature or extent of the damage caused by another major air pollutant from coal-fired plants, the oxides of nitrogen. There has been accumulating evidence that nitrogen dioxide has direct adverse health effects and both NO and NO_2 contribute to the formation of oxidant in photochemical smog [48].

Barrett and Waddell [38] obtained a value of 7.72×10^8 \$/year for the cost of all NO_x air pollution in 1968 but the data upon which this number is based are far less complete than the SO_x data. Using this number we compute a cost of 0.17 mill/kWh as follows:

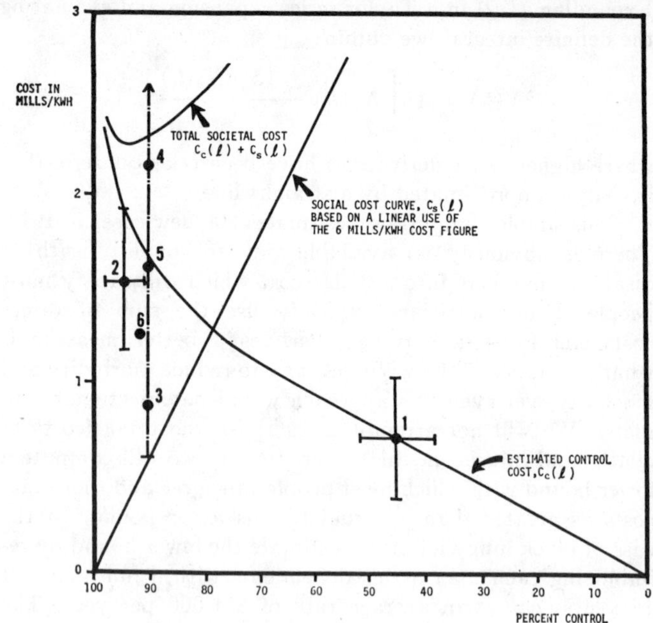

Fig. 6. Estimated curves for $C_c(l)$ and $C_s(l)$ for sulphur oxides. References for the points are as follows: 1) dry limestone injection [45]; 2) wet lime–limestone scrubbing [45]; 3) catalytic oxidation [1], [46]; 4) catalytic oxidation [1]; 5) copper oxide absorption [47]; 6) magnesium oxide scrubbing [1]. Error bars are our own, based on a subjective interpretation of the published results. $C_s(l)$ is based on a linear use of the 6-mills/kWh cost figure cited in the text, and as such probably represents an upper limit.

The nation's air pollution standards are supposed to represent a statement of the relative damage potential of the various pollutants. Of course they were established in the same absence of data that characterizes the Barrett and Waddell work, but they do represent an independent estimate which carries legal weight. Since the SO_x cost numbers are reasonably firm, we can estimate an NO_x cost using the "pindex" method developed by Babcock which weighs the relative effects of different major air pollutants [49]. This method assigns a tolerance factor to each pollutant to determine its relative harmfulness based on the value of its air quality standard as compared with those of the others. The synergistic effect of sulfur oxides when combined with particulates and the synthesis of oxidant from nitrogen oxides and hydrocarbons in the presence of sunlight are also taken into account. We have a sulfur oxides damage cost so that using the "pindex" method we can roughly compute a nitrogen oxides cost. The average ratio of SO_x to NO_x emissions in coal-steam electric generating plants is 5.2 to 1 [4]. If each is the limiting reactant in its particular effect then the pindex method yields a damage ratio of 2.8 to 1. In other words, if the social cost of SO_x damage is 6 mills/kWh then the social cost of NO_x is 2.2 mills/kWh. This result is valid only to the extent that present air pollution standards reflect equivalent damage levels.

There is considerable evidence that the local and regional air pollution damage effects from NO_x are overstated by present air pollution standards. At the same time, the data are so incomplete that we feel confident in arguing that the

$$\frac{(7.72 \times 10^8 \text{ \$/year national } NO_x \text{ damage}) \begin{pmatrix} 0.146 \text{ percent of all } NO_x \text{ which came from} \\ \text{coal-fired steam electric plants in 1968 [7]} \end{pmatrix}}{(0.46 \text{ portion of electric power generated from coal})(1.4 \times 10^{12} \text{ kWh generated in 1968})}.$$

Barrett and Waddell estimate represents a lower bound. The cost of unregulated NO_x emissions clearly lies between 0.17 and 2.2 mills/kWh. Our own estimate is that the actual value lies in the range of 0.3 to 1 mill/kWh.

All large-scale NO_x control technologies with which we are familiar appear to be limited to effecting various changes in the nature and geometry of the combustion process. Data from these processes are plotted as the solid points in Fig. 7. A recent private communication [51] leads us to believe that it may ultimately be possible to control NO_x from utility boilers to 90 percent using NH_3 in a catalytic converter at a cost of 0.16 mill/kWh. This tentative piece of evidence is plotted as the open circle in Fig. 7. If this estimate proves correct, then, as is indicated in Fig. 7, the optimum level of control will fall somewhere between 75 and 90 percent. Total costs with optimum control will fall between 0.16 and 0.26 mill/kWh. And the control costs themselves will lie between 0.07 and 0.1 mill/kWh. It should be obvious that these results may be significantly affected by an improved understanding of the damage effects of NO_x and by future developments in NO_x control technology.

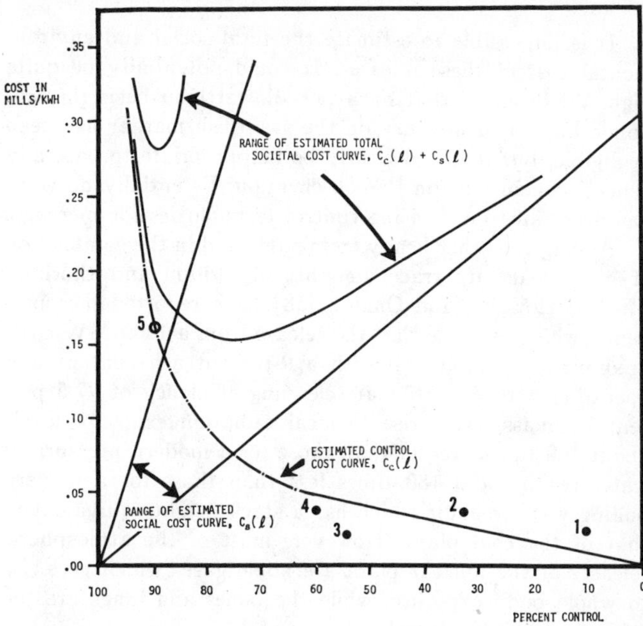

Fig. 7. Estimated curves for $C_c(l)$ and $C_s(l)$ for nitrogen oxides. References for the points are as follows: 1) water injection; 2) flue gas recirculation; 3) low excess air plus flue gas recirculation; 4) low excess air plus two-stage combustion [50]; and 5) estimate for an NH_3 catalytic converter [51].

Particulates

Particulate air pollutants are probably the most serious after sulfur oxides [49]. Coal-fired steam electric plants release over 5 million tons of particulate matter each year or roughly a quarter of all particulate air pollution in the United States [7], [53]. In the size region above about 5 μ, the greatest damage from particulates appears to arise from soiling and visibility effects. Because of the shape of the Junge distribution, these larger particles involve most of the mass of airborne particulates but only a very small fraction of the total particle number. Electrostatic precipitators, centrifugal filters, and mechanical filters, such as bag houses, can easily achieve efficiencies of 95 to 99 percent by mass. However, a 99.7-percent removal by mass in a conventional precipitator frequently corresponds to a removal of only about 30 percent of all particles by number.

Very small and very large particles are effectively blocked from entering the human lung. However, particles in the 0.5 to 5-μ region enter the lung with relative ease. Because even very good precipitators have only a limited effectiveness in the submicron range, a large portion of the lung-entering particles are not blocked. Small particles which enter the lung are of course damaging in their own right but are additionally damaging in that they frequently carry on their surface toxic gases and other unhealthful materials. There are a few proposals to improve the small particle efficiency of precipitators by using the particles as condensation nuclei, thus increasing their diameter. To our knowledge there are not yet field data available on these techniques.

Data adequate to allow an estimate of costs are not available for the general health effects of small particles from power plants. Like smoking and health, these effects are long term and statistically difficult to identify, which is of course not to say that they do not exist.

Control costs for large particles are well known. However, because virtually all sulfur oxides control systems require efficient particle filters, the sulfur oxides control numbers already include these costs. Control costs place a lower limit on the societal cost of particulates at about 0.1 mill/kWh.

The Barrett and Waddell analysis [38] yields a total cost of $\$5.9\times10^9$ for particulates in 1968 which translates to 1.8 mills/kWh.

We can make a simple direct estimate using data reported by Wilson and Minnotte in a U. S. Public Health Service study for Washington, D. C. [54]. During the period 1965–1966, 24 790 tons of particulates were released with a cost of $\$22\times10^6$. We thus compute a cost of 3.4 mills/kWh as follows:

$$\frac{(5 \times 10^6 \text{ tons of coal burned for electricity generation/year})(\$22 \times 10^6 \text{ Washington damage/year})}{(0.46 \text{ portion of electric power from coal})(2 \times 10^{12} \text{ kWh/year})(3.48 \times 10^4 \text{ Washington tons/year})}.$$

A repetition of the "pindex" calculation for particulates taking the SO_x to particulate ratio as 2.8 to 1 [4] yields a cost of 4.3 mills/kWh.

It appears that the social costs of particulate emissions with the modest controls which were in effect in the mid and late 1960's lay somewhere between our estimated control cost of 0.1 mill/kWh and an upper limit of 4.3 mills/kWh. We feel the most probable value is between 2 and 3 mills/kWh.

Heavy Metals

Most coals contain a wide range of metallic trace contaminants which are released by the combustion process [55]. From an environmental point of view the release of several heavy metals appears to pose the greatest problem.

Joensuu [56] has estimated that something like 23 percent of anthropogenic environmental mercury is released by the burning of coal. More recent work by Billings and Matson [57] has shown that most mercury from steam electric stations leaves the plant in the flue gas, with less than 4 percent trapped in conventional electrostatic precipitators or carried in fly ash.

It is impossible to estimate the total social and environmental cost of these releases. It could potentially be quite high. We do know that on a per-kilowatthour basis the economic impact of mercury on the swordfish market has been negligible, but it would clearly be inappropriate to base any general conclusions on this observation. Essentially no work has been done on devising control technologies for mercury.

Among the other heavy metals released in the combustion of most coals are trace amounts of radium and thorium. Martin, Howard, and Oakley [58] have reported measurements which indicate that the release from a 1000-MW coal-fired plant, burning coal with a 9-percent ash content and operating with a particulate cleaning efficiency of 97.5 percent by mass, gives rise to local radiation exposure levels about 400 times greater than those for a modern pressurized water reactor and 180 times less than those for a modern boiling water reactor which has a stack height equivalent to that of the coal plant. However, most of the atmospheric releases of the nuclear plant are noble gases which give rise to whole body exposure, while the bones and lungs are the critical organs for fly ash.

The current consensus of workers in radiation biology appears to be that these levels of exposure do not constitute a significant health hazard. However, this consensus is not unanimous. We are reluctant to assign a cost number but believe that it is small. We are unable to estimate control costs.

Cooling

As long as coal-fired electric plants continue to operate in approximately their current temperature region their thermal efficiency will be limited by classical thermodynamic considerations to something less than about 50 percent. This means that for each new installed kilowatt of electric capacity the environment will be asked to dissipate at least an additional kilowatt of unused heat energy, and in most cases 1.5 kW or more.

In principle this energy need not be entirely "waste." Some central city stations are designed to use such heat for space heating and there are a few plants where the warm water output is used for agricultural or maricultural purposes. However, in the vast majority of cases today this energy is simply dissipated, via local water bodies, or directly, to the atmosphere. We shall examine only the costs of this form of disposal.

It is not immediately clear that heating up a body of water constitutes a societal cost. Indeed some observers have argued that such heating may sometimes be viewed as a benefit and have proposed the term "thermal enrichment." If the mean temperature of a waterway is raised the ecosystem which it supports will be altered. If the resulting new system is in no sense less diverse or less stable, and if the old system is not somehow unique or exceptional, it becomes very difficult to argue that the change represents a loss. However, excessive temperatures, or sudden or extreme fluctuations in temperature, do significantly reduce the degree of diversity in an ecosystem and as such do represent a real cost. Further, losses may be sustained even with low temperature rises if significant killing of small and microsized organisms results from high turbulent shear in the condenser or from the chlorination of water to prevent growth in the condenser plumbing.

TABLE II

Approximate Increase in Costs for Various Cooling Technologies Above Run-of-River Cooling Costs [59]–[61]

Cooling Method	Cost Above Run-of-River Cooling Assuming Workable Site
Cooling pond	0.07 to 0.1 mill/kWh
Natural draft wet cooling tower	0.20 to 0.43 mill/kWh
Forced draft wet cooling tower	0.20 to 0.46 mill/kWh
Dry cooling tower	0.76 to 1.9 mills/kWh

Based on such considerations, it is possible to conceptualize a number of ways to treat the waste heat problem, perhaps including the requirement that plants which significantly raise the temperature of a waterway maintain a standby capability to sustain this higher temperature in the event of a plant shutdown. Such considerations make for interesting academic exercises. But the reality of the situation in the United States is that we are very close to reaching the maximum thermal burden which our inland water systems can sustain without substantial damage. In many areas we have already exceeded this level. We are unable to quantify the damage costs which would result from further use of conventional cooling but the control costs are well known, are relatively low, and are widely thought by neutral observers to be significantly less than the unknown damage costs.

While estimates in the literature vary, Table II contains typical control costs, based on the same capital cost and plant factor values we have assumed previously in this paper. Cooling pond prices are of course very dependent on land availability and such ponds are unworkable in most urban settings. Wet cooling towers are quite acceptable in some parts of the country, but because of fog and icing damage are not an optimal solution in many colder areas. Preliminary studies of fog and icing costs have only recently been initiated and no numbers are yet available [62]. Blowdown waters from wet towers can also pose problems. Dry towers appear to be the "cleanest" solution to this problem but unfortunately firm price estimates for very large towers are unavailable. We estimate average thermal control costs at 0.4–1.0 mill/kWh. We are unable to estimate uncontrolled societal costs with any accuracy but they are probably greater than 1.5 mills/kWh.

Plant Land Use and Esthetic Costs

There is a growing feeling in the United States that land use planning may represent one of our most pressing needs. Land surface, which we have traditionally thought of as a semi-infinite resource, is today becoming painfully scarce. This is especially true of land close to lakes and streams and on the sea shore. Many would argue that by consuming such land, power plants impose a societal cost not reflected in the price paid for the land.

There is also growing concern about the esthetic quality of our environment, a feeling that power plants should be designed to be at least more visibly pleasing . . . or perhaps not even visible.

We are unable to assign costs in either of these areas, because, as in the case of strip-mined land, we do not have an accurate description of the appropriate value orientation functions. However, both problems are important and deserve careful future study.

CONCLUSIONS

We draw the following conclusions.

While the results are incomplete, and not of sufficient accuracy to be the sole basis of hard policy decisions, it is possible, with the present data and level of understanding, to make substantial progress toward quantifying the social costs of coal-fired electric power generation as Table I indicates.

It is clear that if the external costs obtained in this computation are fully internalized they will affect substantially the use patterns of at least the more elastic consumers of electric power. As a result of environmental regulation this process is already under way, and may ultimately lead to significant changes in the levels and shape of electric power price curves [5], [63]–[65].

Significantly more accurate estimates of the social costs of coal-fired electric generation will require large improvements in data, especially in the areas of air pollution and land use costs, as well as some fundamental conceptual progress in techniques for treating value orientation.

ACKNOWLEDGMENT

The authors with to thank the members of the seminar class for their assistance in this work.

Particular thanks are due to P. Schleifer of Gulf Energy and Environmental Systems and Marsha Penner of UCSD.

REFERENCES

[1] "Energy research needs," a report to the National Science Foundation prepared by Resources for the Future in cooperation with the MIT Environmental Lab., Oct. 1971.

[2] G. D. Friedlander, "Power, pollution and the imperiled environment, Part I," *IEEE Spectrum*, vol. 7, pp. 40–50, Nov. 1970.

[3] "Calvert Cliffs Coordinating Committee, Inc. *et al.* versus United States Atomic Energy Commission and United States of America," case no. 24839, D. C. Circuit, July 23, 1971; also: D. Nelkin, *Nuclear Power and its Critics.* Ithaca, N. Y.: Cornell Univ. Press, 1971.

[4] R. G. Ridker, *Economic Costs of Air Pollution.* New York: Praeger, 1967.

[5] C. Cicchetti and W. J. Gillen, "Electricity growth: Economic incentives and environmental quality," presented at the 1973 February Conf. on "Energy: Demand, conservation, and institutional problems" (Mass. Inst. Technol.).

[6] R. T. Anderson, "Simplify power-plant cost calculations," *Power*, pp. 39–41, July 1970.

[7] Federal Power Commission, *The 1970 National Power Survey, Part I.* Washington, D. C.: U. S. Gov. Printing Office, Dec. 1971.

[8] F. R. Kluckhohn and F. L. Strodtbeck, *Variations in Value Orientations.* Evanston, Ill.: Row, Peterson and Co., 1961; also: S. Hood, Ed., *Human Values and Economic Policy.* New York: New York Univ. Press, 1967.

[9] A. C. Worell, *Economics of American Forestry.* New York: Wiley, 1959.

[10] E. A. Nephew, "Healing wounds," *Environment*, vol. 14, pp. 12–21, Jan./Feb. 1972.

[11] J. Meyer, "Cost of reclamation," *Environment*, vol. 14, pp. 14–15, Jan./Feb. 1972.

[12] "Surface mine reclamation: Moraine State Park, Pennsylvania," U. S. Bureau of Mines, Info. Circ. 8456 (also available from U. S. Gov. Printing Office, Washington, D. C.), 1970.

[13] W. Greenburg, "Chewing it up at 200 tons a bite: Strip mining," *Technol. Rev.*, vol. 75, pp. 46–55, Feb. 1973.

[14] Tennessee Valley Authority, private conversation.

[15] "Company replants land torn up by strip mines," *Christian Sci. Moni.*, p. 4, May 4, 1972.

[16] K. P. Davis, *American Forest Management.* New York: McGraw-Hill, 1954.

[17] *Public Land Statistics—1970*, Bureau of Land Management, Department of the Interior. Washington, D. C.: U. S. Gov. Printing Office, 1971.

[18] "Strippable coal reserves of Wyoming: Location, tonnage and characteristics of coal and overburden," U. S. Bureau of Mines, Info. Circ. 8538 (also available from U. S. Gov. Printing Office, Washington, D. C.), 1972.

[19] H. M. Malin, Jr., "Feds eye regulations for strippers," *Environ. Sci. Technol.*, vol. 6, pp. 27–29, Jan. 1972.

[20] H. Perry, "Environmental aspects of coal mining," in *Power Generation and Environmental Change*, D. A. Berkowitz and A. M. Squires, Eds. Cambridge, Mass.: MIT Press, 1971, pp. 317–339.

[21] "Underground coal mining in the United States: Research and development programs," prepared for the Office of Science and Technology by TRW Systems Group, Rep. 13497-6001-R0-00, June 1970.

[22] W. Cochran, "Mine subsidence—Extent and cost of control in a selected area," U. S. Bureau of Mines, Info. Circ. 8507 (also available from U. S. Gov. Printing Office, Washington, D. C.), 1971.

[23] "Acid mine drainage in Appalachia," Appalachian Regional Commission Rep., Washington, D. C., 1969.

[24] J. C. McCartney and R. H. Whaite, "Pennsylvania anthracite refuse: A survey of solid waste from mining and preparation," U. S. Bureau of Mines, Info. Circ. 8409 (also available from U. S. Gov. Printing Office, Washington, D. C.), 1969.

[25] F. C. Andreuzzi, "A method for extinguishing and removing burning coal refuse banks," U. S. Bureau of Mines, Info. Circ. 8485 (also available from U. S. Gov. Printing Office, Washington, D. C.), 1970.

[26] L. M. McNay, "Coal refuse fires: An environmental hazard," U. S. Bureau of Mines, Info. Circ. 8515 (also available from U. S. Government Printing Office, Washington, D. C.), 1971.

[27] "Bureau of Mines environmental action programs for Northern Pennsylvania: Refuse bank removal and subsidence monitoring," The MITRE Corp., McLean, Va., Rep. MTR-6165, Apr. 1972.

[28] "Methods and costs of coal refuse disposal and reclamation," U. S. Bureau of Mines, Info. Circ. 8576 (also available from U. S. Gov. Printing Office, Washington, D. C.), 1973.

[29] W. S. Lainhart *et al.*, "Pneumoconiosis in Appalachian bituminous coal miners," U. S. Public Health Service Bureau of Occupational Safety and Health, U. S. Dep. of Health, Education and Welfare, Cincinnati, Ohio, 1969.

[30] "UMW Welfare and Retirement Fund," hearings before the Subcommittee on Labor of the Committee on Labor and Public Welfare, U. S. Senate, 1970.

[31] *Papers and Proceedings of the National Conference on Medicine and the Federal Coal Mine Health and Safety Act of 1969* (National Conf. on Medicine and the Federal Coal Mine Health and Safety Act), Library of Congress 73-146830, 1970.

[32] M. Bundy, "Recent progress in control of coal workers' pneumoconiosis," presented to the Int. Conf. on Coal Worker's Pneumoconiosis, New York, Sept. 16, 1971.

[33] R. B. Anderson, "A cryogenic supplied air system for continuous miner operators," presented to the 1971 Coal Convention, Pittsburgh, Pa., May 17, 1971.

[34] L. Langlois, "The Cost and Prevention of Coal Worker's Pneumoconiosis," Appalachian Regional Commission Monograph, June 1971.

[35] *Minerals Year Book—1969, Vol. I*, U. S. Bureau of Mines, U. S. Dep. of Interior. Washington, D. C.: U. S. Gov. Printing Office, 1971.

[36] "Coal mine health and safety, Part 2," hearings before the Subcommittee on Labor of the Committee on Labor and Public Welfare, U. S. Senate, 1969.

[37] "Coal mine health and safety, Part 5," hearings before the Subcommittee on Labor of the Committee on Labor and Public Welfare, U. S. Senate, 1969.

[38] L. B. Barrett and T. E. Waddell, "The cost of air pollution damages: A status report," Appendix I–J of the *RECAT Report* (Cumulative regulatory effects on the cost of automotive transportation; prepared for the Office of Sci. Technol., Feb. 28, 1972).

[39] L. B. Lave and E. P. Seskin, "Air pollution and human health," *Science*, vol. 169, pp. 723–733, Aug. 21, 1970.

[40] "The President's 1971 Environmental Program," compiled by the Council on Environmental Quality (available from U. S. Gov. Printing Office, Washington, D. C.), Mar. 1971.

[41] T. K. Sherwood, "Must we breathe sulphur oxide," *Technol. Rev.*, vol. 72, pp. 24–31, Jan. 1970.

[42] S. Ehrlich, "Air pollution control through new combustion processes," *Environ. Sci. Technol.*, vol. 4, pp. 396–400, May 1970.

[43] "Control techniques for sulphur oxide air pollutants," U. S. Public Health Service, National Air Pollution Control Administration, U. S. Dep. of Health, Education and Welfare (available from U. S. Gov. Printing Office, Washington, D. C.), Jan. 1969.

[44] A. V. Slack, "Removing SO_2 from stack gases," *Environ. Sci. Technol.*, vol. 7, pp. 111–119, Feb. 1973.

[45] G. G. McGlamery, H. L. Falkenberry, and A. V. Slack, "Economic factors in recovery of SO_2 from power plant stack gas," *J. APCA*, vol. 21, pp. 9–15, Jan. 1971; also: A. V. Slack, H. L. Falkenberry, and R. E. Harrington, "Sulphur removal from waste gases; lime-limestone scrubbing technology," *J. APCA*, vol. 22, pp. 159–166, Mar. 1972.

[46] A. W. Lemmon, Jr. *et al.*, "A method of comparing costs of alternative SO_2 control techniques for electric power generation facilities," Battelle Memorial Institute, Columbus, Ohio, undated.

[47] D. H. McCrea, A. J. Forney, and J. G. Myers, "Recovery of sulphur from flue gases using a copper oxide absorbent," *J. APCA*, vol. 20, pp. 819–824, Dec. 1970.

[48] "Air quality criteria for nitrogen oxides," Environmental Protection Agency (available from U. S. Gov. Printing Office, Washington, D. C.), Jan. 1971.

[49] L. R. Babcock, "A combined pollution index for measurement of total air pollution," *J. APCA*, vol. 20, pp. 653–659, Oct. 1970.

[50] "Control techniques for nitrogen oxide emissions from stationary sources," Environmental Science Health Service, National Air Pollution Control Administration, U. S. Dep. of Health, Education and Welfare (available from U. S. Gov. Printing Office, Washington, D. C.), p. 5-3, Mar. 1970.

[51] S. S. Penner, Dep. Appl. Mech. and Eng. Sci., Univ. of Calif. at San Diego, private communication, Mar. 1973.

[52] "Air quality standards for particulate matter," National Air Pollution Control Administration, U. S. Dep. of Health, Education and Welfare (available from U. S. Gov. Printing Office, Washington, D. C.), Jan. 1969.

[53] A. E. Vandegrift *et al.*, "Particulate air pollution in the United States," *J. APCA*, vol. 6, pp. 321–328, June 1971.

[54] R. D. Wilson and D. W. Minnotte, "Economic aspects of air pollution," U. S. Public Health Service, National Air Pollution Control Administration, U. S. Dep. of Health, Education and Welfare, Arlington, Va., June 1969.

[55] D. H. Klein and P. Russell, "Heavy metals: Fallout around a power plant," *Environ. Sci. Technol.*, vol. 7, pp. 357–358, Apr. 1973.

[56] O. I. Joensuu, "Fossil fuels as a source of mercury pollution," *Science*, vol. 172, pp. 1027–1028, June 4, 1971.

[57] C. E. Billings and W. R. Matson, "Mercury emissions from coal combustion," *Science*, vol. 176, pp. 1232–1233, June 16, 1972.

[58] J. E. Martin, E. D. Harward, and D. T. Oakley, "Radiation doses from fossil-fuel and nuclear power plants," in *Power Generation and Environmental Change*, D. A. Berkowitz and A. M. Squires, Eds. Cambridge, Mass.: MIT Press, 1971, pp. 107–121.

[59] "A survey of alternative methods for cooling condenser discharge water . . . large scale heat rejection equipment," prepared by Dynatech R/D Co., Water Pollution Control Res. Series, Environmental Protection Agency (available from U. S. Gov. Printing Office, Washington, D. C.), 1969.

[60] K. A. Oleson and R. R. Boyle, "How to cool steam electric plants," *Chem. Eng. Progr.*, vol. 67, pp. 70–76, July 1971.

[61] "Cut pollution at what price," *Elec. World*, vol. 172, pp. 32–33, Jan. 19, 1970.

[62] H. Veldhuizen and J. Ledbetter, "Cooling tower fog: Control and abatement," *J. APCA*, vol. 21, pp. 21–24, Jan. 1971.

[63] D. Chapman and T. Tyrrell, "Alternative assumptions about life style, population, and income growth: Implications for power generation and environmental quality," presented to the Sierra Club Conf. on Power and Public Policy, Johnson City, Vt., Jan. 13–15, 1972.

[64] R. Cohen, Testimony before the public utilities commission of the State of Colorado regarding the investigation and suspension of Colorado P.U.C. No. 5 . . . electric docket No. 705, Denver, Colo., Dec. 21, 1971.

[65] M. J. Roberts, "Economic consequences of energy costs," discussion paper 278, Harvard Inst. Econ. Res., Mar. 1973.

BENEFITS AND COSTS OF SURFACE COAL MINE RECLAMATION IN APPALACHIA

by

Robert A. Bohm, John R. Moore, and F. Schmidt-Bleek*

ENERGY VS. ENVIRONMENT

Coal has long held a position of importance as an energy source in the United States. It fueled the first industrial revolution and was called upon during and immediately after the Second World War to bridge energy supply shortages. While the demand for coal has slipped over the years since then, a slow comeback has been posted since 1961.

Much has been written and said about the renewed need for coal since the present energy supply shortage became imminent during the second half of 1973. As an indication of the rapid change in the position of coal as an energy source, we would like to quote from a report prepared in June of 1973:

> "...Despite the modest resurgence of coal as an energy source, the role which coal will play in the future is difficult to forecast. Basically, one can imagine two scenarios. In the first, coal production will diminish (or at best maintain its present place) in absolute terms and decline in relative terms as a result of a number of factors. Tightening environmental standards applied to the production and consumption of coal could well lead to its substitution by other fuels in an attempt to meet EPA (Environmental Protection Agency) and local (state) regulations at the same time the coal industry is faced with the exhaustion of cheap and readily strippable coal reserves in the East, a reluctance of the Federal and state governments to permit stripping of western coal reserves, and higher costs of underground mining due to stringent enforcement of the 1969 Coal Mine Health and Safety Act. Under such circumstances, the level of coal production and consumption could well fall below 500 million tons per year. A second scenario is much more optimistic. It begins with the picture of continuing balance of payments difficulties making it impossible to meet our rising energy requirements without imported fuels. Political difficulties in the Near East may add to these problems. On the other hand, technological breakthroughs in coal gasification and/or liquefaction coupled with *in situ* processing could mean environmental cost minimization. Continued difficulties with the development of nuclear energy, on the technology side, complicated by the moral issues related to nuclear safety would certainly enhance opportunities for resurgence of the coal industry. Under such a set of circumstances it would not be unthinkable to forecast a demand for coal in the range of 1,500 to 2,000 million tons per year by the year 2000.[1]

In one short year the weight of opinion has clearly shifted dramatically in favor of the second scenario. Now the problem is not one of anticipating the demand for coal as much as determining the ability of the coal industry to supply expanding orders. Realistically, what are the possibilities? We estimate that for 1974, coal production may top the 1973 output by as much as 15 percent. Significant bottlenecks exist, however, in several areas. Bringing new deep mines into production requires two to five years; those planned in light of the present crisis will not contribute significantly to increased output for some time. Existing deep mines may be capable of increasing production by as much as 10 percent per year, provided manpower problems can be overcome and equipment delivered.

The true short-term potential for increased coal production rests with surface extraction. This potential is hindered by two principal factors. First, a severe shortage of equipment exists at this time. Second, surface mining results in dramatic environmental disruption. Large amounts of spoil material must be displaced to reach underlying coal beds. Soil and potentially toxic elements are thereby exposed. Normal hydrology in the mining region is disturbed and flood potential increased. Mined over areas often appear as moonscapes.

---

*The authors are, respectively, Associate Professor of Finance and Associate Director, Appalachian Resources Project; Professor of Economics and Acting Director, Appalachian Resources Project; Associate Professor of Chemistry and Acting Director, The Environment Center; all of the University of Tennessee. The work reported here is supported by a grant from the National Science Foundation/RANN, GI-35137. The authors retain responsibility for any errors in analysis but wish to thank their colleagues G. A. Vaughn, S. L. Carroll, R. E. Shrieves, and J. H. Lord for helpful contributions to this paper.

[1] F. Schmidt-Bleek and J. R. Moore, Benefit/Cost Approach to Decision Making, The Dilemma of Coal Production, (Knoxville, Tennessee, Appalachian Resources Project, 1973) pp. 1-2.

At present, the environmental consequences of surface mining of coal are most acute in the mountainous Appalachian region where contour mining methods are employed. In this area, mining regularly takes place on slopes in excess of twenty degrees. The two polar positions regarding Appalachian surface mining on steep slopes are: (1) outlaw strip mining due to its environmental impact, (2) ignore the environmental impact of strip mining due to the "need" for coal. The intermediate position is to require by law adequate reclamation of surface mined land. The remainder of this paper is devoted to an economic analysis of steep slope surface mine reclamation in Appalachia. The basic question to be answered may be stated as follows: Is the cost of reclaiming steep slope surface mines justified in terms of a diminution of environmental damage.

THE SOCIAL COSTS OF COAL PRODUCTION

The environmental consequences of coal surface mining result in a divergence between the social and private (i.e., market) costs of coal production. This phenomenon results in external effects which are not accurately represented by the selling price of coal.[2] In effect, the market price of coal is lower than the true social cost of production by the value of the externality. This situation is often associated with environmental pollution and, as in the case of coal mining, environmental costs have not been considered as real and quantifiable cost of production in the past.

The traditional economic solution to the problem of externalities is to devise a set of institutional arrangements that will result in balancing private production cost and social cost. For example, if coal mining results in environmental costs which are not reflected in the price of coal, the problem can be corrected by requiring the elimination of factors causing the environmental damage, or alternatively, by compensating those who must bear the cost of the external effect.

Regardless of the course of remedial action undertaken, society as a whole should benefit and resource allocation improve if a significant externality is eliminated. The value of the benefits of any abatement activity, however, is of crucial importance for public policy. Since resources employed to reduce external effects resulting from coal extraction have alternative uses, the benefits resulting from this activity should at least equal the value of the resources utilized in an abatement program. Of course, political realities may result in non-optimization in resource use in the narrow technical sense. Even so, it is imperative to know what has been sacrificed in terms of the usual economic concept of efficiency.

[2] On the general topic of externalities see R. Coase, "The Problem of Social Cost," Journal of Law and Economics, (1960); J. Buchanan and C. Stubblebine, "Externality," Economica, (1962); R. Turvey, "On Divergence Between Social and Private Cost," Economica, (1963). For discussions dealing with mining see D. B. Brooks, "Surface Mine Reclamation and Economic Analysis," Natural Resources Journal, (1966); H. A. Howard, External Diseconomies of Bituminous Coal Surface Mining-A Case Study of Eastern Kentucky, 1960-67, unpublished doctoral dissertation, Indiana University, Bloomington, Ind. 1969; and R. A. Bohm, J. H. Lord and D. A. Patterson, "Market Imperfections, Social Costs of Strip Mining and Policy Alternatives," Rev. Regional Studies (1973).

A BENEFIT/COST FRAMEWORK OF ANALYSIS

A suggested approach to making social decisions in the presence of external environmental effects resulting from coal mining is presented in Figure 1. The framework presented in Figure 1 is constructed of three basic elements: the damage delivery and evaluation system, the abatement delivery and evaluation system, and the benefit-cost assessment and decision system. Application of the overall model will generate data of six basic types as indicated by the letters (A) through (F) in the diagram.

Consider Figure 1 in more detail, beginning with the damage delivery and evaluation system. Given the geologic, hydrologic, and geographic nature of a mining site, the scale and type of activity determines the release of damage agent flows and thus governs damage agent levels in time and space (A). The damage agents, in turn, cause damages to the receptors: humans, animals, plants, and property. The dollar measurement of these damages represents the external effect in the absence of abatement effort and becomes the base from which to compute the benefits associated with various types and levels of abatement effort (B).

Figure 1 also illustrates the abatement delivery and evaluation system. Technological feasibilities coupled with legal constraints and institutional pratices determine abatement efforts. These can be costed out in dollar terms (C). Abatement efforts, however, interdict damage agent flows and produce a decreased level of receptor damage and thus a lower level of external costs. The reduction in external costs when valued before and after abatement activities are undertaken is a measure of the benefit of abatement activity (D).

For each set of parameter values (geologic, hydrologic, geographic, demographic, etc.) three unit cost figures are generated as shown in Figure 1: external costs/unit produced, (B); abatement cost, (C); and external costs/unit produced as modified by abatement efforts (residual external costs), (D). The difference between (B) and (D) is the benefit accruing to a particular abatement strategy. All these costs, along with any external benefits such as differential human capital costs between deep and surface mining associated with accidents and disease (F) are the inputs for the benefit-cost assessment and decision system (E).

In order to implement the benefit-cost framework illustrated by Figure 1, two basic sets of data are required:

1. The external effects of mining activity must be identified and costed out at alternative levels of abatement activity and under various sets of geologic, hydrologic, geographic, and demographic conditions.

2. The cost of alternative abatement strategies must be determined.

MEASUREMENTS OF EXTERNAL COSTS OF COAL PRODUCTION

Geographically, surface mining generates environmental damages at three levels. Inside the coal mining region itself damages occur within the immediate vicinity of the mines as well as in the broader coal mining region taken as a whole. We refer to analysis of the former as direct measurement or

Figure 1

BENEFIT/COST APPROACH TO COST MEASUREMENTS IN COAL MINING

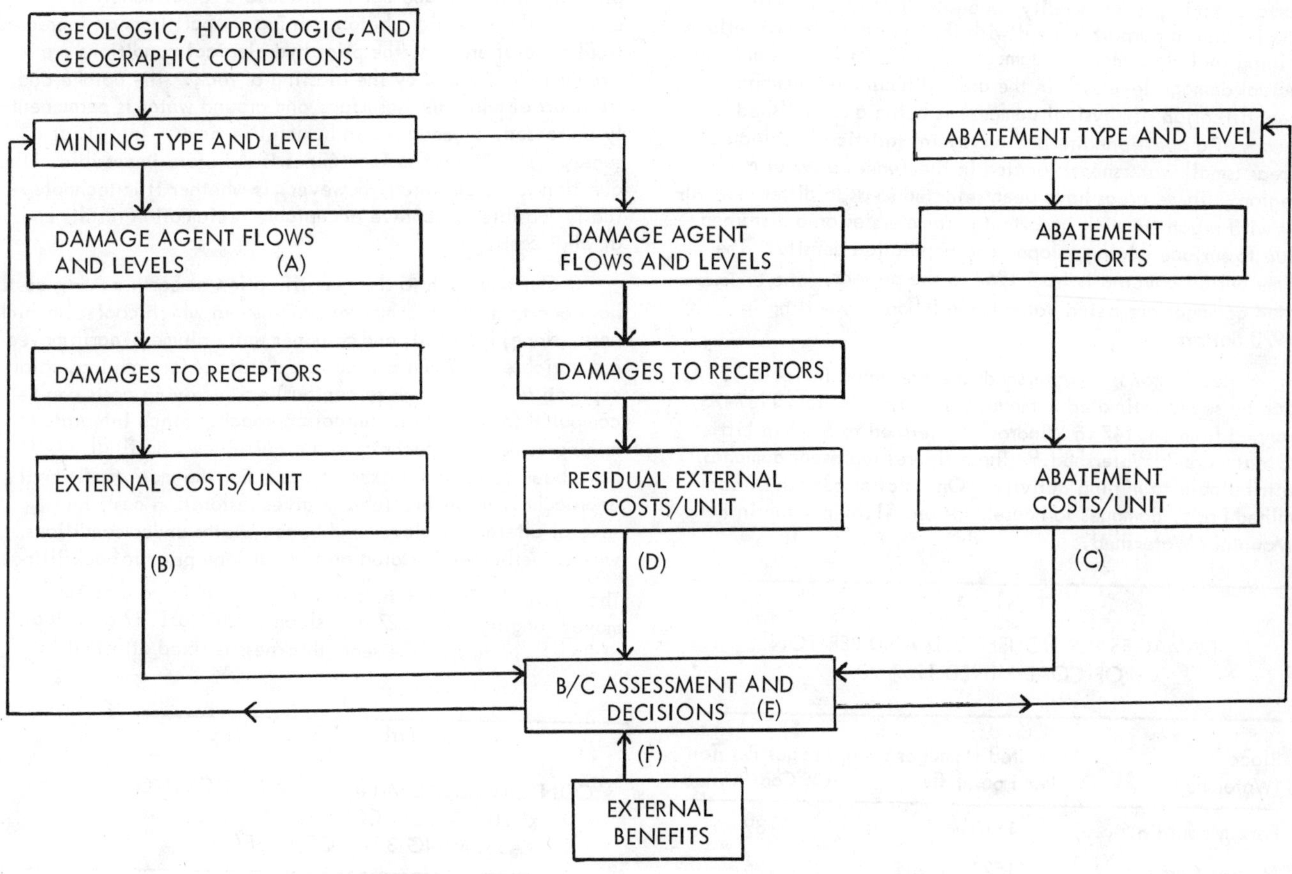

disaggregated analysis and analysis of the latter as aggregate or regional analysis. In addition, damages may occur outside the coal mining region, for example, by the transfer of damage agents downstream via a river system.

At all three levels both tangible and intangible damages may result. Tangible damages are defined as measurable or potentially measurable and may be either short or long term. Short term tangible damages begin to occur simultaneously with mining activity. Short term damage receptors may be classified as: (1) Private Decision Units (e.g., households) (2) Public Property (3) Non-Human Living Things. Long term tangible damage results primarily from the introduction of toxic elements into the environment which generate only low level effects in the short term. Over time, however, factors such as these (e.g., heavy metals, prolonged acidity, etc.) may prove extremely detremental to: (1) Humans (2) Other Living Things (Plants & Animals) (3) Property (Public & Private).

Intangible damages resulting from surface mining are defined as having significant public good characteristics, i.e., they are perceived or received collectively by members of society.[3] In both the long and short term, the principal intangible damage generated by coal surface mining in Appalachia is probably aesthetic damage.

[3]For a discussion of public goods see: R. A. Musgrave, The Theory of Public Finance, (New York: McGraw-Hill Book Company, 1959), Chapter 1; and P. A. Samuleson, "The Pure Theory of Public Expenditures," Review of Economics and Statistics, (November, 1954), pp. 388-89.

In the case of surface mining, short run damages are obviously of most immediate concern. Within this class of damages, damage to decision units (largely households) within the immediate vicinity of the mines is the category where (1) the smallest amount of reliable information is available, and (2) the largest external costs are most likely to be found. Therefore, research has been concentrated in this area in the hope that abatement activity that significantly reduces this form of damage would be economically justified.

Prior to the valuation of damages, they must be enumerated. However, the physical damage function generated by surface mining is not readily apparent.[4] Most available studies which purport to deal with the external costs of surface mining include a menu of damage agents but fail to enumerate actual damage levels.[5] In the case of household damage, quantification of physical damages is being accomplished by means of a questionnaire and survey in statistically selected areas (small watersheds) located in the Tennessee coal mining region. These areas have been selected so as to differ markedly with regard to such important parameters as area disturbed due to surface mining, slope, and population density. The time period covered is from 1967 to the present. The enumerated damages are being valued in a laboratory setting in 1973 dollars.

Results for five watersheds are presented in Table 1. As can be seen, estimated damages per household from 1967-73 ranged from $4,147 in Windrock Watershed to $616 in Little Greasy Creek Watershed.[6] These figures represent damages attributable to mining activity. On a dollar per ton of coal mined basis, damages to households are $1.39 per ton in Fork Mountain Watershed.

Table 1

DAMAGES PER HOUSEHOLD AND PER TON OF COAL MINED 1967-73

Place (Watershed)	($) Estimated Damages Per Household	($) Damages Per Ton Of Coal Mined
Fork Mountain	1618	1.39
Moores Camp	1527	.96
Welch Camp	1467	.85
Windrock	4147	.60
Little Greasy Creek	616	1.24

[4] An interesting discussion of this problem is found in Lester B. Lave, "Air Pollution Damage: Some Difficulties in Estimating the Value of Abatement," in Allen V. Kneese and Blair T. Bower, Environmental Quality Analysis, (Baltimore, Md.: The Johns Hopkins Press, 1972), esp. pp. 214-215.

[5] See Brooks, "Surface Mining Reclamation and Economic Analysis," op.cit.; C. R. Collier, et.al., Influences of Strip Mining on the Hydrologic Environment of Parts of Beaver Creek Basin, Kentucky, 1955-66, Washington, D. C., Department of the Interior, 1970.

[6] Watershed names have been established on the basis of a prominent characteristic of the area studied, for example, a stream, town, or mountain peak. Maps which detail the exact boundaries of the various watersheds under investigation are available from the authors.

MEASUREMENT OF RECLAMATION COSTS

During the preceeding disucssion of Figure 1, it was argued that the assessment of abatement costs/unit ($/ton in the case of coal) is a necessary ingredient for rational decision making. In this section we present a systematic approach to the quantification of two major surface mine reclamation costs: earth moving and revegetation. A few introductory remarks are in order.

The replacing and grading of spoil material on areas from which coal has been extracted and the subsequent replanting of the surface cannot produce a set of conditions equal to the conditions before mining. Apart from an increased likelihood of erosion, the prior hydrological conditions are irreversibly changed by the blasting of rock. The uptake and transport of minerals by surface and ground water is permanently increased, as measured in human life times. In a strict sense, restoration of surface mines is therefore impossible. The question to be answered, however, is whether it is technologically feasible to achieve acceptable reclamation results and at what costs.

Our approach to the quantification of earth moving costs has been to develop computer programs in which coal seam thickness, slope, highwall, and costs per unit volume of earth moved are variables. Given a set of conditions before mining, total restorative (i.e., back to contour) earth moving costs can be computed for a given technical approach. Since integrated mining-restoration techniques are quite new, no final results have been computed for such cases. For the more traditional approaches, however, Table 2 gives restoration costs for several alternative slopes and bench widths under conditions where overburden is stored on the outslope prior to backfilling.[7] The figures in Table 2 assume a cost per cubic yard of earth moved ranging from $.27 at a slope of 20° to $.39 at a slope of 35°.[8] Note that the seam thickness is fixed at 3 feet.

Table 2

CONTOUR COAL MINING EARTH MOVING RESTORATION COSTS IN $/TON, ASSUMING 3 FT. COAL SEAM

Bench Width (Feet)	Slope (Degrees) 20° ($.27)*	25° ($.31)*	30° ($.35)*	35° ($.39)*
20	$0.28	$0.46	$0.70	$1.02
50	$0.71	$1.14	$1.74	$2.55
80	$1.13	$1.83	$2.79	$4.08
110	$1.56	$2.52	$3.83	$5.61
140	$1.98	$3.20	$4.88	$7.14

*Cost per cubic yard of earth moved.

[7] These data, therefore, represent an upper bound on back to contour earth moving costs. For example, costs per ton should be lower when modified block cut techniques are employed.

[8] These figures are based on reported experience in the Tennessee coal mining region.

Inspection of Table 2 shows that earth moving costs are very sensitive to both slope and bench width. Of course, the seam thickness also influences the end-price per ton very strongly. One of the important consequences of these facts is the need to redefine performance bonding for strip mining and make it very site specific. It is not unusual to find mines in Central Appalachia where costs for adequate backfilling run higher than $5,000 per acre.[9]

Revegetation costs presented in Table 3 correspond to costs incurred in Tennessee under actual mine site conditions. We assume the use of 1,000 pounds of mulch per acre, 40 to 80 pounds of seed per acre, 50 pounds of available nitrogen and 100 pounds of available P_2O_5 per acre. As can be seen from Table 3, revegetation costs in the return to contour case range from $.17 to $.33 per ton of coal (at a constant coal seam of 3 feet), depending linearly and almost exclusively upon the slope of the area affected.[10]

Table 3

CONTOUR COAL MINING REVEGETATION COSTS IN $/TON, ASSUMING 3 FOOT COAL SEAM*

Bench Width (Feet)	Slope (Degrees)			
	20°	25°	30°	35°
20	.17	.22	.27	.33
50	.17	.20	.24	.32
80	.17	.21	.26	.32
110	.17	.22	.26	.32
140	.18	.22	.27	.33

*Grass and 1,000 seedlings per acre, mulch and fertilizer

HUMAN CAPITAL COST DIFFERENTIALS

To the extent that deaths, diseases, and injuries are more frequent in deep mining than in surface mining, human capital differentials attributable to these factors may be a partial offset to any environmental damages associated with surface mining. To date it has been found that for 1968 the differences in human capital costs associated with deaths and injuries amounted to between 5 and 10.25 cents per ton depending on assumptions relative to the proper wage rate, discount rate and consumption allowance.[11] Figures for other years during the decade 1960-1970 show little variation from this range. Contrary to the practice of the Bureau of Mines in evaluating time lost due to deaths in terms of the arbitrary allowance of 6000 days as specified by the United States of America Standards Institute, our figures are based on a sample of accident reports on file with the Bureau of Mines which allows computation of age at death which can then be compared with average retirement age compiled from United Mine Workers records. Only recently has the Bureau of Mines used actual time losses for reporting accident severity in cases involving temporary partial disability. Our findings indicate that for the 1960's the Bureau of Mines methods tend to overstate the time losses considerably. For 1968, for example, the Bureau of Mines showed a loss of production due to deaths of 1,650,000 man-days based on the arbitrary measure of time loss, while our figures show a loss of only 1,065,900 days.

The most significant non-accident related cause of disability among bituminous coal miners results from pulmonary disease. We have chosen to estimate pulmonary disease related disability using both a measure of dyspnea (severe shortness of breath) and of pulmonary function. The data used come from the Appalachian Laboratory of Occupational Respiratory Disease's study conducted in 1970-71. We find that each year there are from 900 to 1800 miners newly disabled (both partially and totally) by pulmonary disease. This translates into a lost income stream for 1968 of between 10 and 20.4 cents per ton. Combining this finding with the differential found for deaths and injury, it appears that human capital cost differentials favor surface mining by amounts which ranged (for 1968) between 15 and 30 cents per ton depending on the assumptions made concerning the appropriate measure of income stream losses.

PRELIMINARY BENEFIT/COST ANALYSIS

At this point, it is possible to integrate findings presented thus far on external costs, reclamation costs and human capital costs into a preliminary benefit-cost analysis of surface mine reclamation. Direct benefits of reclamation activity are defined as environmental damages or external costs prevented. For the purposes of this analysis, it will be assumed that 100% of damages to households resulting from surface mine activity will be eliminated by complete backfilling and revegetation. This procedure undoubtedly results in an over estimate of the benefits to be achieved by preventing household damages. In effect, it represents an upper bound on benefits of this type. Table 4 presents the data required to compute benefit-cost ratios for the five watersheds introduced earlier in the discussion. Column 1 of Table 4 reproduces the household damages incurred per ton of coal mined 1967-73 from Table 1.

[9] Most state laws require performance bonding for restoration on a per acre basis. For a 110 foot bench, the data presented in Table 2 show per acre earth moving costs of $3934 at a 25° slope and $5060 at a 30° slope. The 1972 Tennessee surface mine reclamation law, for example, requires only $600 per acre for performance bonding.

[10] By way of contrast, the Council on Environmental Quality (CEQ) "standard condition" calls for $.18 per ton for revegetation. The CEQ figures on backfilling costs are uniformly too low by about a factor of two. See Coal Surface Mining and Reclamation, U.S. Council on Environmental Quality, Washington, D. C., March 1973.

[11] John R. Moore, Human Capital Costs in Deep and Surface Mining of Coal, (Knoxville, Tennessee: Appalachian Resources Project, 1973). Note that the figures reported here represent, for the most part, morbidity figures based on conditions existing prior to the passage of the 1969 Coal Mine Health and Safety Act. Enforcement of new dust standards should result in substantial reduction in the human capital costs of CWP in future years.

Table 4

DATA FOR PRELIMINARY BENEFIT-COST ANALYSIS

Place (Watershed)	(1) Direct Benefits* ($/ton)	(2) Human Capital Differential ($/ton)	(3) Earth Moving Costs** ($/ton)	(4) Revegetation Costs ($/ton)
Fork Mountain	1.39	.15-.30	1.96	.19
Moores Camp	.96	.15-.30	1.26	.15
Welch Camp	.85	.15-.30	1.72	.20
Windrock	.60	.15-.30	1.39	.17
Little Greasy Creek	1.24	.15-.30	1.75	.19

*Equals 100% of household damages due to surface mining.
**Represents maximum earth moving costs for back to contour reclamation.

External benefits are defined as the human capital cost differential observed between deep and surface mining. This differential ranges from $.15 - $.30 as indicated in column 2 of Table 4. Data on human capital cost differentials are being included in the analysis in order to illustrate the potential social cost trade-offs that exist between deep and surface mining. Conceptually, of course, it is not correct to include this factor in a benefit-cost analysis of surface mine reclamation unless all such factors (i.e., both those which favor and thos which do not favor surface mining) are taken into account.[12]

Costs are defined as earth moving and revegetation costs. These data are presented in columns 3 and 4 of Table 4. It will be recalled that earth moving costs are for the case where spoil is stored on the outslope, i.e., these are maximum backfilling costs for back to contour reclamation. It is assumed that reclamation activity takes place concurrently or directly after mining is completed.

Based on the data included in Table 4, three alternative sets of benefit-cost ratios are presented in Table 5. The first set compares household damages to earth moving costs. The sensitivity of the ratio to the inclusion of revegetation costs can be inferred from column B. Finally, in column C a median figure for the human capital cost differential ($.23) is added to direct benefits.

[12] Proceeding in this manner would entail a study of external costs in all coal mining. Such an undertaking is beyond the scope of the present inquiry which deals primarily with the external costs of surface coal mining.

Economic justification of surface mine reclamation requires a benefit-cost ratio greater than one. A benefit-cost ratio greater than one implies that the benefits derived from reclamation activity exceed the costs and that a positive rate of return is being earned on resources employed. All of the benefit-cost ratios reported in Table 5 are less than one. However, all relevant benefits and costs have not been included in the analysis at this time. In terms of the data employed, benefit-cost results to date suggest that total prevention of household damages plus the human capital differential in favor of surface mining off set no more than 84% (Moores Camp Watershed) and perhaps as little as 53% (Windrock Watershed) of the reclamation costs of earth moving and revegetation.

Table 5

PRELIMINARY BENEFIT/COST RATIOS

Place (Watershed)	Benefit/Cost Ratios (A) Col. 1 / Col. 3	(B) Col. 1 / Col. 3+4	(C) Col. 1+2 / Col. 3+4 *
Fork Mountain	.71	.65	.75
Moores Camp	.76	.67	.84
Welch Camp	.71	.64	.56
Windrock	.39	.36	.53
Little Greasy Creek	.43	.45	.76

*Represents maximum backfilling costs (i.e., maximum vertical movement of spoil). Assumes human capital cost differential equals $.23.

CONCLUSION

Continued research on the benefits and costs of surface mine reclamation should significantly alter the benefit-cost ratios presented in Table 5. In the case of reclamation costs, for example, it is anticipated that the introduction of modified block cut techniques will lower earth moving costs considerably. On the benefits side, the analysis to date is clearly incomplete. Further research will have to be undertaken to evaluate additional classes of benefits such as the prevention damage to public property, long term effects and aesthetic costs. Ultimately, however, based on the work completed at this time, it appears that the benefit-cost ratio for surface coal mine reclamation will exceed one.

In conclusion, we would like to point out that the benefit-cost analysis presented in this paper has dealt solely with real economic effects. In the area of benefits evaluation, the external costs of coal surface mining that have been identified are normally defined in the literature as technological externalities.[13] Our benefit-cost analysis is, therefore, an economic efficiency exercise, i.e., a benefit-cost ratio greater than one implies an improvement in the allocation of resources. Distributional considerations such as implicit subsidies to electric power users resulting from environmentally "dirty" coal production have been ignored.[14] To many concerned with the environmental consequences of surface mining, of course, transfer effects of this type are of overriding importance.

[13] See the discussion in Bohm, Lord, and Patterson, "Market Imperfections, Social Costs of Strip Mining and Policy Alternatives, op.cit.

[14] Ibid.

By E. A. Nephew

Healing Wounds

Present surface-mining practices in the United States are devastating large areas of land and are producing great environmental damage. The growing national concern over this problem is evidenced by the number of different surface-mining regulation bills that are currently before Congress. Some of these bills would only slightly alter present strip-mining practices, while others would require a substantially greater degree of land reclamation after completion of the mining, and at least one bill would prohibit future coal stripping altogether.

A program ensuring the full restoration of all lands disturbed by surface mining could be an environmentally acceptable alternative to banning the practice completely. This option has been adopted by the Federal Republic of Germany (West Germany) and several other European countries. To assess the feasibility of this approach to the problem, we visited the Rhineland brown-coal fields of West Germany and studied the land restoration methods employed to reclaim the huge, open-pit lignite (brown coal) mines of that region. Here, we were particularly interested in the German reclamation laws and how well they are enforced. In general, we found that the German land restoration program is highly successful. Many features of the planning, technological, and regulatory procedures used in West Germany to ameliorate the adverse environmental consequences of large-scale surface mining are applicable to strip-mining problems in the United States.

The nature and extent of environmental damage from surface mining in the United States have been documented in several excellent reports.[1] Essentially, two basic factors have contributed to the magnitude of the problem we face today: the rapid development of strip-mining of coal since the beginning of World War II and the greatly increased consumption of coal for electric power generation. Coal consumption for electric power generation increased from 51,474,000 tons in 1940 to 310,312,000 tons in 1969 and, during the same period, the amount of coal that had been either strip-mined or augered (mined by boring into the mountainside) rose from 43,167,000 tons to 213,373,000 tons.[2]

E. A. NEPHEW is a research staff member at the Oak Ridge National Laboratory. He is currently engaged in a National Science Foundation program, "The Environment and Technology Assessment," which includes an investigation of the environmental impacts of electricity production and use.

In coal surface mining, the overburden (the earth and rocks lying above the coal seam) is first removed, and the exposed coal is then extracted. Surface mining conducted in relatively flat terrain is commonly called area stripping. The surface of the land is first scalped of trees, and a deep trench is then cut through the overburden to expose a long strip of the coal bed. The earth, clay, and rock overburden is deposited adjacent to the trench to form a long "spoil" bank. After the coal is removed, a second trench is cut parallel to the first, and the overburden is dumped into the first trench. The process is repeated until all of the coal has been extracted, resulting in a series of parallel, adjacent spoil ridges, which give the mined land the appearance of having been worked over by a giant plow. The final cut produces an open trench bounded by a steep wall called a highwall.

Contour stripping in hilly regions progresses in the same manner except that the process is halted sooner—as the thickness of the overburden becomes too great. Starting at the outcrop of coal along the hillside, a slice of overburden is removed and dumped on the downslope side. The coal is then removed and a second cut is made through the overburden to expose more coal. Finally, when the overburden is too thick for further economical stripping operations, augers as large as seven feet in diameter may be used to drill horizontally some several hundred feet into the mountain to bring out additional coal. Using this combination of stripping and augering, the mining operations proceed along the hillside, leaving a twisting trail of churned earth in their wake. On one side the bench (or shelf) is bounded by a steep, nearly vertical highwall, and on the outside by the mass of spoil material precariously balanced on the downslope of the mountain.

To our knowledge, an accurate survey of the total land area disturbed by the surface mining of coal has never been made. On the basis of data reported by coal producers, Paul Averitt has estimated that, as of January 1, 1970, the cumulative past production of 4.4 billion tons of strip-mined coal resulted in 2,450 square miles of disturbed land in the U.S.[3] He further estimates that the mining of the remaining 128 billion tons of strippable coal in the 0 to 150-foot-thick overburden category would create an area of

Reprinted with permission from *Environment*, vol. 14, pp. 12–21, Jan./Feb. 1972.

The West Germans appear to have developed an exemplary process for exploiting the mineral resources of a region without permanently impairing the quality of their environment.

disturbed land comprising 71,000 square miles. At present coal strip-mining rates, roughly 100 square miles of additional disturbed land are being created each year. We must consider that advanced earth-moving machinery and changing economic conditions will probably make deeper deposits accessible to surface-mining methods. This would expand the strippable coal reserves and the total amount of future land damage.

The damage often extends well beyond the boundaries of the actual mining areas. Where mountains are scarred by contour mining, the whole landscape is rendered ugly even though only a small fraction of the land is disfigured. The destruction of watersheds (catchment areas from which stream waters are drawn) by sedimentation and acid water runoff also extends the harmful effects of strip-mining. (Acid water is water containing high levels of sulfuric or other acids.) According to a recent study,[4] contamination caused by both deep and surface mining has substantially altered the water quality of some 10,500 miles of streams in Appalachia. Acid drainage seriously pollutes about 5,700 miles of streams, reducing or eliminating aquatic life. A recent study by the U.S. Department of the

E. A. Nephew

At left, a wheel excavator of the type used in the Rhineland to mine brown coal. This machine, which weighs 7,400 tons and is 230 feet high and 650 feet long, selectively strips off and saves the top layer of loess, fertile loam which is restored during later reclamation.

Cost of Reclamation

Surface mining for coal has increased sharply in the United States during the past 50 years. Fully 44 percent of the soft (bituminous) coal mined in this country in 1970 came from open surface mines as opposed to deep mines. Little more than 50 years ago, in 1917, only one percent of U.S. soft coal came from surface mines. Strip-mining and its large-scale environmental disruption generally are associated with mining in the Appalachian region, but coal reserves accessible to surface-mining techniques are considerably greater in other areas of the country, according to data compiled by the Center for Science in the Public Interest (CSPI), Washington, D.C. The information was assembled by CSPI Co-Director James B. Sullivan on the basis of material gathered by a team of students from Wells College, Massachusetts Institute of Technology, Thomas More College, and Ohio State University. CSPI is a nonprofit group of scientists and lawyers developing projects in which science can be brought to bear on a varety of public-interest issues.

According to the CSPI data, about 5.2 billion tons of bituminous coal are available for surface mining in Appalachia. By comparison, 7.3 billion tons are available in the Midwest (3.2 billion alone in Illinois), and 650 million tons of bituminous and 20.2 billion tons of subbituminous (lower grade than bituminous but higher than lignite) in Arizona, Montana, New Mexico, and Wyoming. The coal reserves in the four states last mentioned are particularly important since they are low in sulfur content and relatively easy to mine. Strip-mining in the West is expected to increase sharply to meet power needs in that section of the country, as well as to supply the rest of the country with low-sulfur coal that will make it easier to meet new air pollution standards (see "Cloud on the Desert," **Environment,** July/August 1971, and "Coal Rush is On as Strip Mining Spreads Into West," Ben A. Franklin, **New York Times,** August 22, 1971, p. 1).

The accompanying article by E. A. Nephew on strip-mining in West Germany is instructive in at least two respects. First, success in comprehensive recovery of mined minerals and in subsequent reclamation of mined land in West Germany hinges on coordinated advance planning by public, government, and industrial representatives. Such planning has never been done in the United States. Second, the cost to restore mined-out lands to full agricultural productivity ranges from \$3,000 to \$4,500 per acre, a cost comparable to that in several Appalachian projects described by CSPI.

Strip-mining in the United States has been done with such little control and overall planning that

Interior on the environmental effects of strip-mining in Cane Branch Basin, McCreary County, Kentucky, showed a marked decrease in the variety and abundance of invertebrate bottom fauna in Cane Branch and in Hughes Fork downstream from the strip-mined areas.[5] This was caused by alternate deposition and erosion of sediment and the killing of aquatic vegetation by acid water, effectively destroying the stream habitat. Annual erosion losses from freshly strip-mined areas in Appalachia are as high as 27,000 tons per square mile, or up to 1,000 times greater than for undisturbed lands.

Strip-mining is only now beginning on a large scale in the western states, so the extent of damage is not yet clearly defined for climates and terrain different from Appalachia. In addition to increasing the susceptibility of these lands to wind and water erosion, surface mining in the semidesert regions of the Southwest poses the danger of exposing highly saline material to the surface.[6] This overburden material, enriched in salts by the process of percolation of surface water, if brought to the surface, would make it difficult to establish and maintain even sparse vegetation on the disturbed land. Before the environmental impact of surface mining in arid regions can be fully assessed, revegetation methods effective in areas of low average rainfall must be developed, soil conditions must be investigated, and groundwater movements must be determined. (Groundwater is the water beneath the earth's surface, between saturated soil and rock, that supplies wells and springs.) Much work remains to be done. However, it is already clear that strip-mining, whether it is conducted in Appalachia, in the arid regions of the Southwest, or in the harsh climate of the northern coalfields, can seriously affect the natural ecological balance of the region.

Costs of Surface Mining in the U. S.

The true economic and social costs associated with the surface mining of coal have not yet been adequately assessed. Coal can be strip-mined at an average cost of about \$1.50 per ton less than deep-mined coal, largely because a number of important externalities,[7] such as the cost of adequate land reclamation work, are not included in the production costs. For a coal density of 1,800 tons per acre-foot and a recovery factor of 80 percent, the yield of a typical three-foot-thick seam is about 4,300 tons per acre. This means that some \$6,000 per acre could be spent cost comparisons are difficult. State regulations generally require only what is called primary reclamation, in which heavy equipment is used to bury acid-containing materials and to smooth, but not grade, mine-field ridges. The major aim is to alleviate water pollution due to acidic mining wastes. Strip-mine fields that have undergone only primary reclamation still retain long furrows, as though turned up by gigantic plows. Furthermore, the fertile topsoil, saved carefully in the German mining technique, is forever destroyed or buried in the typical American strip-mine field. The field that has undergone primary reclamation is a wasteland on which only the hardiest vegetation can survive. Reclamation practices are somewhat better in Pennsylvania and Ohio, which have stricter standards than other Appalachian states.

The few U.S. studies on the cost of reclamation may be of limited application elsewhere in the country. In a U.S. Bureau of Mines project in Pennsylvania in 1966, the costs of grading and filling a strip-mined area ranged from $912 to $2,770 per acre, depending upon the methods used. Reforestation or planting of forage for grazing could add another $250 per acre to such costs, depending upon the type of planting, according to CSPI estimates. Another study was of 22 different government-sponsored projects to reclaim 178 acres of strip-mined land for a state park in western Pennsylvania. The costs ranged from $420 to nearly $2,600 per acre.

These and other studies must be used cautiously to assess the expense of comprehensive reclamation. The estimates are well within the range of reclamation costs in West Germany, but the German restoration practices are far more extensive than anything that has been done on a large scale in the United States. On the other hand, the German system appears to be profitable in aspects that are neglected in America. The German approach promotes systematic utilization of strip-mined material—coal, fertile topsoil, even sand and gravel. The restored land again provides economic values for agriculture, forestry, or recreation. In the United States, by comparison, strip-mining has turned large areas into economic liabilities. For example, Appalachia must spend millions of dollars each year to control or repair the damage caused by acidic water that drains from strip and deep mines. The Appalachian Regional Commission estimated in 1969 that a 90 percent reduction of acid mine drainage at its source in the region would decrease such costs by $4.2 million for industries, utilities, waterborne commerce, municipal water suppliers, and culvert and bridge authorities. A total of 12 percent of the acid damage is from strip-mine drainage. The meticulously planned approach to strip-mining in West Germany offers many lessons for controlling wasteful mining practices in the United States. J.M.

on land restoration without destroying the competitive position of strip-mined coal with respect to deep-mined coal. In West Germany, the actual costs of restoring mined-out lands to full agricultural productivity range from $3,000 to $4,500 per acre. To the extent that strip-mining is carried out without subsequent restoration of the land, a portion of the true production costs of strip-mined coal is being imposed on neighboring communities and on posterity.

It is easy to find evidence that the burden of external costs associated with strip-mined coal is not fairly borne by the consumer. In some regions of Appalachia, approximately 40 per cent of the maintenance budgets for state and county roads is used to repair damage caused by heavy coal-truck traffic.[8] Land made worthless by strip-mining operations no longer serves as a tax base to provide needed revenues for local community development. Floods and landslides resulting directly from conditions created by coal surface mining destroy both public and private property. Funds that could otherwise provide needed local services are thereby diverted to repair the damages. Poverty, apathy, and blighted social development all too often characterize our coal-producing regions.

The consumption of coal for electric power generation is expected to increase greatly during the coming decades. Perry has estimated that the use of coal for this purpose alone will reach some one billion tons by the year 2000.[9] This would represent nearly a threefold increase over present coal consumption rates for electric power generation. More and larger strip mines may be expected. The problem of meeting the ever-growing energy demands of society without needlessly destroying land, water, and forest resources must somehow be resolved.

In view of the extensive environmental damage already inflicted by coal surface mining, and the anticipated future

The key to the West German success in land restoration lies in detailed advance planning based on the needs of the region as a whole.

E. A. Nephew

In the foreground above, hay is harvested in West Germany on reclaimed land less than two years old. In the background, the adjacent lignite mining operation continues.

Thomas J. Kovarik

A German-made strip-mining machine operated by the Peabody Coal Company 60 miles southwest of Chicago, Illinois.

It has been estimated that, as of January 1, 1970, the cumulative past production of 4.4 billion tons of strip-mined coal resulted in 2,450 square miles of disturbed land in the U.S. At present coal-strip mining rates, roughly 100 square miles of additional land are being damaged each year.

Courier-Journal and Louisville Times

growth of the industry, *the adoption of sound mining and land restoration practices is urgently needed.* The environmental effects of coal surface mining are clearly regional in nature, extending across state boundaries, so that it is difficult for the individual states to deal with them effectively. For this reason, federal mining and land restoration standards that would apply equally to all of the coal-producing states seem highly desirable.

Mining and Restoration in West Germany

The land restoration policies adopted in the Rhineland brown-coal (lignite) fields of West Germany represent one possible way of supplying the energy needed by society while also preserving the environment. The German program for dealing with the social and environmental effects of surface mining affords a valuable reference point in considering the relative merits of various surface-mining control options that have been proposed in the United States. General descriptions of the land restoration practices of the German state of North-Rhine Westphalia are available in the public literature.[10] We have supplemented this information by making an on-site visit to the Rhineland brown-coal fields to gain a firsthand impression of the effectiveness of the land reclamation techniques employed and to obtain a more detailed insight into the regulatory process. The Germans appear to have developed an exemplary process for exploiting the mineral resources of a region without permanently impairing the quality of its environment.

The total West German production of brown coal in 1970 amounted to 108 million tons, of which some 81 million tons were burned in thermal power stations to produce 60 billion kilowatt-hours of electricity. (A kilowatt is equivalent to 1,000 watts; the kilowatt-hour, a common unit of electric power consumption, is the total energy developed by a power of one kilowatt acting for one hour.) This represents 38 percent of all the power generated in the nation's thermal electric power plants in 1970. The importance of brown coal to West German industry, therefore, can hardly be overemphasized. The very existence of such an important domestic energy source exerts a strong stabilizing influence on the economy of the nation, which nevertheless was forced to import 55 per cent of its primary energy during the past year. Thus, it is not possible in West Germany to consider seriously the luxury of banning the surface mining of brown coal. Instead, methods of mining and land restoration had to be developed which would permit continued production of brown coal without incurring serious environmental damage.

Brown-Coal Mining. The Rhineland brown-coal fields lie in flat plains country in the triangle formed by the cities of Aachen, Cologne, and Düsseldorf. Surface mining of brown coal currently encounters overburden thicknesses of up to 550 feet when mining coal from beds 50 to 350 feet thick. The coal bed lies on a slightly inclined plane, near the ground surface in the vicinity of Cologne, but is buried under several hundred yards of overburden near Düsseldorf. For this reason, mining began in the southern portion of the coalfield during the latter half of the nineteenth century and has moved steadily northward, becoming progressively more difficult. The final depth of open-pit mines currently being worked is as much as 900 feet. Such mining required the development of giant wheel excavators and a complex transportation system of conveyor belts and trains to haul away the spoil material and lignite. In 1970, some 243 million cubic yards of spoil were excavated and transported to worked-out mines for use as fill material. In addition to necessitating the moving of earth, each ton of brown coal produced requires pumping about fourteen tons of water out of deep wells to lower the groundwater level. Nearly 2,000 such deep wells have been drilled near the mine pits for this purpose.

The problems of economically moving such massive amounts of material have been solved by introducing large-scale, automated equipment, which increases worker productivity. Excluding maintenance personnel, only two men are needed to operate a 7,400-ton wheel excavator, which has a digging capacity of 130,000 cubic yards per day. Larger machines, weighing 13,000 tons and having a capacity of 260,000 cubic yards per day, have already been ordered and are scheduled to begin operation in late 1975. A 70-mile-long network of conveyor belts moving at speeds up to 12 miles per hour delivers the coal and spoil material to trains to be hauled away. The trains move on some 300 miles of special heavy-duty track, and the locomotives are controlled remotely by radio signal during loading and unloading. These measures have increased the average worker productivity to 64 tons of brown coal per man-day. The productivity at the largest, most modern mine, located at Garsdorf, amounts to 81 tons per man-day, or to a heat equivalent of 22 tons of bituminous coal. The average productivity in U.S. bituminous coal strip mines is 35 tons per man-day.

Land Restoration. Because of the steady northward progression of mining operations during the past 50 years, the various stages of the land restoration process are open to view, spread out in sequential order. At the active mines in the northern and central portions of the brown-coal field, the huge wheel excavators selectively strip off and save the top layer of loess (an extremely fertile type of loam—a soil consisting of a mixture of clay, silt, and sand); remove the remaining sand, gravel, and clay overburden; and extract the loose, black layers of exposed lignite. Immediately to the south, mammoth spreader machines fill the overburden back into mined-out pits while bulldozers level it out in preparation for applying the top layer of loess. Still further southward, the leveled areas are subdivided into five- to ten-acre tracts by loam dikes. These will be filled with loess slurry (a watery mixture) which dries out after several months, leaving behind a three- to six-foot thick top layer of loess. Near Berrenrath, fields of grain and hay are already thriving on land that was restored less than five years ago. The sequence in the forested areas is similar: To the north are newly planted stands of young trees less than five years old, and in the south are recreational and forested areas reclaimed in the 1920s. The latter are nearly indistinguishable from natural forests and are superior to the stands of scrub timber which originally grew there.

Different Conditions. Brown-coal mining in West Germany differs greatly from Appalachian strip-mining in the topography, the type of technology employed, and the degree of government regulation imposed upon the mining industry. In the German lignite fields, excavation is easier because the coal beds are not covered with rock strata as in Appalachia. The terrain is relatively flat, and sulfur-bearing minerals, which produce acid wastes, are not present. Furthermore, the lignite fields are located in a rich agricultural area, providing a strong incentive for restoration of the land after mining is completed. In addition, almost all of the brown-coal resources are located within a single state. This makes it easier for the state to impose adequate land reclamation requirements because significant competition from neighboring states does not exist. As a consequence, nearly all government control of brown-coal surface mining is by the state of North-Rhine Westphalia rather than by the German federal government. Federal laws affecting surface mining in West Germany are general in nature; for example, water quality regulations apply to other industries as well.

Comprehensive Approach. The German land restoration program begins long before the first shovel of brown coal is mined. It begins with detailed plans for the evacuation and relocation of populated settlements and for the restoration of land after the mining operations have ceased. Thus, land-use patterns are proposed and approved far in advance, and the new landscape is planned accordingly—the topography, the water drainage system, lakes, and the designation of areas to be restored for forestry and for agriculture. Such comprehensive early planning allows the mining operations to be tailored to fit the land restoration work which will follow. Modern principles of city planning are used in designing new towns for the displaced people: Residential areas are removed from the main flow of traffic; green areas for recreation are provided; and the towns are more compact than the former unplanned settlements. The basic costs for land reclamation and population resettlement are borne by the mining company. Local and state governments provide supplementary funds to cover the incremental costs of providing better schools, sewer systems, and other community services than existed at the former town site.

This comprehensive approach is based upon an acceptance of the fact that brown-coal mining moves not only coal, but also trees, buildings, people, and the land itself. In most cases, conditions are vastly altered. The state of North-Rhine Westphalia and the lignite mining industry have accepted the responsibility of finding feasible solutions to the entire set of social and environmental problems created by brown-coal surface mining. This approach makes it possible to treat the overall problem as an integral whole rather than dealing with single problems on a piecemeal basis. This approach represents a major change in the philosophy of what constitutes mining. The old concept, which appears to be prevalent in the United States, holds that mining consists merely of extracting minerals from the ground in the quickest, most "economical" manner possible. The new concept includes the qualification that this must be done in a way consistent with the needs of society as a whole. The planning and enforcement methods used in West Germany to assure that this latter goal will be achieved are thus of great interest and relevance.

Government Regulation of Surface Mining

Historical Development. The present form of the brown-coal industry and the adoption of enlightened land restoration practices in Germany emerged gradually over the past several decades. Prior to 1960, four large mining companies dominated the lignite mining industry. In 1960, economic pressures, arising partly from the need to automate the mines, forced the four companies to merge into a single firm, the Rheinische Braunkohlenwerke A.G. mining company. Even before this economic regrouping of the industry, the public became concerned about the large tracts of unreclaimed land left over from World War II. This concern was particularly evident in Cologne, where the proximity of the mining areas made the disturbed lands highly visible to great numbers of people. As a result, new surface-mining control legislation was enacted in 1950 to assure orderly, well-planned mining practices.

On March 11, 1950, the state legislature of North-Rhine Westphalia passed West Germany's first Regional Planning Law. This law, later modified in May 1962, established a Land Planning Commission charged with the responsibility of developing overall guidelines for land use within the region. The main purpose of the commission is to coordinate the diverse social, economic, and industrial activities of the region. With this aim, the commission designates land areas for agriculture, forestry, and industry, and specifies the boundaries of population settlements. It develops long-range plans for transportation networks, the preservation of historic sites, and the construction of recreational facilities to serve the entire region. Later in the same year, on April 4, 1950, the state legislature enacted two additional laws applying specifically to the brown-coal-producing areas of the region. These were The Law for Overall Planning in the Rhineland Brown-Coal Area and another law establishing a community fund to finance land restoration. The first of these laws formed the Brown-Coal Committee, which develops detailed plans for exploiting the lignite resources of the state within the framework of the overall regional planning law.

The basic objective of the Brown-Coal Committee is to ensure that land areas temporarily used for brown-coal mining will not become permanently devalued and made unsuitable for more lasting uses. This means that it is not sufficient merely to prevent the creation of moonscapes by requiring that the land be restored for forestry or agriculture. Rather, in light of the general objectives of the overall regional planning, the land must be restored so that it will harmonize with the social, cultural, and industrial interests of the rest of the region. The Brown-Coal Committee is composed of 27 members especially selected to represent the interest groups affected by mining operations. This broad base provides a coordination of the various conflicting interests long before actual mining activities begin. The committee formulates land restoration requirements based on the future use of the land as defined in the regional planning program.

The Brown-Coal Committee. The composition of the Brown-Coal Committee, as fixed by law, is shown in Figure 1. The primary function of the committee is to review and consider proposals for extending mining operations to new land areas and to make appropriate recommendations to the minister-president of North-Rhine Westphalia. As can be expected in view of the composition of the Brown-Coal Committee, the final recommendation to the state government is based on considerations of overall land use, conflicting local issues, and national energy requirements. The Brown-Coal Committee has gradually emerged as a powerful force defining the conditions under which the brown-coal industry must operate. Its existence subjects the brown-coal industry to public scrutiny and has been instrumental in bringing about the conservation practices of the industry. The Brown-Coal Committee serves as a quasi-public forum where the divergent interests of society can be considered before mining commences. Public hearings and the signature of the state chief executive are required before the recommendations of the committee become legally binding.

The introduction of requirements that a certain portion of the land disturbed by the surface mining of brown coal be restored to agricultural productivity illustrates the importance of such a planning and review body. Although reforestation of disturbed lands has been carried out since the early 1920s, a coalition of agricultural groups within the Brown-Coal Committee became concerned over the destruction of fertile farmland by the mining operations. In the late 1950s, this coalition of agricultural interests, known as the "green front," successfully introduced requirements that the valuable top layer of loess, often fifteen to twenty feet thick, be saved, and that a portion of the land disturbed by surface mining be restored to agricultural productivity.

A break-even point has recently been reached in West Germany because of rising land prices and lowered reclamation costs brought about by the development of more efficient land restoration methods. In the United States, the costs of full land restoration would in most cases greatly exceed the value of the restored land. However, it is interesting to note that reclamation was required in Germany long before it became marginally profitable. Because of land restoration requirements, the rich, thick layer of loess is selectively saved and is now regarded as an important and valuable mineral in its own right. Similarly, commercial exploitation of the sand and gravel contained in the overburden has begun. Thus, the extraction of brown coal is becoming a total mining operation. Of the 53,000 acres of land that have been disturbed by brown-coal surface mining, 33,000 acres have already been restored for forestry, agriculture, and recreational uses. The costs of restoring mined-out lands to full agricultural productivity range from $3,000 to $4,500 per acre.

Interaction Between Planning and Enforcement. The key to the German success in land restoration lies in detailed advance planning based on the needs of the region as a whole. When it is deemed necessary to extend mining operations to new, unopened land areas, the brown-coal mining company submits a proposal containing comprehensive mining and land restoration plans to the Brown-Coal Committee. The committee examines the proposal with respect to regional planning guidelines and hears testimony from technical experts, representatives from the enforcement agency, and the land planning commission. Following committee discussion and review, the original plan may be accepted, modified, or rejected. When final committee approval has been obtained, public hearings are held and the plan is then sent to the titular head of the state land planning commission for adoption.

After the plan has been finally adopted, the state enforcement agency assumes the responsibility of supervising its implementation and assuring that the mining and land restoration activities are carried out in accordance with its stipulated provisions. The mining company is required by law to submit all information which the state enforcement agency needs to carry out its regulatory function. For example, the brown-coal mining company routinely submits aerial survey photographs of its mining and land restoration progress every six months. The planning and enforcement

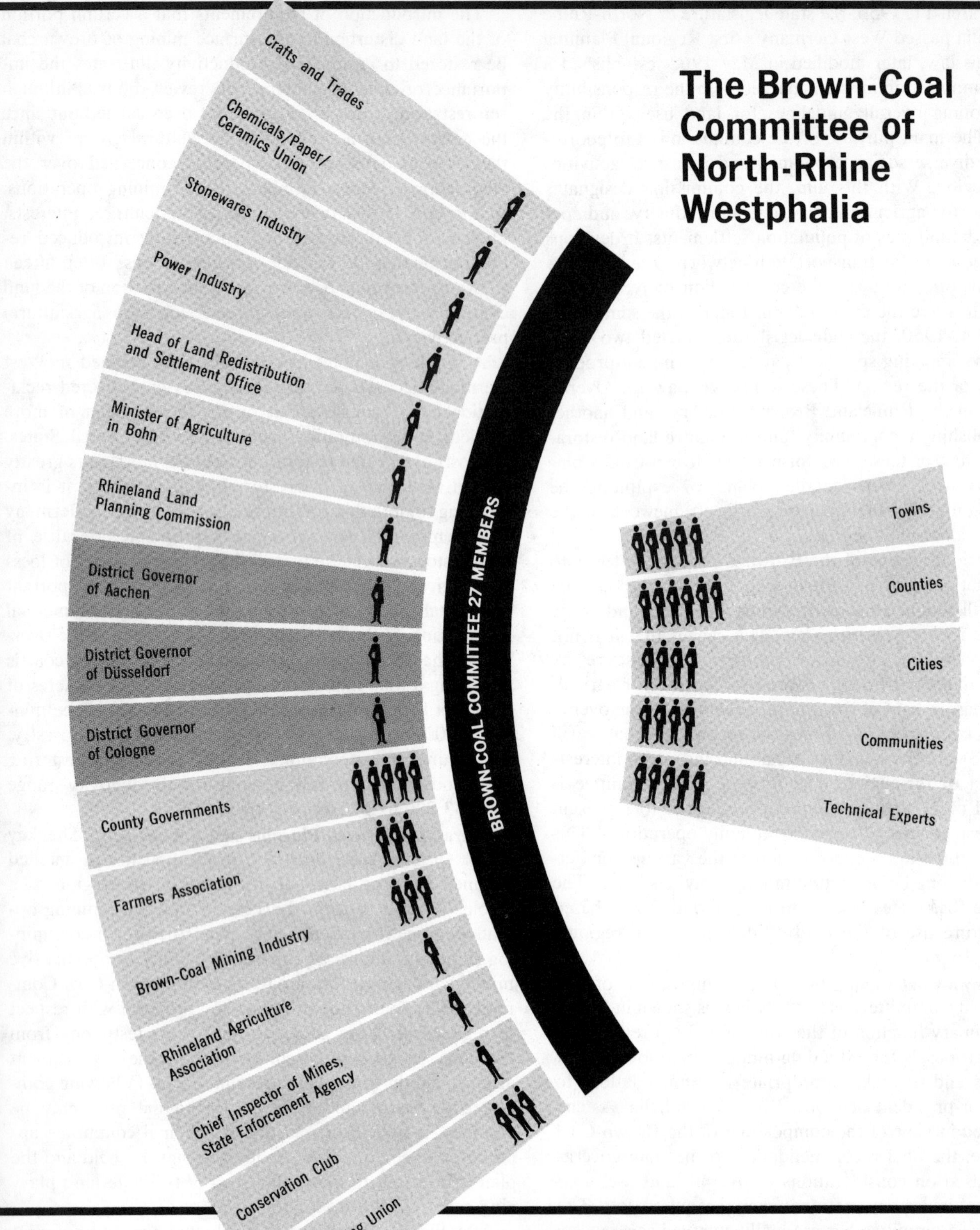

process, with participation of nonmining interests, affords flexibility in resolving the social and environmental problems posed by surface mining. The recommendations of the Brown-Coal Committee function as a living law which changes and adapts to the requirements of specific situations. Since the deliberations are made well in advance of actual mining, sufficient lead time is available for a full consideration of all of the issues and problems.

Application to U. S. Strip-Mining Problems. Some elements of the German surface mining and land reclamation

FIGURE 1

WHAT IT IS

The Rhineland brown-coal fields, which contain 95 percent of the lignite in West Germany, lie in flat plains country in the triangle formed by the cities of Aachen, Cologne, and Düsseldorf in the state of North-Rhine Westphalia. The district governors of these three cities, which are the ones most affected by strip mining operations, share the revolving chairmanship of the Brown-Coal Committee, which was established to ensure that land areas temporarily used for brown-coal mining do not become permanently devalued. The committee is composed of 27 members especially selected to represent the interest groups affected by mining operations. Its primary function is to review proposals for extending mining operations to new land areas and to make appropriate recommendations to the minister-president of North-Rhine Westphalia.

HOW IT WORKS

When it is deemed necessary to extend mining operations to new, unopened land areas, the brown-coal mining company submits a proposal containing comprehensive mining and land restoration plans to the Brown-Coal Committee, which examines the proposal and hears testimony from the chief inspector of mines of the state enforcement agency, the Rhineland land planning commission, any towns, cities, communities, or counties specially affected by the proposal, and any state or private technical experts the committee wishes to call upon. After deliberation and committee approval of the proposal, public hearings are held; the plan is then sent to the minister-president of North-Rhine Westphalia (who is also the head of the land planning commission) for adoption. After the plan has been adopted, the state enforcement agency assumes responsibility for supervising its implementation and assuring that the mining and land restoration activities are carried out in accordance with its stipulated provisions.

techniques are applicable to U.S. strip-mining in spite of important differences in the climate, terrain, and geological features of the mining regions. Wheel excavators of the type used in the Rhineland have already found limited use in North Dakota and Illinois, where they are used to remove soft and unconsolidated overburden. These machines provide continuous operation and can deliver the broken-down overburden by conveyor belt to any point desired. They are especially suitable for separating the fertile layer of topsoil from the remaining overburden and saving it to be used later for land reclamation purposes. The slurry technique of applying topsoil to graded areas being recultivated can almost certainly be applied in some areas of the United States. The considerable amount of basic research which has been performed in West Germany to determine the most suitable trees and plants for revegetation, and the factors affecting their growth rates, may be helpful in our own country. Of primary interest, however, are the institutional arrangements that have been worked out to provide adequate regulation of surface mining and full restoration of the affected lands.

In devising a policy for the United States, consideration should be given to the German experience. The German program has been in effect for some twenty years and has been highly successful in minimizing social dislocations and environmental damage from brown-coal surface mining. The German program embodies four main principles that have contributed greatly to its success. First, the regulation of surface mining is incorporated within an overall regional development plan. Second, a planning body composed of diverse public interests participates in formulating detailed requirements for mining and land restoration long before the actual mining begins. Third, the recommendations of the planning body are reviewed in public hearings. Fourth, an enforcement agency is provided with the necessary powers to enforce the approved plan. The German program offers visible evidence that, with detailed advance planning, striking successes can be achieved in reducing environmental damage from strip-mining at a price easily borne by the consumer. □

NOTES

1. Udall, Stewart L., *Surface Mining and Our Environment—A Special Report to the Nation,* U.S. Department of the Interior, 1967. *Study of Strip and Surface Mining in Appalachia,* U.S. Department of the Interior, 1966. *The Strip Mining of America—Analysis of Surface Coal Mining and The Environment,* Sierra Club, New York, July 1971.

2. *Bituminous Coal Data,* 21st edition, National Coal Association, 1970.

3. Averitt, Paul, *Stripping—Coal Resources of the United States,* Geological Survey Bulletin 1322, Jan. 1, 1970.

4. *Acid Mine Drainage in Appalachia,* A Report by the Appalachian Regional Commission, 1969.

5. Collier, C. R., et al., eds., "Influences of Strip Mining on the Hydrologic Environment of Parts of Beaver Creek Basin, Kentucky, 1955-66," Geological Survey Professional Paper 427-C, 1970.

6. *Fact Summary of the Southwest Power Plants: Ecological and Cultural Effects: Recommended Action,* prepared by Native American Rights Fund, David H. Getches, director, reprinted in the Congressional Record—House of Representatives, May 10, 1971.

7. Van Tassell, Alfred J., ed., *Environmental Side Effects of Rising Industrial Output,* Heath Lexington Books, D. C. Heath and Company, Lexington, Mass., 1970.

8. Vance, Kyle, "Coal Trucks' Damage to Roads is Costing Kentucky Millions," *Louisville Courier-Journal,* Mar. 5, 1971.

9. Perry, Harry, Testimony before the Joint Committee of Atomic Energy, Nov. 4, 1969, in "Environmental Effects of Producing Electric Power."

10. Udall, Stewart L., "Natural Resources Mission in Germany—1966, A Special Report to the President." Ratcliff, J. D., "Transplant in Germany's Heartland," *Readers' Digest,* British edition, Oct. 1968.

> **"What might be highly desirable standards for Chicago, Gary, Indiana, or Birmingham would be an atrocious deterioration of the air quality in Montana."** *(Billings Gazette)*

Bulletin Special Report

The Great Montana Coal Rush

SALLY JACOBSEN

A modern day coal rush, triggered by nationwide demands for clean energy, is underway on the eastern plains of Montana. Underlying the region are 30 billion tons of subbituminous and lignite coal easily recoverable by strip mining. The deposits are part of the Fort Union formation—thought to be the largest coal basin in the world—that extends into Wyoming, the Dakotas and Saskatchewan. The basin contains 40 per cent of the U.S. reserves.

This year an estimated 16 million tons will be strip mined in Montana. By 1980 the figure is expected to reach 75 to 80 million tons. Much of the coal is being shipped to the energy hungry Midwest. In 1972, Commonwealth Edison bought close to seven million tons of coal from Montana and Wyoming mines to fire its generating plants around Chicago.

Reminiscent of gold miners staking their claims, coal companies are rushing to the Big Sky country to acquire leases on the coal. The coal is highly desirable because its low content of sulfur, sodium and ash makes it a cleaner fuel than Appalachian or Illinois coal.

The new bonanza has become an environmental cause celebre among conservationists, federal and state officials, and coal and power people. At issue is the reluctance of the Department of Interior to halt strip mining pending completion of a three-year task force assessment of the environmental, social and economic impacts of development.

In October 1972, the U.S. Senate passed a resolution calling for a temporary moratorium on federal coal leasing in Montana until protective surface mining legislation is enacted. Interior Secretary Rogers C. B. Morton refused to honor the resolution, saying "it would be unwise" to do so "when there is a continuing need for coal."

Until stiff federal legislation is passed, responsibility for regulating mining falls to the state. "Don't make Montana another Appalachia" is a commonly expressed sentiment, referring to the once verdant hills and valleys in West Virginia and Kentucky that now lie wasted after decades of strip mining.

The magnitude of the coal potential was surveyed in the North Central Power Study (NCPS) by the Bureau of Reclamation and 35 power suppliers, including Iowa Electric Light and Power Company, Northern States Power Company (Minnesota), and Idaho Power Company.

The study suggested that 42 generating stations be constructed at the mouths of strip mines in five states to convert the coal to electricity and send it via high voltage transmission lines to the Midwest. The total electrical energy produced from this array of stations would be exceeded only by the total electrical production of the United States or of the USSR, charged the Environmental Defense Fund.

Already one mine-mouth generating plant is under construction at Colstrip, Montana, by Montana Power Company and Puget Sound Power and Light Company.

As reported in the *Billings Gazette*, Nov. 29, 1972, Consolidation Coal Company, a subsidiary of Continental Oil Company, has proposed building a $1 billion coal gasification complex on the Northern Cheyenne reservation. According to the *Gazette*, the company wants to acquire mining rights to a billion tons of coal beneath Indian lands. A spokesman for Consolidation said the gasification proposal was not definite and was "at best tenuous."

Consolidation has leased approximately 1.2 billion tons of coal in Montana. It is presently operating a test pit in the Bull Mountains north of Billings.

Westmoreland Resources—a partnership of Kewanee Oil Co., Morrison-Knudsen Co., Penn-Virginia Corp., and Westmoreland Coal Co.—has mineral leases on 34,000 acres of coal owned by the Crow Indians in Montana. Westmoreland has contracts to ship 76.5 millions tons of coal over a 20-year period beginning March 1, 1974, to Northern States Power Company (Minn.), Interstate Power Company (Iowa), Dairyland Power Cooperative (Wis.), and Wisconsin Power and Light Company. As of January 1973, however, Westmoreland had not begun mining.

Reprinted with permission from *Science and Public Affairs*, Bull. of the Atomic Scientists, vol. 29, pp. 37–42, Apr. 1973.

The deposits lie in the Crow ceded strip, north of the Crow reservation, near Sarpy Creek in southwestern Montana. In 1904 the Crows ceded that portion of land to the federal government. It eventually was homesteaded by settlers who obtained surface but not mineral rights. In 1958, complying with a Supreme Court ruling, Congress returned the mineral rights to the Crows. Before Westmoreland can mine the coal in 1974, it must acquire the surface land lying above the coal deposits. To date, it has purchased or has options to purchase six ranches in the area.

Patchwork Ownership

This patchwork of ownership of surface and mineral rights is typical of the entire region. Early settlers to eastern Montana claimed both surface and mineral rights, but after 1910 the federal government retained the mineral rights. To help pay the construction costs of opening up the West, the government granted alternate sections of public land to the railroads. The Northern Pacific Railroad, now Burlington Northern, which had a route through Billings and Butte, received 44 million acres in land grants. Eventually much of the land was sold, but the railroads kept their valuable mineral rights.

Today, the majority of coal deposits are owned by the federal government and Burlington Northern. The remainder is held by private individuals, the state, and Northern Cheyenne and Crow Indians. In many cases, the surface owner does not own the underlying minerals, which hinders attempts to effectively regulate coal development.

The scene of all this activity is primarily cattle ranching country. Lying in the Great Plains region, eastern Montana is made up of vast expanses of flat land broken by sharply rising buttes and more gently rolling hills. Pink slashes across the buttes are scoria, rock formed by the heat rising from burning coal beds. It is an overt sign of the wealth of thick coal seams beneath the surface.

A Different Life-Style

Towns are few and far apart; their elevation figures often exceed the population. There are still signs that the Old West way of life is not completely dead. A ranch is sometimes called an outfit, and ranchers ride their horses under clear, blue skies that don't seem to end. It is far different from the life-style of the urban areas that need the precious coal for fuel.

Strip mining, the process of tearing back the soil and rock or overburden with giant shovels to expose the coal seams, is going on at Colstrip, 90 miles east of Billings. Peabody Coal Company, a subsidiary of Kennecott Copper, operates its Big Sky Mine near there; and across the road Western Energy Company, a subsidiary of Montana Power Company, operates another.

Peabody mines about 1.5 million tons annually in Montana to fulfill a long-term—about 35 years—contract with Minnesota Power and Light Company. Over the length of the contract, the coal company expects to deliver 100 million tons of coal. Peabody's total reserve holdings in Montana are approximately two billion tons.

Western Energy has approximately 850 million tons of coal reserves in Colstrip alone with contracts to deliver 350 million tons to Northern States Power, Wisconsin Power and Light, Puget Sound Power and Light, and Montana Power.

"Be practical — people always need coal."

Colstrip was built by the Northern Pacific to supply its diesel engines with coal. Spoilbanks—mounds of wastes from pulling back the overburden—on the northwest edge of town are remnants of former mining days. A few cattle graze along the dirt road that runs past the spoilbanks and along the railroad tracks.

Ranching vs. Mining

"Mining and cows don't mix," said rancher Wallace D. McRae. McRae raises quarter horses and Hereford cattle on land south of Colstrip near Rosebud Creek. He owns 27,189 acres and leases 80 acres from the federal government, 640 acres from the state and 960 from private owners. He owns all the minerals beneath his land except iron and coal, which are held by the federal government and Burlington Northern.

Peabody would like to buy his land, but to date he has refused to sell. Peabody bought his neighbor's land to the south, and Western Energy's operations lie to the north of his property.

A thoughtful man, McRae reflected at his ranch on the "psychic wages" of living in Montana. "Everyone here is sacrificing income for psychic wages. I only get a two per cent return on my operation. I made the decision that I'm not interested in the dollars. I wouldn't sell out for $1,000 an acre," said McRae, whose weathered face belies his 36 years.

He thinks the coal company would like to buy his land and lease it back to him for ranching. But, McRae, a third generation rancher, thinks cattle raising and mining are incompatible. Because he does not hold the coal rights, the threat of being forced off the land is real. According to Montana law, mining is considered a public use, and coal companies have the right of eminent domain.

"I can't invest any money in long-range plans. I can't spend money to level and ditch land. I run my outfit on a day-to-day basis," explained McRae.

Although McRae, like other ranchers in the area, knew that minerals underlay the land, it wasn't until the North Central Power Study was released in October 1971 that he realized the potential for development of eastern Montana.

Forty-two sites in five states were named as having strippable coal reserves capable of supporting 1,000 megawatt (mw), or larger, generating units for a 35-year operating life. As proposed, the combined 42 units would supply 200,000 mw of thermal generation. Suggested locations of the generating stations were: Montana, 21; Wyoming, 15; North Dakota, 4; and South Dakota and Colorado, 1 each.

In a letter commenting on the NCPS, the Environmental Defense Fund said: "The NCPP [North Central Power Project] will produce some 403 billion kilowatt hours of electricity per year. Only the total electric energy production of the U.S. or the USSR would now exceed that of NCPP."

The letter charged that the project, even if it met present pollution standards, would emit "far more nitrogen oxides, sulfur dioxide and particulate matter than all sources in New York City and the Los Angeles Air Basin combined."

The first mine-mouth generating plant is being constructed at Colstrip. Two 350 mw units are scheduled to be operating in 1975 and 1976. The station is being built by Montana Power Company and Puget Sound Power and Light Company.

It will consume 90 million tons of coal over its 30-year life span or 3 million tons per year. Coal will be hauled by truck from the nearby Western Energy mine to a crusher and from there to the plant. Approximately 65 surface acres will be disturbed by strip mining each year—2,000 acres over 30 years.

Water for cooling, boiler make-up and scrubbers is to be taken from the Yellowstone River, 30 miles away.

The plant will give off sulfur oxides, particulates, nitrogen oxides, trace elements, water vapor and heat. In the draft environmental impact statement on the plant filed with the state, Montana Power said the "guarantees on the emission control systems are equivalent to 99.5 per cent particulate removal and 39.7 per cent SO_2 (sulfur dioxide) removal."

Construction of the plant has come under sharp attack. The Northern Plains Resource Council, an environmental group with about 300 members, filed suit to halt construction of the plant but the complaint was denied by the 16th Judicial District Court.

What's Good for Chicago

In a full-page advertisement in the *Billings Gazette*, Montana Power said the plants would meet "all federal and state air and water quality requirements." In an editorial Dec. 6, 1972 the *Gazette* replied:

> Whether it [the Colstrip plant] meets federal and state standards for pollution control should not be construed to mean that it will not pollute. . . Both Montana and federal pollution standards are a compromise, not necessarily what is best or most desirable.
>
> What might be highly desirable standards for Chicago, Gary, Indiana, or Birmingham would be an atrocious deterioration of the air quality in Montana.
>
> Dig the coal if we must to meet the energy crisis but do not convert it to power here. Let those who need it pay the air degradation price. Ship the coal until Montanans know their air will be kept clean.

Just as precious as the clear blue skies is the water supply in this semiarid region. Water, in fact, is the key to extensive development. All the coal fields are far from existing surface water sources, according to a report by the Montana Environmental Quality Council.

The NCPS noted that minimizing water use in this area is desirable since it is "relatively expensive." The total amount of water needed for the 42-plant project is estimated at 855,000 acre feet per year. The Environmental Defense Fund pointed

"To alleviate the gas shortage, oil companies plan to gasify coal in increasing amounts. But gasification of coal requires immense amounts of water. . . . Most of the coal for gasification lies in the mountain states where water is scarce and strictly allotted among agricultural and industrial users. Studies now show that most of the water resources of the area will be required to gasify coal in sufficient quantities to supply utilities in the midwest and western states. If current policies, organized by the energy companies, are pursued, this means an industrial change in the mountain states, a diminution of farming, importation of expensive strip mining equipment, decline in employment. In short, a repeat of the process that 100 years ago ruined the land and the people of Appalachia." (James Ridgeway, The Last Play: The Struggle to Monopolize the World's Energy Resources, *1973.)*

out that this is "well over half New York City's annual water requirements of about 1,500,000 acre feet per year."

In April 1972, the Bureau of Reclamation published an appraisal report on building a network of aqueducts in southeastern Montana and northeastern Wyoming to provide water for the generating stations proposed by the NCPS as well as other industrial development of the area. The conclusion of the report was that, according to industry estimates, about 2.6 million acre feet might be required annually for development "that may be attained in less than 30 years." Water would be supplied from three existing reservoirs and nine proposed ones. Already the Bureau of Reclamation has contracts with 14 coal companies to sell over 700,000 acre feet of water from two existing reservoirs.

The study found that as much as 3.2 million acre feet could come from the Yellowstone River and its tributaries.

Needs Disregarded

The Environmental Defense Fund charged that by the year 2000 the water requirements of this area will exceed "by 80 per cent the present municipal and industrial requirements of New York City (population 7,771,000)." It also contended that the mean annual flow of the Yellowstone River would be reduced by 81 per cent.

Speaking before the Western States Water and Power Consumers' Conference, Montana Senator Lee Metcalf said of the NCPS, "I have never seen another resource report that so casually disregarded the water needs of farmers, ranchers and local governments of the entire region."

A smaller version of the NCPS project exists in the Four Corners region of Colorado, Utah, New Mexico and Arizona. Generating stations are fueled by coal from nearby strip mines to supply customers as far away as Los Angeles. By 1977, 14 coal-fired plants are scheduled to be operating in the Southwest project. The first two were started up in 1963 at Farmington, New Mexico.

William L. Bryan, Jr., an environmental activist for Montana, Wyoming and Idaho, said the magnitude of the proposed North Central project makes the Southwest one "look like a popgun."

NCPS Shelved

However, the NCPS has been shelved for the present time. A spokesman for the Bureau of Reclamation in Billings said it was abandoned because the utilities hoped to find a cheaper source of power, wanted "to go it on their own," and didn't want to put the requisite amount of money in transmission lines, preferring instead to ship the coal out of state.

In its place, Secretary of the Interior Morton announced the creation of an interagency federal state task force called the Northern Great Plains Resource Program (NGPRP) to study the environmental and socioeconomic impacts of development in Montana, Wyoming, the Dakotas and Nebraska. A press release said the NGPRP was "an outgrowth of public concern in the region and of prior studies of the region's resources."

The NGPRP will examine not only coal but also other minerals, such as uranium and bentonite, and the oil and gas reserves in the five states. The task force is anticipated to take three years to complete its report. The study will be carried out by the Departments of the Interior and Agriculture, the Environmental Protection Agency and the Old West Regional Commission, a development and planning commission for the five states.

One observer described the federal study as an "end run play to avoid the controversy aroused by the North Central Power Study." He said that he feels attempts will be made to implement the utilities' study with its 42 mine-mouth plants.

Moratorium Requested

While the impacts of development are being studied, the mining will continue. Last fall Montana Senators Mike Mansfield and Lee Metcalf and Utah Senator Frank Moss introduced S. Res. 377 to halt issuance of coal leases on federal lands in Montana until regulatory legislation is passed.

Said Mansfield, "The coal situation is so serious that executive action is imperative. We want action now, until such time that the Congress in cooperation with the States can develop a uniform set of regulations affecting both private and public lands."

After the resolution passed the Senate, Interior Secretary Morton wrote Metcalf that he felt it would be "unwise" to call a moratorium because of the nation's need for coal. He said that existing regulations and "the extensive procedures the Department has developed for analyzing environmental impacts and insuring compliance" with the National Environmental Policy Act (NEPA) of 1969 "should guarantee environmentally acceptable mining where mining is required to meet real energy needs."

Senators Mansfield, Metcalf and Moss replied:

"It is with deep regret that we recognize it [Morton's action] as the refusal of the Department of the Interior, indeed the arrogance of the Executive Branch, to abide by the resolution."

In contrast to Morton's statement, two General Accounting Office (GAO) reports criticized the administration of federal coal leases and of regulations to protect the environment.

One study, requested by a House subcommittee on Conservation and Natural Resources, found that of the prospecting permits and coal leases investigated, the regulations implemented by the Interior Department in 1969 were not being effectively enforced to protect the environment. Both the Bureau of Land Management (BLM), which administers the federal permits and leases, and the Bureau of Indian Affairs (BIA), which oversees Indian lands, were criticized for not preparing environmental impact statements as required by NEPA.

The second study, requested by Senator Metcalf, stated that the most satisfactory reclamation work was being done voluntarily by mine operators or in compliance with state reclamation laws.

In Montana and North Dakota, there are 37 outstanding federal coal leases totaling 52,588 acres; 17 are in Montana. Total production on 12 leases in 1971 was 1.4 million tons. There was no production on 25 leases.

In view of this, Edwin Zaidlicz, state director of the BLM, denied 119 prospecting permit applications over a year and a half period in Montana and North Dakota. Fourteen prospecting permits are valid on 37,544 acres. By his action, Zaidlicz stymied for a time the rush of coal companies to acquire leases. Zaidlicz said in his Billings office that he denied the permits because there was no need to prospect when the known supply of coal under lease was not being developed.

Zaidlicz said he was acting on three premises: the BLM attempts to have orderly development of coal resources; the BLM is required to get a fair market value for the coal; and the BLM must take environmental considerations into account.

The GAO study, requested by Metcalf, said that coal leases were being held for speculative purposes. The report stated that this coal was not being mined because companies hoped that the coal would be more valuable in the future due to "technological breakthroughs in developing methods and processes for converting coal to gaseous, liquid and solid fuels and because of new demands from existing markets and the adoption of new uses for coal."

On February 13, Secretary Morton issued an order rejecting all pending and future applications for prospecting permits "until further notice." This was done, said Morton, "to allow the preparation of a program for the more orderly development of coal resources upon the public lands . . . with proper regard for the protection of the environment."

To lease coal deposits, a company must first obtain a prospecting permit. If commercially-exploitable reserves are located with the permit, the permit holder is issued a federal coal lease to extract the mineral without going through a competitive bidding process. This arrangement is called right of preference, which, said Zaidlicz, is given in recognition of the contribution made in locating valuable resources.

Leases on Indian lands are administered differently. Land is offered on a competitive bid basis with an exclusive prospecting permit, which includes an option to lease the lands at a later date. James F. Canan, area director of the BIA, said that 350,000 acres of land on two Indian reservations in southwestern Montana are under permit. Leases are held by Peabody Coal on 16,000 acres on the Northern Cheyenne reservation. In addition to Westmoreland's holdings on the Crow ceded strip, Shell Oil Company holds leases on 30,000 acres on the Crow reservation.

Canan said that mining companies must submit a mining plan, including proposed reclamation work, to the BIA and the Indian tribal council. The Geological Survey provides technical advice.

Interior Department regulations concerning reclamation of strip-mined land do not apply to

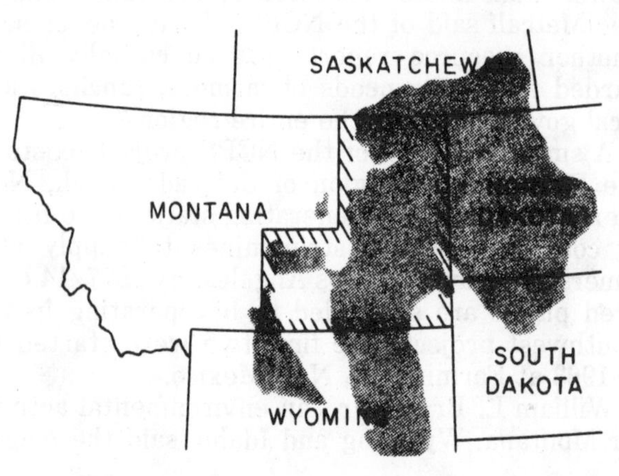

Fort Union Coal Region

surface lands not owned by the federal government.

During the last session of Congress, the Coal Mine Surface Area Protection Act, a surface mining regulatory bill, passed the House, but not the Senate. This act would have covered federal, state, private and Indian lands. Rep. John Melcher, whose congressional district is eastern Montana, is a member of the Committee on Interior and Insular Affairs, which drafted the legislation.

In a statement to the *Bulletin*, Melcher said, "Land must not be strip mined unless it can be reclaimed and is restored to equal or better productivity." He emphasized that mining companies must assume the burden of proof that the land can be reclaimed.

Without this federal legislation, the state is responsible for regulating mining and reclaiming the land. Because of this, the current legislative session is considered to be one of the key sessions in the state's history.

Governor Thomas L. Judge, during the 1972 gubernatorial campaign, listed five means of strengthening strip-mining legislation:

1. Power of selective denial. This would allow the state to deny coal contracts to mine on historical sites, steep slopes or areas where reclamation would be difficult or impossible.

2. Top soil replacement. Coal companies would be required to set aside top soil from the mining site so that it could be replaced after the mining was finished.

3. Higher performance bonds. Corporate surety bonds with the state are required to insure that strip mined land is reclaimed.

4. Annual permits. This would replace the current system of providing almost unlimited contracts. Yearly permits would allow the state to review coal companies' operations.

5. Higher severance tax for the state. This tax is imposed on production of minerals, that is, on the minerals severed from the ground. Judge proposed legislation to create a resource indemnity trust account that would require an annual $25 mining tax plus a tax on the gross value of a product in excess of $5,000. The present tax is a sliding one based on British thermal unit (Btu) rating.

"If we don't have strict laws, then I would support a moratorium [on strip mining] until proper legislation to reclaim the land is passed," said Judge.

A state task force had been appointed by the former Governor, Forrest H. Anderson, in summer 1972 to study the impacts of coal development on the state.

Frank Culver, staff coordinator of the task force, pinpointed five areas of study: air and water quality, fish and wildlife, land use planning, reclamation and leasing authority and water resources.

"We're going to mobilize the talents which exist within the state—universities, state agencies, conservation groups—to define the problem areas and try to determine the alternatives and consequences. For example, we don't know the long-range effects of air emissions from generating plants. We would like to find those out," said Culver.

Five state departments plus the Environmental Quality Council (EQC) are involved in the task force.

In 1971, the Montana legislature passed the Montana Environmental Policy Act (MEPA), patterned after the federal environmental statute. MEPA established a state policy on the environment and a 13-member EQC to oversee it. The Act requires state agencies whose actions affect the environment to file environmental impact statements, similar to federal ones. Sitting on the EQC are representatives of the legislature, the public and the governor's office.

The Council's power lies with its authority to investigate, hold hearings, issue subpoenas and recommend legislation. Although it cannot legally force a state agency to comply with environmental policies, it can bring public pressure to bear on a negligent agency. In its first annual report, the EQC described itself as an "official environment advocate. Of all the organs of state government, the Council is the only one specifically created to speak for the preservation of environmental values," said the report.

Concern for the environment expressed in MEPA was borne out in the new state constitution, proposed by the 1971-72 constitutional convention. A new provision to the constitution established the right to a clean and healthful environment, and a second one requires restoration of the land after removal of a natural resource.

The legislature is expected to create the machinery to carry out the latter directive.

In his office in the capitol at Helena, EQC executive director Fletcher Newby said:

"Our present reclamation law is not adequate because certain requirements and standards that are desirable are not specified in the law. It merely requires that a miner enter into a contract with the Department of State Lands, and they negotiate a reclamation plan. . . . [They have] done a good job negotiating contracts, but that is not insured in the law for the future. We need to amend the present law or need a new law that specifies certain requirements and standards."

Newby explained that the state is trying to get comprehensive control over the rapidly developing coal situation. He said:

"The thing we have to look at in Montana is that national priorities on this picture might be somewhat different from Montana's priorities—not that we are going to build a wall around us. I don't think we'll be allowed that luxury. National priorities are probably going to say that we need some of your coal, and we'll try to keep from ruining your land.

What I fear is that Montana's decisions are going to be made in centers of commerce and government elsewhere."

Part V
Social Issues—Prices, Demand Growth, and Conservation

Until rather recently the conventional wisdom on the subject of prices, demand growth, and conservation was pretty simple. Energy demand will continue to grow at essentially its historical rates, most demand is rather inelastic, that is, it doesn't respond significantly to price changes, and not much one can do in terms of improving the efficiency with which energy gets used will have a significant effect on our overall level of energy consumption.

The last few years have seen a critical reevaluation of these ideas. They have also seen an exploration of such issues as how environmental controls are likely to effect the overall economy and the price of energy, and how energy pricing structures effect the different sectors of American society.

Today these issues are the subject of active research and lively debate. The literature in this field is young, controversial, and growing very rapidly. Much of the best work is still in the form of research reports rather than published papers. The five papers in this last section are selected to provide some indication of the types of investigation now underway. Readers interested in pursuing problems of this sort in depth will find reprint books, secondary sources, and sometimes even the formal technical literature, to be an inadequate vehicle. Seminars, technical meetings, research reports, and face-to-face discussions will provide many of the most useful sources for the next few years.

Economic Consequences of Energy Costs

MARC J. ROBERTS

I. INTRODUCTION AND PERSPECTIVE

THIS paper attempts to present part of the answer to what has recently become a very fashionable set of questions. The questions are, first, what is happening to the balance between supply and demand in the energy sector of our economy? Second, what difference will any foreseeable changes make to our individual and collective lives in the short, medium, and long term? The various time horizons are important. The short-run question involves examining the contention that the nation faces an imminent "energy crisis." The long view implies considering at least briefly whether there are any relevant "limits to growth" imposed by the scarcity of energy resources.

I do not propose to explore these issues in terms of "needs" and "resources," as is now so common in such discussions. Instead I suggest we begin by recognizing that various economizing or resource allocating processes do help to determine the level and pattern of energy use. These operate both nationally and globally and are subject to human choice and control. Such processes come in many forms: free markets, regulated markets, political/administrative allocational schemes, etc. As allocative mechanisms, they assign resources to uses, implicitly or explicitly balancing the costs and benefits of alternative choices. My somewhat more than casual observation suggests that the mechanisms currently in use are neither perfectly responsive nor totally unresponsive to shifts in energy availability and demand. This capacity of the "system" to adjust to "feedback" about its own "state" provides part of the point of departure for what follows. Why inquire into the consequences of a change in some variables unless you believe there will be some of interest?

In other words, the analysis indicated by the title is only worthwhile undertaking if (1) energy costs can be expected to change over the period of interest, and (2) the relevant resource allocating mechanisms are not totally insensitive to those costs.

An interest in the cost of energy amounts to an interest in the "supply curve" for energy. The latter is nothing but a summary of the amount of a good or service available at a given time at a given price (or in a nonmarket context, cost). Thus we are led to ask, what determines the supply curve for energy? What is its shape and position? How will these characteristics change over time? How do changes in the rate of energy production, in technology, in resource availability, and in the prices of other goods and services affect those costs? What will it cost to produce various energy output at any given future time? How do those costs depend upon investment in facilities and new technology in the interim? Further (and the real point of this paper), what effect will variations in energy costs have upon other economic and social variables which we care about?

It is quite uninteresting to ask how much natural combustible hydrocarbon material of one kind or another is contained in the earth's crust. The question of whether a given supply of energy will be "available" at some point in the future simply cannot be answered in a "yes or no" fashion. Instead we must think of supply at a given time and at a cost. And that cost will depend on many characteristics of the system, which in turn will depend on its intervening history.

The unwary should be warned, however, that not all who have discussed these questions will agree to this formulation. The work of Forrester and Meadows *et al.*[1] in particular includes few if any economizing processes. Instead their models are simple machines that blindly use up resources until those resources are gone. In my view this peculiar structural characteristic, along with other difficulties, helps to make their analyses of very little interest scientifically, whatever their popular renown.[2] But discussing that issue further at the moment would lead us too far afield.

Before discussing the impact of energy costs, a prior question is, how can energy costs be expected to change in the future? To suggest part of the answer, I want to briefly examine the role of environmental concerns in increasing energy costs (Section II). Then I will consider what energy demand increases we can expect (Section III) and the extent to which these too might produce cost increases (Section IV). With these very sketchy results in hand, we can then ask the basic question, what difference will all this make? In Section V, I consider what parameters of the system we should care about. The direct and indirect effects on households of energy cost increases and the larger repercussions of such changes are explored in Section VI, VII, and VIII. I will close with some consideration of policy implications in both the short and long run.

This discussion is most emphatically not a "finished product." Nor do I report the results of any major new research.

This is a slightly revised version of a paper presented at the American Association for the Advancement of Science Symp. "Must We Limit Economic Growth?", Washington, D.C., Dec. 29, 1972. A similar version has also appeared as Discussion Paper 278 of the Harvard Institute of Economic Research, Cambridge, Mass.

The author is an Associate Professor of Economics at Harvard University, Cambridge, Mass.

[1] J. Forrester, *World Dynamics.* Cambridge, Mass.: Wright-Allan, 1971; and D. Meadows, *et al.*, *The Limits to Growth.* New York: New American Library, 1972.

[2] See P. Passell, M. J. Roberts, and L. Ross, Untitled review, *New York Times Book Rev.*, p. 1, Apr. 2, 1972; C. Kaysen, "The computer that printed out wolf," *Foreign Affairs*, July 1972; A. Kneese, and R. Ridker, "Predicament of mankind," *Washington Post*, Mar. 2, 1972; and U.S. Department of Health, Education and Welfare, "A report on measurement and the quality of life and the implications for government action of *The Limits to Growth*," Jan. 1973.

It is instead a quick survey, designed to help clarify the issues and to present what is known in a coherent and accessible manner. Much of what I have to say may be well known and perhaps uninteresting to those familiar with these data. But the number of such persons seems small enough to justify the exercise that follows.

II. ENVIRONMENTAL PROTECTION AND RISING ENERGY COSTS

There seems little doubt that in the United States in coming decades measures to protect the environment will bring about some increase in *measured* energy costs. Lower sulphur fuel, smoke stack dust percipitators, stack gas scrubbers, better cooling towers, undergrounding transmission lines, safer tankers, the reclamation of strip-mined land, standby oilspill cleanup capacity: all of these cost money. Recent estimates in the literature imply that sulphur emissions and thermal discharge controls perhaps will each add from less than 10 percent to 25 percent to delivered electricity costs.[3] While the other magnitudes are less easy to determine, altogether it seems we can expect *measured* energy costs in the United States in the short term to go up between 10 and 50 percent for the environmental reasons. The actual outcome depends on what technical progress occurs, on the degree of cleanup chosen by the society, and on the efficiency with which cleanup is undertaken.[4]

It does not seem to be widely appreciated that these *measured short-run cost increases are deceptive in a very important respect. They fail to reflect major improvements in product quality.* We have bought something with these costs increases, a diminished environmental impact from energy production. Presumably we are better off (not worse off) for having spent the money. Why else spend it? In view of this point, the relevant long-run question is about the prospective costs of a given package of energy and environmental outputs. Answering this requires us to look beyond immediate cost changes due to changes in the product mix/product quality of energy industries.

In the long run, we do face a serious additional problem. Each of our many air, land, and water ecosystems will tend to become lower in quality as we increase the amount of waste materials we dump into them. What must we do then when output rises and we want to maintain a given level of ambient quality in the waste-receiving system? We will have to employ a combination of production and waste-control techniques that limits ever more severely the residual material disposed of per unit of output.[5] This will cost more money. *For any given technology, then, the cost of a given level of ambient quality can be expected to increase as production increases.*[6] However, it is not at all clear that costs will in fact go up, exactly because we can expect the technology to change in the interim, especailly if we help the process along.

This is a very crucial point, one which reappears in the context of resource scarcity generally. In general, in the U.S. economy much invention and innovation responds to market incentives: people doing research seek ways to economize on goods whose prices are raising because that is the kind of new technology that will be most valuable (profitable).[7] I would hardly argue for the perfection of the market incentives or administrative mechanisms we currently have at work to aid in developing technologic solutions to the energy-environment problem. But that does not mean nothing will be done.

We cannot be *sure* of what new science and technology will be found or developed. But it requires an immense capacity to ignore the historical record in order to assume that little or no technical change will occur. One only need consider the number of firms seeking to develop devices for stack gas sulphur removal.[8] Similarly, there is substantial evidence that many industrial processes could be run in a very much cleaner manner, but that no one has had the incentive to do this until recently.[9] In sum it is hard to predict how much energy costs will go up in the medium-to-long run due to the need to provide ever-cleaner production as production rises. We can be sure, however, that looking at current technology overstates this cost greatly. Furthermore, it is entirely possible that the costs of environmental protection will *fall* despite these problems. The pace of technical change could outstrip the increasing demand we put on it.

In the area of extraction, given some level of technology, taking into account environmental considerations can change the relative costs of exploiting different deposits. We might find as a result that resources with higher nonenvironmental costs are more attractive on overall social grounds. But here too the increase in money costs really is illusory. *If we are making the right choices, it is really cheaper for the society to pay these direct costs than it is to exploit apparently less expensive reserves in a less controlled manner and pay the nonmarket costs of environmental deterioration.* Those who bemoan the expense of environmental protection are suffering

[3] For a review of a number of these cost estimates, see M. J. Roberts, "Who will pay for cleaner power?" presented at the Sierra Club Conf. Energy and the Environment, Jan. 1972; and P. Sporn, "Possible impacts of environmental standard on electric power availability and costs," in S. Schurr, ed., *Energy, Economic Growth and the Environment.* Baltimore, Md.: The Johns Hopkins Press, 1972, published for Resources for the Future.

[4] For example, scrubbing devices designed to remove sulphur from stack gases are still experimental on large units. Their ultimate costs are unclear and depend in part on prices received for recovered by-products. The costs of lower sulphur fuel will be higher than if higher sulphur content is allowed, etc. See M. J. Roberts, "Who will pay for cleaner power?", presented at the Sierra Club Conf. Energy and the Environment, Jan. 1972.

[5] This assumes that the relevant damage functions have a positive slope with respect to additional discharges. This view is at odds with the apparent logic of the federal air quality standards which appear to imply that damages are a step function, so that variations, which all satisfy the standard, are not of interest. Also some environmental problems do not fit the "residuals" framework, e.g., the aesthetic effects of electricity generating facilities.

[6] This assumes that the total cost of more intensive treatment is higher than the costs of methods that eliminate lower percentage levels of residuals. Marginal costs at the intensive margin could be rising, falling, or constant, as long as they were greater than zero, and this is almost always the case.

[7] J. Schmookler, *Invention and Economic Growth.* Cambridge, Mass.: Harvard Univ. Press. 1966; E. Mansfield *et al.. Research and Innovation in the Modern Corporation.* New York: Norton, 1971.

[8] On the various processes, see A. V. Slack, "Removing SO_2 from stack gases," *Environ. Sci. Technol.*, vol. 7, pp. 110–119, Feb. 1973.

[9] A. V. Kneese, and B. T. Bower, *Managing Water Quality.* Baltimore, Md.: The Johns Hopkins Press, 1968, pp. 44–49, published for Resources for the future.

TABLE I
1965 GNP AND ENERGY CONSUMPTION

Country	GNP per Capita in Dollars	Energy Consumption per Capita Killograms Coal Equivalent	Energy Consumption per Dollar of GNP Coal Equivalent	1950–1965 Energy–GNP Elasticity
United States	3515	9671	2.75	0.81
Canada	2658	8077	3.04	1.13
Sweden	2495	4604	1.85	1.59
West Germany	2195	4625	2.11	0.76
France	2104	3309	1.57	1.00
United Kingdom	1992	5307	2.66	0.62
Australia	1910	4697	2.46	1.21
Finland	1750	2825	1.61	**[a]
East Germany	1562	5534	3.54	0.88
Czechoslovakia	1561	5870	3.76	1.47
Austria	1365	2589	1.89	0.83
USSR	1340	3819	2.85	1.25
Italy	1254	1940	1.55	2.16
Japan	1222	1926	1.58	1.00
Hungary	1094	3188	2.91	1.62
Poland	980	3552	3.62	**[a]
Venezuela	882	3246	3.68	1.61
Trinidad and Tobago	646	3505	5.43	**[a]
South Africa	535	2761	5.16	1.09
World				1.06
Noncommunist developed regions				0.85
Noncommunist underdeveloped regions				1.67
USSR and Eastern Europe				1.23

[a]** indicates unavailable data.
Source: J. Darmstadter *et al.*, *Energy In The World Economy*. Baltimore, Md.: The Johns Hopkins Press, 1971, Tables 16, 17, pp. 34, 37, published for Resources for the Future.

from an accounting myopia, or perhaps self-interest, when they fail to note with equal fervor the costs of nonprotection.

Even the most apparently "limited" and environmentally sensitive resource in the energy sector, land for power plant sites, is not immune from an expansion in supply due to new technology. Fuel cells and solar cells raise the possibility of significantly decreasing the need for large "central station" generating capacity. Advances in transmission technology lower the costs of using more the abundant land that is more distant from metropolitan areas. Developments in cooling towers have already begun to lower the cost penalty of not using especially valuable costal sites.[10] Finally, the possibility of putting plants on offshore barges would directly expand the relevant "land" area available for such facilities.

I am not suggesting that any or all of these particular technical possibilities are necessarily either desirable or feasible on economic, engineering, or environmental grounds. Rather I advance them as examples of the more general process by which supply curves in the economists' language are "shifted outward" by technical change, making the rise in measured energy costs due to environmental protection difficult to forecast.

III. FORESEEABLE PATTERNS OF INCREASING ENERGY DEMAND

What determines the growth of the society's demand for energy? In the simplest sense it is the pattern of growth in the set of goods and services produced by the economy, plus the technology needed to produce those outputs. Hence changes in the rate of growth of GNP, its composition, and the energy input in each sector will all affect energy demand.

Economies vary significantly in their energy intensiveness per unit of output (measured as tons of coal equivalent per dollar of GNP). The United States is not especially energy intensive in this respect compared to countries where high-energy-use industries are relatively more important in the economic structure (see Table I). Interestingly, it is basic materials processing industries that are energy intensive: petroleum refining, basic chemicals, pulp and paper nonferrous metal refining, etc.[11] Of course, the United States has such a high GNP per capita that it has by far the highest energy use per capita, but that is a different issue.

Similarly, economies vary in the relationship they exhibit between growth in output and growth in energy production. In some economies, energy use is increasing at a faster rate than GNP, and in others, at a slower rate. (See Table 1). The more advanced countries are often in the latter category, as the sectoral composition of demand and production shifts away from energy-intensive activities. For the whole postwar period in the United States, the "energy elasticity with GNP" has been slightly less than one.[12] In other words, a 1-percent GNP increase produced a less than 1-percent increase in energy use. Another way to express the same facts is to note that the share of energy in GNP has steadily declined: due both to slower quantity growth and to relative price declines.[13]

On the other hand, since 1966 we have observed energy

[10]R. A. Woodson, "Cooling towers," *Sci. Amer.*, pp. 70–78, May 1971.

[11]H. H. Landsberg, *et al.*, *Resources in America's Future*. Baltimore, Md.: The Johns Hopkins Press, 1963, pp. 203–217, published for Resources for the Future.

[12]J. Darmstadter *et al.*, *Energy In the World Economy*. Baltimore, Md.: The Johns Hopkins Press, 1971, pp. 33–38, published for Resources for the Future.

[13]H. J. Barnett, and C. Morse, Scarcity and Growth. Baltimore, Md.: The Johns Hopkins Press, 1963, pp. 220–223.

growth proceeding at a *faster* rate than GNP in the United States. This reversal appears to be a result of several factors. First, there are some statistical difficulties with the normal figures with respect to the conversion rates employed to calculate total energy use. In addition, the increase in nonfuel use of liquid and solid hydrocarbons is not an insignificant contributor. Increases in air conditioning and an increase in the relatively inefficient use of fuel for electric home heating also appear to play a role. A slowdown in the rate of *increase* of thermal efficiency in fossil fuel power plants also seems relevant. Finally, and most important, has been the slow rate of growth in GNP in recent years.[14]

However, even after the "trend reversal" is taken into account, it seems unlikely that energy growth overall in the economy will continue at levels much above 3.5 to 4.5 percent per year, more or less of the order of magnitude of probable GNP increases. In the period 1950-1971 the annual average rate of increase of GNP in consistent dollars was 3.7 percent.[15] This observation makes it clear that projections of the growth in electric power at 6 to 8 percent per annum assume a continued shift from on-site to central station combustion. Examination of detailed projections for California confirm this point, since much of the expected growth depends upon high forecast rates of conversion of hot water and home heating from other fuels to electricity.[16]

We should be very conscious, however, that all such simple extrapolations are fraught with the possibility of error. We are trying to forecast the behavior of the economy in situations we have not observed. Hence it is difficult at best to conduct such exercises. Perhaps current trends will not continue. What are the structural processes behind postwar developments, and can we expect past relationships to continue to hold into the future?

I want to suggest several reasons for believing that there are some "limits" to the rate of growth and level of demand for energy. Or, more relevantly, there are reasons for believing that these growth rates might well diminish in the United States in the decades ahead. (There are also some contrary indications—when foretelling the future, the world seldom cooperates to simplify the enterprise.) There are two basic facts against which this discussion must be posed. First, as technical change has proceeded, in general each worker's productivity has increased. This has meant both higher incomes and that most people's time has become more valuable, exactly because the opportunity cost of foregone earnings has increased. Second, one clear limit to per capita rates of consumption is the sheer limit of time itself, the number of working hours each day.

These facts as well as others have interacted to produce several relevant developments in economic behavior in the United States recently. First, rates of population growth have decreased noticeably. If parent's time is an important input to a child's upbringing, then rising wage rates have meant that the costs of having children of any given "quality" have increased, leading in turn to fewer children. Other factors no doubt also contributed: the diffusion of contraceptive devices and legal abortions, changing attitudes toward appropriate woman's roles, and a general breakdown of traditional subcultural ties and values. But the implication of such a change is that we can expect slower GNP and energy growth than when the population was growing at 2 percent per annum.[17]

A second potential source of discontinuity is the possibility of shorter work weeks. For the last 40 years Americans have taken the bulk of the benefits of higher productivity in the form of more goods and services instead of increased leisure.[18] There are at least some indications that this may be about to change. One explanation would be that we require income in order to enjoy leisure. Perhaps a significant percentage of Americans are now earning enough to be able to enjoy more leisure and still maintain the income they believe they need to enjoy it. Shorter working hours, with the resulting lower incomes would have an impact much like slower population growth. They would slow the rate of growth of both GNP and energy use.

Third, we do seem to be moving closer and closer to saturation in many of the major energy-using devices in the household sector: refrigerators, washers, dryers, heating, air conditioning, televisions, etc.[19] (Of course, additions to the household capital stock and to household energy use do not always imply a one-for-one increase in net energy consumption. More clothes washing at home means less at laundries and laundromats, etc.) While we cannot discount the possibility of new appliances, existing patterns of diffusion do suggest some foreseeable slowdown in the rate of growth of household energy use. Of course, changing patterns within the energy sector may produce shifts among oil, gas, and electricity, but that is a different matter. And even for *electricity* use, some "drag" on growth rates due to increasing saturation seems a distinct possibility.

Before moving on we should note some contradictory trends. First, changes in the age composition of the population, as we move toward slower population growth, might mean more

[14]A detailed study of these changes is to be found in *Energy Consumption and Gross National Product in the United States*, National Economic Research Assoc., Inc., Mar. 1971.

[15]*Statistical Abstract of the United States 1972*, U.S. Dep. Commerce, Bur. Census, p. 314, Table 510.

[16]R. D. Doctor, *et al.*, *California's Electricity Quandry. III: Slowing the Growth Rate*, Rand Corp., Rep. R-1116-NSF/CSA, Sept. 1972. Compare Table 17, p. 51, with Table 18, p. 52.

[17]Of course, growth in the population is not the same as growth in the labor force. Changes in participation rates (the percentage of any age/sex group in the labor force) mediate the impact of population on the labor input. Lower family sizes may well be accompanied by increased participation by married women, which would tend to counteract any depressing effect in the short run.

[18]Average weekly hours of production workers in manufacturing were 44.2 hours in 1929, and 40.9 hours in (seasonally adjusted) Nov. 1972. See *Business Statistics 1971*, U.S. Dep. Commerce, p. 74; and *Survey of Current Business*, U.S. Dep. Commerce, vol. 52, pp. 5-14, Dec. 1972.

[19]Consider the percentage of households in 1960 and 1971 who owned various energy using devices:

	Washer	Color TV	Black and White TV	Refrigerator	Clothes Dryer	Dishwasher	Air Conditioner
1960	74.5			86.1	17.4	4.9	12.8
1971	71.3	43.3	77.6	83.3	44.5	18.8	31.8

Statistical Abstract of the United States 1972, U.S. Dep. Commerce, Bur. Census, p. 328, Table 537.

TABLE II
RECENT U.S. ENERGY PRICES

	Electricity Component Consumer Price Index (1967 = 100)	Average Revenue for Residential Killowatthours Used	Fuel Oil and Coal Consumer Price Index Annual Averages (1967 = 100)	Consumer Price Index—All Items, Annual Averages (1967 = 100)
1972	119.8 (Sept.)		118.1 (Oct.)	126.6 (Oct.)
1971	113.2	2.19	117.5	121.3
1970	106.2	2.10	110.1	116.3
1969	102.8	2.09	105.6	109.8
1965	99.1	2.25	94.6	94.5
1960	99.8	2.47	89.2	88.7
1955	95.2	2.65		80.2

	Wholesale Price Index All Items 1967 = 100 Annual Averages	WPI-Coal 1967 = 100	WPI-Electric Power 1967 = 100	WPI-Gas 1967 = 100	WPI-Petroleum, refined, 1967 = 100
1972	120.0 (Oct.)	192.4 (Oct.)	123.1 (Oct.)	117.5 (Oct.)	111.5 (Oct.)
1971	113.9	181.8	113.6	108.0	106.8
1970	110.4	150.0	104.8	103.3	101.1
1969	106.5	112.5	102.0	93.1	99.6
1965	96.6	93.4	100.1	92.8	93.8
1960	94.9	95.6	101.2	67.2	95.5
1955	87.8	82.3			92.0

Sources:
Column 1: *Statistical Yearbook of the Electric Utility Industry*, Edison Electric Inst. and *Retail Prices and Indexes of Fuels and Utilities*, U.S. Dep. Labor, Bur. Labor Statistics, monthly, various issues.
Column 2: *Statistical Yearbook of the Electric Utility Industry*, Edison Electric Inst.
Columns 3–9: *Statistical Abstract 1971*, U.S. Dep. Commerce, pp. 339, 335; and *Survey of Current Business*, pp. S-8, S-9, Nov. 1972.

households per capita and less efficient (i.e., greater) energy use per capita. Second, increasing value placed on housewife's time could lead to continued substitution of energy for labor in the home, raising energy use. Recent "new" appliances such as self-cleaning ovens and no-frost refrigerators have often been of this type. Further, the substitution of artificial for natural materials, and plastics and nonferrous metals for iron and steel, could increase energy inputs per unit output. Tighter environmental controls could also have the same effect.

Similarly, increased "suburbanization" and longer commution distances raises energy use. On the other hand, small cars, more mass transit and more multifamily dwellings, have an opposite energy-use-decreasing impact.

Netting all this out, my own "feel" is that the evidence indicates that energy growth could well be somewhat below GNP growth over the next 20 to 50 years, maybe even at or below postwar average elasticity levels. (Of course, unforeseen developments might lead to very different results.) Further, I would not be surprised to also see slower than recent trend increases in measured GNP, due to slower population growth and shorter work weeks. The *overall* energy demand situation thus seems far from catastrophic. On the other hand, short-run regional electricity demand patterns, patterns heavily influenced by internal migration rates, interfuel substitution, and changing patterns of family formation as the "babyboom" comes of age, obviously could present some adjustment difficulties. Indeed we have seen such problems in the last few years. (This situation has also been accelerated in some areas by construction delays due to faulty engineering and labor problems.)

In addition, if costs and prices for energy do increase, this outcome itself can be expected to have a nontrivial depressing impact on the rate of growth of demand. Before discussing that issue, however, we have to ask how supply can be expected to respond to these demand increases and whether prices will in fact rise.

IV. INCREASING RESOUCE SCARCITY AND ENERGY COSTS

What is the real situation with respect to possible impending resource scarcity? Are we running out of fuel inputs, and is that the real origin and nature of our so-called energy crisis? The answer, I suggest below, may surprise some of you, but be forewarned that it depends very much on how one defines the question. Be warned also that much of the discussion in the media appears to reflect the efforts of many special interests who have energetically attempted to define the public dialogue in ways favorable to themselves. And since prophesy, like politics, makes strange bedfellows, we find modern oracles and soothsayers also active here, busy trying to decipher the future in the entrails of birds and the output of computer simulation models.

First of all, what has been happening to energy prices in the United States in recent years? Until literally the last year or two, the clear trend in real prices has been downward (see Table II). This price pattern reflects the fundamental findings of Barnett and Morse: for most of our recent experience, technical change, which has the effect of augmenting resources, has proceeded more rapidly than the demand for resources has increased.[20] The real cost, in terms of capital and labor, of most natural resources, including fuels, in fact *declined* over their period of study. The "limits" receded even faster than our pursuit of them.

Just consider all of the relevant innovations over the last 75 years which lessen the costs of producing a given amount

[20]H. Barnett and C. Morse, *Scarcity and Growth.* Baltimore, Md.: The Johns Hopkins Press, 1963, esp chs. 8–10.

of usable energy: continued development of strip mining and underground mining techniques and equipment; new drilling rigs and techniques, especially for offshore and for deeper water; oil and gas pipelines; unit trains for coal transport; slurry pipelines; oil super tankers; LNG tankers; new oil refining techniques including calaytic cracking; better design, larger scales, higher temperatures and pressures and higher thermal efficiency in both electricity generation and all fuel combustion; better electricity transmission technology, etc.

It is perhaps useful to disentangle briefly the various processes for economizing on resources.[21] First, in part we simply see the *substitution* of capital and labor (especially capital) for new resources in the production of intermediate inputs. For any given technology we can expect this to occur whenever the price of resources increases relative to capital. A steel mill that varies its charge between scrap and primary ingot depending on relative prices is engaging in this kind of activity, since the secondary material was "produced" by the capital and labor of the scrap yard. Second, often such substitutions occur in our economy simultaneous with the introduction of new *conversion technology* which changes the rate at which one can substitute "capital/labor" for "land," and hence makes the substitution more attractive. New and larger boilers that use fuel more efficiently are an example of such a change. Third, there are changes in the *technology of extraction and transportation.* These generally have the effect of augmenting the effective stocks of less costly resources, and often wind up saving labor (and even capital) in the bargain. Finally, *consumers can shift their tastes* to a less resource-intensive product mix (e.g., toward services).

Until the last year or two the effect of these processes, as noted, was to lower energy costs. Recently the situation has become a bit more complicated. First, the costs of environmental controls have begun to affect both electricity prices and the prices of environmentally desirable fuels. The latter effect is in part a transitional disequilibrium situation.[22] Second, unmeasured quality effects show up as part of cost increases due to stricter safety regulations in underground coal mines, again a "quality" improvement in output unnoticed by our statistics.

Recent price changes then are at least in part reflections of quality improvements and transitional phenomena. Perhaps they are completely explicable in these terms. We do not have enough data and experience to say. It certainly is not at all clear, however, that recent price changes are the first taste of those we can expect as energy resources become "scarce." On a worldwide basis, for energy as a whole, it simply is not demonstrable from the data that the historic trend toward ever less costly energy of any given "quality" is being dramatically reversed.

Even if we take a very long run view, the case for ultimate scarcity is equally uncertain. What developments of breeder reactors, fuel cells, fission, hydrogen cycle engines, solar power, etc., might not come about? Who would predict with great confidence that nothing along any of these or other still unexplored lines will be usefully developed? The indisputable facts about the physical finiteness of the earth are simply not very interesting for even long-run policy purposes if they will not be relevant constraints over the next 500 years.

Because world and national energy markets are so thoroughly hedged about with restrictions, controls, and agreements, the argument just reviewed says very little about the short-run prospects for particular U.S. energy prices. Short-run price and supply movements may be quite divorced from long-run equilibrium trends. Further, the argument that we have little to fear for long-run global scarcity does *not* imply that it is impossible for us to have more specific scarcity. In particular, within any given well-explored geographic area, the deposits of a given fuel that are the least expensive to extract with a given technology will, in general, be used first. It is quite possible then that for one or another particular fuel, at a particular time and place, new technology will not outpace the ever more adverse geology encountered as exploitation continues. In such instances the cost of that fuel will tend to rise, and to increase more rapidly the higher the rate of output.

Now what is happening in the U.S. energy sector today? The last-mentioned possibility seems to be true for natural gas and petroleum for the already explored areas of the continental United States and the already explored areas offshore along the Gulf Coast.[23] Natural gas, in addition, is subject to a complicated process of field price regulation by the Federal Power Commission (FPC). It appears that there is a shortage of gas at current prices. A rising demand for natural gas, which is desirable on environmental grounds, and the unwillingness of the FPC to authorize price increases for interstate sales (together possibly with what some have alleged to be shrewd strategic behavior by producers) have combined to bring this imbalance about.[24] Yet such a mismatch between supply and demand in a market with regulated prices is a far cry from an "energy crisis." It is instead a situation produced by the imperfection of our economic institutions, both private and and public.

Onshore continental U.S. oil supplies do likewise seem limited. This has in turn given rise to talk of "crisis" on the part of those who hope to protect the favored position of

[21] H. Barnett and C. Morse, *Scarcity and Growth.* Baltimore, Md.: The Johns Hopkins Press, 1963, ch. 6; and J. Tobin and W. Nordhaus, "Is growth obsolete," unpublished paper prepared for the Nat. Bur. Economic Research Colloquium, San Francisco, Calif., pp. 22–28, Dec. 10, 1970.

[22] Private communications from large oil wholesalers and buyers in the Boston market reveal that prices for low sulphur oil declined from August to October 1972. The same source indicates that current shortages experienced by some companies of low sulphur residual fuel oil along the East Coast are in part due to labor and engineering problems that delayed completion of the desulphurization facilities of one major oil company.

[23] F. M. Fisher, ***Supply and Costs in the U.S. Petroleum Industry.*** Washington, D.C.: Resources for the Future, 1964; K. C. Brown, Ed., ***The Regulation of the National Gas Producing Industry.*** Washington, D.C.: Resources for the Future, 1972.

[24] P. W. MacAvoy, "The regulation-induced shortage of natural gas," *J. Law Econom.*, Apr. 1971; E. W. Erickson and R. M. Spann, "Supply price in a regulated industry: The case of natural gas," *Bell J. Econ. Manage. Sci.*, vol. 2, no. 1, 1971; S. G. Breyer and P. W. MacAvoy, *Energy Regulation by the Federal Power Commission.* Washington, D.C.: Brookings Inst., 1974; J. D. Mazzoon, "The F.P.C. staff's econometric model of natural gas supply in the United States," *Bell J. Econ. Manage. Sci.*, vol 2., no. 1, 1971.

domestic oil production, bolstered as it is by state and federal production controls, favorable federal tax treatment, and a scheme of import quotas to limit foreign competition.[25] These pieces of government protection have come under much attack in recent years.[26] For such producers, talking about an "energy crisis" is a tactic for warding off moves to remove their special privileges. If we might "run short" of energy, they argue, we need to encourage its development; hence we need to preserve quotas, depletion allowances, and so on.

Thus there are some participants in energy markets who urge us to embrace the very contingency "shortage" will presumably produce. They cry for higher prices in order to avoid—yes—higher prices. Sometimes this argument is based on the desirability of energy self-sufficiency and we are urged to provide high domestic prices to avoid the possibility of high import prices. Either way we are being urged to commit suicide to avoid a possible imaginery threat of murder.

Many other economic interests have reason enough to join in promoting the "energy crisis," and many of them have done so. Gas well owners who want higher prices argue that additional exploration incentives are required by the "crisis." Investors in schemes to import liquified natural gas who want licenses to go forward argue that such imports are made necessary by the "crisis." Electric utility executives, unhappy with the obligations of the National Environmental Policy Act, argue for changing that legislation because it is hampering our response to the "crisis." Companies with an investment in Alaskan oil wells distressed at delays over the pipelines and those who want to drill for oil on the outer continental shelf off of the East Coast (to find crude protected by American controls and subsidies) also take the same line.[27]

Regardless of the merits of various positions on these issues, we should note that an imminent and probable worldwide shortage of fuels is alleged only by the most assiduous promoters and apocalyptic prophets of the "energy crisis." On a world basis, is seems to be the case that production plus transportation plus conversion costs for energy in general and oil in particular have continued to fall, right up to the present time.

It is true that recent new agreements between the oil producing countries (OPEC) and the international oil companies to increase host country revenues have recently raised world oil *prices*.[28] As a result, *prices* do diverge substantially from real energy *costs*. Continued successful cartel behavior by OPEC might well further raise the prices (apparent costs) of imported fuels, despite contrary trends in real production plus transportation costs.

Whether the cartel will continue to be successful is another matter. The gains from "cheating" will increase directly with the efficacy of the agreement. Perhaps additional discoveries elsewhere in the world or the forward integration of host countries into refining and marketing will help to "shake up" the structure.[29] U.S. tax policy, too, will have an impact, particularly the fate of the foreign investment tax credit. But in so far as OPEC does continue to be successful, prices will be higher than otherwise.

To sum up, the nature of the energy crisis, short or long run, remains elusive. Monopolistic increases in import price, or higher (apparent) prices for domestic energy, due to expanded environmental and safety outputs, are both possible. But neither is likely to be of "crisis" proportions. For advocates of autarchy (of which I am not one) who believe that the difference between 30-percent importation of oil and 60-percent importation is significant, there is some problem of self-sufficiency.[30] Given the imperfection in our adjustment and allocation mechanisms, there are specific problems to be resolved. For users and regulators of natural gas there are important problems of readjustment to higher costs of production. For the institutions that manage the research on and development of new energy technology, there are important tasks to be undertaken. The process of adjusting our energy system to produce more satisfactory "environmental outputs" will not be an easy one. Indeed, if there are any "crises" in the energy system, they are likely to center around its environmental impact.

We can obviously make very unsatisfactory responses to any or all of these problems. I personally doubt whether together they can usefully be characterized as "the energy crisis." What do we gain by doing so? But enough of preliminaries, what will all this imply for the rest of the system?

V. WHAT PARAMETERS DO WE CARE ABOUT?

Before we can intelligently examine the effects of changing energy costs, we have to know which of those effects are relevant. Economists generally employ a perspective that is individualistic and more-or-less utilitarian. They assume that what matters about a situation is how well off all the individuals in the society perceive themselves as being. From this viewpoint, price increases are undesirable because they mean that people cannot purchase as much as they once could in the way of goods and services. Hence they are presumed to be worse off.[31]

[25]On production controls, see W. E. Lovejoy and P. T. Homan, *Economic Aspects of Oil Conservation Regulation.* Baltimore, Md.: The Johns Hopkins Press, 1967 published for Resources for the Future, esp. ch. 6. On tax treatment, see S. L. McDonald, "Percentage depletion, expensing of intangibles and petroleum conservation," in M. Gaffney, Ed., *Extractive Resources and Taxation.* Madison, Wis. Univ. Wisconsin Press, 1967. On import controls, a useful discussion is *The Oil Import Question*, Cabinet Task Force on Oil Import Control, Washington, D.C., esp. pp. 8-17, 69-78, 1970.

[26]The pressures for reform have been sufficient to lead to small reductions in the percentage depletion allowances on oil wells.

[27]Of course, if the shorter Alaskan pipeline route is taken (to Valdez) instead of a trans-Canada route, it is entirely possible that some significant proportion of the output would find its way into the export market (for Japan), where it would hardly serve to ameliorate any U.S. situation.

[28]For a review and analysis of recent events see M. A. Adleman, "Is the oil shortage real?", *Foreign Policy*, pp. 69-107, winter 1972-1973.

[29]M. A. Adleman, "Is the oil shortage real?", *Foreign Policy*, pp. 69-107, winter 1972-1973; see also T. H. Moran, "New deal or raw deal in raw materials," *Foreign Policy*, pp. 119-134, winter 1971-1972.

[30]The available data suggests approximately the 30-percent level of petroleum import dependence for 1971. See *Statistical Abstract of the United States 1972*, p. 506, Table 820. Any forecast of future imports depends in part on how the oil exporting nations alter export prices, as well as on technical advances in competing fuel sectors.

[31]The argument assumes that individuals act so as to "maximize utility," that is, to put themselves always in a preferred position. There have been many critical discussions of this formulation, for example, see I. M. D. Little, *A Critique of Welfare Economics*, 2nd ed. New York: Oxford Univ. Press, 1958, esp. chs. I-III. A careful, now classic, exposition of the standard model is P. A. Samuelson, *Foundations of Economic Analysis.* Cambridge, Mass.: Harvard Univ. Press, 1948, ch. VIII. Also interesting in this connection is J. Rothenberg, "Consumer choice revisited and the hospitality of freedom of choice," *Amer. Econ. Rev.*, vol. 52, pp. 269-283, May 1962.

While I have a good deal of scepticism as to the exact empirical validity of this "rational model,"[32] the argument does provide one normative focal point that correlates with the "common sense" (prejudices?) prevalent in our society. Of course if you believe that most people in America today are confused, and that they already have "too many" material possessions, this point is of less interest to you.[33]

Evaluating changes in individuals' positions is a bit complex because we might well react differently depending upon how the burden is distributed. Take an extreme case and suppose that (1) only the richest 10 percent or (2) only the poorest 10 percent of the population bore the losses due to increased energy costs. To me, and I suspect to you, these situations are not indistinguishable.

In evaluating the distributive aspects of a policy, economists tend to focus on the percentage change in real income of individuals at various income levels. A policy with equal percentage impact is "proportional." One which leads to a higher percentage decline in the position of higher income groups is "progressive," and the reverse is "regressive."

The conventional view is that regressive measures are "bad," that progressive ones are "good," and that proportional ones are more or less "neutral."[34] You may find, on the other hand, any inequality in dollar burdens "unfair" (and presumably favor regressive taxes). Or you may, like me, find even proportional and mildly progressive taxes unattractive in a society with substantial inequality like our own.

Or you might not be an individualist, but care about national power, prestige, or glory as ends in themselves. Then maintaining fuel self-sufficiency and/or the current value of the foreign exchange rate might be the most important question for you. Here I would demur. I do care about the physical security and political freedom of people (including those not eligible for American passports). However, I find the link between those goals and our fuel import position very complex. Without entering into what would be an extended debate, I am not prepared to accept national fuel self-sufficiency as an important policy goal.

The oil-exporting countries vary in philosophy from feudal monarchies to revolutionary dictatorships. Will they be able to coordinate what would for them be very costly economic actions for political purposes? This contingency seems to me too improbable to spend great amounts to avoid. Invoking the "national security" does not automatically justify any conceivable program.[35] In 1971 the United States satisfied less than 2 percent of its domestic energy requirements and less than 4 percent of its petroleum needs with oil imported from the Middle East and North Africa. Compare this with the position of other industrialized areas! (See Table III.)

TABLE III
1971 ENERGY AND OIL, USE AND IMPORTS IN MILLIONS OF TONS OF OIL EQUIVALENT

Country	Energy Use	Oil Use	Oil Imports	Oil Imports from Middle East and North Africa
United States	1634.1	715.4	198.8	537.0
Canada	145.2	77.0	38.5	13.5
Western Europe	1049.7	652.1	668.0	537.0
Japan	292.3	220.0	231.0	192.8

Source: B.P. Statistical Review of the World Oil Industry. London: 1972.

In 1972 preliminary figures indicate that a bit over 2 percent of our energy came from the areas in question. It seems quite irrational to become so concerned at even modest movements in the direction of a situation other nations have long experienced. After all, Israel has had no trouble finding a willing supplier (Iran) even in recent years.

I will, however, be interested to explore the impact of energy cost changes on trade patterns, because of the implications these might have in turn for the circumstances of individuals worldwide.

In all of this we must not ignore that along with higher energy costs we will also have improvements in the environment. Some of you again may be willing to foresake ethical individualism and assert a value in itself with respect to the state of various ecological systems. Others will be content with considering the effects of policy on the lives of individuals. Stricter or more classical utilitarians will simply ask about the "value" of environmental outputs given our presumably immutable tastes. Less doctrinaire, or more sociologically literate, observers will inquire as well into the impact of both choice process and outcomes on men's tastes, values, and view of themselves.[36]

Well then, what effects can we expect of higher energy prices (even if these are not increases in correctly measured quality-compensated energy costs)?

VI. THE IMPACT OF ENERGY COSTS ON HOUSEHOLDS

It is always nice on these matters to have some data as a basis for discussion. Unfortunately, data here are hard to come by. But Tables IV–VI present some indicative information. The first two indicate electricity expenditures and total household direct energy expenditures (excluding gasoline) for urban families of different sizes and incomes, as derived from a large national sample. Since these survey data are over 10 years old, only the pattern and not the magnitude of the expenditures is relevant. Table VI gives similar data relevant to automobile fuel costs by income class.

What do these data show? They reveal that in our society direct household energy consumption is a "necessity" in a technical sense. That is, households of lower income spend a higher proportion of that income directly on energy.[37] That

[32]See M. J. Roberts, "An unsimple matter of choice," *Saturday Rev.*, pp. 46ff, Jan. 22, 1972, for an informal examination of this issue.

[33]One exposition of this antigrowth position is J. Hardesty *et al.*, "The political economy of environmental destruction," in Johnson and Hardesty, Eds., *Economic Growth versus the Environment.* Belmont, Calif.: Wadsworth, 1971.

[34]A typical textbook exposition of these matters can be found in J. F. Due, *Government Finance*, 4th ed. Homewood, Ill.: Irwin, 1968, ch. 8.

[35]See the discussion in *The Oil Import Question*, Cabinet Task Force on Oil Import Control, Washington, D.C., esp. pp. 124–133, 299–309, 1970.

[36]An intriguing discussion here is G. F. White, "Formation and role of public attitudes," and D. Lowenthal, "Assumptions behind the public attitudes," in H. Jarrett, Ed., *Environmental Quality in a Growing Economy.* Baltimore, Md.: The Johns Hopkins Press, 1966, published for Resources for the Future.

[37]That is, the income elasticity of *expenditure* on energy appears to be discernibly less than one.

TABLE IV
1961 ELECTRICITY EXPENDITURES BY INCOME CLASS AND FAMILY SIZE—URBAN HOUSEHOLDS

	Income in Thousands of Dollars								
Size	1.0–1.9	2.0–2.9	3.0–3.9	4.0–4.9	5.0–5.9	6.0–7.49	7.5–9.9	10.0–14.0	15.0+
1	31.75	34.18	38.33	36.83	43.28	47.04	75.34	93.44	180.00
2	50.14	62.56	70.14	75.17	76.97	97.67	99.96	110.87	153.31
3	40.87	53.11	68.48	79.95	100.37	109.35	111.91	132.17	166.00
4	26.46	64.20	78.56	89.34	107.57	108.97	129.34	123.82	164.92
5	14.48	58.61	97.08	90.30	90.69	119.14	137.32	151.05	171.13
6	77.18	59.51	60.97	77.86	111.68	113.77	142.80	157.56	267.05

Source: Survey of Consumer Expenditures, 1961, U.S. Dep. Labor, Washington, D.C., 1962.
Note: Reported average expenditures have been corrected by dividing each cell by the quantity $(1 - J)$, where J is the percentage of respondents reporting electricity and gas payments jointly.

TABLE V
1961 TOTAL ENERGY EXPENDITURES BY INCOME CLASS AND FAMILY SIZE—URBAN HOUSEHOLDS

	Income in Thousands of Dollars								
Size	1.0–1.9	2.0–2.9	3.0–3.9	4.0–4.9	5.0–5.9	6.0–7.49	7.5–9.9	10.0–14.9	15.0+
1	98.11	95.70	103.12	101.44	95.58	117.31	152.88	207.41	210.00
2	126.73	156.59	170.82	184.67	194.11	208.81	226.63	268.33	463.03
3	106.74	135.31	181.42	178.26	220.48	240.40	261.44	314.24	464.43
4	73.30	182.06	183.92	209.29	238.42	251.00	284.09	311.65	431.32
5	51.33	133.01	198.90	256.18	223.22	285.60	346.60	320.95	377.23
6	166.30	148.89	154.77	185.34	263.74	261.76	346.98	378.96	519.17

Source: Survey of Consumer Expenditures, 1961, U.S. Dep. Labor, Washington, D.C., 1962.

TABLE VI
1961 GASOLINE EXPENDITURES BY INCOME CLASS AND FAMILY SIZE—ALL HOUSEHOLDS

	Income in Thousands of Dollars								
Size	1.0–1.9	2.0–2.9	3.0–3.9	4.0–4.9	5.0–5.9	6.0–7.49	7.5–9.9	10.0–14.9	15.0+
2	42.74	77.72	116.47	167.56	197.90	224.86	253.56	262.56	208.60
3	64.90	94.87	161.67	175.20	219.24	234.15	262.62	331.35	294.00
4	84.27	97.21	178.12	206.48	205.41	249.78	277.45	282.76	258.76
5	68.55	86.59	156.02	194.39	211.83	245.10	265.49	293.96	259.12
6	46.47	99.87	112.96	155.64	194.36	230.59	266.34	307.10	381.50

Source: Survey of Consumer Expenditures, 1961, U.S. Dep. of Labor, Washington, D.C., 1962.
Note: Reported average expenditures have been corrected by dividing each cell by the quantity $(1 - J)$, where J is the percentage of respondents reporting electricity and gas payments jointly.

pattern, however, is heavily influenced by the declining block structure of electricity rates, under which higher consumption of kilowatt hours leads to lower average prices.[38] Thus the relationship of BTU use to income is almost undoubtedly more nearly proportional than the tables suggest. On the other hand, the overall level of regional electricity prices is somewhat related to regional income differentials, which leads to a (probably smaller) offsetting bias in the opposite direction.

What does this pattern imply for the welfare effects of energy price increases? It appears that the direct effects on households of such price rises will be to make the distribution of real income (and probably experienced welfare) in the society somewhat *less* equal, if we consider a proportional impact neutrality. Nontrivial amounts are at issue for many consumers if direct energy costs should double, e.g., a couple of hundred dollars or even more per family. Of course the increase (if any) which we will actually experience might well be very much less; and we will all presumably be getting an improvement in environmental quality for our money.[39]

To put such changes in perspective note that a family of four at $10 000 per year that is now spending $1500 a year on food would have faced an increased cost of over $400 for that consumption bundle given price changes in the last 8 years.[40] This makes it clear that even relatively unfavorable scenarios with respect to energy prices do not mean disaster. Rather they imply events not incomparable to the unpleasant, and to me ethically objectionable, pattern of readjustment we have experienced in recent years. Recent price changes have been accompanied by wage increases for some, but hardly all,

[38]See particularly R. Halvorsen, "Residential electricity demand," Ph.D. dissertation, Harvard Univ., Cambridge, Mass., 1972. Chapters 2, 5, and 6 are especially relevant to the question of income effects on quantities and expenditures.

[39]Of course not everyone will benefit equally. Those who care most about the environment, and whose consumption mix is biased away from environmentally intensive goods, will benefit most from the public provision of environmental services.

[40]In 1964 the food component of the Consumer Price Index was equal to 92.4 (1967 = 100). In Nov. 1972 it was equal to 125.4. See *Economic Report of the President*, Jan. 1972, Table B-45, p. 247; and *Survey of Current Business*, U.S. Dep. Commerce, vol. 52, no. 2, pp. 5–8, Dec. 1972.

TABLE VII
ENERGY PRICE VARIATIONS

City or State	Net Monthly Bill—Residential 500 kWh, (dollars) Sept. 1972	Net Monthly Bill—Residential 40 therms (Nonhouse Heating) (dollars) Sept 1972
Atlanta	10.47	6.24
Baltimore	15.92	7.78
Boston	15.42	12.56
Chicago	15.51	10.68
Cincinnati	13.69	4.41
Cleveland	13.23	5.55
Dallas	13.01	4.04
Detroit	14.60	6.41
Houston	11.95	6.53
Kansas City	13.29	4.70
Los Angeles	13.45	6.18
Milwaukee	14.03	7.13
New York	20.67	10.57
Philadelphia	16.40	8.40
St. Louis	14.11	7.68
San Diego	10.81	6.06
San Francisco	11.59	4.39
Seattle	6.65	8.08
Washington, D.C.	15.17	8.40

	Cost of Fuel Burned by Electric Utilities in 1971 (cents/mill BTU)		Industrial Power, 500-kWh service for 200 000 (dollars) kWh, June 1972
Maine	35.8	New England	3549
Vermont	78.6	Mid Atlantic	3316
Massachusetts	51.6	E.N. Central	3296
New York	50.9	W.N. Central	3147
Pennsylvania	42.5	S. Atlantic	2590
Ohio	35.5	E.S. Central	2520
Indiana	28.7	W.S. Central	2338
Wisconsin	43.2	Mountain	2395
North Dakota	25.6	Pacific	2260
Iowa	35.2		
Nebraska	35.4		
Kansas	26.6		
West Virginia	31.8		
Florida	38.7		
Kentucky	26.0		
Tennessee	32.2		
Louisianna	21.8		
Oklahoma	22.3		
Wyoming	16.2		
New Mexico	18.6		
Nevada	37.6		
California	41.9		
Alaska	66.3 (1969)		

Sources:

Columns 1, 2: ***Retail Prices and Indexes of Fuels and Utilities***, U.S. Dep. Labor, Bur. Labor Statistics, Tables 5 and 6, Sept. 1972.

Column 3: ***Statistical Year book 1972***, Edison Electric Inst., Table 43S.

Column 4: ***Wholesale Prices and Price Indexes***, U.S. Dep. Labor, Bur. Labor Statistics, p. 18, Oct. 1972.

workers, a situation not likely to be fully paralleled if energy prices should continue to rise.

Further perspective is provided by noticing that variations in household energy costs, both between regions of the United States and between the United States and Western Europe, are *very* substantial (see Table VII). Prices in high price locations can be two, three, or even four times the price in low price locations. These observed variations would seem to indicate that consumers can, in the long run, adjust to price changes of the magnitude under discussion without major dislocation.

Additionally, energy consumption depends on price. As prices go up for any one type of energy, or for energy in general, consumers can be expected to use less of it. Such readjustments soften the implicit "tax" burden of price increases. Some recent studies of electricity demand suggest that expenditure actually changes more rapidly than price, when prices change.[41] If this is so, then electricity price increases will, in the long run, be accompanied by consumption declines. Expenditure on electricity will actually decrease as a result. Expenditure on any one type of energy is more price sensitive than spending on energy as a whole. Since such shifts are possible, overall energy prices, and not price for any one fuel or source, are most relevant from a welfare viewpoint.

Electricity and gas prices in most states are set with the ap-

[41] See R. Halvorsen, "Residential electricity demand," Ph.D. dissertation, Harvard Univ., Cambridge, Mass., 1972; and K. P. Anderson, *Residential Demand for Electricity*, Rand Corp., Rep. R-905-NSF, 1972.

proval of the state Public Utility Commission or some equivalent body, usually after a public hearing.[42] Increases in fuel costs for electricity utilities, however, can often be passed on automatically via "fuel cost adjustment" procedures that obviate the rate hearing requirements.

The income distribution effects of energy cost increases will depend significantly on how price structures reflect these rises. How will cost increases be passed on? Will all parts of the rate structure increase by the same absolute amount, or the same percentage amount? Will the occasion be utilized to "flatten" the pattern of declining block rates and will rates for larger purchases increase proportionally more? There have been some attempts of this sort recently.[43] Such changes would ameliorate the adverse redistributive impact somewhat. However, it seems not implausible that the structure of most energy prices will not be affected by the process being discussed. Equal percentage increases is not a bad working hypothesis.

If price increases should turn out to be significant, the distribution effects are not an attractive prospect. Such consequences, in my view, imply that policy makers have a special obligation to not waste resources (i.e., needlessly increase costs) when improving the environmental performance of the energy industries.

VII. INDIRECT IMPACT ON HOUSEHOLDS

Energy price increases for direct household purchases are only part of the picture. In addition, energy price increases will affect the prices of other goods and services. Predicting such results is not simple because the energy input into goods and services itself has both direct and indirect components. When you buy a piece of paper, you do not only pay for the fuel used to produce the paper, you also pay for the energy used to produce the chemical inputs, and the capital equipment, and so on. Given this cost structure, an increase in energy prices will take some time to work itself out. The prices of *existing* capital goods will not go up when energy costs rise. Only new capital equipment will reflect the higher costs. Hence final goods prices will adjust only gradually to increases in their own indirect energy costs.

Furthermore, the cost increases faced by the most energy-intensive industries are a poor guide to the ultimate impact on consumers. The high-energy-use sectors largely do not produce consumer goods (e.g., basic metals, pulp and paper, chemicals, nonmetallic minerals, etc.). Instead the outputs of such sectors often comprise a small percentage of the cost of the goods households actually purchase (e.g., as the cost of the more expensive aluminum in a can containing soda). The cost of consumer goods includes many sales, manufacturing, and distribution expenses, in addition to the cost of energy-intensive raw materials. This structure lessens the indirect impact of energy cost increases on consumer prices. Tripling the cost of an input which amounts to only 5 percent of final prices will only increase final prices by 10 percent even if it is fully "passed on."[44]

These relationships can be explored by means of input-output analysis. An input-output table indicates how much each industry purchases from each other industry, per dollar of final output from the buying sector. Such a table can also be manipulated to indicate direct plus all indirect purchases among industries, including all interactions, no matter how complex.[45] The most recent accurate U.S. tables are for 1963, which separate the U.S. economy into 367 sectors.[46] Less detailed tables separate the economy into 86 sectors. It is, of course, not possible to reproduce such tables here, or even to present the entries relevant to the energy industries.

I can tell you, however, that a reasonably careful eyeball review of these tables located very few industries where direct energy use was over 3 percent of sales. Among the higher *direct* energy users are[47] manufactured ice, 10.1 percent; chemicals, 12.0 percent; synthetic rubber, 7.3 percent; cement, 15.0 percent; primary aluminum, 9.7 percent; lime, 11.1 percent; carbon and graphite, 8.9 percent; air transport, 8.9 percent. In several of these cases, it is clear that important nonenergy use is being made of hydrocarbon inputs.

Further, not many industries have direct plus indirect energy use greater than 10 percent, and relatively few are between 10 and 7.5 percent. Most consumer goods sectors have much lower direct plus indirect requirements. In the less detailed breakdowns, "food and kindred products" has direct plus indirect requirements of 6.5 percent; household appliances, 5.4 percent; motor vehicles and parts, 6.0 percent; etc. None of these sectors has direct energy requirements of as much as 1 percent of shipments.

These data imply that the effects of even major energy price increases will not be an unbearable burden on the consumer. A doubling of energy prices would, in the short run, raise consumer prices for goods except energy by less than 1 percent. This is noticeably less than our current rate of inflation. Over

[42]See for example P. F. Pegrun, *Public Regulation of Business*, revised ed. Homewood, Ill.: Irwin, 1965, ch. 25; or C. Wilcox, *Public Policies Toward Business*, 4th ed. Homewood, Ill.: Irwin, 1971.

[43]See, for example, both the order and opinion of Chairman W. F. Ward and Commissioner L. G. Sculthorp, and the separate opinion of Commissioner W. R. Ralls in Michigan Public Service Comission Case No. U-3910, decided Aug. 18, 1972, which alters the rate structure of Detroit Edison. The issue has prompted a significant staff study by the New York State Department of Public Service. See D. E. Sandler, "The inverted rate structure—An appraisal, part I—Residential usage," New York State Office of Economic Research, No. IX, Feb. 17, 1972.

[44]To simplify the analysis, it is usually assumed that for the products concerned, production occurs at constant returns to scale so that long-run, average, and marginal costs are identical. Then under conditions of perfect competition, the cost increase would be fully "passed on." In a regulatory context, whether they are or not depends, as noted previously, on how the regulatory body behaves. Of course, this is only a partial equilibrium analysis. Final demand can be expected to fall as the result of the price changes. This in turn lowers the derived demand for, and the incomes of, some factors of production. Analyzing the full general equilibrium incidence effects clearly is beyond the scope of this discussion.

[45]Among the many discussions of these models available in the literature are R. E. Kuenne, *The Theory of General Economic Equilibrium*. Princeton, N.J.: Princeton Univ. Press, 1963; W. Leontieff, *The Structure of the American Economy 1919–1939*, 2nd ed. New York: Oxford Univ. Press, and C. Almon, Jr., *The American Economy to 1975*. New York: Harper and Row, 1966.

[46]*Input-Output Structure of the U.S. Economy: 1963*, 3 vols., U.S. Dep. Commerce, Office of Business Economics, Washington, D.C., 1969.

[47]*Input-Output Structure of the U.S. Economy: 1963*, 3 vols., U.S. Dep. Commerce, Office of Business Economics, Washington, D.C., 1969, vol. 2. The figures in the text are the sum of the inputs from the following sectors: 7.00 coal mining; 31.01 petroleum refining and related products, 68.01 electric utilities; 68.02 gas utilities.

the long run (over perhaps many years), when all of the indirect linkages have worked themselves out, the impact on consumer prices of such an increase would be comparable to the *annual* price changes of the worst of recent years. That is not a very attractive prospect. But it hardly constitutes the basis for a major economic crisis, even in conjunction with the direct effects noted previously.

VIII. LARGER EFFECTS OF ENERGY COST INCREASES

So far our attention has been confined to the impact of energy price changes on individual households. But, are there other important effects which that focus does not capture?

One obvious question is the potential impact of these changes on international trade patterns. There are several possible consequences. First suppose we do encourage the use of otherwise noncompetitive domestic energy reserves by maintaining high domestic energy prices. This would injure the international competitiveness of U.S. business. England and West Germany do follow exactly such a policy. And Japan has very high energy prices. But these facts are not really relevant. U.S. firms will be better off if they can obtain energy more cheaply. On the other hand, U.S. exports consist mainly of machinery and other manufactured products (and also agricultural goods). These outputs do not have an especially high energy component. Indirect effects too would be small in most cases, for the same reasons as those just discussed above. Thus any handicap from higher energy prices is likely to be a small one for the most relevant sectors.

Given that environmental and safety concerns may well increase the prices of extraction at home, we will probably find it necessary to import more energy. This would mean that domestic energy prices would come to follow world prices more closely. But price variations due to those circumstances would be suffered by most of our trade competitors as well. Thus such a development probably would not have a major impact on the pattern of international transactions.

More significant are the potential balance-of-payments effects of very high import levels.[48] It may not be easy to afford the low-cost foreign oil which "nature" has chosen to "give" to other nations. To enhance our trade position, we are going to have to start doing better what we do best, that is, exploit advanced technology. Perhaps a lower priority for the aerospace sector (as both NASA and Vietnam diminish in size) will allow us to deploy our skilled technical manpower to the civilian sector, where it is very much needed.[49]

I would argue that in the long run a nation needs to base its trading patterns on its strengths, not on its weaknesses. It seems foolish to compete with the Middle East as a raw materials supplier. But abstract principles will not guarantee acceptable outcomes. We must be prepared to take strong action to improve our trade performance. Otherwise, we could be forced to devalue in the face of large energy import costs. Doing so would raise the costs of imports, including energy, while lowering the export prices of our own goods. This would put domestic consumers in a worse position than they would have been in without the devaluation. However, given the magnitudes involved, the prospect is not a national tragedy. In the 12-month period ending with November 1972 (the last month for which I have data) petroleum products were 7.6 percent of our imports. This compares with 11.6 percent for food products, 14.4 percent for automobiles and parts, 12.6 percent for machinery, and 33.1 percent for manufactured goods and articles.[50]

We do face a general problem of achieving a level of international competitiveness sufficient to preserve domestic standards of living. We must sell goods abroad to raise the foreign exchange to pay for our imports. If we are efficient, the goods we give up will require relatively few inputs. If we are not, the real cost of imports goes up. Suppose balance-of-payments problems do force us to devalue to the point where U.S. reserves become relatively cheap versus imported fuels? Such an outcome would have to be considered a result of the nation's general level of economic inefficiency, as opposed to simply a fuel sector problem.

If we should prove unable to earn our energy keep in world markets, or if our overseas suppliers should prove to be silly enough to price themselves out of our market, then let domestic competition discipline them. The possibility of transitional high prices for 3 to 6 years under such circumstances seems much less unattractive than the certainty of high prices from now until the end of time. Keeping up domestic energy prices, to preserve noncompetitive domestic producers, is not a sound way to provide price discipline for import suppliers.

What should we make of the suggestion that it is not "fair" for the United States to use more than its "share" of world energy resources? Is the United States "draining" or "exploiting" the Middle East by purchasing oil, and is this to be avoided?

I, for one, do not find the political systems of the oil exporting countries uniformly attractive. There probably is some truth to the argument that oil revenues can serve to reinforce the position of the elites that happen to be in power (whether they are traditional or revolutionary). But I believe one is confused if he fails to note how exporting countries *gain* from their exports. All this talk of "fair shares" ignores the devastating impact on many of these "single crop" economies of any major decrease in oil production, sales, and tax revenues. I don't deny both the possibility and the reality of exploitation in international economic relationships. But not all international economic transactions between rich and poor nations are "exploitation." Some are mutually beneficial trade. In any case, if historic trends reassert themselves, resource owning countries might do best to sell resources sooner rather than later. Resource prices may well *decline* with new technology and new discoveries.

[48]See *The Oil Import Question*, Cabinet Task Force on Oil Import Control, Washington, D.C., esp. pp. 43–44, 125. The estimates contained here on potential import dependence are somewhat lower than others that have been advanced.

[49]Almost 50 percent of U.S. exports are in the machinery and equipment category, which includes many advanced technology-intensive sectors, e.g., aircraft, computers, etc. See *Survey of Current Business*, vol. 52, pp. S-22 and S-23, Jan. 1972.

[50]Total imports for the year ending Nov. 1972 were $55.0 billion of which petroleum products were $4.2 billion (*Survey of Current Business*, vol. 52, pp. S-22 and S-23, Jan. 1972.

The "success" of OPEC is also not entirely a "bad thing," despite the attitudes of the usual studies. Its members are poor countries by and large. Should we be happy or sad when they find a way of screwing more foreign exchange out of our relatively rich hides? If you like some of the member governments and not others, how do the international redistributive effects and political effects balance out? It is not an easy question.

Next, will the sort of changes we have been talking about have any impact on the environment? I find very disturbing the possibility that the so-called energy crisis will be used as a club to beat down environmental standards. There are many in the energy industry who would no doubt prefer to proceed with "business as usual," where "usually" one did what one wanted to the environment. I hope that such times are permanently behind us, but one never knows. In particular, there are already very significant political pressures building up to allow drilling to proceed on the East Coast outer continental shelf and for relaxation of the requirements of the National Environmental Policy Act.

I am no expert at political forecasting. It is easier to say what one would like to happen in such cases than it is to predict the outcome. My guess is that the next couple of years are critical. If environmental protection measures remain largely intact, then they will become part of the "woodwork." People will perhaps become used to doing business on these new terms. How long did it take some managers to become accustomed to the idea of labor unions?

I don't want to defend the wisdom of all of our environmental programs. But self-proclaimed well-meaning or responsible critics usually exhibit a notable lack of initiative. Protestations of good faith would be more convincing if, for example, the oil industry lobbyists were ever to be found working for rational and sound environmental protection measures, like the sulphur tax proposals I discuss below. Without such actions, environmentalists will remain justifiably suspicious that industry statements calling for "balanced measures" are employing that term as a codeword for "measures that don't require us to do a hell of a lot."

IX. WHERE DO WE GO FROM HERE?

In discussing these questions, I try hard to separate my roles: as an economist on the one hand and as an advocate on the other. But when one talks about policy, values are necessarily involved. So as a preliminary, let me confess explicitly to being a member of that once-endangered species, the environmentalist, now making a strong comeback.

The situation I have reviewed suggests several policy implications to me. First of all, the distributive implications of increasing energy costs are quite unattractive. This implies an obligation to worry about the efficiency with which we undertake environmental protection. At the moment the environmental movement has something of a "cold war" mentality. No amount of money is too much; no program, however inefficient, should be opposed if it is on the "right side." This willingness to tolerate, indeed encourage, "ecological overkill" just makes the job of "altering our priorities" that much more difficult to accomplish. In my view, given that governments and lower income people, who have much else to spend money on, are paying a large part of the bill, means we should make the bill as small as possible. This is not a plea for low standards or giveaways to established interests, which calls for "responsibility" all too often seem to include. Rather it is a suggestion that the complexity of our choices demands a willingness on our part to entertain complex responses.

Wherever we can alter adverse redistributive consequences, such as by altering rate structures, I would urge my fellow citizens to seize the opportunity and make such changes. Those who lack concern with such "details" do much to invite being categorized as self-serving middle-class reformers. Such people, in an earlier day, tore down and did not replace low-income housing in older urban areas in order to make way for ceremonial plazas and civic centers.[51] The socioeconomic composition of the major environmental groups provides some support to the notion that there is a problem here.

Second, we should make some efforts to have energy prices more accurately reflect the total costs of producing energy, including environmental costs. We should seek ways to inform consumers and to help them make more careful choices about energy use. This means, among other things, instituting effluent fees for residual waste discharges from power plants (e.g., the Aspen-Proxmire proposal to tax sulphur emissions[52]) so that energy prices reflect these costs. It means attempting to create a more economically rational rate structure that more nearly reflects the real costs incurred at various margins of choice. Experiments designed to develop and institute an operational system of peak load pricing, especially for larger customers, are long overdue in this country.[53] Similarly, we should try to develop simple uniform labeling requirements for major home appliances, so that consumers can more intelligently compare expected energy use and operating costs.

Third, it is not obvious that our pattern of energy use reflects any very careful adaptation to real or current costs. I would view any attempt at rationing in the current context with some alarm. But efforts to curtail inefficient energy consumption through better design and improved consumer information seem emminently sensible. Such programs may not have major effects on aggregate energy demand, but they might make a useful contribution to improving intersource choices at the household level. Suppose consumer choice nonetheless proves to be imperfect? Or suppose that the costs (to both government and potential buyers) of diffusing the relevant information are too high to make the effort worthwhile? Then minimum performance standards for appliances,

[51] See E. C. Banfield, *The Unheavenly City*. New York: Little Brown, 1970, ch. 1.

[52] A discussion of various sulphur tax options can be found in *Taxation with Representation*, *The Proposed Tax on Sulphur Emissions*, Washington, D.C., 1972; and *Effluent Charges on Air And Water Pollution*, Environmental Law Inst., Washington, D.C., E.L.I. Monograph 1, pp. 52–63, 1973. The Aspin-Proxmire proposal was contained in H. R. 10890 and S. 3057 in the previous session of Congress. An alternative developed by the administration is presented in *Materials Relating to the Administration's Proposals for Environmental Protection*, U.S. House of Representative, Committee on Ways and Means, Washington, D.C., Feb. 1972.

[53] Such policies have been used abroad for some years. See J. R. Nelson, *Marginal Cost Pricing in Practice*. Englewood Cliffs, N.J.: Prentice-Hall, 1964. esp. chs. 4, 6, and 7.

and tighter building codes with respect to insulation standards, might be considered. After all, we do have auto safety requirements.

Fourth, I would urge us to free up many energy markets. We should remove protectionist and distorting policies in order to allow a more rapid adjustment to the developing situation. I would like to see an end to import controls on oil, an end to oil production controls not essential to very narrow conservation principles, less favorable tax treatment of mineral industries, deregulation of natural gas prices, and perhaps some antitrust actions against the emerging "energy conglomerates."

Fifth, we do need, as most seem to agree, an expanded research and development program with respect to energy conversion technology. How clean and cheap could we make our energy system if we looked harder at new ways of proceeding?

In the environmental area, many problems need to be attacked. Most of them are very complex: strip-mining controls; new federal land leasing policy (both on and offshore); new power plant siting procedures; controls on drilling and transport procedures; better storage of high-level radioactive waste, and so on. Problems of dynamics, uncertainty, irreversibility, and interaction effects tend to arise in most of these cases.[54] The need to grapple with such complicated detail leads to one of my main objections to the way Forrester and Meadows have played with their electronic erector sets. Their diagrams may be pretty, but most of the relevant questions can neither be asked nor answered within the structure of those models. What can you learn about the real problems when all environmental effects, of all kinds, everywhere, are aggregated into one single index of "pollution"?

Yet the whole controversy stirred up by these and related works might ultimately be not without value. It is possible to conceive of American society today as still following a set of decision rules which are rapidly becoming obsolete. Perhaps these do need changing if we are to respond "optimally" to faster technical change and higher levels of material well being. We are becoming the first society in history where physical scarcity will not be in the major concern of the mass of people. How does one live in such a world? We don't have all that much experience with it. Creating new "standard operating procedures" and "mental sets" (i.e., culture) is a difficult business. Perhaps this hoopla will help make people think more seriously about the tradeoffs between industrial use, recreation development, and the preservation of ecosystems in a less disturbed state. Maybe a questioning of the marginal value of material goods is in order. Are we rich enough as a society to now afford other things?

Still, one must remember that if we are rich the world as a whole remains shockingly poor. Is it morally desirable for us to seek leisure and amenity in such a world instead of seeking to create a big production surplus which we can then give away? In handling the problems of energy supply and energy prices we are faced with the classical difficulties of rationally adapting means to ends. Ignoring the prophesy (both of the selfish and silly varieties), we need to begin to construct a careful, balanced, and appropriate energy policy. But the larger problem of how to live justly in a world of such inequality cannot be settled so easily.

[54] M. J. Roberts, "The nature of environmental problems and policy options," presented as the keynote address at the Joint Conf. Technology Governance in Achieving Environmental Quality, New York, N.Y., Sept. 19–22, 1972.

Peak-Load Pricing for Electricity

CHARLES J. CICCHETTI AND WILLIAM J. GILLEN

THIS paper describes a pricing policy for electric power which can do much to alleviate the current shortage both of generating capacity and fuel for power generation. Typically, the price charged for a kilowatt hour of electricity depends on the class of service of the consumer, on the total quantity consumed over the billing period, and, for larger volume users, on the maximum quantity consumed at one time. These factors, purportedly intended to account for variations in cost, in fact mask the principle source of cost variation–diurnal and seasonal load fluctuations. Charging prices which either ignore or average out cost differences between on- and off-peak loads necessarily means that on-peak loads are underpriced, while off-peak loads are overpriced, which leads to overconsumption in the first case and underconsumption in the latter case.

Peak-load or time-differentiated pricing is a type of tariff which varies according to the level of kilowatt demand on a utility system over a daily and seasonal cycle. It is based on the economic theory of marginal cost pricing which postulates that producers should (and in a competitive economy will) expand production so long as consumers are willing to pay the incremental (or marginal) cost of each additional unit of output.

For the electric utilities there are several categories of costs with vastly different characteristics. The price that a consumer should pay for service is the sum of the marginal costs in each category. For example, there are separable costs for the generating plant, the transmission network, and the distribution network.[1] Each of these would be accounted for "at the margin" in an ideal price schedule. Since administrative costs also increase with the complexity of the tariffs, a practically achievable variation of marginal cost pricing is our objective.

The reader should note that the argument for peak-load pricing is usually phrased in terms of savings in capital expenditures. Unless the elasticity and cross elasticity of demand for peak power is zero, less capacity is required to meet demand under peak-load pricing tariffs than under time-uniform tariffs. It is the foregone expenditure in additional capacity which constitutes the bulk of the savings from peak-load pricing. An additional source of savings deserves mention, however, in the days of a shortage of fossil fuels.

A power generating system consists of a mix of plants to serve different types of loads, i.e., peak load, intermediate load, and base load. Each of these types has a different ratio of capital costs to energy (fuel) cost. These vary inversely. High capital costs are associated with relatively low-energy or operating costs, and vice versa. Since peak loads occur only a few hours per year, it makes sense to meet the peak demand with plants that have relatively low capital costs; energy costs will not be significant in any case. Conversely, base loads require plants that economize on fuel, and relatively higher capital costs are acceptable here.

Exactly which mix of plants is appropriate to a particular system will depend on the time distribution of the system load. Where peak-load pricing results in a flattening of the load distribution (without an increase in the total load), fuel savings will result because demands will be met by more energy efficient base-load and intermediate-load plants. Peak-load plants, which squander fuel in the interest of saving on capital investment, will be required less often.

The potential savings in capital investment will overshadow whatever fuel savings may result from peak-load pricing. But the current fuel shortage adds additional impetus for peak-load pricing.

DERIVING THE APPROPRIATE PRICES[2]

Cost data and other information are kept according to accounting formats useful for present pricing strategies, but data deficiencies exist with respect to the information which would be required for the best estimation of appropriate peak-load price differentials. But while refinements in cost and load data collection would be helpful, it is nonetheless possible to demonstrate how existing empirical information may be used to derive a set of time-differentiated prices.

Demand or Capacity Costs

The size of the required generator plant is determined by the level of maximum or peak demand on the utility system.[3] The first approximation to capacity cost is the long-run incremental cost of this capacity, plus desired reserve capacity. This estimate of capacity costs, apportioned among all the peak users according to their on-peak demands, will be the basis of the peak charge.

This work is adapted from *Perspectives on Power: A Study of the Regulation and Pricing of Electric Power* (Cambridge, Mass.: Ballinger,), to be published.

C. J. Cicchetti is an Associate Professor of Economics at the University of Wisconsin, Madison, Wis.

W. J. Gillen is a Consulting Economist for the Environmental Defense Fund, Washington, D.C.

[1] And there is a portion of total costs that does not fit any of these major classifications.

[2] The authors are indebted to D. N. DeSalvia, whose article, "An application of peak load pricing," *J. Business*, vol. 42, p. 458, 1969, we follow closely.

[3] Note that the reference is to *system* peak. There are creatures of utility accounting such as "customer class peak," "customer peak," and others. None of these is relevant here, nor, we would expect, particularly useful elsewhere.

Note, however, that under certain conditions the peak charges may be imposed during certain nonpeak hours. The capacity charge will be set so as to constrain demand to the level of supply at any particular time. Supply, or capacity, is in fact variable. It is not simply the maximum power output of the system at any time during the year. The important point to note is that peak charges are applied to periods which approach or reach *available* capacity and not simply maximum capacity. For this analysis, we assume that available capacity at the time of peak and maximum capacity are the same.

Most electric utilities have a distinct seasonal peak, either winter or summer. In some cases, the maximum winter and summer demands are approximately the same. But by far the greatest variation in demand is over the daily cycle. For purposes of rate design, the peak may be defined as those hours in which the system is operating at or close to maximum available plant capacity. Somewhat arbitrarily, the peak is defined as those hours during the year in which demand exceeds 85 percent of capacity. Let us suppose that on a hypothetical utility system, a peak occurs only during the winter months, between the hours of 4:00 PM and 8:00 PM on weekdays during the months of November through February. This is a total of approximately 340 h during the year. The number of peak hours may of course vary widely among utility systems.

Under peak-load pricing, the total capacity costs of the system would be allocated to consumption during peak hours and no capacity costs are allocated to off-peak consumption.

Transmission Costs

Clearly, transmission facilities requirements are related to peak consumption. But determining the charge to be derived from marginal transmission costs is somewhat more troublesome than capacity charges because there are two geographical sources of variation in cost. First, transmission facilities may serve well-defined geographical areas, presenting capacity limitations apart from those of the whole system. That is, a system may be operating at less than maximum generating capacity, but the transmission lines to a particular area may be strained to the limit. Second, losses do occur as power is transmitted over greater distances, although some lines carry power more efficiently than others. The first source of geographical variation may be ignored in the interest of tariff simplification,[4] especially in the larger integrated grid systems; the better integrated the system, the less able one is to attribute cost causally to a particular area.

In practice, data concerning losses in transmission are recorded only in the form of totals for the whole system, which makes it impossible to precisely calculate the marginal transmission cost. This imperfection must therefore be ignored.

In short, subject to the above qualifications, transmission costs should be allocated in much the same manner as generation costs, although refinement is possible and perhaps desirable.

Energy Costs

The remaining cost element with a temporal character is the energy cost.[5] This is fuel plus an allowance, typically 15 percent, for maintenance. Even a moderate sized utility (or a small interconnected utility) experiences a significant variation in energy costs as demand increases above the base-load level. Energy costs vary because the efficiency of plants varies due to age, location, fuel used, and other factors. The least costly most efficient plants serve the base load. As demand increases, more costly and less efficient plants are brought on line. Further, energy costs may vary for any given unit. The difference between the cost of producing energy on and off peak may vary by a factor of 5 or more.

It may be useful to further divide the nonpeak hours of the year into two or more classes. The number of classes is arbitrary, depending on the design of metering equipment and the desire for easily administered tariffs.[6]

Two nonpeak periods may be designated: "off peak," a fairly continuous load which is required throughout the year; and "full use," which is any level of demand less than peak and greater than, say, 55 percent of maximum demand or whatever the base load is. The remainder is on peak. Energy charges for these periods could be determined from the incremental fuel costs for the particular utility system.

Other Costs

The remaining portion of total costs includes the cost of the distribution network and general administration costs. The appropriate treatment of distribution costs under marginal cost principles is a charge equivalent to a monthly payment on an interest-bearing loan equal to the cost of the distribution facilities required by each consumer to be hooked up to the system. The balance of costs, such as administration, are relatively insignificant and are probably best levied on a per-customer basis.

Data are not now kept in a manner which makes the appropriate assignment of distribution costs possible. Although some effort is made to segregate such costs by category of customer, the failure to distinguish between the distribution costs attributable to various customers constitutes cross subsidization of unknown magnitude.

The prices that would be derived according to the above methodology are interesting—and at first glance startling. Applying this methodology, DeSalvia derived the prices shown in Table I, exclusive of distribution and general administration

TABLE I
PRODUCTION AND TRANSMISSION COSTS (BASED ON EXISTING LOAD PATTERNS)

Daily Period	Winter (¢/kWh)	Summer (¢/kWh)
Peak	11.11	—
Full use	0.46	0.46
Off peak	0.30	0.36

[4]In tariff design, what is possible is not always practical. Tariff complexity imposes its own variety of costs.

[5]As with capacity costs, hydroelectricity is again a special case. The energy costs of hydroelectricity are virtually zero, except for pumped storage hydroelectricity which involves the use of other energy sources, including fossil fuels, during off-peak periods. Pumped storage is of growing significance.

[6]Some meters such as those with time switches will present a limitation on the number of periods. Others, such as remote control frequency sensitive meters, would allow a much greater number of designated periods and therefore prices. The latter are not necessarily more costly than the former.

costs, applicable to a large northeastern utility.[7] The peak-period charge is 24 times the full-use charge and 37 times the off-peak charge. The factor which accounts for the magnitude of the differentials is the brief duration of the peak—348 h/yr, or 4 percent of the total hours.

> The average demand could be serviced with a much smaller physical plant. Thus, a considerable portion of the firm's capacity with its concomitant investment and operating costs is directly attributable to peak consumption.[8]

In practice, price differentials of this size would not be warranted since a drastic change in consumption patterns would result and should be anticipated in the first promulgated set of time-differentiated prices. All customers would seek means of avoiding peak kilowatt-hour consumption. Most of the reduction in consumption would surely be shifted to nonpeak periods. Some would result from more efficient use of that power which necessarily is consumed on peak, and some uses would be dispensed with altogether. Since there is no relevant prior experience with time-differentiated rates in the United States, the net result of the shifts and reductions can only be guessed at. As experience and data accumulated, the elasticities and cross elasticities of peak and off-peak demand could be estimated and used in further efforts to reduce excess investment and resource misallocation.

By making some conservative and not unreasonable assumptions about shifting demand patterns, one can derive a set of prices that are more likely to be appropriate after peak-load pricing has been in effect. DeSalvia assumed a 5-percent decrease in the winter peak, which reduced it to the level of the "full-use" period. He further assumed a 5-percent increase in summer full-use consumption which raised that to the level of winter full use. DeSalvia then assigned a proportion of the capacity costs to some of the 3408 full-use hours. The capacity cost having been spread over so many more hours, the differentials decreased considerably. The prices shown in Table II resulted. These differentials are more realistic; they reveal the extremely high cost (as reflected in cost-determined prices) of sharp daily peaks and excess capacity most of the day and year. And they suggest that even much flattened load curves will support significant price differentials.

TABLE II
PRODUCTION AND TRANSMISSION CHARGES (BASED ON A CHANGED LOAD PATTERN)

Period	Winter (¢/kWh)	Summer (¢/kWh)
Peak	2.04	—
Full use	1.60	1.45
Off peak	0.30	0.30

It is important to note that such prices would apply to all customers. The sole determinants of the electric bill are the time of consumption and the cost of patching into the system. Eliminated are such spurious considerations as what the power is used for, e.g., residential, industrial, or other uses, and the quantity consumed during the month. Should social or political reasons dictate that some customers are to be provided electricity at subsidized rates (religious and rural customers sometimes obtain service at explicitly preferential rates), that subsidy may be applied easily; but inadvertent subsidies such as those provided by present rates would be abolished.

DeSalvia postulated that had his prices been in effect for the time at which he made his calculations, slightly more revenue would have been generated than under the conventional rates.

Let us compare the incentives implicit in present nontime-differentiated rates with those of peak-load pricing. Nontime-differentiated rate schedules do not, by definition, affect the temporal pattern of consumption. Since rates typically decrease as consumption increases, whether kilowatt or kilowatt-hour consumption, the incentive is to expand individual consumption. At one time expanded consumption may have tended to occur in off-peak hours, but consumption expansion now is as likely as not to be on peak. Existing rate structures continue to provide an incentive via lower prices to a customer who exacerbates capacity requirements even though the costs of new capacity are higher than costs associated with existing capacity.

Time-uniform rates carry a portion of capacity costs in all hours, which means that there is, in effect, a penalty associated with off-peak consumption, because in fact there are no appreciable capacity costs off peak. The size of the penalty is usually a several hundred percent surcharge over actual costs, and off-peak consumption is discouraged, albeit to no discernible benefit to producer or consumer.

Peak-load pricing, on the other hand, carries a completely opposite set of incentives. It discourages on-peak consumption, by charging higher on-peak prices, and encourages greater use of excess capacity by charging lower prices off peak.

While peak load pricing has not been implemented in the United States,[9] utility managers have considered problems that might occur if such prices were put into effect. We consider those mentioned most frequently.

The Shifting Peak: As peak-load prices are implemented, changes in the load pattern are to be expected. The original set of peak-load prices would then be wildly inappropriate.

[7]Since his data are a decade old, the prices themselves are useful largely as ratios to one another. The fixed monthly distribution and other costs were allocated on an arbitrary but not unreasonable basis to the various categories of customer. The resultant charges were the following:

residential	$ 4.84
commercial	$ 10.03
farm	$ 16.22
industrial	$165.91
street lighting	$404.67

[8]D. N. DeSalvia, "An application of peak load pricing," *J. Business*, vol. 42, p. 469.

[9]Peak off-peak price differentials have been implemented for such use as water and space heating. This is not a true application of marginal cost pricing principles since it is one-sided, i.e., there is an off-peak discount but no corresponding on-peak premium. Another variation on this theme is "interruptible" rates whereby large consumers receive discounted rates for accepting the possibility that their service may be discontinued on short notice during periods which strain utility capacity. Such "marginalist" pricing schemes in effect here and abroad are discussed in W. G. Shepherd, "Marginal cost pricing in American utilities," *Southern Econom. J.*, vol. 33, p. 58, 1967.

The solution is to redefine the peak and issue revised tariffs.

But there would be a considerable difference between the process presently required to design rates, and the procedure which could be used if peak-load pricing existed. The typical rate case involves determining the appropriate rate of return on investment, the revenue required to meet that target, and the particular rates which will produce that revenue. A host of horrendously tedious subsidiary questions arise and virtually every point is the subject of exhaustive and disputed testimony. The hearings may take weeks, months, or years. But present rate hearings are complicated because almost every issue raised is decided by administrative judgment rather than by objective criteria. Peak-load pricing would significantly reduce the need for rate filings.

First, peak prices would eliminate the existing incentive to consumers to expand demands on peak and thus would reduce the need for capacity additions. Consequently, utilities would not need to petition as often for rate increases to attract investment capital.

Second, costs are rising over time and significant demand growth is in the tail blocks of existing declining block rates which are not fully compensatory. When the need for additional capacity is caused by larger customers as compared to new customers, their contribution to peak demands costs the utility more than the expansion generates in revenue. Peak-load prices, on the other hand, are compensatory at every level of demand. Thus, the need for rate hearings diminishes further.

Finally, disputes between rate payers as to which class shall bear what portion of the additional revenue requirement are eliminated. The same rates apply to all, and these are determined by objective load criteria. Whether a kilowatt hour is used to heat a house, light an office building, or power a drill press is irrelevant to the price charged. It is no longer necessary to distinguish among customers on any basis except the cost of providing them electricity.

Thus, when peaks shift, it would only be necessary to recompute the tariff schedule according to what can be made into a standardized formula. It will not be necessary to engage the whole range of questions that occupy presently styled hearings. These issues will occasionally be faced, but considerably less often.

Repair and Maintenance: It is alleged that repair and maintenance schedules may be disrupted by peak-load pricing. The argument: attempts are currently made to concentrate repair and maintenance during periods when capacity is idle; under peak-load pricing there will be less idle capacity at any one time. Consequently, there will be less opportunity for scheduled outage. The argument implies that higher system load factors are not desirable. It is a curious argument when made on behalf of an industry which has employed selective-load-building advertising, special night rates, Hopkinson rates, Wright rates, Doherty rates, and moral persuasion in an effort to achieve higher load factors. Recall that peak-period pricing requires prices which are highest when demand approaches available plant capacity. If some portion of plant must be taken out of service for repair or maintenance, the period of outage may be defined as peak, although demand at that time does not actually require peak capacity. It should be recognized that unscheduled outage occur regardless of pricing strategies.

Allocation of All Capacity Costs to Peak Period Demand: The argument has been made that it is neither appropriate nor equitable to assign all the capacity costs to the peak period. Generating plants are specifically designed to serve either base load, intermediate load, or peak load. The plants have different capital and running costs depending on the type of load they are intended to serve. The cost of base load plants might then be assigned to all periods. Costs of intermediate-load plants might be assigned to full-use and peak periods, and peaking plant costs to the peak period. Such a cost distribution would reflect the respective contribution of each plant to meeting the distributed load. The argument has a deceptive appeal. It may be answered with reference to the concept of opportunity cost.

Economists postulate that the most accurate way to gauge the cost of anything is to consider what is given up when a resource is used for one purpose which precludes its use for other purposes. For example, if a parcel of land may be used to grow either wheat or corn, then the opportunity cost of growing wheat on the land is the value of the corn which cannot then be grown, and vice versa. In general, opportunity cost of a resource is the value of the resource in its highest (most valuable) alternative use. The opportunity cost of a scenic gorge may be the value of the hydroelectric power which would be produced if a dam were built.

What is the opportunity cost of using generating capacity that would otherwise be standing idle? To what alternative use can it be put? Obviously, none. The appropriate charge then is zero. If a charge greater than zero is levied on off-peak consumption as the objection here proposes, such a penalty would result in underuse. Given that a ceiling is imposed on the revenues of the utility, imposing a charge for using nonpeak capacity would require that less than the opportunity cost be charged for peak consumption. The effect is subsidization of peak consumption.

Is it "fair" to assign all capacity costs to the peak period? If the popular will is to have nonpeak users pay the way for on-peak users, nothing in logic or economic theory can refute that. But an observation is in order. Sooner rather than later, under the present pricing system, the cost to all users, peak and nonpeak alike, will be higher than they otherwise would be.

Stability of Rates: Much is properly made of the desirability of achieving stable electricity prices. The charge considered here is that time-differentiated prices are unstable and impede rational planning and decision making by consumers. But in fact, peak-load pricing is likely to be more stable than existing rate designs. Peak-load prices are not themselves unstable in any meaningful sense. They do vary by time of day; but the pattern of variation is constant. Time-uniform rates, on the other hand, provide incentives which lead to capacity expansion and the need for additional revenues, which is an inherently unstable process. A concern then for stability of electric rates is an argument in favor of peak-load pricing.

Metering: Here is the crux of the matter. Peak-load pricing should be implemented if and only if, and on such a scale, the benefits of less excess capacity and improved fuel economy outweigh the cost of the more sophisticated metering that will be required.[10] Rudimentary and somewhat expensive equipment has been available for some time. To the limited extent to which peak-load pricing has been implemented, the chosen equipment has typically employed watt-hour meters together with a two- or three-stage time switch. More sophisticated devices are in use which use radio controls to switch modes on a meter or to switch from one meter to another. So-called "ripple control" devices transmit messages to points of consumption via the power line itself. Until recently, however, the requisite metering equipment has been either expensive or beset with technical difficulties, either of which limits application on a wide scale.

Recent technical developments have, however, brought the cost of the necessary equipment clearly within the range where it becomes economically feasible to institute time-differentiated pricing. Further, anticipated expenditures for new capacity are of such a magnitude for U.S. electric utilities that even the modest gains from, say, a 5-percent reduction in "requirements" would free those funds for investment in a metering network many times the cost of the existing equipment.

[10]An additional potential exists for reduced meter reading and customer billing costs in integrated computerized processes.

Finally, Vickery has observed that seeking to avoid the cost of metering was

> somewhat as though the operator of a supermarket, chafing at the costs involved in the check-out process, decided to save on these costs by simply weighing each customer's market basket on a scale and charging a uniform price per pound. In a competitive environment such a supermarket would soon be selling only steak and gourmet items while customers went elsewhere for flour, potatoes, and soft drinks. To break even the flat price per pound would have to be pushed up until eventually the store would probably go out of business entirely.[11]

Nor need the metering expense be incurred all at once. Some commercial and industrial customers are already equipped with recording demand meters which may readily be adapted for peak-load pricing. Other large consumers could have such equipment installed at a fraction of even one month's electric bill. The optimal procedure would be to start peak-load pricing where the costs are lowest and the expected benefits greatest. Equipping progressively smaller consumers may proceed as experience, technology, and expected net benefits indicate.

[11]W. Vickery, "Testimony before the New York Public Service Commission," Case 26 402, 1974.

Electricity Demand Growth and the Energy Crisis

An analysis of electricity demand growth projections suggests overestimates in the long run.

Duane Chapman, Timothy Tyrrell, Timothy Mount

In one sense the term energy crisis means simply that the supplies of fuels and power are less than we want, or that they might cost much more in the future.

In another sense, an enlargement of the first, it refers to a tangled web of problems concerning the quality of the environment and the availability, marketing, and growing demand for energy resources. Figure 1 is a diagram of these problems. We can see, for example, that our perception of the need for breeder and fusion reactors is influenced by our understanding of the energy crisis. In turn, our view of this possible crisis is affected by our beliefs about the growth of the demand for energy. If energy prices significantly influence this growth, then our energy system has some interesting circuits. Regulation of surface mines, for example, could reduce the need for the development of fusion power through the mechanism of higher costs and prices and reduced growth rates for demand and supply.

However, most observers today believe that the growth of demands for the various types of energy will be autonomous: Demand will grow independently of any changes in the "factors" box at the top of Fig. 1, or these factors will themselves continue to develop according to past patterns. In this article we focus on the demand for electricity; a generalization to overall energy research problems is suggested in the conclusion.

Observers of electricity demand generally believe that it will continue to increase at constant (or nearly so) compound growth rates for the rest of the century. Table 1 shows predictions made by government, industry, and university researchers. The Federal Power Commission (FPC) estimate is based on the work of six regional advisory committees for the 1970 National Power Survey (*1*). The National Petroleum Council (NPC) projection was prepared by the energy demand task group of the Committee on U.S. Energy Outlook (*2*). The energy demand panel of the Cornell–National Science Foundation (C-NSF) Workshop on Energy and the Environment adopted a modification of the NPC projection as the basis for comparing alternative assumptions (*3*).

Note the similarity in projections. In 1980 the range is only 0.4 of a trillion kilowatt hours (Tkwh). The lowest projection is 89 percent of the highest. The "double-ten" assumption, that electricity demand will double nearly every 10 years, seems to be supported by these projections (*4, 5*). By the end of the century, the demand for electricity is expected to have increased at least six times, to more than 10 Tkwh.

We believe that these projections are generally incorrect. However, before we explain the basis for our conclusion and its possible implication for energy research and development, it is appropriate to note that the kind of projection given in Table 1 has been accurate in the past.

Consider a hypothetical analyst working for a regulatory agency or university in early 1965. He has been asked to project the national electricity requirements for 1969 and 1970. Our analyst favors extrapolated compound growth, and employs that method. The total sales grew 7.35 percent per year for the 5 years from 1959 to 1964. So he predicts that total sales will grow 7.35 percent per year for the next 5 years, to 1.28 Tkwh in 1969 and to 1.37 Tkwh in 1970. For good measure he draws the graph in Fig. 2 (omitting, of course, the actual sales for 1969 and 1970).

Six years pass, a time of war and rebellion, inflation and unemployment, increasing affluence and hardening poverty. In early 1971 he recalls his prediction, and decides to compare it to actual sales. The actual sales were 1.31 Tkwh in 1969 and 1.39 in 1970.

A year later, in early 1972, our analyst is curious about the extension of his projection to 1971. His projection would have been 1.47 Tkwh. The actual sales were 1.47 Tkwh. This kind of experience, repeated at the regional and national level, has reinforced confidence in projections made by assuming compound growth or by extrapolation, or both.

Probably the most accurate 10-year prediction was made in 1960 by the Edison Electric Institute. At that time they predicted that the total sales for 1970 would be 1.31 Tkwh. This estimate was, in fact, too low. It was based

---

D. Chapman and T. Mount are assistant professors at Cornell University, Ithaca, New York 14850, and T. Tyrrell is a statistical programmer at the Oak Ridge National Laboratory, Oak Ridge, Tennessee 37830. This article is based on a statement prepared in June 1972 for the Subcommittee on Science, Research, and Development of the Committee on Science and Astronautics of the House of Representatives.

on an extrapolation of the growth of the gross national product and on a predicted population of 208 million (*6*). The total sales did, in fact, double in the 10 years from 1960 to 1970. Thus, there is historical justification for the type of projections reported in Table 1.

Changes in Causal Factors

It is our opinion that many of the factors influencing the demand for electricity are themselves departing from long-established patterns. Figure 3 shows the post–World War II relationship between the average price of electricity and an overall inflation index. Electricity has become an increasingly better buy. This is apparently true when electricity prices in each consumer class are compared with overall price indexes and the cost of labor, capital, gas, fuel oil, and coal. It seems to be equally true when prices of household electrical appliances and industrial machinery are compared to appropriate overall prices (*7*). All of the changes in relative prices since World War II have pointed toward an increased demand for electricity, and this seems to have been true in all regions of the country.

The information available to us at this time strongly suggests that this pattern will not continue for the rest of the century. The increasing costs involved in environmental protection and decreasing fuel supplies will cause costs and prices to increase. Last year, for the first time since 1946, the deflated average electricity price increased. It rose from 1.66 to 1.69 cents per kilowatt hour in 1971 prices. This marks the beginning of a new period in electricity rates. The concern with the environmental problems noted in Fig. 1 will force continued cost increases in the foreseeable future.

The energy demand panel of the C-NSF workshop reviewed likely cost increases in fuels and environmental protection as well as cost savings from technological improvements. The panel observed, "In general, we conclude that in the course of the next several decades electricity prices (in 1968 dollars) will increase by at least 50 percent." The 1970 National Power Survey estimated a more modest average national cost increase of 19 percent between 1968 and 1990 (*1*, p. I-19-10; *3*, pp. 151–153).

Aubrey J. Wagner, chairman of the Tennessee Valley Authority (TVA), has published estimates of minimum and maximum environmental protection costs for strip mine regulation, mine safety, fly ash and sulfur removal, a sulfur tax, waste heat control, licensing delays, and underground transmission. Minimum environmental protection improvement would mean a one-eighth cost increase. To meet all of the nontransmission requirements he discussed, he stated that (*8*) "TVA would be faced with an approximate doubling of its revenue requirements." Placing underground the high-voltage transmission system of TVA would raise its costs to three times its present revenue.

Factors influencing energy demand

Population
Per capita income
New consumer commodities
Energy intensive production processes
Energy prices

Environmental protection costs

Energy demand growth

Energy crisis

Energy supply growth

Energy supply problems

Sulfur and particulate emission control
Mine safety
Surface mine regulation
Natural gas prices
Oil import quotas
Oil depletion allowances
Offshore oil development
Alaskan oil and gas
Power plant siting
Nuclear power safety
Breeder and fusion development
Inter-energy ownership

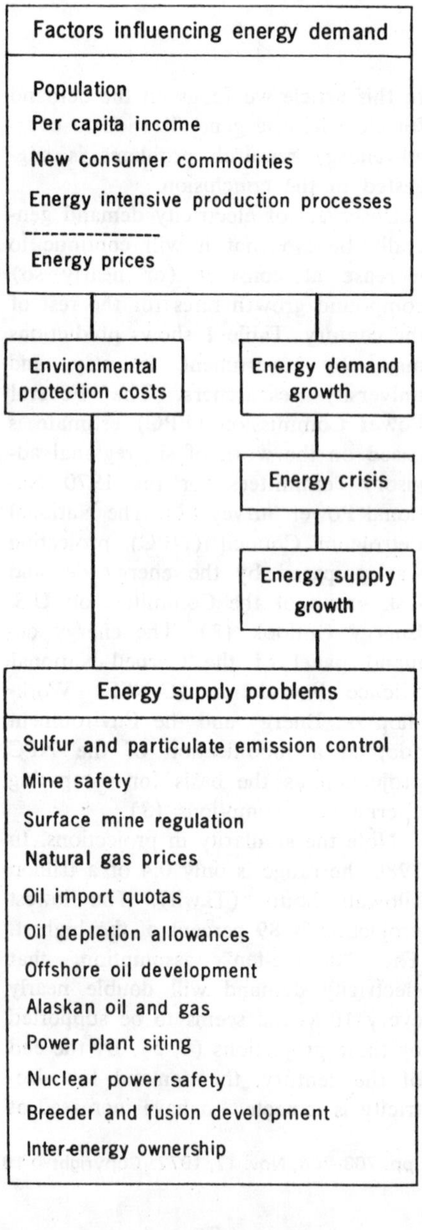

Fig. 1. The energy crisis. Energy supply problems influence energy demand growth through their effect on environmental protection costs. Other demand factors such as population and income influence energy supply problems by their impact on energy requirements.

Analyses of population trends in the studies in Table 1 have lead to the conclusion that past population trends will continue in the foreseeable future. The FPC study groups assumed that the population will continue to grow 1.3 percent per year. In the NPC and C-NSF studies, the Series D projections of the Census Bureau, of 1.1 percent population growth per year, were adopted (*1*, pp. I-3-3, 15; *2*, p. 5; *3*, p. 134). However, there is much uncertainty about future population growth. The recession and new attitudes have combined to drastically lower the fertility rate in the recent past. The birth rate (children born per thousand people) declined for 13 of the 14 months through March of this year. More significantly, the fertility rate (a measure of children born to women of childbearing age) has continued its decline since 1960 and fallen to 2.14 children per woman. The significance of this figure is well known; it is approximately the "zero population growth" fertility rate. At this fertility rate, population growth would slowly decline (except as influenced by immigration and age structure bulges) until stability would be reached in 2035 or 2040 (*9*).

The implication for electricity demand growth is clear. To the extent that past population growth rates continue, the projections in Table 1 are supported. To the extent that the fertility rate declines, the projections will be too high, particularly in the latter part of the century.

Elasticities of Causal Factors

The economic concept of elasticity is useful in describing the magnitude of the influence of causal factors. For particular values of the electricity demand and a causal factor, an elasticity estimate represents the percentage change in the electricity demand associated with a 1 percent change in the causal factor. For example, a commercial price elasticity estimate of −1.5 means that a 1 percent increase in the average price of commercial electricity would, in the long run, cause the demand to be 1.5 percent less than it otherwise would have been.

Wilson (*10*), MacAvoy (*11*), and Halvorsen (*12*) have made quantitative estimates of the influences of causal factors on the demand for electricity. Since each investigator used different definitions of variables and different data bases and statistical techniques, it is difficult to summarize their results (*13*). For illustrative purposes we report eight elasticities from the three studies. Each of those estimates was apparently significant at or beyond the 5 percent level. In these studies residential demand and total demand have price elasticities from −1.1 to −1.3.

Wilson's estimate of a negative income influence of −0.46 is surprising. It may follow from his cross-section study of 77 urban areas in a single year. His explanation is that ". . . federal power projects (and associated low wholesale prices) are concentrated in low-income areas . . . they are totally absent from the high-income northeastern industrial belt from St. Louis to Boston" (*10*, p. 12). Another possible explanation is that one or more omitted factors which are negatively related to demand (such as high fuel oil prices, climate, and appliance prices) are, through geography, positively related to income. In any event, the result is perverse and should be ignored. Halvorsen and MacAvoy examined changes over time as well as geographic differences. Their income elasticity estimates of +0.61 and +0.86, respectively, are likely to be more representative.

MacAvoy reported a population elasticity estimate of +0.91. Since the elasticity demand was defined in terms of consumption per household and per customer by Wilson and Halvorsen, respectively, and since no relationship has been reported between these variables and population, their population elasticity estimate is implicitly +1.00. The elasticity of electricity demand with respect to the price of gas was determined as +0.31 by Wilson and +0.04 by Halvorsen; it was not part of the MacAvoy study.

The major part of our work in the last 2 years has been devoted to numerically evaluating these and other relationships for each consumer class. We are concerned with the stability of these relationships and with the delay or lag in the response of electricity demand to changes in the causal factors. Table 2 shows a summary of the preliminary results of our analysis (*14*).

The small response in the first year (10 or 11 percent) and the length of time necessary for 50 percent of the

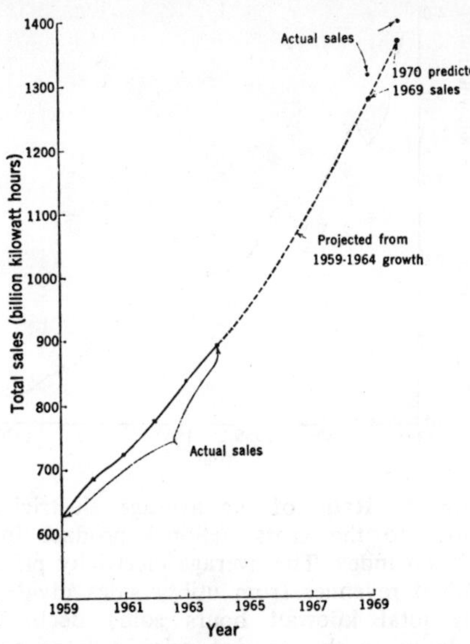

Fig. 2. Projections by 5-year extrapolation of compound growth. The projected sales correspond closely to the actual sales for 1969 and 1970. The extension to 1971 projected sales at 1.47 Tkwh, the actual level for last year.

total response to occur (7 or 8 years) follows from the nature of the processes by which electricity is consumed in residences, commerce, and industry. Electricity is not consumed directly, but through intermediaries such as smelting furnaces, light bulbs, and air conditioning. Changes in preferences for appliances and machinery are reflected in decisions about replacement and new installations, and these decisions in turn affect electricity demand growth.

The estimates in Table 2 suggest a hierarchy of importance of the four factors. The most important factor seems to be the price of electricity, followed by population growth, income, and gas prices in order of decreasing importance. While modification of these estimates is certain, we believe that future amendments will be of small magnitude. The estimates indicate that substantial cost increases (as indicated above) and a reduction in population growth will noticeably lower electricity demand growth in the 1980's and 1990's. Because of the lengthy time period of response, growth reduction in the 1970's might be limited.

Projecting Growth for the Rest of the Century

The growth rate of total electricity demand from 1970 to 1971 was 5.4 percent. This is the lowest rate of growth since 1960, and below the 7.2 percent of the "double-ten" projections. A significant fraction of this growth reduction is probably caused by changes in the causal factors discussed above. The deflated average national price of electricity increased for the first time since World War II: It rose 1.8 percent between 1970 and 1971. The real disposable income per capita increased 3.3 percent per year from 1960 to 1970. The increase from 1970 to the first quarter of this year

Table 1. Selected predictions of electricity demand growth.

Source	Ref.	Electricity demand (Tkwh)					
		1970	1975	1980	1985	1990	2000
Federal Power Commission	(*1*)	1.53		3.07		5.83	
National Petroleum Council	(*2*)	1.59	2.29	3.29	4.54		
Cornell-NSF Workshop	(*3*)	1.57	2.15	2.92	3.96	5.38	10.25

Table 2. Summary of the estimated elasticities of electricity demand associated with electricity prices, income, population, and gas prices. The units of observation are each of the states from 1946 to the present. The average price of electricity is calculated separately for each consumer class. The average price of gas for a particular consumer class, state, and year is calculated by dividing the total revenues of gas utilities for natural gas, mixed gas, manufactured gas, and liquefied petroleum gas by the total sales in therms. Residential electricity and gas prices and per capita personal income are deflated by the Consumer Price Index, and commercial and industrial prices are deflated by the Wholesale Price Index. The percent of response in the first year is equal to 100 $(1-\theta)$ from Eq. 1. The elasticities are derived from coefficients in an equation, which (except for the gas price coefficients) are all significant beyond the 1 percent level. Over 99 percent of the variance of the quantity of electricity is explained by the model for each of the three consumer classes (*14*).

Consumer class	Elasticity				First year response (%)	Years for 50% total response
	Electricity price	Population	Income	Gas price		
Residential	−1.3	+0.9	+0.3	+0.15	10	8
Commercial	−1.5	+1.0	+0.9	+0.15	11	7
Industrial	−1.7	+1.1	+0.5	+0.15	11	7

has been 1.6 percent per year. And, as noted, the fertility rate continues to decline.

It would be incorrect to interpret the statistics for 1971 and early 1972 as defining the quantitative nature of growth for the rest of the century. Rather, these statistics should be seen in a more cautious perspective as indicating that the causal factors influencing electricity growth are changing, and the direction of the change may be toward continued but slower growth.

The methodology of projecting the estimates in Table 1 does not provide quantitative links between population, income, prices, and demand (*15*). (As noted above, this has not been necessary in the past.) In this section we take independent projections of regional and national population, income, and prices and, through the quantitative estimates in Table 2, explicitly relate electricity demand projections to assumptions about causal factors.

One source of estimates of regional populations and per capita income is a study prepared by the Bureau of Economic Analysis (BEA) of the Department of Commerce (*16*). On a national level, population growth is estimated at 1.4 percent per year, gross national product growth at 4 percent per year, and per capita personal income growth at about 2.9 percent per year. Estimates are made for each state and for eight regions. The poorest region, the Southeast, is expected to experience the greatest percentage income increase, but the absolute difference between the poorest region and the richest region (the Midwest) is expected to increase. We may consider an alternative estimate of population growth, that the fertility rates will stay at or below the current level, resulting in a stable population in 2035 or 2040 (*17*).

As discussed above, there is wide variation in analyses of future cost increases. A useful pair of alternatives spans these estimates. Therefore, we use as one assumption the FPC estimate of a 19 percent real cost increase, and as a second assumption a doubling in 30 years. A 13 percent increase in the price of natural gas from 1970 to 1990 is taken from the median projection given in an FPC staff memorandum (*18*).

Thus, we have two alternative cost projections and two alternative population projections. Our assumptions about the growth of gas prices and income are not varied. For comparability, factor projections terminating in 1990 are extrapolated to 2000. Quantitative estimates of the influences of the various factors are taken from Table 2. Projections can be made with this equation:

$$Q_{ijt} = A_{ij}\,(Q_{ijt-1})^{\theta_i}(PE_{ijt})^{\alpha_i} \times (N_{jt})^{\beta_i}(Y_{jt})^{\gamma_i}(PG_{ijt-1})^{\sigma_i} \quad (1)$$

where i is the consumer class; j is the region; t is the year; Q is the demand for electricity; A is a constant; θ is a time response parameter; PE is the average price of electricity; N is the population; Y is the per capita income; PG is the average price of gas; and α, β, γ, and σ are short-run elasticities for electricity price, population, income, and gas price. The units are as described in Table 2.

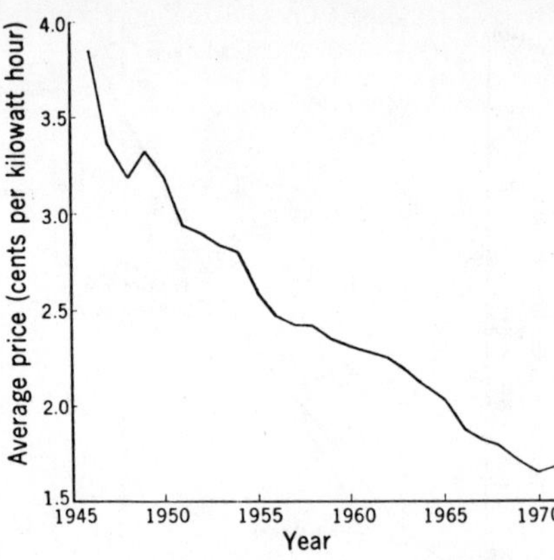

Fig. 3. Ratio of the average electricity price to the gross national product inflation index. The average electricity price (total revenues from utility sales divided by total kilowatt hours sold) declined relative to the implicit price deflator for the gross national product from 1946 to 1970. This real average price inceased in 1971.

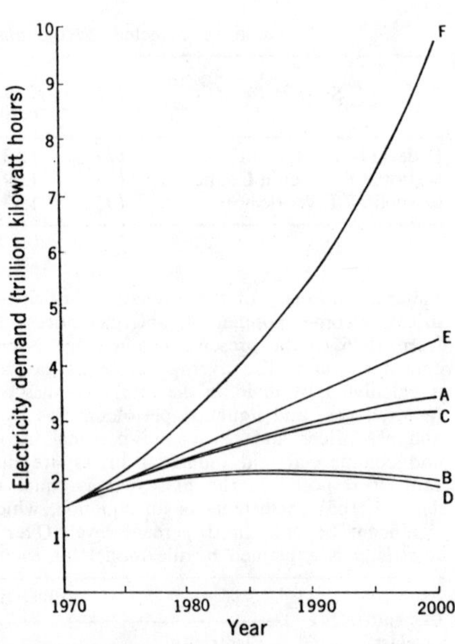

Fig. 4. Electricity demand projections from Table 3. Note the near-term agreement and far-term divergence.

The results of the four combinations of assumptions are shown in Table 3 and Fig. 4 as cases A to D. For case E (the fifth case) we assume that the real average electricity prices for each consumer class and region do not change from their 1970 levels. The detail for case E shown in Table 4 illustrates the framework used in each projection. For case F we use the FPC demand projections and the BEA population projection, and solve for the appropriate price declines. This shows the price declines that might be necessary to cause the growth from 1970 to 1990 shown in Table 1.

Some conclusions based on this analysis are as follows:

1) The near future, say 1975, is not much affected. All projections show that about 2 Tkwh of electricity will be generated in 1975. Supply problems in the next few years will not be eased by likely rate increases and population trends.

2) The population assumption is unimportant for demand growth in the next 20 or 30 years.

3) The generally accepted projections and the estimates of the influence of causal factors are incompatible. In case F, for example, we use the FPC demand in Table 1, the parameter estimates in Table 2, and the BEA population and income projections. In so doing we must reject the FPC price assumptions and instead postulate continued declining prices as indicated.

4) On balance, it seems likely that the projections of demand growth in Table 1 are too high for the 1980's and thereafter.

Implications for Energy Research and Development

This analysis suggests other aspects of our future energy requirements that should be investigated. The results of our work are preliminary and are limited to electricity. Additional studies of the growth of the demand for other forms of energy are desirable. Perhaps future research can clarify the nature of substitution between energy forms as well as the growth of each form and of total energy use.

Table 3. Electricity demand growth and alternative assumptions. BEA, Bureau of Economic Analysis; FPC, Federal Power Commission; ZPG 2035, zero population growth reached in 2035. In the constant price assumption 1970 prices are maintained in each region. In the "double by 2000" assumption, the average price in each region increases annually by 3.33 percent of its 1970 value for 30 years. In case F, the FPC demand projection in Table 1 and the BEA population projections were used, and Eq. 1 was solved for prices. Electricity demand here includes losses of about 9 percent to make the figures comparable with the generation totals in Table 1. A total of 1.53 Tkwh of electricity was generated in 1970.

Case	Population assumption	Electricity price assumption	Electricity demand (Tkwh)			
			1975	1980	1990	2000
A	BEA	FPC	1.98	2.38	3.01	3.45
B	BEA	Double by 2000	1.88	2.07	2.11	2.01
C	ZPG 2035	FPC	1.98	2.37	2.95	3.29
D	ZPG 2035	Double by 2000	1.88	2.05	2.07	1.91
E	BEA	Constant	2.02	2.54	3.56	4.56
F	BEA	*	2.14	3.05	5.66	9.89

* Average prices decline 24 percent from 1970 to 1980, and 12 percent each 10 years thereafter until 2000.

Most observers other than ourselves believe that electricity use per capita will at least quadruple by the end of the century. Per capita generation was 7500 kwh in 1970, and the C-NSF projection is for 35,900 kwh per capita in 2000. This presumably implies substantial growth in the use of metal products in construction and of electric heating, air conditioning, automobiles, plastics, concrete, packaging, paper, drugs, fertilizers, pesticides, chemicals, and other p[...] nected with an intensiv[...] tricity. It is desirable [...] material living standard[...] fected if growth is less[...] projected, and whether [...] the future has any sig[...] welfare of low-income groups.

Our analysis indicates that if increased environmental protection costs are passed on to consumers in the form of higher prices, then the rate of growth of demand is reduced. Thus, further research into the energy supply problems outlined in Fig. 1 is linked to better predictions of future energy demand growth.

Finally, a question is raised about technological research and development. If energy demand growth in the 1980's and 1990's is less than expected at present, will the relative importance of research on present energy sources

Table 4. Detailed estimates of electricity demand (in million kilowatt hours) for case E. The BEA population projection and constant price assumption are used (see Table 3). Total demand includes other uses, as for subways, street lighting, and so forth. The estimated transmission losses (average about 9 percent) are added to the demand in order to derive the generating requirements.

Area	Electricity demand (Mkwh)						
	1970	1975	1980	1985	1990	1995	2000
			Residential demand				
New England	20,900.0	31,733.7	42,246.4	51,858.4	60,529.6	68,495.2	75,994.9
Mideast	69,146.0	100,351.0	129,657.1	155,928.6	179,365.1	200,764.2	220,864.0
Great Lakes	79,687.0	110,721.4	139,981.8	166,671.1	191,007.4	213,706.6	235,384.2
Plains	35,339.0	49,549.8	62,558.2	74,049.1	84,223.4	93,471.7	102,123.9
Southeast	129,124.0	202,016.5	272,694.1	336,848.4	394,286.7	446,574.9	495,391.5
Southwest	40,127.0	61,098.3	81,273.6	99,686.9	116,378.9	131,778.9	146,326.1
Rocky Mountains	9,652.0	12,842.5	15,795.9	18,572.1	21,249.3	23,866.1	26,454.3
Far West	63,820.0	81,279.3	98,599.3	115,467.0	131,860.8	147,990.2	164,024.0
United States	447,795.0	649,592.4	842,806.1	1,019,081.6	1,178,901.0	1,326,647.0	1,466,544.0
			Commercial demand				
New England	14,643.0	22,546.0	30,840.3	39,148.1	47,410.6	55,799.5	64,453.8
Mideast	57,696.0	79,827.4	103,129.0	126,737.3	150,476.4	174,782.6	199,969.4
Great Lakes	53,911.0	80,329.4	108,139.0	136,017.4	163,702.7	191,792.2	220,749.2
Plains	21,406.0	30,375.9	39,663.0	48,934.9	58,153.1	67,499.6	77,109.3
Southeast	63,556.0	96,842.8	132,157.8	167,914.3	203,815.9	240,472.1	278,392.0
Southwest	33,628.0	47,667.9	62,652.6	78,141.6	94,053.6	110,588.8	127,900.1
Rocky Mountains	10,356.0	14,273.2	18,474.5	22,925.2	27,639.5	32,644.4	37,960.4
Far West	57,554.0	78,612.2	102,300.1	127,479.1	153,682.2	181,191.9	210,218.7
United States	312,750.0	450,474.8	597,356.1	747,297.6	898,933.8	1,054,771.0	1,216,752.0
			Industrial demand				
New England	18,161.0	20,136.2	22,897.8	26,100.4	29,578.1	33,285.7	37,194.1
Mideast	94,108.0	107,519.6	123,566.9	141,013.1	159,293.4	178,356.0	198,167.6
Great Lakes	123,395.0	127,440.1	139,361.6	155,343.8	173,705.3	193,840.3	215,382.5
Plains	30,703.0	38,085.8	45,549.8	52,938.6	60,241.7	67,563.8	74,978.0
Southeast	160,003.0	193,058.1	229,185.7	267,079.3	306,206.7	346,648.6	388,467.9
Southwest	50,853.0	69,864.1	88,833.6	107,463.4	125,841.4	144,249.6	162,912.2
Rocky Mountains	16,642.0	17,973.4	20,065.0	22,706.8	25,779.2	29,180.9	32,849.0
Far West	78,657.0	90,713.8	106,143.1	123,604.7	142,426.2	162,457.5	183,594.9
United States	572,522.0	664,791.1	775,603.4	896,249.8	1,023,071.9	1,155,582.0	1,293,544.0
			Total demand				
New England	55,261.4	76,573.9	98,768.0	120,502.9	141,506.3	162,150.2	182,794.4
Mideast	233,765.1	304,384.5	377,021.4	448,252.3	517,504.7	586,029.1	654,884.1
Great Lakes	267,272.7	331,230.6	402,981.6	476,353.5	549,552.0	622,312.6	698,376.5
Plains	90,421.2	122,023.8	152,795.1	181,903.8	209,507.2	236,305.3	262,854.4
Southeast	365,732.3	510,118.4	657,496.9	800,400.0	937,768.7	1,071,942.0	1,205,254.0
Southwest	129,966.1	186,311.3	242,768.4	297,559.4	350,733.7	403,241.8	455,935.3
Rocky Mountains	38,262.6	47,073.0	56,726.1	67,029.1	77,953.3	89,461.8	101,543.2
Far West	210,632.6	263,887.4	323,315.8	385,978.0	450,651.4	517,696.4	587,403.0
United States	1,391,312.0	1,841,601.0	2,311,872.0	2,777,978.0	3,235,174.0	3,690,136.0	4,149,043.0

ared with new energy sources be nged? We have no answer to this ifficult question.

Perhaps the most appropriate conclusion is to observe that energy demand growth is partly a matter of choice. Our decisions about the quality of our natural environment, our material standards of living, and equity will influence our demand for energy and will, in turn, be affected by our use of energy.

References and Notes

1. U.S. Federal Power Commission, *The 1970 National Power Survey* (Government Printing Office, Washington, D.C., 1971), part I, p. I-3-15.
2. Committee on U.S. Energy Outlook, *U.S. Energy Outlook* (National Petroleum Council, Washington, D.C., 1971), vol. 2, pp. 7, 10, 11. The NPC projections for electricity generation were actually reported for 1970, 1975, 1980, and 1985 as 16 : 695; 23 : 525; 32 : 996; and 44 : 363 quadrillion British thermal units, respectively. The heat rate was estimated to decline 7 percent from 1970 to 1985. By taking the average heat rate of 10,508 Btu/kwh for 1970 reported in the *Statistical Year Book of the Electrical Utility Industry for 1970* (Edison Electric Institute, New York, 1971) and calculating as indicated, the NPC projections in Table 1 are obtained. These projections grow at 7.2 percent annually, as assumed by the NPC. The amount of electricity generated by the utilities in 1970 was (on a preliminary reporting basis) 1.53 Tkwh. The NPC estimate for 1970 may have been prepared before this figure was known.
3. U.S. Congress, Senate Committee on Interior and Insular Affairs, *Summary Report of the Cornell Workshop on Energy and the Environment (Sponsored by the National Science Foundation)* (Print, 92nd Congr., 2nd Sess., May 1972), p. 137.
4. In a major pollution control study (*5*, p. 96) it was concluded that "We . . . concur in the Federal Power Commission forecast that electricity consumption will continue to grow at a doubling rate of every ten years (i.e., at 7.2 percent per year) during the 1970's."
5. U.S. Council on Environmental Quality, Department of Commerce, and Environmental Protection Agency, *The Economic Impact of Pollution Control* (Government Printing Office, Washington, D.C., 1972).
6. U.S. Congress, Senate Select Committee on National Water Resources, *Water Resources Activities in the United States: Electric Power in Relation to the Nation's Water Resources* (Print, 86th Congr., 2nd Sess., January 1960), p. 17. This prediction was brought to our attention by P. Auer and H. J. Young.
7. The wholesale electrical machinery price index has declined relative to overall wholesale prices since 1959.
8. A. J. Wagner, *Public Util. Fortn.* **89** (No. 13), 27 (1972).
9. J. Rosenthal, "Birth rate drop is accelerating," *New York Times*, 24 May 1972; "Population growth in U.S. sharply off," *ibid.*, 5 November 1971.
10. J. W. Wilson, *Quart. Rev. Econ. Bus.* **11**, 7 (1971).
11. P. W. MacAvoy, *Economic Strategy for Developing Nuclear Breeder Reactors* (M.I.T. Press, Cambridge, Mass., 1969).
12. R. Halvorsen, *Sierra Club Conference on Power and Public Policy* (Public Resources, Inc., Burlington, Vt., 1972).
13. Wilson (*10*) worked with the average annual residential consumption per household (in kilowatt hours) for 77 cities in (apparently) 1966. MacAvoy (*11*) studied the total added electrical capacity (in megawatts) for nine regions over three 4-year periods. Halvorsen (*12*) examined the annual residential consumption per customer (in kilowatt hours), by states, in the period from 1961 to 1969.
14. We work with data from 1946 to the present for each state, region, and consumer class. Various functional forms, variables, and dynamic models are compared. Details for our analysis and a more comprehensive review of other studies are discussed elsewhere. T. Mount, D. Chapman, T. J. Tyrrell, in preparation; also papers presented at the meeting of the American Association for the Advancement of Science in Philadelphia, 1971; at (*3*); and at (*12*).
15. In the studies cited in Table 1 various opinions are offered about the competitiveness of electricity prices as a significant influence on electricity demand growth (*1*, p. I-1-14; *2*, vol. 1, p. 8, vol. 2, p. 2; *3*, pp. 134, 155, 156). In *The Economic Impact of Pollution Control* (*5*, p. 97) it was stated that "We assume for purposes of this report that the demand for electricity is relatively inelastic."
16. R. E. Graham, Jr., H. C. Degraff, E. A. Trott, Jr., *Surv. Curr. Bus.* **52** (No. 4), 22 (1972).
17. More precisely, we consider that the fertility rate is low enough so that the sum of births and new immigrants less the number of deaths will be zero by 2035 or 2040.
18. P. Kline, "Projections of electricity consumption in the United States 1970–1990" (unpublished) (Federal Power Commission, Washington, D.C., 1971), p. 14.

19. We appreciate the assistance of E. Fleming, J. Baldwin, and J. M. Ostro, and the helpful comments of the referees and editor. Supported by the National Science Foundation through the Atomic Energy Commission, Union Carbide Corporation, Oak Ridge National Laboratory and Cornell University.

Efficiency of Energy Use in the United States

Transportation, space heating, and air conditioning provide opportunities for large energy savings.

Eric Hirst and John C. Moyers

Conflicts between the demand for energy and environmental quality goals can be resolved in several ways. The two most important are (i) development and use of pollution control technologies and of improved energy-conversion technologies and (ii) the improvement in efficiency of energy use. Increased efficiency of energy use would help to slow energy growth rates, thereby relieving pressure on scarce energy resources and reducing environmental problems associated with energy production, conversion, and use.

Between 1950 and 1970, U.S. consumption of energy resources (coal, oil, natural gas, falling water, and uranium) doubled (*1*), with an average annual growth rate of 3.5 percent—more than twice the population growth rate.

Energy resources are used for many purposes in the United States (*2*) (Table 1). In 1970, transportation of people and freight consumed 25 percent of total energy, primarily as petroleum. Space heating of homes and commercial establishments was the second largest end-use, consuming an additional 18 percent. Industrial uses of energy [process steam, direct heat, electric drive, fuels used as raw materials (*3*), and electrolytic processes] accounted for 42 percent. The remaining 15 percent was used by the commercial and residential sectors for water heating, air conditioning, refrigeration, cooking, lighting, operation of small appliances, and other miscellaneous purposes.

The authors are research staff members in the Oak Ridge National Laboratory–National Science Foundation environmental program, Oak Ridge National Laboratory, Oak Ridge, Tennessee 37830. The work reported here was sponsored by the National Science Foundation RANN program under Union Carbide Corporation contract with the U.S. Atomic Energy Commission.

During the 1960's, the percentage of energy consumed for electric drive, raw materials, air conditioning, refrigeration, and electrolytic processes increased relative to the total. Air conditioning showed the largest relative growth, increasing its share of total energy use by 81 percent, while the other uses noted increased their shares of the total by less than 10 percent in this period.

The growth in energy consumption by air conditioners, refrigerators, electric drive, and electrolytic processes—coupled with the substitution of electricity for direct fossil fuel combustion for some space and water heating, cooking, and industrial heat—accounts for the rapid growth in electricity consumption. Between 1960 and 1970, while consumption of primary energy (*1*) grew by 51 percent, the use of electricity (*4*) grew by 104 percent. The increasing use of electricity relative to the primary fuels is an important factor accounting for energy growth rates because of the inherently low efficiency of electricity generation, transmission, and distribution which averaged 30 percent during this decade (*1, 4*). In 1970, electrical generation (*1*) accounted for 24 percent of energy resource consumption as compared to 19 percent in 1960.

Industry, the largest energy user, includes manufacturing; mining; and agriculture, forestry, and fisheries. Six manufacturers—of primary metals; of chemicals; of petroleum and coal; of stone, clay, and glass; of paper; and of food—account for half of industrial energy consumption (*5*), equivalent to 20 percent of the total energy budget.

Energy consumption is determined by at least three factors: population, affluence, and efficiency of use. In this article we describe three areas in which energy-efficiency improvements (the third factor) might be particularly important: (i) transportation of people and freight, (ii) space heating, and (iii) space cooling (air conditioning).

Energy efficiency varies considerably among the different passenger and freight transport modes. Shifts from energy-intensive modes (airplanes, trucks, automobiles) to energy-efficient modes (boats, pipelines, trains, buses) could significantly reduce energy consumption. Increasing the amount of building insulation could reduce both space-heating and air-conditioning energy consumption in homes and save money for the homeowner. Energy consumption for air conditioning could be greatly reduced through the use of units that are more energy efficient.

Transportation

Transportation of people and goods consumed 16,500 trillion British thermal units (*6*) in 1970 (25 percent of total energy consumption) (*1*). Energy requirements for transportation increased by 89 percent between 1950 and 1970, an average annual growth rate of 3.2 percent.

Increases in transportation energy consumption (*7*) are due to (i) growth in traffic levels, (ii) shifts toward the use of less energy-efficient transport modes, and (iii) declines in energy efficiency for individual modes. Energy intensiveness, the inverse of energy efficiency, is expressed here as British thermal units per ton-mile for freight and as British thermal units per passenger-mile for passenger traffic.

Table 2 shows approximate values (*8*) for energy consumption and average revenue in 1970 for intercity freight modes; the large range in energy efficiency among modes is noteworthy. Pipelines and waterways (barges and boats) are very efficient; however, they are limited in the kinds of materials they can transport and in the flexibility of their pickup and delivery points. Railroads are slightly less efficient than pipelines. Trucks, which are faster and more flexible than the preceding three modes, are, with respect to energy, only one-fourth as efficient as railroads. Airplanes, the fastest mode, are only 1/60 as efficient as trains.

The variation in freight prices shown in Table 2 closely parallels the variation in energy intensiveness. The increased prices of the less efficient modes reflect their greater speed, flexibility, and reliability.

Table 3 gives approximate 1970 energy and price data for various passenger modes (*8*). For intercity passenger traffic, trains and buses are the most efficient modes. Cars are less than one-half as efficient as buses, and airplanes are only one-fifth as efficient as buses.

For urban passenger traffic, mass transit systems (of which about 60 percent are bus systems) are more than twice as energy efficient as automobiles. Walking and bicycling are an order of magnitude more efficient than autos, on the basis of energy consumption to produce food. Urban values of efficiency for cars and buses are much lower than intercity values because of poorer vehicle performance (fewer miles per gallon) and poorer utilization (fewer passengers per vehicle).

Passenger transport prices are also shown in Table 3. The correlation between energy intensiveness and price, while positive, is not as strong as for freight transport. Again, the differences in price reflect the increased values of the more energy-intensive modes.

The transportation scenario for 1970 shown in Table 4 gives energy savings that may be possible through increased use of more efficient modes. The first calculation uses the actual 1970 transportation patterns. The scenario—entirely speculative—indicates the potential energy savings that could have occurred through shifts to more efficient transport modes. In this hypothetical scenario, half the freight traffic carried by truck and by aiplane is assumed to have been carried by rail; half the intercity passenger traffic carried by airplane and one-third the traffic carried by car are assumed to have been carried by bus and train; and half the urban automobile traffic is assumed to have been carried by bus. The load factors (percentage of transport capacity utilized) and prices are assumed to be the same for both calculations. The scenario ignores several factors that might inhibit shifts to energy-efficient transport modes, such as existing land-use patterns, capital costs, changes in energy efficiency within a given mode, substitutability among modes, new technologies, transportation ownership patterns, and other institutional arrangements.

The hypothetical scenario requires only 78 percent as much energy to move the same traffic as does the actual calculation. This savings of 2800 trillion Btu is equal to 4 percent of the total 1970 energy budget. The scenario also results in a total transportation cost that is $19 billion less than the actual 1970 cost (a 12 percent reduction). The dollar savings (which includes the energy saved) must be balanced against any losses in speed, comfort, and flexibility resulting from a shift to energy-efficient modes.

To some extent, the current mix of transport modes is optimal, chosen in response to a variety of factors. However, noninternalized social costs, such as noise and air pollution and various government activities (regulation, subsidization, research), may tend to distort the mix, and, therefore, present modal patterns may not be socially optimal.

Present trends in modal mix are determined by personal preference, private economics, convenience, speed, reliability, and government policy. Emerging factors such as fuel scarcities, rising energy prices, dependence on petroleum imports, urban land-use problems, and environmental quality considerations may provide incentives to shift transportation patterns toward greater energy efficiency.

Table 1. End-uses of energy in the United States.

Item	1960* (%)	1970† (%)
Transportation	25.2	24.7
Space heating	18.5	17.7
Process steam	17.8	16.4
Direct heat	12.9	11.0
Electric drive	7.4	8.1
Raw materials	5.2	5.6
Water heating	4.0	4.0
Air conditioning	1.6	2.9
Refrigeration	2.1	2.3
Cooking	1.5	1.2
Electrolytic processes	1.1	1.2
Other‡	2.7	4.9

* Data for 1960 obtained from Stanford Research Institute (SRI) (*2*). † Estimates for 1970 obtained by extrapolating changes in energy-use patterns from SRI data. ‡ Includes clothes drying, small appliances, lighting, and other miscellaneous energy uses.

Table 2. Energy and price data for intercity freight transport.

Mode	Energy (Btu/ ton-mile)	Price (cents/ ton-mile)
Pipeline	450	0.27
Railroad	670	1.4
Waterway	680	0.30
Truck	2,800	7.5
Airplane	42,000	21.9

Table 3. Energy and price data for passenger transport.

Mode	Energy (Btu/passenger-mile)	Price (cents/passenger-mile)
	*Intercity**	
Bus	1600	3.6
Railroad	2900	4.0
Automobile	3400	4.0
Airplane	8400	6.0
	Urban†	
Mass transit	3800	8.3
Automobile	8100	9.6

* Load factors (percentage of transport capacity utilized) for intercity travel are about: bus, 45 percent; railroad, 35 percent; automobile, 48 percent; and airplane, 50 percent. † Load factors for urban travel are about: mass transit, 20 percent; and automobile, 28 percent.

Space Heating

The largest single energy-consuming function in the home is space heating. In an average all-electric home in a moderate climate, space heating uses over half the energy delivered to the home; in gas- or oil-heated homes, the fraction is probably larger because the importance of thermal insulation has not been stressed where these fuels are used.

The nearest approach to a national standard for thermal insulation in residential construction is "Minimum Property Standards (MPS) for One and Two Living Units," issued by the Federal Housing Administration (FHA). In June 1971, FHA revised the MPS to require more insulation, with the stated objectives of reducing air pollution and fuel consumption.

A recent study (*9*) estimated the value of different amounts of thermal insulation in terms both of dollar savings to the homeowner and of reduction in energy consumption. Hypothetical model homes (1800 square feet) were placed in three climatic regions, each representing one-third of the U.S. population. The three regions were represented by Atlanta, New York, and Minneapolis.

As an example of the findings of the study, Table 5 presents the results applicable to a New York residence, including the insulation requirements of the unrevised and the revised MPS, the insulation that yields the maximum economic benefit to the homeowner, and the monetary and energy savings that result in each case. The net monetary savings are given after recovery of the cost of the insulation installation, and would be realized each year of the lifetime of the home. A mortgage interest rate of 7 percent was assumed.

The revised MPS provide appreciable savings in energy consumption and in the cost of heating a residence, although more insulation is needed to minimize the long-term cost to the homeowner. A further increase in insulation requirements would increase both dollar and energy savings.

The total energy consumption of the United States (*1*) in 1970 was 67,000 trillion Btu, and about 11 percent was devoted to residential space heating and 7 percent to commercial space heating (*2*). Table 5 shows reductions in energy required for space heating of 49 percent for gas-heated homes and 47 percent for electric-heated homes in the New York area by going from the MPS-required insulation in 1970 to the economically optimum amount of insulation. The nationwide average reductions are 43 percent for gas-heated homes and 41 percent for electric-heated homes. An average savings of 42 percent, applied to the space heating energy requirements for all residential units (single family and apartment, gas and electric), would have amounted to 3100 trillion Btu in 1970 (4.6 percent of total energy consumption). The energy savings are somewhat understated—as insulation is added, the heat from lights, stoves, refrigerators, and other appliances becomes a significant part of the total heat required. The use of additional insulation also reduces the energy consumption for air conditioning as discussed later.

Electrical resistance heating is more wasteful of primary energy than is direct combustion heating. The average efficiency for electric power plants (*1*) in the United States is about 33 percent, and the efficiency (*4*) of transmitting and distributing the power to the customer is about 91 percent. The end-use efficiency of electrical resistance heating is 100 percent; so the overall efficiency is approximately 30 percent. Thus, for every unit of heat delivered in the home, 3.3 units of heat must be extracted from the fuel at the power plant. Conversely, the end-use efficiency of gas- or oil-burning home heating systems is about 60 percent (claimed values range from 40 to 80 percent), meaning that 1.7 units of heat must be extracted from the fuel for each unit delivered to the living area of the home. Therefore, the electrically heated home requires about twice as much fuel per unit of heat as the gas- or oil-heated home, assuming equivalent insulation.

The debate about whether gas, oil, or electric-resistance space heating is better from a conservation point of view may soon be moot because of the shortage of natural gas and petroleum. The use of electricity generated by nuclear plants for this purpose can be argued to be a more prudent use of resources than is the combustion of natural gas or oil for its energy content. Heating by coal-generated electricity may also be preferable to heating by gas or oil in that a plentiful resource is used and dwindling resources are conserved.

The use of electrical heat pumps could equalize the positions of electric-, oil-, and gas-heating systems from a fuel conservation standpoint. The heat pump delivers about 2 units of heat energy for each unit of electric energy that it consumes. Therefore, only 1.7 units of fuel energy would be required at the power plant for each unit of delivered heat, essentially the same as that required for fueling a home furnace.

Heat pumps are not initially expensive when installed in conjunction with central air conditioning; the basic equipment and air handling systems are the same for both heating and cooling. A major impediment to their widespread use has been high maintenance cost associated with equipment failure. Several manufacturers of heat pumps have carried out extensive programs to improve component reliability that, if successful, should improve acceptance by homeowners.

Table 4. Actual and hypothetical energy consumption patterns for transportation in 1970.

	Total traffic	Percentage of total traffic: Air	Truck	Rail	Waterway and pipeline	Auto	Bus*	Total energy (10^{12} Btu)	Total cost (10^9 \$)
Intercity freight traffic									
Actual	2210†	0.2	19	35	46			2400	45
Hypothetical	2210	0.1	9	44	46			1900	33
Intercity passenger traffic									
Actual	1120‡	10		1		87	2	4300	47
Hypothetical	1120	5		12		58	25	3500	45
Urban passenger traffic									
Actual	710‡					97	3	5700	68
Hypothetical	710					49	51	4200	63
Totals									
Actual								12,400	160
Hypothetical								9600	141

* Intercity bus or urban mass transit. † Billion ton-miles. ‡ Billion passenger-miles.

Table 5. Comparison of insulation requirements and monetary and energy savings for a New York residence.

Insulation specification	Unrevised MPS*: Gas	Unrevised MPS*: Electric	Revised MPS*: Gas	Revised MPS*: Electric	Economic optimum: Gas	Economic optimum: Electric
Wall insulation thickness (inches)	0	1⅞	1⅞	1⅞	3½	3½
Ceiling insulation thickness (inches)	1⅞	1⅞	3½	3½	3½	6
Floor insulation	No	No	Yes	Yes	Yes	Yes
Storm windows	No	No	No	No	Yes	Yes
Monetary savings (\$/yr)	0	0	28	75	32	155
Reduction of energy consumption (%)	0	0	29	19	49	47

* Minimum property standards (MPS) for one and two living units.

Space Cooling

In all-electric homes, air conditioning ranks third as a major energy-consuming function, behind space heating and water heating. Air conditioning is particularly important because it contributes to or is the cause of the annual peak load that occurs in the summertime for many utility systems.

In addition to reducing the energy required for space heating, the ample use of thermal insulation reduces the energy required for air conditioning. In the New York case, use of the economically optimum amount of insulation results in a reduction of the electricity consumed for air conditioning of 26 percent for the gas home or 18 percent for the electric home, compared to the 1970 MPS-compliance homes.

The popularity of room air conditioners is evidenced by an exponential sales growth with a doubling time of 5 years over the past decade; almost 6 million were sold in 1970. The strong growth in sales is expected to continue since industry statistics show a market saturation of only about 40 percent.

There are about 1400 models of room air conditioners available on the market today, sold under 52 different brand names (*10*). A characteristic of the machines that varies widely but is not normally advertised is the efficiency with which energy is converted to cooling. Efficiency ranges from 4.7 to 12.2 Btu per watt-hour. Thus the least efficient machine consumes 2.6 times as much electricity per unit of cooling as the most efficient one. Figure 1 shows the efficiencies of all units having ratings up to 24,000 Btu per hour, as listed in (*10*).

From an economic point of view, the purchaser should select the particular model of air conditioner that provides the needed cooling capacity and the lowest total cost (capital, maintenance, operation) over the unit's lifetime. Because of the large number of models available and the general ignorance of the fact that such a range of efficiencies exists, the most economical choice is not likely to be made. An industry-sponsored certification program requires that the cooling rating and wattage input be listed on the nameplate of each unit, providing the basic information required for determining efficiency. However, the nameplate is often hard to locate and does not state the efficiency explicitly.

The magnitude of possible savings that would result from buying a more efficient unit is illustrated by the following case. Of the 90 models with a capacity of 10,000 Btu per hour, the lowest efficiency model draws 2100 watts and the highest efficiency model draws 880 watts. In Washington, D.C., the average room air conditioner operates about 800 hours per year. The low-efficiency unit would use 976 kilowatt-hours more electricity each year than the high-efficiency unit. At 1.8 cents per kilowatt-hour, the operating cost would increase by $17.57 per year. The air conditioner could be expected to have a life of 10 years. If the purchaser operates on a credit card economy, with an 18 percent interest rate, he would be economically justified in paying up to $79 more for the high-efficiency unit. If his interest rate were 6 percent, an additional purchase price of $130 would be justified.

In the above example, the two units were assumed to operate the same number of hours per year. However, many of the low-priced, low-efficiency units are not equipped with thermostats. As a result, they may operate almost continuously, with a lower-than-desired room temperature. This compounds the inefficiency and, in addition, shortens the lifetime of the units.

In addition to the probable economic advantage to the consumer, an improvement in the average efficiency of room air conditioners would result in appreciable reductions in the nation's energy consumption and required generating capacity. If the size distribution of all existing room units is that for the 1970 sales, the average efficiency (*10*) is 6 Btu per watt-hour, and the average annual operating time is 886 hours per year, then the nation's room air conditioners consumed 39.4 billion kilowatt-hours during 1970. On the same basis, the connected load was 44,500 megawatts, and the annual equivalent coal consumption was 18.9 million tons. If the assumed efficiency is changed to 10 Btu per watt-hour, the annual power consumption would have been 23.6 billion kilowatt-hours, a reduction of 15.8 billion kilowatt-hours. The connected load would have decreased to 26,700 megawatts, a reduction of 17,800 megawatts. The annual coal consumption for room air conditioners would have been 11.3 million tons, a reduction of 7.6 million tons, or at a typical strip mine yield of 5000 tons per acre, a reduction in stripped area of 1500 acres in 1970.

Other Potential Energy Savings

Energy-efficiency improvements can be effected for other end-uses of energy besides the three considered here. Improved appliance design could increase the energy efficiency of hot-water heaters, stoves, and refrigerators. The use of solar energy for residential space and water heating is technologically feasible and might some day be economically feasible. Alternatively, waste heat from air conditioners could be used for water heating. Improved design or elimination of gas pilot lights and elimination of gas yard lights would also provide energy savings (*11*). Increased energy efficiency within homes would tend to reduce summer air-conditioning loads.

In the commercial sector, energy savings in space heating and cooling such as those described earlier are possible. In addition, the use of total energy systems (on-site generation of electricity and the use of waste heat for space and water heating and absorption air conditioning) would increase the overall energy efficiency of commercial operations.

Commercial lighting accounts for about 10 percent of total electricity consumption (*12*). Some architects claim that currently recommended lighting levels can be reduced without danger to eyesight or worker performance (*13*). Such reduction would save energy directly and by reducing air-conditioning loads. Alternatively, waste heat from lighting can be circulated in winter for space heating and shunted outdoors in summer to reduce air-conditioning loads.

Changes in building design practices might effect energy savings (*13*). Such changes could include use of less glass and of windows that open for circulation of outside air.

Waste heat and low temperature steam from electric power plants may be useful for certain industries and for space heating in urban districts (*14*). This thermal energy (about 8 percent of energy consumption in 1970) (*15*) could be used for industrial process steam, space heating, water heating, and air conditioning in a carefully planned urban complex.

The manufacture of a few basic materials accounts for a large fraction of industrial energy consumption. Increased recycle of energy-intensive materials such as aluminum, steel, and paper would save energy. Savings could also come from lower production of certain materials. For example, the production of packaging materials (paper, metal, glass, plastic, wood) requires about 4 percent of the total energy budget. In general, it may be possible to design products and choose materials to decrease the use of packaging and to reduce energy costs per unit of production.

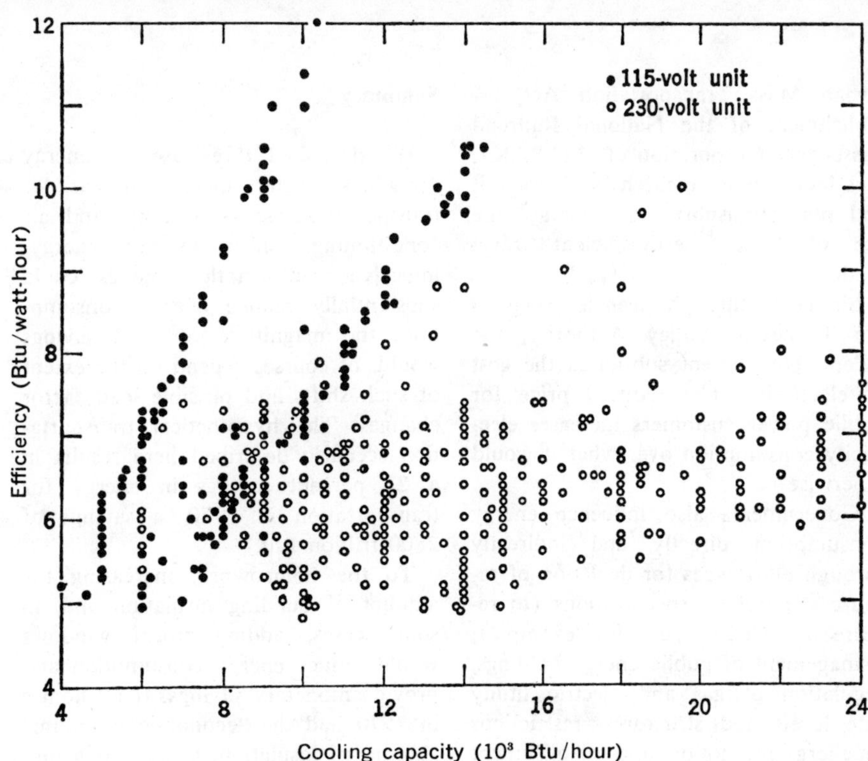

Fig. 1. Efficiency of room air conditioners as a function of unit size.

Implementation

Changes in *energy prices*, both levels and rate structures, would influence decisions concerning capital versus life costs, and this would affect the use of energy-conserving technologies. *Public education* to increase awareness of energy problems might heighten consumer sensitivity toward personal energy consumption. Various local, state, and federal *government policies* exist that, directly and indirectly, influence the efficiency of energy use. These three routes are not independent; in particular, government policies could affect prices or public education (or both) on energy use.

One major factor that promotes energy consumption is the low price of energy. A typical family in the United States spends about 5 percent of its annual budget on electricity, gas, and gasoline. The cost of fuels and electricity to manufacturers is about 1.5 percent of the value of their total shipments. Because the price of energy is low relative to other costs, efficient use of energy has not been of great importance in the economy. Not only are fuel prices low, but historically they have declined relative to other prices.

The downward trend in the relative price of energy has begun to reverse because of the growing scarcity of fuels, increasing costs of both money and energy-conversion facilities (power plants, petroleum refineries), and the need to internalize social costs of energy production and use. The impact of rising energy prices on demand is difficult to assess. According to one source (*16*):

> . . . In the absence of any information, we assume a long-run price elasticity of demand of — 0.5 (meaning that in the long-run a doubling of energy prices will reduce demand by a factor of the square root of 2, namely to about 70 percent of what it would have been otherwise).

The factors cited above (fuel scarcity, rising costs, environmental constraints) are likely to influence energy price structures as well as levels. If these factors tend to increase energy prices uniformly (per Btu delivered), then energy price structures will become flatter; that is, the percentage difference in price between the first and last unit purchased by a customer will be less than that under existing rate structures. The impact of such rate structure changes on the demand for energy is unknown, and research is needed.

Increases in the price of energy should decrease the quantity demanded and this is likely to encourage more efficient use of energy. For example, if the price of gasoline rises, there will probably be a shift to the use of smaller cars and perhaps to the use of public transportation systems.

Public education programs may slow energy demand. As Americans understand better the environmental problems associated with energy production and use, they may voluntarily decrease their personal energy-consumption growth rates. Experiences in New York City and in Sweden with energy-conservation advertising programs showed that the public is willing and able to conserve energy, at least during short-term emergencies.

Consumers can be educated about the energy consumption of various appliances. The energy-efficiency data for air conditioners presented here are probably not familiar to most prospective buyers of air conditioners. If consumers understood energy and dollar costs of low-efficiency units, perhaps they would opt for more expensive, high-efficiency units to save money over the lifetime of the unit and also to reduce environmental impacts. Recently, at least two air-conditioner manufacturers began marketing campaigns that stress energy efficiency. Some electric utilities have also begun to urge their customers to use electricity conservatively and efficiently.

Public education can be achieved through government publications or government regulation, for example, by requiring labels on appliances which state the energy efficiency and provide estimates of operating costs. Advertisements for energy-consuming equipment might be required to state the energy efficiency.

Federal policies, reflected in research expenditures, construction of facilities, taxes and subsidies, influence energy consumption. For example, the federal government spends several billion dollars annually on highway, airway, and airport construction, but nothing is spent for railway and railroad construction. Until recently, federal transportation research and development funds were allocated almost exclusively to air and highway travel. Passage of the

Urban Mass Transportation Act, establishment of the National Railroad Passenger Corporation (AMTRAK), plus increases in research funds for rail and mass transport may increase the use of these energy-efficient travel modes.

Similarly, through agencies such as the Tennessee Valley Authority, the federal government subsidizes the cost of electricity. The reduced price for public power customers increases electricity consumption over what it would otherwise be.

Governments also influence energy consumption directly and indirectly through allowances for depletion of resources, purchase specifications (to require recycled paper, for example), management of public energy holdings, regulation of gas and electric utility rate levels and structures, restrictions on energy promotion, and establishment of minimum energy performance standards for appliances and housing.

The federal government spends about $0.5 billion a year on research and development for civilian energy, of which the vast majority is devoted to energy supply technologies (*16*):

> . . . Until recently only severely limited funds were available for developing a detailed understanding of the ways in which the nation uses energy. . . . The recently instituted Research Applied to National Needs (RANN) Directorate of the National Science Foundation . . . has been supporting research directed toward developing a detailed understanding of the way in which the country utilizes energy. . . . This program also seeks to examine the options for meeting the needs of society at reduced energy and environmental costs.

Perhaps new research on energy use will reveal additional ways to reduce energy growth rates.

Summary

We described three uses of energy for which greater efficiency is feasible: transportation, space heating, and air conditioning. Shifts to less energy-intensive transportation modes could substantially reduce energy consumption; the magnitude of such savings would, of course, depend on the extent of such shifts and possible load factor changes. The hypothetical transportation scenario described here results in a 22 percent savings in energy for transportation in 1970, a savings of 2800 trillion Btu.

To the homeowner, increasing the amount of building insulation and, in some cases, adding storm windows would reduce energy consumption and provide monetary savings. If all homes in 1970 had the "economic optimum" amount of insulation, energy consumption for residential heating would have been 42 percent less than if the homes were insulated to meet the pre-1971 FHA standards, a savings of 3100 trillion Btu.

Increased utilization of energy-efficient air conditioners and of building insulation would provide significant energy savings and help to reduce peak power demands during the summer. A 67 percent increase in energy efficiency for room air conditioners would have saved 15.8 billion kilowatt-hours in 1970.

In conclusion, it is possible—from an engineering point of view—to effect considerable energy savings in the United States. Increases in the efficiency of energy use would provide desired end results with smaller energy inputs. Such measures will not reduce the *level* of energy consumption, but they could slow energy growth *rates*.

References and Notes

1. Bureau of Mines, *U.S. Energy Use at New High in 1971* (News Release, 31 March 1972).
2. Stanford Research Institute, *Patterns of Energy Consumption in the United States* (Menlo Park, Calif., November 1971).
3. In this article all fuels used as raw materials are charged to the industrial sector, although fuels are also used as feedstocks by the commercial and transportation sectors.
4. Edison Electric Institute, *Statistical Yearbook of the Electric Utility Industry for 1970* (Edison Electric Institute, New York, 1971).
5. U.S. Bureau of the Census, *1967 Census of Manufactures, Fuels and Electric Energy Consumed* (MC67 (S)-4, Government Printing Office, Washington, D.C., 1971).
6. Conversion factors are: from British thermal units to joules (1055), from miles to meters (1609), from inches to meters (0.0254), from acres to square meters (4047), and from tons to kilograms (907).
7. E. Hirst, *Energy Consumption for Transportation in the U.S.* (Oak Ridge National Laboratory Report ORNL-NSF-EP-15, Oak Ridge, Tenn., 1972); R. A. Rice, "System energy as a factor in considering future transportation," presented at American Society of Mechanical Engineers annual meeting, December 1970).
8. Energy efficiency and unit revenue values for 1970 are computed in E. Hirst, *Energy Intensiveness of Passenger and Freight Transport Modes: 1950–1970* (Oak Ridge National Laboratory Report ORNL-NSF-EP-44, Oak Ridge, Tenn., 1973).
9. J. C. Moyers, *The Value of Thermal Insulation in Residential Construction: Economics and the Conservation of Energy* (Oak Ridge National Laboratory Report ORNL-NSF-EP-9, Oak Ridge, Tenn., December 1971).
10. Association of Home Appliance Manufacturers, *1971 Directory of Certified Room Air Conditioners*, 15 June 1971.
11. Hittman Associates, *Residential Energy Consumption—Phase I Report*, No. HUD-HAI-1, (Columbia, Md., March 1972).
12. C. M. Crysler, General Electric Company, private communication.
13. R. G. Stein, "Architecture and energy," presented at the annual meeting of the American Association for the Advancement of Science, Philadelphia, 29 December 1971.
14. A. J. Miller *et al.*, *Use of Steam-Electric Power Plants to Provide Thermal Energy to Urban Areas* (Oak Ridge National Laboratory Report ORNL-HUD-14, Oak Ridge, Tenn., 1971).
15. R. M. Jimeson and G. G. Adkins, "Factors in waste heat disposal associated with power generation," presented at the American Institute of Chemical Engineers national meeting, Houston, Texas, March 1971).
16. National Science Foundation RANN Program, *Summary Report of the Cornell Workshop on Energy and the Environment, 22 to 24 February 1972*, Senate Committee on Interior and Insular Affairs, No. 92-23, May 1972 (Government Printing Office, Washington, D.C., 1972).

Energy Conservation through Effective Utilization

Energy consumption could be reduced by improved efficiency of utilization in buildings and in industry.

Charles A. Berg

There are indications that the demand for energy in the United States will soon outstrip both power generating capacity and fuel supply.

The basic problems in energy supply can be divided as follows. In the immediate future (1972 to 1980) the most important problem appears to be inadequate power generating capacity. In the distant future (the year 2000 and beyond) the basic problem is availability of fuel or of energy in another form, such as solar or geothermal energy. In the intermediate time range (1972 to 2000) the conservation of energy by means which do not damage the functioning of the economy could well be the most important consideration.

There are two main approaches to solving the problem of providing sufficient energy for future needs: either the supply of energy can be increased, or the demand for energy can be reduced. However, these approaches are not independent of each other. For example, a decrease in the demand for energy caused by curtailing industrial electrolytic processing could adversely affect the capacity to increase the energy supply by causing shortages of electrical conductor material. Such interactions between supply and demand must be considered and evaluated in planning how to meet the overall energy needs of the nation.

In this article I discuss various ways in which the demand for energy could be decreased, focusing not so much on discouraging demand by increasing prices (the belt-tightening approach) as on reducing energy consumption by improved efficiency of energy utilization in buildings service and in industry. In the long run, effective actions to moderate demand will probably consist of a combination of the belt-tightening approach and improvements in efficiency of energy utilization.

Efforts to improve efficiency are essentially technological in nature. The implementation of technological improvements in energy-consuming processes would probably require, or at least be greatly facilitated by, appropriate price tax, loan, and regulatory policies, especially as these pertain to new building construction and industrial plant equipment.

The actions that might be taken to moderate demand should begin to take effect in the intermediate time range (1980 to 2000); as a matter of principle, actions proposed to take effect during this time period should be reviewed for possible conflict with solutions of long-term problems of fuel and energy supply. For example, there are frequent proposals that local combustion of fuels should be used to drive power generation equipment and that reject heat should be utilized locally. Such proposals should be considered in light of the fact that centralized combustion facilities, for example, large power plants, can operate at higher combustion efficiency than small low-cost plants appropriate for installation in individual buildings. Also, large power plants can usually be more efficiently operated and maintained. Thus short-term measures to relieve demand on power generation via local combustion could, in the long run, result in poor use of fuel on a national basis.

The complementary relationship between efforts to increase energy supply and efforts to improve efficiency of utilization merit specific attention. The increasing national demand for energy reflects, in large part, influences which are not subject to immediate control, such as increasing population. There appears to be no way to meet the future needs of society without increasing the national capacity to supply energy. However, I will show that many of the ways in which energy is now used to satisfy the needs of society are not particularly effective. Large amounts of energy are allowed to "leak" out of the national energy system at the point of consumption, and available techniques for utilization of reject heat and heat wasted in energy consuming processes are seldom applied. Faced with a developing shortage of nonpolluting fuels and with the recognition that fuels of all types are a nonrenewable resource, it is appropriate that we give serious attention to improving the effectiveness with which energy is used, as well as to improving the national capacity to supply energy. If the effectiveness of energy utilization were not to be improved, then newly developed supplies of energy would be permitted to escape utilization through presently accepted "leaks." Surplus energy, with all of its economic implications, could have to be produced merely to supply these leaks. Thus, improvement of the effectiveness of energy utilization is seen here to be necessary both as a measure for conservation of natural resources and as a measure for economic optimization of investments in energy.

Technological efforts to moderate demand through improved effectiveness of utilization should consist of a combination of short-range measures, such as the upgrading of housing insulation or furnace performance, and long-range measures including the application of thermal management techniques to industrial processes, improved building design, and the institution of new technological means of improving the efficiency of energy utilization in devices.

The author is deputy director for engineering at the Institute for Applied Technology, U.S. Department of Commerce, National Bureau of Standards, Washington, D.C. 20234.

Present Uses of Energy in the United States

The possibility that demands for energy can be moderated by improving the effectiveness of energy utilization without affecting the expected output of present processes can be evaluated by determining (i) how much energy is consumed in various sectors of society, and (ii) how much of the energy consumed could be saved by the use of more efficient practices. The term "waste energy" will be used here to mean energy which need not be wasted were presently available technology applied. The possibility of further reductions of energy requirements for existing processes, through the development of new technology, is an important subject that I discuss later.

Considerable data are available about energy consumption and are well summarized in a report by the Stanford Research Institute (SRI) (*1*). However, data required for the determination of waste energy are known for only a few sectors of the economy.

Data from SRI indicate that, of the total national energy consumption (NEC), 19.2 percent is used in residential building services, 14.4 percent in commercial building services, 41.2 percent in industrial processes, and 25.2 percent is used in transportation. Considerable amounts of data concerning the effectiveness of energy utilization are available for commercial and residential building services, but few such data are available for industrial processes.

Energy Conservation by Improved Thermal Performance of Structures

The major uses for energy in buildings are space heating, air conditioning, and hot water heating (Table 1) (*2*). Space heating of residences accounts for 11 percent of the total national energy consumption, while space heating of commercial occupancies represents an additional 6.9 percent of that total. Air conditioning in commercial and residential buildings represents 2.5 percent of the total national energy consumption (*1*). The leaks that affect the effectiveness of heating and air conditioning in buildings are essentially the same, the major sources of heat loss or heat gain being inadequate insulation, excessive ventilation, high rates of air infiltration from outside, and excessive fenestration. To estimate the present effectiveness of building insulation and ventilation one may note that the Federal Housing Administration (FHA) minimum property standards of 1965 permitted heat losses of 2000 British thermal units per thousand cubic feet–degree day (*3*) in residences. The property standards required by Housing and Urban Development (HUD) Operation Breakthrough in 1970 reduced this figure to 1500 and the newly implemented FHA minimum property standards (1972) require heat losses to be less than 1000 Btu per thousand cubic feet–degree day. The reduction in energy consumption implied by these standards is to be achieved largely by thermal insulation and control of air infiltration.

The 1972 FHA minimum property standards, of course, bear upon new construction. Because few buildings are designed to exceed the standard requirements one may assume that most of the residential buildings in use today may consume approximately 40 percent more energy for heating and air conditioning than they would, had they been insulated and sealed in accordance with present-day minimum property standards. In fact, in certain areas of the country, whole residential neighborhoods were built prior to the advent of either FHA or housing insulation, and consist of buildings with little or no insulation and with very high air infiltration rates. These neighborhoods are, as a rule, located in areas of high heating requirements. In these neighborhoods the fuel consumption for heating is at least twice as great as would be required if insulation and infiltration control required by modern standards were applied. Heat losses in these neighborhoods therefore represent an especially significant leak in the national energy consumption system (Fig. 1) (*4*).

Sample field observations indicate that the state of insulation and draft sealing in existing commercial buildings is not significantly different from that in existing residences. Similar savings of space heating fuel—approximately 40 percent—may be assumed to be attainable through insulation and draft control in commercial buildings.

The infiltration of outside air accounts for approximately 25 to 50 percent of the heating and cooling requirements of individual buildings, depending upon the type of insulation installed (*5*). Present construction practices yield infiltration rates which exceed by a factor of 4 the average ventilation requirements of typical buildings (*2*, *6*). Special areas, such as toilet facilities, kitchens, and conference rooms where heavy smoking may occur, have high ventilation requirements when in use. However, in most buildings the high ventilation rates required for use of these areas are maintained all day.

Reduction of infiltration to currently accepted levels could be assumed to yield a 10 to 20 percent reduction in national fuel requirements for space heating and air conditioning. Control of ventilation in critical areas so that high rates were supplied only when required could provide an additional relief of up to 5 percent in total fuel requirements.

Future standards for insulation, ventilation, and infiltration may offer even greater potential for saving energy. Engineers studying building insulation estimate that it will be technologically and economically feasible to reduce heat losses from buildings to approximately 700 Btu per thousand cubic feet–degree day, through the use of insulation. If these estimates prove to be correct, it would be feasible to reduce total energy requirements of buildings by more than 50 percent through well-designed insulation and careful control of ventilation.

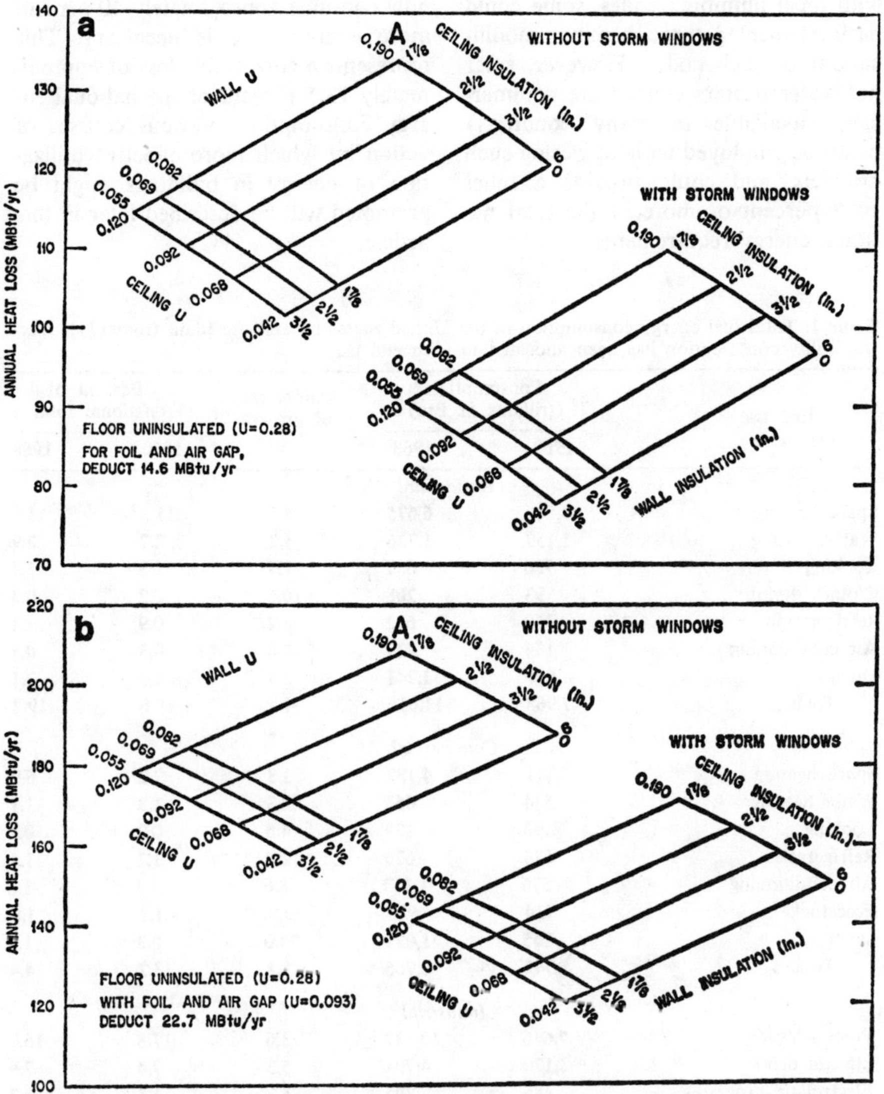

Fig. 1. The annual heat loss from model homes with various weights of insulation in (a) New York, and (b) Minneapolis. A quantitative estimate of the savings in energy for home heating, which are technologically feasible, may be obtained as follows (*4*). The upper point, *A*, in (a) and (b) may be assumed to represent the approximate state of insulation and storm window sealing in approximately 90 percent of housing built prior to issuance of minimum property standards. The heat losses from these houses can be reduced by approximately 45 percent by application of heavy ceiling insulation, side wall insulation, and installation of storm windows. Thus, it would not be unreasonable to assume that if installation of insulation and storm windows on housing units now in service were feasible, the present national demand for fuel consumed in space heating of residences could be reduced by approximately 40 percent (1 MBtu = 10^6 Btu; U = Btu/ft^2·hr·°F). [From Moyers (*4*)]

Heating and Air Conditioning Equipment

The efficiency of heating and air conditioning equipment is especially sensitive to the percent of full load at which the equipment is operated and to how well it is maintained. Heating equipment for buildings, including the home furnace as sold, is typically 75 percent efficient when run at full load. However, the full load capacity of the equipment is seldom needed, and the equipment is most often operated intermittently in which mode it is much less efficient. In addition, small accumulations of soot on boiler surfaces and other minor unattended items of maintenance continuously reduce the efficiency of heating equipment during its lifetime. If one takes the few field data now available together with what is known about the effects of unattended maintenance and intermittent operation upon combustion apparatus in heating plants, it would appear reasonable to estimate that the actual efficiency of heating equipment in the field is 50 percent or less, with units functioning at efficiencies as low as 35 percent not being uncommon [for example, see (*7*)].

The efficiency of air conditioning equipment varies widely. Air conditioners of the same rated output may differ by a factor of 2 in their power requirements. Thus, substantial savings in energy could be realized through inclusion of energy consumption in the criteria for selection of equipment, and by diligent maintenance of equipment.

Illumination in Buildings

Illumination of residences and commercial buildings accounts for 1.5 percent of the national energy consumption (*1*). In office buildings in the United States, the illumination provided often exceeds, by as much as a factor of 2, the amount of illumination used in similar European buildings, and there is no concrete evidence that the increased illumination is of any benefit to the building occupants. Also, much greater use could be made of daylight in office buildings and residences. While the heat generated by illumination may lighten the heating load during the colder part of the year, during the time when indoor space is cooled by air conditioning systems, increased illumination imposes double energy costs upon building operations. Design techniques to permit greater use of daylight and optional use of artificial light exist, and could be more widely applied. Further research to ascertain the most beneficial amounts of illumination to building occupants would be of great value in efforts to reduce excessive energy consumption in buildings.

Hot Water Heating in Buildings

Hot water heating merits special consideration in energy conservation efforts (*2*). Once the hot water is used, the water, with the energy it contains, literally goes down the drain. Furthermore, as shown in Table 1, hot water heating accounts for approximately 4 percent of the total national energy consumption. When viewed in this context, the energy lost in hot water heating is indeed significant. A number of ways to recapture the heat in expended hot water, by using heat exchangers on drains, have been suggested. Although most such proposals are in conflict with local plumbing codes, some could be implemented through minor modifications of such codes. However, solar hot water heaters (which are commercially available in many countries) could be employed without raising such conflicts, and could provide a relief of 2 percent or more of the total national energy requirements.

Other Means of Saving Energy in Buildings

Other components of building operations contribute additional small amounts to national energy consumption (for example, cooking 2.2 percent, clothes drying 0.3 percent) (*1*). Although extensive data on the effectiveness of energy utilization in appliances are not available, sample observations indicate that energy is used with no greater effectiveness than in other building services. Improved design of appliances could significantly enhance their efficiency of energy utilization.

The data cited indicate that buildings now consume approximately 40 percent more energy than is necessary. This represents a correctable loss of approximately 13.5 percent of the national energy consumption. Various courses of action by which more effective utilization of energy in buildings might be promoted will be described later in this article.

Table 1. Total fuel energy consumption in the United States by end use [data from (*1*)]. Electric utility consumption has been allocated to each end use.

End use	Consumption (trillions of Btu) 1960	Consumption (trillions of Btu) 1968	Annual rate of growth (%)	Percent of national total 1960	Percent of national total 1968
Residential					
Space heating	4,848	6,675	4.1	11.3	11.0
Water heating	1,159	1,736	5.2	2.7	2.9
Cooking	556	637	1.7	1.3	1.1
Clothes drying	93	208	10.6	0.2	0.3
Refrigeration	369	692	8.2	0.9	1.1
Air conditioning	134	427	15.6	0.3	0.7
Other	809	1,241	5.5	1.9	2.1
Total	7,968	11,616	4.8	18.6	19.2
Commercial					
Space heating	3,111	4,182	3.8	7.2	6.9
Water heating	544	653	2.3	1.3	1.1
Cooking	93	139	4.5	0.2	0.2
Refrigeration	534	670	2.9	1.2	1.1
Air conditioning	576	1,113	8.6	1.3	1.8
Feedstock	734	984	3.7	1.7	1.6
Other	145	1,025	28.0	0.3	1.7
Total	5,742	8,766	5.4	13.2	14.4
Industrial					
Process steam	7,646	10,132	3.6	17.8	16.7
Electric drive	3,170	4,794	5.3	7.4	7.9
Electrolytic processes	486	705	4.8	1.1	1.2
Direct heat	5,550	6,929	2.8	12.9	11.5
Feedstock	1,370	2,202	6.1	3.2	3.6
Other	118	198	6.7	0.3	0.3
Total	18,340	24,960	3.9	42.7	41.2
Transportation					
Fuel	10,873	15,038	4.1	25.2	24.9
Raw materials	141	146	0.4	0.3	0.3
Total	11,014	15,184	4.1	25.5	25.2
National total	43,064	60,526	4.3	100.0	100.0

Thermal Effectiveness of Industrial Processes

Although data pertaining to the effectiveness of energy utilization in industry are less extensive than those available for building services, there is reason to believe that it would not be unreasonable to assume that energy savings of approximately 30 percent might be realized by applying already developed energy conservation techniques to industrial processes. The effectiveness with which energy is used in industry varies greatly, depending upon the nature of the industry and the size of the plant. In industries such as electric power generation and chemical refining, the nature of the industry is to convert the energy content of fuel to some more readily salable form. As a rule, optimal design of large plants in such industries is based on both initial costs and operating costs, especially fuel costs. Thus, the design of piping systems to minimize pumping costs and the design of pipe insulation to optimize the trade-off between heat loss and total costs of insulation operation and maintenance are common practices, and are representative of the consideration given to effective energy utilization in large plants for power generation and in other similar industries (*8*).

In other industries, including machine manufacturing, materials processing, and metal forming, the role of energy as an essential ingredient seems to be less clearly recognized, perhaps because energy costs have not been a major part of overall costs of operation, and the effectiveness with which individual items of plant equipment use energy has not been a major concern. Indeed, in some instances where energy costs have been taken into account in selection of plant equipment and plant design, industry has found that because of the prevailing low price of energy it has been cheaper to permit a leak of energy than to modify or replace inefficient equipment. The assumption that this rule applies broadly appears not to be justifiable. Moreover, with the prospect of substantial increases in fuel prices by 1985 (*9*), industrial concern for effective utilization of energy may be expected to increase sharply. In particular, the management of small plants, to which effective energy utilization seems now to be an item of small concern, may be expected to take a much greater interest in obtaining more effective use of the energy. It is appropriate to point out that price is not the only factor influencing industrial concern for effective energy utilization. In some districts of the country, gas suppliers have assigned industries fuel quotas which may not be exceeded. Already one can observe an intense effort on the part of the affected industries to improve the effectiveness of energy utilization in their plants.

That energy can be saved in many industrial operations is indicated by many sales offices of large volume gas suppliers having representatives assigned to advise industry how to use less fuel to conduct their operations. The success of these recently instituted programs has yet to be measured, but one may presume that they will prove beneficial.

Certain examples of improved equipment merit attention here. Gas fired vacuum furnaces have recently been developed for industry. Through the use of well-designed vacuum insulation, heat pipe technology, and modern heat transfer and combustion techniques, these furnaces operate with 25 percent of the total fuel consumption of previous vacuum furnaces (*10*). Other studies of the effectiveness of industrial energy utilization indicate that application of available heat recovery devices (for example, heat wheels) and thermal management techniques could yield net energy savings of 30 percent or more in typical industrial operations (*11*).

Surveys of energy utilization in steel making and in related industrial operations indicate that fuel savings of as much as 39 percent could be realized in operation of certain items of equipment and that average fuel savings of 25 percent or more could be realized by the application of current techniques of waste heat management and up-to-date equipment design to the industry as a whole (*12*).

Interviews with engineering consultants and plant supervisors in a variety of industries have indicated that effectiveness of energy utilization has been of sufficiently little concern in the past that there is ample opportunity for improvement. Approximately 30 percent of the energy used in industrial processes could be saved through the application of existing techniques that are economically justifiable at today's fuel prices. Predicted increases in fuel prices are expected to make energy conservation measures even more attractive to industry in the near future. The invention of more efficient devices, more efficient processes (for example, cement making, refining, chemical processing), and especially the institution of a methodology for the utilization of waste heat in plants may be expected to yield further energy savings in industry, beyond the estimated 30 percent quoted above.

Summary of the Problem

From all the data cited, it is evident that approximately one-quarter of the total national energy consumption may escape effective use because of correctable leaks at the point of utilization. A major reason for this is that at the point of energy utilization, economic justification of energy consuming equipment tends to be governed by initial costs (*13*). Thus, one often finds that high energy consumption has been designed into devices and buildings in order to reduce initial costs. Whatever technological steps might be taken to increase the effectiveness of energy utilization must be coupled with steps to induce a change in the methods of economic justification of building and equipment purchases. If the purchaser can be alerted to the significance of the lifetime operating costs of buildings and energy utilizing equipment, as well as the initial costs, the technological possibilities to enhance the effectiveness of energy utilization in buildings and industrial processes might then be brought to field implementation.

Means to Promote Effective Energy Utilization

I now suggest three ways to approach the problem of improving the effectiveness of energy utilization. The first approach focuses upon improved effectiveness of use of present fuels; the results from such an approach could take effect in a relatively short time (about 2 to 5 years). The second approach focuses upon the utilization of unused energy sources and fuels; results from this approach would require new programs and might be expected to take effect in about 5 to 10 years. In the third approach, energy utilization is considered in the broader context of the energy invested in materials and manufactured goods. Results from this approach would require new technological and economic studies and might take effect in about 10 to 20 years.

Improved Effectiveness of Use of Present Fuels in Current Applications

Building design. Two basic branches of activity are required in building design. First, design criteria and standards for energy conservation in new construction are required. Second, a technology for upgrading the thermal performance of existing structures is required. Even with present high rates of construction, approximately half of the buildings in service in the year 2000 will have been built before 1973 (*14*).

To improve energy utilization in buildings of the future, technological attention must be given to insulation, draft sealing, ventilation, proper selection and maintenance of equipment, envelope design, fenestration design, and illumination. The optimization of trade-offs between increased capital outlay and decreased operating costs over the life of a building must be determined with reliable estimates of fuel price increases being taken into account (*9*). Field data to establish the economic benefits of energy conservation measures must be compiled and made known. Preliminary data (*15*) indicate that, at current fuel prices, increased capital expenditures on thermal upgrading of existing structures (for example, installation of additional insulation) can pay off in approximately 5 years. If one assumes that fuel prices will increase, the payoff time may realistically be assumed to be substantially less than 5 years. Although this is encouraging, conclusive data must be compiled and made known before one can expect the building purchaser to embrace the theory that it is to his financial benefit to invest in energy conserving aspects of buildings.

Thermal upgrading of existing buildings entails technological, economic, and social considerations. Materials and techniques to permit inexpensive, reliable, attractive and safe (for example, fireproof) insulation of existing buildings require development. In addition, some criteria for estimating the expected life of existing buildings must be developed; this involves social as well as technological considerations. Certain neighborhoods of older buildings undoubtedly should not be torn down and replaced with new construction even though modern technology could offer physical improvements. The social effects of demolition and reconstruction would prove unacceptable.

Finally, performance standards by which the effectiveness of energy utilization in buildings can be judged must be established. To evaluate the performance of buildings, test methods must first be devised for measuring heat transmission from buildings, for determining ventilation and infiltration rates, and for determining the effectiveness of building equipment; standard duty cycles in accordance with which building systems can be tested must be established; and means of interpreting test results in terms of effectiveness of energy utilization must be developed. It will then be possible to demonstrate that the field practices advocated actually do lead to more effective use of energy.

The possibility of improving the effectiveness of energy utilization in buildings on a national scale, through federal standards maintained by FHA and the Veterans Administration (VA), and through incentives (for example, home improvement loans) is immense. Thirty-seven percent of the construction in the United States is either built for the federal government or financially assisted by the federal government, and the influence of federal regulations in construction extends well beyond this sector. Properly coupled technological and economic efforts by regulatory agencies could have a powerful influence upon improving the effectiveness of energy utilization.

Industrial processes. In the study of energy utilization in industry, the real efficiency of industrial processes must first be determined. More data must be obtained that will indicate precisely what improvements in effectiveness of energy utilization are technologically feasible and economically justifiable. Programs to distribute information on the technological and economic aspects of improved effectiveness of energy use should be developed on a national scale and methodologies of energy conservation should be demonstrated. For example, techniques of waste heat management by means of heat recovery devices, the application of efficient heat transfer devices such as heat pipes, and coupling between presently independent items of plant equipment, should be demonstrated. As the concern for energy conservation increases, federal laboratories may be able to contribute directly to innovation in industrial processes by developing certain generic processes useful to industry. For example, if effective "air slides" using hot gases of combustion as the fluidizing media could be developed, these could enhance a number of materials processing operations. The federal government may be able to facilitate invention and innovation in industrial processes by providing assistance in questions of patent policy and related matters.

Fuel efficiency and maintenance of total energy systems. A total energy system is one in which electric power for a small complex of buildings is generated locally, and the reject heat is used to provide comfort conditioning and hot water for the dwelling units. Total energy has been enthusiastically embraced by some as a means for effective fuel utilization. Indeed, the promise of reducing the total fuel requirements of building complexes by 25 to 50 percent has been shown to be possible in principle. However, reliable field data to establish the feasibility of total energy systems must be obtained, and technological problems bearing upon the effectiveness of such systems, such as their maintainability and fuel efficiency, must be studied.

Total energy plants are, of necessity, small installations, and, as a rule, the effective temperature of combustion of a small plant is less than that in a large central power station. In large stations, one can afford to make use of heat recovery equipment or topping cycles to attain high combustion temperatures and the efficiencies of power generation associated with them (*16*). To alleviate local thermal pollution from large plants, use might be made of "bottoming cycles" employing low temperature working fluids, but such cycles have yet to be developed (*17*). In small plants the use of such equipment is not economically justifiable. By using efficient central station power generators and employing heat pumps to provide comfort conditioning, it is possible, in principle, to attain greater effectiveness of fuel consumption than by using small scale total energy plants with their limited thermal efficiencies.

The combination of efficient central power generation with local heat pumps permits flexibility in the choice of fuels for power generation. This may be an important consideration, especially in connection with future shifts toward nuclear power. Local applications of nuclear power, as in a total energy system, seem not to be feasible for many reasons. The wisdom of long-range investments in a system, such as total energy, which is both dependent upon fossil fuels and of limited thermal efficiency must be carefully studied even though the total energy concept appears to offer certain advantages in the short range.

Central power stations, with their high effective temperatures of combustion and consequent high thermal efficiencies, also have certain disadvantages. High temperature combustion produces, in addition to thermal pollution, large amounts of other pollutants, particularly nitrogen-oxygen compounds. While the combination of central power and heat pumps appears extremely promising, certain environmental questions about large aerial densities of heat pumps still have to be resolved (for example, where to obtain or reject heat without upsetting the local environment). Efficient, easily maintainable heat pumps have to be developed.

A broad spectrum of fossil fuels exists, some of which are low energy fuels which will not yield high temperature combustion. If it may be assumed that a sufficient quantity of the low energy fuels will be available for some reasonably long period compared with the useful life of small scale total energy power-generation equipment, then it may be appropriate to plan total energy systems to use low energy fuels while reserving high energy fuels for central stations. Compatibility of such total energy equipment with expansions of the national energy system through central station power plants using nonfossil fuels may not pose serious problems if the supply of low energy fuel will survive the equipment. However, the supply of such fuels is not well determined at present; the technology for utilizing such fuels in power generation needs development; and, in any event, supplies of all fossil fuels, including low energy fuels, are exhaustible.

Small scale power units of 500 kilowatts or less are important to total energy planning, for these are the units which can power small complexes of, say, 300 residences or less. However, a survey of total energy systems in the field shows that these small plants have a very high rate of failure (*18*) that is largely attributable to faulty maintenance. For comparison, one may consider that very high quality reciprocating engines in aircraft provide 2000 hours of service between major overhauls; a high quality turbojet engine provides approximately 7000 hours of service between overhauls; a very high quality natural gas-fired reciprocating engine will provide as much as 10,000 hours of service between major overhauls. In the period between major overhauls, numerous minor overhauls are commonly required (for example, the "top" overhaul of reciprocating engines). The engines cited above are typical of the prime movers required in small total energy plants. At best one can expect slightly more than 1 year (8600 hours) of continuous service from these prime movers before a major overhaul will be required. In the interim, several minor overhaul procedures may be required. This poses severe maintenance and management problems for small total energy plants, where overhaul of the prime mover requires temporary shutdown or substantial reduction of power generation.

Cost of maintenance is an especially important aspect of small power plants. Public utility companies now sell electrical power at an average price of approximately 3 cents per kilowatt hour. Some utility companies estimate that their maintenance costs are approximately 1 percent of sales prices, or approximately 2.5/100 to 3/100 cent per kilowatt hour (*19*). The level of maintenance costs for total energy units is 3/10 cent per kilowatt hour (*20*). To examine what this maintenance cost means, let us consider a plant with a 200-kw capacity operating on the average of 50 percent of full load to supply electrical power for a complex of 100 residences. In a day, the plant will produce 2400 kwh of electrical energy. The maximum maintenance expenditure which can be justified for such a plant is $75 per day. At today's labor prices this amount is barely sufficient to support the salary and benefits of one skilled mechanic. The equipment in a total energy plant is technologically advanced (for example, reciprocating prime movers, heat transfer apparatus, air conditioning equipment, electrical generators, switching apparatus, and controls) and requires a wider range of skills for maintenance than one can hope to secure through direct employment of staff. The possibility of prime movers, or other items of equipment, being leased from large companies having effective maintenance staffs should be investigated; the maintenance of small power plants requires very careful management planning. All these considerations indicate that there are fundamental unresolved technological questions upon which the evaluation of total energy systems and other matters of energy planning depend.

Utilization of Unused Energy Sources: Solar Energy

Solar energy at the point of utilization. The use of solar energy for space heating, air conditioning, and hot water heating is one of the extremely attractive possibilities for conservation of nonrenewable energy resources. The annual incidence of solar energy on average buildings in the United States is six to ten times the amount required to heat the buildings (*7*). Solar energy is not only an unused and renewable source of energy, but during the cooling season unused incident solar energy imposes a high load on air conditioning equipment which consumes energy from nonrenewable sources. It would, therefore, be appropriate to mount efforts to utilize solar energy in local applications for building services.

An important consideration in such applications is that most of the energy required by buildings is low temperature heat. For example, space heating requires air at approximately 28°C and water heating temperatures are commonly 60° to 65°C. These temperatures are below the reject heat temperatures of most steam power plants. At present, combustion of high energy fuels, such as natural gas or fuel oil, is used to provide this low temperature heat. But in this practice the capacity of the fuel to produce work, which is the precious commodity of energy, is permanently lost. The utilization of solar energy would, therefore, be an important means for conservation.

In addition, solar energy could be used to provide air conditioning. Absorption refrigeration equipment, which appears particularly attractive for solar power, is now very low in efficiency, but the prospects of substantial improvements in efficiency through application of modern heat transfer technology are promising. The chief advantages of absorption equipment are that it requires very little power (mechanical compression is replaced by chemical effects, so that pumping for circulation is the only work required) and it is easy to maintain. These advantages weigh heavily in favor of development of solar-powered absorption machinery of improved efficiency.

By using what is known today, it would be technologically feasible to apply solar energy to space heating and water heating on a national scale; approximately 50 percent of these energy requirements (representing approximately 11 percent of the national energy consumption) could be met through local application of solar energy. To do this, maintenance-free solar equipment which can be manufactured cheaply would have to be produced. The basic scientific requirements, such as the design of collectors that can optimize collection efficiency, have already been satisfied and can be incorporated into designs for maintainability and low cost (*21*, *22*).

In the most sophisticated solar energy devices in use today (the solid-state devices for direct conversion of sunlight to electricity in space exploration applications), two thirds of the cost of each unit is represented by the supporting frame or case (*23*). While much research has gone into improving the efficiency of the electronics and the special materials (for example, selective absorbers) used in solar devices, rather little has been done on designing the prosaic components of solar equipment (such as the cases) so that they can be manufactured cheaply. And yet, it is the cost of such components which largely determines the cost of the solar equipment, upon which public acceptance and economic justifiability of local solar energy applications ultimately depends. A major effort to address the technological problems of designing solar equipment for ease of manufacture and simplicity of maintenance is urgently needed.

Systems design is an additional important technological aspect of local use of solar energy which requires intensive study. Regardless of the price of fuel within the foreseeable future, the economic optimization of solar energy systems for buildings will require some "booster" heating or air conditioning capacity which utilizes other energy sources; the collection and storage facilities necessary to provide all building services by solar energy are simply too expensive to justify (*24*). Economic justification of a solar energy system depends upon the trade-offs between initial capital requirements of solar devices and operating costs (for example, fuel) of conventional equipment (*25*). At present there is no alternative to one's installing a full size stock item home furnace as the "booster" in an experimental solar home. But this means that the capital costs of the solar heating system are simply an addition to the capital costs of a nonsolar building. The design of integrated solar energy systems for buildings, in which appropriately small booster equipment —with sufficient capacity to "boost" but without surplus capacity which mostly remains idle—could have an immensely favorable influence upon the economic justification of solar energy in building services. For example, the combination of solar power with absorption refrigeration machinery, designed with reroutable circulation, could provide solar-powered air conditioning in summer and solar-powered building heating—the absorption machine being used as a heat pump—in the winter. Heat could be obtained from the intermediate temperature station of the absorption device for hot water heating all through the year. Booster capacity in this case could be provided by low capacity heaters applied to the distillation chamber of the absorption device. In such a system substantial capital savings appear to be possible, especially if current expectations for improved effectiveness of absorption devices are realized. In addition, such a system could provide for control of humidity as well as temperature, through solar energy. This would be a significant attraction to the home owner or building operator.

The major obstacles to local application of solar energy are cultural and institutional. Solar collectors on roofs appear strange and impose certain constraints upon building style and orientation. Reliable data on maintenance requirements and measured performance of solar energy equipment in actual field service are lacking. In general, solar energy appears to the building buyer, the financier, and the building constructor as an interesting but unproved idea. Testing and evaluation of field equipment can provide the information with which the institutional and cultural obstacles to the realization of these benefits might be overcome.

Solar energy and building design. The use of solar energy to provide low temperature heat for buildings should be coupled with efforts to improve the thermal performance of the buildings themselves. Estimates in the literature pertaining to the extent to which solar energy can be used to provide space heating are generally based upon the assumption that the basic thermal design of buildings will remain conventional (*25*). Because proper thermal design of buildings could reduce the energy required to provide space heating by 40 percent or more compared with energy required by buildings of conventional design it should be possible to build solar homes which derive substantially more than 50 percent of their space heating requirements from the sun. The precise amount of solar heating that could be achieved in a suitably designed home depends upon economic trade-offs between costs of solar energy, storage facilities, and costs of additional insulation, draft sealing, and double glazing, for example. Detailed study of this problem should be made to provide a basis for design.

Some economic aspects of solar energy in local application. One may consider the costs of utilizing solar energy to provide low temperature heat in local applications, such as domestic hot water heating, in at least two ways. First, the cost to the consumer of installing a solar device may be compared with the total costs which the consumer would have to pay for fuel or electric power were the solar device not to be installed (*26*). A second way to consider the costs of solar energy is to compare the net capital outlay required to effect a reduction in demand upon fuels currently used to supply energy with the costs of increasing the capacity of the present national energy supply system, it being assumed that continued expansion is indeed possible. The former consideration has been treated extensively in the literature of solar research (*25*, *27*). The latter aspect of solar energy appears not to have been treated previously, and I will consider it here. I reemphasize that I consider the local application of solar energy to provide low temperature heat.

Hot water heating by solar energy. At present, domestic hot water heating accounts for approximately 3 percent of the national energy consumption (Table 1). Hot water heating by solar energy has been exploited abroad (*28*, *29*). Given the climatic conditions of the United States, it should be feasible to provide at least 50 percent of this demand for energy via solar devices (*28*). Thus, one is dealing with a potential reduction of 1.5 percent of the national requirement for fuels.

Solar collectors of 1 square meter surface area and having the capacity to retain an average of 3.5 kwh per day of solar energy in the form of low-temperature heat have been demonstrated (*7*, *21*, *30*). At present, such a collector suitable for domestic hot water heating can be produced for approximately $18 per square meter. Through the application of modern materials and manufacturing techniques, it should be possible to reduce this cost to $15 per square meter, or less. At the former figure, the capital cost of collecting 1 kwh per year of solar energy, in the form of low-temperature heat, is 1.4 cents. To make use of solar energy for hot water heating in a typical residence, the addition of a collector may be sufficient. For solar space heating, an energy storage facility, which may double the price of the system, would be required. To see what the costs estimated above would mean to the individual householder, consider that a typical dwelling in the United States uses approximately 10,000 kwh per year for hot water heating. A collector to provide half of this energy annually would be approximately 4 square meters in area and would cost approximately $76. The cost of implementing solar domestic hot water heating in all the 60 million dwelling units of the United States, to effect a 1.5 percent reduction in demand upon fuels, would be approximately $4.5 billion. Thus, we may take a figure of $3 billion as an approximate estimate of the capital requirements of reducing national requirement for fuel by 1 percent, through implementation of solar energy to provide low-temperature heat.

To estimate the capital costs of increasing the capacity of the national energy supply by 1 percent to supply low-temperature heat in building services, one must note first that most of the low-temperature heat used in buildings is provided by combustion of natural gas. Thus, expansion of energy supply capacity to meet growth in building services will require an increase in the supply of gas, or the conversion of domestic equipment for utilization of more readily available fuels such as coal. The costs of converting domestic or industrial equipment are very high. It would not be unreasonable to take $200 as the cost of converting a gas-fired residential space heating plant to permit combustion of residual oil or coal. In addition, in some regions it may not be desirable, or even feasible, to convert equipment for combustion of other types of fuels (*31*). Thus, in a discussion of the need to increase the capacity of the national energy supply to provide low-temperature heat in building services, it is reasonable to consider the possibility of expanding the national capacity to supply gas. Because domestic supplies of natural gas are severely strained, expansion of capacity to supply gas would probably entail the importation of liquefied gas. Current estimates for the capital costs of gas liquefaction plants indicate that a plant capable of delivering 100 million cubic feet of gas per day will cost between $200 and $300 million to build (*32*).

If one takes the heating value of the gas (about 1000 Btu per cubic foot) as the basis for estimating the cost of such a plant, one finds that the capital costs of gas liquefaction are approximately $160 to $240 per kilowatt. If one assumes that such a plant can be operated for 8000 hours per year, the capital costs of gas liquefaction may be estimated as 2 to 3 cents to increase the capacity of the national energy supply by 1 kwh per year. These figures reflect only the costs of liquefaction. In addition, one must include costs of increasing the transportation system (for example, refrigerated tankers), storage capacity (refrigerated tanks), and distribution network (pipelines) for this form of fuel, as well as the capitalization of wells to provide raw petroleum for liquefaction and the energy required for transporting the fuel to the point of utilization. Taking all considerations into account, it would appear reasonable to estimate the costs of increasing gas supply capacity through liquefaction as being approximately 5 cents per kilowatt hour per year. Thus, to increase the present capacity of the national energy supply by 1 percent, through gas liquefaction, would cost approximately $10 billion.

The fastest growing mode of space heating has been electric heat (*1*). To estimate the costs of increasing the capacity of the national electrical energy supply, one may note that a modern power plant costs between $200 and $300 per kilowatt to build. In addition to building the plant, one must provide additional fuel and additional electrical distribution capacity for the plant. If one assumes that these additional capital requirements can be met by $100 per kilowatt of plant capacity, the net cost of increasing electrical power capacity may be taken to be approximately $400 per kilowatt. If one further assumes that newly constructed plants may run, on the average, at 65 percent of peak capacity, the capital cost of increasing the capacity of national electrical supply by 1 kwh per year is found to be approximately 8 cents. To increase the capacity of the national energy supply by 1 percent through expansion of electrical power would, therefore, cost approximately $16 billion.

The difference in capital requirements for reducing demand for fuels by 1 percent via utilization of solar energy for low-temperature heat ($3 billion) and the costs to increase present supply capacity by 1 percent (as much as $16 billion), may be debatable in certain details, and, in any event, requires further study. I have tried, not to offer a definitive estimate of these costs, but rather to provide a preliminary estimate to assist one in judging whether local utilization of solar energy for low-temperature heat offers an effective investment opportunity, in the context of national energy planning. Based upon the data provided, one may conclude that it does.

One of the reasons why the use of solar energy for low-temperature heat would cost less than a corresponding expansion of the capacity of the national energy supply is that the latter requires energy in high quality form (*33*). High-quality energy, such as fuels for high-temperature combustion or electrical energy, is readily convertible to work but is generally expensive to generate. Low-quality forms of energy, such as low-temperature heat, are usually inexpensive to generate; in fact, low-quality energy is often discarded.

Investment in the use of solar energy can be furthered two ways: (i) by developing effective, reliable, and inexpensive solar collectors, and (ii) by establishing suitable government construction regulations; in particular, the FHA and VA could adopt standards under which domestic solar equipment might qualify for residential construction and loan support.

Utilization of Unused Energy Sources: Incineration

The use of solid waste as fuel has attracted favorable attention of many technologists. Solid waste products are known to have heating values varying from one half that of fuel oil (paper) to as much as that of fuel oil (consumer plastics). In addition, solid waste represents a form of unexploited fuel which will probably remain in relative abundance for some time to come. It has been estimated that by the year 1990 the heating content of collected urban refuse could be used to generate as much as 35,000 megawatts of electrical power (*34*). However, before the apparently rich fuel resources of solid refuse can be put to use, a number of technological problems must be solved. These include the design of combustion plants (in this instance, incinerators) to use widely varying fuels (for example, waste paper, consumer plastics) and the selection of materials to tolerate some of the highly corrosive products of combustion of solid refuse. The supply of solid refuse in a given area may fluctuate seasonally so that incineration plants must be planned to operate effectively with widely varying fuel loadings or equipment must be designed for sorting refuse according to its combustion properties. Electrical generating plants or district heating systems that can use the heat generated by such plants must also be developed before incineration can become a source of useful heat. In addition, techniques must be found for controlling the potentially polluting emissions from incinerator combustion chambers.

It is evident that the incineration of solid refuse can provide useful heat and constitutes a field of technology in itself. The federal government will undoubtedy be substantially involved in incinerator construction in the coming years, but at present there is no identifiable resource, either within the federal government or private industry, to provide the technological basis for construction standards, environmental regulations, or rules for qualification for federal support which may apply to incineration plants. Such a facility would appear to be called for.

Conservation of Energy Invested in Materials

The largest single consumer of energy in the United States is industry. Here I consider the possibility of conserving energy by improving the products in which the energy of processing is invested. I will describe three specific examples.

The use of materials as dictated by design standards. In building construction, plumbing, and several other areas, the codes governing design are overly conservative for most applications. This simplifies design procedures, but leads to excessive use of materials, which in turn requires excessive use of energy. The possibility that more accurate design standards can be devised, which would permit construction, plumbing, and manufacturing operations to proceed without excessive use of material and without the functionality or safety of the product being reduced, merits detailed study. For example, the size of air-vent piping used in plumbing systems is chosen so that it can satisfy the needs of toilet systems in large apartment complexes or office buildings. An air-vent pipe of one-fifth the conventional size (with correspondingly smaller investment in energy of manufacture) would be adequate for most residences. This and other examples of excessive materials should be studied in the context of conservation both of energy and of natural resources. The economic implications of shifting industrial emphasis from areas of materials production to areas of effective materials utilization, should be considered as an integral part of such a study.

Durable as opposed to disposable goods. The disposable goods to which the public has become accustomed are widely recognized to constitute a drain on natural resources in general, and energy in particular. However, while reusable glass milk bottles, for example, are beneficial to society in connection with conservation of resources, they may constitute a hazard to the householder. Field interviews with physicians have revealed that glass milk bottles, which are wet, slippery, and heavy when removed from the refrigerator, often slip from one's grasp, fall and shatter, producing shards that can inflict serious wounds. The seriousness of the accidents was compounded by the fact that the spilled milk made footing slippery and led at times to the accident victim's falling on the glass shards. With the advent of paper milk cartons the frequency of this type of accident appears to have diminished. However, conclusive evidence is not available at present. The object of citing the glass milk bottle here is just to point out that important ramifications of any major change, such as converting from durable to disposable goods, exist and require study. For example, the design of glass containers for both reusability and safety should be studied if an effective conversion from disposable to durable containers is to be proposed.

Maintainability of machines. The manufacturing of machines is one of the largest components of U.S. industry. The possibility that the average life of machines might be extended through design for durability, careful utilization of durable materials and, especially, design for effective maintenance should be considered. If the average life of machines and other manufactured goods could be extended by, say 25 percent, then the energy requirements of the manufacturing industry might be reduced by a corresponding fraction. The technological prerequisites for such an alteration of design and manufacturing have yet to be satisfied. Moreover, the social and economic ramifications of such a step require careful study. Nevertheless, the possibility of more careful designs being used for the production of more durable goods in which energy materials are invested with greater effectiveness than at present appears to be an attractive possible measure for the conservation of natural resources.

It is evident that a technological field concerned with maintenance procedures is urgently needed. The quality of machine maintenance at present could be substantially improved by the incorporation of existing techniques of monitoring performance into field practice; methods for the detection of impending failure or malfunction could be developed through application of existing techniques. By such means, unexpected costly and dangerous failures could be avoided, the useful life of equipment could be extended, and natural resources could be utilized more effectively. Machine maintenance, as it is practiced today, does not constitute a technology; the disparity between existing and developable techniques on the one hand, and the techniques used in field practice on the other, is simply too great.

Conclusions and Recommendations

The ineffective utilization of energy in buildings and industrial processes constitutes a major component of the energy problems in the United States. Not only could the effectiveness of energy utilization be improved, but such improvement appears to be justifiable economically, especially when the costs of the alternative of expanding the national capacity to supply increasing energy demands are considered. The measures to improve effectiveness of energy utilization are basically technological in nature. However, at present, there does not exist an identifiable technological field concerned with energy conservation through effective utilization. Although techniques for this purpose exist and others can be developed, extant techniques have not been integrated and applied in rational field practice, and there is no disciplinary framework within which further developments might be made. Appropriate measures should be undertaken on a national scale to create and implement a technology for energy conservation through more effective utilization.

References and Notes

1. Stanford Research Institute, *Patterns of Energy Consumption in the United States* (Menlo Park, Calif., November 1971; prepared for the Office of Science and Technology, Washington, D.C., January 1972), p. 6.
2. American Society of Heating, Refrigerating, and Air-Conditioning Engineers, *Handbook of Fundamentals* (New York, 1972), p. 337.
3. Quantities expressed as British thermal units per thousand cubic feet—degree day represent the heating requirements of a building relative to its size and the severity of the climate in which it serves. Conversion factors are: from British thermal units to joules, 1055; cubic feet to cubic meters, 2.83×10^{-2}.
4. J. C. Moyers, *The Value of Thermal Insulation in Residential Construction: Economics and Conservation of Energy* (Oak Ridge National Laboratory, Report ORNL-NSF-EP-9, Oak Ridge, Tenn., December 1971), p. 28.
5. American Society of Heating, Refrigeration, and Air-Conditioning, *Engineers Handbook of Fundamentals* (New York, 1972), pp. 381–383.
6. ———, *ibid.*, p. 421; National Association of Home Builders Research Foundation, *Insulation Manual* (Rockville, Md., 1971).
7. H. C. Hottel and T. B. Howard, *New Energy Technology—Some Facts and Assessments* (M.I.T. Press, Cambridge, Mass., 1971).
8. Even in large power plants much heat is lost. In a typical modern power plant approximately two-thirds of the heating value of the fuel consumed must be rejected to the atmosphere. This reject heat is not "waste" in the present context; the rejection of this heat is required by the second law of thermodynamics. But, by siting a plant near a consumer, much of this heat could be put to use in waste processing, water purification, space heating, or air conditioning, for example.
9. U.S. Department of Commerce, *The Energy Crisis: An Analysis* (Washington, D.C., April 1972).
10. P. K. Shefsiek and L. J. Lazaridis, *Nat. Gas Res. Technol.*, in press.
11. G. A. Maier, "Practical Means of Conserving Energy Today in the Residential, Commercial and Industrial Market" (University of Pittsburgh School of Engineering Library, Pittsburgh, 1971); D. P. Gregory, *A Techno-Economic Study of the Cost-Effectiveness of Methods of Conserving the Use of Energy* (Institute of Gas Technology, Chicago, 1971); R. B. Rosenberg, *The Future of Industrial Sales* (Institute of Gas Technology, Chicago, 1972).
12. J. D. Nesbitt, *Improving the Utilization of Natural Gas in Major Steel Mill Applications* (Institute of Gas Technology, Chicago, 1972).
13. It is not intended to imply here that quality of performance is not considered in acquisition of industrial equipment, but rather that of two devices which yield the same product, the cheaper will tend to be preferred, irrespective of energy consumption. Those industrial accounting systems in which energy requirements are carried as overhead appear to reinforce this tendency.
14. U.S. Department of Housing and Urban Development, International Brief, January 1971.
15. National Mineral World Insulation Association, Impact of Improved Thermal Performance in Conserving Energy (National Bureau of Standards, Washington, D.C., April 1972), p. 35.
16. A topping cycle is an additional power generation plant which receives heat at the temperature of combustion, and rejects heat at the maximum temperature required by the main power plant. The topping cycle utilizes the temperature drop between the combustion chamber and the boiler of the plant, to generate power.
17. In nighttime power generation, it would be possible to reject heat at subatmospheric temperatures, through radiative techniques, and thus avert local thermal overloading of the atmosphere.
18. P. R. Achenbach, J. B. Coble, B. C. Cadoff, T. Kasuda, *A Feasibility Study of Total Energy Systems for Breakthrough Housing Sites* (National Bureau of Standards Report 10 402, Washington, D.C., August 1971), Appendix A.
19. K. Boer, Institute for Energy Conversion, University of Delaware, personal communication.
20. Few small units can justify spending more than 3 cents per kilowatt hour for maintenance; few can be maintained for less.
21. H. Buchberg, O. A. LaLude, D. K. Edwards, *Solar Energy J. Solar Energy Sci. Eng.* **13**, 193 (1972).
22. H. C. Hottel and A. Whillier, "Evaluation of flat-plate solar-collector performance," in *International Conference on the Uses of Solar Energy Proceedings* (University of Arizona Press, Tucson, 1958); H. Tabor, *Bull. Res. Counc. Israel* **5C**, No. 1 (1955).
23. Committee report, *Solar Cells, Outlook for Improved Efficiency* (National Academy of Sciences, Washington, D.C., 1972), p. 3.
24. A. Whillier, "Solar house heating—a panel," in *International Conference on Uses of Solar Energy* (Univ. of Arizona Press, Tucson, 1958).
25. R. A. Tybout and G. O. G. Löf, *Nat. Resourc. J.* **10**, 268 (1970).
26. In this comparison it is assumed that the consumer will own and maintain the solar device, but such an arrangement may be neither necessary nor desirable.
27. H. C. Hottel, "Residential uses of solar energy," in *International Conference on Uses of Solar Energy* (Univ. of Arizona Press, Tucson, 1958); G. Pheijel and B. Lindström, *New Sources of Energy, United Nations Conference, Rome, 21 to 31 August 1961*, p. 207–223.
28. D. N. W. Chinnery, *CSIR (S. Afr. Counc. Sci. Ind. Res.) Res. Rep. No. 248*, pp. 1–79 (1967).
29. S. J. Richards and D. N. W. Chinnery, *ibid., No. 237*, pp. 1–26 (1967).
30. The average cited here was obtained over a period during which cloudy and sunny weather obtained.
31. For example, in certain industrial areas local efforts to decrease air pollution started with conversion of both industrial domestic combustion equipment from coal and oil to gas. It would be difficult to justify reconversion at this point. The same argument applies to domestic heating equipment.
32. Such a plant will also be able to supply a large quantity of residual fuel oil to those installations capable of using it. However, for the purposes of estimating the costs of supplying energy for domestic consumption, I compare the cost of the plant to its capacity to supply gas, which is its principal function.
33. The thermodynamic notion of quality (or more precisely, availability) is a measure of the extent to which the form of energy can be converted to work.
34. "Refuse-fueled power station," *Technol. Rev.* May 1972, p. 62.

Appendix I Some Useful Numbers

PROPERTIES OF SOLID AND LIQUID FUELS

Fuel	Approximate Energy Content per Unit Mass: 10^3 Btu/lb	10^3 cal/g	Approximate Specific Gravity[a]
Oak	~7.9	~4.4	~0.83
Pine	~8.7	~4.8	~0.48
Bituminous coal	7.9 to 14.8	4.4 to 8.2	1.27 to 1.45
Anthracite coal	9.0 to 14.1	5.0 to 7.8	1.4 to 1.7
Crude oil	18.8 to 19.5	10.1 to 10.8	0.81 to 0.98
Fuel oil	18.0 to 19.4	10.0 to 10.8	~0.94
Gasoline	20.0 to 21.0	11.1 to 11.7	0.72 to 0.74
Ethyl alcohol	12.8 to 13.2	7.1 to 7.3	0.79

[a]The density of water is ~1 g/cm^3 or ~62.4 lb/ft^3.

PROPERTIES OF GASEOUS FUELS

Fuel	Approximate Energy Content of Gas at STP: 10^3 Btu/lb	10^3 cal/g	Gaseous Specific Gravity (air = 1[a])	Liquid Specific Gravity (water = 1[b])	Heat of Fusion (cal/g)
Methane ($CH_4 \simeq$ natural gas)	23.7	13.2	0.55	0.466	14.5
Propane (C_3H_8)	21.7	12.1	1.55	0.501	19.11
n-butane (C_4H_{10})	21.3	11.8	2.08	0.579	19.18
Hydrogen	52	29	0.07	0.07	13.8

[a]The density of air is 1.29 kg/m^3 (at 0°C, 1 atm) or 8×10^{-3} lb/ft^3.
[b]The density of water is ~1 g/cm^3 or 62.4 lb/ft^3.

The energy content of some nuclear fuels is as follows:

1) U^{235} at 200 MeV/fission yields 3.5×10^{10} Btu/lb for complete fission of all atoms. In a conventional burner reactor 1 lb of enriched uranium fuel yields about 3.2×10^8 Btu/lb;
2) deuterium-deuterium fusion reaction produces 24.8 MeV/fusion, which, if water is used as the fuel source, translates to 1.8×10^6 Btu/gal.

APPROXIMATE ENERGY EQUIVALENCES

	Coal (tons)	Oil (bbls)	Natural Gas (10^3 ft^2)	Gasoline (gals)	U^{238} Fuel (lbs)	H_2O for D-D Fusion (gals)	Solar Flux (m^2 · h)	Btu
Coal (1 ton) (2000 lb)	1	4.3	25	200	0.08	14.3	8.3×10^3	25×10^6
Oil (1 bbl)	0.23	1	5.8	45.5	0.02	3.3	2×10^3	5.8×10^6
Natural gas (10^3 ft^3)	0.04	0.17	1	7.9	0.003	0.56	300	10^6
Automotive gasoline (1 gal)	0.005	0.022	0.127	1	4×10^{-4}	0.07	47.3	127×10^3
Enriched uranium fuel (1 lb)	12.8	55	320	2500	1	167	10^5	320×10^6
Water used for deuterium-deuterium fusion (1 U.S. gal)	0.07	0.31	1.8	14.3	0.006	1	600	1.8×10^6
Solar flux (1 m^2 for 1 h)	1.2×10^{-4}	5.1×10^{-4}	3×10^{-3}	0.024	9×10^{-6}	1.7×10^{-3}	1	3×10^3

	Btu	erg	ft · lb	hp · h	J	cal	kWh	eV	MeV	kg	amu
1 Btu	1	1.055×10^{10}	777.9	3.929×10^{-4}	1055	252.0	2.930×10^{-4}	6.585×10^{21}	6.585×10^{15}	1.174×10^{-14}	7.074×10^{12}
1 erg	9.481×10^{-11}	1	7.376×10^{-8}	3.725×10^{-14}	10^{-7}	2.389×10^{-8}	2.778×10^{-14}	6.242×10^{11}	6.242×10^{5}	1.113×10^{-24}	670.5
1 ft · lb	1.285×10^{-3}	1.356×10^{7}	1	5.051×10^{-7}	1.356	0.3239	3.766×10^{-7}	8.464×10^{18}	8.464×10^{12}	1.509×10^{-17}	9.092×10^{9}
1 hp · h	2545	2.685×10^{13}	1.980×10^{6}	1	2.685×10^{6}	6.414×10^{5}	0.7457	1.676×10^{25}	1.676×10^{19}	2.988×10^{-11}	1.800×10^{16}
1 J	9.481×10^{-4}	10^{7}	0.7376	3.725×10^{-7}	1	0.2389	2.778×10^{-7}	6.242×10^{18}	6.242×10^{12}	1.113×10^{-17}	6.705×10^{9}
1 cal	3.968×10^{-3}	4.186×10^{7}	3.087	1.559×10^{-6}	4.186	1	1.163×10^{-6}	2.613×10^{19}	2.613×10^{13}	4.659×10^{17}	2.807×10^{10}
1 kWh	3413	3.6×10^{13}	2.655×10^{6}	1.341	3.6×10^{6}	8.601×10^{5}	1	2.247×10^{25}	2.270×10^{19}	4.007×10^{-11}	2.414×10^{16}
1 eV	1.519×10^{-22}	1.602×10^{-12}	1.182×10^{-19}	5.967×10^{-26}	1.602×10^{-19}	3.827×10^{-20}	4.450×10^{-26}	1	10^{-6}	1.783×10^{-38}	1.074×10^{-9}
1 MeV	1.519×10^{-16}	1.602×10^{-6}	1.182×10^{-13}	5.967×10^{-20}	1.602×10^{-13}	3.827×10^{-14}	4.450×10^{-20}	10^{6}	1	1.783×10^{-30}	1.074×10^{-3}
1 kg	8.521×10^{13}	8.987×10^{23}	6.629×10^{16}	3.348×10^{10}	8.987×10^{16}	2.147×10^{16}	2.497×10^{10}	5.610×10^{35}	5.610×10^{29}	1	6.025×10^{26}
1 amu	1.415×10^{-13}	1.492×10^{-3}	1.100×10^{-10}	5.558×10^{-17}	1.492×10^{-10}	3.564×10^{-11}	4.145×10^{-17}	9.31×10^{8}	931.0	1.660×10^{-27}	1

Notes:
The electronvolt (eV) is the kinetic energy an electron gains from being accelerated through the potential difference of 1 V in an electric field. The MeV is the kinetic energy it gains from being accelerated through a million-volt potential difference.

The last two items in this table are not properly energy units but are included for convenience. They arise from the relativistic mass-energy equivalence formula $E = mc^2$ and represent the energy released if a kilogram or atomic mass unit (amu) is destroyed completely.

Again, care should be used when employing this table.

Source: R. Resnick and D. Halliday, *Physics for Students of Science and Engineering*. New York: Wiley, 1960.

Some other energy units are as follows:

1 kcal = 1000 cal

1 J = 1 W · s = 1 N · m = 10^7 erg

1 erg = 1 cm · dyne

1 mkgf = 9.807 J

1 $Q = 10^{18}$ Btu.

POWER

	Btu/h	ft · lb/min	ft · lb/s	hp	cal/s	kW	W
1 Btu/h	1	12.97	0.2161	3.929×10^{-4}	7.000×10^{-2}	2.930×10^{-4}	0.2930
1 ft · lb/min	7.713×10^{-2}	1	1.667×10^{-2}	3.030×10^{-5}	5.399×10^{-3}	2.260×10^{-5}	2.260×10^{-2}
1 ft · lb/s	4.628	60	1	1.818×10^{-3}	0.3239	1.356×10^{-3}	1.356
1 hp	2545	3.3×10^{4}	550	1	178.2	0.7457	745.7
1 cal/s	14.29	1.852×10^{2}	3.087	5.613×10^{-3}	1	4.186×10^{-3}	4.186
1 kW	3413	4.425×10^{4}	737.6	1.341	238.9	1	1000
1 W	3.413	44.25	0.7376	1.341×10^{-3}	0.2389	0.001	1

Source: R. Resnick and D. Halliday, *Physics for Students of Science and Engineering*. New York: Wiley, 1960.

Weights, measures, and other conversion factors and constants frequently encountered in energy problems as follows:

1 short ton = 2000 lb = 907.2 kg

1 long ton = 2240 lb = 1016 kg

1 metric ton = 2205 lb = 1000 kg

1 U.S. fl gal = 231 in^3 = 0.134 ft^3 = 4.23×10^{-3} m^3 = 4.23 l (liters)

1 bbl (barrel) = 42 gal = 5.62 ft^3 = 0.259 m^3 (Warning: In the United States the barrel is often defined as 31.5 gal, but it is the 42 U.S. fluid gallon barrel that is used in connection with international petroleum transactions.)

1 mi^2 = 640 acres; 1 acre = 43560 ft^2 ; 1 m^2 = 10.76 ft^2

1 ha (hectare) = 1.0×10^4 m^2 = 2.471 acres

1 yr = 8760 h

1 Å (Angstrom) = 10^{-10} m

1 μm (micron) = 10^{-6} m

First Bohr orbit radius in hydrogen atom = 0.528×10^{-10} m.

Air pollutant concentration conversions (25°C, 760 mmHg) are as follows:

1 ppm CH_4 = 655 μg/m^3

1 ppm NO = 1230 μg/m^3

1 ppm NO_2 = 1880 μg/m^3

1 ppm O_3 = 1960 μg/m^3

1 ppm SO_2 = 2617 μg/m^3.

U.S. AMBIENT AIR QUALITY STANDARDS

Pollutant	Primary (Health) (3 years to reach)	Secondary (Welfare) (no deadline)
Particulates		
annual mean	75	60
maximum 24-h concentration	260	150
Sulfur oxides		
annual mean	80	–
maximum 24-h concentration	365	260
maximum 3-h concentration	–	1300
Carbon monoxide		
maximum 8-h concentration	10	10
maximum 1-h concentration	40	40
Photochemical oxidants		
maximum 1-h concentration	160	160
Hydrocarbons		
maximum 3-h concentration, 6 to 9 A.M.	160	160
Nitrogen dioxide		
annual mean	100	100

Note: Concentrations for pollutants are in micrograms per cubic meter except for carbon monoxide, which is in milligrams per cubic meter.

U.S. NEW SOURCE PERFORMANCE STANDARDS FOR FOSSIL FUEL-FIRED STEAM GENERATORS

Pollutant	Emmission Standard ($lb/10^6$ Btu)
Particulates	0.1
Sulfur dioxide	
liquid fuel	0.8
solid fuel	1.2
Nitrogen oxides	
gaseous fuel	0.2
liquid fuel	0.3
solid fuel	0.7
Visible emissions	#1 Ringel. or 20 percent opacity

Some of the abbreviations commonly used in the energy field are as follows:

AEC Atomic Energy Commission
AEPI American Electric Power Institute
AGA American Gas Association
API American Petroleum Institute
BPA Bonneville Power Administration
BWR boiling water reactor
DCF discounted cash flow
DWT deadweight ton
ECCS emergency core cooling system
EPA Environmental Protection Agency
EPRI Electric Power Research Institute
ERDA Energy Research and Development Agency
FBR fast breeder reactor
FEA Federal Energy Administration
FEO Federal Energy Office
FPC Federal Power Commission
GNP gross national product
HTGR high-temperature gas-cooled reactor
LMFBR liquid-metal fast-breeder reactor
LNG liquefied natural gas
LPG liquefied petroleum gas
LWR light-water reactor
MB/D thousand barrels per day
MCF thousand cubic feet
MHD magnetohydrodynamics
MMB/D million barrels per day
MMCF million cubic feet
NGL natural gas liquids
NSF National Science Foundation
OCR Office of Coal Research (U.S. Department of the Interior)
OCS outer continental shelf
OIP oil in place
OPEC Organization of Petroleum Exporting Countries
PAD Petroleum Administration for Defense
PGC Potential Gas Committee
PWR pressurized water reactor
R/P reserves per production ratio
TCF trillion cubic feet
TVA Tennessee Valley Authority
USGS U.S. Geological Survey
VLCC very large crude carriers.

Appendix II
Some Energy Bibliographies

Most of the IEEE Press reprint books are accompanied by a topical bibliography. However, because the energy literature has become so vast, it was decided that in the case of this book a list of some of the available bibliographies would be more appropriate.

A large volume of material of a bibliographic nature has been produced by the various energy-related committees of the U.S. Congress during the past several years. A good starting place is

1) Publication List, "National fuels energy policy study" (S. Res. 45), U.S. Senate Committee on Interior and Insular Affairs, Washington, D.C.
2) Selected Reading on the Fuels and Energy Crisis, 92nd Congress, 2nd Session. Available from the U.S. Government Printing Office, Washington, D.C., 1972.

In the area of energy-related research, there are several general bibliographies now available, including

1) NSF-RANN Energy Abstracts; a monthly abstract journal of energy research, published monthly by Oak Ridge National Labs.,

   Miriam P. Guthrie, Ed.
   P. O. Box X
   Oak Ridge, Tenn. 37830

2) "An inventory of energy research," a report prepared for the National Science Foundation by Booz, Allen and Hamilton, Inc., Washington, D.C., Oct. 15, 1971

or its sequel,

3) An inventory of energy research," prepared for the Task Force on Energy of the Subcommittee on Science, Research and Development of the Committee on Science and Astronautics, U.S. House of Representatives, by Oak Ridge National Labs. Mar. 1972. Available from the U.S. Government Printing Office, Washington, D.C.
4) D. R. Limaye, R. Ciliano, J. R. Sharko, "Quantitative energy studies and models: A state of the art review," prepared by Decision Sciences Corp. for CEQ, Mar. 1973. (See especially Appendixes II and III which contain an annotated bibliography and model summary). Available from the National Technical Information Service (NTIS).
5) "Bibliography of R&D research reports" (Socioeconomic Environmental Studies Ser.), U.S. Environmental Protection Agency, Publ. EPA 600/5-73-002, July 1973.

The NTIS maintains a large collection of technical reports and will provide copies for a charge. Bibliographic citations on all of their material are maintained as a computerized data bank. They can perform special bibliographic searches on request. They also perform standard searches and produce bibliographies in areas of current interest. Several bibliographies in energy are available. For information write

National Technical Information Service
U.S. Department of Commerce
Springfield, Va. 22151

Various federal agencies publish bibliographies. For example, the Atomic Energy Commission has several, including

1) "Books on atomic energy for adults and children," Div. Technical Information, U.S. Atomic Energy Commission.

Two useful collections from the Environmental Protection Agency are

1) "EPA reports bibliography," U.S. Environmental Protection Agency, Rep. EPA-LIB-73-01, July 1973.
2) "Air pollution aspects of emission sources: Electric power production—A bibliography with abstracts," Environmental Protection Agency, Publ. AP-96, May 1971.

The U.S. Government Printing Office also produces lists of its publications. Four examples are

1) "Mines: Explosives, fuel, gasoline, gas, petroleum, minerals," Superintendent of Documents, U.S. Government Printing Office, Washington, D.C., price list PL 58, Nov. 1970.
2) "Atomic energy and civil defense," Superintendent of Documents, U.S. Government Printing Office, Washington, D.C., price list PL 84, Dec. 1970.
3) "Irrigation, drainage and water power," Superintendent of Documents, U.S. Government Printing Office, Washington, D.C., price list PL 42, Dec. 1970.
4) "Ecology," Superintendent of Documents, U.S. Government Printing Office, Washington, D.C., price list PL 88, Feb. 1972.

Five more general, bibliographic lists are the following:

1) R. H. Romer, "Resource letter ERPEE-1 on energy: Resources, production, and environmental effects," *Amer. J. Phys.*, vol. 40, p. 805, June 1972.
2) *Science for Society: A bibliography*, 3rd ed. Washington, D.C.: Amer. Assoc. Advancement of Science, 1972. Prepared by H. T. Bansum; previously by J. A. Moore.
3) L. K. Caldwell, H. S. Kibbey, and T. A. Siddigi, Eds., *Science Technology and Public Policy*, vol. 3, School of Public and Environmental Affairs, Univ. Indiana, Bloomington, 1972 (this bibliography is annotated).
4) "Science bibliography of energy," *Science*, vol. 184, p. 386, Apr. 19, 1974.
5) "Bibliographies," in *Energy and Power*. San Francisco: Freeman, 1971, pp. 139–140. (Scientific American reprint book.)

Appendix III

Energy and Public Policy: An Introductory Bibliography

K. N. LEE

Whether in obscurity or on the front page, energy has been—and will continue to be—a staple of governmental policies, and a matter of long-range social importance. The literature suggested in this bibliography has two immediate aims: first, to complement the more technical material presented elsewhere in this book; second, to serve as a basic primer of the *recent* precursors of the energy crisis of the early 1970's.

As will be clear after leafing through the book, technology and public policy cannot be cleanly subdivided. The articles by Russell Train and Allen Kneese reprinted here, for instance, arguably belong to a "policy" collection, not a technological reader. So be it; the world which uses energy does not respect the distinction between technology and policy either. Some subject matter divisions are useful, nonetheless. In the following I have cut things five ways:

1) *Background*, including materials useful in setting the social context of energy as a policy problem;

2) *Fossil fuels;*

3) *Nuclear energy;*

4) *Conservation, use, and demand management;*

5) *Institutional redesign* and proposals for the future, including materials dealing with the outlines of the long-term social future.

Despite this categorial neatness, of course, energy policy remains a turbulent, rapidly changing field (indeed, the bibliography, although representative, is by no means comprehensive). Some major trends have begun to emerge—some of them only tentatively. Four trends are worth pointing out:

1) Major *federal reorganization* efforts, already underway, are likely to lead to a Department of Energy and Natural Resources, together with an Energy Research and Development Agency drawn from the existing Atomic Energy Commission research establishment. Far and away the most intractable part of this reorganization, however, will be the reshuffling of Congressional responsibilities, especially those of the Joint Committee on Atomic Energy, long a legislative powerhouse.

2) *State* and even *local government* will come to play larger and more explicit roles in day-to-day energy use policies. During the recent fuel shortages, state governments became the locus of what rationing procedures were necessary; similar efforts are probably in the offing.

3) *Research and development* expenditures for energy independence are highly likely, although it may well turn out that they are, from an economic standpoint, wasteful of scarce public resources.

4) *Environmental protection* is probably a permanent part of the energy development agenda. Although some of this environmental concern will be window dressing, a lot will not. In particular, the extensive use of *technology assessment* as a methodological perspective in analyzing new technology is quite likely.

These trends, however, are based upon the assumption that business roughly as usual will prevail. More narrowly, the continued stress on environmental protection is probably dependent on the continued possibilities of moderate, but real, economic growth. Whether real growth and the political maneuvering room it promises will show up remains, of course, to be seen.

As long as they do, the readings listed here—which share these assumptions—will remain useful as approaches to the many-headed problem of energy policy.

The author is with the Program in Social Management of Technology and the Department of Political Science, University of Washington, Seattle, Wash.

1) Background

Committee on Interior and Insular Affairs, U.S. Senate, *Federal Energy Organization*, a staff analysis, serial no. 93-6. Washington: Committee Print, 1973. A summary of the organizational maze of federal energy policy.

A. Downs, "Up and down with ecology—The 'issue-attention cycle'", *Public Interest*, no. 28, pp. 38–50, 1972. An interesting speculation on the dynamics of political attention.

R. J. Dubos, "Humanizing the earth," *Science*, vol. 179, pp. 769-772, 1973. A vision of man's intervention into the natural system which is provocatively optimistic.

M. Edel, *Economies and the Environment.* Englewood Cliffs, N.J.: Prentice-Hall, 1973. An introductory text which discusses auto use as an environmental problem in useful detail.

A. L. Hammond, W. D. Metz, and T. W. Maugh, II, *Energy and the Future.* Washington: American Association for the Advancement of Science, 1973. A useful compilation of future technological options.

J. P. Holdren and P. Herrera, *Energy.* San Francisco: Sierra

Club, 1972. A lucid introduction for the nontechnical reader.

T. LaPorte and D. Metlay, *They Watch and Wonder—The Public's Attitudes Toward Technology: A Survey*, working paper no. 6, Institute of Governmental Studies. Berkeley: Univ. California, 1973. Has data on public opinions regarding energy technologies.

D. B. Luten, "The economic geography of energy," *Energy and Power.* San Francisco: Freeman, 1971, pp. 109-17. A comparative view, too often neglected.

V. E. McKelvey, "Mineral resource estimates and public policy," *Amer. Sci.*, vol. 60, pp. 32-40, 1972. The economic analysis of resource scarcity—a useful comparison with Hubbert's piece, collected here.

R. G. Ridker, "Resource and environmental consequences of population growth in the United States . . . A summary," in *Commission on Population Growth and the American Future, Population, Resources, and the Environment*, Ridker, Ed. Washington: Government Printing Office, 1972, pp. 19-33. The global growth scene.

M. J. Roberts, "Is there an energy crisis?," *Public Interest*, no. 31, pp. 17-37, 1973. A cogent summary of the skeptical view.

John Walsh, "Britain and energy policy: Problems of interdependence," *Science*, vol. 180, pp. 1343-1347, 1973. A comparative view.

2) Fossil Fuels

M. A. Adelman, "Is the oil shortage real?—Oil companies as OPEC tax-collectors," *Foreign Policy*, no. 9, pp. 76-114, 1972. This and the articles by Akins and Amuzegar lay out the main lines of the best known energy policy issue, petroleum from the Middle East. Adelman's economic view is shared, at a more general level, by Marc Roberts and by Theodore Moran. The political-accommodation school of thinking represented by Akins (currently U.S. Ambassador to Saudi Arabia) and Amuzegar (an Iranian Ambassador to the United States) is not ignorant of market realities, but it does take a freer view of the range of human bargaining behavior. Both Akins and Amuzegar stress the importance of *secure supply*, while the economists argue that the most secure supply is to be found in a free market, unfettered by high-priced (and unenforceable) sweetheart deals.

J. E. Akins, "The oil crisis: This time the wolf is here," *Foreign Affairs*, vol. 51, pp. 462-490, 1973.

J. Amuzegar, "The oil story: Facts, fiction and fair play," *Foreign Affairs*, vol. 51, pp. 676-689, 1973.

L. J. Carter, "Rio Blanco: Stimulating gas and conflict in Colorado," *Science*, vol. 180, pp. 844-848, 1973. The problems of developing natural gas in the Rockies are briefly sketched here.

Committee on Interior and Insular Affairs, U.S. Senate, "Coal surface mining and reclamation," background paper by the Council on Environmental Quality, serial no. 93-8. Washington: Committee Print, 1973. A comprehensive regulatory scheme for strip mining is outlined.

Committee on Interior and Insular Affairs, U.S. Senate, "Factors affecting the use of coal in present and future energy markets," background paper by the Congressional Research Service, serial no. 93-9. Washington: Committee Print, 1973. An economic review of the factors influencing choice of fuel; increasingly dated, given recent rises of petroleum prices.

A. T. Demaree, "Aramco is a lesson in the management of chaos," *Fortune*, p. 58ff, Feb. 1974. A look at the managerial style of oil companies in the field. With the upsurge of interest in energy, *Fortune* has begun to carry articles on energy industries, many of which provide interesting journalistic insights into industrial decisions.

J. W. Devanney, III, "Key issues in offshore oil," *Technol. Rev.*, pp. 21-25, Jan. 1974. A brief look at the Georges Bank study, an economic and environmental survey of oil resources off the New England coast.

R. Gillette, "Western coal: Does the debate follow irreversible commitment?," *Science*, vol. 182, pp. 456-458, 1973. The problem is lack of water—a problem intelligent policy analysis would have spotted.

R. Gillette, "Synthetic fuels: Will government lend the oil industry a hand?," *Science*, vol. 183, pp. 641-643, 1974. The puzzles of governmental support of commercial energy development.

M. I. Goldman, "Red black gold," *Foreign Policy*, no. 8, pp. 138-148, 1972. Oil in the Soviet Union.

W. J. Mead, "The system of government subsidies to the oil industry," *Natural Resources J.*, vol. 10, pp. 113-125, 1970. The analysis is aging—the oil import quota has been seriously modified—but the attitude toward governmental distortion of the market is not.

T. H. Moran, "New deal or raw deal in raw materials," *Foreign Policy*, no. 5, pp. 119-134, 1971-1972. An interesting speculative essay on the political economy of industrial technology.

A. E. Keir Nash, D. E. Mann, and P. G. Olsen, *Oil Pollution and the Public Interest: A Study of the Santa Barbara Oil Spill.* Berkeley: Institute of Governmental Studies, 1972. The legal struggles over the Santa Barbara oil spill.

C. T. Rand, "The Arabian fantasy," *Harper's*, pp. 42-54, Jan. 1974. A brief—but somewhat breathless—recapitulation of the international oil debate (see Adelman, above).

3) Nuclear Energy

G. Buchan, "Institutional design for energy systems/environmental decision-making," unpublished paper presented at M.I.T. Conference on Energy: Demand, Conservation and Institutional Problems, Feb. 1973. A perspective from decision theory on power-plant siting.

R. Gillette, articles on reactor safety in *Science*, beginning with vol. 172, p. 918, 1971, and ending with vol. 178, p. 482, 1972. A prize-winning series of investigative reports on AEC research on the safety of water reactors, the principal type used in power generation. The regulation versus promotion issue in AEC policy is given technical form here.

H. P. Green, "Nuclear safety and the public interest," *Nuclear News*, vol. 15, no. 9, pp. 75-78, 1972. An articulate critique.

Holmes and Narver, Inc., *California Power Plant Siting*

Study, 1973. Prepared for Resources Agency, State of California, and U.S. Atomic Energy Commission. Executive summary, vol. 1. Land use considerations make inland power plant sites, far from valuable coastline land, attractive despite higher operating costs.

D. Nelkin, *Nuclear Power and Its Critics.* Ithaca: Cornell Univ. Press, 1971. A well-written case study.

W. H. Rodgers, "Sitting power plants in Washington State," *Washington Law Rev.*, vol. 47, pp. 9-33, 1971. A discussion of the Washington State power plant siting process, a pioneering effort in "one-stop" siting.

C. Starr, "Realities of the energy crisis," *Science and Public Affairs*, pp. 15-20, Sept. 1973. The relationship between the energy crisis and nuclear power by one of its strongest supporters.

J. Walsh, 1) "Vermont: A small state faces up to a dilemma over development," *Science*, vol. 173, pp. 895-897, 1971; 2) "Vermont: A power deficit raises pressure for new plants," *Science*, vol. 173, pp. 1110-1115, 1971; 3) "Vermont: Forced to figure in the big power picture," *Science*, vol. 174, pp. 44-47, 1971. The dilemmas of economic development and nuclear energy development.

4) Conservation, Use, and Demand Management

B. Commoner, testimony in *Fuel and Energy Resources*, part II, pp. 565-604, 1972. Hearings before the House Interior and Insular Affairs Committee, Apr. 1972. Serial no. 92-42. The energy-environment tradeoff put in its harshest terms.

W. K. Foell and J. E. Rushton, "Energy use in Wisconsin," working paper 4, Institute of Environmental Studies, Univ. Wisconsin, 1972. Like the study by Romer *et al.*, cited below, a state-level survey of energy usage, useful for comparative purposes.

B. Hannon, "System energy and recycling: A study of the beverage industry," unpublished paper, presented at AAAS Meeting, Philadelphia, Pa., Dec. 1972. Center for Advanced Computation, Univ. Illinois, 1971. An interesting use of input-output economic methods to depict energy flows and their economic implications.

B. Hannon, "Options for energy conservation," *Technol. Rev.*, pp. 24-31, Feb. 1974. Application of Hannon's input-output work to conservation possibilities.

R. A. Herendeen, "An energy input-output matrix for the United States, 1963: User's guide," CAC no. 69. Center for Advanced Computation, Univ. Illinois, 1973. Description and documentation of the Hannon group's input-output work.

RAND Corporation, *California's Electricity Quandary*, 3 vols., Santa Monica, 1972. An in-depth policy study for the State of California. The conclusions should be taken as indicative rather than settled; there has been sharp disagreement from the electric utilities in the state.

W. A. Reardon, "An input/output analysis of energy use changes from 1947 to 1958 and 1958 to 1963," Res. Rep., Battelle Pacific Northwest Lab., June 1972. The earliest use of input-output methods for energy analysis. Does not have the detail of Hannon's work, but it does permit longitudinal comparisons over 20 years. Things have changed remarkably little, all in all.

H. F. Romer, S. H. Flajser, and C. H. Martin, *Energy Profile of the State of Washington*, Institute for Environmental Studies, Univ. Washington, 1973. A state-level survey for Washington, a somewhat atypical state.

5) Institutional Redesign

P. H. Abelson, in a series of editorials in *Science* during late 1973 and 1974, has taken up a number of issues in energy policy. Abelson's views are technologically optimistic and cautiously conservative about the utility of governmental policy as a tool of society-wide change.

Special issue on "The Energy Crisis: Reality or Myth?," *Annals Amer. Acad. Political and Social Sci.*, vol. 410, Nov. 1973. A compilation of articles by distinguished social scientists.

J. Cameron, "Growth is a fighting word in Colorado's mountain wonderland," *Fortune*, pp. 148-159, 212-218, Oct. 1973. Growth requires energy. The struggle in Colorado seems to be a harbinger of conflicts to come.

J. D. Carroll, "Participatory technology," *Science*, vol. 171, pp. 647-653, 1971. A sophisticated and optimistic analysis of the possibilities for public participation in the design and location of technical facilities. Carroll's optimism will need to be redeemed in full if the scale of capital investment in energy processing facilities is to be achieved.

L. J. Carter, "Environmental law (I): Maturing field for lawyers and scientists," *Science*, vol. 179, pp. 1205-1209, 1973; "Environmental law (II): A strategic weapon against degradation?," *Science*, vol. 179, pp. 1310, 1312, 1350, 1973. As the most powerful organized group criticizing and forcing change upon the energy industries (and the AEC), environmental lawyers bulk large in future planning—and uncertainties.

L. J. Carter, "Land use law (I): Congress on verge of a modest beginning," *Science*, vol. 182, pp. 691-697, 1973; "Land use law (II): Florida is a major testing ground," *Science*, vol. 182, pp. 902-908, 1973. Likely to be a major testing ground for environmental politics, land use legislation highlights the growing resurgence of state and local governments in energy and environmental policy.

Special issue on "The No-Growth Society," *Daedalus*, Fall 1973. A collection of articles on the implications of a stable world, mainly by social scientists.

E. E. David, Jr., "Energy: A strategy of diversity," *Technol. Rev.*, pp. 26-31, June 1973. By the former Presidential Science Adviser—a proposal once followed by the White House before the energy crisis and Senator Henry Jackson's massive spending plans hit the front pages.

D. H. Davis, *Energy Politics.* New York: St. Martin's, 1974. A political science text with a usefully historical view and helpful analytic framework.

Energy Policy Project, *A Time to Choose.* Cambridge, Mass.: Ballinger, 1974. The summary report of the Ford Foundation-sponsored policy research group. A provocative discussion—including the possibility that *economic* growth may not have to be sacrificed.

R. W. Fri, "Facing up to pollution controls," *Harvard Business Rev.*, p. 26ff, Mar.-Apr. 1974. By the former Deputy Administrator of the Environmental Protection Agency, who argues that good energy sense makes good environmental sense—makes good business sense.

R. G. Noll, *Reforming Regulation.* Washington: Brookings Institution, 1971. Originally written as a critique of federal reorganization plans, this perceptive analysis of regulatory agencies is important reading for those harboring hopes for governmental intervention in energy supply. Noll is not pessimistic—but, given the record of regulatory agencies, caution *is* indicated.

L. H. Olsen, "The energy crisis and the balance of payments," *Science and Public Affairs*, pp. 26-29, Mar. 1974. The chief economist of First National City Bank notes that the flow of money to the Arabs is not the whole of our foreign trade.

D. J. Rose, "Energy policy in the U.S.," *Sci. Amer.*, pp. 20-29, Jan. 1974. A trenchant analysis by a broad-ranging nuclear engineer.

L. Winner, "On criticizing technology," *Public Policy*, pp. 35-59, Winter 1972. Important qualifications to the promise of technology assessment.

E. J. Woodhouse, "Revisioning the future of the third world: An ecological perspective on development," *World Politics*, pp. 1-33, Oct. 1972. For the less-developed countries, the energy crisis is bad news.

Author Index

Subject Index

Editor's Biography

M. Granger Morgan (S'62–M'71) was born on March 17, 1941. He concentrated in physics and graduated cum laude in general studies from Harvard College, Cambridge, Mass., in 1963. He received an experimental M.S. degree in astronomy and space science from Cornell University, Ithaca, N.Y., in 1965. During 1965 he was a graduate student in modern Latin American history at the University of California, Berkeley. He received the Ph.D. degree from the Department of Applied Physics and Information Sciences, University of California, San Diego, in 1969 with a dissertation entitled "A Laboratory Model for Radio Star Scintillation and Other Diffraction Phenomena."

From 1969 to 1972 he was employed at the University of California, San Diego, where he created and for three years directed the Computer Jobs Through Training Project, a motivational and job training project for the disadvantaged. In addition, he served as a Lecturer (1970–1971) and as an Acting Assistant Professor (1971–1972) in the Department of Applied Physics and Information Sciences, where he taught and did research on problems in the technology-society area. From 1972 to 1974 he served as a Program Director in the NSF Division of Computer Research, where he assisted in the design and development of research support programs in the computers and society area. He spent the summer of 1974 as a Visiting Scientist at Brookhaven National Laboratories doing research on the external costs of energy production. He currently holds a joint appointment in Electrical Engineering and Engineering and Public Affairs at Carnegie-Mellon University, Pittsburgh, Pa., where he teaches, does research on technology-society problems involving energy and information technology, and is responsible for the development of a graduate program in Engineering and Public Affairs.